THEORY OF BESSEL FUNCTIONS

W. B. F.

A TREATISE ON THE THEORY OF BESSEL FUNCTIONS

BY

G. N. WATSON

SECOND EDITION

CAMBRIDGE
AT THE UNIVERSITY PRESS
1966

PUBLISHED BY
THE SYNDICS OF THE CAMBRIDGE UNIVERSITY PRESS

Bentley House, 200 Euston Road, London, N.W. 1
American Branch: 32 East 57th Street, New York, N.Y. 10022
West African Office: P.M.B. 5181, Ibadan, Nigeria

First Edition	1922
Second Edition	1944
Reprinted	1952
	1958
	1962
	1966
First paperback edition	1966

First printed in Great Britain at the University Press, Cambridge
Reprinted by lithography in Great Britain by
Hazell Watson & Viney Ltd, Aylesbury, Bucks

PREFACE

THIS book has been designed with two objects in view. The first is the development of applications of the fundamental processes of the theory of functions of complex variables. For this purpose Bessel functions are admirably adapted; while they offer at the same time a rather wider scope for the application of parts of the theory of functions of a real variable than is provided by trigonometrical functions in the theory of Fourier series.

The second object is the compilation of a collection of results which would be of value to the increasing number of Mathematicians and Physicists who encounter Bessel functions in the course of their researches. The existence of such a collection seems to be demanded by the greater abstruseness of properties of Bessel functions (especially of functions of large order) which have been required in recent years in various problems of Mathematical Physics.

While my endeavour has been to give an account of the theory of Bessel functions which a Pure Mathematician would regard as fairly complete, I have consequently also endeavoured to include all formulae, whether general or special, which, although without theoretical interest, are likely to be required in practical applications; and such results are given, so far as possible, in a form appropriate for these purposes. The breadth of these aims, combined with the necessity for keeping the size of the book within bounds, has made it necessary to be as concise as is compatible with intelligibility.

Since the book is, for the most part, a development of the theory of functions as expounded in the *Course of Modern Analysis* by Professor Whittaker and myself, it has been convenient to regard that treatise as a standard work of reference for general theorems, rather than to refer the reader to original sources.

It is desirable to draw attention here to the function which I have regarded as the canonical function of the second kind, namely the function which was defined by Weber and used subsequently by Schläfli, by Graf and Gubler and by Nielsen. For historical and sentimental reasons it would have been pleasing to have felt justified in using Hankel's function of the second kind; but three considerations prevented this. The first is the necessity for standardizing the function of the second kind; and, in my opinion, the authority of the group of mathematicians who use Weber's function has greater weight than the authority of the mathematicians who use any other one function of the second kind. The second is the parallelism which the use of Weber's function exhibits between the two kinds of Bessel functions and the two kinds (cosine and sine) of trigonometrical functions. The third is the existence of the device by which interpolation is made possible in Tables I and III at the end of Chapter XX, which seems to make the use of Weber's function inevitable in numerical work.

It has been my policy to give, in connexion with each section, references to any memoirs or treatises in which the results of the section have been previously enunciated; but it is not to be inferred that proofs given in this book are necessarily those given in any of the sources cited. The bibliography at the end of the book has been made as complete as possible, though doubtless omissions will be found in it. While I do not profess to have inserted every memoir in which Bessel functions are mentioned, I have not consciously omitted any memoir containing an original contribution, however slight to the theory of the functions; with regard to the related topic of Riccati's equation, I have been eclectic to the extent of inserting only those memoirs which seemed to be relevant to the general scheme.

In the case of an analytical treatise such as this, it is probably useless to hope that no mistakes, clerical or other, have remained undetected; but the number of such mistakes has been considerably diminished by the criticisms and the vigilance of my colleagues Mr C. T. Preece and Mr T. A. Lumsden, whose labours to remove errors and obscurities have been of the greatest value. To these gentlemen and to the staff of the University Press, who have given every assistance, with unfailing patience, in a work of great typographical complexity, I offer my grateful thanks.

G. N. W.

August 21, 1922.

PREFACE TO THE SECOND EDITION

To incorporate in this work the discoveries of the last twenty years would necessitate the rewriting of at least Chapters XII—XIX; my interest in Bessel functions, however, has waned since 1922, and I am consequently not prepared to undertake such a task to the detriment of my other activities. In the preparation of this new edition I have therefore limited myself to the correction of minor errors and misprints and to the emendation of a few assertions (such as those about the unproven character of Bourget's hypothesis) which, though they may have been true in 1922, would have been definitely false had they been made in 1941.

My thanks are due to many friends for their kindness in informing me of errors which they had noticed; in particular, I cannot miss this opportunity of expressing my gratitude to Professor J. R. Wilton for the vigilance which he must have exercised in the compilation of his list of corrigenda.

G. N. W.

March 31, 1941.

CONTENTS

To stand upon every point, and go over things at large, and to be curious in particulars, belongeth to the first author of the story: but to use brevity, and avoid much labouring of the work, is to be granted to him that will make an abridgement.

2 MACCABEES ii. 30, 31.

CHAPTER I

BESSEL FUNCTIONS BEFORE 1826

1·1. *Riccati's differential equation.*

The theory of Bessel functions is intimately connected with the theory of a certain type of differential equation of the first order, known as Riccati's equation. In fact a Bessel function is usually defined as a particular solution of a linear differential equation of the second order (known as Bessel's equation) which is derived from Riccati's equation by an elementary transformation.

The earliest appearance in Analysis of an equation of Riccati's type occurs in a paper* on curves which was published by John Bernoulli in 1694. In this paper Bernoulli gives, as an example, an equation of this type and states that he has not solved it†.

In various letters‡ to Leibniz, written between 1697 and 1704, James Bernoulli refers to the equation, which he gives in the form

$$dy = yy\,dx + xx\,dx,$$

and states, more than once, his inability to solve it. Thus he writes (Jan. 27, 1697): "Vellem porro ex Te scire num et hanc tentaveris $dy = yy\,dx + xx\,dx$. Ego in mille formas transmutavi, sed operam meam improbum Problema perpetuo lusit." Five years later he succeeded in reducing the equation to a linear equation of the second order and wrote§ to Leibniz (Nov. 15, 1702): "Qua occasione recordor aequationes alias memoratae $dy = yy\,dx + x^2 dx$ in qua nunquam separare potui indeterminatas a se invicem, sicut aequatio maneret simpliciter differentialis: sed separavi illas reducendo aequationem ad hanc differentio-differentialem‖ $ddy : y = - x^2\,dx^2$."

When this discovery had been made, it was a simple step to solve the last equation in series, and so to obtain the solution of the equation of the first order as the quotient of two power-series.

* *Acta Eruditorum publicata Lipsiae*, 1694, pp. 435—437.

† "Esto proposita aequatio differentialis haec $x^2dx + y^2dx = a^2dy$ quae an per separationem indeterminatarum construi possit nondum tentavi" (p. 436).

‡ See *Leibnizens gesamellte Werke*, Dritte Folge (Mathematik), III. (Halle, 1855), pp. 50—87.

§ *Ibid.* p. 65. Bernoulli's procedure was, effectively, to take a new variable u defined by the formula

$$-\frac{1}{u}\frac{du}{dx} = y$$

in the equation $dy/dx = x^2 + y^2$, and then to replace u by y.

‖ The connexion between this equation and a special form of Bessel's equation will be seen in § 4·3.

And, in fact, this form of the solution was communicated to Leibniz by James Bernoulli within a year (Oct. 3, 1703) in the following terms*:

"Reduco autem aequationem $dy = yy\,dx + xx\,dx$ ad fractionem cujus uterque terminus per seriem exprimitur, ita

$$y = \frac{\dfrac{x^3}{3} - \dfrac{x^7}{3.4.7} + \dfrac{x^{11}}{3.4.7.8.11} - \dfrac{x^{15}}{3.4.7.8.11.12.15} + \dfrac{x^{19}}{3.4.7.8.11.12.15.16.19} - \text{etc.}}{1 - \dfrac{x^4}{3.4} + \dfrac{x^8}{3.4.7.8} - \dfrac{x^{12}}{3.4.7.8.11.12} + \dfrac{x^{16}}{3.4.7.8.11.12.15.16} - \text{etc.}}$$

quae series quidem actuali divisione in unam conflari possunt, sed in qua ratio progressionis non tam facile patescat, scil.

$$y = \frac{x^3}{3} + \frac{x^7}{3.3.7} + \frac{2x^{11}}{3.3.3.7.11} + \frac{13x^{15}}{3.3.3.3.5.7.7.11} + \text{etc."}$$

Of course, at that time, mathematicians concentrated their energy, so far as differential equations were concerned, on obtaining solutions *in finite terms*, and consequently James Bernoulli seems to have received hardly the full credit to which his discovery entitled him. Thus, twenty-two years later, the paper†, in which Count Riccati first referred to an equation of the type which now bears his name, was followed by a note‡ by Daniel Bernoulli in which it was stated that the solution of the equation§

$$ax^n\,dx + uu\,dx = b\,du$$

was a hitherto unsolved problem. The note ended with an announcement in an anagram of the solution: "Solutio problematis ab Ill. Riccato proposito characteribus occultis involuta 24*a*, 6*b*, 6*c*, 8*d*, 33*e*, 5*f*, 2*g*, 4*h*, 33*i*, 6*l*, 21*m*, 26*n*, 16*o*, 8*p*, 5*q*, 17*r*, 16*s*, 25*t*, 32*u*, 5*x*, 3*y*, +, −, ——, ±, =, 4, 2, 1."

The anagram appears never to have been solved; but Bernoulli published his solution‖ of the problem about a year after the publication of the anagram. The solution consists of the determination of a set of values of n, namely $-4m/(2m \pm 1)$, where m is any integer, for any one of which the equation is soluble in finite terms; the details of this solution will be given in §§ 4·1, 4·11.

The prominence given to the work of Riccati by Daniel Bernoulli, combined with the fact that Riccati's equation was of a slightly more general type than

* See *Leibnizens gesamellte Werke*, Dritte Folge (Mathematik), III. (Halle, 1855), p. 75.

† *Acta Eruditorum*, *Suppl.* VIII. (1724), pp. 66—73. The form in which Riccati took the equation was

$$x^m dq = du + uu\,dx : q,$$

where $q = x^n$.

‡ *Ibid.* pp. 73—75. Daniel Bernoulli mentioned that solutions had been obtained by three other members of his family—John, Nicholas and the younger Nicholas.

§ The reader should observe that the substitution

$$u = -\frac{b}{z}\frac{dz}{dx}$$

gives rise to an equation which is easily soluble in series.

‖ *Exercitationes quaedam mathematicae* (Venice, 1724), pp. 77—80; *Acta Eruditorum*, 1725, pp. 465—473.

John Bernoulli's equation* has resulted in the name of Riccati being associated not only with the equation which he discussed without solving, but also with a still more general type of equation.

It is now customary to give the name† *Riccati's generalised equation* to any equation of the form

$$\frac{dy}{dx} = P + Qy + Ry^2,$$

where P, Q, R are given functions of x.

It is supposed that neither P nor R is identically zero. If $R=0$, the equation is linear; if $P=0$, the equation is reducible to the linear form by taking $1/y$ as a new variable.

The last equation was studied by Euler‡; it is reducible to the general linear equation of the second order, and this equation is sometimes reducible to Bessel's equation by an elementary transformation (cf. §§ 3·1, 4·3, 4·31).

Mention should be made here of two memoirs by Euler. In the first§ it is proved that, when a particular integral y_1 of Riccati's generalised equation is known, the equation is reducible to a linear equation of the first order by replacing y by $y_1 + 1/u$, and so the general solution can be effected by two quadratures. It is also shewn (*ibid.* p. 59) that, if two particular solutions are known, the equation can be integrated completely by a single quadrature; and this result is also to be found in the second|| of the two papers. A brief discussion of these theorems will be given in Chapter IV.

1·2. *Daniel Bernoulli's mechanical problem.*

In 1738 Daniel Bernoulli published a memoir¶ containing enunciations of a number of theorems on the oscillations of heavy chains. The eighth** of these is as follows: "*De figura catenae uniformiter oscillantis.* Sit catena AC uniformiter gravis et perfecte flexilis suspensa de puncto A, eaque oscillationes facere uniformes intelligatur: pervenerit catena in situm AMF; fueritque longitudo catenae $= l$: longitudo cujuscunque partis $FM = x$, sumatur n ejus valoris†† ut fit

$$1 - \frac{l}{n} + \frac{ll}{4nn} - \frac{l^3}{4 \,.\, 9n^3} + \frac{l^4}{4 \,.\, 9 \,.\, 16n^4} - \frac{l^5}{4 \,.\, 9 \,.\, 16 \,.\, 25n^5} + \text{etc.} = 0.$$

* See James Bernoulli, *Opera Omnia*, II. (Geneva, 1744), pp. 1054—1057; it is stated that the point of Riccati's problem is the determination of a solution in finite terms, and a solution which resembles the solution by Daniel Bernoulli is given.

† The term 'Riccati's equation' was used by D'Alembert, *Hist. de l'Acad. R. des Sci. de Berlin*, XIX. (1763), [published 1770], p. 242.

‡ *Institutiones Calculi Integralis*, II. (Petersburg, 1769), § 831, pp. 88—89. In connexion with the reduction, see James Bernoulli's letter to Leibniz already quoted.

§ *Novi Comm. Acad. Petrop.* VIII. (1760—1761), [published 1763], p. 32.

|| *Ibid.* IX. (1762—1763), [published 1764], pp. 163—164.

¶ "Theoremata de oscillationibus corporum filo flexili connexorum et catenae verticaliter suspensae," *Comm. Acad. Sci. Imp. Petrop.* VI. (1732—3), [published 1738], pp. 108—122.

** *Loc. cit.* p. 116.

†† The length of the simple equivalent pendulum is n.

Ponatur porro distantia extremi puncti F ab linea verticali = 1, dico fore distantiam puncti ubicunque assumpti M ab eadem linea verticali aequalem

$$1 - \frac{x}{n} + \frac{xx}{4nn} - \frac{x^3}{4.9n^3} + \frac{x^4}{4.9.16n^4} - \frac{x^5}{4.9.16.25n^5} + \text{etc.''}$$

He goes on to say: "Invenitur brevissimo calculo n = proxime 0·691 l.... Habet autem littera n infinitos valores alios."

The last series is now described as a Bessel function* of order zero and argument $2\sqrt{(x/n)}$; and the last quotation states that this function has an infinite number of zeros.

Bernoulli published† proofs of his theorems soon afterwards; in theorem VIII, he obtained the equation of motion by considering the forces acting on the portion FM of length x. The equation of motion was also obtained by Euler‡ many years later from a consideration of the forces acting on an element of the chain.

The following is the substance of Euler's investigation:

Let ρ be the line density of the chain (supposed uniform) and let T be the tension at height x above the lowest point of the chain in its undisturbed position. The motion being transversal, we obtain the equation $\delta T = g\rho\,\delta x$ by resolving vertically for an element of chain of length δx. The integral of the equation is $T = g\rho x$.

The horizontal component of the tension is, effectively, $T(dy/dx)$ where y is the (horizontal) displacement of the element; and so the equation of motion is

$$\rho\,\delta x \frac{d^2y}{dt^2} = \delta\left(T\frac{dy}{dx}\right).$$

If we substitute for T and proceed to the limit, we find that

$$\frac{d^2y}{dt^2} = g\frac{d}{dx}\left(x\frac{dy}{dx}\right).$$

If f is the length of the simple equivalent pendulum for any one normal vibration, we write

$$y = A\Pi\left(\frac{x}{f}\right)\sin\left(\zeta + t\sqrt{\frac{g}{f}}\right),$$

where A and ζ are constants; and then $\Pi(x/f)$ is a solution of the equation

$$\frac{d}{dx}\left(x\frac{dv}{dx}\right) + \frac{v}{f} = 0.$$

If $x/f = u$, we obtain the solution in the form of Bernoulli's series, namely

$$v = 1 - \frac{u}{1} + \frac{u^2}{1.4} - \frac{u^3}{1.4.9} + \frac{u^4}{1.4.9.16} - \ldots.$$

* On the Continent, the functions are usually called *cylinder functions*, or, occasionally, *functions of Fourier-Bessel*, after Heine, *Journal für Math.* LXIX. (1868), p. 128; see also *Math. Ann.* III. (1871), pp. 609—610.

† *Comm. Acad. Petrop.* VII. (1734—5), [published 1740], pp. 162—179.

‡ *Acta Acad. Petrop.* V. pars 1 (Mathematica), (1781), [published 1784], pp. 157—177. Euler took the weight of length e of the chain to be E, and he defined g to be the measure of the distance (not twice the distance) fallen by a particle from rest under gravity in a second. Euler's notation has been followed in the text apart from the significance of g and the introduction of ρ and δ (for d).

The general solution of the equation is then shewn to be $Dv + Cv \int^u \frac{du}{uv^2}$, where C and D are constants. Since y is finite when $x=0$, C must be zero.

If a is the whole length of the chain, $y=0$ when $x=a$, and so the equation to determine f is

$$1 - \frac{a}{1.f} + \frac{a^2}{1.4f^2} - \frac{a^3}{1.4.9f^3} + \ldots = 0.$$

By an extremely ingenious analysis, which will be given fully in Chapter XV, Euler proceeded to shew that the three smallest roots of the equation in a/f are 1·445795, 7·6658 and 18·63. [More accurate values are 1·4457965, 7·6178156 and 18·7217517.]

In the memoir* immediately following this investigation Euler obtained the general solution (in the form of series) of the equation $\frac{d}{du}\left(u\frac{dv}{du}\right) + v = 0$, but his statement of the law of formation of successive coefficients is rather incomplete. The law of formation had, however, been stated in his *Institutiones Calculi Integralis*†, II. (Petersburg, 1769), § 977, pp. 233–235.

1·3. *Euler's mechanical problem.*

The vibrations of a stretched membrane were investigated by Euler‡ in 1764. He arrived at the equation

$$\frac{1}{e^2}\frac{d^2z}{dt^2} = \frac{d^2z}{dr^2} + \frac{1}{r}\frac{dz}{dr} + \frac{1}{r^2}\frac{d^2z}{d\phi^2},$$

where z is the transverse displacement at time t at the point whose polar coordinates are (r, ϕ); and e is a constant depending on the density and tension of the membrane.

To obtain a normal solution he wrote

$$z = u \sin(\alpha t + A)\sin(\beta\phi + B),$$

where α, A, β, B are constants and u is a function of r; and the result of substitution of this value of z is the differential equation

$$\frac{d^2u}{dr^2} + \frac{1}{r}\frac{du}{dr} + \left(\frac{\alpha^2}{e^2} - \frac{\beta^2}{r^2}\right)u = 0.$$

The solution of this equation which is finite at the origin is given on p. 256 of Euler's memoir; it is

$$u = r^\beta\left\{1 - \frac{\alpha^2 r^2}{2(n+1)e^2} + \frac{\alpha^4 r^4}{2.4(n+1)(n+3)e^4} - \ldots\right\},$$

where n has been written§ in place of $2\beta + 1$.

This differential equation is now known as Bessel's equation for functions of order β; and β may have‖ any of the values 0, 1, 2, ...

Save for an omitted constant factor the series is now called a Bessel coefficient of order β and argument $\alpha r/e$. The periods of vibration, $2\pi/\alpha$, of a

* *Acta Acad. Petrop.* v. pars 1 (Mathematica), (1781), [published 1784], pp. 178—190.

† See also §§ 935, 936 (p. 187 *et seq.*) for the solution of an associated equation which will be discussed in § 3·52.

‡ *Novi Comm. Acad. Petrop.* x. (1764), [published 1766], pp. 243—260.

§ The reason why Euler made this change of notation is not obvious.

‖ If β were not an integer, the displacement would not be a one-valued function of position, in view of the factor $\sin(\beta\phi + B)$.

circular membrane of radius a with a fixed boundary* are to be determined from the consideration that u vanishes when $r = a$.

This investigation by Euler contains the earliest appearance in Analysis of a Bessel coefficient of general integral order.

1·4. *The researches of Lagrange, Carlini and Laplace.*

Only a few years after Euler had arrived at the general Bessel coefficient in his researches on vibrating membranes, the functions reappeared, in an astronomical problem. It was shewn by Lagrange† in 1770 that, in the elliptic motion of a planet about the sun at the focus attracting according to the law of the inverse square, the relations between the radius vector r, the mean anomaly M and the eccentric anomaly E, which assume the forms

$$M = E - \epsilon \sin E, \quad r = a(1 - \epsilon \cos E),$$

give rise to the expansions

$$E = M + \sum_{n=1}^{\infty} A_n \sin nM, \quad \frac{r}{a} = 1 + \tfrac{1}{2}\epsilon^2 + \sum_{n=1}^{\infty} B_n \cos nM,$$

in which a and ϵ are the semi-major axis and the eccentricity of the orbit, and

$$A_n = 2 \sum_{m=0}^{\infty} \frac{(-)^m\, n^{n+2m-1}\, \epsilon^{n+2m}}{2^{n+2m}\, m!\,(n+m)!}, \quad B_n = -2 \sum_{m=0}^{\infty} \frac{(-)^m\,(n+2m)\,.\,n^{n+2m-2}\,\epsilon^{n+2m}}{2^{n+2m}\, m!\,(n+m)!}.$$

Lagrange gave these expressions for $n = 1, 2, 3$. The object of the expansions is to obtain expressions for the eccentric anomaly and the radius vector in terms of the time.

In modern notation these formulae are written

$$A_n = 2J_n(n\epsilon)/n, \quad B_n = -2(\epsilon/n)\, J_n{}'(n\epsilon).$$

It was noted by Poisson, *Connaissance des Tems*, 1836 [published 1833], p. 6 that

$$B_n = -\frac{\epsilon}{n}\frac{dA_n}{d\epsilon};$$

a memoir by Lefort, *Journal de Math.* XI. (1846), pp. 142—152, in which an error made by Poisson is corrected, should also be consulted.

A remarkable investigation of the approximate value of A_n when n is large and $0 < \epsilon < 1$ is due to Carlini‡; though the analysis is not rigorous (and it would be difficult to make it rigorous) it is of sufficient interest for a brief account of it to be given here.

* Cf. Bourget, *Ann. Sci. de l'École norm. sup.* III. (1866), pp. 55—95, and Chree, *Quarterly Journal.* XXI. (1886), p. 298.

† *Hist. de l'Acad. R. des Sci. de Berlin*, XXV. (1769), [published 1771], pp. 204—233. [*Oeuvres*, III. (1869), pp. 113—138.]

‡ *Ricerche sulla convergenza della serie che serva alla soluzione del problema di Keplero* (Milan, 1817). This work was translated into German by Jacobi, *Astr. Nach.* XXX. (1850), col. 197—254 [*Werke*, VII. (1891), pp. 189—245]. See also two papers by Scheibner dated 1856, reprinted in *Math. Ann.* XVII. (1880), pp. 531—544, 545—560.

It is easy to shew that A_n is a solution of the differential equation

$$\epsilon^2 \frac{d^2 A_n}{d\epsilon^2} + \epsilon \frac{dA_n}{d\epsilon} - n^2 (1-\epsilon^2) A_n = 0.$$

Define u by the formula $A_n = 2n^{n-1} e^{\int u d\epsilon}/n!$ and then

$$\epsilon^2 \left(\frac{du}{d\epsilon} + u^2\right) + \epsilon u - n^2 (1-\epsilon^2) = 0.$$

Hence when n is large either u or u^2 or $du/d\epsilon$ must be large.

If $u = O(n^a)$ we should expect u^2 and $du/d\epsilon$ to be $O(n^{2a})$ and $O(n^a)$ respectively; and on considering the highest powers of n in the various terms of the last differential equation, we find that $a=1$. It is consequently *assumed* that u admits of an expansion in descending powers of n in the form

$$u = nu_0 + u_1 + u_2/n + \ldots,$$

where $u_0, u_1, u_2, \ldots$ are independent of n.

On substituting this series in the differential equation of the first order and equating to zero the coefficients of the various powers of n, we find that

$$u_0{}^2 = (1-\epsilon^2)/\epsilon^2, \quad \epsilon(u_0' + 2u_0 u_1) + u_0 = 0, \ldots$$

where $u_0' = du_0/d\epsilon$; so that $u_0 = \pm \frac{\sqrt{(1-\epsilon^2)}}{\epsilon}$, $u_1 = \frac{\frac{1}{2}\epsilon}{1-\epsilon^2}$, and therefore

$$\int u d\epsilon = n \left\{\log \frac{\epsilon}{1 \pm \sqrt{(1-\epsilon^2)}} \pm \sqrt{(1-\epsilon^2)} \mp 1\right\} - \tfrac{1}{4} \log (1-\epsilon^2) + \ldots,$$

and, since the value of A_n shews that $\int u d\epsilon \sim n \log \frac{1}{2}\epsilon$ when ϵ is small, the upper sign must be taken and no constant of integration is to be added.

From Stirling's formula it now follows at once that

$$A_n \sim \frac{\epsilon^n \exp\{n\sqrt{(1-\epsilon^2)}\}}{\sqrt{(\frac{1}{2}\pi)} \cdot n^{\frac{3}{2}} (1-\epsilon^2)^{\frac{1}{4}} \{1+\sqrt{(1-\epsilon^2)}\}^n},$$

and this is the result obtained by Carlini. This method of approximation has been carried much further by Meissel (see § 8·11), while Cauchy* has also discussed approximate formulae for A_n in the case of comets moving in nearly parabolic orbits (see § 8·42), for which Carlini's approximation is obviously inadequate.

The investigation of which an account has just been given is much more plausible than the arguments employed by Laplace† to establish the corresponding approximation for B_n.

The investigation given by Laplace is quite rigorous and the method which he uses is of considerable importance when the value of B_n is modified by taking all the coefficients in the series to be positive—or, alternatively, by supposing that ϵ is a pure imaginary. But Laplace goes on to argue that an approximation established in the case of purely imaginary variables may be used 'sans crainte' in the case of real variables. To anyone who is acquainted with the modern theory of asymptotic series, the fallacious character of such reasoning will be evident.

* *Comptes Rendus*, XXXVIII. (1854), pp. 990—993.

† *Mécanique Céleste*, supplément, t. V. [first published 1827]. *Oeuvres*, V. (Paris, 1882), pp. 486—489.

The earlier portion of Laplace's investigation is based on the principle that, in the case of a series of positive terms in which the terms steadily increase up to a certain point and then steadily decrease, the order of magnitude of the sum of the series may frequently be obtained from a consideration of the order of magnitude of the greatest term of the series.

For other and more recent applications of this principle, see Stokes, *Proc. Camb. Phil. Soc.* VI. (1889), pp. 362—366 [*Math. and Phys. Papers*, V. (1905), pp. 221—225], and Hardy, *Proc. London Math. Soc.* (2) II. (1905), pp. 332—339; *Messenger*, XXXIV. (1905), pp. 97—101. A statement of the principle was given by Borel, *Acta Mathematica*, XX. (1897), pp. 393—394.

The following exposition of the principle applied to the example considered by Laplace may not be without interest:

The series considered is

$$B_n^{(1)} = 2 \sum_{m=0}^{\infty} \frac{(n+2m)\, n^{n+2m-2}\, \epsilon^{n+2m}}{2^{n+2m}\, m!\, (n+m)!},$$

in which n is large and ϵ has a fixed positive value. The greatest term is that for which $m=\mu$, where μ is the greatest integer such that

$$4\mu\,(n+\mu)\,(n+2\mu-2) \leqslant (n+2\mu)\, n^2\epsilon^2,$$

and so μ is approximately equal to

$$\tfrac{1}{2}n\,\{\sqrt{(1+\epsilon^2)}-1\}+\tfrac{1}{2}\epsilon^2/(1+\epsilon^2).$$

Now, if u_m denotes the general term in $B_n^{(1)}$, it is easy to verify by Stirling's theorem that, to a first approximation, $\dfrac{u_{\mu\pm t}}{u_\mu} \sim q^{t^2}$, where

$$\log q = -2\sqrt{(1+\epsilon^2)}/(n\epsilon^2).$$

Hence

$$B_n^{(1)} \sim u_\mu\,\{1+2q+2q^4+2q^9+\ldots\}$$
$$\sim 2u_\mu \sqrt{\{\pi/(1-q)\}},$$

since* q is nearly equal to 1.

Now, by Stirling's theorem,

$$u_\mu \sim \frac{\epsilon^{n-1}\exp\{n\sqrt{(1+\epsilon^2)}\}}{\pi n^2\,\{1+\sqrt{(1+\epsilon^2)}\}^n},$$

and so

$$B_n^{(1)} \sim \left\{\frac{2\sqrt{(1+\epsilon^2)}}{\pi n^3}\right\}^{\frac{1}{2}} \frac{\epsilon^n \exp\{n\sqrt{(1+\epsilon^2)}\}}{\{1+\sqrt{(1+\epsilon^2)}\}^n}.$$

The inference which Laplace drew from this result is that

$$B_n \sim -\left(\frac{2\sqrt{(1-\epsilon^2)}}{\pi n^3}\right)^{\frac{1}{2}} \frac{\epsilon^n \exp\{n\sqrt{(1-\epsilon^2)}\}}{\{1+\sqrt{(1-\epsilon^2)}\}^n}.$$

This approximate formula happens to be valid when $\epsilon<1$ (though the reason for this restriction is not apparent, apart from the fact that it is obviously necessary), but it is difficult to prove it without using the methods of contour

* The formula $1+2\sum_{t=0}^{\infty} q^{t^2} \sim \sqrt{\{\pi/(1-q)\}}$ may be inferred from general theorems on series; cf. Bromwich, *Theory of Infinite Series*, § 51. It is also a consequence of Jacobi's transformation formula in the theory of elliptic functions,

$$\vartheta_3(0\,|\,\tau) = (-i\tau)^{-\frac{1}{2}}\,\vartheta_3(0\,|\,-\tau^{-1});$$

see *Modern Analysis*, § 21·51.

integration (cf. § 8·31). Laplace seems to have been dubious as to the validity of his inference because, immediately after his statement about real and imaginary variables, he mentioned, by way of confirmation, that he had another proof; but the latter proof does not appear to be extant.

1·5. *The researches of Fourier.*

In 1822 appeared the classical treatise by Fourier*, *La Théorie analytique de la Chaleur*; in this work Bessel functions of order zero occur in the discussion of the symmetrical motion of heat in a solid circular cylinder. It is shewn by Fourier (§§ 118—120) that the temperature v, at time t, at distance x from the axis of the cylinder, satisfies the equation

$$\frac{dv}{dt} = \frac{K}{CD}\left(\frac{d^2v}{dx^2} + \frac{1}{x}\frac{dv}{dx}\right),$$

where K, C, D denote respectively the Thermal Conductivity, Specific Heat and Density of the material of the cylinder; and he obtained the solution

$$v = e^{-mt}\left\{1 - \frac{gx^2}{2^2} + \frac{g^2x^4}{2^2 . 4^2} - \frac{g^3x^6}{2^2 . 4^2 . 6^2} + \dots\right\},$$

where $g = mCD/K$ and m has to be so chosen that

$$hv + K\,(dv/dx) = 0$$

at the boundary of the cylinder, where h is the External Conductivity.

Fourier proceeded to give a proof (§§ 307—309) by Rolle's theorem that the equation to determine the values of m has† an infinity of real roots and no complex roots. His proof is slightly incomplete because he assumes that certain theorems which have been proved for polynomials are true of integral functions; the defect is not difficult to remedy, and a memoir by Hurwitz‡ has the object of making Fourier's demonstration quite rigorous.

It should also be mentioned that Fourier discovered the continued fraction formula (§ 313) for the quotient of a Bessel function of order zero and its derivate; generalisations of this formula will be discussed in §§ 5·6, 9·65. Another formula given by Fourier, namely

$$1 - \frac{\alpha^2}{2^2} + \frac{\alpha^4}{2^2 . 4^2} - \frac{\alpha^6}{2^2 . 4^2 . 6^2} + \dots = \frac{1}{\pi}\int_0^\pi \cos(\alpha \sin x)\,dx,$$

had been proved some years earlier by Parseval§; it is a special case of what are now known as Bessel's and Poisson's integrals (§§ 2·2, 2·3).

* The greater part of Fourier's researches was contained in a memoir deposited in the archives of the French Institute on Sept. 28, 1811, and crowned on Jan. 6, 1812. This memoir is to be found in the *Mém. de l'Acad. des Sci.*, IV. (1819), [published 1824], pp. 185—555; V. (1820), [published 1826], pp. 153—246.

† This is a generalisation of Bernoulli's statement quoted in § 1·2.

‡ *Math. Ann.* XXXIII. (1889), pp. 246—266.

§ *Mém. des savans étrangers*, I. (1805), pp. 639—648. This paper also contains the formal statement of the theorem on Fourier constants which is sometimes called Parseval's theorem; another paper by this little known writer, *Mém. des savans étrangers*, I. (1805), pp. 379—398, contains a general solution of Laplace's equation in a form involving arbitrary functions.

The expansion of an arbitrary function into a series of Bessel functions of order zero was also examined by Fourier (§§ 314—320); he gave the formula for the general coefficient in the expansion as a definite integral.

The validity of Fourier's expansion was examined much more recently by Hankel, *Math. Ann.* VIII. (1875), pp. 471—494; Schläfli, *Math. Ann.* X. (1876), pp. 137—142; Dini, *Serie di Fourier*, I. (Pisa, 1880), pp. 246—269; Hobson, *Proc. London Math. Soc.* (2) VII. (1909), pp. 359—388; and Young, *Proc. London Math. Soc.* (2) XVIII. (1920), pp. 163—200. This expansion will be dealt with in Chapter XVIII.

1·6. *The researches of Poisson.*

The unsymmetrical motions of heat in a solid sphere and also in a solid cylinder were investigated by Poisson* in a lengthy memoir published in 1823.

In the problem of the sphere†, he obtained the equation

$$\frac{d^2 R}{dr^2} - \frac{n(n+1)}{r^2} R = -\rho^2 R,$$

where r denotes the distance from the centre, ρ is a constant, n is a positive integer (zero included), and R is that factor of the temperature, in a normal mode, which is a function of the radius vector. It was shewn by Poisson that a solution of the equation is

$$r^{n+1} \int_0^\pi \cos(r\rho \cos\omega) \sin^{2n+1}\omega \, d\omega$$

and he discussed the cases $n = 0, 1, 2$ in detail. It will appear subsequently (§ 3·3) that the definite integral is (save for a factor) a Bessel function of order $n + \frac{1}{2}$.

In the problem of the cylinder (*ibid.* p. 340 *et seq.*) the analogous integral is

$$\lambda^n \int_0^\pi \cos(h\lambda \cos\omega) \sin^{2n}\omega \, d\omega,$$

where $n = 0, 1, 2, \ldots$ and λ is the distance from the axis of the cylinder. The integral is now known as Poisson's integral (§ 2·3).

In the case $n = 0$, an important approximate formula for the last integral and its derivate was obtained by Poisson (*ibid.*, pp. 350—352) when the variable is large; the following is the substance of his investigation:

Let‡ $J_0(k) = \frac{1}{\pi} \int_0^\pi \cos(k\cos\omega)\, d\omega, \quad J_0'(k) = -\frac{1}{\pi} \int_0^\pi \cos\omega \sin(k\cos\omega)\, d\omega.$

Then $J_0(k)$ is a solution of the equation

$$\frac{d^2(y\sqrt{k})}{dk^2} + \left(1 + \frac{1}{4k^2}\right) y\sqrt{k} = 0.$$

* *Journal de l'École R. Polytechnique*, XII. (cahier 19), (1823), pp. 249—403.

† *Ibid.* p. 300 *et seq.* The equation was also studied by Plana, *Mem. della R. Accad. delle Sci. di Torino*, XXV. (1821), pp. 532—534, and has since been studied by numerous writers, some of whom are mentioned in § 4·3. See also Poisson, *La Théorie Mathématique de la Chaleur* (Paris, 1835), pp. 366, 369.

‡ See also Röhrs, *Proc. London Math. Soc.* v. (1874), pp. 136—137. The notation $J_0(k)$ was not used by Poisson.

When k is large, $1/(4k^2)$ may be neglected in comparison with unity and so we may expect that $J_0(k)\surd k$ is approximately of the form $A\cos k+B\sin k$ where A and B are constants.

To determine A and B observe that

$$\cos k\,.\,J_0(k)-\sin k\,.\,J_0'(k)=\frac{1}{\pi}\int_0^\pi\{\cos^2\tfrac{1}{2}\omega\cos(2k\sin^2\tfrac{1}{2}\omega)+\sin^2\tfrac{1}{2}\omega\cos(2k\cos^2\tfrac{1}{2}\omega)\}\,d\omega.$$

Write $\pi-\omega$ for ω in the latter half of the integral and then

$$\cos k\,.\,J_0(k)-\sin k\,.\,J_0'(k)=\frac{2}{\pi}\int_0^\pi\cos^2\tfrac{1}{2}\omega\cos(2k\sin^2\tfrac{1}{2}\omega)\,d\omega$$

$$=\frac{2\surd 2}{\pi\surd k}\int_0^{\surd(2k)}\left(1-\frac{x^2}{2k}\right)^{\frac{1}{2}}\cos x^2\,dx,$$

and similarly $$\sin k\,.\,J_0(k)+\cos k\,.\,J_0'(k)=\frac{2\surd 2}{\pi\surd k}\int_0^{\surd(2k)}\left(1-\frac{x^2}{2k}\right)^{\frac{1}{2}}\sin x^2\,dx.$$

But $$\lim_{k\to\infty}\int_0^{\surd(2k)}\left(1-\frac{x^2}{2k}\right)^{\frac{1}{2}}\begin{matrix}\cos\\ \sin\end{matrix}\,x^2\,.\,dx=\int_0^\infty\begin{matrix}\cos\\ \sin\end{matrix}\,x^2\,.\,dx=\tfrac{1}{2}\surd(\tfrac{1}{2}\pi),$$

by a well known formula*.

[Note. It is not easy to prove rigorously that the passage to the limit is permissible; the simplest procedure is to appeal to Bromwich's integral form of Tannery's theorem, Bromwich, *Theory of Infinite Series*, § 174.]

It follows that

$$\begin{cases}\cos k\,.\,J_0(k)-\sin k\,.\,J_0'(k)=\dfrac{1}{\surd(\pi k)}(1+\epsilon_k),\\[2mm] \sin k\,.\,J_0(k)+\cos k\,.\,J_0'(k)=\dfrac{1}{\surd(\pi k)}(1+\eta_k),\end{cases}$$

where $\epsilon_k\to 0$ and $\eta_k\to 0$ as $k\to\infty$; and therefore

$$\begin{cases}J_0(k)=\dfrac{1}{\surd(\pi k)}[(1+\epsilon_k)\cos k+(1+\eta_k)\sin k],\\[2mm] J_0'(k)=\dfrac{1}{\surd(\pi k)}[-(1+\epsilon_k)\sin k+(1+\eta_k)\cos k].\end{cases}$$

It was then *assumed* by Poisson that $J_0(k)$ is expressible in the form

$$\frac{1}{\surd(\pi k)}\left[\left(A+\frac{A'}{k}+\frac{A''}{k^2}+\dots\right)\cos k+\left(B+\frac{B'}{k}+\frac{B''}{k^2}+\dots\right)\sin k\right],$$

where $A=B=1$. The series are, however, not convergent but asymptotic, and the validity of this expansion was not established, until nearly forty years later, when it was investigated by Lipschitz, *Journal für Math.* LVI. (1859), pp. 189—196.

The result of formally operating on the expansion assumed by Poisson for the function $J_0(k)\surd(\pi k)$ with the operator $\dfrac{d^2}{dk^2}+1+\dfrac{1}{4k^2}$ is

$$-\cos k\left[\frac{2\,.\,1\,.\,B'-\frac{1}{4}A}{k^2}+\frac{2\,.\,2B''-(1\,.\,2+\frac{1}{4})A'}{k^3}+\frac{2\,.\,3B'''-(2\,.\,3+\frac{1}{4})A''}{k^4}+\dots\right]$$

$$+\sin k\left[\frac{2\,.\,1\,.\,A'+\frac{1}{4}B}{k^2}+\frac{2\,.\,2A''+(1\,.\,2+\frac{1}{4})B'}{k^3}+\frac{2\,.\,3A'''+(2\,.\,3+\frac{1}{4})B''}{k^4}+\dots\right],$$

* Cf. Watson, *Complex Integration and Cauchy's Theorem* (Camb. Math. Tracts, no. 15, 1914), p. 71, for a proof of these results by using contour integrals.

and so, by equating to zero the various coefficients, we find that

$$\begin{cases} A' = -\frac{1}{8}B, & A'' = -\frac{9}{2.8^2}A, \quad A''' = \frac{9.25}{2.3.8^3}B, \ldots \\ B' = \frac{1}{8}A, & B'' = -\frac{9}{2.8^2}B, \quad B''' = -\frac{9.25}{2.3.8^3}A, \ldots \end{cases}$$

and hence the expansion of Poisson's integral is

$$\int_0^\pi \cos(k\cos\omega)\,d\omega \sim \left(\frac{\pi}{k}\right)^{\frac{1}{2}}\left[\left(1-\frac{1}{8k}-\frac{9}{2.8^2k^2}+\frac{9.25}{2.3.8^3k^3}+\ldots\right)\cos k \right.$$
$$\left. +\left(1+\frac{1}{8k}-\frac{9}{2.8^2k^2}-\frac{9.25}{2.3.8^3k^3}+\ldots\right)\sin k\right].$$

But, since the series on the right are not convergent, the researches of Lipschitz and subsequent writers are a necessary preliminary to the investigation of the significance of the latter portion of Poisson's investigation.

It should be mentioned that an explicit formula for the general term in the expansion was first given by W. R. Hamilton, *Trans. R. Irish Acad.* XIX. (1843), p. 313; his result was expressed thus:

$$\frac{1}{\pi}\int_0^\pi \cos(2\beta\sin\alpha)\,d\alpha = \frac{1}{\sqrt{(\pi\beta)}}\sum_{n=0}^{\infty}[0]^{-n}([-\tfrac{1}{2}]^n)^2(4\beta)^{-n}\cos(2\beta-\tfrac{1}{2}n\pi-\tfrac{1}{4}\pi),$$

and he described the expansion as semi-convergent; the expressions $[0]^{-n}$ and $[-\frac{1}{2}]^n$ are to be interpreted as $1/n!$ and $(-\frac{1}{2})(-\frac{3}{2})\ldots(-n+\frac{1}{2})$.

A result of some importance, which was obtained by Poisson in a subsequent memoir*, is that the general solution of the equation

$$\frac{d^2y}{dx^2}+\frac{y}{4x^2}=h^2y$$

is
$$y = Ax^{\frac{1}{2}}\int_0^\pi e^{-hx\cos\omega}\,d\omega + Bx^{\frac{1}{2}}\int_0^\pi e^{-hx\cos\omega}\log(x\sin^2\omega)\,d\omega,$$

where A and B are constants.

It follows at once that the general solution of the equation

$$\frac{d^2y}{dx^2}+\frac{1}{x}\frac{dy}{dx}-h^2y=0$$

is
$$y = A\int_0^\pi e^{-hx\cos\omega}\,d\omega + B\int_0^\pi e^{-hx\cos\omega}\log(x\sin^2\omega)\,d\omega.$$

This result was quoted by Stokes† as a known theorem in 1850, and it is likely that he derived his knowledge of it from the integral given in Poisson's memoir; but the fact that the integral is substantially due to Poisson has been sometimes overlooked‡.

* *Journal de l'École R. Polytechnique*, XII. (cahier 19), (1823), p. 476. The corresponding general integral of an associated partial differential equation was given in an earlier memoir, *ibid*, p. 227.

† *Camb. Phil. Trans.* IX. (1856), p. [38], [*Math. and Phys. Papers*, III. (1901), p. 42].

‡ See *Encyclopédie des Sci. Math.* II. 28 (§ 53), p. 213.

1·7. *The researches of Bessel.*

The memoir* in which Bessel examined in detail the functions which now bear his name was written in 1824, but in an earlier memoir† he had shewn that the expansion of the radius vector in planetary motion is

$$\frac{r}{a} = 1 + \tfrac{1}{2}\epsilon^2 + \sum_{n=1}^{\infty} B_n \cos nM,$$

where
$$B_n = -\frac{\epsilon}{n\pi}\int_0^{2\pi} \sin u \sin (nu - n\epsilon \sin u)\, du;$$

this expression for B_n should be compared with the series given in § 1·4.

In the memoir of 1824 Bessel investigated systematically the function I_k^h defined by the integral‡

$$I_k^h = \frac{1}{2\pi}\int_0^{2\pi} \cos (hu - k \sin u)\, du.$$

He took h to be an integer and obtained many of the results which will be given in detail in Chapter II. Bessel's integral is not adapted for defining the function which is most worth study when h is not an integer (see § 10·1); the function which is of most interest for non-integral values of h is not I_k^h but the function defined by Lommel which will be studied in Chapter III.

After the time of Bessel investigations on the functions became so numerous that it seems convenient at this stage to abandon the chronological account and to develop the theory in a systematic and logical order.

An historical account of researches from the time of Fourier to 1858 has been compiled by Wagner, *Bern Mittheilungen*, 1894, pp. 204—266; a briefer account of the early history was given by Maggi, *Atti della R. Accad. dei Lincei*, (*Transunti*), (3) IV. (1880), pp. 259—263.

* *Berliner Abh.* 1824 [published 1826], pp. 1—52. The date of this memoir, "Untersuchung des Theils der planetarischen Störungen, welcher aus der Bewegung der Sonne entsteht," is Jan. 29, 1824.

† *Berliner Abh.* 1816—17 [published 1819], pp. 49—55.

‡ This integral occurs in the expansion of the eccentric anomaly; with the notation of § 1·4,

$$nA_n = 2I^n_{n\epsilon},$$

a formula given by Poisson, *Connaissance des Tems*, 1825 [published 1822], p. 383.

CHAPTER II

THE BESSEL COEFFICIENTS

2·1. *The definition of the Bessel coefficients.*

The object of this chapter is the discussion of the fundamental properties of a set of functions known as *Bessel coefficients.* There are several ways of defining these functions; the method which will be adopted in this work is to define them as the coefficients in a certain expansion. This procedure is due to Schlömilch*, who derived many properties of the functions from his definition, and proved incidentally that the functions thus defined are equal to the definite integrals by which they had previously been defined by Bessel†. It should, however, be mentioned that the converse theorem that Bessel's integrals are equal to the coefficients in the expansion, was discovered by Hansen‡ fourteen years before the publication of Schlömilch's memoir. Some similar results had been published in 1836 by Jacobi (§ 2·22).

The generating function of the Bessel coefficients is

$$e^{\frac{1}{2}z\left(t-\frac{1}{t}\right)}.$$

It will be shewn that this function can be developed into a Laurent series, *qua* function of t; the coefficient of t^n in the expansion is called the *Bessel coefficient of argument z and order n,* and it is denoted by the symbol $J_n(z)$, so that

$$(1) \qquad e^{\frac{1}{2}z\left(t-\frac{1}{t}\right)} = \sum_{n=-\infty}^{\infty} t^n J_n(z).$$

To establish this development, observe that $e^{\frac{1}{2}zt}$ can be expanded into an absolutely convergent series of ascending powers of t; and for all values of t, with the exception of zero, $e^{-\frac{1}{2}z/t}$ can be expanded into an absolutely convergent series of descending powers of t. When these series are multiplied together, their product is an absolutely convergent series, and so it may be arranged according to powers of t; that is to say, we have an expansion of the form (1), which is valid for all values of z and t, $t = 0$ excepted.

* *Zeitschrift für Math. und Phys.* II. (1857), pp. 137—165. For a somewhat similar expansion, namely that of $e^{z\cos\theta}$, see Frullani, *Mem. Soc. Ital. (Modena)*, XVIII. (1820), p. 503. It must be pointed out that Schlömilch, following Hansen, denoted by $J_{z,n}$ what we now write as $J_n(2z)$; but the definition given in the text is now universally adopted. Traces of Hansen's notation are to be found elsewhere, e.g. Schläfli, *Math. Ann.* III. (1871), p. 148.

† *Berliner Abh.* 1824 [published 1826], p. 22.

‡ *Ermittelung der Absoluten Störungen in Ellipsen von beliebiger Excentricität und Neigung,* I. theil, [Schriften der Sternwarte Seeburg: Gotha, 1843], p. 106. See also the French translation, *Mémoire sur la détermination des perturbations absolues* (Paris, 1845), p. 100, and *Leipziger Abh.* II. (1855), pp. 250—251.

If in (1) we write $-1/t$ for t, we get

$$e^{\frac{1}{2}z(-1/t+t)} = \sum_{n=-\infty}^{\infty} (-t)^{-n} J_n(z)$$

$$= \sum_{n=-\infty}^{\infty} (-t)^{n} J_{-n}(z),$$

on replacing n by $-n$. Since the Laurent expansion of a function is unique*, a comparison of this formula with (1) shews that

$$(2) \qquad J_{-n}(z) = (-)^n J_n(z),$$

where n is any integer – a formula derived by Bessel from his definition of $J_n(z)$ as an integral.

From (2) it is evident that (1) may be written in the form

$$(3) \qquad e^{\frac{1}{2}z(t-1/t)} = J_0(z) + \sum_{n=1}^{\infty} \{t^n + (-)^n t^{-n}\} J_n(z).$$

A summary of elementary results concerning $J_n(z)$ has been given by Hall, *The Analyst*, I. (1874), pp. 81—84, and an account of elementary applications of these functions to problems of Mathematical Physics has been compiled by Harris, *American Journal of Math.* XXXIV. (1912), pp. 391—420.

The function of order unity has been encountered by Turrière, *Nouv. Ann. de Math.* (4) IX. (1909), pp. 433—441, in connexion with the steepest curves on the surface $z=y(5x^2-y^4)$.

2·11. *The ascending series for $J_n(z)$.*

An explicit expression for $J_n(z)$ in the form of an ascending series of powers of z is obtainable by considering the series for $\exp(\frac{1}{2}zt)$ and $\exp(-\frac{1}{2}z/t)$, thus

$$\exp\{\tfrac{1}{2}z(t-1/t)\} = \sum_{r=0}^{\infty} \frac{(\frac{1}{2}z)^r t^r}{r!} \sum_{m=0}^{\infty} \frac{(-\frac{1}{2}z)^m t^{-m}}{m!}.$$

When n is a positive integer or zero, the only term of the first series on the right which, when associated with the general term of the second series gives rise to a term involving t^n is the term for which $r = n + m$; and, since $n \geqslant 0$, there is always one term for which r has this value. On associating these terms for all the values of m, we see that the coefficient of t^n in the product is

$$\sum_{m=0}^{\infty} \frac{(\frac{1}{2}z)^{n+m}}{(n+m)!} \frac{(-\frac{1}{2}z)^m}{m!}.$$

We therefore have the result

$$(1) \qquad J_n(z) = \sum_{m=0}^{\infty} \frac{(-)^m (\frac{1}{2}z)^{n+2m}}{m!\,(n+m)!},$$

* For, if not, zero could be expanded into a Laurent series in t, in which some of the coefficients (say, in particular, that of t^m) were not zero. If we then multiplied the expansion by t^{-m-1} and integrated it round a circle with centre at the origin, we should obtain a contradiction. This result was noticed by Cauchy, *Comptes Rendus*, XIII. (1841), p. 911.

where n is a positive integer or zero. The first few terms of the series are given by the formula

$$(2)\qquad J_n(z) = \frac{z^n}{2^n \,.\, n!}\left\{1 - \frac{z^2}{2^2 \,.\, 1 \,.\, (n+1)} + \frac{z^4}{2^4 \,.\, 1 \,.\, 2 \,.\, (n+1)(n+2)} - \ldots\right\}.$$

In particular

$$(3)\qquad J_0(z) = 1 - \frac{z^2}{2^2} + \frac{z^4}{2^2 \,.\, 4^2} - \frac{z^6}{2^2 \,.\, 4^2 \,.\, 6^2} + \ldots.$$

To obtain the Bessel coefficients of negative order, we select the terms involving t^{-n} in the product of the series representing $\exp(\tfrac{1}{2}zt)$ and $\exp(-\tfrac{1}{2}z/t)$, where n is still a positive integer. The term of the second series which, when associated with the general term of the first series gives rise to a term in t^{-n} is the term for which $m = n + r$; and so we have

$$J_{-n}(z) = \sum_{r=0}^{\infty} \frac{(\tfrac{1}{2}z)^r}{r!} \frac{(-\tfrac{1}{2}z)^{n+r}}{(n+r)!},$$

whence we evidently obtain anew the formula § 2·1 (2), namely

$$J_{-n}(z) = (-)^n J_n(z).$$

It is to be observed that, in the series (1), the ratio of the $(m+1)$th term to the mth term is $-\tfrac{1}{4}z^2/\{m(n+m)\}$, and this tends to zero as $m \to \infty$, for *all* values of z and n. By D'Alembert's ratio test for convergence, it follows that the series representing $J_n(z)$ is convergent for all values of z and n, and so it is an *integral function* of z when $n = 0, \pm 1, \pm 2, \pm 3, \ldots$.

It will appear later (§ 4·73) that $J_n(z)$ is not an algebraic function of z and so it is a transcendental function; moreover, it is not an elementary transcendent, that is to say it is not expressible as a finite combination of exponential, logarithmic and algebraic functions operated on by signs of indefinite integration.

From (1) we can obtain two useful inequalities, which are of some importance (cf. Chapter XVI) in the discussion of series whose general term is a multiple of a Bessel coefficient.

Whether z be real or complex, we have

$$|J_n(z)| \leqslant |\tfrac{1}{2}z|^n \sum_{m=0}^{\infty} \frac{|\tfrac{1}{2}z|^{2m}}{m!\,(n+m)!}$$

$$\leqslant \frac{|\tfrac{1}{2}z|^n}{n!} \sum_{m=0}^{\infty} \frac{|\tfrac{1}{2}z|^{2m}}{m!\,(n+1)^m},$$

and so, when $n \geqslant 0$, we have

$$(4)\qquad |J_n(z)| \leqslant \frac{|\tfrac{1}{2}z|^n}{n!} \exp\left(\frac{\tfrac{1}{4}|z|^2}{n+1}\right) \leqslant \frac{|\tfrac{1}{2}z|^n}{n!} \exp(\tfrac{1}{4}|z|^2).$$

This result was given in substance by Cauchy, *Comptes Rendus*, XIII. (1841), pp. 687, 854; a similar but weaker inequality, namely

$$|J_n(z)| \leqslant \frac{|\tfrac{1}{2}z|^n}{n!} \exp(|z|^2),$$

was given by Neumann, *Theorie der Bessel'schen Functionen* (Leipzig, 1867), p. 27.

By considering all the terms of the series for $J_n(z)$ except the first, it is found that

(5) $$J_n(z) = \frac{(\frac{1}{2}z)^n}{n!}(1+\theta),$$

where $$|\theta| \leqslant \exp\left(\frac{\frac{1}{4}|z|^2}{n+1}\right) - 1 \leqslant \frac{\exp(\frac{1}{4}|z|^2) - 1}{n+1}.$$

It should be observed that the series on the right in § 2·1 (1) converges uniformly in any bounded domain of the variables z and t which does not contain the origin in the t-plane. For if δ, Δ and R are positive constants and if

$$\delta \leqslant |t| \leqslant \Delta, \quad |z| \leqslant |R|,$$

the terms in the expansion of $\exp(\frac{1}{2}zt)\exp(\frac{1}{2}z/t)$ do not exceed in absolute value the corresponding terms of the product $\exp(\frac{1}{2}R\Delta)\exp(\frac{1}{2}R/\delta)$, and the uniformity of the convergence follows from the test of Weierstrass. Similar considerations apply to the series obtained by term-by-term differentiations of the expansion $\Sigma t^n J_n(z)$, whether the differentiations be performed with respect to z or t or both z and t.

2·12. *The recurrence formulae.*

The equations*

(1) $$J_{n-1}(z) + J_{n+1}(z) = \frac{2n}{z} J_n(z),$$

(2) $$J_{n-1}(z) - J_{n+1}(z) = 2J_n'(z),$$

which connect three contiguous functions are useful in constructing Tables of Bessel coefficients; they are known as *recurrence formulae.*

To prove the former, differentiate the fundamental expansion of § 2·1, namely

$$e^{\frac{1}{2}z(t-1/t)} = \sum_{n=-\infty}^{\infty} t^n J_n(z),$$

with respect to t; we get

$$\tfrac{1}{2}z(1+1/t^2)\, e^{\frac{1}{2}z(t-1/t)} = \sum_{n=-\infty}^{\infty} nt^{n-1} J_n(z),$$

so that

$$\tfrac{1}{2}z(1+1/t^2) \sum_{n=-\infty}^{\infty} t^n J_n(z) \equiv \sum_{n=-\infty}^{\infty} nt^{n-1} J_n(z).$$

If the expression on the left is arranged in powers of t and coefficients of t^{n-1} are equated in the two Laurent series, which are identically equal, it is evident that

$$\tfrac{1}{2}z\{J_{n-1}(z) + J_{n+1}(z)\} = nJ_n(z),$$

which is the first of the formulae†.

* Throughout the work primes are used to denote the derivate of a function with respect to its argument.

† Differentiations are permissible because (§ 2·11) the resulting series are uniformly convergent. The equating of coefficients is permissible because Laurent expansions are unique.

Again, differentiate the fundamental expansion with respect to z; and then

$$\tfrac{1}{2}(t-1/t)\,e^{\frac{1}{2}z(t-1/t)} = \sum_{n=-\infty}^{\infty} t^n J_n'(z),$$

so that

$$\tfrac{1}{2}(t-1/t) \sum_{n=-\infty}^{\infty} t^n J_n(z) \equiv \sum_{n=-\infty}^{\infty} t^n J_n'(z).$$

By equating coefficients of t^n on either side of this identity we obtain formula (2) immediately.

The results of adding and subtracting (1) and (2) are

(3) $$zJ_n'(z) + nJ_n(z) = zJ_{n-1}(z),$$

(4) $$zJ_n'(z) - nJ_n(z) = -zJ_{n+1}(z).$$

These are equivalent to

(5) $$\frac{d}{dz}\{z^n J_n(z)\} = z^n J_{n-1}(z),$$

(6) $$\frac{d}{dz}\{z^{-n} J_n(z)\} = -z^{-n} J_{n+1}(z).$$

In the case $n=0$, (1) is trivial while the other formulae reduce to

(7) $$J_0'(z) = -J_1(z).$$

The formulae (1) and (4) from which the others may be derived were discovered by Bessel, *Berliner Abh.* 1824, [1826], pp. 31, 35. The method of proof given here is due to Schlömilch, *Zeitschrift für Math. und Phys.* II. (1857), p. 138. Schlömilch proved (1) in this manner, but he obtained (2) by direct differentiation of the series for $J_n(z)$.

A formula which Schlömilch derived (*ibid.* p. 143) from (2) is

(8) $$2^r \frac{d^r J_n(z)}{dz^r} = \sum_{m=0}^{r} (-)^m\, {}_rC_m \,.\, J_{n-r+2m}(z),$$

where ${}_rC_m$ is a binomial coefficient.

By obvious inductions from (5) and (6), we have

(9) $$\left(\frac{d}{z\,dz}\right)^m \{z^n J_n(z)\} = z^{n-m} J_{n-m}(z),$$

(10) $$\left(\frac{d}{z\,dz}\right)^m \{z^{-n} J_n(z)\} = (-)^m z^{-n-m} J_{n+m}(z),$$

where n is any integer and m is any positive integer. The formula (10) is due to Bessel (*ibid.* p. 34).

As an example of the results of this section observe that

$$\begin{aligned} zJ_1(z) &= 4J_2(z) - zJ_3(z) \\ &= 4J_2(z) - 8J_4(z) + zJ_5(z) \\ &= \ldots\ldots\ldots\ldots\ldots\ldots \\ &= 4\sum_{n=1}^{N} (-)^{n-1} nJ_{2n}(z) + (-)^N z J_{2N+1}(z) \\ &= 4\sum_{n=1}^{\infty} (-)^{n-1} nJ_{2n}(z), \end{aligned}$$

since $zJ_{2N+1}(z) \to 0$ as $N \to \infty$, by § 2·11 (4).

The expansion thus obtained,

(11) $$zJ_1(z) = 4 \sum_{n=1}^{\infty} (-)^{n-1} n J_{2n}(z),$$

is useful in the developments of Neumann's theory of Bessel functions (§ 3·57).

2·13. *The differential equation satisfied by $J_n(z)$.*

When the formulae § 2·12 (5) and (6) are written in the forms

$$\frac{d}{dz}\{z^n J_n(z)\} = z^n J_{n-1}(z), \quad \frac{d}{dz}\{z^{1-n} J_{n-1}(z)\} = -z^{1-n} J_n(z),$$

the result of eliminating $J_{n-1}(z)$ is seen to be

$$\frac{d}{dz}\left[z^{1-2n}\frac{d}{dz}\{z^n J_n(z)\}\right] = -z^{1-n} J_n(z),$$

that is to say

$$\frac{d}{dz}\left\{z^{1-n}\frac{dJ_n(z)}{dz} + nz^{-n}J_n(z)\right\} = -z^{1-n}J_n(z),$$

and so we have Bessel's differential equation*

(1) $$z^2\frac{d^2J_n(z)}{dz^2} + z\frac{dJ_n(z)}{dz} + (z^2 - n^2)J_n(z) = 0.$$

The analysis is simplified by using the operator ϑ defined as $z(d/dz)$.

Thus the recurrence formulae are

$$(\vartheta + n)J_n(z) = zJ_{n-1}(z), \quad (\vartheta - n + 1)J_{n-1}(z) = -zJ_n(z),$$

and so

$$(\vartheta - n + 1)\{z^{-1}(\vartheta + n)J_n(z)\} = -zJ_n(z),$$

that is

$$z^{-1}(\vartheta - n)(\vartheta + n)J_n(z) = -zJ_n(z),$$

and the equation

$$(\vartheta^2 - n^2)J_n(z) = -z^2J_n(z)$$

reduces at once to Bessel's equation.

Corollary. The same differential equation is obtained if $J_{n+1}(z)$ is eliminated from the formulae

$$(\vartheta + n + 1)J_{n+1}(z) = zJ_n(z), \quad (\vartheta - n)J_n(z) = -zJ_{n+1}(z).$$

2·2. *Bessel's integral for the Bessel coefficients.*

We shall now prove that

(1) $$J_n(z) = \frac{1}{2\pi}\int_0^{2\pi}\cos(n\theta - z\sin\theta)\,d\theta.$$

This equation was taken by Bessel† as the definition of $J_n(z)$, and he derived the other properties of the functions from this definition.

* *Berliner Abh.* 1824 [published 1826], p. 34; see also Frullani, *Mem. Soc. Ital.* (*Modena*), XVIII. (1820), p. 504.

† *Ibid.* pp. 22 and 35.

It is frequently convenient to modify (1) by bisecting the range of integration and writing $2\pi - \theta$ for θ in the latter part. This procedure gives

(2) $$J_n(z) = \frac{1}{\pi}\int_0^\pi \cos(n\theta - z\sin\theta)\,d\theta.$$

Since the integrand has period 2π, the first equation may be transformed into

(3) $$J_n(z) = \frac{1}{2\pi}\int_\alpha^{2\pi+\alpha} \cos(n\theta - z\sin\theta)\,d\theta,$$

where α is any angle.

To prove (1), multiply the fundamental expansion of § 2·1 (1) by t^{-n-1} and integrate* round a contour which encircles the origin once counterclockwise. We thus get

$$\frac{1}{2\pi i}\int^{(0+)} t^{-n-1}\, e^{\frac{1}{2}z(t-1/t)}\,dt = \sum_{m=-\infty}^{\infty} \frac{J_m(z)}{2\pi i}\int^{(0+)} t^{m-n-1}\,dt.$$

The integrals on the right all vanish except the one for which $m = n$; and so we obtain the formula

(4) $$J_n(z) = \frac{1}{2\pi i}\int^{(0+)} t^{-n-1}\, e^{\frac{1}{2}z(t-1/t)}\,dt.$$

Take the contour to be a circle of unit radius and write $t = e^{-i\theta}$, so that θ may be taken to decrease from $2\pi + \alpha$ to α. It is thus found that

(5) $$J_n(z) = \frac{1}{2\pi}\int_\alpha^{2\pi+\alpha} e^{i(n\theta - z\sin\theta)}\,d\theta,$$

a result given by Hansen† in the case $\alpha = 0$.

In this equation take $\alpha = -\pi$, bisect the range of integration and, in the former part, replace θ by $-\theta$. This procedure gives

$$J_n(z) = \frac{1}{2\pi}\int_0^\pi \{e^{i(n\theta - z\sin\theta)} + e^{-i(n\theta - z\sin\theta)}\}\,d\theta,$$

and equation (2), from which (1) may be deduced, is now obvious.

Various modifications of Bessel's integral are obtainable by writing

$$J_n(z) = \frac{1}{\pi}\int_0^\pi \cos n\theta\cos(z\sin\theta)\,d\theta + \frac{1}{\pi}\int_0^\pi \sin n\theta\sin(z\sin\theta)\,d\theta.$$

If θ be replaced by $\pi - \theta$ in these two integrals, the former changes sign when n is odd, the latter when n is even, the other being unaffected in each case; and therefore

(6) $$\left.\begin{aligned} J_n(z) &= \frac{1}{\pi}\int_0^\pi \sin n\theta\sin(z\sin\theta)\,d\theta \\ &= \frac{2}{\pi}\int_0^{\frac{1}{2}\pi} \sin n\theta\sin(z\sin\theta)\,d\theta \end{aligned}\right\} \quad (n \text{ odd}),$$

* Term-by-term integration is permitted because the expansion is uniformly convergent on the contour. It is convenient to use the symbol $\int^{(a+)}$ to denote integration round a contour encircling the point a once counterclockwise.

† *Ermittelung der absoluten Störungen* (Gotha, 1843), p. 105.

$$(7) \qquad \left.\begin{aligned} J_n(z) &= \frac{1}{\pi}\int_0^{\pi} \cos n\theta \cos(z\sin\theta)\,d\theta \\ &= \frac{2}{\pi}\int_0^{\frac{1}{2}\pi} \cos n\theta \cos(z\sin\theta)\,d\theta \end{aligned}\right\} \qquad (n \text{ even}).$$

If θ be replaced by $\frac{1}{2}\pi - \eta$ in the latter parts of (6) and (7), it is found that

$$(8) \qquad J_n(z) = \frac{2}{\pi}(-)^{\frac{1}{2}(n-1)}\int_0^{\frac{1}{2}\pi} \cos n\eta \sin(z\cos\eta)\,d\eta \qquad (n \text{ odd}),$$

$$(9) \qquad J_n(z) = \frac{2}{\pi}(-)^{\frac{1}{2}n}\int_0^{\frac{1}{2}\pi} \cos n\eta \cos(z\cos\eta)\,d\eta \qquad (n \text{ even}).$$

The last two results are due substantially to Jacobi*.

[NOTE. It was shewn by Parseval, *Mém. des savans étrangers*, I. (1805), pp. 639—648, that

$$1 - \frac{a^2}{2^2} + \frac{a^4}{2^2 . 4^2} - \frac{a^6}{2^2 . 4^2 . 6^2} + \ldots = \frac{1}{\pi}\int_0^{\pi} \cos(a\sin x)\,dx,$$

and so, in the special case in which $n=0$, (2) will be described as *Parseval's integral*. It will be seen in § 2·3 that two integral representations of $J_n(z)$, namely Bessel's integral and Poisson's integral become identical when $n=0$, so a special name for this case is justified.]

The reader will find it interesting to obtain (after Bessel) the formulae § 2·12 (1) and § 2·12 (4) from Bessel's integral.

2·21. *Modifications of Parseval's integral.*

Two formulae involving definite integrals which are closely connected with Parseval's integral formula are worth notice. The first, namely

$$(1) \qquad J_0\{\sqrt{(z^2-y^2)}\} = \frac{1}{\pi}\int_0^{\pi} e^{y\cos\theta}\cos(z\sin\theta)\,d\theta,$$

is due to Bessel†. The simplest method of proving it is to write the expression on the right in the form

$$\frac{1}{2\pi}\int_{-\pi}^{\pi} e^{y\cos\theta + iz\sin\theta}\,d\theta,$$

expand in powers of $y\cos\theta + iz\sin\theta$ and use the formulae

$$\int_{-\pi}^{\pi} (y\cos\theta + iz\sin\theta)^{2n+1}\,d\theta = 0, \quad \int_{-\pi}^{\pi} (y\cos\theta + iz\sin\theta)^{2n}\,d\theta = \frac{2\Gamma(n+\frac{1}{2})\,\Gamma(\frac{1}{2})}{\Gamma(n+1)}(y^2-z^2)^n;$$

the formula then follows without difficulty.

The other definite integral, due to Catalan‡, namely

$$(2) \qquad J_0(2i\sqrt{z}) = \frac{1}{\pi}\int_0^{\pi} e^{(1+z)\cos\theta}\cos\{(1-z)\sin\theta\}\,d\theta,$$

is a special case of (1) obtained by substituting $1-z$ and $1+z$ for z and y respectively.

* *Journal für Math.* XV. (1836), pp. 12—13. [*Ges. Math. Werke*, VI. (1891), pp. 100—102]; the integrals actually given by Jacobi had limits 0 and π with factors $1/\pi$ replacing the factors $2/\pi$. See also Anger, *Neueste Schriften der Naturf. Ges. in Danzig*, V. (1855), p. 1, and Cauchy, *Comptes Rendus*, XXXVIII. (1854), pp. 910—913.

† *Berliner Abh.*, 1824 [published 1826], p. 37. See also Anger, *Neueste Schriften der Naturf. Ges. in Danzig*, V. (1855), p. 10, and Lommel, *Zeitschrift für Math. und Phys.* XV. (1870), p. 151.

‡ *Bulletin de l'Acad. R. de Belgique*, (2) XLI. (1876), p. 938.

Catalan's integral may be established independently by using the formula

$$\frac{1}{m!} = \frac{1}{2\pi i}\int^{(0+)} t^{-m-1} e^t\, dt,$$

so that

$$J_0(2i\sqrt{z}) = \sum_{m=0}^{\infty} \frac{z^m}{(m!)^2} = \frac{1}{2\pi i}\sum_{m=0}^{\infty} \frac{z^m}{m!}\int^{(0+)} t^{-m-1} e^t\, dt$$

$$= \frac{1}{2\pi i}\int^{(0+)} \exp\left(t + \frac{z}{t}\right)\frac{dt}{t} = \frac{1}{2\pi}\int_{-\pi}^{\pi} \exp\{e^{i\theta} + ze^{-i\theta}\}\, d\theta,$$

by taking the contour to be a unit circle; the result then follows by bisecting the range of integration.

2·22. *Jacobi's expansions in series of Bessel coefficients.*

Two series, which are closely connected with Bessel's integral, were discovered by Jacobi*. The simplest method of obtaining them is to write $t = \pm e^{i\theta}$ in the fundamental expansion § 2·1 (3). We thus get

$$e^{\pm iz\sin\theta} = J_0(z) + \sum_{n=1}^{\infty} (\pm 1)^n \{e^{ni\theta} + (-)^n e^{-ni\theta}\} J_n(z)$$

$$= J_0(z) + 2\sum_{n=1}^{\infty} J_{2n}(z)\cos 2n\theta \pm 2i\sum_{n=0}^{\infty} J_{2n+1}(z)\sin(2n+1)\theta.$$

On adding and subtracting the two results which are combined in this formula, we find

$$(1)\qquad \cos(z\sin\theta) = J_0(z) + 2\sum_{n=1}^{\infty} J_{2n}(z)\cos 2n\theta,$$

$$(2)\qquad \sin(z\sin\theta) = 2\sum_{n=0}^{\infty} J_{2n+1}(z)\sin(2n+1)\theta.$$

Write $\frac{1}{2}\pi - \eta$ for θ, and we get

$$(3)\qquad \cos(z\cos\eta) = J_0(z) + 2\sum_{n=1}^{\infty} (-)^n J_{2n}(z)\cos 2n\eta,$$

$$(4)\qquad \sin(z\cos\eta) = 2\sum_{n=0}^{\infty} (-)^n J_{2n+1}(z)\cos(2n+1)\eta.$$

The results (3) and (4) were given by Jacobi, while the others were obtained later by Anger†. Jacobi's procedure was to expand $\cos(z\cos\eta)$ and $\sin(z\cos\eta)$ into a series of cosines of multiples of η, and use Fourier's rule to obtain the coefficients in the form of integrals which are seen to be associated with Bessel's integrals.

In view of the fact that the first terms in (1) and (3) are not formed according to the same law as the other terms, it is convenient to introduce *Neumann's factor*‡ ϵ_n, which is defined to be equal to 2 when n is not zero, and to be equal to 1 when n is zero. The employment of this factor, which

* *Journal für Math.* xv. (1836), p. 12. [*Ges. Math. Werke*, vi. (1891), p. 101.]

† *Neueste Schriften der Naturf. Ges. in Danzig*, v. (1855), p. 2.

‡ Neumann, *Theorie der Bessel'schen Functionen* (Leipzig, 1867), p. 7.

will be of frequent occurrence in the sequel, enables us to write (1) and (2) in the compact forms:

$$\cos(z\sin\theta) = \sum_{n=0}^{\infty} \epsilon_{2n} J_{2n}(z)\cos 2n\theta, \tag{5}$$

$$\sin(z\sin\theta) = \sum_{n=0}^{\infty} \epsilon_{2n+1} J_{2n+1}(z)\sin(2n+1)\theta. \tag{6}$$

If we put $\theta = 0$ in (5), we find

$$1 = \sum_{n=0}^{\infty} \epsilon_{2n} J_{2n}(z). \tag{7}$$

If we differentiate (5) and (6) any number of times before putting $\theta = 0$, we obtain expressions for various polynomials as series of Bessel coefficients. We shall, however, use a slightly different method subsequently (§ 2·7) to prove that z^m is expansible into a series of Bessel coefficients when m is any positive integer. It is then obvious that any polynomial is thus expansible. This is a special case of an expansion theorem, due to Neumann, which will be investigated in Chapter XVI.

For the present, we will merely notice that, if (6) be differentiated once before θ is put equal to 0, there results

$$z = \sum_{n=0}^{\infty} \epsilon_{2n+1}(2n+1) J_{2n+1}(z), \tag{8}$$

while, if θ be put equal to $\frac{1}{2}\pi$ after two differentiations of (5) and (6), then

$$z\sin z = 2\{2^2 J_2(z) - 4^2 J_4(z) + 6^2 J_6(z) - \dots\}, \tag{9}$$

$$z\cos z = 2\{1^2 J_1(z) - 3^2 J_3(z) + 5^2 J_5(z) - \dots\}. \tag{10}$$

These results are due to Lommel*.

NOTE. The expression $\exp\{\frac{1}{2}z(t-1/t)\}$ introduced in § 2·1 is not a generating function in the strict sense. The generating function† associated with $\epsilon_n J_n(z)$ is $\sum_{n=0}^{\infty} \epsilon_n t^n J_n(z)$.

If this expression be called S, by using the recurrence formula § 2·12 (2), we have

$$\frac{dS}{dz} = \frac{1}{2}\left(t - \frac{1}{t}\right) S + \frac{1}{2}\left(t + \frac{1}{t}\right) J_0(z).$$

If we solve this differential equation we get

$$S = e^{\frac{1}{2}z(t-1/t)} + \frac{1}{2}\left(t + \frac{1}{t}\right) e^{\frac{1}{2}z(t-1/t)} \int_0^z e^{-\frac{1}{2}z(t-1/t)} J_0(z)\,dz. \tag{11}$$

A result equivalent to this was given by Brenke, *Bull. American Math. Soc.* XVI. (1910), pp. 225—230.

* *Studien über die Bessel'schen Functionen* (Leipzig, 1868), p. 41.

† It will be seen in Chapter XVI. that this is a form of "Lommel's function of two variables."

2·3. *Poisson's integral for the Bessel coefficients.*

Shortly before the appearance of Bessel's memoir on planetary perturbations, Poisson had published an important work on the Conduction of Heat*, in the course of which he investigated integrals of the types†

$$\int_0^\pi \cos(z\cos\theta)\sin^{2n+1}\theta\,d\theta,\quad \int_0^\pi \cos(z\cos\theta)\sin^{2n}\theta\,d\theta,$$

where n is a positive integer or zero. He proved that these integrals are solutions of certain differential equations‡ and gave the investigation, which has already been reproduced in § 1·6, to determine an approximation to the latter integral when z is large and positive, in the special case $n = 0$.

We shall now prove that

$$(1)\qquad J_n(z) = \frac{z^n}{1.3.5\ldots(2n-1)\pi}\int_0^\pi \cos(z\cos\theta)\sin^{2n}\theta\,d\theta$$

$$= \frac{(\tfrac{1}{2}z)^n}{\Gamma(n+\tfrac{1}{2})\,\Gamma(\tfrac{1}{2})}\int_0^\pi \cos(z\cos\theta)\sin^{2n}\theta\,d\theta;$$

and, in view of the importance of Poisson's researches, it seems appropriate to describe the expressions on the right§ as *Poisson's integrals* for $J_n(z)$. In the case $n = 0$, Poisson's integral reduces to Parseval's integral (§ 2·2).

It is easy to prove that the expressions under consideration are equal to $J_n(z)$; for, if we expand the integrand in powers of z and then integrate term-by-term‖, we have

$$\frac{1}{\pi}\int_0^\pi \cos(z\cos\theta)\sin^{2n}\theta\,d\theta = \frac{1}{\pi}\sum_{m=0}^\infty \frac{(-)^m z^{2m}}{(2m)!}\int_0^\pi \cos^{2m}\theta\sin^{2n}\theta\,d\theta$$

$$= \sum_{m=0}^\infty \frac{(-)^m z^{2m}}{(2m)!}\cdot\frac{1.3.5\ldots(2n-1).1.3.5\ldots(2m-1)}{2.4.6\ldots(2n+2m)}$$

$$= 1.3.5\ldots(2n-1)\sum_{m=0}^\infty \frac{(-)^m z^{2m}}{2^{n+2m}\,m!\,(n+m)!},$$

and the result is obvious.

* *Journal de l'École R. Polytechnique*, XII. (cahier 19), (1823), pp. 249—403.

† *Ibid.* p. 293, *et seq.*; p. 340, *et seq.* Integrals equivalent to them had previously been examined by Euler, *Inst. Calc. Int.* II. (Petersburg, 1769), Ch. X. § 1036, but Poisson's forms are more elegant, and his study of them is more systematic. See also § 3·3.

‡ E.g. on p. 300, he proved that, if

$$R = r^{n+1}\int_0^\pi \cos(r\rho\cos\omega)\sin^{2n+1}\omega\,d\omega,$$

then R satisfies the differential equation

$$\frac{d^2R}{dr^2} - \frac{n(n+1)}{r^2}R = -\rho^2 R.$$

§ Nielsen, *Handbuch der Theorie der Cylinderfunktionen* (Leipzig, 1904), p. 51, calls them *Bessel's second integral*, but the above nomenclature seems preferable.

‖ The series to be integrated is obviously uniformly convergent; the procedure adopted is due to Poisson, *ibid.* pp. 314, 340.

Poisson also observed* that

$$\int_0^\pi e^{iz\cos\theta}\sin^{2n}\theta\,d\theta = \int_0^\pi \cos(z\cos\theta)\sin^{2n}\theta\,d\theta;$$

this is evident when we consider the arithmetic mean of the integral on the left and the integral derived from it by replacing θ by $\pi-\theta$.

We thus get

$$J_n(z) = \frac{(\frac{1}{2}z)^n}{\Gamma(n+\frac{1}{2})\Gamma(\frac{1}{2})}\int_0^\pi e^{iz\cos\theta}\sin^{2n}\theta\,d\theta. \tag{2}$$

A slight modification of this formula, namely

$$J_n(z) = \frac{(\frac{1}{2}z)^n}{\Gamma(n+\frac{1}{2})\Gamma(\frac{1}{2})}\int_{-1}^1 e^{izt}(1-t^2)^{n-\frac{1}{2}}\,dt, \tag{3}$$

has suggested important developments (cf. § 6·1) in the theory of Bessel functions.

It should also be noticed that

$$\begin{aligned}\int_0^\pi \cos(z\cos\theta)\sin^{2n}\theta\,d\theta &= 2\int_0^{\frac{1}{2}\pi}\cos(z\cos\theta)\sin^{2n}\theta\,d\theta \\ &= 2\int_0^{\frac{1}{2}\pi}\cos(z\sin\theta)\cos^{2n}\theta\,d\theta,\end{aligned} \tag{4}$$

and each of these expressions gives rise to a modified form of Poisson's integral.

An interesting application of Bessel's and Poisson's integrals was obtained by Lommel† who multiplied the formula

$$\cos 2n\theta = \sum_{m=0}^{n}(-)^m \frac{4n^2\{4n^2-2^2\}\ldots\{4n^2-(2m-2)^2\}}{(2m)!}\sin^{2m}\theta$$

by $\cos(z\cos\theta)$ and integrated. It thus follows that

$$J_{2n}(z) = (-)^n\sum_{m=0}^{n}(-)^m\frac{4n^2\{4n^2-2^2\}\ldots\{4n^2-(2m-2)^2\}}{2^m\,.\,m!}\frac{J_m(z)}{z^m}. \tag{5}$$

2·31. *Bessel's investigation of Poisson's integral.*

The proof, that $J_n(z)$ is equal to Poisson's integral, which was given by Bessel‡, is somewhat elaborate; it is substantially as follows:

It is seen on differentiation that

$$\begin{aligned}&\frac{d}{d\theta}\left[\cos\theta\sin^{2n-1}\theta\cos(z\cos\theta) - \frac{z}{2n+1}\sin^{2n+1}\theta\sin(z\cos\theta)\right] \\ &\quad= \left[(2n-1)\sin^{2n-2}\theta - 2n\sin^{2n}\theta + \frac{z^2}{2n+1}\sin^{2n+2}\theta\right]\cos(z\cos\theta),\end{aligned}$$

* Poisson actually made the statement (p. 293) concerning the integral which contains $\sin^{2n+1}\theta$; but, as he points out on p. 340, odd powers may be replaced by even powers throughout his analysis.

† *Studien über die Bessel'schen Functionen* (Leipzig, 1868), p. 30.

‡ *Berliner Abh.* 1824 [published 1826], pp. 36—37. Jacobi, *Journal für Math.* xv. (1836), p. 13, [*Ges. Math. Werke*, vi. (1891), p. 102], when giving his proof (§ 2·32) of Poisson's integral formula, objected to the artificial character of Bessel's demonstration.

and hence, on integration, when $n \geqslant 1$,

$$(2n-1)\int_0^\pi \cos(z\cos\theta)\sin^{2n-2}\theta\,d\theta - 2n\int_0^\pi \cos(z\cos\theta)\sin^{2n}\theta\,d\theta + \frac{z^2}{2n+1}\int_0^\pi \cos(z\cos\theta)\sin^{2n+2}\theta\,d\theta = 0.$$

If now we write

$$\frac{(\frac{1}{2}z)^n}{\Gamma(n+\frac{1}{2})\,\Gamma(\frac{1}{2})}\int_0^\pi \cos(z\cos\theta)\sin^{2n}\theta\,d\theta \equiv \phi(n),$$

the last formula shews that

$$z\phi(n-1) - 2n\phi(n) + z\phi(n+1) = 0,$$

so that $\phi(n)$ and $J_n(z)$ satisfy the same recurrence formula.

But, by using Bessel's integral, it is evident that

$$\phi(0) = J_0(z),$$

$$\phi(1) = \frac{z}{\pi}\int_0^\pi \cos(z\cos\theta)\sin^2\theta\,d\theta = -\frac{1}{\pi}\int_0^\pi \frac{d}{d\theta}\left\{\sin(z\cos\theta)\right\}\sin\theta\,d\theta$$

$$= \frac{1}{\pi}\int_0^\pi \sin(z\cos\theta)\cos\theta\,d\theta = -J_0'(z) = J_1(z),$$

and so, by induction from the recurrence formula, we have

$$\phi(n) = J_n(z),$$

when $n = 0, 1, 2, 3, \ldots$.

2·32. *Jacobi's investigation of Poisson's integral.*

The problem of the direct transformation of Poisson's integral into Bessel's integral was successfully attacked by Jacobi*; this method necessitates the use of Jacobi's transformation formula

$$\frac{d^{n-1}\sin^{2n-1}\theta}{d\mu^{n-1}} = (-)^{n-1}\frac{1.3.5\ldots(2n-1)}{n}\sin n\theta,$$

where $\mu \equiv \cos\theta$. We shall assume this formula for the moment, and, since no simple direct proof of it seems to have been previously published, we shall give an account of various proofs in §§ 2·321—2·323.

If we observe that the first $n-1$ derivates of $(1-\mu^2)^{n-\frac{1}{2}}$, with respect to μ, vanish when $\mu = \pm 1$, it is evident that, by n partial integrations, we have

$$z^n\int_0^\pi \cos(z\cos\theta)\sin^{2n}\theta\,d\theta = z^n\int_{-1}^1 \cos(z\mu).(1-\mu^2)^{n-\frac{1}{2}}\,d\mu$$

$$= (-)^n\int_{-1}^1 \cos(z\mu - \tfrac{1}{2}n\pi)\frac{d^n(1-\mu^2)^{n-\frac{1}{2}}}{d\mu^n}\,d\mu.$$

* *Journal für Math.* xv. (1836), pp. 12—13. [*Ges. Math. Werke*, vi. (1891), pp. 101—102.] See also *Journal de Math.* i. (1836), pp. 195—196.

If we now use Jacobi's formula, this becomes

$$-\frac{1.3.5\ldots(2n-1)}{n}\int_{-1}^{1}\cos(z\mu-\tfrac{1}{2}n\pi)\,\frac{d\sin n\theta}{d\mu}\,d\mu$$
$$=1.3.5\ldots(2n-1)\int_0^\pi \cos(z\cos\theta-\tfrac{1}{2}n\pi)\cos n\theta\,d\theta$$
$$=1.3.5\ldots(2n-1)\,\pi J_n(z),$$

by Jacobi's modification of § 2·2 (8) and (9), since $\cos(z\cos\theta-\frac{1}{2}n\pi)$ is equal to $(-)^{\frac{1}{2}n}\cos(z\cos\theta)$ or $(-)^{\frac{1}{2}(n-1)}\sin(z\cos\theta)$ according as n is even or odd; and this establishes the transformation.

2·321. *Proofs of Jacobi's transformation.*

Jacobi's proof of the transformation formula used in § 2·32 consisted in deriving it as a special case of a formula due to Lacroix*; but the proof which Lacroix gave of his formula is open to objection in that it involves the use of infinite series to obtain a result of an elementary character. A proof, based on the theory of linear differential equations, was discovered by Liouville, *Journal de Math.* VI. (1841), pp. 69—73; this proof will be given in § 2·322. Two years after Liouville, an interesting symbolic proof was published by Boole, *Camb. Math. Journal*, III. (1843), pp. 216—224. An elementary proof by induction was given by Grunert, *Archiv der Math. und Phys.* IV. (1844), pp. 104—109. This proof consists in shewing that, if

$$\Theta_n=\frac{d^{n-1}(1-\mu^2)^{n-\frac{1}{2}}}{d\mu^{n-1}},$$

then

$$\Theta_{n+1}=(1-\mu^2)\frac{d\Theta_n}{d\mu}-2n\mu\Theta_n-n(n-1)\int_1^\mu\Theta_n\,d\mu,$$

and that $(-)^{n-1}\,1.3.5\ldots(2n-1)(\sin n\theta)/n$ satisfies the same recurrence formula.

Other proofs of this character have been given by Todhunter, *Differential Calculus* (London 1871), Ch. XXVIII., and Crawford†, *Proc. Edinburgh Math. Soc.* XX. (1902), pp. 11—15, but all these proofs involve complicated algebra.

A proof depending on the use of contour integration is due to Schläfli, *Ann. di Mat.* (2) V. (1873), pp. 201—202. The contour integrals are of the type used in establishing Lagrange's expansion; and in § 2·323 we shall give the modification of Schläfli's proof, in which the use of contour integrals is replaced by a use of Lagrange's expansion.

To prove Jacobi's formula, differentiate by Leibniz' theorem, thus:

$$\frac{(-)^{n-1}n}{1.3.5\ldots(2n-1)}\frac{d^{n-1}}{d\mu^{n-1}}\{(1-\mu)^{n-\frac{1}{2}}(1+\mu)^{n-\frac{1}{2}}\}$$
$$=\frac{n}{2^{n-1}}\sum_{m=0}^{n-1}(-)^m\,{}_{n-1}C_m\,\frac{(n-\frac{1}{2})(n-\frac{3}{2})\ldots(n-m+\frac{1}{2})}{\frac{3}{2}\cdot\frac{5}{2}\ldots(m+\frac{1}{2})}(1-\mu)^{m+\frac{1}{2}}(1+\mu)^{n-m-\frac{1}{2}}$$
$$=\sum_{m=0}^{n-1}(-)^m\,{}_{2n}C_{2m+1}(\sin\tfrac{1}{2}\theta)^{2m+1}(\cos\tfrac{1}{2}\theta)^{2n-2m-1}$$
$$=\sin(2n\times\tfrac{1}{2}\theta),$$

and this is the transformation required‡.

* *Traité du Calc. Diff.* I. (Paris, 1810, 2nd edition), pp. 182—183. See also a note written by Catalan in 1868, *Mém. de la Soc. R. des Sci. de Liège*, (2) XII. (1885), pp. 312—316.

† Crawford attributes the formula to Rodrigues, possibly in consequence of an incorrect statement by Frenet, *Recueil d'Exercices* (Paris, 1866), p. 93, that it is given in Rodrigues' dissertation, *Corresp. sur l'École R. Polytechnique*, III. (1814—1816), pp. 361—385.

‡ I owe this proof to Mr C. T. Preece.

2·322. *Liouville's proof of Jacobi's transformation.*

The proof given by Liouville of Jacobi's formula is as follows:

Let $y=(1-\mu^2)^{n-\frac{1}{2}}$ and let D be written for $d/d\mu$; then obviously

$$(1-\mu^2)\,Dy+(2n-1)\,\mu y=0.$$

Differentiate this equation n times; and then

$$(1-\mu^2)\,D^{n+1}y-\mu\,D^n y+n^2 D^{n-1}y=0:$$

but

$$(1-\mu^2)\,D^2-\mu D=\sin\theta\,\frac{d}{d\theta}\left(\frac{1}{\sin\theta}\,\frac{d}{d\theta}\right)+\cot\theta\,\frac{d}{d\theta}=\frac{d^2}{d\theta^2},$$

so that

$$\left(\frac{d^2}{d\theta^2}+n^2\right)D^{n-1}y=0.$$

Hence

$$D^{n-1}y=A\sin n\theta+B\cos n\theta,$$

where A and B are constants; since $D^{n-1}y$ is obviously an odd function of θ, B is zero. To determine A compare the coefficients of θ in the expansions of $D^{n-1}y$ and $A\sin n\theta$ in ascending powers of θ. The term involving θ in $D^{n-1}y$ is easily seen to be

$$\left(\frac{d}{-\theta\,d\theta}\right)^{n-1}\theta^{2n-1}=(-)^{n-1}(2n-1)(2n-3)\ldots3.1.\theta,$$

so that

$$nA=(-)^{n-1}\,1.3.5\ldots(2n-1),$$

and thence we have the result, namely

$$\frac{d^{n-1}\sin^{2n-1}\theta}{d\mu^{n-1}}=(-)^{n-1}\,\frac{1.3.5\ldots(2n-1)}{n}\sin n\theta.$$

2·323. *Schläfli's proof of Jacobi's transformation.*

We first recall Lagrange's expansion, which is that, if $z=\mu+hf(z)$, then

$$\phi(z)=\phi(\mu)+\sum_{n=1}^{\infty}\frac{h^n}{n!}\,\frac{d^{n-1}}{d\mu^{n-1}}\,[\{f(\mu)\}^n\,\phi'(\mu)],$$

so that

$$\phi'(z)\,\frac{\partial z}{\partial\mu}=\sum_{n=0}^{\infty}\frac{h^n}{n!}\,\frac{d^n}{d\mu^n}\,[\{f(\mu)\}^n\,\phi'(\mu)],$$

subject to the usual conditions of convergence*.

Now take

$$f(z)\equiv-\tfrac{1}{2}(1-z^2),\quad \phi'(z)\equiv\sqrt{(1-z^2)},$$

it being supposed that $\phi'(z)$ reduces to $\sqrt{(1-\mu^2)}$, i.e. to $\sin\theta$ when $h\to0$.

The singularities of z *qua* function of h are at $h=e^{\pm i\theta}$; and so, when θ is real, the expansion of $\sqrt{(1-z^2)}$ in powers of h is convergent when both $|h|$ and $|z|$ are less than unity.

Now

$$z=\{1-\sqrt{(1-2\mu h+h^2)}\}/h,$$

and so

$$\frac{\partial z}{\partial\mu}=\frac{1}{\sqrt{(1-2\mu h+h^2)}},\quad \sqrt{(1-z^2)}=\frac{(1-he^{-i\theta})^{\frac{1}{2}}-(1-he^{i\theta})^{\frac{1}{2}}}{hi}.$$

Hence it follows that $\dfrac{(-)^{n-1}}{2^{n-1}.(n-1)!}\,\dfrac{d^{n-1}\sin^{2n-1}\theta}{d\mu^{n-1}}$ is the coefficient of h^{n-1} in the expansion of $\sqrt{(1-z^2)}.(\partial z/\partial\mu)$ in powers of h. But it is evident that

$$\sqrt{(1-z^2)}.\frac{\partial z}{\partial\mu}=\frac{(1-he^{i\theta})^{-\frac{1}{2}}-(1-he^{-i\theta})^{-\frac{1}{2}}}{hi}=\sum_{n=1}^{\infty}\frac{1.3.5\ldots(2n-1)}{2.4.6\ldots(2n)}.\frac{e^{ni\theta}-e^{-ni\theta}}{i}\,h^{n-1},$$

and a consideration of the coefficient of h^{n-1} in the last expression establishes the truth of Jacobi's formula.

* Cf. *Modern Analysis*, § 7·32.

2·33. *An application of Jacobi's transformation.*

The formal expansion

$$\int_0^\pi f(\cos x)\cos nx\,dx = \int_0^\pi \sum_{m=0}^{\infty} (-)^m \alpha_m f^{(n+2m)}(\cos x)\,dx,$$

in which α_m is the coefficient of t^{n+2m} in the expansion of $J_n(t)/J_0(t)$ in ascending powers of t, has been studied by Jacobi*. To establish it, integrate the expression on the left n times by parts; it transforms (§ 2·32) into

$$\frac{1}{1.3.5\ldots(2n-1)}\int_0^\pi f^{(n)}(\cos x)\sin^{2n} x\,dx,$$

and, when $\sin^{2n} x$ is replaced by a series of cosines of multiples of x, this becomes

$$\frac{1}{2.4.6\ldots(2n)}\int_0^\pi f^{(n)}(\cos x)\left[1 - \frac{2n}{n+1}\cos 2x + \frac{2n(n-1)}{(n+1)(n+2)}\cos 4x - \ldots\right]dx.$$

We now integrate $f^{(n)}(\cos x)\cos 2x$, $f^{(n)}(\cos x)\cos 4x$, ... by parts, and by continual repetitions of this process, we evidently arrive at a formal expansion of the type stated. When $f(\cos x)$ is a polynomial in $\cos x$, the process obviously terminates and the transformation is certainly valid.

To determine the values of the coefficients α_m in the expansion

$$\int_0^\pi f(\cos x)\cos nx\,dx = \int_0^\pi \sum_{m=0}^{\infty} (-)^m \alpha_m f^{(n+2m)}(\cos x)\,dx$$

thus obtained, write

$$f(\cos x) \equiv (-)^{\frac{1}{2}n}\cos(t\cos x),\quad (-)^{\frac{1}{2}(n-1)}\sin(t\cos x),$$

according as n is even or odd, and we deduce from § 2·2 (8) and (9) that

$$J_n(t) = \sum_{m=0}^{\infty} (-)^m \alpha_m t^{n+2m}\{(-)^m J_0(t)\},$$

so that α_m has the value stated.

It has been stated that the expansion is valid when $f(\cos x)$ is a polynomial in $\cos x$; it can, however, be established when $f(\cos x)$ is merely restricted to be an integral function of $\cos x$, say

$$\sum_{n=0}^{\infty} \frac{b_n \cos^n x}{n!},$$

provided that $\overline{\lim_{n\to\infty}} \sqrt[n]{|b_n|}$ is less than the smallest positive root of the equation $J_0(t) = 0$; the investigation of this will not be given since it seems to be of no practical importance.

* *Journal für Math.* xv. (1836), pp. 25—26 [*Ges. Math. Werke*, vi. (1891), pp. 117—118]. See also Jacobi, *Astr. Nach.* xxviii. (1849), col. 94 [*Ges. Math. Werke*, vii. (1891), p. 174].

2·4. *The addition formula for the Bessel coefficients.*

The Bessel coefficients possess an *addition formula* by which $J_n(y+z)$ may be expressed in terms of Bessel coefficients of y and z. This formula, which was first given by Neumann* and Lommel†, is

$$J_n(y+z) = \sum_{m=-\infty}^{\infty} J_m(y)\, J_{n-m}(z). \tag{1}$$

The simplest way of proving this result is from the formula § 2·2 (4), which gives

$$\begin{aligned} J_n(y+z) &= \frac{1}{2\pi i}\int^{(0+)} t^{-n-1} e^{\frac{1}{2}(y+z)(t-1/t)}\, dt \\ &= \frac{1}{2\pi i}\int^{(0+)} \sum_{m=-\infty}^{\infty} t^{m-n-1} J_m(y)\, e^{\frac{1}{2}z(t-1/t)}\, dt \\ &= \frac{1}{2\pi i}\sum_{m=-\infty}^{\infty} J_m(y)\int^{(0+)} t^{m-n-1} e^{\frac{1}{2}z(t-1/t)}\, dt \\ &= \sum_{m=-\infty}^{\infty} J_m(y)\, J_{n-m}(z), \end{aligned}$$

on changing the order of summation and integration in the third line of the analysis; and this is the result to be established.

Numerous generalisations of this expansion will be given in Chapter XI.

2·5. *Hansen's series of squares and products of Bessel coefficients.*

Special cases of Neumann's addition formula were given by Hansen‡ as early as 1843. The first system of formulae is obtainable by squaring the fundamental expansion § 2·1 (1), so that

$$e^{z(t-1/t)} = \left\{\sum_{r=-\infty}^{\infty} t^r J_r(z)\right\}\left\{\sum_{m=-\infty}^{\infty} t^m J_m(z)\right\}.$$

By expressing the product on the right as a Laurent series in t, and equating the coefficient of t^n in the result to the coefficient of t^n in the Laurent expansion of the expression on the left, we find that

$$J_n(2z) = \sum_{r=-\infty}^{\infty} J_r(z)\, J_{n-r}(z).$$

In particular, taking $n = 0$, we have§

$$J_0(2z) = J_0^2(z) + 2\sum_{r=1}^{\infty} (-)^r J_r^2(z) = \sum_{r=0}^{\infty} (-)^r \epsilon_r J_r^2(z). \tag{1}$$

* *Theorie der Bessel'schen Functionen* (Leipzig, 1867), p. 40.

† *Studien über die Bessel'schen Functionen* (Leipzig, 1868), pp. 26—27; see also Schläfli, *Math. Ann.* III. (1871), pp. 135—137.

‡ *Ermittelung der absoluten Störungen* (Gotha, 1843), p. 107 *et seq.* Hansen did not give (4), and he gave only the special case of (2) in which $n=1$. The more general formulae are due to Lommel, *Studien über die Bessel'schen Functionen* (Leipzig, 1868), p. 33.

§ For brevity, $J_n^2(z)$ is written in place of $\{J_n(z)\}^2$.

From the general formula we find that

(2) $$J_n(2z) = \sum_{r=0}^{n} J_r(z)\,J_{n-r}(z) + 2\sum_{r=1}^{\infty} (-)^r J_r(z)\,J_{n+r}(z),$$

when the Bessel coefficients of negative order are removed by using § 2·1 (2).

Similarly, since

$$\left\{\sum_{r=-\infty}^{\infty} t^r J_r(z)\right\}\left\{\sum_{m=-\infty}^{\infty} (-)^m t^m J_m(z)\right\}$$
$$= \exp\{\tfrac{1}{2}z(t-1/t)\}\exp\{\tfrac{1}{2}z(-t+1/t)\}$$
$$= 1,$$

it follows that

(3) $$J_0^2(z) + 2\sum_{r=1}^{\infty} J_r^2(z) = 1,$$

(4) $$\sum_{r=0}^{2n} (-)^r J_r(z)\,J_{2n-r}(z) + 2\sum_{r=1}^{\infty} J_r(z)\,J_{2n+r}(z) = 0.$$

Equation (4) is derived by considering the coefficient of t^{2n} in the Laurent expansion; the result of considering the coefficient of t^{2n+1} is nugatory.

A very important consequence of (3), namely that, when x is real,

(5) $$|J_0(x)| \leqslant 1, \quad |J_r(x)| \leqslant 1/\sqrt{2},$$

where $r = 1, 2, 3, \ldots$, was noticed by Hansen.

2·6. *Neumann's integral for* $J_n^2(z)$.

It is evident from § 2·2 (5) that

$$J_n(z) = \frac{1}{2\pi}\int_{-\pi}^{\pi} e^{i(n\theta - z\sin\theta)}\,d\theta,$$

and so

$$J_n^2(z) = \frac{1}{4\pi^2}\int_{-\pi}^{\pi}\int_{-\pi}^{\pi} e^{ni(\theta+\phi)}\,e^{-iz(\sin\theta+\sin\phi)}\,d\theta\,d\phi.$$

To reduce this double integral to a single integral take new variables defined by the equations

$$\theta - \phi = 2\chi, \quad \theta + \phi = 2\psi,$$

so that

$$\frac{\partial(\theta, \phi)}{\partial(\chi, \psi)} = 2.$$

It follows that

$$J_n^2(z) = \frac{1}{2\pi^2}\iint e^{2ni\psi}\,e^{-2iz\sin\psi\cos\chi}\,d\chi\,d\psi,$$

where the field of integration is the square for which

$$-\pi \leqslant \chi - \psi \leqslant \pi, \quad -\pi \leqslant \chi + \psi \leqslant \pi.$$

Since the integrand is unaffected if both χ and ψ are increased by π, or if χ is increased by π while ψ is simultaneously decreased by π, the field of integration may evidently be taken to be the rectangle for which

$$0 \leqslant \chi \leqslant \pi, \quad -\pi \leqslant \psi \leqslant \pi.$$

Hence

$$J_n^2(z) = \frac{1}{2\pi^2}\int_0^\pi \int_{-\pi}^\pi e^{2ni\psi - 2iz\sin\psi\cos\chi}\, d\psi\, d\chi$$

$$= \frac{1}{\pi}\int_0^\pi J_{2n}(2z\cos\chi)\, d\chi.$$

If we replace χ by $\frac{1}{2}\pi \mp \theta$, according as χ is acute or obtuse, we obtain the result

(1) $$J_n^2(z) = \frac{2}{\pi}\int_0^{\frac{1}{2}\pi} J_{2n}(2z\sin\theta)\, d\theta.$$

This formula may obviously be written in the form

(2) $$J_n^2(z) = \frac{1}{\pi}\int_0^\pi J_{2n}(2z\sin\theta)\, d\theta,$$

which is the result actually given by Neumann*. It was derived by him by some elaborate transformations from the addition-theorem which will be given in § 11·2. The proof which has just been given is suggested by the proof of that addition-theorem which was published by Graf and Gubler†.

We obtain a different form of the integral if we perform the integration with respect to χ instead of with respect to ψ. This procedure gives

$$J_n^2(z) = \frac{1}{2\pi}\int_{-\pi}^\pi J_0(2z\sin\psi)\, e^{2ni\psi}\, d\psi,$$

so that

(3) $$J_n^2(z) = \frac{1}{2\pi}\int_{-\pi}^\pi J_0(2z\sin\psi)\cos 2n\psi\, d\psi$$

$$= \frac{1}{\pi}\int_0^\pi J_0(2z\sin\psi)\cos 2n\psi\, d\psi,$$

a result which Schläfli‡ attributed to Neumann.

2·61. *Neumann's series for $J_n^2(z)$.*

By taking the formula § 2·6 (1), expanding the Bessel coefficient on the right in powers of z and then integrating term-by-term, Neumann§ shewed that

$$J_n^2(z) = \frac{1}{\pi}\int_0^\pi \sum_{m=0}^\infty \frac{(-)^m z^{2n+2m}\sin^{2n+2m}\theta}{m!\,(2n+m)!}\, d\theta$$

$$= \sum_{m=0}^\infty \frac{(-)^m (2n+2m)!\,(\frac{1}{2}z)^{2n+2m}}{m!\,(2n+m)!\,\{(n+m)!\}^2}.$$

* *Theorie der Bessel'schen Functionen* (Leipzig, 1867), p. 70.

† *Einleitung in die Theorie der Bessel'schen Funktionen*, II. (Bern, 1900), pp. 81—85.

‡ The formula is an immediate consequence of equation 16 on p. 69 of Neumann's treatise.

§ *Math. Ann.* III. (1871), p. 603. The memoir, in which this result was given, was first published in the *Leipziger Berichte*, XXI. (1869), pp. 221—256.

This result was written by Neumann in the form

$$(1)\quad J_n^2(z) = \frac{(\frac{1}{2}z)^{2n}}{(n!)^2}\left[1 - \frac{T_2 z^2}{1\,(2n+1)} + \frac{T_4 z^4}{1\,.\,2\,.\,(2n+1)(2n+2)} - \dots\right],$$

where

$$(2)\quad \begin{cases} T_2 = \dfrac{2n+1}{2n+2}, \\ T_4 = \dfrac{(2n+1)(2n+3)}{(2n+2)(2n+4)}, \\ T_6 = \dfrac{(2n+1)(2n+3)(2n+5)}{(2n+2)(2n+4)(2n+6)}, \\ \dots\dots\dots\dots\dots\dots \end{cases}$$

This expansion is a special case of a more general expansion (due to Schläfli) for the product of any two Bessel functions as a series of powers with comparatively simple coefficients (§ 5·41).

2·7. *Schlömilch's expansion of z^m in a series of Bessel coefficients.*

We shall now obtain the result which was foreshadowed in § 2·22 concerning the expansibility of z^m in a series of Bessel coefficients, where m is any positive integer. The result for $m = 0$ has already been given in § 2·22 (7).

In the results § 2·22 (1) and (2) substitute for $\cos 2n\theta$ and $\sin(2n+1)\theta$ their expansions in powers of $\sin^2\theta$. These expansions are*

$$\cos 2n\theta = \sum_{s=0}^{n} (-)^s \frac{n\,.\,(n+s-1)!}{(n-s)!\,(2s)!}(2\sin\theta)^{2s},$$

$$\sin(2n+1)\theta = \frac{1}{2}\sum_{s=0}^{n} (-)^s \frac{(2n+1)\,.\,(n+s)!}{(n-s)!\,(2s+1)!}(2\sin\theta)^{2s+1}.$$

The results of substitution are

$$\begin{cases} \cos(z\sin\theta) = J_0(z) + 2\displaystyle\sum_{n=1}^{\infty} J_{2n}(z)\left\{\sum_{s=0}^{n} (-)^s \frac{n\,.\,(n+s-1)!}{(n-s)!\,(2s)!}(2\sin\theta)^{2s}\right\}, \\ \sin(z\sin\theta) = \displaystyle\sum_{n=0}^{\infty} J_{2n+1}(z)\left\{\sum_{s=0}^{n} (-)^s \frac{(2n+1)\,.\,(n+s)!}{(n-s)!\,(2s+1)!}(2\sin\theta)^{2s+1}\right\}. \end{cases}$$

If we rearrange the series on the right as power series in $\sin\theta$ (assuming that it is permissible to do so), we have

$$\begin{cases} \cos(z\sin\theta) = \left\{J_0(z) + 2\displaystyle\sum_{n=1}^{\infty} J_{2n}(z)\right\} + \displaystyle\sum_{s=1}^{\infty} \frac{(-)^s(2\sin\theta)^{2s}}{(2s)!}\left\{\sum_{n=s}^{\infty} \frac{2n\,.\,(n+s-1)!}{(n-s)!} J_{2n}(z)\right\}, \\ \sin(z\sin\theta) = \displaystyle\sum_{s=0}^{\infty} \frac{(-)^s(2\sin\theta)^{2s+1}}{(2s+1)!}\left\{\sum_{n=s}^{\infty} \frac{(2n+1)\,.\,(n+s)!}{(n-s)!} J_{2n+1}(z)\right\}. \end{cases}$$

* Cf. Hobson, *Plane Trigonometry* (1918), §§ 80, 82.

If we expand the left-hand sides in powers of $\sin\theta$ and equate coefficients, we find that

$$\left\{\begin{aligned} 1 &= J_0(z) + 2\sum_{n=1}^{\infty} J_{2n}(z), \\ (\tfrac{1}{2}z)^{2s} &= \sum_{n=s}^{\infty} \frac{2n\,.\,(n+s-1)!}{(n-s)!} J_{2n}(z), \qquad (s=1, 2, 3, \ldots) \\ (\tfrac{1}{2}z)^{2s+1} &= \sum_{n=s}^{\infty} \frac{(2n+1)\,.\,(n+s)!}{(n-s)!} J_{2n+1}(z). \qquad (s=0, 1, 2, \ldots) \end{aligned}\right.$$

The first of these is the result already obtained; the others may be combined into the single formula

$$(1) \qquad (\tfrac{1}{2}z)^m = \sum_{n=0}^{\infty} \frac{(m+2n)\,.\,(m+n-1)!}{n!} J_{m+2n}(z). \qquad (m=1, 2, 3, \ldots)$$

The particular cases of (1) for which $m = 1, 2, 3$, were given by Schlömilch*. He also shewed how to obtain the general formula which was given explicitly some years later by Neumann† and Lommel‡.

The rearrangement of the double series now needs justification; the rearrangement is permissible if we can establish the absolute convergence of the double series.

If we make use of the inequalities

$$|J_{2n+1}(z)| \leqslant \frac{|\tfrac{1}{2}z|^{2n+1}}{(2n+1)!} \exp(\tfrac{1}{4}|z|^2), \qquad \frac{(n+s)!}{(2n)!} \leqslant 1, \quad (n \geqslant s),$$

in connexion with the series for $\sin(z\sin\theta)$ we see that

$$\begin{aligned} \sum_{s=0}^{\infty} \frac{|2\sin\theta|^{2s+1}}{(2s+1)!} \sum_{n=s}^{\infty} \frac{(2n+1)\,.\,(n+s)!}{(n-s)!} |J_{2n+1}(z)| &\leqslant \sum_{s=0}^{\infty} \frac{|2\sin\theta|^{2s+1}}{(2s+1)!} \sum_{n=s}^{\infty} \frac{|\tfrac{1}{2}z|^{2n+1}}{(n-s)!} \exp(\tfrac{1}{4}|z|^2) \\ &= \sum_{s=0}^{\infty} \frac{|2\sin\theta|^{2s+1}}{(2s+1)!} |\tfrac{1}{2}z|^{2s+1} \exp(\tfrac{1}{2}|z|^2) \\ &= \sinh(|z\sin\theta|)\exp(\tfrac{1}{2}|z|^2), \end{aligned}$$

and so the series of moduli is convergent. The series for $\cos(z\sin\theta)$ may be treated in a similar manner.

The somewhat elaborate analysis which has just been given is avoided in Lommel's proof by induction, but this proof suffers from the fact that it is supposed that the form of the expansion is known and merely needs verification. If, following Lommel, we assume that

$$(\tfrac{1}{2}z)^m = \sum_{n=0}^{\infty} \frac{(m+2n)\,.\,(m+n-1)!}{n!} J_{m+2n}(z),$$

* *Zeitschrift für Math. und Phys.* II. (1857), pp. 140—141.

† *Theorie der Bessel'schen Functionen* (Leipzig, 1867), p. 38.

‡ *Studien über die Bessel'schen Functionen* (Leipzig, 1868), pp. 35—36. Lommel's investigation is given later in this section.

[which has been proved in § 2·22 (8) in the special case $m = 1$], we have

$$(\tfrac{1}{2}z)^{m+1} = \sum_{n=0}^{\infty} \frac{(m+2n).(m+n-1)!}{n!} \tfrac{1}{2}z\, J_{m+2n}(z)$$

$$= \tfrac{1}{2}z \,.\, m!\, J_m(z) + \sum_{n=1}^{\infty} \left\{ \frac{(m+n)!}{n!} + \frac{(m+n-1)!}{(n-1)!} \right\} \tfrac{1}{2}z\, J_{m+2n}(z)$$

$$= \sum_{n=0}^{\infty} \frac{(m+n)!}{n!} \{\tfrac{1}{2}z\, J_{m+2n}(z) + \tfrac{1}{2}z\, J_{m+2n+2}(z)\}$$

$$= \sum_{n=0}^{\infty} \frac{(m+1+2n).(m+n)!}{n!} J_{m+1+2n}(z).$$

Since $(m+n)!\, J_{m+2n}(z)/n! \to 0$ as $n \to \infty$, the rearrangement in the third line of the analysis is permissible. It is obvious from this result that the induction holds for $m = 2, 3, 4, \ldots$.

An extremely elegant proof of the expansion, due to A. C. Dixon*, is as follows:—

Let t be a complex variable and let u be defined by the equation $u = \dfrac{2t}{1-t^2}$, so that when t describes a small circuit round the origin (inside the circle $|t| = 1$), u does the same.

We then have

$$(\tfrac{1}{2}z)^m = \frac{m!}{2^{m+1}\pi i} \int^{(0+)} u^{m-1} \exp(z/u)\, du$$

$$= \frac{m!}{2^{m+1}\pi i} \int^{(0+)} \exp\{-\tfrac{1}{2}z(t - 1/t)\} \,.\, \frac{1}{m} \frac{du^m}{dt}\, dt$$

$$= \frac{1}{2\pi i} \int^{(0+)} \exp\{-\tfrac{1}{2}z(t-1/t)\} \sum_{n=0}^{\infty} \frac{(m+2n).(m+n-1)!}{n!} t^{m+2n-1}\, dt$$

$$= \sum_{n=0}^{\infty} \frac{(m+2n).(m+n-1)!}{n!} J_{m+2n}(z),$$

when we calculate the sum of the residues at the origin for the last integral; the interchange of the order of summation and integration is permitted because the series converges uniformly on the contour; and the required result is obtained.

[NOTE. When m is zero, $\dfrac{1}{m}\dfrac{du^m}{dt}$ has to be replaced by $\dfrac{d \log u}{dt}$.]

2·71. *Schlömilch's expansions of the type* $\Sigma n^p J_n(z)$.

The formulae

(1) $$\sum_{n=1}^{\infty} (2n)^{2p} J_{2n}(z) = \sum_{m=0}^{p} P_{2m}^{(2p)} z^{2m},$$

(2) $$\sum_{n=0}^{\infty} (2n+1)^{2p+1} J_{2n+1}(z) = \sum_{m=0}^{p} P_{2m+1}^{(2p+1)} z^{2m+1},$$

in which p is any positive integer [zero included in (2) but not in (1)] and $P_m^{(p)}$ is a numerical coefficient, are evidently very closely connected with the results of § 2·7. The formulae

* *Messenger*, XXXII. (1903), p. 8; a proof on the same lines for the case $m=1$ had been previously given by Kapteyn, *Nieuw Archief voor Wiskunde*, XX. (1893), p. 120.

were obtained by Schlömilch, *Zeitschrift für Math. und Phys.* II. (1857), p. 141, and he gave, as the value of $P_m^{(p)}$,

(3) $$P_m^{(p)} = \sum_{k=0}^{<\frac{1}{2}m} \frac{(-)^k {}_mC_k (m-2k)^p}{2^m . m!},$$

where ${}_mC_k$ is a binomial coefficient and the last term of the summation is that for which k is $\frac{1}{2}m - 1$ or $\frac{1}{2}(m-1)$. To prove the first formula, take the equation § 2·22 (1), differentiate $2p$ times with respect to θ, and then make θ equal to zero. It is thus found that

$$2(-)^p \sum_{n=1}^{\infty} (2n)^{2p} J_{2n}(z) = \left[\frac{d^{2p} \cos(z \sin\theta)}{d\theta^{2p}}\right]_{\theta=0} = \left[\frac{d^{2p}}{d\theta^{2p}} \sum_{m=0}^{\infty} \frac{(-)^m z^{2m} \sin^{2m}\theta}{(2m)!}\right]_{\theta=0}.$$

The terms of the series for which $m > p$, when expanded in ascending powers of θ, contain no term in θ^{2p}, and so it is sufficient to evaluate

$$\left[\frac{d^{2p}}{d\theta^{2p}} \sum_{m=0}^{p} \frac{(-)^m z^{2m} \sin^{2m}\theta}{(2m)!}\right]_{\theta=0} = \sum_{m=0}^{p} \frac{(-)^m z^{2m}}{(2m)!} \sum_{k=0}^{2m} \left[\frac{d^{2p}}{d\theta^{2p}} \left\{\frac{(-)^k {}_{2m}C_k e^{(2m-2k)i\theta}}{(2i)^{2m}}\right\}\right]_{\theta=0}$$

$$= (-)^p \sum_{m=0}^{p} \frac{(\frac{1}{2}z)^{2m}}{(2m)!} \sum_{k=0}^{2m} (-)^k {}_{2m}C_k (2m-2k)^{2p}$$

$$= 2(-)^p \sum_{m=0}^{p} z^{2m} P_{2m}^{(2p)},$$

since terms equidistant from the beginning and the end of the summation with respect to k are equal. The truth of equation (1) is now evident, and equation (2) is proved in a similar manner from § 2·22 (2).

The reader will easily establish the following special cases, which were stated by Schlömilch:

(4) $$\begin{cases} 1^3 J_1(z) + 3^3 J_3(z) + 5^3 J_5(z) + \ldots = \frac{1}{2}(z + z^3), \\ 2^2 J_2(z) + 4^2 J_4(z) + 6^2 J_6(z) + \ldots = \frac{1}{2}z^2, \\ 2.3.4\, J_3(z) + 4.5.6\, J_5(z) + 6.7.8\, J_7(z) + \ldots = \frac{1}{2}z^3. \end{cases}$$

2·72. *Neumann's expansion of z^{2m} as a series of squares of Bessel coefficients.*

From Schlömilch's expansion (§ 2·7) of z^{2m} as a series of Bessel coefficients of even order, it is easy to derive an expansion of z^{2m} as a series of squares of Bessel coefficients, by using Neumann's integral given in § 2·6.

Thus, if we take the expansion

$$(z \sin\theta)^{2m} = \sum_{n=0}^{\infty} \frac{(2m+2n).(2m+n-1)!}{n!} J_{2m+2n}(2z \sin\theta),$$

and integrate with respect to θ, we find that

$$\frac{z^{2m}}{\pi} \int_0^{\pi} \sin^{2m}\theta \, d\theta = \sum_{n=0}^{\infty} \frac{(2m+2n).(2m+n-1)!}{n!} J^2_{m+n}(z),$$

so that (when $m > 0$)

(1) $$(\tfrac{1}{2}z)^{2m} = \frac{(m!)^2}{(2m)!} \sum_{n=0}^{\infty} \frac{(2m+2n).(2m+n-1)!}{n!} J^2_{m+n}(z).$$

This result was given by Neumann*. An alternative form is

$$(2)\qquad (\tfrac{1}{2}z)^{2m} = \frac{(m!)^2}{(2m)!} \sum_{n=m}^{\infty} n\epsilon_n \frac{\Gamma(n+m)}{\Gamma(n-m+1)} J_n^2(z),$$

and this is true when $m = 0$, for it then reduces to Hansen's formula of § 2·5.

As special cases, we have

$$(3)\qquad \begin{cases} z^2 = \dfrac{1}{2} \displaystyle\sum_{n=1}^{\infty} \epsilon_n \,.\, 4n^2 J_n^2(z), \\ z^4 = \dfrac{1\,.\,2}{3\,.\,4} \displaystyle\sum_{n=2}^{\infty} \epsilon_n\; 4n^2(4n^2-2^2) J_n^2(z), \\ z^6 = \dfrac{1\,.\,2\,.\,3}{4\,.\,5\,.\,6} \displaystyle\sum_{n=3}^{\infty} \epsilon_n \,.\, 4n^2(4n^2-2^2)(4n^2-4^2) J_n^2(z), \\ \ldots\ldots\ldots\ldots\ldots\ldots\ldots\ldots\ldots\ldots\ldots\ldots \end{cases}$$

If we differentiate (1), use § 2·12 (2) and then rearrange, it is readily found that

$$(4)\qquad (\tfrac{1}{2}z)^{2m-1} = \frac{m!\,(m-1)!}{(2m-1)!} \sum_{n=0}^{\infty} \frac{(2m+2n-1)\,.\,(2m+n-2)!}{n!} J_{m+n-1}(z) J_{m+n}(z),$$

an expansion whose existence was indicated by Neumann.

* *Leipziger Berichte*, XXI. (1869), p. 226. [*Math. Ann.* III. (1871), p. 585.]

CHAPTER III

BESSEL FUNCTIONS

3·1. *The generalisation of Bessel's differential equation.*

The Bessel coefficients, which were discussed in Chapter II, are functions of two variables, z and n, of which z is unrestricted but n has hitherto been required to be an integer. We shall now generalise these functions so as to have functions of two unrestricted (complex) variables.

This generalisation was effected by Lommel *, whose definition of a *Bessel function* was effected by a generalisation of Poisson's integral; in the course of his analysis he shewed that the function, so defined, is a solution of the linear differential equation which is to be discussed in this section. Lommel's definition of the Bessel function $J_\nu(z)$ of argument z and order ν was†

$$J_\nu(z) = \frac{(\frac{1}{2}z)^\nu}{\Gamma(\nu+\frac{1}{2})\,\Gamma(\frac{1}{2})}\int_0^\pi \cos(z\cos\theta)\sin^{2\nu}\theta\,d\theta,$$

and the integral on the right is convergent for general complex values of ν for which $R(\nu)$ exceeds $-\frac{1}{2}$. Lommel apparently contemplated only real values of ν, the extension to complex values being effected by Hankel‡; functions of order less than $-\frac{1}{2}$ were defined by Lommel by means of an extension of the recurrence formulae of § 2·12.

The reader will observe, on comparing § 3·3 with § 1·6 that Plana and Poisson had investigated Bessel functions whose order is half of an odd integer nearly half a century before the publication of Lommel's treatise.

We shall now replace the integer n which occurs in Bessel's differential equation by an unrestricted (real or complex) number§ ν, and then define a *Bessel function* of order ν to be a certain solution of this equation; it is of course desirable to select such a solution as reduces to $J_n(z)$ when ν assumes the integral value n.

We shall therefore discuss solutions of the differential equation

$$z^2\frac{d^2y}{dz^2} + z\frac{dy}{dz} + (z^2-\nu^2)\,y = 0, \tag{1}$$

which will be called *Bessel's equation for functions of order* ν.

* *Studien über die Bessel'schen Functionen* (Leipzig, 1868), p. 1.

† Integrals resembling this (with ν not necessarily an integer) were studied by Duhamel, *Cours d'Analyse*, II. (Paris, 1840), pp. 118—121.

‡ *Math. Ann.* I. (1869), p. 469.

§ Following Lommel, we use the symbols ν, μ to denote unrestricted numbers, the symbols n, m being reserved for integers. This distinction is customary on the Continent, though it has not yet come into general use in this country. It has the obvious advantage of shewing at a glance whether a result is true for unrestricted functions or for functions of integral order only.

Let us now construct a solution of (1) which is valid near the origin; the form assumed for such a solution is a series of ascending powers of z, say

$$y = \sum_{m=0}^{\infty} c_m z^{\alpha+m},$$

where the index α and the coefficients c_m are to be determined, with the proviso that c_0 is not zero.

For brevity the differential operator which occurs in (1) will be called ∇_ν, so that

$$\text{(2)} \qquad \nabla_\nu \equiv z^2 \frac{d^2}{dz^2} + z\frac{d}{dz} + z^2 - \nu^2.$$

It is easy to see that*

$$\nabla_\nu \sum_{m=0}^{\infty} c_m z^{\alpha+m} = \sum_{m=0}^{\infty} c_m \{(\alpha+m)^2 - \nu^2\} z^{\alpha+m} + \sum_{m=0}^{\infty} c_m z^{\alpha+m+2}.$$

The expression on the right reduces to the first term of the first series, namely $c_0(\alpha^2 - \nu^2) z^\alpha$, if we choose the coefficients c_m so that the coefficients of corresponding powers of z in the two series on the right cancel.

This choice gives the system of equations

$$\text{(3)} \qquad \left\{\begin{array}{ll} c_1\{(\alpha+1)^2 - \nu^2\} & = 0 \\ c_2\{(\alpha+2)^2 - \nu^2\} + c_0 & = 0 \\ c_3\{(\alpha+3)^2 - \nu^2\} + c_1 & = 0 \\ \dots\dots\dots\dots & \\ c_m\{(\alpha+m)^2 - \nu^2\} + c_{m-2} & = 0 \\ \dots\dots\dots\dots & \end{array}\right.$$

If, then, these equations are satisfied, we have

$$\text{(4)} \qquad \nabla_\nu \sum_{m=0}^{\infty} c_m z^{\alpha+m} = c_0(\alpha^2 - \nu^2) z^\alpha.$$

From this result, it is evident that the postulated series can be a solution of (1) only if $\alpha = \pm\nu$; for c_0 is not zero, and z^α vanishes only for exceptional values of z.

Now consider the mth equation in the system (3) when $m > 1$. It can be written in the form

$$c_m(\alpha - \nu + m)(\alpha + \nu + m) + c_{m-2} = 0,$$

and so it determines c_m in terms of c_{m-2} for all values of m greater than 1 unless $\alpha - \nu$ or $\alpha + \nu$ is a negative integer, that is, unless -2ν is a negative integer (when $\alpha = -\nu$) or unless 2ν is a negative integer (when $\alpha = \nu$).

We disregard these exceptional values of ν for the moment (see §§ 3·11, 3·5), and then $(\alpha+m)^2 - \nu^2$ does not vanish when $m = 1, 2, 3, \dots$. It now

* When the constants α and c_m have been determined by the following analysis, the series obtained by formal processes is easily seen to be convergent and differentiable, so that the formal procedure actually produces a solution of the differential equation.

follows from the equations (3) that $c_1 = c_3 = c_5 = \dots = 0$, and that c_{2m} is expressible in terms of c_0 by the equation

$$c_{2m} = \frac{(-)^m c_0}{(\alpha - \nu + 2)(\alpha - \nu + 4) \dots (\alpha - \nu + 2m)(\alpha + \nu + 2)(\alpha + \nu + 4) \dots (\alpha + \nu + 2m)}.$$

The system of equations (3) is now satisfied; and, if we take $\alpha = \nu$, we see from (4) that

$$c_0 z^\nu \left[1 + \sum_{m=1}^{\infty} \frac{(-)^m (\frac{1}{2}z)^{2m}}{m!(\nu+1)(\nu+2)\dots(\nu+m)}\right] \tag{5}$$

is a formal solution of equation (1). If we take $\alpha = -\nu$, we obtain a second formal solution

$$c_0' z^{-\nu} \left[1 + \sum_{m=1}^{\infty} \frac{(-)^m (\frac{1}{2}z)^{2m}}{m!(-\nu+1)(-\nu+2)\dots(-\nu+m)}\right]. \tag{6}$$

In the latter, c_0' has been written in place of c_0, because the procedure of obtaining (6) can evidently be carried out without reference to the existence of (5), so that the constants c_0 and c_0' are independent.

Any values independent of z may be assigned to the constants c_0 and c_0'; but, in view of the desirability of obtaining solutions reducible to $J_n(z)$ when $\nu \to n$, we define them by the formulae*

$$c_0 = \frac{1}{2^\nu \Gamma(\nu+1)}, \qquad c_0' = \frac{1}{2^{-\nu}\Gamma(-\nu+1)}. \tag{7}$$

The series (5) and (6) may now be written

$$\sum_{m=0}^{\infty} \frac{(-)^m (\frac{1}{2}z)^{\nu+2m}}{m!\,\Gamma(\nu+m+1)}, \qquad \sum_{m=0}^{\infty} \frac{(-)^m (\frac{1}{2}z)^{-\nu+2m}}{m!\,\Gamma(-\nu+m+1)}.$$

In the circumstances considered, namely when 2ν is not an integer, these series of powers converge for all values of z, ($z = 0$ excepted) and so term-by-term differentiations are permissible. The operations involved in the analysis† by which they were obtained are consequently legitimate, and so we have obtained two solutions of equation (1).

The first of the two series defines a function called a *Bessel function* of *order* ν and *argument* z, of the *first kind*‡; and the function is denoted by the symbol $J_\nu(z)$. Since ν is unrestricted (apart from the condition that, for the present, 2ν is not an integer), the second series is evidently $J_{-\nu}(z)$.

Accordingly, the function $J_\nu(z)$ *is defined by the equation*

$$J_\nu(z) = \sum_{m=0}^{\infty} \frac{(-)^m (\frac{1}{2}z)^{\nu+2m}}{m!\,\Gamma(\nu+m+1)}. \tag{8}$$

It is evident from § 2·11 that this definition continues to hold when ν is a *positive* integer (zero included), a Bessel function of integral order being identical with a Bessel coefficient.

* For properties of the Gamma-function, see *Modern Analysis*, ch. XII.

† Which, up to the present, has been purely formal.

‡ Functions of the second and third kinds are defined in §§ 3·5, 3·54, 3·57, 3·6.

An interesting symbolic solution of Bessel's equation has been given by Cotter* in the form

$$[1+z^{\nu}D^{-1}z^{-2\nu-1}D^{-1}z^{\nu+1}]^{-1}(Az^{\nu}+Bz^{-\nu}),$$

where $D\equiv d/dz$ while A and B are constants. This may be derived by writing successively

$$[D(zD-2\nu)+z]\,z^{\nu}y=0,$$

$$[zD-2\nu+D^{-1}z]\,z^{\nu}y=-2\nu B,$$

$$zD(z^{-\nu}y)+z^{-2\nu}D^{-1}z^{\nu+1}y=-2\nu Bz^{-2\nu},$$

$$z^{-\nu}y+D^{-1}z^{-2\nu-1}D^{-1}z^{\nu+1}y=A+Bz^{-2\nu},$$

which gives Cotter's result.

3·11. *Functions whose order is half of an odd integer.*

In § 3·1, two cases of Bessel's generalised equation were temporarily omitted from consideration, namely (i) when ν is half of an odd integer, (ii) when ν is an integer†. It will now be shewn that case (i) may be included in the general theory for unrestricted values of ν.

When ν is half of an odd integer, let

$$\nu^2=(r+\tfrac{1}{2})^2,$$

where r is a positive integer or zero.

If we take $\alpha=r+\frac{1}{2}$ in the analysis of § 3·1, we find that

(1) $$\begin{cases} c_1 . 1\,(2r+2) &=0, \\ c_m . m\,(m+2r+1)+c_{m-2}=0, \end{cases} \qquad (m>1)$$

and so

(2) $$c_{2m}=\frac{(-)^m c_0}{2 . 4 \ldots (2m) . (2r+3)(2r+5)\ldots(2r+2m+1)},$$

which is the value of c_{2m} given by § 3·1 when α and ν are replaced by $r+\frac{1}{2}$. If we take

$$c_0=\frac{1}{2^{r+\frac{1}{2}}\,\Gamma(r+\frac{3}{2})},$$

we obtain the solution

$$\sum_{m=0}^{\infty}\frac{(-)^m(\frac{1}{2}z)^{r+\frac{1}{2}+2m}}{m!\,\Gamma(r+m+\frac{3}{2})},$$

which is naturally denoted by the symbol $J_{r+\frac{1}{2}}(z)$, so that the definition of § 3·1 (8) is still valid.

If, however, we take $\alpha=-r-\frac{1}{2}$, the equations which determine c_m become

(3) $$\begin{cases} c_1 . 1\,(-2r) &=0, \\ c_m . m\,(m-1-2r)+c_{m-2}=0. \end{cases} \qquad (m>1).$$

As before, $c_1, c_3, \ldots, c_{2r-1}$ are all zero, but the equation to determine c_{2r+1} is

$$0 . c_{2r+1}+c_{2r-1}=0,$$

and this equation is satisfied by an arbitrary *value of* c_{2r+1}; when $m>r$, c_{2m+1} is defined by the equation

$$c_{2m+1}=\frac{(-)^{m-r}\,c_{2r+1}}{(2r+3)(2r+5)\ldots(2m+1) . 2 . 4 \ldots (2m-2r)}.$$

* *Proc. R. Irish. Acad.* XXVII. (A), (1909), pp. 157—161.

† The cases combine to form the case in which 2ν is an integer.

If $J_\nu(z)$ be defined by § 3·1 (8) when $\nu = -r - \frac{1}{2}$, the solution now constructed is*

$$c_0\, 2^{-r-\frac{1}{2}}\, \Gamma(\tfrac{1}{2} - r)\, J_{-r-\frac{1}{2}}(z) + c_{2r+1}\, 2^{r+\frac{1}{2}}\, \Gamma(r + \tfrac{3}{2})\, J_{r+\frac{1}{2}}(z).$$

It follows that no modification in the definition of $J_\nu(z)$ is necessary when $\nu = \pm(r + \frac{1}{2})$; the real peculiarity of the solution in this case is that the negative root of the indicial equation gives rise to a series containing *two* arbitrary constants, c_0 and c_{2r+1}, i.e. to the general solution of the differential equation.

3·12. *A fundamental system of solutions of Bessel's equation.*

It is well known that, if y_1 and y_2 are two solutions of a linear differential equation of the second order, and if y_1' and y_2' denote their derivates with respect to the independent variable, then the solutions are linearly independent if the *Wronskian determinant*†

$$\begin{vmatrix} y_1 & y_2 \\ y_1' & y_2' \end{vmatrix}$$

does not vanish identically; and if the Wronskian does vanish identically, then, either one of the two solutions vanishes identically, or else the ratio of the two solutions is a constant.

If the Wronskian does not vanish identically, then *any* solution of the differential equation is expressible in the form $c_1 y_1 + c_2 y_2$ where c_1 and c_2 are constants depending on the particular solution under consideration; the solutions y_1 and y_2 are then said to form a fundamental system.

For brevity the Wronskian of y_1 and y_2 will be written in the forms

$$\mathfrak{W}_z\{y_1, y_2\}, \qquad \mathfrak{W}\{y_1, y_2\},$$

the former being used when it is necessary to specify the independent variable.

We now proceed to evaluate

$$\mathfrak{W}\{J_\nu(z), J_{-\nu}(z)\}.$$

If we multiply the equations

$$\nabla_\nu J_{-\nu}(z) = 0, \qquad \nabla_\nu J_\nu(z) = 0$$

by $J_\nu(z)$, $J_{-\nu}(z)$ respectively and subtract the results, we obtain an equation which may be written in the form

$$\frac{d}{dz}[z\, \mathfrak{W}\{J_\nu(z), J_{-\nu}(z)\}] = 0,$$

* In connexion with series representing this solution, see Plana, *Mem. della R. Accad. delle Sci. di Torino*, XXVI. (1821), pp. 519—538.

† For references to theorems concerning Wronskians, see *Encyclopédie des Sci. Math.* II. 16 (§ 23), p. 109. Proofs of the theorems quoted in the text are given by Forsyth, *Treatise on Differential Equations* (1914), §§ 72—74.

and hence, on integration,

(1) $$\mathfrak{W}\{J_\nu(z), J_{-\nu}(z)\} = \frac{C}{z},$$

where C is a determinate constant.

To evaluate C, we observe that, *when ν is not an integer*, and $|z|$ is small, we have

$$J_\nu(z) = \frac{(\frac{1}{2}z)^\nu}{\Gamma(\nu+1)}\{1 + O(z^2)\},\quad J_\nu'(z) = \frac{(\frac{1}{2}z)^{\nu-1}}{2\Gamma(\nu)}\{1 + O(z^2)\},$$

with similar expressions for $J_{-\nu}(z)$ and $J'_{-\nu}(z)$; and hence

$$J_\nu(z)\,J'_{-\nu}(z) - J_{-\nu}(z)\,J_\nu'(z) = \frac{1}{z}\left\{\frac{1}{\Gamma(\nu+1)\,\Gamma(-\nu)} - \frac{1}{\Gamma(\nu)\,\Gamma(-\nu+1)}\right\} + O(z)$$

$$= -\frac{2\sin\nu\pi}{\pi z} + O(z).$$

If we compare this result with (1), it is evident that the expression on the right which is $O(z)$ must vanish, and so*

(2) $$\mathfrak{W}\{J_\nu(z), J_{-\nu}(z)\} = -\frac{2\sin\nu\pi}{\pi z}.$$

Since $\sin\nu\pi$ is not zero (because ν is not an integer), the functions $J_\nu(z)$, $J_{-\nu}(z)$ form a fundamental system of solutions of equation § 3·1 (1).

When ν is an integer, n, we have seen that, with the definition of § 2·1 (2),

$$J_{-n}(z) = (-)^n J_n(z);$$

and when ν is made equal to $-n$ in § 3·1 (8), we find that

$$J_{-n}(z) = \sum_{m=0}^{\infty} \frac{(-)^m (\frac{1}{2}z)^{-n+2m}}{m!\,\Gamma(-n+m+1)}.$$

Since the first n terms of the last series vanish, the series is easily reduced to $(-)^n J_n(z)$, so that the two definitions of $J_{-n}(z)$ are equivalent, and the functions $J_n(z)$, $J_{-n}(z)$ do *not* form a fundamental system of solutions of Bessel's equation for functions of order n. The determination of a fundamental system in this case will be investigated in § 3·63.

To sum up, the function $J_\nu(z)$ is defined, for all values of ν, by the expansion of § 3·1 (8); and $J_\nu(z)$, so defined, is always a solution of the equation $\nabla_\nu y = 0$. When ν is not an integer, a fundamental system of solutions of this equation is formed by the functions $J_\nu(z)$ and $J_{-\nu}(z)$.

A generalisation of the Bessel function has been effected by F. H. Jackson in his researches on "basic numbers." Briefly, a basic number $[n]$ is defined as $\dfrac{p^n-1}{p-1}$, where p is the *base*, and the basic Gamma function $\Gamma_p(\nu)$ is defined to satisfy the recurrence formula $\Gamma_p(\nu+1) = [\nu].\Gamma_p(\nu)$.

The basic Bessel function is then defined by replacing the numbers which occur in the series for the Bessel function by basic numbers. It has been shewn that very many theorems

* This result is due to Lommel, *Math. Ann.* IV. (1871), p. 104. He derived the value of C by making $z \to \infty$ and using the approximate formulae which will be investigated in Chapter VII.

concerning Bessel functions have their analogues in the theory of basic Bessel functions, but the discussion of these analogues is outside the scope of this work. Jackson's main results are to be found in a series of papers, *Proc. Edinburgh Math. Soc.* XXI. (1903), pp. 65—72; XXII. (1904), pp. 80—85; *Proc. Royal Soc. Edinburgh*, XXV. (1904), pp. 273—276; *Trans. Royal Soc. Edinburgh*, XLI. (1905), pp. 1—28, 105—118, 399—408; *Proc. London Math. Soc.* (2) I. (1904), pp. 361—366; (2) II. (1905), pp. 192—220; (2) III. (1905), pp. 1—23.

The more obvious generalisation of the Bessel function, obtained by increasing the number of sets of factors in the denominators of the terms of the series, will be dealt with in § 4·4. In connexion with this generalisation see Cailler, *Mém. de la Soc. de Phys. de Genève*, XXXIV. (1905), p. 354; another generalisation, in the shape of Bessel functions of two variables, has been dealt with by Whittaker, *Math. Ann.* LVII. (1903), p. 351, and Pérès, *Comptes Rendus*, CLXI. (1915), pp. 168—170.

3·13. *General properties of $J_\nu(z)$.*

The series which defines $J_\nu(z)$ converges absolutely and uniformly* in any closed domain of values of z [the origin not being a point of the domain when $R(\nu) < 0$], and in any bounded domain of values of ν.

For, when $|\nu| \leqslant N$ and $|z| \leqslant \Delta$, the test ratio for this series is

$$\left|\frac{-\frac{1}{4}z^2}{m(\nu+m)}\right| \leqslant \frac{\frac{1}{4}\Delta^2}{m(m-N)} < 1,$$

whenever m is taken to be greater than the positive root of the equation

$$m^2 - mN - \tfrac{1}{4}\Delta^2 = 0.$$

This choice of m being independent of ν and z, the result stated follows from the test of Weierstrass.

Hence† $J_\nu(z)$ is an analytic function of z for all values of z ($z = 0$ possibly being excepted) and it is an analytic function of ν for all values of ν.

An important consequence of this theorem is that term-by-term differentiations and integrations (with respect to z or ν) of the series for $J_\nu(z)$ are permissible.

An inequality due to Nielsen ‡ should be noticed here, namely

(1) $$J_\nu(z) = \frac{(\frac{1}{2}z)^\nu}{\Gamma(\nu+1)}(1+\theta),$$

where $$|\theta| < \exp\left\{\frac{\frac{1}{4}|z|^2}{|\nu_0+1|}\right\} - 1,$$

and $|\nu_0+1|$ is the smallest of the numbers $|\nu+1|, |\nu+2|, |\nu+3|, \ldots$.

This result may be proved in exactly the same way as § 2·11 (5); it should be compared with the inequalities which will be given in § 3·3.

Finally, the function z^ν, which is a factor of $J_\nu(z)$, needs precise specifica-

* Bromwich, *Theory of Infinite Series*, § 82.

† *Modern Analysis*, § 5·3.

‡ *Math. Ann.* LII. (1899), p. 230; *Nyt Tidsskrift*, IX. B (1898), p. 73; see also *Math. Ann.* LV. (1902), p. 494.

tion. We define it to be $\exp(\nu \log z)$ where the *phase* (or *argument*) of z is given its principal value so that

$$-\pi < \arg z \leqslant \pi.$$

When it is necessary to "continue" the function $J_\nu(z)$ outside this range of values of $\arg z$, explicit mention will be made of the process to be carried out.

3·2. *The recurrence formulae for* $J_\nu(z)$.

Lommel's generalisations* of the recurrence formulae for the Bessel coefficients (§ 2·12) are as follows:

(1) $$J_{\nu-1}(z) + J_{\nu+1}(z) = \frac{2\nu}{z} J_\nu(z)$$

(2) $$J_{\nu-1}(z) - J_{\nu+1}(z) = 2J_\nu'(z),$$

(3) $$zJ_\nu'(z) + \nu J_\nu(z) = zJ_{\nu-1}(z),$$

(4) $$zJ_\nu'(z) - \nu J_\nu(z) = -zJ_{\nu+1}(z).$$

These are of precisely the same form as the results of § 2·12, the only difference being the substitution of the unrestricted number ν for the integer n.

To prove them, we observe first that

$$\begin{aligned}\frac{d}{dz}\{z^\nu J_\nu(z)\} &= \frac{d}{dz}\sum_{m=0}^{\infty}\frac{(-)^m z^{2\nu+2m}}{2^{\nu+2m}\,.\,m!\,\Gamma(\nu+m+1)}\\ &= \sum_{m=0}^{\infty}\frac{(-)^m z^{2\nu-1+2m}}{2^{\nu-1+2m}\,.\,m!\,\Gamma(\nu+m)}\\ &= z^\nu J_{\nu-1}(z).\end{aligned}$$

When we differentiate out the product on the left, we at once obtain (3). In like manner,

$$\begin{aligned}\frac{d}{dz}\{z^{-\nu} J_\nu(z)\} &= \frac{d}{dz}\sum_{m=0}^{\infty}\frac{(-)^m z^{2m}}{2^{\nu+2m}\,.\,m!\,\Gamma(\nu+m+1)}\\ &= \sum_{m=1}^{\infty}\frac{(-)^m z^{2m-1}}{2^{\nu+2m-1}\,.\,(m-1)!\,\Gamma(\nu+m+1)}\\ &= \sum_{m=0}^{\infty}\frac{(-)^{m+1} z^{2m+1}}{2^{\nu+2m+1}\,.\,m!\,\Gamma(\nu+m+2)}\\ &= -z^{-\nu}J_{\nu+1}(z),\end{aligned}$$

whence (4) is obvious; and (2) and (1) may be obtained by adding and subtracting (3) and (4).

* *Stüdien über die Bessel'schen Functionen* (Leipzig, 1868), pp. 2, 6, 7. Formula (3) was given when ν is half of an odd integer by Plana, *Mem. della R. Accad. delle Sci. di Torino*, XXVI. (1821), p. 533.

We can now obtain the generalised formulae

(5) $$\left(\frac{d}{z\,dz}\right)^m \{z^\nu J_\nu(z)\} = z^{\nu-m} J_{\nu-m}(z),$$

(6) $$\left(\frac{d}{z\,dz}\right)^m \{z^{-\nu} J_\nu(z)\} = (-)^m z^{-\nu-m} J_{\nu+m}(z)$$

by repeated differentiations, when m is any positive integer.

Lommel obtained all these results from his generalisation of Poisson's integral which has been described in § 3·1.

The formula (1) has been extensively used* in the construction of Tables of Bessel functions.

By expressing $J_{\nu-1}(z)$ and $J_{1-\nu}(z)$ in terms of $J_{\pm\nu}(z)$ and $J'_{\pm\nu}(z)$ by (3) and (4), we can derive Lommel's formula†

(7) $$J_\nu(z) J_{1-\nu}(z) + J_{-\nu}(z) J_{\nu-1}(z) = \frac{2\sin\nu\pi}{\pi z}$$

from formula (2) of § 3·12.

An interesting consequence of (1) and (2) is that, if $Q_\nu(z) \equiv J_\nu^2(z)$, then

(8) $$Q_{\nu-1}(z) - Q_{\nu+1}(z) = \frac{2\nu}{z} Q_\nu'(z);$$

this formula was discovered by Lommel, who derived various consequences of it, *Studien über die Bessel'schen Functionen* (Leipzig, 1868), pp. 48 *et seq.* See also Neumann, *Math. Ann.* III. (1871), p. 600.

3·21. *Bessel functions of complex order.*

The real and imaginary parts of the function $J_{\nu+i\mu}(x)$, where ν, μ and x are real, have been discussed in some detail by Lommel‡, and his results were subsequently extended by Bôcher§.

In particular, after defining the real functions $K_{\nu,\mu}(x)$ and $S_{\nu,\mu}(x)$ by the equation‖

$$J_{\nu+i\mu}(x) = \frac{(\frac{1}{2}x)^{\nu+i\mu}}{\Gamma(\nu+i\mu+\frac{1}{2})\,\Gamma(\frac{1}{2})}\{K_{\nu,\mu}(x) + iS_{\nu,\mu}(x)\},$$

Lommel obtained the results

(1) $$\frac{d^2}{dx^2}\{K_{\nu,\mu}(x) \pm iS_{\nu,\mu}(x)\} + \{K_{\nu,\mu}(x) \pm iS_{\nu,\mu}(x)\} + \frac{2(\nu \pm i\mu)+1}{x}\frac{d}{dx}\{K_{\nu,\mu}(x) \pm iS_{\nu,\mu}(x)\} = 0,$$

(2) $$K_{\nu+1,\mu}(x) = K_{\nu,\mu}(x) + K''_{\nu,\mu}(x),$$

(3) $$S_{\nu+1,\mu}(x) = S_{\nu,\mu}(x) + S''_{\nu,\mu}(x),$$

* See, e.g. Lommel, *Münchener Abh.* xv. (1884—1886), pp. 644—647.

† *Math. Ann.* IV. (1871), p. 105. Some associated formulae are given in § 3·63.

‡ *Math. Ann.* III. (1871), pp. 481—486.

§ *Annals of Math.* VI. (1892), pp. 137—160.

‖ The reason for inserting the factor on the right is apparent from formulae which will be established in § 3·3.

with numerous other formulae of like character. These results seem to be of no great importance, and consequently we merely refer the reader to the memoirs in which they were published.

In the special case in which $\nu = 0$, Bessel's equation becomes

$$z^2 \frac{d^2y}{dz^2} + z\frac{dy}{dz} + (z^2 + \mu^2)y = 0;$$

solutions of this equation in the form of series were given by Boole* many years ago.

3·3. *Lommel's expression of $J_\nu(z)$ by an integral of Poisson's type.*

We shall now shew that, when $R(\nu) > -\frac{1}{2}$, then

$$(1) \qquad J_\nu(z) = \frac{(\frac{1}{2}z)^\nu}{\Gamma(\nu+\frac{1}{2})\,\Gamma(\frac{1}{2})}\int_0^\pi \cos(z\cos\theta)\sin^{2\nu}\theta\, d\theta.$$

It was proved by Poisson† that, when 2ν is a positive integer (zero included), the expression on the right is a solution of Bessel's equation; and this expression was adopted by Lommel‡ as the definition of $J_\nu(z)$ for positive values of $\nu + \frac{1}{2}$.

Lommel subsequently proved that the function, so defined, is a solution of Bessel's generalised equation and that it satisfies the recurrence formulae of § 3·2; and he then defined $J_\nu(z)$ for values of ν in the intervals $(-\frac{1}{2}, -\frac{3}{2})$, $(-\frac{3}{2}, -\frac{5}{2})$, $(-\frac{5}{2}, -\frac{7}{2})$, ... by successive applications of § 3·2 (1).

To deduce (1) from the definition of $J_\nu(z)$ adopted in this work, we transform the general term of the series for $J_\nu(z)$ in the following manner:

$$\frac{(-)^m(\frac{1}{2}z)^{\nu+2m}}{m!\,\Gamma(\nu+m+1)} = \frac{(-)^m(\frac{1}{2}z)^\nu}{\Gamma(\nu+\frac{1}{2})\,\Gamma(\frac{1}{2})}\cdot\frac{z^{2m}}{(2m)!}\cdot\frac{\Gamma(\nu+\frac{1}{2})\,\Gamma(m+\frac{1}{2})}{\Gamma(\nu+m+1)}$$

$$= \frac{(-)^m(\frac{1}{2}z)^\nu}{\Gamma(\nu+\frac{1}{2})\,\Gamma(\frac{1}{2})}\cdot\frac{z^{2m}}{(2m)!}\int_0^1 t^{\nu-\frac{1}{2}}(1-t)^{m-\frac{1}{2}}\,dt,$$

provided that $R(\nu) > -\frac{1}{2}$.

Now when $R(\nu) \geqslant \frac{1}{2}$, the series

$$\sum_{m=1}^{\infty} \frac{(-)^m z^{2m}}{(2m)!} t^{\nu-\frac{1}{2}}(1-t)^{m-\frac{1}{2}}$$

converges uniformly with respect to t throughout the interval (0, 1), and so it may be integrated term-by-term; on adding to the result the term for which

* *Phil. Trans. of the Royal Soc.* 1844, p. 239. See also a question set in the Mathematical Tripos, 1894.

† *Journal de l'École R. Polytechnique*, XII. (cahier 19), (1823), pp. 300 *et seq.*, 340 *et seq.* Strictly speaking, Poisson shewed that, when 2ν is an odd integer, the expression on the right multiplied by $\sqrt{z}$ is a solution of the equation derived from Bessel's equation by the appropriate change of dependent variable.

‡ *Studien über die Bessel'schen Functionen* (Leipzig, 1868), pp. 1 *et seq.*

$m=0$, namely $\int_0^1 t^{\nu-\frac{1}{2}}(1-t)^{-\frac{1}{2}}dt$, which is convergent, we find that, when $R(\nu)\geqslant\frac{1}{2}$,

$$J_\nu(z)=\frac{(\frac{1}{2}z)^\nu}{\Gamma(\nu+\frac{1}{2})\,\Gamma(\frac{1}{2})}\int_0^1 t^{\nu-\frac{1}{2}}\left\{\sum_{m=0}^{\infty}\frac{(-)^m z^{2m}(1-t)^{m-\frac{1}{2}}}{(2m)!}\right\}dt,$$

whence the result stated follows by making the substitution $t=\sin^2\theta$ and using the fact that the integrand is unaffected by writing $\pi-\theta$ in place of θ.

When $-\frac{1}{2}<R(\nu)<\frac{1}{2}$, the analysis necessary to establish the last equation is a little more elaborate. The simplest procedure seems to be to take the series with the first two terms omitted and integrate by parts, thus

$$\begin{aligned}\sum_{m=2}^{\infty}\frac{(-)^m z^{2m}}{(2m)!}\int_0^1 t^{\nu-\frac{1}{2}}(1-t)^{m-\frac{1}{2}}dt &= \sum_{m=2}^{\infty}\frac{m-\frac{1}{2}}{\nu+\frac{1}{2}}\cdot\frac{(-)^m z^{2m}}{(2m)!}\int_0^1 t^{\nu+\frac{1}{2}}(1-t)^{m-\frac{3}{2}}dt\\ &=\int_0^1\frac{t^{\nu+\frac{1}{2}}}{\nu+\frac{1}{2}}\left\{\sum_{m=2}^{\infty}\frac{(-)^m(m-\frac{1}{2})z^{2m}}{(2m)!}(1-t)^{m-\frac{3}{2}}\right\}dt\\ &=-\int_0^1\frac{t^{\nu+\frac{1}{2}}}{\nu+\frac{1}{2}}\frac{d}{dt}\left\{\sum_{m=2}^{\infty}\frac{(-)^m z^{2m}}{(2m)!}(1-t)^{m-\frac{1}{2}}\right\}dt\\ &=\int_0^1 t^{\nu-\frac{1}{2}}\left\{\sum_{m=2}^{\infty}\frac{(-)^m z^{2m}}{(2m)!}(1-t)^{m-\frac{1}{2}}\right\}dt,\end{aligned}$$

on integrating by parts a second time. The interchange of the order of summation and integration in the second line of analysis is permissible on account of the uniformity of convergence of the series. On adding the integrals corresponding to the terms $m=0$, $m=1$ (which are convergent), we obtain the desired result.

It follows that, when $R(\nu)>-\frac{1}{2}$, then

$$J_\nu(z)=\frac{(\frac{1}{2}z)^\nu}{\Gamma(\nu+\frac{1}{2})\,\Gamma(\frac{1}{2})}\int_0^1 t^{\nu-\frac{1}{2}}(1-t)^{-\frac{1}{2}}\cos\{z(1-t)^{\frac{1}{2}}\}\,dt.$$

Obvious transformations of this result, in addition to (1), are the following:

(2) $$J_\nu(z)=\frac{2(\frac{1}{2}z)^\nu}{\Gamma(\nu+\frac{1}{2})\,\Gamma(\frac{1}{2})}\int_0^1(1-t^2)^{\nu-\frac{1}{2}}\cos(zt)\,dt,$$

(3) $$J_\nu(z)=\frac{(\frac{1}{2}z)^\nu}{\Gamma(\nu+\frac{1}{2})\,\Gamma(\frac{1}{2})}\int_{-1}^1(1-t^2)^{\nu-\frac{1}{2}}\cos(zt)\,dt,$$

(4) $$J_\nu(z)=\frac{(\frac{1}{2}z)^\nu}{\Gamma(\nu+\frac{1}{2})\,\Gamma(\frac{1}{2})}\int_{-1}^1(1-t^2)^{\nu-\frac{1}{2}}e^{izt}\,dt,$$

(5) $$J_\nu(z)=\frac{2(\frac{1}{2}z)^\nu}{\Gamma(\nu+\frac{1}{2})\,\Gamma(\frac{1}{2})}\int_0^{\frac{1}{2}\pi}\cos(z\cos\theta)\sin^{2\nu}\theta\,d\theta,$$

(6) $$J_\nu(z)=\frac{(\frac{1}{2}z)^\nu}{\Gamma(\nu+\frac{1}{2})\,\Gamma(\frac{1}{2})}\int_0^{\pi}e^{iz\cos\theta}\sin^{2\nu}\theta\,d\theta.$$

The formula obtained by a partial integration of (5), namely

(7) $$J_\nu(z)=\frac{(2\nu-1)\,.\,(\frac{1}{2}z)^{\nu-1}}{\Gamma(\nu+\frac{1}{2})\,\Gamma(\frac{1}{2})}\int_0^{\frac{1}{2}\pi}\sin(z\cos\theta)\sin^{2\nu-2}\theta\cos\theta\,d\theta,$$

is sometimes useful; it is valid only when $R(\nu)>\frac{1}{2}$.

An expansion involving Bernoullian polynomials has been obtained from (4) by Nielsen* with the help of the expansion

$$e^{a\xi}=(1-e^{-a})\sum_{n=0}^{\infty}\frac{a^{n-1}\phi'_{n+1}(\xi+1)}{(n+1)!},$$

in which $\phi_n(\xi)$ denotes the nth Bernoullian polynomial and $a=izt$.

[NOTE. Integrals of the type (3) were studied before Poisson by Plana, *Mem. della R. Accad. delle Sci. di Torino*, XXVI. (1821), pp. 519—538, and subsequently by Kummer, *Journal für Math.* XII. (1834), pp. 144—147; Lobatto, *Journal für Math.* XVII. (1837), pp. 363—371; and Duhamel, *Cours d'Analyse*, II. (Paris, 1840), pp. 118—121.

A function, substantially equivalent to $J_\nu(z)$, defined by the equation

$$J(\mu, x)=\int_0^1 (1-v^2)^\mu \cos vx\,.\,dv,$$

was investigated by Lommel, *Archiv der Math. und Phys.* XXXVII. (1861), pp. 349—360. The converse problem of obtaining the differential equation satisfied by

$$z^\lambda\int_\alpha^\beta e^{zv}(v-\alpha)^{\mu-1}(v-\beta)^{\nu-1}dv$$

was also discussed by Lommel, *Archiv der Math. und Phys.* XL. (1863), pp. 101—126. In connexion with this integral see also Euler, *Inst. Calc. Int.* II. (Petersburg, 1769), § 1036, and Petzval, *Integration der linearen Differentialgleichungen* (Vienna, 1851), p. 48.]

3·31. *Inequalities derived from Poisson's integral.*

From § 3·3 (6) it follows that, if ν be real and greater than $-\frac{1}{2}$, then

$$(1)\qquad |J_\nu(z)|\leqslant\frac{|(\frac{1}{2}z)^\nu|}{\Gamma(\nu+\frac{1}{2})\,\Gamma(\frac{1}{2})}\int_0^\pi \exp|I(z)|\sin^{2\nu}\theta\,d\theta$$

$$=\frac{|(\frac{1}{2}z)^\nu|}{\Gamma(\nu+1)}\exp|I(z)|.$$

By using the recurrence formulae § 3·2 (1) and (4), we deduce in a similar manner that

$$(2)\qquad |J_\nu(z)|\leqslant\frac{|(\frac{1}{2}z)^\nu|}{|\Gamma(\nu+1)|}\left\{1+\frac{|(\frac{1}{2}z)^2|}{|(\nu+1)(\nu+2)|}\right\}\exp|I(z)|\qquad(-\tfrac{3}{2}<\nu\leqslant-\tfrac{1}{2}),$$

$$(3)\qquad |J_\nu'(z)|\leqslant\frac{|(\frac{1}{2}z)^{\nu-1}|}{2\,|\Gamma(\nu)|}\left\{1+\frac{\frac{1}{2}|z^2|}{|\nu(\nu+1)|}\right\}\exp|I(z)|\qquad(\nu>-\tfrac{1}{2}).$$

By using the expression† $\{2/(\pi z)\}^{\frac{1}{2}}\cos z$ for $J_{-\frac{1}{2}}(z)$ it may be shewn that (1) is valid when $\nu=-\frac{1}{2}$.

These inequalities should be compared with the less stringent inequalities obtained in § 3·13. When ν is complex, inequalities of a more complicated character can be obtained in the same manner, but they are of no great importance.

* *Math. Ann.* LIX. (1904), p. 108. The notation used in the text is that given in *Modern Analysis*, § 7·2; Nielsen uses a different notation.

† The reader should have no difficulty in verifying this result. A formal proof of a more general theorem will be given in § 3·4.

3·32. *Gegenbauer's generalisation of Poisson's integral.*

The integral formula

$$(1)\qquad J_{\nu+n}(z)=\frac{(-i)^n\,\Gamma(2\nu)\,.\,n!\,(\tfrac{1}{2}z)^\nu}{\Gamma(\nu+\tfrac{1}{2})\,\Gamma(\tfrac{1}{2})\,\Gamma(2\nu+n)}\int_0^\pi e^{iz\cos\theta}\sin^{2\nu}\theta\,.\,C_n^\nu(\cos\theta)\,d\theta,$$

in which $C_n^\nu(t)$ is the coefficient of α^n in the expansion of $(1-2\alpha t+\alpha^2)^{-\nu}$ in ascending powers of α, is due to Gegenbauer*; the formula is valid when $R(\nu)>-\tfrac{1}{2}$ and n is any of the integers 0, 1, 2, ... When $n=0$, it obviously reduces to Poisson's integral.

In the special case in which $\nu=\tfrac{1}{2}$, the integral assumes the form

$$(2)\qquad J_{n+\frac{1}{2}}(z)=(-i)^n\left(\frac{z}{2\pi}\right)^{\frac{1}{2}}\int_0^\pi e^{iz\cos\theta}P_n(\cos\theta)\sin\theta\,d\theta;$$

this equation has been the subject of detailed study by Whittaker†.

To prove Gegenbauer's formula, we take Poisson's integral in the form

$$J_{\nu+n}(z)=\frac{(\tfrac{1}{2}z)^{\nu+n}}{\Gamma(\nu+n+\tfrac{1}{2})\,\Gamma(\tfrac{1}{2})}\int_{-1}^1 e^{izt}(1-t^2)^{\nu+n-\frac{1}{2}}\,dt,$$

and integrate n times by parts; the result is

$$J_{\nu+n}(z)=\frac{(\tfrac{1}{2}z)^\nu}{(-2i)^n\,\Gamma(\nu+n+\tfrac{1}{2})\,\Gamma(\tfrac{1}{2})}\int_{-1}^1 e^{izt}\left\{\frac{d^n(1-t^2)^{\nu+n-\frac{1}{2}}}{dt^n}\right\}dt.$$

Now it is known that‡

$$\frac{d^n(1-t^2)^{\nu+n-\frac{1}{2}}}{dt^n}=\frac{(-2)^n\,n!\,\Gamma(\nu+n+\tfrac{1}{2})\,\Gamma(2\nu)}{\Gamma(\nu+\tfrac{1}{2})\,\Gamma(2\nu+n)}(1-t^2)^{\nu-\frac{1}{2}}\,C_n^\nu(t),$$

whence we have

$$(3)\qquad J_{\nu+n}(z)=\frac{(-i)^n\,.\,\Gamma(2\nu)\,.\,n!\,(\tfrac{1}{2}z)^\nu}{\Gamma(\nu+\tfrac{1}{2})\,\Gamma(\tfrac{1}{2})\,\Gamma(2\nu+n)}\int_{-1}^1 e^{izt}(1-t^2)^{\nu-\frac{1}{2}}C_n^\nu(t)\,dt,$$

and Gegenbauer's result is evident.

A symbolic form of Gegenbauer's equation is

$$(4)\qquad J_{\nu+n}(z)=\frac{(\tfrac{1}{2}z)^\nu\,\Gamma(2\nu)\,.\,n!}{\Gamma(2\nu+n)}(-i)^n C_n^\nu\left(\frac{d}{idz}\right)\{(\tfrac{1}{2}z)^{-\nu}J_\nu(z)\};$$

this was given by Rayleigh§ in the special case $\nu=\tfrac{1}{2}$.

The reader will find it instructive to establish (3) by induction with the aid of the recurrence formula

$$(n+1)\,C_{n+1}^\nu(t)=(2\nu+n)\,tC_n^\nu(t)-(1-t^2)\frac{dC_n^\nu(t)}{dt}.$$

* *Wiener Sitzungsberichte*, LXVII. (2), (1873), p. 203; LXX. (2), (1875), p. 15. See also Bauer, *Münchener Sitzungsberichte*, v. (1875), p. 262, and O. A. Smith, *Giornale di Mat.* (2) XII. (1905), pp. 365—373. The function $C_n^\nu(t)$ has been extensively studied by Gegenbauer in a series of memoirs in the *Wiener Sitzungsberichte*; some of the more important results obtained by him are given in *Modern Analysis*, § 15·8.

† *Proc. London Math. Soc.* XXXV. (1903), pp. 198–206. See §§ 6·17, 10·5.

‡ Cf. *Modern Analysis*, § 15·8.

§ *Proc. London Math. Soc.* IV. (1873), pp. 100, 263.

A formula which is a kind of converse of (4), namely*

$$(5)\qquad P_\nu^{-\mu}\left(\frac{z}{\sqrt{(\rho^2+z^2)}}\right)=\frac{\Gamma(\nu-\mu+1)}{\Gamma(\nu+1)\,(\rho^2+z^2)^{\frac{1}{2}\nu}}\,J_\mu\left(\rho\frac{\partial}{\partial z}\right)z^\nu,$$

in which $P_\nu^{-\mu}$ denotes a generalised Legendre function, is due to Filon, *Phil. Mag.* (6) VI. (1903), p. 198; the proof of this formula is left to the reader.

3·33. *Gegenbauer's double integral of Poisson's type.*

It has been shewn by Gegenbauer† that, when $R(\nu)>0$,

$$(1)\qquad J_\nu(\varpi)=\frac{(\tfrac{1}{2}\varpi)^\nu}{\pi\Gamma(\nu)}\int_0^\pi\int_0^\pi \exp\left[iZ\cos\theta-iz(\cos\phi\cos\theta+\sin\phi\sin\theta\cos\psi)\right]\sin^{2\nu-1}\psi\sin^{2\nu}\theta\,d\psi\,d\theta,$$

where $\varpi^2=Z^2+z^2-2Zz\cos\phi$ and Z, z, ϕ are unrestricted (complex) variables. This result was originally obtained by Gegenbauer by applying elaborate integral transformations to certain addition formulae which will be discussed in Chapter XI. It is possible, however, to obtain the formula in a quite natural manner by means of transformations of a type used in the geometry of the sphere‡.

After noticing that, when $z=0$, the formula reduces to a result which is an obvious consequence of Poisson's integral, namely

$$J_\nu(Z)=\frac{(\tfrac{1}{2}Z)^\nu}{\pi\,\Gamma(\nu)}\int_0^\pi e^{iZ\cos\theta}\sin^{2\nu}\theta\,.\int_0^\pi\sin^{2\nu-1}\psi\,d\psi,$$

we proceed to regard ψ and θ as longitude and colatitude of a point on a unit sphere; we denote the direction-cosines of the vector from the centre to this point by (l, m, n) and the element of surface at the point by $d\omega$.

We then transform Poisson's integral by making a cyclical interchange of the coordinate axes in the following manner§:

$$\begin{aligned}J_\nu(\varpi)&=\frac{(\tfrac{1}{2}\varpi)^\nu}{\pi\Gamma(\nu)}\int_0^\pi\int_0^\pi e^{i\varpi\cos\theta}\sin^{2\nu}\theta\sin^{2\nu-1}\psi\,d\theta\,d\psi\\&=\frac{(\tfrac{1}{2}\varpi)^\nu}{\pi\Gamma(\nu)}\iint_{m\geqslant 0}e^{i\varpi n}\,m^{2\nu-1}\,d\omega\\&=\frac{(\tfrac{1}{2}\varpi)^\nu}{\pi\Gamma(\nu)}\iint_{n\geqslant 0}e^{i\varpi l}\,n^{2\nu-1}\,d\omega\\&=\frac{(\tfrac{1}{2}\varpi)^\nu}{\pi\Gamma(\nu)}\int_0^{\frac{1}{2}\pi}\int_0^{2\pi}e^{i\varpi\sin\theta\cos\psi}\cos^{2\nu-1}\theta\sin\theta\,d\psi\,d\theta.\end{aligned}$$

* It is supposed that

$$\frac{\partial^\mu z^\nu}{\partial z^\mu}=\frac{\Gamma(\nu+1)\,z^{\nu-\mu}}{\Gamma(\nu-\mu+1)}.$$

† *Wiener Sitzungsberichte*, LXXIV. (2), (1877), pp. 128—129.

‡ This method is effective in proving numerous formulae of which analytical proofs were given by Gegenbauer; and it seems not unlikely that he discovered these formulae by the method in question; cf. §§ 12·12, 12·14. The device is used by Beltrami, *Lombardo Rendiconti*, (2) XIII. (1880), p. 328, for a rather different purpose.

§ The symbol $\iint_{m\geqslant 0}$ means that the integration extends over the surface of the hemisphere on which m is positive.

Now the integrand is an integral periodic function of ψ, and so the limits of integration with respect to ψ may be taken to be α and $\alpha + 2\pi$, where α is an arbitrary (complex) number. This follows from Cauchy's theorem.

We thus get

$$J_\nu(\varpi) = \frac{(\frac{1}{2}\varpi)^\nu}{\pi\Gamma(\nu)} \int_0^{\frac{1}{2}\pi} \int_\alpha^{\alpha+2\pi} e^{i\varpi \sin\theta \cos\psi} \cos^{2\nu-1}\theta \sin\theta \, d\psi \, d\theta$$

$$= \frac{(\frac{1}{2}\varpi)^\nu}{\pi\Gamma(\nu)} \int_0^{\frac{1}{2}\pi} \int_0^{2\pi} e^{i\varpi \sin\theta \cos(\psi+\alpha)} \cos^{2\nu-1}\theta \sin\theta \, d\psi \, d\theta.$$

We now define α by the pair of equations

$$\varpi \cos\alpha = Z - z\cos\phi, \quad \varpi \sin\alpha = z \sin\phi,$$

so that

$$J_\nu(\varpi) = \frac{(\frac{1}{2}\varpi)^\nu}{\pi\Gamma(\nu)} \int_0^{\frac{1}{2}\pi} \int_0^{2\pi} \exp\left[i(Z - z\cos\phi)\sin\theta\cos\psi - iz\sin\phi\sin\psi\sin\theta\right] \cos^{2\nu-1}\theta \sin\theta \, d\psi \, d\theta.$$

The only difference between this formula and the formula

$$J_\nu(\varpi) = \frac{(\frac{1}{2}\varpi)^\nu}{\pi\Gamma(\nu)} \int_0^{\frac{1}{2}\pi} \int_0^{2\pi} \exp\left[i\varpi\sin\theta\cos\psi\right] \cos^{2\nu-1}\theta \sin\theta \, d\psi \, d\theta$$

is in the form of the exponential factor; and we now retrace the steps of the analysis with the modified form of the exponential factor. When the steps are retraced the successive exponents are

$$i(Z - z\cos\phi)\, l - iz\sin\phi \,.\, m,$$
$$i(Z - z\cos\phi)\, n - iz\sin\phi \,.\, l,$$
$$i(Z - z\cos\phi)\cos\theta - iz\sin\phi\cos\psi\sin\theta.$$

The last expression is

$$iZ\cos\theta - iz(\cos\phi\cos\theta + \sin\phi\sin\theta\cos\psi),$$

so that the result of retracing the steps is

$$\frac{(\frac{1}{2}\varpi)^\nu}{\pi\Gamma(\nu)} \int_0^\pi \int_0^\pi \exp\left[iZ\cos\theta - iz(\cos\phi\cos\theta + \sin\phi\sin\theta\cos\psi)\right] \sin^{2\nu-1}\psi \sin^{2\nu}\theta \, d\psi \, d\theta,$$

and consequently Gegenbauer's formula is established.

[NOTE. The device of using transformations of polar coordinates, after the manner of this section, to evaluate definite integrals seems to be due to Legendre, *Mém. de l'Acad. des Sci.*, 1789, p. 372, and Poisson, *Mém. de l'Acad. des Sci.* III. (1818), p. 126.]

3·4. *The expression of $J_{\pm(n+\frac{1}{2})}(z)$ in finite terms.*

We shall now deduce from Poisson's integral the important theorem that, *when ν is half of an odd integer, the function $J_\nu(z)$ is expressible in finite terms by means of algebraic and trigonometrical functions of z.*

It will appear later (§ 4·74) that, when ν has not such a value, then $J_\nu(z)$ is not so expressible; but of course this converse theorem is of a much more recondite character than the theorem which is now about to be proved.

[NOTE. Solutions in finite terms of differential equations associated with $J_{n+\frac{1}{2}}(z)$ were obtained by various early writers; it was observed by Euler, *Misc. Taurinensia*, III. (1762—1765), p. 76 that a solution of the equation for $e^{iz} J_{n+\frac{1}{2}}(z)$ is expressible in finite terms; while the equation satisfied by $z^{\frac{1}{2}} J_{n+\frac{1}{2}}(z)$ was solved in finite terms by Laplace, *Conn. des Tems*, 1823 [1820], pp. 245—257 and *Mécanique Céleste*, V. (Paris, 1825), pp. 82—84; by Plana, *Mem. della R. Accad. delle Sci. di Torino*, XXVI. (1821), pp. 533—534; by Paoli, *Mem. di Mat. e di Fis. (Modena)*, XX. (1828), pp. 183—188; and also by Stokes in 1850, *Trans. Camb. Phil. Soc.* IX. (1856), p. 187 [*Math. and Phys. Papers*, II. (1883), p. 356]. The investigation which will now be given is based on the work of Lommel, *Studien über die Bessel'schen Functionen* (Leipzig, 1868), pp. 51—56.]

It is convenient to restrict n to be a *positive* integer (zero included), and then, by § 3·3 (4),

$$\begin{aligned} J_{n+\frac{1}{2}}(z) &= \frac{(\frac{1}{2}z)^{n+\frac{1}{2}}}{n!\sqrt{\pi}} \int_{-1}^{1} e^{izt}(1-t^2)^n\,dt \\ &= \frac{(\frac{1}{2}z)^{n+\frac{1}{2}}}{n!\sqrt{\pi}} \left[-e^{izt} \sum_{r=0}^{2n} \frac{i^{r+1}}{z^{r+1}} \frac{d^r(1-t^2)^n}{dt^r} \right]_{-1}^{1}, \end{aligned}$$

when we integrate by parts $2n+1$ times; since $(1-t^2)^n$ is a polynomial of degree $2n$, the process then terminates.

To simplify the last expression we observe that if $d^r(1-t^2)^n/dt^r$ be calculated from Leibniz' theorem by writing $(1-t^2)^n \equiv (1-t)^n(1+t)^n$, the only term which does not vanish at the upper limit arises from differentiating n times the factor $(1-t)^n$, and therefore from differentiating the other factor $r-n$ times; so that we need consider only the terms for which $r \geqslant n$.

Hence
$$\left[\frac{d^r(1-t^2)^n}{dt^r}\right]_{t=1} = (-)^n \cdot {}_rC_n \cdot n! \frac{n!\,2^{2n-r}}{(2n-r)!}$$

and similarly
$$\left[\frac{d^r(1-t^2)^n}{dt^r}\right]_{t=-1} = (-)^{r-n} \cdot {}_rC_n \cdot n! \frac{n!\,2^{2n-r}}{(2n-r)!}.$$

It follows that

$$\begin{aligned} J_{n+\frac{1}{2}}(z) = \frac{(\frac{1}{2}z)^{n+\frac{1}{2}}}{\sqrt{\pi}} \Bigg[&(-)^{n+1} e^{iz} \sum_{r=n}^{2n} \frac{i^{r+1}\,2^{2n-r} \cdot r!}{z^{r+1} \cdot (r-n)!\,(2n-r)!} \\ &+ (-)^{n+1} e^{-iz} \sum_{r=n}^{2n} \frac{(-i)^{r+1}\,2^{2n-r} \cdot r!}{z^{r+1} \cdot (r-n)!\,(2n-r)!} \Bigg], \end{aligned}$$

and hence

$$(1) \quad J_{n+\frac{1}{2}}(z) = \frac{1}{\sqrt{(2\pi z)}} \left[e^{iz} \sum_{r=0}^{n} \frac{i^{r-n-1}(n+r)!}{r!\,(n-r)!\,(2z)^r} + e^{-iz} \sum_{r=0}^{n} \frac{(-i)^{r-n-1}(n+r)!}{r!\,(n-r)!\,(2z)^r} \right]$$

This result may be written in the form*

$$\begin{aligned} (2) \quad J_{n+\frac{1}{2}}(z) = \left(\frac{2}{\pi z}\right)^{\frac{1}{2}} \Bigg[&\sin(z - \tfrac{1}{2}n\pi) \sum_{r=0}^{\leqslant \frac{1}{2}n} \frac{(-)^r \cdot (n+2r)!}{(2r)!\,(n-2r)!\,(2z)^{2r}} \\ &+ \cos(z - \tfrac{1}{2}n\pi) \sum_{r=0}^{\leqslant \frac{1}{2}(n-1)} \frac{(-)^r \cdot (n+2r+1)!}{(2r+1)!\,(n-2r-1)!\,(2z)^{2r+1}} \Bigg]. \end{aligned}$$

* A compact method of obtaining this formula is given by de la Vallée Poussin, *Ann. de la Soc. Sci. de Bruxelles*, XXIX. (1905), pp. 140—143.

In particular we have

$$(3) \qquad J_{\frac{1}{2}}(z) = \left(\frac{2}{\pi z}\right)^{\frac{1}{2}} \sin z, \quad J_{\frac{3}{2}}(z) = \left(\frac{2}{\pi z}\right)^{\frac{1}{2}} \left(\frac{\sin z}{z} - \cos z\right);$$

the former of these results is also obvious from the power series for $J_{\frac{1}{2}}(z)$.

Again, from the recurrence formula we have

$$J_{n+\frac{1}{2}}(z) = (-)^n z^{n+\frac{1}{2}} \left(\frac{d}{z\,dz}\right)^n \{z^{-\frac{1}{2}} J_{\frac{1}{2}}(z)\},$$

and hence, from (1),

$$e^{iz} \sum_{r=0}^{n} \frac{i^{r-n} \cdot (n+r)!}{r!\,(n-r)!\,(2z)^r} - e^{-iz} \sum_{r=0}^{n} \frac{(-i)^{r-n} \cdot (n+r)!}{r!\,(n-r)!\,(2z)^r}$$
$$\equiv (-)^n z^{n+1} \left(\frac{d}{z\,dz}\right)^n \frac{e^{iz}}{z} - (-)^n z^{n+1} \left(\frac{d}{z\,dz}\right)^n \frac{e^{-iz}}{z}.$$

But, obviously, by induction we can express

$$z^{n+1} \left(\frac{d}{z\,dz}\right)^n \frac{e^{\pm iz}}{z}$$

as a polynomial in $1/z$ multiplied by $e^{\pm iz}$, *and so we must have*

$$e^{\pm iz} \sum_{r=0}^{n} \frac{(\pm i)^{r-n} (n+r)!}{r!\,(n-r)!\,(2z)^r} \equiv (-)^n z^{n+1} \left(\frac{d}{z\,dz}\right)^n \frac{e^{\pm iz}}{z};$$

for, if not, the preceding identity would lead to a result of the form

$$e^{iz} \phi_1(z) - e^{-iz} \phi_2(z) \equiv 0,$$

where $\phi_1(z)$ and $\phi_2(z)$ are polynomials in $1/z$; and such an identity is obviously impossible*.

Hence it follows that†

$$e^{iz} \sum_{r=0}^{n} \frac{i^{r-n} \cdot (n+r)!}{r!\,(n-r)!\,(2z)^r} + e^{-iz} \sum_{r=0}^{n} \frac{(-i)^{r-n} \cdot (n+r)!}{r!\,(n-r)!\,(2z)^r}$$
$$= (-)^n z^{n+1} \left(\frac{d}{z\,dz}\right)^n \frac{e^{iz} + e^{-iz}}{z}$$
$$= (-)^n (2\pi)^{\frac{1}{2}} z^{n+1} \left(\frac{d}{z\,dz}\right)^n \{z^{-\frac{1}{2}} J_{-\frac{1}{2}}(z)\}$$
$$= (-)^n (2\pi z)^{\frac{1}{2}} J_{-n-\frac{1}{2}}(z).$$

Consequently

$$(4) \quad J_{-n-\frac{1}{2}}(z) = \frac{1}{\sqrt{(2\pi z)}} \left[e^{iz} \sum_{r=0}^{n} \frac{i^{r+n} \cdot (n+r)!}{r!\,(n-r)!\,(2z)^r} + e^{-iz} \sum_{r=0}^{n} \frac{(-i)^{r+n} \cdot (n+r)!}{r!\,(n-r)!\,(2z)^r} \right]$$

* Cf. Hobson, *Squaring the Circle* (Cambridge, 1913), p. 51.

† From the series

$$J_{-\frac{1}{2}}(z) = \frac{(\frac{1}{2}z)^{-\frac{1}{2}}}{\Gamma(\frac{1}{2})} \sum_{m=0}^{\infty} \frac{(-)^m (\frac{1}{2}z)^{2m}}{m!\,\frac{1}{2} \cdot \frac{3}{2} \ldots (m-\frac{1}{2})},$$

it is obvious that

$$J_{-\frac{1}{2}}(z) = \left(\frac{2}{\pi z}\right)^{\frac{1}{2}} \cos z.$$

and hence

$$(5)\quad J_{-n-\frac{1}{2}}(z)=\left(\frac{2}{\pi z}\right)^{\frac{1}{2}}\left[\cos(z+\tfrac{1}{2}n\pi)\sum_{r=0}^{\leqslant\frac{1}{2}n}\frac{(-)^r.(n+2r)!}{(2r)!(n-2r)!(2z)^{2r}}\right.$$
$$\left.-\sin(z+\tfrac{1}{2}n\pi)\sum_{r=0}^{\leqslant\frac{1}{2}(n-1)}\frac{(-)^r.(n+2r+1)!}{(2r+1)!(n-2r-1)!(2z)^{2r+1}}\right].$$

In particular, we have

$$(6)\qquad J_{-\frac{1}{2}}(z)=\left(\frac{2}{\pi z}\right)^{\frac{1}{2}}\cos z,\quad J_{-\frac{3}{2}}(z)=\left(\frac{2}{\pi z}\right)^{\frac{1}{2}}\left(-\frac{\cos z}{z}-\sin z\right).$$

We have now expressed in finite terms any Bessel function, whose order is half of an odd integer, by means of algebraic and trigonometrical functions.

The explicit expression of a number of these functions can be written down from numerical results contained in a letter from Hermite to Gordan, *Journal für Math.* LXVI. (1873), pp. 303—311.

3·41. *Notations for functions whose order is half of an odd integer.*

Functions of the types $J_{\pm(n+\frac{1}{2})}(z)$ occur with such frequency in various branches of Mathematical Physics that various writers have found it desirable to denote them by a special functional symbol. Unfortunately no common notation has been agreed upon and none of the many existing notations can be said to predominate over the others. Consequently, apart from the summary which will now be given, the notations in question will not be used in this work.

In his researches on vibrating spheres surrounded by a gas, Stokes, *Phil. Trans. of the Royal Soc.* CLVIII. (1868), p. 451 [*Math. and Phys. Papers*, IV. (1904), p. 306], made use of the series

$$1+\frac{n(n+1)}{2.imr}+\frac{(n-1)n(n+1)(n+2)}{2.4.(imr)^2}+\ldots,$$

which is annihilated by the operator

$$\frac{d^2}{dr^2}-2im\frac{d}{dr}-\frac{n(n+1)}{r^2}.$$

This series Stokes denoted by the symbol $f_n(r)$ and he wrote

$$r\psi_n=S_n e^{-imr}f_n(r)+S_n' e^{imr}f_n(-r),$$

where S_n and S_n' are zonal surface harmonics; so that ψ_n is annihilated by the total operator

$$\frac{d^2}{dr^2}+\frac{2}{r}\frac{d}{dr}+m^2-\frac{n(n+1)}{r^2},$$

and by the partial operator

$$\frac{\partial^2}{\partial r^2}+\frac{2}{r}\frac{\partial}{\partial r}+\frac{1}{r^2\sin\theta}\frac{\partial}{\partial\theta}\left\{\sin\theta\frac{\partial}{\partial\theta}\right\}+m^2.$$

In this notation Stokes was followed by Rayleigh, *Proc. London Math. Soc.* IV. (1873), pp. 93—103, 253—283, and again *Proc. Royal Soc.* LXXII. (1903), pp. 40—41 [*Scientific Papers*, V. (1912), pp. 112—114], apart from the comparatively trivial change that Rayleigh would have written $f_n(imr)$ where Stokes wrote $f_n(r)$.

In order to obtain a solution finite at the origin, Rayleigh found it necessary to take $S_n' \equiv (-)^{n+1} S_n$ in the course of his analysis, and then

$$\psi_n = (-i)^{n+1} m S_n \left(\frac{2\pi}{mr}\right)^{\frac{1}{2}} J_{n+\frac{1}{2}}(mr).$$

It follows from § 3·4 that $$\frac{e^{-ir} f_n(ir)}{r^{n+1}} = \left(i \frac{\partial}{r\partial r}\right)^n \frac{e^{-ir}}{r},$$

and that $$J_{n+\frac{1}{2}}(r) = \frac{1}{\surd(2\pi r)} [e^{-ir} i^{n+1} f_n(ir) + e^{ir} i^{-n-1} f_n(-ir)].$$

In order to have a simple notation for the combinations of the types $e^{\mp ir} f_n(\pm ir)$ which are required for solutions finite at the origin, Lamb found it convenient to write

$$\psi_n(z) = 1 - \frac{z^2}{2(2n+3)} + \frac{z^4}{2 \cdot 4 \cdot (2n+3)(2n+5)} - \dots,$$

in his earlier papers, *Proc. London Math. Soc.* XIII. (1882), pp. 51—66; 189—212; XV. (1884), pp. 139—149; XVI. (1885), pp. 27—43; *Phil. Trans. of the Royal Soc.* CLXXIV. (1883), pp. 519—549; and he was followed by Rayleigh, *Proc. Royal Soc.* LXXVII. A, (1906), pp. 486—499 [*Scientific Papers,* V. (1912), pp. 300—312], and by Love*, *Proc. London Math. Soc.* XXX. (1899), pp. 308—321.

With this notation it is evident that

$$\psi_n(z) = \frac{\Gamma(n+\frac{3}{2})}{(\frac{1}{2}z)^{n+\frac{1}{2}}} J_{n+\frac{1}{2}}(z) = (-)^n 1 \cdot 3 \cdot 5 \dots (2n+1) \cdot \left(\frac{d}{zdz}\right)^n \frac{\sin z}{z}.$$

Subsequently, however, Lamb found it convenient to modify this notation, and accordingly in his treatise on Hydrodynamics and also *Proc. London Math. Soc.* XXXII. (1901), pp. 11—20, 120—150 he used the notation†

$$\psi_n(z) = \frac{1}{1 \cdot 3 \cdot 5 \dots (2n+1)} \left[1 - \frac{z^2}{2(2n+3)} + \frac{z^4}{2 \cdot 4(2n+3)(2n+5)} - \dots\right],$$

and he also wrote $$f_n(z) = \left(-\frac{d}{zdz}\right)^n \frac{e^{-iz}}{z} = \Psi_n(z) - i\psi_n(z),$$

so that $$\Psi_n(z) = \frac{(-)^n \surd(\frac{1}{2}\pi) \cdot J_{-n-\frac{1}{2}}(z)}{z^{n+\frac{1}{2}}}, \quad \psi_n(z) = \frac{\surd(\frac{1}{2}\pi) \cdot J_{n+\frac{1}{2}}(z)}{z^{n+\frac{1}{2}}},$$

while Rayleigh, *Phil. Trans. of the Royal Soc.* CCIII. A, (1904), pp. 87—110 [*Scientific Papers,* V. (1912) pp. 149—161] found it convenient to replace the symbol $f_n(z)$ by $\chi_n(z)$. Love, *Phil. Trans. of the Royal Soc.* CCXV. A, (1915), p. 112 omitted the factor $(-)^n$ and wrote

$$E_n(z) = \left(\frac{d}{zdz}\right)^n \frac{e^{-iz}}{z}, \quad \psi_n(z) = \left(\frac{d}{zdz}\right)^n \frac{\sin z}{z},$$

while yet another notation has been used by Sommerfeld, *Ann. der Physik und Chemie,* (4) XXVIII. (1909), pp. 665—736, and two of his pupils, namely March, *Ann. der Physik und Chemie,* (4) XXXVII. (1912), p. 29 and Rybczyński, *Ann. der Physik und Chemie,* (4) XLI. (1913), p. 191; this notation is

$$\psi_n(z) = (\tfrac{1}{2}\pi z)^{\frac{1}{2}} J_{n+\frac{1}{2}}(z) = z^{n+1}\left(-\frac{d}{zdz}\right)^n \frac{\sin z}{z},$$

$$\zeta_n(z) = (\tfrac{1}{2}\pi z)^{\frac{1}{2}} [J_{n+\frac{1}{2}}(z) + (-)^n i J_{-n-\frac{1}{2}}(z)],$$

and it is certainly the best adapted for the investigation on electric waves which was the subject of their researches.

* In this paper Love defined the function $E_n(z)$ as $(-)^n \cdot 1 \cdot 3 \dots (2n-1) \left(\frac{d}{z\,dz}\right)^n \frac{e^{-iz}}{z}$, but, as stated, he modified the definition in his later work.

† This is nearer the notation used by Heine, *Handbuch der Kugelfunctionen,* I. (Berlin, 1878), p. 82; except that Heine defined $\psi_n(z)$ to be twice the expression on the right in his treatise, but not in his memoir, *Journal für Math.* LXIX. (1869), pp. 128—141.

Sommerfeld's notation is a slightly modified form of the notation used by L. Lorenz, who used v_n and $v_n + (-)^n iw_n$ in place of ψ_n and ζ_n; see his memoir on reflexion and refraction of light, *K. Danske Videnskabernes Selskabs Skrifter*, (6) VI. (1890), [*Oeuvres scientifiques*, I. (1898), pp. 405—502.]

3·5. *A second solution of Bessel's equation for functions of integral order.*

It has been seen (§ 3·12) that, whenever ν is not an integer, a fundamental system of solutions of Bessel's equation for functions of order ν is formed by the pair of functions $J_\nu(z)$ and $J_{-\nu}(z)$. When ν is an integer $(=n)$, this is no longer the case, on account of the relation $J_{-n}(z) = (-)^n J_n(z)$.

It is therefore necessary to obtain a solution of Bessel's equation which is linearly independent of $J_n(z)$; and the combination of this solution with $J_n(z)$ will give a fundamental system of solutions.

The solution which will now be constructed was obtained by Hankel*; the full details of the analysis involved in the construction were first published by Bôcher†.

An alternative method of constructing Hankel's solution was discovered by Forsyth; his procedure is based on the general method of Frobenius, *Journal für Math.* LXXVI. (1874), pp. 214—235, for dealing with any linear differential equation. Forsyth's solution was contained in his lectures on differential equations delivered in Cambridge in 1894, and it has since been published in his *Theory of Differential Equations*, IV. (Cambridge, 1902), pp. 101—102, and in his *Treatise on Differential Equations* (London, 1903 and 1914), Chapter VI. note 1.

It is evident that, if ν be unrestricted, and if n be any integer (positive, negative or zero), the function

$$J_\nu(z) - (-)^n J_{-\nu}(z)$$

is a solution of Bessel's equation for functions of order ν; and this function vanishes when $\nu = n$.

Consequently, so long as $\nu \neq n$, the function

$$\frac{J_\nu(z) - (-)^n J_{-\nu}(z)}{\nu - n}$$

is also a solution of Bessel's equation for functions of order ν; and this function assumes an undetermined form‡ when $\nu = n$.

We shall now evaluate

$$\lim_{\nu \to n} \frac{J_\nu(z) - (-)^n J_{-\nu}(z)}{\nu - n},$$

and we shall shew that it is a solution of Bessel's equation for functions of

* *Math. Ann.* I. (1869), pp 469—472.

† *Annals of Math.* VI. (1892), pp. 85—90. See also Niemöller, *Zeitschrift für Math. und Phys.* XXV. (1880), pp. 65—71

‡ The essence of Hankel's investigation is the construction of an expression which satisfies the equation when ν is not an integer, which assumes an undetermined form when ν is equal to the integer n and which has a limit when $\nu \to n$.

order n and that it is linearly independent of $J_n(z)$; so that it may be taken to be the second solution required*.

It is evident that

$$\frac{J_\nu(z)-(-)^n J_{-\nu}(z)}{\nu-n}=\frac{J_\nu(z)-J_n(z)}{\nu-n}-(-)^n\frac{J_{-\nu}(z)-J_{-n}(z)}{\nu-n}$$

$$\to\left[\frac{\partial J_\nu(z)}{\partial\nu}-(-)^n\frac{\partial J_{-\nu}(z)}{\partial\nu}\right]_{\nu=n},$$

as $\nu\to n$, since both of the differential coefficients exist†.

Hence

$$\lim_{\nu\to n}\frac{J_\nu(z)-(-)^n J_{-\nu}(z)}{\nu-n}$$

exists; it is called a Bessel function *of the second kind* of order n.

To distinguish it from other functions which are also called functions of the second kind it may be described as *Hankel's function*. Following Hankel, we shall denote it by the symbol‡ $\mathbf{Y}_n(z)$ so that

$$\text{(1)}\qquad \mathbf{Y}_n(z)=\lim_{\nu\to n}\left[\frac{J_\nu(z)-(-)^n J_{-\nu}(z)}{\nu-n}\right],$$

and also

$$\text{(2)}\qquad \mathbf{Y}_n(z)=\left[\frac{\partial J_\nu(z)}{\partial\nu}-(-)^n\frac{\partial J_{-\nu}(z)}{\partial\nu}\right]_{\nu=n}.$$

It has now to be shewn that $\mathbf{Y}_n(z)$ is a solution of Bessel's equation.

Since the two functions $J_{\pm\nu}(z)$ are analytic functions of both z and ν, the order of performing partial differentiations on $J_{\pm\nu}(z)$ with respect to z and ν is a matter of indifference§. Hence the result of differentiating the pair of equations

$$\nabla_\nu J_{\pm\nu}(z)=0$$

with respect to ν may be written

$$z^2\frac{d^2}{dz^2}\frac{\partial J_{\pm\nu}(z)}{\partial\nu}+z\frac{d}{dz}\frac{\partial J_{\pm\nu}(z)}{\partial\nu}+(z^2-\nu^2)\frac{\partial J_{\pm\nu}(z)}{\partial\nu}-2\nu J_{\pm\nu}(z)=0.$$

When we combine the results contained in this formula, we find that

$$\nabla_\nu\left[\frac{\partial J_\nu(z)}{\partial\nu}-(-)^n\frac{\partial J_{-\nu}(z)}{\partial\nu}\right]=2\nu\{J_\nu(z)-(-)^n J_{-\nu}(z)\},$$

* The reader will realise that, given a solution of a differential equation, it is not obvious that a limiting form of this solution is a solution of the corresponding limiting form of the equation.

† See § 3·1. It is conventional to write differentiations with respect to z as total differential coefficients while differentiations with respect to ν are written as partial differential coefficients. Of course, in many parts of the theory, variations in ν are not contemplated.

‡ The symbol $Y_n(z)$, which was actually used by Hankel, is used in this work to denote a function equal to $1/\pi$ times Hankel's function (§ 3·54).

§ See, e.g. Hobson, *Functions of a Real Variable* (1921), §§ 312, 313.

so that

$$\nabla_n\left[\frac{\partial J_\nu(z)}{\partial\nu}-(-)^n\frac{\partial J_{-\nu}(z)}{\partial\nu}\right]=(\nu^2-n^2)\left[\frac{\partial J_\nu(z)}{\partial\nu}-(-)^n\frac{\partial J_{-\nu}(z)}{\partial\nu}\right]$$
$$+2\nu\{J_\nu(z)-(-)^n J_{-\nu}(z)\}.$$

Now make $\nu\to n$. All the expressions in the last equation are continuous functions of ν, and so we have

$$\nabla_n\left[\frac{\partial J_\nu(z)}{\partial\nu}-(-)^n\frac{\partial J_{-\nu}(z)}{\partial\nu}\right]_{\nu=n}=0,$$

where ν is to be made equal to n immediately after the differentiations with respect to ν have been performed. We have therefore proved that

(3) $$\nabla_n \mathbf{Y}_n(z)=0,$$

so that $\mathbf{Y}_n(z)$ *is a solution of Bessel's equation for functions of order* n.

It is to be noticed that

$$\mathbf{Y}_{-n}(z)=\lim_{\nu\to -n}\frac{J_\nu(z)-(-)^{-n}J_{-\nu}(z)}{\nu+n}$$
$$=\lim_{\mu\to n}\frac{J_{-\mu}(z)-(-)^n J_\mu(z)}{-\mu+n},$$

whence follows a result substantially due to Lommel*,

(4) $$\mathbf{Y}_{-n}(z)=(-)^n\mathbf{Y}_n(z).$$

Again,

$$\mathbf{Y}_0(z)=\left[\frac{\partial J_\nu(z)}{\partial\nu}\right]_{\nu=0}-\left[\frac{\partial J_{-\nu}(z)}{\partial\nu}\right]_{\nu=0},$$

while, because $J_\nu(z)$ is a monogenic function of ν at $\nu=0$, we have

$$\left[\frac{\partial J_{-\nu}(z)}{\partial\nu}\right]_{\nu=0}=\left[\frac{\partial J_\nu(z)}{\partial(-\nu)}\right]_{\nu=0}=-\left[\frac{\partial J_\nu(z)}{\partial\nu}\right]_{\nu=0},$$

and hence it follows that

(5) $$\mathbf{Y}_0(z)=2\left[\frac{\partial J_\nu(z)}{\partial\nu}\right]_{\nu=0}.$$

A result equivalent to this was given by Duhamel† as early as 1840.

3·51. *The expansion of* $\mathbf{Y}_0(z)$ *in an ascending series.*

Before considering the expansion of the general function $\mathbf{Y}_n(z)$, it is convenient to examine the function of order zero because the analysis is simpler and the resulting expansion is more compact.

We use the formula just obtained,

$$\mathbf{Y}_0(z)=2\left[\frac{\partial}{\partial\nu}\left\{\sum_{m=0}^{\infty}\frac{(-)^m(\tfrac12 z)^{\nu+2m}}{m!\,\Gamma(\nu+m+1)}\right\}\right]_{\nu=0},$$

* *Studien über die Bessel'schen Functionen* (Leipzig, 1868), p. 87. Lommel actually proved this result for what is sometimes called *Neumann's function* of the second kind. See § 3·58 (8).

† *Cours d'Analyse*, II. (Paris, 1840), pp. 122—124.

and the result of term-by-term differentiation is

$$\mathbf{Y}_0(z) = 2\left[\sum_{m=0}^{\infty} \frac{(-)^m (\tfrac{1}{2}z)^{\nu+2m}}{m!\,\Gamma(\nu+m+1)}\left\{\log(\tfrac{1}{2}z) - \frac{\partial}{\partial\nu}\log\Gamma(\nu+m+1)\right\}\right]_{\nu=0}$$

$$= 2\sum_{m=0}^{\infty} \frac{(-)^m (\tfrac{1}{2}z)^{2m}}{(m!)^2}\{\log\tfrac{1}{2}z - \psi(m+1)\},$$

where ψ denotes, as is customary, the logarithmic derivate of the Gamma-function*.

Since $0 < \psi(m+1) < m$ when $m = 1, 2, 3, \ldots$ the convergence of the series for $\mathbf{Y}_0(z)$ may be established by using D'Alembert's ratio-test for the series in which $\psi(m+1)$ is replaced by m. The convergence is also an immediate consequence of a general theorem concerning analytic functions. See *Modern Analysis*, § 5·3.

The following forms of the expansion are to be noticed:

(1) $$\mathbf{Y}_0(z) = 2\sum_{m=0}^{\infty} \frac{(-)^m (\tfrac{1}{2}z)^{2m}}{(m!)^2}\{\log(\tfrac{1}{2}z) - \psi(m+1)\},$$

(2) $$\mathbf{Y}_0(z) = 2\left[\log(\tfrac{1}{2}z)\,.\,J_0(z) - \sum_{m=0}^{\infty} \frac{(-)^m (\tfrac{1}{2}z)^{2m}}{(m!)^2}\psi(m+1)\right],$$

(3) $$\mathbf{Y}_0(z) = 2\{\gamma + \log(\tfrac{1}{2}z)\}\,J_0(z) - 2\sum_{m=1}^{\infty} \frac{(-)^m (\tfrac{1}{2}z)^{2m}}{(m!)^2}\left\{\frac{1}{1} + \frac{1}{2} + \ldots + \frac{1}{m}\right\}.$$

The reader will observe that

$$\tfrac{1}{2}\mathbf{Y}_0(z) + (\log 2 - \gamma)\,J_0(z)$$

is a solution of Bessel's equation for functions of order zero. The expansion of this function is

$$(\log z)\sum_{m=0}^{\infty} \frac{(-)^m (\tfrac{1}{2}z)^{2m}}{(m!)^2} - \sum_{m=1}^{\infty} \frac{(-)^m (\tfrac{1}{2}z)^{2m}}{(m!)^2}\left\{\frac{1}{1} + \frac{1}{2} + \ldots + \frac{1}{m}\right\}.$$

This function was adopted as the canonical function of the second kind of order zero by Neumann, *Theorie der Bessel'schen Functionen* (Leipzig, 1867), pp. 42—44; see § 3·57.

But the series was obtained as a solution of Bessel's equation, long before, by Euler †. Euler's result in his own notation is that the general solution of the equation

$$xx\,\partial\partial y + x\,\partial x\,\partial y + gx^n y\,\partial x^2 = 0$$

is
$$y = \frac{2Ag}{n^3}x^n - \frac{6Ag^2}{1\,.\,8n^5}x^{2n} + \frac{22Ag^3}{1\,.\,8\,.\,27n^7}x^{3n} - \frac{100Ag^4}{1\,.\,8\,.\,27\,.\,64n^9}x^{4n} + \text{etc.}$$

$$+ A\left(1 - \frac{g}{nn}x^n + \frac{g^2}{1\,.\,4n^4}x^{2n} - \frac{g^3}{1\,.\,4\,.\,9n^6}x^{3n} + \frac{g^4}{1\,.\,4\,.\,9\,.\,16n^8}x^{4n} - \text{etc.}\right)lx$$

$$+ a - \frac{ag}{nn}x^n + \frac{ag^2}{1\,.\,4n^4}x^{2n} - \frac{ag^3}{1\,.\,4\,.\,9n^6}x^{3n} + \frac{ag^4}{1\,.\,4\,.\,9\,.\,16n^8}x^{4n} - \text{etc.},$$

* *Modern Analysis*, Ch. XII. It is to be remembered that, when m is a positive integer, then

$$\psi(1) = -\gamma, \quad \psi(m+1) = \frac{1}{1} + \frac{1}{2} + \ldots + \frac{1}{m} - \gamma,$$

where γ denotes Euler's constant, 0·5772157... .

† *Inst. Calc. Int.* II. (Petersburg, 1769), § 977, pp. 233—235. See also *Acta Acad. Petrop.* v. (1781) [published 1784], pars I. Mathematica, pp. 186—190.

where A and a are arbitrary constants. He gave the following law to determine successive numerators in the first line:

$$6=3.2-1.0, \qquad 22=5.6-4.2, \qquad 100=7.22-9.6,$$

$$548=9.100-16.22, \qquad 3528=11.548-25.100 \text{ etc.}$$

If
$$2\left(\frac{1}{1}+\frac{1}{2}+\ldots+\frac{1}{m}\right)=\frac{\sigma_m}{m!},$$

this law is evidently expressed by the formula

$$\sigma_{m+1}=(2m+1)\,\sigma_m-m^2\,\sigma_{m-1}.$$

3·52. *The expansion of $\mathbf{Y}_n(z)$ in an ascending series and the definition of $\mathfrak{J}_\nu(z)$.*

We shall now obtain Hankel's* expansion of the more general function $\mathbf{Y}_n(z)$, where n is any positive integer. [Cf. equation (4) of § 3·5.]

It is clear that

$$\frac{\partial J_\nu(z)}{\partial\nu}=\sum_{m=0}^{\infty}\frac{\partial}{\partial\nu}\left\{\frac{(-)^m(\tfrac{1}{2}z)^{\nu+2m}}{m!\,\Gamma(\nu+m+1)}\right\}$$

$$=\sum_{m=0}^{\infty}\frac{(-)^m(\tfrac{1}{2}z)^{\nu+2m}}{m!\,\Gamma(\nu+m+1)}\{\log(\tfrac{1}{2}z)-\psi(\nu+m+1)\}$$

$$\to\sum_{m=0}^{\infty}\frac{(-)^m(\tfrac{1}{2}z)^{n+2m}}{m!\,(n+m)!}\{\log(\tfrac{1}{2}z)-\psi(n+m+1)\},$$

when $\nu\to n$, where n is a positive integer. That is to say

$$(1)\qquad \left[\frac{\partial J_\nu(z)}{\partial\nu}\right]_{\nu=n}=\{\gamma+\log(\tfrac{1}{2}z)\}\,J_n(z)-\sum_{m=0}^{\infty}\frac{(-)^m(\tfrac{1}{2}z)^{n+2m}}{m!\,(n+m)!}\left\{\frac{1}{1}+\frac{1}{2}+\ldots+\frac{1}{n+m}\right\}.$$

The evaluation of $[\partial J_{-\nu}(z)/\partial\nu]_{\nu=n}$ is a little more tedious because of the pole of $\psi(-\nu+m+1)$ at $\nu=n$ in the terms for which $m=0, 1, 2, \ldots, n-1$. We break the series for $J_{-\nu}(z)$ into two parts, thus

$$J_{-\nu}(z)=\sum_{m=0}^{n-1}\frac{(-)^m(\tfrac{1}{2}z)^{-\nu+2m}}{m!\,\Gamma(-\nu+m+1)}+\sum_{m=n}^{\infty}\frac{(-)^m(\tfrac{1}{2}z)^{-\nu+2m}}{m!\,\Gamma(-\nu+m+1)},$$

and in the former part we replace

$$\frac{1}{\Gamma(-\nu+m+1)}\quad\text{by}\quad\frac{\Gamma(\nu-m)\sin(\nu-m)\pi}{\pi}.$$

Now, when $0\leqslant m<n$,

$$\left[\frac{\partial}{\partial\nu}\left\{\frac{(\tfrac{1}{2}z)^{-\nu+2m}\,\Gamma(\nu-m)\sin(\nu-m)\pi}{\pi}\right\}\right]_{\nu=n}$$

$$=[(\tfrac{1}{2}z)^{-\nu+2m}\,\Gamma(\nu-m)$$
$$\{\pi^{-1}\psi(\nu-m)\sin(\nu-m)\pi+\cos(\nu-m)\pi-\pi^{-1}\log(\tfrac{1}{2}z)\sin(\nu-m)\pi\}]_{\nu=n}$$

$$=(\tfrac{1}{2}z)^{-n+2m}\,\Gamma(n-m)\cos(n-m)\pi.$$

* *Math. Ann.* I. (1869), p. 471.

Hence

$$\left[\frac{\partial J_{-\nu}(z)}{\partial \nu}\right]_{\nu=n} = \sum_{m=0}^{n-1} \frac{(-)^n \Gamma(n-m)(\tfrac{1}{2}z)^{-n+2m}}{m!}$$
$$+ \sum_{m=n}^{\infty} \frac{(-)^m (\tfrac{1}{2}z)^{-n+2m}}{m!\,(-n+m)!}\{-\log(\tfrac{1}{2}z)+\psi(-n+m+1)\},$$

that is to say

$$\left[\frac{\partial J_{-\nu}(z)}{\partial \nu}\right]_{\nu=n} = (-)^n \sum_{m=0}^{n-1} \frac{(n-m-1)!}{m!}(\tfrac{1}{2}z)^{-n+2m} + (-)^{n-1} \sum_{m=0}^{\infty} \frac{(-)^m (\tfrac{1}{2}z)^{n+2m}}{m!\,(n+m)!} \times \{\log(\tfrac{1}{2}z)-\psi(m+1)\}, \tag{2}$$

when we replace m by $n+m$ in the second series.

On combining (1) and (2) we have Hankel's formula, namely

$$\begin{aligned}\mathbf{Y}_n(z) &= -\sum_{m=0}^{n-1} \frac{(n-m-1)!}{m!}(\tfrac{1}{2}z)^{-n+2m} + \sum_{m=0}^{\infty} \frac{(-)^m (\tfrac{1}{2}z)^{n+2m}}{m!\,(n+m)!} \\ &\qquad \times \{2\log(\tfrac{1}{2}z)-\psi(m+1)-\psi(n+m+1)\} \\ &= 2\{\gamma+\log(\tfrac{1}{2}z)\}J_n(z) - (\tfrac{1}{2}z)^{-n}\sum_{m=0}^{n-1}\frac{(n-m-1)!}{m!}(\tfrac{1}{2}z)^{2m} \\ &\quad - \sum_{m=0}^{\infty}\frac{(-)^m(\tfrac{1}{2}z)^{n+2m}}{m!\,(n+m)!}\left\{\frac{1}{1}+\frac{1}{2}+\dots+\frac{1}{m}+\frac{1}{1}+\frac{1}{2}+\dots+\frac{1}{n+m}\right\}.\end{aligned} \tag{3}$$

In the first term ($m=0$) of the last summation, the expression in { } is

$$\frac{1}{1}+\frac{1}{2}+\dots+\frac{1}{n}.$$

It is frequently convenient (following Lommel*) to write

$$\mathfrak{J}_\nu(z) = \frac{\partial J_\nu(z)}{\partial \nu} - J_\nu(z)\log z, \tag{4}$$

so that

$$\mathfrak{J}_\nu(z) = -\sum_{m=0}^{\infty}\frac{(-)^m(\tfrac{1}{2}z)^{\nu+2m}}{m!\,\Gamma(\nu+m+1)}\{\log 2+\psi(\nu+m+1)\}; \tag{5}$$

when ν is a negative integer, $\mathfrak{J}_\nu(z)$ is defined by the limit of the expression on the right.

We thus have

$$\mathbf{Y}_n(z) = 2J_n(z)\log z + \mathfrak{J}_n(z) + (-)^n\mathfrak{J}_{-n}(z). \tag{6}$$

The complete solution of $x\dfrac{d^2y}{dx^2}+ay=0$ was given in the form of a series (part of which contained a logarithmic factor) by Euler, *Inst. Calc. Int.* II. (Petersburg, 1769), §§ 935, 936; solutions of this equation are

$$x^{\frac{1}{2}}J_1(2a^{\frac{1}{2}}x^{\frac{1}{2}}), \qquad x^{\frac{1}{2}}\mathbf{Y}_1(2a^{\frac{1}{2}}x^{\frac{1}{2}}).$$

Euler also gave (*ibid.* §§ 937, 938) the complete solution of $x^{\frac{3}{2}}\dfrac{d^2y}{dx^2}+ay=0$; solutions of this equation are

$$x^{\frac{1}{2}}J_2(4a^{\frac{1}{2}}x^{\frac{1}{4}}), \qquad x^{\frac{1}{2}}\mathbf{Y}_2(4a^{\frac{1}{2}}x^{\frac{1}{4}}).$$

* *Studien über die Bessel'schen Functionen* (Leipzig, 1868), p. 77.

3·53. *The definition of* $\mathbf{Y}_\nu(z)$.

Hitherto the function of the second kind has been defined only when its order is an integer. The definition which was adopted by Hankel* for *unrestricted* values of ν (integral values of 2ν excepted) is

$$\mathbf{Y}_\nu(z) = 2\pi e^{\nu\pi i} \frac{J_\nu(z)\cos\nu\pi - J_{-\nu}(z)}{\sin 2\nu\pi}. \tag{1}$$

This definition fails both when ν is an integer and when ν is half of an odd integer, because of the vanishing of $\sin 2\nu\pi$. The failure is complete in the latter case; but, in the former case, the function is defined by the limit of the expression on the right and it is easy to reconcile this definition with the definition of § 3·5.

To prove this statement, observe that

$$\begin{aligned}\lim_{\nu\to n} \mathbf{Y}_\nu(z) &= \lim_{\nu\to n}\left[\frac{\pi e^{\nu\pi i}}{\cos\nu\pi}\cdot\frac{\nu-n}{\sin\nu\pi}\cdot\frac{J_\nu(z)\cos\nu\pi - J_{-\nu}(z)}{\nu-n}\right]\\ &= (-)^n \lim_{\nu\to n}\left[\frac{J_\nu(z)\cos\nu\pi - J_{-\nu}(z)}{\nu-n}\right]\\ &= \mathbf{Y}_n(z) + \lim_{\nu\to n}\left[\frac{(-)^n\cos\nu\pi - 1}{\nu-n} J_\nu(z)\right]\\ &= \mathbf{Y}_n(z),\end{aligned}$$

and so we have proved that

$$\lim_{\nu\to n} \mathbf{Y}_\nu(z) = \mathbf{Y}_n(z). \tag{2}$$

It is now evident that $\mathbf{Y}_\nu(z)$, defined either by (1) or by the limiting form of that equation, is a solution of Bessel's equation for functions of order ν both when (i) ν has *any* value for which 2ν is not an integer, and when (ii) ν is an integer: the latter result follows from equation (2) combined with § 3·5 (3).

The function $\mathbf{Y}_\nu(z)$, defined in this way, is called a Bessel function *of the second kind* (of Hankel's type) of order ν; and the definition fails only when $\nu + \frac{1}{2}$ is an integer.

NOTE. The reader should be careful to observe that, in spite of the change of form, the function $\mathbf{Y}_\nu(z)$, *qua* function of ν, is *continuous* at $\nu = n$, except when z is zero; and, in fact, $J_\nu(z)$ and $\mathbf{Y}_\nu(z)$ approach their limits $J_n(z)$ and $\mathbf{Y}_n(z)$, as $\nu \to n$, uniformly with respect to z, except in the neighbourhood of $z = 0$, where n is any integer, positive or negative.

3·54. *The Weber-Schläfli function of the second kind.*

The definition of the function of the second kind which was given by Hankel (§ 3·53) was modified slightly by Weber† and Schläfli‡ in order to avoid the inconveniences produced by the failure of the definition when the order of the function is half of an odd integer.

* *Math. Ann.* I. (1869), p. 472.

† *Journal für Math.* LXXVI. (1873), p. 9; *Math. Ann.* VI. (1873), p. 148. These papers are dated Sept. 1872 and Oct. 1872 respectively. In a paper written a few months before these, *Journal für Math.* LXXV. (1873), pp. 75—105, dated May 1872, Weber had used Neumann's function of the second kind (see §§ 3·57, 3·58).

‡ *Ann. di Mat.* (2) VII. (1875), p. 17; this paper is dated Oct. 4, 1872.

The function which was adopted by Weber as the canonical function of the second kind is expressible in terms of functions of the first kind by the formula*

$$\frac{J_\nu(z)\cos\nu\pi - J_{-\nu}(z)}{\sin\nu\pi}$$

(or the limit of this, when ν is an integer).

Schläfli, however, inserted a factor $\frac{1}{2}\pi$; and he denoted his function by the symbol K, so that, with his definition,

$$K_\nu(z) = \tfrac{1}{2}\pi\,\frac{J_\nu(z)\cos\nu\pi - J_{-\nu}(z)}{\sin\nu\pi}.$$

Subsequent writers, however, have usually omitted this factor $\frac{1}{2}\pi$, e.g. Graf and Gubler in their treatise†, and also Nielsen, so that these writers work with Weber's function.

The symbol K is, however, used largely in this country, especially by Physicists, to denote a completely different type of Bessel function (§ 3·7), and so it is advisable to use a different notation. The procedure which seems to produce least confusion is *to use the symbol $Y_\nu(z)$ to denote Weber's function*, after the manner of Nielsen‡, and to adopt this as the canonical function of the second kind, save in rare instances when the use of Hankel's function of integral order saves the insertion of the number π in certain formulae.

We thus have

$$(1) \qquad Y_\nu(z) = \frac{J_\nu(z)\cos\nu\pi - J_{-\nu}(z)}{\sin\nu\pi} = \frac{\cos\nu\pi}{\pi e^{\nu\pi i}}\mathbf{Y}_\nu(z),$$

$$(2) \qquad Y_n(z) = \lim_{\nu\to n}\frac{J_\nu(z)\cos\nu\pi - J_{-\nu}(z)}{\sin\nu\pi} = \frac{1}{\pi}\mathbf{Y}_n(z).$$

[NOTE. Schläfli's function has been used by Bôcher, *Annals of Math.* VI. (1892), pp. 85—90, and by McMahon, *Annals of Math.* VIII. (1894), pp. 57—61; IX. (1895), pp. 23—30. Schafheitlin and Heaviside use Weber's function with the sign changed, so that the function which we (with Nielsen) denote by $Y_\nu(z)$ is written as $-Y_\nu(z)$ by Schafheitlin§ and (when $\nu = n$) as $-G_n(z)$ by Heaviside||.

Gray and Mathews sometimes¶ use Weber's function, and they denote it by the symbol Y_n.

* Weber's definition was by an integral (see § 6·1) which is equal to this expression; the expression (with the factor $\frac{1}{2}\pi$ inserted) was actually given by Schläfli.

† *Einleitung in die Theorie der Bessel'schen Funktionen*, I. (Bern, 1898), p. 34 *et seq.*

‡ Nielsen, as in the case of other functions, writes the number indicating the order as an index, thus $Y^\nu(z)$, *Handbuch der Theorie der Cylinderfunktionen* (Leipzig, 1904), p. 11. There are obvious objections to such a notation, and we reserve it for the obsolete function used by Neumann (§ 3·58).

§ See, e.g. *Journal für Math.* CXIV. (1895), pp. 31—44, and other papers; also *Die Theorie der Bessel'schen Funktionen* (Leipzig, 1908).

|| *Proc. Royal Soc.* LIV. (1893), p. 138, and *Electromagnetic Theory*, II. (London, 1899), p. 255; a change in sign has been made from his *Electrical Papers*, II. (London, 1892), p. 445.

¶ *A Treatise on Bessel Functions* (London, 1895), pp. 65—66.

Lommel, in his later work, used Neumann's function of the second kind (see § 3·57), but in his *Studien über die Bessel'schen Functionen* (Leipzig, 1868), pp. 85—86, he used the function

$$\tfrac{1}{2}\pi Y_n(z) + \{\psi(n+\tfrac{1}{2}) + \log 2\} J_n(z),$$

where $Y_n(z)$ is the function of Weber. One disadvantage of this function is that the presence of the term $\psi(n+\tfrac{1}{2})$ makes the recurrence formulae for the function much more complicated; see Julius, *Archives Néerlandaises*, XXVIII. (1895), pp. 221—225, in this connexion.]

3·55. *Heine's definition of the function of the second kind.*

The definition given by Heine* of the function of the second kind possesses some advantages from the aspect of the theory of Legendre functions; it enables certain generalisations of Mehler's formula (§ 5·71), namely

$$\lim_{n\to\infty} P_n(\cos\theta/n) = J_0(\theta),$$

to be expressed in a compact form. The function, which Heine denoted by the symbol $K_n(z)$, is expressible in terms of the canonical functions, and it is equal to $-\tfrac{1}{2}\pi Y_n(z)$ and to $-\tfrac{1}{2}\mathbf{Y}_n(z)$; the function consequently differs only in sign from the function originally used by Schläfli.

The use of Heine's function seems to have died out on the Continent many years ago; the function was occasionally used by Gray and Mathews in their treatise†, and they term it $G_n(z)$. In this form the function has been extensively tabulated first by Aldis‡ and Airey §, and subsequently in *British Association Reports*, 1913, 1914 and 1916.

This revival of the use of Heine's function seems distinctly unfortunate, both on account of the existing multiplicity of functions of the second kind and also on account of the fact (which will become more apparent in Chapters VI and VII) that the relations between the functions $J_n(z)$ and $Y_n(z)$ present many points of resemblance to the relations between the cosine and sine; so that the adoption‖ of $J_n(z)$ and $G_n(z)$ as canonical functions is comparable to the use of $\cos z$ and $-\tfrac{1}{2}\pi\sin z$ as canonical functions. It must also be pointed out that the symbol $G_n(z)$ has been used in senses other than that just explained by at least two writers, namely Heaviside, *Proc. Royal Soc.* LIV. (1893), p. 138 (as was stated in § 3·54), and Dougall, *Proc. Edinburgh Math. Soc.* XVIII. (1900), p. 36.

NOTE. An error in sign on p. 245 of Heine's treatise has been pointed out by Morton, *Nature*, LXIII. (1901), p. 29; the error is equivalent to a change in the sign of γ in formula § 3·51 (3) *supra*. It was also stated by Morton that this error had apparently been copied by various other writers, including (as had been previously noticed by Gray ¶) J. J. Thomson, *Recent Researches in Electricity and Magnetism* (Oxford, 1893), p. 263. A further error

* *Handbuch der Kugelfunctionen*, I. (Berlin, 1878), pp. 185—248.

† *A Treatise on Bessel Functions* (London, 1895), pp. 91, 147, 242.

‡ *Proc. Royal Soc.* LXVI. (1900), pp. 32—43.

§ *Phil. Mag.* (6) XXII. (1911), pp. 658—663.

‖ From the historical point of view there is something to be said for using Hankel's function, and also for using Neumann's function; but Heine's function, being more modern than either, has not even this in its favour.

¶ *Nature*, XLIX. (1894), p. 359.

noticed by Morton in Thomson's work seems to be due to a most confusing notation employed by Heine; for on p. 245 of his treatise Heine uses the symbol K_0 to denote the function called $-\frac{1}{2}\pi Y_0$ in this work, while on p. 248 the same symbol K_0 denotes $-\frac{1}{2}\pi(Y_0 - iJ_0)$.

3·56. *Recurrence formulae for $Y_\nu(z)$ and $\mathbf{Y}_\nu(z)$.*

The recurrence formulae which are satisfied by $Y_\nu(z)$ are of the same form as those which are satisfied by $J_\nu(z)$; they are consequently as follows:

$$(1)\qquad Y_{\nu-1}(z) + Y_{\nu+1}(z) = \frac{2\nu}{z} Y_\nu(z),$$

$$(2)\qquad Y_{\nu-1}(z) - Y_{\nu+1}(z) = 2Y_\nu'(z),$$

$$(3)\qquad zY_\nu'(z) + \nu Y_\nu(z) = zY_{\nu-1}(z),$$

$$(4)\qquad zY_\nu'(z) - \nu Y_\nu(z) = -zY_{\nu+1}(z),$$

and in these formulae the function Y may be replaced throughout by the function $\mathbf{Y}$.

To prove them we take § 3·2 (3) and (4) in the forms

$$\frac{d}{dz}\{z^\nu J_\nu(z)\} = z^\nu J_{\nu-1}(z), \qquad \frac{d}{dz}\{z^\nu J_{-\nu}(z)\} = -z^\nu J_{-\nu+1}(z);$$

if we multiply these by $\cot \nu\pi$ and $\operatorname{cosec} \nu\pi$, and then subtract, we have

$$\frac{d}{dz}\{z^\nu Y_\nu(z)\} = z^\nu Y_{\nu-1}(z),$$

whence (3) follows at once. Equation (4) is derived in a similar manner from the formulae

$$\frac{d}{dz}\{z^{-\nu} J_\nu(z)\} = -z^{-\nu} J_{\nu+1}(z), \qquad \frac{d}{dz}\{z^{-\nu} J_{-\nu}(z)\} = z^{-\nu} J_{-\nu-1}(z).$$

By addition and subtraction of (3) and (4) we obtain (2) and (1).

The formulae are, so far, proved on the hypothesis that ν is not an integer; but since $Y_\nu(z)$ and its derivatives are continuous functions of ν, the result of proceeding to the limit when ν tends to an integral value n, is simply to replace ν by n.

Again, the effect of multiplying the four equations by $\pi e^{\nu\pi i}\sec \nu\pi$, which is equal to $\pi e^{(\nu\pm1)\pi i}\sec(\nu\pm1)\pi$, is to replace the functions Y by the functions $\mathbf{Y}$ throughout.

In the case of functions of integral order, these formulae were given by Lommel, *Studien über die Bessel'schen Functionen* (Leipzig, 1868), p. 87. The reader will find it instructive to establish them for such functions directly from the series of § 3·52.

Neumann's investigation connected with the formula (4) will be discussed in § 3·58.

3·57. *Neumann's function of the second kind.*

The function which Neumann* adopted as the canonical function of the second kind possesses the advantage that it is represented more simply by integrals of Poisson's type than the functions of the second kind which have been hitherto discussed; but this is its only merit.

We first define the function *of order zero*†, which will be called $Y^{(0)}(z)$.

The second solution of Bessel's equation for functions of order zero being known to contain logarithms, Neumann assumed as a solution the expression

$$J_0(z)\log z + w,$$

where w is a function of z to be determined.

If this expression is to be annihilated by ∇_0, we must have

$$\nabla_0 w = -\nabla_0\{J_0(z)\log z\}$$
$$= -2z\,J_0'(z).$$

But, by § 2·12 (11),

$$-2z\,J_0'(z) = 2z\,J_1(z) = 8\sum_{n=1}^{\infty}(-)^{n-1}n\,J_{2n}(z);$$

and so, since $\nabla_0 J_{2n}(z) = 4n^2 J_{2n}(z)$, we have

$$\nabla_0 w = 2\sum_{n=1}^{\infty}(-)^{n-1}\nabla_0 J_{2n}(z)/n$$
$$= 2\nabla_0\sum_{n=1}^{\infty}(-)^{n-1}J_{2n}(z)/n;$$

the change of the order of the operations Σ and ∇_0 is easily justified.

Hence a possible value for w is

$$2\sum_{n=1}^{\infty}(-)^{n-1}J_{2n}(z)/n,$$

and therefore Neumann's function $Y^{(0)}(z)$, defined by the equation

$$(1)\qquad Y^{(0)}(z) = J_0(z)\log z + 2\sum_{n=1}^{\infty}(-)^{n-1}\frac{J_{2n}(z)}{n},$$

is a solution of Bessel's equation for functions of order zero.

Since $w \to 0$ as $z \to 0$, (the series for w being an analytic function of z near the origin), it is evident that $J_0(z)$ and $Y^{(0)}(z)$ form a fundamental system of solutions, and hence $\mathbf{Y}_0(z)$ is expressible as a linear combination of $J_0(z)$ and $Y^{(0)}(z)$; a comparison of the behaviours of the three functions near the origin shews that the relation connecting them is

$$(2)\qquad Y^{(0)}(z) = \tfrac{1}{2}\mathbf{Y}_0(z) + (\log 2 - \gamma)\,J_0(z).$$

* *Theorie der Bessel'schen Functionen* (Leipzig, 1867), pp. 42—44. Neumann calls this function *Bessel's associated function*, and he describes another function, $O_n(z)$, as the function of the second kind (§ 9·1). But, because $O_n(z)$ is not a solution of Bessel's equation, this description is undesirable and it has not survived.

† Neumann's function is distinguished from the Weber-Schläfli function by the position of the suffix which indicates the order.

3·571. *The integral of Poisson's type for $Y^{(0)}(z)$.*

It was shewn by Poisson* that

$$\int_0^{\pi} e^{ix\cos\omega} \log(x \sin^2\omega)\, d\omega$$

is a solution of Bessel's equation for functions of order zero and argument x; and subsequently Stokes obtained an expression of the integral in the form of an ascending series (see § 3·572).

The associated integral

$$\frac{2}{\pi}\int_0^{\frac{1}{2}\pi} \cos(z\sin\theta) . \log(4z\cos^2\theta)\, d\theta$$

was identified by Neumann† with the function $Y^{(0)}(z)$; and the analysis by which he obtained this result is of sufficient interest to be given here, with some slight modifications in matters of detail.

From § 2·2 (9) we have

$$\frac{(-)^n J_{2n}(z)}{n} = \frac{2}{n\pi}\int_0^{\frac{1}{2}\pi} \cos(z\cos\theta)\cos 2n\theta\, d\theta,$$

and so, if we assume that the order of summation and integration can be changed, we deduce that

$$2\sum_{n=1}^{\infty} \frac{(-)^n J_{2n}(z)}{n} = \frac{4}{\pi}\int_0^{\frac{1}{2}\pi} \cos(z\cos\theta) \sum_{n=1}^{\infty} \frac{\cos 2n\theta}{n}\, d\theta$$

$$= -\frac{2}{\pi}\int_0^{\frac{1}{2}\pi} \cos(z\cos\theta) . \log(4\sin^2\theta)\, d\theta;$$

from this result combined with Parseval's integral (§ 2·2) and the definition of $Y^{(0)}(z)$, we at once obtain the formula

(1) $$Y^{(0)}(z) = \frac{2}{\pi}\int_0^{\frac{1}{2}\pi} \cos(z\cos\theta) . \log(4z\sin^2\theta)\, d\theta,$$

from which Neumann's result is obvious.

The change of the order of summation and integration has now to be examined, because $\Sigma n^{-1}\cos 2n\theta$ is non-uniformly convergent near $\theta = 0$. To overcome this difficulty we observe that, since $\Sigma(-)^n J_{2n}(z)/n$ is convergent, it follows from Abel's theorem ‡ that

$$\sum_{n=1}^{\infty} (-)^n J_{2n}(z)/n = \lim_{a\to 1-0} \sum_{n=1}^{\infty} (-)^n a^n J_{2n}(z)/n = \lim_{a\to 1-0} \frac{2}{\pi}\sum_{n=1}^{\infty}\int_0^{\frac{1}{2}\pi} \cos(z\cos\theta)\frac{a^n\cos 2n\theta}{n}\, d\theta.$$

* *Journal de l'École R. Polytechnique*, XII. (cahier 19), (1823), p. 476. The solution of an associated partial differential equation had been given earlier (*ibid.* p. 227). See also Duhamel, *Cours d'Analyse*, II. (Paris, 1840), pp. 122—124, and Spitzer, *Zeitschrift für Math. und Phys.* II. (1857), pp. 165—170.

† *Theorie der Bessel'schen Functionen* (Leipzig, 1867), pp. 45—49. See also Niemöller, *Zeitschrift für Math. und Phys.* XXV. (1880), pp. 65—71.

‡ Cf. Bromwich, *Theory of Infinite Series*, § 51.

Now, since a is less than 1, $\Sigma\,(a^n \cos 2n\theta)/n$ does converge uniformly throughout the range of integration (by comparison with Σa^n), and so the interchange is permissible; that is to say

$$\frac{2}{\pi}\sum_{n=1}^{\infty}\int_0^{\frac{1}{2}\pi}\cos(z\cos\theta)\frac{a^n\cos 2n\theta}{n}\,d\theta=\frac{2}{\pi}\int_0^{\frac{1}{2}\pi}\cos(z\cos\theta)\sum_{n=1}^{\infty}\frac{a^n\cos 2n\theta}{n}\,d\theta$$
$$=-\frac{1}{\pi}\int_0^{\frac{1}{2}\pi}\cos(z\cos\theta)\log(1-2a\cos 2\theta+a^2)\,d\theta.$$

Hence we have

$$\sum_{n=1}^{\infty}\frac{(-)^n J_{2n}(z)}{n}=-\lim_{a\to 1-0}\frac{1}{\pi}\int_0^{\frac{1}{2}\pi}\cos(z\cos\theta)\log(1-2a\cos 2\theta+a^2)\,d\theta.$$

We now proceed to shew that*

$$\lim_{a\to 1-0}\int_0^{\frac{1}{2}\pi}\cos(z\cos\theta)\{\log(1-2a\cos 2\theta+a^2)-\log(4a\sin^2\theta)\}\,d\theta=0.$$

It is evident that $\quad 1-2a\cos 2\theta+a^2-4a\sin^2\theta=(1-a)^2\geqslant 0,$

and so $\quad \log(1-2a\cos 2\theta+a^2)\geqslant\log(4a\sin^2\theta).$

Hence, if A be the upper bound† of $|\cos(z\cos\theta)|$ when $0\leqslant\theta\leqslant\frac{1}{2}\pi$, we have

$$\left|\int_0^{\frac{1}{2}\pi}\cos(z\cos\theta)\{\log(1-2a\cos 2\theta+a^2)-\log(4a\sin^2\theta)\}\,d\theta\right|$$
$$\leqslant A\int_0^{\frac{1}{2}\pi}\{\log(1-2a\cos 2\theta+a^2)-\log(4a\sin^2\theta)\}\,d\theta$$
$$=A\int_0^{\frac{1}{2}\pi}\left\{-2\sum_{n=1}^{\infty}\frac{a^n\cos 2n\theta}{n}+\log(1/a)-2\log(2\sin\theta)\right\}d\theta$$
$$=\tfrac{1}{2}\pi A\log(1/a),$$

term-by-term integration being permissible since $a<1$. Hence, when $a<1$,

$$\left|\int_0^{\frac{1}{2}\pi}\cos(z\cos\theta)\{\log(1-2a\cos 2\theta+a^2)-\log(4a\sin^2\theta)\}\,d\theta\right|\leqslant\tfrac{1}{2}\pi A\log(1/a)\to 0,$$

as $a\to 1-0$; and this is the result to be proved.

Consequently

$$\sum_{n=1}^{\infty}\frac{(-)^n J_{2n}(z)}{n}=-\lim_{a\to 1-0}\frac{1}{\pi}\int_0^{\frac{1}{2}\pi}\cos(z\cos\theta)\,.\log(4a\sin^2\theta)\,d\theta$$
$$=-\frac{1}{\pi}\int_0^{\frac{1}{2}\pi}\cos(z\cos\theta)\,.\log(4\sin^2\theta)\,d\theta,$$

and the interchange is finally justified.

The reader will find it interesting to deduce this result from Poisson's integral for $J_\nu(z)$ combined with § 3·5 (5).

3·572. *Stokes' series for the Poisson-Neumann integral.*

The differential equation considered by Stokes‡ in 1850 was $\dfrac{d^2y}{dz^2}+\dfrac{1}{z}\dfrac{dy}{dz}-m^2y=0$, where m is a constant. This is Bessel's equation for functions of order zero and argument imz. Stokes stated (presumably with reference to Poisson) that it was known that the general solution was

$$y=\int_0^{\frac{1}{2}\pi}\{C+D\log(z\sin^2\theta)\}\cosh(mz\cos\theta)\,d\theta.$$

* The value of this limit was assumed by Neumann.

† If z is real, $A=1$; if not, $A\leqslant\exp\{|I(z)|\}$.

‡ *Trans. Camb. Phil. Soc.* IX. (1856), p. [38]. [*Mathematical and Physical Papers*, III. (1901), p. 42.]

It is easy to see that, with Neumann's notation, the value of the expression on the right is

$$\tfrac{1}{2}\pi\{C-D\log(4im)\}J_0(imz)+\tfrac{1}{2}\pi D Y^{(0)}(imz).$$

The expression was expanded into a series by Stokes; it is equal to

$$\tfrac{1}{2}\pi(C+D\log z)J_0(imz)+2D\sum_{n=0}^{\infty}\frac{m^{2n}z^{2n}}{(2n)!}\int_0^{\frac{1}{2}\pi}\cos^{2n}\theta\log\sin\theta\,d\theta,$$

and, by integrating by parts, Stokes obtained a recurrence formula from which it may be deduced that

$$-\int_0^{\frac{1}{2}\pi}\cos^{2n}\theta\log\sin\theta\,d\theta=\frac{(2n)!}{2^{2n}.(n!)^2}\left\{\tfrac{1}{2}\pi\log 2+\tfrac{1}{4}\pi\left(\frac{1}{1}+\frac{1}{2}+\ldots+\frac{1}{n}\right)\right\}.$$

3·58. *Neumann's definition of* $Y^{(n)}(z)$.

The Bessel function of the second kind, of integral order n, was defined by Neumann* in terms of $Y^{(0)}(z)$ by induction from the formula

$$z\frac{dY^{(n)}(z)}{dz}-nY^{(n)}(z)=-zY^{(n+1)}(z), \tag{1}$$

which is a recurrence formula of the same type as § 2·12 (4). It is evident from this equation that

$$Y^{(n)}(z)=(-z)^n\left(\frac{d}{zdz}\right)^n Y^{(0)}(z). \tag{2}$$

Now $Y^{(0)}(z)$ satisfies the equation

$$z^2\left(\frac{d}{zdz}\right)^2 Y^{(0)}(z)+2\left(\frac{d}{zdz}\right)Y^{(0)}(z)+Y^{(0)}(z)=0;$$

and, if we apply the operator† $\dfrac{d}{zdz}$ to this equation n times, and use Leibniz' theorem, we get

$$z^2\left(\frac{d}{zdz}\right)^{n+2}Y^{(0)}(z)+(2n+2)\left(\frac{d}{zdz}\right)^{n+1}Y^{(0)}(z)+\left(\frac{d}{zdz}\right)^n Y^{(0)}(z)=0, \tag{3}$$

and so

$$z^2\left(\frac{d}{zdz}\right)^2\{z^{-n}Y^{(n)}(z)\}+(2n+2)\left(\frac{d}{zdz}\right)\{z^{-n}Y^{(n)}(z)\}+z^{-n}Y^{(n)}(z)=0.$$

This equation is at once reducible to

$$\nabla_n Y^{(n)}(z)=0, \tag{4}$$

and so $Y^{(n)}(z)$ is a solution of Bessel's equation for functions of order n.

Again, (3) may be written in the form

$$z\frac{d}{dz}\{-z^{-n-1}Y^{(n+1)}(z)\}-(2n+2)z^{-n-1}Y^{(n+1)}(z)+z^{-n}Y^{(n)}(z)=0,$$

* *Theorie der Bessel'schen Functionen* (Leipzig, 1867), p. 51. The function is undefined when its order is not an integer.

† The analysis is simplified by taking $\tfrac{1}{2}z^2=\zeta$, so that

$$\frac{d}{z\,dz}=\frac{d}{d\zeta}.$$

so that

$$\frac{dY^{(n+1)}(z)}{dz} + \frac{n+1}{z} Y^{(n+1)}(z) - Y^{(n)}(z) = 0,$$

whence we obtain another recurrence formula

(5) $$z\frac{dY^{(n)}(z)}{dz} + nY^{(n)}(z) = zY^{(n-1)}(z).$$

When we combine (1) with (5) we at once deduce the other recurrence formulae

(6) $$Y^{(n-1)}(z) + Y^{(n+1)}(z) = \frac{2n}{z} Y^{(n)}(z),$$

(7) $$Y^{(n-1)}(z) - Y^{(n+1)}(z) = 2\frac{dY^{(n)}(z)}{dz}.$$

Consequently $Y^{(n)}(z)$ satisfies the same recurrence formulae as $J_n(z)$, $Y_n(z)$ and $\mathbf{Y}_n(z)$. It follows from § 3·57 (2) that

(8) $$Y^{(n)}(z) = \tfrac{1}{2}\pi Y_n(z) + (\log 2 - \gamma) J_n(z)$$
$$= \tfrac{1}{2}\mathbf{Y}_n(z) + (\log 2 - \gamma) J_n(z).$$

A solution of the equation $\nabla_n(y)=0$ in the form of a definite integral, which reduces to the integral of § 3·571 when $n=0$, has been constructed by Spitzer, *Zeitschrift für Math. und Phys.* III. (1858), pp. 244–246; cf. § 3·583.

3·581. *Neumann's expansion of* $Y^{(n)}(z)$.

The generalisation of the formula § 3·57 (1) has been given by Neumann*; it is

(1) $$Y^{(n)}(z) = J_n(z)\{\log z - s_n\} - \sum_{m=0}^{n-1} \frac{2^{n-m-1}.\, n!}{(n-m).\, m!} \frac{J_m(z)}{z^{n-m}} + \sum_{m=1}^{\infty} \frac{(-)^{m-1}(n+2m)}{m(n+m)} J_{n+2m}(z),$$

where $s_n = \frac{1}{1} + \frac{1}{2} + \frac{1}{3} + \ldots + \frac{1}{n}$, $s_0 = 0$.

To establish this result, we first define the functions $L_n(z)$ and $U_n(z)$ by the equations

(2) $$L_n(z) = J_n(z)\log z - \sum_{m=0}^{n-1} \frac{2^{n-m-1}.\, n!}{(n-m).\, m!} \frac{J_m(z)}{z^{n-m}},$$

(3) $$U_n(z) = s_n J_n(z) + \sum_{m=1}^{\infty} \frac{(-)^m (n+2m)}{m(n+m)} J_{n+2m}(z),$$

so that $Y^{(0)}(z) = L_0(z) - U_0(z)$.

We shall prove that $L_n(z)$ and $U_n(z)$ satisfy the recurrence formulae

(4) $L_{n+1}(z) = -L_n'(z) + (n/z)L_n(z),\quad U_{n+1}(z) = -U_n'(z) + (n/z)U_n(z),$

and then (1) will be evident by induction from § 3·58 (2).

* *Theorie der Bessel'schen Functionen* (Leipzig, 1867), p. 52. See also Lommel, *Studien über die Bessel'schen Functionen* (Leipzig, 1868), pp. 82—84; Otti, *Bern Mittheilungen*, 1898, pp. 34—35; and Haentzschel, *Zeitschrift für Math. und Phys.* XXXI. (1886), pp. 25—33.

It is evident that

$$\frac{d}{dz}\left\{\frac{L_n(z)}{z^n}\right\} = \log z \frac{d}{dz}\left\{\frac{J_n(z)}{z^n}\right\} + \frac{J_n(z)}{z^{n+1}} - \sum_{m=0}^{n-1} \frac{2^{n-m-1} \cdot n!}{(n-m) \cdot m!} \frac{d}{dz}\left\{\frac{J_m(z)}{z^{2n-m}}\right\}$$

$$= \frac{1}{z^n}\left[-J_{n+1}(z)\log z + \frac{J_n(z)}{z} + \sum_{m=0}^{n-1} \frac{2^{n-m-1} \cdot n!}{(n-m) \cdot m!}\left\{2(n-m)\frac{J_m(z)}{z^{n-m+1}} + \frac{J_{m+1}(z)}{z^{n-m}}\right\}\right]$$

$$= \frac{1}{z^n}\left[-J_{n+1}(z)\log z + \frac{(n+1)J_n(z)}{z} + \sum_{m=0}^{n-1} \frac{2^{n-m} \cdot n!}{m!}\left\{1 + \frac{m}{n-m+1}\right\}\frac{J_m(z)}{z^{n-m+1}}\right]$$

$$= -\frac{L_{n+1}(z)}{z^n},$$

and the first part of (4) is proved. To prove the second part, we have

$$\frac{d}{dz}\left\{\frac{U_n(z)}{z^n}\right\} = s_n \frac{d}{dz}\left\{\frac{J_n(z)}{z^n}\right\} + \sum_{m=1}^{\infty} \frac{(-)^m (n+2m)}{m(n+m)} \frac{d}{dz}\left\{\frac{J_{n+2m}(z)}{z^n}\right\}$$

$$= -s_n \frac{J_{n+1}(z)}{z^n} + \frac{1}{z^n}\sum_{m=1}^{\infty} \frac{(-)^m}{m(n+m)}\{mJ_{n+2m-1}(z) - (n+m)J_{n+2m+1}(z)\}$$

$$= -s_{n+1}\frac{J_{n+1}(z)}{z^n} - \frac{1}{z^n}\sum_{m=1}^{\infty}(-)^m J_{n+2m+1}(z)\left\{\frac{1}{m} + \frac{1}{n+m+1}\right\}$$

$$= -\frac{U_{n+1}(z)}{z^n},$$

and the second part of (4) is proved. It follows from § 3·58 (2) that

$$\frac{Y^{(n+1)}(z) - L_{n+1}(z) + U_{n+1}(z)}{z^n} = -\frac{d}{dz}\left\{\frac{Y^{(n)}(z) - L_n(z) + U_n(z)}{z^n}\right\},$$

and since the expression on the right vanishes when $n = 0$, it is evident by induction that it vanishes for all integral values of n. Hence

$$Y^{(n)}(z) = L_n(z) - U_n(z),$$

and the truth of equation (1) is therefore established.

3·582. *The power series for $U_n(z)$.*

The function $U_n(z)$, which was defined in § 3·581 (3) as a series of Bessel coefficients, has been expressed by Schläfli* as a power series with simple coefficients, namely

(1) $$U_n(z) = \sum_{m=0}^{\infty} \frac{(-)^m (\tfrac{1}{2}z)^{n+2m}}{m!(n+m)!} s_{n+m}.$$

To establish this result, observe that it is true when $n = 0$ by § 3·51 (3) and § 3·57 (1); and that, by straightforward differentiation, the expression on the right satisfies the same recurrence formula as that of § 3·581 (4) for $U_n(z)$; equation (1) is then evident by induction.

NOTE. It will be found interesting to establish this result by evaluating the coefficient of $(\tfrac{1}{2}z)^{n+2m}$ in the expansion on the right of § 3·581 (3).

* *Math. Ann.* III. (1871), pp. 146—147.

The reader will now easily prove the following formulae:

(2) $$\mathfrak{Y}_n(z) = \{\gamma - \log 2\} J_n(z) - U_n(z),$$

(3) $$Y^{(n)}(z) = L_n(z) + \mathfrak{Y}_n(z) + \{\log 2 - \gamma\} J_n(z),$$

(4) $$\tfrac{1}{2}\pi Y_n(z) = L_n(z) + \mathfrak{Y}_n(z).$$

3·583. *The integral of Poisson's type for $Y^{(n)}(z)$.*

The Poisson-Neumann formula of § 3·571 for $Y^{(0)}(z)$ was generalised by Lommel, *Studien über die Bessel'schen Functionen* (Leipzig, 1868), p. 86, with a notation rather different from Neumann's; to obtain Lommel's result in Neumann's notation, we first observe that, by differentiation of Poisson's integral for $J_\nu(z)$, we have

$$\frac{\partial J_\nu(z)}{\partial \nu} - J_\nu(z)\log z = \frac{2(\frac{1}{2}z)^\nu}{\Gamma(\nu+\frac{1}{2})\Gamma(\frac{1}{2})}\int_0^{\frac{1}{2}\pi} \cos(z\sin\theta)\cos^{2\nu}\theta\,\{\log(\tfrac{1}{2}\cos^2\theta) - \psi(\nu+\tfrac{1}{2})\}\,d\theta,$$

and so, from § 3·582 (3),

$$Y^{(n)}(z) = \frac{2(\frac{1}{2}z)^n}{\Gamma(n+\frac{1}{2})\Gamma(\frac{1}{2})}\int_0^{\frac{1}{2}\pi} \cos(z\sin\theta)\cos^{2n}\theta\,\{\log\cos^2\theta - \psi(n+\tfrac{1}{2}) - \gamma\}\,d\theta + L_n(z),$$

and hence, since $\psi(\frac{1}{2}) = \psi(1) - 2\log 2 = -\gamma - 2\log 2$, we have the formula

(1) $$Y^{(n)}(z) = \frac{2(\frac{1}{2}z)^n}{\Gamma(n+\frac{1}{2})\Gamma(\frac{1}{2})}\int_0^{\frac{1}{2}\pi} \cos(z\sin\theta)\cos^{2n}\theta\log(4\cos^2\theta)\,d\theta - \{\psi(n+\tfrac{1}{2}) - \psi(\tfrac{1}{2})\}J_n(z) + L_n(z),$$

in which it is to be remembered that $L_n(z)$ is expressible as a finite combination of Bessel coefficients and powers of z.

3·6. *Functions of the third kind.*

In numerous developments of the theory of Bessel functions, especially those which are based on Hankel's researches (Chapters VI and VII) on integral representations and asymptotic expansions of $J_\nu(z)$ and $Y_\nu(z)$, two combinations of Bessel functions, namely $J_\nu(z) \pm iY_\nu(z)$, are of frequent occurrence. The combinations also present themselves in the theory of "Bessel functions of purely imaginary argument" (§ 3·7).

It has consequently seemed desirable to Nielsen* to regard the pair of functions $J_\nu(z) \pm iY_\nu(z)$ as standard solutions of Bessel's equation, and he describes them as *functions of the third kind*; and, in honour of Hankel, Nielsen denotes them by the symbol H. The two functions of the third kind are defined by the equations†

(1) $$H_\nu^{(1)}(z) = J_\nu(z) + iY_\nu(z), \quad H_\nu^{(2)}(z) = J_\nu(z) - iY_\nu(z).$$

From these definitions, combined with § 3·54 (1), we have

(2) $$H_\nu^{(1)}(z) = \frac{J_{-\nu}(z) - e^{-\nu\pi i}J_\nu(z)}{i\sin\nu\pi}, \quad H_\nu^{(2)}(z) = \frac{J_{-\nu}(z) - e^{\nu\pi i}J_\nu(z)}{-i\sin\nu\pi}.$$

When ν is an integer, the right-hand sides are to be replaced by their limits.

Since $J_\nu(z)$ and $Y_\nu(z)$ satisfy the same recurrence formulae (§§ 3·2, 3·56), in which the functions enter linearly, and since the functions of the third kind

* *Ofversigt over det K. Danske Videnskabernes Selskabs Forhandlinger*, 1902, p. 125. *Handbuch der Theorie der Cylinderfunktionen* (Leipzig, 1904), p. 16.

† Nielsen uses the symbols $H_1^\nu(z)$, $H_2^\nu(z)$.

are linear functions (with constant coefficients) of $J_\nu(z)$ and $Y_\nu(z)$, it follows that these same recurrence formulae are satisfied by functions of the third kind.

Hence we can at once write down the following formulae:

(3) $H^{(1)}_{\nu-1}(z)+H^{(1)}_{\nu+1}(z)=\frac{2\nu}{z}H^{(1)}_\nu(z)$, $\quad H^{(2)}_{\nu-1}(z)+H^{(2)}_{\nu+1}(z)=\frac{2\nu}{z}H^{(2)}_\nu(z)$,

(4) $H^{(1)}_{\nu-1}(z)-H^{(1)}_{\nu+1}(z)=2\frac{dH^{(1)}_\nu(z)}{dz}$, $\quad H^{(2)}_{\nu-1}(z)-H^{(2)}_{\nu+1}(z)=2\frac{dH^{(2)}_\nu(z)}{dz}$,

(5) $z\frac{dH^{(1)}_\nu(z)}{dz}+\nu H^{(1)}_\nu(z)=zH^{(1)}_{\nu-1}(z)$, $\quad z\frac{dH^{(2)}_\nu(z)}{dz}+\nu H^{(2)}_\nu(z)=zH^{(2)}_{\nu-1}(z)$,

(6) $z\frac{dH^{(1)}_\nu(z)}{dz}-\nu H^{(1)}_\nu(z)=-zH^{(1)}_{\nu+1}(z)$, $\quad z\frac{dH^{(2)}_\nu(z)}{dz}-\nu H^{(2)}_\nu(z)=-zH^{(2)}_{\nu+1}(z)$,

(7) $\frac{dH^{(1)}_0(z)}{dz}=-H^{(1)}_1(z)$, $\quad \frac{dH^{(2)}_0(z)}{dz}=-H^{(2)}_1(z)$,

(8) $\nabla_\nu H^{(1)}_\nu(z)=0$, $\quad \nabla_\nu H^{(2)}_\nu(z)=0$,

(9) $\left(\frac{d}{zdz}\right)^m\{z^\nu H^{(1)}_\nu(z)\}=z^{\nu-m}H^{(1)}_{\nu-m}(z)$, $\quad \left(\frac{d}{zdz}\right)^m\{z^\nu H^{(2)}_\nu(z)\}=z^{\nu-m}H^{(2)}_{\nu-m}(z)$,

(10) $\left(\frac{d}{zdz}\right)^m\left\{\frac{H^{(1)}_\nu(z)}{z^\nu}\right\}=(-)^m\frac{H^{(1)}_{\nu+m}(z)}{z^{\nu+m}}$, $\quad \left(\frac{d}{zdz}\right)^m\left\{\frac{H^{(2)}_\nu(z)}{z^\nu}\right\}=(-)^m\frac{H^{(2)}_{\nu+m}(z)}{z^{\nu+m}}$.

NOTE. Rayleigh on several occasions, *e.g. Phil. Mag.* (5) XLIII. (1897), p. 266; (6) XIV. (1907), pp. 350—359 [*Scientific Papers*, IV. (1904), p. 290; V. (1912), pp. 410—418], has used the symbol $D_n(z)$ to denote the function which Nielsen calls $\frac{1}{2}\pi i H^{(2)}_n(z)$.

3·61. *Relations connecting the three kinds of Bessel functions.*

It is easy to obtain the following set of formulae, which express each function in terms of functions of the other two kinds. The reader will observe that some of the formulae are simply the definitions of the functions on the left.

(1) $$J_\nu(z)=\frac{H^{(1)}_\nu(z)+H^{(2)}_\nu(z)}{2}=\frac{Y_{-\nu}(z)-Y_\nu(z)\cos\nu\pi}{\sin\nu\pi},$$

(2) $$J_{-\nu}(z)=\frac{e^{\nu\pi i}H^{(1)}_\nu(z)+e^{-\nu\pi i}H^{(2)}_\nu(z)}{2}=\frac{Y_{-\nu}(z)\cos\nu\pi-Y_\nu(z)}{\sin\nu\pi},$$

(3) $$Y_\nu(z)=\frac{J_\nu(z)\cos\nu\pi-J_{-\nu}(z)}{\sin\nu\pi}=\frac{H^{(1)}_\nu(z)-H^{(2)}_\nu(z)}{2i},$$

(4) $$Y_{-\nu}(z)=\frac{J_\nu(z)-J_{-\nu}(z)\cos\nu\pi}{\sin\nu\pi}=\frac{e^{\nu\pi i}H^{(1)}_\nu(z)-e^{-\nu\pi i}H^{(2)}_\nu(z)}{2i},$$

(5) $$H^{(1)}_\nu(z)=\frac{J_{-\nu}(z)-e^{-\nu\pi i}J_\nu(z)}{i\sin\nu\pi}=\frac{Y_{-\nu}(z)-e^{-\nu\pi i}Y_\nu(z)}{\sin\nu\pi},$$

(6) $$H^{(2)}_\nu(z)=\frac{e^{\nu\pi i}J_\nu(z)-J_{-\nu}(z)}{i\sin\nu\pi}=\frac{Y_{-\nu}(z)-e^{\nu\pi i}Y_\nu(z)}{\sin\nu\pi}.$$

From (5) and (6) it is obvious that

(7) $$H^{(1)}_{-\nu}(z)=e^{\nu\pi i}H^{(1)}_\nu(z),\quad H^{(2)}_{-\nu}(z)=e^{-\nu\pi i}H^{(2)}_\nu(z).$$

3·62. *Bessel functions with argument $-z$ and $ze^{m\pi i}$.*

Since Bessel's equation is unaltered if z is replaced by $-z$, we must expect the functions $J_{\pm\nu}(-z)$ to be solutions of the equation satisfied by $J_{\pm\nu}(z)$.

To avoid the slight difficulty produced by supposing that the phases of both of the complex variables z and $-z$ have their principal values*, we shall construct Bessel functions of argument $ze^{m\pi i}$, where m is any integer, arg z has its principal value, and it is supposed that

$$\arg(ze^{m\pi i}) = m\pi + \arg z.$$

Since $J_\nu(z)/z^\nu$ is definable as a one-valued function, it is obviously convenient to assume that, when the phase of z is unrestricted, $J_\nu(z)$ is to be defined by the same convention as that by which z^ν is defined; and accordingly we have the equations

(1) $$J_\nu(ze^{m\pi i}) = e^{m\nu\pi i} J_\nu(z),$$

(2) $$J_{-\nu}(ze^{m\pi i}) = e^{-m\nu\pi i} J_{-\nu}(z).$$

The functions of the second and third kinds will now be defined for all values of the argument by means of the equations § 3·54 (1), § 3·6 (1); and then the construction of the following set of formulae is an easy matter:

(3) $$Y_\nu(ze^{m\pi i}) = e^{-m\nu\pi i} Y_\nu(z) + 2i \sin m\nu\pi \cot \nu\pi\, J_\nu(z),$$

(4) $$Y_{-\nu}(ze^{m\pi i}) = e^{-m\nu\pi i} Y_{-\nu}(z) + 2i \sin m\nu\pi \operatorname{cosec} \nu\pi\, J_\nu(z),$$

(5) $$H_\nu^{(1)}(ze^{m\pi i}) = e^{-m\nu\pi i} H_\nu^{(1)}(z) - 2e^{-\nu\pi i} \frac{\sin m\nu\pi}{\sin \nu\pi} J_\nu(z)$$
$$= \frac{\sin(1-m)\nu\pi}{\sin\nu\pi} H_\nu^{(1)}(z) - e^{-\nu\pi i} \frac{\sin m\nu\pi}{\sin\nu\pi} H_\nu^{(2)}(z),$$

(6) $$H_\nu^{(2)}(ze^{m\pi i}) = e^{-m\nu\pi i} H_\nu^{(2)}(z) + 2e^{\nu\pi i} \frac{\sin m\nu\pi}{\sin \nu\pi} J_\nu(z)$$
$$= \frac{\sin(1+m)\nu\pi}{\sin\nu\pi} H_\nu^{(2)}(z) + e^{\nu\pi i} \frac{\sin m\nu\pi}{\sin\nu\pi} H_\nu^{(1)}(z).$$

Of these results, (3) was given by Hankel, *Math. Ann.* VIII. (1875), p. 454, in the special case when $m=1$ and ν is an integer. Formulae equivalent to (5) and (6) were obtained by Weber, *Math. Ann.* XXXVII. (1890), pp. 411, 412, when $m=1$; see § 6·11. And a memoir by Graf, *Zeitschrift für Math. und Phys.* XXXVIII. (1893), pp. 115—120, contains the general formulae.

3·63. *Fundamental systems of solutions of Bessel's equation.*

It has been seen (§ 3·12) that $J_\nu(z)$ and $J_{-\nu}(z)$ form a fundamental system of solutions of Bessel's equation when, and only when, ν is not an integer. We shall now examine the Wronskians of other pairs of solutions with a view to determining fundamental systems in the critical case when ν is an integer.

* For $\operatorname{Arg}(-z) = \operatorname{Arg} z \mp \pi$, according as $I(z) \gtrless 0$.

It is clear from § 3·54 (1) that

$$\mathfrak{W}\{J_\nu(z), Y_\nu(z)\} = -\operatorname{cosec}\nu\pi \,.\, \mathfrak{W}\{J_\nu(z), J_{-\nu}(z)\}$$
$$= \frac{2}{\pi z}.$$

This result is established on the hypothesis that ν is not an integer; but considerations of continuity shew that

(1) $$\mathfrak{W}\{J_\nu(z), Y_\nu(z)\} = 2/(\pi z),$$

whether ν be an integer or not. *Hence $J_\nu(z)$ and $Y_\nu(z)$ always form a fundamental system of solutions.*

It is easy to deduce that

(2) $$\mathfrak{W}\{J_\nu(z), \mathbf{Y}_\nu(z)\} = \frac{2e^{\nu\pi i}}{z\cos\nu\pi},$$

and, in particular*,

(3) $$\mathfrak{W}\{J_n(z), \mathbf{Y}_n(z)\} = 2/z.$$

When we express the functions of the third kind in terms of $J_\nu(z)$ and $Y_\nu(z)$, it is found that

(4) $$\mathfrak{W}\{H_\nu^{(1)}(z), H_\nu^{(2)}(z)\} = -2i\,\mathfrak{W}\{J_\nu(z), Y_\nu(z)\} = -4i/(\pi z),$$

so that the functions of the third kind also form a fundamental system of solutions for all values of ν.

Various formulae connected with (1) and (3) have been given by Basset, *Proc. London Math. Soc.* XXI. (1889), p. 55; they are readily obtainable by expressing successive differential coefficients of $J_\nu(z)$ and $Y_\nu(z)$ in terms of $J_\nu(z)$, $J_\nu'(z)$, and $Y_\nu(z)$, $Y_\nu'(z)$ by repeated differentiations of Bessel's equation. Basset's results (of which the earlier ones are frequently required in physical problems) are expressed in the notation used in this work by the following formulae:

(5) $$J_\nu(z)\,Y_\nu''(z) - Y_\nu(z)\,J_\nu''(z) = -\frac{2}{\pi z^2},$$

(6) $$J_\nu'(z)\,Y_\nu''(z) - Y_\nu'(z)\,J_\nu''(z) = \frac{2}{\pi z}\left(1 - \frac{\nu^2}{z^2}\right),$$

(7) $$J_\nu(z)\,Y_\nu'''(z) - Y_\nu(z)\,J_\nu'''(z) = \frac{2}{\pi z}\left(\frac{\nu^2+2}{z^2} - 1\right),$$

(8) $$J_\nu'(z)\,Y_\nu'''(z) - Y_\nu'(z)\,J_\nu'''(z) = \frac{2}{\pi z^2}\left(\frac{3\nu^2}{z^2} - 1\right),$$

(9) $$J_\nu''(z)\,Y_\nu'''(z) - Y_\nu''(z)\,J_\nu'''(z) = \frac{2}{\pi z}\left(1 - \frac{2\nu^2+1}{z^2} + \frac{\nu^4-\nu^2}{z^4}\right),$$

(10) $$J_\nu(z)\,Y_\nu^{(\mathrm{iv})}(z) - Y_\nu(z)\,J_\nu^{(\mathrm{iv})}(z) = \frac{4}{\pi z^2}\left(1 - \frac{3\nu^2+3}{z^2}\right),$$

(11) $$J_\nu'(z)\,Y_\nu^{(\mathrm{iv})}(z) - Y_\nu'(z)\,J_\nu^{(\mathrm{iv})}(z) = -\frac{2}{\pi z}\left(\frac{\nu^4+11\nu^2}{z^4} - \frac{2\nu^2+3}{z^2} + 1\right).$$

Throughout these formulae Y_ν may be replaced by $J_{-\nu}$ if the expressions on the right are multiplied by $-\sin\nu\pi$; and J_ν, Y_ν may be replaced by $H_\nu^{(1)}$, $H_\nu^{(2)}$ throughout if the expressions on the right are multiplied by $-2i$.

* Cf. Lommel, *Math. Ann.* IV. (1871), p. 106, and Hankel, *Math. Ann.* VIII. (1875), p. 457.

An associated formula, due to Lommel*, *Math. Ann.* IV. (1871), p. 106, and Hankel, *Math. Ann.* VIII. (1875), p. 458, is

(12) $$J_\nu(z)\,Y_{\nu+1}(z) - J_{\nu+1}(z)\,Y_\nu(z) = -\frac{2}{\pi z}.$$

This is proved in the same way as § 3·2 (7).

3·7. *Bessel functions of purely imaginary argument.*

The differential equation

(1) $$z^2\frac{d^2y}{dz^2} + z\frac{dy}{dz} - (z^2+\nu^2)\,y = 0,$$

which differs from Bessel's equation only in the coefficient of y, is of frequent occurrence in problems of Mathematical Physics; in such problems, it is usually desirable to present the solution in a real form, and the fundamental systems $J_\nu(iz)$ and $J_{-\nu}(iz)$ or $J_\nu(iz)$ and $Y_\nu(iz)$ are unsuited for this purpose.

However the function $e^{-\frac{1}{2}\nu\pi i}J_\nu(iz)$ is a real function of z which is a solution of the equation. It is customary to denote it by the symbol $I_\nu(z)$ so that

(2) $$I_\nu(z) = \sum_{m=0}^{\infty} \frac{(\tfrac{1}{2}z)^{\nu+2m}}{m!\,\Gamma(\nu+m+1)}.$$

When z is regarded as a complex variable, it is usually convenient to define its phase, not with reference to the principal value of arg iz, as the consideration of the function $J_\nu(iz)$ would suggest, but with reference to the principal value of arg z, so that

$$\begin{cases} I_\nu(z) = e^{-\frac{1}{2}\nu\pi i}J_\nu(ze^{\frac{1}{2}\pi i}), & (-\pi < \arg z \leqslant \tfrac{1}{2}\pi), \\ I_\nu(z) = e^{\frac{3}{2}\nu\pi i}J_\nu(ze^{-\frac{3}{2}\pi i}), & (\tfrac{1}{2}\pi < \arg z \leqslant \pi). \end{cases}$$

The introduction of the symbol $I_\nu(z)$ to denote "the function of imaginary argument" is due to Basset† and it is now in common use. It should be mentioned that four years before the publication of Basset's work, Nicolas‡ had suggested the use of the symbol $F_\nu(z)$, but this notation has not been used by other writers.

The relative positions of Pure and Applied Mathematics on the Continent as compared with this country are remarkably illustrated by the fact that, in Nielsen's standard treatise§, neither the function $I_\nu(z)$, nor the second solution $K_\nu(z)$, which will be defined immediately, is even mentioned, in spite of their importance in physical applications.

The function $I_{-\nu}(z)$ is also a solution of (1), and it is easy to prove (cf. § 3·12) that

(3) $$\mathfrak{W}\{I_\nu(z),\, I_{-\nu}(z)\} = -\frac{2\sin\nu\pi}{\pi z}.$$

* Lommel gave the corresponding formula for Neumann's function of the second kind.

† *Proc. Camb. Phil. Soc.* VI. (1889), p. 11. [This paper was first published in 1886.] Basset, in this paper, defined the function of integral order to be $i^{+n}J_n(iz)$, but he subsequently changed it, in his *Hydrodynamics*, II. (Cambridge, 1888), p. 17, to that given in the text. The more recent definition is now universally used.

‡ *Ann. Sci. de l'École norm. sup.* (2) XI. (1882), supplément, p. 17.

§ *Handbuch der Theorie der Cylinderfunktionen* (Leipzig, 1904).

It follows that, when ν is not an integer, the functions $I_\nu(z)$ and $I_{-\nu}(z)$ form a fundamental system of solutions of equation (1).

In the case of functions of integral order, a second solution has to be constructed by the methods of §§ 3·5—3·54.

The function $K_n(z)$, which will be adopted throughout this work as the second solution, is defined by the equation

$$K_n(z) = \lim_{\nu \to n} \frac{(-)^n}{2} \left[\frac{I_{-\nu}(z) - I_\nu(z)}{\nu - n} \right]. \tag{4}$$

An equivalent definition (cf. § 3·5) is

$$K_n(z) = \frac{(-)^n}{2} \left[\frac{\partial I_{-\nu}(z)}{\partial \nu} - \frac{\partial I_\nu(z)}{\partial \nu} \right]_{\nu=n}. \tag{5}$$

It may be verified, by the methods of § 3·5, that $K_n(z)$ is a solution of (1) when the order ν is equal to n.

The function $K_\nu(z)$ has been defined, for unrestricted values of ν, by Macdonald*, by the equation

$$K_\nu(z) = \tfrac{1}{2}\pi \frac{I_{-\nu}(z) - I_\nu(z)}{\sin \nu\pi}, \tag{6}$$

and, with this definition, it may be verified that

$$K_n(z) = \lim_{\nu \to n} K_\nu(z). \tag{7}$$

It is easy to deduce from (6) that

$$K_\nu(z) = \tfrac{1}{2}\pi i e^{\frac{1}{2}\nu\pi i} H_\nu^{(1)}(iz) = \tfrac{1}{2}\pi i e^{-\frac{1}{2}\nu\pi i} H_{-\nu}^{(1)}(iz). \tag{8}$$

The physical importance of the function $K_\nu(z)$ lies in the fact that it is a solution of equation (1) which tends exponentially to zero as $z \to \infty$ through positive values. This fundamental property of the function will be established in § 7·23.

The definition of $K_n(z)$ is due to Basset, *Proc. Camb. Phil. Soc.* VI. (1889), p. 11, and his definition is equivalent to that given by equations (4) and (5); the infinite integrals by which he actually defined the function will be discussed in §§ 6·14, 6·15. Basset subsequently modified his definition of the function in his *Hydrodynamics*, II. (Cambridge, 1888), pp. 18—19, and his final definition is equivalent to $\frac{1}{2^{n+1}}\left[\frac{\partial I_{-\nu}(z)}{\partial \nu} - \frac{\partial I_\nu(z)}{\partial \nu}\right]_{\nu=n}$.

In order to obtain a function which satisfies the same recurrence formulae as $I_\nu(z)$, Gray and Mathews in their work, *A Treatise on Bessel Functions* (London, 1895), p. 67, omit the factor $1/2^n$, so that their definition is equivalent to

$$\frac{1}{2}\left[\frac{\partial I_{-\nu}(z)}{\partial \nu} - \frac{\partial I_\nu(z)}{\partial \nu}\right]_{\nu=n}.$$

The only simple extension of this definition to functions of unrestricted order is by the formula

$$\mathbf{K}_\nu(z) \equiv \tfrac{1}{2}\pi \cot \nu\pi \{I_{-\nu}(z) - I_\nu(z)\},$$

* *Proc. London Math. Soc.* XXX. (1899), p. 167.

(cf. *Modern Analysis*, § 17·71) but this function suffers from the serious disadvantage that it vanishes whenever 2ν is an odd integer. Consequently in this work, Macdonald's function will be used although it has the disadvantage of not satisfying the same recurrence formulae as $I_\nu(z)$.

An inspection of formula (8) shews that it would have been advantageous if a factor $\frac{1}{2}\pi$ had been omitted from the definition of $K_\nu(z)$; but in view of the existence of extensive tables of Macdonald's function it is now inadvisable to make the change, and the presence of the factor is not so undesirable as the presence of the corresponding factor in Schläfli's function (§ 3·54) because linear combinations of $I_\nu(z)$ and $K_\nu(z)$ are not of common occurrence.

3·71. *Formulae connected with $I_\nu(z)$ and $K_\nu(z)$.*

We shall now give various formulae for $I_\nu(z)$ and $K_\nu(z)$ analogous to those constructed in §§ 3·2—3·6 for the ordinary Bessel functions. The proofs of the formulae are left to the reader.

(1) $$I_{\nu-1}(z) - I_{\nu+1}(z) = \frac{2\nu}{z} I_\nu(z), \quad K_{\nu-1}(z) - K_{\nu+1}(z) = -\frac{2\nu}{z} K_\nu(z),$$

(2) $$I_{\nu-1}(z) + I_{\nu+1}(z) = 2I_\nu'(z), \quad K_{\nu-1}(z) + K_{\nu+1}(z) = -2K_\nu'(z),$$

(3) $$zI_\nu'(z) + \nu I_\nu(z) = zI_{\nu-1}(z), \quad zK_\nu'(z) + \nu K_\nu(z) = -zK_{\nu-1}(z),$$

(4) $$zI_\nu'(z) - \nu I_\nu(z) = zI_{\nu+1}(z), \quad zK_\nu'(z) - \nu K_\nu(z) = -zK_{\nu+1}(z),$$

(5) $$\left(\frac{d}{z\,dz}\right)^m \{z^\nu I_\nu(z)\} = z^{\nu-m} I_{\nu-m}(z), \quad \left(\frac{d}{z\,dz}\right)^m \{z^\nu K_\nu(z)\} = (-)^m z^{\nu-m} K_{\nu-m}(z),$$

(6) $$\left(\frac{d}{z\,dz}\right)^m \left\{\frac{I_\nu(z)}{z^\nu}\right\} = \frac{I_{\nu+m}(z)}{z^{\nu+m}}, \quad \left(\frac{d}{z\,dz}\right)^m \left\{\frac{K_\nu(z)}{z^\nu}\right\} = (-)^m \frac{K_{\nu+m}(z)}{z^{\nu+m}},$$

(7) $$I_0'(z) = I_1(z), \qquad K_0'(z) = -K_1(z),$$

(8) $$I_{-n}(z) = I_n(z), \qquad K_{-\nu}(z) = K_\nu(z).$$

The following integral formulae are valid only when $R(\nu + \frac{1}{2}) > 0$:

(9) $$I_\nu(z) = \frac{(\frac{1}{2}z)^\nu}{\Gamma(\nu+\frac{1}{2})\,\Gamma(\frac{1}{2})} \int_0^\pi \cosh(z\cos\theta)\sin^{2\nu}\theta\,d\theta$$

$$= \frac{(\frac{1}{2}z)^\nu}{\Gamma(\nu+\frac{1}{2})\,\Gamma(\frac{1}{2})} \int_{-1}^1 (1-t^2)^{\nu-\frac{1}{2}} \cosh(zt)\,dt$$

$$= \frac{(\frac{1}{2}z)^\nu}{\Gamma(\nu+\frac{1}{2})\,\Gamma(\frac{1}{2})} \int_{-1}^1 (1-t^2)^{\nu-\frac{1}{2}} e^{\pm zt}\,dt$$

$$= \frac{(\frac{1}{2}z)^\nu}{\Gamma(\nu+\frac{1}{2})\,\Gamma(\frac{1}{2})} \int_0^\pi e^{\pm z\cos\theta}\sin^{2\nu}\theta\,d\theta$$

$$= \frac{2(\frac{1}{2}z)^\nu}{\Gamma(\nu+\frac{1}{2})\,\Gamma(\frac{1}{2})} \int_0^{\frac{1}{2}\pi} \cosh(z\cos\theta)\sin^{2\nu}\theta\,d\theta$$

$$= \frac{2(\frac{1}{2}z)^\nu}{\Gamma(\nu+\frac{1}{2})\,\Gamma(\frac{1}{2})} \int_0^1 (1-t^2)^{\nu-\frac{1}{2}} \cosh(zt)\,dt.$$

These results are due to Basset. We also have

(10) $$I_{n+\frac{1}{2}}(z) = \frac{1}{\sqrt{(2\pi z)}}\left[e^z \sum_{r=0}^{n} \frac{(-)^r (n+r)!}{r!\,(n-r)!\,(2z)^r} + (-)^{n+1} e^{-z} \sum_{r=0}^{n} \frac{(n+r)!}{r!\,(n-r)!\,(2z)^r}\right],$$

(11) $$I_{-(n+\frac{1}{2})}(z) = \frac{1}{\sqrt{(2\pi z)}}\left[e^z \sum_{r=0}^{n} \frac{(-)^r (n+r)!}{r!\,(n-r)!\,(2z)^r} + (-)^n e^{-z} \sum_{r=0}^{n} \frac{(n+r)!}{r!\,(n-r)!\,(2z)^r}\right],$$

(12) $$K_{n+\frac{1}{2}}(z) = \left(\frac{\pi}{2z}\right)^{\frac{1}{2}} e^{-z} \sum_{r=0}^{n} \frac{(n+r)!}{r!\,(n-r)!\,(2z)^r},$$

(13) $$K_{\frac{1}{2}}(z) = \left(\frac{\pi}{2z}\right)^{\frac{1}{2}} e^{-z},$$

(14) $$K_0(z) = -\log(\tfrac{1}{2}z)\,.\,I_0(z) + \sum_{m=0}^{\infty} \frac{(\frac{1}{2}z)^{2m}}{(m!)^2}\,\psi(m+1),$$

(15) $$K_n(z) = \frac{1}{2}\sum_{m=0}^{n-1} \frac{(-)^m (n-m-1)!}{m!\,(\frac{1}{2}z)^{n-2m}} + (-)^{n+1} \sum_{m=0}^{\infty} \frac{(\frac{1}{2}z)^{n+2m}}{m!\,(n+m)!}\{\log(\tfrac{1}{2}z) - \tfrac{1}{2}\psi(m+1) - \tfrac{1}{2}\psi(n+m+1)\},$$

(16) $$K_0(z) = -\frac{1}{\pi}\int_0^{\pi} e^{z\cos\theta}\{\log(2z\sin^2\theta) + \gamma\}\,d\theta,$$

(17) $$I_\nu(ze^{m\pi i}) = e^{m\nu\pi i}\,I_\nu(z),$$

(18) $$K_\nu(ze^{m\pi i}) = e^{-m\nu\pi i}\,K_\nu(z) - \pi i\,\frac{\sin m\nu\pi}{\sin \nu\pi}\,I_\nu(z),$$

(19) $$\mathfrak{W}\{I_\nu(z),\ K_\nu(z)\} = -1/z,$$

(20) $$I_\nu(z)\,K_{\nu+1}(z) + I_{\nu+1}(z)\,K_\nu(z) = 1/z.$$

The integral involved in (16) has been discussed by Stokes (cf. § 3·572).

The integrals involved in (9) and the series in (14) were discussed by Riemann in his memoir "Zur Theorie der Nobili'schen Farbenringe," *Ann. der Physik und Chemie*, (2) XCV. (1855), pp. 130—139, in the special case in which $\nu=0$; he also discussed the ascending power series for $I_0(z)$.

The recurrence formulae have been given by Basset, *Proc. Camb. Phil. Soc.* VI. (1889), pp. 2—19; by Macdonald, *Proc. London Math. Soc.* XXIX. (1899), pp. 110—115; and by Aichi, *Proc. Phys. Math. Soc. of Japan*, (3) II. (1920), pp. 8—19.

Functions of this type whose order is half an odd integer, as in equations (10) and (12), were used by Hertz in his Berlin Dissertation, 1880 [*Ges. Werke*, I. (1895), pp. 77—91]; and he added yet another notation to those described in § 3·41.

3·8. *Thomson's functions* ber (z) *and* bei (z) *and their generalisations.*

A class of functions which occurs in certain electrical problems consists of Bessel functions whose arguments have their phases equal to $\frac{1}{4}\pi$ or $\frac{3}{4}\pi$.

The functions of order zero were first examined by W. Thomson*; they may be defined by the equation†

$$\text{(1)} \qquad \operatorname{ber}(x) + i \operatorname{bei}(x) = J_0(xi\sqrt{i}) = I_0(x\sqrt{i}),$$

where x is real, and ber and bei denote real functions. For complex arguments we adopt the definitions expressed by the formulae

$$\text{(2)} \qquad \operatorname{ber}(z) \pm i \operatorname{bei}(z) = J_0(zi\sqrt{\pm i}) = I_0(z\sqrt{\pm i}).$$

Hence we have

$$\text{(3)} \qquad \operatorname{ber}(z) = 1 - \frac{(\frac{1}{2}z)^4}{(2!)^2} + \frac{(\frac{1}{2}z)^8}{(4!)^2} - \dots,$$

$$\text{(4)} \qquad \operatorname{bei}(z) = \frac{(\frac{1}{2}z)^2}{(1!)^2} - \frac{(\frac{1}{2}z)^6}{(3!)^2} + \frac{(\frac{1}{2}z)^{10}}{(5!)^2} - \dots.$$

Extensions of these definitions to functions of any order of the first, second and third kinds have been effected by Russell‡ and Whitehead§.

The functions of the second kind of order zero were defined by Russell by a pair of equations resembling (2), the function I_0 being replaced by the function K_0, thus

$$\text{(5)} \qquad \operatorname{ker}(z) \pm i \operatorname{kei}(z) = K_0(z\sqrt{\pm i}).$$

Functions of unrestricted order ν were defined by Whitehead with reference to Bessel functions of the first and third kinds, thus

$$\text{(6)} \qquad \operatorname{ber}_\nu(z) \pm i \operatorname{bei}_\nu(z) = J_\nu(ze^{\pm \frac{3}{4}\pi i}),$$

$$\text{(7)} \qquad \operatorname{her}_\nu(z) \pm i \operatorname{hei}_\nu(z) = H_\nu^{(1)}(ze^{\pm \frac{3}{4}\pi i}).$$

It will be observed that‖

$$\text{(8)} \qquad \operatorname{ker}(z) = -\tfrac{1}{2}\pi \operatorname{hei}(z), \quad \operatorname{kei}(z) = \tfrac{1}{2}\pi \operatorname{her}(z),$$

in consequence of § 3·7 (8).

The following series, due to Russell, are obtainable without difficulty:

$$\text{(9)} \qquad \operatorname{ker}(z) = -\log(\tfrac{1}{2}z) \,.\, \operatorname{ber}(z) + \tfrac{1}{4}\pi \operatorname{bei}(z) + \sum_{m=0}^{\infty} \frac{(-)^m (\frac{1}{2}z)^{4m}}{\{(2m)!\}^2} \psi(2m+1),$$

* Presidential Address to the Institute of Electrical Engineers, 1889. [*Math. and Phys. Papers*, III. (1890), p. 492.]

† In the case of functions of zero order, it is customary to omit the suffix which indicates the order.

‡ *Phil. Mag.* (6) XVII. (1909), pp. 524—552.

§ *Quarterly Journal*, XLII. (1911), pp. 316—342.

‖ Integrals equal to ker (z) and kei (z) occur in a memoir by Hertz, *Ann. der Physik und Chemie*, (3) XXII. (1884), p. 450 [*Ges. Werke*, I. (1895), p. 289].

(10) $$\operatorname{kei}(z) = -\log(\tfrac{1}{2}z) \,.\, \operatorname{bei}(z) - \tfrac{1}{4}\pi \operatorname{ber}(z) + \sum_{m=0}^{\infty} \frac{(-)^m (\tfrac{1}{2}z)^{4m+2}}{\{(2m+1)!\}^2} \psi(2m+2).$$

It has also been observed by Russell that the first few terms of the expansion of $\operatorname{ber}^2(z) + \operatorname{bei}^2(z)$ have simple coefficients, thus

(11) $$\operatorname{ber}^2(z) + \operatorname{bei}^2(z) = 1 + \frac{(\tfrac{1}{2}z)^4}{2!} + \frac{(\tfrac{1}{2}z)^8}{4\,.\,4!} + \frac{(\tfrac{1}{2}z)^{12}}{6^2\,.\,6!} + \frac{(\tfrac{1}{2}z)^{16}}{8^2\,.\,9!} + \dots,$$

but this result had previously been obtained, with a different notation, by Nielsen (cf. § 5·41); the coefficient of $(\tfrac{1}{2}z)^{4m}$ in the expansion on the right is $1/[(m!)^2 \,.\, (2m)!]$.

Numerous expansions involving squares and products of the general functions have been obtained by Russell; for such formulae the reader is referred to Russell's memoir and also to a paper by Savidge*.

Formulae analogous to the results of §§ 3·61, 3·62 have been discussed by Whitehead; it is sufficient to quote the following here:

(12) $$\operatorname{ber}_{-\nu}(z) = \cos\nu\pi \,.\, \operatorname{ber}_\nu(z) - \sin\nu\pi \,.\, [\operatorname{hei}_\nu(z) - \operatorname{bei}_\nu(z)],$$

(13) $$\operatorname{bei}_{-\nu}(z) = \cos\nu\pi \,.\, \operatorname{bei}_\nu(z) + \sin\nu\pi \,.\, [\operatorname{her}_\nu(z) - \operatorname{ber}_\nu(z)],$$

(14) $$\operatorname{her}_{-\nu}(z) = \cos\nu\pi \,.\, \operatorname{her}_\nu(z) - \sin\nu\pi \,.\, \operatorname{hei}_\nu(z),$$

(15) $$\operatorname{hei}_{-\nu}(z) = \sin\nu\pi \,.\, \operatorname{her}_\nu(z) + \cos\nu\pi \,.\, \operatorname{hei}_\nu(z).$$

The reader will be able to construct the recurrence formulae which have been worked out at length by Whitehead.

The functions of order unity have recently been examined in some detail by B. A. Smith†.

3·9. *The definition of cylinder functions.*

Various writers, especially Sonine‡ and Nielsen§, have studied the general theory of analytic functions of two variables $\mathscr{C}_\nu(z)$ which satisfy the pair of recurrence formulae

(1) $$\mathscr{C}_{\nu-1}(z) + \mathscr{C}_{\nu+1}(z) = \frac{2\nu}{z}\mathscr{C}_\nu(z),$$

(2) $$\mathscr{C}_{\nu-1}(z) - \mathscr{C}_{\nu+1}(z) = 2\mathscr{C}_\nu'(z),$$

in which z and ν are unrestricted complex variables. These recurrence formulae are satisfied by each of the three kinds of Bessel functions.

Functions which satisfy only one of the two formulae are also discussed by Sonine in his elaborate memoir; a brief account of his researches will be given in Chapter x.

* *Phil. Mag.* (6) XIX. (1910), pp. 49—58.

† *Proc. American Soc. of Civil Engineers*, XLVI. (1920), pp. 375—425.

‡ *Math. Ann.* XVI. (1880), pp. 1—80.

§ *Handbuch der Theorie der Cylinderfunktionen* (Leipzig, 1904), pp. 1, 42 *et seq.*

Following Sonine we shall call any function $\mathscr{C}_\nu(z)$, which satisfies *both* of the formulae, a *cylinder function*. It will now be shewn that cylinder functions are expressible in terms of Bessel functions.

When we combine the formulae (1) and (2), we find that

(3) $$z\mathscr{C}_\nu'(z) + \nu\mathscr{C}_\nu(z) = z\mathscr{C}_{\nu-1}(z),$$

(4) $$z\mathscr{C}_\nu'(z) - \nu\mathscr{C}_\nu(z) = -z\mathscr{C}_{\nu+1}(z),$$

and so, if ϑ be written for $z(d/dz)$, we deduce that

(5) $$(\vartheta + \nu)\mathscr{C}_\nu(z) = z\mathscr{C}_{\nu-1}(z),$$

(6) $$(\vartheta - \nu)\mathscr{C}_\nu(z) = -z\mathscr{C}_{\nu+1}(z).$$

It follows that

$$\begin{aligned}(\vartheta^2 - \nu^2)\mathscr{C}_\nu(z) &= (\vartheta - \nu)\{z\mathscr{C}_{\nu-1}(z)\}\\ &= z(\vartheta - \nu + 1)\mathscr{C}_{\nu-1}(z)\\ &= -z^2\mathscr{C}_\nu(z),\end{aligned}$$

that is to say

(7) $$\nabla_\nu \mathscr{C}_\nu(z) = 0.$$

Hence $$\mathscr{C}_\nu(z) = a_\nu J_\nu(z) + b_\nu Y_\nu(z),$$

where a_ν and b_ν are independent of z, though they may depend on ν. When we substitute in (3) we find that

$$a_\nu J_{\nu-1}(z) + b_\nu Y_{\nu-1}(z) \equiv a_{\nu-1} J_{\nu-1}(z) + b_{\nu-1} Y_{\nu-1}(z),$$

and so, since $J_{\nu-1}(z)/Y_{\nu-1}(z)$ is not independent of z, we must have

$$a_\nu = a_{\nu-1}, \quad b_\nu = b_{\nu-1}.$$

Hence a_ν and b_ν *must be* periodic functions of ν with period unity; and, conversely, if they are such functions of ν, it is easy to see that both (1) and (2) are satisfied.

Hence the general solution of (1) and (2) is

(8) $$\mathscr{C}_\nu(z) = \varpi_1(\nu) J_\nu(z) + \varpi_2(\nu) Y_\nu(z),$$

where $\varpi_1(\nu)$ and $\varpi_2(\nu)$ are arbitrary periodic functions of ν with period unity. It may be observed that an equivalent solution is

(9) $$\mathscr{C}_\nu(z) = \varpi_3(\nu) H_\nu^{(1)}(z) + \varpi_4(\nu) H_\nu^{(2)}(z).$$

A difference equation, which is more general than (1), has been examined by Barnes, *Messenger*, XXXIV. (1905), pp. 52—71; in certain circumstances the solution is expressible by Bessel functions, though it usually involves hypergeometric functions.

NOTE. The name *cylinder function* is used by Nielsen to denote $J_\nu(z)$, $Y_\nu(z)$, $H_\nu^{(1)}(z)$ and $H_\nu^{(2)}(z)$ as well as the more general functions discussed in this section. This procedure is in accordance with the principle laid down by Mittag-Leffler that it is, in general, undesirable to associate functions with the names of particular mathematicians.

The name cylinder function is derived from the fact that normal solutions of Laplace's equation in cylindrical coordinates are

$$e^{\pm kz} J_m(k\rho) \begin{matrix}\cos\\ \sin\end{matrix} m\phi$$

(cf. § 4·8 and *Modern Analysis*, § 18·5).

Some writers*, following Heine† who called $J_n(z)$ a *Fourier-Bessel function*, call $J_n(z)$ a *Fourier function*.

Although Bessel coefficients of any order were used long before the time of Bessel (cf. §§ 1·3, 1·4), it seems desirable to associate Bessel's name with them, not only because it has become generally customary to do so, but also because of the great advance made by Bessel on the work of his predecessors in the invention of a simple and compact notation for the functions.

Bessel's name was associated with the functions by Jacobi, *Journal für Math.* XV. (1836), p. 13 [*Ges. Math. Werke*, VI. (1891), p. 101]. "Transcendentium $I_k{}^i$ naturam variosque usus in determinandis integralibus definitis exposuit ill. Bessel in commentatione celeberrima."

A more recent controversy on the name to be applied to the functions is to be found in a series of letters in *Nature*, LX. (1899), pp. 101, 149, 174; LXXXI. (1909), p. 68.

* E.g. Nicolas, *Ann. Sci. de l'École norm. sup.* (2) XI. (1882), supplément.

† *Journal für Math.* LXIX. (1868), p. 128. Heine also seems to be responsible for the term cylinder function.

CHAPTER IV

DIFFERENTIAL EQUATIONS

4·1. *Daniel Bernoulli's solution of Riccati's equation.*

The solution given by Bernoulli* of the equation

(1) $$\frac{dy}{dz} = az^n + by^2$$

consisted in shewing that when the index n has any of the values

$$0;\quad -\tfrac{4}{1}, -\tfrac{4}{3};\quad -\tfrac{8}{3}, -\tfrac{8}{5};\quad -\tfrac{12}{5}, -\tfrac{12}{7};\quad -\tfrac{16}{7}, -\tfrac{16}{9};\quad \ldots,$$

while a and b have any constant values†, then the equation is soluble by means of algebraic, exponential and logarithmic functions. The values of n just given are comprised in the formula

(2) $$n = -\frac{4m}{2m \pm 1},$$

where m is zero or a positive integer.

Bernoulli's method of solution is as follows: If n be called the index of the equation, it is first proved that the general equation‡ of index n is transformable into the general equation of index N, where

(3) $$N = -\frac{n}{n+1};$$

and it is also proved that the general equation of index n is transformable into the general equation of index ν, where

(4) $$\nu = -n - 4.$$

The Riccati equation of index zero is obviously integrable, because the variables are separable. Hence, by (4), the equation of index -4 is integrable. Hence by (3), the equation of index $-\tfrac{4}{3}$ is integrable. If this process be continued by using the transformations (3) and (4) alternately, we arrive at the set of soluble cases given above, and it is easy to see that these cases are comprised in the general formula (2).

* *Exercitationes quaedam mathematicae* (Venice, 1724), pp. 77—80; *Acta Eruditorum*, 1725, pp. 473—475. The notation used by Bernoulli has been slightly modified; and in this analysis n is not restricted to be an integer.

† It is assumed that neither a nor b is zero. If either were zero the variables would obviously be separable.

‡ That is, the equation in which a and b have arbitrary values.

4·11. *Daniel Bernoulli's transformations of Riccati's equation.*

Now that the outlines of Bernoulli's procedure have been indicated, we proceed to give the analysis by which the requisite transformations are effected.

Take § 4·1 (1) as the standard equation of index n and make the substitutions

$$\frac{z^{n+1}}{n+1} = Z, \quad y = -\frac{1}{Y}.$$

[NOTE. The substitutions are possible because -1 is not included among the values of n. The factor $n+1$ in the denominator was not inserted by Bernoulli; the effect of its presence is that the transformed equation is more simple than if it were omitted.]

The equation becomes

$$\frac{1}{Y^2}\frac{dY}{dZ} = a + \frac{b}{Y^2 z^n},$$

that is

$$\frac{dY}{dZ} = b\,(n+1)^N Z^N + a Y^2,$$

where $N = -n/(n+1)$; and this is the general equation of index N.

Again in § 4·1 (1) make the substitutions

$$z = \frac{1}{\zeta}, \quad y = -\frac{\zeta}{b} - \eta\zeta^2.$$

The equation becomes

$$\frac{d\eta}{d\zeta} = a\zeta^\nu + b\eta^2,$$

where $\nu = -n-4$; and this is the general equation of index ν.

The transformations described in § 4·1 are therefore effected, and so the equation is soluble in the cases stated. But this procedure does not give the solution in a compact form.

4·12. *The limiting form of Riccati's equation, with index* -2.

When the processes described in §§ 4·1, 4·11 are continually applied to Riccati's equation, the value to which the index tends, when $m \to \infty$ in § 4·1 (2), is -2. The equation with index -2 is consequently not soluble by a finite number of transformations of the types hitherto under consideration.

To solve the equation with index -2, namely

$$\frac{dy}{dz} = \frac{a}{z^2} + by^2, \tag{1}$$

write $y = v/z$, and the equation becomes

$$z\frac{dv}{dz} = a + v + bv^2;$$

and this is an equation with the variables separable.

Hence, in this limiting case, Riccati's equation is still soluble by the use of elementary functions.

This solution was implicitly given by Euler, *Inst. Calc. Int.* II. (Petersburg, 1769), § 933, p. 185. If we write (cf. § 4·14) $y=-\frac{1}{b\eta}\frac{d\eta}{dz}$, the equation which determines η is

$$\frac{d^2\eta}{dz^2}+\frac{ab\eta}{z^2}=0,$$

which is homogeneous, and consequently it is immediately soluble.

Euler does not seem to mention the limiting case of Riccati's equation explicitly, although he gave both the solution of the homogeneous linear equation and the transformation which connects any equation of Riccati's type with a linear equation.

It will appear subsequently (§§ 4·7—4·75) that the *only* cases in which Riccati's equation is soluble in finite terms are the cases which have now been examined; that is to say, those in which the index has one of the values

$$0;\quad -\tfrac{4}{1},\ -\tfrac{4}{3};\quad -\tfrac{8}{3},\ -\tfrac{8}{5};\quad \dots,\ -2,$$

and also the trivial cases in which a or b (or both) is zero.

This converse theorem, due to Liouville, is, of course, much more recondite than Bernoulli's theorem that the equation is soluble in the specified cases.

4·13. *Euler's solution of Riccati's equation.*

A practical method of constructing a solution of Riccati's equation in the soluble cases was devised by Euler*, and this method (with some slight changes in notation), will now be explained.

First transform Riccati's equation, § 4·1 (1), by taking new variables and constants as follows:

(1) $$y=-\eta/b,\quad ab=-c^2,\quad n=2q-2;$$

the transformed equation is

(2) $$\frac{d\eta}{dz}+\eta^2-c^2z^{2q-2}=0;$$

and the soluble cases are those in which $1/q$ is an odd integer.

Define a new variable w by the equation

(3) $$\eta=cz^{q-1}+\frac{1}{w}\frac{dw}{dz},$$

so that the equation in w is

(4) $$\frac{d^2w}{dz^2}+2cz^{q-1}\frac{dw}{dz}+(q-1)\,cz^{q-2}w=0.$$

A solution in series of the last equation is

$$w=z^{-\frac{1}{2}(q-1)}\sum_{r=0}^{\infty}A_rz^{-qr},$$

provided that

$$\frac{A_{r+1}}{A_r}=\frac{(2qr+q+1)(2qr+q-1)}{8qc\,(r+1)},$$

* *Nov. Comm. Acad. Petrop.* VIII. (1760—1761) [1763], pp. 3—63; and IX. (1762—1763) [1764], pp. 154—169.

and so the series terminates with the term $A_m z^{-qm}$ if q has either of the values $\pm 1/(2m+1)$; and this procedure gives the solution* examined by Bernoulli.

The general solution of Riccati's equation, which is not obvious by this method, was given explicitly by Hargreave, *Quarterly Journal*, VII. (1866), pp. 256—258, but Hargreave's form of the solution was unnecessarily complicated; two years later Cayley, *Phil. Mag.* (4) XXXVI. (1868), pp. 348—351 [*Collected Papers*, VII. (1894), pp. 9—12], gave the general solution in a form which closely resembles Euler's particular solution, the chief difference between the two solutions being the reversal of the order of the terms of the series involved.

Cayley used a slightly simpler form of the equation than (2), because he took constant multiples of *both* variables in Riccati's equation in such a way as to reduce it to

$$\frac{d\eta}{d\zeta}+\eta^2-\zeta^{2q-2}=0. \tag{5}$$

4·14. *Cayley's general solution of Riccati's equation.*

We have just seen that Riccati's equation is reducible to the form

$$\frac{d\eta}{dz}+\eta^2-c^2z^{2q-2}=0,$$

given in § 4·13 (2); and we shall now explain Cayley's† method of solving this equation, which is to be regarded as a canonical form of Riccati's equation.

When we make the substitution‡ $\eta = d(\log v)/dz$, the equation becomes

$$\frac{d^2v}{dz^2}-c^2z^{2q-2}v=0; \tag{1}$$

and, if U_1 and U_2 are a fundamental system of solutions of this equation, the general solution of the canonical form of Riccati's equation is

$$\eta=\frac{C_1U_1'+C_2U_2'}{C_1U_1+C_2U_2}, \tag{2}$$

where C_1 and C_2 are arbitrary constants and primes denote differentiations with respect to z.

To express U_1 and U_2 in a finite form, we write

$$v = w \exp(cz^q/q),$$

so that the equation satisfied by w is § 4·13 (4). A solution of this equation in w proceeding in *ascending* powers of z^q is

$$1-\frac{q-1}{q(q-1)}cz^q+\frac{(q-1)(3q-1)}{q(q-1)\,2q(2q-1)}c^2z^{2q} \\ -\frac{(q-1)(3q-1)(5q-1)}{q(q-1)\,2q(2q-1)\,3q(3q-1)}c^3z^{3q}+\dots,$$

and we take U_1 to be $\exp(cz^q/q)$ multiplied by this series.

* When the index n of the Riccati equation is -2, equation (4) is homogeneous.

† *Phil. Mag.* (4) XXXVI. (1868), pp. 348—351 [*Collected Papers*, VII. (1894), pp. 9—12]. Cf. also the memoirs by Euler which were cited in § 4·13.

‡ This is, of course, the substitution used in 1702 by James Bernoulli; cf. § 1·1.

Now equation (1) is unaffected by changing the sign of c, and so we take

$$U_1,\ U_2 = \exp(\pm cz^q/q)\left[1 \mp \frac{q-1}{q(q-1)}cz^q + \frac{(q-1)(3q-1)}{q(q-1)\,2q\,(2q-1)}c^2z^{2q} \mp \frac{(q-1)(3q-1)(5q-1)}{q(q-1)\,2q\,(2q-1)\,3q\,(3q-1)}c^3z^{3q} + \dots\right],$$

and *both* of these series terminate when q is the reciprocal of an odd positive integer. Since the ratio $U_1 : U_2$ is the exponential function $\exp(2cz^q/q)$ multiplied by an algebraic function of z^q, it cannot be a constant; and so U_1, U_2 form a fundamental system of solutions of (1).

If q were the reciprocal of an odd negative integer, we should write equation (1) in the form

$$\frac{d^2(v/z)}{d(1/z)^2} - c^2(1/z)^{-2q-2}(v/z) = 0,$$

whence it follows that

$$\eta = \frac{d}{dz}\log(\gamma_1 V_1 + \gamma_2 V_2),$$

where γ_1 and γ_2 are constants, and

$$V_1,\ V_2 = z\exp(\mp cz^q/q)\left[1 \pm \frac{q+1}{q(q+1)}cz^q + \frac{(q+1)(3q+1)}{q(q+1)\,2q\,(2q+1)}c^2z^{2q} \pm \dots\right].$$

The series which have now been obtained will be examined in much greater detail in §§ 4·4—4·42.

The reader should have no difficulty in constructing the following solutions of Riccati's equation, when it is soluble in finite terms.

	Equation	Values of U_1, U_2
(i)	$(d\eta/dz)+\eta^2=1$	$\exp(\pm z)$
(ii)	$(d\eta/dz)+\eta^2=z^{-4/3}$	$(1\mp 3z^{1/3})\exp(\pm 3z^{1/3})$
(iii)	$(d\eta/dz)+\eta^2=z^{-8/5}$	$(1\mp 5z^{1/5}+\frac{25}{3}z^{2/5})\exp(\pm 5z^{1/5})$
	....	....

	Equation	Values of V_1, V_2
(i)	$(d\eta/dz)+\eta^2=z^{-4}$	$z\exp(\pm 1/z)$
(ii)	$(d\eta/dz)+\eta^2=z^{-8/3}$	$z(1\mp 3z^{-1/3})\exp(\pm 3z^{-1/3})$
(iii)	$(d\eta/dz)+\eta^2=z^{-12/5}$	$z(1\mp 5z^{-1/5}+\frac{25}{3}z^{-2/5})\exp(\pm 5z^{-1/5})$
	....	....

It is to be noticed that the series U_1, U_2 (or V_1, V_2, as the case may be) are supposed to terminate with the term before the first term which has a zero factor in the numerator; see § 4·42 and Glaisher, *Phil. Trans. of the Royal Soc.* CLXXII. (1881), p. 773.

Among the writers who have studied equation (1) are Kummer, *Journal für Math.* XII. (1834), pp. 144—147, Lobatto, *Journal für Math.* XVII. (1837), pp. 363—371, Glaisher (in the memoir to which reference has just been made), and Suchar, *Bull. de la Soc. Math. de France*, XXXII. (1904), pp. 103—116; for other references see § 4·3.

The reader will observe that when $q=0$, the equation (1) is homogeneous and immediately soluble; and that the second order equation solved by James Bernoulli (§ 1·1) is obtainable by taking $q=2$ in (1), and so it is not included among the soluble cases.

4·15. *Schläfli's canonical form of Riccati's equation.*

The form of Riccati's equation which was examined by Schläfli* was

$$\frac{du}{dt} = t^a - t^{-a-1}u^2. \tag{1}$$

This is easily reduced to the form of § 4·13 (2) by taking $-t^{-a}/a$ as a new independent variable.

To solve the equation, Schläfli wrote

$$u = t^{a+1}\frac{d \log y}{dt},$$

and arrived at the equation

$$t\frac{d^2y}{dt^2} + (a+1)\frac{dy}{dt} - y = 0.$$

If†

$$F(a, t) \equiv \sum_{m=0}^{\infty} \frac{t^m}{m!\,\Gamma(a+m+1)},$$

the general solution of the equation in y is

$$y = c_1 F(a, t) + c_2 t^{-a} F(-a, t).$$

The solution of (1) is then

$$u = \frac{c_1 t^{a+1} F(a+1, t) + c_2 F(-a-1, t)}{c_1 F(a, t) + c_2 t^{-a} F(-a, t)}.$$

The connexion between Riccati's equation and Bessel's equation is thus rendered evident; but a somewhat tedious investigation is necessary (§ 4·43) to exhibit the connexion between Cayley's solution and Schläfli's solution.

NOTE. The function $\phi : z$, defined as the series

$$1 + \frac{a}{z} + \frac{1}{2}\cdot\frac{a^2}{z(z+1)} + \frac{1}{2\,.\,3}\cdot\frac{a^3}{z(z+1)(z+2)} + \dots,$$

which is evidently expressible in terms of Schläfli's function, was used by Legendre, *Éléments de Géométrie* (Paris, 1802), note 4, in the course of his proof that π is irrational.

Later the function was studied (with a different notation) by Clifford; see a posthumous fragment in his *Math. Papers* (London, 1882), pp. 346—349.

* *Ann. di Mat.* (2) I. (1868), p. 232. The reader will see that James Bernoulli's solution in series (§ 1·1) is to be associated with Schläfli's solution rather than with Cayley's solution.

† This notation should be compared with the notation of § 4·4.

It is obvious that $$J_\nu(z) = (\tfrac{1}{2}z)^\nu F(\nu, -\tfrac{1}{4}z^2),$$

and it has recently been suggested* that, because the Schläfli-Clifford notation simplifies the analysis in the discussion of certain problems on the stability of vertical wires under gravity, the standard notation for Bessel functions should be abandoned in favour of a notation resembling the notation used by Schläfli-Clifford:—a procedure which seems comparable to a proposal to replace the ordinary tables of trigonometrical functions by tables of the functions

$$\sum_{n=0}^{\infty} \frac{x^n}{(2n)!}, \quad \sum_{n=0}^{\infty} \frac{x^n}{(2n+1)!}.$$

4·16. *Miscellaneous researches on Riccati's equation.*

A solution of Riccati's equation, which involves definite integrals, was given by Murphy, *Trans. Camb. Phil. Soc.* III. (1830), pp. 440—443. The equation which he considered is

$$\frac{du}{dt} + Au^2 = Bt^m,$$

and, if a be written for $1/(m+2)$ and $A^{-1} d(\log y)/dt$ for u, his solution (when $ABa^2 = 1$) is

$$y = \tfrac{1}{2}t \int_{-1}^{1} h^{-1} [\phi(h) \exp(t^{1/a}/h) + \phi(1/h) \exp(ht^{1/a})]\, dh,$$

where $$\phi(h) = e^h h^{-a} \int_0^h e^{-h} h^{a-1}\, dh = \sum_{n=0}^{\infty} \frac{h^n}{a(a+1)(a+2)\ldots(a+n)}.$$

If $1/h$ be written for h in the second part of the integral, then the last expression given for y reduces to $\pi i t$ multiplied by the residue at the origin of $h^{-1}\phi(h)\exp(t^{1/a}/h)$, and the connexion between Murphy's solution and Schläfli's solution (§ 4·15) is evident.

An investigation was published by Challis, *Quarterly Journal*, VII. (1866), pp. 51—53, which shewed how to connect two equations of the type of § 4·13 (2), namely

$$\frac{d\eta}{dz} + \eta^2 = c^2 z^{2q-2},$$

in one of which $1/q$ is an odd positive integer, and in the other it is an odd negative integer. This investigation is to be associated with the discovery of the two types of solution given in § 4·14.

The equation $$\frac{du}{dz} + \frac{au}{z} + bz^n u^2 - cz^m = 0,$$

which is easily transformed into an equation of Riccati's type by taking z^{n-a+1} and $z^a u$ as new variables, was investigated by Rawson, *Messenger*, VII. (1878), pp. 69—72. He transformed it into the equation

$$\frac{dy}{dz} - \frac{a+a}{z} y + bz^{m-a} y^2 - cz^{n+a} = 0,$$

by taking $bu = cz^a/y$; two such equations are called *cognate* Riccati equations. A somewhat similar equation was reduced to Riccati's type by Brassinne, *Journal de Math.* XVI. (1851), pp. 255—256.

The connexions between the various types of equations which different writers have adopted as canonical forms of Riccati's equation have been set out in a paper by Greenhill, *Quarterly Journal*, XVI. (1879), pp. 294—298.

* Greenhill, *Engineering*, CVII. (1919), p. 334; *Phil. Mag.* (6) XXXVIII. (1919), pp. 501—528; see also *Engineering*, CIX. (1920), p. 851.

The reader should also consult a short paper by Siacci, *Napoli Rendiconti*, (3) VII. (1901), pp. 139—143. And a monograph on Riccati's equation, which apparently contains the majority of the results of this chapter, has been produced by Feldblum, *Warschau Univ. Nach.* 1898, nos. 5, 7, and 1899, no. 4.

4·2. *The generalised Riccati equation.*

An obvious generalisation of the equation discussed in § 4·1 is

$$\frac{dy}{dz} = P + Qy + Ry^2, \tag{1}$$

where P, Q, R are any given functions of z. This equation was investigated by Euler*. It is supposed that neither P nor R is identically zero; for, if either P or R is zero, the equation is easily integrable by quadratures.

It was pointed out by Eneström, *Encyclopédie des Sci. Math.* II. 16, § 10, p. 75, that a special equation of this type namely

$$nxx\,dx - nyy\,dx + xx\,dy = xy\,dx$$

was studied by Manfredius, *De constructione aequationum differentialum primi gradus* (Bologna, 1707), p. 167. "Sed tamèn haec eadem aequatio non apparet quomodo construibilis sit, neque enìm videmus quomodò illam integremus, nec quomodo indeterminatas ab invicèm separemus."

The equation (1) is easily reduced to the linear equation of the second order, by taking a new dependent variable u defined by the equation†

$$y = -\frac{1}{R}\frac{d\log u}{dz}. \tag{2}$$

The equation then becomes

$$\frac{d^2u}{dz^2} - \left\{Q + \frac{1}{R}\frac{dR}{dz}\right\}\frac{du}{dz} + PRu = 0. \tag{3}$$

Conversely, if in the general linear equation of the second order,

$$p_0\frac{d^2u}{dz^2} + p_1\frac{du}{dz} + p_2 u = 0, \tag{4}$$

(where p_0, p_1, p_2 are given functions of z), we write

$$u = e^{\int y\,dz}, \tag{5}$$

the equation to determine y is

$$\frac{dy}{dz} = -\frac{p_2}{p_0} - \frac{p_1}{p_0}y - y^2, \tag{6}$$

which is of the same type as (1). The complete equivalence of the generalised Riccati equation with the linear equation of the second order is consequently established.

The equations of this section have been examined by Anisimov, *Warschau Univ. Nach.* 1896, pp. 1—33. [*Jahrbuch über die Fortschritte der Math.* 1896, p. 256.]

* *Nov. Comm. Acad. Petrop.* VIII. (1760—1761) [1763], p. 32; see also a short paper by W. W. Johnson, *Ann. of Math.* III. (1887), pp. 112—115.

† This is the generalisation of James Bernoulli's substitution (§ 1·1). See also Euler, *Inst. Calc. Int.* II. (Petersburg, 1769), §§ 831, 852, pp. 88, 104.

4·21. *Euler's theorems concerning the generalised Riccati equation.*

It has been shewn by Euler* that, if a particular solution of the generalised Riccati equation is known, the general solution can be obtained by two quadratures; if two particular solutions are known the general solution is obtainable by a single quadrature†. And it follows from theorems discovered by Weyr and Picard that, if three particular solutions are known, the general solution can be effected without a quadrature.

To prove the first result, let y_0 be a particular solution of

$$\frac{dy}{dz} = P + Qy + Ry^2,$$

and write $y = y_0 + 1/v$. The equation in v is

$$\frac{dv}{dz} + (Q + 2Ry_0)\, v + R = 0,$$

of which the solution is

$$v \exp\{\textstyle\int (Q + 2Ry_0)\, dz\} + \int R \exp\{\int (Q + 2Ry_0)\, dz\} \,.\, dz = 0,$$

and, since $v = 1/(y - y_0)$, the truth of the first theorem is manifest.

To prove the second, let y_0 and y_1 be two particular solutions, and write

$$w = \frac{y - y_0}{y - y_1}.$$

The result of substituting $(y_1 w - y_0)/(w - 1)$ for y in the equation is

$$\frac{y_0 - y_1}{(w-1)^2}\frac{dw}{dz} + \frac{w}{w-1}\frac{dy_1}{dz} - \frac{1}{w-1}\frac{dy_0}{dz} = P + Q\,\frac{y_1 w - y_0}{w-1} + R\left(\frac{y_1 w - y_0}{w-1}\right)^2,$$

and, when we substitute for (dy_1/dz) and (dy_0/dz) the values $P + Qy_1 + Ry_1^2$ and $P + Qy_0 + Ry_0^2$, the last equation is reduced to

$$\frac{1}{w}\frac{dw}{dz} = Ry_0 - Ry_1,$$

so that
$$w = c \exp\{\textstyle\int (Ry_0 - Ry_1)\, dz\},$$

where c is the constant of integration. Hence, from the equation defining w, we see that y is expressed as a function involving a single quadrature.

To prove the third result, let y_0 and y_1 be the solutions already specified, let y_2 be a third solution, and let c' be the value to be assigned to c to make y reduce to y_2. Then

$$\frac{y - y_0}{y - y_1} = \frac{c}{c'} \cdot \frac{y_2 - y_0}{y_2 - y_1},$$

and this is the integral in a form free from quadratures.

* *Nov. Comm. Acad. Petrop.* VIII. (1760—1761) [1763], p. 32.

† *Ibid.* p. 59, and IX. (1762—1763) [1764], pp. 163—164. See also Minding, *Journal für Math.* XL. (1850), p. 361.

It follows that the general solution is expressible in the form

$$y = \frac{Cf_1(z) + f_2(z)}{Cf_3(z) + f_4(z)}.$$

Hence it is evident that, if y_1, y_2, y_3, y_4 be *any* four solutions, obtained by giving C the values C_1, C_2, C_3, C_4 respectively, *then the cross-ratio*

$$\frac{(y_1 - y_2)(y_3 - y_4)}{(y_1 - y_4)(y_3 - y_2)}$$

is independent of z; for it is equal to

$$\frac{(C_1 - C_2)(C_3 - C_4)}{(C_1 - C_4)(C_3 - C_2)}.$$

In spite of the obvious character of this theorem, it does not seem to have been noticed until some forty years ago*.

Other properties of the generalised Riccati equation may be derived from properties of the corresponding linear equation (§ 4·2). Thus Raffy† has given two methods of reducing the Riccati equation to the canonical form

$$\frac{du}{d\xi} + u^2 = F(\xi);$$

these correspond to the methods of reducing a linear equation to its normal form by changes of the dependent and independent variables respectively.

Various properties of the solution of Riccati's equation in which P, Q, R are *rational* functions have been obtained by C. J. D. Hill, *Journal für Math.* XXV. (1843), pp. 23—37; Autonne, *Comptes Rendus*, XCVI. (1883), pp. 1354—1356; CXXVIII. (1899), pp. 410—412; and Jamet, *Comptes Rendus de l'Assoc. Française (Ajaccio)*, (1901), pp. 207—228; *Ann. de la Fac. des Sci. de Marseille*, XII. (1902), pp. 1—21.

The behaviour of the solution near singularities of P, Q, R has been studied by Falkenhagen, *Nieuw Archief voor Wiskunde*, (2) VI. (1905), pp. 209—248.

The equation of the *second* order whose primitive is of the type

$$y = \frac{c_1\eta_1 + c_2\eta_2 + c_3\eta_3}{c_1\zeta_1 + c_2\zeta_2 + c_3\zeta_3},$$

where c_1, c_2, c_3 are constants of integration (which is an obvious generalisation of the primitive of the Riccati equation), has been studied by Vessiot, *Ann. de la Fac. des Sci. de Toulouse*, IX. (1895), no. 6 and by Wallenburg, *Journal für Math.* CXXI. (1900), pp. 210—217; and *Comptes Rendus*, CXXXVII. (1903), pp. 1033—1035.

* Weyr, *Abh. böhm. Ges. Wiss.* (6) VIII. (1875—1876), *Math. Mem.* I. p. 30; Picard, *Ann. Sci. de l'École norm. sup.* (2) VI. (1877), pp. 342—343. Picard's thesis, in which the result is contained, is devoted to the theory of surfaces and twisted curves—a theory in which Riccati's equation has various applications.

† *Nouv. Ann. de Math.* (4) II. (1902), pp. 529—545.

4·3. *Various transformations of Bessel's equation.*

The equations which we are now about to investigate are derived from Bessel's equation by elementary transformations of the dependent and independent variables.

The first type which we shall consider is*

(1) $$\frac{d^2u}{dz^2} - c^2u = \frac{p(p+1)}{z^2}u,$$

where c is an unrestricted constant. The equation is of frequent occurrence in physical investigations, and, in such problems, p is usually an integer.

The equation has been encountered in the Theory of Conduction of Heat and the Theory of Sound by Poisson, *Journal de l'École Polytechnique*, XII. (cahier 19), (1823), pp. 249—403; Stokes, *Phil. Trans. of the Royal Soc.* 1868, pp. 447—464 [*Phil. Mag.* (4) XXXVI. (1868), pp. 401—421, *Math. and Phys. Papers*, IV. (1904), pp. 299—324]; Rayleigh, *Proc. London Math. Soc.* IV. (1873), pp. 93—103, 253—283 [*Scientific Papers*, I. (1899), pp. 138, 139]. The special equation in which $p=2$ occurs in the Theory of the Figure of the Earth; see Ellis, *Camb. Math. Journal*, II. (1841), pp. 169—177, 193—201.

Since equation (1) may be written in the form

$$z^2\frac{d^2(uz^{-\frac{1}{2}})}{dz^2} + z\frac{d(uz^{-\frac{1}{2}})}{dz} - \{c^2z^2 + (p+\tfrac{1}{2})^2\}\,.\,uz^{-\frac{1}{2}} = 0,$$

its general solution is

(2) $$u = z^{\frac{1}{2}}\mathscr{C}_{p+\frac{1}{2}}(ciz).$$

Consequently the equation is equivalent to Bessel's equation when p is unrestricted, and no advantage is to be gained by studying equations of the form (1) rather than Bessel's equation. But, when p is an integer, the solutions of (1) are expressible "in finite terms†" (cf. § 3·4), and it is then frequently desirable to regard (1) as a canonical form. The relations between various types of solutions of (1) will be examined in detail in §§ 4·41—4·43.

The second type of equation is derived from (1) by a transformation of the dependent variable which makes the indicial equation have a zero root. The roots of the indicial equation of (1) are $p+1$ and $-p$, and so we write $u = vz^{-p}$; we are thus led to the equation

(3) $$\frac{d^2v}{dz^2} - \frac{2p}{z}\frac{dv}{dz} - c^2v = 0,$$

of which the general solution is

(4) $$v = z^{p+\frac{1}{2}}\mathscr{C}_{p+\frac{1}{2}}(ciz).$$

* See Plana, *Mem. della R. Accad. delle Sci. di Torino*, XXVI. (1821), pp. 519—538, and Paoli, *Mem. di Mat. e di Fis. della Soc. Italiana delle Sci.* XX. (1828), pp. 183—188.

† This was known to Plana, who studied equations (1) and (5) in the paper to which reference has just been made.

Equation (3), which has been studied in detail by Bach, *Ann. Sci. de l'École norm. sup.* (2) III. (1874), pp. 47—68, occurs in certain physical investigations; see L. Lorenz, *Ann. der Physik und Chemie*, (2) XX. (1883), pp. 1—21 [*Oeuvres Scientifiques*, I. (1898), pp. 371—396]; and Lamb, *Hydrodynamics* (Cambridge, 1906), §§ 287—291. Solutions of equation (3) in the form of continued fractions (cf. §§ 5·6, 9·65) have been examined by Catalan, *Bulletin de l'Acad. R. de Belgique*, (2) XXXI. (1871), pp. 68—73. See also Le Paige, *ibid.* (2) XLI. (1876), pp. 1011—1016, 935—939.

Next, we derive from (3), by a change of independent variable, an equation in its normal form. We write $z = \zeta^q/q$, where $q = 1/(2p+1)$, the equation then becomes

$$\frac{d^2v}{d\zeta^2} - c^2\zeta^{2q-2}v = 0, \tag{5}$$

and its solution is

$$v = (\zeta^q/q)^{1/(2q)}\mathscr{C}_{1/(2q)}(ci\zeta^q/q). \tag{6}$$

When a constant factor is absorbed into the symbol $\mathscr{C}$, the solution may be taken to be

$$\zeta^{\frac{1}{2}}\mathscr{C}_{1/(2q)}(ci\zeta^q/q).$$

Equation (5), which has already been encountered in § 4·14, has been studied by Plana, *Mem. della R. Accad. delle Sci. di Torino*, XXVI. (1821), pp. 519—538; Cayley, *Phil. Mag.* (4) XXXVI. (1868), pp. 348—351 [*Collected Papers*, VII. (1894), pp. 9—12]; and Lommel, *Studien über die Bessel'schen Functionen* (Leipzig, 1868), pp. 112—118.

The system of equations which has now been constructed has been discussed systematically by Glaisher*, whose important memoir contains an interesting account of the researches of earlier writers.

The equations have been studied from a different aspect by Haentzschel† who regarded them as degenerate forms of Lamé's equations in which both of the invariants g_2 and g_3 are zero.

The following papers by Glaisher should also be consulted: *Phil. Mag.* (4) XLIII. (1872), pp. 433—438; *Messenger*, VIII. (1879), pp. 20—23; *Proc. London Math. Soc.* IX. (1878), pp. 197—202.

It may be noted that the forms of equation (1) used by various writers are as follows:

$$\frac{d^2y}{dx^2} + y = \frac{k(k+1)}{x^2}y, \qquad \text{(Plana)},$$

$$\frac{d^2R}{dr^2} - \rho^2R = \frac{n(n+1)}{r^2}R, \qquad \text{(Poisson)},$$

$$\frac{d^2u}{dx^2} - a^2u = \frac{p(p+1)}{x^2}u. \qquad \text{(Glaisher)}.$$

Equation (5) has been encountered by Greenhill‡ in his researches on the stability of a vertical pole of variable cross-section, under the action of gravity. When the cross-section is constant, the special equation in which $q = \frac{3}{2}$ is obtained, and the solution of it leads to Bessel functions of order $\pm\frac{1}{3}$.

* *Phil. Trans. of the Royal Soc.* CLXXII. (1881), pp. 759—828; see also a paper by Curtis, *Cambridge and Dublin Math. Journal*, IX. (1854), pp. 272—290.

† *Zeitschrift für Math. und Phys.* XXXI. (1886), pp. 25—33.

‡ *Proc. Camb. Phil. Soc.* IV. (1883), pp. 65—73.

4·31. *Lommel's transformations of Bessel's equation.*

Various types of transformations of Bessel's equation were examined by Lommel on two occasions; his earlier researches* were of a somewhat special type, the later† were much more general.

In the earlier investigation, after observing that the general solution of

$$\frac{d^2y}{dz^2} - \frac{2\nu-1}{z}\frac{dy}{dz} + y = 0 \tag{1}$$

is

$$y = z^\nu \mathscr{C}_\nu(z), \tag{2}$$

Lommel proceeded by direct transformations to construct the equation whose general solution is $z^{\beta\nu-\alpha}\mathscr{C}_\nu(\gamma z^\beta)$, where α, β, γ are constants. His result, which it will be sufficient to quote, is that the general solution of

$$z^2\frac{d^2u}{dz^2} + (2\alpha - 2\beta\nu + 1)\, z\frac{du}{dz} + \{\beta^2\gamma^2 z^{2\beta} + \alpha(\alpha - 2\beta\nu)\}\, u = 0 \tag{3}$$

is

$$u = z^{\beta\nu-\alpha}\mathscr{C}_\nu(\gamma z^\beta). \tag{4}$$

When $\beta=0$, the general solution of (3) degenerates into

$$u = z^{-\alpha}(c_1 + c_2 \log z);$$

and when $\gamma=0$, it degenerates into

$$u = z^{-\alpha}(c_1 + c_2 z^{2\beta\nu})$$

unless $\beta\nu$ is zero.

The solution of (3) was given explicitly by Lommel in numerous special cases. It will be sufficient to quote the following for reference:

$$\frac{d^2u}{dz^2} + \frac{1}{z}\frac{du}{dz} + 4\left(z^2 - \frac{\nu^2}{z^2}\right)u = 0; \qquad u = \mathscr{C}_\nu(z^2). \tag{5}$$

$$z\frac{d^2u}{dz^2} + \frac{du}{dz} + \frac{1}{4}\left(1 - \frac{\nu^2}{z}\right)u = 0; \qquad u = \mathscr{C}_\nu(\sqrt{z}). \tag{6}$$

$$z\frac{d^2u}{dz^2} + (1-\nu)\frac{du}{dz} + \frac{1}{4}u = 0; \qquad u = z^{\frac{1}{2}\nu}\mathscr{C}_\nu(\sqrt{z}). \tag{7}$$

$$z\frac{d^2u}{dz^2} + (1-\nu)\frac{du}{dz} - \frac{1}{4}u = 0; \qquad u = z^{\frac{1}{2}\nu}\mathscr{C}_\nu(i\sqrt{z}). \tag{8}$$

$$\frac{d^2u}{dz^2} + \beta^2\gamma^2 z^{2\beta-2}u = 0; \qquad u = z^{\frac{1}{2}}\mathscr{C}_{1/(2\beta)}(\gamma z^\beta). \tag{9}$$

$$z^{\frac{1}{2}}\frac{d^2u}{dz^2} \pm u = 0; \qquad u = z^{\frac{1}{2}}\mathscr{C}_{\frac{2}{3}}(\tfrac{4}{3}z^{\frac{3}{4}}), \quad z^{\frac{1}{2}}\mathscr{C}_{\frac{2}{3}}(\tfrac{4}{3}iz^{\frac{3}{4}}). \tag{10}$$

$$\frac{d^2u}{dz^2} \pm zu = 0; \qquad u = z^{\frac{1}{2}}\mathscr{C}_{\frac{1}{3}}(\tfrac{2}{3}z^{\frac{3}{2}}), \quad z^{\frac{1}{2}}\mathscr{C}_{\frac{1}{3}}(\tfrac{2}{3}iz^{\frac{3}{2}}). \tag{11}$$

An account of Stokes' researches on the solutions of equation (11) will be given in §§ 6·4, 10·2.

* *Studien über die Bessel'schen Functionen* (Leipzig, 1868), pp. 98—120; *Math. Ann.* III. (1871), pp. 475—487.

† *Math. Ann.* XIV. (1879), pp. 510—536.

Lommel's later researches appeared at about the same time as a memoir by Pearson*, and several results are common to the two papers. Lommel's procedure was to simplify the equation†

$$\frac{d^2\{y/\chi(z)\}}{d\{\psi(z)\}^2}-\frac{2\nu-1}{\psi(z)}\frac{d\{y/\chi(z)\}}{d\psi(z)}+\frac{y}{\chi(z)}=0,$$

of which the solution is (§ 4·3)

(12) $$y=\chi(z)\{\psi(z)\}^{\nu}\mathscr{C}_\nu\{\psi(z)\}.$$

On reduction the equation becomes

(13) $$\frac{d^2y}{dz^2}-\left\{\frac{\psi''(z)}{\psi'(z)}+(2\nu-1)\frac{\psi'(z)}{\psi(z)}+2\frac{\chi'(z)}{\chi(z)}\right\}\frac{dy}{dz}$$
$$+\left[\left\{\frac{\psi''(z)}{\psi'(z)}+(2\nu-1)\frac{\psi'(z)}{\psi(z)}+2\frac{\chi'(z)}{\chi(z)}\right\}\frac{\chi'(z)}{\chi(z)}-\frac{\chi''(z)}{\chi(z)}+\{\psi'(z)\}^2\right]y=0.$$

Now define the function $\phi(z)$ by the equation

$$\frac{\phi'(z)}{\phi(z)}=\frac{\psi''(z)}{\psi'(z)}+(2\nu-1)\frac{\psi'(z)}{\psi(z)}+2\frac{\chi'(z)}{\chi(z)}.$$

It will be adequate to take

(14) $$\phi(z)=\psi'(z)\{\chi(z)\}^2\{\psi(z)\}^{2\nu-1}.$$

If we eliminate $\chi(z)$, it is apparent that the general solution of

(15) $$\frac{d^2y}{dz^2}-\frac{\phi'(z)}{\phi(z)}\frac{dy}{dz}+\left[\frac{3}{4}\left\{\frac{\phi'(z)}{\phi(z)}\right\}^2-\frac{1}{2}\frac{\phi''(z)}{\phi(z)}\right.$$
$$\left.-\frac{3}{4}\left\{\frac{\psi''(z)}{\psi'(z)}\right\}^2+\frac{1}{2}\frac{\psi'''(z)}{\psi'(z)}+\{\psi^2(z)-\nu^2+\tfrac{1}{4}\}\left\{\frac{\psi'(z)}{\psi(z)}\right\}^2\right]y=0$$

is

(16) $$y=\sqrt{\left\{\frac{\phi(z)\,\psi(z)}{\psi'(z)}\right\}}\,.\,\mathscr{C}_\nu\{\psi(z)\}.$$

As a special case, if we take $\phi(z)\equiv 1$, it is seen that the general solution of

(17) $$\frac{d^2y}{dz^2}+\left[\frac{1}{2}\frac{\psi'''(z)}{\psi'(z)}-\frac{3}{4}\left\{\frac{\psi''(z)}{\psi'(z)}\right\}^2+\{\psi^2(z)-\nu^2+\tfrac{1}{4}\}\left\{\frac{\psi'(z)}{\psi(z)}\right\}^2\right]y=0$$

is

(18) $$y=\sqrt{\{\psi(z)/\psi'(z)\}}\,.\,\mathscr{C}_\nu\{\psi(z)\}.$$

Next, returning to (13), we take $\chi(z)\equiv\{\psi(z)\}^{\mu-\nu}$, and we find that the general solution of

(19) $$\frac{d^2y}{dz^2}-\left\{\frac{\psi''(z)}{\psi'(z)}+(2\mu-1)\frac{\psi'(z)}{\psi(z)}\right\}\frac{dy}{dz}+\{\mu^2-\nu^2+\psi^2(z)\}\left\{\frac{\psi'(z)}{\psi(z)}\right\}^2y=0$$

is

(20) $$y=\{\psi(z)\}^{\mu}\mathscr{C}_\nu\{\psi(z)\}.$$

* *Messenger*, IX. (1880), pp. 127—131. † The functions $\chi(z)$ and $\psi(z)$ are arbitrary.

The following are special cases of (17):

(21) $$\frac{d^2y}{dz^2}+(e^{2z}-\nu^2)\,y=0\,;\quad y=\mathscr{C}_\nu(e^z),$$

(22) $$\frac{d^2y}{dz^2}+\frac{e^{2/z}-\nu^2}{z^4}\,y=0\,;\quad y=z\mathscr{C}_\nu(e^{1/z}).$$

The independent researches of Pearson proceeded on very similar lines except that he started from Bessel's equation instead of from the modified form of it. The reader will find many special cases of equation (17) worked out in his paper.

A partial differential equation closely connected with (7) and (8), namely

$$z\,\frac{\partial^2 u}{\partial z^2}+(1+\nu)\frac{\partial u}{\partial z}-\mu\,\frac{\partial u}{\partial t}=0,$$

has been investigated by Kepinski, *Math. Ann.* LXI. (1906), pp. 397—405, and Myller-Lebedeff, *Math. Ann.* LXVI. (1909), pp. 325—330. The reader may verify that Kepinski's formula

$$u=\frac{z^{-\frac{1}{2}\nu}}{t}\int_0^\infty \exp\left\{\frac{-\mu z-\mu w}{t}\right\}\mathscr{C}_\nu\left(\frac{2i\mu\sqrt{(zw)}}{t}\right)f(w)\,dw$$

is a solution, when $f(w)$ denotes an arbitrary function of w.

The special case of the equation when $\nu=-1$ was also investigated by Kepinski, *Bull. int. de l'Acad. des Sci. de Cracovie*, 1905, pp. 198—205.

4·32. *Malmstén's differential equation.*

Twenty years before Lommel published his researches on transformations of Bessel's equation, Malmstén* investigated conditions for the integrability in finite terms of the equation

(1) $$\frac{d^2y}{dz^2}+\frac{r}{z}\frac{dy}{dz}=\left(Az^m+\frac{s}{z^2}\right)y$$

which is obviously a generalisation of Bessel's equation; and it is a special case of § 4·31 (15).

To reduce the equation, Malmstén chose new variables defined by the formulae

$$z=\zeta^q,\quad y=u\zeta^{pq},$$

where p and q are constants to be suitably chosen.

The transformed equation is

$$\frac{d^2u}{d\zeta^2}+(2pq-q+1+qr)\frac{1}{\zeta}\frac{du}{d\zeta}=\left[Aq^2\zeta^{(m+2)q-2}+\frac{sq^2-pq\,(pq+rq-q)}{\zeta^2}\right]u.$$

We choose p and q so that this may reduce to the equation of § 4·3 (1) considered by Plana, and therefore we take

$$2pq-q+1+qr=0,\quad (m+2)\,q=2,$$

so that $p=-\frac{1}{2}r-\frac{1}{4}m$.

The equation then reduces to

$$\frac{d^2u}{d\zeta^2}=\left[\frac{4A}{(m+2)^2}+\frac{q^2\{4s+(1-r)^2\}-1}{4\zeta^2}\right]u.$$

* *Camb. and Dublin Math. Journal*, v. (1850), pp. 180—182. The case in which $s=0$ had been previously considered by Malmstén, *Journal für Math.* XXXIX. (1850), pp. 108—115.

By § 4·3 this is integrable in finite terms if

$$\tfrac{1}{4}q^2\{4s+(1-r)^2\}-\tfrac{1}{4}=n(n+1),$$

where n is an integer; so that

(2) $$m+2=\pm\frac{\sqrt{\{4s+(1-r)^2\}}}{n+\frac{1}{2}}.$$

The equation is also obviously integrable in the trivial cases $A=0$ and $m=-2$.

4·4. *The notation of Pochhammer for series of hypergeometric type.*

A compact notation, invented by Pochhammer* and modified by Barnes†, is convenient for expressing the series which are to be investigated. We shall write now and subsequently

$$(\alpha)_n=\alpha(\alpha+1)(\alpha+2)\dots(\alpha+n-1),\qquad (\alpha)_0=1.$$

The notation which will be used is, in general,

$${}_pF_q(\alpha_1,\alpha_2,\dots,\alpha_p;\ \rho_1,\rho_2,\dots,\rho_q;\ z)=\sum_{n=0}^{\infty}\frac{(\alpha_1)_n(\alpha_2)_n\dots(\alpha_p)_n}{n!\,(\rho_1)_n(\rho_2)_n\dots(\rho_q)_n}z^n.$$

In particular,

$${}_1F_1(\alpha;\ \rho;\ z)=1+\frac{(\alpha)_1}{1!\,(\rho)_1}z+\frac{(\alpha)_2}{2!\,(\rho)_2}z^2+\frac{(\alpha)_3}{3!\,(\rho)_3}z^3+\dots$$

$$=\sum_{n=0}^{\infty}\frac{(\alpha)_n}{n!\,(\rho)_n}z^n,$$

$${}_0F_1(\rho;\ z)=\sum_{n=0}^{\infty}\frac{z^n}{n!\,(\rho)_n},$$

$${}_2F_1(\alpha,\beta;\ \rho;\ z)=\sum_{n=0}^{\infty}\frac{(\alpha)_n(\beta)_n}{n!\,(\rho)_n}z^n.$$

The functions defined by the first three series are called generalised hypergeometric functions.

It may be noted here that the function ${}_1F_1(\alpha;\ \rho;\ z)$ is a solution of the differential equation

$$z\frac{d^2y}{dz^2}+(\rho-z)\frac{dy}{dz}-\alpha y=0,$$

and, when ρ is not an integer, an independent solution of this equation is

$$z^{1-\rho}\,.\,{}_1F_1(\alpha-\rho+1;\ 2-\rho;\ z).$$

It is evident that

$$J_\nu(z)=\frac{(\frac{1}{2}z)^\nu}{\Gamma(\nu+1)}\cdot{}_0F_1(\nu+1;\ -\tfrac{1}{4}z^2).$$

Various integral representations of functions of the types ${}_1F_1$, ${}_0F_2$, ${}_0F_3$ have been studied by Pochhammer, *Math. Ann.* XLI. (1893), pp. 174—178, 197—218.

* *Math. Ann.* XXXVI. (1890), p. 84; XXXVIII. (1891), pp. 227, 586, 587. Cf. § 4·15.

† *Proc. London Math. Soc.* (2) V. (1907), p. 60. The modification due to Barnes is the insertion of the suffixes p and q before and after the F to render evident the number of sets of factors.

4·41. *Various solutions in series.*

We shall now examine various solutions of the equation

$$\frac{d^2u}{dz^2} - c^2u = \frac{p(p+1)}{z^2}u,$$

and obtain relations between them, which will for the most part be expressed in Pochhammer's notation.

It is supposed for the present that p is not a positive integer or zero, and, equally, since the equation is unaltered by replacing p by $-p-1$, it is supposed that p is not a negative integer.

It is already known (§ 4·3) that the general solution* is $z^{\frac{1}{2}}\mathscr{C}_{p+\frac{1}{2}}(ciz)$, and this gives rise to the special solutions

$$z^{p+1}\,.\,{}_0F_1(p+\tfrac{3}{2};\ \tfrac{1}{4}c^2z^2);\quad z^{-p}\,.\,{}_0F_1(\tfrac{1}{2}-p;\ \tfrac{1}{4}c^2z^2).$$

The equation may be written in the forms

$$\frac{d^2(ue^{\mp cz})}{dz^2} \pm 2c\frac{d(ue^{\mp cz})}{dz} = \frac{p(p+1)}{z^2}(ue^{\mp cz}),$$

which are suggested by the fact that the functions $e^{\pm cz}$ are solutions of the original equation with the right-hand side suppressed.

When ϑ is written for $z(d/dz)$, the last pair of equations become

$$(\vartheta - p - 1)(\vartheta + p)\,.\,(ue^{\mp cz}) \pm 2cz\vartheta(ue^{\mp cz}) = 0.$$

When we solve these in series we are led to the following four expressions for u:

$$z^{p+1}e^{cz}\,.\,{}_1F_1(p+1;\ 2p+2;\ -2cz);\qquad z^{-p}e^{cz}\,.\,{}_1F_1(-p;\ -2p;\ -2cz);$$

$$z^{p+1}e^{-cz}\,.\,{}_1F_1(p+1;\ 2p+2;\ 2cz);\qquad z^{-p}e^{-cz}\,.\,{}_1F_1(-p;\ -2p;\ 2cz).$$

Now, by direct multiplication of series, the two expressions on the left are expansible in ascending series involving $z^{p+1}, z^{p+2}, z^{p+3}, \ldots$. And the expressions on the right similarly involve $z^{-p}, z^{1-p}, z^{2-p}, \ldots$. Since none of the two sets of powers are the same when $2p$ is not an integer, we must have

$$(1)\qquad e^{cz}\,.\,{}_1F_1(p+1;\ 2p+2;\ -2cz) = e^{-cz}\,.\,{}_1F_1(p+1;\ 2p+2;\ 2cz) = {}_0F_1(p+\tfrac{3}{2};\ \tfrac{1}{4}c^2z^2),$$

$$(2)\qquad e^{cz}\,.\,{}_1F_1(-p;\ -2p;\ -2cz) = e^{-cz}\,.\,{}_1F_1(-p;\ -2p;\ 2cz) = {}_0F_1(\tfrac{1}{2}-p;\ \tfrac{1}{4}c^2z^2).$$

These formulae are due to Kummer†. When (1) has been proved for general values of p, the truth of (2) is obvious on replacing p by $-p-1$ in (1).

We now have to consider the cases when $2p$ is an integer.

* It follows from § 3·1 that a special investigation is also necessary when p is half of an odd integer.

† *Journal für Math.* xv. (1836), pp. 138—141.

When p has any of the values $\frac{1}{2}, \frac{3}{2}, \frac{5}{2}, \ldots$, the solutions which contain z^{-p} as a factor have to be replaced by series involving logarithms (§§ 3·51, 3·52), and there is only one solution which involves only powers of z. By the previous reasoning, equation (1) still holds.

When p has any of the values 0, 1, 2, ... a comparison of the lowest powers of z involved in the solutions shews that (1) still holds; but it is not obvious that there are no relations of the form

$$\begin{aligned} z^{-p}{}_0F_1(\tfrac{1}{2}-p;\ \tfrac{1}{4}c^2z^2) &= z^{-p}e^{cz}{}_1F_1(-p;\ -2p;\ -2cz) + k_1 z^{p+1}{}_0F_1(p+\tfrac{3}{2};\ \tfrac{1}{4}c^2z^2) \\ &= z^{-p}e^{-cz}{}_1F_1(-p;\ -2p;\ 2cz) + k_2 z^{p+1}{}_0F_1(p+\tfrac{3}{2};\ \tfrac{1}{4}c^2z^2), \end{aligned}$$

where k_1, k_2 are constants which are not zero.

We shall consequently have to give an independent investigation of (1) and (2) which depends on direct multiplication of series.

NOTE. In addition to Kummer's researches, the reader should consult the investigations of the series by Cayley, *Phil. Mag.* (4) XXXVI. (1868), pp. 348—351 [*Collected Papers*, VII. (1894), pp. 9—12] and Glaisher, *Phil. Mag.* (4) XLIII. (1872), pp. 433—438; *Phil. Trans. of the Royal Soc.* CLXXII. (1881), pp. 759—828.

4·42. *Relations between the solutions in series.*

The equation

$$e^{cz}{}_1F_1(p+1;\ 2p+2;\ -2cz) = e^{-cz}{}_1F_1(p+1;\ 2p+2;\ 2cz),$$

which forms part of equation (1) of § 4·41, is a particular case of the more general formula due to Kummer*

(1) $$ {}_1F_1(\alpha;\ \rho;\ \zeta) = e^{\zeta}{}_1F_1(\rho-\alpha;\ \rho;\ -\zeta),$$

which holds for all values of α and ρ subject to certain conventions (which will be stated presently) which have to be made when α and ρ are negative integers.

We first suppose that ρ is not a negative integer and then the coefficient of ζ^n in the expansion of the product of the series for e^{ζ} and ${}_1F_1(\rho-\alpha;\ \rho;\ -\zeta)$ is

$$\begin{aligned} \sum_{m=0}^{n} \frac{(-)^m}{(n-m)!}\frac{(\rho-\alpha)_m}{m!\,(\rho)_m} &= \frac{(-)^n}{n!\,(\rho)_n}\sum_{m=0}^{n} {}_nC_m\,.\,(\rho-\alpha)_m(1-\rho-n)_{n-m} \\ &= \frac{(-)^n}{n!\,(\rho)_n}\,.\,(1-\alpha-n)_n \\ &= \frac{(\alpha)_n}{n!\,(\rho)_n}, \end{aligned}$$

if we first use Vandermonde's theorem† and then reverse the order of the factors in the numerator; and the last expression is the coefficient of ζ^n in ${}_1F_1(\alpha;\ \rho;\ \zeta)$. The result required is therefore established when α and ρ have general complex values‡.

* *Journal für Math.* XV. (1836), pp. 138—141; see also Bach, *Ann. Sci. de l'École norm. sup.* (2) III. (1874), p. 55.

† See, e.g. Chrystal, *Algebra*, II. (1900), p. 9.

‡ Another proof depending on the theory of contour integration has been given by Barnes, *Trans. Camb. Phil. Soc.* XX. (1908), pp. 254—257.

When ρ is a negative integer, equation (1) is obviously meaningless unless also α is a negative integer and $|\alpha|<|\rho|$. The interpretation of (1) in these circumstances will be derived by an appropriate limiting process.

First let α be a negative integer $(=-N)$ and let ρ not be an integer, so that the preceding analysis is valid. The series ${}_1F_1(-N;\ \rho;\ \zeta)$ is now a terminating series, while ${}_1F_1(\rho+N;\ \rho;\ -\zeta)$ is an infinite series which consists of $N+1$ terms followed by terms in which the earlier factors $\rho+N$, $\rho+N+1, \rho+N+2, \ldots$ in the sequences in the numerators can be cancelled with the later factors of the sequences $\rho, \rho+1, \rho+2, \ldots$ in the denominators.

When these factors have been cancelled, the series for ${}_1F_1(-N;\ \rho;\ \zeta)$ and ${}_1F_1(\rho+N;\ \rho;\ -\zeta)$ are both continuous functions of ρ near $\rho=-M$, where M is any of the integers $N, N+1, N+2, \ldots$.

Hence we may proceed to the limit when $\rho\to-M$, and the limiting form of (1) may then be written*

$$(2)\qquad {}_1F_1(-N;\ -M;\ \zeta)\urcorner = e^{\zeta}\,{}_1F_1(N-M;\ -M;\ -\zeta)\rceil,$$

in which the symbol $\urcorner$ means that the series is to stop at the term in ζ^N, i.e. the last term in which the numerator does not contain a zero factor, while the symbol $\rceil$ means that the series is to proceed normally as far as the term in ζ^{M-N}, and then it is to continue with terms in $\zeta^{M+1}, \zeta^{M+2}, \ldots$, the vanishing factors in numerator and denominator being cancelled as though their ratio were one of equality.

With this convention, it is easy to see that

$$(3)\qquad {}_1F_1(-N;\ -M;\ \zeta)\rceil = {}_1F_1(-N;\ -M;\ \zeta)\urcorner + (-)^{M-N}\frac{N!\,(M-N)!}{M!\,(M+1)!}\zeta^{M+1}\,{}_1F_1(M-N+1;\ M+2;\ \zeta).$$

When we replace N by $M-N$ and ζ by $-\zeta$, we have

$$(4)\qquad {}_1F_1(N-M;\ -M;\ -\zeta)\rceil = {}_1F_1(N-M;\ -M;\ -\zeta)\urcorner + (-)^{N}\frac{N!\,(M-N)!}{M!\,(M+1)!}(-\zeta)^{M+1}\,{}_1F_1(N+1;\ M+2;\ -\zeta).$$

As an ordinary case of (1) we have

$${}_1F_1(M-N+1;\ M+2;\ \zeta) = e^{\zeta}\,{}_1F_1(N+1;\ M+2;\ -\zeta),$$

and from this result combined with (2), (3) and (4) we deduce that

$$(5)\qquad {}_1F_1(-N;\ -M;\ \zeta)\rceil = e^{\zeta}\,{}_1F_1(N-M;\ -M;\ -\zeta)\urcorner.$$

This could have been derived directly from (1) by giving $\rho-\alpha$ (instead of α) an integral value, and then making ρ tend to its limit.

* Cf. Cayley, *Messenger* (old series), v. (1871), pp. 77—82 [*Collected Papers*, VIII. (1895), pp. 458—462], and Glaisher, *Messenger*, VIII. (1879), pp. 20—23.

We next examine the equation

$$(6) \qquad e^{cz}\,{}_1F_1(p+1;\ 2p+2;\ -2cz) = {}_0F_1(p+\tfrac{3}{2};\ \tfrac{1}{4}c^2z^2),$$

which forms the remainder of equation (1) in § 4·41, and which is also due to Kummer*.

If we suppose that $2p$ is not a negative integer, the coefficient of $(cz)^n$ in the product of the series on the left in (6) is

$$\sum_{m=0}^{n} \frac{(-2)^m}{(n-m)!}\,\frac{(p+1)_m}{m!\,(2p+2)_m} = \frac{(-)^n}{(2p+2)_n}\sum_{m=0}^{n} 2^m\,\frac{(p+1)_m\,(-n-2p-1)_{n-m}}{m!\,(n-m)!}.$$

Now $\dfrac{1}{n!}\sum\limits_{m=0}^{n} 2^m\,.\,{}_nC_m\,(p+1)_m\,(-n-2p-1)_{n-m}$ is the coefficient of t^n in the expansion of $(1-2t)^{-p-1}(1-t)^{n+2p+1}$, and so it is equal to

$$\frac{1}{2\pi i}\int^{(0+)} (1-2t)^{-p-1}(1-t)^{n+2p+1}\,t^{-n-1}\,dt = \frac{1}{2\pi i}\int^{(0+)} (1-u^2)^{-p-1}u^{-n-1}\,du,$$

where $u = t/(1-t)$ and the contours enclose the origin but no other singularities of the integrands. By expanding the integrand in ascending powers of u, we see that the integral is zero if n is odd, but it is equal to $\dfrac{(p+1)_{\frac{1}{2}n}}{(\frac{1}{2}n)!}$ when n is even.

Hence it follows that

$$\begin{aligned} e^{cz}\,{}_1F_1(p+1;\ 2p+2;\ -2cz) &= \sum_{n=0}^{\infty} \frac{(cz)^{2n}\,.\,(p+1)_n}{(2p+2)_{2n}\,.\,n!} \\ &= \sum_{n=0}^{\infty} \frac{(cz)^{2n}}{2^{2n}\,.\,n!\,(p+\frac{3}{2})_n}, \end{aligned}$$

and this is the result to be proved.

When we make p tend to the value of a negative integer, $-N$, we find by the same limiting process as before that

$$\begin{aligned} \lim_{p\to -N} {}_1F_1(p+1;\ 2p+2;\ -2cz) &= {}_1F_1(1-N;\ 2-2N;\ -2cz)\urcorner \\ &+ \frac{(-)^{N-1}(N-1)!\,N!}{(2N-2)!\,(2N)!}(-2cz)^{2N-1}\,.\,{}_1F_1(N;\ 2N;\ -2cz). \end{aligned}$$

It follows that

$$\begin{aligned} {}_0F_1(\tfrac{3}{2}-N;\ \tfrac{1}{4}c^2z^2) &= e^{cz}\,.\,{}_1F_1(1-N;\ 2-2N;\ -2cz)\urcorner \\ &+ \frac{(-)^N(N-1)!\,N!}{(2N-2)!\,(2N)!}(2cz)^{2N-1}\,e^{cz}\,.\,{}_1F_1(N;\ 2N;\ -2cz). \end{aligned}$$

If we change the signs of c and z throughout and add the results so obtained, we find that

$$\begin{aligned} (7) \qquad 2\,.\,{}_0F_1(\tfrac{3}{2}-N;\ \tfrac{1}{4}c^2z^2) &= e^{cz}\,.\,{}_1F_1(1-N;\ 2-2N;\ -2cz)\urcorner \\ &+ e^{-cz}\,.\,{}_1F_1(1-N;\ 2-2N;\ 2cz)\urcorner, \end{aligned}$$

* *Journal für Math.* xv. (1836), pp. 138—141. In connexion with the proof given here, see Barnes, *Trans. Camb. Phil. Soc.* xx. (1908), p. 272.

the other terms on the right cancelling by a use of equation (1) This is, of course, the expression for $J_{-N+\frac{1}{2}}(icz)$ in finite terms with a different notation.

For Barnes' proof of Kummer's formulae, by the methods of contour integration, see § 6·5.

4·43. *Sharpe's differential equation.*

The equation

$$z\frac{d^2y}{dz^2}+\frac{dy}{dz}+(z+A)\,y=0, \tag{1}$$

which is a generalisation of Bessel's equation for functions of order zero, occurs in the theory of the reflexion of sound by a paraboloid. It has been investigated by Sharpe*, who has shewn that the integral which reduces to unity at the origin is

$$y=C\int_0^{\frac{1}{2}\pi}\cos\left(z\cos\theta+A\log\cot\tfrac{1}{2}\theta\right)d\theta, \tag{2}$$

where

$$1=C\int_0^{\frac{1}{2}\pi}\cos\left(A\log\cot\tfrac{1}{2}\theta\right)d\theta. \tag{3}$$

This is the appropriate modification of Parseval's integral (§ 2·3). To investigate its convergence write $\cos\theta=\tanh\phi$, and it becomes

$$y=C\int_0^{\infty}\frac{\cos\left(A\phi+z\tanh\phi\right)}{\cosh\phi}\,d\phi. \tag{4}$$

It is easy to see from this form of the integral that it converges for (complex) values of A for which $|I(A)|<1$, and†

$$C=\frac{2}{\pi}\cosh\tfrac{1}{2}\pi A.$$

The integral has been investigated in great detail by Sharpe and he has given elaborate rules for calculating successive coefficients in the expansion of y in powers of z.

A simple form of the solution (which was not given by Sharpe) is

$$y=e^{\pm iz}\,{}_1F_1\left(\tfrac{1}{2}\mp\tfrac{1}{2}iA\,;\,1\,;\,\mp 2iz\right).$$

The reader should have no difficulty in verifying this result.

* *Messenger*, x. (1881), pp. 174—185; xii. (1884), pp. 66—79; *Proc. Camb. Phil. Soc.* x. (1900), pp. 101—136.

† See, e.g. Watson, *Complex Integration and Cauchy's Theorem* (1914), pp. 64—65.

4·5. *Equations of order higher than the second.*

The construction of a differential equation of any order, which is soluble by means of Bessel functions, has been effected by Lommel*; its possibility depends on the fact that cylinder functions exist for which the quotient $\mathscr{C}_\nu(z)/\mathscr{C}_{-\nu}(z)$ is independent of z.

Each of the functions $J_n(z)$ and $Y_n(z)$, of integral order, possesses this property [§§ 2·31, 3·5]; and the functions of the third kind $H_\nu^{(1)}(z)$, $H_\nu^{(2)}(z)$ possess it (§ 3·61), whether ν is an integer or not.

Now when § 3·9 (5) is written in the form

$$\frac{d^m}{dz^m} z^{\frac{1}{2}\nu} \mathscr{C}_\nu(\gamma\sqrt{z}) = (\tfrac{1}{2}\gamma)^m z^{\frac{1}{2}(\nu-m)} \mathscr{C}_{\nu-m}(\gamma\sqrt{z}), \tag{1}$$

the cylinder function on the right is of order $-\nu$ if $m = 2\nu$.

This is the case either (i) if ν is an integer, n, and $m = 2n$, or (ii) if $\nu = n + \frac{1}{2}$ and $m = 2n + 1$.

Hence if $\mathscr{C}_n$ denotes either J_n or Y_n, we have

$$\frac{d^{2n}\{z^{\frac{1}{2}n}\mathscr{C}_n(\gamma\sqrt{z})\}}{dz^{2n}} = (\tfrac{1}{2}\gamma)^{2n} z^{-\frac{1}{2}n} \mathscr{C}_{-n}(\gamma\sqrt{z}).$$

From this equation we obtain Lommel's result that the functions $z^{\frac{1}{2}n}J_n(\gamma\sqrt{z})$, $z^{\frac{1}{2}n}Y_n(\gamma\sqrt{z})$ are solutions of†

$$\frac{d^{2n}y}{dz^{2n}} = \frac{(\frac{1}{2}c)^{2n}y}{z^n}, \tag{2}$$

where γ has any value such that $\gamma^{2n} = (-)^n c^{2n}$, so that

$$\gamma = ic\exp(r\pi i/n). \qquad (r = 0, 1, 2, \ldots, n-1)$$

By giving γ all possible values we obtain $2n$ solutions of (2), and these form a fundamental system.

Next, if $\mathscr{C}_{n+\frac{1}{2}}$ denotes $H^{(1)}{}_{n+\frac{1}{2}}$, we have $\mathscr{C}_{-(n+\frac{1}{2})} = e^{(n+\frac{1}{2})\pi i}\mathscr{C}_{n+\frac{1}{2}}$, so that

$$\frac{d^{2n+1}\{z^{\frac{1}{2}n+\frac{1}{4}}\mathscr{C}_{n+\frac{1}{2}}(\gamma\sqrt{z})\}}{dz^{2n+1}} = e^{(n+\frac{1}{2})\pi i}(\tfrac{1}{2}\gamma)^{2n+1} z^{-\frac{1}{2}n-\frac{1}{4}}\mathscr{C}_{n+\frac{1}{2}}(\gamma\sqrt{z}),$$

and hence $z^{\frac{1}{2}n+\frac{1}{4}}H^{(1)}{}_{n+\frac{1}{2}}(\gamma\sqrt{z})$ is a solution of

$$\frac{d^{2n+1}y}{dz^{2n+1}} = \frac{(\frac{1}{2}c)^{2n+1}y}{z^{n+\frac{1}{2}}}, \tag{3}$$

where γ has any value such that $\gamma^{2n+1} = c^{2n+1}e^{-(n+\frac{1}{2})\pi i}$, so that

$$\gamma = -ic\exp\{r\pi i/(n+\tfrac{1}{2})\}, \qquad (r = 0, 1, 2, \ldots, 2n)$$

and the solutions so obtained form a fundamental system.

* *Studien über die Bessel'schen Functionen* (Leipzig, 1868), p. 120; *Math. Ann.* II. (1870), pp. 624—635.

† The more general equation

$$\frac{d^k y}{dz^k} = az^m y$$

has been discussed by Molins, *Mém. de l'Acad. des Sci. de Toulouse*, (7) VIII. (1876), pp. 167—189.

For some applications of these results, see Forsyth, *Quarterly Journal*, XIX. (1883), pp. 317—320.

In view of (1), which holds when m is an integer, Lommel, *Math. Ann.* II. (1870), p. 635, has suggested an interpretation of a "fractional differential coefficient." Thus he would interpret $\left(\frac{d}{dz}\right)^{\frac{1}{2}} \exp(\pm\gamma\sqrt{z})$ to mean $\mathscr{C}_0(\gamma i\sqrt{z})$. The idea has been developed at some length by Heaviside in various papers.

Lommel's formulae may be generalised by considering equation (3) of § 4·31, after writing it in the form

$$(\vartheta+\alpha)(\vartheta+\alpha-2\beta\nu)\,u=-\beta^2\gamma^2z^{2\beta}u,$$

the solution of the equation being $u=z^{\beta\nu-\alpha}\mathscr{C}_\nu(\gamma z^\beta)$. For it is easy to verify by induction that, with this value of u,

$$\prod_{r=0}^{n-1}(\vartheta+\alpha-2r\beta)(\vartheta+\alpha-2\beta\nu-2r\beta)\,u=(-)^n\beta^{2n}c^{2n}z^{2n\beta}u,$$

and so solutions of

(4) $$\prod_{r=0}^{n-1}(\vartheta+\alpha-2r\beta)(\vartheta+\alpha-2\beta\nu-2r\beta)\,u=(-)^n\beta^{2n}c^{2n}z^{2n\beta}u$$

are of the form $$u=z^{\beta\nu-\alpha}\mathscr{C}_\nu(\gamma z^\beta),$$

where $$\gamma=c\exp(r\pi i/n).\qquad(r=0,1,\ldots,n-1)$$

By giving γ these values, we obtain $2n$ solutions which form a fundamental system.

In the special case in which $n=2$, equation (4) reduces to

$$(\vartheta+\alpha)(\vartheta+\alpha-2\beta)(\vartheta+\alpha-2\beta\nu)(\vartheta+\alpha-2\beta\nu-2\beta)\,u=\beta^4c^4z^{4\beta}u.$$

This equation resembles an equation which has been encountered by Nicholson* in the investigation of the shapes of Sponge Spicules, namely

(5) $$\frac{d^2}{dz^2}\left\{z^{4\mu}\frac{d^2u}{dz^2}\right\}=z^{2\mu}u,$$

that is to say $$\vartheta(\vartheta-1)(\vartheta+4\mu-2)(\vartheta+4\mu-3)\,u=z^{4-2\mu}u.$$

If we identify this with the special form of (4) we obtain the following four distinct sets of values for α, β, μ, ν:

α	β	μ	ν
0	1	0	$\frac{1}{2}$
2	$\frac{1}{2}$	1	2
$\frac{6}{5}$	$\frac{3}{5}$	$\frac{4}{5}$	$\frac{5}{6}$
-1	$-\frac{1}{2}$	3	10

* *Proc. Royal Soc.* XCIII. A (1917), pp. 506—519. See also Dendy and Nicholson, *Proc. Royal Soc.* LXXXIX. B (1917), pp. 573—587; the special cases of (5) in which $\mu=0$ or 1 had been solved previously by Kirchhoff, *Berliner Monatsberichte*, 1879, pp. 815—828. [*Ann. der Physik und Chemie*, (3) X. (1880), pp. 501—512.]

These four cases give the following equations and their solutions:

$$(6)\qquad \frac{d^4 u}{dz^4}=u; \qquad u=z^{\frac{1}{2}}\{\mathscr{C}_{\frac{1}{2}}(z)+\overline{\mathscr{C}}_{\frac{1}{2}}(iz)\},$$

$$(7)\qquad \frac{d^2}{dz^2}\left\{z^4\frac{d^2u}{dz^2}\right\}=z^2u; \qquad u=z^{-1}\{\mathscr{C}_2(2\sqrt{z})+\overline{\mathscr{C}}_2(2i\sqrt{z})\},$$

$$(8)\qquad \frac{d^2}{dz^2}\left\{z^{\frac{16}{5}}\frac{d^2u}{dz^2}\right\}=z^{\frac{8}{5}}u; \qquad u=z^{-\frac{7}{10}}\{\mathscr{C}_{\frac{5}{8}}(\tfrac{5}{3}z^{\frac{3}{5}})+\overline{\mathscr{C}}_{\frac{5}{8}}(\tfrac{5}{3}iz^{\frac{3}{5}})\},$$

$$(9)\qquad \frac{d^2}{dz^2}\left\{z^{12}\frac{d^2u}{dz^2}\right\}=z^6u; \qquad u=z^{-4}\{\mathscr{C}_{10}(2z^{-\frac{1}{2}})+\overline{\mathscr{C}}_{10}(2iz^{-\frac{1}{2}})\}.$$

These seem to be the only equations of Nicholson's type which are soluble with the aid of Bessel functions; in the case $\mu=2$, the equation (5) is homogeneous. Nicholson's general equation is associated with the function

$${}_0F_3\left(\frac{3-2\mu}{4-2\mu},\ \frac{2+2\mu}{4-2\mu},\ \frac{1+2\mu}{4-2\mu};\ \frac{z^{4-2\mu}}{(4-2\mu)^4}\right).$$

4·6. *Symbolic solutions of differential equations.*

Numerous mathematicians have given solutions of the equation § 4·3 (1) namely

$$(1)\qquad \frac{d^2u}{dz^2}-c^2u=\frac{p\,(p+1)}{z^2}\,u,$$

in symbolic forms, when p is a positive integer (zero included). These forms are intimately connected with the recurrence formulae for Bessel functions.

It has been seen (§ 4·3) that the general solution of the equation is

$$z^{\frac{1}{2}}\mathscr{C}_{p+\frac{1}{2}}(ciz);$$

and from the recurrence formula § 3·9 (6) we have

$$z^{\frac{1}{2}}\mathscr{C}_{p+\frac{1}{2}}(ciz)=(-ci)^p z^{p+1}\left(\frac{d}{z\,dz}\right)^p\{z^{-\frac{1}{2}}\mathscr{C}_{\frac{1}{2}}(ciz)\}.$$

Since any cylinder function of the form $\mathscr{C}_{\frac{1}{2}}(ciz)$ is expressible as

$$(\alpha e^{cz}+\beta e^{-cz})/\sqrt{z},$$

where α and β are constants, it follows that the general solution of (1) may be written

$$(2)\qquad u=z^{p+1}\left(\frac{d}{z\,dz}\right)^p\frac{\alpha e^{cz}+\beta e^{-cz}}{z}.$$

A modification of this, due to Glaisher*, is

$$(3)\qquad u=z^{p+1}\left(\frac{d}{z\,dz}\right)^{p+1}(\alpha' e^{cz}+\beta' e^{-cz}),$$

where $\alpha'=\alpha/c$, $\beta'=-\beta/c$. This may be seen by differentiating $\alpha' e^{cz}+\beta' e^{-cz}$ once.

* *Phil. Trans. of the Royal Soc.* CLXXII. (1881), p. 813. It was remarked by Glaisher that equation (3) is substantially given by Earnshaw, *Partial Differential Equations* (London, 1871), p. 92. See also Glaisher, *Quarterly Journal*, XI. (1871), p. 269, formula (9), and p. 270.

NOTE. A result equivalent to (2) was set by Gaskin as a problem* in the Senate House Examination, 1839; and a proof was published by Leslie Ellis, *Camb. Math. Journal*, II. (1841), pp. 193—195, and also by Donkin, *Phil. Trans. of the Royal Soc.* CXLVII. (1857), pp. 43—57. In the question as set by Gaskin, the sign of c^2 was changed, so that the solution involved circular functions instead of exponential functions.

Next we shall prove the symbolic theorem, due to Glaisher†, that

$$(4)\qquad z^{p+1}\left(\frac{d}{z\,dz}\right)^p = \frac{1}{z^{p+1}}\left[\left(z^3\frac{d}{dz}\right)^p \frac{1}{z^{2p-2}}\right].$$

In operating on a function with the operator on the right, it is supposed that the function is multiplied by $1/z^{2p-2}$ before the application of the operators $z^3(d/dz)$.

It is convenient to write

$$z = e^\theta,\quad z\frac{d}{dz} = \vartheta,$$

and then to use the symbolic formula

$$(5)\qquad f(\vartheta).(e^{a\theta}Z) = e^{a\theta}.f(\vartheta+a)Z,$$

in which a is a constant and Z is any function of z.

The proof of this formula presents no special difficulties when $f(\vartheta)$ is a polynomial in ϑ, as is the case in the present investigation. See, e.g. Forsyth, *Treatise on Differential Equations* (1914), § 33.

It is easy to see from (5) that

$$\begin{aligned} z^{p+1}\left(\frac{d}{z\,dz}\right)^p &= e^{(p+1)\theta}(e^{-2\theta}\vartheta)^p \\ &= e^{(p+1)\theta}\{(e^{-2\theta}\vartheta).(e^{-2\theta}\vartheta).(e^{-2\theta}\vartheta)\dots(e^{-2\theta}\vartheta)\} \\ &= e^{(1-p)\theta}(\vartheta-2p+2)(\vartheta-2p+4)(\vartheta-2p+6)\dots\vartheta, \end{aligned}$$

when we bring the successive functions $e^{-2\theta}$ (beginning with those on the left) past the operators one at a time, by repeated applications of (5).

We now reverse the order‡ of the operators in the last result, and by a reversal of the previous procedure we get

$$\begin{aligned} z^{p+1}\left(\frac{d}{z\,dz}\right)^p &= e^{(1-p)\theta}\vartheta(\vartheta-2)(\vartheta-4)\dots(\vartheta-2p+2) \\ &= e^{(p-1)\theta}[(\vartheta+2p-2)(\vartheta+2p-4)\dots(\vartheta+2)\vartheta.e^{-(2p-2)\theta}] \\ &= e^{-(p+1)\theta}[(e^{2\theta}\vartheta)(e^{2\theta}\vartheta)\dots(e^{2\theta}\vartheta).e^{-(2p-2)\theta}] \\ &= \frac{1}{z^{p+1}}\left[\left(z^3\frac{d}{dz}\right)^p\frac{1}{z^{2p-2}}\right], \end{aligned}$$

* The problem was the second part of question 8, Tuesday afternoon, Jan. 8, 1839; see the *Cambridge University Calendar*, 1839, p. 319.

† *Nouvelle Corr. Math.* II. (1876), pp. 240—243, 349—350; and *Phil. Trans. of the Royal Soc.* CLXXII. (1881), pp. 803—805.

‡ It was remarked by Cayley, *Quarterly Journal*, XII. (1872), p. 132, in a footnote to a paper by Glaisher, that differential operators of the form $z^{a+1}\dfrac{d}{dz}z^{-a}$, i.e. $\vartheta-a$, obey the commutative law.

and this is the result to be proved. If we replace p by $p+1$, we find that

(6) $$z^{p+1}\left(\frac{d}{z\,dz}\right)^{p+1} = \frac{1}{z^{p+3}}\left[\left(z^3\frac{d}{dz}\right)^{p+1}\frac{1}{z^{2p}}\right].$$

When we transform (2) and (3) with the aid of (4) and (6), we see that the general solution of (1) is expressible in the following forms:

(7) $$u = \frac{1}{z^{p+1}}\left(z^3\frac{d}{dz}\right)^p \frac{\alpha e^{cz}+\beta e^{-cz}}{z^{2p-1}},$$

(8) $$u = \frac{1}{z^{p+3}}\left(z^3\frac{d}{dz}\right)^{p+1} \frac{\alpha' e^{cz}+\beta' e^{-cz}}{z^{2p}}.$$

The solutions of the equation

$$\frac{d^2v}{dz^2} - \frac{2p}{z}\frac{dv}{dz} - c^2v = 0,$$

[(3) of § 4·3], which correspond to (2), (3), (7) and (8) are

(9) $$v = z^{2p+1}\left(\frac{d}{z\,dz}\right)^p \frac{\alpha e^{cz}+\beta e^{-cz}}{z},$$

(10) $$v = z^{2p+1}\left(\frac{d}{z\,dz}\right)^{p+1} (\alpha' e^{cz}+\beta' e^{-cz}),$$

(11) $$v = \frac{1}{z}\left(z^3\frac{d}{dz}\right)^p \frac{\alpha e^{cz}+\beta e^{-cz}}{z^{2p-1}},$$

(12) $$v = \frac{1}{z^3}\left(z^3\frac{d}{dz}\right)^{p+1} \frac{\alpha' e^{cz}+\beta' e^{-cz}}{z^{2p}}.$$

A different and more direct method of obtaining (7) is due to Boole, *Phil. Trans. of the Royal Soc.* 1844, pp. 251, 252; *Treatise on Differential Equations* (London, 1872), ch. XVII. pp. 423—425; see also Curtis, *Cambridge and Dublin Math. Journal*, IX. (1854), p. 281. The solution (9) was first given by Leslie Ellis, *Camb. Math. Journal*, II. (1841), pp. 169, 193, and Lebesgue, *Journal de Math.* XI. (1846), p. 338; developments in series were obtained from it by Bach, *Ann. Sci. de l'École norm. sup.* (2) III. (1874), p. 61.

Similar symbolic solutions for the equation $\dfrac{d^2v}{dz^2} - c^2z^{2q-2}v = 0$ were discussed by Fields, *John Hopkins University Circulars*, VI. (1886—7), p. 29.

A transformation of the solution (9), due to Williamson, *Phil. Mag.* (4) XI. (1856), pp. 364—371, is

(13) $$v = c^{2p}\left(\frac{\partial}{\partial c}\cdot\frac{1}{c}\right)^p (\alpha e^{cz}+\beta e^{-cz}).$$

This is derived from the equivalence of the operators $\dfrac{1}{c}\dfrac{\partial}{\partial z}$, $\dfrac{1}{z}\dfrac{\partial}{\partial c}$, when they operate on functions of cz.

We thus obtain the equivalence of the following operators

$$z^{2p+1}\left[\left(\frac{\partial}{z\,\partial z}\right)^p\frac{1}{z}\right] = (cz)^{2p+1}\left[\left(\frac{\partial}{cz\,.\,c\partial z}\right)^p\frac{1}{cz}\right]$$
$$= (cz)^{2p+1}\left[\left(\frac{\partial}{cz^2\,\partial c}\right)^p\frac{1}{cz}\right] = c^{2p+1}\left[\left(\frac{\partial}{c\,\partial c}\right)^p\cdot\frac{1}{c}\right],$$

it being supposed that the operators operate on a function of cz; and Williamson's formula is then manifest.

4·7. *Liouville's classification of elementary transcendental functions.*

Before we give a proof of Liouville's general theorem (which was mentioned in § 4·12) concerning the impossibility of solving Riccati's equation "in finite terms" except in the classical cases discovered by Daniel Bernoulli (and the limiting form of index −2), we shall give an account of Liouville's* theory of a class of functions known as *elementary transcendental functions*; and we shall introduce a convenient notation for handling such functions.

For brevity we write†

$$l_1(z) \equiv l(z) \equiv \log z, \qquad l_2(z) = l(l(z)), \qquad l_3(z) = l(l_2(z)), \qquad \ldots,$$

$$e_1(z) \equiv e(z) \equiv e^z, \qquad e_2(z) = e(e(z)), \qquad e_3(z) = e(e_2(z)), \qquad \ldots,$$

$$\varsigma_1 f(z) = \varsigma f(z) = \int f(z)\, dz, \quad \varsigma_2 f(z) = \varsigma\{\varsigma f(z)\}, \quad \varsigma_3 f(z) = \varsigma\{\varsigma_2 f(z)\}, \quad \ldots.$$

A function of z is then said to be an *elementary transcendental function*‡ if it is expressible as an algebraic function of z and of functions of the types $l_r \phi(z)$, $e_r \psi(z)$, $\varsigma_r \chi(z)$, where the auxiliary functions $\phi(z)$, $\psi(z)$, $\chi(z)$ are expressible in terms of z and of a second set of auxiliary functions, and so on; provided that there exists a finite number n, such that the nth set of auxiliary functions are all algebraic functions of z.

The *order* of an elementary transcendental function of z is then defined inductively as follows:

(I) Any algebraic function of z is of order zero§.

(II) If $f_r(z)$ denotes any function of order r, then any algebraic function of functions of the types

$$l f_r(z), \quad e f_r(z), \quad \varsigma f_r(z), \quad f_r(z), \quad f_{r-1}(z), \ldots f_0(z)$$

(into which at least one of the first three enters) is said to be of order $r+1$.

(III) Any function is supposed to be expressed as a function of the lowest possible order. Thus $el f_r(z)$ is to be replaced by $f_r(z)$, and it is a function of order r, not of order $r+2$.

In connexion with this and the following sections, the reader should study Hardy, *Orders of Infinity* (Camb. Math. Tracts, no. 12, 1910). The functions discussed by Hardy were of a slightly more restricted character than those now under consideration, since, for his purposes, the symbol ς is not required, and also, for his purposes, it is convenient to postulate the reality of the functions which he investigates.

It may be noted that Liouville did not study properties of the symbol ς in detail, but merely remarked that it had many properties akin to those of the symbol l.

* *Journal de Math.* II. (1837), pp. 56—105; III. (1838), pp. 523—547; IV. (1839), pp. 423—456.

† It is supposed that the integrals are all indefinite.

‡ "Une fonction finie explicite."

§ For the purposes of this investigation, *irrational* powers of z, such as z^π, of course must not be regarded as algebraic functions.

4·71. *Liouville's first theorem* concerning linear differential equations.*

The investigation of the character of the solution of the equation

$$\frac{d^2u}{dz^2} = u\chi(z), \tag{1}$$

in which $\chi(z)$ is a transcendant of order† n, has been made by Liouville, who has established the following theorem:

If equation (1) *has a solution which is a transcendant of order* $m+1$, *where* $m \geqslant n$, *then* either *there exists a solution of the equation which is of order*‡ n, or else *there exists a solution*, u_1, *of the equation expressible in the form*

$$u_1 = \phi_\mu(z) \,.\, ef_\mu(z), \tag{2}$$

where $f_\mu(z)$ *is of order* μ, *and the order of* $\phi_\mu(z)$ *does not exceed* μ, *and* μ *is such that* $n \leqslant \mu \leqslant m$.

If the equation (1) has a solution of order $m+1$, let it be $f_{m+1}(z)$; then $f_{m+1}(z)$ is an algebraic function of one or more functions of the types $lf_m(z)$, $\varsigma f_m(z)$, $ef_m(z)$ as well as (possibly) of functions whose order does not exceed m. Let us concentrate our attention on a particular function of one of the three types, and let it be called θ, ϑ or Θ according to its type.

(I) We shall first shew how to prove that, if (1) has a solution of order $m+1$, then a solution can be constructed which does not involve functions of the types θ and ϑ.

For, if possible, let $f_{m+1}(z) = F(z, \theta)$, where F is an algebraic function of θ; and any function of z (other than θ itself) of order $m+1$ which occurs in F is algebraically independent of θ.

Then it is easy to shew that

$$\frac{d^2F}{dz^2} - F\,.\,\chi(z) \equiv \frac{\partial^2 F}{\partial z^2} + \frac{2}{f_m(z)}\frac{df_m(z)}{dz}\frac{\partial^2 F}{\partial\theta\,\partial z}$$
$$+\left\{\frac{1}{f_m(z)}\frac{df_m(z)}{dz}\right\}^2\frac{\partial^2 F}{\partial\theta^2} + \left[\frac{d}{dz}\left\{\frac{1}{f_m(z)}\frac{df_m(z)}{dz}\right\}\right]\frac{\partial F}{\partial\theta} - F\,.\,\chi(z), \tag{3}$$

it being supposed that z and θ are the independent variables in performing the partial differentiations.

The expression on the right in (3) is an algebraic function of θ which vanishes identically when θ is replaced by $lf_m(z)$. Hence it must vanish *identically* for all values of θ; for if it did not, the result of equating it to zero would express $lf_m(z)$ as an algebraic function of transcendants whose orders do not exceed m together with transcendants of order $m+1$ which are, *ex hypothesi*, algebraically independent of θ.

* *Journal de Math.* IV. (1839), pp. 435—442.

† This phrase is used as an abbreviation of "elementary transcendental function of order n."

‡ Null solutions are disregarded; if u were of order *less* than n, then $\frac{1}{u}\frac{d^2u}{dz^2}$ would be of order less than n, which is contrary to hypothesis.

In particular, the expression on the right of (3) vanishes when θ is replaced by $\theta + c$, where c is an arbitrary constant; and when this change is made the expression on the left of (3) changes into

$$\frac{d^2F(z, \theta + c)}{dz^2} - F(z, \theta + c) . \chi(z),$$

which is therefore zero. That is to say

$$(4) \qquad \frac{d^2F(z, \theta + c)}{dz^2} - F(z, \theta + c) . \chi(z) = 0.$$

When we differentiate (4) partially with regard to c, we find that

$$\frac{\partial F(z, \theta + c)}{\partial c}, \quad \frac{\partial^2 F(z, \theta + c)}{\partial c^2}, \quad \ldots$$

are solutions of (1) for *all* values of c independent of z. If we put $c = 0$ after performing the differentiations, these expressions become

$$\frac{\partial F(z, \theta)}{\partial \theta}, \quad \frac{\partial^2 F(z, \theta)}{\partial \theta^2}, \quad \ldots,$$

which are consequently solutions of (1). For brevity they will be called F_θ, $F_{\theta\theta}$, ...

Now either F and F_θ form a fundamental system of solutions of (1) or they do not.

If they do not, we must have*

$$F_\theta = AF,$$

where A is independent both of z and θ. On integration we find that

$$F = \Phi e^{A\theta},$$

where Φ involves transcendants (of order not exceeding $m + 1$) which are algebraically independent of θ. But this is impossible because $e^{A\theta}$ is not an algebraic function of θ; *and therefore F and F_θ form a fundamental system of solutions of* (1).

Hence $F_{\theta\theta}$ is expressible in terms of F and F_θ by an equation of the form

$$F_{\theta\theta} = AF_\theta + BF,$$

where A and B are constants. Now this may be regarded as a linear equation in θ (with constant coefficients) and its solution is

$$F = \Phi_1 e^{\alpha\theta} + \Phi_2 e^{\beta\theta} \quad \text{or} \quad F = e^{\alpha\theta}\{\Phi_1 + \Phi_2\theta\},$$

where Φ_1 and Φ_2 are functions of the same nature as Φ, while α and β are the roots of the equation

$$x^2 - Ax - B = 0.$$

The only value of F which is an algebraic function of θ is obtained when $\alpha = \beta = 0$; *and then F is a linear function of θ.*

Similarly, if $f_{m+1}(z)$ involves a function of the type ϑ, we can prove that it must be a linear function of ϑ.

* Since F *must* involve θ, $F\theta$ cannot be identically zero.

It follows that, in so far as $f_{m+1}(z)$ involves functions of the types θ and ϑ, it involves them linearly, so that we may write

$$f_{m+1}(z) = \Sigma\theta_1(z)\,\theta_2(z)\ldots\theta_p(z)\,.\,\vartheta_1(z)\,\vartheta_2(z)\ldots\vartheta_q(z)\,.\,\psi_{p,q}(z),$$

where the functions $\psi_{p,q}(z)$ are of order $m+1$ at most, and the only functions of order $m+1$ involved in them are of the type Θ.

Take any one of the terms in $f_{m+1}(z)$ which is of the highest degree, *qua* function of $\theta_1, \theta_2, \ldots \vartheta_1, \vartheta_2, \ldots$, and let it be

$$\theta_1(z)\,\theta_2(z)\ldots\theta_P(z)\,.\,\vartheta_1(z)\ldots\vartheta_Q(z)\,.\,\psi_{P,Q}(z).$$

Then, by arguments resembling those previously used, it follows that

$$\left[\frac{\partial}{\partial\theta_1}\frac{\partial}{\partial\theta_2}\cdots\frac{\partial}{\partial\theta_P}\cdot\frac{\partial}{\partial\vartheta_1}\frac{\partial}{\partial\vartheta_2}\cdots\frac{\partial}{\partial\vartheta_Q}\right]f_{m+1}(z)$$

is a solution of (1); i.e. $\psi_{P,Q}(z)$ is a solution of (1).

But $\psi_{P,Q}(z)$ is *either* a function of order not exceeding m, *or else* it is a function of order $m+1$ which involves functions of the type Θ and not of the types θ and ϑ.

In the former case, we repeat the process of reduction to functions of lower order, and in the latter case we see that some solution of the equation is an algebraic function of functions of the type Θ.

We have therefore proved that, if (1) has a solution which is a transcendant of order greater than n, then *either* it has a solution of order n *or else* it has a solution which is an algebraic function of functions of the type $ef_\mu(z)$ and $\phi_\mu(z)$, where $f_\mu(z)$ is of order μ and $\phi_\mu(z)$ is of an order which does not exceed μ.

(II) We shall next prove that, whenever (1) has a solution which is a transcendant of order greater than n, then it has a solution which involves the transcendant $ef_\mu(z)$ only in having a power of it as a factor.

We concentrate our attention on a particular transcendant Θ of the form $ef_\mu(z)$, and then the postulated solution may be written in the form $G(z, \Theta)$, where G is an algebraic function of Θ; and any function (other than Θ itself) of order $\mu+1$ which occurs in G is algebraically independent of Θ.

Then it is easy to shew that

$$(5)\qquad \frac{d^2G}{dz^2} - G\,.\,\chi(z) = \frac{\partial^2 G}{\partial z^2} + 2\Theta f_\mu'(z)\frac{\partial^2 G}{\partial z\,\partial\Theta} + \{\Theta f_\mu'(z)\}^2\frac{\partial^2 G}{\partial\Theta^2}$$
$$+\,\Theta\,[f_\mu''(z) + \{f_\mu'(z)\}^2]\frac{\partial G}{\partial\Theta} - G\,.\,\chi(z).$$

The expression on the right is an algebraic function of Θ which vanishes when Θ is replaced by $ef_\mu(z)$, and so it vanishes identically, by the arguments used in (I). In particular it vanishes when Θ is replaced by $c\,\Theta$, where c is independent of z. But its value is then

$$\frac{d^2G(z, c\,\Theta)}{dz^2} - G(z, c\,\Theta)\,.\,\chi(z),$$

so that

$$\text{(6)} \qquad \frac{d^2G(z, c\Theta)}{dz^2} - G(z, c\Theta)\,.\,\chi(z) = 0.$$

When we differentiate this with regard to c, we find that

$$\frac{\partial G(z, c\Theta)}{\partial c}, \quad \frac{\partial^2 G(z, c\Theta)}{\partial c^2}, \quad \ldots$$

are solutions of (1) for *all* values of c independent of z. If we put $c = 1$, these expressions become

$$\Theta\frac{\partial G(z, \Theta)}{\partial \Theta}, \quad \Theta^2\frac{\partial^2 G(z, \Theta)}{\partial \Theta^2}, \quad \ldots.$$

Hence, by the reasoning used in (I), we have $\Theta G_\Theta = AG$ *or else*

$$\Theta^2 G_{\Theta\Theta} = A\Theta G_\Theta + BG,$$

where A and B are constants.

In the former case $G = \Phi\Theta^A$, and in the latter G has one of the values

$$\Phi_1\Theta^\gamma + \Phi_2\Theta^\delta \quad or \quad \Theta^\gamma\{\Phi_1 + \Phi_2 \log\Theta\} = \Theta^\gamma\{\Phi_1 + \Phi_2 f_\mu(z)\},$$

where Φ, Φ_1, Φ_2 are functions of z of order $\mu + 1$ at most, any functions of order $\mu + 1$ which are involved being algebraically independent of Θ; while γ and δ are the roots of the equation

$$x(x-1) - Ax - B = 0.$$

In any case, G either contains Θ only by a factor which is a power of Θ or else G is the sum of two expressions which contain Θ only in that manner. In the latter case*,

$$G(z, c\Theta) - c^\delta G(z, \Theta)$$

is a solution of (1) which contains Θ only by a factor which is a power of Θ.

By repetitions of this procedure, we see that, if $\Theta_1, \Theta_2, \ldots \Theta_r$ are all the transcendants of order $\mu + 1$ which occur in the postulated solution, we can derive from that solution a sequence of solutions of which the sth contains $\Theta_1, \Theta_2, \ldots \Theta_s$ only by factors which are powers of $\Theta_1, \Theta_2, \ldots \Theta_s$; and the rth member of the sequence consequently consists of a product of powers of $\Theta_1, \Theta_2, \ldots \Theta_r$ multiplied by a transcendant which is of order μ at most; this solution is of the form

$$\phi_\mu(z) \exp\left\{\sum_{s=1}^{r} \gamma_s \log\Theta_s\right\},$$

which is of the form $\phi_\mu(z)\,.\,ef_\mu(z)$.

* If Φ_1 is not identically zero; if it is, then $\Phi_2\Theta^\delta$ is a solution of the specified type.

4·72. *Liouville's second theorem concerning linear differential equations.*

We have just seen that, if the equation

(1) $$\frac{d^2u}{dz^2} = u\chi(z)$$

[in which $\chi(z)$ is of order n] has a solution which is an elementary transcendant of order greater than n, then it must have a solution of the form

$$\phi_\mu(z)\, ef_\mu(z),$$

where $\mu \geqslant n$. If the equation has more than one solution of this type, let a solution for which μ has the smallest value be chosen, and let it be called u_1. Liouville's theorem, which we shall now prove, is that, *for this solution, the order of* $d(\log u_1)/dz$ *is equal to* n.

Let

$$\frac{d\log u_1}{dz} \equiv t,$$

and then t is of order μ at most; let the order of t be N, where $N \leqslant \mu$.

If $N = n$, the theorem required is proved. If $N > n$, then the equation satisfied by t, namely

(2) $$\frac{dt}{dz} + t^2 = \chi(z),$$

has a solution whose order N is greater than n.

Now t is an algebraic function of at least one transcendant of the types $lf_{N-1}(z)$, $\varsigma f_{N-1}(z)$, $ef_{N-1}(z)$ and (possibly) of transcendants whose order does not exceed $N-1$. We call the first three transcendants θ, ϑ, Θ respectively.

If t contains more than one transcendant of the type θ, we concentrate our attention on a particular function of this type, and we write

$$t = F(z, \theta).$$

By arguments resembling those used in § 4·71, we find that, if $N > n$, then

$$F(z, \theta + c)$$

is also a solution of (2). The corresponding solution of (1) is

$$\exp \int F(z, \theta + c)\, dz,$$

and this is a solution for all values of c independent of z. Hence, by differentiation with respect to c, we find that the function u_2 defined as

$$\left[\frac{\partial}{\partial c}\{\exp \int F(z, \theta + c)\, dz\}\right]_{c=0}$$

is also a solution of (1); and we have

$$u_2 = u_1 \int F_\theta\, dz,$$

so that

$$u_1 \frac{du_2}{dz} - u_2 \frac{du_1}{dz} = u_1^2 F_\theta.$$

But the Wronskian of any two solutions of (1) is a constant*; and so

$$u_1^2 F_\theta = C,$$

where C is a constant.

If $C=0$, F is independent of θ, which is contrary to hypothesis; so $C \neq 0$, and

$$u_1 = \sqrt{(C/F_\theta)}.$$

Hence u_1 is an algebraic function of θ; and similarly it is an algebraic function of all the functions of the types θ and ϑ which occur in t.

Next consider any function of the type Θ which occurs in t; we write

$$t = G(z, \Theta),$$

and, by arguments resembling those used in § 4·71 and those used earlier in this section, we find that the function u_3 defined as

$$\frac{\partial}{\partial c}\left[\exp \int G(z, c\Theta)\, dz\right]_{c=1}$$

is a solution of (1); and we have

$$u_3 = u_1 \int \Theta G_\Theta \, dz,$$

so that

$$u_1 \frac{du_3}{dz} - u_3 \frac{du_1}{dz} = u_1^2 \Theta G_\Theta .$$

This Wronskian is a constant, C_1, and so

$$u_1 = \sqrt{\{C_1/(\Theta G_\Theta)\}}.$$

Consequently u_1 is an algebraic function, not only of all the transcendants of the types θ and ϑ, but also of those of type Θ which occur in t; and therefore u_1 is of order N. This is contrary to the hypothesis that u_1 is of order $\mu+1$, where $\mu \geqslant N$, if $N > n$.

The contradiction shews that N cannot be greater than n; hence the order of $d(\log u_1)/dz$ is n. And this is the theorem to be established.

4·73. *Liouville's theorem† that Bessel's equation has no algebraic integral.*

We shall now shew that the equation

$$z^2 \frac{d^2y}{dz^2} + z\frac{dy}{dz} + (z^2 - \nu^2)\, y = 0$$

has no integral (other than a null-function) which is an algebraic function of z.

We first reduce the equation to its normal form

$$\text{(1)} \qquad \frac{d^2u}{dz^2} + \left\{1 - \frac{p(p+1)}{z^2}\right\} u = 0,$$

by writing

$$y = uz^{-\frac{1}{2}}, \quad p = \pm \nu - \tfrac{1}{2}.$$

* See e.g. Forsyth, *Treatise on Differential Equations* (1914), § 65.

† *Journal de Math.* IV. (1839), pp. 429—435; VI. (1841), pp. 4—7. Liouville's first investigation was concerned with the general case in which $\chi(z)$ is any polynomial; the application (with various modifications) to Bessel's equation was given in his later paper, *Journal de Math.* VI. (1841), pp. 1—13, 36.

This is of the form

$$\frac{d^2u}{dz^2} = u\chi(z),$$

where

(2) $$\chi(z) = \frac{p(p+1)}{z^2} - 1.$$

If possible, let Bessel's equation have an algebraic integral; then (1) also has an algebraic integral. Let the equation which expresses this integral, u, as an algebraic function of z be

(3) $$\mathcal{A}(u, z) = 0,$$

where $\mathcal{A}$ is a polynomial both in u and in z; and it is supposed that $\mathcal{A}$ is irreducible*.

Since u is a solution of (1) we have

(4) $$\mathcal{A}_{uu}\mathcal{A}_z^2 - 2\mathcal{A}_{uz}\mathcal{A}_u\mathcal{A}_z + \mathcal{A}_{zz}\mathcal{A}_u^2 + \mathcal{A}_u^3 u\chi(z) = 0.$$

The equations (3) and (4) have a common root, and hence *all* the roots of (3) satisfy (4).

For, if not, the left-hand sides of (3) and (4) (*qua* functions of u) would have a highest common factor other than $\mathcal{A}$ itself, and this would be a polynomial in u and in z. Hence $\mathcal{A}$ would be reducible, which is contrary to hypothesis.

Let all the roots of (3) be $u_1, u_2, \ldots u_M$. Then, if s is any positive integer,

$$u_1^s + u_2^s + \ldots + u_M^s$$

is a rational function of z; and there is at least one value of s not exceeding M for which this sum is not zero†.

Let any such value of s be taken, and let

$$W_0 = \sum_{m=1}^{M} u_m^s.$$

Also let $$W_r = s(s-1)\ldots(s-r+1)\sum_{m=1}^{M} u_m^{s-r}\left(\frac{du_m}{dz}\right)^r,$$

where $r = 1, 2, \ldots s$. Since $u_1, u_2, \ldots u_M$ are all solutions‡ of (1), it is easy to prove that

(5) $$\frac{dW_0}{dz} = W_1,$$

(6) $$\frac{dW_r}{dz} = W_{r+1} + r(s-r+1)\chi(z)W_{r-1}, \quad (r = 1, 2, \ldots s-1)$$

(7) $$\frac{dW_s}{dz} = s\chi(z)W_{s-1}.$$

* That is to say, $\mathcal{A}$ has no factors which are polynomials in u or in z or in both u and z.

† If not, all the roots of (3) would be zero.

‡ Because (4) is satisfied by all the roots of (3), *qua* equation in u.

Since W_0 is a rational function of z, it is expressible in partial fractions, so that

$$W_0 = \sum_{n=-\kappa}^{\lambda} A_n z^n + \sum_{n,q} \frac{B_{n,q}}{(z-a_q)^n},$$

where A_n and $B_{n,q}$ are constants, κ and λ are integers, n assumes positive integral values only in the last summation and $a_q \neq 0$.

Let the highest power of $1/(z-a_q)$ which occurs in W_0 be $1/(z-a_q)^P$.

It follows by an easy induction from (5) and (6) that the highest power of $1/(z-a_q)$ in W_r is $1/(z-a_q)^{P+r}$, where $r=1, 2, \ldots s$.

Hence there is a higher power on the left of (7) than on the right. This contradiction shews that there are no terms of the type $B_{n,q}(z-a_q)^{-n}$ in W_0 and so

$$W_0 = \sum_{n=-\kappa}^{\lambda} A_n z^n.$$

We may now assume that $A_\lambda \neq 0$, because this expression for W_0 must have a last term if it does not vanish identically.

From (5) and (6) it is easy to see that the terms of highest degree in z which occur in $W_0, W_1, W_2, W_3, \ldots$ are*

$$A_\lambda z^\lambda, \quad \lambda A_\lambda z^{\lambda-1}, \quad A_\lambda s z^\lambda, \quad \lambda A_\lambda (3s-2) z^{\lambda-1}, \ldots.$$

By a simple induction it is possible to shew that the term of highest degree in W_{2r} is

$$A_\lambda z^\lambda . 1 . 3 \ldots (2r-1) . s(s-2) \ldots (s-2r+2).$$

An induction of a more complicated nature is then necessary to shew that the term of highest degree in W_{2r+1} is

$$\lambda A_\lambda z^{\lambda-1}\, 2 . 4 \ldots (2r) . (s-1)(s-3) \ldots (s-2r+1) . {}_2F_1(\tfrac{1}{2}, -\tfrac{1}{2}s; \tfrac{1}{2}-\tfrac{1}{2}s; 1)_{r+1},$$

where the suffix $r+1$ indicates that the first $r+1$ terms only of the hypergeometric series are to be taken.

If s is odd, the terms of highest degree on the left and right of (7) are of degrees $\lambda-2$ and λ respectively, which is impossible. Hence W_0 vanishes whenever s is odd.

When s is even, the result of equating coefficients of $z^{\lambda-1}$ in (7) is

$$\lambda A_\lambda . s! = -\lambda A_\lambda . s!\, {}_2F_1(\tfrac{1}{2}, -\tfrac{1}{2}s; \tfrac{1}{2}-\tfrac{1}{2}s; 1)_{\frac{1}{2}s}.$$

That is to say

$$\lambda A_\lambda . s!\, {}_2F_1(\tfrac{1}{2}, -\tfrac{1}{2}s; \tfrac{1}{2}-\tfrac{1}{2}s; 1) = 0,$$

and so, by Vandermonde's theorem,

$$\lambda A_\lambda . s! \frac{2 . 4 . 6 \ldots s}{1 . 3 . 5 \ldots (s-1)} = 0.$$

The expression on the left vanishes only when λ is zero†.

* It is to be remembered that the term of highest degree in $\chi(z)$ is -1.

† The analysis given by Liouville, *Journal de Math.* VI. (1841), p. 7, seems to fail at this point, because he apparently overlooked the possibility of λ vanishing. The failure seems inevitable in view of the fact that $J^2_{n+\frac{1}{2}}(z) + J^2_{-n-\frac{1}{2}}(z)$ is an algebraic function of z, by § 3·4. The subsequent part of the proof given here is based on a suggestion made by Liouville, *Journal de Math.* IV. (1839), p. 435; see also Genocchi, *Mem. Accad. delle Sci. di Torino*, XXIII. (1866), pp. 299—362; *Comptes Rendus*, LXXXV. (1877), pp. 391—394.

We have therefore proved that, when s is odd, W_0 vanishes, and that, when s is even, W_0 is expressible in the form

$$\sum_{n=0}^{\kappa s} A_{n,s}\, z^{-n},$$

where $A_{0,s}$ does not vanish.

From Newton's theorem which expresses the coefficients in an equation in terms of the sums of powers of the roots, it appears that M must be even, and that the equation $\mathscr{A}(u, z) = 0$ is expressible in the form

$$u^M + \sum_{r=1}^{\frac{1}{2}M} u^{M-2r}\, \mathfrak{P}_r(1/z) = 0, \tag{8}$$

where the functions $\mathfrak{P}_r$ are polynomials in $1/z$.

When we solve (8) in a series of ascending powers of $1/z$, we find that each of the branches of u is expressible in the form

$$\sum_{m=0}^{\infty} c_m z^{-m/n},$$

where n is a positive integer and, in the case of one branch at least, c_0 does not vanish because the constant terms in the functions $\mathfrak{P}_r$ are not all zero. And the series which are of the form

$$\sum_{m=0}^{\infty} c_m z^{-m/n}$$

are convergent* for all sufficiently large values of z.

When we substitute the series into the left-hand side of (1), we find that the coefficient of the constant term in the result is c_0, and so, for every branch, c_0 must be zero, contrary to what has just been proved. The contradiction thus obtained shews that Bessel's equation has no algebraic integral.

4·74. *On the impossibility of integrating Bessel's equation in finite terms.*

We are now in a position to prove Liouville's theorem† that Bessel's equation for functions of order ν has no solution (except a null-function) which is expressible in finite terms by means of elementary transcendental functions, if 2ν is not an odd integer.

As in § 4·73, we reduce Bessel's equation to its normal form

$$\frac{d^2u}{dz^2} = u\chi(z), \tag{1}$$

where $\chi(z) = -1 + p(p+1)/z^2$ and $p = \pm\nu - \frac{1}{2}$.

Now write $d(\log u)/dz = t$, and we have

$$\frac{dt}{dz} + t^2 + 1 - \frac{p(p+1)}{z^2} = 0. \tag{2}$$

* Goursat, *Cours d'Analyse*, II. (Paris, 1911), pp. 273—281. Many treatises tacitly assume the convergence of a series derived in this manner from an algebraic equation.

† *Journal de Math.* VI. (1841), pp. 1—13, 36.

Since $\chi(z)$ is of order zero, it follows from § 4·72 that, if Bessel's equation has an integral expressible in finite terms, then (2) must have a solution which is of order zero, i.e. it must have an algebraic integral.

If (2) has an algebraic integral, let the equation which expresses this integral, t, as an algebraic function of z, be

$$\mathscr{A}(t, z) = 0, \tag{3}$$

where $\mathscr{A}$ is an irreducible polynomial in t and z.

Since t is a solution of (2), we have

$$\mathscr{A}_z + \{\chi(z) - t^2\}\mathscr{A}_t = 0. \tag{4}$$

As in the corresponding analysis of § 4·73, all the branches of t satisfy (4).

First suppose that there are more than two branches of t, and let three of them be called t_1, t_2, t_3, the corresponding values of u (defined as $\exp \int t\,dz$) being u_1, u_2, u_3. These functions are all solutions of (1) and so the Wronskians

$$u_2 \frac{du_3}{dz} - u_3 \frac{du_2}{dz}, \quad u_3 \frac{du_1}{dz} - u_1 \frac{du_3}{dz}, \quad u_1 \frac{du_2}{dz} - u_2 \frac{du_1}{dz}$$

are constants, which will be called C_1, C_2, C_3.

Now it is easy to verify that

$$C_1 = u_2 \frac{du_3}{dz} - u_3 \frac{du_2}{dz} = u_2 u_3 (t_3 - t_2);$$

and $t_3 - t_2$ is not zero, because, if it were zero, the equation (3) would have a pair of equal roots, and would therefore be reducible.

Hence $C_1 \neq 0$, and so

$$u_2 u_3 = C_1/(t_3 - t_2).$$

Therefore $u_2 u_3$ (and similarly $u_3 u_1$ and $u_1 u_2$) is an algebraic function of z.

But
$$u_1 = \sqrt{\frac{u_3 u_1 \,.\, u_1 u_2}{u_2 u_3}},$$

and therefore u_1 is an algebraic function of z. This, as we have seen in § 4·73, cannot be the case, and so t has not more than two branches.

Next suppose that t has two branches, so that $\mathscr{A}(t, z)$ is quadratic in t. Let the branches be $U \pm \sqrt{V}$, where U and V are rational functions of z. By substituting in (2) we find that

$$\begin{cases} U' + U^2 + V = \chi(z), \\ V' + 4UV = 0. \end{cases} \tag{5}$$

Let V be factorised so that

$$V = Az^{\lambda} \Pi (z - a_q)^{\kappa_q},$$

where A is constant, κ_q and λ are integers, and κ_q and a_q are not zero.

From the second member of (5) it follows that

$$U = -\frac{\lambda}{4z} - \sum_q \frac{\kappa_q}{4(z-a_q)},$$

and then by substituting into the first member of (5) we have

$$(6) \qquad \frac{\lambda}{4z^2} + \sum_q \frac{\kappa_q}{4(z-a_q)^2} + \left\{\frac{\lambda}{4z} + \sum \frac{\kappa_q}{4(z-a_q)}\right\}^2 + Az^\lambda \prod_q (z-a_q)^{\kappa_q} - \chi(z) \equiv 0.$$

Now consider the principal part of the expression on the left near a_q. It is evident that none of the numbers κ_q can be less than -2, and, if any one of them is greater than -2 it must satisfy the equation

$$\kappa_q^2 + 4\kappa_q = 0,$$

so that κ_q is 0 or -4, which are both excluded from consideration. Hence all the numbers κ_q are equal to -2.

Again, if we consider the principal part near ∞, we see that the highest power in V must cancel with the -1 in $\chi(z)$, so that $\lambda = -\sum_q \kappa_q$.

It follows that $\sqrt{V}$ is rational, and consequently $\mathscr{A}(t, z)$ is reducible, which is contrary to hypothesis.

Hence t cannot have as many as two branches and so it must be rational.

Accordingly, let the expression for t in partial fractions be

$$t = \sum_{n=-\kappa}^{\lambda} A_n z^n + \sum_{n,q} \frac{B_{n,q}}{(z-a_q)^n},$$

where A_n and $B_{n,q}$ are constants, κ and λ are integers, n assumes positive values only in the last summation and $a_q \neq 0$.

If we substitute this value of t in (2) we find that

$$\sum_{n=-\kappa}^{\lambda} nA_n z^{n-1} - \sum_{n,q} \frac{nB_{n,q}}{(z-a_q)^{n+1}} + \left\{\sum_{n=-\kappa}^{\lambda} A_n z^n + \sum \frac{B_{n,q}}{(z-a_q)^n}\right\}^2 + 1 - \frac{p(p+1)}{z^2} \equiv 0.$$

If we consider the principal part of the left-hand side near a_q we see that $1/(z-a_q)$ cannot occur in t to a higher power than the first and that

$$B_{1,q} - B_{1,q}^2 = 0,$$

so that

$$B_{1,q} = 1.$$

Similarly, if we consider the principal parts near 0 and ∞, we find that

$$\kappa = 1, \; (A_{-1})^2 - A_{-1} = p(p+1); \quad \lambda = 0, \; A_0^2 = -1.$$

Since $p = \pm\nu - \frac{1}{2}$, we may take $A_{-1} = -p$ without loss of generality.

It then follows that

$$u = z^{-p} e^{\pm iz} \prod_q (z-a_q).$$

Accordingly, if we replace u by $z^{-p} e^{\pm iz} w$ in (1), we see that the equation

$$(7) \qquad \frac{d^2w}{dz^2} + 2\left(\pm i - \frac{p}{z}\right)\frac{dw}{dz} \mp \frac{2ip}{z} w = 0$$

must have a solution which is a polynomial in z, and the constant term in this polynomial does not vanish.

When we substitute $\Sigma\, c_m z^m$ for w in (7) we find that the relation connecting successive coefficients is

$$m(m-2p-1)\,c_m \pm 2ic_{m-1}(m-p-1)=0,$$

and so the series for w cannot terminate unless $m-p-1$ can vanish, i.e. unless p is zero or a positive integer.

Hence the hypothesis that Bessel's equation is soluble in finite terms leads of necessity to the consequence that one of the numbers $\pm\nu-\frac{1}{2}$ is zero or a positive integer; and this is the case if, and only if, 2ν is an odd integer.

Conversely we have seen (§ 3·4) that, when 2ν is an odd integer, Bessel's equation actually possesses a fundamental system of solutions expressible in finite terms. The investigation of the solubility of the equation is therefore complete.

Some applications of this theorem to equations of the types discussed in § 4·3 have been recorded by Lebesgue, *Journal de Math.* XI. (1846), pp. 338—340.

4·75. *On the impossibility of integrating Riccati's equation in finite terms.*

By means of the result just obtained, we can discuss Riccati's equation

$$\frac{dy}{dz}=az^n+by^2$$

with a view to proving that it is, in general, not integrable in finite terms.

It has been seen (§ 4·21) that the equation is reducible to

$$\frac{d^2u}{d\zeta^2}-c^2\,\zeta^{2q-2}\,u=0,$$

where $n=2q-2$; and, by § 4·3, the last equation is reducible to Bessel's equation for functions of order $1/(2q)$ unless $q=0$.

Hence the only possible cases in which Riccati's equation is integrable in finite terms are those in which q is zero or $1/q$ is an odd integer; and these are precisely the cases in which n is equal to -2 or to

$$-\frac{4m}{2m\pm 1}. \qquad (m=0,\ 1,\ 2,\ \ldots)$$

Consequently the only cases in which Riccati's equation is integrable in finite terms are the classical cases discovered by Daniel Bernoulli (cf. § 4·11) and the limiting case discussed after the manner of Euler in § 4·12.

This theorem was proved by Liouville, *Journal de Math.* VI. (1841), pp. 1—13. It seems impossible to establish it by any method which is appreciably more brief than the analysis used in the preceding sections.

4·8. *Solutions of Laplace's equation.*

The first appearance in analysis of the general Bessel coefficient has been seen (§ 1·3) to be in connexion with an equation, equivalent to Laplace's equation, which occurs in the problem of the vibrations of a circular membrane.

We shall now shew how Bessel coefficients arise in a natural manner from Whittaker's* solution of Laplace's equation

$$\frac{\partial^2 V}{\partial x^2} + \frac{\partial^2 V}{\partial y^2} + \frac{\partial^2 V}{\partial z^2} = 0. \tag{1}$$

The solution in question is

$$V = \int_{-\pi}^{\pi} f(z + ix\cos u + iy\sin u,\ u)\, du, \tag{2}$$

in which f denotes an arbitrary function of the two variables involved.

In particular, a solution is

$$\int_{-\pi}^{\pi} e^{k(z+ix\cos u+iy\sin u)} \cos mu\, du,$$

in which k is any constant and m is any integer.

If we take cylindrical-polar coordinates, defined by the equations

$$x = \rho\cos\phi, \quad y = \rho\sin\phi,$$

this solution becomes

$$\begin{aligned} e^{kz}\int_{-\pi}^{\pi} e^{ik\rho\cos(u-\phi)} \cos mu\, du &= e^{kz}\int_{-\pi}^{\pi} e^{ik\rho\cos v}\cos m(v+\phi)\, dv, \\ &= 2e^{kz}\int_{0}^{\pi} e^{ik\rho\cos v}\cos mv\cos m\phi\, dv, \\ &= 2\pi i^m e^{kz}\cos m\phi \,.\, J_m(k\rho), \end{aligned}$$

by § 2·2. In like manner a solution is

$$\int_{-\pi}^{\pi} e^{k(z+ix\cos u+iy\sin u)} \sin mu\, du,$$

and this is equal to $2\pi i^m e^{kz}\sin m\phi \,.\, J_m(k\rho)$. Both of these solutions are analytic near the origin.

Again, if Laplace's equation be transformed† to cylindrical-polar coordinates, it is found to become

$$\frac{\partial^2 V}{\partial \rho^2} + \frac{1}{\rho}\frac{\partial V}{\partial \rho} + \frac{1}{\rho^2}\frac{\partial^2 V}{\partial \phi^2} + \frac{\partial^2 V}{\partial z^2} = 0;$$

* *Monthly Notices of the R. A. S.* LXII. (1902), pp. 617—620; *Math. Ann.* LVII. (1902), pp. 333—341.

† The simplest method of effecting the transformation is by using Green's theorem. See W. Thomson, *Camb. Math. Journal*, IV. (1845), pp. 33—42.

and a normal solution of this equation of which e^{kz} is a factor must be such that

$$\frac{1}{V}\frac{\partial^2 V}{\partial \phi^2}$$

is independent of ϕ, and, if the solution is to be one-valued, it must be equal to $-m^2$ where m is an integer. Consequently the function of ρ which is a factor of V must be annihilated by

$$\frac{d^2}{d\rho^2}+\frac{1}{\rho}\frac{d}{d\rho}+\left(k^2-\frac{m^2}{\rho^2}\right),$$

and therefore it must be a multiple of $J_m(k\rho)$ if it is to be analytic along the line $\rho=0$.

We thus obtain anew the solutions

$$e^{kz}\begin{matrix}\cos\\ \sin\end{matrix} m\phi \,.\, J_m(k\rho).$$

These solutions have been derived by Hobson* from the solution $e^{kz}J_0(k\rho)$ by Clerk Maxwell's method of differentiating harmonics with respect to axes.

Another solution of Laplace's equation involving Bessel functions has been obtained by Hobson (*ibid.* p. 447) from the equation in cylindrical-polar coordinates by regarding $\partial/\partial z$ as a symbolic operator. The solution so obtained is

$$\begin{matrix}\cos\\ \sin\end{matrix} m\phi \,.\, \mathscr{C}_m\left(\rho\frac{d}{dz}\right)f(z),$$

where $f(z)$ is an arbitrary function; but the interpretation of this solution when $\mathscr{C}_m$ involves a function of the second kind is open to question. Other solutions involving a Bessel function of an operator acting on an arbitrary function have been given by Hobson, *Proc. London Math. Soc.* XXIV. (1893), pp. 55—67; XXVI. (1895), pp. 492—494.

4·81. *Solutions of the equations of wave motions.*

We shall now examine the equation of wave motions

$$\text{(1)} \qquad \frac{\partial^2 V}{\partial x^2}+\frac{\partial^2 V}{\partial y^2}+\frac{\partial^2 V}{\partial z^2}=\frac{1}{c^2}\frac{\partial^2 V}{\partial t^2},$$

in which t represents the time and c the velocity of propagation of the waves, from the same aspect.

Whittaker's† solution of this equation is

$$\text{(2)} \qquad V=\int_{-\pi}^{\pi}\int_0^{\pi} f(x\sin u\cos v+y\sin u\sin v+z\cos u+ct,\ u,\ v)\,du\,dv,$$

where f denotes an arbitrary function of the three variables involved.

In particular, a solution is

$$V=\int_{-\pi}^{\pi}\int_0^{\pi} e^{ik(x\sin u\cos v+y\sin u\sin v+z\cos u+ct)}F(u,v)\,du\,dv,$$

where F denotes an arbitrary function of u and v.

* *Proc. London Math. Soc.* XXII. (1892), pp. 431—449.

† *Math. Ann.* LVII. (1902), pp. 342—345. See also Havelock, *Proc. London Math. Soc.* (2) II. (1904), pp. 122—137, and Watson, *Messenger*, XXXVI. (1907), pp. 98—106.

The physical importance of this particular solution lies in the fact that it is the general solution in which the waves all have the same frequency kc.

Now let the polar coordinates of (x, y, z) be (r, θ, ϕ), and let (ω, ψ) be the angular coordinates of the direction (u, v) referred to new axes for which the polar axis is the direction (θ, ϕ) and the plane $\psi = 0$ passes through the z-axis. The well-known formulae of spherical trigonometry then shew that

$$\cos\omega = \cos\theta\cos u + \sin\theta\sin u\cos(v-\phi),$$

$$\sin u\sin(v-\phi) = \sin\omega\sin\psi.$$

Now take the arbitrary function $F(u, v)$ to be $S_n(u, v)\sin u$, where S_n denotes a surface harmonic in (u, v) of degree n; we may then write

$$S_n(u, v) = \bar{S}_n(\theta, \phi;\ \omega, \psi),$$

where $\bar{S}_n$ is a surface harmonic* in (ω, ψ) of degree n.

We thus get the solution

$$V_n = e^{ikct}\int_{-\pi}^{\pi}\int_0^{\pi} e^{ikr\cos\omega}\,\bar{S}_n(\theta, \phi;\ \omega, \psi)\sin\omega\, d\omega\, d\psi.$$

Since $\bar{S}_n$ is a surface harmonic of degree n in (ω, ψ), we may write

$$\begin{aligned}\bar{S}_n(\theta, \phi;\ \omega, \psi) &= A_n(\theta, \phi)\,.\,P_n(\cos\omega)\\ &+\sum_{m=1}^{n}\{A_n^{(m)}(\theta, \phi)\cos m\psi + B_n^{(m)}(\theta, \phi)\sin m\psi\}\,P_n^m(\cos\omega),\end{aligned}$$

where $A_n(\theta, \phi)$, $A_n^{(m)}(\theta, \phi)$ and $B_n^{(m)}(\theta, \phi)$ are independent of ω and ψ.

Performing the integration with respect to ψ, we get

$$\begin{aligned}V_n &= 2\pi e^{ikct} A_n(\theta, \phi)\int_0^{\pi} e^{ikr\cos\omega} P_n(\cos\omega)\sin\omega\, d\omega\\ &= (2\pi)^{\frac{3}{2}}\, i^n e^{ikct}\,\frac{J_{n+\frac{1}{2}}(kr)}{\sqrt{(kr)}}\, A_n(\theta, \phi)\end{aligned}$$

by § 3·32.

Now the equation of wave motions is unaffected if we multiply x, y, z and t by the same constant factor, i.e. if we multiply r and t by the same constant factor, leaving θ and ϕ unaltered; so that $A_n(\theta, \phi)$ may be taken to be independent† of the constant k which multiplies r and t.

Hence $\lim\limits_{k\to 0}(k^{-n}V_n)$ is a solution of the equation of wave motions, that is to say, $r^n A_n(\theta, \phi)$ is a solution (independent of t) of the equation of wave motions, and is consequently a solution of Laplace's equation. Hence $A_n(\theta, \phi)$

* This follows from the fact that Laplace's operator is an invariant for changes of rectangular axes.

† This is otherwise obvious, because S_n may be taken independent of k.

is a surface harmonic of degree n. If we assume it to be permissible to take $A_n(\theta, \phi)$ to be any such harmonic, *we obtain the result that*

$$e^{ikct} r^{-\frac{1}{2}} J_{n+\frac{1}{2}}(kr) P_n^m(\cos\theta) \overset{\cos}{\underset{\sin}{\cdot}} m\phi$$

*is a solution of the equation of wave motions**; and the motion represented by this solution has frequency kc.

To justify the assumption that $A_n(\theta, \phi)$ may be any surface harmonic of degree n, we construct the normal solution of the equation of wave motions

$$\frac{\partial}{\partial r}\left(r^2 \frac{\partial V}{\partial r}\right) + \frac{1}{\sin\theta}\frac{\partial}{\partial\theta}\left(\sin\theta \frac{\partial V}{\partial\theta}\right) + \frac{1}{\sin^2\theta}\frac{\partial^2 V}{\partial\phi^2} = \frac{r^2}{c^2}\frac{\partial^2 V}{\partial t^2},$$

which has factors of the form $e^{ikct} \overset{\cos}{\underset{\sin}{\cdot}} m\phi$. The factor which involves θ must then be of the form $P_n^m(\cos\theta)$; and the factor which involves r is annihilated by the operator

$$\frac{d}{dr}\left(r^2 \frac{d}{dr}\right) - n(n+1) + k^2 r^2,$$

so that if this factor is to be analytic at the origin it must be a multiple of $J_{n+\frac{1}{2}}(kr)/\sqrt{r}$.

4·82. *Theorems derived from solutions of the equations of Mathematical Physics.*

It is possible to prove (or, at any rate, to render probable) theorems concerning Bessel functions by a comparison of various solutions of Laplace's equation or of the equation of wave motions.

Thus, if we take the function

$$e^{kz} J_0\{k\sqrt{(\rho^2 + a^2 - 2a\rho\cos\phi)}\},$$

by making a change of origin to the point $(a, 0, 0)$, we see that it is a solution of Laplace's equation in cylindrical-polar coordinates. This solution has e^{kz} as a factor and it is analytic at all points of space. It is therefore natural to expect it to be expansible in the form

$$e^{kz}\left[A_0 J_0(k\rho) + 2\sum_{m=1}^{\infty}(A_m \cos m\phi + B_m \sin m\phi) J_m(k\rho)\right].$$

Assuming the possibility of this expansion, we observe that the function under consideration is an even function of ϕ, and so $B_m = 0$; and, from the symmetry in ρ and a, A_m is of the form $c_m J_m(ka)$, where c_m is independent of ρ and a.

We thus get

$$J_0\{k\sqrt{(\rho^2 + a^2 - 2a\rho\cos\phi)}\} = \sum_{m=0}^{\infty} \epsilon_m c_m J_m(k\rho) J_m(ka)\cos m\phi.$$

If we expand both sides in powers of ρ, a and $\cos\phi$, and compare the coefficients of $(k^2\rho a\cos\phi)^m$, we get

$$c_m = 1,$$

* Cf. Bryan, *Nature*, LXXX. (1909), p. 309.

and so we are led to the expansion*

$$J_0\{k\sqrt{(\rho^2 + a^2 - 2a\rho\cos\phi)}\} = \sum_{m=0}^{\infty} \epsilon_m J_m(k\rho) J_m(ka)\cos m\phi,$$

of which a more formal proof will be given in § 11·2.

Again, if we take $e^{ik(ct+z)}$, which is a solution of the equation of wave motions, and which represents a wave moving in the direction of the axis of z from $+\infty$ to $-\infty$ with frequency kc and wave-length $2\pi/k$, we expect this expression to be expansible† in the form

$$\left(\frac{2\pi}{kr}\right)^{\frac{1}{2}} e^{ikct} \sum_{n=0}^{\infty} c_n i^n J_{n+\frac{1}{2}}(kr) P_n(\cos\theta),$$

where c_n is a constant; so that

$$e^{ikr\cos\theta} = \left(\frac{2\pi}{kr}\right)^{\frac{1}{2}} \sum_{n=0}^{\infty} c_n i^n J_{n+\frac{1}{2}}(kr) P_n(\cos\theta).$$

If we compare the coefficients of $(kr\cos\theta)^n$ on each side, we find that

$$\frac{i^n}{n!} = (2\pi)^{\frac{1}{2}} \frac{c_n i^n}{2^{n+\frac{1}{2}}\Gamma(n+\frac{3}{2})} \cdot \frac{(2n)!}{2^n . (n!)^2},$$

and so $c_n = n + \frac{1}{2}$; we are thus led to the expansion‡

$$e^{ikr\cos\theta} = \left(\frac{2\pi}{kr}\right)^{\frac{1}{2}} \sum_{n=0}^{\infty} (n+\tfrac{1}{2})\, i^n J_{n+\frac{1}{2}}(kr) P_n(\cos\theta),$$

of which a more formal proof will be given in § 11·5.

4·83. *Solutions of the wave equation in space of p dimensions.*

The analysis just explained has been extended by Hobson§ to the case of the equation

$$\frac{\partial^2 V}{\partial x_1^2} + \frac{\partial^2 V}{\partial x_2^2} + \ldots + \frac{\partial^2 V}{\partial x_p^2} = \frac{1}{c^2}\frac{\partial^2 V}{\partial t^2}.$$

A normal solution of this equation of frequency kc which is expressible as a function of r and t only, where

$$r = \sqrt{(x_1^2 + x_2^2 + \ldots + x_p^2)},$$

must be annihilated by the operator

$$\frac{\partial^2}{\partial r^2} + \frac{p-1}{r}\frac{\partial}{\partial r} + k^2$$

and so such a solution, containing a time-factor e^{ikct}, must be of the form

$$e^{ikct}\, \mathscr{C}_{\frac{1}{2}(p-2)}(kr)/(kr)^{\frac{1}{2}(p-2)}.$$

* This is due to Neumann, *Theorie der Bessel'schen Functionen* (Leipzig, 1867), pp. 59—65.

† The tesseral harmonics do not occur because the function is symmetrical about the axis of z.

‡ This expansion is due to Bauer, *Journal für Math.* LVI. (1859), pp. 104, 106.

§ *Proc. London Math. Soc.* XXV. (1894), pp. 49—75.

Hobson describes the quotient $\mathscr{C}_{\frac{1}{2}(p-2)}(kr)/(kr)^{\frac{1}{2}(p-2)}$ as a cylinder function of *rank p*; such a function may be written in the form

$$\mathscr{C}\,(kr\,|\,p).$$

By using this notation combined with the concept of p-dimensional space, Hobson succeeded in proving a number of theorems for cylinder functions of integral order and of order equal to half an odd integer simultaneously.

As an example of such theorems we shall consider an expansion for

$$J\,\{k\,\sqrt{(r^2+a^2-2ar\cos\phi)}\,|\,p\},$$

where it is convenient to regard ϕ as being connected with x_p by the equation $x_p = r\cos\phi$. This function multiplied by e^{ikct} is a solution of the wave equation, and when we write $\rho = r\sin\phi$, it is expressible as a function of ρ, ϕ, t and of no other coordinates.

Hence

$$e^{ikct}J\,\{k\,\sqrt{(r^2+a^2-2ar\cos\phi)}\,|\,p\}$$

is annihilated by the operator

$$\frac{\partial^2}{\partial\rho^2}+\frac{p-2}{\rho}\frac{\partial}{\partial\rho}+\frac{\partial^2}{\partial x_p{}^2}+k^2,$$

that is to say, by the operator

$$\frac{\partial^2}{\partial r^2}+\frac{p-1}{r}\frac{\partial}{\partial r}+\frac{(p-2)\cos\phi}{r^2\sin\phi}\frac{\partial}{\partial\phi}+\frac{1}{r^2}\frac{\partial^2}{\partial\phi^2}+k^2.$$

Now normal functions which are annihilated by this operator are of the form

$$\frac{\mathscr{C}_{n+\frac{1}{2}p-1}(kr)}{(kr)^{\frac{1}{2}p-1}}P_n(\cos\phi\,|\,p),$$

where $P_n(\cos\phi\,|\,p)$ is the coefficient* of α^n in the expansion of

$$(1-2\alpha\cos\phi+\alpha^2)^{1-\frac{1}{2}p}.$$

By the reasoning used in § 4·82, we infer that

$$J\,\{k\,\sqrt{(r^2+a^2-2ar\cos\phi)}\,|\,p\}$$
$$=\frac{1}{(ka)^{\frac{1}{2}p-1}(kr)^{\frac{1}{2}p-1}}\sum_{n=0}^{\infty}A_nJ_{n+\frac{1}{2}p-1}(kr)\,J_{n+\frac{1}{2}p-1}(ka)\,P_n(\cos\phi\,|\,p).$$

Now expand all the Bessel functions and equate the coefficients of $(k^2ar\cos\phi)^n$ on each side; we find that

$$\frac{2^n}{2^{2n+\frac{1}{2}p-1}\,n!\,\Gamma(n+\frac{1}{2}p)}=\frac{A_n}{\{2^{n+\frac{1}{2}p-1}\,\Gamma(n+\frac{1}{2}p)\}^2}\cdot\frac{2^n\,\Gamma(n+\frac{1}{2}p-1)}{n!\,\Gamma(\frac{1}{2}p-1)},$$

so that $A_n = 2^{\frac{1}{2}p-1}(n+\frac{1}{2}p-1)\,\Gamma(\frac{1}{2}p-1)$.

* So that, in Gegenbauer's notation,

$$P_n(\cos\phi\,|\,p)\equiv C_n^{\frac{1}{2}p-1}(\cos\phi).$$

We thus obtain the expansion

$$\frac{J_{\frac{1}{2}p-1}\{k\sqrt{(r^2+a^2-2ar\cos\phi)}\}}{(r^2+a^2-2ar\cos\phi)^{\frac{1}{2}p-\frac{1}{2}}}$$

$$=\frac{2^{\frac{1}{2}p-1}\,\Gamma(\frac{1}{2}p-1)}{(kar)^{\frac{1}{2}p-1}}\sum_{n=0}^{\infty}(n+\tfrac{1}{2}p-1)\,J_{n+\frac{1}{2}p-1}(kr)\,J_{n+\frac{1}{2}p-1}(ka)\,C_n^{\frac{1}{2}p-1}(\cos\phi).$$

An analytical proof of this expansion, which holds for Bessel functions of all orders (though the proof given here is valid only when p is an integer), will be given in § 11·4.

4·84. *Bateman's solutions of the generalised equation of wave motions.*

Two systems of normal solutions of the equation

$$\frac{\partial^2 V}{\partial x_1^2}+\frac{\partial^2 V}{\partial x_2^2}+\frac{\partial^2 V}{\partial x_3^2}+\frac{\partial^2 V}{\partial x_4^2}=\frac{1}{c^2}\frac{\partial^2 V}{\partial t^2} \tag{1}$$

have been investigated by Bateman*, who also established a connexion between the two systems.

If we take new variables ρ, σ, χ, ψ defined by the equations

$$x_1=\rho\cos\chi,\quad x_3=\sigma\cos\psi,$$

$$x_2=\rho\sin\chi,\quad x_4=\sigma\sin\psi,$$

the equation transforms into

$$\left\{\frac{\partial^2 V}{\partial\rho^2}+\frac{1}{\rho}\frac{\partial V}{\partial\rho}+\frac{1}{\rho^2}\frac{\partial^2 V}{\partial\chi^2}\right\}+\left\{\frac{\partial^2 V}{\partial\sigma^2}+\frac{1}{\sigma}\frac{\partial V}{\partial\sigma}+\frac{1}{\sigma^2}\frac{\partial^2 V}{\partial\psi^2}\right\}=\frac{1}{c^2}\frac{\partial^2 V}{\partial t^2}. \tag{2}$$

A normal solution of this equation with frequency kc is

$$J_\mu(k\rho\cos\Phi)\,J_\nu(k\sigma\sin\Phi)\,e^{i(\mu\chi+\nu\psi+kct)},$$

where Φ is any constant.

Further, if we write

$$\rho=r\cos\phi,\quad \sigma=r\sin\phi,$$

so that (r, χ, ψ, ϕ) form a system of polar coordinates, equation (2) transforms into

$$\frac{\partial^2 V}{\partial r^2}+\frac{3}{r}\frac{\partial V}{\partial r}+\frac{1}{r^2}\frac{\partial^2 V}{\partial\phi^2}+\frac{\cot\phi-\tan\phi}{r^2}\frac{\partial V}{\partial\phi}$$
$$+\frac{1}{r^2\cos^2\phi}\frac{\partial^2 V}{\partial\chi^2}+\frac{1}{r^2\sin^2\phi}\frac{\partial^2 V}{\partial\psi^2}=\frac{1}{c^2}\frac{\partial^2 V}{\partial t^2}. \tag{3}$$

Now normal solutions of this equation which have $e^{i(\mu\chi+\nu\psi+kct)}$ as a factor are annihilated by the operator

$$\frac{\partial^2}{\partial r^2}+\frac{3}{r}\frac{\partial}{\partial r}+k^2+\frac{1}{r^2}\left\{\frac{\partial^2}{\partial\phi^2}+(\cot\phi-\tan\phi)\frac{\partial}{\partial\phi}-\frac{\mu^2}{\cos^2\phi}-\frac{\nu^2}{\sin^2\phi}\right\},$$

* *Messenger*, XXXIII. (1904), pp. 182—188; *Proc. London Math. Soc.* (2) III. (1905), pp. 111—123.

and since such solutions are expressible as the product of a function of r and a function of ϕ they must be annihilated by each of the operators

$$\frac{\partial^2}{\partial r^2}+\frac{3}{r}\frac{\partial}{\partial r}+k^2-\frac{4\lambda(\lambda+1)}{r^2},$$

$$\frac{\partial^2}{\partial\phi^2}+(\cot\phi-\tan\phi)\frac{\partial}{\partial\phi}+4\lambda(\lambda+1)-\frac{\mu^2}{\cos^2\phi}-\frac{\nu^2}{\sin^2\phi},$$

where λ is a constant whose value depends on the particular solution under consideration. The normal solutions so obtained are now easily verified to be of the form

$$(kr)^{-1}J_{2\lambda+1}(kr)\cos^\mu\phi\sin^\nu\phi$$
$$\times{}_2F_1\left(\frac{\mu+\nu}{2}-\lambda,\ \frac{\mu+\nu}{2}+\lambda+1;\ \nu+1;\ \sin^2\phi\right)e^{i(\mu\chi+\nu\psi+kct)}.$$

It is therefore suggested that

$$J_\mu(kr\cos\phi\cos\Phi)\,J_\nu(kr\sin\phi\sin\Phi)$$

is expressible in the form

$$\sum_\lambda a_\lambda(kr)^{-1}J_{2\lambda+1}(kr)\cos^\mu\phi\sin^\nu\phi\,.\,{}_2F_1\left(\frac{\mu+\nu}{2}-\lambda,\ \frac{\mu+\nu}{2}+\lambda+1;\ \nu+1;\ \sin^2\phi\right),$$

where the summation extends over various values of λ, and the coefficients a_λ depend on λ and Φ, but not on r or ϕ. By symmetry it is clear that

$$a_\lambda=b_\lambda\cos^\mu\Phi\sin^\nu\Phi\,.\,{}_2F_1\left(\frac{\mu+\nu}{2}-\lambda,\ \frac{\mu+\nu}{2}+\lambda+1;\ \nu+1;\ \sin^2\Phi\right),$$

where b_λ is independent of Φ.

It is not difficult to see that

$$\lambda=\tfrac{1}{2}(\mu+\nu)+n,\qquad (n=0,1,2,\ldots)$$

and Bateman has proved that

$$b_\lambda=2(-)^n(\mu+\nu+2n+1)\frac{\Gamma(\mu+\nu+n+1)\,\Gamma(\nu+n+1)}{n!\ \Gamma(\mu+n+1)\{\Gamma(\nu+1)\}^2}.$$

We shall not give Bateman's proof, which is based on the theory of linear differential equations, but later (§ 11·6) we shall establish the expansion of $J_\mu(kr\cos\phi\cos\Phi)\,J_\nu(kr\sin\phi\sin\Phi)$ by a direct transformation.

CHAPTER V

MISCELLANEOUS PROPERTIES OF BESSEL FUNCTIONS

5·1. *Indefinite integrals containing a single Bessel function.*

In this chapter we shall discuss some properties of Bessel functions which have not found a place in the two preceding chapters, and which have but one feature in common, namely that they are all obtainable by processes of a definitely elementary character.

We shall first evaluate some indefinite integrals.

The recurrence formulae § 3·9 (5) and (6) at once lead to the results

$$\int^z z^{\nu+1}\mathscr{C}_\nu(z)\,dz = z^{\nu+1}\mathscr{C}_{\nu+1}(z), \tag{1}$$

$$\int^z z^{-\nu+1}\mathscr{C}_\nu(z)\,dz = -z^{-\nu+1}\mathscr{C}_{\nu-1}(z). \tag{2}$$

To generalise these formulae, consider

$$\int^z z^{\nu+1} f(z)\,\mathscr{C}_\nu(z)\,dz;$$

let this integral be equal to

$$z^{\nu+1}\{A(z)\,\mathscr{C}_\nu(z) + B(z)\,\mathscr{C}_{\nu+1}(z)\},$$

where $A(z)$ and $B(z)$ are to be determined.

The result of differentiation is that

$$z^{\nu+1} f(z)\,\mathscr{C}_\nu(z) \equiv z^{\nu+1}\left\{A'(z)\,\mathscr{C}_\nu(z) + A(z)\frac{2\nu+1}{z}\mathscr{C}_\nu(z) - A(z)\,\mathscr{C}_{\nu+1}(z)\right\}$$
$$+ z^{\nu+1}\{B'(z)\,\mathscr{C}_{\nu+1}(z) + B(z)\,\mathscr{C}_\nu(z)\}.$$

In order that $A(z)$ and $B(z)$ may not depend on the cylinder function, we take $A(z) \equiv B'(z)$, and then

$$f(z) \equiv A'(z) + \frac{2\nu+1}{z}A(z) + B(z).$$

Hence it follows that

$$\int^z z^{\nu+1}\left\{B''(z) + \frac{2\nu+1}{z}B'(z) + B(z)\right\}\mathscr{C}_\nu(z)\,dz \tag{3}$$
$$= z^{\nu+1}\{B'(z)\,\mathscr{C}_\nu(z) + B(z)\,\mathscr{C}_{\nu+1}(z)\}.$$

This result was obtained by Sonine, *Math. Ann.* XVI. (1880), p. 30, though an equivalent formula (with a different notation) had been obtained previously by Lommel, *Studien über die Bessel'schen Functionen* (Leipzig, 1868), p. 70. Some developments of formula (3) are due to Nielsen, *Nyt Tidsskrift*, IX. (1898), pp. 73—83 and *Ann. di Mat.* (3) VI. (1901), pp. 43—46.

For some associated integrals which involve the functions ber and bei, see Whitehead, *Quarterly Journal*, XLII. (1911), pp. 338—340.

The following reduction formula, which is an obvious consequence of (3), should be noted:

$$(4) \qquad \int^z z^{\mu+1}\mathscr{C}_\nu(z)\,dz = -(\mu^2-\nu^2)\int^z z^{\mu-1}\mathscr{C}_\nu(z)\,dz + \left[z^{\mu+1}\mathscr{C}_{\nu+1}(z) + (\mu-\nu)\,z^\mu\mathscr{C}_\nu(z)\right].$$

5·11. *Lommel's integrals containing two cylinder functions.*

The simplest integrals which contain two Bessel functions are those derived from the Wronskian formula of § 3·12 (2), namely

$$J_\nu(z)\,J'_{-\nu}(z) - J_{-\nu}(z)J_\nu'(z) = -\frac{2\sin\nu\pi}{\pi z},$$

which gives

$$(1) \qquad \int^z \frac{dz}{z\,J_\nu^2(z)} = -\frac{\pi}{2\sin\nu\pi}\,\frac{J_{-\nu}(z)}{J_\nu(z)},$$

$$(2) \qquad \int^z \frac{dz}{z\,J_\nu(z)\,J_{-\nu}(z)} = -\frac{\pi}{2\sin\nu\pi}\log\frac{J_{-\nu}(z)}{J_\nu(z)},$$

and similarly, from § 3·63 (1),

$$(3) \qquad \int^z \frac{dz}{zJ_\nu^2(z)} = \frac{\pi}{2}\,\frac{Y_\nu(z)}{J_\nu(z)},$$

$$(4) \qquad \int^z \frac{dz}{zJ_\nu(z)\,Y_\nu(z)} = \frac{\pi}{2}\log\frac{Y_\nu(z)}{J_\nu(z)},$$

$$(5) \qquad \int^z \frac{dz}{zY_\nu^2(z)} = -\frac{\pi}{2}\,\frac{J_\nu(z)}{Y_\nu(z)}.$$

The reader should have no difficulty in evaluating the similar integrals which contain any two cylinder functions of the same order in the denominator. The formulae actually given are due to Lommel, *Math. Ann.* IV. (1871), pp. 103—116. The reader should compare (3) with the result due to Euler which was quoted in § 1·2.

Some more interesting results, also due to Lommel*, are obtained from generalisations of Bessel's equation.

It is at once verified by differentiation that, if y and η satisfy the equations

$$\frac{d^2y}{dz^2} + Py = 0, \qquad \frac{d^2\eta}{dz^2} + Q\eta = 0,$$

then

$$\int^z (P-Q)\,y\eta\,dz = y\frac{d\eta}{dz} - \eta\frac{dy}{dz}.$$

* *Math. Ann.* XIV. (1879), pp. 520—536.

Now apply this result to any two equations of the type of § 4·31 (17). If $\mathscr{C}_\mu$, $\overline{\mathscr{C}}_\nu$ denote any two cylinder functions of orders μ and ν respectively, we have

$$(6) \quad \sqrt{\left\{\frac{\phi(z)\psi(z)}{\phi'(z)\psi'(z)}\right\}}\cdot\left[\mathscr{C}_\mu\{\phi(z)\}\frac{d\overline{\mathscr{C}}_\nu\{\psi(z)\}}{dz} - \overline{\mathscr{C}}_\nu\{\psi(z)\}\frac{d\mathscr{C}_\mu\{\phi(z)\}}{dz}\right.$$
$$\left. - \frac{1}{2}\left\{\frac{\phi'(z)}{\phi(z)} - \frac{\phi''(z)}{\phi'(z)} - \frac{\psi'(z)}{\psi(z)} + \frac{\psi''(z)}{\psi'(z)}\right\}\mathscr{C}_\mu\{\phi(z)\}\overline{\mathscr{C}}_\nu\{\psi(z)\}\right]$$
$$= \int^z \left[\frac{\phi'''(z)}{2\phi'(z)} - \frac{3}{4}\left\{\frac{\phi''(z)}{\phi'(z)}\right\}^2 + \{\phi^2(z) - \mu^2 + \tfrac{1}{4}\}\left\{\frac{\phi'(z)}{\phi(z)}\right\}^2\right.$$
$$\left. - \frac{\psi'''(z)}{2\psi'(z)} + \frac{3}{4}\left\{\frac{\psi''(z)}{\psi'(z)}\right\}^2 - \{\psi^2(z) - \nu^2 + \tfrac{1}{4}\}\left\{\frac{\psi'(z)}{\psi(z)}\right\}^2\right]$$
$$\times \mathscr{C}_\mu\{\phi(z)\}\overline{\mathscr{C}}_\nu\{\psi(z)\}\cdot\sqrt{\left\{\frac{\phi(z)\psi(z)}{\phi(z)\psi'(z)}\right\}}\,dz,$$

where $\phi(z)$ and $\psi(z)$ are arbitrary functions of z.

This formula is too general to be of practical use. As a special case, take $\phi(z)$ and $\psi(z)$ to be multiples of z, say kz and lz. It is then found that

$$(7) \quad \int^z \left\{(k^2 - l^2)z - \frac{\mu^2 - \nu^2}{z}\right\}\mathscr{C}_\mu(kz)\overline{\mathscr{C}}_\nu(lz)\,dz$$
$$= z\left\{\mathscr{C}_\mu(kz)\frac{d\overline{\mathscr{C}}_\nu(lz)}{dz} - \overline{\mathscr{C}}_\nu(lz)\frac{d\mathscr{C}_\mu(kz)}{dz}\right\}$$
$$= z\{k\mathscr{C}_{\mu+1}(kz)\overline{\mathscr{C}}_\nu(lz) - l\mathscr{C}_\mu(kz)\overline{\mathscr{C}}_{\nu+1}(lz)\} - (\mu - \nu)\mathscr{C}_\mu(kz)\overline{\mathscr{C}}_\nu(lz).$$

The expression on the left simplifies still further in two special cases (i) $\mu = \nu$, (ii) $k = l$.

If we take $\mu = \nu$, it is found that

$$(8) \quad \int^z z\,\mathscr{C}_\mu(kz)\overline{\mathscr{C}}_\mu(lz)\,dz = \frac{z\{k\mathscr{C}_{\mu+1}(kz)\overline{\mathscr{C}}_\mu(lz) - l\mathscr{C}_\mu(kz)\overline{\mathscr{C}}_{\mu+1}(lz)\}}{k^2 - l^2}.$$

This formula may be verified by differentiating the expression on the right. It becomes nugatory when $k = l$, for the denominator is then zero, while the numerator is a constant.

If this constant is omitted, an application of l'Hospital's rule shews that, when $l \to k$,

$$(9) \quad \int^z z\mathscr{C}_\mu(kz)\overline{\mathscr{C}}_\mu(kz)\,dz = -\frac{z}{2k}\{kz\,\mathscr{C}_{\mu+1}(kz)\overline{\mathscr{C}}_\mu'(kz)$$
$$- kz\mathscr{C}_\mu(kz)\overline{\mathscr{C}}'_{\mu+1}(kz) - \mathscr{C}_\mu(kz)\overline{\mathscr{C}}_{\mu+1}(kz)\}.$$

The result of using recurrence formulae to remove the derivates on the right of (9) is

$$(10) \quad \int^z z\mathscr{C}_\mu(kz)\overline{\mathscr{C}}_\mu(kz)\,dz = \tfrac{1}{4}z^2\{2\mathscr{C}_\mu(kz)\overline{\mathscr{C}}_\mu(kz) - \mathscr{C}_{\mu-1}(kz)\overline{\mathscr{C}}_{\mu+1}(kz)$$
$$- \mathscr{C}_{\mu+1}(kz)\overline{\mathscr{C}}_{\mu-1}(kz)\}.$$

Special cases of these formulae are:

$$(11)\qquad \int^z z\,\mathscr{C}_\mu^2(kz)\,dz = \tfrac{1}{2}z^2\{\mathscr{C}_\mu^2(kz) - \mathscr{C}_{\mu-1}(kz)\,\mathscr{C}_{\mu+1}(kz)\}$$
$$= \tfrac{1}{2}z^2\left\{\left(1-\frac{\mu^2}{k^2z^2}\right)\mathscr{C}_\mu^2(kz) + \mathscr{C}_\mu'^2(kz)\right\},$$

$$(12)\qquad \int^z z\,\mathscr{C}_\mu(kz)\,\overline{\mathscr{C}}_{-\mu}(kz)\,dz = \tfrac{1}{4}z^2\{2\mathscr{C}_\mu(kz)\,\overline{\mathscr{C}}_{-\mu}(kz) + \mathscr{C}_{\mu-1}(kz)\,\overline{\mathscr{C}}_{-\mu-1}(kz)$$
$$+ \mathscr{C}_{\mu+1}(kz)\,\overline{\mathscr{C}}_{-\mu+1}(kz)\},$$

the latter equation being obtained by regarding $e^{-\mu\pi i}\overline{\mathscr{C}}_\mu(kz)$ as a cylinder function of order $-\mu$.

To obtain a different class of elementary integrals take $k=l$ in (7) and it is found that

$$(13)\qquad \int^z \mathscr{C}_\mu(kz)\,\overline{\mathscr{C}}_\nu(kz)\,\frac{dz}{z} = -\frac{kz\{\mathscr{C}_{\mu+1}(kz)\,\overline{\mathscr{C}}_\nu(kz) - \mathscr{C}_\mu(kz)\,\overline{\mathscr{C}}_{\nu+1}(kz)\}}{\mu^2-\nu^2}$$
$$+\frac{\mathscr{C}_\mu(kz)\,\overline{\mathscr{C}}_\nu(kz)}{\mu+\nu}.$$

The result of making $\nu \to \mu$ in this formula is

$$(14)\qquad \int^z \mathscr{C}_\mu(kz)\,\overline{\mathscr{C}}_\mu(kz)\,\frac{dz}{z} = \frac{kz}{2\mu}\left\{\mathscr{C}_{\mu+1}(kz)\,\frac{\partial\overline{\mathscr{C}}_\mu(kz)}{\partial\mu}\right.$$
$$\left.-\,\mathscr{C}_\mu(kz)\,\frac{\partial\overline{\mathscr{C}}_{\mu+1}(kz)}{\partial\mu}\right\} + \frac{\mathscr{C}_\mu(kz)\,\overline{\mathscr{C}}_\mu(kz)}{2\mu}.$$

The last equation is also readily obtainable by multiplying the equations

$$\nabla_\mu\mathscr{C}_\mu(z) = 0,\quad \nabla_\mu\frac{\partial\overline{\mathscr{C}}_\mu(z)}{\partial\mu} = 2\mu\overline{\mathscr{C}}_\mu(z)$$

by $\dfrac{1}{z}\dfrac{\partial\overline{\mathscr{C}}_\mu(z)}{\partial\mu}$, $\dfrac{1}{z}\mathscr{C}_\mu(z)$ respectively, subtracting and integrating, and then replacing z by kz.

As a special case we have

$$(15)\qquad \int^z J_\mu^2(kz)\,\frac{dz}{z} = \frac{kz}{2\mu}\{J_{\mu+1}(kz)\,\mathfrak{J}_\mu(kz) - J_\mu(kz)\,\mathfrak{J}_{\mu+1}(kz)\} + \frac{1}{2\mu}J_\mu^2(kz).$$

An alternative method of obtaining this result will be given immediately.

Results equivalent to (11) are as old as Fourier's treatise, *La Théorie Analytique de la Chaleur* (Paris, 1822), §§ 318—319, in the case of functions of order zero; but none of the other formulae of this section seem to have been discovered before the publication of Lommel's memoir.

Various special cases of the formulae have been worked out in detail by Marcolongo, *Napoli Rendiconti*, (2) III. (1889), pp. 91—99 and by Chessin, *Trans. Acad. Sci. of St Louis*, XII. (1902), pp. 99—108.

5·12. *Indefinite integrals containing two cylinder functions; Lommel's second method.*

An alternative method has been given by Lommel* for evaluating some of the integrals just discussed. By this method their values are obtained in a form more suitable for numerical computation.

The method consists in adding the two results

$$\frac{d}{dz}\{z^\rho \mathscr{C}_\mu(z)\overline{\mathscr{C}}_\nu(z)\} = -z^\rho\{\mathscr{C}_\mu(z)\overline{\mathscr{C}}_{\nu+1}(z) + \mathscr{C}_{\mu+1}(z)\overline{\mathscr{C}}_\nu(z)\} + (\rho+\mu+\nu)z^{\rho-1}\mathscr{C}_\mu(z)\overline{\mathscr{C}}_\nu(z),$$

$$\frac{d}{dz}\{z^\rho \mathscr{C}_{\mu+1}(z)\overline{\mathscr{C}}_{\nu+1}(z)\} = z^\rho\{\mathscr{C}_\mu(z)\overline{\mathscr{C}}_{\nu+1}(z) + \mathscr{C}_{\mu+1}(z)\overline{\mathscr{C}}_\nu(z)\} + (\rho-\mu-\nu-2)z^{\rho-1}\mathscr{C}_{\mu+1}(z)\overline{\mathscr{C}}_{\nu+1}(z),$$

so that

$$(\rho+\mu+\nu)\int^z z^{\rho-1}\mathscr{C}_\mu(z)\overline{\mathscr{C}}_\nu(z)\,dz + (\rho-\mu-\nu-2)\int^z z^{\rho-1}\mathscr{C}_{\mu+1}(z)\overline{\mathscr{C}}_{\nu+1}(z)\,dz = z^\rho\{\mathscr{C}_\mu(z)\overline{\mathscr{C}}_\nu(z) + \mathscr{C}_{\mu+1}(z)\overline{\mathscr{C}}_{\nu+1}(z)\},$$

and then giving special values to ρ.

Thus we have

(1) $$\int^z z^{-\mu-\nu-1}\mathscr{C}_{\mu+1}(z)\overline{\mathscr{C}}_{\nu+1}(z)\,dz = -\frac{z^{-\mu-\nu}}{2(\mu+\nu+1)}\{\mathscr{C}_\mu(z)\overline{\mathscr{C}}_\nu(z) + \mathscr{C}_{\mu+1}(z)\overline{\mathscr{C}}_{\nu+1}(z)\},$$

(2) $$\int^z z^{\mu+\nu+1}\mathscr{C}_\mu(z)\overline{\mathscr{C}}_\nu(z)\,dz = \frac{z^{\mu+\nu+2}}{2(\mu+\nu+1)}\{\mathscr{C}_\mu(z)\overline{\mathscr{C}}_\nu(z) + \mathscr{C}_{\mu+1}(z)\overline{\mathscr{C}}_{\nu+1}(z)\}.$$

As special cases of these

(3) $$\int^z z^{-2\mu-1}\mathscr{C}^2_{\mu+1}(z)\,dz = -\frac{z^{-2\mu}}{4\mu+2}\{\mathscr{C}_\mu^2(z) + \mathscr{C}^2_{\mu+1}(z)\},$$

(4) $$\int^z z^{2\mu+1}\mathscr{C}_\mu^2(z)\,dz = \frac{z^{2\mu+2}}{4\mu+2}\{\mathscr{C}_\mu^2(z) + \mathscr{C}^2_{\mu+1}(z)\}.$$

Again, if ρ be made zero, it is found that

$$(\mu+\nu)\int^z \mathscr{C}_\mu(z)\overline{\mathscr{C}}_\nu(z)\frac{dz}{z} - (\mu+\nu+2)\int^z \mathscr{C}_{\mu+1}(z)\overline{\mathscr{C}}_{\nu+1}(z)\frac{dz}{z} = \mathscr{C}_\mu(z)\overline{\mathscr{C}}_\nu(z) + \mathscr{C}_{\mu+1}(z)\overline{\mathscr{C}}_{\nu+1}(z),$$

so that, by summing formulae of this type, we get

(5) $$(\mu+\nu)\int^z \mathscr{C}_\mu(z)\overline{\mathscr{C}}_\nu(z)\frac{dz}{z} - (\mu+\nu+2n)\int^z \mathscr{C}_{\mu+n}(z)\overline{\mathscr{C}}_{\nu+n}(z)\frac{dz}{z} = \mathscr{C}_\mu(z)\overline{\mathscr{C}}_\nu(z) + 2\sum_{m=1}^{n-1}\mathscr{C}_{\mu+m}(z)\overline{\mathscr{C}}_{\nu+m}(z) + \mathscr{C}_{\mu+n}(z)\overline{\mathscr{C}}_{\nu+n}(z).$$

* *Math. Ann.* XIV. (1879), pp. 530—536.

In particular, if $\mu=\nu=0$,

$$(6)\qquad \int^z \mathscr{C}_n(z)\,\overline{\mathscr{C}}_n(z)\,\frac{dz}{z}$$
$$= -\frac{1}{2n}\left[\mathscr{C}_0(z)\,\overline{\mathscr{C}}_0(z) + 2\sum_{m=1}^{n-1}\mathscr{C}_m(z)\,\overline{\mathscr{C}}_m(z) + \mathscr{C}_n(z)\,\overline{\mathscr{C}}_n(z)\right],$$

where $n=1, 2, 3, \ldots$. But there seems to be no simple formula for

$$\int \mathscr{C}_0(z)\,\overline{\mathscr{C}}_0(z)\,\frac{dz}{z}.$$

For a special case of (1) see Rayleigh, *Phil. Mag.* (5) XI. (1881), p. 217. [*Scientific Papers*, I. (1899), p. 516.]

5·13. *Sonine's integrals containing two cylinder functions.*

The analysis of § 5·1 has been extended by Sonine, *Math. Ann.* XVI. (1880), pp. 30—33, to the discussion of conditions that

$$\int f(z)\,\mathscr{C}_\mu\{\phi(z)\}\,\overline{\mathscr{C}}_\nu\{\psi(z)\}\,dz$$

may be expressible in the form

$$A(z)\,\mathscr{C}_\mu\{\phi(z)\}\,\overline{\mathscr{C}}_\nu\{\psi(z)\} + B(z)\,\mathscr{C}_{\mu+1}\{\phi(z)\}\,\overline{\mathscr{C}}_\nu\{\psi(z)\}$$
$$+ C(z)\,\mathscr{C}_\mu\{\phi(z)\}\,\overline{\mathscr{C}}_{\nu+1}\{\psi(z)\} + D(z)\,\mathscr{C}_{\mu+1}\{\phi(z)\}\,\overline{\mathscr{C}}_{\nu+1}\{\psi(z)\},$$

but the results are too complicated and not sufficiently important to justify their insertion here.

5·14. *Schafheitlin's reduction formula.*

A reduction formula for

$$\int^z z^\mu\,\mathscr{C}_\nu^2(z)\,dz,$$

which is a natural extension of the formula § 5·1 (4), has been discovered by Schafheitlin* and applied by him to discuss the rate of change of the zeros of $\mathscr{C}_\nu(z)$ as ν varies (§ 15·6).

To obtain the formula we observe that

$$\int^z z^\mu(z^2-\nu^2)\,\mathscr{C}_\nu^2(z)\,dz$$
$$= -\int^z z^\mu\,\mathscr{C}_\nu(z)\left\{z^2\frac{d^2}{dz^2} + z\frac{d}{dz}\right\}\mathscr{C}_\nu(z)\,dz$$
$$= \left[-z^{\mu+2}\,\mathscr{C}_\nu(z)\,\mathscr{C}_\nu'(z)\right] + \int^z \left\{z^{\mu+2}\,\mathscr{C}_\nu'^2(z) + (\mu+1)\,z^{\mu+1}\,\mathscr{C}_\nu(z)\,\mathscr{C}_\nu'(z)\right\}dz.$$

Now, by a partial integration,

$$(\mu+3)\int^z z^{\mu+2}\,\mathscr{C}_\nu'^2(z)\,dz = \left[z^{\mu+3}\,\mathscr{C}_\nu'^2(z)\right]$$
$$+ 2\int^z z^{\mu+1}\,\mathscr{C}_\nu'(z)\left\{z\,\mathscr{C}_\nu'(z) + (z^2-\nu^2)\,\mathscr{C}_\nu(z)\right\}dz,$$

* *Berliner Sitzungsberichte*, V. (1906), p. 88.

and so

$$(\mu+1)\int^z z^{\mu+2}\mathscr{C}_\nu'^2(z)\,dz = [z^{\mu+3}\mathscr{C}_\nu'^2(z)] + 2\int^z z^{\mu+1}(z^2-\nu^2)\mathscr{C}_\nu(z)\mathscr{C}_\nu'(z)$$

Hence, on substitution,

$$(\mu+1)\int^z z^\mu(z^2-\nu^2)\mathscr{C}_\nu^2(z)\,dz$$
$$= [z^{\mu+3}\mathscr{C}_\nu'^2(z) - (\mu+1)z^{\mu+2}\mathscr{C}_\nu(z)\mathscr{C}_\nu'(z)]$$
$$+ 2\int^z z^{\mu+3}\mathscr{C}_\nu(z)\mathscr{C}_\nu'(z)\,dz + \{(\mu+1)^2-2\nu^2\}\int^z z^{\mu+1}\mathscr{C}_\nu(z)\mathscr{C}_\nu'(z$$
$$= [z^{\mu+3}\mathscr{C}_\nu'^2(z) - (\mu+1)z^{\mu+2}\mathscr{C}_\nu(z)\mathscr{C}_\nu'(z) + z^{\mu+3}\mathscr{C}_\nu^2(z)$$
$$+ \{\tfrac{1}{2}(\mu+1)^2-\nu^2\}z^{\mu+1}\mathscr{C}_\nu^2(z)$$
$$-(\mu+3)\int^z z^{\mu+2}\mathscr{C}_\nu^2(z) - (\mu+1)\{\tfrac{1}{2}(\mu+1)^2-\nu^2\}\int^z z^\mu\mathscr{C}_\nu^2(z)\,dz$$

By rearranging we find that

$$(\mu+2)\int^z z^{\mu+2}\mathscr{C}_\nu^2(z)\,dz = (\mu+1)\{\nu^2-\tfrac{1}{4}(\mu+1)^2\}\int^z z^\mu\mathscr{C}_\nu^2(z)\,dz$$
$$+\tfrac{1}{2}[z^{\mu+1}\{z\mathscr{C}_\nu'(z)-\tfrac{1}{2}(\mu+1)\mathscr{C}_\nu(z)\}^2 + z^{\mu+1}\{z^2-\nu^2+\tfrac{1}{4}(\mu+1)^2\}\mathscr{C}_\nu^2($$

and this is the reduction formula in question.

5·2. *Expansions in series of Bessel functions.*

We shall now discuss some of the simplest expansions of the type tained for $(\frac{1}{2}z)^m$ in § 2·7. The general theory of such expansions is reser for Chapter XVI.

The result of § 2·7 at once suggests the possibility of the expansion

$$(1)\qquad (\tfrac{1}{2}z)^\mu = \sum_{n=0}^\infty \frac{(\mu+2n)\,\Gamma(\mu+n)}{n!}J_{\mu+2n}(z),$$

which is due to Gegenbauer* and is valid when μ is not a negative intege

To establish the expansion, observe that

$$\sum_{n=0}^\infty \frac{(\mu+2n)\,\Gamma(\mu+n)}{n!}(\tfrac{1}{2}z)^{-\mu}J_{\mu+2n}(z)$$

is a series of analytic functions which converges uniformly throughout a bounded domain of the z-plane (cf. § 3·13); and since

$$\frac{d}{dz}\{(\tfrac{1}{2}z)^{-\mu}J_{\mu+2n}(z)\} = \frac{(\frac{1}{2}z)^{-\mu}}{\mu+2n}\{nJ_{\mu+2n-1}(z)-(\mu+n)J_{\mu+2n+1}(z)\},$$

it is evident that the derivate of the series now under consideration is

$$(\tfrac{1}{2}z)^{-\mu}\left[\sum_{n=0}^\infty \frac{\Gamma(\mu+n)}{n!}nJ_{\mu+2n-1}(z) - \sum_{n=0}^\infty\frac{\Gamma(\mu+n+1)}{n!}J_{\mu+2n+1}(z)\right]=0,$$

* *Wiener Sitzungsberichte*, LXXIV. (2), (1877), pp. 124—130.

and so the sum is a constant. When we make $z \to 0$, we see that the constant is unity; that is to say

$$\sum_{n=0}^{\infty} \frac{(\mu+2n)\,\Gamma(\mu+n)}{n!} (\tfrac{1}{2}z)^{-\mu} J_{\mu+2n}(z) = 1,$$

and the required result is established.

The reader will find that it is not difficult to verify that when the expansion on the right in (1) is rearranged in powers of z, all the coefficients except that of z^{μ} vanish; but this is a crude method of proving the result.

5·21. *The expansion of a Bessel function as a series of Bessel functions.*

The expansion

(1) $$(\tfrac{1}{2}z)^{-\nu} J_\nu(z) = (\tfrac{1}{2}z)^{-\mu}\,\Gamma(\nu+1-\mu) \times \sum_{n=0}^{\infty} \frac{(\mu+2n)\,\Gamma(\mu+n)}{n!\,\Gamma(\nu+1-\mu-n)\,\Gamma(\nu+n+1)} J_{\mu+2n}(z)$$

is a generalisation of a formula proved by Sonine* when the difference $\nu-\mu$ is a positive integer; it is valid when μ, ν and $\nu-\mu$ are not negative integers.

It is most easily obtained by expanding each power of z in the expansion of $(\tfrac{1}{2}z)^{\mu-\nu} J_\nu(z)$ with the aid of § 5·2, and rearranging the resulting double series, which is easily seen to be absolutely convergent.

It is thus found that

$$\begin{aligned}
(\tfrac{1}{2}z)^{\mu-\nu} J_\nu(z) &= \sum_{m=0}^{\infty} \frac{(-)^m (\tfrac{1}{2}z)^{\mu+2m}}{m!\,\Gamma(\nu+m+1)} \\
&= \sum_{m=0}^{\infty} \frac{(-)^m}{m!\,\Gamma(\nu+m+1)} \sum_{p=0}^{\infty} \frac{(\mu+2m+2p)\,\Gamma(\mu+2m+p)}{p!} J_{\mu+2m+2p}(z) \\
&= \sum_{m=0}^{\infty} \frac{(-)^m}{m!\,\Gamma(\nu+m+1)} \sum_{n=m}^{\infty} \frac{(\mu+2n)\,\Gamma(\mu+m+n)}{(n-m)!} J_{\mu+2n}(z) \\
&= \sum_{n=0}^{\infty} \left\{ \sum_{m=0}^{n} \frac{(-)^m\,\Gamma(\mu+m+n)}{m!\,(n-m)!\,\Gamma(\nu+m+1)} \right\} (\mu+2n) J_{\mu+2n}(z) \\
&= \sum_{n=0} \frac{\Gamma(\mu+n)\,\Gamma(\nu+1-\mu)}{n!\,\Gamma(\nu+1-\mu-n)\,\Gamma(\nu+n+1)} \cdot (\mu+2n) J_{\mu+2n}(z),
\end{aligned}$$

by Vandermonde's theorem; and the result is established.

If we put $\nu = \mu + m$, we find that

(2) $$(\tfrac{1}{2}z)^{-m} J_{\mu+m}(z) = \sum_{n=0}^{m} {}_mC_n \frac{(\mu+2n)\,\Gamma(\mu+n)}{\Gamma(\mu+m+n+1)} J_{\mu+2n}(z),$$

which is Sonine's form of the result, and is readily proved by induction.

* *Math. Ann.* XVI. (1880), p. 22.

By a slight modification of the analysis, we may prove that, if k is any constant,

$$(3)\quad (\tfrac{1}{2}kz)^{\mu-\nu} J_\nu(kz) = k^\mu \sum_{n=0}^{\infty} \frac{\Gamma(\mu+n)}{n!\,\Gamma(\nu+1)} \times {}_2F_1(\mu+n, -n; \nu+1; k^2)\,(\mu+2n)\,J_{\mu+2n}(z).$$

This formula will be required in establishing some more general expansions in § 11·6.

5·22. *Lommel's expansions of* $(z+h)^{\pm\frac{1}{2}\nu} J_\nu\{\sqrt{(z+h)}\}$.

It is evident that $(z+h)^{-\frac{1}{2}\nu} J_\nu\{\sqrt{(z+h)}\}$, *qua* function of $z+h$, is analytic for all values of the variable, and consequently, by Taylor's theorem combined with § 3·21 (6), we have

$$(1)\quad (z+h)^{-\frac{1}{2}\nu} J_\nu\{\sqrt{(z+h)}\} = \sum_{m=0}^{\infty} \frac{h^m}{m!} \frac{d^m}{dz^m} \{z^{-\frac{1}{2}\nu} J_\nu(\sqrt{z})\}$$
$$= \sum_{m=0}^{\infty} \frac{(-\frac{1}{2}h)^m}{m!} z^{-\frac{1}{2}(\nu+m)} J_{\nu+m}(\sqrt{z}).$$

Again, $(z+h)^{\frac{1}{2}\nu} J_\nu\{\sqrt{(z+h)}\}$ is analytic except when $z+h=0$; and so, provided that $|h|<|z|$, we have

$$(2)\quad (z+h)^{\frac{1}{2}\nu} J_\nu\{\sqrt{(z+h)}\} = \sum_{m=0}^{\infty} \frac{h^m}{m!} \frac{d^m}{dz^m} \{z^{\frac{1}{2}\nu} J_\nu(\sqrt{z})\}$$
$$= \sum_{m=0}^{\infty} \frac{(\frac{1}{2}h)^m}{m!} z^{\frac{1}{2}(\nu-m)} J_{\nu-m}(\sqrt{z}).$$

These formulae are due to Lommel*. If we take $\nu=-\frac{1}{2}$ in (1) and $\nu=\frac{1}{2}$ in (2) we deduce from § 3·4, after making some slight changes in notation,

$$(3)\quad \left(\frac{2}{\pi z}\right)^{\frac{1}{2}} \cos\sqrt{(z^2-2zt)} = \sum_{m=0}^{\infty} \frac{t^m}{m!} J_{m-\frac{1}{2}}(z),$$

$$(4)\quad \left(\frac{2}{\pi z}\right)^{\frac{1}{2}} \sin\sqrt{(z^2+2zt)} = \sum_{m=0}^{\infty} \frac{t^m}{m!} J_{\frac{1}{2}-m}(z),$$

equation (4) being true only when $|t|<\frac{1}{2}|z|$. These formulae are due to Glaisher†, who regarded the left-hand sides as the generating functions associated with the functions whose order is half of an odd integer, just as $\exp\{\frac{1}{2}z(t-1/t)\}$ is the generating function associated with the Bessel coefficients.

Proofs of (3) and (4) by direct expansion of the right-hand sides have been given by Glaisher; the algebra involved in investigations of this nature is somewhat formidable.

* *Studien über die Bessel'schen Functionen* (Leipzig, 1868), pp. 11—16. Formula (1) was given by Bessel, *Berliner Abh.* 1824 [1826], p. 35, for the Bessel coefficients.

† *Quarterly Journal*, XII. (1873), p. 136; *British Association Report*, 1878, pp. 469—470. *Phil. Trans. of the Royal Soc.* CLXXII. (1881), pp. 774—781, 813.

We shall now enumerate various modifications of (1) and (2).

In (1) replace z and h by z^2 and kz^2, and then

(5) $$J_\nu\{z\sqrt{(1+k)}\} = (1+k)^{\frac{1}{2}\nu} \sum_{m=0}^{\infty} \frac{(-)^m (\frac{1}{2}kz)^m}{m!} J_{\nu+m}(z),$$

and, in particular,

(6) $$J_\nu(z\sqrt{2}) = 2^{\frac{1}{2}\nu} \sum_{m=0}^{\infty} \frac{(-)^m (\frac{1}{2}z)^m}{m!} J_{\nu+m}(z).$$

If we divide (5) by $(1+k)^{\frac{1}{2}\nu}$ and then make $k \to -1$, we find that

(7) $$\frac{(\frac{1}{2}z)^\nu}{\Gamma(\nu+1)} = \sum_{m=0}^{\infty} \frac{(\frac{1}{2}z)^m}{m!} J_{\nu+m}(z).$$

In like manner, from (2),

(8) $$J_\nu\{z\sqrt{(1+k)}\} = (1+k)^{-\frac{1}{2}\nu} \sum_{m=0}^{\infty} \frac{(\frac{1}{2}kz)^m}{m!} J_{\nu-m}(z),$$

provided that $|k| < 1$.

If we make $k \to -1+0$, we find, by Abel's theorem,

$$\lim_{k\to -1+0} [(1+k)^{\frac{1}{2}\nu} J_\nu\{z\sqrt{(1+k)}\}] = \sum_{m=0}^{\infty} \frac{(-)^m (\frac{1}{2}z)^m}{m!} J_{\nu-m}(z),$$

provided that the series on the right is convergent. The convergence is obvious when ν is an integer. If ν is not an integer, then, for large values of m,

$$\frac{(-)^m (\frac{1}{2}z)^m}{m!} J_{\nu-m}(z) = \frac{(-)^m (\frac{1}{2}z)^\nu \Gamma(m-\nu)}{\pi . m!} \sin(m-\nu)\pi\{1+O(1/m)\}$$

$$= -\frac{(\frac{1}{2}z)^\nu \sin\nu\pi}{\pi m^{\nu+1}}\{1+O(1/m)\}.$$

Hence the condition for convergence is $R(\nu) > 0$, and if the condition is satisfied, the convergence is absolute. Consequently, when $R(\nu) > 0$, and also when ν is *any* integer,

(9) $$\sum_{m=0}^{\infty} \frac{(-)^m (\frac{1}{2}z)^m}{m!} J_{\nu-m}(z) = 0.$$

In like manner, if $R(\nu) > -1$, and also when ν is *any* integer, we have

(10) $$J_\nu(z\sqrt{2}) = 2^{-\frac{1}{2}\nu} \sum_{m=0}^{\infty} \frac{(\frac{1}{2}z)^m}{m!} J_{\nu-m}(z).$$

It should be observed that functions of the second kind may be substituted for functions of the first kind in (1), (2), (5) and (8) provided that $|h| < |z|$ and $|k| < 1$; so that

(11) $$(z+h)^{-\frac{1}{2}\nu} Y_\nu\{\sqrt{(z+h)}\} = \sum_{m=0}^{\infty} \frac{(-\frac{1}{2}h)^m}{m!} z^{-\frac{1}{2}(\nu+m)} Y_{\nu+m}(\sqrt{z}),$$

(12) $$(z+h)^{\frac{1}{2}\nu} Y_\nu\{\sqrt{(z+h)}\} = \sum_{m=0}^{\infty} \frac{(\frac{1}{2}h)^m}{m!} z^{\frac{1}{2}(\nu-m)} Y_{\nu-m}(\sqrt{z}),$$

(13) $$Y_\nu\{z\sqrt{(1+k)}\} = (1+k)^{\frac{1}{2}\nu} \sum_{m=0}^{\infty} \frac{(-)^m (\frac{1}{2}kz)^m}{m!} Y_{\nu+m}(z),$$

(14) $$Y_\nu\{z\sqrt{(1+k)}\} = (1+k)^{-\frac{1}{2}\nu} \sum_{m=0}^{\infty} \frac{(\frac{1}{2}kz)^m}{m!} Y_{\nu-m}(z).$$

These may be proved by expressing the functions of the second kind as a linear combination of functions of the first kind; by proceeding to the limit when ν tends to an integral value, we see that they hold for functions of integral order.

By combining (11)—(14) with the corresponding results for functions of the first kind, we see that we may substitute the symbol $\mathscr{C}$ for the symbol Y throughout.

These last formulae were noted by Lommel, *Studien*, p. 87. Numerous generalisations of them will be given in Chapter XI. It has been observed by Airey, *Phil. Mag.* (6) XXXVI. (1918), pp. 234—242, that they are of some use in calculations connected with zeros of Bessel functions.

When we combine (5) and (13), and then replace $\sqrt{(1+k)}$ by λ, we find that, when $|\lambda^2-1|<1$,

$$\mathscr{C}_\nu(\lambda z)=\lambda^\nu \sum_{m=0}^{\infty} \frac{(-)^m(\lambda^2-1)^m(\tfrac{1}{2}z)^m}{m!}\mathscr{C}_{\nu+m}(z), \tag{15}$$

and, in particular, when λ is unrestricted,

$$J_\nu(\lambda z)=\lambda^\nu \sum_{m=0}^{\infty} \frac{(-)^m(\lambda^2-1)^m(\tfrac{1}{2}z)^m}{m!}J_{\nu+m}(z). \tag{16}$$

These two results are frequently described* as *multiplication theorems* for Bessel functions.

It may be observed that the result of treating (14) in the same way as (8) is that (when ν is taken equal to an integer n)

$$-(n-1)!\,(2/z)^n=\pi \sum_{m=0}^{\infty} \frac{(-\tfrac{1}{2}z)^m}{m!} Y_{n-m}(z). \tag{17}$$

An alternative proof of the multiplication formula has been given by Böhmer, *Berliner Sitzungsberichte*, XIII. (1913), p. 35, with the aid of the methods of complex integration; see also Nielsen, *Math. Ann.* LIX. (1904), p. 108, and (for numerous extensions of the formulae) Wagner, *Bern Mittheilungen*, 1895, pp. 115—119; 1896, pp. 53—60.

[NOTE. A special case of formula (1), namely that in which $\nu=1$, was discovered by Lommel seven years before the publication of his treatise; see *Archiv der Math.* XXXVII. (1861), p. 356.

His method consisted in taking the integral

$$\frac{1}{2\pi}\iint \cos(\xi r\cos\theta+\eta r\sin\theta)\,d\xi\,d\eta$$

over the area of the circle $\xi^2+\eta^2=1$, and evaluating it by two different methods.

The result of integrating with respect to η is

$$\begin{aligned}\frac{1}{2\pi}\int_{-1}^{1}\Big[\sin(\xi r\cos\theta+\eta r\sin\theta)\Big]_{-\sqrt{(1-\xi^2)}}^{\sqrt{(1-\xi^2)}}\frac{d\xi}{r\sin\theta}&\\
=\frac{1}{\pi}\int_{-1}^{1}\cos(\xi r\cos\theta)\sin\{\sqrt{(1-\xi^2)}\,.\,r\sin\theta\}\frac{d\xi}{r\sin\theta}&\\
=\frac{1}{\pi}\sum_{m=0}^{\infty}\frac{(-)^m(r\sin\theta)^{2m}}{(2m+1)!}\int_{-1}^{1}\cos(\xi r\cos\theta)\,.\,(1-\xi^2)^{m+\frac{1}{2}}\,d\xi&\\
=\sum_{m=0}^{\infty}\frac{(-)^m(\tfrac{1}{2}r\sin\theta)^{2m}}{m!}\frac{J_{m+1}(r\cos\theta)}{(r\cos\theta)^{m+1}},&\end{aligned}$$

* See, e.g. Schafheitlin, *Die Theorie der Besselschen Funktionen* (Leipzig, 1908), p. 83.

and the result of changing to polar coordinates (ρ, ϕ) is

$$\frac{1}{2\pi}\int_{-\pi}^{\pi}\int_0^1 \cos\{r\rho\cos(\phi-\theta)\}\,\rho\,d\rho\,d\phi = \frac{1}{2\pi}\int_{-\pi}^{\pi}\int_0^1 \cos(r\rho\cos\phi)\,\rho d\rho d\phi$$

$$=\frac{1}{2\pi}\iint \cos(\xi r)\,d\xi\,d\eta = \frac{1}{\pi}\int_{-1}^{1}(1-\xi^2)^{\frac{1}{2}}\cos(\xi r)\,d\xi = J_1(r)/r.$$

If we compare these equations we obtain (1) in the case $\nu=1$ with z and h replaced by $r^2\cos^2\theta$ and $r^2\sin^2\theta$.]

5·23. *The expansion of a Bessel function as a series of Bessel functions.*

From formula § 5·22 (7), Lommel has deduced an interesting series of Bessel functions which represents any given Bessel function.

If μ and ν are unequal, and μ is not a negative integer, we have

$$J_\nu(z) = \sum_{n=0}^{\infty}\frac{(-)^n(\tfrac{1}{2}z)^{\nu-\mu+n}}{n!\,\Gamma(\nu+n+1)}\cdot(\tfrac{1}{2}z)^{\mu+n}$$

$$=\sum_{n=0}^{\infty}\frac{(-)^n(\tfrac{1}{2}z)^{\nu-\mu+n}}{n!\,\Gamma(\nu+n+1)}\cdot\Gamma(\mu+n+1)\sum_{p=0}^{\infty}\frac{(\tfrac{1}{2}z)^p}{p!}J_{\mu+n+p}(z).$$

The repeated series is absolutely convergent; consequently we may rearrange it by replacing p by $m-n$, and then we have

$$J_\nu(z) = \sum_{m=0}^{\infty}(\tfrac{1}{2}z)^{\nu-\mu+m}J_{\mu+m}(z)\left\{\sum_{n=0}^{m}\frac{(-)^n\,\Gamma(\mu+n+1)}{n!\,(m-n)!\,\Gamma(\nu+n+1)}\right\},$$

and hence, by Vandermonde's theorem,

$$(1)\qquad J_\nu(z) = \frac{\Gamma(\mu+1)}{\Gamma(\nu-\mu)}\sum_{m=0}^{\infty}\frac{\Gamma(\nu-\mu+m)}{\Gamma(\nu+m+1)}\cdot\frac{(\tfrac{1}{2}z)^{\nu-\mu+m}}{m!}J_{\mu+m}(z).$$

This formula was given by Lommel, *Studien über die Bessel'schen Functionen* (Leipzig, 1868), pp. 22—23, in the special case $\mu=0$; by differentiating with respect to ν and then putting $\nu=0$, it is found that

$$(2)\quad \tfrac{1}{2}\pi Y_0(z) = J_0(z)\log(\tfrac{1}{2}z) - \frac{\Gamma(\mu+1)}{\Gamma(-\mu)}\sum_{m=0}^{\infty}\frac{\Gamma(m-\mu)}{(m!)^2}(\tfrac{1}{2}z)^{m-\mu}J_{\mu+m}(z) \times\{\psi(m+1)+\psi(-\mu)-\psi(m-\mu)\},$$

and, when $\mu=0$, we have Lommel's formula

$$(3)\qquad \tfrac{1}{2}\pi Y_0(z) = J_0(z)\,.\,\{\gamma+\log(\tfrac{1}{2}z)\} + \sum_{m=1}^{\infty}\frac{(\tfrac{1}{2}z)^m J_m(z)}{m\,.\,m!}.$$

This should be compared with Neumann's expansion given in § 3·57.

5·3. *An addition formula for Bessel functions.*

An extension of the formula of § 2·4 to Bessel functions of any order is

$$(1)\qquad J_\nu(z+t) = \sum_{m=-\infty}^{\infty}J_{\nu-m}(t)\,J_m(z),$$

where $|z|<|t|$, ν being unrestricted. This formula is due to Schläfli*; and the similar but more general formula

$$(2)\qquad \mathscr{C}_\nu(z+t) = \sum_{m=-\infty}^{\infty}\mathscr{C}_{\nu-m}(t)\,J_m(z)$$

is due to Sonine†.

* *Math. Ann.* III. (1871), pp. 135—137.

† *Ibid.* XVI. (1880), pp. 7—8.

It will first be shewn that the series on the right of (1) is a uniformly convergent series of analytic functions of both z and t when

$$|z| \leqslant r, \quad R \leqslant |t| \leqslant \Delta,$$

where r, R, Δ are unequal positive numbers in ascending order of magnitude.

When m is large and positive, $J_{\nu-m}(t) J_m(z)$ is comparable with

$$\sin \nu\pi \,.\, (\tfrac{1}{2}R)^\nu \,.\, (r/R)^m \frac{\Gamma(m-\nu)}{m!}$$

and the convergence of the series is comparable with that of the binomial series for $(1 - r/R)^\nu$. When m is large and negative $(= -n)$, the general term is comparable with

$$\frac{(-)^n (\tfrac{1}{2}\Delta)^\nu (\tfrac{1}{4}\Delta r)^n}{\Gamma(\nu+n+1) \,.\, n!}$$

and the uniformity of the convergence follows for both sets of values of m by the test of Weierstrass.

Term-by-term differentiation is consequently permissible*, so that

$$\begin{aligned}\left(\frac{\partial}{\partial t} - \frac{\partial}{\partial z}\right) \sum_{m=-\infty}^{\infty} J_{\nu-m}(t) J_m(z) &= \sum_{m=-\infty}^{\infty} \{J'_{\nu-m}(t) J_m(z) - J_{\nu-m}(t) J'_m(z)\} \\ &= \frac{1}{2} \sum_{m=-\infty}^{\infty} \{J_{\nu-m-1}(t) - J_{\nu-m+1}(t)\} J_m(z) \\ &\quad - \frac{1}{2} \sum_{m=-\infty}^{\infty} J_{\nu-m}(t) \{J_{m-1}(z) - J_{m+1}(z)\},\end{aligned}$$

and it is seen, on rearrangement, that all the terms on the right cancel, so that

$$\left(\frac{\partial}{\partial t} - \frac{\partial}{\partial z}\right) \sum_{m=-\infty}^{\infty} J_{\nu-m}(t) J_m(z) = 0.$$

Hence, when $|z| < |t|$, the series $\sum_{m=-\infty}^{\infty} J_{\nu-m}(t) J_m(z)$ is an analytic function of z and t *which is expressible as a function of* $z+t$ *only*, since its derivates with respect to z and t are identically equal. If this function be called $F(z+t)$, then

$$F(z+t) \equiv \sum_{m=-\infty}^{\infty} J_{\nu-m}(t) J_m(z).$$

If we put $z = 0$, we see that $F(t) \equiv J_\nu(t)$, and the truth of (1) becomes evident.

Again, if the signs of ν and m in (1) be changed, we have

$$J_{-\nu}(z+t) = \sum_{m=-\infty}^{\infty} (-)^m J_{-\nu+m}(t) J_m(z),$$

and when this result is combined with (1), we see that

(3) $$Y_\nu(z+t) = \sum_{m=-\infty}^{\infty} Y_{\nu-m}(t) J_m(z).$$

* Cf. *Modern Analysis*, § 5·3.

When this is combined with (1), equation (2) becomes evident.

The reader will readily prove by the same method that, when $|z| < |t|$,

(4) $$J_\nu(t-z) = \sum_{m=-\infty}^{\infty} J_{\nu+m}(t) J_m(z),$$

(5) $$\mathscr{C}_\nu(t-z) = \sum_{m=-\infty}^{\infty} \mathscr{C}_{\nu+m}(t) J_m(z),$$

(6) $$Y_\nu(t-z) = \sum_{m=-\infty}^{\infty} Y_{\nu+m}(t) J_m(z).$$

Of these results, (3) was given by Lommel, *Studien über die Bessel'schen Functionen* (Leipzig, 1868), when ν is an integer; while (4), (5) and (6) were given* explicitly by Graf, *Math. Ann.* XLIII. (1893), pp. 141—142. Various generalisations of these formulae will be given in Chapter XI.

5·4. *Products of Bessel functions.*

The ascending series for the product $J_\mu(z) J_\nu(z)$ has been given by various writers; the expansion is sometimes stated to be due to Schönholzer†, who published it in 1877, but it had, in fact, been previously published (in 1870) by Schläfli‡. More recently the product has been examined by Orr§, while Nicholson|| has given expansions (cf. § 5·42) for products of the forms

$$J_\mu(z) Y_n(z) \text{ and } Y_m(z) Y_n(z).$$

In the present section we shall construct the differential equation satisfied by the product of two Bessel functions, and solve it in series. We shall then (§ 5·41) obtain the expansion anew by direct multiplication of series.

Given two differential equations in their normal forms

$$\frac{d^2v}{dz^2} + Iv = 0, \qquad \frac{d^2w}{dz^2} + Jw = 0,$$

if y denotes the product vw, we have

$$\begin{aligned} y'' &= v''w + 2v'w' + vw'' \\ &= -(I+J)y + 2v'w', \end{aligned}$$

where primes indicate differentiations with respect to z.

* See also Epstein, *Die vier Rechnungsoperationen mit Bessel'schen Functionen* (Bern, 1894), [*Jahrbuch über die Fortschritte der Math.* 1893—1894, pp. 845—846].

† *Ueber die Auswerthung bestimmter Integrale mit Hülfe von Veränderungen des Integrationsweges* (Bern, 1877), p. 13. The authorities who attribute the expansion to Schönholzer include Graf and Gubler, *Einleitung in die Theorie der Bessel'schen Funktionen*, II. (Bern, 1900), pp. 85—87, and Nielsen, *Ann. Sci. de l'École norm. sup.* (3) XVIII. (1901), p. 50; *Handbuch der Theorie der Cylinderfunktionen* (Leipzig, 1904), p. 20. According to Nielsen, *Nouv. Ann. de Math.* (4) II. (1902), p. 396, Meissel obtained some series for products in the *Iserlohn Programm*, 1862.

‡ *Math. Ann.* III. (1871), pp. 141—142. A trivial defect in Schläfli's proof is that he uses a contour integral which (as he points out) converges only when $R(\mu+\nu+1)>0$.

§ *Proc. Camb. Phil. Soc.* X. (1900), pp. 93—100.

|| *Quarterly Journal*, XLIII. (1912), pp. 78—100.

It follows that $$\frac{d}{dz}\{y'' + (I+J)\,y\} = 2v''w' + 2v'w''$$
$$= -2Ivw' - 2Jv'w$$

and hence $$y''' + 2(I+J)y' + (I'+J')\,y = (I-J)(v'w - vw').$$

Hence, in the special case when $I = J$, y satisfies the equation

(1) $$y''' + 4Iy' + 2I'y = 0;$$

but, if $I \neq J$, it is easy to shew by differentiation that

(2) $$\frac{d}{dz}\left\{\frac{y''' + 2(I+J)\,y' + (I'+J')\,y}{I-J}\right\} = -(I-J)\,y.$$

This is the form of the differential equation used by Orr; in connexion with (1), s Appell, *Comptes Rendus*, XCI. (1880), pp. 211—214.

To apply these results to Bessel's equation, the equation has to be reduc to a normal form; both Orr and Nicholson effect the reduction by takir $z^{\frac{1}{2}}\mathscr{C}_\nu(z)$ as a new dependent variable, but, for purposes of solution in series, is simpler to take a new independent variable by writing

$$z = e^\theta, \quad z\frac{d}{dz} = \frac{d}{d\theta} = \vartheta,$$

so that $$\frac{d^2 J_\nu(z)}{d\theta^2} + (e^{2\theta} - \nu^2)\,J_\nu(z) = 0.$$

Hence the equation satisfied by $J_\mu(z)\,J_\nu(z)$, when $\mu^2 \neq \nu^2$, is

$$\frac{d}{d\theta}\left\{\frac{d^3y}{d\theta^3} + 2(2e^{2\theta} - \mu^2 - \nu^2)\frac{dy}{d\theta} + 4e^{2\theta}y\right\} + (\mu^2 - \nu^2)^2\,y = 0,$$

that is to say

(3) $$[\vartheta^4 - 2(\mu^2+\nu^2)\,\vartheta^2 + (\mu^2-\nu^2)^2]\,y + 4e^{2\theta}(\vartheta+1)(\vartheta+2)\,y = 0,$$

and the equation satisfied by $J_\nu(z)\,J_{\pm\nu}(z)$ is

(4) $$\vartheta(\vartheta^2 - 4\nu^2)\,y + 4e^{2\theta}(\vartheta+1)\,y = 0.$$

Solutions in series of (3) are

$$z^\alpha \sum_{m=0}^{\infty} (-)^m c_m z^{2m},$$

where $\alpha = \pm\mu\pm\nu$ and

$$c_m = \frac{4(\alpha+2m-1)(\alpha+2m)\,c_{m-1}}{(\alpha+\mu+\nu+2m)(\alpha+\mu-\nu+2m)(\alpha-\mu+\nu+2m)(\alpha-\mu-\nu+2m)}.$$

If we take $\alpha = \mu+\nu$ and

$$c_0 = \frac{1}{2^{\mu+\nu}\,\Gamma(\mu+1)\,\Gamma(\nu+1)},$$

we obtain the series

$$\sum_{m=0}^{\infty} \frac{(-)^m (\frac{1}{2}z)^{\mu+\nu+2m}\,\Gamma(\mu+\nu+2m+1)}{m!\,\Gamma(\mu+\nu+m+1)\,\Gamma(\mu+m+1)\,\Gamma(\nu+m+1)},$$

and the other series which are solutions of (3) are obtained by changing t signs of either μ or ν or both μ and ν.

By considering the powers of z which occur in the product $J_\mu(z) J_\nu(z)$ it is easy to infer that, if 2μ, 2ν and $2(\mu+\nu)$ are not negative integers and if $\mu^2 \neq \nu^2$, then

$$(5)\quad J_\mu(z) J_\nu(z) = \sum_{m=0}^{\infty} \frac{(-)^m (\tfrac{1}{2}z)^{\mu+\nu+2m}\, \Gamma(\mu+\nu+2m+1)}{m!\, \Gamma(\mu+\nu+m+1)\, \Gamma(\mu+m+1)\, \Gamma(\nu+m+1)}.$$

In like manner, by solving (4) in series, we find that, when 2ν is not a negative integer, then

$$(6)\qquad J_\nu^2(z) = \sum_{m=0}^{\infty} \frac{(-)^m (\tfrac{1}{2}z)^{2\nu+2m}\, \Gamma(2\nu+2m+1)}{m!\, \Gamma(2\nu+m+1)\, \{\Gamma(\nu+m+1)\}^2},$$

and, when ν is not a negative integer, then

$$(7)\qquad J_\nu(z) J_{-\nu}(z) = \sum_{m=0}^{\infty} \frac{(-)^m (\tfrac{1}{2}z)^{2m}\, (2m)!}{(m!)^2\, \Gamma(\nu+m+1)\, \Gamma(-\nu+m+1)}.$$

By reasoning which resembles that given in § 4·42, it may be shewn that (6) holds when ν is half of an odd negative integer, provided that the quotient $\Gamma(2\nu+2m+1)/\Gamma(2\nu+m+1)$ is replaced by the product $(2\nu+m+1)_m$.

5·41. *Products of series representing Bessel functions.*

It is easy to obtain the results of § 5·4 by direct multiplication of series. This method has the advantage that special investigations, for the cases in which $\mu^2 = \nu^2$ and those in which $\mu+\nu$ is a negative integer, are superfluous.

The coefficient of $(-)^m (\tfrac{1}{2}z)^{\mu+\nu+2m}$ in the product of the two absolutely convergent series

$$\sum_{m=0}^{\infty} \frac{(-)^m (\tfrac{1}{2}z)^{\mu+2m}}{m!\, \Gamma(\mu+m+1)} \times \sum_{n=0}^{\infty} \frac{(-)^n (\tfrac{1}{2}z)^{\nu+2n}}{n!\, \Gamma(\nu+n+1)}$$

is
$$\sum_{n=0}^{m} \frac{1}{n!\, \Gamma(\nu+n+1)\,.\,(m-n)!\, \Gamma(\mu+m-n+1)}$$
$$= \frac{(-)^m}{m!\, \Gamma(\mu+m+1)\, \Gamma(\nu+m+1)} \sum_{n=0}^{m} {}_mC_n\,.\,(-\nu-m)_{m-n}\,.\,(-\mu-m)_n$$
$$= \frac{(-)^m (-\mu-\nu-2m)_m}{m!\, \Gamma(\mu+m+1)\, \Gamma(\nu+m+1)}$$
$$= \frac{(\mu+\nu+m+1)_m}{m!\, \Gamma(\mu+m+1)\, \Gamma(\nu+m+1)},$$

when Vandermonde's theorem is used to sum the finite series.

Hence, for *all* values of μ and ν,

$$(1)\qquad J_\mu(z) J_\nu(z) = \sum_{m=0}^{\infty} \frac{(-)^m (\tfrac{1}{2}z)^{\mu+\nu+2m} (\mu+\nu+m+1)_m}{m!\, \Gamma(\mu+m+1)\, \Gamma(\nu+m+1)},$$

and this formula comprises the formulae (5), (6) and (7) of § 5·4.

This obvious mode of procedure does not seem to have been noticed by any of the earlier writers; it was given by Nielsen, *Math. Ann.* LII. (1899), p. 228.

The series for $J_0(z)\cos z$ and $J_0(z)\sin z$ were obtained by Bessel, *Berliner Abh.* 1824, [1826], pp. 38—39, and the corresponding results for $J_\nu(z)\cos z$ and $J_\nu(z)\sin z$ were deduced from Poisson's integral by Lommel, *Studien über die Bessel'schen Functionen* (Leipzig, 1868), pp. 16—18. Some deductions concerning the functions ber and bei have been made by Whitehead, *Quarterly Journal*, XLII. (1911), p. 342.

More generally, if we multiply the series for $J_\mu(az)$ and $J_\nu(bz)$, we obtain an expansion in which the coefficient of $(-)^m a^\mu b^\nu (\frac{1}{2}z)^{\mu+\nu+2m}$ is

$$\sum_{n=0}^{m} \frac{a^{2m-2n} b^{2n}}{n!\,\Gamma(\nu+n+1).(m-n)!\,\Gamma(\mu+m-n+1)}$$
$$= \frac{a^{2m}\,{}_2F_1(-m, -\mu-m;\ \nu+1;\ b^2/a^2)}{m!\,\Gamma(\mu+m+1)\,\Gamma(\nu+1)},$$

and so

$$(2)\quad J_\mu(az)\,J_\nu(bz) = \frac{(\frac{1}{2}az)^\mu (\frac{1}{2}bz)^\nu}{\Gamma(\nu+1)} \times \sum_{m=0}^{\infty} \frac{(-)^m (\frac{1}{2}az)^{2m}\,{}_2F_1(-m, -\mu-m;\ \nu+1;\ b^2/a^2)}{m!\,\Gamma(\mu+m+1)},$$

and this result can be simplified whenever the hypergeometric series is expressible in a compact form.

One case of reduction is the case $b = a$, which has already been discussed; another is the case $b = ia$, provided that $\mu^2 = \nu^2$.

In this case we use the formula*

$$F(\alpha, \beta;\ \alpha-\beta+1;\ -1) = \frac{\Gamma(\alpha-\beta+1)\,\Gamma(\frac{1}{2})}{2^\alpha\,\Gamma(\frac{1}{2}\alpha+\frac{1}{2})\,\Gamma(\frac{1}{2}\alpha-\beta+1)}$$

and then we see that

$$(3)\quad J_\nu(az)\,I_\nu(az) = \sum_{m=0}^{\infty} \frac{(-)^m (\frac{1}{2}az)^{2\nu+2m} \cos\frac{1}{2}m\pi}{\Gamma(\frac{1}{2}m+1)\,\Gamma(\nu+\frac{1}{2}m+1)\,\Gamma(\nu+m+1)}$$
$$= \sum_{m=0}^{\infty} \frac{(-)^m (\frac{1}{2}az)^{2\nu+4m}}{m!\,\Gamma(\nu+m+1)\,\Gamma(\nu+2m+1)},$$

$$(4)\quad J_{-\nu}(az)\,I_\nu(az) = \sum_{m=0}^{\infty} \frac{(-)^m (\frac{1}{2}az)^{2m} \cos(\frac{1}{2}\nu-\frac{1}{2}m)\pi}{m!\,\Gamma(\frac{1}{2}\nu+\frac{1}{2}m+1)\,\Gamma(-\frac{1}{2}\nu+\frac{1}{2}m+1)},$$

$$(5)\quad J_\nu(az)\,I_{-\nu}(az) = \sum_{m=0}^{\infty} \frac{(-)^m (\frac{1}{2}az)^{2m} \cos(\frac{1}{2}\nu+\frac{1}{2}m)\pi}{m!\,\Gamma(\frac{1}{2}\nu+\frac{1}{2}m+1)\,\Gamma(-\frac{1}{2}\nu+\frac{1}{2}m+1)}.$$

If we take $a = e^{\frac{1}{4}\pi i}$ in (3) we find that

$$(6)\quad \mathrm{ber}_\nu^2(z) + \mathrm{bei}_\nu^2(z) = \sum_{m=0}^{\infty} \frac{(\frac{1}{2}z)^{2\nu+4m}}{m!\,\Gamma(\nu+m+1)\,\Gamma(\nu+2m+1)},$$

an expansion of which the leading terms were given in § 3·8.

* Cf. Kummer, *Journal für Math.* XV. (1836), p. 78, formula (53).

The formulae (3), (4), (5) were discovered by Nielsen, *Atti della R. Accad. dei Lincei*, (5) xv. (1906), pp. 490—497 and *Monatshefte für Math. und Phys.* xix. (1908), pp. 164—170, from a consideration of the differential equation satisfied by $J_\nu(az) J_{\pm\nu}(bz)$.

Some series have been given, *Quarterly Journal*, xli. (1910), p. 55, for products of the types $J_\nu^3(z)$ and $J_\nu^2(z) J_{-\nu}(z)$, but they are too cumbrous to be of any importance.

By giving μ the special values $\pm\frac{1}{2}$ in (2), it is easy to prove that

$$(7)\qquad e^{z\cos\theta} J_{\nu-\frac{1}{2}}(z\sin\theta) = \frac{\Gamma(\nu)}{\Gamma(\frac{1}{2})}(2\sin\theta)^{\nu-\frac{1}{2}} \sum_{n=0}^{\infty} \frac{z^{\nu+n-\frac{1}{2}}}{\Gamma(2\nu+n)} C_n^\nu(\cos\theta).$$

The special case of this formula in which 2ν is an integer has been given by Hobson*.

5·42. *Products involving Bessel functions of the second kind.*

The series for the products $J_\mu(z) Y_n(z)$, $J_m(z) Y_n(z)$, and $Y_m(z) Y_n(z)$ have been the subject of detailed study by Nicholson†; the following is an outline of his analysis with some modifications.

We have

$$\pi J_\mu(z) Y_n(z) = \frac{\partial}{\partial\nu}\{J_\mu(z) J_\nu(z)\} - (-)^n \frac{\partial}{\partial\nu}\{J_\mu(z) J_{-\nu}(z)\},$$

where ν is to be made equal to n after the differentiations have been performed. Now

$$\frac{\partial}{\partial\nu}\{J_\mu(z) J_\nu(z)\} = \log(\tfrac{1}{2}z) . J_\mu(z) J_\nu(z)$$
$$+ \sum_{r=0}^{\infty}\left[\frac{(-)^r (\frac{1}{2}z)^{\mu+\nu+2r}\,\Gamma(\mu+\nu+2r+1)}{r!\,\Gamma(\mu+\nu+r+1)\,\Gamma(\mu+r+1)\,\Gamma(\nu+r+1)}\right.$$
$$\left.\times\{\psi(\mu+\nu+2r+1) - \psi(\mu+\nu+r+1) - \psi(\nu+r+1)\}\right]$$

and

$$\frac{\partial}{\partial\nu}\{J_\mu(z) J_{-\nu}(z)\} = -\log(\tfrac{1}{2}z) . J_\mu(z) J_{-\nu}(z)$$
$$-\sum_{r=0}^{\infty}\left[\frac{(-)^r (\frac{1}{2}z)^{\mu-\nu+2r}\,\Gamma(\mu-\nu+2r+1)}{r!\,\Gamma(\mu-\nu+r+1)\,\Gamma(\mu+r+1)\,\Gamma(-\nu+r+1)}\right.$$
$$\left.\times\{\psi(\mu-\nu+2r+1) - \psi(\mu-\nu+r+1) - \psi(-\nu+r+1)\}\right].$$

We divide the last series into two parts, $\sum_{r=0}^{n-1}$ and $\sum_{r=n}^{\infty}$. In the former part we have

$$\lim_{\nu\to n} \frac{\psi(-\nu+r+1)}{\Gamma(-\nu+r+1)} = (-)^{r-n} . (n-r-1)!,$$

* *Proc. London Math. Soc.* xxv. (1894), p. 66; see also Cailler, *Mém. de la Soc. de Phys. de Genève*, xxxiv. (1902—1905), p. 316.

† *Quarterly Journal*, xliii. (1912), pp. 78—100. The expansion of $J_0(z) Y_0(z)$ had been given previously by Nielsen, *Handbuch der Theorie der Cylinderfunktionen* (Leipzig, 1904), p. 21.

while in the latter part there is no undetermined form to be evaluated. V
r is replaced in this part by $n+r$, it is seen that

$$(1) \qquad \pi J_\mu(z)\, Y_n(z) = -\sum_{r=0}^{n-1} \frac{(\tfrac{1}{2}z)^{\mu-n+2r}(\mu-n+r+1)_r\,(n-r-1)!}{r!\,\Gamma(\mu+r+1)}$$

$$+\sum_{r=0}^{\infty} \frac{(-)^r (\tfrac{1}{2}z)^{\mu+n+2r}(\mu+n+r+1)_r}{r!\,(n+r)!\,\Gamma(\mu+r+1)}$$

$$\times \{2\log(\tfrac{1}{2}z) + 2\psi(\mu+n+2r+1)$$

$$-\psi(\mu+n+r+1) - \psi(\mu+r+1)$$

$$-\psi(n+r+1) - \psi(r+1)\}.$$

The expression on the right is a continuous function of μ at $\mu = m$ w
$m = 0, 1, 2, \ldots$, and so the series for $\pi J_m(z)\, Y_n(z)$ is obtained by replaci
by m on the right in (1).

The series for $Y_m(z)\, Y_n(z)$ can be calculated by constructing series f

$$\left[\frac{\partial^2 J_{\pm\mu}(z)\, J_{\pm\nu}(z)}{\partial\mu\,\partial\nu}\right]_{\mu=m,\ \nu=n}$$

in a similar manner. The details of the analysis, which is extremely labor
have been given by Nicholson, and will not be repeated here.

5·43. *The integral for $J_\mu(z)\, J_\nu(z)$.*

A generalisation of Neumann's integral (§ 2·6) for $J_n^2(z)$ is obtainabl
applying the formula*

$$\int_0^{\frac{1}{2}\pi} \cos^{\mu+\nu+2m}\theta\cos(\mu-\nu)\theta\, d\theta = \frac{\pi\,\Gamma(\mu+\nu+2m+1)}{2^{\mu+\nu+2m+1}\,\Gamma(\mu+m+1)\,\Gamma(\nu+m+1)}$$

to the result of § 5·41; the integral has this value when $m = 0, 1, 2, \ldots$, prov
that $R(\mu+\nu) > -1$.

It is then evident that

$$J_\mu(z)\, J_\nu(z) = \frac{2}{\pi}\sum_{m=0}^{\infty}\int_0^{\frac{1}{2}\pi} \frac{(-)^m z^{\mu+\nu+2m}\cos^{\mu+\nu+2m}\theta}{m!\,\Gamma(\mu+\nu+m+1)}\cos(\mu-\nu)\theta\, d\theta,$$

so that when $R(\mu+\nu) > -1$,

$$(1) \qquad J_\mu(z)\, J_\nu(z) = \frac{2}{\pi}\int_0^{\frac{1}{2}\pi} J_{\mu+\nu}(2z\cos\theta)\cos(\mu-\nu)\theta\, d\theta;$$

the change of the order of summation and integration presents no ser
difficulty.

* This formula is due to Cauchy; for a proof by contour integration, see *Modern Ana*
p. 263.

If n be a positive integer and $R(\mu - n) > -1$, then

$$J_\mu(z) J_n(z) = \frac{2(-)^n}{\pi} \int_0^{\frac{1}{2}\pi} J_{\mu-n}(2z\cos\theta)\cos(\mu+n)\theta\, d\theta, \tag{2}$$

and this formula is also true if μ and n are *both* integers, but are otherwise unrestricted.

Formula (1) was given by Schläfli, *Math. Ann.* III. (1871), p. 142, when $\mu \pm \nu$ are both integers; the general formula is due to Gegenbauer, *Wiener Sitzungsberichte*, CXI. (2*a*), (1902), p. 567.

5·5. *The expansion of* $(\frac{1}{2}z)^{\mu+\nu}$ *as a series of products.*

A natural generalisation of the formulae of Neumann (§ 2·7) and Gegenbauer (§ 5·2) is that

$$(\tfrac{1}{2}z)^{\mu+\nu} = \frac{\Gamma(\mu+1)\Gamma(\nu+1)}{\Gamma(\mu+\nu+1)} \times \sum_{m=0}^{\infty} \frac{(\mu+\nu+2m)\Gamma(\mu+\nu+m)}{m!} J_{\mu+m}(z) J_{\nu+m}(z). \tag{1}$$

The formula is true if μ and ν are not negative integers, but the following proof applies only if $R(\mu+\nu+1) > -1$.

From § 5·2 we have

$$(z\cos\theta)^{\mu+\nu} = \sum_{m=0}^{\infty} \frac{(\mu+\nu+2m)\Gamma(\mu+\nu+m)}{m!} J_{\mu+\nu+2m}(2z\cos\theta).$$

If we multiply by $\cos(\mu-\nu)\theta$ and integrate, it is clear from § 5·43 that

$$\frac{2z^{\mu+\nu}}{\pi} \int_0^{\frac{1}{2}\pi} \cos^{\mu+\nu}\theta \cos(\mu-\nu)\theta\, d\theta = \sum_{m=0}^{\infty} \frac{(\mu+\nu+2m)\Gamma(\mu+\nu+m)}{m!} \times J_{\mu+m}(z) J_{\nu+m}(z),$$

and the result follows by evaluating the integral on the left; for other values of μ and ν the result may be established by analytic continuation.

The formula is at once deducible from formulae given by Gegenbauer, *Wiener Sitzungsberichte*, LXXV. (2), (1877), p. 220.

5·51. *Lommel's series of squares of Bessel functions.*

An expansion derived by Lommel* from the formula

$$\frac{2\nu}{z}\frac{dJ_\nu^2(z)}{dz} = J^2_{\nu-1}(z) - J^2_{\nu+1}(z)$$

is

$$J^2_{\nu-1}(z) = \sum_{n=0}^{\infty} \frac{2(\nu+2n)}{z} \frac{dJ^2_{\nu+2n}(z)}{dz},$$

so that

$$\frac{1}{2}\int^z z J^2_{\nu-1}(z)\, dz = \left[\sum_{n=0}^{\infty} (\nu+2n) J^2_{\nu+2n}(z)\right].$$

* The results of this section will be found in *Math. Ann.* II. (1870), pp. 632—633; XIV. (1878), p. 532; *Münchener Abh.* XV. (1886), pp. 548—549.

Hence, by § 5·11 (11), we have

(1) $$\tfrac{1}{4}z^2\{J^2_{\nu-1}(z) - J_{\nu-2}(z)\,J_\nu(z)\} = \sum_{n=0}^{\infty}(\nu+2n)\,J^2_{\nu+2n}(z),$$

on taking zero as the lower limit when $R(\nu) > 0$; by adding on terms at the beginning of the series, it may be seen that the restriction $R(\nu) > 0$ is superfluous.

If we take in turn $\nu = \tfrac{1}{2}$, $\nu = \tfrac{3}{2}$, and add and subtract the results so obtained, we have (§ 3·4)

(2) $$\frac{z}{\pi} = \sum_{n=0}^{\infty}(n+\tfrac{1}{2})\,J^2_{n+\frac{1}{2}}(z),$$

(3) $$\frac{\sin 2z}{2\pi} = \sum_{n=0}^{\infty}(-)^n(n+\tfrac{1}{2})\,J^2_{n+\frac{1}{2}}(z),$$

while, by taking $\nu = 1$, we see that

(4) $$\tfrac{1}{4}z^2\{J_0^2(z) + J_1^2(z)\} = \sum_{n=0}^{\infty}(2n+1)\,J^2_{2n+1}(z).$$

Another formula of the same type is derived by differentiating the series

$$\sum_{n=0}^{\infty}\epsilon_n\,J^2_{\nu+n}(z);$$

for it is evident that

$$\begin{aligned}\frac{d}{dz}\sum_{n=0}^{\infty}\epsilon_n\,J^2_{\nu+n}(z) &= 2\sum_{n=0}^{\infty}\epsilon_n\,J_{\nu+n}(z)\,J'_{\nu+n}(z)\\ &= \sum_{n=0}^{\infty}\epsilon_n\,J_{\nu+n}(z)\{J_{\nu+n-1}(z) - J_{\nu+n+1}(z)\}\\ &= J_\nu(z)\{J_{\nu-1}(z) + J_{\nu+1}(z)\}\\ &= 2\nu\,J_\nu^2(z)/z,\end{aligned}$$

and so, when $R(\nu) > 0$, we obtain a modification of Hansen's formula (§ 2·5), namely

(5) $$\sum_{n=0}^{\infty}\epsilon_n\,J^2_{\nu+n}(z) = 2\nu\int_0^z J_\nu^2(t)\frac{dt}{t}.$$

An important consequence of this formula, namely the value of an upper bound for $|J_\nu(x)|$, will be given in § 13·42.

By taking $\nu = \tfrac{1}{2}$, it is found that

$$\begin{aligned}\sum_{n=0}^{\infty}\epsilon_n\,J^2_{n+\frac{1}{2}}(z) &= \frac{2}{\pi}\int_0^z \sin^2 t\,\frac{dt}{t^2}\\ &= \left[-\frac{2}{\pi}\frac{\sin^2 t}{t}\right]_0^z + \frac{2}{\pi}\int_0^z \sin 2t\,\frac{dt}{t}\end{aligned}$$

and so

(6) $$\sum_{n=0}^{\infty}J^2_{n+\frac{1}{2}}(z) = \frac{1}{\pi}Si(2z),$$

where, as usual, the symbol Si denotes the "sine integral." This result is given by Lommel in the third of the memoirs to which reference has been made.

5·6. *Continued fraction formulae.*

Expressions for quotients of Bessel functions as continued fractions are deducible immediately from the recurrence formula given by § 3·2 (1); thus, if the formula be written

$$\frac{J_\nu(z)}{J_{\nu-1}(z)} = \cfrac{\tfrac{1}{2}z/\nu}{1 - \cfrac{\tfrac{1}{2}z\, J_{\nu+1}(z)/\nu}{J_\nu(z)}},$$

it is at once apparent that

$$(1) \quad \frac{J_\nu(z)}{J_{\nu-1}(z)} = \frac{\tfrac{1}{2}z/\nu}{1\,-}\;\frac{\tfrac{1}{4}z^2/\{\nu(\nu+1)\}}{1\,-}\;\frac{\tfrac{1}{4}z^2/\{(\nu+1)(\nu+2)\}}{1}\;-\ldots$$
$$-\frac{\tfrac{1}{4}z^2/\{(\nu+m-1)(\nu+m)\}}{1\,-}\;\frac{\tfrac{1}{2}z\, J_{\nu+m+1}(z)/(\nu+m)}{J_{\nu+m}(z)}.$$

This formula is easily transformed into

$$(2) \quad \frac{J_\nu(z)}{J_{\nu-1}(z)} = \frac{1}{2\nu/z\,-}\;\frac{1}{2(\nu+1)/z\,-}\;\ldots\,-\frac{1}{2(\nu+m)/z\,-}\;\frac{J_{\nu+m+1}(z)}{J_{\nu+m}(z)}.$$

These results are true for general values of ν; (1) was discovered by Bessel* for integral values of ν. An equivalent result, due to Schlömilch†, is that, if $Q_\nu(z) = J_{\nu+1}(z)/\{\tfrac{1}{2}z\, J_\nu(z)\}$, then

$$(3) \quad Q_\nu(z) = \frac{1}{\nu+1\,-}\;\frac{\tfrac{1}{4}z^2}{\nu+2\,-}\;\frac{\tfrac{1}{4}z^2}{\nu+3\,-}\;\ldots\,-\frac{\tfrac{1}{4}z^2}{\nu+m\,-}\;\frac{z^2 Q_{\nu+m}(z)}{4}.$$

Other formulae, given by Lommel‡, are

$$(4) \quad \frac{J_{\nu+1}(z)}{J_\nu(z)} = \frac{z}{2(\nu+1)\,-}\;\frac{z^2}{2(\nu+2)\,-}\;\frac{z^2}{2(\nu+3)\,-}\;\ldots\,-\frac{z^2}{2(\nu+m)\,-}\;\frac{z\, J_{\nu+m+1}(z)}{J_{\nu+m}(z)},$$

$$(5) \quad \frac{J_{\nu+2}(z)}{J_\nu(z)} = -1 + \frac{2(\nu+1)}{2(\nu+1)\,-}\;\frac{z^2}{2(\nu+2)\,-}\;\ldots\,-\frac{z^2}{2(\nu+m)\,-}\;\frac{z\, J_{\nu+m+1}(z)}{J_{\nu+m}(z)}.$$

The Bessel functions in all these formulae may obviously be replaced by any cylinder functions.

It was assumed by Bessel that, when $m \to \infty$, the last quotient may be neglected, so that

$$(6) \quad \frac{J_\nu(z)}{J_{\nu-1}(z)} = \frac{\tfrac{1}{2}z/\nu}{1\,-}\;\frac{\tfrac{1}{4}z^2/\{\nu(\nu+1)\}}{1\,-}\;\frac{\tfrac{1}{4}z^2/\{(\nu+1)(\nu+2)\}}{1}\;-\ldots.$$

* *Berliner Abh.* (1824), [1826], p. 31. Formula (2) seems not to have been given by the earlier writers; see *Encyclopédie des Sci. Math.* II. 28, § 58, p. 217. A slightly different form is used by Graf, *Ann. di Mat.* (2) XXIII. (1895), p. 47.

† *Zeitschrift für Math. und Phys.* II. (1857), p. 142; Schlömilch considered integral values of ν only.

‡ *Studien über die Bessel'schen Functionen* (Leipzig, 1868), p. 5; see also Spitzer, *Archiv der Math. und Phys.* XXX. (1858), p. 332, and Günther, *Archiv der Math. und Phys.* LVI. (1874), pp. 292—297.

It is not obvious that this assumption is justifiable, though it happens t[o] be so, and a rigorous proof of the expansion of a quotient of Bessel function[s] into an infinite continued fraction will be given in § 9·65 with the help of th[e] theory of "Lommel's polynomials."

[NOTE. The reason why the assumption is not obviously correct is that, even thoug[h] the fraction p_m/q_m tends to a limit as $m\to\infty$, it is *not* necessarily the case that $\dfrac{a_m p_m + p_{m+1}}{a_m q_m + q_{m+1}}$ tends to that limit; this may be seen by taking

$$p_m = m + \sin m, \quad q_m = m, \quad a_m = -1.]$$

The reader will find an elaborate discussion on the representation of $J_\nu(z)/J_{\nu-1}(z)$ as continued fraction in a memoir* by Perron, *Münchener Sitzungsberichte*, XXXVII. (1907 pp. 483—504; solutions of Riccati's equation, depending on such a representation, hav[e] been considered by Wilton, *Quarterly Journal*, XLVI. (1915), pp. 320—323. The connexio[n] between continued fractions of the types considered in this section and the relations con[-]necting contiguous hypergeometric functions has been noticed by Heine, *Journal fü[r] Math.* LVII. (1860), pp. 231—247 and Christoffel, *Journal für Math.* LVIII. (1861 pp. 90—92.

5·7. *Hansen's expression for $J_\nu(z)$ as a limit of a hypergeometric functio[n]*

It was stated by Hansen† that

$$(1) \qquad J_\nu(z) = \lim_{\substack{\lambda\to\infty \\ \mu\to\infty}} \frac{(\frac{1}{2}z)^\nu}{\Gamma(\nu+1)} \cdot {}_2F_1\left(\lambda, \mu; \nu+1; -\frac{z^2}{4\lambda\mu}\right).$$

We shall prove this result for general (complex) values of ν and z when λ an[d] μ tend to infinity through complex values.

If $\lambda = 1/\delta$, $\mu = 1/\eta$, the $(m+1)$th term of the expansion on the right is

$$\frac{(-)^m (\frac{1}{2}z)^{\nu+2m}}{m!\,\Gamma(\nu+m+1)} \prod_{r=1}^{m-1} [(1+r\delta)(1+r\eta)].$$

This is a continuous function of δ and η; and, if δ_0, η_0 are arbitrary positiv[e] numbers (less than $2|z|^{-1}$), the series of which it is the $(m+1)$th term con[-]verges uniformly with respect to δ and η whenever both $|\delta| \leqslant \delta_0$ and $|\eta| \leqslant \eta_0$[.] For the term in question is numerically less than the modulus of the $(m+1)$t[h] term of the (absolutely convergent) expansion of

$$\frac{(\frac{1}{2}z)^\nu}{\Gamma(\nu+1)} \cdot {}_2F_1(1/\delta_0, 1/\eta_0; \nu+1; \tfrac{1}{4}z^2\delta_0\eta_0),$$

and the uniformity of the convergence follows from the test of Weierstrass.

Since the convergence is uniform, the sum of the terms is a continuo[us]

* This memoir is the subject of a paper by Nielsen, *Münchener Sitzungsberichte*, XXXVI[II.] (1908), pp. 85—88.

† *Leipziger Abh.* II. (1855), p. 252; see also a Halberstadt dissertation by F. Neumann, 190[9] [*Jahrbuch über die Fortschritte der Math.* 1909, p. 575.]

function of both the variables (δ, η) at $(0, 0)$, and so the limit of the series is the sum of the limits of the individual terms; that is to say

$$\lim_{\substack{\delta \to 0 \\ \eta \to 0}} \frac{(\frac{1}{2}z)^\nu}{\Gamma(\nu+1)} \cdot {}_2F_1\left(\frac{1}{\delta}, \frac{1}{\eta}; \nu+1; -\frac{z^2 \delta\eta}{4}\right) = \sum_{m=0}^{\infty} \frac{(-)^m (\frac{1}{2}z)^{\nu+2m}}{m!\,\Gamma(\nu+m+1)},$$

and this is the result stated.

5·71. *Bessel functions as limits of Legendre functions.*

It is well known that solutions of Laplace's equation, which are analytic near the origin and which are appropriate for the discussion of physical problems connected with a sphere, may be conveniently expressed as linear combinations of functions of the type

$$r^n P_n(\cos\theta), \quad r^n P_n^m(\cos\theta) \begin{matrix}\cos \\ \sin\end{matrix} m\phi;$$

these are *normal solutions* of Laplace's equation when referred to polar coordinates (r, θ, ϕ).

Now consider the nature of the structure of spheres, cones and planes associated with polar coordinates in a region of space at a great distance from the origin near the axis of harmonics. The spheres approximate to planes and the cones approximate to cylinders, and the structure resembles the structure associated with cylindrical-polar coordinates; and normal solutions of Laplace's equation referred to such coordinates are of the form (§ 4·8)

$$e^{\pm kz} J_m(k\rho) \begin{matrix}\cos \\ \sin\end{matrix} m\phi.$$

It is therefore to be expected that, when r and n are large* while θ is small in such a way that $r \sin\theta$ (i.e. ρ) remains bounded, the Legendre function should approximate to a Bessel function; in other words, we must expect Bessel functions to be expressible as limits of Legendre functions.

The actual formulae by which Bessel functions are so expressed are, in effect, special cases of Hansen's limit.

The most important formula of this type is

(1) $$\lim_{n \to \infty} P_n\left(\cos\frac{z}{n}\right) = J_0(z).$$

This result, which seems to have been known to Neumann† in 1862, has been investigated by Mehler, *Journal für Math.* LXVIII. (1868), p. 140; *Math. Ann.* V. (1872), pp. 136, 141—144; Heine, *Journal für Math.* LXIX. (1869), p. 130; Rayleigh, *Proc. London Math. Soc.* IX. (1878), pp. 61—64; *Proc. Royal Soc.* XCII. A, (1916), pp. 433—437 [*Scientific Papers*, I. (1899), pp. 338—341; VI. (1920), pp. 393—397]; and Giuliani, *Giorn. di Mat.* XXII. (1884), pp. 236—239. The result has been extended to generalised Legendre functions by Heine and Rayleigh.

It has usually been assumed that n tends to infinity through integral values in proving (1); but it is easier to prove it when n tends to infinity as a continuous real variable.

* If n were not large, the approximate formula for $P_n^m(\cos\theta)$ would be $(\sin^m\theta)/m!$.

† Cf. *Journal für Math.* LXII. (1863), pp. 36—49.

We take Murphy's formula

$$P_n(\cos z/n) = {}_2F_1(-n, n+1;\ 1;\ \sin^2 \tfrac{1}{2}z/n);$$

and the reasoning of the preceding section is applicable with the slight modification that we use the inequality

$$|\sin(\tfrac{1}{2}z/n)| \leqslant \tfrac{3}{5}|(z/n)|,$$

when $|z| \leqslant 2|n|$, and then we can compare the two series

$${}_2F_1(-n, n+1;\ 1;\ \sin^2 \tfrac{1}{2}z/n), \quad {}_2F_1(1/\delta_0, 1/\delta_0+1;\ 1;\ \tfrac{9}{25}\delta_0^2|z|^2),$$

where δ_0 is an arbitrary positive number less than $\frac{5}{3}|z|^{-1}$ and the comparison is made when $|n| > 1/\delta_0$. The details of the proof may now be left to the reader.

When n is restricted to be a positive integer, the series for $P_n(\cos z/n)$ terminates, and it is convenient to appeal to Tannery's theorem* to complete the proof. This fact was first noticed by Giuliani; the earlier writers took for granted the permissibility of the passage to the limit.

In the case of generalised Legendre functions (of *unrestricted* order m), the definition depends on whether the argument of the functions is between $+1$ and -1 or not; for real values of x (between 0 and π) we have

$$P_n^{-m}\left(\cos\frac{x}{n}\right) = \frac{\tan^m(\frac{1}{2}x/n)}{\Gamma(m+1)}\, {}_2F_1(-n, n+1;\ m+1;\ \sin^2 \tfrac{1}{2}x/n),$$

so that

$$\lim_{n\to\infty} n^m P_n^{-m}\left(\cos\frac{x}{n}\right) = J_m(x), \tag{2}$$

but otherwise, we have

$$P_n^{-m}\left(\cosh\frac{z}{n}\right) = \frac{\tanh^m(\frac{1}{2}z/n)}{\Gamma(m+1)}\, {}_2F_1(-n, n+1;\ m+1;\ -\sinh^2 \tfrac{1}{2}z/n),$$

so that

$$\lim_{n\to\infty} n^m P_n^{-m}\left(\cosh\frac{z}{n}\right) = I_m(z). \tag{3}$$

The corresponding formula for functions of the second kind may be deduced from the equation which expresses† Q_n^m in terms of P_n^m and P_n^{-m}; it is

$$\lim_{n\to\infty}\left[\frac{n^{-m}\sin n\pi}{\sin(m+n)\pi}\, Q_n^m\left(\cosh\frac{z}{n}\right)\right] = K_m(z). \tag{4}$$

This formula has been given (with a different notation) by Heine‡; it is most easily proved by substituting the integral of Laplace's type for the Legendre function, proceeding to the limit and using formula (5) of § 6·22.

* Cf. Bromwich, *Theory of Infinite Series*, § 49.

† Cf. Barnes, *Quarterly Journal*, XXXIX. (1908), p. 109; the equation is

$$2\Gamma(-m-n)\,\frac{\sin m\pi \sin n\pi}{\pi^2}\, Q_n^m = \frac{P_n^{-m}}{\Gamma(1-m+n)} - \frac{P_n^m}{\Gamma(1+m+n)}$$

in Barnes' notation, which is adopted in this work.

‡ *Journal für Math.* LXIX. (1868), p. 131.

Another formula, slightly different from those just discussed, is

$$\lim_{n\to\infty} P_n\left(\frac{n^2+z^2}{n^2-z^2}\right) = I_0(2z); \tag{5}$$

this is due to Laurent*, and it may be proved by using the second of Murphy's formulae, namely

$$P_n(\cos\theta) = \cos^n \tfrac{1}{2}\theta \, . \, {}_2F_1(-n, -n; 1; -\tan^2 \tfrac{1}{2}\theta).$$

[NOTE. The existence of the formulae of this section must be emphasized because it used to be generally believed that there was no connexion between Legendre functions and Bessel functions. Thus it was stated by Todhunter in his *Elementary Treatise on Laplace's Functions, Lamé's Functions and Bessel's Functions* (London, 1875), p. vi, that "these [i.e. Bessel functions] are not connected with the main subject of this book."]

5·72. *Integrals associated with Mehler's formula.*

A completely different method of establishing the formulae of the last section was given by Mehler and also, later, by Rayleigh; this method depends on a use of Laplace's integral, thus

$$P_n(\cos\theta) = \frac{1}{\pi}\int_0^\pi (\cos\theta + i\sin\theta\cos\phi)^n\,d\phi$$

$$= \frac{1}{\pi}\int_0^\pi e^{n\log(\cos\theta + i\sin\theta\cos\phi)}\,d\phi.$$

Since

$$n\log\{\cos(z/n) + i\sin(z/n)\cos\phi\} \to iz\cos\phi$$

uniformly as $n \to \infty$ when $0 \leqslant \phi \leqslant \pi$, we have at once

$$\lim_{n\to\infty} P_n(\cos z/n) = \frac{1}{\pi}\int_0^\pi e^{iz\cos\phi}\,d\phi = J_0(z).$$

Heine† and de Ball‡ have made similar passages to the limit with integrals of Laplace's type for Legendre functions. In this way Heine has *defined* Bessel functions of the second and third kinds; reference will be made to his results in § 6·22 when we deal with integral representations of $Y_\nu(z)$.

Mehler has also given a proof of his formula by using the Mehler-Dirichlet integral

$$P_n(\cos\theta) = \frac{2}{\pi}\int_0^\theta \frac{\cos(n+\frac{1}{2})\phi\,d\phi}{\sqrt{\{2(\cos\phi - \cos\theta)\}}}.$$

If $n\phi = \psi$, it may be shewn that

$$P_n(\cos z/n) \to \frac{2}{\pi}\int_0^z \frac{\cos\psi\,d\psi}{\sqrt{(z^2-\psi^2)}},$$

but the passage to the limit presents some little difficulty because the integral is an improper integral.

Various formulae have been given recently which exhibit the way in which

* *Journal de Math.* (3) I. (1875), pp. 384—385; the formula actually given by Laurent is erroneous on account of an arithmetical error.

† *Journal für Math.* LXIX. (1868), p. 131. See also Sharpe, *Quarterly Journal*, XXIV. (1890), pp. 383—386.

‡ *Astr. Nach.* CXXVIII. (1891), col. 1—4.

the Legendre function approaches its limit as its degree tends to infin[ity]. Thus, a formal expansion due to Macdonald* is

$$(1)\quad P_n^{-m}(\cos\theta) = (n+\tfrac{1}{2})^{-m}(\cos\tfrac{1}{2}\theta)^{-m}[J_m(x) + \sin^2\tfrac{1}{2}\theta\,\{\tfrac{1}{6}x\,J_{m+3}(x) - J_{m+2}(x) + \tfrac{1}{2}x^{-1}J_{m+1}(x)\} + \ldots]$$

where $x = (2n+1)\sin\frac{1}{2}\theta$.

Other formulae, which exhibit an upper limit for the error due to replac[ing] a Legendre function of large degree by a Bessel function, are†

$$(2)\quad P_n(\cos\eta) \pm i\pi^{-1}Q_n(\cos\eta) = \sqrt{(\sec\eta)}\,.\,e^{\pm(n+\frac{1}{2})i(\eta-\tan\eta)}[J_0\{(n+\tfrac{1}{2})\tan\eta\} \pm iY_0\{(n+\tfrac{1}{2})\tan\eta\}] + \frac{4\theta_1\sqrt{(\sec\eta)}}{\pi\{R(n)+\frac{1}{2}\}}$$

$$(3)\quad P_n(\cosh\xi) = \left(\frac{\xi}{\sinh\xi}\right)^{\frac{1}{2}} I_0(n\xi) + \frac{\frac{2}{5}\theta_2\xi\sinh n\xi}{6^{\frac{3}{2}}R(n)},$$

$$(4)\quad Q_n(\cosh\xi) = e^{-(n+\frac{1}{2})(\xi-\tanh\xi)}\sqrt{(\operatorname{sech}\xi)}\,.\,K_0\{(n+\tfrac{1}{2})\tanh\xi\} + \frac{\frac{2}{3}\theta_3\sqrt{(\operatorname{sech}\xi)}\,.\,e^{-(n+\frac{1}{2})\xi}}{R(n)+\frac{1}{2}},$$

where, in (2), $0 \leqslant \eta < \frac{1}{2}\pi$, and, in (3) and (4), $\xi \geqslant 0$; the numbers θ_1, θ_2, θ_3 [are] less than unity in absolute magnitude, and n may be complex provided th[at] its real part is positive. But the proof of these results is too lengthy to [be] given here.

5·73. *The formulae of Olbricht.*

The fact that a Bessel function is expressible by Hansen's formula as [a] limit of a hypergeometric function has led Olbricht‡ to investigate metho[ds] by which Bessel's equation is expressible as a confluent form of equatio[ns] associated with Riemann's P-functions.

If we take the equation

$$\frac{d^2y}{dz^2} + \frac{2\mu+1}{z}\frac{dy}{dz} + \left(1 - \frac{\nu^2-\mu^2}{z^2}\right)y = 0,$$

of which a fundamental system of solutions is the pair of functions

$$z^{-\mu}J_\nu(z),\quad z^{-\mu}Y_\nu(z),$$

and compare the equation with the equation defined by the scheme

$$P\begin{Bmatrix} a, & b, & c, & \\ \alpha, & \beta, & \gamma, & z \\ \alpha', & \beta', & \gamma', & \end{Bmatrix},$$

* *Proc. London Math. Soc.* (2) XIII. (1914), pp. 220—221; some associated results had be[en] obtained previously by the same writer, *Proc. London Math. Soc.* XXXI. (1899), p. 269.

† Watson, *Trans. Camb. Phil. Soc.* XXII. (1918), pp. 277—308; *Messenger*, XLVII. (191[8]) pp. 151—160.

‡ *Nova Acta Caes.-Leop.-Acad. (Halle)*, 1888, pp. 1—48.

namely

$$\frac{d^2y}{dz^2}+\left\{\frac{1-\alpha-\alpha'}{z-a}+\frac{1-\beta-\beta'}{z-b}+\frac{1-\gamma-\gamma'}{z-c}\right\}\frac{dy}{dz}$$
$$+\left\{\frac{\alpha\alpha'(a-b)(a-c)}{z-a}+\frac{\beta\beta'(b-c)(b-a)}{z-b}+\frac{\gamma\gamma'(c-a)(c-b)}{z-c}\right\}$$
$$\times\frac{y}{(z-a)(z-b)(z-c)}=0,$$

we see that the latter reduces to the former if

$$a=0,\quad \alpha=\nu-\mu,\quad \alpha'=-\nu-\mu,$$

while b, c, β, β', γ, γ' tend to infinity in such a way that $\beta+\beta'$ and $\gamma+\gamma'$ remain finite (their sum being $2\mu+1$) while $\beta\beta'=\gamma\gamma'=\frac{1}{4}b^2$ and $b+c=0$.

We thus obtain the scheme

$$\lim_{\beta\to\infty} P\begin{Bmatrix} 0, & 2i\beta, & -2i\beta, & \\ \nu-\mu, & \beta, & \gamma, & z \\ -\nu-\mu, & -\beta, & \gamma', & \end{Bmatrix},$$

where $\gamma,\gamma'=\mu+\frac{1}{2}\pm\sqrt{\{(\mu+\frac{1}{2})^2+\beta^2\}}$.

Another similar scheme is

$$\lim_{\beta\to\infty} P\begin{Bmatrix} 0, & i\beta, & \infty, & \\ \nu-\mu, & \beta, & \gamma, & z \\ -\nu-\mu, & -\beta, & \gamma', & \end{Bmatrix}$$

with the same values of γ and γ' as before.

A scheme for $J_\nu(z)$ derived directly from Hansen's formula is

$$\lim_{\left(\substack{\alpha\to\infty\\ \beta\to\infty}\right)} P\begin{Bmatrix} 0, & \infty, & -4\alpha\beta, & \\ \frac{1}{2}\nu, & \alpha-\frac{1}{2}\nu, & 0, & z^2 \\ -\frac{1}{2}\nu, & \beta-\frac{1}{2}\nu, & \nu+1-\alpha-\beta. & \end{Bmatrix}.$$

Olbricht has given other schemes but they are of no great importance and those which have now been constructed will be sufficient examples.

NOTE. It has been observed by Haentzschel, *Zeitschrift für Math. und Phys.* XXXI. (1886), p. 31, that the equation

$$\frac{d^2y}{du^2}=\left\{\frac{\nu^2-\frac{1}{4}}{u^2}-h^2\right\}y,$$

whose solution (§ 4·3) is $u^{\frac{1}{2}}\mathscr{C}_\nu(hu)$, may be derived by confluence from Lamé's equation

$$\frac{d^2y}{du^2}=[(\nu^2-\tfrac{1}{4})\{\wp(u)-e_1\}-h^2]\,y,$$

when the invariants g_2 and g_3 of the Weierstrassian elliptic function are made to tend to zero.

CHAPTER VI

INTEGRAL REPRESENTATIONS OF BESSEL FUNCTIONS

6·1. *Generalisations of Poisson's integral.*

In this chapter we shall study various contour integrals associated with Poisson's integral (§§ 2·3, 3·3) and Bessel's integral (§ 2·2). By suitable choices of the contour of integration, large numbers of elegant formulae can be obtained which express Bessel functions as definite integrals. The contour integrals will also be applied in Chapters VII and VIII to obtain approximate formulae and asymptotic expansions for $J_\nu(z)$ when z or ν is large.

It happens that the applications of Poisson's integral are of a more elementary character than the applications of Bessel's integral, and accordingly we shall now study integrals of Poisson's type, deferring the study of integrals of Bessel's type to § 6·2. The investigation of generalisations of Poisson's integral which we shall now give is due in substance to Hankel*.

The simplest of the formulae of § 3·3 is § 3·3 (4), since this formula contains a single exponential under the integral sign, while the other formulae contain circular functions, which are expressible in terms of two exponentials. We shall therefore examine the circumstances in which contour integrals of the type

$$z^\nu \int_a^b e^{izt}\, T dt$$

are solutions of Bessel's equation; it is supposed that T is a function of t but not of z, and that the end-points, a and b, are complex numbers independent of z.

The result of operating on the integral with Bessel's differential operator ∇_ν, defined in § 3·1, is as follows:

$$\nabla_\nu \left\{ z^\nu \int_a^b e^{izt}\, T dt \right\} = z^{\nu+2} \int_a^b e^{izt}\, T(1-t^2)\, dt + (2\nu+1)\, iz^{\nu+1} \int_a^b e^{izt}\, Tt dt$$
$$= iz^{\nu+1} \left[e^{izt}\, T(t^2-1) \right]_a^b + iz^{\nu+1} \int_a^b e^{izt} \left[(2\nu+1)\, Tt - \frac{d}{dt}\{T(t^2-1)\} \right] dt,$$

* *Math. Ann.* I. (1869), pp. 473—485. The discussion of the corresponding integrals for $I_\nu(z)$ and $K_\nu(z)$ is due to Schläfli, *Ann. di Mat.* (2) I. (1868), pp. 232—242, though Schläfli's results are expressed in the notation explained in § 4·15. The integrals have also been examined in great detail by Gubler, *Zürich Vierteljahrsschrift*, XXXIII. (1888), pp. 147—172, and, from the aspect of the theory of the linear differential equations which they satisfy, by Graf, *Math. Ann.* XLV. (1894), pp. 235—262; LVI. (1903), pp. 432—444. See also de la Vallée Poussin, *Ann. de la Soc. Sci. de Bruxelles*, XXIX. (1905), pp. 140—143.

by a partial integration. Accordingly we obtain a solution of Bessel's equation if T, a, b are so chosen that

$$\frac{d}{dt}\{T(t^2-1)\} = (2\nu+1)\,Tt, \qquad \Big[e^{izt}\,T(t^2-1)\Big]_a^b = 0.$$

The former of these equations shews that T is a constant multiple of $(t^2-1)^{\nu-\frac{1}{2}}$, and the latter shews that we may choose the path of integration, either so that it is a closed circuit such that $e^{izt}(t^2-1)^{\nu+\frac{1}{2}}$ returns to its initial value after t has described the circuit, or so that $e^{izt}(t^2-1)^{\nu+\frac{1}{2}}$ vanishes at each limit.

A contour of the first type is a figure-of-eight passing round the point $t=1$ counter-clockwise and round $t=-1$ clockwise. And, if we suppose temporarily that the real part of z is positive, a contour of the second type is one which starts from $+\infty i$ and returns there after encircling both the points -1, $+1$ counter-clockwise (Fig. 1 and Fig. 2). If we take $a, b = \pm 1$, it is

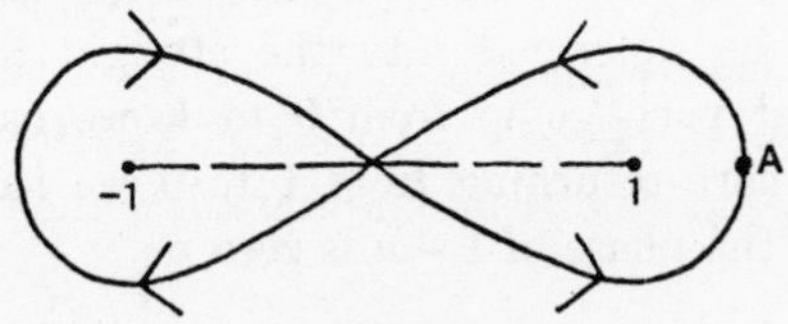

Fig. 1.

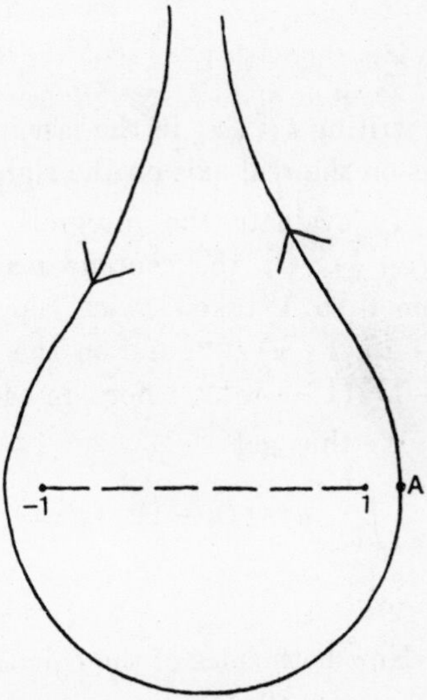

Fig. 2.

necessary to suppose that $R(\nu+\frac{1}{2})>0$, and we merely obtain Poisson's integral.

To make the many-valued function $(t^2-1)^{\nu-\frac{1}{2}}$ definite*, we take the phases of $t-1$ and $t+1$ to vanish at the point A where the contours cross the real axis on the right of $t=1$.

We therefore proceed to examine the contour integrals

$$z^\nu \int_A^{(1+,\,-1-)} e^{izt}(t^2-1)^{\nu-\frac{1}{2}}\,dt, \qquad z^\nu \int_{+\infty i}^{(-1+,\,1+)} e^{izt}(t^2-1)^{\nu-\frac{1}{2}}\,dt.$$

* It is supposed that ν has not one of the values $\frac{1}{2}, \frac{3}{2}, \frac{5}{2}, \ldots$; for then the integrands are analytic at ± 1, and both integrals vanish, by Cauchy's theorem.

It is to be observed that, when $R(z) > 0$, both integrals are converg[ent] and differentiations under the integral sign are permissible. Also, b[oth] integrals are analytic functions of ν for all values of ν.

In order to express the first integral in terms of Bessel functions, [we] expand the integrand in powers of z, the resulting series being unifor[mly] convergent with respect to t on the contour. It follows that

$$z^\nu \int_A^{(1+,\,-1-)} e^{izt} (t^2-1)^{\nu-\frac{1}{2}}\, dt = \sum_{m=0}^{\infty} \frac{i^m z^{\nu+m}}{m!} \int_A^{(1+,\,-1-)} t^m (t^2-1)^{\nu-\frac{1}{2}}\, dt.$$

Now $t^m (t^2-1)^{\nu-\frac{1}{2}}$ is an even or an odd function of t according as m is e[ven] or odd; and so, taking the contour to be symmetrical with respect to the orig[in] we see that the alternate terms of the series on the right vanish, and we [are] then left with the equation

$$z^\nu \int_A^{(1+,\,-1-)} e^{izt} (t^2-1)^{\nu-\frac{1}{2}}\, dt = 2 \sum_{m=0}^{\infty} \frac{(-)^m z^{\nu+2m}}{(2m)!} \int_0^{(1+)} t^{2m} (t^2-1)^{\nu-\frac{1}{2}}\, dt$$

$$= \sum_{m=0}^{\infty} \frac{(-)^m z^{\nu+2m}}{(2m)!} \int_0^{(1+)} u^{m-\frac{1}{2}} (u-1)^{\nu-\frac{1}{2}}\, du,$$

on writing $t = \sqrt{u}$; in the last integral the phases of u and $u-1$ vanish wh[en] u is on the real axis on the right of $u=1$.

To evaluate the integrals on the right, we assume temporarily th[at] $R(\nu+\frac{1}{2}) > 0$; the contour may then be deformed into the straight li[ne] from 0 to 1 taken twice; on the first part, going from 0 to 1, we ha[ve] $u-1 = (1-u)\, e^{-\pi i}$, and on the second part, returning from 1 to 0, we ha[ve] $u-1 = (1-u)\, e^{+\pi i}$, where, in each case, the phase of $1-u$ is zero.

We thus get

$$\int_0^{(1+)} u^{m-\frac{1}{2}} (u-1)^{\nu-\frac{1}{2}}\, du = \left\{e^{-(\nu-\frac{1}{2})\pi i} - e^{(\nu-\frac{1}{2})\pi i}\right\} \int_0^1 u^{m-\frac{1}{2}} (1-u)^{\nu-\frac{1}{2}}\, du$$

$$= 2i \cos \nu\pi \frac{\Gamma(m+\frac{1}{2})\,\Gamma(\nu+\frac{1}{2})}{\Gamma(m+\nu+1)}.$$

Now both sides of the equation

$$\int_0^{(1+)} u^{m-\frac{1}{2}} (u-1)^{\nu-\frac{1}{2}}\, du = 2i \cos \nu\pi \frac{\Gamma(m+\frac{1}{2})\,\Gamma(\nu+\frac{1}{2})}{\Gamma(m+\nu+1)}$$

are analytic functions of ν for all values of ν; and so, by the general theory [of] analytic continuation*, this result, which has been proved when $R(\nu+\frac{1}{2}) >$ [0,] persists for *all* values of ν.

* *Modern Analysis*, § 5·5. The reader will also find it possible to obtain the result, wh[en] $R(\nu+\frac{1}{2}) < 0$, by repeatedly using the recurrence formula

$$\int_0^{(1+)} u^{m-\frac{1}{2}} (u-1)^{\nu+n-\frac{1}{2}}\, du = -\frac{m+\nu+n+1}{\nu+n+\frac{1}{2}} \int_0^{(1+)} u^{m-\frac{1}{2}} (u-1)^{\nu+n+\frac{1}{2}}\, du,$$

which is obtained by integrating the formula

$$\frac{d}{du}\{u^{m+\frac{1}{2}} (u-1)^{\nu+n+\frac{1}{2}}\} = (m+\nu+n+1)\, u^{m-\frac{1}{2}} (u-1)^{\nu+n+\frac{1}{2}} + (\nu+n+\tfrac{1}{2})\, u^{m-\frac{1}{2}} (u-1)^{\nu+n-\frac{1}{2}};$$

the integral is then expressed in terms of an integral of the same type in which the exponent [of] $u-1$ has a positive real part.

Hence, for all* values of ν,

$$z^\nu \int_A^{(1+,-1-)} e^{izt}(t^2-1)^{\nu-\frac{1}{2}}\,dt = 2i\cos\nu\pi\,.\,\Gamma(\nu+\tfrac{1}{2})\sum_{m=0}^{\infty}\frac{(-)^m z^{\nu+2m}\,\Gamma(m+\frac{1}{2})}{(2m)!\,\Gamma(\nu+m+1)}$$
$$= 2^{\nu+1} i\,\Gamma(\tfrac{1}{2})\,\Gamma(\nu+\tfrac{1}{2})\cos\nu\pi\,.\,J_\nu(z).$$

Therefore, if $\nu+\frac{1}{2}$ is not a positive integer,

$$(1)\qquad J_\nu(z) = \frac{\Gamma(\frac{1}{2}-\nu)\,.\,(\frac{1}{2}z)^\nu}{2\pi i\,\Gamma(\frac{1}{2})}\int_A^{(1+,-1-)} e^{izt}(t^2-1)^{\nu-\frac{1}{2}}\,dt,$$

and this is Hankel's generalisation of Poisson's integral.

Next let us consider the second type of contour. Take the contour to lie wholly outside the circle $|t|=1$, and then $(t^2-1)^{\nu-\frac{1}{2}}$ is expansible in a series of descending powers of t, uniformly convergent on the contour; thus we have

$$(t^2-1)^{\nu-\frac{1}{2}} = \sum_{m=0}^{\infty}\frac{\Gamma(\frac{1}{2}-\nu+m)}{m!\,\Gamma(\frac{1}{2}-\nu)}\,t^{2\nu-1-2m},$$

and in the series the phase of t lies between $-\frac{3}{2}\pi$ and $+\frac{1}{2}\pi$.

Assuming† the permissibility of integrating term-by-term, we have

$$z^\nu \int_{\infty i}^{(-1+,1+)} e^{izt}(t^2-1)^{\nu-\frac{1}{2}}\,dt = \sum_{m=0}^{\infty}\frac{z^\nu\,\Gamma(\frac{1}{2}-\nu+m)}{m!\,\Gamma(\frac{1}{2}-\nu)}\int_{\infty i}^{(-1+,1+)} t^{2\nu-1-2m}e^{izt}\,dt.$$

But

$$\int_{\infty i}^{(-1+,1+)} t^{2\nu-1-2m}e^{izt}\,dt = (-)^{m+1}e^{-\nu\pi i}z^{2m-2\nu}\int_{\infty\exp i\alpha}^{(0+)}(-u)^{2\nu-1-2m}e^{-u}\,du,$$

where α is the phase of z (between $\pm\frac{1}{2}\pi$); and, by a well-known formula‡, the last integral equals $-2\pi i/\Gamma(2m-2\nu+1)$.

Hence

$$z^\nu \int_{\infty i}^{(-1+,1+)} e^{izt}(t^2-1)^{\nu-\frac{1}{2}}\,dt = \sum_{m=0}^{\infty}\frac{2\pi i\,(-)^m e^{-\nu\pi i}z^{-\nu+2m}\,\Gamma(\frac{1}{2}-\nu+m)}{m!\,\Gamma(\frac{1}{2}-\nu)\,\Gamma(2m-2\nu+1)}$$
$$= \frac{2^{\nu+1}\pi i\,e^{-\nu\pi i}\,\Gamma(\frac{1}{2})}{\Gamma(\frac{1}{2}-\nu)}J_{-\nu}(z),$$

when we use the duplication formula§ to express $\Gamma(2m-2\nu+1)$ in terms of $\Gamma(\frac{1}{2}-\nu+m)$ and $\Gamma(-\nu+m+1)$.

* If $\nu-\frac{1}{2}$ is a negative integer, the simplest way of evaluating the integral is to calculate the residue of the integrand at $u=1$.

† To justify the term-by-term integration, observe that $\int_{\infty i}^{(-1+,1+)}|e^{izt}\,dt|$ is convergent; let its value be K. Since the expansion of $(t^2-1)^{\nu-\frac{1}{2}}$ converges uniformly, it follows that, when we are given a positive number ϵ, we can find an integer M_0 independent of t, such that the remainder after M terms of the expansion does not exceed ϵ/K in absolute value when $M\geqslant M_0$. We then have at once

$$\left|\int_{\infty i}^{(-1+,1+)} e^{izt}(t^2-1)^{\nu-\frac{1}{2}}\,dt - \sum_{m=0}^{M-1}\frac{\Gamma(\frac{1}{2}-\nu+m)}{m!\,\Gamma(\frac{1}{2}-\nu)}\int_{\infty i}^{(-1+,1+)} e^{izt}\,t^{2\nu-1-2m}\,dt\right|$$
$$< \epsilon K^{-1}\int_{\infty i}^{(-1+,1+)}|e^{izt}\,dt| = \epsilon,$$

and the required result follows from the definition of the sum of an infinite series.

‡ Cf. *Modern Analysis*, § 12·22. § Cf. *Modern Analysis*, § 12·15.

Thus, when $R(z) > 0$ and $\nu + \frac{1}{2}$ is not a positive integer,

(2) $$J_{-\nu}(z) = \frac{\Gamma(\frac{1}{2} - \nu)\, e^{\nu\pi i} (\frac{1}{2} z)^\nu}{2\pi i\, \Gamma(\frac{1}{2})} \int_{\infty i}^{(-1+, 1+)} e^{izt} (t^2 - 1)^{\nu - \frac{1}{2}}\, dt.$$

This equation was also obtained by Hankel.

Next consider

$$\int_{\infty i \exp(-i\omega)}^{(-1+, 1+)} e^{izt} (t^2 - 1)^{\nu - \frac{1}{2}}\, dt,$$

where ω is an acute angle, positive or negative. This integral defines a function of z which is analytic when

$$-\tfrac{1}{2}\pi + \omega < \arg z < \tfrac{1}{2}\pi + \omega;$$

and, if z is subject to the further condition that $|\arg z| < \frac{1}{2}\pi$, the contour can be deformed into the second of the two contours just considered. Hence the analytic continuation of $J_{-\nu}(z)$ can be defined by the new integral over an extended range of values of arg z; so that we have

(3) $$J_{-\nu}(z) = \frac{\Gamma(\frac{1}{2} - \nu)\, e^{\nu\pi i} (\frac{1}{2} z)^\nu}{2\pi i\, \Gamma(\frac{1}{2})} \int_{\infty i \exp(-i\omega)}^{(-1+, 1+)} e^{izt} (t^2 - 1)^{\nu - \frac{1}{2}}\, dt,$$

where arg z has any value between $-\frac{1}{2}\pi + \omega$ and $\frac{1}{2}\pi + \omega$.

By giving ω a suitable value*, we can obtain a representation of $J_{-\nu}(z)$ for any assigned value of arg z between $-\pi$ and π.

When $R(z) > 0$ and $R(\nu + \frac{1}{2}) > 0$ we may take the contour to be that shewn in Fig. 3,

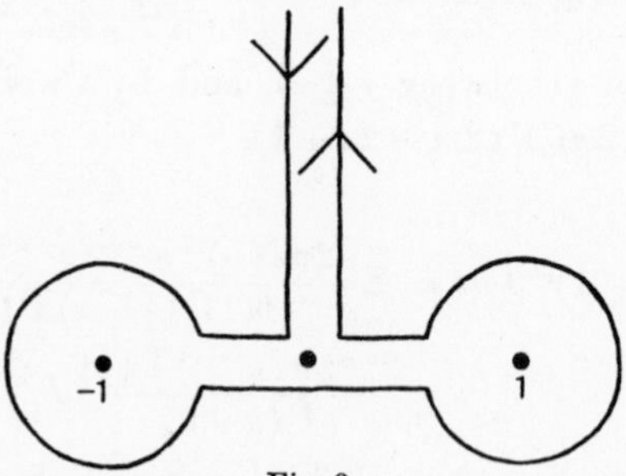

Fig. 3.

in which it is supposed that the radii of the circles are ultimately made indefinitely small. By taking each straight line in the contour separately, we get

$$J_{-\nu}(z) = \frac{\Gamma(\frac{1}{2} - \nu)\, e^{\nu\pi i} (\frac{1}{2} z)^\nu}{2\pi i\, \Gamma(\frac{1}{2})} \Big[\int_{\infty i}^{0} e^{izt} e^{-3\pi i(\nu - \frac{1}{2})} (1 - t^2)^{\nu - \frac{1}{2}}\, dt$$
$$+ \int_{0}^{-1} e^{izt} e^{-3\pi i(\nu - \frac{1}{2})} (1 - t^2)^{\nu - \frac{1}{2}}\, dt$$
$$+ \int_{-1}^{1} e^{izt} e^{-\pi i(\nu - \frac{1}{2})} (1 - t^2)^{\nu - \frac{1}{2}}\, dt$$
$$+ \int_{1}^{0} e^{izt} e^{\pi i(\nu - \frac{1}{2})} (1 - t^2)^{\nu - \frac{1}{2}}\, dt$$
$$+ \int_{0}^{\infty i} e^{izt} e^{\pi i(\nu - \frac{1}{2})} (1 - t^2)^{\nu - \frac{1}{2}}\, dt \Big].$$

* If $|\omega|$ be increased in a series of stages to an appropriate value (greater than $\frac{1}{2}\pi$), a representation of $J_{-\nu}(z)$ valid for any preassigned value of arg z may be obtained.

On bisecting the third path of integration and replacing t in the various integrals by it, $-t$, $\pm t$, t, it respectively, we obtain a formula for $J_{-\nu}(z)$, due to Gubler*, which corresponds to Poisson's integral for $J_\nu(z)$; the formula is

$$(4)\quad J_{-\nu}(z)=\frac{2(\frac{1}{2}z)^\nu}{\Gamma(\nu+\frac{1}{2})\Gamma(\frac{1}{2})}\left[\sin\nu\pi\int_0^\infty e^{-zt}(1+t^2)^{\nu-\frac{1}{2}}dt+\int_0^1\cos(zt+\nu\pi).(1-t^2)^{\nu-\frac{1}{2}}dt\right],$$

and, if this be combined with Poisson's integral, it is found that

$$(5)\qquad Y_\nu(z)=\frac{2(\frac{1}{2}z)^\nu}{\Gamma(\nu+\frac{1}{2})\Gamma(\frac{1}{2})}\left[\int_0^1\sin(zt).(1-t^2)^{\nu-\frac{1}{2}}dt-\int_0^\infty e^{-zt}(1+t^2)^{\nu-\frac{1}{2}}dt\right],$$

a formula which was also discovered by Gubler, though it had been previously stated by Weber† in the case of integral values of ν.

After what has gone before the reader should have no difficulty in obtaining a formula closely connected with (1), namely

$$(6)\qquad J_\nu(z)=\frac{\Gamma(\frac{1}{2}-\nu).(\frac{1}{2}z)^\nu}{\pi i\,\Gamma(\frac{1}{2})}\int_0^{(1+)}(t^2-1)^{\nu-\frac{1}{2}}\cos(zt).dt,$$

in which it is supposed that the phase of t^2-1 vanishes when t is on the real axis on the right of $t=1$.

6·11. *Modifications of Hankel's contour integrals.*

Taking $R(z)>0$, let us modify the two contours of § 6·1 into the contours shewn in Figs. 4 and 5 respectively.

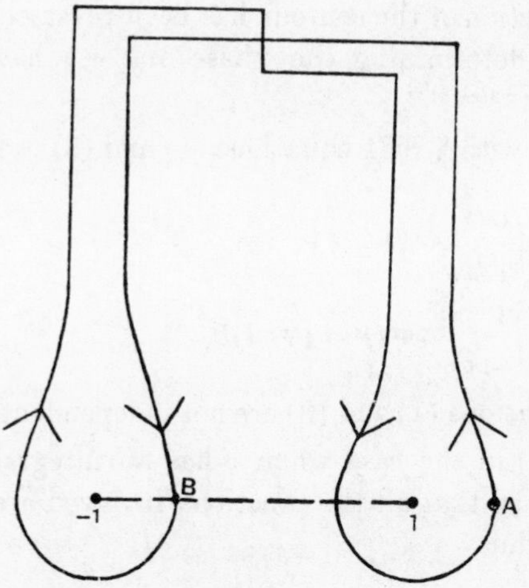

Fig. 4.

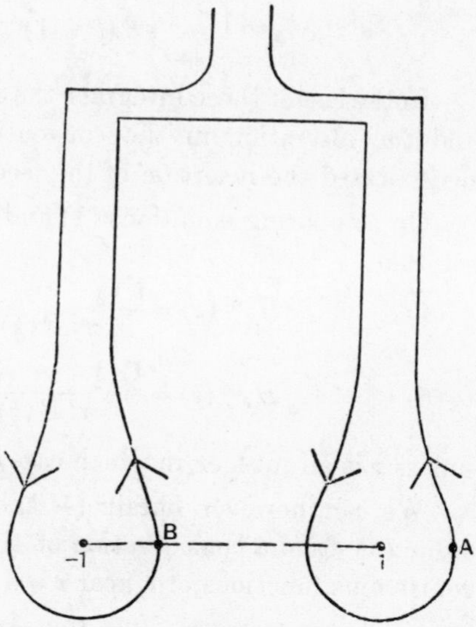

Fig. 5.

By making those portions of the contours which are parallel to the real

* *Zürich Vierteljahrsschrift*, XXXIII. (1888), p. 159. See also Graf, *Zeitschrift für Math. und Phys.* XXXVIII. (1893), p. 115.

† *Journal für Math.* LXXVI. (1873), p. 9. Cf. Hayashi, *Nyt Tidsskrift for Math.* XXIII. B, (1912), pp. 86—90. The formula was examined in the case $\nu=0$ by Escherich, *Monatshefte für Math. und Phys.* III. (1892), pp. 142, 234.

axis move off to infinity (so that the integrals along them tend to zero), we obtain the two following formulae:

$$(1)\quad J_\nu(z) = \frac{\Gamma(\tfrac{1}{2}-\nu).(\tfrac{1}{2}z)^\nu}{2\pi i\,\Gamma(\tfrac{1}{2})} \times \left[\int_{1+\infty i}^{(1+)} e^{izt}(t^2-1)^{\nu-\frac{1}{2}}\,dt + \int_{-1+\infty i}^{(-1-)} e^{izt}(t^2-1)^{\nu-\frac{1}{2}}\,dt\right],$$

$$(2)\quad J_{-\nu}(z) = \frac{\Gamma(\tfrac{1}{2}-\nu)\,e^{\nu\pi i}(\tfrac{1}{2}z)^\nu}{2\pi i\,\Gamma(\tfrac{1}{2})} \times \left[\int_{1+\infty i}^{(1+)} e^{izt}(t^2-1)^{\nu-\frac{1}{2}}\,dt + \int_{-1+\infty i}^{(-1+)} e^{izt}(t^2-1)^{\nu-\frac{1}{2}}\,dt\right].$$

In the first result the many-valued functions are to be interpreted by taking the phase of t^2-1 to be 0 at A *and to be* $+\pi$ *at* B, while in the second the phase of t^2-1 is 0 at A *and is* $-\pi$ *at* B.

To avoid confusion it is desirable to have the phase of t^2-1 interpreted in the same way in both formulae; and when it is supposed that the phase of t^2-1 is $+\pi$ at B, the formula (1) is of course unaltered, while (2) is replaced by

$$(3)\quad J_{-\nu}(z) = \frac{\Gamma(\tfrac{1}{2}-\nu).(\tfrac{1}{2}z)^\nu}{2\pi i\,\Gamma(\tfrac{1}{2})} \times \left[e^{\nu\pi i}\int_{1+\infty i}^{(1+)} e^{izt}(t^2-1)^{\nu-\frac{1}{2}}\,dt + e^{-\nu\pi i}\int_{-1+\infty i}^{(-1-)} e^{izt}(t^2-1)^{\nu-\frac{1}{2}}\,dt\right].$$

In the last of these integrals, the direction of the contour has been reversed and the alteration in the convention determining the phase of t^2-1 has necessitated the insertion of the factor $e^{-2(\nu-\frac{1}{2})\pi i}$.

On comparing equations (1) and (3) with § 3·61 equations (1) and (2), we see that

$$(4)\qquad H_\nu^{(1)}(z) = \frac{\Gamma(\tfrac{1}{2}-\nu).(\tfrac{1}{2}z)^\nu}{\pi i\,\Gamma(\tfrac{1}{2})}\int_{1+\infty i}^{(1+)} e^{izt}(t^2-1)^{\nu-\frac{1}{2}}\,dt,$$

$$(5)\qquad H_\nu^{(2)}(z) = \frac{\Gamma(\tfrac{1}{2}-\nu).(\tfrac{1}{2}z)^\nu}{\pi i\,\Gamma(\tfrac{1}{2})}\int_{-1+\infty i}^{(-1-)} e^{izt}(t^2-1)^{\nu-\frac{1}{2}}\,dt,$$

unless ν is an integer, in which case equations (1) and (3) are not independent.

We can, however, obtain (4) and (5) in the case when ν has an integral value (n), from a consideration of the fact that all the functions involved are continuous functions of ν near $\nu = n$. Thus

$$\begin{aligned} H_n^{(1)}(z) &= \lim_{\nu\to n} H_\nu^{(1)}(z) \\ &= \lim_{\nu\to n}\frac{\Gamma(\tfrac{1}{2}-\nu).(\tfrac{1}{2}z)^\nu}{\pi i\,\Gamma(\tfrac{1}{2})}\int_{1+\infty i}^{(1+)} e^{izt}(t^2-1)^{\nu-\frac{1}{2}}\,dt \\ &= \frac{\Gamma(\tfrac{1}{2}-n).(\tfrac{1}{2}z)^n}{\pi i\,\Gamma(\tfrac{1}{2})}\int_{1+\infty i}^{(1+)} e^{izt}(t^2-1)^{n-\frac{1}{2}}\,dt, \end{aligned}$$

and similarly for $H_n^{(2)}(z)$.

As in the corresponding analysis of § 6·1, the ranges of validity of (4) and (5) may be extended by swinging round the contours and using the theory of analytic continuation.

Thus, if $-\frac{1}{2}\pi < \omega < \frac{3}{2}\pi$, we have

$$(6) \qquad H_\nu^{(1)}(z) = \frac{\Gamma(\frac{1}{2}-\nu)\,.\,(\frac{1}{2}z)^\nu}{\pi i\,\Gamma(\frac{1}{2})} \int_{\infty i \exp(-i\omega)}^{(1+)} e^{izt}(t^2-1)^{\nu-\frac{1}{2}}\,dt;$$

while, if $-\frac{3}{2}\pi < \omega < \frac{1}{2}\pi$, we have

$$(7) \qquad H_\nu^{(2)}(z) = \frac{\Gamma(\frac{1}{2}-\nu)\,.\,(\frac{1}{2}z)^\nu}{\pi i\,\Gamma(\frac{1}{2})} \int_{\infty i \exp(-i\omega)}^{(-1-)} e^{izt}(t^2-1)^{\nu-\frac{1}{2}}\,dt,$$

provided that, in both (6) and (7), the phase of z lies between $-\frac{1}{2}\pi + \omega$ and $\frac{1}{2}\pi + \omega$.

Representations are thus obtained of $H_\nu^{(1)}(z)$ when $\arg z$ has any value between $-\pi$ and 2π, and of $H_\nu^{(2)}(z)$ when $\arg z$ has any value between -2π and π.

If ω be increased beyond the limits stated, it is necessary to make the contours coil round the singular points of the integrand, and numerical errors are liable to occur in the interpretation of the integrals unless great care is taken. Weber, however, has adopted this procedure, *Math. Ann.* XXXVII. (1890), pp. 411—412, to determine the formulae of § 3·62 connecting $H_\nu^{(1)}(-z)$, $H_\nu^{(2)}(-z)$ with $H_\nu^{(1)}(z)$, $H_\nu^{(2)}(z)$.

NOTE. The formula $2iY_\nu(z) = H_\nu^{(1)}(z) - H_\nu^{(2)}(z)$ makes it possible to express $Y_\nu(z)$ in terms of loop integrals, and in this manner Hankel obtained the series of § 3·52 for $\mathbf{Y}_n(z)$; this investigation will not be reproduced in view of the greater simplicity of Hankel's other method which has been described in § 3·52.

6·12. *Integral representations of functions of the third kind.*

In the formula § 6·11 (6) suppose that the phase of z has any given value between $-\pi$ and 2π, and define β by the equation

$$\arg z = \omega + \beta,$$

so that $-\frac{1}{2}\pi < \beta < \frac{1}{2}\pi$.

Then we shall write

$$t - 1 = e^{-\frac{1}{2}\pi i} z^{-1}(-u),$$

so that the phase of $-u$ increases from $-\pi + \beta$ to $\pi + \beta$ as t describes the contour; and it follows immediately that

$$(1) \qquad H_\nu^{(1)}(z) = \frac{i\,\Gamma(\frac{1}{2}-\nu)\,e^{i(z-\frac{1}{2}\nu\pi-\frac{1}{4}\pi)}}{\pi\sqrt{(2\pi z)}} \int_{\infty \exp i\beta}^{(0+)} e^{-u}(-u)^{\nu-\frac{1}{2}}\left(1+\frac{iu}{2z}\right)^{\nu-\frac{1}{2}} du,$$

where the phase of $1 + \frac{1}{2}iu/z$ has its principal value. Again, if β be a given acute angle (positive or negative), this formula affords a representation of $H_\nu^{(1)}(z)$ valid over the sector of the z-plane in which

$$-\tfrac{1}{2}\pi + \beta < \arg z < \tfrac{3}{2}\pi + \beta.$$

Similarly*, from § 6·11 (7),

$$(2)\qquad H_\nu^{(2)}(z)=\frac{i\Gamma(\frac{1}{2}-\nu)\,e^{-i(z-\frac{1}{2}\nu\pi-\frac{1}{4}\pi)}}{\pi\sqrt{(2\pi z)}}\int_{\infty\exp i\beta}^{(0+)} e^{-u}(-u)^{\nu-\frac{1}{2}}\left(1-\frac{iu}{2z}\right)^{\nu-\frac{1}{2}}du,$$

where β is any acute angle (positive or negative) and

$$-\tfrac{3}{2}\pi+\beta<\arg z<\tfrac{1}{2}\pi+\beta.$$

Since†, by § 3·61 (7), $H_{-\nu}^{(1)}(z)=e^{\nu\pi i}H_\nu^{(1)}(z)$, it follows that we lose nothing by restricting ν so that $R(\nu+\frac{1}{2})>0$; and it is then permissible to deform the contours into the line joining the origin to $\infty\exp i\beta$, taken twice; for the integrals taken round a small circle (with centre at the origin) tend to zero with the radius of the circle‡.

On deforming the contour of (1) in the specified manner, we find that

$$(3)\qquad H_\nu^{(1)}(z)=\left(\frac{2}{\pi z}\right)^{\frac{1}{2}}\frac{e^{i(z-\frac{1}{2}\nu\pi-\frac{1}{4}\pi)}}{\Gamma(\nu+\frac{1}{2})}\int_0^{\infty\exp i\beta} e^{-u}u^{\nu-\frac{1}{2}}\left(1+\frac{iu}{2z}\right)^{\nu-\frac{1}{2}}du,$$

where β may be any acute angle (positive or negative) and

$$R(\nu+\tfrac{1}{2})>0,\quad -\tfrac{1}{2}\pi+\beta<\arg z<\tfrac{3}{2}\pi+\beta.$$

In like manner, from (2),

$$(4)\qquad H_\nu^{(2)}(z)=\left(\frac{2}{\pi z}\right)^{\frac{1}{2}}\frac{e^{-i(z-\frac{1}{2}\nu\pi-\frac{1}{4}\pi)}}{\Gamma(\nu+\frac{1}{2})}\int_0^{\infty\exp i\beta} e^{-u}u^{\nu-\frac{1}{2}}\left(1-\frac{iu}{2z}\right)^{\nu-\frac{1}{2}}du,$$

where β may be any acute angle (positive or negative) and

$$R(\nu+\tfrac{1}{2})>0,\quad -\tfrac{3}{2}\pi+\beta<\arg z<\tfrac{1}{2}\pi+\beta.$$

The results (3) and (4) have not yet been proved when 2ν is an odd positive integer. But in view of the continuity near $\nu=n+\frac{1}{2}$ of the functions involved (where $n=0, 1, 2, \ldots$) it follows, as in the somewhat similar work of § 6·11, that (3) and (4) are true when $\nu=\frac{1}{2}, \frac{3}{2}, \frac{5}{2}, \ldots$. The results may also be obtained for such values of ν by expanding the integrands in terminating series of descending powers of z, and integrating term-by-term; the formulae so obtained are easily reconciled with the equations of § 3·4.

The general formulae (3) and (4) are of fundamental importance in the discussion of asymptotic expansions of $J_{\pm\nu}(z)$ for large values of $|z|$. These applications of the formulae will be dealt with in Chapter VII.

A useful modification of the formulae is due to Schafheitlin§. If we take $\arg z=\beta$ (so that $\arg z$ is restricted to be an acute angle), and then write $u=2z\cot\theta$, it follows that

$$(5)\qquad H_\nu^{(1)}(z)=\frac{2^{\nu+1}z^\nu}{i\Gamma(\nu+\frac{1}{2})\Gamma(\frac{1}{2})}\int_0^{\frac{1}{2}\pi}\frac{\cos^{\nu-\frac{1}{2}}\theta\,.\,e^{i(z-\nu\theta+\frac{1}{2}\theta)}}{\sin^{2\nu+1}\theta}e^{-2z\cot\theta}\,d\theta,$$

$$(6)\qquad H_\nu^{(2)}(z)=-\frac{2^{\nu+1}z^\nu}{i\Gamma(\nu+\frac{1}{2})\Gamma(\frac{1}{2})}\int_0^{\frac{1}{2}\pi}\frac{\cos^{\nu-\frac{1}{2}}\theta\,.\,e^{-i(z-\nu\theta+\frac{1}{2}\theta)}}{\sin^{2\nu+1}\theta}e^{-2z\cot\theta}\,d\theta,$$

* To obtain this formula, write

$$t+1=e^{-\frac{1}{2}\pi i}z^{-1}(-u),\quad t-1=2e^{\pi i}(1-\tfrac{1}{2}iu/z).$$

† There seems to be no simple direct proof that

$$\Gamma(\tfrac{1}{2}-\nu)\int_{\infty\exp i\beta}^{(0+)} e^{-u}(-u)^{\nu-\frac{1}{2}}\left(1+\frac{iu}{2z}\right)^{\nu-\frac{1}{2}}du$$

is an even function of ν.

‡ Cf. *Modern Analysis*, § 12·22.

§ *Journal für Math.* CXIV. (1894), pp. 31—44.

and hence that

$$(7)\qquad J_\nu(z)=\frac{2^{\nu+1}z^\nu}{\Gamma(\nu+\frac{1}{2})\Gamma(\frac{1}{2})}\int_0^{\frac{1}{2}\pi}\frac{\cos^{\nu-\frac{1}{2}}\theta\,.\sin(z-\nu\theta+\frac{1}{2}\theta)}{\sin^{2\nu+1}\theta}e^{-2z\cot\theta}\,d\theta,$$

$$(8)\qquad Y_\nu(z)=-\frac{2^{\nu+1}z^\nu}{\Gamma(\nu+\frac{1}{2})\Gamma(\frac{1}{2})}\int_0^{\frac{1}{2}\pi}\frac{\cos^{\nu-\frac{1}{2}}\theta\,.\cos(z-\nu\theta+\frac{1}{2}\theta)}{\sin^{2\nu+1}\theta}e^{-2z\cot\theta}\,d\theta.$$

These formulae, which are of course valid only when $R(\nu+\frac{1}{2})>0$, were applied by Schafheitlin to obtain properties of the zeros of Bessel functions (§§ 15·32—15·35). They were obtained by him from the consideration that the expressions on the right are solutions of Bessel's equation which behave in the appropriate manner near the origin.

The integral $\int_0^\infty e^{-uz}u^{\nu-\frac{1}{2}}(1+u)^{\mu-\frac{1}{2}}\,du$, which is reducible to integrals of the types occurring in (3) and (4) when $\mu=\nu$, has been studied in some detail by Nielsen, *Math. Ann.* LIX. (1904), pp. 89—102.

The integrals of this section are also discussed from the aspect of the theory of asymptotic solutions of differential equations by Brajtzew, *Warschau Polyt. Inst. Nach.* 1902, nos. 1, 2 [*Jahrbuch über die Fortschritte der Math.* 1903, pp. 575—577].

6·13. *The generalised Mehler-Sonine integrals.*

Some elegant definite integrals may be obtained to represent Bessel functions of a positive variable of a suitably restricted order. To construct them, observe that, when z is positive ($=x$) and the real part of ν *is less than* $\frac{1}{2}$, it is permissible to take $\omega=\frac{1}{2}\pi$ in § 6·11 (6) and to take $\omega=-\frac{1}{2}\pi$ in § 6·11 (7), so that the contours are those shewn in Fig. 6. When, in addition, the real part of ν is greater than $-\frac{1}{2}$, it is permissible to deform the contours (after the manner of § 6·12) so that the first contour consists of the real axis from $+1$ to $+\infty$ taken twice while the second contour consists of the real axis from -1 to $-\infty$ taken twice.

Fig. 6.

We thus obtain the formulae

$$\begin{cases} H_\nu^{(1)}(x)=\dfrac{\Gamma(\frac{1}{2}-\nu)\,.\,(\frac{1}{2}x)^\nu}{\pi i\,\Gamma(\frac{1}{2})}(1-e^{-2(\nu-\frac{1}{2})\pi i})\displaystyle\int_1^\infty e^{ixt}(t^2-1)^{\nu-\frac{1}{2}}\,dt, \\ H_\nu^{(2)}(x)=-\dfrac{\Gamma(\frac{1}{2}-\nu)\,.\,(\frac{1}{2}x)^\nu}{\pi i\,\Gamma(\frac{1}{2})}(1-e^{2(\nu-\frac{1}{2})\pi i})\displaystyle\int_1^\infty e^{-ixt}(t^2-1)^{\nu-\frac{1}{2}}\,dt, \end{cases}$$

the second being derived from § 6·11 (7) by replacing t by $-t$.

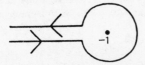

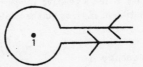

In these formulae replace ν by $-\nu$ and use the transformation formulae given by § 3·61 (7). It follows that, when $x>0$ and $-\frac{1}{2}<R(\nu)<\frac{1}{2}$, then

(1) $$H_\nu^{(1)}(x)=\frac{2}{i\Gamma(\frac{1}{2}-\nu)\,\Gamma(\frac{1}{2}).(\frac{1}{2}x)^\nu}\int_1^\infty \frac{e^{ixt}\,dt}{(t^2-1)^{\nu+\frac{1}{2}}},$$

(2) $$H_\nu^{(2)}(x)=-\frac{2}{i\Gamma(\frac{1}{2}-\nu)\,\Gamma(\frac{1}{2}).(\frac{1}{2}x)^\nu}\int_1^\infty \frac{e^{-ixt}\,dt}{(t^2-1)^{\nu+\frac{1}{2}}},$$

so that

(3) $$J_\nu(x)=\frac{2}{\Gamma(\frac{1}{2}-\nu)\,\Gamma(\frac{1}{2}).(\frac{1}{2}x)^\nu}\int_1^\infty \frac{\sin(xt).\,dt}{(t^2-1)^{\nu+\frac{1}{2}}},$$

(4) $$Y_\nu(x)=-\frac{2}{\Gamma(\frac{1}{2}-\nu)\,\Gamma(\frac{1}{2}).(\frac{1}{2}x)^\nu}\int_1^\infty \frac{\cos(xt).\,dt}{(t^2-1)^{\nu+\frac{1}{2}}}.$$

Of these results, (3) was given by Mehler, *Math. Ann.* v. (1872), p. 142, in the special case $\nu=0$, while Sonine, *Math. Ann.* XVI. (1880), p. 39, gave both (3) and (4) in the same special case. Other generalisations of the Mehler-Sonine integrals will be given in § 6·21.

6·14. *Symbolic formulae due to Hargreave and Macdonald.*

When $R(z)>0$ and $R(\nu+\frac{1}{2})>0$, it is evident from formula § 6·11 (6) that

$$H_\nu^{(1)}(z)=\frac{(\frac{1}{2}z)^\nu}{\Gamma(\nu+\frac{1}{2})\,\Gamma(\frac{1}{2})}\int_{1+\infty i}^1 e^{izt}(1-t^2)^{\nu-\frac{1}{2}}\,dt,$$

where the phase of $1-t^2$ lies between 0 and $-\frac{1}{2}\pi$.

If D denotes (d/dz) and f is any polynomial, then

$$f(it).\,e^{izt}=f(D)\,e^{izt},$$

and so, *when* $\nu+\frac{1}{2}$ *is a positive integer*, we have

$$\begin{aligned}H_\nu^{(1)}(z)&=\frac{(\frac{1}{2}z)^\nu}{\Gamma(\nu+\frac{1}{2})\,\Gamma(\frac{1}{2})}\int_{1+\infty i}^1 \{(1+D^2)^{\nu-\frac{1}{2}}e^{izt}\}\,dt\\&=\frac{(\frac{1}{2}z)^\nu}{\Gamma(\nu+\frac{1}{2})\,\Gamma(\frac{1}{2})}(1+D^2)^{\nu-\frac{1}{2}}\int_{1+\infty i}^1 e^{izt}\,dt\\&=\frac{(\frac{1}{2}z)^\nu}{\Gamma(\nu+\frac{1}{2})\,\Gamma(\frac{1}{2})}(1+D^2)^{\nu-\frac{1}{2}}\frac{\sin z-i\cos z}{z}.\end{aligned}$$

When $\nu+\frac{1}{2}$ is not a positive integer, the last expression may be regarded as a symbolic representation of $H_\nu^{(1)}(z)$, on the understanding that $f(D)(e^{\pm iz}/z)$ is to be interpreted as

$$i\int_{\pm1+\infty i}^{\pm1} e^{izt} f(it)\,dt.$$

Consequently

(1) $$H_\nu^{(1)}(z)=\frac{(\frac{1}{2}z)^\nu}{\Gamma(\nu+\frac{1}{2})\,\Gamma(\frac{1}{2})}(1+D^2)^{\nu-\frac{1}{2}}\frac{\sin z-i\cos z}{z},$$

and similarly

(2) $$H_\nu^{(2)}(z)=\frac{(\frac{1}{2}z)^\nu}{\Gamma(\nu+\frac{1}{2})\,\Gamma(\frac{1}{2})}(1+D^2)^{\nu-\frac{1}{2}}\frac{\sin z+i\cos z}{z},$$

so that

(3) $$J_\nu(z)=\frac{(\frac{1}{2}z)^\nu}{\Gamma(\nu+\frac{1}{2})\,\Gamma(\frac{1}{2})}(1+D^2)^{\nu-\frac{1}{2}}\frac{\sin z}{z},$$

(4) $$Y_\nu(z)=-\frac{(\frac{1}{2}z)^\nu}{\Gamma(\nu+\frac{1}{2})\,\Gamma(\frac{1}{2})}(1+D^2)^{\nu-\frac{1}{2}}\frac{\cos z}{z}.$$

The series obtained from (4) by expanding in ascending powers of D does not converge unless it terminates; the series obtained in a similar manner from (3) converges only when $R(\nu) > \frac{1}{2}$.

The expressions on the right of (3) and (4), with constant factors omitted, were given by Hargreave, *Phil. Trans. of the Royal Soc.* 1848, p. 36 as solutions of Bessel's equation. The exact formulae are due to Macdonald, *Proc. London Math. Soc.* XXIX. (1898), p. 114.

An associated formula, valid for all values of ν, is derivable from § 6·11 (4). If n is any positive integer, we see from the equation in question that

$$H_\nu^{(1)}(z) = \frac{\Gamma(\frac{1}{2}-\nu).(\frac{1}{2}z)^\nu}{\pi i \Gamma(\frac{1}{2})} \int_{1+\infty i}^{(1+)} (t^2-1)^{\nu-n-\frac{1}{2}} (-)^n . (1+D^2)^n e^{izt}\, dt$$

$$= \frac{(-)^n \Gamma(\frac{1}{2}-\nu).(\frac{1}{2}z)^\nu}{\pi i \Gamma(\frac{1}{2})} (1+D^2)^n \int_{1+\infty i}^{(1+)} (t^2-1)^{\nu-n-\frac{1}{2}} e^{izt}\, dt,$$

so that

(5) $$H_\nu^{(1)}(z) = \frac{\Gamma(\nu+\frac{1}{2}-n).(\frac{1}{2}z)^\nu}{\Gamma(\nu+\frac{1}{2})} (1+D^2)^n \{(\tfrac{1}{2}z)^{n-\nu} H_{\nu-n}^{(1)}(z)\}.$$

A similar equation holds for the other function of the third kind, and so

(6) $$\mathscr{C}_\nu(z) = \frac{\Gamma(\nu+\frac{1}{2}-n).(\frac{1}{2}z)^\nu}{\Gamma(\nu+\frac{1}{2})} (1+D^2)^n \{(\tfrac{1}{2}z)^{n-\nu} \mathscr{C}_{\nu-n}(z)\}.$$

This result, proved when $R(z) > 0$, is easily extended to all values of z by the theory of analytic continuation; it was discovered by Sonine, *Math. Ann.* XVI. (1880), p. 66, when $\nu = n$, and used by Steinthal, *Quarterly Journal*, XVIII. (1882), p. 338 when $\nu = n+\frac{1}{2}$; in the case when $\nu = n+\frac{1}{2}$ the result was given slightly earlier (without the use of the notation of Bessel functions) by Glaisher, *Proc. Camb. Phil. Soc.* III. (1880), pp. 269—271. A proof based on arguments of a physical character has been given by Havelock, *Proc. London Math. Soc.* (2) II. (1904), pp. 124—125.

6·15. *Schläfli's* integrals of Poisson's type for $I_\nu(z)$ and $K_\nu(z)$.*

If we take $\omega = \frac{1}{2}\pi$ in § 6·1 (3) and then replace z by iz, we find that, when $|\arg z| < \frac{1}{2}\pi$,

(1) $$I_{-\nu}(z) = \frac{\Gamma(\frac{1}{2}-\nu)\, e^{2\nu\pi i} (\frac{1}{2}z)^\nu}{2\pi i \Gamma(\frac{1}{2})} \int_\infty^{(1+,-1+)} e^{-zt} (t^2-1)^{\nu-\frac{1}{2}}\, dt;$$

and the phase of t^2-1 at the point where t crosses the negative real axis is -2π.

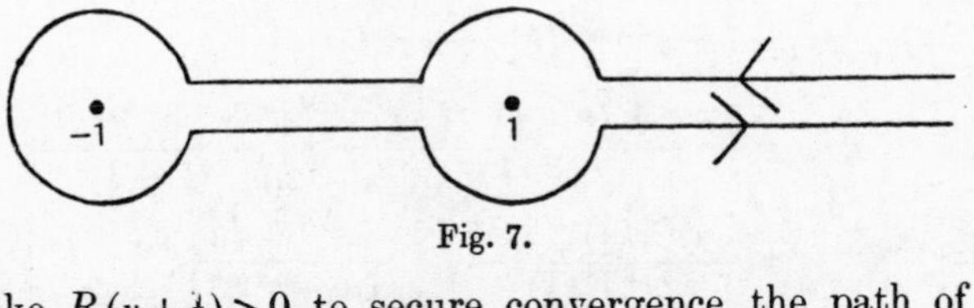

Fig. 7.

If we take $R(\nu+\frac{1}{2}) > 0$ to secure convergence, the path of integration may be taken to be the contour of Fig. 7, in which the radii of the circles may be made to tend to zero. We thus find the formula†

$$I_{-\nu}(z) = \frac{\Gamma(\frac{1}{2}-\nu)\, e^{2\nu\pi i} (\frac{1}{2}z)^\nu}{2\pi i \Gamma(\frac{1}{2})} \left[(1-e^{-4\nu\pi i}) \int_1^\infty e^{-zt} (t^2-1)^{\nu-\frac{1}{2}}\, dt \right.$$
$$\left. + i(e^{-\nu\pi i} + e^{-3\nu\pi i}) \int_{-1}^1 e^{-zt} (1-t^2)^{\nu-\frac{1}{2}}\, dt \right],$$

* *Ann. di Mat.* (2) I. (1868), pp. 239—241. Schläfli obtained the results (1) and (2) directly by the method of § 6·1.

† Cf. Serret, *Journal de Math.* IX. (1844), p. 204.

in which the phases of t^2-1 and of $1-t^2$ are both zero. Now, from § 3·71 (9), we have

$$(2)\qquad I_\nu(z)=\frac{(\tfrac{1}{2}z)^\nu}{\Gamma(\nu+\tfrac{1}{2})\,\Gamma(\tfrac{1}{2})}\int_{-1}^{1}e^{-zt}(1-t^2)^{\nu-\frac{1}{2}}\,dt,$$

and so

$$(3)\qquad I_{-\nu}(z)-I_\nu(z)=\frac{\Gamma(\tfrac{1}{2}-\nu)\sin 2\nu\pi\,.\,(\tfrac{1}{2}z)^\nu}{\pi\Gamma(\tfrac{1}{2})}\int_1^\infty e^{-zt}(t^2-1)^{\nu-\frac{1}{2}}\,dt,$$

that is to say*

$$(4)\qquad K_\nu(z)=\frac{\Gamma(\tfrac{1}{2})\,.\,(\tfrac{1}{2}z)^\nu}{\Gamma(\nu+\tfrac{1}{2})}\int_1^\infty e^{-zt}(t^2-1)^{\nu-\frac{1}{2}}\,dt,$$

whence we obtain the formula

$$(5)\qquad K_\nu(z)=\frac{\Gamma(\tfrac{1}{2})\,.\,(\tfrac{1}{2}z)^\nu}{\Gamma(\nu+\tfrac{1}{2})}\int_0^\infty e^{-z\cosh\theta}\sinh^{2\nu}\theta\,d\theta,$$

a result set by Hobson as a problem in the Mathematical Tripos, 1898. The formulae are all valid when $R(\nu+\tfrac{1}{2})>0$ and $|\arg z|<\tfrac{1}{2}\pi$. The reader will find it instructive to obtain (4) directly from § 6·11 (6).

6·16. *Basset's integral for $K_\nu(xz)$.*

When x is positive and z is a complex number subject to the condition $|\arg z|<\tfrac{1}{2}\pi$, the integral for $H_{-\nu}^{(1)}(xze^{\frac{1}{2}\pi i})$ derived from § 6·11 (6) may be written in the form

$$H_{-\nu}^{(1)}(xze^{\frac{1}{2}\pi i})=\frac{\Gamma(\nu+\tfrac{1}{2})\,.\,(\tfrac{1}{2}ixz)^{-\nu}}{\pi i\,\Gamma(\tfrac{1}{2})}\int_{+\infty}^{(1+)}\frac{e^{-xzt}\,dt}{(t^2-1)^{\nu+\frac{1}{2}}}.$$

Now, when $R(\nu)\geqslant-\tfrac{1}{2}$, the integral, taken round arcs of a circle from ρ to $\rho e^{\pm\frac{1}{2}\pi i-i\arg z}$, tends to zero as $\rho\to\infty$, by Jordan's lemma. Hence, by Cauchy's theorem, the path of integration may be opened out until it becomes the line on which $R(zt)=0$. If then we write $zt=iu$, the phase of $-(u^2/z^2)-1$ is $-\pi$ at the origin in the u-plane.

It then follows from § 3·7 (8) that

$$\begin{aligned}K_\nu(xz)&=\tfrac{1}{2}\pi i e^{-\frac{1}{2}\nu\pi i}H_{-\nu}^{(1)}(xze^{\frac{1}{2}\pi i})\\&=\frac{e^{-\nu\pi i}\,\Gamma(\nu+\tfrac{1}{2})\,.\,(\tfrac{1}{2}xz)^{-\nu}}{2\Gamma(\tfrac{1}{2})}\int_{\infty i/z}^{-\infty i/z}\frac{e^{-xzt}\,dt}{(t^2-1)^{\nu+\frac{1}{2}}}\\&=\frac{\Gamma(\nu+\tfrac{1}{2})\,.\,(2z)^\nu}{2x^\nu\,\Gamma(\tfrac{1}{2})}\int_{-\infty}^{\infty}\frac{e^{-ixu}\,du}{(u^2+z^2)^{\nu+\frac{1}{2}}},\end{aligned}$$

and so we have Basset's formula

$$(1)\qquad K_\nu(xz)=\frac{\Gamma(\nu+\tfrac{1}{2})\,.\,(2z)^\nu}{x^\nu\,\Gamma(\tfrac{1}{2})}\int_0^\infty\frac{\cos xu\,.\,du}{(u^2+z^2)^{\nu+\frac{1}{2}}},$$

valid when $R(\nu+\tfrac{1}{2})\geqslant 0$, $x>0$, $|\arg z|<\tfrac{1}{2}\pi$ The formula was obtained by Basset†, for integral values of ν only, by regarding $K_0(x)$ as the limit of

* The integral on the right was examined in the case $\nu=0$ by Riemann, *Ann. der Physik und Chemie*, (2) xcv. (1855), pp. 130—139.

† *Proc. Camb. Phil. Soc.* VI. (1889), p. 11; *Hydrodynamics*, II. (Cambridge, 1888), p. 19.

a Legendre function of the second kind and expressing it by the corresponding limit of the integral of Laplace's type (*Modern Analysis*, § 15·33). The formula for $K_n(xz)$ is obtainable by repeated applications of the operator $\dfrac{d}{z\,dz}$.

Basset also investigated a similar formula for $I_\nu(xz)$, but there is an error in his result.

The integral on the right in (1) was studied by numerous mathematicians before Basset. Among these investigators were Poisson (see § 6·32), *Journal de l'École Polytechnique*, IX. (1813), pp. 239—241; Catalan, *Journal de Math.* V. (1840), pp. 110—114 (reprinted with some corrections, *Mém. de la Soc. R. des Sci. de Liège*, (2) XII. (1885), pp. 26—31); and Serret, *Journal de Math.* VIII. (1843), pp. 20, 21; IX. (1844), pp. 193—210; Schlömilch, *Analytischen Studien*, II. (Leipzig, 1848), pp. 96—97. These writers evaluated the integral in finite terms when $\nu+\frac{1}{2}$ is a positive integer.

Other writers who must be mentioned are Malmstén, *K. Svenska V. Akad. Handl.* LXII. (1841), pp. 65—74 (see § 7·23); Svanberg, *Nova Acta Reg. Soc. Sci. Upsala*, X. (1832), p. 232; Leslie Ellis, *Trans. Camb. Phil. Soc.* VIII. (1849), pp. 213—215; Enneper, *Math. Ann.* VI. (1873), pp. 360—365; Glaisher, *Phil. Trans. of the Royal Soc.* CLXXII. (1881), pp. 792—815; J. J. Thomson, *Quarterly Journal*, XVIII. (1882), pp. 377—381; Coates, *Quarterly Journal*, XX. (1885), pp. 250—260; and Oltramare, *Comptes Rendus de l'Assoc. Française*, XXIV. (1895), part II. pp. 167—171.

The last named writer proved by contour integration that

$$\int_0^\infty \frac{\cos xu\,.\,du}{(u^2+z^2)^n} = \frac{(-)^{n-1}\pi}{2z^{2n-1}\,.\,(n-1)!}\left[\frac{d^{n-1}}{dp^{n-1}}\left(\frac{e^{-xz\sqrt{p}}}{\sqrt{p}}\right)\right]_{p=1}$$
$$= \frac{(-)^{n-1}\pi}{z^{2n-1}\,.\,(n-1)!}\left[\frac{d^{n-1}}{dp^{n-1}}\frac{e^{-xzp}}{(1+p)^n}\right]_{p=1}.$$

The former of these results may be obtained by differentiating the equation

$$\int_0^\infty \frac{\cos xu\,.\,du}{u^2+z^2p} = \frac{\pi e^{-xz\sqrt{p}}}{2z\sqrt{p}},$$

and the latter is then obtainable by using Lagrange's expansion.

6·17. *Whittaker's* generalisations of Hankel's integrals.*

Formulae of the type contained in § 3·32 suggest that solutions of Bessel's equation should be constructed in the form

$$z^{\frac{1}{2}}\int_a^b e^{izt}\,T\,dt.$$

It may be shewn by the methods of § 6·1 that

$$\nabla_\nu\left\{z^{\frac{1}{2}}\int_a^b e^{izt}\,T\,dt\right\} = \left[z^{\frac{1}{2}}e^{izt}(1-t^2)\left\{\frac{dT}{dt}-izT\right\}\right]_a^b$$
$$- z^{\frac{1}{2}}\int_a^b e^{izt}\left\{(1-t^2)\frac{d^2T}{dt^2} - 2t\frac{dT}{dt} + (\nu^2-\tfrac{1}{4})\,T\right\}dt,$$

and so the integral is a solution if T is a solution of Legendre's equation for functions of order $\nu-\frac{1}{2}$ and the values of the integrated part are the same at each end of the contour.

* *Proc. London Math. Soc.* XXXV. (1903), pp. 198—206.

If T be taken to be the Legendre function $Q_{\nu-\frac{1}{2}}(t)$, the contour may start and end at $+\infty i \exp(-i\omega)$, where ω is an acute angle (positive or negative) provided that z satisfies the inequalities

$$-\tfrac{1}{2}\pi + \omega < \arg z < \tfrac{1}{2}\pi + \omega.$$

If T be taken to be $P_{\nu-\frac{1}{2}}(t)$, the same contour is possible; but the logarithmic singularity of $P_{\nu-\frac{1}{2}}(t)$ at $t=-1$ (when $\nu-\frac{1}{2}$ is not an integer) makes it impossible to take the line joining -1 to 1 as a contour except in the special case considered in § 3·32; for a detailed discussion of the integral in the general case, see § 10·5.

We now proceed to take various contours in detail.

First consider

$$z^{\frac{1}{2}} \int_{\infty i \exp(-i\omega)}^{(-1+,\,1+)} e^{izt}\, Q_{\nu-\frac{1}{2}}(t)\, dt,$$

where the phase of t is zero at the point on the right of $t=1$ at which the contour crosses the real axis. Take the contour to lie wholly outside the circle $|t|=1$ and expand $Q_{\nu-\frac{1}{2}}(t)$ in descending powers of t. It is thus found, as in the similar analysis of § 6·1, that

(1) $$J_\nu(z) = \frac{(\frac{1}{2}z)^{\frac{1}{2}}\, e^{-\frac{1}{2}(\nu+\frac{1}{2})\pi i}}{\pi\, \Gamma(\frac{1}{2})} \int_{\infty i \exp(-i\omega)}^{(-1+,\,1+)} e^{izt}\, Q_{\nu-\frac{1}{2}}(t)\, dt,$$

and therefore

(2) $$J_{-\nu}(z) = \frac{(\frac{1}{2}z)^{\frac{1}{2}}\, e^{\frac{1}{2}(\nu-\frac{1}{2})\pi i}}{\pi\, \Gamma(\frac{1}{2})} \int_{\infty i \exp(-i\omega)}^{(-1+,\,1+)} e^{izt}\, Q_{-\nu-\frac{1}{2}}(t)\, dt.$$

If we combine these formulae and use the relation* connecting the two kinds of Legendre functions, we find that

(3) $$H_\nu^{(2)}(z) = \frac{(\frac{1}{2}z)^{\frac{1}{2}}\, e^{\frac{1}{2}(\nu+\frac{1}{2})\pi i}}{\pi\, \Gamma(\frac{1}{2}) \cos \nu\pi} \int_{\infty i \exp(-i\omega)}^{(-1+,\,1+)} e^{izt}\, P_{\nu-\frac{1}{2}}(t)\, dt.$$

Again, consider

$$z^{\frac{1}{2}} \int_{\infty i \exp(-i\omega)}^{(1+)} e^{izt}\, Q_{\nu-\frac{1}{2}}(t)\, dt;$$

this is a solution of Bessel's equation, and, if the contour be taken to lie on the right of the line $R(t)=a$, it is clear that the integral is $O\{z^{\frac{1}{2}} \exp(-a|z|)\}$ as $z \to +\infty i$. Hence the integral is a multiple of $H_\nu^{(1)}(z)$. Similarly by making $z \to -\infty i$, we find that

$$z^{\frac{1}{2}} \int_{\infty i \exp(-i\omega)}^{(-1+)} e^{izt}\, Q_{\nu-\frac{1}{2}}(t)\, dt$$

* The relation, discovered by Schläfli, is

$$P_n(z) = \frac{\tan n\pi}{\pi} \{Q_n(z) - Q_{-n-1}(z)\};$$

cf. Hobson, *Phil. Trans. of the Royal Soc.* CLXXXVII. (1896), p. 461.

is a multiple of $H_\nu^{(2)}(z)$. From a consideration of (1) it is then clear that

$$(4) \qquad H_\nu^{(1)}(z) = \frac{(2z)^{\frac{1}{2}} e^{-\frac{1}{2}(\nu+\frac{1}{2})\pi i}}{\pi\,\Gamma(\frac{1}{2})} \int_{\infty i \exp(-i\omega)}^{(1+)} e^{izt} Q_{\nu-\frac{1}{2}}(t)\,dt,$$

$$(5) \qquad H_\nu^{(2)}(z) = \frac{(2z)^{\frac{1}{2}} e^{-\frac{1}{2}(\nu+\frac{1}{2})\pi i}}{\pi\,\Gamma(\frac{1}{2})} \int_{\infty i \exp(-i\omega)}^{(-1+)} e^{izt} Q_{\nu-\frac{1}{2}}(t)\,dt,$$

and hence, by § 3·61 combined with Schläfli's relation,

$$(6) \qquad H_\nu^{(2)}(z) = \frac{(\frac{1}{2}z)^{\frac{1}{2}} e^{\frac{1}{2}(\nu+\frac{1}{2})\pi i}}{\pi\,\Gamma(\frac{1}{2})\cos\nu\pi} \int_{\infty i \exp(-i\omega)}^{(-1+)} e^{izt} P_{\nu-\frac{1}{2}}(t)\,dt;$$

this is also obvious from (3).

The integral which differs from (6) only by encircling the point $+1$ instead of -1 is zero since the integrand is analytic inside such a contour.

In (5) and (6), $\arg(t+1)$ vanishes where the contour crosses the real axis on the right of -1, and, in (5), $\arg(t-1)$ is $-\pi$ at that point.

6·2. *Generalisations of Bessel's integral.*

We shall next examine various representations of Bessel functions by a system of definite integrals and contour integrals due to Sonine* and Schläfli†. The fundamental formula which will be obtained is easily reduced to Bessel's integral in the case of functions whose order is an integer.

We take Hankel's well-known generalisation‡ of the second Eulerian integral

$$\frac{1}{\Gamma(\nu+m+1)} = \frac{1}{2\pi i}\int_{-\infty}^{(0+)} t^{-\nu-m-1} e^t\,dt,$$

in which the phase of t increases from $-\pi$ to π as t describes the contour, and then

$$\sum_{m=0}^{\infty} \frac{(-)^m (\frac{1}{2}z)^{\nu+2m}}{m!\,\Gamma(\nu+m+1)} = \frac{(\frac{1}{2}z)^\nu}{2\pi i} \sum_{m=0}^{\infty} \int_{-\infty}^{(0+)} \frac{(-)^m(\frac{1}{2}z)^{2m}}{m!} t^{-\nu-m-1} e^t\,dt.$$

Consider the function obtained by interchanging the signs of summation and integration on the right; it is

$$\int_{-\infty}^{(0+)} t^{-\nu-1} \exp\left\{t - \frac{z^2}{4t}\right\} dt.$$

This is an analytic function of z for all values of z, and, when expanded in ascending powers of z by Maclaurin's theorem, the coefficients may be obtained by differentiating with regard to z *under the integral sign* and making z zero after the differentiations§. Hence

$$\int_{-\infty}^{(0+)} t^{-\nu-1} \exp\left\{t - \frac{z^2}{4t}\right\} dt = \sum_{m=0}^{\infty} \frac{(-)^m(\frac{1}{2}z)^{2m}}{m!} \int_{-\infty}^{(0+)} t^{-\nu-m-1} e^t\,dt,$$

* *Mathematical Collection*, v. (Moscow, 1870); *Math. Ann.* XVI. (1880), pp. 9—29.

† *Ann. di Mat.* (2) v. (1873), p. 204. His memoir, *Math. Ann.* III. (1871), pp. 134—149, should also be consulted. In addition, see Graf, *Math. Ann.* LVI. (1903), pp. 423—432, and Chessin, *Johns Hopkins University Circulars*, XIV. (1895), pp. 20—21.

‡ Cf. *Modern Analysis*, § 12·22.

§ Cf. *Modern Analysis*, §§ 5·32, 4·44.

and so we have at once

$$(1)\qquad J_\nu(z) = \frac{(\frac{1}{2}z)^\nu}{2\pi i}\int_{-\infty}^{(0+)} t^{-\nu-1}\exp\left\{t - \frac{z^2}{4t}\right\} dt.$$

This result, which was discovered by Schläfli, was rediscovered by Sonine; and the latter writer was the first to point out its importance.

When $|\arg z| < \frac{1}{2}\pi$, we may swing round the contour about the origin until it passes to infinity in a direction making an angle $\arg z$ with the negative real axis.

On writing $t = \frac{1}{2}zu$, we then find that, when $|\arg z| < \frac{1}{2}\pi$,

$$(2)\qquad J_\nu(z) = \frac{1}{2\pi i}\int_{-\infty}^{(0+)} u^{-\nu-1}\exp\left\{\tfrac{1}{2}z\left(u - \frac{1}{u}\right)\right\} du.$$

This form was given in Sonine's earlier paper (p. 335).

Again, writing $u = e^w$, we have

$$(3)\qquad J_\nu(z) = \frac{1}{2\pi i}\int_{\infty-\pi i}^{\infty+\pi i} e^{z\sinh w - \nu w}\, dw,$$

valid when $|\arg z| < \frac{1}{2}\pi$. This is the first of the results obtained by Schläfli.

In this formula take the contour to consist of three sides of a rectangle, as in Fig. 8, with vertices at $\infty - \pi i$, $-\pi i$, πi and $\infty + \pi i$.

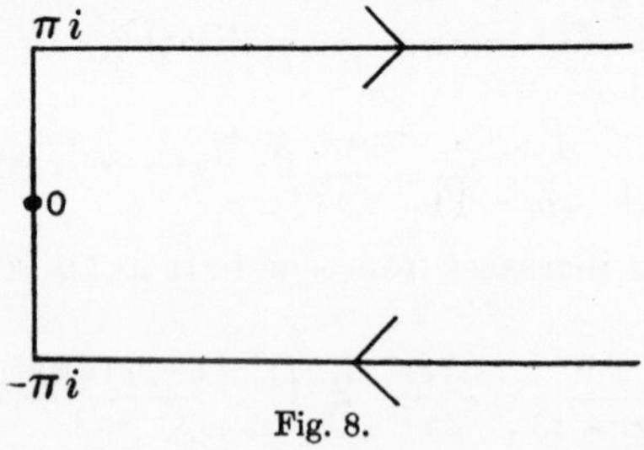

Fig. 8.

If we write $t \mp \pi i$ for w on the sides parallel to the real axis and $\pm i\theta$ for w on the lines joining 0 to $\pm\pi i$, we get *Schläfli's generalisation of Bessel's integral*

$$(4)\qquad J_\nu(z) = \frac{1}{\pi}\int_0^\pi \cos(\nu\theta - z\sin\theta)\,d\theta - \frac{\sin\nu\pi}{\pi}\int_0^\infty e^{-\nu t - z\sinh t}\,dt,$$

valid when $|\arg z| < \frac{1}{2}\pi$.

If we make $\arg z \to \pm\frac{1}{2}\pi$, the first integral on the right is continuous and, if $R(\nu) > 0$, so also is the second, and $J_\nu(z)$ is known to be continuous. So (4) is still true when z is a pure imaginary if $R(\nu)$ is positive.

The integrals just discussed were examined methodically by Sonine in his second memoir; in that memoir he obtained numerous definite integrals by appropriate modifications of the contour. For example, if ψ be an acute angle (positive or negative) and if

$$-\tfrac{1}{2}\pi + \psi < \arg z < \tfrac{1}{2}\pi - \psi,$$

the contour in (3) may be replaced by one which goes from $\infty - (\pi - \psi) i$ to $\infty + (\pi + \psi) i$. By taking the contour to be three sides of a rectangle with corners at $\infty - (\pi - \psi) i$, $-(\pi - \psi) i$, $(\pi + \psi) i$, and $\infty + (\pi + \psi) i$, we obtain, as a modification of (4),

$$(5) \qquad J_\nu(z) = \frac{e^{-\nu i\psi}}{\pi} \int_0^\pi e^{iz \sin\psi \cos\theta} \cos(\nu\theta - z\cos\psi \sin\theta)\, d\theta \\ - \frac{e^{-\nu i\psi} \sin\nu\pi}{\pi} \int_0^\infty e^{-z\sinh(t+i\psi) - \nu t}\, dt.$$

Again, if we take ψ to be an angle between 0 and π, the contour in (3) may be replaced by one which passes from $\infty - (\frac{1}{2}\pi + \psi) i$ to $\infty + (\frac{1}{2}\pi + \psi) i$, and so we find that

$$(6) \qquad J_\nu(z) = \frac{1}{\pi}\int_0^{\frac{1}{2}\pi+\psi} \cos(\nu\theta - z\sin\theta)\, d\theta \\ + \frac{1}{\pi}\int_0^\infty e^{-z\sinh t \sin\psi - \nu t} \sin(z\cosh t \cos\psi - \tfrac{1}{2}\nu\pi - \nu\psi)\, dt,$$

provided that $|\arg z|$ is less than both ψ and $\pi - \psi$.

When $R(\nu) > 0$ and z is positive $(= x)$, we may take $\psi = 0$ in the last formula, and get*

$$(7) \qquad J_\nu(x) = \frac{1}{\pi}\int_0^{\frac{1}{2}\pi} \cos(\nu\theta - x\sin\theta)\, d\theta + \frac{1}{\pi}\int_0^\infty e^{-\nu t} \sin(x\cosh t - \tfrac{1}{2}\nu\pi)\, dt.$$

Another important formula, derived from (1), is obtained by spreading out the contour until it is parallel to the imaginary axis on the right of the origin; by Jordan's lemma this is permissible if $R(\nu) > -1$, and we then obtain the formula

$$(8) \qquad J_\nu(z) = \frac{(\frac{1}{2}z)^\nu}{2\pi i} \int_{c-\infty i}^{c+\infty i} t^{-\nu-1} \exp\left\{t - \frac{z^2}{4t}\right\} dt,$$

in which c may have any positive value; this integral is the basis of many of Sonine's investigations.

Integrals which resemble those given in this section are of importance in the investigation of the diffraction of light by a prism; see Carslaw, *Proc. London Math. Soc.* XXX. (1899), pp. 121—161; W. H. Jackson, *Proc. London Math. Soc.* (2) I. (1904), pp. 393—414; Whipple, *Proc. London Math. Soc.* (2) XVI. (1917), pp. 94—111.

6·21. *Integrals which represent functions of the second and third kinds.*

If we substitute Schläfli's integral § 6·2 (4) for both of the Bessel functions on the right of the equation

$$Y_\nu(z) = J_\nu(z)\cot\nu\pi - J_{-\nu}(z)\operatorname{cosec}\nu\pi,$$

we find that

$$\pi Y_\nu(z) = \cot\nu\pi \int_0^\pi \cos(\nu\theta - z\sin\theta)\, d\theta - \operatorname{cosec}\nu\pi \int_0^\pi \cos(\nu\theta + z\sin\theta)\, d\theta \\ - \cos\nu\pi \int_0^\infty e^{-\nu t - z\sinh t}\, dt - \int_0^\infty e^{\nu t - z\sinh t}\, dt.$$

* Cf. Gubler, *Math. Ann.* XLIX. (1897), pp. 583—584.

Replace θ by $\pi - \theta$ in the second integral on the right, and it is found on reduction that

$$(1) \quad Y_\nu(z) = \frac{1}{\pi}\int_0^\pi \sin(z\sin\theta - \nu\theta)\,d\theta - \frac{1}{\pi}\int_0^\infty (e^{\nu t} + e^{-\nu t}\cos\nu\pi)\,e^{-z\sinh t}\,dt,$$

a formula, practically discovered by Schläfli (who actually gave the corresponding formula for Neumann's function), which is valid when $|\arg z| < \frac{1}{2}\pi$.

By means of this result we can evaluate

$$\frac{1}{\pi i}\int_{-\infty}^{\infty+\pi i} e^{z\sinh w - \nu w}\,dw,$$

when $|\arg z| < \frac{1}{2}\pi$; for we take the contour to be rectilinear, as in Fig. 9, and

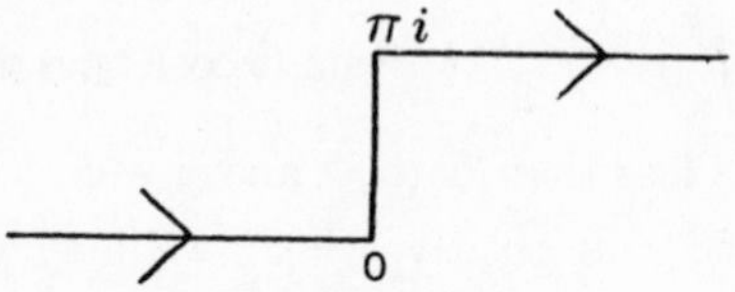

Fig. 9.

write $-t$, $i\theta$, $t + \pi i$ for w on the three parts of the contour; we then see that the expression is equal to

$$\frac{1}{\pi i}\int_0^\infty e^{\nu t - z\sinh t}\,dt + \frac{1}{\pi}\int_0^\pi e^{i(z\sin\theta - \nu\theta)}\,d\theta + \frac{e^{-\nu\pi i}}{\pi i}\int_0^\infty e^{-\nu t - z\sinh t}\,dt,$$

and this is equal to $J_\nu(z) + iY_\nu(z)$ from formula (1) combined with § 6·2 (4). Hence, when $|\arg z| < \frac{1}{2}\pi$, we have

$$(2) \qquad H_\nu^{(1)}(z) = \frac{1}{\pi i}\int_{-\infty}^{\infty+\pi i} e^{z\sinh w - \nu w}\,dw,$$

$$(3) \qquad H_\nu^{(2)}(z) = -\frac{1}{\pi i}\int_{-\infty}^{\infty-\pi i} e^{z\sinh w - \nu w}\,dw.$$

Formulae equivalent to these were discovered by Sommerfeld, *Math. Ann.* XLVII. (1896), pp. 327—357. The only difference between these formulae and Sommerfeld's is a rotation of the contours through a right angle, with a corresponding change in the parametric variable; see also Hopf and Sommerfeld, *Archiv der Math. und Phys.* (3) XVIII. (1911), pp. 1—16.

By an obvious change of variable we may write (2) and (3) in the forms

$$(4) \qquad H_\nu^{(1)}(z) = \frac{1}{\pi i}\int_0^{\infty\exp\pi i} u^{-\nu-1}\exp\left\{\tfrac{1}{2}z\left(u - \frac{1}{u}\right)\right\}du,$$

$$(5) \qquad H_\nu^{(2)}(z) = -\frac{1}{\pi i}\int_0^{\infty\exp(-\pi i)} u^{-\nu-1}\exp\left\{\tfrac{1}{2}z\left(u - \frac{1}{u}\right)\right\}du;$$

the contours are those shewn in Fig. 10, emerging from the origin and then bending round to the left and right respectively; results equivalent to these were discovered by Schläfli.

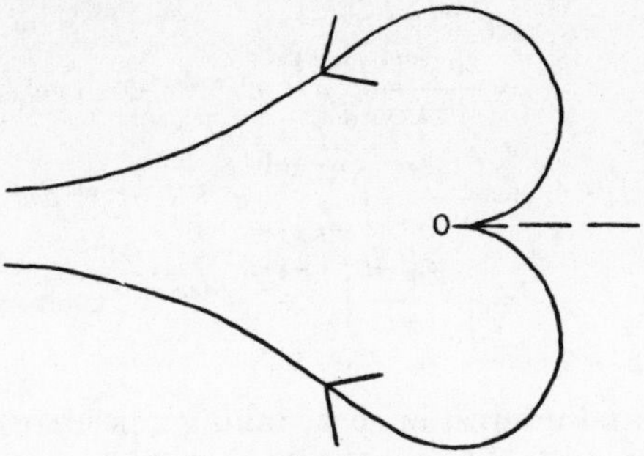

Fig. 10.

[NOTE. There is no difficulty in proving these results for integral values of ν, in view of the continuity of the functions involved; cf. § 6·11.]

We proceed to modify the contours involved in (4) and (5) to obtain the analytic continuations of the functions on the left.

If ω is an angle between $-\pi$ and π such that $|\omega - \arg z| < \frac{1}{2}\pi$, we have

(6) $$H_\nu^{(1)}(z) = \frac{1}{\pi i}\int_{0 \exp i\omega}^{\infty \exp(\pi - \omega) i} u^{-\nu-1} \exp\left\{\tfrac{1}{2}z\left(u - \frac{1}{u}\right)\right\} du,$$

and

(7) $$H_\nu^{(2)}(z) = -\frac{1}{\pi i}\int_{0 \exp i\omega}^{\infty \exp(-\pi - \omega) i} u^{-\nu-1} \exp\left\{\tfrac{1}{2}z\left(u - \frac{1}{u}\right)\right\} du,$$

the contours being those shewn in Fig. 11 and Fig. 12; and these formulae give the analytic continuations of the functions on the left over the range of

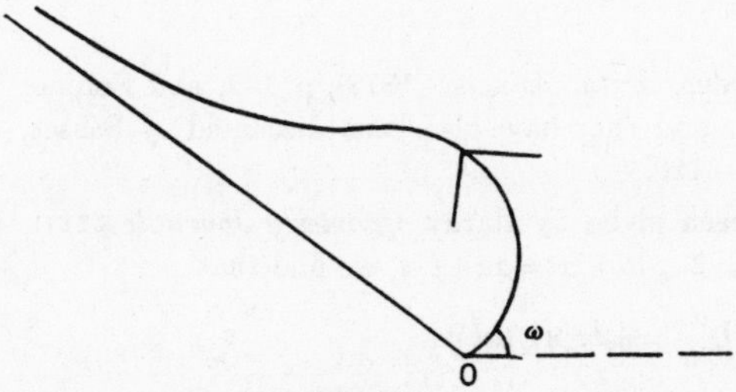

Fig. 11.

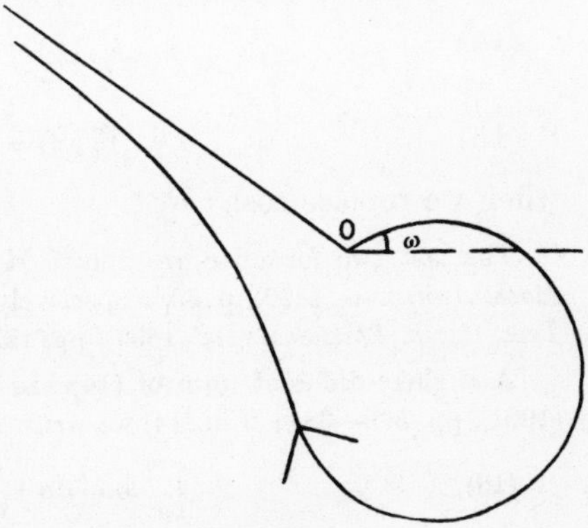

Fig. 12.

values of z for which $\omega - \frac{1}{2}\pi < \arg z < \omega + \frac{1}{2}\pi$; and ω may have any value between* $-\pi$ and π.

* If $|\omega|$ were increased beyond these limits, difficulties would arise in the interpretation of the phase of u.

Modifications of (2) and (3) are obtained by replacing w by $w \pm \frac{1}{2}\pi i$; it is thus found that*

(8)
$$H_\nu^{(1)}(z) = \frac{e^{-\frac{1}{2}\nu\pi i}}{\pi i}\int_{-\infty-\frac{1}{2}\pi i}^{\infty+\frac{1}{2}\pi i} e^{iz\cosh w - \nu w}\,dw$$
$$= \frac{2e^{-\frac{1}{2}\nu\pi i}}{\pi i}\int_0^{\infty+\frac{1}{2}\pi i} e^{iz\cosh w}\cosh \nu w\,.\,dw,$$

(9)
$$H_\nu^{(2)}(z) = -\frac{e^{\frac{1}{2}\nu\pi i}}{\pi i}\int_{-\infty+\frac{1}{2}\pi i}^{\infty-\frac{1}{2}\pi i} e^{-iz\cosh w - \nu w}\,dw$$
$$= -\frac{2e^{\frac{1}{2}\nu\pi i}}{\pi i}\int_0^{\infty-\frac{1}{2}\pi i} e^{-iz\cosh w}\cosh \nu w\,.\,dw,$$

provided that $|\arg z| < \frac{1}{2}\pi$.

Formulae of special interest arise by taking z positive $(=x)$ in (6) and (7) and $-1 < R(\nu) < 1$. A double application of Jordan's lemma (to circles of large and small radius respectively) shews that, in such circumstances, we may take $\omega = \frac{1}{2}\pi$ in (6) and $\omega = -\frac{1}{2}\pi$ in (7). It is thus clear, if u be replaced by $\pm ie^t$, that

(10)
$$H_\nu^{(1)}(x) = \frac{e^{-\frac{1}{2}\nu\pi i}}{\pi i}\int_{-\infty}^{\infty} e^{ix\cosh t - \nu t}\,dt = \frac{2e^{-\frac{1}{2}\nu\pi i}}{\pi i}\int_0^\infty e^{ix\cosh t}\cosh \nu t\,.\,dt,$$

(11)
$$H_\nu^{(2)}(x) = -\frac{e^{\frac{1}{2}\nu\pi i}}{\pi i}\int_{-\infty}^{\infty} e^{-ix\cosh t - \nu t}\,dt = -\frac{2e^{\frac{1}{2}\nu\pi i}}{\pi i}\int_0^\infty e^{-ix\cosh t}\cosh \nu t\,.\,dt,$$

and hence, when $x > 0$ and $-1 < R(\nu) < 1$, we have

(12)
$$J_\nu(x) = \frac{2}{\pi}\int_0^\infty \sin(x\cosh t - \tfrac{1}{2}\nu\pi)\,.\,\cosh \nu t\,.\,dt,$$

(13)
$$Y_\nu(x) = -\frac{2}{\pi}\int_0^\infty \cos(x\cosh t - \tfrac{1}{2}\nu\pi)\,.\,\cosh \nu t\,.\,dt;$$

and, in particular (cf. § 6·13),

(14)
$$J_0(x) = \frac{2}{\pi}\int_1^\infty \frac{\sin xt\,.\,dt}{\sqrt{(t^2-1)}},$$

(15)
$$Y_0(x) = -\frac{2}{\pi}\int_1^\infty \frac{\cos xt\,.\,dt}{\sqrt{(t^2-1)}},$$

when we replace $\cosh t$ by t.

The last two formulae are due to Mehler, *Math. Ann.* V. (1872), p. 142, and Sonine, *Math. Ann.* XVI. (1880), p. 39, respectively; and they have also been discussed by Basset, *Proc. Camb. Phil. Soc.* VIII. (1895), pp. 122—128.

A slightly different form of (14) has been given by Hardy, *Quarterly Journal*, XXXII. (1901), pp. 369—384; if in (14) we write $x = 2\sqrt{(ab)}$, $xt = au + b/u$, we find that

(16)
$$\int_0^\infty \sin\left(au + \frac{b}{u}\right)\frac{du}{u} = \pi J_0\{2\sqrt{(ab)}\}.$$

NOTE. The reader will find it instructive to obtain (14) from the formula

$$P_n(\cos\theta) = \frac{2}{\pi}\int_\theta^\pi \frac{\sin(n+\frac{1}{2})\phi}{\sqrt{\{2(\cos\theta - \cos\phi)\}}}\,d\phi$$

combined with the formula § 5·71 (1). This was Mehler's original method.

* Cf. Coates, *Quarterly Journal*, XXI. (1886), pp. 183—192.

6·22. *Integrals representing $I_\nu(z)$ and $K_\nu(z)$.*

The modifications of the previous analysis which are involved in the discussion of $I_\nu(z)$ and $K_\nu(z)$ are of sufficient interest to be given fully; they are due to Schläfli*, though he expressed his results mainly in terms of the function $F(a, t)$ of § 4·15.

The analysis of § 6·2 is easily modified so as to prove that

(1) $$I_\nu(z)=\frac{(\frac{1}{2}z)^\nu}{2\pi i}\int_{-\infty}^{(0+)} t^{-\nu-1}\exp\left(t+\frac{z^2}{4t}\right)dt,$$

and hence, when $|\arg z|<\frac{1}{2}\pi$,

(2) $$I_\nu(z)=\frac{1}{2\pi i}\int_{-\infty}^{(0+)} u^{-\nu-1}\exp\left\{\tfrac{1}{2}z\left(u+\frac{1}{u}\right)\right\}du,$$

(3) $$I_\nu(z)=\frac{1}{2\pi i}\int_{\infty-\pi i}^{\infty+\pi i} e^{z\cosh w-\nu w}\,dw.$$

The formulae (2) and (3) are valid when $\arg z=\pm\frac{1}{2}\pi$ if $R(\nu)>0$.

If in (3) the contour is taken to be three sides of a rectangle with corners at $\infty-\pi i$, $-\pi i$, πi, $\infty+\pi i$, it is found that

(4) $$I_\nu(z)=\frac{1}{\pi}\int_0^\pi e^{z\cos\theta}\cos\nu\theta\,d\theta-\frac{\sin\nu\pi}{\pi}\int_0^\infty e^{-z\cosh t-\nu t}\,dt,$$

so that

$$I_{-\nu}(z)-I_\nu(z)=\frac{2\sin\nu\pi}{\pi}\int_0^\infty e^{-z\cosh t}\cosh\nu t\,.\,dt,$$

and hence, when $|\arg z|<\frac{1}{2}\pi$,

(5) $$K_\nu(z)=\int_0^\infty e^{-z\cosh t}\cosh\nu t\,.\,dt,$$

a formula obtained by Schläfli† by means of somewhat elaborate transformations.

From the results just obtained, we can evaluate

$$\frac{1}{2\pi i}\int_{-\infty-\pi i}^{\infty+\pi i} e^{z\cosh w-\nu w}\,dw$$

when $|\arg z|<\frac{1}{2}\pi$. For it is easily seen that

$$\begin{aligned}\frac{1}{2\pi i}\int_{-\infty-\pi i}^{\infty+\pi i} e^{z\cosh w-\nu w}\,dw&=\frac{1}{2\pi i}\left\{\int_{-\infty-\pi i}^{\infty-\pi i}+\int_{\infty-\pi i}^{\infty+\pi i}\right\}e^{z\cosh w-\nu w}\,dw\\&=\frac{1}{2\pi i}\int_{-\infty-\pi i}^{\infty-\pi i}e^{z\cosh w-\nu w}\,dw+I_\nu(z)\\&=\frac{e^{\nu\pi i}}{2\pi i}\int_{-\infty}^{\infty}e^{-z\cosh t-\nu t}\,dt+I_\nu(z)\\&=\frac{e^{\nu\pi i}}{2i\sin\nu\pi}\{I_{-\nu}(z)-I_\nu(z)\}+I_\nu(z),\end{aligned}$$

* *Ann. di Mat.* (2) v. (1873), pp. 199—205.

† *Ann. di Mat.* (2) v. (1873), pp. 199—201; this formula was used by Heine, *Journal für Math.* LXIX. (1868), p. 131, as the definition to which reference was made in § 5·72.

and hence

(6) $$\frac{1}{2\pi i}\int_{-\infty-\pi i}^{\infty+\pi i} e^{z\cosh w-\nu w}\,dw = \frac{e^{\nu\pi i} I_{-\nu}(z) - e^{-\nu\pi i} I_\nu(z)}{2i\sin\nu\pi}.$$

Again, we may write (5) in the form

(7) $$K_\nu(z) = \frac{1}{2}\int_{-\infty}^{\infty} e^{-z\cosh t-\nu t}\,dt,$$

and hence, by the processes used in § 6·21,

(8) $$K_\nu(z) = \frac{1}{2}\int_{0\exp i\omega}^{\infty\exp(-i\omega)} u^{-\nu-1}\exp\left\{-\tfrac{1}{2}z\left(u+\frac{1}{u}\right)\right\}du,$$

when $-\pi < \omega < \pi$ and $-\frac{1}{2}\pi + \omega < \arg z < \frac{1}{2}\pi + \omega$.

Similarly

(9) $$e^{\nu\pi i} I_{-\nu}(z) - e^{-\nu\pi i} I_\nu(z) = \frac{\sin\nu\pi}{\pi}\int_{0\exp(-\pi+\omega)i}^{\infty\exp(\pi-\omega)i} u^{-\nu-1}\exp\left\{\tfrac{1}{2}z\left(u+\frac{1}{u}\right)\right\}du;$$

this is valid when $0 < \omega < 2\pi$ and $-\frac{1}{2}\pi + \omega < \arg z < \frac{1}{2}\pi + \omega$.

The contours for the formulae (8) and (9) are shewn in Figs. 13 and 14 respectively.

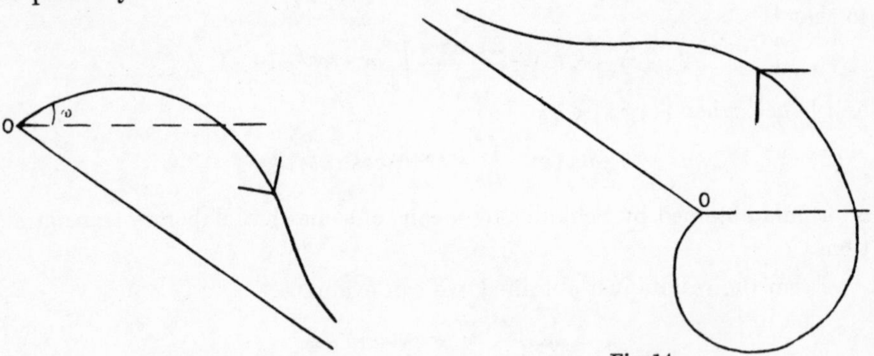

Fig. 13. Fig. 14.

Further, when z is positive $(= x)$ and $-1 < R(\nu) < 1$, the path of integration in (8) may be swung round until it becomes the positive half of the imaginary axis; it is thus found that

$$K_\nu(x) = \tfrac{1}{2}e^{-\frac{1}{2}\nu\pi i}\int_0^\infty v^{-\nu-1}\exp\left\{-\tfrac{1}{2}ix\left(v-\frac{1}{v}\right)\right\}dv,$$

so that

(10) $$K_\nu(x) = \tfrac{1}{2}e^{-\frac{1}{2}\nu\pi i}\int_{-\infty}^{\infty} e^{-ix\sinh t-\nu t}\,dt,$$

and, on changing the sign of ν,

(11) $$K_\nu(x) = \tfrac{1}{2}e^{\frac{1}{2}\nu\pi i}\int_{-\infty}^{\infty} e^{-ix\sinh t+\nu t}\,dt.$$

From these results we see that

(12) $$2\cos\tfrac{1}{2}\nu\pi \,.\, K_\nu(x) = \int_{-\infty}^{\infty} e^{-ix\sinh t}\cosh \nu t \,.\, dt,$$

so that

(13) $$K_\nu(x) = \frac{1}{\cos\frac{1}{2}\nu\pi}\int_0^\infty \cos(x\sinh t)\cosh\nu t\,.\,dt,$$

and these formulae are all valid when $x > 0$ and $-1 < R(\nu) < 1$.

In particular

(14) $$K_0(x) = \int_0^\infty \cos(x\sinh t)\,dt = \int_0^\infty \frac{\cos(xt)\,dt}{\sqrt{(t^2+1)}},$$

a result obtained by Mehler* in 1870.

It may be observed that if, in (7), we make the substitution $\frac{1}{2}ze^t = \tau$, we find that

(15) $$K_\nu(z) = \tfrac{1}{2}(\tfrac{1}{2}z)^\nu \int_0^\infty \exp\left\{-\tau - \frac{z^2}{4\tau}\right\}\frac{d\tau}{\tau^{\nu+1}},$$

provided that $R(z^2) > 0$. The integral on the right has been studied by numerous mathematicians, among whom may be mentioned Poisson, *Journal de l'École Polytechnique*, IX. (cahier 16), 1813, p. 237; Glaisher, *British Association Report*, 1872, pp. 15—17; *Proc. Camb. Phil. Soc.* III. (1880), pp. 5—12; and Kapteyn, *Bull. des Sci. Math.* (2) XVI. (1892), pp. 41—44. The integrals in which ν has the special values $\frac{1}{2}$ and $\frac{3}{2}$ were discussed by Euler, *Inst. Calc. Int.* IV. (Petersburg, 1794), p. 415; and, when ν is half of an odd integer, the integral has been evaluated by Legendre, *Exercices de Calcul Intégral*, I. (Paris, 1811), p. 366; Cauchy, *Exercices des Math.* (Paris, 1826), pp. 54—56; and Schlömilch, *Journal für Math.* XXXIII. (1846), pp. 268—280. The integral in which the limits of integration are arbitrary has been examined by Binet, *Comptes Rendus*, XII. (1841), pp. 958—962.

6·23. *Hardy's formulae for integrals of Du Bois Reymond's type.*

The integrals

$$\int_0^\infty \sin t\,.\,\sin\frac{x^2}{t}\,.\,t^{\nu-1}\,dt, \quad \int_0^\infty \cos t\,.\,\cos\frac{x^2}{t}\,.\,t^{\nu-1}\,dt,$$

in which $x > 0$, $-1 < R(\nu) < 1$, have been examined by Hardy† as examples of Du Bois Reymond's integrals

$$\int_0^\infty f(t) \begin{matrix}\sin\\ \cos\end{matrix} t\,.\,t^{\nu-1}\,dt,$$

in which $f(t)$ oscillates rapidly as $t \to 0$. By constructing a differential equation of the fourth order, Hardy succeeded in expressing them in terms of Bessel functions; but a simpler way of evaluating them is to make use of the results of §§ 6·21, 6·22.

* *Math. Ann.* XVIII. (1881), p. 182.

† *Messenger*, XL. (1911), pp. 44—51.

If we replace t by xe^t, it is clear that

$$\int_0^\infty \sin t \sin \frac{x^2}{t} . t^{\nu-1}\, dt = x^\nu \int_{-\infty}^\infty \sin(xe^t)\sin(xe^{-t}) . e^{\nu t}\, dt$$

$$= -\tfrac{1}{4}x^\nu \int_{-\infty}^\infty \{e^{2ix\cosh t} + e^{-2ix\cosh t} - e^{2ix\sinh t} - e^{-2ix\sinh t}\} e^{\nu t}\, dt$$

$$= -\tfrac{1}{4}x^\nu [\pi i e^{-\frac{1}{2}\nu\pi i} H_{-\nu}^{(1)}(2x) - \pi i e^{\frac{1}{2}\nu\pi i} H_{-\nu}^{(2)}(2x)$$
$$- 2e^{-\frac{1}{2}\nu\pi i} K_\nu(2x) - 2e^{\frac{1}{2}\nu\pi i} K_{-\nu}(2x)],$$

and hence we have

(1) $$\int_0^\infty \sin t \sin \frac{x^2}{t} . t^{\nu-1} dt = \frac{\pi x^\nu}{4\sin\frac{1}{2}\nu\pi}[J_\nu(2x) - J_{-\nu}(2x) + I_{-\nu}(2x) - I_\nu(2x)],$$

and similarly

(2) $$\int_0^\infty \cos t \cos \frac{x^2}{t} . t^{\nu-1}\, dt = \frac{\pi x^\nu}{4\sin\frac{1}{2}\nu\pi}[J_{-\nu}(2x) - J_\nu(2x) + I_{-\nu}(2x) - I_\nu(2x)].$$

When ν has the special value zero, these formulae become

(3) $$\int_0^\infty \sin t \sin \frac{x^2}{t} . \frac{dt}{t} = \tfrac{1}{2}\pi Y_0(2x) + K_0(2x),$$

(4) $$\int_0^\infty \cos t \cos \frac{x^2}{t} . \frac{dt}{t} = -\tfrac{1}{2}\pi Y_0(2x) + K_0(2x).$$

6·24. *Theisinger's extension of Bessel's integral.*

A curious extension of Jacobi's formulae of § 2·2 has been obtained in the case of $J_0(x)$ and $J_1(x)$ by Theisinger, *Monatshefte für Math. und Phys.* XXIV. (1913), pp. 337—341; we shall now give a generalisation of Theisinger's formula which is valid for functions of order ν where $-\frac{1}{2} < \nu < \frac{1}{2}$.

If a is any positive number*, it is obvious from Poisson's integral that

$$J_\nu(x) = \frac{2(\frac{1}{2}x)^\nu}{\Gamma(\nu+\frac{1}{2})\Gamma(\frac{1}{2})}\int_0^{\frac{1}{2}\pi} e^{-ax\sin\theta}\cos(x\cos\theta)\sin^{2\nu}\theta\, d\theta$$
$$+ \frac{2(\frac{1}{2}x)^\nu}{\Gamma(\nu+\frac{1}{2})\Gamma(\frac{1}{2})}\int_0^{\frac{1}{2}\pi}(1 - e^{-ax\sin\theta})\cos(x\cos\theta)\sin^{2\nu}\theta\, d\theta.$$

Now $$2\int_0^{\frac{1}{2}\pi}(1 - e^{-ax\sin\theta})\cos(x\cos\theta)\sin^{2\nu}\theta\, d\theta$$

$$= \int_0^\pi \frac{1 - e^{-ax\sin\theta}}{\sinh(x\sin\theta)}\sinh(x\sin\theta - ix\cos\theta)\sin^{2\nu}\theta\, d\theta$$

$$= -i\int_{-1}^1 \frac{1 - \exp\{\frac{1}{2}axi(z - 1/z)\}}{\sinh\{\frac{1}{2}xi(z - 1/z)\}}\sinh(-ixz) . \left(\frac{z - 1/z}{2i}\right)^{2\nu}\frac{dz}{z},$$

where the contour passes above the origin. Take the contour to be the real axis with an indentation at the origin, and write $z = \pm\tan\frac{1}{2}\phi$ on the two parts of the contour; we thus find that the last expression is equal to

$$-i\int_0^{\frac{1}{2}\pi}\frac{1 - \exp(-axi\cot\phi)}{\sin(x\cot\phi)}\sin(x\tan\tfrac{1}{2}\phi) . e^{\nu\pi i}\cot^{2\nu}\phi\,\frac{d\phi}{\sin\phi}$$
$$+ i\int_0^{\frac{1}{2}\pi}\frac{1 - \exp(axi\cot\phi)}{\sin(x\cot\phi)}\sin(x\tan\tfrac{1}{2}\phi) . e^{-\nu\pi i}\cot^{2\nu}\phi\,\frac{d\phi}{\sin\phi}$$
$$= 4\int_0^{\frac{1}{2}\pi}\sin(\tfrac{1}{2}ax\cot\phi)\cos(\tfrac{1}{2}ax\cot\phi - \nu\pi)\frac{\sin(x\tan\frac{1}{2}\phi)}{\sin(x\cot\phi)}\cot^{2\nu}\phi\,\frac{d\phi}{\sin\phi},$$

* In Theisinger's analysis, a is an even integer.

and therefore

$$(1)\quad \frac{\Gamma(\nu+\frac{1}{2})\,\Gamma(\frac{1}{2})}{2(\frac{1}{2}x)^\nu}J_\nu(x)=\int_0^{\frac{1}{2}\pi}e^{-ax\sin\theta}\cos(x\cos\theta)\sin^{2\nu}\theta\,d\theta$$
$$+2\int_0^{\frac{1}{2}\pi}\sin(\tfrac{1}{2}ax\cot\phi)\cos(\tfrac{1}{2}ax\cot\phi-\nu\pi)\frac{\sin(x\tan\frac{1}{2}\phi)}{\sin(x\cot\phi)}\cot^{2\nu}\phi\frac{d\phi}{\sin\phi}.$$

The transformation fails when $\nu \geqslant \frac{1}{2}$, because the integral round the indentation does not tend to zero with the radius of the indentation. The form given by Theisinger in the case $\nu=1$ differs from (1) because he works with § 3·3 (7) which gives

$$(2)\quad \frac{\Gamma(\nu-\frac{1}{2})\,\Gamma(\frac{1}{2})}{2(\frac{1}{2}x)^{\nu-1}}J_\nu(x)=\int_0^{\frac{1}{2}\pi}e^{-ax\sin\theta}\sin(x\cos\theta)\sin^{2\nu-2}\theta\cos\theta\,d\theta$$
$$+4\int_0^{\frac{1}{2}\pi}\sin(\tfrac{1}{2}ax\cot\phi)\cos(\tfrac{1}{2}ax\cot\phi-\nu\pi)\frac{\sin^2(\frac{1}{2}x\tan\frac{1}{2}\phi)}{\sin(x\cot\phi)}\cot^{2\nu-2}\phi\frac{d\phi}{\sin^2\phi},$$

provided that $\frac{1}{2}<\nu<\frac{3}{2}$.

6·3. *The equivalence of the integral representations of $K_\nu(z)$.*

Three different types of integrals which represent $K_\nu(z)$ have now been obtained in §§ 6·15 (4), 6·22 (5) and 6·16 (1), namely

$$K_\nu(z)=\frac{\Gamma(\frac{1}{2}).(\frac{1}{2}z)^\nu}{\Gamma(\nu+\frac{1}{2})}\int_1^\infty e^{-zt}(t^2-1)^{\nu-\frac{1}{2}}\,dt$$
$$=\int_0^\infty e^{-z\cosh t}\cosh\nu t\,.\,dt,$$
$$K_\nu(xz)=\frac{\Gamma(\nu+\frac{1}{2}).(2z)^\nu}{x^\nu\,\Gamma(\frac{1}{2})}\int_0^\infty\frac{\cos xu\,.\,du}{(u^2+z^2)^{\nu+\frac{1}{2}}}.$$

The equality of the first and second was directly demonstrated in 1871 by Schläfli*; but Poisson proved the equivalence of the second and third as early as 1813, while Malmstén gave a less direct proof of the equivalence of the second and third in 1841. We proceed to describe the three transformations in question.

6·31. *Schläfli's transformation.*

We first give an abstract of the analysis used by Schläfli, *Ann. di Mat.* (2) v. (1873), pp. 199—201, to prove the relation

$$\frac{\Gamma(\frac{1}{2}).(\frac{1}{2}z)^\nu}{\Gamma(\nu+\frac{1}{2})}\int_1^\infty e^{-zt}(t^2-1)^{\nu-\frac{1}{2}}\,dt=\int_0^\infty e^{-z\cosh\theta}\cosh\nu\theta\,d\theta$$

which arises from a comparison of two of the integral representations of $K_\nu(z)$, and which may be established by analysis resembling that of § 2·323.

We have, of course, to suppose that $R(z)>0$ to secure convergence, and it is convenient at first to take† $-\frac{1}{2}<R(\nu)<1$.

* An earlier proof is due to Kummer, *Journal für Math.* XVII. (1837), pp. 228—242, but it is much more elaborate than Schläfli's investigation.

† The result is established for larger values of $R(\nu)$ either by the theory of analytic continuation or by the use of recurrence formulae.

Now define S by the equation

$$S=\int_1^x (t^2-1)^{\nu-\frac{1}{2}}\frac{dt}{(x-t)^\nu},$$

where $x\geqslant 1$; and then, if $t=x-(x-1)u$, we have

$$S=(x+1)^{\nu-\frac{1}{2}}(x-1)^{\frac{1}{2}}\int_0^1 u^{-\nu}(1-u)^{\nu-\frac{1}{2}}\left\{1-\frac{u(x-1)}{x+1}\right\}^{\nu-\frac{1}{2}}du$$

$$=\frac{\Gamma(1-\nu)\,\Gamma(\nu+\frac{1}{2})}{2\nu\Gamma(\frac{1}{2})}(x+1)^\nu\left[\left\{1+\sqrt{\frac{x-1}{x+1}}\right\}^{2\nu}-\left\{1-\sqrt{\frac{x-1}{x+1}}\right\}^{2\nu}\right],$$

on expanding the last factor of the integrand in powers of u and integrating term-by-term.

Replacing x by $\cosh\theta$, we see that

$$\int_1^{\cosh\theta}(t^2-1)^{\nu-\frac{1}{2}}\frac{dt}{(\cosh\theta-t)^\nu}=2^\nu\frac{\Gamma(1-\nu)\,\Gamma(\nu+\frac{1}{2})}{\nu\,\Gamma(\frac{1}{2})}\sinh\nu\theta,$$

so that, by a partial integration,

$$\frac{2^\nu\,\Gamma(\nu+\frac{1}{2})}{\Gamma(\frac{1}{2})}\int_0^\infty e^{-z\cosh\theta}\cosh\nu\theta\,d\theta=\frac{2^\nu z\Gamma(\nu+\frac{1}{2})}{\nu\Gamma(\frac{1}{2})}\int_0^\infty e^{-z\cosh\theta}\sinh\theta\sinh\nu\theta\,d\theta$$

$$=\frac{z}{\Gamma(1-\nu)}\int_0^\infty e^{-z\cosh\theta}S\sinh\theta\,d\theta$$

$$=\frac{z}{\Gamma(1-\nu)}\int_1^\infty\int_1^x e^{-zx}(t^2-1)^{\nu-\frac{1}{2}}\frac{dt\,dx}{(x-t)^\nu}$$

$$=\frac{z}{\Gamma(1-\nu)}\int_1^\infty\int_t^\infty e^{-zx}(t^2-1)^{\nu-\frac{1}{2}}\frac{dx\,dt}{(x-t)^\nu}$$

$$=\frac{z}{\Gamma(1-\nu)}\int_1^\infty\int_0^\infty e^{-z(t+u)}(t^2-1)^{\nu-\frac{1}{2}}\frac{du\,dt}{u^\nu}$$

$$=z^\nu\int_1^\infty e^{-zt}(t^2-1)^{\nu-\frac{1}{2}}dt;$$

the inversion of the order of the integrations presents no great theoretical difficulty, and the transformation is established.

6·32. *Poisson's transformation.*

The direct proof that

$$\frac{1}{2}\int_{-\infty}^\infty e^{-xz\cosh t-\nu t}\,dt=\frac{\Gamma(\nu+\frac{1}{2})\,.\,(2z)^\nu}{x^\nu\,\Gamma(\frac{1}{2})}\int_0^\infty\frac{\cos(xu)\,du}{(u^2+z^2)^{\nu+\frac{1}{2}}}$$

is due to Poisson*, *Journal de l'École Polytechnique*, IX. (1813), pp. 239—241. The equation is true when $|\arg z|<\frac{1}{2}\pi$, $x>0$ and $R(\nu)>-\frac{1}{2}$, but it is convenient to assume in the course of the proof that $R(\nu)>\frac{1}{2}$ and $|\arg z|<\frac{1}{4}\pi$, and to derive the result for other values of z and ν by an appeal to recurrence formulae and the theory of analytic continuation.

If we replace t by a new variable defined by the equation $v=x^\nu e^{-\nu t}$, we see that it is sufficient to prove that

$$\Gamma(\nu+\tfrac{1}{2})\int_0^\infty\frac{\cos(xu)\,du}{(u^2+z^2)^{\nu+\frac{1}{2}}}=\frac{\frac{1}{2}\Gamma(\frac{1}{2})}{\nu(2z)^\nu}\int_0^\infty\exp\{-\tfrac{1}{2}z(v^{1/\nu}+x^2v^{-1/\nu})\}\,dv.$$

* See also Paoli, *Mem. di Mat. e di Fis. della Soc. Italiana delle Sci.* XX. (1828), p. 172.

Now the expression on the left is equal to

$$\int_0^\infty \int_0^\infty \frac{t^{\nu-\frac{1}{2}} e^{-t} \cos xu}{(u^2+z^2)^{\nu+\frac{1}{2}}} dt\, du = \int_0^\infty \int_0^\infty s^{\nu-\frac{1}{2}} \exp\{-s(u^2+z^2)\} \cos xu \,.\, ds\, du$$

$$= \int_0^\infty \int_0^\infty \{\exp(-su^2) \cos xu \,.\, du\} \,.\, s^{\nu-\frac{1}{2}} \exp(-sz^2)\, ds,$$

when we write $t=s(u^2+z^2)$ and change the order of the integrations*.

Now
$$\int_0^\infty \exp(-su^2) \cos xu \,.\, du = \tfrac{1}{2}\Gamma(\tfrac{1}{2})\, s^{-\frac{1}{2}} \exp(-\tfrac{1}{4}x^2/s),$$

and so we have

$$\Gamma(\nu+\tfrac{1}{2}) \int_0^\infty \frac{\cos(xu)\, du}{(u^2+z^2)^{\nu+\frac{1}{2}}} = \tfrac{1}{2}\Gamma(\tfrac{1}{2}) \int_0^\infty s^{\nu-1} \exp\{-sz^2 - \tfrac{1}{4}x^2/s\}\, ds$$

$$= \frac{\frac{1}{2}\Gamma(\frac{1}{2})}{\nu(2z)^\nu} \int_0^\infty \exp\{-\tfrac{1}{2}z(v^{1/\nu} + x^2 v^{-1/\nu})\}\, dv,$$

which establishes the result.

[NOTE. It is evident that $s=\frac{1}{2}xe^{-t}/z=\frac{1}{2}v^{1/\nu}/z$. The only reason for modifying

$$\frac{1}{2}\int_{-\infty}^\infty e^{-xz\cosh t - \nu t}\, dt$$

by taking v as a parametric variable is to obtain an integral which is ostensibly of the same form as the integral actually investigated by Poisson; with his notation the integral is

$$\int_0^\infty \exp(-x^n - a^2 x^{-n})\, dx.]$$

6·33. *Malmstén's transformation.*

The method employed by Malmstén† in proving that, when $R(z)>0$ and $R(\nu)>-\frac{1}{2}$, then

$$\frac{\Gamma(\nu+\frac{1}{2})}{(\frac{1}{2}x)^\nu \Gamma(\frac{1}{2})} \int_0^\infty \frac{\cos(xu)\, du}{(u^2+z^2)^{\nu+\frac{1}{2}}} = \frac{\Gamma(\frac{1}{2})(\frac{1}{2}x)^\nu}{\Gamma(\nu+\frac{1}{2})} \int_1^\infty e^{-xzt}(t^2-1)^{\nu-\frac{1}{2}}\, dt,$$

is not so direct as the analysis of §§ 6·31, 6·32, inasmuch as it involves an appeal to the theory of linear differential equations. It is first shewn by Malmstén that the three expressions

$$\int_0^\infty \frac{\cos(xu)\, du}{(u^2+z^2)^{\nu+\frac{1}{2}}}, \quad \int_x^\infty e^{-zt}(t^2-x^2)^{\nu-\frac{1}{2}}\, dt, \quad \int_{-x}^x e^{-zt}(x^2-t^2)^{\nu-\frac{1}{2}}\, dt,$$

qua functions of x, are annihilated‡ by the operator

$$x\frac{d^2}{dx^2} - (2\nu-1)\frac{d}{dx} - xz^2,$$

and that as $x \to +\infty$, the third is $O(e^{xz})$ while the first and second are bounded, provided that $R(\nu)>0$. It follows that the second and third expressions form a fundamental system of solutions of the equation

$$x\frac{d^2y}{dx^2} - (2\nu-1)\frac{dy}{dx} - xz^2 y = 0,$$

* Cf. Bromwich, *Theory of Infinite Series*, § 177.

† *K. Svenska V. Akad. Handl.* LXII. (1841), pp. 65—74.

‡ The reader should have no difficulty in supplying a proof of this.

and the first is consequently a linear combination of the second and third. In view of the unboundedness of the third as $x \to +\infty$, it is obvious that the first must be a constant multiple of the second so that

$$\int_0^\infty \frac{(\cos xu)\,du}{(u^2+z^2)^{\nu+\frac{1}{2}}} = C\int_x^\infty e^{-zt}(t^2-x^2)^{\nu-\frac{1}{2}}\,dt,$$

where C is independent of x. To determine C, make $x \to 0$ and then

$$\int_0^\infty \frac{du}{(u^2+z^2)^{\nu+\frac{1}{2}}} = C\int_0^\infty e^{-zt}\,t^{2\nu-1}\,dt,$$

so that

$$\frac{\Gamma(\nu)\,\Gamma(\frac{1}{2})}{2z^{2\nu}\,\Gamma(\nu+\frac{1}{2})} = \frac{C\Gamma(2\nu)}{z^{2\nu}},$$

and the required transformation follows, when $R(\nu)>0$, if we use the duplication formula for the Gamma function.

An immediate consequence of Malmstén's transformation is that $\int_0^\infty \frac{\cos xu\,.\,du}{(u^2+z^2)^n}$ is expressible in finite terms; for it is equal to

$$\frac{\pi}{2^{2n-1}\{(n-1)!\}^2}\int_x^\infty e^{-zt}(t^2-x^2)^{n-1}\,dt$$

$$=\frac{\pi e^{-xz}}{2^{2n-1}\{(n-1)!\}^2}\int_0^\infty e^{-zv}v^{n-1}(2x+v)^{n-1}\,dv$$

$$=\frac{\pi e^{-xz}}{(2z)^{2n-1}(n-1)!}\sum_{m=0}^{n-1}\frac{(2xz)^m(2n-m-1)!}{m!\,(n-m-1)!}.$$

This method of evaluating the integral is simpler than a method given by Catalan, *Journal de Math.* v. (1840), pp. 110—114; and his investigation is not rigorous in all its stages. The transformation is discussed by Serret, *Journal de Math.* VIII. (1843), pp. 20, 21; IX. (1844), pp. 193—216; see also Cayley, *Journal de Math.* XII. (1847), p. 236 [*Collected Papers*, I. (1889), p. 313.]

6·4. *Airy's integral.*

The integral

$$\int_0^\infty \cos(t^3 \pm xt)\,dt$$

which appeared in the researches of Airy* "On the Intensity of Light in the neighbourhood of a Caustic" is a member of a class of integrals which are expressible in terms of Bessel functions. The integral was tabulated by Airy by quadratures, but the process was excessively laborious. Later, De Morgan† obtained a series in ascending powers of x by a process which needs justification either by Stokes' transformation (which will be explained immediately) or by the use of Hardy's theory of generalised integrals‡.

* *Trans. Camb. Phil. Soc.* VI. (1838), pp. 379—402. Airy used the form

$$\int_0^\infty \cos\tfrac{1}{2}\pi(w^3-mw)\,dw,$$

but this is easily reduced to the integral given above.

† The result was communicated to Airy on March 11, 1848; see *Trans. Camb. Phil. Soc.* VIII. (1849), pp. 595—599.

‡ *Quarterly Journal*, XXXV. (1904), pp. 22—66; *Trans. Camb. Phil. Soc.* XXI. (1912), pp. 1—48.

Stokes observed* that the integral satisfies the differential equation

$$\frac{d^2v}{dx^2} \pm \tfrac{1}{3}xv = 0,$$

and he also obtained the asymptotic expansions of the integral for large values of x, both positive and negative.

As was observed by Stokes (*loc. cit.* p. 187), this differential equation can be reduced to Bessel's equation; cf. § 4·3 (5) with $2q = 3$. The expression of Airy's integral in terms of Bessel functions of orders† $\pm \frac{1}{3}$ was published first in a little-known paper by Wirtinger, *Berichte des natur.-med. Vereins in Innsbruck*, XXIII. (1897), pp. 7—15, and later by Nicholson, *Phil. Mag.* (6) XVII. (1909), pp. 6—17.

Subsequently Hardy, *Quarterly Journal*, XLI. (1910), pp. 226—240, pointed out the connexion between Airy's integral and the integrals discussed in §§ 6·21, 6·22, and he then examined various generalisations of Airy's integral (§§ 10·2—10·22).

To evaluate Airy's integral‡, we observe that it may be written in the form

$$\frac{1}{2}\int_{-\infty}^{\infty} \exp(it^3 \pm ixt)\, dt.$$

Now consider this integrand taken along two arcs of a circle of radius ρ with centre at the origin, the arcs terminating at ρ, $\rho e^{\frac{1}{6}\pi i}$ and $\rho e^{\frac{5}{6}\pi i}$, $\rho e^{\pi i}$ respectively. The integrals along these arcs tend to zero as $\rho \to \infty$, by Jordan's lemma, and hence, by Cauchy's theorem, we obtain Stokes' transformation

$$\int_0^\infty \cos(t^3 \pm xt)\, dt = \frac{1}{2}\int_{\infty \exp \frac{5}{6}\pi i}^{\infty \exp \frac{1}{6}\pi i} \exp(it^3 \pm ixt)\, dt$$

$$= \frac{1}{2}\int_0^\infty \{e^{\frac{1}{6}\pi i} \exp(-\tau^3 \pm e^{\frac{2}{3}\pi i} x\tau) + e^{-\frac{1}{6}\pi i} \exp(-\tau^3 \pm e^{-\frac{2}{3}\pi i} x\tau)\}\, d\tau;$$

the contour of the second integral consists of two rays emerging from the origin and the third integral is obtained by writing $\tau e^{\frac{1}{6}\pi i}$, $\tau e^{\frac{5}{6}\pi i}$ for t on these rays.

Now, since the resulting series are convergent, it may be shewn that§

$$\int_0^\infty \exp(-\tau^3 \pm e^{\pm\frac{2}{3}\pi i} x\tau)\, d\tau = \sum_{m=0}^{\infty} \frac{(\pm)^m e^{\pm\frac{2}{3}m\pi i} x^m}{m!} \int_0^\infty \tau^m \exp(-\tau^3)\, d\tau,$$

* *Trans. Camb. Phil. Soc.* IX. (1856), pp. 166—187. [*Math. and Phys. Papers*, II. (1883), pp. 329—349.] See also Stokes' letter of May 12, 1848, to Airy, *Sir G. G. Stokes, Memoir and Scientific Correspondence*, II. (Cambridge, 1907), pp. 159—160.

† For other occurrences of these functions, see Rayleigh, *Phil. Mag.* (6) XXVIII. (1914), pp. 609—619; XXX. (1915), pp. 329—338 [*Scientific Papers*, VI. (1920), pp. 266—275; 341—349] on stability of motion of a viscous fluid; also Weyl, *Math. Ann.* LXVIII. (1910), p. 267, and, for approximate formulae, § 8·43 *infra*.

‡ The integral is convergent. Cf. Hardy, *loc. cit.* p. 228, or de la Vallée Poussin, *Ann. de la Soc. Sci. de Bruxelles*, XVI. (1892), pp. 150—180.

§ Bromwich, *Theory of Infinite Series*, § 176.

and so

$$\int_0^\infty \cos(t^3 \pm xt)\,dt = \sum_{m=0}^\infty \frac{(\pm x)^m \cos(\frac{2}{3}m + \frac{1}{6})\pi}{m!} \int_0^\infty \tau^m \exp(-\tau^3)\,d\tau$$

$$= \frac{1}{3}\sum_{m=0}^\infty (\pm x)^m \sin \tfrac{2}{3}(m+1)\pi \,.\, \Gamma(\tfrac{1}{3}m + \tfrac{1}{3})/m!$$

$$= \tfrac{1}{3}\pi \left[\sum_{m=0}^\infty \frac{(\pm\frac{1}{3}x)^{3m}}{m!\,\Gamma(m+\frac{2}{3})} \mp \tfrac{1}{3}x \sum_{m=0}^\infty \frac{(\pm\frac{1}{3}x)^{3m}}{m!\,\Gamma(m+\frac{4}{3})}\right]$$

This is the result obtained by De Morgan. When the series on the right are expressed in terms of Bessel functions, we obtain the formulae (in which x is to be taken to be positive) due to Wirtinger and Nicholson:

$$(1) \qquad \int_0^\infty \cos(t^3 - xt)\,dt = \tfrac{1}{3}\pi\sqrt{(\tfrac{1}{3}x)}\,.\left[J_{-\frac{1}{3}}\left(\frac{2x\sqrt{x}}{3\sqrt{3}}\right) + J_{\frac{1}{3}}\left(\frac{2x\sqrt{x}}{3\sqrt{3}}\right)\right]$$

$$(2) \qquad \int_0^\infty \cos(t^3 + xt)\,dt = \tfrac{1}{3}\pi\sqrt{(\tfrac{1}{3}x)}\,.\left[I_{-\frac{1}{3}}\left(\frac{2x\sqrt{x}}{3\sqrt{3}}\right) - I_{\frac{1}{3}}\left(\frac{2x\sqrt{x}}{3\sqrt{3}}\right)\right]$$

$$= \frac{\sqrt{x}}{3} K_{\frac{1}{3}}\left(\frac{2x\sqrt{x}}{3\sqrt{3}}\right).$$

6·5. *Barnes' integral representations of Bessel functions.*

By using integrals of a type introduced by Pincherle* and Mellin†, Barnes‡ has obtained representations of Bessel functions which render possible an easy proof of Kummer's formula of § 4·42.

Let us consider the residue of

$$-\Gamma(2m - s)\,.\,(iz)^s$$

at $s = 2m + r$, where $r = 0, 1, 2, \ldots$. This residue is $(-)^r (iz)^{2m+r}/r!$, so the sum of the residues is $(-)^m z^{2m} e^{-iz}$.

Hence, by Cauchy's theorem,

$$J_\nu(z)\,e^{-iz} = -\frac{(\frac{1}{2}z)^\nu}{2\pi i}\sum_{m=0}^\infty \int_\infty^{(0+)} \frac{\Gamma(2m-s)\,.\,(iz)^s}{2^{2m}\,.\,m!\,\Gamma(\nu+m+1)}\,ds,$$

if the contour encloses the points $0, 1, 2, \ldots$. It may be verified, by using Stirling's formula that the integrals are convergent.

Now suppose that $R(\nu) > -\frac{1}{2}$, and choose the contour so that, on it, $R(\nu + s) > -\frac{1}{2}$. When this last condition is satisfied the series

$$\sum_{m=0}^\infty \frac{\Gamma(2m-s)}{2^{2m}\,.\,m!\,\Gamma(\nu+m+1)}$$

is convergent and equal to

$$\frac{\Gamma(-s)}{\Gamma(\nu+1)}\,.\,{}_2F_1(-\tfrac{1}{2}s, \tfrac{1}{2}-\tfrac{1}{2}s;\ \nu+1;\ 1) = \frac{\Gamma(-s)\,\Gamma(\nu+s+\frac{1}{2})}{\Gamma(\nu+\frac{1}{2}s+\frac{1}{2})\,\Gamma(\nu+\frac{1}{2}s+1)},$$

* *Rend. del R. Istituto Lombardo*, (2) XIX. (1886), pp. 559—562; *Atti della R. Accad. dei Lincei*, ser. 4, *Rendiconti*, IV. (1888), pp. 694—700, 792—799.

† Mellin has given a summary of his researches, *Math. Ann.* LXVIII. (1910), pp. 305—337.

‡ *Camb. Phil. Trans.* XX. (1908), pp. 270—279. For a bibliography of researches on integrals of this type, see Barnes, *Proc. London Math. Soc.* (2) V. (1907), pp. 59—65.

by the well-known formula due to Gauss. If therefore we change the order of summation and integration* we have

$$J_\nu(z)\,e^{-iz} = -\frac{(\frac{1}{2}z)^\nu}{2\pi i}\int_\infty^{(0+)} \frac{\Gamma(-s)\,\Gamma(\nu+s+\frac{1}{2})\,.\,(iz)^s\,ds}{\Gamma(\nu+\frac{1}{2}s+\frac{1}{2})\,\Gamma(\nu+\frac{1}{2}s+1)}.$$

The only poles of the integrand inside the contour are at 0, 1, 2, ... When we calculate the sum of the residues at these poles, we find that

$$J_\nu(z)\,e^{-iz} = (\tfrac{1}{2}z)^\nu \sum_{m=0}^{\infty} \frac{\Gamma(\nu+m+\frac{1}{2})\,.\,(-iz)^m}{m!\,\Gamma(\nu+\frac{1}{2}m+\frac{1}{2})\,\Gamma(\nu+\frac{1}{2}m+1)},$$

so that

$$\text{(1)}\qquad J_\nu(z)\,e^{-iz} = \frac{(\frac{1}{2}z)^\nu}{\Gamma(\nu+1)}\,{}_1F_1(\nu+\tfrac{1}{2};\ 2\nu+1;\ -2iz),$$

which is Kummer's relation. In like manner, we find that

$$\text{(2)}\qquad J_\nu(z)\,e^{iz} = \frac{(\frac{1}{2}z)^\nu}{\Gamma(\nu+1)}\,{}_1F_1(\nu+\tfrac{1}{2};\ 2\nu+1;\ 2iz).$$

These formulae, proved when $R(\nu) > -\frac{1}{2}$, are relations connecting functions of ν which are analytic for all values of ν, and so, by the theory of analytic continuation, they are universally true.

It is also possible to represent Bessel functions by integrals in which no exponential factor is involved. To do this, we consider the function

$$\Gamma(-\nu-s)\,\Gamma(-s)\,(\tfrac{1}{2}iz)^{\nu+2s},$$

qua function of s. It has poles at the points

$$s = 0, 1, 2, \ldots;\ -\nu, -\nu+1, -\nu+2, \ldots.$$

The residue at $s = m$ is

$$\frac{\pi i^\nu}{\sin\nu\pi}\cdot\frac{(-)^m\,(\frac{1}{2}z)^{\nu+2m}}{m!\,\Gamma(\nu+m+1)},$$

while the residue at $s = -\nu+m$ is

$$-\frac{\pi i^{-\nu}}{\sin\nu\pi}\cdot\frac{(-)^m\,(\frac{1}{2}z)^{-\nu+2m}}{m!\,\Gamma(\nu+m+1)},$$

so that

$$\text{(3)}\qquad \pi e^{-\frac{1}{2}(\nu+1)\pi i}\,H_\nu^{(2)}(z) = -\frac{1}{2\pi i}\int \Gamma(-\nu-s)\,\Gamma(-s)\,(\tfrac{1}{2}iz)^{\nu+2s}\,ds,$$

and, in like manner,

$$\text{(4)}\qquad \pi e^{\frac{1}{2}(\nu+1)\pi i}\,H_\nu^{(1)}(z) = -\frac{1}{2\pi i}\int \Gamma(-\nu-s)\,\Gamma(-s)\,(-\tfrac{1}{2}iz)^{\nu+2s}\,ds,$$

where the contours start from and return to $+\infty$ after encircling the poles of the integrand counter-clockwise. When $|\arg iz| < \frac{1}{2}\pi$ in (3) or $|\arg(-iz)| < \frac{1}{2}\pi$ in (4) the contours may be opened out, so as to start from ∞i and end at $-\infty i$. If we reverse the directions of the contours we find that

$$\text{(5)}\qquad \pi e^{-\frac{1}{2}(\nu+1)\pi i}\,H_\nu^{(2)}(z) = \frac{1}{2\pi i}\int_{-c-\infty i}^{-c+\infty i} \Gamma(-\nu-s)\,\Gamma(-s)\,(\tfrac{1}{2}iz)^{\nu+2s}\,ds,$$

* Cf. Bromwich, *Theory of Infinite Series*, § 176.

provided that $|\arg iz| < \frac{1}{2}\pi$; and

(6) $$\pi e^{\frac{1}{2}(\nu+1)\pi i} H_\nu^{(1)}(z) = \frac{1}{2\pi i}\int_{-c-\infty i}^{-c+\infty i} \Gamma(-\nu-s)\,\Gamma(-s)\,(-\tfrac{1}{2}iz)^{\nu+2s}\,ds,$$

provided that $|\arg(-iz)| < \frac{1}{2}\pi$; and, in each integral, c is any positive number exceeding $R(\nu)$ and the path of integration is parallel to the imaginary axis.

There is an integral resembling these which represents the function of the first kind of order ν, but it converges only when $R(\nu) > 0$ and the argument of the function is positive. The integral in question is

(7) $$J_\nu(x) = \frac{1}{2\pi i}\int_{-\infty i}^{\infty i} \frac{\Gamma(-s)\,(\frac{1}{2}x)^{\nu+2s}}{\Gamma(\nu+s+1)}\,ds;$$

and it is obtained in the same way as the preceding integrals; the reader will notice that, when $|s|$ is large on the contour, the integrand is $O(|s|^{-\nu-1})$.

6·51. *Barnes' representations of functions of the third kind.*

By using the duplication formula for the Gamma function we may write the results just obtained in the form

(1) $$J_\nu(z)\,e^{\mp iz} = \frac{(2z)^\nu}{2i\sqrt{\pi}}\int_\infty^{(0+)} \frac{\Gamma(\nu+s+\frac{1}{2})\,.\,(\pm 2iz)^s\,ds}{\Gamma(s+1)\,\Gamma(2\nu+s+1)\sin s\pi}.$$

Consider now the integral

$$-\frac{(2z)^\nu}{2i\sqrt{\pi}}\int_{-\infty i}^{\infty i} \Gamma(-s)\,\Gamma(-2\nu-s)\,\Gamma(\nu+s+\tfrac{1}{2})\,.\,(2iz)^s\,ds,$$

in which the integrand differs from the integrand in (1) by a factor which is periodic in s. It is to be supposed temporarily that 2ν is not an integer and that the path of integration is so drawn that the sequences of poles $0, 1, 2, \ldots$; $-2\nu, 1-2\nu, 2-2\nu, \ldots$ lie on the right of the contour while the sequence of poles $-\nu-\frac{1}{2}, -\nu-\frac{3}{2}, -\nu-\frac{5}{2}, \ldots$ lies on the left of the contour. In the first place, we shall shew that, if $|\arg iz| < \frac{3}{2}\pi$, the integral taken round a semicircle of radius ρ on the right of the imaginary axis tends to zero as $\rho \to \infty$; for, if $s = \rho e^{i\theta}$, we have

$$s\Gamma(-s)\,\Gamma(-2\nu-s)\,\Gamma(\nu+s+\tfrac{1}{2}).(2iz)^s = \frac{\pi^2\,\Gamma(\nu+s+\frac{1}{2})\,.\,(2iz)^s}{\Gamma(s)\,\Gamma(2\nu+s+1)\sin s\pi\,\sin(2\nu+s)\pi},$$

and, by Stirling's formula,

$$\log\frac{\Gamma(\nu+s+\frac{1}{2})\,.\,(2iz)^s}{\Gamma(s+1)\,\Gamma(2\nu+s+1)}$$
$$\sim \rho e^{i\theta}\log(2iz) - (\nu+\rho e^{i\theta})(\log\rho + i\theta) + \rho e^{i\theta} - \tfrac{1}{2}\log(2\pi);$$

and the real part of this tends to $-\infty$ when $-\frac{1}{2}\pi < \theta < \frac{1}{2}\pi$, because the dominant term is $-\rho\cos\theta\log\rho$. When θ is nearly equal to $\pm\frac{1}{2}\pi$, $|\sin s\pi|$ is comparable with $\frac{1}{2}\exp\{\rho\pi|\sin\theta|\}$ and the dominant term in the real part of the logarithm of s times the integrand is

$$\rho\cos\theta\log|2z| - \rho\sin\theta\,.\arg 2iz - \rho\cos\theta\log\rho + \rho\theta\sin\theta + \rho\cos\theta - 2\rho\pi|\sin\theta|,$$

and this tends to $-\infty$ as $\rho\to\infty$ if $|\arg iz| < \frac{3}{2}\pi$.

Hence s times the integrand tends to zero all along the semicircle, and so the integral round the semicircle tends to zero if the semicircle is drawn so as to pass between (and not through) the poles of the integrand.

It follows from Cauchy's theorem that, when $|\arg iz| < \frac{3}{2}\pi$ and 2ν is not an integer, then

$$-\frac{(2z)^\nu}{2i\sqrt{\pi}}\int_{-\infty i}^{\infty i}\Gamma(-s)\,\Gamma(-2\nu-s)\,\Gamma(\nu+s+\tfrac{1}{2}).(2iz)^s\,ds$$

may be calculated by evaluating the residues at the poles on the right of the contour.

The residues of

$$\Gamma(-s)\,\Gamma(-2\nu-s)\,\Gamma(\nu+s+\tfrac{1}{2}).(2iz)^s$$

at $s=m$ and $s=-2\nu+m$ are respectively

$$\frac{\pi}{\sin 2\nu\pi}\,\frac{\Gamma(\nu+m+\frac{1}{2}).(2iz)^m}{m!\,\Gamma(2\nu+m+1)},\quad -\frac{\pi}{\sin 2\nu\pi}\,\frac{\Gamma(-\nu+m+\frac{1}{2}).(2iz)^{-2\nu+m}}{m!\,\Gamma(-2\nu+m+1)}$$

and hence

$$\begin{aligned}-\frac{(2z)^\nu}{2i\sqrt{\pi}}&\int_{-\infty i}^{\infty i}\Gamma(-s)\,\Gamma(-2\nu-s)\,\Gamma(\nu+s+\tfrac{1}{2}).(2iz)^s\,ds\\ &=\frac{(2z)^\nu\pi^{\frac{3}{2}}}{\sin 2\nu\pi}\,\frac{\Gamma(\nu+\frac{1}{2})}{\Gamma(2\nu+1)}\,.\,{}_1F_1(\nu+\tfrac{1}{2};\ 2\nu+1;\ 2iz)\\ &\qquad-\frac{e^{-\nu\pi i}(2z)^{-\nu}\pi^{\frac{3}{2}}}{\sin 2\nu\pi}\,\frac{\Gamma(\frac{1}{2}-\nu)}{\Gamma(1-2\nu)}\,.\,{}_1F_1(\tfrac{1}{2}-\nu;\ 1-2\nu;\ 2iz)\\ &=\frac{\pi^2 e^{iz}}{\sin 2\nu\pi}\{J_\nu(z)-e^{-\nu\pi i}J_{-\nu}(z)\}.\end{aligned}$$

It follows that, when $|\arg iz| < \frac{3}{2}\pi$,

$$\text{(2)}\qquad H_\nu^{(2)}(z)=\frac{e^{-i(z-\nu\pi)}\cos(\nu\pi).(2z)^\nu}{\pi^{\frac{5}{2}}}\times\int_{-\infty i}^{\infty i}\Gamma(-s)\,\Gamma(-2\nu-s)\,\Gamma(\nu+s+\tfrac{1}{2}).(2iz)^s\,ds,$$

and similarly, when $|\arg(-iz)| < \frac{3}{2}\pi$,

$$\text{(3)}\qquad H_\nu^{(1)}(z)=-\frac{e^{i(z-\nu\pi)}\cos(\nu\pi).(2z)^\nu}{\pi^{\frac{5}{2}}}\times\int_{-\infty i}^{\infty i}\Gamma(-s)\,\Gamma(-2\nu-s)\,\Gamma(\nu+s+\tfrac{1}{2}).(-2iz)^s\,ds.$$

The restriction that ν is not to be an integer may be removed in the usual manner by a limiting process, but the restriction that 2ν must not be an odd integer cannot be removed, since then poles which must be on the right of the contour would have to coincide with poles which must be on the left.

CHAPTER VII

ASYMPTOTIC EXPANSIONS OF BESSEL FUNCTIONS

7·1. *Approximate formulae for* $J_\nu(z)$.

In Chapter III various representations of Bessel functions were obtained in the form of series of ascending powers of the argument z, multiplied in some cases by $\log z$. These series are well adapted for numerical computation when z^2 is not large compared with $4(\nu+1)$, $4(\nu+2)$, $4(\nu+3)$, ..., since the series converge fairly rapidly for such values of z. But, when $|z|$ is large, the series converge slowly, and an inspection of their initial terms affords no clue to the approximate values of $J_\nu(z)$ and $Y_\nu(z)$. There is one exception to this statement; when $\nu+\frac{1}{2}$ is an integer which is not large, the expressions for $J_{\pm\nu}(z)$ in finite terms (§ 3·4) enable the functions to be calculated without difficulty.

The object of this chapter is the determination of formulae which render possible the calculation of the values of a fundamental system of solutions of Bessel's equation when z is large.

There are really two aspects of the problem to be considered; the investigation when ν is large is very different from the investigation when ν is not large. The former investigation is, in every respect, of a more recondite character than the latter, and it is postponed until Chapter VIII.

It must, however, be mentioned that the first step towards the solution of the more recondite problem was made by Carlini* some years before Poisson's† investigation of the behaviour of $J_0(x)$, for large positive values of x, was published.

The formal expansion obtained by Poisson was

$$J_0(x) = \left(\frac{2}{\pi x}\right)^{\frac{1}{2}} \left[\cos(x-\tfrac{1}{4}\pi)\,.\left\{1 - \frac{1^2\,.\,3^2}{2!\,(8x)^2} + \frac{1^2\,.\,3^2\,.\,5^2\,.\,7^2}{4!\,(8x)^4} - \ldots\right\}\right.$$
$$\left. + \sin(x-\tfrac{1}{4}\pi)\,.\left\{\frac{1^2}{1!\,8x} - \frac{1^2\,.\,3^2\,.\,5^2}{3!\,(8x)^3} + \ldots\right\}\right],$$

when x is large and positive. But, since the series on the right are not convergent, and since Poisson gave no investigation of the remainders in the series, his analysis (apart from his method of obtaining the dominant term) is to be regarded as suggestive and ingenious rather than convincing.

* *Ricerche sulla convergenza della serie che serva alla soluzione del problema di Keplero* (Milan, 1817). An account of these investigations has already been given in § 1·4.

† *Journal de l'École Polytechnique*, XII. (cahier 19), (1823), pp. 350—352; see § 1·6. An investigation of $J_\nu(x)$ similar to Poisson's investigation of $J_0(x)$ has been constructed by Gray and Mathews, *A Treatise on Bessel Functions* (London, 1895), pp. 34—38.

It will be seen in the course of this chapter that Poisson's series are *asymptotic*; this has been proved by Lipschitz, Hankel, Schläfli, Weber, Stieltjes and Barnes.

It must be mentioned that Poisson merely indicated the law of formation of successive terms of the series without giving an explicit expression for the general term; such an expression was actually obtained by W. R. Hamilton* (cf. § 1·6).

The analogous formal expansion for $J_1(x)$ is due to Hansen†; and a few years later, Jacobi‡ obtained the more general formula which is now usually written in the form

$$J_n(x) \sim \left(\frac{2}{\pi x}\right)^{\frac{1}{2}} \left[\cos\left(x - \tfrac{1}{2}n\pi - \tfrac{1}{4}\pi\right) \times \left\{1 - \frac{(4n^2-1^2)(4n^2-3^2)}{2!\,(8x)^2} + \frac{(4n^2-1^2)(4n^2-3^2)(4n^2-5^2)(4n^2-7^2)}{4!\,(8x)^4} - \dots\right\} - \sin\left(x - \tfrac{1}{2}n\pi - \tfrac{1}{4}\pi\right).\left\{\frac{4n^2-1^2}{1!\,8x} - \frac{(4n^2-1^2)(4n^2-3^2)(4n^2-5^2)}{3!\,(8x)^3} + \dots\right\}\right].$$

These expansions for $J_0(x)$ and $J_1(x)$ were used by Hansen for purposes of numerical computation, and a comparison of the results so obtained for isolated values of x with the results obtained from the ascending series led Hansen to infer that the expansions, although not convergent, could safely be used for purposes of computation §.

Two years before the publication of Jacobi's expansion, Plana|| had discovered a method of transforming Parseval's integral which placed the expansion of $J_0(x)$ on a much more satisfactory basis¶. His work was followed by the researches of Lipschitz**, who gave the first rigorous investigation of the asymptotic expansion of $J_0(z)$ with the aid of the theory of contour integration; Lipschitz also briefly indicated how his results could be applied to $J_n(z)$.

The general formulae for $J_\nu(z)$ and $Y_\nu(z)$, where ν has any assigned (complex) value and z is large and complex, were obtained in the great memoir by Hankel††, written in 1868.

* Some information concerning W. R. Hamilton's researches will be found in *Sir George Gabriel Stokes, Memoir and Scientific Correspondence*, I. (Cambridge, 1907), pp. 130—135.

† *Ermittelung der absoluten Störungen* [*Schriften der Sternwarte Seeburg*]. (Gotha, 1843), pp. 119—123.

‡ *Astr. Nach.* XXVIII. (1849), col. 94. [*Ges. Math. Werke*, VII. (1891), p. 174.] Jacobi's result is obtained by making the substitutions

$$\sqrt{2}\,.\cos\left(x - \tfrac{1}{2}n\pi - \tfrac{1}{4}\pi\right) = (-1)^{\frac{1}{2}n(n+1)}\cos x + (-1)^{\frac{1}{2}n(n-1)}\sin x,$$
$$\sqrt{2}\,.\sin\left(x - \tfrac{1}{2}n\pi - \tfrac{1}{4}\pi\right) = (-1)^{\frac{1}{2}n(n+1)}\sin x - (-1)^{\frac{1}{2}n(n-1)}\cos x,$$

in the form quoted.

§ See a note by Niemöller, *Zeitschrift für Math. und Phys.* XXV. (1880), pp. 44—48.

|| *Mem. della R. Accad. delle Sci. di Torino*, (2) X. (1849), pp. 275—292.

¶ Analysis of Plana's type was used to obtain the asymptotic expansions of $J_\nu(z)$ and $Y_\nu(z)$ by McMahon, *Annals of Math.* VIII. (1894), pp. 57—61.

** *Journal für Math.* LVI. (1859), pp. 189—196.

†† *Math. Ann.* I. (1869), pp. 467—501.

The general character of the formula for $Y_n(z)$ had been indicated by Lommel, *Studien über die Bessel'schen Functionen* (Leipzig, 1868), just before the publication of Hankel's memoir; and the researches of Weber, *Math. Ann.* VI. (1873), pp. 146—149 must also be mentioned.

The asymptotic expansion of $K_\nu(z)$ was investigated (and proved to be asymptotic) at an early date by Kummer*; this result was reproduced, with the addition of the corresponding formula for $I_\nu(z)$, by Kirchhoff†; and a little-known paper by Malmstén‡ also contains an investigation of the asymptotic expansion of $K_\nu(z)$.

A close study of the remainders in the asymptotic expansions of $J_0(x)$, $Y_0(x)$, $I_0(x)$ and $K_0'(x)$ has been made by Stieltjes, *Ann. Sci. de l'École norm. sup.* (3) III. (1886), pp. 233—252, and parts of his analysis have been extended by Callandreau, *Bull. des Sci. Math.* (2) XIV. (1890), pp. 110—114, to include functions of any integral order; while results concerning the remainders when the variables are complex have been obtained by Weber, *Math. Ann.* XXXVII. (1890), pp. 404—416.

The expansions have also been investigated by Adamoff§, *Petersburg Ann. Inst. polyt.* 1906, pp. 239—265, and by Valewink|| in a Haarlem dissertation, 1905.

Investigations concerning asymptotic expansions of $J_\nu(z)$ and $Y_\nu(z)$, when $|z|$ is large while ν is fixed, seem to be most simply carried out with the aid of integrals of Poisson's type. But Schläfli¶ has shewn that a large number of results are obtainable by a peculiar treatment of integrals of Bessel's type, while, more recently, Barnes** has discussed the asymptotic expansions by means of the Pincherle-Mellin integrals, involving gamma-functions, which were examined in §§ 6·5, 6·51.

7·2. *Asymptotic expansions of $H_\nu^{(1)}(z)$ and $H_\nu^{(2)}(z)$ after Hankel.*

We shall now obtain the asymptotic expansions of the functions of the third kind, valid for large values of $|z|$; the analysis, apart from some slight modifications, will follow that given by Hankel††.

Take the formula § 6·12 (3), namely

$$H_\nu^{(1)}(z) = \left(\frac{2}{\pi z}\right)^{\frac{1}{2}} \frac{e^{i(z-\frac{1}{2}\nu\pi-\frac{1}{4}\pi)}}{\Gamma(\nu+\frac{1}{2})} \int_0^{\infty \exp i\beta} e^{-u} u^{\nu-\frac{1}{2}} \left(1+\frac{iu}{2z}\right)^{\nu-\frac{1}{2}} du,$$

valid when $-\frac{1}{2}\pi < \beta < \frac{1}{2}\pi$ and $-\frac{1}{2}\pi + \beta < \arg z < \frac{3}{2}\pi + \beta$, provided that $R(\nu+\frac{1}{2}) > 0$.

The expansion of the factor $(1+\frac{1}{2}iu/z)^{\nu-\frac{1}{2}}$ in descending powers of z is

$$1 + \frac{(\nu-\frac{1}{2})\, iu}{2z} + \frac{(\nu-\frac{1}{2})(\nu-\frac{3}{2}) . (iu)^2}{2 . 4 . z^2} + \dots;$$

* *Journal für Math.* XVII. (1837), pp. 228—242. † *Ibid.* XLVIII. (1854), pp. 348—376.
‡ *K. Svenska V. Akad. Handl.* LXII. (1841), pp. 65—74.
§ See the *Jahrbuch über die Fortschritte der Math.* 1907, p. 492.
|| *Ibid.* 1905, p. 328. ¶ *Ann. di Mat.* (2) VI. (1875), pp. 1—20.
** *Trans. Camb. Phil. Soc.* XX. (1908), pp. 270—279.
†† *Math. Ann.* I. (1869), pp. 491—495.

but since this expansion is not convergent all along the path of integration, we shall replace it by a finite number of terms plus a remainder.

For all positive integral values of p, we have*

$$\left(1+\frac{iu}{2z}\right)^{\nu-\frac{1}{2}} = \sum_{m=0}^{p-1} \frac{(\frac{1}{2}-\nu)_m}{m!}\left(\frac{u}{2iz}\right)^m + \frac{(\frac{1}{2}-\nu)_p}{(p-1)!}\left(\frac{u}{2iz}\right)^p \int_0^1 (1-t)^{p-1}\left(1-\frac{ut}{2iz}\right)^{\nu-p-\frac{1}{2}} dt.$$

It is convenient to take p so large that $R(\nu-p-\frac{1}{2}) \leqslant 0$; and we then choose any positive angle δ which satisfies the inequalities

$$|\beta| \leqslant \tfrac{1}{2}\pi - \delta, \quad |\arg z - (\tfrac{1}{2}\pi + \beta)| \leqslant \pi - \delta.$$

The effect of this choice is that, when δ is given, z is restricted so that

$$-\pi + 2\delta \leqslant \arg z \leqslant 2\pi - 2\delta.$$

When the choice has been made, then

$$\left|1-\frac{ut}{2iz}\right| \geqslant \sin\delta, \quad \left|\arg\left(1-\frac{ut}{2iz}\right)\right| < \pi,$$

for the values of t and u under consideration, and so

$$\left|\left(1-\frac{ut}{2iz}\right)^{\nu-p-\frac{1}{2}}\right| \leqslant e^{\pi|I(\nu)|}(\sin\delta)^{R(\nu-p-\frac{1}{2})} = A_p,$$

say, where A_p is independent of z.

On substituting its expansion for $(1+\frac{1}{2}iu/z)^{\nu-\frac{1}{2}}$ and integrating term-by-term, we find that

$$H_\nu^{(1)}(z) = \left(\frac{2}{\pi z}\right)^{\frac{1}{2}} e^{i(z-\frac{1}{2}\nu\pi-\frac{1}{4}\pi)}\left[\sum_{m=0}^{p-1}\frac{(\frac{1}{2}-\nu)_m \,.\, \Gamma(\nu+m+\frac{1}{2})}{m!\,\Gamma(\nu+\frac{1}{2})\,.\,(2iz)^m} + R_p^{(1)}\right],$$

where

$$|R_p^{(1)}| \leqslant \frac{A_p}{(p-1)!}\left|\frac{(\frac{1}{2}-\nu)_p}{\Gamma(\nu+\frac{1}{2})(2iz)^p}\left\{\int_0^1 (1-t)^{p-1}dt \int_0^{\infty \exp i\beta} e^{-u}|u^{\nu+p-\frac{1}{2}}du|\right\}\right|$$
$$= B_p \,.\, |z|^{-p},$$

where B_p is a function of ν, p and δ which is independent of z.

Hence, when $R(\nu-p-\frac{1}{2}) < 0$ and $R(\nu+\frac{1}{2}) > 0$, we have

$$(1) \qquad H_\nu^{(1)}(z) = \left(\frac{2}{\pi z}\right)^{\frac{1}{2}} e^{i(z-\frac{1}{2}\nu\pi-\frac{1}{4}\pi)}\left[\sum_{m=0}^{p-1}\frac{(\frac{1}{2}-\nu)_m(\frac{1}{2}+\nu)_m}{m!\,(2iz)^m} + O(z^{-p})\right],$$

when z is such that $-\pi+2\delta \leqslant \arg z \leqslant 2\pi - 2\delta$, δ being any positive acute angle; and the symbol O is the Bachmann-Landau symbol which denotes a function of the order of magnitude† of z^{-p} as $|z| \to \infty$.

The formula (1) is also valid when $R(\nu-p-\frac{1}{2}) > 0$; this may be seen by

* Cf. *Modern Analysis*, § 5·41. The use of this form of the binomial expansion seems to be due to Graf and Gubler, *Einleitung in die Theorie der Bessel'schen Funktionen*, I. (Bern, 1898), pp. 86—87. Cf. Whittaker, *Modern Analysis* (Cambridge, 1902), § 161; Gibson, *Proc. Edinburgh Math. Soc.* XXXVIII. (1920), pp. 6—9; and MacRobert, *ibid.* pp. 10—19.

† Cf. *Modern Analysis*, § 2·1.

supposing that $R(\nu - p - \frac{1}{2}) > 0$ and then taking an integer q so large that $R(\nu - q - \frac{1}{2}) < 0$; if the expression which is contained in [ ] in (1) is then rewritten with q in place of p throughout, it may be expressed as p terms followed by $q - p + 1$ terms each of which is $O(z^{-p})$ or $o(z^{-p})$; and the sum of these $q - p + 1$ terms is therefore $O(z^{-p})$.

In a similar manner (by changing the sign of i throughout the previous work) we can deduce from § 6·12 (4) that

$$(2) \qquad H_\nu^{(2)}(z) = \left(\frac{2}{\pi z}\right)^{\frac{1}{2}} e^{-i(z - \frac{1}{2}\nu\pi - \frac{1}{4}\pi)} \left[\sum_{m=0}^{p-1} \frac{(\frac{1}{2} - \nu)_m (\frac{1}{2} + \nu)_m}{m!(-2iz)^m} + O(z^{-p})\right],$$

provided that $R(\nu + \frac{1}{2}) > 0$ and that the domain of values of z is now given by the inequalities

$$-2\pi + 2\delta \leqslant \arg z \leqslant \pi - 2\delta.$$

If, following Hankel, we write

$$(\nu, m) = (-)^m \frac{(\frac{1}{2} - \nu)_m (\frac{1}{2} + \nu)_m}{m!} = \frac{\Gamma(\nu + m + \frac{1}{2})}{m!\,\Gamma(\nu - m + \frac{1}{2})}$$

$$= \frac{\{4\nu^2 - 1^2\}\{4\nu^2 - 3^2\} \dots \{4\nu^2 - (2m-1)^2\}}{2^{2m} . m!},$$

these expansions become

$$(3) \qquad H_\nu^{(1)}(z) = \left(\frac{2}{\pi z}\right)^{\frac{1}{2}} e^{i(z - \frac{1}{2}\nu\pi - \frac{1}{4}\pi)} \left[\sum_{m=0}^{p-1} \frac{(-)^m . (\nu, m)}{(2iz)^m} + O(z^{-p})\right],$$

$$(4) \qquad H_\nu^{(2)}(z) = \left(\frac{2}{\pi z}\right)^{\frac{1}{2}} e^{-i(z - \frac{1}{2}\nu\pi - \frac{1}{4}\pi)} \left[\sum_{m=0}^{p-1} \frac{(\nu, m)}{(2iz)^m} + O(z^{-p})\right].$$

For brevity we write these equations thus:

$$(5) \qquad H_\nu^{(1)}(z) \sim \left(\frac{2}{\pi z}\right)^{\frac{1}{2}} e^{i(z - \frac{1}{2}\nu\pi - \frac{1}{4}\pi)} \sum_{m=0}^{\infty} \frac{(-)^m . (\nu, m)}{(2iz)^m},$$

$$(6) \qquad H_\nu^{(2)}(z) \sim \left(\frac{2}{\pi z}\right)^{\frac{1}{2}} e^{-i(z - \frac{1}{2}\nu\pi - \frac{1}{4}\pi)} \sum_{m=0}^{\infty} \frac{(\nu, m)}{(2iz)^m}.$$

Since (ν, m) is an even function of ν, it follows from the formulae of § 3·61 (7), which connect functions of the third kind of order ν with the corresponding functions of order $-\nu$, that the restriction that the real part of ν exceeds $-\frac{1}{2}$ is unnecessary. So the formulae (1)—(6) are valid for all values of ν, when z is confined to one or other of two sectors of angle just less than 3π.

In the notation of generalised hypergeometric functions, the expansions are

$$(7) \qquad H_\nu^{(1)}(z) \sim \left(\frac{2}{\pi z}\right)^{\frac{1}{2}} e^{i(z - \frac{1}{2}\nu\pi - \frac{1}{4}\pi)} . {}_2F_0\left(\tfrac{1}{2} + \nu, \tfrac{1}{2} - \nu; \frac{1}{2iz}\right),$$

$$(8) \qquad H_\nu^{(2)}(z) \sim \left(\frac{2}{\pi z}\right)^{\frac{1}{2}} e^{-i(z - \frac{1}{2}\nu\pi - \frac{1}{4}\pi)} . {}_2F_0\left(\tfrac{1}{2} + \nu, \tfrac{1}{2} - \nu; -\frac{1}{2iz}\right),$$

of which (7) is valid when $-\pi < \arg z < 2\pi$, and (8) when $-2\pi < \arg z < \pi$.

7·21. *Asymptotic expansions of $J_\nu(z)$, $J_{-\nu}(z)$ and $Y_\nu(z)$.*

If we combine the formulae of § 7·2, we deduce from the formulae of § 3·61 (which express Bessel functions of the first and second kinds in terms of functions of the third kind) that

(1) $$J_\nu(z) \sim \left(\frac{2}{\pi z}\right)^{\frac{1}{2}} \left[\cos(z - \tfrac{1}{2}\nu\pi - \tfrac{1}{4}\pi) \,.\, \sum_{m=0}^{\infty} \frac{(-)^m \,.\, (\nu, 2m)}{(2z)^{2m}} - \sin(z - \tfrac{1}{2}\nu\pi - \tfrac{1}{4}\pi) \,.\, \sum_{m=0}^{\infty} \frac{(-)^m \,.\, (\nu, 2m+1)}{(2z)^{2m+1}}\right],$$

(2) $$Y_\nu(z) \sim \left(\frac{2}{\pi z}\right)^{\frac{1}{2}} \left[\sin(z - \tfrac{1}{2}\nu\pi - \tfrac{1}{4}\pi) \,.\, \sum_{m=0}^{\infty} \frac{(-)^m \,.\, (\nu, 2m)}{(2z)^{2m}} + \cos(z - \tfrac{1}{2}\nu\pi - \tfrac{1}{4}\pi) \,.\, \sum_{m=0}^{\infty} \frac{(-)^m \,.\, (\nu, 2m+1)}{(2z)^{2m+1}}\right],$$

(3) $$J_{-\nu}(z) \sim \left(\frac{2}{\pi z}\right)^{\frac{1}{2}} \left[\cos(z + \tfrac{1}{2}\nu\pi - \tfrac{1}{4}\pi) \,.\, \sum_{m=0}^{\infty} \frac{(-)^m \,.\, (\nu, 2m)}{(2z)^{2m}} - \sin(z + \tfrac{1}{2}\nu\pi - \tfrac{1}{4}\pi) \,.\, \sum_{m=0}^{\infty} \frac{(-)^m \,.\, (\nu, 2m+1)}{(2z)^{2m+1}}\right],$$

(4) $$Y_{-\nu}(z) \sim \left(\frac{2}{\pi z}\right)^{\frac{1}{2}} \left[\sin(z + \tfrac{1}{2}\nu\pi - \tfrac{1}{4}\pi) \,.\, \sum_{m=0}^{\infty} \frac{(-)^m \,.\, (\nu, 2m)}{(2z)^{2m}} + \cos(z + \tfrac{1}{2}\nu\pi - \tfrac{1}{4}\pi) \,.\, \sum_{m=0}^{\infty} \frac{(-)^m \,.\, (\nu, 2m+1)}{(2z)^{2m+1}}\right],$$

and (in the case of functions of *integral order n only*),

(5) $$\mathbf{Y}_n(z) \sim \left(\frac{2\pi}{z}\right)^{\frac{1}{2}} \left[\sin(z - \tfrac{1}{2}n\pi - \tfrac{1}{4}\pi) \,.\, \sum_{m=0}^{\infty} \frac{(-)^m \,.\, (n, 2m)}{(2z)^{2m}} + \cos(z - \tfrac{1}{2}n\pi - \tfrac{1}{4}\pi) \,.\, \sum_{m=0}^{\infty} \frac{(-)^m \,.\, (n, 2m+1)}{(2z)^{2m+1}}\right].$$

These formulae are all valid for large values of $|z|$ provided that $|\arg z| < \pi$; and the error due to stopping at any term is obviously of the order of magnitude of that term multiplied by $1/z$. Actually, however, this factor $1/z$ may be replaced by $1/z^2$; this may be seen by taking the expansions of $H_\nu^{(1)}(z)$ and $H_\nu^{(2)}(z)$ to two terms further than the last term required in the particular combination with which we have to deal.

As has been seen in § 7·2, the integrals which are dealt with when $R(\nu) > -\frac{1}{2}$ represent $H_\nu^{(1)}(z)$ and $H_\nu^{(2)}(z)$, but, when $R(\nu) \leqslant -\frac{1}{2}$, the integrals from which the asymptotic expansions are derived *are those which represent $H^{(1)}_{-\nu}(z)$ and $H^{(2)}_{-\nu}(z)$.* This difference in the mode of treatment of $J_\nu(z)$ and $Y_\nu(z)$ for such values of ν seems to have led some writers to think* that formula (1) is not valid unless $R(\nu) > -\frac{1}{2}$.

* Cf. Sheppard, *Quarterly Journal*, XXIII. (1889), p. 223; Searle, *Quarterly Journal*, XXXIX. (1908), p. 60. The error appears to have originated from Todhunter, *An Elementary Treatise on Laplace's Functions, Lamé's Functions and Bessel's Functions* (London, 1875), pp. 312—313.

The asymptotic expansion of $J_0(z)$ was obtained by Lipschitz* by integrating $e^{izt}(1-t^2)^{-\frac{1}{2}}$ round a rectangle (indented at ± 1) with corners at ± 1 and $\pm 1+\infty i$. Cauchy's theorem gives at once

$$\int_{-1}^{1} e^{izt}(1-t^2)^{-\frac{1}{2}}\,dt + e^{\frac{3}{4}\pi i}\int_0^\infty e^{(i-u)z}u^{-\frac{1}{2}}(2+iu)^{-\frac{1}{2}}\,du$$
$$- e^{\frac{1}{4}\pi i}\int_0^\infty e^{-(i+u)z}u^{-\frac{1}{2}}(2-iu)^{-\frac{1}{2}}\,du = 0,$$

and the analysis then proceeds on the lines already given; but in order to obtain asymptotic expansions of a *pair* of solutions of Bessel's equation it seems necessary to use a method which involves at some stage the loop integrals discussed in Chapter VI.

It may be convenient to note explicitly the initial terms in the expansions involved in equations (1)—(4); they are as follows:

$$\sum_{m=0}^\infty \frac{(-)^m.(\nu, 2m)}{(2z)^{2m}} = 1 - \frac{(4\nu^2-1^2)(4\nu^2-3^2)}{2!(8z)^2}$$
$$+ \frac{(4\nu^2-1^2)(4\nu^2-3^2)(4\nu^2-5^2)(4\nu^2-7^2)}{4!(8z)^4} - \dots,$$

$$\sum_{m=0}^\infty \frac{(-)^m.(\nu, 2m+1)}{(2z)^{2m+1}} = \frac{4\nu^2-1^2}{1!8z} - \frac{(4\nu^2-1^2)(4\nu^2-3^2)(4\nu^2-5^2)}{3!(8z)^3} + \dots.$$

The reader should notice that

$$J_\nu^2(z) + J_{\nu+1}^2(z) \sim 2/(\pi z),$$

a formula given by Lommel, *Studien*, p. 67.

NOTE. The method by which Lommel endeavoured to obtain the asymptotic expansion of $Y_n(z)$ in his *Studien*, pp. 93—97, was by differentiating the expansions of $J_{\pm\nu}(z)$ with respect to ν; but of course it is now known that the term-by-term differentiation of an asymptotic expansion with respect to a parameter raises various theoretical difficulties. It should be noticed that Lommel's later work, *Math. Ann.* IV. (1871), p. 103, is free from the algebraical errors which occur in his earlier work. These errors have been enumerated by Julius, *Archives Néerlandaises*, XXVIII. (1895), pp. 221—225. The asymptotic expansions of $J_n(z)$ and $Y_n(z)$ have also been studied by McMahon, *Ann. of Math.* VIII. (1894), pp. 57—61, and Kapteyn, *Monatshefte für Math. und Phys.* XIV. (1903), pp. 275—282.

A novel application of these asymptotic expansions has been discovered in recent years: they are of some importance in the analytic theory of the divisors of numbers. In such investigations the dominant terms of the expansions are adequate for the purpose in view. This fact combined with the consideration that the theory of Bessel functions forms only a trivial part of the investigations in question has made it seem desirable merely to mention the work of Voronoi† and Wigert‡ and the more recent papers by Hardy§.

* *Journal für Math.* LVI. (1859), pp. 189—196.

† *Ann. Sci. de l'École norm. sup.* (3) XXI. (1904), pp. 207—268, 459—534; *Verh. des Int. Math. Kongresses in Heidelberg*, 1904, pp. 241—245.

‡ *Acta Mathematica*, XXXVII. (1914), pp. 113—140.

§ *Quarterly Journal*, XLVI. (1915), pp. 263—283; *Proc. London Math. Soc.* (2) XV. (1916), pp. 1—25.

7·22. *Stokes' phenomenon.*

The formula § 7·21 (1) for $J_\nu(z)$ was established for values of z such that $|\arg z| < \pi$. If we took $\arg z$ to lie between 0 and 2π (so that $\arg ze^{-\pi i}$ lies, between $-\pi$ and π) we should consequently have

$$J_\nu(z) = e^{\nu\pi i} J_\nu(ze^{-\pi i})$$
$$\sim e^{\nu\pi i}\left(\frac{2}{\pi ze^{-\pi i}}\right)^{\frac{1}{2}}\left[\cos(ze^{-\pi i} - \tfrac{1}{2}\nu\pi - \tfrac{1}{4}\pi)\sum_{m=0}^{\infty}\frac{(-)^m.(\nu, 2m)}{(2ze^{-\pi i})^{2m}}\right.$$
$$\left. - \sin(ze^{-\pi i} - \tfrac{1}{2}\nu\pi - \tfrac{1}{4}\pi)\sum_{m=0}^{\infty}\frac{(-)^m.(\nu, 2m+1)}{(2ze^{-\pi i})^{2m+1}}\right],$$

so that, when $0 < \arg z < 2\pi$,

$$J_\nu(z) \sim e^{(\nu+\frac{1}{2})\pi i}\left(\frac{2}{\pi z}\right)^{\frac{1}{2}}\left[\cos(z + \tfrac{1}{2}\nu\pi + \tfrac{1}{4}\pi)\sum_{m=0}^{\infty}\frac{(-)^m.(\nu, 2m)}{(2z)^{2m}}\right.$$
$$\left. - \sin(z + \tfrac{1}{2}\nu\pi + \tfrac{1}{4}\pi)\sum_{m=0}^{\infty}\frac{(-)^m.(\nu, 2m+1)}{(2z)^{2m+1}}\right],$$

and this expansion is superficially quite different from the expansion of § 7·21 (1). We shall now make a close examination of this change.

The expansions of § 7·21 are derived from the formula

$$J_\nu(z) = \tfrac{1}{2}\{H_\nu^{(1)}(z) + H_\nu^{(2)}(z)\},$$

and throughout the sector in which $-\pi < \arg z < 2\pi$, the function $H_\nu^{(1)}(z)$ has the asymptotic expansion

$$H_\nu^{(1)}(z) \sim \left(\frac{2}{\pi z}\right)^{\frac{1}{2}} e^{i(z-\frac{1}{2}\nu\pi-\frac{1}{4}\pi)}\sum_{m=0}^{\infty}\frac{(-)^m.(\nu, m)}{(2iz)^m}.$$

The corresponding expansion for $H_\nu^{(2)}(z)$, namely

$$(1)\qquad H_\nu^{(2)}(z) \sim \left(\frac{2}{\pi z}\right)^{\frac{1}{2}} e^{-i(z-\frac{1}{2}\nu\pi-\frac{1}{4}\pi)}\sum_{m=0}^{\infty}\frac{(\nu, m)}{(2iz)^m},$$

is, however, valid for the sector $-2\pi < \arg z < \pi$. To obtain an expansion valid for the sector $0 < \arg z < 2\pi$ we use the formula of § 3·62 (6), namely

$$H_\nu^{(2)}(z) = 2\cos\nu\pi . H_\nu^{(2)}(ze^{-\pi i}) + e^{\nu\pi i} H_\nu^{(1)}(ze^{-\pi i}),$$

and this gives

$$(2)\qquad H_\nu^{(2)}(z) \sim \left(\frac{2}{\pi z}\right)^{\frac{1}{2}} e^{-i(z-\frac{1}{2}\nu\pi-\frac{1}{4}\pi)}\sum_{m=0}^{\infty}\frac{(\nu, m)}{(2iz)^m}$$
$$+ 2\cos\nu\pi . \left(\frac{2}{\pi z}\right)^{\frac{1}{2}} e^{i(z+\frac{1}{2}\nu\pi+\frac{3}{4}\pi)}\sum_{m=0}^{\infty}\frac{(-)^m.(\nu, m)}{(2iz)^m}.$$

The expansions (1) and (2) are both valid when $0 < \arg z < \pi$; now the difference between them has the asymptotic expansion

$$2\cos\nu\pi . \left(\frac{2}{\pi z}\right)^{\frac{1}{2}} e^{i(z+\frac{1}{2}\nu\pi+\frac{3}{4}\pi)}\sum_{m=0}^{\infty}\frac{(-)^m.(\nu, m)}{(2iz)^m},$$

and, on account of the factor e^{iz} which multiplies the series, this expression is of *lower order of magnitude* (when $|z|$ is large) than the error due to stopping at *any definite* term of the expansion (1); for this error is $O(e^{-iz} z^{-p-\frac{1}{2}})$ when we stop at the pth term. Hence the discrepancy between (1) and (2), which occurs when $0 < \arg z < \pi$, is only apparent, since the series in (1) has to be used in conjunction with its remainder.

Generally we have

$$J_\nu(z) \sim c_1 \left(\frac{2}{\pi z}\right)^{\frac{1}{2}} e^{i(z-\frac{1}{2}\nu\pi-\frac{1}{4}\pi)} \sum_{m=0}^{\infty} \frac{(-)^m . (\nu, m)}{(2iz)^m} + c_2 \left(\frac{2}{\pi z}\right)^{\frac{1}{2}} e^{-i(z-\frac{1}{2}\nu\pi-\frac{1}{4}\pi)} \sum_{m=0}^{\infty} \frac{(\nu, m)}{(2iz)^m},$$

where the constants c_1, c_2 have values which depend on the domain of values assigned to $\arg z$. And, if $\arg z$ is continually increased (or decreased) while $|z|$ is unaltered, the values of c_1 and c_2 have to be changed abruptly at various stages, the change in either constant being made when the function which multiplies it is negligible compared with the function multiplying the other constant. That is to say, changes in c_1 occur when $I(z)$ is positive, while changes in c_2 occur when $I(z)$ is negative.

It is not difficult to prove that the values to be assigned to the constants c_1 and c_2 are as follows:

$$c_1 = \tfrac{1}{2} e^{2p(\nu+\frac{1}{2})\pi i}, \quad c_2 = \tfrac{1}{2} e^{2p(\nu+\frac{1}{2})\pi i}, \quad [(2p-1)\pi < \arg z < (2p+1)\pi],$$

$$c_1 = \tfrac{1}{2} e^{2(p+1)(\nu+\frac{1}{2})\pi i}, \quad c_2 = \tfrac{1}{2} e^{2p(\nu+\frac{1}{2})\pi i}, \quad [2p\pi < \arg z < (2p+2)\pi],$$

where p is any integer, positive or negative.

This phenomenon of the *discontinuity of the constants* was discovered by Stokes and was discussed by him in a series of papers. It is a phenomenon which is not confined to Bessel functions, and it is characteristic of integral functions which possess asymptotic expansions of a simple type*.

The fact that the constants involved in the asymptotic expansion of the analytic function $J_\nu(z)$ are discontinuous was discovered by Stokes in (March?) 1857, and the discovery was apparently one of those which are made at three o'clock in the morning. See *Sir George Gabriel Stokes, Memoir and Scientific Correspondence*, I. (Cambridge, 1907), p. 62. The papers in which Stokes published his discovery are the following†: *Trans. Camb. Phil. Soc.* X. (1864), pp. 106—128; XI. (1871), pp. 412—425; *Acta Math.* XXVI. (1902), pp. 393—397. [*Math. and Phys. Papers*, IV. (1904), pp. 77—109; 283—298; V. (1905), pp. 283—287.] The third of these seems to have been the last paper written by Stokes.

7·23. *Asymptotic expansions of $I_\nu(z)$ and $K_\nu(z)$.*

The formula § 7·2 (5) combined with equation § 3·7 (8), which connects $K_\nu(z)$ and $H_\nu^{(1)}(iz)$, shews at once that

$$(1) \qquad K_\nu(z) \sim \left(\frac{\pi}{2z}\right)^{\frac{1}{2}} e^{-z} \sum_{m=0}^{\infty} \frac{(\nu, m)}{(2z)^m}$$

$$\sim \left(\frac{\pi}{2z}\right)^{\frac{1}{2}} e^{-z} \left[1 + \frac{4\nu^2 - 1^2}{1!\,8z} + \frac{(4\nu^2-1^2)(4\nu^2-3^2)}{2!\,(8z)^2} + \dots\right],$$

* Cf. Bromwich, *Theory of Infinite Series*, § 133.

† Stokes illustrated the change with the aid of Bessel functions whose orders are 0 and $\pm\frac{1}{3}$, the latter being those associated with Airy's integral (§ 6·4).

when $|\arg z| < \frac{3}{2}\pi$. And the formula $I_\nu(z) = e^{\frac{1}{2}\nu\pi i} J_\nu(e^{-\frac{1}{2}\pi i} z)$ shews that

$$(2) \qquad I_\nu(z) \sim \frac{e^z}{(2\pi z)^{\frac{1}{2}}} \sum_{m=0}^{\infty} \frac{(-)^m (\nu, m)}{(2z)^m} + \frac{e^{-z+(\nu+\frac{1}{2})\pi i}}{(2\pi z)^{\frac{1}{2}}} \sum_{m=0}^{\infty} \frac{(\nu, m)}{(2z)^m},$$

provided that $-\frac{1}{2}\pi < \arg z < \frac{3}{2}\pi$.

On the other hand, the formula $I_\nu(z) = e^{-\frac{1}{2}\nu\pi i} J_\nu(e^{\frac{1}{2}\pi i} z)$ shews that

$$(3) \qquad I_\nu(z) \sim \frac{e^z}{(2\pi z)^{\frac{1}{2}}} \sum_{m=0}^{\infty} \frac{(-)^m (\nu, m)}{(2z)^m} + \frac{e^{-z-(\nu+\frac{1}{2})\pi i}}{(2\pi z)^{\frac{1}{2}}} \sum_{m=0}^{\infty} \frac{(\nu, m)}{(2z)^m},$$

provided that $-\frac{3}{2}\pi < \arg z < \frac{1}{2}\pi$.

The apparent discrepancy between (2) and (3) when z has a value for which $\arg z$ lies between $-\frac{1}{2}\pi$ and $\frac{1}{2}\pi$ is, of course, an example of Stokes' phenomenon which has just been investigated.

The formulae of this section were stated explicitly by Kummer, *Journal für Math.* XVII. (1837), pp. 228—242, and Kirchhoff, *Journal für Math.* XLVIII. (1854), pp. 348—376, except that, in (2) and (3), the negligible second series is omitted. The object of the retention of the negligible series is to make (1) and (3) formally consistent with § 3·7 (6).

The formulae are also given by Riemann, *Ann. der Physik und Chemie*, (2) XCV. (1855), p. 135, when $\nu = 0$. Proofs are to be found on pp. 496—498 of Hankel's memoir.

A number of extremely interesting symbolic investigations of the formulae are to be found in Heaviside's* papers, but it is difficult to decide how valuable such researches are to be considered when modern standards of rigour are adopted.

A remarkable memoir is due to Malmstén†, in which the formula

$$\int_0^\infty \frac{\cos ax \,.\, dx}{(1+x^2)^{n+1}} \sim \frac{\pi e^{-a}}{2^{2n+1} \,.\, n!}$$
$$\times [(2a)^n + {}_nC_1 (n+1) \,.\, (2a)^{n-1} + {}_nC_2 (n+1)(n+2) \,.\, (2a)^{n-2} + \ldots]$$

is obtained (cf. § 6·3). This formula is written symbolically in the form

$$\int_0^\infty \frac{\cos ax \,.\, dx}{(1+x^2)^{n+1}} = \frac{\pi e^{-a}}{2^{2n+1} \,.\, n!} \left\{ 2a + \frac{1}{[n]^{-1}} \right\}^n,$$

the [ ] denoting that $[n]^{-m}$ is to be replaced by $(n)_{-m}$, and this, in Malmstén's notation, means

$$1/\{(n+1)(n+2)\ldots(n+m)\}.$$

It will be observed that this notation is different from the notation of § 4·4.

7·24. *The asymptotic expansions of* ber (z) *and* bei (z).

From the formulae obtained in §§ 7·21, 7·23, the asymptotic expansions of Thomson's functions ber (z) and bei (z), and of their generalisations, may be written down without difficulty. The formulae for functions of any order have been given by Whitehead‡, but, on account of their complexity, they will not

* *Proc. Royal Soc.* LII. (1893), pp. 504—529; *Electromagnetic Theory*, II. (London, 1899). My thanks are due to Dr Bromwich for bringing to my notice the results contained in the latter work.

† *K. Svenska V. Akad. Handl.* LXII. (1841), pp. 65—74.

‡ *Quarterly Journal*, XLII. (1911), pp. 329—338.

be quoted here. The functions of zero order had been examined previously by Russell*; he found it convenient to deal with the logarithms† of the functions of the third kind which are involved, and his formulae may be written as follows:

(1) $$\begin{cases}\operatorname{ber}(z)\\ \operatorname{bei}(z)\end{cases} = \frac{\exp\alpha(z)}{\sqrt{(2\pi z)}} \begin{matrix}\cos\\ \sin\end{matrix}\, \beta(z),$$

(2) $$\begin{cases}\operatorname{ker}(z)\\ \operatorname{kei}(z)\end{cases} = \frac{\exp\alpha(-z)}{\sqrt{(2z/\pi)}} \begin{matrix}\cos\\ \sin\end{matrix}\, \beta(-z),$$

where

$$\alpha(z) \sim \frac{z}{\sqrt{2}} + \frac{1}{8z\sqrt{2}} - \frac{25}{384z^3\sqrt{2}} - \frac{13}{128z^4} - \ldots,$$

$$\beta(z) \sim \frac{z}{\sqrt{2}} - \frac{\pi}{8} - \frac{1}{8z\sqrt{2}} - \frac{1}{16z^2} - \frac{25}{384z^3\sqrt{2}} + \ldots.$$

The ranges of validity of the formulae are $|\arg z| < \frac{1}{4}\pi$ in the case of (1) and $|\arg z| < \frac{5}{4}\pi$ in the case of (2).

These results have been expressed in a modified form by Savidge, *Phil. Mag.* (6) XIX. (1910), p. 51.

7·25. *Hadamard's modification of the asymptotic expansions.*

A result which is of considerable theoretical importance is due to Hadamard‡; he has shewn that it is possible to modify the various asymptotic expansions, *so that they become convergent series together with a negligible remainder term.* The formulae will be stated for real values of the variables, but the reader should have no difficulty in making the modifications appropriate to complex variables.

We take first the case of $I_\nu(x)$ when $\nu > -\frac{1}{2}$. When we replace $\sin\frac{1}{2}\theta$ by u, we have

$$\begin{aligned} I_\nu(x) &= \frac{(\frac{1}{2}x)^\nu}{\Gamma(\nu+\frac{1}{2})\Gamma(\frac{1}{2})}\int_0^\pi e^{x\cos\theta}\sin^{2\nu}\theta\, d\theta \\ &= \frac{2(2x)^\nu e^x}{\Gamma(\nu+\frac{1}{2})\Gamma(\frac{1}{2})}\int_0^1 \exp(-2u^2x)\,.\,u^{2\nu}(1-u^2)^{\nu-\frac{1}{2}}\,du \\ &= \frac{2(2x)^\nu e^x}{\Gamma(\nu+\frac{1}{2})\Gamma(\frac{1}{2})}\sum_{m=0}^\infty \frac{(\frac{1}{2}-\nu)_m}{m!}\int_0^1 u^{2\nu+2m}\exp(-2u^2x)\,du, \end{aligned}$$

the last result being valid because the series of integrals is convergent.

We may write this equation in the form

(1) $$\begin{aligned} I_\nu(x) &= \frac{e^x}{\Gamma(\nu+\frac{1}{2})\Gamma(\frac{1}{2})\sqrt{(2x)}}\sum_{m=0}^\infty \frac{(\frac{1}{2}-\nu)_m}{m!(2x)^m}\int_0^{2x} t^{\nu+m-\frac{1}{2}}e^{-t}\,dt \\ &= \frac{e^x}{\sqrt{(2\pi x)}}\sum_{m=0}^\infty \frac{(\frac{1}{2}-\nu)_m\,.\,\gamma(\nu+m+\frac{1}{2}, 2x)}{\Gamma(\nu+\frac{1}{2})\,.\,m!(2x)^m}, \end{aligned}$$

where γ denotes the "incomplete Gamma-function" of Legendre§.

* *Phil. Mag.* (6) XVII. (1909), pp. 531, 537.

† Cf. the similar procedure due to Meissel, which will be explained in § 8·11.

‡ *Bull. de la Soc. Math. de France*, XXXVI. (1908), pp. 77—85.

§ Cf. *Modern Analysis*, § 16·2.

For large values of x, the difference between

$$\frac{\gamma(\nu+n+\frac{1}{2}, 2x)}{\Gamma(\nu+\frac{1}{2})} \text{ and } (\tfrac{1}{2}+\nu)_n$$

is $O(x^{\nu+n+\frac{1}{2}}e^{-2x})$ which is $o(1)$ for each integral value of n.

In the case of the ordinary Bessel functions, we take the expression for the function of the third kind

$$\begin{aligned} H_\nu^{(1)}(x) &= \left(\frac{2}{\pi x}\right)^{\frac{1}{2}} \frac{e^{i(x-\frac{1}{2}\nu\pi-\frac{1}{4}\pi)}}{\Gamma(\nu+\frac{1}{2})} \int_0^\infty e^{-u} u^{\nu-\frac{1}{2}} \left(1+\frac{iu}{2x}\right)^{\nu-\frac{1}{2}} du \\ &= \left(\frac{2}{\pi x}\right)^{\frac{1}{2}} \frac{e^{i(x-\frac{1}{2}\nu\pi-\frac{1}{4}\pi)}}{\Gamma(\nu+\frac{1}{2})} \int_0^{2x} e^{-u} u^{\nu-\frac{1}{2}} \left(1+\frac{iu}{2x}\right)^{\nu-\frac{1}{2}} du + O(x^\nu e^{-2x}) \\ &= \left(\frac{2}{\pi x}\right)^{\frac{1}{2}} \frac{e^{i(x-\frac{1}{2}\nu\pi-\frac{1}{4}\pi)}}{\Gamma(\nu+\frac{1}{2})} \sum_{m=0}^{\infty} \frac{(-i)^m(\frac{1}{2}-\nu)_m}{m!(2x)^m} \int_0^{2x} u^{\nu+m-\frac{1}{2}} e^{-u} du + O(x^\nu e^{-2x}), \end{aligned}$$

so that

$$(2) \quad H_\nu^{(1)}(x) = \left(\frac{2}{\pi x}\right)^{\frac{1}{2}} e^{i(x-\frac{1}{2}\nu\pi-\frac{1}{4}\pi)} \sum_{m=0}^{\infty} \frac{(\frac{1}{2}-\nu)_m \gamma(\nu+m+\frac{1}{2}, 2x)}{\Gamma(\nu+\frac{1}{2}) . m!(2ix)^m} + O(x^\nu e^{-2x}),$$

and similarly

$$(3) \quad H_\nu^{(2)}(x) = \left(\frac{2}{\pi x}\right)^{\frac{1}{2}} e^{-i(x-\frac{1}{2}\nu\pi-\frac{1}{4}\pi)} \sum_{m=0}^{\infty} \frac{(\frac{1}{2}-\nu)_m \gamma(\nu+m+\frac{1}{2}, 2x)}{\Gamma(\nu+\frac{1}{2}) . m!(-2ix)^m} + O(x^\nu e^{-2x}).$$

From these results it is easy to derive convergent series for the functions of the first and second kinds.

Hadamard gave the formulae for functions of order zero only; but the extension to functions of any order exceeding $-\frac{1}{2}$ is obvious.

7·3. *Formulae for the remainders in the asymptotic expansions.*

In § 7·2 we gave an investigation which shewed that the remainders in the asymptotic expansions of $H_\nu^{(1)}(z)$ and $H_\nu^{(2)}(z)$ are of the same order of magnitude as the first terms neglected. In the case of functions of the first and second kinds, it is easy to obtain a more exact and rather remarkable theorem to the effect that when ν is real* and x is positive the remainders after a certain stage in the asymptotic expansions of $J_{\pm\nu}(x)$ and $Y_{\pm\nu}(x)$ are *numerically less* than the first terms neglected, and, by a slightly more recondite investigation (§ 7·32), it can be proved that the remainders are *of the same sign* as the first terms neglected.

Let us write

$$\frac{1}{2\Gamma(\nu+\frac{1}{2})} \int_0^\infty e^{-u} u^{\nu-\frac{1}{2}} \left\{\left(1+\frac{iu}{2x}\right)^{\nu-\frac{1}{2}} + \left(1-\frac{iu}{2x}\right)^{\nu-\frac{1}{2}}\right\} du = P(x, \nu),$$

$$\frac{1}{2i\Gamma(\nu+\frac{1}{2})} \int_0^\infty e^{-u} u^{\nu-\frac{1}{2}} \left\{\left(1+\frac{iu}{2x}\right)^{\nu-\frac{1}{2}} - \left(1-\frac{iu}{2x}\right)^{\nu-\frac{1}{2}}\right\} du = Q(x, \nu),$$

* We may take $\nu \geqslant 0$ without losing generality.

so that

(1) $$J_{\pm\nu}(x) = \left(\frac{2}{\pi x}\right)^{\frac{1}{2}} [\cos(x \mp \tfrac{1}{2}\nu\pi - \tfrac{1}{4}\pi) P(x, \nu) - \sin(x \mp \tfrac{1}{2}\nu\pi - \tfrac{1}{4}\pi) Q(x, \nu)],$$

(2) $$Y_{\pm\nu}(x) = \left(\frac{2}{\pi x}\right)^{\frac{1}{2}} [\sin(x \mp \tfrac{1}{2}\nu\pi - \tfrac{1}{4}\pi) P(x, \nu) + \cos(x \mp \tfrac{1}{2}\nu\pi - \tfrac{1}{4}\pi) Q(x, \nu)].$$

Now $I(\nu) = 0$ and, in the analysis of § 7·2, we may take δ to be $\frac{1}{2}\pi$ since the variables are real, and so $A_{2p} = 1$.

It follows that, if p be taken so large that $2p \geqslant \nu - \frac{1}{2}$, there exists a number θ, not exceeding unity in absolute value, such that

$$\left(1 \pm \frac{iu}{2x}\right)^{\nu - \frac{1}{2}} = \sum_{m=0}^{2p-1} \frac{(\frac{1}{2} - \nu)_m}{m!} \left(\frac{\pm u}{2ix}\right)^m + \frac{\theta . (\frac{1}{2} - \nu)_{2p}}{(2p)!} \left(\frac{\pm u}{2ix}\right)^{2p},$$

and, on adding the results combined in this formula, we have*

$$\left(1 + \frac{iu}{2x}\right)^{\nu - \frac{1}{2}} + \left(1 - \frac{iu}{2x}\right)^{\nu - \frac{1}{2}} = 2 \sum_{m=0}^{p-1} \frac{(\frac{1}{2} - \nu)_{2m}}{(2m)!} \left(\frac{u}{2ix}\right)^{2m} + \frac{2\theta_0 . (\frac{1}{2} - \nu)_{2p}}{(2p)!} \left(\frac{u}{2x}\right)^{2p},$$

where $|\theta_0| \leqslant 1$; and, since θ_0 is obviously real, $-1 \leqslant \theta_0 \leqslant 1$.

It follows on integration that

$$P(x, \nu) = \sum_{m=0}^{p-1} \frac{(-)^m . (\frac{1}{2} - \nu)_{2m} (\frac{1}{2} + \nu)_{2m}}{(2m)! (2x)^{2m}} + \frac{(\frac{1}{2} - \nu)_{2p}}{(2p)! (2x)^{2p} \Gamma(\nu + \frac{1}{2})} \int_0^\infty \theta_0 e^{-u} u^{\nu + 2p - \frac{1}{2}} du,$$

and since

$$\left| \int_0^\infty \theta_0 e^{-u} u^{\nu + 2p - \frac{1}{2}} du \right| \leqslant \int_0^\infty e^{-u} u^{\nu + 2p - \frac{1}{2}} du = \Gamma(\nu + 2p + \tfrac{1}{2}),$$

we see that the remainder after p terms in the expansion of $P(x, \nu)$ does not exceed the $(p + 1)$th term in absolute value, provided that $2p > \nu - \frac{1}{2}$.

From the formula

$$\left(1 \pm \frac{iu}{2x}\right)^{\nu - \frac{1}{2}} = \sum_{m=0}^{2p} \frac{(\frac{1}{2} - \nu)_m}{m!} \left(\frac{\pm u}{2ix}\right)^m + \frac{\theta . (\frac{1}{2} - \nu)_{2p+1}}{(2p+1)!} \left(\frac{\pm u}{2ix}\right)^{2p+1},$$

we find in a similar manner that the remainder after p terms in the expansion of $Q(x, \nu)$ does not exceed the $(p + 1)$th term in absolute value, provided that $2p \geqslant \nu - \frac{3}{2}$.

These results were given by Hankel, *Math. Ann.* I. (1869), pp. 491—494, and were reproduced by Gray and Mathews in their *Treatise on Bessel Functions* (London, 1895), p. 70, but small inaccuracies have been pointed out in both investigations by Orr, *Trans. Camb. Phil. Soc.* XVII. (1899), pp. 172—180.

In the case of $K_\nu(x)$ we have the formula

$$K_\nu(x) = \left(\frac{\pi}{2x}\right)^{\frac{1}{2}} \frac{e^{-x}}{\Gamma(\nu + \frac{1}{2})} \int_0^\infty e^{-u} u^{\nu - \frac{1}{2}} \left(1 + \frac{u}{2x}\right)^{\nu - \frac{1}{2}} du,$$

* This result was obtained in a rather different manner by Lipschitz, *Journal für Math.* LVI. (1859), pp. 189—196.

and

$$\left(1+\frac{u}{2x}\right)^{\nu-\frac{1}{2}}=\sum_{m=0}^{p-1}\frac{(-)^m.(\frac{1}{2}-\nu)_m}{m!}\left(\frac{u}{2x}\right)^m$$
$$+\frac{(-)^p(\frac{1}{2}-\nu)_p}{(p-1)!}\left(\frac{u}{2x}\right)^p\int_0^1(1-t)^{p-1}\left(1+\frac{ut}{2x}\right)^{\nu-p-\frac{1}{2}}dt,$$

and, when $p \geqslant \nu-\frac{1}{2}$, the last term may be written

$$\frac{(-)^p.(\frac{1}{2}-\nu)_p}{(p-1)!}\left(\frac{u}{2x}\right)^p\theta_1\int_0^1(1-t)^{p-1}\,dt,$$

where $0<\theta_1\leqslant 1$, and so, on integration,

$$K_\nu(x)=\left(\frac{\pi}{2x}\right)^{\frac{1}{2}}e^{-x}\left[\sum_{m=0}^{p-1}\frac{(\nu, m)}{(2x)^m}+\theta_2\frac{(\nu, p)}{(2x)^p}\right],$$

where $0\leqslant\theta_2\leqslant 1$ when $p\geqslant\nu-\frac{1}{2}$.

This is a more exact result than those obtained for $P(x,\nu)$ and $Q(x,\nu)$ by the same methods; the reason why the greater exactness is secured is, of course, the fact that $(1+\frac{1}{2}ut/x)^{\nu-p-\frac{1}{2}}$ is positive and does not oscillate in sign after the manner of $(1+\frac{1}{2}iut/x)^{\nu-p-\frac{1}{2}}\pm(1-\frac{1}{2}iut/x)^{\nu-p-\frac{1}{2}}$.

7·31. *The researches of Stieltjes on $J_0(x)$, $Y_0(x)$ and $K_0(x)$.*

The results of § 7·3 were put into a more precise form by Stieltjes*, who proved not only that the remainders in the asymptotic expansions associated with $J_0(x)$, $Y_0(x)$ and $K_0(x)$ are numerically less than the first terms neglected, but also that the remainders *have the same sign* as those terms.

Stieltjes also examined $I_0(x)$, but his result is complicated and we shall not reproduce it†. It is only to be expected that $I_0(x)$ is intractable because in the dominant expansion the terms all have the same sign whereas in the other three asymptotic expansions the terms alternate in sign.

It is evident from the definitions of § 7·3 that

$$P(x,0)=\frac{\sqrt{x}}{2\sqrt{\pi}}\int_0^\infty e^{-ux}u^{-\frac{1}{2}}\{(1+\tfrac{1}{2}iu)^{-\frac{1}{2}}+(1-\tfrac{1}{2}iu)^{-\frac{1}{2}}\}\,du,$$

$$Q(x,0)=\frac{\sqrt{x}}{2i\sqrt{\pi}}\int_0^\infty e^{-ux}u^{-\frac{1}{2}}\{(1+\tfrac{1}{2}iu)^{-\frac{1}{2}}-(1-\tfrac{1}{2}iu)^{-\frac{1}{2}}\}\,du.$$

In these formulae replace $(1\pm\frac{1}{2}iu)^{-\frac{1}{2}}$ by

$$\frac{2}{\pi}\int_0^{\frac{1}{2}\pi}\frac{d\phi}{1\pm\frac{1}{2}iu\sin^2\phi}.$$

* *Ann. Sci. de l'École norm. sup.* (3) III. (1886), pp. 233—252.

† The function $I_\nu(x)$ has also been examined by Schafheitlin, *Jahresbericht der Deutschen Math.-Vereinigung*, XIX. (1910), pp. 120—129, but he appears to use Lagrange's form for the remainder in Taylor's theorem when it is inapplicable.

It is then evident that

$$(1+\tfrac{1}{2}iu)^{-\frac{1}{2}}+(1-\tfrac{1}{2}iu)^{-\frac{1}{2}}=\frac{4}{\pi}\int_0^{\frac{1}{2}\pi}\frac{d\phi}{1+\tfrac{1}{4}u^2\sin^4\phi}$$

$$=\frac{4}{\pi}\int_0^{\frac{1}{2}\pi}\{1-\tfrac{1}{4}u^2\sin^4\phi+\ldots+(-)^{p-1}(\tfrac{1}{4}u^2\sin^4\phi)^{p-1}$$
$$+(-)^p(\tfrac{1}{4}u^2\sin^4\phi)^p/(1+\tfrac{1}{4}u^2\sin^4\phi)\}\,d\phi,$$

where p is any positive integer (zero included).

Now, obviously,

$$\int_0^{\frac{1}{2}\pi}\frac{(\tfrac{1}{4}u^2\sin^4\phi)^p}{1+\tfrac{1}{4}u^2\sin^2\phi}\,d\phi=\theta\int_0^{\frac{1}{2}\pi}(\tfrac{1}{4}u^2\sin^4\phi)^p\,d\phi,$$

where θ *lies between* 0 *and* 1; and hence

$$\tfrac{1}{2}\{(1+\tfrac{1}{2}iu)^{-\frac{1}{2}}+(1-\tfrac{1}{2}iu)^{-\frac{1}{2}}\}=1-\frac{1.3}{2\,!}(\tfrac{1}{4}u)^2+\frac{1.3.5.7}{4\,!}(\tfrac{1}{4}u)^4-\ldots$$
$$+(-)^{p-1}\frac{1.3.5\ldots(4p-5)}{(2p-2)\,!}(\tfrac{1}{4}u)^{2p-2}+(-)^p\,\theta\,\frac{1.3.5\ldots(4p-1)}{(2p)\,!}(\tfrac{1}{4}u)^{2p}.$$

If we multiply by the positive function $e^{-ux}u^{-\frac{1}{2}}$ and integrate, it is evident that

$$(1)\quad P(x,0)=1-\frac{1^2.3^2}{2\,!\,(8x)^2}+\ldots+(-)^{p-1}\frac{1^2.3^2.5^2\ldots(4p-5)^2}{(2p-2)\,!\,(8x)^{2p-2}}$$
$$+(-)^p\,\theta_1\,\frac{1^2.3^2.5^2\ldots(4p-1)^2}{(2p)\,!\,(8x)^{2p}},$$

where $0<\theta_1<1$, and p is any positive integer (zero included); and this is the result which had to be proved for $P(x,0)$.

Similarly, from the formula

$$i\{(1+\tfrac{1}{2}iu)^{-\frac{1}{2}}-(1-\tfrac{1}{2}iu)^{-\frac{1}{2}}\}=\frac{4}{\pi}\int_0^{\frac{1}{2}\pi}\frac{\tfrac{1}{2}u\sin^2\phi\,d\phi}{1+\tfrac{1}{4}u^2\sin^2\phi},$$

we find that

$$(2)\quad Q(x,0)=-\frac{1^2}{1\,!\,8x}+\frac{1^2.3^2.5^2}{3\,!\,(8x)^3}-\ldots+(-)^p\frac{1^2.3^2.5^2\ldots(4p-3)^2}{(2p-1)\,!\,(8x)^{2p-1}}$$
$$+(-)^{p+1}\,\theta_2\,\frac{1^2.3^2.5^2\ldots(4p+1)^2}{(2p+1)\,!\,(8x)^{2p+1}},$$

where $0<\theta_2<1$, and p is any positive integer (zero included); and this is the result which had to be proved for $Q(x,0)$.

In the case of $K_0(x)$, Stieltjes took the formula

$$K_0(x)=\frac{e^{-x}}{\sqrt{2}}\int_0^\infty e^{-ux}u^{-\frac{1}{2}}(1+\tfrac{1}{2}u)^{-\frac{1}{2}}\,du,$$

and replaced $(1+\tfrac{1}{2}u)^{-\frac{1}{2}}$ by $\dfrac{2}{\pi}\displaystyle\int_0^{\frac{1}{2}\pi}\frac{d\phi}{1+\tfrac{1}{2}u\sin^2\phi}$; the procedure then follows the method just explained, and gives again the result of § 7·3.

By an ingenious device, Callandreau* succeeded in applying the result of Stieltjes to obtain the corresponding results for functions of any *integral* order; but we shall now explain a method which is effective in obtaining the precise results for functions of any *real* order.

* *Bull. des Sci. Math.* (2) XIV. (1890), pp. 110—114.

7·32. *The signs of the remainders in the asymptotic expansions associated with $J_\nu(x)$ and $Y_\nu(x)$.*

It has already been seen that $J_\nu(x)$ and $Y_\nu(x)$ are expressible in terms of two functions $P(x, \nu)$ and $Q(x, \nu)$ which have asymptotic expansions of a simpler type. We shall now extend the result of Stieltjes (§ 7·31) so as to shew that for *any real value** of the order ν, the remainder after p terms of the expansion of $P(x, \nu)$ is *of the same sign as* (in addition to being numerically less than†) the $(p+1)$th term provided that $2p > \nu - \frac{1}{2}$: a corresponding result holds for $Q(x, \nu)$ when $2p > \nu - \frac{3}{2}$. The restrictions which these conditions lay on p enable the theorem to be stated in the following manner:

In the oscillatory parts of the series for $P(x, \nu)$ and $Q(x, \nu)$, the remainders are of the same sign as, and numerically less than, the first terms neglected.

By a slight modification of the formulae of § 7·3, we have

$$P(x, \nu) = \frac{x^{\nu+\frac{1}{2}}}{2\Gamma(\nu+\frac{1}{2})}\int_0^\infty e^{-ux} u^{\nu-\frac{1}{2}} \{(1+\tfrac{1}{2}iu)^{\nu-\frac{1}{2}} + (1-\tfrac{1}{2}iu)^{\nu-\frac{1}{2}}\}\, du,$$

$$Q(x, \nu) = \frac{x^{\nu+\frac{1}{2}}}{2i\Gamma(\nu+\frac{1}{2})}\int_0^\infty e^{-ux} u^{\nu-\frac{1}{2}} \{(1+\tfrac{1}{2}iu)^{\nu-\frac{1}{2}} - (1-\tfrac{1}{2}iu)^{\nu-\frac{1}{2}}\}\, du,$$

and, exactly as in § 7·3, we may shew that

$$\tfrac{1}{2}\{(1+\tfrac{1}{2}iu)^{\nu-\frac{1}{2}} + (1-\tfrac{1}{2}iu)^{\nu-\frac{1}{2}}\} = \sum_{m=0}^{p-1} \frac{(-)^m . (\frac{1}{2}-\nu)_{2m} (\frac{1}{2}u)^{2m}}{(2m)!}$$
$$+ \frac{(-)^p (\frac{1}{2}-\nu)_{2p-1} (\frac{1}{2}u)^{2p-1}}{(2p-2)!}\int_0^1 (1-t)^{2p-2}\, \tfrac{1}{2}i\, \{(1+\tfrac{1}{2}iut)^{\nu-2p+\frac{1}{2}} - (1-\tfrac{1}{2}iut)^{\nu-2p+\frac{1}{2}}\}\, dt.$$

The reader will see that we can establish the theorem if we can prove that, when $2p > \nu - \frac{1}{2}$, the last term on the right is of fixed sign and its sign is that of

$$(-)^p . (\tfrac{1}{2}-\nu)_{2p} (\tfrac{1}{2}u)^{2p}/(2p)!.$$

It is clearly sufficient to shew that

$$\frac{1}{2p-\nu-\frac{1}{2}}\int_0^1 (1-t)^{2p-2}\, \tfrac{1}{2}i\, \{(1+\tfrac{1}{2}iut)^{\nu-2p+\frac{1}{2}} - (1-\tfrac{1}{2}iut)^{\nu-2p+\frac{1}{2}}\}\, dt$$

is positive. Now this expression is equal to‡

$$\frac{1}{(2p-\nu-\frac{1}{2})\,\Gamma(2p-\nu-\frac{1}{2})}\int_0^1 (1-t)^{2p-2}\, \tfrac{1}{2}i\, .\int_0^\infty \lambda^{2p-\nu-\frac{3}{2}} \{e^{-\lambda(1+\frac{1}{2}iut)} - e^{-\lambda(1-\frac{1}{2}iut)}\}\, d\lambda\, dt$$
$$= \frac{1}{\Gamma(2p-\nu+\frac{1}{2})}\int_0^1\int_0^\infty (1-t)^{2p-2} \lambda^{2p-\nu-\frac{3}{2}} \sin(\tfrac{1}{2}\lambda ut) . e^{-\lambda}\, d\lambda\, dt$$
$$= \frac{1}{\Gamma(2p-\nu+\frac{1}{2})}\int_0^\infty \lambda^{2p-\nu-\frac{3}{2}} e^{-\lambda} \int_0^1 (1-t)^{2p-2} \sin(\tfrac{1}{2}\lambda ut)\, dt\, d\lambda.$$

* As in § 7·3 we may take $\nu \geqslant 0$ without loss of generality.

† This has already been proved in § 7·3.

‡ Since $|\sin(\frac{1}{2}\lambda ut)| \leqslant \frac{1}{2}\lambda ut$, the condition $2p > \nu - \frac{1}{2}$ secures the absolute convergence of the infinite integral.

Now $(1-t)^{2p-2}$ is a monotonic decreasing function of t; and hence, by the second mean-value theorem, a number ξ, between 0 and 1, exists such that

$$\int_0^1 (1-t)^{2p-2} \sin(\tfrac{1}{2}\lambda ut)\, dt = \int_0^\xi \sin(\tfrac{1}{2}\lambda ut)\, dt \geqslant 0.$$

Since $\Gamma(2p-\nu+\frac{1}{2})$ is positive, we have succeeded in transforming

$$\frac{1}{2p-\nu-\frac{1}{2}} \int_0^1 (1-t)^{2p-2} \tfrac{1}{2}i \{(1+\tfrac{1}{2}iut)^{\nu-2p+\frac{1}{2}} - (1-\tfrac{1}{2}iut)^{\nu-2p+\frac{1}{2}}\}\, dt$$

into an infinite integral in which the integrand is *positive*, and so the expression under consideration is positive.

That is to say,

$$\tfrac{1}{2}\{(1+\tfrac{1}{2}iu)^{\nu-\frac{1}{2}} + (1-\tfrac{1}{2}iu)^{\nu-\frac{1}{2}}\}$$
$$= \sum_{m=0}^{p-1} \frac{(-)^m . (\frac{1}{2}-\nu)_{2m} (\frac{1}{2}u)^{2m}}{(2m)!} + \theta \frac{(-)^p . (\frac{1}{2}-\nu)_{2p} (\frac{1}{2}u)^{2p}}{(2p)!},$$

where $\theta \geqslant 0$ when $2p > \nu - \frac{1}{2}$. And it has already been seen (§ 7·3) that in these circumstances $|\theta| \leqslant 1$. *Consequently* $0 \leqslant \theta \leqslant 1$; and then, on multiplying the last equation by $e^{-ux} u^{\nu-\frac{1}{2}}$ and integrating, we at once obtain the property stated for $P(x, \nu)$.

The corresponding property for $Q(x, \nu)$ follows from the equation

$$\frac{1}{2i}\{(1+\tfrac{1}{2}iu)^{\nu-\frac{1}{2}} - (1-\tfrac{1}{2}iu)^{\nu-\frac{1}{2}}\} = -\sum_{m=0}^{p-1} \frac{(-)^m . (\frac{1}{2}-\nu)_{2m+1} (\frac{1}{2}u)^{2m+1}}{(2m+1)!}$$
$$+ \frac{(-)^{p+1} . (\frac{1}{2}-\nu)_{2p} (\frac{1}{2}u)^{2p}}{(2p-1)!} \int_0^1 (1-t)^{2p-1} \tfrac{1}{2}i \{(1+\tfrac{1}{2}iut)^{\nu-2p-\frac{1}{2}} - (1-\tfrac{1}{2}iut)^{\nu-2p-\frac{1}{2}}\}\, dt;$$

the details of the analysis will easily be supplied by the reader.

NOTE. The analysis fails when $-\frac{1}{2} < \nu < \frac{1}{2}$ if we take $p=0$, but then the phase of $(1 \pm \frac{1}{2}iu)^{\nu-\frac{1}{2}}$ lies between 0 and $\pm\frac{1}{2}(\nu-\frac{1}{2})\pi$, and so $\frac{1}{2}\{(1+\frac{1}{2}iu)^{\nu-\frac{1}{2}} + (1-\frac{1}{2}iu)^{\nu-\frac{1}{2}}\}$ has the same sign as unity, and, in like manner, $\frac{1}{2}\{(1+\frac{1}{2}iu)^{\nu-\frac{1}{2}} - (1-\frac{1}{2}iu)^{\nu-\frac{1}{2}}\}/i$ has the same sign as $\frac{1}{2}(\nu-\frac{1}{2})u$, and hence $P(x, \nu)$ and $Q(x, \nu)$ have the same sign as the first terms in their expansions, so the conclusions are still true; and the conclusion is true for $Q(x, \nu)$ when $\frac{1}{2} < \nu < \frac{3}{2}$ if $p=0$.

7·33. *Weber's formulae for the remainders in the expansions of functions of the third kind.*

Some inequalities which are satisfied by the remainders in the asymptotic expansions of $H_\nu^{(1)}(z)$ and $H_\nu^{(2)}(z)$ have been given by Weber*. These inequalities owe their importance to the fact that they are true whether z and ν are real or complex. In the investigations which we shall give it will be supposed for simplicity that ν is real, though it will be obvious that modifications of detail only are adequate to make the mode of analysis applicable to complex values

* *Math. Ann.* XXXVII. (1890), pp. 404—416.

of ν. There is no further loss of generality in assuming that $\nu \geqslant 0$, $R(z) \geqslant 0$. We shall write $|z| = r$, and, since *large* values of $|z|$ are primarily under consideration, we shall suppose that $2r \geqslant \nu - \frac{1}{2}$.

If $\nu - \frac{1}{2} > 0$, we have*, by § 6·12 (3),

$$\begin{aligned} |H_\nu^{(1)}(z)| &= \left| \left(\frac{2}{\pi z}\right)^{\frac{1}{2}} \frac{e^{i(z-\frac{1}{2}\nu\pi-\frac{1}{4}\pi)}}{\Gamma(\nu+\frac{1}{2})} \int_0^\infty e^{-u} u^{\nu-\frac{1}{2}} \left(1+\frac{iu}{2z}\right)^{\nu-\frac{1}{2}} du \right| \\ &\leqslant \left| \left(\frac{2}{\pi z}\right)^{\frac{1}{2}} \frac{e^{i(z-\frac{1}{2}\nu\pi-\frac{1}{4}\pi)}}{\Gamma(\nu+\frac{1}{2})} \right| . \int_0^\infty e^{-u} u^{\nu-\frac{1}{2}} \left(1+\frac{u}{2r}\right)^{\nu-\frac{1}{2}} du \\ &\leqslant \left| \left(\frac{2}{\pi z}\right)^{\frac{1}{2}} \frac{e^{i(z-\frac{1}{2}\nu\pi-\frac{1}{4}\pi)}}{\Gamma(\nu+\frac{1}{2})} \right| . \int_0^\infty \exp\left\{-u\left(1-\frac{\nu-\frac{1}{2}}{2r}\right)\right\} . u^{\nu-\frac{1}{2}} du \\ &= \left| \left(\frac{2}{\pi z}\right)^{\frac{1}{2}} e^{i(z-\frac{1}{2}\nu\pi-\frac{1}{4}\pi)} \right| . \left(1-\frac{\nu-\frac{1}{2}}{2r}\right)^{-\nu-\frac{1}{2}} \end{aligned}$$

If $0 \leqslant \nu < \frac{1}{2}$, we use the recurrence formula

$$H_\nu^{(1)}(z) = (2/z)(\nu+1) H_{\nu+1}^{(1)}(z) - H_{\nu+2}^{(1)}(z)$$

and apply the inequality just obtained to each of the functions on the right.

It is thus found that

(1) $$H_\nu^{(1)}(z) \leqslant G \left| (\tfrac{1}{2}\pi z)^{-\frac{1}{2}} e^{i(z-\frac{1}{2}\nu\pi-\frac{1}{4}\pi)} \right|,$$

and similarly

(2) $$H_\nu^{(2)}(z) \leqslant G \left| (\tfrac{1}{2}\pi z)^{-\frac{1}{2}} e^{-i(z-\frac{1}{2}\nu\pi-\frac{1}{4}\pi)} \right|,$$

where†

(3) $$\begin{cases} G = \left(1-\dfrac{\nu-\frac{1}{2}}{2r}\right)^{-\nu-\frac{1}{2}}, & (\nu > \tfrac{1}{2}) \\ G = \left(1-\dfrac{\nu+\frac{3}{2}}{2r}\right)^{-\nu-\frac{5}{2}} \left(1+\dfrac{2\nu+2}{r}\right). & (\nu < \tfrac{1}{2}) \end{cases}$$

The results may be called Weber's crude inequalities satisfied by $H_\nu^{(1)}(z)$ and $H_\nu^{(2)}(z)$. By an elegant piece of analysis, Weber succeeded in deducing more refined inequalities from them in the following manner:

Take the first p terms of the series involved in Hankel's two expansions and denote them by the symbols $\Sigma_\nu^{(1)}(z;\,p)$, $\Sigma_\nu^{(2)}(z;\,p)$, so that

$$\Sigma_\nu^{(1)}(z;\,p) = \sum_{m=0}^{p-1} \frac{(-)^m . (\nu, m)}{(2iz)^m}, \quad \Sigma_\nu^{(2)}(z;\,p) = \sum_{m=0}^{p-1} \frac{(\nu, m)}{(2iz)^m}.$$

It is easy to verify that

$$\left[\frac{d^2}{dz^2} + 2i\frac{d}{dz} + \frac{\frac{1}{4}-\nu^2}{z^2}\right] \Sigma_\nu^{(1)}(z;\,p) = \frac{-p.(\nu,p)}{z^2(-2iz)^{p-1}}.$$

We regard this as an equation to be solved by the method of variation of parameters; we thus find that

$$\Sigma_\nu^{(1)}(z;\,p) = (\tfrac{1}{2}\pi z)^{\frac{1}{2}} e^{-i(z-\frac{1}{2}\nu\pi-\frac{1}{4}\pi)} \{A(z) H_\nu^{(1)}(z) + B(z) H_\nu^{(2)}(z)\},$$

* In the third line of analysis the inequality $e^x \geqslant 1 + x$ $(x \geqslant 0)$ has been used.

† When $\nu < \frac{1}{2}$ we take $2r > \nu + \frac{3}{2}$.

where $A(z)$ and $B(z)$ are functions of z so chosen that

$$\begin{cases} A'(z)\,H_\nu^{(1)}(z) + B'(z)\,H_\nu^{(2)}(z) = 0, \\ A'(z)\dfrac{d}{dz}H_\nu^{(1)}(z) + B'(z)\dfrac{d}{dz}H_\nu^{(2)}(z) = -(\tfrac{1}{2}\pi z)^{-\frac{1}{2}}\,e^{i(z-\frac{1}{2}\nu\pi-\frac{1}{4}\pi)}\dfrac{p\,.\,(\nu, p)}{z^2(-2iz)^{p-1}}. \end{cases}$$

It follows that

$$A'(z) = \tfrac{1}{2}\pi\,(\tfrac{1}{2}\pi z)^{-\frac{1}{2}}\,e^{i(z-\frac{1}{2}\nu\pi-\frac{1}{4}\pi)}\,\frac{p\,.\,(\nu, p)}{(-2iz)^p}\,H_\nu^{(2)}(z),$$

and so

$$A(z) = A - \tfrac{1}{2}\pi\int_0^\infty \{\tfrac{1}{2}\pi(z+t)\}^{-\frac{1}{2}}\,e^{i(z+t-\frac{1}{2}\nu\pi-\frac{1}{4}\pi)}\,\frac{p\,.\,(\nu, p)}{\{-2i(z+t)\}^p}\,H_\nu^{(2)}(z+t)\,dt,$$

where A is a constant.

We obtain a similar expression for $B(z)$, and hence it follows that

$$\Sigma_\nu^{(1)}(z;\,p) = \{AH_\nu^{(1)}(z) + BH_\nu^{(2)}(z)\}\,(\tfrac{1}{2}\pi z)^{\frac{1}{2}}\,e^{-i(z-\frac{1}{2}\nu\pi-\frac{1}{4}\pi)}$$
$$-\tfrac{1}{2}\pi p\,.\,(\nu, p)\int_0^\infty\left(\frac{z}{z+t}\right)^{\frac{1}{2}}\frac{H_\nu^{(1)}(z)\,H_\nu^{(2)}(z+t) - H_\nu^{(2)}(z)\,H_\nu^{(1)}(z+t)}{\{-2i(z+t)\}^p}\,dt.$$

By considering the behaviour of both sides of this equation as $z \to +\infty$, it is not difficult to see that $A = 1$ and $B = 0$.

Hence we may write Hankel's formulae in the forms

$$\begin{cases} H_\nu^{(1)}(z) = \left(\dfrac{2}{\pi z}\right)^{\frac{1}{2}} e^{i(z-\frac{1}{2}\nu\pi-\frac{1}{4}\pi)}\,\{\Sigma_\nu^{(1)}(z;\,p) + R_p^{(1)}\}, \\ H_\nu^{(2)}(z) = \left(\dfrac{2}{\pi z}\right)^{\frac{1}{2}} e^{-i(z-\frac{1}{2}\nu\pi-\frac{1}{4}\pi)}\,\{\Sigma_\nu^{(2)}(z;\,p) + R_p^{(2)}\}, \end{cases}$$

where the remainder $R_p^{(1)}$ may be defined by the equation

$$R_p^{(1)} = \tfrac{1}{2}\pi p\,.\,(\nu, p)\int_0^\infty\left(\frac{z}{z+t}\right)^{\frac{1}{2}}\frac{H_\nu^{(1)}(z)\,H_\nu^{(2)}(z+t) - H_\nu^{(2)}(z)\,H_\nu^{(1)}(z+t)}{\{-2i(z+t)\}^p}\,dt.$$

Since $R(z) \geqslant 0$, we have $|z+t| \geqslant \sqrt{(r^2+t^2)}$, and so, by using the crude inequalities, we see that the modulus of the last integrand does not exceed

$$2^{2-p}\,\pi^{-1}\,G^2\,.\,(r^2+t^2)^{-\frac{1}{2}(p+1)}.$$

Hence

$$|R_p^{(1)}| \leqslant 2^{1-p}\,G^2 p\,|(\nu, p)|\int_0^\infty (r^2+t^2)^{-\frac{1}{2}(p+1)}\,dt,$$

and so, when $p \geqslant 1$, we have

$$(4) \qquad |R_p^{(1)}| \leqslant 2G^2\,|(\nu, p)|\,\frac{\Gamma(\frac{1}{2})\,\Gamma(\frac{1}{2}p+1)}{\Gamma(\frac{1}{2}p+\frac{1}{2})\,.\,|(2z)^p|},$$

and similarly

$$(5) \qquad |R_p^{(2)}| \leqslant 2G^2\,|(\nu, p)|\,\frac{\Gamma(\frac{1}{2})\,\Gamma(\frac{1}{2}p+1)}{\Gamma(\frac{1}{2}p+\frac{1}{2})\,.\,|(2z)^p|}.$$

These are the results obtained by Weber; and it will be observed that in the analysis no hypothesis has been made concerning the relative values of ν and p; in this respect Weber's results differ from the results obtained by other writers.

7·34. *Approximations to remainders in the asymptotic expansions.*

When the argument of a Bessel function is not very large*, the asymptotic expansion is not well adapted for numerical computation because the smallest term in it (with the remainder after the smallest term) is not particularly small; at the same time the argument may be sufficiently large for the ascending series to converge very slowly.

An ingenious method for meeting these numerical difficulties was devised by Stieltjes†; we shall explain the method in detail as applied to the function $K_0(x)$ and state the results which were obtained by Stieltjes by applying the method to $J_0(x)$ and $Y_0(x)$.

We apply the transformation indicated in § 7·31 to the formula § 6·15 (4), so that

$$\begin{aligned} K_0(x) &= \frac{e^{-x}}{\sqrt{2}} \int_0^\infty \frac{e^{-xu}\,du}{u^{\frac{1}{2}}(1+\frac{1}{2}u)^{\frac{1}{2}}} \\ &= \frac{e^{-x}\sqrt{2}}{\pi} \int_0^\infty \int_0^{\frac{1}{2}\pi} \frac{e^{-xu}\,d\theta\,du}{u^{\frac{1}{2}}(1+\frac{1}{2}u\sin^2\theta)} \\ &= \frac{e^{-x}\sqrt{2}}{\pi}\left[\sum_{m=0}^{p-1}\int_0^\infty\int_0^{\frac{1}{2}\pi}\frac{e^{-xu}}{u^{\frac{1}{2}}}(-\tfrac{1}{2}u\sin^2\theta)^m\,d\theta\,du\right. \\ &\qquad \left. + \int_0^\infty\int_0^{\frac{1}{2}\pi}\frac{e^{-xu}(-\frac{1}{2}u\sin^2\theta)^p}{u^{\frac{1}{2}}(1+\frac{1}{2}u\sin^2\theta)}\,d\theta\,du\right]. \end{aligned}$$

That is to say,

(1) $$K_0(x) = \left(\frac{\pi}{2x}\right)^{\frac{1}{2}} e^{-x}\left[\sum_{m=0}^{p-1}\frac{(0, m)}{(2x)^m} + (-)^p R_p\right],$$

where

$$R_p = \frac{x^{\frac{1}{2}}}{\pi^{\frac{3}{2}}}\int_0^\infty\int_0^\pi \frac{e^{-xu}(\frac{1}{2}u\sin^2\theta)^p}{u^{\frac{1}{2}}(1+\frac{1}{2}u\sin^2\theta)}\,d\theta\,du.$$

Now the value of m for which $(0, m)/(2x)^m$ is least is nearly equal to $2x$ when x is large; accordingly, in order to consider the remainder after the smallest term of the series for $K_0(x)$, we choose p so that

$$x = \tfrac{1}{2}p + \sigma,$$

where σ is numerically less than unity; and then

$$R_p = \frac{x^{\frac{1}{2}}}{\pi^{\frac{3}{2}}}\int_0^\infty\int_0^\pi (\tfrac{1}{2}ue^{-\frac{1}{2}u}\sin^2\theta)^p\,\frac{e^{-\sigma u}}{u^{\frac{1}{2}}(1+\frac{1}{2}u\sin^2\theta)}\,d\theta\,du.$$

Now, as u increases from 0 to ∞, $\frac{1}{2}ue^{-\frac{1}{2}u}$ increases from 0 up to a maximum e^{-1} (when $u = 2$) and then decreases to zero; so we write

$$\tfrac{1}{2}ue^{-\frac{1}{2}u} = e^{-1-\xi^2},$$

where ξ increases from $-\infty$ to ∞, and, for similar reasons, we write

$$\sin^2\theta = e^{-\eta^2}.$$

* The range of values of x under contemplation for the functions $J_0(x)$, $Y_0(x)$ and $K_0(x)$ is from about 4 to about 10.

† *Ann. Sci. de l'École norm. sup.* (3) III. (1886), pp. 241—252.

The domain of integration becomes the whole of the (ξ, η) plane; and it is found that

$$R_p = \frac{2x^{\frac{1}{2}} e^{-2x}}{\pi^{\frac{3}{2}}} \int_{-\infty}^{\infty} \int_{-\infty}^{\infty} e^{-p(\xi^2+\eta^2)} \left\{ \sum_{r,s=0}^{\infty} a_{r,s} \xi^r \eta^s \right\} d\xi d\eta,$$

where

$$a_{0,0} = \tfrac{1}{2}, \qquad a_{2,0} = 2\sigma^2 - \tfrac{1}{24}, \qquad a_{0,2} = \tfrac{1}{8},$$

by some rather tedious arithmetic. It follows* that the dominant terms of the asymptotic expansion of R_p for large values of p are

$$R_p \sim 2\left(\frac{x}{\pi}\right)^{\frac{1}{2}} \frac{e^{-2x}}{p} \left[a_{0,0} + \frac{a_{0,2} + a_{2,0}}{2p} + \dots \right],$$

so that

(2) $$R_p \sim 2\left(\frac{x}{\pi}\right)^{\frac{1}{2}} \frac{e^{-2x}}{p} \left[\frac{1}{2} + \frac{\sigma^2 + \frac{1}{24}}{p} + \dots \right].$$

It is easy to verify by Stirling's theorem that

$$\frac{(0, p)}{(2x)^p} \sim (-)^p \cdot 2\left(\frac{x}{\pi}\right)^{\frac{1}{2}} \frac{e^{-2x}}{p},$$

so that the error due to stopping at one of the smallest terms is roughly half of the first term omitted.

In like manner Stieltjes proved that, if $P(x, 0)$ and $Q(x, 0)$ are defined as in § 7·3, then

(3) $$P(x, 0) = \sum_{m=0}^{p-1} \frac{(-)^m \cdot (0, 2m)}{(2x)^{2m}} + (-)^p R_p^{(P)},$$

(4) $$Q(x, 0) = \sum_{m=0}^{p-1} \frac{(-)^m \cdot (0, 2m+1)}{(2x)^{2m+1}} + (-)^p R_p^{(Q)},$$

where

(5) $$R_p^{(P)} \sim \left(\frac{x}{\pi}\right)^{\frac{1}{2}} \frac{e^{-2x}}{p} \left[\frac{1}{2} + \frac{\frac{1}{2}\tau^2 + \frac{1}{4}\tau + \frac{1}{12}}{p} + \dots \right],$$

(6) $$R_p^{(Q)} \sim \left(\frac{x}{\pi}\right)^{\frac{1}{2}} \frac{e^{-2x}}{p} \left[\frac{1}{2} + \frac{\frac{1}{2}\tau^2 - \frac{1}{4}\tau - \frac{1}{6}}{p} + \dots \right],$$

provided that p is chosen so as to be nearly equal to x, and τ is defined to be $x - p$.

Results of this character are useful for tabulating Bessel functions in the critical range; some similar formulae have been actually used for that purpose by Airey, *Archiv der Math. und Phys.* (3) XX. (1913), pp. 240—244; (3) XXII. (1914), pp. 30—43; and *British Association Reports*, 1913, 1914.

It would be of some interest to extend the results, which Stieltjes has established for Bessel functions of zero order (as well as for the logarithmic integral and some other functions), to Bessel functions of arbitrary order.

* Cf. Bromwich, *Theory of Infinite Series*, §§ 133, 137, and 174, or the lemma which will be proved in § 8·3.

7·35. *Deductions from Schafheitlin's integrals.*

If we replace u by $2\tan\theta$ in the formulae of § 7·32, we deduce that

$$P(x,\nu)=\frac{(2x)^{\nu+\frac{1}{2}}}{\Gamma(\nu+\frac{1}{2})}\int_0^{\frac{1}{2}\pi}\frac{\sin^{\nu-\frac{1}{2}}\theta\cos(\nu-\frac{1}{2})\theta}{\cos^{2\nu+1}\theta}e^{-2x\tan\theta}\,d\theta,$$

$$Q(x,\nu)=\frac{(2x)^{\nu+\frac{1}{2}}}{\Gamma(\nu+\frac{1}{2})}\int_0^{\frac{1}{2}\pi}\frac{\sin^{\nu-\frac{1}{2}}\theta\sin(\nu-\frac{1}{2})\theta}{\cos^{2\nu+1}\theta}e^{-2x\tan\theta}\,d\theta,$$

which resemble Schafheitlin's integrals of § 6·12.

It is obvious from these results that

$$P(x,\nu)>0, \qquad (-\tfrac{1}{2}<\nu<\tfrac{3}{2})$$

$$Q(x,\nu)>0, \qquad (\tfrac{1}{2}<\nu<\tfrac{5}{2})$$

$$Q(x,\nu)<0. \qquad (-\tfrac{1}{2}<\nu<\tfrac{1}{2})$$

An interesting consequence of these results is that we can prove that

$$Q(x,\nu)/P(x,\nu)$$

is an increasing function of x when $-\frac{1}{2}<\nu<\frac{1}{2}$ and that it is a decreasing function of x when $\frac{1}{2}<\nu<\frac{5}{2}$.

For we have

$$\begin{aligned}Q'(x,\nu)\,P(x,\nu)&-P'(x,\nu)\,Q(x,\nu)\\&=2\left\{\frac{(2x)^{\nu+\frac{1}{2}}}{\Gamma(\nu+\frac{1}{2})}\right\}^2\int_0^{\frac{1}{2}\pi}\int_0^{\frac{1}{2}\pi}e^{-2x(\tan\theta+\tan\phi)}F(\theta,\phi)\,d\theta\,d\phi,\end{aligned}$$

where

$$F(\theta,\phi)=\frac{(\sin\theta\sin\phi)^{\nu-\frac{1}{2}}}{(\cos\theta\cos\phi)^{2\nu+1}}(\tan\theta-\tan\phi)\cos(\nu-\tfrac{1}{2})\theta\sin(\nu-\tfrac{1}{2})\phi,$$

so that

$$F(\theta,\phi)+F(\phi,\theta)=\frac{(\sin\theta\sin\phi)^{\nu-\frac{1}{2}}}{(\cos\theta\cos\phi)^{2\nu+1}}(\tan\theta-\tan\phi)\sin(\tfrac{1}{2}-\nu)(\theta-\phi).$$

If we interchange the parametric variables θ, ϕ in the double integral and add the results so obtained we see that, when $-\frac{1}{2}<\nu<\frac{5}{2}$, the double integral has the same sign as $\frac{1}{2}-\nu$; and this proves the result.

7·4. *Schläfli's investigation of the asymptotic expansions of Bessel functions.*

In a memoir which seems hardly to have received the recognition which its importance deserves, Schläfli* has given a very elegant but somewhat elaborate investigation of the asymptotic expansions of the functions of the third kind.

The integral formulae from which he derived these expansions are generalisations of Bessel's integral; although Bessel's integral is not so well adapted as Poisson's integral for constructing the asymptotic expansion of

* *Ann. di Mat.* (2) VI. (1875), pp. 1—20. The only standard work on Bessel functions in which the importance of this memoir is recognised is the treatise by Graf and Gubler.

$J_\nu(z)$ when z is large and ν is fixed, yet Schläfli's method not only succeeds in obtaining the expansion, but also it expresses the remainders in a neat and compact form.

Schläfli's procedure consisted in taking integrals of the type

$$\frac{1}{2\pi i}\int u^{-\nu-1}\exp\left\{\pm\tfrac{1}{2}re^{i\alpha}\left(u+\frac{1}{u}\right)\right\}du,$$

and selecting the contour of integration in such a way that, on it, *the phase* of*

$$\tfrac{1}{2}re^{i\alpha}(u-2+1/u)$$

is constant. He took two contours, the constants for the respective contours being 0 and π; and it is supposed that r is positive and α is real.

(I) Let us first take the phase to be π; write

$$u=1+\rho e^{i\theta},$$

where ρ is positive and θ is real, and then

$$re^{i\alpha}\rho^2 e^{2i\theta}/(1+\rho e^{i\theta})$$

is negative, and is consequently equal to its conjugate complex.

Hence we have

$$\rho=-\frac{\sin(\alpha+2\theta)}{\sin(\alpha+\theta)},\qquad u=-\frac{\sin\theta}{\sin(\alpha+\theta)}e^{i(2\theta+\alpha)}. \tag{1}$$

Next choose a new parametric variable ϕ such that

$$\phi=2\theta+\alpha-\pi,$$

and then

$$u=\frac{\cos\frac{1}{2}(\alpha-\phi)}{\cos\frac{1}{2}(\alpha+\phi)}e^{i\phi},\qquad re^{i\alpha}\frac{(u-1)^2}{u}=\frac{-r\sin^2\phi}{\cos\frac{1}{2}(\alpha-\phi)\cos\frac{1}{2}(\alpha+\phi)}. \tag{2}$$

Now, as ϕ varies from $-(\pi-\alpha)$ to $(\pi-\alpha)$, u traces out a contour emerging from the origin at an angle $-(\pi-\alpha)$ with the positive real axis and passing to infinity at an angle $(\pi-\alpha)$ with the positive real axis, *provided that* $0<\alpha<2\pi$.

If this restriction is not laid on α the contour passes to infinity more than once.

We shall now lay this restriction on α; and then the contour is of the type specified for formula § 6·22 (9), provided that we give ω and $\arg z$ the same value α, as is permissible.

It follows that

$$\frac{e^{\nu\pi i}I_{-\nu}(re^{i\alpha})-e^{-\nu\pi i}I_\nu(re^{i\alpha})}{2i\sin\nu\pi}=\frac{1}{2\pi i}\int_{-(\pi-\alpha)}^{\pi-\alpha}u^{-\nu-1}\exp\left\{\tfrac{1}{2}re^{i\alpha}\left(u+\frac{1}{u}\right)\right\}\cdot\frac{du}{d\phi}\,d\phi,$$

where u is defined in terms of ϕ by equation (2).

* The reader will find it interesting to compare the general methods of this section with the "method of steepest descents" which is applied to obtain various asymptotic expansions in Chapter VIII.

Changing the sign of ϕ is equivalent to replacing u by $1/u$, and so, replacing the expression on the left by its value as a function of the third kind, we have

$$(3)\quad e^{\frac{1}{2}\nu\pi i} H_\nu^{(1)}(re^{i(\alpha-\frac{1}{2}\pi)}) = \frac{1}{\pi i}\int_0^{\pi-\alpha}(u^{-\nu}+u^{\nu})\exp\left\{\tfrac{1}{2}re^{i\alpha}\left(u+\frac{1}{u}\right)\right\}.\frac{d\log u}{d\phi}\,d\phi.$$

From (2) it follows that $-re^{i\alpha}(u-1)^2/u$ increases steadily* from 0 to $+\infty$ as ϕ varies monotonically from 0 to $\pi-\alpha$; and, if we write

$$-re^{i\alpha}(u-1)^2/u = t,$$

so that t is positive when u is on the contour, we have

$$\frac{du}{u} = \frac{dt}{-re^{i\alpha}(u-1/u)} = \frac{dt}{e^{-\frac{1}{2}(\pi-\alpha)i}(u^{\frac{1}{2}}+u^{-\frac{1}{2}})\sqrt{(rt)}},$$

the range of values of $\arg u$ being less than π.

Next, by Cauchy's theorem,

$$u^\nu + u^{-\nu} = \frac{1}{2\pi i}\int^{(u+,1/u+)}\zeta^{\nu-\frac{1}{2}}\left\{\frac{u^{\frac{1}{2}}}{\zeta-u}+\frac{u^{-\frac{1}{2}}}{\zeta-1/u}\right\}d\zeta;$$

it is convenient to take the point $\zeta=1$ inside the contour, but $\zeta=0$ must be outside the contour because the origin is a branch-point.

It follows that

$$(4)\qquad u^\nu + u^{-\nu} = \frac{u^{\frac{1}{2}}+u^{-\frac{1}{2}}}{2\pi i}\int^{(u+,1/u+)}\frac{\zeta^{\nu-\frac{1}{2}}(\zeta-1)\,d\zeta}{(\zeta-1)^2+\zeta t/(re^{i\alpha})}.$$

Hence

$$(5)\quad H_\nu^{(1)}(re^{i(\alpha-\frac{1}{2}\pi)}) = \frac{e^{-\frac{1}{2}\nu\pi i}\exp(re^{i\alpha})}{2\pi^2 i\, r^{\frac{1}{2}} e^{\frac{1}{2}i\alpha}}\int_0^\infty\int^{(u+,1/u+,1+)}\frac{e^{-\frac{1}{2}t}\,t^{-\frac{1}{2}}\,\zeta^{\nu-\frac{1}{2}}(\zeta-1)\,d\zeta dt}{(\zeta-1)^2+\zeta t/(re^{i\alpha})}.$$

Now it is evident that

$$\frac{1}{(\zeta-1)^2+\zeta t/(re^{i\alpha})} = \sum_{m=0}^{p-1}\frac{(-)^m\zeta^m t^m}{(\zeta-1)^{2m+2}(re^{i\alpha})^m} + \frac{(-)^p\zeta^p t^p}{(\zeta-1)^{2p}(re^{i\alpha})^p\{(\zeta-1)^2+\zeta t/(re^{i\alpha})\}},$$

where p is any positive integer (zero included). It will be convenient subsequently to suppose that p exceeds both $R(\nu-\frac{1}{2})$ and $R(-\nu-\frac{1}{2})$.

On making this substitution in the last integrand and observing that

$$\frac{1}{2\pi i}\int^{(u+,1/u+,1+)}\zeta^{\nu+m-\frac{1}{2}}(\zeta-1)^{-2m-1}d\zeta = \frac{\Gamma(\nu+m+\frac{1}{2})}{\Gamma(\nu-m+\frac{1}{2}).(2m!)} = \frac{m!\,(\nu,m)}{(2m)!},$$

(with the notation of § 7·2), we deduce that

$$(6)\quad H_\nu^{(1)}(re^{i(\alpha-\frac{1}{2}\pi)}) = \left(\frac{2}{\pi re^{i\alpha}}\right)^{\frac{1}{2}}\exp(re^{i\alpha}-\tfrac{1}{2}\nu\pi i)\left[\sum_{m=0}^{p-1}\frac{(-)^m.(\nu,m)}{(2re^{i\alpha})^m}+R_p^{(1)}\right],$$

where

$$R_p^{(1)} = \frac{(-)^p}{2\pi i\sqrt{(2\pi)}}\int_0^\infty\int^{(u+,1/u+,1+)}\frac{e^{-\frac{1}{2}t}\,t^{p-\frac{1}{2}}\,\zeta^{\nu+p-\frac{1}{2}}\,d\zeta dt}{(\zeta-1)^{2p-1}(re^{i\alpha})^p\{(\zeta-1)^2+\zeta t/(re^{i\alpha})\}}.$$

* Since $$\frac{d}{d\phi}\frac{\sin^2\phi}{\cos\alpha+\cos\phi} = \frac{\sin\phi\,(1+2\cos\alpha\cos\phi+\cos^2\phi)}{(\cos\alpha+\cos\phi)^2}.$$

First consider

$$\frac{1}{2\pi i}\int^{(u+,\,1/u+,\,1+)} \frac{\zeta^{\nu+p-\frac{1}{2}}\,d\zeta}{(\zeta-1)^{2p-1}\{(\zeta-1)^2+\zeta t/(re^{ia})\}}.$$

When p is so large that it exceeds both $R(\nu-\frac{1}{2})$ and $R(-\nu-\frac{1}{2})$, we take the contour to be as shewn in Fig. 15; and when the radii of the large and small circles tend to ∞ and 0 respectively the integrals along them tend to zero. If now we write

$$\zeta = e^{\pm\pi i}(1-x)/x$$

on the two rays (which are all that survives of the contour), we find that

$$\frac{1}{2\pi i}\int^{(u+,\,1/u+,\,1+)} \frac{\zeta^{\nu+p-\frac{1}{2}}\,d\zeta}{(\zeta-1)^{2p-1}\{(\zeta-1)^2+\zeta t/(re^{ia})\}}$$
$$= \frac{(-)^p\cos\nu\pi}{\pi}\int_0^1 \frac{x^{p-\nu-\frac{1}{2}}(1-x)^{p+\nu-\frac{1}{2}}\,dx}{1-tx(1-x)/(re^{ia})}.$$

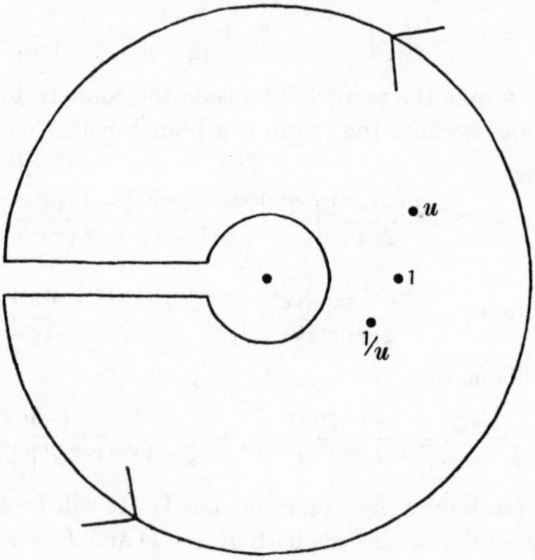

Fig. 15.

Now the numerator of the integrand is positive (when ν is real), and the modulus of the denominator is never less than 1 when $\frac{1}{2}\pi < \alpha < \frac{3}{2}\pi$; for other values of α it is never less than $|\sin\alpha|$.

Therefore

$$|R_p^{(1)}| \leqslant \frac{\theta_0\,|\cos\nu\pi|}{\pi r^p\sqrt{(2\pi)}}\int_0^\infty\int_0^1 e^{-\frac{1}{2}t}\,t^{p-\frac{1}{2}}\,x^{p-\nu-\frac{1}{2}}(1-x)^{p+\nu-\frac{1}{2}}\,dx\,dt = \theta_0\,|(\nu,p)|\div(2r)^p,$$

where $|\theta_0|$ is 1 or $|\operatorname{cosec}\alpha|$ according as $\cos\alpha$ is negative or positive. When ν is complex, it is easy to see that

$$(7)\qquad |R_p^{(1)}| \leqslant \left|\frac{\cos\nu\pi}{\cos R(\nu\pi)}\right|\frac{\theta_0\,|(R(\nu),p)|}{(2r)^p}.$$

Hence, finally, when $-\frac{1}{2}\pi < \arg z < \frac{3}{2}\pi$,

$$(8) \qquad H_\nu^{(1)}(z) = \left(\frac{2}{\pi z}\right)^{\frac{1}{2}} e^{i(z-\frac{1}{2}\nu\pi-\frac{1}{4}\pi)} \left[\sum_{m=0}^{p-1} \frac{(-)^m . (\nu, m)}{(2iz)^m} + \theta_1 \frac{(-)^p . (\nu, p)}{(2iz)^p}\right],$$

where $|\theta_1|$ does not exceed 1 or $|\sec(\arg z)|$ according as $I(z)$ is positive or negative, provided that ν is real and $p + \frac{1}{2} > |\nu|$. When ν is complex, the modified form of the remainder given by (7) has to be used.

Since $R\{1 - tx(1-x)/(re^{i\alpha})\} \geqslant 0$ when $R(e^{-i\alpha}) \leqslant 0$, we see that, in (8), θ_1 has its real part positive when ν is real and $I(z) \geqslant 0$.

If z be replaced by iz in (8) we find that, when $|\arg z| < \pi$,

$$(9) \qquad K_\nu(z) = \left(\frac{\pi}{2z}\right)^{\frac{1}{2}} e^{-z} \left[\sum_{m=0}^{p-1} \frac{(\nu, m)}{(2z)^m} + \theta_2 \frac{(\nu, p)}{(2z)^p}\right],$$

and, when ν is real,

(i) $R(\theta_2) \geqslant 0$ and $|\theta_2| < 1$, if $R(z) \geqslant 0$,

(ii) $|\theta_2| < |\operatorname{cosec}(\arg z)|$, if $R(z) < 0$.

The modifications necessary for complex values of ν are left to the reader.

(II) We next discuss the consequences of taking the phase of

$$\tfrac{1}{2} re^{i\alpha}(u - 2 + 1/u)$$

to be zero. As before, we write

$$u = 1 + \rho e^{i\theta},$$

and then $re^{i\alpha}\rho^2 e^{2i\theta}/(1+\rho e^{i\theta})$ is positive, and therefore equal to its conjugate complex, so that we obtain anew equation (1). We then diverge from the preceding analysis by writing

$$\phi = -(2\theta + \alpha)$$

so that

$$(10) \qquad u = \frac{\sin\frac{1}{2}(\alpha+\phi)}{\sin\frac{1}{2}(\alpha-\phi)} e^{-i\phi}, \quad re^{i\alpha}\frac{(u-1)^2}{u} = \frac{r\sin^2\phi}{\sin\frac{1}{2}(\alpha-\phi)\sin\frac{1}{2}(\alpha+\phi)}.$$

Now, as ϕ varies from $-\alpha$ to α, u traces out a contour emerging from the origin at an angle α with the positive real axis and passing to infinity at an angle $-\alpha$ with the positive real axis, *provided that α lies between $-\pi$ and π.* The contour is then of the type specified for formula § 6·22 (8) if, as is permissible, we give ω and $\arg z$ the same value α.

It follows that, when $-\pi < \alpha < \pi$,

$$K_\nu(re^{i\alpha}) = \tfrac{1}{2}\cos\nu\pi \int_{-\alpha}^{\alpha} u^{-\nu-1} \exp\left\{-\tfrac{1}{2}re^{i\alpha}\left(u + \frac{1}{u}\right)\right\} \frac{du}{d\phi}\, d\phi,$$

where u is defined as a function of ϕ by equation (10); and therefore

$$(11) \qquad H_\nu^{(2)}(re^{i(\alpha-\frac{1}{2}\pi)}) = \frac{ie^{\frac{1}{2}\nu\pi i}}{\pi} \int_0^{\alpha} (u^{-\nu} + u^{\nu}) \exp\left\{-\tfrac{1}{2}re^{i\alpha}\left(u + \frac{1}{u}\right)\right\} \frac{d\log u}{d\phi}\, d\phi,$$

and hence, if now

$$t = re^{ia}(u-1)^2/u,$$

we find that

$$(12)\qquad H_\nu^{(2)}(re^{i(\alpha-\frac{1}{2}\pi)}) = \frac{e^{\frac{1}{2}\nu\pi i}\exp(-re^{ia})}{2\pi^2 r^{\frac{1}{2}} e^{\frac{1}{2}ia}} \int_0^\infty \int^{(u+,1/u+)} \frac{e^{-\frac{1}{2}t}\, t^{-\frac{1}{2}}\, \zeta^{\nu-\frac{1}{2}}(\zeta-1)\, d\zeta\, dt}{(\zeta-1)^2 - \zeta t/(re^{ia})}.$$

We have consequently expressed a second solution of Bessel's equation in a form from which its asymptotic expansion can be deduced; and the analysis proceeds as in the case of $H_\nu^{(1)}(z)$, the final result being that, when

$$-\tfrac{3}{2}\pi < \arg z < \tfrac{1}{2}\pi,$$

$$(13)\qquad H_\nu^{(2)}(z) = \left(\frac{2}{\pi z}\right)^{\frac{1}{2}} e^{-i(z-\frac{1}{2}\nu\pi-\frac{1}{4}\pi)} \left[\sum_{m=0}^{p-1} \frac{(\nu, m)}{(2iz)^m} + \theta_2 \frac{(\nu, p)}{(2iz)^p}\right],$$

where $|\theta_2|$ does not exceed 1 or $|\sec \arg z|$ according as $I(z) \leqslant 0$ or $I(z) \geqslant 0$, provided that ν is real and $p + \frac{1}{2} > |\nu|$; and $R(\theta_2) \geqslant 0$ when $I(z) \leqslant 0$. If ν is complex the form of the remainder has to be modified, just as in the case of (8).

It should be observed that, since the integrands in (3) and (11) are even functions of ν, it is unnecessary in this investigation to suppose that $R(\nu)$ must exceed $-\frac{1}{2}$, as was necessary in investigations based on integrals of Poisson's type.

7·5. *Barnes' investigation* of asymptotic expansions of Bessel functions.*

The asymptotic expansions of functions of the third kind follow immediately from Barnes' formulae which were obtained in §§ 6·5, 6·51. Let us consider

$$\int_{-\infty i-\nu-p}^{\infty i-\nu-p} \Gamma(-s)\,\Gamma(-2\nu-s)\,\Gamma(\nu+s+\tfrac{1}{2})\,(-2iz)^s\, ds$$

$$= (-2iz)^{-\nu-p} \int_{-\infty i}^{\infty i} \Gamma(-s+\nu+p)\,\Gamma(-s-\nu+p)\,\Gamma(s-p+\tfrac{1}{2})\,(-2iz)^s\, ds.$$

If $|\arg(-iz)| \leqslant \frac{3}{2}\pi - \delta$, we have

$$\left|\int_{-\infty i}^{\infty i} \Gamma(-s+\nu+p)\,\Gamma(-s-\nu+p)\,\Gamma(s-p+\tfrac{1}{2})\,(-2iz)^s\, ds\right|$$

$$\leqslant \int_{-\infty i}^{\infty i} |\Gamma(-s+\nu+p)\,\Gamma(-s-\nu+p)\,\Gamma(s-p+\tfrac{1}{2})\, e^{(\frac{3}{2}\pi-\delta)|s|}\, ds|,$$

and the last integral is convergent and so the first integral of all is

$$O\{(-2iz)^{-\nu-p}\}.$$

But, by the arguments of § 6·51, the first integral is $-2\pi i$ times the sum of the residues at the poles on the right of the contour, and so it is equal to $-\pi^{\frac{3}{2}} H_\nu^{(1)}(z)/[e^{i(z-\nu\pi)}\cos\nu\pi\,(2z)^\nu]$ plus $-2\pi i$ times the sum of the residues at $s = -\nu-\frac{1}{2}, -\nu-\frac{3}{2}, \ldots, -\nu-p+\frac{1}{2}$. The residue at $-\nu-m-\frac{1}{2}$ is

$$\frac{(-)^m\,\Gamma(\nu+m+\frac{1}{2})\,\Gamma(-\nu+m+\frac{1}{2})}{m!\,(-2iz)^{\nu+m+\frac{1}{2}}},$$

* *Trans. Camb. Phil. Soc.* xx. (1908), pp. 273—279.

and so, when $|\arg(-iz)| \leqslant \frac{3}{2}\pi - \delta$

$$H_\nu^{(1)}(z) = -\frac{e^{i(z-\nu\pi)}\cos(\nu\pi).(2z)^\nu}{\pi^{\frac{3}{2}}}$$
$$\left[2\pi i \sum_{m=0}^{p-1} \frac{(-)^m\,\Gamma(\nu+m+\frac{1}{2})\,\Gamma(-\nu+m+\frac{1}{2})}{m!\,(-2iz)^{\nu+m+\frac{1}{2}}} + O\{(-2iz)^{-\nu-p}\}\right]$$
$$= \left(\frac{2}{\pi z}\right)^{\frac{1}{2}} e^{i(z-\frac{1}{2}\nu\pi-\frac{1}{4}\pi)} \left[\sum_{m=0}^{p-1} \frac{(\nu, m)}{(-2iz)^m} + O(z^{\frac{1}{2}-p})\right],$$

and this is equivalent to the result obtained in § 7·2. The investigation of $H_\nu^{(2)}(z)$ may be constructed by replacing i by $-i$ throughout.

The reader should notice that, although the determination of the *order of magnitude* of the remainders by this method is transparently simple, it is not possible to obtain concrete formulae, concerning the magnitude and the sign of the remainders, which are ultimately supplied by the methods which have been previously considered.

7·51. *Asymptotic expansions of products of Bessel functions.*

It does not seem possible to obtain asymptotic expansions of the four products $J_{\pm\mu}(z)\,J_{\pm\nu}(z)$ in which the coefficients have simple forms, even when $\mu = \nu$. The reason for this is that the products $H_\mu^{(1)}(z)\,H_\nu^{(1)}(z)$ and $H_\mu^{(2)}(z)\,H_\nu^{(2)}(z)$ have asymptotic expansions for which no simple expression exists for the general term; the leading terms in the two expansions are

$$\frac{2e^{\pm 2iz \mp \frac{1}{2}(\mu+\nu+1)\pi i}}{\pi z}\left\{1 \mp \frac{2\mu^2 + 2\nu^2 - 1}{4iz} + \ldots\right\}.$$

The products $H_\mu^{(1)}(z)\,H_\nu^{(2)}(z)$ and $H_\mu^{(2)}(z)\,H_\nu^{(1)}(z)$, however, do possess simple asymptotic expansions; and from them we can deduce asymptotic expansions for

$$J_\mu(z)\,J_\nu(z) + Y_\mu(z)\,Y_\nu(z)$$

and for

$$J_\mu(z)\,Y_\nu(z) - Y_\mu(z)\,J_\nu(z).$$

The simplest way of constructing the expansions is by Barnes' method, just explained in § 7·5. A consideration of series of the type obtained in § 5·41 suggests that we should examine the integral

$$\frac{1}{2\pi i}\int_{-\infty i}^{\infty i} \Gamma(2s+1)\,\Gamma\left(\frac{\mu+\nu}{2}-s\right)\Gamma\left(\frac{\mu-\nu}{2}-s\right)\Gamma\left(\frac{\nu-\mu}{2}-s\right)\Gamma\left(-\frac{\mu+\nu}{2}-s\right)(\tfrac{1}{2}iz)^{2s}\,ds;$$

the contour is to be chosen so that the poles of $\Gamma(2s+1)$ lie on the left of the contour and the poles of the other four Gamma functions lie on the right of the contour; and it is temporarily supposed that μ, ν and $\mu \pm \nu$ are not integers, so that the integrand has no double poles. The integral is convergent provided that $|\arg(iz)| < \frac{3}{2}\pi$.

First evaluate the integral by swinging round the contour to enclose the sequences of poles which lie to the right of the original contour; the expression is equal to minus the sum of the residues at these poles, and the residue at $m+\frac{1}{2}(\mu+\nu)$ is

$$-\frac{\pi^3 e^{\frac{1}{2}(\mu+\nu)\pi i}}{\sin\mu\pi\,.\,\sin\nu\pi\,.\,\sin(\mu+\nu)\pi}\cdot\frac{\Gamma(\mu+\nu+2m+1)\,.\,(-)^m(\frac{1}{2}z)^{\mu+\nu+2m}}{m!\,\Gamma(\mu+m+1)\,\Gamma(\nu+m+1)\,\Gamma(\mu+\nu+m+1)}.$$

It follows that

$$\frac{1}{2\pi i}\int_{-\infty i}^{\infty i}\Gamma(2s+1)\,\Gamma\left(\frac{\mu+\nu}{2}-s\right)\Gamma\left(\frac{\mu-\nu}{2}-s\right)$$

$$\times\,\Gamma\left(\frac{\nu-\mu}{2}-s\right)\Gamma\left(-\frac{\mu+\nu}{2}-s\right)(\tfrac{1}{2}iz)^{2s}\,ds$$

$$=\frac{-\pi^3}{\sin\mu\pi\sin\nu\pi}\left\{\frac{e^{\frac{1}{2}(\mu+\nu)\pi i}J_\mu(z)\,J_\nu(z)}{\sin(\mu+\nu)\pi}-\frac{e^{\frac{1}{2}(\mu-\nu)\pi i}J_\mu(z)\,J_{-\nu}(z)}{\sin(\mu-\nu)\pi}\right.$$

$$\left.-\frac{e^{\frac{1}{2}(\nu-\mu)\pi}J_{-\mu}(z)\,J_\nu(z)}{\sin(\nu-\mu)\pi}-\frac{e^{-\frac{1}{2}(\mu+\nu)\pi i}J_{-\mu}(z)\,J_{-\nu}(z)}{\sin(\mu+\nu)\pi}\right\}$$

$$=\frac{\pi^3 e^{-\frac{1}{2}(\mu+\nu)\pi i}}{\sin(\mu+\nu)\pi}\{J_\mu(z)\,J_\nu(z)+Y_\mu(z)\,Y_\nu(z)\}$$

$$-\frac{\pi^3 e^{-\frac{1}{2}(\mu+\nu)\pi i}\{2\cos\mu\pi\cos\nu\pi+i\sin(\mu+\nu)\pi\}}{\sin(\mu+\nu)\pi\sin(\mu-\nu)\pi}\{J_\mu(z)\,Y_\nu(z)-Y_\mu(z)\,J_\nu(z)\}$$

$$=\frac{\pi^3}{2\sin\frac{1}{2}(\mu+\nu)\pi}[\{J_\mu(z)\,J_\nu(z)+Y_\mu(z)\,Y_\nu(z)\}$$

$$-\cot\tfrac{1}{2}(\mu-\nu)\pi\,\{J_\mu(z)\,Y_\nu(z)-Y_\mu(z)\,J_\nu(z)\}]$$

$$-\frac{\pi^3 i}{2\cos\frac{1}{2}(\mu+\nu)\pi}[\{J_\mu(z)\,J_\nu(z)+Y_\mu(z)\,Y_\nu(z)\}$$

$$+\tan\tfrac{1}{2}(\mu-\nu)\pi\,\{J_\mu(z)\,Y_\nu(z)-Y_\mu(z)\,J_\nu(z)\}].$$

By writing $-i$ for i throughout the analysis we deduce that, if both $|\arg iz|$ and $|\arg(-iz)|$ are less than $\frac{3}{2}\pi$, i.e. if $|\arg z|<\pi$, then

$$(1)\qquad\frac{\pi^3}{2\sin\frac{1}{2}(\mu+\nu)\pi}[\{J_\mu(z)\,J_\nu(z)+Y_\mu(z)\,Y_\nu(z)\}$$

$$-\cot\tfrac{1}{2}(\mu-\nu)\pi\,.\,\{J_\mu(z)\,Y_\nu(z)-Y_\mu(z)\,J_\nu(z)\}]$$

$$=\frac{1}{2\pi i}\int_{-\infty i}^{\infty i}\Gamma(2s+1)\,\Gamma\left(\frac{\mu+\nu}{2}-s\right)\Gamma\left(\frac{\mu-\nu}{2}-s\right)$$

$$\times\,\Gamma\left(\frac{\nu-\mu}{2}-s\right)\Gamma\left(-\frac{\mu+\nu}{2}-s\right)\cos s\pi\,.\,(\tfrac{1}{2}z)^{2s}\,ds,$$

and

$$(2)\qquad \frac{\pi^3}{2\cos\frac{1}{2}(\mu+\nu)\pi}[\{J_\mu(z)J_\nu(z)+Y_\mu(z)Y_\nu(z)\} + \tan\tfrac{1}{2}(\mu-\nu)\pi.\{J_\mu(z)Y_\nu(z)-Y_\mu(z)J_\nu(z)\}]$$

$$= -\frac{1}{2\pi i}\int_{-\infty i}^{\infty i}\Gamma(2s+1)\Gamma\left(\frac{\mu+\nu}{2}-s\right)\Gamma\left(\frac{\mu-\nu}{2}-s\right) \times\Gamma\left(\frac{\nu-\mu}{2}-s\right)\Gamma\left(-\frac{\mu+\nu}{2}-s\right)\sin s\pi.(\tfrac{1}{2}z)^{2s}\,ds.$$

These results hold for all values of μ and ν (whether integers or not) provided that, in the case of the former $\mu+\nu$ and $\mu-\nu$ are not even integers, and, in the case of the latter $\mu+\nu$ and $\mu-\nu$ are not odd integers.

We now obtain the asymptotic expansions of the functions on the left of (1) and (2) after the manner of § 7·5.

We first take p to be an integer so large that the only poles of the integrands on the left of the line $R(s)=-p-\frac{1}{4}$ are poles of $\Gamma(2s+1)$; and then

$$\int_{-\infty i}^{\infty i} - \int_{-\infty i-p-\frac{1}{4}}^{\infty i-p-\frac{1}{4}}$$

(when either integrand is inserted) is equal to $2\pi i$ times the sum of the residues at the poles between the contours. Since

$$\int_{-\infty i-p-\frac{1}{4}}^{\infty i-p-\frac{1}{4}} f(s)(\tfrac{1}{2}z)^{2s}\,ds = O(z^{-2p-\frac{1}{2}}),$$

we deduce that the asymptotic expansions, when $|\arg z|<\pi$, are

$$(3)\quad [J_\mu(z)J_\nu(z)+Y_\mu(z)Y_\nu(z)]-\cot\tfrac{1}{2}(\mu-\nu)\pi.[J_\mu(z)Y_\nu(z)-J_\nu(z)Y_\mu(z)]$$

$$=\frac{\sin\dfrac{\mu+\nu}{2}\pi}{\pi^3}\times$$

$$\left[\sum_{m=0}^{p-1}\frac{(-)^m\Gamma\left(\frac{\mu+\nu}{2}+m+1\right)\Gamma\left(\frac{\mu-\nu}{2}+m+1\right)\Gamma\left(\frac{\nu-\mu}{2}+m+1\right)\Gamma\left(-\frac{\mu+\nu}{2}+m+1\right)}{(2m+1)!\,(\frac{1}{2}z)^{2m+2}} + O\left(\frac{1}{z^{2p+\frac{1}{2}}}\right)\right]$$

$$\sim\frac{\mu^2-\nu^2}{\pi z^2\sin\frac{1}{2}(\mu-\nu)\pi}\cdot {}_4F_1\left(\frac{\mu+\nu}{2}+1,\ \frac{\mu-\nu}{2}+1,\ \frac{\nu-\mu}{2}+1,\ 1-\frac{\mu+\nu}{2};\ \frac{3}{2};\ -\frac{1}{z^2}\right)$$

and

$$(4)\quad [J_\mu(z)J_\nu(z)+Y_\mu(z)Y_\nu(z)]+\tan\tfrac{1}{2}(\mu-\nu)\pi.[J_\mu(z)Y_\nu(z)-J_\nu(z)Y_\mu(z)]$$

$$\sim\frac{2}{\pi z\cos\frac{1}{2}(\mu-\nu)\pi}\cdot {}_4F_1\left(\frac{\mu+\nu+1}{2},\ \frac{\mu-\nu+1}{2},\ \frac{\nu-\mu+1}{2},\ \frac{1-\mu-\nu}{2};\ \frac{1}{2};\ -\frac{1}{z^2}\right).$$

In the special case when $\mu = \nu$, the last formula reduces to

$$(5)\qquad J_\nu^2(z) + Y_\nu^2(z) \sim \frac{2}{\pi z} \sum_{m=0}^{\infty} \{1 . 3 . 5 \ldots (2m-1)\} \frac{(\nu, m)}{2^m z^{2m}},$$

and, in particular,

$$(6)\qquad J_0^2(z) + Y_0^2(z) \sim \frac{2}{\pi z} \sum_{m=0}^{\infty} \frac{(-)^m \{1 . 3 . 5 \ldots (2m-1)\}^4}{(2m)!\,(2z)^{2m}}.$$

Formula (5) seems to have been discovered by Lorenz, *K. Danske Vidensk. Selskabs Skrifter*, (6) VI. (1890). [*Oeuvres scientifiques*, I. (1898), p. 435], while the more general formulae (3) and (4) were stated by Orr, *Proc. Camb. Phil. Soc.* X. (1900), p. 99. A proof of (5) which depends on transformations of repeated integrals was given by Nielsen, *Handbuch der Theorie der Cylinderfunktionen* (Leipzig, 1904), pp. 245—247; the expansion (5) is, however, attributed to Walter Gregory by A. Lodge, *British Association Report*, 1906, pp. 494—498.

It is not easy to estimate exactly the magnitude or the sign of the remainder after any number of terms in these asymptotic expansions when this method is used. An alternative method of obtaining the asymptotic expansion of $J_\nu^2(z) + Y_\nu^2(z)$ will be given in § 13·75, and it will then be possible to form such an estimate.

CHAPTER VIII

BESSEL FUNCTIONS OF LARGE ORDER

8·1. *Bessel functions of large order.*

The subject of this chapter is the investigation of descriptive properties, including approximate formulae, complete asymptotic expansions, and inequalities of various types connected with Bessel functions; and the properties which will be examined are of primary importance *when the orders of the functions concerned are large*, though many of the results happen to be true for functions of all positive orders.

We shall first obtain results which are of a purely formal character, associated with Carlini's formula (§ 1·4). Next, we shall obtain certain approximate formulae with the aid of Kelvin's* "principle of stationary phase." And finally, we shall examine the contour integrals discovered by Debye†; these will be employed firstly to obtain asymptotic expansions when the variables concerned are real, secondly, to obtain numerous inequalities of varying degrees of importance, and thirdly, to obtain asymptotic expansions of Bessel functions in which both the order and the argument are complex.

In dealing with the function $J_\nu(x)$, in which ν and x are positive, it is found that the problems under consideration have to be divided into three classes, according as x/ν is less than, nearly equal to, or greater than unity. Similar sub-divisions also have to be made in the corresponding theorems concerned with complex variables.

The trivial problem of determining the asymptotic expansion of $J_\nu(z)$, when ν is large and z is fixed, may be noticed here.

It is evident, by applying Stirling's theorem to the expansion of § 3·1, that

$$J_\nu(z) \sim \exp\{\nu + \nu\log(\tfrac{1}{2}z) - (\nu+\tfrac{1}{2})\log\nu\}\,.\left[c_0 + \frac{c_1}{\nu} + \frac{c_2}{\nu^2} + \ldots\right],$$

where $c_0 = 1/\sqrt{(2\pi)}$; this result has been pointed out by Horn, *Math. Ann.* LII. (1899), p. 359.

[NOTE. For physical applications of approximate formulae for functions of large order, the following writers may be consulted: Macdonald, *Proc. Royal Soc.* LXXI. (1903), pp. 251—258; LXXII. (1904), pp. 59—68; XC. A (1914), pp. 50—61; *Phil. Trans. of the Royal Soc.* CCX. A (1910), pp. 113—144; Debye, *Ann. der Physik und Chemie*, (4) XXX. (1909), pp. 57—136; March, *Ann. der Physik und Chemie*, (4) XXXVII. (1912), pp. 29—50; Rybczyński, *Ann. der*

* *Phil. Mag.* (5) XXIII. (1887), pp. 252—255. [*Math. and Phys. Papers*, IV. (1910), pp. 303—306.] In connexion with the principle, see Stokes, *Trans. Camb. Phil. Soc.* IX. (1856), p. 175 footnote. [*Math. and Phys. Papers*, II. (1883), p. 341.]

† *Math. Ann.* LXVII. (1909), pp. 535—558; *Münchener Sitzungsberichte*, XL. [5], (1910).

Physik und Chemie, (4) XLI. (1913), pp. 191—208; Nicholson, *Phil. Mag.* (6) XIX. (1910), pp. 516—537; Love, *Phil. Trans. of the Royal Soc.* CCXV. A (1915), pp. 105—131; Watson, *Proc. Royal Soc.* XCV. A (1918), pp. 83—99, 546—563. The works quoted all deal with the problem of the propagation of electric waves over the surface of the earth, and are largely concerned with attempts to reconcile theoretical with experimental results.]

8·11. *Meissel's first extension of Carlini's formula.*

The approximation (§ 1·4) obtained by Carlini is the first term of the asymptotic expansion of a Bessel function of large order; subsequent terms in the expansion were formally calculated by Meissel, *Astr. Nach.* CXXIX. (1892), col. 281—284, in the following manner:

It is clear that Bessel's equation may be written

$$z^2 \frac{d^2 J_\nu(\nu z)}{dz^2} + z \frac{dJ_\nu(\nu z)}{dz} - \nu^2(1-z^2) J_\nu(\nu z) = 0; \tag{1}$$

if we define a function $u(z)$ by the equation

$$J_\nu(\nu z) = \frac{\nu^\nu}{\Gamma(\nu+1)} \exp\left\{\int^z u(z)\, dz\right\},$$

then equation (1) transforms into

$$z^2[u'(z) + \{u(z)\}^2] + zu(z) - \nu^2(1-z^2) = 0. \tag{2}$$

If now we *assume* that, for large values of ν, $u(z)$ is expansible in a series of descending powers of ν, thus

$$u(z) = \nu u_0 + u_1 + u_2/\nu + u_3/\nu^2 + \ldots,$$

where $u_0, u_1, u_2, u_3, \ldots$ denote functions of z which are independent of ν, by substituting in (2) and equating to zero the coefficients of the various powers of ν on the left, we find that

$$u_0 = \{\sqrt{(1-z^2)}\}/z, \quad u_1 = \frac{z}{2(1-z^2)}, \quad u_2 = -\frac{4z+z^3}{8(1-z^2)^{\frac{5}{2}}},$$

$$u_3 = \frac{4z + 10z^3 + z^5}{8(1-z^2)^4}, \quad u_4 = -\frac{64z + 560z^3 + 456z^5 + 25z^7}{128(1-z^2)^{\frac{11}{2}}},$$

$$u_5 = \frac{16z + 368z^3 + 924z^5 + 374z^7 + 13z^9}{32(1-z^2)^7},$$

....

Hence, on integration, it is found (cf. § 1·4) that

$$\begin{aligned}\int^z u(z)\,dz = \nu\left\{\log\frac{z}{1+\sqrt{(1-z^2)}} + \sqrt{(1-z^2)} - 1\right\} - \tfrac{1}{4}\log(1-z^2) \\ - \frac{1}{24\nu}\left\{\frac{2+3z^2}{(1-z^2)^{\frac{3}{2}}} - 2\right\} + \frac{4z^2+z^4}{16\nu^2(1-z^2)^3} \\ + \frac{1}{5760\nu^3}\left\{\frac{16 - 1512z^2 - 3654z^4 - 375z^6}{(1-z^2)^{\frac{9}{2}}} - 16\right\} \\ + \frac{32z^2 + 288z^4 + 232z^6 + 13z^8}{128\nu^4(1-z^2)^6} + \ldots.\end{aligned}$$

Hence we have Meissel's formula

$$(3)\qquad J_\nu(\nu z) = \frac{(\nu z)^\nu \exp\{\nu\sqrt{(1-z^2)}\} \,.\, \exp(-V_\nu)}{e^\nu \Gamma(\nu+1)(1-z^2)^{\frac{1}{4}}\{1+\sqrt{(1-z^2)}\}^\nu},$$

where

$$(4)\qquad V_\nu = \frac{1}{24\nu}\left\{\frac{2+3z^2}{(1-z^2)^{\frac{3}{2}}} - 2\right\} - \frac{4z^2+z^4}{16\nu^2(1-z^2)^3}$$
$$- \frac{1}{5760\nu^3}\left\{\frac{16-1512z^2-3654z^4-375z^6}{(1-z^2)^{\frac{9}{2}}} - 16\right\}$$
$$- \frac{32z^2+288z^4+232z^6+13z^8}{128\nu^4(1-z^2)^6} + \dots .$$

It will appear in § 8·4 that the expression given for V_ν is the sum of the four dominant terms of an asymptotic expansion which is certainly valid when z lies between 0 and 1 and ν is large.

It is stated by Graf and Gubler* that the first approximation derivable from (3), namely

$$J_\nu(\nu z) \sim \frac{z^\nu \exp\{\nu\sqrt{(1-z^2)}\}}{(2\pi\nu)^{\frac{1}{2}}(1-z^2)^{\frac{1}{4}}\{1+\sqrt{(1-z^2)}\}^\nu},$$

is due to Duhamel; but a search for the formula in Duhamel's writings has not been successful, and it seems certain that, even if it had been discovered by Duhamel, his discovery would have been subsequent to Carlini's.

NOTE. The reader should observe that (3) may also be written in the form

$$(5)\qquad J_\nu(\nu \operatorname{sech} a) = \frac{\exp\{-\nu(a-\tanh a)\} \,.\, \exp(-W_\nu)}{\sqrt{(2\pi\nu\tanh a)}},$$

where

$$(6)\qquad W_\nu = \frac{\coth^3 a}{24\nu}(2+3\operatorname{sech}^2 a) - \frac{\coth^6 a}{16\nu^2}(4\operatorname{sech}^2 a + \operatorname{sech}^4 a)$$
$$- \frac{\coth^9 a}{5760\nu^3}(16 - 1512\operatorname{sech}^2 a - 3654\operatorname{sech}^4 a - 375\operatorname{sech}^6 a)$$
$$- \frac{\coth^{12} a}{128\nu^4}(32\operatorname{sech}^2 a + 288\operatorname{sech}^4 a + 232\operatorname{sech}^6 a + 13\operatorname{sech}^8 a)$$
$$+ \dots .$$

8·12. *Meissel's second expansion.*

The expansion obtained in § 8·11 obviously fails to represent $J_\nu(\nu z)$ when z is real and greater than unity; for such values of z, Meissel† obtained two formal solutions of Bessel's equation; and, if we write $z = \sec\beta$, the reader will see, by making some modifications in § 8·11 (5), that these solutions may be written in the form

$$\sqrt{\left(\frac{2\cot\beta}{\nu\pi}\right)} \exp\{-P_\nu \pm iQ_\nu\},$$

* *Einleitung in die Theorie der Bessel'schen Funktionen*, I. (Bern, 1898), p. 102.

† *Astr. Nach.* CXXX. (1892), col. 363—368.

where*

$$(1)\quad P_\nu = \frac{\cot^6\beta}{16\nu^2}(4\sec^2\beta + \sec^4\beta)$$
$$- \frac{\cot^{12}\beta}{128\nu^4}(32\sec^2\beta + 288\sec^4\beta + 232\sec^6\beta + 13\sec^8\beta)$$
$$+ \frac{\cot^{18}\beta}{3072\nu^6}(768\sec^2\beta + 41280\sec^4\beta + 14884\sec^6\beta + 17493\sec^8\beta + 4242\sec^{10}\beta + 103\sec^{12}\beta)$$
$$+ \ldots,$$

$$(2)\quad Q_\nu = \nu(\tan\beta - \beta) - \frac{\cot^3\beta}{24\nu}(2 + 3\sec^2\beta)$$
$$- \frac{\cot^9\beta}{5760\nu^3}(16 - 1512\sec^2\beta - 3654\sec^4\beta - 375\sec^6\beta)$$
$$- \frac{\cot^{15}\beta}{322560\nu^5}(256 + 78720\sec^2\beta + 1891200\sec^4\beta + 4744640\sec^6\beta + 1914210\sec^8\beta + 67599\sec^{10}\beta)$$
$$+ \ldots.$$

To determine $J_\nu(\nu\sec\beta)$ in terms of these expansions, we take β to tend to $\frac{1}{2}\pi$, and compare the results so obtained with the expansions of Hankel's type given in § 7·21; we see that, as $\beta \to \frac{1}{2}\pi$,

$$P_\nu \to 0, \quad Q_\nu \sim \nu(\sec\beta - \tfrac{1}{2}\pi),$$

and we infer that

$$(3)\qquad H_\nu^{(1)}(\nu\sec\beta) = \sqrt{\left(\frac{2\cot\beta}{\nu\pi}\right)} \cdot e^{-P_\nu + iQ_\nu - \frac{1}{4}\pi i},$$

$$(4)\qquad H_\nu^{(2)}(\nu\sec\beta) = \sqrt{\left(\frac{2\cot\beta}{\nu\pi}\right)} \cdot e^{-P_\nu - iQ_\nu + \frac{1}{4}\pi i}.$$

It follows that

$$(5)\qquad J_\nu(\nu\sec\beta) = \sqrt{\left(\frac{2\cot\beta}{\nu\pi}\right)} \cdot e^{-P_\nu}\cos(Q_\nu - \tfrac{1}{4}\pi),$$

$$(6)\qquad Y_\nu(\nu\sec\beta) = \sqrt{\left(\frac{2\cot\beta}{\nu\pi}\right)} \cdot e^{-P_\nu}\sin(Q_\nu - \tfrac{1}{4}\pi),$$

where P_ν and Q_ν are defined by (1) and (2). It will appear subsequently (§ 8·41) that these formulae are actually asymptotic expansions of $J_\nu(\nu\sec\beta)$ and $Y_\nu(\nu\sec\beta)$ when ν is large and β is any assigned acute angle.

Formulae which are valid for small values of β, i.e. asymptotic expansions of $J_\nu(z)$ and $Y_\nu(z)$ which are valid when z and ν are both large and are nearly equal, cannot easily be obtained by this method; but it will be seen in § 8·2 that, for such values of the variables, approximations can be obtained by rigorous methods from Schläfli's extension of Bessel's integral.

* The reader will observe that the approximation has been carried two stages further than in the corresponding analysis of § 8·11.

NOTE. The dominant terms in the expansions (5) and (6), which may be written in the form

(7) $$J_\nu(x) = M_\nu \cos(Q_\nu - \tfrac{1}{4}\pi), \quad Y_\nu(x) = M_\nu \sin(Q_\nu - \tfrac{1}{4}\pi),$$

where $$M_\nu \sim \left(\frac{2}{\pi\sqrt{(x^2-\nu^2)}}\right)^{\frac{1}{2}},$$

$$Q_\nu \sim \sqrt{(x^2-\nu^2)} - \tfrac{1}{2}\nu\pi + \nu \arcsin(\nu/x),$$

had been obtained two years before the publication of Meissel's paper by L. Lorenz in a memoir on Physical Optics, *K. Danske Videnskabernes Selskabs Skrifter*, (6) VI. (1890). [*Oeuvres Scientifiques*, I. (1898), pp. 421—436.]

The procedure of Lorenz was to take for granted that, as a consequence of the result which has been proved in § 7·51,

$$M_\nu^2 \sim \frac{2}{\pi x}\left[1 + \frac{1}{2}\cdot\frac{\nu^2-\frac{1}{4}}{x^2} + \frac{1\,.\,3}{2\,.\,4}\frac{(\nu^2-\frac{1}{4}).(\nu^2-\frac{9}{4})}{x^4} + \ldots\right]$$
$$\sim \frac{2}{\pi x}\left[1 - \frac{\nu^2-\frac{1}{4}}{x^2}\right]^{-\frac{1}{2}},$$

and then to use the exact equation

(8) $$\frac{dQ_\nu}{dx} = \frac{2}{\pi x M_\nu^2},$$

which is easily deduced from the Wronskian formula of § 3·63 (1), to prove that

$$Q_\nu = x - \tfrac{1}{2}\nu\pi - \int_x^\infty \left\{\frac{2}{\pi x M_\nu^2} - 1\right\} dx,$$

whence the approximation stated for Q_ν follows without difficulty.

Subsequent researches on the lines laid down by Lorenz are due to Macdonald, *Phil. Trans. of the Royal Soc.* CCX. A (1910), pp. 131—144, and Nicholson, *Phil. Mag.* (6) XIV. (1907), pp. 697—707; (6) XIX. (1910), pp. 228—249; 516—537; *Proc. London Math. Soc.* (2) IX. (1911), pp. 67—80; (2) XI. (1913), pp. 104—126. A result concerning $Q_{\nu+1} - Q_\nu$, which is closely connected with (8), has been published by A. Lodge, *British Association Report*, 1906, pp. 494—498.

8·2. *The principle of stationary phase. Bessel functions of equal order and argument.*

The principle of stationary phase was formally enunciated by Kelvin* in connexion with a problem of Hydrodynamics, though the essence of the principle is to be found in some much earlier work by Stokes† on Airy's integral (§ 6·4) and Parseval's integral (§ 2·2), and also in a posthumous paper by Riemann‡.

The problem which Kelvin propounded was to find an approximate expression for the integral

$$u = \frac{1}{2\pi}\int_0^\infty \cos[m\{x - tf(m)\}]\,dm,$$

which expresses the effect at place and time (x, t) of an impulsive disturbance at place and time $(0, 0)$, when $f(m)$ is the velocity of propagation of two-dimensional waves in water corresponding to a wave-length $2\pi/m$. The principle of interference set forth by Stokes

* *Phil. Mag.* (5) XXIII. (1887), pp. 252—255. [*Math. and Phys. Papers*, IV. (1910), pp. 303—306.]

† *Camb. Phil. Trans.* IX. (1856), pp. 175, 183. [*Math. and Phys. Papers*, II. (1883), pp. 341, 351.]

‡ *Ges. Math. Werke* (Leipzig, 1876), pp. 400—406.

and Rayleigh in their treatment of group-velocity and wave-velocity suggested to Kelvin that, for large values of $x-tf(m)$, the parts of the integral outside the range $(\mu-a, \mu+a)$ of values of m are negligible on account of interference if μ is a value (or the value) of m which makes

$$\frac{d}{dm}[m\{x-tf(m)\}]=0.$$

In the range $(\mu-a, \mu+a)$, the expression $m\{x-tf(m)\}$ is then replaced by the first three terms of its expansion by Taylor's theorem, namely

$$\mu\{x-tf(\mu)\}+0.(m-\mu)-\tfrac{1}{2}t\{\mu f''(\mu)+2f'(\mu)\}(m-\mu)^2,$$

and it is found that, if*

$$m-\mu=\frac{\sigma\sqrt{2}}{\sqrt{[-t\{\mu f''(\mu)+2f'(\mu)\}]}},$$

then

$$u\sim\frac{\int_{-\infty}^{\infty}\cos\{t\mu^2 f'(\mu)+\sigma^2\}\,d\sigma}{\pi\sqrt{[-2t\{\mu f''(\mu)+2f'(\mu)\}]}}$$

$$=\frac{\cos\{t\mu^2 f'(\mu)+\tfrac{1}{4}\pi\}}{\sqrt{[-2\pi t\{\mu f''(\mu)+2f'(\mu)\}]}}.$$

In the last integral the limits for σ, which are large even though a be small, have been replaced by $-\infty$ and $+\infty$.

It will be seen from the foregoing analysis that Kelvin's principle is, effectively, that *in the case of the integral of a rapidly oscillating function, the important part of the integral is due to that part of the range of integration near which the phase of the trigonometrical function involved is stationary*†.

It has subsequently been noticed‡ that it is possible to give a formal mathematical proof of Kelvin's principle, for a large class of oscillating functions, by using Bromwich's generalisation§ of an integral formula due to Dirichlet.

The form of Bromwich's theorem which will be adequate for the applications of the principle to Bessel functions is as follows:

Let $F(x)$ be a function of x which has limited total fluctuation when $x\geqslant 0$; let γ be a function of ν which is such that $\nu\gamma\to\infty$ as $\nu\to\infty$. Then, if $-1<\mu<1$,

$$\nu^{\mu}\int_0^{\gamma}x^{\mu-1}F(x)\sin\nu x\,.\,dx\to F(+0)\int_0^{\infty}t^{\mu-1}\sin t\,.\,dt=F(+0)\,\Gamma(\mu)\sin\tfrac{1}{2}\mu\pi;$$

and, if $0<\mu<1$, the sines *may be replaced by* cosines *throughout.*

The method which has just been explained will now be used to obtain an

* This is the appropriate substitution when $m\{x-tf(m)\}$ has a minimum at $m=\mu$; for a maximum the sign of the expression under the radical is changed.

† A persistent search reveals traces of the use of the principle in the writings of Cauchy. See e.g. equation (119) in note 16 of his *Théorie de la propagation des Ondes*, crowned Sept. 1815, *Mém. présentés par divers savants*, I. (1827). [*Oeuvres*, (1) I. (1882), p. 230.]

‡ *Proc. Camb. Phil. Soc.* XIX. (1918), pp. 49—55.

§ Bromwich, *Theory of Infinite Series*, § 174.

approximate formula for $J_\nu(\nu)$ when ν is large and positive. This formula, which was discovered by Cauchy*, is

$$(1) \qquad J_\nu(\nu) \sim \frac{\Gamma(\frac{1}{3})}{2^{\frac{2}{3}} \cdot 3^{\frac{1}{6}} \pi \nu^{\frac{1}{3}}}.$$

This formula has been investigated by means of the principle of stationary phase, comparatively recently, by Nicholson, *Phil. Mag.* (6) XVI. (1909), pp. 276—277, and Rayleigh, *Phil. Mag.* (6) XX. (1910), pp. 1001—1004 [*Scientific Papers*, V. (1912), pp. 617—620]; see also Watson, *Proc. Camb. Phil. Soc.* XIX. (1918), pp. 42—48.

From § 6·2 (4) it is evident that

$$J_\nu(\nu) = \frac{1}{\pi}\int_0^\pi \cos\{\nu(\theta - \sin\theta)\}\, d\theta - \frac{\sin\nu\pi}{\pi}\int_0^\infty e^{-\nu(t+\sinh t)}\, dt,$$

and obviously

$$\left|\frac{\sin\nu\pi}{\pi}\int_0^\infty e^{-\nu(t+\sinh t)}\, dt\right| \leqslant \frac{1}{\pi}\int_0^\infty e^{-2\nu t}\, dt = O(1/\nu).$$

Hence

$$J_\nu(\nu) = \frac{1}{\pi}\int_0^\pi \cos\{\nu(\theta - \sin\theta)\}\, d\theta + O(1/\nu).$$

Now let $\phi = \theta - \sin\theta$, and then

$$\int_0^\pi \cos\{\nu(\theta - \sin\theta)\}\, d\theta = \int_0^\pi \frac{\cos\nu\phi}{1 - \cos\theta}\, d\phi.$$

But

$$\lim_{\theta\to 0} \frac{\phi^{\frac{2}{3}}}{1 - \cos\theta} = \frac{2}{6^{\frac{2}{3}}},$$

and hence, *if* $\phi^{\frac{2}{3}}/(1 - \cos\theta)$ *has limited total fluctuation in the interval* $(0, \pi)$, *it follows from Bromwich's theorem that*

$$\int_0^\pi \frac{\cos\nu\phi}{1 - \cos\theta}\, d\phi \sim \frac{2}{6^{\frac{2}{3}}}\int_0^\infty \phi^{-\frac{2}{3}} \cos\nu\phi\, d\phi$$

$$= \frac{2}{6^{\frac{2}{3}} \nu^{\frac{1}{3}}} \Gamma(\tfrac{1}{3}) \cos\tfrac{1}{6}\pi,$$

and then (1) follows at once.

It still has to be proved that $\phi^{\frac{2}{3}}/(1 - \cos\theta)$ has limited total fluctuation; to establish this result we observe that

$$\frac{d}{d\theta}\left\{\frac{\phi^{\frac{2}{3}}}{1 - \cos\theta}\right\} = \frac{\phi^{-\frac{1}{3}} \sin\theta\, g(\theta)}{3(1 - \cos\theta)^2},$$

where

$$g(\theta) = \frac{2(1 - \cos\theta)^2}{\sin\theta} - 3(\theta - \sin\theta),$$

so that

$$g(0) = 0, \quad g(\pi - 0) = +\infty,$$

$$g'(\theta) = (1 - \cos\theta)^2/(1 + \cos\theta) \geqslant 0,$$

and therefore, by integration, $g(\theta) \geqslant 0$ when $0 \leqslant \theta \leqslant \pi$. Consequently $\phi^{\frac{2}{3}}/(1 - \cos\theta)$ is monotonic and it is obviously bounded. The result required is therefore proved.

* *Comptes Rendus*, XXXVIII. (1854), p. 993. [*Oeuvres*, (1) XII. (1900), p. 663.] A proof by Cauchy's methods will be given in § 8·21.

By means of some tedious integrations by parts*, it is possible to obtain a second approximation, namely

(2) $$J_\nu(\nu) = \frac{\Gamma(\frac{1}{3})}{2^{\frac{2}{3}} \cdot 3^{\frac{1}{6}} \cdot \pi \nu^{\frac{1}{3}}} - \frac{3^{\frac{5}{6}} \Gamma(\frac{5}{3})}{2^{\frac{1}{3}} \cdot 140 \pi \nu^{\frac{5}{3}}} + o(\nu^{-\frac{5}{3}}),$$

and it may also be proved that

(3) $$J_\nu'(\nu) = \frac{3^{\frac{1}{6}} \Gamma(\frac{2}{3})}{2^{\frac{1}{3}} \pi \nu^{\frac{2}{3}}} + o(\nu^{-\frac{2}{3}});$$

an associated formula is

(4) $$Y_\nu(\nu) \sim -\frac{3^{\frac{1}{3}} \Gamma(\frac{1}{3})}{2^{\frac{2}{3}} \pi \nu^{\frac{1}{3}}}.$$

The asymptotic expansions, of which these results give the dominant terms, will be investigated with the aid of more powerful analytical machinery in § 8·42.

8·21. *Meissel's third expansion.*

The integral just discussed has been used by Cauchy† and Meissel‡ to obtain the formal asymptotic expansion of $J_n(n)$ when n is a large integer. It will now be explained how this expansion was obtained by Cauchy and (in a more complete form) by Meissel; the theoretical justification of the processes employed will be investigated in § 8·42.

Taking the formula

$$J_n(n) = \frac{1}{\pi} \int_0^\pi \cos\{n(\theta - \sin\theta)\}\, d\theta,$$

let us write $\theta - \sin\theta = \frac{1}{6}t^3$; it then follows that, for sufficiently small values of t,

$$\theta = t + \tfrac{1}{60}t^3 + \tfrac{1}{1400}t^5 + \ldots = \sum_{m=0}^{\infty} \lambda_m t^{2m+1};$$

and $\lambda_0 = 1,\quad \lambda_1 = \frac{1}{60},\quad \lambda_2 = \frac{1}{1400},\quad \lambda_3 = \frac{1}{25200},\quad \lambda_4 = \frac{43}{17248000},$

$\lambda_5 = \frac{1213}{7207200000},\quad \ldots.$

It follows that

$$J_n(n) = \frac{1}{\pi} \int_0^\pi \left\{ \sum_{m=0}^{\infty} (2m+1) \lambda_m t^{2m} \right\} \cos(\tfrac{1}{6}nt^3) \frac{dt}{d\theta}\, d\theta.$$

When n is large, $\frac{1}{6}nt^3$ is large at the upper limit, and Meissel inferred that

$$J_n(n) \sim \frac{1}{\pi} \sum_{m=0}^{\infty} (2m+1) \lambda_m \cdot G \int_0^\infty t^{2m} \cos(\tfrac{1}{6}nt^3)\, dt,$$

* See *Proc. Camb. Phil. Soc.* XIX. (1918), pp. 42—48.

† *Comptes Rendus*, XXXVIII. (1854), pp. 990—993, 1104—1107. [*Oeuvres*, (1) XII. (1900), pp. 161—164, 167—170.]

‡ *Astr. Nach.* CXXVII. (1891), col. 359—362; CXXVIII. (1891), col. 145—154. Concerning formula (1), Meissel stated "Schon vor dreissig Jahren war ich zu folgenden Formel gelangt."

where G is the sign indicating a "generalised integral" (§ 6·4); and hence, by integrating term-by-term and using Euler's formula, Meissel deduced that

$$(1) \qquad J_n(n) \sim \frac{1}{\pi} \sum_{m=0}^{\infty} \lambda_m \Gamma\left(\tfrac{2}{3}m + \tfrac{4}{3}\right) \left(\frac{6}{n}\right)^{\frac{2}{3}m+\frac{1}{3}} \cos\left(\tfrac{1}{3}m + \tfrac{1}{6}\right)\pi.$$

Meissel also gave an approximation for λ_m, valid when m is large; and this approximation exhibits the divergent character of the expansion (1).

The approximation is obtainable by the theory developed in the memoir of Darboux, "Sur l'approximation des fonctions de très grands nombres," *Journal de Math.* (3) IV. (1878), pp. 5—56, 377—416.

We consider the singularities of θ *qua* function of t; the singularities (where θ fails to be monogenic) are the points at which $\theta = 2r\pi$ and $t = (12r\pi)^{\frac{1}{3}}$, where $r = \pm 1, \pm 2, \pm 3, \ldots$; and near* $t = \pm(12\pi)^{\frac{1}{3}}$ the dominant terms in the expansion of θ are

$$\pm 2\pi \mp (36\pi)^{\frac{1}{3}} \left\{1 \mp \frac{t}{(12\pi)^{\frac{1}{3}}}\right\}^{\frac{1}{3}}.$$

By the theory of Darboux, an approximation to λ_m is the sum of the coefficients of t^{2m+1} in the expansions of the two functions comprised in the last formula; that is to say that

$$\lambda_m \sim 2 \,.\, (36\pi)^{\frac{1}{3}} \frac{\frac{1}{3} \cdot \frac{2}{3} \cdot \frac{5}{3} \ldots (2m - \frac{1}{3})}{(2m+1)!\,(12\pi)^{\frac{2}{3}m+\frac{1}{3}}}$$

$$= \frac{2\,\Gamma(2m + \frac{2}{3})}{3^{\frac{2}{3}}\,\Gamma(\frac{2}{3})\,\Gamma(2m+2) \,.\, (12\pi)^{\frac{2}{3}m}},$$

and so, by Stirling's formula,

$$(2) \qquad \lambda_m \sim \frac{1}{(18)^{\frac{1}{3}}\,\Gamma(\frac{2}{3})\,(m+\frac{1}{3})^{\frac{4}{3}}\,(12\pi)^{\frac{2}{3}m}}.$$

This is Meissel's approximation; an approximation of the same character was obtained by Cauchy, *loc. cit.*, p. 1106.

8·22. *The application of Kelvin's principle to $J_\nu(\nu \sec \beta)$.*

The principle of stationary phase has been applied by Rayleigh† to obtain an approximate formula for $J_\nu(\nu \sec \beta)$ where β is a fixed positive acute angle, and ν is large‡.

As in § 8·2 we have

$$J_\nu(\nu \sec \beta) = \frac{1}{\pi} \int_0^\pi \cos\{\nu(\theta - \sec\beta \sin\theta)\}\, d\theta + O(1/\nu),$$

and $\theta - \sec\beta\sin\theta$ is stationary (a minimum) when $\theta = \beta$.

Write $\theta - \sec\beta\sin\theta = \beta - \tan\beta + \phi$, so that ϕ decreases to zero as θ increases from 0 to β and then increases as θ increases from β to π.

* These are the singularities which are nearest to the origin.

† *Phil. Mag.* (6) XX. (1910), p. 1004. [*Scientific Papers*, V. (1912), p. 620.]

‡ See also Macdonald, *Phil. Trans. of the Royal Soc.* CCX. A (1910), pp. 131—144; and *Proc. Royal Soc.* LXXI. (1903), pp. 251—258; LXXII. (1904), pp. 59—68.

Now

$$\int_0^\pi \cos\{\nu(\theta - \sec\beta\sin\theta)\}\,d\theta$$
$$= \left[\int_{\tan\beta-\beta}^0 + \int_0^{\pi-\beta+\tan\beta}\right] \cos\{\nu(\phi+\beta-\tan\beta)\}\frac{d\theta}{d\phi}\,d\phi,$$

and $\phi^{\frac{1}{2}}\left|\frac{d\theta}{d\phi}\right| \to \frac{1}{\sqrt{(2\tan\beta)}}$ as $\theta \to \beta$.

Hence, *if $\phi^{\frac{1}{2}}(d\theta/d\phi)$ has limited total fluctuation in the range $0 \leqslant \theta \leqslant \pi$, it follows from Bromwich's theorem that*

$$\int_0^\pi \cos\{\nu(\theta - \sec\beta\sin\theta)\}\,d\theta \sim 2\int_0^\infty \cos\{\nu(\phi+\beta-\tan\beta)\}\frac{d\phi}{\sqrt{(2\phi\tan\beta)}}$$
$$= \sqrt{\left(\frac{2\pi}{\nu\tan\beta}\right)}.\cos\{\nu(\tan\beta-\beta)-\tfrac{1}{4}\pi\},$$

and so

(1) $$J_\nu(\nu\sec\beta) \sim \frac{\cos\{\nu(\tan\beta-\beta)-\frac{1}{4}\pi\}}{\sqrt{(\frac{1}{2}\nu\pi\tan\beta)}}.$$

The formula

(2) $$Y_\nu(\nu\sec\beta) \sim \frac{\sin\{\nu(\tan\beta-\beta)-\frac{1}{4}\pi\}}{\sqrt{(\frac{1}{2}\nu\pi\tan\beta)}}$$

is derived in a similar manner from § 6·21 (1).

The reader will observe that these are the dominant terms in Meissel's expansions § 8·12 (5), (6).

To complete the rigorous proof of these formulae we have to shew that $\phi^{\frac{1}{2}}(d\theta/d\phi)$ has limited total fluctuation.

Now the square of this function, namely $\phi\,(d\theta/d\phi)^2$, is equal to

$$\frac{\theta - \sec\beta\sin\theta - \beta + \tan\beta}{(1-\sec\beta\cos\theta)^2} \equiv h(\theta),$$

say. But

$$h'(\theta) = \frac{\cos\beta\,\mathrm{cosec}\,\theta\,(1-\sec\beta\cos\theta)^2 - 2(\theta - \sec\beta\sin\theta - \beta + \tan\beta)}{\cos\beta\,\mathrm{cosec}\,\theta\,(1-\sec\beta\cos\theta)^3}.$$

The numerator, $k(\theta)$, of this fraction has the differential coefficient

$$-\cos\beta\cos\theta\,\mathrm{cosec}^2\,\theta\,(1-\sec\beta\cos\theta)^2,$$

and so $k(\theta)$ decreases steadily as θ increases from 0 to $\frac{1}{2}\pi$, and then increases steadily as θ increases from $\frac{1}{2}\pi$ to π; since $k(\theta)=0$ when $\theta=\beta<\frac{1}{2}\pi$, it follows that $h'(\theta)\leqslant 0$ when $0\leqslant\theta\leqslant\beta$ and $h'(\theta)$ changes sign once (from negative to positive) in the range $\beta\leqslant\theta\leqslant\pi$.

Hence $|\sqrt{h(\theta)}|$ is monotonic (and decreasing) when $0\leqslant\theta\leqslant\beta$, and it has one stationary point (a minimum) in the range $\beta<\theta<\pi$; since $|\sqrt{h(\theta)}|$ is bounded and continuous when $0\leqslant\theta\leqslant\pi$ it consequently has limited total fluctuation when $0\leqslant\theta\leqslant\pi$, as had to be proved.

8·3. *The method of steepest descents.*

A development of the theory of contour integration, called the method of steepest descents*, has been applied by Debye† to obtain integral representations of Bessel functions of large order from which asymptotic expansions are readily deduced. If, in general, we consider the integral

$$\int e^{\nu f(w)} \phi(w)\, dw,$$

in which $|\nu|$ is supposed to be large, the contour is chosen so that it passes through a point w_0 at which $f'(w)$ vanishes; and the whole of the contour is then determined by the assumption that the imaginary part of $f(w)$ is to be constant on it, so that the equation of the contour may be written in the form

$$If(w) = If(w_0).$$

To obtain a geometrical conception of the contour, let $w = u + iv$, where u, v are real; and draw the surface such that the three coordinates of any point on it are

$$u, \ v, \ Rf(w).$$

If $Rf(w) = z$, and if the z-axis be supposed to be vertical, the surface has no absolute maxima or minima except where $f(w)$ fails to be monogenic; for, at all other points,

$$\frac{\partial^2 z}{\partial u^2} + \frac{\partial^2 z}{\partial v^2} = 0.$$

The points $[u_0, v_0, Rf(w_0)]$ are saddle points, or passes, on the surface, so that the contour of integration is the plan of a curve on the surface which goes through one of the passes on the surface. This curve possesses a further property derived from the equation of the contour; for the rate of change of $f(w)$, at any given value of w, has a definite modulus, since $f(w)$ is supposed to be monogenic; and since $If(w)$ does not change as w traverses the contour, it follows that $Rf(w)$ must change as rapidly as possible; that is to say, that the curve is characterised by the property that its direction, at any point of it, is so chosen that it is the steepest curve through that point and on the surface.

It may happen that we have a freedom of choice in selecting a pass and then in selecting a contour through that pass; our choice is to be determined from the consideration that the curve must *descend* on both sides of the pass; for if the curve ascended, $Rf(w)$ would tend to $+\infty$ (except in very special cases) as w left the pass, and then the integral would diverge if $R(\nu) > 0$.

* French "Méthode du Col," German "Methode der Sattelpunkte."

† *Math. Ann.* LXVII. (1909), pp. 535—558; *Münchener Sitzungsberichte*, XL. [5], (1910). The method is to be traced to a posthumous paper by Riemann, *Werke*, p. 405; and it has recently been applied to obtain asymptotic expansions of a variety of functions.

The contour has now been selected* so that the integrand does not oscillate rapidly on it; and so we may expect that an approximate value of the integral will be determined from a consideration of the integrand in the neighbourhood of the pass: from the physical point of view, we have evaded the interference effects (cf. § 8·2) which occur with any other type of contour.

The mode of derivation of asymptotic expansions from the integral will be seen clearly from the special functions which will be studied in §§ 8·4—8·43, 8·6, 8·61; but it is convenient to enunciate at this stage a lemma† which will be useful subsequently in proving that the expansions which will be obtained are asymptotic in the sense of Poincaré.

LEMMA. *Let $F(\tau)$ be analytic when $|\tau| \leqslant a+\delta$, where $a>0$, $\delta>0$; and let*

$$F(\tau)=\sum_{m=1}^{\infty} a_m \tau^{(m/r)-1},$$

when $|\tau| \leqslant a$, r being positive; also, let $|F(\tau)|<Ke^{b\tau}$, where K and b are positive numbers independent of τ, when τ is positive and $\tau \geqslant a$. Then the asymptotic expansion

$$\int_0^{\infty} e^{-\nu\tau} F(\tau)\, d\tau \sim \sum_{m=1}^{\infty} a_m \Gamma(m/r)\, \nu^{-m/r}$$

is valid in the sense of Poincaré when $|\nu|$ is sufficiently large and

$$|\arg \nu| \leqslant \tfrac{1}{2}\pi-\Delta,$$

where Δ is an arbitrary positive number.

It is evident that, if M be any fixed integer, a constant K_1 can be found such that

$$\left|F(\tau)-\sum_{m=1}^{M-1} a_m \tau^{(m/r)-1}\right| \leqslant K_1 \tau^{(M/r)-1} e^{b\tau}$$

whenever $\tau \geqslant 0$ whether $\tau \leqslant a$ or $\tau \geqslant a$; and therefore

$$\int_0^{\infty} e^{-\nu\tau} F(\tau)\, d\tau=\sum_{m=1}^{M-1} \int_0^{\infty} e^{-\nu\tau} a_m \tau^{(m/r)-1}\, d\tau+R_M,$$

where

$$\begin{aligned} |R_M| &\leqslant \int_0^{\infty} |e^{-\nu\tau}| \,.\, K_1 \tau^{(M/r)-1} e^{b\tau}\, d\tau \\ &= K_1 \Gamma(M/r)/\{R(\nu)-b\}^{M/r} \\ &= O(\nu^{-M/r}), \end{aligned}$$

provided that $R(\nu)>b$, which is the case when $|\nu|>b \operatorname{cosec} \Delta$. The analysis remains valid even when b is a function of ν such that $R(\nu)-b$ is not small compared with ν. We have therefore proved that

$$\int_0^{\infty} e^{-\nu\tau} F(\tau)\, d\tau=\sum_{m=1}^{M-1} a_m \Gamma(m/r)\, \nu^{-m/r}+O(\nu^{-M/r}),$$

and so the lemma is established.

* For an account of researches in which the contour is the real axis see pp. 1343—1350 of Burkhardt's article in the *Encyclopädie der Math. Wiss.* II. 1 (1916).

† Cf. *Proc. London Math. Soc.* (2) XVII. (1918), p. 133.

8·31. *The construction of Debye's contours* when the variables are real.*

It has been seen in §§ 6·2, 6·21 that the various types of functions associated with $J_\nu(x)$ can be represented by integrals of the form

$$\int e^{x \sinh w - \nu w}\, dw$$

taken along suitable contours. On the hypothesis that ν and x are positive, we shall now examine whether any of the contours appropriate for the method of steepest descents are of the types investigated in §§ 6·2, 6·21.

In accordance with the principles of the method of steepest descents, as explained in § 8·3, we have first to find the stationary points of

$$x \sinh w - \nu w,$$

qua function of w, i.e. we have to solve the equation

(1) $$x \cosh w - \nu = 0;$$

and it is at once seen that we shall have three distinct cases to consider, in which x/ν is less than, greater than, or equal to 1, respectively. We consider these three cases in turn.

(I) When $x/\nu < 1$, we can find a positive number α such that

(2) $$x = \nu \operatorname{sech} \alpha,$$

and then the complete solution of (1) is

$$w = \pm \alpha + 2n\pi i.$$

It will be sufficient to confine our attention to the stationary points† $\pm \alpha$; at these points the imaginary part of $x \sinh w - \nu w$ is zero, and so the equation of the contour to be discussed is

$$I(x \sinh w - \nu w) = 0.$$

Write $w \equiv u + iv$, where u, v are real, and this equation becomes

$$\cosh u \sin v - v \cosh \alpha = 0,$$

so that $v = 0$, or

(3) $$\cosh u = \frac{v \cosh \alpha}{\sin v}.$$

The contour $v = 0$ gives a divergent integral. We therefore consider the contour given by equation (3). To values of v between 0 and π, correspond pairs of values of u which are equal but opposite in sign; and as v increases from 0 to π, the positive value of u steadily increases from α to $+\infty$.

* The contours investigated in this section are those which were discussed in Debye's earlier paper, *Math. Ann.* LXVII. (1909), pp. 535—558, except that their orientation is different; cf. § 6·21.

† The effect of taking stationary points other than $\pm \alpha$ would be to translate the contour parallel to the imaginary axis.

The equation is unaltered by changing the sign of v and so the contour is symmetrical with regard to the axes; the shape of the part of the contour between $v=-\pi$ and $v=\pi$ is shewn in Fig. 16.

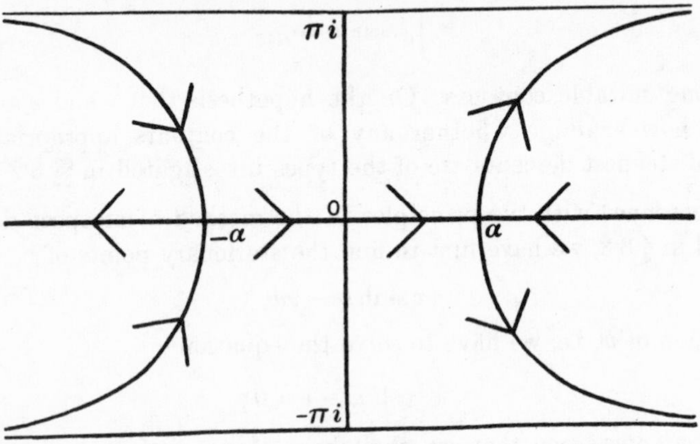

Fig. 16.

If
$$\tau = \sinh\alpha - \alpha\cosh\alpha - (\sinh w - w\cosh\alpha),$$
it is easy to verify that τ (which is real on the curves shewn in the figure) increases in the directions indicated by the arrows.

As w travels along the contour from $\infty - \pi i$ to $\infty + \pi i$, τ decreases from $+\infty$ to 0 and then increases to $+\infty$; and since, by § 6·2 (3),
$$J_\nu(x) = \frac{1}{2\pi i}\int_{\infty-\pi i}^{\infty+\pi i} e^{x\sinh w - \nu w}\,dw,$$
we have obtained a curve from which we can derive information concerning $J_\nu(x)$ when x and ν are large and $x/\nu < 1$. The detailed discussion of the integral will be given subsequently in §§ 8·4, 8·5.

The contours from $-\infty$ to $\infty \pm \pi i$ give information concerning a second solution of Bessel's equation; but this problem is complicated by Stokes' phenomenon, on account of the two stationary points on the contour.

(II) When $x/\nu > 1$, we can find a positive acute angle β such that

(4) $$x = \nu\sec\beta,$$

and the relevant stationary points, which are now roots of the equation
$$\cosh w - \cos\beta = 0,$$
are
$$w = \pm i\beta.$$
When we take the stationary point $i\beta$, the contour which we obtain is
$$I(\sinh w - w\cos\beta) = \sin\beta - \beta\cos\beta,$$
so that, replacing w by $u+iv$, the equation of the contour is

(5) $$\cosh u = \frac{\sin\beta + (v-\beta)\cos\beta}{\sin v}$$

Now, for values of v between 0 and π, the function

$$\sin\beta + (v-\beta)\cos\beta - \sin v$$

has one minimum ($v = \beta$) at which the value of the function is zero; for other values of v between 0 and π,

$$\sin\beta + (v-\beta)\cos\beta > \sin v.$$

Hence, for values of v between 0 and π, equation (5) gives two real values of u (equal but opposite in sign), and these coincide only when $v = \beta$. They are infinite when v is 0 or π.

The shape of the curves given by equation (5) is as shewn in the upper half of Fig. 17; and if

$$\tau = i(\sin\beta - \beta\cos\beta) - (\sinh w - w\cos\beta),$$

it is easy to verify that τ (which is real on the curves) increases in the directions indicated by the arrows. As w travels along the contour from $-\infty$ to $\infty + \pi i$, τ decreases from $+\infty$ to 0 and then increases to $+\infty$ and so we

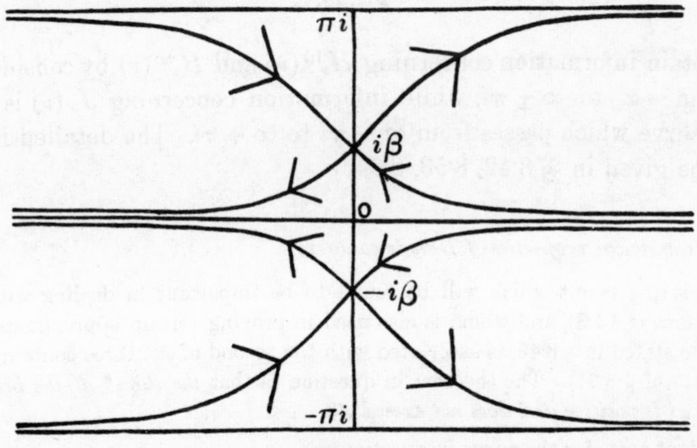

Fig. 17.

have obtained a curve from which [§ 6·21 (4)] we can derive information concerning $H_\nu^{(1)}(x)$ when x and ν are large and $x/\nu > 1$. The detailed discussion of the integral will be given in §§ 8·41, 15·8.

If we had taken the stationary point $-i\beta$, we should have obtained the curves shewn in the lower half of Fig. 17, and the curve going from $-\infty$ to $\infty - \pi i$ gives an integral associated with $H_\nu^{(2)}(x)$; this also will be discussed in § 8·41. The two integrals now obtained form a fundamental system of solutions of Bessel's equation, so that there is a marked distinction between the case $x/\nu < 1$ and the case $x/\nu > 1$.

(III) The case in which $\nu = x$ may be derived as a limiting case either from (I) or from (II) by taking α or β equal to 0. The curves now to be considered are $v = 0$ and

$$\cosh u = v/\sin v, \tag{6}$$

and they are shewn in Fig. 18.

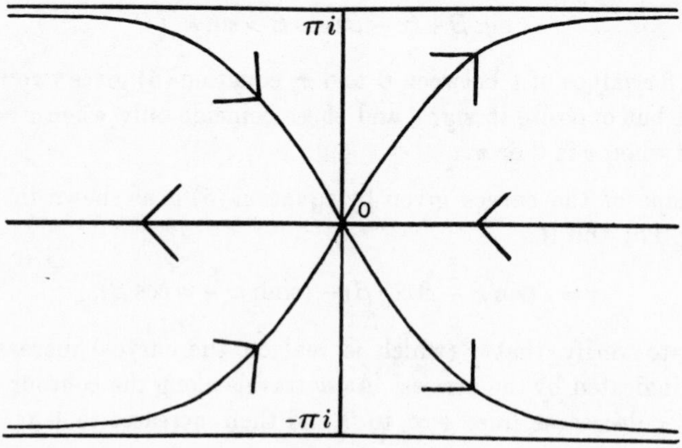

Fig. 18.

We obtain information concerning $H_\nu^{(1)}(\nu)$ and $H_\nu^{(2)}(\nu)$ by considering the curves from $-\infty$ to $\infty \pm \pi i$, while information concerning $J_\nu(\nu)$ is obtained from the curve which passes from $\infty - \pi i$ to $\infty + \pi i$. The detailed investigation will be given in §§ 8·42, 8·53, 8·54.

8·32. *Geometrical properties of Debye's contours.*

An interesting result which will be found to be important in dealing with zeros of Bessel functions (§ 15·8), and which is also used in proving certain approximate formulae which will be stated in § 8·43, is associated with the second of the three contours just discussed (Fig. 17 of § 8·31). The theorem in question is that *the slope* of the branch from* $-\infty$ *to* $\infty + \pi i$ *is positive and does not exceed* $\sqrt{3}$.

It is evident that, for the curve in question,

$$\sinh u \frac{du}{dv} = \frac{\sin(v-\beta) - (v-\beta)\cos v \cos\beta}{\sin^2 v}.$$

But $\sin(v-\beta)\sec v - (v-\beta)\cos\beta$ has the positive derivative $\cos\beta\tan^2 v$, and hence it follows that

$$\sin(v-\beta) - (v-\beta)\cos v\cos\beta$$

has the same sign† as $v-\beta$. Therefore since $v-\beta$ and u are both positive or both negative for the curve under consideration, dv/du is positive.

* *Proc. Camb. Phil. Soc.* XIX. (1918), p. 105. Since, in the limiting case (Fig. 18) in which $\beta = 0$, the slope is 0 on the left of the origin and is $\sqrt{3}$ immediately on the right of the origin, no better results of this type exist.

† This is obvious from a figure.

Again, to prove that dv/du does not exceed $\sqrt{3}$, we write

$$\psi(v) \equiv \frac{\sin\beta + (v-\beta)\cos\beta}{\sin v},$$

and then it is sufficient to prove that

$$3\psi'^2(v) - \psi^2(v) + 1 \geqslant 0.$$

Now the expression on the left (which vanishes when $v=\beta$) has the derivate

$$2\psi'(v)[3\psi''(v) - \psi(v)]$$
$$= \frac{4\psi'(v)}{\sin^3 v}[(v-\beta)\{\sin^2 v + 3\cos^2 v\}\cos\beta + \sin^2 v \sin\beta - 3\cos v \sin(v-\beta)].$$

But
$$(v-\beta)\cos\beta + \frac{\sin^2 v\sin\beta - 3\cos v\sin(v-\beta)}{\sin^2 v + 3\cos^2 v}$$

has the positive derivate $\dfrac{4\sin^4 v\cos\beta}{(\sin^2 v + 3\cos^2 v)^2}$, and so, since it is positive when $v=0$, it is positive when $0<v<\pi$. Therefore, since $\psi'(v)$ has the same sign as $v-\beta$, it follows that

$$2\psi'(v)[3\psi''(v) - \psi(v)]$$

has the same sign as $v-\beta$, and consequently

$$3\psi'^2(v) - \psi^2(v) + 1$$

has $v=\beta$ for its only minimum between $v=0$ and $v=\pi$; and therefore it is not negative. This proves the result stated.

8·4. *The asymptotic expansion* of $J_\nu(\nu\,\text{sech}\,\alpha)$.*

From the results obtained in § 8·31 we shall now obtain the asymptotic expansion of the function of the first kind in which the argument is less than the order, both being large and positive.

We retain the notation of § 8·31 (I); and it is clear that, corresponding to any positive value of τ, there are two values of w, which will be called w_1 and w_2; the values of w_1 and w_2 differ only in the sign of their imaginary part, and it will be supposed that

$$I(w_1) > 0, \qquad I(w_2) < 0.$$

We then have

$$J_\nu(\nu\,\text{sech}\,\alpha) = \frac{e^{\nu(\tanh\alpha - \alpha)}}{2\pi i}\int_0^\infty e^{-x\tau}\left\{\frac{dw_1}{d\tau} - \frac{dw_2}{d\tau}\right\}d\tau,$$

where $x = \nu\,\text{sech}\,\alpha$.

Next we discuss the expansions of w_1 and w_2 in ascending powers of τ. Since τ and $d\tau/dw$ vanish when $w=\alpha$, it follows that the expansion of τ in powers of $w-\alpha$ begins with a term in $(w-\alpha)^2$; by reverting this expansion, we obtain expansions of the form

$$w_1 - \alpha = \sum_{m=0}^\infty \frac{a_m}{m+1}\tau^{\frac{1}{2}(m+1)}, \quad w_2 - \alpha = \sum_{m=0}^\infty (-)^{m+1}\frac{a_m}{m+1}\tau^{\frac{1}{2}(m+1)},$$

* The asymptotic expansions contained in this section and in §§ 8·41, 8·42 were established by Debye, *Math. Ann.* LXVII. (1909), pp. 535—558.

and, by Lagrange's theorem, these expansions are valid for sufficiently small values of $|\tau|$. Moreover

$$a_m = \frac{1}{2\pi i}\int^{(0+,0+)} \left(\frac{dw_1}{d\tau}\right) \frac{d\tau}{\tau^{\frac{1}{2}(m+1)}}$$
$$= \frac{1}{2\pi i}\int^{(\alpha+)} \frac{dw}{\tau^{\frac{1}{2}(m+1)}}.$$

The double circuit in the τ-plane is necessary in order to dispose of the fractional powers of τ; and a single circuit round α in the w-plane corresponds to a double circuit round the origin in the τ-plane. From the last contour integral it follows that a_m is the coefficient of $1/(w-\alpha)$ in the expansion of $\tau^{-\frac{1}{2}(m+1)}$ in ascending powers of $w-\alpha$; we are thus enabled to calculate the coefficients a_m.

Write $w - \alpha = W$ and we have

$$\tau = -\sinh\alpha\,(\cosh W - 1) - \cosh\alpha\,(\sinh W - W)$$
$$= W^2 (c_0 + c_1 W + c_2 W^2 + \ldots),$$

where $c_0 = -\frac{1}{2}\sinh\alpha$, $c_1 = -\frac{1}{6}\cosh\alpha$, $c_2 = -\frac{1}{24}\sinh\alpha$, ... Therefore a_m is the coefficient of W^m in the expansion of $\{c_0 + c_1 W + c_2 W^2 + \ldots\}^{-\frac{1}{2}(m+1)}$.

The coefficients in this expansion will be called $a_0(m)$, $a_1(m)$, $a_2(m)$, ..., and so we have

$$(1)\quad \begin{cases} a_0(m) = c_0^{-\frac{1}{2}(m+1)}, \\ a_1(m) = c_0^{-\frac{1}{2}(m+1)}\left\{-\dfrac{m+1}{2\,.\,1!}\cdot\dfrac{c_1}{c_0}\right\}, \\ a_2(m) = c_0^{-\frac{1}{2}(m+1)}\left\{-\dfrac{m+1}{2\,.\,1!}\cdot\dfrac{c_2}{c_0} + \dfrac{(m+1)(m+3)}{2^2\,.\,2!}\cdot\dfrac{c_1^2}{c_0^2}\right\}, \\ a_3(m) = c_0^{-\frac{1}{2}(m+1)}\left\{-\dfrac{m+1}{2\,.\,1!}\cdot\dfrac{c_3}{c_0} + \dfrac{(m+1)(m+3)}{2^2\,.\,2!}\cdot\dfrac{2c_1c_2}{c_0^2} \right. \\ \qquad\qquad \left. - \dfrac{(m+1)(m+3)(m+5)}{2^3\,.\,3!}\cdot\dfrac{c_1^3}{c_0^3}\right\}, \\ a_4(m) = c_0^{-\frac{1}{2}(m+1)}\left\{-\dfrac{m+1}{2\,.\,1!}\cdot\dfrac{c_4}{c_0} + \dfrac{(m+1)(m+3)}{2^2\,.\,2!}\left(\dfrac{2c_1c_3}{c_0^2} + \dfrac{c_2^2}{c_0^2}\right) \right. \\ \qquad\qquad - \dfrac{(m+1)(m+3)(m+5)}{2^3\,.\,3!}\cdot\dfrac{3c_1^2c_2}{c_0^3} \\ \qquad\qquad \left. + \dfrac{(m+1)(m+3)(m+5)(m+7)}{2^4\,.\,4!}\cdot\dfrac{c_1^4}{c_0^4}\right\}, \\ \ldots\ldots\ldots\ldots\ldots\ldots\ldots\ldots \end{cases}$$

On substitution we find that

$$(2)\quad \begin{cases} a_0 = a_0(0) = +(-\frac{1}{2}\sinh\alpha)^{-\frac{1}{2}}, \\ a_1 = a_1(1) = -(-\frac{1}{2}\sinh\alpha)^{-1}\{\frac{1}{3}\coth\alpha\}, \\ a_2 = a_2(2) = -(-\frac{1}{2}\sinh\alpha)^{-\frac{3}{2}}\{\frac{1}{8} - \frac{5}{24}\coth^2\alpha\}, \\ a_3 = a_3(3) = -(-\frac{1}{2}\sinh\alpha)^{-2}\{\frac{2}{15}\coth\alpha - \frac{4}{27}\coth^3\alpha\}, \\ a_4 = a_4(4) = +(-\frac{1}{2}\sinh\alpha)^{-\frac{5}{2}}\{\frac{3}{128} - \frac{77}{576}\coth^2\alpha + \frac{385}{3456}\coth^4\alpha\}, \\ \ldots\ldots\ldots\ldots\ldots\ldots\ldots\ldots \end{cases}$$

Now
$$\frac{d(w_1 - w_2)}{d\tau} = \sum_{m=0}^{\infty} a_{2m}\,\tau^{m-\frac{1}{2}}$$
when $|\tau|$ is sufficiently small; and since
$$\frac{d\tau}{dw} = \cosh\alpha - \cosh w,$$
it follows that $d(w_1 - w_2)/d\tau$ tends to zero as τ tends to $+\infty$.

Hence the conditions stated in the lemma of § 8·3 are satisfied, *and so*
$$\int_0^\infty e^{-x\tau}\left\{\frac{dw_1}{d\tau} - \frac{dw_2}{d\tau}\right\} d\tau$$
has the asymptotic expansion
$$\sum_{m=0}^{\infty} \frac{a_{2m}\,\Gamma(m+\frac{1}{2})}{x^{m+\frac{1}{2}}}$$
when x is large.

Since $\arg\{(w_1 - \alpha)/\tau^{\frac{1}{2}}\} \to \frac{1}{2}\pi$ as $\tau \to 0$, it follows that, in (2), the phase of a_0 has to be interpreted by the convention $\arg a_0 = +\frac{1}{2}\pi$, and hence

(3) $$J_\nu(\nu \operatorname{sech}\alpha) \sim \frac{e^{\nu(\tanh\alpha - \alpha)}}{\sqrt{(2\pi\nu\tanh\alpha)}} \sum_{m=0}^{\infty} \frac{\Gamma(m+\frac{1}{2})}{\Gamma(\frac{1}{2})} \cdot \frac{A_m}{(\frac{1}{2}\nu\tanh\alpha)^m},$$

where

(4) $$\begin{cases} A_0 = 1, \quad A_1 = \frac{1}{8} - \frac{5}{24}\coth^2\alpha, \\ A_2 = \frac{3}{128} - \frac{77}{576}\coth^2\alpha + \frac{385}{3456}\coth^4\alpha, \\ \ldots\ldots\ldots\ldots\ldots\ldots\ldots\ldots\ldots\ldots\ldots\ldots \end{cases}$$

The formula (3) gives the asymptotic expansion of $J_\nu(\nu\operatorname{sech}\alpha)$ valid when α is any fixed positive number and ν is large and positive.

The corresponding expansion for the function of the second kind, obtained by taking a contour from $-\infty$ to $\infty \pm \pi i$, is

(5) $$Y_\nu(\nu\operatorname{sech}\alpha) \sim -\frac{e^{\nu(\alpha - \tanh\alpha)}}{\sqrt{(\frac{1}{2}\pi\nu\tanh\alpha)}} \sum_{m=0}^{\infty} \frac{\Gamma(m+\frac{1}{2})}{\Gamma(\frac{1}{2})} \cdot \frac{(-)^m A_m}{(\frac{1}{2}\nu\tanh\alpha)^m}.$$

The position of the singularities of $d(w_1 - w_2)/d\tau$, *qua* function of the complex variable τ, should be noted. These singularities correspond to the points where w fails to be a monogenic function of τ, i.e. the points where $d\tau/dw$ vanishes. Hence the singularities correspond to the values $\pm\alpha + 2n\pi i$ of w, so they are the points where
$$\tau = 2n\pi i\cosh\alpha, \quad \tau = 2(\sinh\alpha - \alpha\cosh\alpha) + 2n\pi i\cosh\alpha,$$
and n assumes all integral values.

It is convenient to obtain a formula for $dw/d\tau$ in the form of a contour integral; if (w_0, τ_0) be a pair of corresponding values of (w, τ), then, by Cauchy's theorem,
$$\left(\frac{dw}{d\tau}\right)_0 = \frac{1}{2\pi i}\int^{(\tau_0+)} \frac{dw}{d\tau}\,\frac{d\tau}{\tau - \tau_0} = \frac{1}{2\pi i}\int^{(w_0+)} \frac{dw}{\tau - \tau_0},$$
where the contour includes no point (except w_0) at which τ has the value τ_0.

8·41. *The asymptotic expansions of $J_\nu(\nu \sec\beta)$ and $Y_\nu(\nu\sec\beta)$.*

In § 8·4 we obtained the asymptotic expansion of a Bessel function in which the argument was *less* than the order, both being large; we shall now obtain the asymptotic expansions of a fundamental system of solutions of Bessel's equation when the argument is *greater* than the order, both being large.

We retain the notation of § 8·31 (II); it is clear that, corresponding to any positive value of τ, there are two values of w lying on the contour which passes from $-\infty$ to $\infty + \pi i$; these values will be called w_1 and w_2, and it will be supposed that

$$R(w_1) > 0, \quad R(w_2) < 0.$$

We then have

$$H_\nu^{(1)}(\nu\sec\beta) = \frac{e^{\nu i(\tan\beta-\beta)}}{\pi i}\int_0^\infty e^{-x\tau}\left\{\frac{dw_1}{d\tau} - \frac{dw_2}{d\tau}\right\}d\tau,$$

where $x = \nu\sec\beta$. The analysis now proceeds exactly on the lines of § 8·4 except that α is replaced throughout by $i\beta$, and the Bessel function is of the third kind.

It is thus found that

$$\int_0^\infty e^{-x\tau}\left\{\frac{dw_1}{d\tau} - \frac{dw_2}{d\tau}\right\}d\tau \sim \sum_{m=0}^\infty \frac{a_{2m}\,\Gamma(m+\tfrac{1}{2})}{x^{m+\frac{1}{2}}}.$$

To determine the phase of a_0, that is of $(-\tfrac{1}{2}i\sin\beta)^{-\frac{1}{2}}$, we observe that $\arg\{(w_1 - i\beta)/\tau\} \to +\tfrac{1}{4}\pi$ as $\tau \to 0$, and so

$$a_0 = e^{\frac{1}{4}\pi i}/\sqrt{(\tfrac{1}{2}\sin\beta)}.$$

Consequently

$$(1) \qquad H_\nu^{(1)}(\nu\sec\beta) \sim \frac{e^{\nu i(\tan\beta-\beta)-\frac{1}{4}\pi i}}{\sqrt{(\tfrac{1}{2}\nu\pi\tan\beta)}}\sum_{m=0}^\infty \frac{\Gamma(m+\tfrac{1}{2})}{\Gamma(\tfrac{1}{2})}\cdot\frac{A_m}{(\tfrac{1}{2}\nu i\tan\beta)^m}.$$

In like manner, by taking as contour the reflexion of the preceding contour in the real axis of the w-plane, we find that

$$(2) \qquad H_\nu^{(2)}(\nu\sec\beta) \sim \frac{e^{-\nu i(\tan\beta-\beta)+\frac{1}{4}\pi i}}{\sqrt{(\tfrac{1}{2}\nu\pi\tan\beta)}}\sum_{m=0}^\infty \frac{\Gamma(m+\tfrac{1}{2})}{\Gamma(\tfrac{1}{2})}\cdot\frac{A_m}{(-\tfrac{1}{2}\nu i\tan\beta)^m}.$$

In these formulae, which are valid when β is a fixed positive acute angle and ν is large and positive, we have to make the substitutions:

$$(3) \qquad \begin{cases} A_0 = 1, \quad A_1 = \tfrac{1}{8} + \tfrac{5}{24}\cot^2\beta, \\ A_2 = \tfrac{3}{128} + \tfrac{77}{576}\cot^2\beta + \tfrac{385}{3456}\cot^4\beta, \\ \dots\dots\dots\dots\dots\dots\dots\dots\dots\dots\dots\dots \end{cases}$$

If we combine (1) and (2), we find that

$$(4) \qquad J_\nu(\nu\sec\beta) \sim$$

$$\left(\frac{2}{\nu\pi\tan\beta}\right)^{\frac{1}{2}}\left[\cos(\nu\tan\beta - \nu\beta - \tfrac{1}{4}\pi)\sum_{m=0}^\infty \frac{(-)^m\,\Gamma(2m+\tfrac{1}{2})}{\Gamma(\tfrac{1}{2})}\cdot\frac{A_{2m}}{(\tfrac{1}{2}\nu\tan\beta)^{2m}}\right.$$
$$\left. + \sin(\nu\tan\beta - \nu\beta - \tfrac{1}{4}\pi)\sum_{m=0}^\infty \frac{(-)^m\,\Gamma(2m+\tfrac{3}{2})}{\Gamma(\tfrac{1}{2})}\cdot\frac{A_{2m+1}}{(\tfrac{1}{2}\nu\tan\beta)^{2m+1}}\right],$$

(5) $Y_\nu(\nu \sec\beta) \sim$

$$\left(\frac{2}{\nu\pi\tan\beta}\right)^{\frac{1}{2}}\left[\sin(\nu\tan\beta-\nu\beta-\tfrac{1}{4}\pi)\sum_{m=0}^{\infty}\frac{(-)^m\,\Gamma(2m+\frac{1}{2})}{\Gamma(\frac{1}{2})}\cdot\frac{A_{2m}}{(\frac{1}{2}\nu\tan\beta)^{2m}}\right.$$
$$\left.-\cos(\nu\tan\beta-\nu\beta-\tfrac{1}{4}\pi)\sum_{m=0}^{\infty}\frac{(-)^m\,\Gamma(2m+\frac{3}{2})}{\Gamma(\frac{1}{2})}\cdot\frac{A_{2m+1}}{(\frac{1}{2}\nu\tan\beta)^{2m+1}}\right].$$

The dominant terms in these expansions are those obtained by the principle of stationary phase in § 8·21.

8·42. *Asymptotic expansions of Bessel functions whose order and argument are nearly equal.*

The formulae which have been established in §§ 8·4, 8·41 obviously fail to give adequate approximations when α (or β) is small, that is when the argument and order of the Bessel function concerned are nearly equal. It is, however, possible to use the same method for determining asymptotic expansions in these circumstances, and it happens that *no complications arise by supposing the variables to be complex.*

Accordingly we shall discuss the functions

$$H_\nu^{(1)}(z), \quad H_\nu^{(2)}(z),$$

where z and ν are complex numbers of large modulus, such that $|z-\nu|$ is not large. It will appear that it is necessary to assume that $z-\nu = o(z^{\frac{1}{3}})$, in order that the terms of low rank in the expansions may be small.

We shall write

$$\nu = z(1-\epsilon),$$

and it is convenient to suppose temporarily that

$$|\arg z| < \tfrac{1}{2}\pi.$$

We then have

$$(1) \qquad H_\nu^{(1)}(z) = \frac{1}{\pi i}\int_{-\infty}^{\infty+\pi i}\exp\{z(\sinh w - w) + z\epsilon w\}\,dw,$$

where the contour is that shewn in Fig. 18; on this contour $\sinh w - w$ is real and negative.

We write

$$\tau = w - \sinh w,$$

and the values of w corresponding to any positive value of τ will be called w_1 and w_2, of which w_1 is a complex number with a positive real part, and w_2 is a real negative number.

We then have

$$(2) \qquad H_\nu^{(1)}(z) = \frac{1}{\pi i}\int_0^\infty e^{-z\tau}\left\{\exp(z\epsilon w_1)\frac{dw_1}{d\tau} - \exp(z\epsilon w_2)\frac{dw_2}{d\tau}\right\}d\tau.$$

The expansion of τ in powers of w begins with a term in w^3, and hence we obtain expansions of the form

$$\begin{cases} \exp(z\epsilon w_1)\dfrac{dw_1}{d\tau} = \tau^{-\frac{2}{3}} \sum\limits_{m=0}^{\infty} b_m \tau^{\frac{1}{3}m}, \\ \exp(z\epsilon w_2)\dfrac{dw_2}{d\tau} = \tau^{-\frac{2}{3}} \sum\limits_{m=0}^{\infty} e^{\frac{2}{3}(m+1)\pi i} b_m \tau^{\frac{1}{3}m}, \end{cases}$$

and these are valid when $|\tau|$ is sufficiently small.

To determine the coefficients b_m we observe that

$$b_m = \frac{1}{6\pi i}\int^{(0+,0+,0+)} \exp(z\epsilon w_1) \cdot \left(\frac{dw_1}{d\tau}\right)\frac{d\tau}{\tau^{\frac{1}{3}(m+1)}}$$

$$= \frac{1}{6\pi i}\int^{(0+)} \exp(z\epsilon w)\frac{dw}{(w-\sinh w)^{\frac{1}{3}(m+1)}}.$$

As in the analogous investigation of § 8·4, a single circuit in the τ-plane is inadequate, and the triple circuit is necessary to dispose of the fractional powers of τ; a triple circuit round the origin in the τ-plane corresponds to a single circuit in the w-plane.

It follows that b_m is equal to $\frac{1}{3}e^{\frac{1}{3}(m+1)\pi i}$ multiplied by the coefficient of w^m in the expansion of

$$\exp(z\epsilon w) \cdot \{(\sinh w - w)/w^3\}^{-\frac{1}{3}(m+1)}.$$

The coefficients in this expansion will be called $b_0(m), b_1(m), b_2(m), \ldots$ so that

$$b_m = \tfrac{1}{3}e^{\frac{1}{3}(m+1)\pi i} b_m(m).$$

It is easy to shew that

$$(3)\quad \begin{cases} b_0(m) = 6^{\frac{1}{3}(m+1)}, \\ b_1(m) = 6^{\frac{1}{3}(m+1)}\epsilon z, \\ b_2(m) = 6^{\frac{1}{3}(m+1)}\left\{\dfrac{\epsilon^2 z^2}{2} - \dfrac{m+1}{60}\right\}, \\ b_3(m) = 6^{\frac{1}{3}(m+1)}\left\{\dfrac{\epsilon^3 z^3}{6} - \dfrac{(m+1)\epsilon z}{60}\right\}, \\ b_4(m) = 6^{\frac{1}{3}(m+1)}\left\{\dfrac{\epsilon^4 z^4}{24} - \dfrac{(m+1)\epsilon^2 z^2}{120} + \dfrac{(m+1)(7m+8)}{50400}\right\}. \\ \ldots\ldots\ldots\ldots\ldots\ldots\ldots\ldots\ldots\ldots \end{cases}$$

For brevity we write

$$b_m(m) = 6^{\frac{1}{3}(m+1)} B_m(\epsilon z),$$

so that*

$$(4)\quad \begin{cases} B_0(\epsilon z) = 1, & B_1(\epsilon z) = \epsilon z, \\ B_2(\epsilon z) = \tfrac{1}{2}\epsilon^2 z^2 - \tfrac{1}{20}, & B_3(\epsilon z) = \tfrac{1}{6}\epsilon^3 z^3 - \tfrac{1}{15}\epsilon z, \\ B_4(\epsilon z) = \tfrac{1}{24}\epsilon^4 z^4 - \tfrac{1}{24}\epsilon^2 z^2 + \tfrac{1}{280}, & \\ B_5(\epsilon z) = \tfrac{1}{120}\epsilon^5 z^5 - \tfrac{1}{60}\epsilon^3 z^3 + \tfrac{43}{8400}\epsilon z, & \\ \dots\dots\dots\dots & \end{cases}$$

$$[B_6(0) = -\tfrac{1}{3600},\quad B_8(0) = \tfrac{387}{17248000},\quad B_{10}(0) = -\tfrac{1213}{655200000}.]$$

We then have

$$\begin{cases} \exp(z\epsilon w_1)\dfrac{dw_1}{d\tau} = \tfrac{1}{3}\tau^{-\frac{2}{3}} \sum\limits_{m=0}^{\infty} e^{\frac{1}{3}(m+1)\pi i}\, 6^{\frac{1}{3}(m+1)}\, B_m(\epsilon z)\, \tau^{\frac{1}{3}m}, \\ \exp(z\epsilon w_2)\dfrac{dw_2}{d\tau} = \tfrac{1}{3}\tau^{-\frac{2}{3}} \sum\limits_{m=0}^{\infty} e^{(m+1)\pi i}\, 6^{\frac{1}{3}(m+1)}\, B_m(\epsilon z)\, \tau^{\frac{1}{3}m}, \end{cases}$$

and $[\exp(z\epsilon w).(dw/d\tau)]$ satisfies the conditions of the lemma of § 8·3.

It follows from the lemma of § 8·3 that

$$(5)\qquad H_\nu^{(1)}(z) \sim -\frac{2}{3\pi}\sum_{m=0}^{\infty} e^{\frac{2}{3}(m+1)\pi i}\, B_m(\epsilon z)\sin\tfrac{1}{3}(m+1)\pi\,.\,\frac{\Gamma(\frac{1}{3}m+\frac{1}{3})}{(\frac{1}{6}z)^{\frac{1}{3}(m+1)}},$$

and similarly

$$(6)\qquad H_\nu^{(2)}(z) \sim -\frac{2}{3\pi}\sum_{m=0}^{\infty} e^{-\frac{2}{3}(m+1)\pi i}\, B_m(\epsilon z)\sin\tfrac{1}{3}(m+1)\pi\,.\,\frac{\Gamma(\frac{1}{3}m+\frac{1}{3})}{(\frac{1}{6}z)^{\frac{1}{3}(m+1)}}.$$

We deduce at once that

$$(7)\qquad J_\nu(z) \sim \frac{1}{3\pi}\sum_{m=0}^{\infty} B_m(\epsilon z)\sin\tfrac{1}{3}(m+1)\pi\,.\,\frac{\Gamma(\frac{1}{3}m+\frac{1}{3})}{(\frac{1}{6}z)^{\frac{1}{3}(m+1)}},$$

$$(8)\qquad Y_\nu(z) \sim -\frac{2}{3\pi}\sum_{m=0}^{\infty} (-)^m B_m(\epsilon z)\sin^2\tfrac{1}{3}(m+1)\pi\,.\,\frac{\Gamma(\frac{1}{3}m+\frac{1}{3})}{(\frac{1}{6}z)^{\frac{1}{3}(m+1)}}.$$

From the Cauchy-Meissel formula § 8·21 (2), it is to be inferred that, when m is large,

$$(9)\qquad B_{2m}(0) \sim \frac{(-)^m(\frac{2}{3})^{\frac{2}{3}}}{\Gamma(\frac{2}{3}).(m+\frac{1}{3})^{\frac{1}{3}}(12\pi)^{\frac{2}{3}m}},$$

but there seems to be no very simple approximate formula for $B_m(\epsilon z)$.

The dominant terms in (7) were obtained by Meissel, in a *Kiel Programm* †, 1892; and some similar results, which seem to resemble those stated in § 8·43, were obtained by Koppe in a *Berlin Programm* ‡, 1899. The dominant terms in (8) as well as in (7) were also investigated by Nicholson, *Phil. Mag.* (6) XVI. (1908), pp. 271—279, shortly before the appearance of Debye's memoir.

* The values of $B_0(0)$, $B_2(0)$, ... $B_{10}(0)$ were given by Meissel, *Astr. Nach.* CXXVII. (1891), col. 359—362; apart from the use of the contours Meissel's analysis (cf. § 8·21) is substantially the same as the analysis given in this section. The object of using the methods of contour integration is to evade the difficulties produced by using generalised integrals.

The values of $B_6(\epsilon z)$, $B_7(\epsilon z)$ and $B_9(\epsilon z)$ will be found in a paper by Airey, *Phil. Mag.* (6) XXXI. (1916), p. 524.

† See the *Jahrbuch über die Fortschritte der Math.* 1892, pp. 476—478.

‡ See the *Jahrbuch über die Fortschritte der Math* 1899, pp. 420, 421.

We next consider the extent to which the condition $|\arg z| < \frac{1}{2}\pi$, which has so far been imposed on formulae (5)—(8), is removable.

The singularities of the integrand in (2), *qua* function of τ, are the values of τ for which w_1 (or w_2) fails to be a monogenic function of τ, so that the singularities are the values of τ corresponding to those values of w for which

$$d\tau/dw = 0.$$

They are therefore the points

$$\tau = 2n\pi i,$$

where n assumes all integral values.

It is consequently permissible to swing the contour through any angle η less than a right angle (either positively or negatively), and we then obtain the analytic continuation of $H_\nu^{(1)}(z)$ or $H_\nu^{(2)}(z)$ over the range $-\frac{1}{2}\pi - \eta < \arg z < \frac{1}{2}\pi - \eta$. By giving η suitable values, we thus find that the expansions (5)—(8) are valid over the extended region

$$-\pi < \arg z < \pi.$$

If we confine our attention to real variables, we see that the solution of the problem is not quite complete; we have determined asymptotic expansions of $J_\nu(x)$ valid when x and ν are large and (i) $x/\nu < 1$, (ii) $x/\nu > 1$, (iii) $|x - \nu|$ not large compared with $x^{\frac{1}{3}}$. But there are transitional regions between (i) and (iii) and also between (ii) and (iii), and in these transitional regions x/ν is nearly equal to 1 while $|x - \nu|$ is large. In these transitional regions simple expansions (involving elementary functions only in each term) do not exist. But important approximate formulae have been discovered by Nicholson, which involve Bessel functions of orders $\pm\frac{1}{3}$. Formulae of this type will now be investigated.

8·43. *Approximate formulae valid in the transitional regions.*

The failure of the formulae of §§ 8·4—8·42 in the transitional regions led Nicholson* to investigate second approximations to Bessel's integral in the following manner:

In the case of functions of integral order n,

$$J_n(x) = \frac{1}{\pi}\int_0^\pi \cos(n\theta - x\sin\theta)\,d\theta,$$

and, when x and n are nearly equal (both being large), it follows from Kelvin's principle of stationary phase (§ 8·2) that the important part of the path of integration is the part on which θ is small; now, on this part of the path, $\sin\theta$ is approximately equal to $\theta - \frac{1}{6}\theta^3$. It is inferred that, for the values of x and n under consideration,

$$J_n(x) \sim \frac{1}{\pi}\int_0^\pi \cos(n\theta - x\theta + \tfrac{1}{6}x\theta^3)\,d\theta$$

$$\sim \frac{1}{\pi}\int_0^\infty \cos(n\theta - x\theta + \tfrac{1}{6}x\theta^3)\,d\theta,$$

* *Phil. Mag.* (6) XIX. (1910), pp. 247—249; see also Emde, *Archiv der Math. und Phys.* (3) XXIV. (1916), pp. 239—250.

and the last expression is one of Airy's integrals (§ 6·4). It follows that, when $x < n$,

(1) $$J_n(x) \sim \frac{1}{\pi}\left\{\frac{2(n-x)}{3x}\right\}^{\frac{1}{2}} K_{\frac{1}{3}}\left(\frac{2^{\frac{3}{2}}(n-x)^{\frac{3}{2}}}{3x^{\frac{1}{2}}}\right),$$

and, when $x > n$,

(2) $$J_n(x) \sim \frac{1}{3}\left\{\frac{2(x-n)}{x}\right\}^{\frac{1}{2}} \{J_{-\frac{1}{3}} + J_{\frac{1}{3}}\},$$

where the arguments of the Bessel functions on the right are $\frac{1}{3}\{2(x-n)\}^{\frac{3}{2}}/x^{\frac{1}{2}}$.

The corresponding formula for $Y_n(x)$ when $x > n$ was also found by Nicholson; with the notation employed in this work it is

(3) $$Y_n(x) \sim -\left\{\frac{2(x-n)}{3x}\right\}^{\frac{1}{2}} \{J_{-\frac{1}{3}} - J_{\frac{1}{3}}\}.$$

The chief disadvantage of these formulae is that it seems impossible to determine, by rigorous methods, their domains of validity and the order of magnitude of the errors introduced in using them.

With a view to remedying this defect, Watson* examined Debye's integrals, and discovered a method which is theoretically simple (though actually it is very laborious), by means of which formulae analogous to Nicholson's are obtained together with an upper limit for the errors involved.

The method employed is the following:

Debye's integral for a Bessel function whose order ν exceeds its argument $x\,(\equiv \nu \operatorname{sech} \alpha)$ may be written in the form†

$$J_\nu(\nu \operatorname{sech} \alpha) = \frac{e^{\nu(\tanh \alpha - \alpha)}}{2\pi i}\int_{\infty - \pi i}^{\infty + \pi i} e^{-x\tau}\, dw,$$

where $$\tau = -\sinh \alpha (\cosh w - 1) - \cosh \alpha (\sinh w - w),$$
the contour being chosen so that τ is positive on it.

If τ is expanded in ascending powers of w, Carlini's formula is obtained when we approximate by neglecting all powers of w save the lowest, $-\frac{1}{2}w^2 \sinh \alpha$; and when $\alpha = 0$, Cauchy's formula of § 8 2 (1) is similarly obtained by neglecting all powers of w save the lowest, $-\frac{1}{6}w^3$.

These considerations suggest that it is desirable to examine whether the first two terms, namely

$$-\tfrac{1}{2}w^2 \sinh \alpha - \tfrac{1}{6}w^3 \cosh \alpha,$$

may not give an approximation valid throughout the first transitional region.

The integral which we shall investigate is therefore

$$\int e^{-x\tau}\, dW,$$

where $$\tau = -\tfrac{1}{2}W^2 \sinh \alpha - \tfrac{1}{6}W^3 \cosh \alpha,$$

* *Proc. Camb. Phil. Soc.* XIX. (1918), pp. 96—110.

† This is deducible from § 8·31 by making a change of origin in the w-plane.

and the contour in the plane of the complex variable W is so chosen that τ is positive on it. If $W \equiv U + iV$, this contour is the right-hand branch of the hyperbola

$$U \tanh \alpha + \tfrac{1}{2}U^2 = \tfrac{1}{6}V^2,$$

and this curve has contact of the third order with Debye's contour at the origin.

It therefore has to be shewn that an approximation to

$$\int_{\infty - \pi i}^{\infty + \pi i} e^{-x\tau}\, dw \quad \text{is} \quad \int_{\infty \exp(-\frac{1}{3}\pi i)}^{\infty \exp(\frac{1}{3}\pi i)} e^{-x\tau}\, dW.$$

These integrals differ by

$$\left\{\int_\infty^0 + \int_0^\infty\right\} e^{-x\tau} \left\{\frac{dw}{d\tau} - \frac{dW}{d\tau}\right\} d\tau,$$

and so the problem is reduced to the determination of an upper bound for $|\{d(w-W)/d\tau\}|$. And it has been proved, by exceedingly heavy analysis which will not be reproduced here, that

$$\left|\frac{d(w-W)}{d\tau}\right| < 3\pi \operatorname{sech} \alpha,$$

and so

$$\left|\left\{\int_\infty^0 + \int_0^\infty\right\} e^{-x\tau} \left\{\frac{dw}{d\tau} - \frac{dW}{d\tau}\right\} d\tau\right| < \frac{6\pi}{\nu}.$$

Hence

$$\frac{1}{2\pi i}\int_{\infty - \pi i}^{\infty + \pi i} e^{-x\tau}\, dw = \frac{1}{2\pi i}\int_{\infty \exp(-\frac{1}{3}\pi i)}^{\infty \exp(\frac{1}{3}\pi i)} e^{-x\tau}\, dW + \frac{3\theta_1}{\nu},$$

where $|\theta_1| < 1$.

To evaluate the integral on the right (which is of the type discussed in § 6·4), modify the contour into two lines starting from the point at which $W = -\tanh \alpha$ and making angles $\pm \frac{1}{3}\pi$ with the real axis.

If we write $W = -\tanh \alpha + \xi e^{\pm \frac{1}{3}\pi i}$ on the respective rays, the integral becomes

$$e^{\frac{1}{3}\pi i} \exp(\tfrac{1}{3}\nu \tanh^3 \alpha) \int_0^\infty \exp\{-\tfrac{1}{6}\nu\xi^3 - \tfrac{1}{2}\nu\xi e^{\frac{1}{3}\pi i} \tanh^2 \alpha\}\, d\xi$$

$$- e^{-\frac{1}{3}\pi i} \exp(\tfrac{1}{3}\nu \tanh^3 \alpha) \int_0^\infty \exp\{-\tfrac{1}{6}\nu\xi^3 - \tfrac{1}{2}\nu\xi e^{-\frac{1}{3}\pi i} \tanh^2 \alpha\}\, d\xi.$$

Expand the integrands in powers of $\tanh^2 \alpha$ and integrate term-by-term—a procedure which is easily justified—and we get on reduction

$$\tfrac{2}{3}\pi i \tanh \alpha \exp(\tfrac{1}{3}\nu \tanh^3 \alpha) . [I_{-\frac{1}{3}}(\tfrac{1}{3}\nu \tanh^3 \alpha) - I_{\frac{1}{3}}(\tfrac{1}{3}\nu \tanh^3 \alpha)],$$

and hence we obtain the formula

$$(4) \quad J_\nu(\nu \operatorname{sech} \alpha) = \frac{\tanh \alpha}{\pi\sqrt{3}} \exp\{\nu(\tanh \alpha + \tfrac{1}{3}\tanh^3 \alpha - \alpha)\} K_{\frac{1}{3}}(\tfrac{1}{3}\nu \tanh^3 \alpha) + 3\theta_1 \nu^{-1} \exp\{\nu(\tanh \alpha - \alpha)\},$$

where $|\theta_1| < 1$. This is the more precise form of Nicholson's approximation (1).

It can be shewn that, whether $\frac{1}{3}\nu \tanh^3 \alpha$ be small, of a moderate size, or large, the error is of a smaller order of magnitude (when ν is large) than the approximation given by the first term on the right.

Next we take the case in which the order ν is less than the argument $x\,(\equiv \nu \sec \beta)$.

We then have

$$H_\nu^{(1)}(\nu \sec \beta) = \frac{e^{\nu i(\tan\beta - \beta)}}{\pi i} \int_{-\infty - i\beta}^{\infty + i(\pi - \beta)} e^{-x\tau}\, dw,$$

where $$\tau = -i \sin \beta (\cosh w - 1) - \cos \beta (\sinh w - w),$$

the contour being so chosen that τ is positive on it.

The process of reasoning already employed leads us to consider the integral

$$\int e^{-x\tau}\, dW,$$

where $$\tau = -\tfrac{1}{2} i W^2 \sin \beta - \tfrac{1}{6} W^3 \cos \beta,$$

and the contour in the plane of the complex variable W is such that τ is positive on it. If $W \equiv U + iV$, this contour is the branch of the cubic

$$(U^2 - V^2) \tan \beta + \tfrac{1}{3} V (3U^2 - V^2) = 0$$

which passes from $-\infty - i \tan \beta$ through the origin to $\infty \exp \frac{1}{3}\pi i$.

It therefore has to be shewn that an approximation to

$$\int_{-\infty - i\beta}^{\infty + i(\pi-\beta)} e^{-x\tau}\, dw \quad \text{is} \quad \int_{-\infty - i\tan\beta}^{\infty \exp\frac{1}{3}\pi i} e^{-x\tau}\, dW.$$

The difference of these integrals is

$$\left\{ \int_\infty^0 + \int_0^\infty \right\} e^{-x\tau} \left\{ \frac{dw}{d\tau} - \frac{dW}{d\tau} \right\} d\tau,$$

and it has been proved that, when* $\beta \leqslant \frac{1}{4}\pi$, then

$$\left| \frac{d(w - W)}{d\tau} \right| < 12\pi \sec \beta.$$

Hence it follows that

$$\frac{1}{\pi i} \int_{-\infty - i\beta}^{\infty + i(\pi-\beta)} e^{-x\tau}\, dw = \frac{1}{\pi i} \int_{-\infty - i\tan\beta}^{\infty \exp\frac{1}{3}\pi i} e^{-x\tau}\, dW + \frac{24\theta'}{\nu},$$

where $|\theta'| < 1$.

To evaluate the integral on the right, modify the contour into two lines meeting at $W = -i \tan \beta$ and inclined at angles $\frac{1}{3}\pi$ and π respectively to the real axis. On these lines, write

$$W = -i \tan \beta - \xi, \quad -i \tan \beta + \xi e^{\frac{1}{3}\pi i},$$

* The important values of β are, of course, *small* values. If β is not small, Debye's formulae of § 8·41 yield effective approximations. The geometrical property of Debye's contour which was proved in § 8·32 is used in the proof of the theorem quoted.

expand the integrands in powers of $\tan^2\beta$, integrate term-by-term, and it is found that

$$\int_{-\infty - i\tan\beta}^{\infty \exp\frac{1}{3}\pi i} e^{-x\tau}\, dW = \tfrac{2}{3}\pi i \tan\beta \exp\left(-\tfrac{1}{3}\nu i \tan^3\beta\right) \times \left[e^{-\frac{1}{3}\pi i} J_{-\frac{1}{3}}\left(\tfrac{1}{3}\nu\tan^3\beta\right) + e^{\frac{1}{3}\pi i} J_{\frac{1}{3}}\left(\tfrac{1}{3}\nu\tan^3\beta\right)\right]$$

$$= \frac{\pi e^{\frac{2}{3}\pi i}\tan\beta}{\sqrt{3}} \exp\left(-\tfrac{1}{3}\nu i\tan^3\beta\right) H_{\frac{1}{3}}^{(1)}\left(\tfrac{1}{3}\nu\tan^3\beta\right).$$

On equating real and imaginary parts, it is at once found that

$$(5)\quad J_\nu(\nu\sec\beta) = \tfrac{1}{3}\tan\beta\cos\{\nu(\tan\beta - \tfrac{1}{3}\tan^3\beta - \beta)\}.[J_{-\frac{1}{3}} + J_{\frac{1}{3}}] + 3^{-\frac{1}{2}}\tan\beta\sin\{\nu(\tan\beta - \tfrac{1}{3}\tan^3\beta - \beta)\}.[J_{-\frac{1}{3}} - J_{\frac{1}{3}}] + 24\theta_2/\nu,$$

$$(6)\quad Y_\nu(\nu\sec\beta) = \tfrac{1}{3}\tan\beta\sin\{\nu(\tan\beta - \tfrac{1}{3}\tan^3\beta - \beta)\}.[J_{-\frac{1}{3}} + J_{\frac{1}{3}}] - 3^{-\frac{1}{2}}\tan\beta\cos\{\nu(\tan\beta - \tfrac{1}{3}\tan^3\beta - \beta)\}.[J_{-\frac{1}{3}} - J_{\frac{1}{3}}] + 24\theta_3/\nu,$$

where the argument of each of the Bessel functions $J_{\pm\frac{1}{3}}$ on the right is $\frac{1}{3}\nu\tan^3\beta$; and $|\theta_2|$ and $|\theta_3|$ are both less than 1. These are the more precise forms of Nicholson's formulae (2) and (3); and they give effective approximations except near the zeros of the dominant terms on the right.

It is highly probable that the upper limits obtained for the errors are largely in excess of the actual values of the errors.

8·5. *Descriptive properties* of $J_\nu(\nu x)$ when $0 < x \leq 1$.*

The contour integral, which was obtained in § 8·31 (I) to represent $J_\nu(\nu\,\text{sech}\,\alpha)$ was shewn in § 8·4 to yield an asymptotic expansion of the function. But the contour integral is really of much greater importance than has hitherto appeared; for an integral is an exact representation of a function, whereas an asymptotic expansion can only give, at best, an approximate representation. And the contour integral (together with the limiting form of it when $x = 1$) is peculiarly well adapted for giving interesting information concerning $J_\nu(\nu x)$ when ν is positive.

In the contour integral take ν to be positive and write

$$w = \log\{re^{i\theta}\},$$

so that $u = \log r,\ v = \theta$.

With the contour selected,

$$x\sinh w - w$$

is equal to its conjugate complex, and the path of integration is its own reflexion in the real axis. Hence

$$J_\nu(\nu x) = \frac{1}{2\pi i}\int_{\infty-\pi i}^{\infty+\pi i} e^{\nu(x\sinh w - w)}\, dw$$

$$= \frac{1}{\pi}\int_0^\pi e^{\nu(x\sinh w - w)}\, dv.$$

* The results of this section are investigated in rather greater detail in *Proc. London Math. Soc.* (2) XVI. (1917), pp. 150—174.

Changing the notation, we find that the equation of the contour is

$$r + \frac{1}{r} = \frac{2\theta}{x \sin\theta},$$

so that

$$r = \frac{\theta}{x \sin\theta}\left\{1 + \sqrt{\left(1 - \frac{x^2 \sin^2\theta}{\theta^2}\right)}\right\},$$

and, when this substitution is made for r, the value of $(w - x \sinh w)$ is

$$\log \frac{\theta + \sqrt{(\theta^2 - x^2 \sin^2\theta)}}{x \sin\theta} - \cot\theta \,.\, \sqrt{(\theta^2 - x^2 \sin^2\theta)}.$$

This last expression will invariably be denoted by the symbol* $F(\theta, x)$, so that

(1) $$J_\nu(\nu x) = \frac{1}{\pi}\int_0^\pi e^{-\nu F(\theta, x)}\, d\theta,$$

and by differentiating under the integral sign (a procedure which is easily justified) it is found that

(2) $$J_\nu'(\nu x) = \frac{1}{\pi}\int_0^\pi e^{-\nu F(\theta, x)} \frac{\theta - x^2 \sin\theta\cos\theta}{x\sqrt{(\theta^2 - x^2\sin^2\theta)}}\, d\theta.$$

This is also easily deduced from the equation

$$J_\nu'(\nu x) = \frac{1}{2\pi i}\int_{\infty - \pi i}^{\infty + \pi i} e^{\nu(x \sinh w - w)} \sinh w\, dw.$$

Before proceeding to obtain further results concerning Bessel functions, it is convenient to set on record various properties† of $F(\theta, x)$. The reader will easily verify that

(3) $$\frac{\partial}{\partial\theta} F(\theta, x) = \frac{(1 - \theta\cot\theta)^2}{\sqrt{(\theta^2 - x^2\sin^2\theta)}} + \sqrt{(\theta^2 - x^2\sin^2\theta)} \geqslant 0,$$

(4) $$\frac{\partial}{\partial x} F(0, x) = -\frac{\sqrt{(1 - x^2)}}{x} \leqslant 0,$$

so that

(5) $$F(\theta, x) \geqslant F(0, x) \geqslant F(0, 1) = 0;$$

and also

(6) $$\frac{\partial}{\partial x} F(\theta, x) = -\frac{\theta - x^2\sin\theta\cos\theta}{x\sqrt{(\theta^2 - x^2\sin^2\theta)}} \leqslant 0.$$

Next we shall establish the more abstruse property

(7) $$F(\theta, x) \geqslant F(0, x) + \tfrac{1}{2}(\theta^2 - x^2\sin^2\theta)/\sqrt{(1 + x^2)}.$$

To prove it, we shall first shew that

$$g(\theta, x) \equiv \frac{\theta - x^2\sin\theta\cos\theta}{\sqrt{(\theta^2 - x^2\sin^2\theta)}} \leqslant \sqrt{(1 + x^2)}.$$

* This function will not be confused with Schläfli's function defined in § 4·15.

† It is supposed throughout the following analysis that $0 < x \leqslant 1$, $0 \leqslant \theta \leqslant \pi$.

It is clear that

$$\begin{cases} g(0, x) = \sqrt{(1 - x^2)} < \sqrt{(1 + x^2)}, \\ g(\pi, x) = 1 \qquad\qquad < \sqrt{(1 + x^2)}, \end{cases}$$

so that, if $g(\theta, x)$, *qua* function of θ, attained its greatest value at 0 or π, that value would be less than $\sqrt{(1 + x^2)}$. If, however, $g(\theta, x)$ attained its greatest value when θ had a value θ_0 between 0 and π, then

$$\frac{1 - x^2 \cos 2\theta_0}{(\theta_0^2 - x^2 \sin^2 \theta_0)^{\frac{1}{2}}} - \frac{(\theta_0 - x^2 \sin \theta_0 \cos \theta_0)^2}{(\theta_0^2 - x^2 \sin^2 \theta_0)^{\frac{3}{2}}} = 0,$$

and therefore

$$g(\theta, x) \leqslant g(\theta_0, x) = \sqrt{(1 - x^2 \cos 2\theta_0)} \leqslant \sqrt{(1 + x^2)},$$

so that, no matter where $g(\theta, x)$ attains its greatest value, that value does not exceed $\sqrt{(1 + x^2)}$.

Hence

$$\frac{\partial F(\theta, x)}{\partial \theta} \geqslant \sqrt{(\theta^2 - x^2 \sin^2 \theta)} \geqslant \frac{\theta - x^2 \sin \theta \cos \theta}{\sqrt{(1 + x^2)}}$$

and so

$$F(\theta, x) - F(0, x) = \int_0^\theta \frac{\partial F(\theta, x)}{\partial \theta} d\theta \geqslant \int_0^\theta \frac{\theta - x^2 \sin \theta \cos \theta}{\sqrt{(1 + x^2)}} d\theta,$$

whence (7) follows at once.

Another, but simpler, inequality of the same type is

(8) $$F(\theta, x) \geqslant F(0, x) + \tfrac{1}{2} \theta^2 \sqrt{(1 - x^2)}.$$

To prove this, observe that

$$\frac{\partial F(\theta, x)}{\partial \theta} \geqslant \sqrt{(\theta^2 - x^2 \sin^2 \theta)} \geqslant \theta \sqrt{(1 - x^2)},$$

and integrate; then the inequality is obvious.

From these results we are now in a position to obtain theorems concerning $J_\nu(\nu x)$ and $J_\nu'(\nu x)$ *qua* functions of ν.

Thus, since

$$\frac{\partial J_\nu(\nu x)}{\partial \nu} = -\frac{1}{\pi} \int_0^\pi F(\theta, x) e^{-\nu F(\theta, x)} d\theta \leqslant 0,$$

the integrand being positive by (5), it follows that *$J_\nu(\nu x)$ is a positive decreasing function of ν*; in like manner, *$J_\nu'(\nu x)$ is a positive decreasing function of ν.*

Also, since

$$\frac{\partial \{e^{\nu F(0, x)} J_\nu(\nu x)\}}{\partial \nu} = -\frac{1}{\pi} \int_0^\pi \{F(\theta, x) - F(0, x)\} e^{\nu F(0, x) - \nu F(\theta, x)} d\theta \leqslant 0,$$

the integrand being positive by (5), it follows that *$e^{\nu F(0, x)} J_\nu(\nu x)$ is a decreasing function of ν; and so also, similarly, is $e^{\nu F(0, x)} J_\nu'(\nu x)$.*

Again, from (8) we have

$$J_\nu(\nu x) \leqslant \frac{e^{-\nu F(0, x)}}{\pi} \int_0^\pi \exp\{-\tfrac{1}{2}\nu\theta^2 \sqrt{(1-x^2)}\}\, d\theta$$

$$< \frac{e^{-\nu F(0, x)}}{\pi} \int_0^\infty \exp\{-\tfrac{1}{2}\nu\theta^2 \sqrt{(1-x^2)}\}\, d\theta,$$

so that

(9) $$J_\nu(\nu x) \leqslant \frac{e^{-\nu F(0, x)}}{(1-x^2)^{\frac{1}{4}} \sqrt{(2\pi\nu)}}.$$

The last expression is easily reduced to Carlini's approximate expression (§§ 1·4, 8·11) for $J_\nu(\nu x)$; and so Carlini's expression is always in error *by excess*, for all* positive values of ν.

The corresponding result for $J_\nu'(\nu x)$ is derived from (7). Write

$$\theta^2 - x^2 \sin^2\theta \equiv G(\theta, x),$$

and replace $G(\theta, x)$ by G for brevity.

Then

$$2xJ_\nu'(\nu x) = \frac{1}{\pi}\int_0^\pi e^{-\nu F(\theta, x)} \frac{dG(\theta, x)}{d\theta} \{G(\theta, x)\}^{-\frac{1}{2}}\, d\theta$$

$$\leqslant \frac{e^{-\nu F(0, x)}}{\pi} \int_0^{\pi^2} \exp\{-\tfrac{1}{2}\nu G/\sqrt{(1+x^2)}\} \,.\, G^{-\frac{1}{2}}\, dG$$

$$\leqslant \frac{e^{-\nu F(0, x)}}{\pi} \int_0^\infty \exp\{-\tfrac{1}{2}\nu G/\sqrt{(1+x^2)}\} \,.\, G^{-\frac{1}{2}}\, dG,$$

and so

(10) $$xJ_\nu'(\nu x) \leqslant e^{-\nu F(0, x)} (1+x^2)^{\frac{1}{4}}/\sqrt{(2\pi\nu)}.$$

The absence of the factor $\sqrt[4]{(1-x^2)}$ from the denominator is remarkable.

It is possible to prove the formula†

(11) $$\int_0^x J_\nu(\nu t)\, dt \sim \frac{x^{\nu+1}\exp\{\nu\sqrt{(1-x^2)}\}}{(2\pi\nu^3)^{\frac{1}{2}}(1-x^2)^{\frac{3}{4}}\{1+\sqrt{(1-x^2)}\}^\nu}$$

in a very similar manner.

This concludes the results which we shall establish concerning a single Bessel function whose argument is less than its order.

8·51. *Lemma concerning $F(\theta, x)$.*

We shall now prove the lemma that, when $0 \leqslant x \leqslant 1$ and $0 \leqslant \theta \leqslant \pi$, then

(1) $$\frac{dF(\theta, x)}{d\theta} - \{F(\theta, x) - F(0, x)\}\frac{\theta - x^2\sin\theta\cos\theta}{\theta^2 - x^2\sin^2\theta} \geqslant 0.$$

The lemma will be used immediately to prove an important theorem concerning the rate of increase of $J_\nu(\nu x)$.

* It is evident from Debye's expansion that the expression is in error by excess for *sufficiently large* values of ν.

† Cf. *Proc. London Math. Soc.* (2) XVI. (1917), p. 157.

If $\sqrt{(\theta^2 - x^2 \sin^2 \theta)} \equiv H(\theta, x)$, we shall first prove that

$$\frac{dF(\theta, x)}{d\theta} \Big/ \frac{dH(\theta, x)}{d\theta}$$

is a non-decreasing function of θ; that is to say that

$$\frac{(1 - \theta \cot \theta)^2 + \theta^2 - x^2 \sin^2 \theta}{\theta - x^2 \sin \theta \cos \theta}$$

is a non-decreasing function of θ.

The differential coefficient of this last function of θ is

$$(\theta - x^2 \sin \theta \cos \theta)^{-2} [(\theta^2 \operatorname{cosec}^2 \theta - 1 - \tfrac{1}{3} \sin^2 \theta)(1 - x^2) \\ + 2(\theta^2 \operatorname{cosec}^2 \theta - \theta^3 \cot \theta \operatorname{cosec}^2 \theta - \tfrac{1}{3} \sin^2 \theta)(1 - x^2) \\ + 2x^2 (1 - \theta \cot \theta)(\theta \operatorname{cosec} \theta - \cos \theta)^2 + \sin^2 \theta (1 - x^2)^2],$$

and every group of terms in this expression is positive (or zero) in consequence of elementary trigonometrical inequalities.

To establish the trigonometrical inequalities, we first observe that, when $0 \leqslant \theta \leqslant \pi$,

$$\text{(i)} \quad \theta + \sin \theta \cos \theta - 2\theta^{-1} \sin^2 \theta \geqslant 0,$$
$$\text{(ii)} \quad \theta + \sin \theta \cos \theta - 2\theta^2 \cot \theta \geqslant 0,$$
$$\text{(iii)} \quad \sin \theta - \theta \cos \theta - \tfrac{1}{3} \sin^3 \theta \geqslant 0,$$

because the expressions on the left vanish when $\theta = 0$ and have the positive differential coefficients

$$\text{(i)} \ 2(\cos \theta - \theta^{-1} \sin \theta)^2, \quad \text{(ii)} \ 2(\cos \theta - \theta \operatorname{cosec} \theta)^2, \quad \text{(iii)} \ \sin \theta (\theta - \sin \theta \cos \theta),$$

and then

$$\theta^2 \operatorname{cosec}^2 \theta - \theta^3 \cot \theta \operatorname{cosec}^2 \theta - \tfrac{1}{3} \sin^2 \theta \\ = (\theta^2 \operatorname{cosec}^2 \theta - 1)(1 - \theta \cot \theta) + \operatorname{cosec} \theta (\sin \theta - \theta \cos \theta - \tfrac{1}{3} \sin^3 \theta) \geqslant 0,$$

$$\theta^2 \operatorname{cosec}^2 \theta - 1 - \tfrac{1}{3} \sin^2 \theta \\ = \theta \operatorname{cosec}^2 \theta (\theta + \sin \theta \cos \theta - 2\theta^{-1} \sin^2 \theta) + \operatorname{cosec} \theta (\sin \theta - \theta \cos \theta - \tfrac{1}{3} \sin^3 \theta) \geqslant 0,$$

so that the inequalities are proved.

It has consequently been shewn that

$$\frac{d}{d\theta} \left\{ \frac{F'}{H'} \right\} \geqslant 0,$$

where the variables are understood to be θ and x, and primes denote differentiations with regard to θ. It is now obvious that

$$\frac{d}{d\theta} \left\{ \frac{F'H}{H'} - F \right\} \equiv H \frac{d}{d\theta} \left\{ \frac{F'}{H'} \right\} \geqslant 0,$$

and, if we integrate this inequality from 0 to θ, we get

$$\left[\frac{F'H}{H'} - F \right]_0^\theta \geqslant 0.$$

Since F' and H/H' vanish when $\theta = 0$, this inequality is equivalent to

$$\frac{F'(\theta, x) H(\theta, x)}{H'(\theta, x)} - F(\theta, x) + F(0, x) \geqslant 0,$$

and the truth of the lemma becomes obvious when we substitute the value of $H(\theta, x)$ in the last inequality.

8·52. *The monotonic property of $J_\nu(\nu x)/J_\nu(\nu)$.*

We shall now prove a theorem of some importance, to the effect that, if x is fixed, and $0 \leqslant x \leqslant 1$, *then $J_\nu(\nu x)/J_\nu(\nu)$ is a non-increasing function of* ν, when ν is positive.

[The actual proof of the theorem will be valid only when $\delta \leqslant x \leqslant 1$, (where δ is an arbitrarily small positive number), since some expressions introduced in the proof contain an x in their denominators; but the theorem is obvious when $0 \leqslant x \leqslant \delta$ since $e^{\nu F(0,x)} J_\nu(\nu x)$ and $e^{-\nu F(0,x)}/J_\nu(\nu)$ are non-increasing functions of ν when x is sufficiently small; moreover, as will be seen in Chapter XVII, the theorem owes its real importance to the fact that it is true for values of x *in the neighbourhood of unity*.]

It will first be shewn that

$$(1) \qquad J_\nu(\nu x)\frac{\partial^2 J_\nu(\nu x)}{\partial\nu\,\partial x} - \frac{\partial J_\nu(\nu x)}{\partial x}\frac{\partial J_\nu(\nu x)}{\partial\nu} \geqslant 0.$$

To establish this result, we observe that, with the usual notation,

$$J_\nu(\nu x) = \frac{1}{\pi}\int_0^\pi e^{-\nu F(\psi, x)}\,d\psi,$$

$$\frac{\partial J_\nu(\nu x)}{\partial x} = \frac{\nu}{2\pi x}\int_0^\pi \{G(\theta, x)\}^{-\frac{1}{2}}\frac{dG(\theta, x)}{d\theta}e^{-\nu F(\theta, x)}\,d\theta,$$

and, when we differentiate under the integral sign,

$$\frac{\partial J_\nu(\nu x)}{\partial\nu} = -\frac{1}{\pi}\int_0^\pi F(\psi, x)\,e^{-\nu F(\psi, x)}\,d\psi,$$

$$\begin{aligned}\frac{\partial^2 J_\nu(\nu x)}{\partial\nu\,\partial x} &= \frac{1}{2\pi x}\int_0^\pi \{G(\theta, x)\}^{-\frac{1}{2}}\frac{dG(\theta, x)}{d\theta}e^{-\nu F(\theta, x)}\,d\theta \\ &\quad - \frac{\nu}{2\pi x}\int_0^\pi F(\theta, x)\{G(\theta, x)\}^{-\frac{1}{2}}\frac{dG(\theta, x)}{d\theta}e^{-\nu F(\theta, x)}\,d\theta \\ &= \frac{\nu}{\pi x}\int_0^\pi\left[\{G(\theta, x)\}^{\frac{1}{2}}\frac{dF(\theta, x)}{d\theta} - \tfrac{1}{2}F(\theta, x)\{G(\theta, x)\}^{-\frac{1}{2}}\frac{dG(\theta, x)}{d\theta}\right] \\ &\qquad\qquad \times e^{-\nu F(\theta, x)}\,d\theta,\end{aligned}$$

if we integrate by parts the former of the two integrals.

Hence it follows that

$$J_\nu(\nu x)\frac{\partial^2 J_\nu(\nu x)}{\partial\nu\,\partial x} - \frac{\partial J_\nu(\nu x)}{\partial x}\frac{\partial J_\nu(\nu x)}{\partial\nu} = \frac{\nu}{2\pi^2 x}\int_0^\pi\int_0^\pi \Omega(\theta, \psi)\,e^{-\nu\{F(\theta, x)+F(\psi, x)\}}\,d\theta\,d\psi,$$

where

$$\begin{aligned}\Omega(\theta, \psi) &= 2\{G(\theta, x)\}^{\frac{1}{2}}\left[\frac{dF(\theta, x)}{d\theta} + \frac{F(\psi, x) - F(\theta, x)}{2G(\theta, x)}\frac{dG(\theta, x)}{d\theta}\right] \\ &\geqslant 2\{G(\theta, x)\}^{\frac{1}{2}}\left[\frac{dF(\theta, x)}{d\theta} + \frac{F(0, x) - F(\theta, x)}{2G(\theta, x)}\frac{dG(\theta, x)}{d\theta}\right] \\ &\geqslant 0,\end{aligned}$$

by using the inequality $F(\psi, x) \geqslant F(0, x)$ combined with the theorem of § 8·51.

Since $\Omega(\theta, \psi)$ is not negative, the repeated integral cannot be negative; that is to say, we have proved that

$$J_\nu(\nu x)\frac{\partial^2 J_\nu(\nu x)}{\partial\nu\partial x} - \frac{\partial J_\nu(\nu x)}{\partial x}\frac{\partial J_\nu(\nu x)}{\partial\nu} \geqslant 0,$$

so that

$$\frac{\partial}{\partial x}\left\{\frac{\partial J_\nu(\nu x)}{\partial\nu}\bigg/ J_\nu(\nu x)\right\} \geqslant 0.$$

Integrating this inequality between the limits x and 1, we get

$$\left[\frac{\partial J_\nu(\nu x)}{\partial\nu}\bigg/ J_\nu(\nu x)\right]_x^1 \geqslant 0,$$

so that

$$\frac{\partial J_\nu(\nu x)}{\partial\nu}\bigg/ J_\nu(\nu x) \leqslant \frac{\partial J_\nu(\nu)}{\partial\nu}\bigg/ J_\nu(\nu).$$

Since $J_\nu(\nu x)$ and $J_\nu(\nu)$ are both positive, this inequality may be written in the form

(2) $$\frac{\partial}{\partial\nu}\{J_\nu(\nu x)/J_\nu(\nu)\} \leqslant 0,$$

and this exhibits the result which was to be proved, namely that $J_\nu(\nu x)/J_\nu(\nu)$ is a non-increasing function of ν.

8·53. *Properties of $J_\nu(\nu)$ and $J_\nu'(\nu)$.*

If, for brevity, we write $F(\theta)$ in place of $F(\theta, 1)$, so that

(1) $$F(\theta) \equiv \log\frac{\theta + \sqrt{(\theta^2 - \sin^2\theta)}}{\sin\theta} - \cot\theta\sqrt{(\theta^2 - \sin^2\theta)},$$

the formulae* for $J_\nu(\nu)$ and $J_\nu'(\nu)$ are

(2) $$J_\nu(\nu) = \frac{1}{\pi}\int_0^\pi e^{-\nu F(\theta)}\,d\theta, \quad J_\nu'(\nu) = \frac{1}{\pi}\int_0^\pi \frac{\theta - \sin\theta\cos\theta}{\sqrt{(\theta^2 - \sin^2\theta)}}\,e^{-\nu F(\theta)}\,d\theta.$$

The first term in the expansion of $F(\theta)$ in ascending powers of θ is $4\theta^3/(9\sqrt{3})$; and we shall prove a series of inequalities leading up to the result that $F(\theta)/\theta^3$ is a non-decreasing function of θ.

We shall first shew that

$$\frac{d}{d\theta}\left\{\frac{F'(\theta)}{\theta^2}\right\} \geqslant 0.$$

To prove this we observe that

$$\frac{F'(\theta)}{\theta^2} = \frac{\{(1 - \theta\cot\theta)/\theta^2\}^2}{\theta^{-2}\sqrt{(\theta^2 - \sin^2\theta)}} + \theta^{-2}\sqrt{(\theta^2 - \sin^2\theta)},$$

* It is to be understood that $J_\nu'(\nu)$ means the value of $dJ_\nu(x)/dx$ when x has the particular value ν.

and that

$$\frac{d}{d\theta}\left\{\frac{1-\theta\cot\theta}{\theta^2}\right\}=\frac{\theta^2\operatorname{cosec}^2\theta+\theta\cot\theta-2}{\theta^3},$$

$$\frac{d}{d\theta}\left\{\frac{\sqrt{(\theta^2-\sin^2\theta)}}{\theta^2}\right\}=-\frac{(\theta^2\operatorname{cosec}^2\theta+\theta\cot\theta-2)\sin^2\theta}{\theta^3\sqrt{(\theta^2-\sin^2\theta)}}.$$

Hence it follows that

$$\frac{d}{d\theta}\left\{\frac{F'(\theta)}{\theta^2}\right\}=\frac{\theta-\sin\theta\cos\theta}{\theta^3(\theta^2-\sin^2\theta)^{\frac{3}{2}}}(\theta^2\operatorname{cosec}^2\theta+\theta\cot\theta-2)\times(\theta+\sin\theta\cos\theta-2\theta^2\cot\theta)$$

$$\geqslant 0,$$

by inequalities proved in § 8·51.

Consequently

(3) $$\theta F''(\theta)-2F'(\theta)\geqslant 0,$$

that is to say $$\frac{d}{d\theta}\{\theta F'(\theta)-3F(\theta)\}\geqslant 0.$$

If we integrate this inequality from 0 to θ we get

(4) $$\theta F'(\theta)-3F(\theta)\geqslant 0,$$

and this is the condition that $F(\theta)/\theta^3$ should be a non-decreasing function of θ. It follows that

$$\frac{F(\theta, 1)}{\theta^3}\geqslant\lim_{\theta\to 0}\frac{F(\theta)}{\theta^3}=\frac{4}{9\sqrt{3}},$$

and therefore

$$J_\nu(\nu)<\frac{1}{\pi}\int_0^\pi\exp\left\{-\frac{4\nu\theta^3}{9\sqrt{3}}\right\}d\theta$$

$$<\frac{1}{\pi}\int_0^\infty\exp\left\{-\frac{4\nu\theta^3}{9\sqrt{3}}\right\}d\theta$$

$$=\frac{\Gamma(\frac{1}{3})}{2^{\frac{2}{3}}3^{\frac{1}{6}}\pi\nu^{\frac{1}{3}}},$$

so that Cauchy's approximation for $J_\nu(\nu)$ is always in error *by excess.*

An inequality which will be required subsequently is

(5) $$2(\theta^2-\sin^2\theta)F''(\theta)-3(\theta-\sin\theta\cos\theta)F(\theta)\geqslant 0.$$

The truth of this may be seen by writing the expression on the left in the form

$$(\theta^2-2\sin^2\theta+\theta\sin\theta\cos\theta)F'(\theta)+(\theta-\sin\theta\cos\theta)\{\theta F'(\theta)-3F(\theta)\},$$

in which each group of terms is positive (cf. § 8·51).

[NOTE. A formula resembling those which have just been established is

(6) $$\int_0^1 J_\nu(\nu t)\,dt\sim\frac{1}{3\nu}-\frac{2^{\frac{2}{3}}}{3^{\frac{1}{3}}.5\Gamma(\frac{1}{3})\nu^{\frac{5}{3}}},$$

see *Phil. Mag.* (6) XXXV. (1918), pp. 364—370.]

8·54. *Monotonic properties of $J_\nu(\nu)$ and $J_\nu'(\nu)$.*

It has already been seen (§ 8·5) that the functions $J_\nu(\nu)$ and $J_\nu'(\nu)$ are decreasing functions of ν. It will now be shewn that *both* $\nu^{\frac{1}{3}} J_\nu(\nu)$ *and* $\nu^{\frac{2}{3}} J_\nu'(\nu)$ *are steadily increasing* functions of* ν.

To prove the first result we observe that

$$\frac{d\{\nu^{\frac{1}{3}} J_\nu(\nu)\}}{d\nu} = \frac{\nu^{-\frac{2}{3}}}{3\pi}\int_0^\pi e^{-\nu F(\theta)}\,d\theta - \frac{\nu^{\frac{1}{3}}}{\pi}\int_0^\pi F(\theta)\, e^{-\nu F(\theta)}\,d\theta$$

$$= \frac{\nu^{-\frac{2}{3}}}{3\pi}\Big[\theta e^{-\nu F(\theta)}\Big]_0^\pi + \frac{\nu^{\frac{1}{3}}}{3\pi}\int_0^\pi \{\theta F'(\theta) - 3F(\theta)\}\, e^{-\nu F(\theta)}\,d\theta$$

$$> 0,$$

since the integrated part vanishes at each limit and (§ 8·53) the integrand is positive.

Hence $\nu^{\frac{1}{3}} J_\nu(\nu)$ is an increasing function of ν; and therefore

(1) $$\nu^{\frac{1}{3}} J_\nu(\nu) < \lim_{\nu\to\infty}\{\nu^{\frac{1}{3}} J_\nu(\nu)\} = \Gamma(\tfrac{1}{3})/(2^{\frac{2}{3}} 3^{\frac{1}{6}}\pi) = 0{\cdot}44731.$$

In connexion with this result it may be noted that

$$J_1(1) = 0{\cdot}44005, \quad 2J_8(8) = 0{\cdot}44691.$$

To prove the second result, by following the same method we find that

$$\frac{d\{\nu^{\frac{2}{3}} J_\nu'(\nu)\}}{d\nu} = \frac{2\nu^{-\frac{1}{3}}}{3\pi}\Big[e^{-\nu F(\theta)}\sqrt{(\theta^2-\sin^2\theta)}\Big]_0^\pi$$

$$+ \frac{\nu^{\frac{2}{3}}}{3\pi}\int_0^\pi \left\{2F'(\theta)\sqrt{(\theta^2-\sin^2\theta)} - \frac{3(\theta-\sin\theta\cos\theta)}{\sqrt{(\theta^2-\sin^2\theta)}}F(\theta)\right\} e^{-\nu F(\theta)}\,d\theta$$

$$> 0,$$

by § 8·53 (5), and so $\nu^{\frac{2}{3}} J_\nu'(\nu)$ is an increasing function of ν.

Hence

(2) $$\nu^{\frac{2}{3}} J_\nu'(\nu) < \lim_{\nu\to\infty}\{\nu^{\frac{2}{3}} J_\nu'(\nu)\} = 3^{\frac{1}{6}}\,\Gamma(\tfrac{2}{3})/(2^{\frac{1}{3}}\pi) = 0{\cdot}41085.$$

It is to be noted that

$$J_1'(1) = 0{\cdot}32515, \quad 4J_8'(8) = 0{\cdot}38854.$$

8·55. *The monotonic property of* $\nu^{\frac{1}{3}} J_\nu'(\nu)/J_\nu(\nu)$.

A theorem which is slightly more recondite than the theorems just proved is that *the quotient*

$$\{\nu^{\frac{2}{3}} J_\nu'(\nu)\} \div \{\nu^{\frac{1}{3}} J_\nu(\nu)\}$$

is a steadily increasing function of ν.

* It is not possible to deduce these monotonic properties from the asymptotic expansions. If, as $\nu\to\infty$, $f(\nu)\sim\phi(\nu)$, and if $\phi(\nu)$ is monotonic, nothing can be inferred concerning monotonic properties of $f(\nu)$ in the absence of further information concerning $f(\nu)$.

To prove this result we use the integrals already mentioned in §§ 8·53, 8·54 for the four functions

$$\nu^{\frac{1}{3}} J_\nu(\nu), \quad \nu^{\frac{2}{3}} J_\nu'(\nu), \quad \frac{d\{\nu^{\frac{1}{3}} J_\nu(\nu)\}}{d\nu}, \quad \frac{d\{\nu^{\frac{2}{3}} J_\nu'(\nu)\}}{d\nu}.$$

Taking the parametric variable in the first and third integrals to be ψ in place of θ, we find that

$$\{\nu^{\frac{1}{3}} J_\nu(\nu)\}^2 \frac{d}{d\nu}\left\{\frac{\nu^{\frac{2}{3}} J_\nu'(\nu)}{\nu^{\frac{1}{3}} J_\nu(\nu)}\right\} = \frac{\nu}{\pi^2}\int_0^\pi\int_0^\pi \Omega_1(\theta, \psi)\, e^{-\nu\{F(\theta)+F(\psi)\}}\, d\theta\, d\psi,$$

where

$$\Omega_1(\theta, \psi) = \tfrac{2}{3} F'(\theta)\sqrt{(\theta^2 - \sin^2\theta)} - \frac{\theta - \sin\theta\cos\theta}{\sqrt{(\theta^2 - \sin^2\theta)}} F(\theta)$$

$$- \frac{\theta - \sin\theta\cos\theta}{\sqrt{(\theta^2 - \sin^2\theta)}}\{\tfrac{1}{3}\psi F'(\psi) - F(\psi)\}$$

$$= \frac{2\sqrt{(\theta^2 - \sin^2\theta)}}{\theta}\{\tfrac{1}{3}\theta F'(\theta) - F(\theta)\} - \frac{\theta - \sin\theta\cos\theta}{\sqrt{(\theta^2 - \sin^2\theta)}}\{\tfrac{1}{3}\psi F'(\psi) - F(\psi)\}$$

$$+ \frac{F(\theta)}{\theta\sqrt{(\theta^2 - \sin^2\theta)}}\{\theta^2 + \theta\sin\theta\cos\theta - 2\sin^2\theta\}$$

$$\geqslant \frac{2\sqrt{(\theta^2 - \sin^2\theta)}}{\theta}\{\tfrac{1}{3}\theta F'(\theta) - F(\theta)\} - \frac{\theta - \sin\theta\cos\theta}{\sqrt{(\theta^2 - \sin^2\theta)}}\{\tfrac{1}{3}\psi F'(\psi) - F(\psi)\},$$

by § 8·51. The function $\Omega_1(\theta, \psi)$ does not seem to be essentially positive (cf. § 8·52); to overcome this difficulty, interchange the parametric variables θ and ψ, when it will be found that

$$2\{\nu^{\frac{1}{3}} J_\nu(\nu)\}^2 \frac{d}{d\nu}\left\{\frac{\nu^{\frac{2}{3}} J_\nu'(\nu)}{\nu^{\frac{1}{3}} J_\nu(\nu)}\right\} = \frac{\nu}{\pi^2}\int_0^\pi\int_0^\pi \{\Omega_1(\theta, \psi) + \Omega_1(\psi, \theta)\}\, e^{-\nu\{F(\theta)+F(\psi)\}}\, d\theta\, d\psi.$$

Now, from the inequality just proved,

$$\Omega_1(\theta, \psi) + \Omega_1(\psi, \theta)$$

$$\geqslant \left\{\frac{2\sqrt{(\theta^2 - \sin^2\theta)}}{\theta} - \frac{2\sqrt{(\psi^2 - \sin^2\psi)}}{\psi}\right\}\left[\{\tfrac{1}{3}\theta F'(\theta) - F(\theta)\} - \{\tfrac{1}{3}\psi F'(\psi) - F(\psi)\}\right]$$

$$+ \frac{\psi^2 + \psi\sin\psi\cos\psi - 2\sin^2\psi}{\psi\sqrt{(\psi^2 - \sin^2\psi)}}\{\tfrac{1}{3}\theta F'(\theta) - F(\theta)\}$$

$$+ \frac{\theta^2 + \theta\sin\theta\cos\theta - 2\sin^2\theta}{\theta\sqrt{(\theta^2 - \sin^2\theta)}}\{\tfrac{1}{3}\psi F'(\psi) - F(\psi)\}.$$

Since $\theta^{-1}\sqrt{(\theta^2 - \sin^2\theta)}$ and $\frac{1}{3}\theta F'(\theta) - F(\theta)$ are both (§ 8·53) increasing functions of θ, the factors of the first term in the sum on the right are both positive or both negative; and, by §§ 8·51, 8·53, the second and third terms are both positive. Hence $\Omega_1(\theta, \psi) + \Omega_1(\psi, \theta)$ is positive, and therefore

$$(1) \qquad \frac{d}{d\nu}\left\{\frac{\nu^{\frac{1}{3}} J_\nu'(\nu)}{J_\nu(\nu)}\right\} \geqslant 0,$$

which establishes the result stated.

8·6. *Asymptotic expansions of Bessel functions of large complex order.*

The results obtained (§§ 8·31—8·42) by Debye in connexion with $J_\nu(x)$ and $Y_\nu(x)$ where ν and x are large and positive were subsequently extended* to the case of complex variables. In the following investigation, which is, in some respects, more detailed than Debye's memoir, we shall obtain asymptotic expansions associated with $J_\nu(z)$ when ν and z are large and complex.

It will first be supposed that $|\arg z| < \frac{1}{2}\pi$, and we shall write

$$\nu = z\cosh\gamma = z\cosh(\alpha + i\beta),$$

where α and β are real and γ is complex. There is a one-one correspondence between $\alpha + i\beta$ and ν/z if we suppose that β is restricted to lie *between*† 0 and π, while α may have any real value. This restriction prevents z/ν from lying between -1 and 1, but this case has already (§ 8·4) been investigated.

The integrals to be investigated are

$$H_\nu^{(1)}(z) = \frac{1}{\pi i}\int_{-\infty}^{\infty+\pi i} e^{-zf(w)}\,dw,$$

$$H_\nu^{(2)}(z) = -\frac{1}{\pi i}\int_{-\infty}^{\infty-\pi i} e^{-zf(w)}\,dw = -\frac{1}{\pi i}\int_{-\infty+\pi i}^{\infty} e^{zf(w)}\,dw,$$

where $f(w) \equiv w\cosh\gamma - \sinh w$.

A stationary point of the integrand is at γ, and we shall therefore investigate the curve whose equation is

$$If(w) = If(\gamma).$$

If we replace w by $u + iv$, this equation may be written in the form

$$(v-\beta)\cosh\alpha\cos\beta + (u-\alpha)\sinh\alpha\sin\beta - \cosh u\sin v + \cosh\alpha\sin\beta = 0.$$

The shape of the curve near (α, β) is

$$\{(u-\alpha)^2 - (v-\beta)^2\}\cosh\alpha\sin\beta + 2(u-\alpha)(v-\beta)\sinh\alpha\cos\beta = 0,$$

so the slopes of the two branches through that point are

$$\tfrac{1}{4}\pi + \tfrac{1}{2}\arctan(\tanh\alpha\cot\beta),$$

$$-\tfrac{1}{4}\pi + \tfrac{1}{2}\arctan(\tanh\alpha\cot\beta),$$

where the arc tan denotes an acute angle, positive or negative; $Rf(w)$ increases as w moves away from γ on the first branch, while it decreases as w moves away from γ on the second branch. The increase (or decrease) is steady, and $Rf(w)$ tends to $+\infty$ (or $-\infty$) as w moves off to infinity unless the curve has a second double-point‡.

* *Münchener Sitzungsberichte*, XL. [5], (1910); the asymptotic expansions of $I_\nu(x)$ and $K_\nu(x)$ were stated explicitly by Nicholson, *Phil. Mag.* (6) XX. (1910), pp. 938—943.

† That is to say $0 < \beta < \pi$.

‡ As will be seen later, this is the exceptional case.

If (i) and (ii) denote the whole of the contours of which portions are marked with those numbers in Fig. 19, we shall write

$$S_\nu^{(1)}(z)=\frac{1}{\pi i}\int_{(i)} e^{-zf(w)}\,dw,\qquad S_\nu^{(2)}(z)=-\frac{1}{\pi i}\int_{(ii)} e^{zf(w)}\,dw,$$

and by analysis identical with that of § 8·41 (except that $i\beta$ is to be replaced by γ), it is found that the asymptotic expansions of $S_\nu^{(1)}(z)$ and $S_\nu^{(2)}(z)$ are given by the formulae

$$(1)\qquad S_\nu^{(1)}(z)\sim\frac{e^{\nu(\tanh\gamma-\gamma)-\frac{1}{4}\pi i}}{\sqrt{(-\frac{1}{2}\nu\pi i\tanh\gamma)}}\sum_{m=0}^{\infty}\frac{\Gamma(m+\frac{1}{2})}{\Gamma(\frac{1}{2})}\cdot\frac{A_m}{(\frac{1}{2}\nu\tanh\gamma)^m},$$

$$(2)\qquad S_\nu^{(2)}(z)\sim\frac{e^{-\nu(\tanh\gamma-\gamma)+\frac{1}{4}\pi i}}{\sqrt{(-\frac{1}{2}\nu\pi i\tanh\gamma)}}\sum_{m=0}^{\infty}\frac{\Gamma(m+\frac{1}{2})}{\Gamma(\frac{1}{2})}\cdot\frac{A_m}{(-\frac{1}{2}\nu\tanh\gamma)^m},$$

where $\qquad \arg(-\frac{1}{2}\nu\pi i\tanh\gamma)=\arg z+\arg(-i\sinh\gamma),$

and the value of $\arg(-i\sinh\gamma)$ which lies between $-\frac{1}{2}\pi$ and $\frac{1}{2}\pi$ is to be taken.

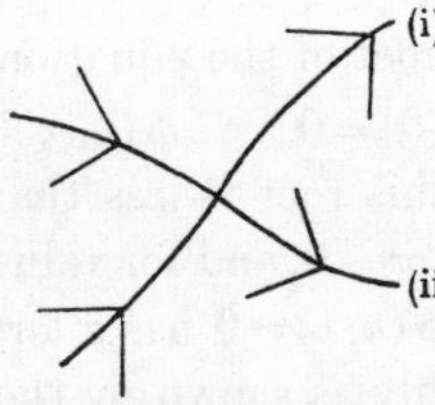

Fig. 19.

The values of $A_0, A_1, A_2, \ldots$ are

$$(3)\qquad \begin{cases} A_0=1,\qquad A_1=\frac{1}{8}-\frac{5}{24}\coth^2\gamma,\\ A_2=\frac{3}{128}-\frac{77}{576}\coth^2\gamma+\frac{385}{3456}\coth^4\gamma,\\ \ldots\ldots\ldots\ldots\ldots\ldots\ldots\ldots\ldots\ldots\ldots\end{cases}$$

It remains to express $H_\nu^{(1)}(z)$ and $H_\nu^{(2)}(z)$ in terms of $S_\nu^{(1)}(z)$ and $S_\nu^{(2)}(z)$; and to do this an intensive study of the curve on which

$$If(w)=If(\gamma)$$

is necessary.

8·61. *The form of Debye's contours when the variables are complex.*

The equation of the curve introduced in the last section is

$$(1)\qquad (v-\beta)\cosh\alpha\cos\beta+(u-\alpha)\sinh\alpha\sin\beta \\ -\cosh u\sin v+\cosh\alpha\sin\beta=0,$$

where (u, v) are current Cartesian coordinates and $0<\beta<\pi$.

Since the equation is unaltered by a change of sign in both u and α, we shall first study the case in which $\alpha\geqslant 0$; and since the equation is unaltered when $\pi-v$ and $\pi-\beta$ are written for v and β, we shall also at first suppose that $0<\beta\leqslant\frac{1}{2}\pi$, though many of the results which will be proved when β is an acute angle are still true when β is an obtuse angle.

For brevity, the expression on the left in (1) will be called $\phi(u, v)$. Since

$$\frac{\partial \phi(u, v)}{\partial u} = \sinh \alpha \sin \beta - \sinh u \sin v,$$

it follows that, when v is given, $\partial\phi/\partial u$ vanishes for only one value of u, and so the equation in u,

$$\phi(u, v) = 0,$$

has at most two real roots; and one of these is infinite whenever v is a multiple of π.

When $0 < v < \pi$, we have*

$$\phi(-\infty, v) = -\infty, \qquad \phi(+\infty, v) = -\infty,$$
$$\phi(\alpha, v) = \cosh \alpha \{(v - \beta) \cos \beta - \sin v + \sin \beta\} \geqslant 0,$$

and so one root of the equation in u,

$$\phi(u, v) = 0,$$

is less than α and the other is greater than α, both becoming equal when $v = \beta$.

By considering the finite root of the equations

$$\phi(u, 0) = 0, \qquad \phi(u, \pi) = 0,$$

it is seen that, in each case, this root is less than α, so the larger root tends to $+\infty$ as v tends to $+0$ or to $\pi - 0$, and for values of v just less than 0 or just greater than π the equation $\phi(u, v) = 0$ has a large negative root. The shape of the curve is therefore roughly as shewn by the continuous lines in Fig. 20.

Next consider the configuration when v lies between 0 and $-\pi$.

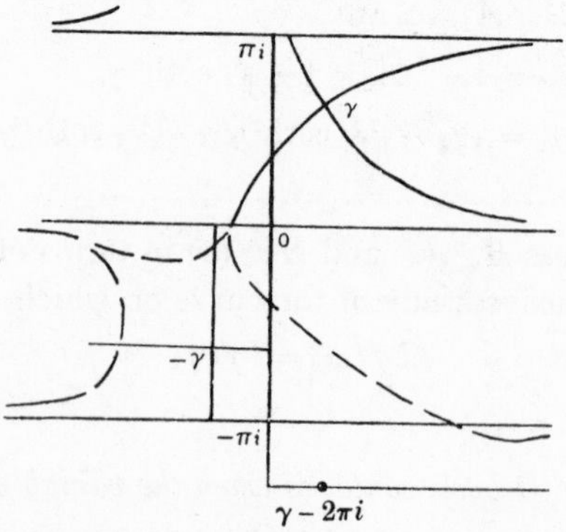

Fig. 20.

When v is $-\beta$, $\partial\phi(u, v)/\partial u$ vanishes at $u = -\alpha$, and hence $\phi(u, -\beta)$ has a minimum value

$$2 \cosh \alpha \sin \beta (1 - \beta \cot \beta - \alpha \tanh \alpha)$$

at $u = -\alpha$. There are now two cases to consider according as

$$1 - \beta \cot \beta - \alpha \tanh \alpha$$

is (I) positive or (II) negative.

* Since $\partial\phi(\alpha, v)/\partial v = \cosh \alpha (\cos \beta - \cos v)$, and this has the same sign as $v - \beta$, $\phi(\alpha, v)$ has a *minimum* value zero at $v = \beta$.

The domains of values of the complex $\gamma \equiv \alpha + i\beta$ for which

$$1 - \beta \cot \beta - \alpha \tanh \alpha$$

is positive (in the strip $0 \leqslant \beta \leqslant \pi$) are numbered 1, 4, 5 in Fig. 21; in the domains numbered 2, 3, $6a$, $6b$, $7a$, $7b$ the expression is negative; the corresponding domains for the complex $\nu/z \equiv \cosh(\alpha + i\beta)$ have the same numbers in Fig. 22.

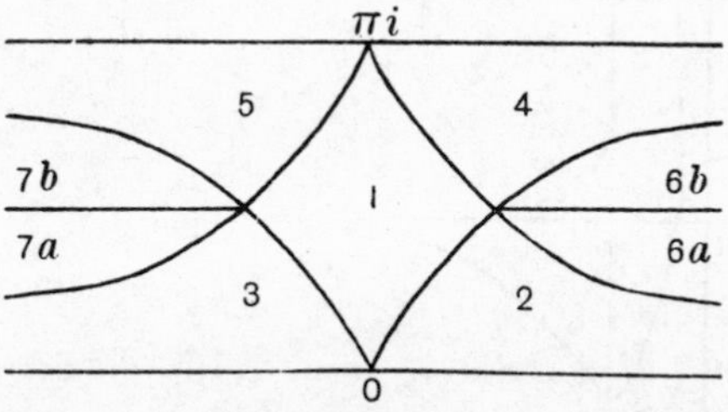

Fig. 21.

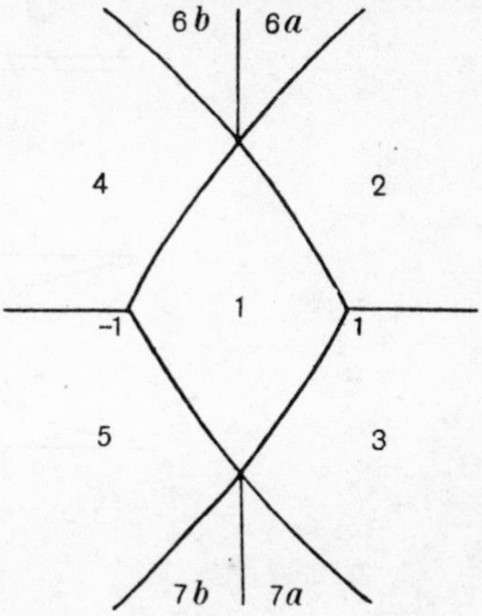

Fig. 22.

(I) When $1 - \beta \cot \beta - \alpha \tanh \alpha$ is positive, $\phi(u, -\beta)$ is essentially positive, so that the curve never crosses the line $v = -\beta$. The only possibility therefore is that the curve after crossing the real axis goes off to $-\infty$ as shewn by the upper dotted curve in Fig. 20.

(II) When $1 - \beta \cot \beta - \alpha \tanh \alpha$ is negative, the equation $\phi(-\alpha, v) = 0$ has no real root between 0 and $\beta - 2\pi$, for

$$\partial \phi(-\alpha, v)/\partial v = \cosh \alpha (\cos \beta - \cos v).$$

Therefore $\phi(-\alpha, v)$ has a single maximum at $-\beta$, and its value there is negative, so that $\phi(-\alpha, v)$ is negative when v lies between 0 and $\beta - 2\pi$.

Also $\phi(u, \beta - 2\pi)$ has a maximum at $u = \alpha$, and its value there is negative, so that the curve $\phi(u, v) = 0$ does not cross $v = \beta - 2\pi$; hence, after crossing the real axis, the curve must pass off to $\infty - \pi i$, as shewn by the dotted curve on the right of Fig. 20.

This completes the discussion of the part of the curve associated with $S_\nu^{(1)}(z)$ when $\alpha > 0$, $0 < \beta \leqslant \frac{1}{2}\pi$.

Next we have to consider what happens to the curve after crossing the line $v = +\pi$.

Since $$\phi(\alpha, v) = \cosh \alpha \{(v - \beta) \cos \beta - \sin v + \sin \beta\},$$

and the expression on the right is positive when $v \geqslant \beta$, the curve never crosses the line $u = \alpha$; also

$$\phi(u, n\pi) = (u - \alpha) \sinh \alpha \sin \beta + (n\pi - \beta) \cosh \alpha \cos \beta + \cosh \alpha \sin \beta,$$

and this is positive when $u > \alpha$, so that the parts of the curve which go off to infinity on the right must lie as shewn in the north-east corner of Fig. 23.

When
$$1 - \alpha \tanh \alpha + (\pi - \beta) \cot \beta > 0,$$
i.e. when (α, β) lies in any of the domains numbered 1, 2 and 3 in Fig. 21, it is found that the curve does not cross $v = 2\pi - \beta$, and so the curve after crossing $v = \pi$ passes off to $-\infty + \pi i$ as shewn in Fig. 23 by a broken curve.

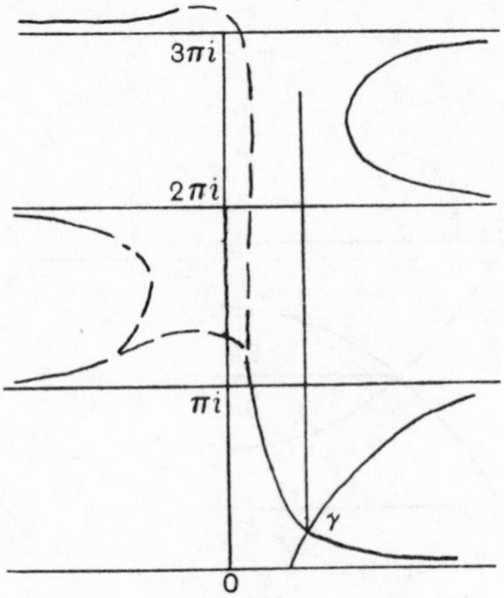

Fig. 23.

We now have to consider what happens when (α, β) lies in the domain numbered $6a$ in Fig. 21. In such circumstances
$$1 - \alpha \tanh \alpha + (\pi - \beta) \cot \beta < 0;$$
and $\phi(-\alpha, v)$ has a maximum at $v = 2\pi - \beta$, the value of $\phi(-\alpha, 2\pi - \beta)$ being negative. The curve, after crossing $v = \pi$, consequently remains on the right of $u = -\alpha$ until it has got above $v = 2\pi - \beta$.

Now $\phi(-\alpha, v)$ is increasing in the intervals
$$(\beta, 2\pi - \beta), \quad (2\pi + \beta, 4\pi - \beta), \quad (4\pi + \beta, 6\pi - \beta), \ldots;$$
let the first of these intervals in which it becomes positive be
$$(2M\pi + \beta, 2M\pi + 2\pi - \beta).$$

Then $\phi(u, 2M\pi + 2\pi - \beta)$ has a minimum at $u = -\alpha$, at which its value is positive, and so the curve cannot cross the line $v = 2M\pi + 2\pi - \beta$; it must therefore go off to infinity on the left, and consequently goes to
$$-\infty + (2M + 1)\pi i;$$
it cannot go to infinity lower than this, for then the complete curve would meet a horizontal line in more than two points.

When (α, β) is in $6a$, the curve consequently goes to infinity at

$$-\infty + (2M+1)\pi i,$$

where M is the smallest integer for which $1 - \alpha \tanh \alpha + \{(M+1)\pi - \beta\} \cot \beta$ is positive.

We can now construct a table of values of the end-points of the contours for $S_\nu^{(1)}(z)$ and $S_\nu^{(2)}(z)$, and thence we can express these integrals in terms of $H_\nu^{(1)}(z)$ and $H_\nu^{(2)}(z)$ when (α, β) lies in the domains numbered 1, 2 and $6a$ in Fig. 21; and by suitable reflexions we obtain their values for the rest of the complete strip in which $0 < \beta < \pi$. The reader should observe that, so far as the domain 1 is concerned, it does not matter whether β is acute or obtuse.

If M is the smallest integer for which

$$1 - \alpha \tanh \alpha + \{(M+1)\pi - \beta\} \cot \beta$$

is positive when $\cot \beta$ is positive, and if N is the smallest integer for which

$$1 - \alpha \tanh \alpha - (N\pi + \beta) \cot \beta$$

is positive when $\cot \beta$ is negative, the tables of values of $S_\nu^{(1)}(z)$ and $S_\nu^{(2)}(z)$ are as follows:

Regions	End-points	$S_\nu^{(1)}(z)$
1, 3, 4	$-\infty,\ \infty + \pi i$	$H_\nu^{(1)}(z)$
2, $6a$	$\infty - \pi i,\ \infty + \pi i$	$2J_\nu(z)$
5, $7b$	$-\infty,\ -\infty + 2\pi i$	$2e^{-\nu\pi i} J_{-\nu}(z)$
$6b$	$-\infty - 2N\pi i,\ \infty + \pi i$	$e^{N\nu\pi i} H_\nu^{(1)}(ze^{-N\pi i})$
$7a$	$-\infty,\ \infty + (2M+1)\pi i$	$e^{-M\nu\pi i} H_\nu^{(1)}(ze^{-M\pi i})$

Regions	End-points	$S_\nu^{(2)}(z)$
1, 2, 5	$-\infty + \pi i,\ \infty$	$H_\nu^{(2)}(z)$
3, $7a$	$-\infty + \pi i,\ -\infty - \pi i$	$2J_\nu(z)$
4, $6b$	$\infty + 2\pi i,\ \infty$	$2e^{\nu\pi i} J_{-\nu}(z)$
$6a$	$-\infty + (2M+1)\pi i,\ \infty$	$e^{M\nu\pi i} H_\nu^{(2)}(ze^{M\pi i})$
$7b$	$-\infty + \pi i,\ \infty - 2N\pi i$	$e^{-N\nu\pi i} H_\nu^{(2)}(ze^{N\pi i})$

From these tables asymptotic expansions of any fundamental system of solutions of Bessel's equation can be constructed when ν and z are both arbitrarily large complex numbers, the real part of z being positive. The range of validity of the expansions can be extended to a somewhat wider range of values of $\arg z$ by means of the device used in § 8·42.

The reader will find it interesting to prove that, in the critical case $\beta=\frac{1}{2}\pi$, the contours pass from $-\infty$ to $\infty+\pi i$ and from $-\infty+\pi i$ to ∞, so that the expansions appropriate to the region 1 are valid.

NOTE. The differences between the formulae for the regions 6*a* and 6*b* and also for the regions 7*a* and 7*b* appear to have been overlooked by Debye, and by Watson, *Proc. Royal Soc.* XCV. A, (1918), p. 91.

8·7. *Kapteyn's inequality for* $J_n(nz)$.

An extension of Carlini's formula (§§ 8·11, 8·5) to Bessel coefficients in which the argument is complex has been effected by Kapteyn* who has shewn that, when z has any value, real or complex, for which z^2-1 is not a real positive number†, then

$$|J_n(nz)| \leqslant \left|\frac{z^n \exp\{n\sqrt{(1-z^2)}\}}{\{1+\sqrt{(1-z^2)}\}^n}\right|. \tag{1}$$

This formula is less precise than Carlini's formula because the factor $(2\pi n)^{\frac{1}{2}}(1-z^2)^{\frac{1}{4}}$ does not appear in the denominator on the right, but nevertheless the inequality is sufficiently powerful for the purposes for which it is required‡.

To obtain the inequality, consider the integral formula

$$J_n(nz)=\frac{1}{2\pi i}\int^{(0+)} t^{-n-1}\exp\{\tfrac{1}{2}nz(t-1/t)\}\,dt,$$

in which the contour is a circle of radius e^u, where u is a positive number to be chosen subsequently.

If we write $t=e^{u+i\theta}$, we get

$$J_n(nz)=\frac{1}{2\pi}\int_{-\pi}^{\pi}\exp\left[n\{\tfrac{1}{2}z(e^u e^{i\theta}-e^{-u}e^{-i\theta})-u-i\theta\}\right]d\theta.$$

Now, if M be the maximum value of

$$|\exp\{\tfrac{1}{2}z(e^u e^{i\theta}-e^{-u}e^{-i\theta})-u-i\theta\}|$$

on the contour, it is clear that

$$|J_n(nz)|\leqslant M^n.$$

But if $z=\rho e^{i\alpha}$, where ρ is positive and α is real, then the real part of

$$\tfrac{1}{2}z(e^u e^{i\theta}-e^{-u}e^{-i\theta})-u-i\theta$$

is

$$\tfrac{1}{2}\rho\{e^u\cos(\alpha+\theta)-e^{-u}\cos(\alpha-\theta)\}-u,$$

and this attains its maximum value when

$$\tan\theta=-\coth u\tan\alpha,$$

and its value is then

$$\rho\sqrt{(\sinh^2 u+\sin^2\alpha)}-u.$$

* *Ann. Sci. de l'École norm sup.* (3) x. (1893), pp. 91—120.

† Since both sides of (1) are continuous when z approaches the real axis it follows that the inequality is still true when z^2-1 is positive: for such values of z, either sign may be given to the radicals according to the way in which z approaches the cuts.

‡ See Chapter XVII.

Hence, for all positive values of u,

$$|J_n(n\rho e^{i\alpha})| \leqslant \exp[n\rho\sqrt{(\sinh^2 u + \sin^2\alpha)} - nu].$$

We now choose u so that the expression on the right may be as small as possible in order to get the strongest inequality attainable by this method.

The expression

$$\rho\sqrt{(\sinh^2 u + \sin^2\alpha)} - u$$

has a minimum, *qua* function of u, when u is chosen to be the positive root of the equation*

$$\frac{\sinh u \cosh u}{\sqrt{(\sinh^2 u + \sin^2\alpha)}} = \frac{1}{\rho}.$$

With this choice of u it may be proved that

$$2\sqrt{(1-z^2)} \,.\, \sinh u \cosh u = \pm(\cosh 2u - e^{2i\alpha}),$$

and, by taking z to be real, it is clear that the positive sign must be taken in the ambiguity. Hence

$$2\{1+\sqrt{(1-z^2)}\}\sinh u\cosh u = e^{2u} - e^{2i\alpha},$$

and so

$$\begin{aligned}\log\left|\frac{z\exp\sqrt{(1-z^2)}}{1+\sqrt{(1-z^2)}}\right| &= \log\frac{2\sqrt{(\sinh^2 u+\sin^2\alpha)}\,.\,|\exp\sqrt{(1-z^2)}|}{|e^{2u}-e^{2i\alpha}|}\\ &= R\sqrt{(1-z^2)} - u\\ &= \frac{\sinh^2 u + \sin^2\alpha}{\sinh u\cosh u} - u\\ &= \rho\sqrt{(\sinh^2 u + \sin^2\alpha)} - u,\end{aligned}$$

and it is now clear that

$$|J_n(nz)| \leqslant \left|\left\{\frac{z\exp\sqrt{(1-z^2)}}{1+\sqrt{(1-z^2)}}\right\}^n\right|.$$

An interesting consequence of this inequality is that $|J_n(nz)| \leqslant 1$ so long as both $|z| \leqslant 1$ and

$$\left|\frac{z\exp\sqrt{(1-z^2)}}{1+\sqrt{(1-z^2)}}\right| \leqslant 1.$$

To construct the domain in which the last inequality is satisfied, write as before $z = \rho e^{i\alpha}$, and define u by the equation

$$\frac{\sinh u\cosh u}{\sqrt{(\sinh^2 u + \sin^2\alpha)}} = \frac{1}{\rho}.$$

The previous analysis shews at once that, when

$$\left|\frac{z\exp\sqrt{(1-z^2)}}{1+\sqrt{(1-z^2)}}\right| = 1,$$

then

$$\rho\sqrt{(\sinh^2 u + \sin^2\alpha)} - u = 0.$$

* This equation is a quadratic in $\sinh^2 u$ with one positive root.

It follows that

$$\rho^2 = \frac{2u}{\sinh 2u}, \qquad \sin^2 \alpha = \sinh u\,(u \cosh u - \sinh u).$$

As u increases from 0 to 1·1997 ..., $\sin^2 \alpha$ increases from 0 to 1 and ρ decreases from 1 to* 0·6627434 .... It is then clear that $\left|\frac{z \exp \sqrt{(1-z^2)}}{1+\sqrt{(1-z^2)}}\right| \leqslant 1$ inside and on the boundary of an oval curve containing the origin. This curve

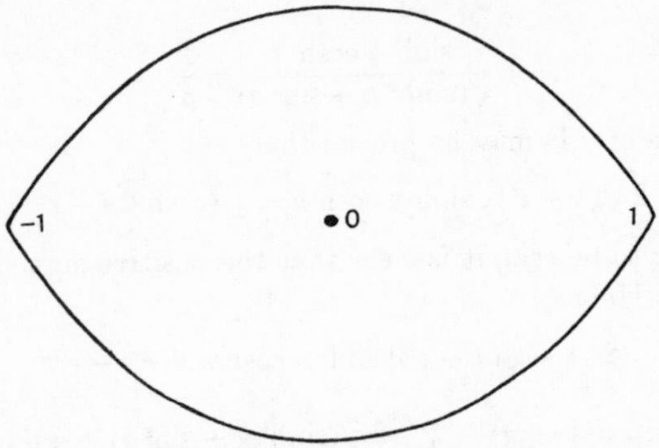

Fig. 24. The domain in which $|J_n(nz)|$ certainly does not exceed unity.

is shewn in Fig. 24; it will prove to be of considerable importance in the theory of Kapteyn series (Chapter XVII).

When the order of the Bessel function is positive but not restricted to be an integer we take the contour of integration to be a circle of radius e^u terminated by two rays inclined $\pm\pi - \text{arc tan}\,(\coth u \tan a)$ to the real axis. If we take $|t| = e^v$ on these rays, we get

$$|J_\nu(\nu z)| \leqslant M^\nu + \left|\frac{\sin \nu\pi}{\pi}\right| \left|\int_u^\infty \exp\left[-\tfrac{1}{2}\nu\rho \frac{\cosh(u+v) - \cos 2a \cosh(v-u)}{\sqrt{(\sinh^2 u + \sin^2 a)}} - \nu v\right] dv\right.$$

$$\leqslant M^\nu \left\{1 + \left|\frac{\sin \nu\pi}{\pi}\right| \int_u^\infty \exp\{-\nu(v-u)\}\, dv\right\}$$

and so

$$|J_\nu(\nu z)| \leqslant \left\{1 + \left|\frac{\sin \nu\pi}{\nu\pi}\right|\right\} \cdot \left|\left\{\frac{z \exp \sqrt{(1-z^2)}}{1+\sqrt{(1-z^2)}}\right\}^\nu\right|.$$

* This value is given by Plummer, *Dynamical Astronomy* (Cambridge, 1918), p. 47.

CHAPTER IX

POLYNOMIALS ASSOCIATED WITH BESSEL FUNCTIONS

9·1. *The definition of Neumann's polynomial $O_n(t)$.*

The object of this chapter is the discussion of certain polynomials which occur in various types of investigations connected with Bessel functions.

The first of these polynomials to appear in analysis occurs in Neumann's* investigation of the problem of expanding an arbitrary analytic function $f(z)$ into a series of the form $\Sigma a_n J_n(z)$. The function $O_n(t)$, which is now usually called *Neumann's polynomial*, is defined as the coefficient of $\epsilon_n J_n(z)$ in the expansion of $1/(t-z)$ as a series of Bessel coefficients†, so that

$$(1) \qquad \frac{1}{t-z} \equiv J_0(z)\,O_0(t) + 2J_1(z)\,O_1(t) + 2J_2(z)\,O_2(t) + \dots$$

$$= \sum_{n=0}^{\infty} \epsilon_n J_n(z)\,O_n(t).$$

From this definition we shall derive an explicit expression for the function, and it will then appear that the expansion (1) is valid whenever $|z| < |t|$. In order to obtain this expression, assume that $|z| < |t|$ and, after expanding $1/(t-z)$ in ascending powers of z, substitute Schlömilch's series of Bessel coefficients (§ 2·7) for each power of z.

This procedure gives

$$\frac{1}{t-z} = \frac{1}{t} + \sum_{s=1}^{\infty} \frac{z^s}{t^{s+1}}$$

$$= \frac{1}{t} \sum_{m=0}^{\infty} \epsilon_{2m} J_{2m}(z) + \sum_{s=1}^{\infty} \frac{2^s}{t^{s+1}} \left\{ \sum_{m=0}^{\infty} \frac{(s+2m).(s+m-1)!}{m!} J_{s+2m}(z) \right\}.$$

Assuming for the moment that the repeated series is absolutely convergent‡,

* *Theorie der Bessel'schen Functionen* (Leipzig, 1867), pp. 8—15, 33; see also *Journal für Math.* LXVII. (1867), pp. 310—314. Neumann's procedure, after assuming the expansion (1), is to derive the differential equation which will be given subsequently (§ 9·12) and to solve it in series.

† In anticipation of § 16·11, we observe that the expansion of an arbitrary function is obtained by substituting for $1/(t-z)$ in the formula

$$f(z) = \frac{1}{2\pi i} \int^{(z+)} \frac{f(t)\,dt}{t-z}.$$

‡ Cf. Pincherle's rather more general investigation, *Rendiconti R. Ist. Lombardo*, (2) XV. (1882), pp. 224—225.

we effect a rearrangement by replacing s by $n-2m$, and the rearranged series is a series of Bessel coefficients; we thus get

$$\frac{1}{t-z}=\frac{1}{t}\sum_{m=0}^{\infty}\epsilon_{2m}J_{2m}(z)+\sum_{n=1}^{\infty}\epsilon_n\left\{\sum_{m=0}^{\leqslant\frac{1}{2}(n-1)}\frac{2^{n-2m-1}}{t^{n-2m+1}}\cdot\frac{n.(n-m-1)!}{m!}\right\}J_n(z)$$

$$=\frac{1}{t}J_0(z)+\sum_{n=1}^{\infty}\epsilon_n\left\{\sum_{m=0}^{\leqslant\frac{1}{2}n}\frac{2^{n-2m-1}n.(n-m-1)!}{m!\,t^{n-2m+1}}\right\}J_n(z).$$

Accordingly the functions $O_n(t)$ are defined by the equations

(2) $$O_n(t)=\frac{1}{4}\sum_{m=0}^{\leqslant\frac{1}{2}n}\frac{n.(n-m-1)!}{m!\,(\frac{1}{2}t)^{n-2m+1}},\qquad (n\geqslant 1)$$

(3) $$O_0(t)=1/t.$$

It is easy to see that

(4) $$\epsilon_n O_n(t)=\frac{2^n.n!}{t^{n+1}}\left\{1+\frac{t^2}{2(2n-2)}+\frac{t^4}{2.4.(2n-2)(2n-4)}+\dots\right\},$$

and the series terminates before there is any possibility of a denominator factor being zero or negative.

We have now to consider the permissibility of rearranging the repeated series for $1/(t-z)$. A sufficient condition is that the series

$$\sum_{s=1}^{\infty}\frac{2^s}{|t|^{s+1}}\left\{\sum_{m=0}^{\infty}\frac{(s+2m).(s+m-1)!}{m!}|J_{s+2m}(z)|\right\}$$

should be convergent. To prove that this is actually the case, we observe that, by § 2·11 (4), we have

$$\sum_{m=0}^{\infty}\frac{(s+2m).(s+m-1)!}{m!}|J_{s+2m}(z)|\leqslant\sum_{m=0}^{\infty}\frac{(s+m-1)!\,(\frac{1}{2}|z|)^{s+2m}}{m!\,(s+2m-1)!}\exp(\tfrac{1}{4}|z|^2)$$

$$\leqslant\sum_{m=0}^{\infty}(\tfrac{1}{2}|z|)^{s+2m}\{\exp(\tfrac{1}{4}|z|^2)\}/(2m)!$$

$$\leqslant(\tfrac{1}{2}|z|)^s\exp(\tfrac{1}{2}|z|^2).$$

Hence

$$\sum_{s=1}^{\infty}\frac{2^s}{|t|^{s+1}}\left\{\sum_{m=0}^{\infty}\frac{(s+2m).(s+m-1)!}{m!}|J_{s+2m}(z)|\right\}\leqslant\sum_{s=1}^{\infty}\frac{|z|^s}{|t|^{s+1}}\exp(\tfrac{1}{2}|z|^2)$$

$$=\frac{|z|\exp(\frac{1}{2}|z|^2)}{|t|(|t|-|z|)}.$$

The absolute convergence of the repeated series is therefore established under the hypothesis that $|z|<|t|$. And so the expansion (1) is valid when $|z|<|t|$, and the coefficients of the Bessel functions in the expansion are defined by (2) and (3).

It is also easy to establish the uniformity of the convergence of the expansion (1) throughout the regions $|t|\geqslant R$, $|z|\leqslant r$, where $R>r>0$.

When these inequalities are satisfied, the sum of the moduli of the terms does not exceed

$$\sum_{s=0}^{\infty} \frac{2^s}{R^{s+1}} \left\{ \sum_{m=0}^{\infty} \frac{(s+2m) \, . \, (s+m-1)!}{m!} \cdot \frac{(\frac{1}{2}r)^{s+2m} \exp(\frac{1}{4}r^2)}{(s+2m)!} \right\} \leqslant \frac{\exp(\frac{1}{2}r^2)}{R-r}.$$

Since the expression on the right is independent of z and t, the uniformity of the convergence follows from the test of Weierstrass.

The function $O_n(t)$ was called by Neumann *a Bessel function of the second kind**; but this term is now used (cf. §§ 3·53, 3·54) to describe a certain solution of Bessel's equation, and so it has become obsolete as a description of Neumann's function. The function $O_n(t)$ is a polynomial of degree $n+1$ in $1/t$, and it is usually called *Neumann's polynomial of order n*.

If the order of the terms in Neumann's polynomial is reversed by writing $\frac{1}{2}n - m$ or $\frac{1}{2}(n-1) - m$ for m in (2), according as n is even or odd, it is at once found that

$$(5) \qquad O_n(t) = \frac{1}{4} \sum_{m=0}^{\frac{1}{2}n} \frac{n \, . \, (\frac{1}{2}n + m - 1)!}{(\frac{1}{2}n - m)! \, (\frac{1}{2}t)^{2m+1}} \qquad (n \text{ even})$$

$$= \frac{1}{t} + \frac{n^2}{t^3} + \frac{n^2(n^2-2^2)}{t^5} + \frac{n^2(n^2-2^2)(n^2-4^2)}{t^7} + \dots,$$

$$(6) \qquad O_n(t) = \frac{1}{4} \sum_{m=0}^{\frac{1}{2}(n-1)} \frac{n \, . \, (\frac{1}{2}n + m - \frac{1}{2})!}{(\frac{1}{2}n - m - \frac{1}{2})! \, (\frac{1}{2}t)^{2m+2}} \qquad (n \text{ odd})$$

$$= \frac{n}{t^2} + \frac{n(n^2-1^2)}{t^4} + \frac{n(n^2-1^2)(n^2-3^2)}{t^6} + \dots.$$

These results may be combined in the formula

$$(7) \qquad O_n(t) = \frac{1}{4} \sum_{m=0}^{n} \frac{n \, \Gamma(\frac{1}{2}n + \frac{1}{2}m) \cos^2 \frac{1}{2}(m \pm n)\pi}{\Gamma(\frac{1}{2}n - \frac{1}{2}m + 1) \, . \, (\frac{1}{2}t)^{m+1}}.$$

The equations (5), (6) and (7) were given by Neumann.

By the methods of § 2·11, it is easily proved that

$$(8) \qquad |\epsilon_n O_n(t)| \leqslant \tfrac{1}{2} \, . \, (n!) \, . \, (\tfrac{1}{2}|t|)^{-n-1} \exp(\tfrac{1}{4}|t|^2),$$

$$(9) \qquad \epsilon_n O_n(t) = \tfrac{1}{2} \, . \, (n!) \, . \, (\tfrac{1}{2}t)^{-n-1}(1+\theta), \qquad (n > 1)$$

where $$|\theta| \leqslant [\exp(\tfrac{1}{4}|t|^2) - 1]/(2n-2).$$

From these formulae it follows that the series $\Sigma a_n J_n(z) O_n(t)$ is convergent whenever the series $\Sigma a_n (z/t)^n$ is absolutely convergent; and, when z is outside the circle of convergence of the latter series, $a_n J_n(z) O_n(t)$ does not tend to zero as $n \to \infty$, and so the former series does not converge. Again, it is easy to prove that, as $n \to \infty$,

$$\epsilon_n J_n(z) O_n(t) = \frac{z^n}{t^{n+1}} \left\{ 1 - \frac{z^2 - t^2}{4n} + O(n^{-2}) \right\},$$

* By analogy with the Legendre function of the second kind, $Q_n(t)$, which is such that

$$\frac{1}{t-z} = \sum_{n=0}^{\infty} (2n+1) P_n(z) Q_n(t).$$

Cf. *Modern Analysis*, § 15·4.

and hence it may be shewn* that the points on the circle of convergence at which either series converges† are identical with the points on the circle at which the other series is convergent. It may also be proved that, if either series is uniformly convergent in any domains of values of z and t, so also is the other series.

Since the series on the right of (1) is a uniformly convergent series of analytic functions when $|z|<|t|$, it follows by differentiation‡ that

$$\frac{(-)^q \cdot (p+q)!}{(t-z)^{p+q+1}} = \sum_{n=0}^{\infty} \epsilon_n \frac{d^p J_n(z)}{dz^p} \frac{d^q O_n(t)}{dt^q}, \tag{10}$$

where p, q are any positive integers (zero included).

It may be convenient to place on record the following expressions:

$$O_0(t) = 1/t, \qquad O_1(t) = 1/t^2,$$

$$O_2(t) = 1/t + 4/t^3, \qquad O_3(t) = 3/t^2 + 24/t^4,$$

$$O_4(t) = 1/t + 16/t^3 + 192/t^5, \qquad O_5(t) = 5/t^2 + 120/t^4 + 1920/t^6.$$

The coefficients in the polynomial $O_n(t)$, for $n=0, 1, 2, \ldots 15$, have been calculated by Otti, *Bern Mittheilungen*, 1898, pp. 4, 5.

9·11. *The recurrence formulae satisfied by $O_n(t)$.*

We shall now obtain the formulae

$$(n-1)\, O_{n+1}(t) + (n+1)\, O_{n-1}(t) - \frac{2(n^2-1)}{t} O_n(t) = \frac{2n \sin^2 \frac{1}{2}n\pi}{t}, \quad (n \geqslant 1) \tag{1}$$

$$O_{n-1}(t) - O_{n+1}(t) = 2O_n'(t), \quad (n \geqslant 1) \tag{2}$$

$$-O_1(t) = O_0'(t) \tag{3}$$

The first of these was stated by Schläfli, *Math. Ann.* III. (1871), p. 137, and proved by Gegenbauer, *Wiener Sitzungsberichte*, LXV. (2), (1872), pp. 33—35, but the other two were proved some years earlier by Neumann, *Theorie der Bessel'schen Functionen* (Leipzig, 1867), p. 21.

Since early proofs consisted merely of a verification, we shall not repeat them, but give in their place an investigation by which the recurrence formulae are derived in a natural manner from the corresponding formulae for Bessel coefficients.

Taking $|z|<|t|$, observe that, by § 9·1 (1) and § 2·22 (7),

$$(t-z) \sum_{n=0}^{\infty} \epsilon_n J_n(z)\, O_n(t) = 1 = \sum_{n=0}^{\infty} \epsilon_n \cos^2 \tfrac{1}{2}n\pi \,.\, J_n(z),$$

* It is sufficient to use the theorems that, if Σb_n is convergent, so also is $\Sigma b_n/n$, and that then $\Sigma b_n/n^2$ is absolutely convergent.

† This was pointed out by Pincherle, *Bologna Memorie*, (4) III. (1881—2), p. 160.

‡ Cf. *Modern Analysis*, § 5·33.

and hence

$$z \sum_{n=0}^{\infty} \epsilon_n J_n(z)\, O_n(t) = \sum_{n=0}^{\infty} \epsilon_n J_n(z)\, \{tO_n(t) - \cos^2 \tfrac{1}{2} n\pi\}$$
$$= \sum_{n=1} \epsilon_n J_n(z)\, \{tO_n(t) - \cos^2 \tfrac{1}{2} n\pi\},$$

since $tO_0(t) = 1$. If now we use the recurrence formula for $J_n(z)$ to modify the expression on the right, we get

$$\sum_{n=0}^{\infty} \epsilon_n J_n(z)\, O_n(t) = \sum_{n=1}^{\infty} \{J_{n-1}(z) + J_{n+1}(z)\}\, \{tO_n(t) - \cos^2 \tfrac{1}{2} n\pi\}/n.$$

If we notice that $J_{n+1}(z)\{tO_n(t) - \cos^2 \tfrac{1}{2} n\pi\}/n$ tends to zero as $n \to \infty$, it is clear on rearrangement that

$$J_0(z)\{O_0(t) - tO_1(t)\} + J_1(z)\{2O_1(t) - \tfrac{1}{2} tO_2(t) + \tfrac{1}{2}\}$$
$$+ \sum_{n=2}^{\infty} J_n(z) \left\{ 2O_n(t) - \frac{tO_{n+1}(t)}{n+1} - \frac{tO_{n-1}(t)}{n-1} + \frac{2n \sin^2 \frac{1}{2} n\pi}{n^2 - 1} \right\} = 0.$$

Now regard z as a variable, while t remains constant; if the coefficients of all the Bessel functions on the left do not vanish, the first term which does not vanish can be made to exceed the sum of all the others in absolute value, by taking $|z|$ sufficiently small. Hence *all* the coefficients vanish identically* and, from this result, formula (1) is obvious.

To prove (2) and (3) observe that

$$\left(\frac{\partial}{\partial t} + \frac{\partial}{\partial z}\right) \frac{1}{t - z} = 0,$$

and so, $|z|$ being less than $|t|$, we have

$$\sum_{n=0}^{\infty} \epsilon_n J_n(z)\, O_n'(t) + \sum_{n=0}^{\infty} \epsilon_n J_n'(z)\, O_n(t) = 0.$$

By rearranging the series on the left we find that

$$\sum_{n=0}^{\infty} \epsilon_n J_n(z)\, O_n'(t) = J_1(z)\, O_0(t) - \sum_{n=1}^{\infty} \{J_{n-1}(z) - J_{n+1}(z)\}\, O_n(t)$$
$$= -J_0(z)\, O_1(t) - \sum_{n=1}^{\infty} J_n(z)\, \{O_{n+1}(t) - O_{n-1}(t)\},$$

that is to say,

$$J_0(z)\{O_0'(t) + O_1(t)\} + \sum_{n=1}^{\infty} J_n(z)\, \{2O_n'(t) + O_{n+1}(t) - O_{n-1}(t)\} \equiv 0.$$

On equating to zero the coefficient of $J_n(z)$ on the left, just as in the proof of (1), we obtain (2) and (3).

* This is the argument used to prove that, if a convergent power series vanishes identically, then all its coefficients vanish (cf. *Modern Analysis*, § 3·73). The argument is valid here because the various series of Bessel coefficients converge uniformly throughout a domain containing $z = 0$.

By combining (1) and (2) we at once obtain the equivalent formulae

(4) $$ntO_{n-1}(t) - (n^2-1)\,O_n(t) = (n-1)\,tO_n'(t) + n\sin^2\tfrac{1}{2}n\pi,$$

(5) $$ntO_{n+1}(t) - (n^2-1)\,O_n(t) = -(n+1)\,tO_n'(t) + n\sin^2\tfrac{1}{2}n\pi.$$

If ϑ be written for $t\,(d/dt)$, these formulae become

(6) $$(n-1)(\vartheta+n+1)\,O_n(t) = n\,\{tO_{n-1}(t) - \sin^2\tfrac{1}{2}n\pi\},$$

(7) $$(n+1)(\vartheta-n+1)\,O_n(t) = -n\,\{tO_{n+1}(t) - \sin^2\tfrac{1}{2}n\pi\}.$$

The Neumann polynomial of *negative* integral order was defined by Schläfli* by the equation

(8) $$O_{-n}(t) = (-)^n\,O_n(t).$$

With this definition the formulae (1)—(7) are valid for all integral values of n.

9·12. *The differential equation† satisfied by $O_n(t)$.*

From the recurrence formulae § 9·11 (6) and (7), it is clear that

$$\begin{aligned}(\vartheta+n+1)(\vartheta-n+1)\,O_n(t) &= \frac{1}{n+1}(\vartheta+n+1)\,\{-ntO_{n+1}(t) + n\sin^2\tfrac{1}{2}n\pi\}\\ &= -\frac{nt}{n+1}(\vartheta+n+2)\,O_{n+1}(t) + n\sin^2\tfrac{1}{2}n\pi\\ &= -t\,\{tO_n(t) - \cos^2\tfrac{1}{2}n\pi\} + n\sin^2\tfrac{1}{2}n\pi,\end{aligned}$$

and consequently $O_n(t)$ satisfies the differential equation

$$(\vartheta+1)^2\,O_n(t) + (t^2-n^2)\,O_n(t) = t\cos^2\tfrac{1}{2}n\pi + n\sin^2\tfrac{1}{2}n\pi.$$

It follows that the general solution of the differential equation

(1) $$\frac{d^2y}{dt^2} + \frac{3}{t}\frac{dy}{dt} + \left(1 - \frac{n^2-1}{t^2}\right)y = \frac{\cos^2\frac{1}{2}n\pi}{t} + \frac{n\sin^2\frac{1}{2}n\pi}{t^2}$$

is

$$y = O_n(t) + t^{-1}\mathscr{C}_n(t),$$

and so the only solution of (1) which is expressible as a terminating series is $O_n(t)$.

It is sometimes convenient to write (1) in the form

(2) $$\frac{d^2y}{dt^2} + \frac{3}{t}\frac{dy}{dt} + \left(1 - \frac{n^2-1}{t^2}\right)y = g_n(t),$$

where

(3) $$g_n(t) = \begin{cases}1/t, & (n \text{ even})\\ n/t^2. & (n \text{ odd})\end{cases}$$

* *Math. Ann.* III. (1871), p. 138.

† Neumann, *Theorie der Bessel'schen Functionen* (Leipzig, 1867), p. 13; *Journal für Math.* LXVII. (1867), p. 314.

Another method of constructing the differential equation is to observe that

$$\left\{z^2\frac{\partial^2}{\partial z^2}+z\frac{\partial}{\partial z}+z^2\right\}\sum_{n=0}^{\infty}\epsilon_n J_n(z)\,O_n(t)=\sum_{n=0}^{\infty}\epsilon_n n^2 J_n(z)\,O_n(t),$$

and so

$$\begin{aligned}\sum_{n=0}^{\infty}\epsilon_n n^2 J_n(z)\,O_n(t)&=\left\{z^2\frac{\partial^2}{\partial z^2}+z\frac{\partial}{\partial z}+z^2\right\}\frac{1}{t-z}\\&=\frac{2z^2}{(t-z)^3}+\frac{z}{(t-z)^2}+\frac{z^2}{t-z}\\&=\left\{t^2\frac{\partial^2}{\partial t^2}+3t\frac{\partial}{\partial t}+1+t^2\right\}\frac{1}{t-z}-t-z.\end{aligned}$$

Now

$$1=\sum_{n=0}^{\infty}\epsilon_{2n}J_{2n}(z),\quad z=\sum_{n=0}^{\infty}\epsilon_{2n+1}(2n+1)\,J_{2n+1}(z),$$

and hence

$$t+z=t^2\sum_{n=0}^{\infty}\epsilon_n g_n(t)\,J_n(z).$$

Therefore

$$\sum_{n=0}^{\infty}\epsilon_n J_n(z)\left[\left\{t^2\frac{\partial^2}{\partial t^2}+3t\frac{\partial}{\partial t}+1+t^2-n^2\right\}O_n(t)-t^2 g_n(t)\right]\equiv 0.$$

On equating to zero the coefficient of $J_n(z)$ on the left-hand side of this identity, just as in § 9·11, we obtain at once the differential equation satisfied by $O_n(t)$.

9·13. *Neumann's contour integrals associated with $O_n(z)$.*

It has been shewn by Neumann* that, if C be any closed contour,

(1) $$\int_C O_m(z)\,O_n(z)\,dz=0, \qquad (m=n \text{ and } m\neq n)$$

(2) $$\int_C J_m(z)\,O_n(z)\,dz=0, \qquad (m^2\neq n^2)$$

(3) $$\int_C J_n(z)\,O_n(z)\,dz=2\pi ik/\epsilon_n,$$

where k is the excess of the number of positive circuits of the contour round the origin over the number of negative circuits.

The first result is obvious from Cauchy's theorem, because the only singularity of $O_m(z)\,O_n(z)$ is at the origin, and the residue there is zero.

The third result follows in a similar manner; the only pole of the integrand is a simple pole at the origin, and the residue at this point is $1/\epsilon_n$.

To prove the second result, multiply the equations

$$\nabla_m J_m(z)=0,\qquad \nabla_n\{zO_n(z)\}=z^3 g_n(z)$$

by $zO_n(z)$ and $J_m(z)$ respectively, and subtract. If $U(z)$ be written in place of

$$J_m(z)\frac{d\{zO_n(z)\}}{dz}-zO_n(z)\frac{dJ_m(z)}{dz},$$

the result of subtracting assumes the form

$$z^2U'(z)+zU(z)+(m^2-n^2)\,zJ_m(z)\,O_n(z)=z^3 g_n(z)\,J_m(z),$$

* *Theorie der Bessel'schen Functionen* (Leipzig, 1867), p. 19.

and hence

$$[zU(z)]_C + (m^2 - n^2)\int_C J_m(z)\, O_n(z)\, dz = \int_C z^2 g_n(z)\, J_m(z)\, dz.$$

The integrated part vanishes because $U(z)$ is one-valued, and the integral on the right vanishes because the integrand is analytic for all values of z; and hence we deduce (2) when $m^2 \neq n^2$.

Two corollaries, due to Schläfli, *Math. Ann.* III. (1871), p. 138, are that

(4) $$\frac{1}{\pi i}\int^{(0+)} J_n(x+y)\, O_m(y)\, dy = J_{n-m}(x) + (-)^m J_{n+m}(x),$$

(5) $$\frac{1}{\pi i}\int^{(-x+)} O_m(x+y)\, J_n(y)\, dy = J_{m-n}(x) + (-)^n J_{m+n}(x).$$

The first is obtained by applying (2) and (3) to the formula § 2·4 (1), namely

$$J_n(x+y) = \sum_{p=-\infty}^{\infty} J_{n+p}(x)\, J_{-p}(y),$$

and the second follows by making an obvious change of variable.

9·14. *Neumann's integral for $O_n(z)$.*

It was stated by Neumann* that

(1) $$O_n(z) = \int_0^\infty \frac{\{u + \sqrt{(u^2+z^2)}\}^n + \{u - \sqrt{(u^2+z^2)}\}^n}{2z^{n+1}}\, e^{-u}\, du.$$

We shall now prove by induction the equivalent formula

(2) $$O_n(z) = \tfrac{1}{2}\int_0^{\infty \exp i\alpha} [\{t + \sqrt{(1+t^2)}\}^n + \{t - \sqrt{(1+t^2)}\}^n]\, e^{-zt}\, dt,$$

where α is any angle such that $|\alpha + \arg z| < \tfrac{1}{2}\pi$; on writing $t = u/z$, the truth of (1) will then be manifest.

A modification of equation (2) is

(3) $$O_n(z) = \tfrac{1}{2}\int_0^{\infty + i\alpha} \{e^{n\theta} + (-)^n e^{-n\theta}\}\, e^{-z\sinh\theta} \cosh\theta\, d\theta.$$

To prove (2) we observe that

$$O_0(z) = \int_0^{\infty \exp i\alpha} e^{-zt}\, dt, \quad O_1(z) = \int_0^{\infty \exp i\alpha} t e^{-zt}\, dt;$$

and so, by using the recurrence formula § 9·11 (2), it follows that we may write

$$O_n(z) = \int_0^{\infty \exp i\alpha} \phi_n(t)\, e^{-zt}\, dt,$$

where

(4) $$\phi_{n+1}(t) - 2t\phi_n(t) - \phi_{n-1}(t) = 0,$$

and

(5) $$\phi_0(t) = 1, \quad \phi_1(t) = t.$$

* *Theorie der Bessel'schen Functionen* (Leipzig, 1867), p. 16; *Journal für Math.* LXVII. (1867), p. 312.

The solution of the difference equation (4) is

$$\phi_n(t) = A\,\{t + \sqrt{(t^2+1)}\}^n + B\,\{t - \sqrt{(1+t^2)}\}^n,$$

where A and B are independent of n, though they might be functions of t. The conditions (5) shew, however, that $A = B = \frac{1}{2}$; and the formula (2) is established.

This proof was given in a symbolic form by Sonine*, who wrote $\phi_n(D) . (1/z)$ where we have written $\int_0^{\infty \exp ia} \phi_n(t)\, e^{-zt}\, dt$, D standing for (d/dz).

A completely different investigation of this result is due to Kapteyn†, whose analysis is based on the expansion of § 9·1 (1), which we now write in the form

$$\frac{1}{z-\zeta} = \sum_{n=0}^{\infty} \epsilon_n J_n(\zeta)\, O_n(z).$$

When $|\zeta| < |z|$, we have

$$\frac{1}{z-\zeta} = \frac{1}{z}\int_0^{\infty} \exp\left\{\frac{\zeta u}{z} - u\right\} du$$

$$= \frac{1}{z}\int_0^{\infty} \left\{\sum_{n=-\infty}^{\infty} \rho^n J_n(\zeta)\right\} e^{-u}\, du,$$

if ρ be so chosen that

$$\frac{u}{z} = \frac{1}{2}\left(\rho - \frac{1}{\rho}\right).$$

It follows that

$$\frac{1}{z-\zeta} = \int_0^{\infty}\left[\sum_{n=-\infty}^{\infty} \frac{\{u+\sqrt{(u^2+z^2)}\}^n}{z^{n+1}} . J_n(\zeta)\right] e^{-u}\, du.$$

We shall now shew that the interchange of summation and integration is justifiable; it will be sufficient to shew that, for any given values of ζ and z (such that $|\zeta| < |z|$),

$$\sum_{n=N+1}^{M} \int_0^{\infty} \left|\frac{\{u \pm \sqrt{(u^2+z^2)}\}^n}{z^n} J_n(\zeta)\right| e^{-u}\, du$$

can be made arbitrarily small by taking N sufficiently large‡; now

$$|u \pm \sqrt{(u^2+z^2)}| \leqslant 2\,(u + |z|),$$

and so $$|J_n(\zeta)| \int_0^{\infty} \left|\frac{\{u \pm \sqrt{(u^2+z^2)}\}^n}{z^n}\right| e^{-u}\, du \leqslant \frac{|J_n(\zeta)|}{|z|^n}\int_0^{\infty} 2^n (u + |z|)^n e^{-u}\, du$$

$$= \frac{|J_n(\zeta)|}{|z|^n}\int_{|z|}^{\infty} 2^n t^n e^{-t+|z|}\, dt$$

$$\leqslant |(\zeta/z)^n| \exp\{|z| + \tfrac{1}{4}|\zeta|^2\}.$$

* *Math. Ann.* XVI. (1880), p. 7. For a similar symbolic investigation see § 6·14 *supra*.

† *Ann. Sci. de l'École norm. sup.* (3) x. (1893), p. 108.

‡ Cf. Bromwich, *Theory of Infinite Series*, § 176.

Therefore, since $|\zeta| < |z|$, we have

$$\sum_{n=N+1}^{M} \int_0^\infty \left| \frac{\{u \pm \sqrt{(u^2+z^2)}\}^n}{z^n} J_n(z) \right| e^{-u} du \leqslant \frac{|\zeta^{N+1}| \exp\{|z| + \frac{1}{4}|\zeta|^2\}}{|z^N| \{|z| - |\zeta|\}},$$

and the expression on the left can be made arbitrarily small by taking N sufficiently large when z and ζ are fixed.

Hence, when $|\zeta| < |z|$, we have

$$\frac{1}{z-\zeta} = \sum_{n=-\infty}^{\infty} J_n(\zeta) \int_0^\infty \frac{\{u + \sqrt{(u^2+z^2)}\}^n}{z^{n+1}} e^{-u} du$$

$$= \sum_{n=0}^{\infty} \epsilon_n J_n(\zeta) O_n(z),$$

where $O_n(z)$ is *defined* by the equation

$$O_n(z) = \int_0^\infty \frac{\{u + \sqrt{(u^2+z^2)}\}^n + \{u - \sqrt{(u^2+z^2)}\}^n}{2z^{n+1}} e^{-u} du;$$

and it is easy to see that $O_n(z)$, so defined, is a polynomial in $1/z$ of degree $n+1$.

When the integrand is expanded* in powers of z and integrated term by term, it is easy to reconcile this definition of $O_n(z)$ with the formula § 9·1 (4).

9·15. *Sonine's investigation of Neumann's integral.*

An extremely interesting and suggestive investigation of a general type of expansion of $1/(\alpha - z)$ is due to Sonine†; from this general expansion, Neumann's formula (§ 9·1) with the integral of § 9·14 can be derived without difficulty. Sonine's general theorem is as follows:

Let $\psi(w)$ be an arbitrary function of w; and, if $\psi(w) = x$, let $w = \Psi(x)$, so that Ψ is the function inverse to ψ.

Let Z_n and A_n be defined by the equations‡

$$Z_n = \frac{1}{2\pi i} \int^{(0+)} e^{z\psi(w)} \frac{dw}{w^{n+1}}, \quad A_n = \int_0^\infty e^{-\alpha x} \{\Psi(x)\}^n dx.$$

Then
$$\frac{1}{\alpha - z} = \sum_{n=-\infty}^{\infty} Z_n A_n,$$

it being assumed that the series on the right is convergent.

Suppose that for any given positive value of x, $|w| > |\Psi(x)|$ on a closed curve C surrounding the origin and the point z, and $|w| < |\Psi(x)|$ on a closed

* Cf. Hobson, *Plane Trigonometry* (1918), § 264.

† *Mathematical Collection* (Moscow), v (1870), pp. 323—382. Sonine's notation has been modified slightly, but the symbols ψ and Ψ are his.

‡ This is connected with Laplace's transformation. See Burkhardt, *Encyclopädie der Math. Wiss.* II. (*Analysis*) (1916), pp. 781—784.

curve c surrounding the origin but not enclosing the point z. Then

$$\sum_{n=-\infty}^{\infty} Z_n A_n = \frac{1}{2\pi i}\sum_{n=0}^{\infty}\int_0^\infty \int_C e^{z\psi(w)-ax}\frac{\{\phi(x)\}^n}{w^{n+1}}\,dw\,dx$$
$$+\frac{1}{2\pi i}\sum_{n=0}^{\infty}\int_0^\infty \int_c e^{z\psi(w)-ax}\frac{w^n}{\{\phi(x)\}^{n+1}}\,dw\,dx$$
$$=\frac{1}{2\pi i}\int_0^\infty\left\{\int_C-\int_c\right\}e^{z\psi(w)-ax}\frac{dw\,dx}{w-\phi(x)}$$
$$=\frac{1}{2\pi i}\int_0^\infty\int^{\phi(x)+} e^{z\psi(w)-ax}\frac{dw\,dx}{w-\phi(x)}$$
$$=\int_0^\infty e^{(z-a)x}\,dx = 1/(\alpha-z),$$

provided that $R(z)<R(\alpha)$; and the result is established if it is assumed that the various transformations are permissible.

In order to obtain Neumann's expansion, take

$$\psi(w)=\tfrac{1}{2}(w-1/w),\quad \phi(x)=x\pm\sqrt{(x^2+1)},$$

and then

$$\frac{1}{\alpha-z}=\sum_{n=-\infty}^{\infty}J_n(z)A_n$$
$$=\sum_{n=0}^{\infty}\tfrac{1}{2}\epsilon_n J_n(z)\{A_n+(-)^n A_{-n}\}.$$

Since

$$A_n+(-)^n A_{-n}=\int_0^\infty e^{-ax}[\{x\pm\sqrt{(x^2+1)}\}^n+(-)^n\{x\pm\sqrt{(x^2+1)}\}^{-n}]\,dx,$$

we at once obtain Neumann's integral.

Sonine notes (p. 328) that

$$J_n(z)\sim(\tfrac{1}{2}z)^n/n!,\quad \epsilon_n O_n(a)\sim n!\,(\tfrac{1}{2}a)^{-n},$$

so that the expansion of $1/(a-z)$ converges when $|z|<|a|$; and in the later part of his memoir he gives further applications of his general expansion.

9·16. *The generating function of $O_n(z)$.*

The series $\sum_{n=0}^{\infty}(-)^n\epsilon_n t^n O_n(z)$, which is a generating function associated with $O_n(z)$, does not converge for any value of t except zero. Kapteyn*, however, has "summed" the series after the method of Borel, in the following manner:

$$\sum_{n=0}^{\infty}(-)^n\epsilon_n t^n O_n(z)=\frac{1}{z}+\sum_{n=0}^{\infty}\sum_{m=0}^{n}\frac{n.(n+m-1)!\,t^{2n}}{(n-m)!\,(\frac{1}{2}z)^{2m+1}}$$
$$-\sum_{n=0}^{\infty}\sum_{m=0}^{n}\frac{(n+\frac{1}{2}).(n+m)!\,t^{2n+1}}{(n-m)!\,(\frac{1}{2}z)^{2m+2}}$$
$$=\frac{1}{z}\frac{1+t^2}{1-t^2}+\sum_{m=1}^{\infty}\sum_{n=m}^{\infty}\frac{n.(n+m-1)!\,t^{2n}}{(n-m)!\,(\frac{1}{2}z)^{2m+1}}$$
$$-\sum_{m=0}^{\infty}\sum_{n=m}^{\infty}\frac{(n+\frac{1}{2}).(n+m)!\,t^{2n+1}}{(n-m)!\,(\frac{1}{2}z)^{2m+2}}$$
$$=\frac{1}{2}\sum_{m=0}^{\infty}\frac{(2m)!}{(\frac{1}{2}z)^{2m+1}}\frac{t^{2m}(1+t^2)}{(1-t^2)^{2m+1}}-\frac{1}{2}\sum_{m=0}^{\infty}\frac{(2m+1)!}{(\frac{1}{2}z)^{2m+2}}\frac{t^{2m+1}(1+t^2)}{(1-t^2)^{2m+2}}$$
$$=\frac{1+t^2}{z(1-t^2)}\sum_{m=0}^{\infty}\frac{(-)^m.m!\,t^m}{\{\frac{1}{2}z(1-t^2)\}^m}.$$

* *Nieuw Archief voor Wiskunde* (2), VI. (1905), pp. 49—55.

The Borel-sum associated with this series is $(1+t^2)\int_0^\infty \frac{e^{-u}\,du}{(1-t^2)\,z+2tu}$, and this integral is convergent so long as $(1-t^2)\,z/t$ is not negative.

There is no great difficulty in verifying that the series $\sum_{n=0}^{\infty}(-)^n \epsilon_n t^n O_n(z)$ is an asymptotic expansion of the integral for small positive values of t when $|\arg z|<\pi$, and so the integral may be regarded as the generating function of $O_n(z)$. Kapteyn has built up much of the theory of Neumann's function from this result.

9·17. *The inequality of Kapteyn's type for $O_n(nz)$.*

It is possible to deduce from Neumann's integral an inequality satisfied by $O_n(nz)$ which closely resembles the inequality satisfied by $J_n(nz)$ obtained in § 8·7.

We have

$$O_n(nz)=\frac{1}{2z^{n+1}}\int_0^\infty [\{w+\sqrt{(w^2+z^2)}\}^n+\{w-\sqrt{(w^2+z^2)}\}^n]\,e^{-nw}\,dw,$$

the path of integration being a contour in the w-plane, and so

$$|O_n(nz)|\leqslant\frac{1}{|z|^{n+1}}\int_0^\infty |[\{w\pm\sqrt{(w^2+z^2)}\}\,e^{-w}]|^n\,.\,|dw|,$$

where that value of the radical is taken which gives the integrand with the greater modulus.

Now the stationary point of

$$\{w+\sqrt{(w^2+z^2)}\}\,e^{-w}$$

is $\sqrt{(1-z^2)}$, and so

$$(1)\qquad |O_n(nz)|\leqslant\frac{1}{|z|^{n+1}}\left\{\frac{1+\sqrt{(1-z^2)}}{\exp\sqrt{(1-z^2)}}\right\}^{n-1}\int|\{w+\sqrt{(w^2+z^2)}\}\,e^{-w}|\,.\,|dw|,$$

where the path of integration is one for which the integrand is greatest at the stationary point.

If a surface of the type indicated in § 8·3 is constructed over the w-plane, the stationary point is the only pass on the surface; and both $w=0$ and $w=+\infty$ are at a lower level than the pass if

$$(2)\qquad \left|\frac{z\exp\sqrt{(1-z^2)}}{1+\sqrt{(1-z^2)}}\right|<1.$$

Hence, since a contour joining the origin to infinity can be drawn when (2) is satisfied, and since the integral involved in (1) is convergent with this contour, it follows that, throughout the domain in which (2) is satisfied, the inequality

$$(3)\qquad O_n(nz)<\frac{A}{|z|^2}\cdot\left|\frac{1+\sqrt{(1-z^2)}}{z\exp\sqrt{(1-z^2)}}\right|^{n-1}$$

is satisfied for some constant value of A; and this is an inequality of the same character as the inequality of § 8·7.

9·2. *Gegenbauer's generalisation* of Neumann's polynomial.*

If we expand $z^\nu/(t-z)$ in ascending powers of z and replace each power of z by the expansion as a series of Bessel functions given in § 5·2, we find on rearrangement that

$$\begin{aligned}\frac{z^\nu}{t-z} &= \sum_{s=0}^{\infty} \frac{z^{\nu+s}}{t^{s+1}} \\ &= \sum_{s=0}^{\infty} \frac{2^{\nu+s}}{t^{s+1}} \left\{ \sum_{m=0}^{\infty} \frac{(\nu+s+2m) \,.\, \Gamma(\nu+s+m)}{m!} J_{\nu+s+2m}(z) \right\} \\ &= \sum_{n=0}^{\infty} \left\{ \sum_{m=0}^{\leqslant \frac{1}{2}n} \frac{2^{\nu+n-2m}}{t^{n-2m+1}} \,.\, \frac{(\nu+n) \,.\, \Gamma(\nu+n-m)}{m!} J_{\nu+n}(z) \right\};\end{aligned}$$

the rearrangement has been effected by replacing s by $n-2m$, and it presents no greater theoretical difficulties than the corresponding rearrangement in § 9·1.

We are thus led to consider Gegenbauer's polynomial $A_{n,\nu}(t)$, defined by the equation

$$(1) \qquad A_{n,\nu}(t) = \frac{2^{\nu+n}(\nu+n)}{t^{n+1}} \sum_{m=0}^{\leqslant \frac{1}{2}n} \frac{\Gamma(\nu+n-m)}{m!} (\tfrac{1}{2}t)^{2m};$$

this definition is valid whenever ν is not zero or a negative integer; and when $|z|<|t|$, we have

$$(2) \qquad \frac{z^\nu}{t-z} = \sum_{n=0}^{\infty} A_{n,\nu}(t) J_{\nu+n}(z).$$

The reader should have no difficulty in proving the following recurrence formulae:

$$(3) \quad (\nu+n-1) A_{n+1,\nu}(t) + (\nu+n+1) A_{n-1,\nu}(t) - \frac{2\{(\nu+n)^2-1\}}{t} A_{n,\nu}(t)$$
$$= \frac{2^\nu(\nu+n)\{(\nu+n)^2-1\}\Gamma(\nu+\frac{1}{2}n-\frac{1}{2})}{t\,\Gamma(\frac{1}{2}n+\frac{3}{2})} \sin^2 \tfrac{1}{2}n\pi,$$

$$(4) \qquad \frac{2\nu+n-1}{\nu+n-1} A_{n-1,\nu}(t) - \frac{n+1}{\nu+n+1} A_{n+1,\nu}(t) = 2A'_{n,\nu}(t),$$

$$(5) \quad (\nu+n) tA_{n-1,\nu}(t) - (n+1)(\nu+n-1) A_{n,\nu}(t)$$
$$= (\nu+n-1) tA'_{n,\nu}(t) + \frac{2^\nu(\nu+n)(\nu+n-1)\Gamma(\nu+\frac{1}{2}n-\frac{1}{2})}{\Gamma(\frac{1}{2}n+\frac{1}{2})} \sin^2 \tfrac{1}{2}n\pi,$$

$$(6) \quad (\nu+n) tA_{n+1,\nu}(t) - (\nu+n+1)(2\nu+n-1) A_{n,\nu}(t)$$
$$= -(\nu+n+1) tA'_{n,\nu}(t) + \frac{2^\nu(\nu+n)(\nu+n+1)\Gamma(\nu+\frac{1}{2}n+\frac{1}{2})}{\Gamma(\frac{1}{2}n+\frac{3}{2})} \sin^2 \tfrac{1}{2}n\pi,$$

$$(7) \qquad A_{0,\nu}(t) = 2^\nu \Gamma(\nu+1)/t.$$

* *Wiener Sitzungsberichte*, LXXIV. (2), (1877), pp. 124—130.

The differential equation of which $A_{n,\nu}(t)$ is a solution is

$$\text{(8)} \qquad \frac{d^2y}{dt^2} + \frac{3-2\nu}{t}\frac{dy}{dt} + \left\{1 - \frac{(n+1)(2\nu+n-1)}{t^2}\right\} y = g_{n,\nu}(t),$$

where

$$\text{(9)} \qquad g_{n,\nu}(t) = \frac{2^\nu(\nu+n)\,\Gamma(\nu+\tfrac{1}{2}n)}{\Gamma(\tfrac{1}{2}n+1)\,t}\cos^2\tfrac{1}{2}n\pi + \frac{2^{\nu+1}(\nu+n)\,\Gamma(\nu+\tfrac{1}{2}n+\tfrac{1}{2})}{\Gamma(\tfrac{1}{2}n+\tfrac{1}{2})\,t^2}\sin^2\tfrac{1}{2}n\pi.$$

The general solution of (8) is $A_{n,\nu}(t) + t^{\nu-1}\mathscr{C}_{\nu+n}(t)$.

Of these results, (3), (4), (8) and (9) are due to Gegenbauer; and he also proved that

$$\text{(10)} \qquad \frac{1}{2\pi i}\int^{(0+)} A_{n,\nu}(t)\,e^{izt}\,dt = 2^\nu i^n\,\Gamma(\nu)\,.\,(\nu+n)\,C_n^\nu(z),$$

where $C_n^\nu(z)$ is the coefficient of α^n in the expansion of $(1-2\alpha z+\alpha^2)^{-\nu}$; this formula is easily proved by calculating the residue of $(izt)^m A_{n,\nu}(t)$ at the origin.

The corresponding formula for Neumann's polynomial is

$$\text{(11)} \qquad \frac{1}{2\pi i}\int^{(0+)} O_n(t)\,e^{izt}\,dt = i^n\cos\{n\arccos z\}.$$

The following formulae may also be mentioned:

$$\text{(12)} \qquad \int_C A_{m,\nu}(z)\,A_{n,\nu}(z)\,dz = 0, \qquad (m = n \text{ and } m \neq n)$$

$$\text{(13)} \qquad \int_C z^{-\nu} J_{\nu+m}(z)\,A_{n,\nu}(z)\,dz = 0, \qquad (m^2 \neq n^2)$$

$$\text{(14)} \qquad \int_C z^{-\nu} J_{\nu+n}(z)\,A_{n,\nu}(z)\,dz = 2\pi ik,$$

where C is any closed contour, $n = 0, 1, 2, \ldots$, and k is the excess of the number of positive circuits over the number of negative circuits of C round the origin.

The first and third of these last results are proved by the method of § 9·13; the second is derived from the equations

$$\nabla_{\nu+m} J_{\nu+m}(z) = 0, \quad \nabla_{\nu+n}\{z^{1-\nu}A_{n,\nu}(z)\} = z^{3-\nu}g_{n,\nu}(z),$$

whence we find that

$$(m-n)(2\nu+m+n)\int_C z^{-\nu}J_{\nu+m}(z)\,A_{n,\nu}(z)\,dz = \int_C z^{2-\nu}g_{n,\nu}(z)\,J_{\nu+m}(z)\,dz = 0.$$

9·3. *Schläfli's polynomial $S_n(t)$.*

A polynomial closely connected with Neumann's polynomial $O_n(t)$ was investigated by Schläfli. In view of the greater simplicity of some of its properties, it is frequently convenient to use it rather than Neumann's polynomial.

Schläfli's definition* of the polynomial is

(1) $$S_n(t) = \sum_{m=0}^{<\frac{1}{2}n} \frac{(n-m-1)!}{m!} (\tfrac{1}{2}t)^{-n+2m}, \qquad (n \geqslant 1)$$

(2) $$S_0(t) = 0.$$

On comparing (1) with § 9·1 (2), we see at once that

(3) $$\tfrac{1}{2}n\, S_n(t) = tO_n(t) - \cos^2 \tfrac{1}{2}n\pi.$$

If we substitute for the functions $O_n(t)$ in the recurrence formulae § 9·11 (1) and (2), we find from the former that

(4) $$S_{n+1}(t) + S_{n-1}(t) - 2nt^{-1} S_n(t) = 4t^{-1} \cos^2 \tfrac{1}{2}n\pi,$$

and from the latter,

$$\tfrac{1}{2}(n-1) S_{n-1}(t) - \tfrac{1}{2}(n+1) S_{n+1}(t) = nS_n'(t) - nt^{-1} S_n(t) - 2t^{-1} \cos^2 \tfrac{1}{2}n\pi.$$

If we multiply this by 2 and add the result to (4), we get

(5) $$S_{n-1}(t) - S_{n+1}(t) = 2S_n'(t).$$

The formulae (4) and (5) may, of course, be proved by elementary algebra by using the definition of $S_n(t)$, without appealing to the properties of Neumann's polynomial.

The definition of Schläfli's polynomial of negative order is

(6) $$S_{-n}(t) = (-)^{n+1} S_n(t),$$

and, with this definition, (4) and (5) are true for all integral values of n.

The interesting formula, pointed out by Schläfli,

(7) $$S_{n-1}(t) + S_{n+1}(t) = 4O_n(t),$$

is easily derived from (3) and (4).

Other forms of the recurrence formulae which may be derived from (4) and (5) are

(8) $$tS_{n-1}(t) - nS_n(t) - tS_n'(t) = 2\cos^2 \tfrac{1}{2}n\pi,$$

(9) $$tS_{n+1}(t) - nS_n(t) + tS_n'(t) = 2\cos^2 \tfrac{1}{2}n\pi.$$

If we write ϑ for $t\,(d/dt)$, these formulae become

(10) $$(\vartheta + n) S_n(t) = tS_{n-1}(t) - 2\cos^2 \tfrac{1}{2}n\pi,$$

(11) $$(\vartheta - n) S_n(t) = -\, tS_{n+1}(t) + 2\cos^2 \tfrac{1}{2}n\pi.$$

It follows that

$$\begin{aligned}(\vartheta^2 - n^2) S_n(t) &= t(\vartheta + 1 - n) S_{n-1}(t) + 2n\cos^2 \tfrac{1}{2}n\pi \\ &= -\, t^2 S_n(t) + 2t\sin^2 \tfrac{1}{2}n\pi + 2n\cos^2 \tfrac{1}{2}n\pi,\end{aligned}$$

and so $S_n(t)$ is a solution of the differential equation

(12) $$t^2 \frac{d^2y}{dt^2} + t\frac{dy}{dt} + (t^2 - n^2)\, y = 2t\sin^2 \tfrac{1}{2}n\pi + 2n\cos^2 \tfrac{1}{2}n\pi.$$

* *Math. Ann.* III. (1871), p. 138.

It may be convenient to place on record the following expressions :

$$S_1(t)=2/t, \qquad S_2(t)=4/t^2,$$
$$S_3(t)=2/t+16/t^3, \qquad S_4(t)=8/t^2+96/t^4,$$
$$S_5(t)=2/t+48/t^3+768/t^5, \qquad S_6(t)=12/t^2+384/t^4+7680/t^6.$$

The general descending series, given explicitly by Otti, are

$$(13) \qquad S_n(t)=\sum_{m=1}^{\frac{1}{2}n}\frac{(\frac{1}{2}n+m-1)!}{(\frac{1}{2}n-m)!\,(\frac{1}{2}t)^{2m}} \qquad (n \text{ even})$$
$$=\frac{2n}{t^2}+\frac{2n(n^2-2^2)}{t^4}+\frac{2n(n^2-2^2)(n^2-4^2)}{t^6}+\dots,$$

$$(14) \qquad S_n(t)=\sum_{m=0}^{\frac{1}{2}(n-1)}\frac{(\frac{1}{2}n+m-\frac{1}{2})!}{(\frac{1}{2}n-m-\frac{1}{2})!\,(\frac{1}{2}t)^{2m+1}} \qquad (n \text{ odd})$$
$$=\frac{2}{t}+\frac{2(n^2-1^2)}{t^3}+\frac{2(n^2-1^2)(n^2-3^2)}{t^5}+\dots.$$

The coefficients in the polynomial $S_n(t)$, for $n=1, 2, \dots 12$, have been calculated by Otti, *Bern Mittheilungen*, 1898, pp. 13—14; Otti's formulae are reproduced (with some obvious errors) by Graf and Gubler, *Einleitung in die Theorie der Bessel'schen Funktionen*, II. (Bern, 1900), p. 24.

9·31. *Formulae connecting the polynomials of Neumann and Schläfli.*

We have already encountered two formulae connecting the polynomials of Neumann and Schläfli, namely

$$\tfrac{1}{2}n\,S_n(t)=tO_n(t)-\cos^2\tfrac{1}{2}n\pi,$$
$$S_{n-1}(t)+S_{n+1}(t)=4O_n(t),$$

of which the former is an immediate consequence of the definitions of the functions, and the latter follows from the recurrence formulae. A number of other formulae connecting the two functions are due to Crelier*; they are easily derivable from the formulae already obtained, and we shall now discuss the more important of them.

When we eliminate $\cos^2\frac{1}{2}n\pi$ from § 9·3 (3) and either § 9·3 (8) or (9), we find that

$$(1) \qquad S_{n-1}(t)-S_n'(t)=2O_n(t),$$
$$(2) \qquad S_{n+1}(t)+S_n'(t)=2O_n(t).$$

Next, on summing equations of the type § 9·3 (5), we find that

$$(3) \qquad S_n(t)=-2\sum_{m=0}^{<(\frac{1}{2}n-1)}S'_{n-2m-1}(t)+\sin^2\tfrac{1}{2}n\pi\,.\,S_1(t),$$

and hence

$$(4) \qquad S_n(t)+S_{n-1}(t)=-2\sum_{m=0}^{n-2}S'_{n-m-1}(t)+S_1(t).$$

* *Comptes Rendus*, cxxv. (1897), pp. 421—423, 860—863; *Bern Mittheilungen*, 1897, pp. 61—96.

Again from § 9·3 (7) and (5) we have

$$\begin{aligned}4\{O_{n-1}(t)+O_{n+1}(t)\} &= S_{n-2}(t)+2S_n(t)+S_{n+2}(t)\\ &= \{S_{n-2}(t)-S_n(t)\}-\{S_n(t)-S_{n+2}(t)\}\,4S_n(t)\\ &= 2S'_{n-1}(t)-2S'_{n+1}(t)+4S_n(t)\\ &= 4S_n''(t)+4S_n(t),\end{aligned}$$

so that

(5) $$S_n''(t)+S_n(t)=O_{n-1}(t)+O_{n+1}(t).$$

This is the most interesting of the formulae obtained by Crelier.

Again, on summing formulae of the type of § 9·11 (2), we find that

(6) $$O_n(t)=-2\sum_{m=0}^{<\frac{1}{2}(n-1)} O'_{n-2m-1}(t)+\sin^2\tfrac{1}{2}n\pi\,.\,O_1(t)+\cos^2\tfrac{1}{2}n\pi\,.\,O_0(t),$$

and hence

(7) $$O_n(t)+O_{n-1}(t)=-2\sum_{m=0}^{n-2} O'_{n-m-1}(t)+O_1(t)+O_0(t).$$

9·32. *Graf's expression of $S_n(z)$ as a sum.*

The peculiar summatory formula

(1) $$S_n(z)=\pi\sum_{m=-n}^{n}\{J_n(z)\,Y_m(z)-J_m(z)\,Y_n(z)\}$$

was stated by Graf* in 1893, the proof being supplied later in Graf and Gubler's treatise†. This formula is most readily proved by induction; it is obviously true when $n=0$, and also, by § 3·63 (12), when $n=1$. If now the sum on the right be denoted temporarily by $\phi_n(z)$, it is clear that

$$\begin{aligned}&\phi_{n+1}(z)+\phi_{n-1}(z)-(2n/z)\,\phi_n(z)\\ &\quad=\pi J_{n+1}(z)\sum_{m=-n-1}^{n+1} Y_m(z)-\pi\,Y_{n+1}(z)\sum_{m=-n-1}^{n+1} J_m(z)\\ &\qquad+\pi J_{n-1}(z)\sum_{m=-n+1}^{n-1} Y_m(z)-\pi Y_{n-1}(z)\sum_{m=-n+1}^{n-1} J_m(z)\\ &\qquad-(2n\pi/z)\,J_n(z)\sum_{m=-n}^{n} Y_m(z)+(2n\pi/z)\,Y_n(z)\sum_{m=-n}^{n} J_m(z).\end{aligned}$$

Now modify the summations on the right by suppressing or inserting terms at the beginning and end so that all the summations run from $-n$ to n; and we then see that the complete coefficients of the sums $\Sigma J_m(z)$ and $\Sigma Y_m(z)$ vanish. It follows that

$$\begin{aligned}&\phi_{n+1}(z)+\phi_{n-1}(z)-(2n/z)\,\phi_n(z)\\ &\quad=\pi J_{n+1}(z)\{Y_{n+1}(z)+Y_{-n-1}(z)\}-\pi\,Y_{n+1}(z)\{J_{n+1}(z)+J_{-n-1}(z)\}\\ &\qquad-\pi J_{n-1}(z)\{Y_n(z)+Y_{-n}(z)\}+\pi Y_{n-1}(z)\{J_n(z)+J_{-n}(z)\}\\ &\quad=-\pi\{1+(-1)^n\}\{J_{n-1}(z)\,Y_n(z)-J_n(z)\,Y_{n-1}(z)\}\\ &\quad=4z^{-1}\cos^2\tfrac{1}{2}n\pi,\end{aligned}$$

by § 3·63 (12); and so $\phi_n(z)$ satisfies the recurrence formula which is satisfied by $S_n(z)$, and the induction that $\phi_n(z)=S_n(z)$ is evident.

* *Math. Ann.* XLIII. (1893), p. 138.

† *Einleitung in die Theorie der Bessel'schen Funktionen*, II. (Bern, 1900), pp. 34—41.

9·33. *Crelier's integral for $S_n(z)$.*

If we take the formula § 9·14 (2), namely

$$O_n(z) = \tfrac{1}{2}\int_0^{\infty \exp i\alpha} [\{t + \sqrt{(1+t^2)}\}^n + \{t - \sqrt{(1+t^2)}\}^n]\, e^{-zt}\, dt,$$

and integrate by parts, we find that

$$O_n(z) = \frac{\tfrac{1}{2}\{1 + (-1)^n\}}{z} + \frac{1}{2z}\int_0^{\infty \exp i\alpha} e^{-zt}\frac{d}{dt}[\{t + \sqrt{(1+t^2)}\}^n + \{t - \sqrt{(1+t^2)}\}^n]\, dt$$

$$= \frac{\cos^2 \tfrac{1}{2}n\pi}{z} + \frac{n}{2z}\int_0^{\infty \exp i\alpha} \frac{\{t + \sqrt{(1+t^2)}\}^n - \{t - \sqrt{(1+t^2)}\}^n}{\sqrt{(1+t^2)}}\, e^{-zt}\, dt.$$

Hence it follows that

(1) $$S_n(z) = \int_0^{\infty \exp i\alpha} \frac{\{t + \sqrt{(1+t^2)}\}^n - \{t - \sqrt{(1+t^2)}\}^n}{\sqrt{(1+t^2)}}\, e^{-zt}\, dt.$$

This equation, which was given by Schläfli, *Math. Ann.* III. (1871), p. 146, in the form

(2) $$S_n(z) = \int_0^{\infty + i\alpha} \{e^{n\theta} - (-)^n e^{-n\theta}\}\, e^{-z\sinh\theta}\, d\theta,$$

is fundamental in Crelier's researches*, of which we shall now give an outline.

We write temporarily

$$T_n \equiv \{t + \sqrt{(1+t^2)}\}^n - \{t - \sqrt{(1+t^2)}\}^n,$$

and then

$$T_{n+1} - 2t T_n - T_{n-1} = 0,$$

so that

$$\frac{T_{n+1}}{T_n} = 2t + \frac{1}{T_n/T_{n-1}},$$

and therefore

$$\frac{T_{n+1}}{T_n} = 2t + \frac{1}{2t+}\,\frac{1}{2t+}\,\dots + \frac{1}{2t},$$

the continued fraction having n elements. It follows that T_{n+1}/T_n is the quotient of two simple *continuants*† so that

$$\frac{T_{n+1}}{T_n} = \frac{K(2t, 2t, \dots, 2t)_n}{K(2t, 2t, \dots, 2t)_{n-1}},$$

the suffixes n, $n-1$ denoting the number of elements in the continuants.

It follows that‡ $T_n/K(2t)_{n-1}$ is independent of n; and since

$$T_1 = 2\sqrt{(1+t^2)}, \quad K(2t)_0 = 1,$$

we have

$$T_n = 2\sqrt{(1+t^2)}\,.\,K(2t)_{n-1},$$

* *Comptes Rendus*, CXXV. (1897), pp. 421—423, 860—863; *Bern Mittheilungen*, 1897, pp. 61—96.

† Chrystal, *Algebra*, II. (1900), pp. 494—502.

‡ Since all the elements of the continuant are the same, the continuant may be expressed by this abbreviated notation.

and hence

(3) $$S_n(z) = 2\int_0^{\infty \exp i\alpha} K(2t)_{n-1}\, e^{-zt}\, dt.$$

From this result it is possible to obtain all the recurrence formulae for $S_n(z)$ by using properties of continuants.

9·34. *Schläfli's expansion of $S_n(t+z)$ as a series of Bessel coefficients.*

We shall now obtain the result due to Schläfli* that, when $|z| < |t|$, $S_n(t+z)$ can be expanded in the form

(1) $$S_n(t+z) = \sum_{m=-\infty}^{\infty} S_{n-m}(t)\, J_m(z).$$

The simplest method of establishing this formula for positive values of n is by induction†. It is evidently true when $n = 0$, for then both sides vanish; when $n = 1$, the expression on the right is equal to

$$\begin{aligned} S_1(t)\, J_0(z) &+ \sum_{m=1}^{\infty} \{S_{1-m}(t)\, J_m(z) + S_{m+1}(t)\, J_{-m}(z)\} \\ &= 2O_0(t)\, J_0(-z) + \sum_{m=1}^{\infty} \{S_{m-1}(t) + S_{m+1}(t)\}\, J_m(-z) \\ &= 2\sum_{m=0}^{\infty} \epsilon_m O_m(t)\, J_m(-z) \\ &= 2/(t+z) = S_1(t+z), \end{aligned}$$

by § 9·1 (1) and § 9·3 (7).

Now, if we assume the truth of (1) for Schläfli's polynomials of orders 0, 1, 2, ... n, we have

$$\begin{aligned} S_{n+1}(t+z) &= S_{n-1}(t+z) - 2S_n'(t+z) \\ &= \sum_{m=-\infty}^{\infty} S_{n-m-1}(t)\, J_m(z) - 2\sum_{m=-\infty}^{\infty} S'_{n-m}(t)\, J_m(z) \\ &= \sum_{m=-\infty}^{\infty} \{S_{n-m-1}(t) - 2S'_{n-m}(t)\}\, J_m(z) \\ &= \sum_{m=-\infty}^{\infty} S_{n+1-m}(t)\, J_m(z), \end{aligned}$$

and the induction is established; to obtain the second line in the analysis, we have used the obvious result that

$$S_n'(t+z) = \frac{\partial}{\partial t} S_n(t+z).$$

* *Math. Ann.* III. (1871), pp. 139—141; the examination of the convergence of the series is left to the reader (cf. § 9·1).

† The extension to negative values of n follows on the proof for positive values, by § 9·3 (6).

The expansion was obtained by Schläfli by expanding every term on the right of (1) in ascending powers of z and descending powers of t. The investigation given here is due to Sonine, *Math. Ann.* XVI. (1880), p. 7; Sonine's investigation was concerned with a more general class of functions than Schläfli's polynomial, known as *hemi-cylindrical functions* (§ 10·8).

When we make use of equation § 9·3 (7), it is clear that, when $|z| < |t|$,

$$O_n(t+z) = \sum_{m=-\infty}^{\infty} O_{n-m}(t) J_m(z). \tag{2}$$

This was proved directly by Gegenbauer, *Wiener Sitzungsberichte*, LXVI. (2), (1872), pp. 220—223, who expanded $O_n(t+z)$ in ascending powers of z by Taylor's theorem, used the obvious formula [cf. § 9·11 (2)]

$$2^p \frac{d^p O_n(t)}{dt^p} = \sum_{m=0}^{p} (-)^m {}_pC_m \cdot O_{n-p+2m}(t), \tag{3}$$

and rearranged the resulting double series.

It is easy to deduce Graf's* results (valid when $|z| < |t|$),

$$S_n(t-z) = \sum_{m=-\infty}^{\infty} S_{n+m}(t) J_m(z), \tag{4}$$

$$O_n(t-z) = \sum_{m=-\infty}^{\infty} O_{n+m}(t) J_m(z). \tag{5}$$

9·4. *The definition of Neumann's polynomial $\Omega_n(t)$.*

The problem of expanding an arbitrary even analytic function into a series of squares of Bessel coefficients was suggested to Neumann† by the formulae of § 2·72, which express any even power of z as a series of this type.

The preliminary expansion, corresponding to the expansion of $1/(t-z)$ given in § 9·1, is the expansion of $1/(t^2-z^2)$; and the function $\Omega_n(t)$ will be defined as the coefficient of $\epsilon_n J_n^2(z)$ in the expansion of $1/(t^2-z^2)$, so that

$$\frac{1}{t^2-z^2} = J_0^2(z)\,\Omega_0(t) + 2J_1^2(z)\,\Omega_1(t) + 2J_2^2(z)\,\Omega_2(t) + \dots \tag{1}$$

$$= \sum_{n=0}^{\infty} \epsilon_n J_n^2(z)\,\Omega_n(t).$$

To obtain an explicit expression for $\Omega_n(t)$, take $|z| < |t|$, and, after expanding $1/(t^2-z^2)$ in ascending powers of z, substitute for each power of z the

* *Math. Ann.* XLIII. (1893), pp. 141—142; see also Epstein, *Die vier Rechnungsoperationen mit Bessel'schen Functionen* (Bern, 1894). [*Jahrbuch über die Fortschritte der Math.* 1893—1894, pp. 845—846.]

† *Leipziger Berichte*, XXI. (1869), pp. 221—256. [*Math. Ann.* III. (1871), pp. 581—610.]

series of squares of Bessel coefficients given by Neumann (§ 2·72). As in § 9·1, we have

$$\begin{aligned}\frac{1}{t^2-z^2} &= \sum_{s=0}^{\infty} \frac{z^{2s}}{t^{2s+2}} \\ &= \frac{1}{t^2}\sum_{m=0}^{\infty} \epsilon_m J_m^2(z) + \sum_{s=1}^{\infty} \frac{2^{2s}}{t^{2s+2}} \cdot \frac{(s!)^2}{(2s)!}\left\{\sum_{m=0}^{\infty} \frac{(2s+2m).(2s+m-1)!}{m!} J^2_{s+m}(z)\right\} \\ &= \frac{1}{t^2}\sum_{n=0}^{\infty} \epsilon_n J_n^2(z) + \sum_{n=1}^{\infty}\sum_{s=1}^{n} \frac{2^{2s}}{t^{2s+2}} \frac{(s!)^2}{(2s)!} n\epsilon_n \frac{(n+s-1)!}{(n-s)!} J_n^2(z),\end{aligned}$$

when we rearrange the series by writing $n-s$ for m; this rearrangement presents no greater theoretical difficulties than the corresponding rearrangement in § 9·1.

Accordingly the function $\Omega_n(t)$ is defined by the equations

$$\Omega_n(t) = \frac{1}{4}\sum_{s=0}^{n} \frac{n.(n+s-1)!(s!)^2}{(n-s)!(2s)!(\tfrac{1}{2}t)^{2s+2}}, \qquad (n \geqslant 1) \tag{2}$$

$$\Omega_0(t) = 1/t^2. \tag{3}$$

On reversing the order of the terms in (2) we find that

$$\Omega_n(t) = \frac{1}{4}\sum_{m=0}^{n} \frac{n.(2n-m-1)!\{(n-m)!\}^2}{m!(2n-2m)!(\tfrac{1}{2}t)^{2n-2m+2}}, \qquad (n \geqslant 1) \tag{4}$$

while, if (2) be written out in full, it assumes the form

$$\Omega_n(t) = \frac{1}{t^2} + \frac{1}{2}\frac{4n^2}{t^4} + \frac{1.2}{3.4}\frac{4n^2(4n^2-2^2)}{t^6} + \frac{1.2.3}{4.5.6}\frac{4n^2(4n^2-2^2)(4n^2-4^2)}{t^8} + \ldots. \tag{5}$$

Also

$$\epsilon_n\Omega_n(t) = \frac{2^{2n}(n!)^2}{t^{2n+2}}\left\{1 + \frac{\Theta_2 t^2}{1.(2n-1)} + \frac{\Theta_4 t^4}{1.2.(2n-1)(2n-2)} + \ldots + \frac{\Theta_{2n} t^{2n}}{1.2\ldots n.(2n-1)(2n-2)\ldots n}\right\}, \tag{6}$$

where

$$\left\{\begin{aligned} &\Theta_2 = \frac{2n-1}{2n}, \quad \Theta_4 = \frac{(2n-1)(2n-3)}{2n(2n-2)}, \\ &\Theta_6 = \frac{(2n-1)(2n-3)(2n-5)}{2n(2n-2)(2n-4)}, \\ &\ldots\ldots\ldots\ldots\ldots\ldots\ldots\ldots, \\ &\Theta_{2n} = \frac{(2n-1)(2n-3)\ldots 1}{2n(2n-2)\ldots 2}. \end{aligned}\right. \tag{7}$$

Since $0 < \Theta_{2n} < 1$, it is easy to shew by the methods of § 2·11 that

$$|\epsilon_n\Omega_n(t)| \leqslant 2^{2n}|t|^{-2n-2}(n!)^2\exp(|t|^2), \tag{8}$$

and, when $n > 0$,

$$\epsilon_n\Omega_n(t) = 2^{2n}t^{-2n-2}(n!)^2(1+\theta), \tag{9}$$

where $\qquad |\theta| \leqslant \{\exp|t|^2 - 1\}/(2n-1).$

By reasoning similar to that given at the end of § 9·1, it is easy to shew that the domains of convergence of the series $\Sigma a_n J_n^2(z)\,\Omega_n(t)$ and $\Sigma a_n (z/t)^{2n}$ are the same.

The reader should have no difficulty in verifying the curious formula, due to Kapteyn*,

$$\Omega_n(t) = -2\frac{d}{dt}\int_0^{\frac{1}{2}\pi} O_{2n}\left(\frac{2t}{\sin 2\theta}\right) d\theta. \tag{10}$$

9·41. *The recurrence formulae for* $\Omega_n(t)$.

The formulae corresponding to § 9·11 (2) and (3) are

$$\frac{2}{t}\Omega_n'(t) = \frac{\Omega_{n-1}(t)}{n-1} - \frac{\Omega_{n+1}(t)}{n+1} - \frac{2\Omega_0(t)}{n^2-1}, \qquad (n \geqslant 2) \tag{1}$$

$$(2/t)\,\Omega_1'(t) = \tfrac{1}{2}\Omega_0(t) - \tfrac{1}{2}\Omega_2(t), \tag{2}$$

$$(2/t)\,\Omega_0'(t) = -2\Omega_1(t) + 2\Omega_0(t). \tag{3}$$

There seems to be no simple analogue of § 9·11 (1). The method by which Neumann† obtained these formulae is that described in § 9·11.

Take the fundamental expansion § 9·4 (1), and observe that

$$\frac{dJ_n^2(z)}{dz} = \frac{z}{2n}\{J^2_{n-1}(z) - J^2_{n+1}(z)\},$$

and that, by Hansen's expansion of § 2·5,

$$2J_0(z)\,J_0'(z) = -z\sum_{n=1}^{\infty}\{J^2_{n-1}(z) - J^2_{n+1}(z)\}/n.$$

We find by differentiations with regard to t, and with regard to z, that

$$-2t/(t^2-z^2)^2 = \sum_{n=0}^{\infty}\epsilon_n J_n^2(z)\,\Omega_n'(t),$$

$$\begin{aligned}2z/(t^2-z^2)^2 &= 2J_0(z)\,J_0'(z)\,\Omega_0(t) + z\sum_{n=1}^{\infty}\{J^2_{n-1}(z) - J^2_{n+1}(z)\}\,\Omega_n(t)/n \\ &= z\sum_{n=1}^{\infty}\{J^2_{n-1}(z) - J^2_{n+1}(z)\}\,\{\Omega_n(t) - \Omega_0(t)\}/n.\end{aligned}$$

On comparing these results, it is clear that

$$t^{-1}\sum_{n=0}^{\infty}\epsilon_n J_n^2(z)\,\Omega_n'(t) + \sum_{n=1}^{\infty}\{J^2_{n-1}(z) - J^2_{n+1}(z)\}\,.\,\{\Omega_n(t) - \Omega_0(t)\}/n = 0.$$

On selecting the coefficient of $J_n^2(z)$ on the left and equating it to zero (cf. § 9·1), we at once obtain the three stated formulae.

* *Ann. Sci. de l'École norm. sup.* (3) x. (1893), p. 111.

† *Leipziger Berichte*, xxi. (1869), p. 251. [*Math. Ann.* iii. (1871), p. 606.]

9·5. *Gegenbauer's generalisation of Neumann's polynomial $\Omega_n(t)$.*

If we expand $z^{\mu+\nu}/(t-z)$ in ascending powers of z and replace each power of z by its expansion as a series of products of Bessel functions given in § 5·5, we find on rearrangement (by replacing s by $n-2m$) that

$$\frac{z^{\mu+\nu}}{t-z}=\sum_{s=0}^{\infty}\frac{z^{\mu+\nu+s}}{t^{s+1}}$$

$$=\sum_{s=0}^{\infty}\frac{2^{\mu+\nu+s}}{t^{s+1}}\left\{\sum_{m=0}^{\infty}\frac{\Gamma(\mu+\frac{1}{2}s+1)\,\Gamma(\nu+\frac{1}{2}s+1)}{\Gamma(\mu+\nu+s+1)}\frac{(\mu+\nu+s+2m)\,\Gamma(\mu+\nu+s+m)}{m!}\right.$$

$$\left.\times J_{\mu+\frac{1}{2}s+m}(z)\,J_{\nu+\frac{1}{2}s+m}(z)\right\}$$

$$=\sum_{n=0}^{\infty}\left\{\sum_{m=0}^{\leqslant\frac{1}{2}n}\frac{2^{\mu+\nu+n-2m}}{t^{n-2m+1}}\right.$$

$$\times\frac{(\mu+\nu+n)\,\Gamma(\mu+\frac{1}{2}n-m+1)\,\Gamma(\nu+\frac{1}{2}n-m+1)\,\Gamma(\mu+\nu+n-m)}{m!\,\Gamma(\mu+\nu+n-2m+1)}$$

$$\left.\times J_{\mu+\frac{1}{2}n}(z)\,J_{\nu+\frac{1}{2}n}(z)\right\};$$

it is supposed that $|z|<|t|$, and then the rearrangement presents no greater theoretical difficulties than the corresponding rearrangement of § 9·1.

We consequently are led to consider the polynomial $B_{n;\,\mu,\nu}(t)$, defined by the equation

$$(1)\quad B_{n;\,\mu,\nu}(t)=\frac{2^{\mu+\nu+n}(\mu+\nu+n)}{t^{n+1}}$$
$$\times\sum_{m=0}^{\leqslant\frac{1}{2}n}\frac{\Gamma(\mu+\frac{1}{2}n-m+1)\,\Gamma(\nu+\frac{1}{2}n-m+1)\,\Gamma(\mu+\nu+n-m)}{m!\,\Gamma(\mu+\nu+n-2m+1)}(\tfrac{1}{2}t)^{2m}.$$

This polynomial was investigated by Gegenbauer*; it satisfies various recurrence formulae, none of which are of a simple character.

It may be noted that

$$(2)\qquad B_{2n;\,0,0}(t)=\epsilon_n t\,\Omega_n(t).$$

The following generalisations of Gegenbauer's formulae are worth placing on record. They are obtained by expanding the Bessel functions in ascending series and calculating the residues.

$$(3)\qquad \frac{1}{2\pi i}\int^{(0+)}t^{-\nu}J_\nu(2t\sin\phi)\,B_{2n+1;\,\mu,\nu}(t)\,dt=0.$$

$$(4)\quad \frac{1}{2\pi i}\int^{(0+)}t^{-\nu}J_\nu(2t\sin\phi)\,B_{2n;\,\mu,\nu}(t)\,dt$$
$$=\frac{2^{\mu+\nu}(\mu+\nu+2n)\,\Gamma(\mu+1)\,\Gamma(\mu+\nu+n)\sin^\nu\phi}{n!\,\Gamma(\mu+\nu+1)}$$
$$\times{}_3F_2(-n,\mu+1,\mu+\nu+n;\ \tfrac{1}{2}\mu+\tfrac{1}{2}\nu+\tfrac{1}{2},\ \tfrac{1}{2}\mu+\tfrac{1}{2}\nu+1;\ \sin^2\phi).$$

In the special case in which $\mu=\nu$, this reduces to

$$(5)\quad \frac{1}{2\pi i}\int^{(0+)}t^{-\nu}J_\nu(2t\sin\phi)\,B_{2n;\,\nu,\nu}(t)\,dt=2^{2\nu}(\nu+n)\,\Gamma(\nu)\sin^\nu\phi\,C_n^\nu(\cos2\phi).$$

This formula may be still further specialised by taking ϕ equal to $\frac{1}{4}\pi$ or $\frac{1}{2}\pi$.

* *Wiener Sitzungsberichte*, LXXV. (2), (1877), pp. 218—222.

9·6. *The genesis of Lommel's* polynomial $R_{m,\nu}(z)$.*

The recurrence formula

$$J_{\nu+1}(z) = (2\nu/z) J_\nu(z) - J_{\nu-1}(z)$$

may obviously be used to express $J_{\nu+m}(z)$ linearly in terms of $J_\nu(z)$ and $J_{\nu-1}(z)$; and the coefficients in this linear relation are polynomials in $1/z$ which are known as *Lommel's polynomials.* We proceed to shew how to obtain explicit expressions for them.

The result of eliminating $J_{\nu+1}(z), J_{\nu+2}(z), \ldots J_{\nu+m-1}(z)$ from the system of equations

$$J_{\nu+p+1}(z) - \{2(\nu+p)/z\} J_{\nu+p}(z) + J_{\nu+p-1}(z) = 0, \qquad (p = 0, 1, \ldots m-1)$$

is easily seen to be

$$\begin{vmatrix} J_{\nu+m}(z), & -2z^{-1}(\nu+m-1), & 1, & \ldots\ldots & 0 \\ 0, & 1, & -2z^{-1}(\nu+m-2), & \ldots\ldots & 0 \\ 0, & 0, & 1, & \ldots\ldots & 0 \\ \cdots & \cdots & \cdots & \cdots & \cdots \\ \cdots & \cdots & \cdots & \cdots & \cdots \\ 0, & 0, & 0, & \ldots\ldots & 1 \\ J_\nu(z), & 0, & 0, & \ldots\ldots & -2z^{-1}(\nu+1) \\ J_{\nu-1}(z) - (2\nu/z) J_\nu(z), & 0, & 0, & \ldots\ldots & 1 \end{vmatrix} = 0.$$

By expanding in cofactors of the first column, we see that the cofactor of $J_{\nu+m}(z)$ is unity; and the cofactor of $(-)^{m-1} J_\nu(z)$ is

$$\begin{vmatrix} -2z^{-1}(\nu+m-1), & 1, & 0, & \ldots\ldots & 0, & 0 \\ 1, & -2z^{-1}(\nu+m-2), & 1, & \ldots\ldots & 0, & 0 \\ 0, & 1, & -2z^{-1}(\nu+m-3), & \ldots\ldots & 0, & 0 \\ \cdots & \cdots & \cdots & \cdots & \cdots & \cdots \\ \cdots & \cdots & \cdots & \cdots & \cdots & \cdots \\ 0, & 0, & 0, & \ldots\ldots & -2z^{-1}(\nu+1), & 1 \\ 0, & 0, & 0, & \ldots\ldots & 1, & -2z^{-1}\nu \end{vmatrix}$$

The cofactor of $(-)^{m-1} J_{\nu-1}(z)$ is this determinant modified by the suppression of the last row and column.

The cofactor of $(-)^{m-1} J_\nu(z)$ is denoted by the symbol $(-)^m R_{m,\nu}(z)$; and $R_{m,\nu}(z)$, thus defined, is called Lommel's polynomial. It is of degree m in $1/z$ and it is also of degree m in ν.

The effect of suppressing the last row and column of the determinant by which $R_{m,\nu}(z)$ is defined is equivalent to increasing ν and diminishing m by unity; and so the cofactor of $(-)^{m-1} J_{\nu-1}(z)$ is $(-)^{m-1} R_{m-1,\nu+1}(z)$.

Hence it follows that

$$J_{\nu+m}(z) - J_\nu(z) R_{m,\nu}(z) + J_{\nu-1}(z) R_{m-1,\nu+1}(z) = 0,$$

* *Math. Ann.* IV. (1871), pp. 108—116.

that is to say

$$(1) \qquad J_{\nu+m}(z) = J_\nu(z)\, R_{m,\nu}(z) - J_{\nu-1}(z)\, R_{m-1,\nu+1}(z).$$

It is easy to see that* $R_{m,\nu}(z)$ is the numerator of the last convergent of the continued fraction

$$2z^{-1}(\nu+m-1) - \frac{1}{2z^{-1}(\nu+m-2)-}\,\frac{1}{2z^{-1}(\nu+m-3)-}\cdots\frac{1}{-2z^{-1}\nu}.$$

The function $R_{m,\nu}(z)$ was defined by Lommel by means of equation (1). He then derived an explicit expression for the coefficients in the polynomial by a somewhat elaborate induction; it is, however, simpler to determine the coefficients by using the series for the product of two Bessel functions in the way which will be explained in § 9·61.

It had been observed by Bessel, *Berliner Abh.*, 1824, p. 32, that, in consequence of the recurrence formulae, polynomials $A_{n-1}(z)$, $B_{n-1}(z)$ exist such that

$$J_n(z) = -\frac{2.4\ldots(2n-2)}{z^{n-1}}\{A_{n-1}(z)\,J_0(z) - B_{n-1}(z)\,J_1(z)\},$$

where [cf. § 9·62 (8)]

$$A_{n-1}(z)\,B_n(z) - A_n(z)\,B_{n-1}(z) = \frac{-z^{2n-1}}{2^2.4^2\ldots(2n-2)^2\,2n}.$$

It should be noticed that Graf† and Crelier‡ use a notation which differs from Lommel's notation; they write equation (1) in the form

$$J_{a+m}(x) = P_m^{a-1}(x)\,J_a(x) - P_{m-1}^{a}(x)\,J_{a-1}(x).$$

9·61. *The series for Lommel's polynomial.*

It is easy to see that $(-)^m J_{-\nu-m}(z)$, *qua* function of the integer m, satisfies the same recurrence formulae as $J_{\nu+m}(z)$; and hence the analysis of § 9·6 also shews that

$$(1) \qquad (-)^m J_{-\nu-m}(z) = J_{-\nu}(z)\, R_{m,\nu}(z) + J_{-\nu+1}(z)\, R_{m-1,\nu+1}(z).$$

Multiply this equation by $J_{\nu-1}(z)$ and § 9·6 (1) by $J_{-\nu+1}(z)$, and add the results. It follows that

$$(2) \qquad J_{\nu+m}(z)\,J_{-\nu+1}(z) + (-)^m J_{-\nu-m}(z)\,J_{\nu-1}(z)$$
$$= R_{m,\nu}(z)\{J_\nu(z)\,J_{-\nu+1}(z) + J_{-\nu}(z)\,J_{\nu-1}(z)\}$$
$$= \frac{2\sin\nu\pi}{\pi z} R_{m,\nu}(z),$$

* Cf. Chrystal, *Algebra*, II. (1900), p. 502.

† *Ann. di Mat.* (2) XXIII. (1895), pp. 45—65; *Einleitung in die Theorie der Bessel'schen Funktionen*, II. (Bern, 1900), pp. 98—109.

‡ *Ann. di Mat.* (2) XXIV. (1896), pp. 131—163.

by § 3·2 (7). But, by § 5·41, we have

$$\begin{cases} J_{\nu+m}(z)\,J_{-\nu+1}(z) = \sum_{p=0}^{\infty} \dfrac{(-)^p\,(m+p+2)_p\,(\frac{1}{2}z)^{m+2p+1}}{p!\,\Gamma(\nu+m+p+1)\,\Gamma(-\nu+2+p)}, \\ J_{-\nu-m}(z)\,J_{\nu-1}(z) = \sum_{n=0}^{\infty} \dfrac{(-)^n\,(-m+n)_n\,(\frac{1}{2}z)^{-m+2n-1}}{n!\,\Gamma(-\nu-m+n+1)\,\Gamma(\nu+n)} \\ \qquad = \sum_{n=0}^{m} \dfrac{(-)^n\,(-m+n)_n\,(\frac{1}{2}z)^{-m+2n-1}}{n!\,\Gamma(-\nu-m+n+1)\,\Gamma(\nu+n)} \\ \qquad + \sum_{p=0}^{\infty} \dfrac{(-)^{m+p+1}\,(p+1)_{m+p+1}\,(\frac{1}{2}z)^{m+2p+1}}{(m+p+1)!\,\Gamma(-\nu+2+p)\,\Gamma(\nu+m+p+1)}, \end{cases}$$

when we replace n in the last summation by $m+p+1$. Now it is clear that

$$\frac{(p+1)_{m+p+1}}{(m+p+1)!} = \frac{(m+2p+1)!}{p!\,(m+p+1)!} = \frac{(m+p+2)_p}{p!},$$

and so, when we combine the series for the products of the Bessel functions, we find that

$$\frac{2\sin\nu\pi}{\pi z} R_{m,\nu}(z) = \sum_{n=0}^{m} \frac{(-)^{m+n}\,(-m+n)_n\,(\frac{1}{2}z)^{-m+2n-1}}{n!\,\Gamma(-\nu-m+n+1)\,\Gamma(\nu+n)}$$
$$= \frac{\sin\nu\pi}{\pi} \sum_{n=0}^{\leqslant\frac{1}{2}m} \frac{(-)^n\,(m-n)!\,\Gamma(\nu+m-n)\,(\frac{1}{2}z)^{-m+2n-1}}{n!\,(m-2n)!\,\Gamma(\nu+n)};$$

the terms for which $n > \frac{1}{2}m$ vanish on account of the presence of the factor $(-m+n)_n$ in the numerator.

When ν is not an integer, we infer that

(3) $$R_{m,\nu}(z) = \sum_{n=0}^{\leqslant\frac{1}{2}m} \frac{(-)^n\,(m-n)!\,\Gamma(\nu+m-n)\,(\frac{1}{2}z)^{-m+2n}}{n!\,(m-2n)!\,\Gamma(\nu+n)}$$
$$= \sum_{n=0}^{\leqslant\frac{1}{2}m} (-)^n\,{}_{m-n}C_n \frac{\Gamma(\nu+m-n)}{\Gamma(\nu+n)} (\tfrac{1}{2}z)^{-m+2n}$$

But the original definition of $R_{m,\nu}(z)$, by means of a determinant, shews that $R_{m,\nu}(z)$ is a continuous function of ν for *all* values of ν, integral or not; and so, by an obvious limiting process, we infer that (3) is a valid expression for $R_{m,\nu}(z)$ even when ν is an integer. When ν is a negative integer it may be necessary to replace the quotient

$$\frac{\Gamma(\nu+m-n)}{\Gamma(\nu+n)} \text{ by } (-)^m \frac{\Gamma(-\nu-n+1)}{\Gamma(-\nu-m+n+1)}$$

in part of the series.

The series (3) was given by Lommel, *Math. Ann.* IV. (1871), pp. 108—111; an equivalent result, in a different notation, had, however, been published by him ten years earlier, *Archiv der Math. und Phys.* XXXVII. (1861), pp. 354—355.

An interesting result, depending on the equivalence of the quotients just mentioned, was first noticed by Graf*, namely that

(4) $$R_{m,\nu}(z) = (-)^m R_{m,-\nu-m+1}(z).$$

* *Ann. di Mat.* (2) XXIII. (1895), p. 56.

In the notation of Pochhammer (cf. §§ 4·4, 4·42), we have

(5) $R_{m,\nu}(z) = (\nu)_m (\tfrac{1}{2}z)^{-m} \,.\, {}_2F_3(\tfrac{1}{2} - \tfrac{1}{2}m, -\tfrac{1}{2}m;\ \nu, -m, 1-\nu-m;\ -z^2).$

Since $R_{m,\nu}(z)/z$ is a linear combination of products of cylinder functions of orders $\nu + m$ and $\nu - 1$, it follows from § 5·4 that it is annihilated by the operator

$$[\vartheta^4 - 2\{(\nu+m)^2 + (\nu-1)^2\}\vartheta^2 + \{(\nu+m)^2 - (\nu-1)^2\}^2] + 4z^2(\vartheta^2 + 3\vartheta + 2);$$

where $\vartheta \equiv z(d/dz)$; and so $R_{m,\nu}(z)$ is a solution of the differential equation

(6) $$[(\vartheta + m)(\vartheta + 2\nu + m - 2)(\vartheta - 2\nu - m)(\vartheta - m - 2)]\,y + 4z^2\vartheta(\vartheta+1)\,y = 0.$$

An equation equivalent to this was stated by Hurwitz, *Math. Ann.* XXXIII. (1889), p. 251; and a lengthy proof of it was given by Nielsen, *Ann. di Mat.* (3) VI. (1901), pp. 332—334; a simple proof, differing from the proof just given, may be obtained from formula (5).

9·62. *Various properties of Lommel's polynomial.*

We proceed to enumerate some theorems concerning $R_{m,\nu}(z)$, which were published by Lommel in his memoir of 1871.

In the first place, § 9·6 (1) holds if the Bessel functions are replaced by any other functions satisfying the same recurrence formulae; and, in particular,

(1) $$\mathscr{Y}_{\nu+m}(z) = \mathscr{Y}_\nu(z)\,R_{m,\nu}(z) - \mathscr{Y}_{\nu-1}(z)\,R_{m-1,\nu+1}(z),$$

whence it follows that

(2) $$\mathscr{Y}_{\nu+m}(z)\,J_{\nu-1}(z) - J_{\nu+m}(z)\,\mathscr{Y}_{\nu-1}(z)$$
$$= R_{m,\nu}(z)\{\mathscr{Y}_\nu(z)\,J_{\nu-1}(z) - J_\nu(z)\,\mathscr{Y}_{\nu-1}(z)\} = -2R_{m,\nu}(z)/(\pi z).$$

Next, in § 9·61 (2), take m to be an even integer; replace m by $2m$, and ν by $\nu - m$. The equation then becomes

(3) $$J_{\nu+m}(z)J_{m+1-\nu}(z) + J_{-\nu-m}(z)J_{-m-1+\nu}(z) = 2(-)^m \sin\nu\pi \,.\, R_{2m,\nu-m}(z)/(\pi z),$$

and, in the special case $\nu = \tfrac{1}{2}$, we get

(4) $$J^2_{m+\frac{1}{2}}(z) + J^2_{-m-\frac{1}{2}}(z) = 2(-)^m R_{2m,\frac{1}{2}-m}(z)/(\pi z),$$

that is to say

(5) $$J^2_{m+\frac{1}{2}}(z) + J^2_{-m-\frac{1}{2}}(z) = \frac{2}{\pi z}\sum_{n=0}^{m} \frac{(2z)^{2n-2m}(2m-n)!\,(2m-2n)!}{\{(m-n)!\}^2 \,.\, n!}.$$

This is the special case of the asymptotic expansion of § 7·51 when the order is half of an odd integer.

In particular, we have

(6) $$\begin{cases} J^2_{\frac{1}{2}}(z) + J^2_{-\frac{1}{2}}(z) = \dfrac{2}{\pi z}, \\ J^2_{\frac{3}{2}}(z) + J^2_{-\frac{3}{2}}(z) = \dfrac{2}{\pi z}\left(1 + \dfrac{1}{z^2}\right), \\ J^2_{\frac{5}{2}}(z) + J^2_{-\frac{5}{2}}(z) = \dfrac{2}{\pi z}\left(1 + \dfrac{3}{z^2} + \dfrac{9}{z^4}\right), \\ J^2_{\frac{7}{2}}(z) + J^2_{-\frac{7}{2}}(z) = \dfrac{2}{\pi z}\left(1 + \dfrac{6}{z^2} + \dfrac{45}{z^4} + \dfrac{225}{z^6}\right). \end{cases}$$

Formula (5) was published in 1870 by Lommel*, who derived it at that time by a direct multiplication of the expansions (§ 3·4)

$$J_{m+\frac{1}{2}}(z) \mp (-)^m iJ_{-m-\frac{1}{2}}(z) = \left(\frac{2}{\pi z}\right)^{\frac{1}{2}} (\mp i)^{m+1} e^{\pm iz} \sum_{r=0}^{m} \frac{(m-r+1)_{2r}\,(\pm i)^r}{2^r \,.\, r!\, z^r},$$

followed by a somewhat lengthy induction to determine the coefficients in the product.

As special cases of § 9·6 (1) and § 9·61 (1), we have

$$(7) \quad \begin{cases} J_{m+\frac{1}{2}}(z) = \left(\dfrac{2}{\pi z}\right)^{\frac{1}{2}} \sin z \,.\, R_{m,\frac{1}{2}}(z) - \left(\dfrac{2}{\pi z}\right)^{\frac{1}{2}} \cos z \,.\, R_{m-1,\frac{3}{2}}(z), \\ (-)^m J_{-m-\frac{1}{2}}(z) = \left(\dfrac{2}{\pi z}\right)^{\frac{1}{2}} \cos z \,.\, R_{m,\frac{1}{2}}(z) + \left(\dfrac{2}{\pi z}\right)^{\frac{1}{2}} \sin z \,.\, R_{m-1,\frac{3}{2}}(z). \end{cases}$$

By squaring and adding we deduce from (4) that†

$$(8) \qquad R^2_{m,\frac{1}{2}}(z) + R^2_{m-1,\frac{3}{2}}(z) = (-)^m R_{2m,\frac{1}{2}-m}(z).$$

Finally, if, in § 9·61 (2), we replace m by the odd integer $2m+1$ and then replace ν by $\nu - m$, we get

$$(9) \quad J_{\nu+m+1}(z)\, J_{-\nu+m+1}(z) - J_{-\nu-m-1}(z)\, J_{\nu-m-1}(z) = 2(-)^m \sin \nu\pi\, R_{2m+1,\nu-m}(z)/(\pi z).$$

An interesting result, pointed out by Nielsen, *Ann. di Mat.* (3) v. (1901), p. 23, is that if we have any identity of the type $\sum_{m=0}^{n} \cdot f_m(z)\, J_{\nu+m}(z) \equiv 0$, where the functions $f_m(z)$ are algebraic in z, we can at once infer the two identities

$$\sum_{m=0}^{n} f_m(z)\, R_{m,\nu}(z) \equiv 0, \qquad \sum_{m=0}^{n} f_m(z)\, R_{m-1,\nu+1}(z) \equiv 0,$$

by writing the postulated identity in the form

$$\sum_{m=0}^{n} f_m(z)\, \{J_\nu(z)\, R_{m,\nu}(z) - J_{\nu-1}(z)\, R_{m-1,\nu+1}(z)\} \equiv 0,$$

and observing that, by § 4·74 combined with § 3·2 (3), the quotient $J_{\nu-1}(z)/J_\nu(z)$ is not an algebraic function. Nielsen points out in this memoir, and its sequel, *ibid.* (3) VI. (1901), pp. 331—340, that this result leads to many interesting expansions in series of Lommel's polynomials; some of these formulae will be found in his *Handbuch der Theorie der Cylinderfunktionen* (Leipzig, 1904), but they do not seem to be of sufficient practical importance to justify their insertion here.

9·63. *Recurrence formulae for Lommel's polynomial.*

In the fundamental formula

$$J_{\nu+m}(z) = J_\nu(z)\, R_{m,\nu}(z) - J_{\nu-1}(z)\, R_{m-1,\nu+1}(z),$$

replace m and ν by $m+1$ and $\nu - 1$; on comparing the two expressions for $J_{\nu+m}(z)$, we see that

$$J_{\nu-1}(z)\, \{R_{m-1,\nu+1}(z) + R_{m+1,\nu-1}(z)\} = \{J_\nu(z) + J_{\nu-2}(z)\}\, R_{m,\nu}(z).$$

* *Math. Ann.* II. (1870), pp. 627—632.

† This result was obtained by Lommel, *Math. Ann.* IV. (1871), pp. 115—116.

Divide by $J_{\nu-1}(z)$, which is not identically zero, and it is apparent that

(1) $$R_{m-1,\nu+1}(z)+R_{m+1,\nu-1}(z)=\frac{2(\nu-1)}{z}R_{m,\nu}(z).$$

To obtain another recurrence formula, we replace m in § 9·62 (2) by $m+1$ and $m-1$, and use the recurrence formula connecting Bessel functions of orders $\nu+m-1$, $\nu+m$ and $\nu+m+1$; it is then seen that

(2) $$R_{m-1,\nu}(z)+R_{m+1,\nu}(z)=\frac{2(\nu+m)}{z}R_{m,\nu}(z),$$

and hence, by combining (1) and (2),

(3) $$R_{m-1,\nu}(z)+R_{m+1,\nu}(z)-R_{m-1,\nu+1}(z)-R_{m+1,\nu-1}(z)=\frac{2(m+1)}{z}R_{m,\nu}(z).$$

Again, write § 9·62 (2) in the form

$$\frac{2}{\pi}z^{-m-2}R_{m,\nu}(z)=\{z^{-\nu-m}J_{\nu+m}(z)\}\{z^{\nu-1}Y_{\nu-1}(z)\}-\{z^{-\nu-m}Y_{\nu+m}(z)\}\{z^{\nu-1}J_{\nu-1}(z)\},$$

and differentiate it. We deduce that

(4) $$R'_{m,\nu}(z)=\frac{m+2}{z}R_{m,\nu}(z)+R_{m+1,\nu-1}(z)-R_{m+1,\nu}(z),$$

and so, by (3), (1) and (2),

(5) $$R'_{m,\nu}(z)=-\frac{m}{z}R_{m,\nu}(z)+R_{m-1,\nu}(z)-R_{m-1,\nu+1}(z),$$

(6) $$R'_{m,\nu}(z)=\frac{2\nu+m}{z}R_{m,\nu}(z)-R_{m-1,\nu+1}(z)-R_{m+1,\nu}(z),$$

(7) $$R'_{m,\nu}(z)=-\frac{2\nu+m-2}{z}R_{m,\nu}(z)+R_{m+1,\nu-1}(z)+R_{m-1,\nu}(z).$$

The majority of these formulae were given by Lommel, *Math. Ann.* IV. (1871), pp. 113—116, but (6) is due to Nielsen, *Ann. di Mat.* (3) VI. (1901), p. 332; formula (2) has been used by Porter, *Annals of Math.* (2) III. (1901), p. 66, in discussing the zeros of $R_{m,\nu}(z)$.

It is evident that (2) may be used to *define* $R_{m,\nu}(z)$, when the parameter m is zero or a negative integer; thus, if (2) is to hold for all integral values of m, we find in succession from the formulae

$$R_{2,\nu}(z)=\frac{4\nu(\nu+1)}{z^2}-1,\quad R_{1,\nu}(z)=\frac{2\nu}{z},$$

that

(8) $$R_{0,\nu}(z)=1,\quad R_{-1,\nu}(z)=0,\quad R_{-2,\nu}(z)=-1,$$

and hence generally, by induction,

(9) $$R_{-m,\nu}(z)=(-)^{m-1}R_{m-2,2-\nu}(z).$$

This formula was given by Graf, *Ann. di Mat.* (2) XXIII. (1895), p. 59.

If we compare (9) with Graf's other formula, § 9·61 (4), we find that

(10) $$R_{m,\nu}(z)=(-)^{m-1}R_{-m-2,2-\nu}(z)=-R_{-m-2,\nu+m+1}(z)=(-)^m R_{m,-\nu-m+1}(z).$$

When the functions of negative parameter are defined by equation (9), all the formulae (1)—(7) are true for negative as well as positive values of m.

9·64. *Three-term relations connecting Lommel polynomials.*

It is possible to deduce from the recurrence formulae a class of relations which has been discussed by Crelier*. The relations were obtained by Crelier from the theory of continued fractions.

First observe that § 9·63 (2) shews that $J_{\nu+m}(z)$ and $R_{m,\nu}(z)$, *qua* functions of m, satisfy precisely the same recurrence formula connecting three contiguous functions; and so a repetition of the arguments of § 9·6 (modified by replacing the Bessel functions by the appropriate Lommel polynomials) shews that

(1) $$R_{m+n,\nu}(z) = R_{m,\nu}(z)\, R_{n,\nu+m}(z) - R_{m-1,\nu}(z)\, R_{n-1,\nu+m+1}(z).$$

Next in § 9·63 (2) replace m by $m-1$ and ν by $\nu+1$, and eliminate $2(m+\nu)/z$ from the two equations; it is then seen that

$$R_{m,\nu}(z)\, R_{m,\nu+1}(z) - R_{m+1,\nu}(z)\, R_{m-1,\nu+1}(z)$$
$$= R_{m-1,\nu}(z)\, R_{m-1,\nu+1}(z) - R_{m,\nu}(z)\, R_{m-2,\nu+1}(z),$$

and so the value of the function on the left is unaffected by changing m into $m-1$. It is consequently independent of m; and, since its value when $m=0$ is unity, we have Crelier's formula

(2) $$R_{m,\nu}(z)\, R_{m,\nu+1}(z) - R_{m+1,\nu}(z)\, R_{m-1,\nu+1}(z) = 1,$$

a result essentially due to Bessel (cf. § 9·6) in the special case $\nu = 0$.

More generally, if in § 9·63 (2) we had replaced m by $m-n$ and ν by $\nu+n$, we should have similarly found that

$$R_{m,\nu}(z)\, R_{m-n+1,\nu+n}(z) - R_{m+1,\nu}(z)\, R_{m-n,\nu+n}(z)$$
$$= R_{m-1,\nu}(z)\, R_{m-n,\nu+n}(z) - R_{m,\nu}(z)\, R_{m-n-1,\nu+n}(z),$$

and so the value of the function on the left is unaffected by changing m into $m-1$. It is consequently independent of m; and since its value when $m=n$ is $R_{n-1,\nu}(z)$, we find from § 9·63 (10) that

(3) $$R_{m,\nu}(z)\, R_{m-n+1,\nu+n}(z) - R_{m+1,\nu}(z)\, R_{m-n,\nu+n}(z) = R_{n-1,\nu}(z),$$

a result given in a different form by Lommel†.

Replace m and n by $m-1$ and $n+1$ in this equation, and it is found that

(4) $$R_{m-1,\nu}(z)\, R_{m-n-1,\nu+n+1}(z) - R_{m,\nu}(z)\, R_{m-n-2,\nu+n+1}(z) = R_{n,\nu}(z).$$

If we rewrite this equation with p in place of n and eliminate $R_{m-1,\nu}(z)$ between the two equations, we see that

$$R_{n,\nu}(z)\, R_{m-p-1,\nu+p+1}(z) - R_{p,\nu}(z)\, R_{m-n-1,\nu+n+1}(z)$$
$$= R_{m,\nu}(z)\left[R_{m-p-2,\nu+p+1}(z)\, R_{m-n-1,\nu+n+1}(z) - R_{m-n-2,\nu+n+1}(z)\, R_{m-p-1,\nu+p+1}(z)\right]$$
$$= R_{m,\nu}(z)\, R_{n-p-1,\nu+p+1}(z),$$

by (3). If we transform the second factor of each term by means of § 9·63 (10), we obtain Crelier's result (*loc. cit.* p. 143),

(5) $$R_{n,\nu}(z)\, R_{p-m-1,\nu+m+1}(z) - R_{p,\nu}(z)\, R_{n-m-1,\nu+m+1}(z)$$
$$= R_{m,\nu}(z)\, R_{p-n-1,\nu+n+1}(z).$$

* *Ann. di Mat.* (2) XXIV. (1896), p. 136 *et seq.*

† *Math. Ann.* IV. (1871), p. 115.

This is the most general linear relation of the types considered by Crelier; it connects any three polynomials $R_{m,\nu}(z)$, $R_{n,\nu}(z)$, $R_{p,\nu}(z)$ which have the same parameter ν and the same argument z. The formula may be written more symmetrically

$$(6)\quad R_{n,\nu}(z)\, R_{p-m-1,\nu+m+1}(z) + R_{p,\nu}(z)\, R_{m-n-1,\nu+n+1}(z) + R_{m,\nu}(z)\, R_{n-p-1,\nu+p+1}(z) = 0,$$

that is to say

$$(7)\quad \sum_{m,n,p} R_{n,\nu}(z)\, R_{p-m-1,\nu+m+1}(z) = 0.$$

A similar result may be obtained which connects any three Bessel functions whose orders differ by integers. If we eliminate $J_{\nu+m-1}(z)$ between the equations*

$$\begin{cases} J_{\nu+n}(z) = J_{\nu+m}(z)\, R_{n-m,\nu+m}(z) - J_{\nu+m-1}(z)\, R_{n-m-1,\nu+m+1}(z), \\ J_{\nu+p}(z) = J_{\nu+m}(z)\, R_{p-m,\nu+m}(z) - J_{\nu+m-1}(z)\, R_{p-m-1,\nu+m+1}(z), \end{cases}$$

we find that

$$\begin{aligned} J_{\nu+n}(z)\, R_{p-m-1,\nu+m+1}(z) &- J_{\nu+p}(z)\, R_{n-m-1,\nu+m+1}(z) \\ &= J_{\nu+m}(z)\,[R_{n-m,\nu+m}(z)\, R_{p-m-1,\nu+m+1}(z) - R_{p-m,\nu+m}(z)\, R_{n-m-1,\nu+m+1}(z)] \\ &= J_{\nu+m}(z)\, R_{p-n-1,\nu+n+1}(z); \end{aligned}$$

the last expression is obtained from a special case of (5) derived by replacing m, n, p, ν by $0, n-m, p-m, \nu+m$ respectively.

It follows that

$$(8)\quad \sum_{m,n,p} J_{\nu+n}(z)\, R_{p-m-1,\nu+m+1}(z) = 0,$$

and obviously we can prove the more general equation

$$(9)\quad \sum_{m,n,p} \mathscr{C}_{\nu+n}(z)\, R_{p-m-1,\nu+m+1}(z) = 0,$$

where $\mathscr{C}$ denotes any cylinder function.

The last two formulae seem never to have been previously stated explicitly, though Graf and Gubler hint at the existence of such equations, *Einleitung in die Theorie der Bessel'schen Funktionen*, II. (Bern, 1900), pp. 108, 109.

[NOTE. If we eliminate $J_{\nu-1}(z)$ from the equations

$$\begin{cases} J_{\nu+m}(z) = J_\nu(z)\, R_{m,\nu}(z) - J_{\nu-1}(z)\, R_{m-1,\nu+1}(z), \\ J_{\nu+m-1}(z) = J_\nu(z)\, R_{m-1,\nu}(z) - J_{\nu-1}(z)\, R_{m-2,\nu+1}(z), \end{cases}$$

and use (2) to simplify the resulting equation, we find that

$$J_\nu(z) = -J_{\nu+m}(z)\, R_{m-2,\nu+1}(z) + J_{\nu+m-1}(z)\, R_{m-1,\nu+1}(z),$$

and so, replacing ν by $\nu-m$, we have

$$J_{\nu-m}(z) = -J_\nu(z)\, R_{m-2,\nu-m+1}(z) + J_{\nu-1}(z)\, R_{m-1,\nu-m+1}(z).$$

By using § 9·63 (10), we deduce that

$$J_{\nu-m}(z) = J_\nu(z)\, R_{-m,\nu}(z) - J_{\nu-1}(z)\, R_{-m-1,\nu+1}(z),$$

that is to say that the equation § 9·6 (1), which has hitherto been considered only for positive values of the parameter m, is still true for negative values.]

* It is supposed temporarily that m is the smallest of the integers m, n, p; but since the final result is symmetrical, this restriction may be removed. See also the note at the end of the section.

9·65. *Hurwitz' limit of a Lommel polynomial.*

We shall now prove that

(1) $$\lim_{m\to\infty} \frac{(\frac{1}{2}z)^{\nu+m} R_{m,\nu+1}(z)}{\Gamma(\nu+m+1)} = J_\nu(z).$$

This result was applied by Hurwitz, *Math. Ann.* XXXIII. (1889), pp. 250—252, to discuss the reality of the zeros of $J_\nu(z)$ when ν has an assigned real value (§ 15·27). It has also been examined by Graf, *Ann. di Mat.* (2) XXIII. (1895), pp. 49—52, and by Crelier, *Bern Mittheilungen*, 1897, pp. 92—96.

From § 9·61 (3) we have

$$\frac{(\frac{1}{2}z)^{\nu+m} R_{m,\nu+1}(z)}{\Gamma(\nu+m+1)} = \sum_{n=0}^{\leqslant \frac{1}{2}m} \frac{(-)^n (\frac{1}{2}z)^{\nu+2n}}{n!\,\Gamma(\nu+n+1)} \cdot \frac{(m-n)!\,\Gamma(\nu+m-n+1)}{(m-2n)!\,\Gamma(\nu+m+1)}.$$

Now write

$$\frac{(m-n)!\,\Gamma(\nu+m-n+1)}{(m-2n)!\,\Gamma(\nu+m+1)} \equiv \theta(m, n),$$

so that

$$\theta(m, n) = \frac{(m-n)(m-n-1)\dots(m-2n+1)}{(\nu+m)(\nu+m-1)\dots(\nu+m-n+1)}.$$

If now N be the greatest integer contained in $|\nu|$, then each factor in the numerator of $\theta(m, n)$ is numerically less than the corresponding factor in the denominator, provided that $n > N$.

Hence, when $n > N$, and $m > 2N$,

$$|\theta(m, n)| < 1,$$

while, when n has any *fixed* value,

$$\lim_{m\to\infty} \theta(m, n) = 1.$$

Since $$\sum_{n=0}^{\infty} \frac{(-)^n (\frac{1}{2}z)^{\nu+2n}}{n!\,\Gamma(\nu+n+1)}$$

is absolutely convergent, it follows from Tannery's theorem* that

$$\lim_{m\to\infty} \sum_{n=0}^{\leqslant \frac{1}{2}m} \frac{(-)^n (\frac{1}{2}z)^{\nu+2n}}{n!\,\Gamma(\nu+n+1)} \theta(m, n) = \sum_{n=0}^{\infty} \frac{(-)^n (\frac{1}{2}z)^{\nu+2n}}{n!\,\Gamma(\nu+n+1)},$$

and the theorem of Hurwitz is established.

Again, since the convergence of $\sum_{n=0}^{\infty} \frac{(-)^n (\frac{1}{2}z)^{\nu+2n}}{n!\,\Gamma(\nu+n+1)}$ is uniform in any bounded domain† of values of z (by the test due to Weierstrass), it follows that the convergence of

$$(\tfrac{1}{2}z)^{\nu+m} R_{m,\nu+1}(z)/\Gamma(\nu+m+1)$$

to its limit is also uniform in any bounded domain of values of z.

* Cf. Bromwich, *Theory of Infinite Series*, § 49.

† An arbitrarily small region of which the origin is an internal point must obviously be excluded from this domain when $R(\nu) \leqslant 0$.

From the theorem of Hurwitz it is easy to derive an infinite continued fraction for $J_{\nu-1}(z)/J_\nu(z)$. For, when $J_\nu(z)\neq 0$, we have

$$\frac{J_{\nu-1}(z)}{J_\nu(z)}=\lim_{m\to\infty}\left[\frac{R_{m+1,\nu}(z)}{R_{m,\nu+1}(z)}\right]$$
$$=\lim_{m\to\infty}\left[2\nu z^{-1}-1\Big/\frac{R_{m,\nu+1}(z)}{R_{m-1,\nu+2}(z)}\right],$$

by § 9·63 (1). On carrying out the process of reduction and noticing that

$$\frac{R_{2,\nu+m-1}(z)}{R_{1,\nu+m}(z)}=2(\nu+m-1)z^{-1}-\frac{1}{2z^{-1}(\nu+m)},$$

we find that

$$\frac{R_{m+1,\nu}(z)}{R_{m,\nu+1}(z)}=2\nu z^{-1}-\frac{1}{2(\nu+1)z^{-1}-}\,\frac{1}{2(\nu+2)z^{-1}-}\,\dots\,\frac{1}{-2(\nu+m)z^{-1}},$$

and hence

(2) $$\frac{J_{\nu-1}(z)}{J_\nu(z)}=2\nu z^{-1}-\frac{1}{2(\nu+1)z^{-1}-}\,\frac{1}{2(\nu+2)z^{-1}-}\dots$$

This procedure avoids the necessity of proving directly that, when $m\to\infty$, the last element of the continued fraction

$$\frac{J_{\nu-1}(z)}{J_\nu(z)}=2\nu z^{-1}-\frac{1}{2(\nu+1)z^{-1}-}\dots\frac{1}{-2(\nu+m)z^{-1}-}\,\frac{J_{\nu+m+1}(z)}{J_{\nu+m}(z)}$$

may be neglected; the method is due to Graf, *Ann. di Mat.* (2) XXIII. (1895), p. 52.

9·7. *The modified notation for Lommel polynomials.*

In order to discuss properties of the zeros of Lommel polynomials, it is convenient to follow Hurwitz by making a change in the notation, for the reason that Lommel polynomials contain only alternate powers of the variable.

Accordingly we define the modified Lommel polynomial $g_{m,\nu}(z)$ by the equation*

(1) $$g_{m,\nu}(z)=\sum_{n=0}^{\leq\frac{1}{2}m}{}_{m-n}C_n\,\frac{(-)^n\,\Gamma(\nu+m-n+1)\,z^n}{\Gamma(\nu+n+1)},$$

so that

(2) $$R_{m,\nu+1}(z)=(\tfrac{1}{2}z)^{-m}\,g_{m,\nu}(\tfrac{1}{4}z^2).$$

By making the requisite changes in notation in §§ 9·63, 9·64, the reader will easily obtain the following formulae:

(3) $$g_{m+1,\nu}(z)=(\nu+m+1)\,g_{m,\nu}(z)-zg_{m-1,\nu}(z),\qquad [\S\,9{\cdot}63\,(2)]$$

(4) $$g_{m+1,\nu-1}(z)=\nu g_{m,\nu}(z)-zg_{m-1,\nu+1}(z),\qquad [\S\,9{\cdot}63\,(1)]$$

(5) $$\frac{1}{z^{\nu-1}}\frac{d}{dz}\{z^\nu g_{m,\nu}(z)\}=zg_{m-1,\nu}(z)+g_{m+1,\nu-1}(z),\qquad [\S\,9{\cdot}63\,(7)]$$

(6) $$z^{m+2}\frac{d}{dz}\{z^{-m-1}\,g_{m,\nu}(z)\}=g_{m+1,\nu-1}(z)-g_{m+1,\nu}(z),\qquad [\S\,9{\cdot}63\,(4)]$$

(7) $$g_{m,\nu}(z)\,g_{m+1,\nu+1}(z)-g_{m+2,\nu}(z)\,g_{m-1,\nu+1}(z)=z^m\,g_{0,\nu}(z)\,g_{1,\nu+m+1}(z).$$

[A special case of § 9·64 (5).]

* This notation differs in unimportant details from the notation used by Hurwitz.

These results will be required in the sequel; it will not be necessary to write down the analogues of all the other formulae of §§ 9·6—9·64.

The result of eliminating alternate functions from the system (3) is of some importance. The eliminant is

$$(\nu + m)\, g_{m+2,\nu}(z) = c_m(z)\, g_{m,\nu}(z) - (\nu + m + 2)\, z^2 g_{m-2,\nu}(z),$$

where $$c_m(z) \equiv (\nu + m + 1)\{(\nu + m)(\nu + m + 2) - 2z\}.$$

We thus obtain the set of equations:

$$(8)\quad \begin{cases} (\nu+2)\, g_{4,\nu}(z) = c_2(z)\, g_{2,\nu}(z) - (\nu+4)\, z^2 g_{0,\nu}(z), \\ (\nu+4)\, g_{6,\nu}(z) = c_4(z)\, g_{4,\nu}(z) - (\nu+6)\, z^2 g_{2,\nu}(z), \\ \dots\dots\dots\dots\dots\dots\dots\dots\dots\dots \\ (\nu+2s)\, g_{2s+2,\nu}(z) = c_{2s}(z)\, g_{2s,\nu}(z) - (\nu+2s+2)\, z^2 g_{2s-2,\nu}(z), \\ \dots\dots\dots\dots\dots\dots\dots\dots\dots\dots \\ (\nu+2m-2)\, g_{2m,\nu}(z) = c_{2m-2}(z)\, g_{2m-2,\nu}(z) - (\nu+2m)\, z^2 g_{2m-4,\nu}(z). \end{cases}$$

9·71. *The reality of the zeros of $g_{2m,\nu}(z)$ when ν exceeds -2.*

We shall now give Hurwitz' proof of his theorem* that *when $\nu > -2$, the zeros of $g_{2m,\nu}(z)$ are all real; and also that they are all positive, except when $-1 > \nu > -2$, in which case one of them is negative.*

After observing that $g_{2m,\nu}(z)$ is a polynomial in z of degree m, we shall shew that the set of functions $g_{2m,\nu}(z)$, $g_{2m-2,\nu}(z)$, ... $g_{2,\nu}(z)$, $g_{0,\nu}(z)$ form a set of Sturm's functions. Sufficient conditions for this to be the case are (i) the existence of the set of relations § 9·7 (8), combined with (ii) the theorem that the real zeros of $g_{2m-2,\nu}(z)$ alternate with those of $g_{2m,\nu}(z)$.

To prove that the zeros alternate, it is sufficient to prove that the quotient $g_{2m,\nu}(z)/g_{2m-2,\nu}(z)$ is a monotonic function of the real variable z, except at the zeros of the denominator, where the quotient is discontinuous.

We have $$g^2_{2m-2,\nu}(z)\frac{d}{dz}\left\{\frac{g_{2m,\nu}(z)}{g_{2m-2,\nu}(z)}\right\} = -\mathfrak{W}_{2m,2m-2},$$

where $$\mathfrak{W}_{r,s} = g_{r,\nu}(z)\, g'_{s,\nu}(z) - g_{s,\nu}(z)\, g'_{r,\nu}(z);$$

and from § 9·7 (3) it follows that

$$\begin{cases} \mathfrak{W}_{2m,2m-2} = g^2_{2m-2,\nu}(z) + (\nu+2m)\,\mathfrak{W}_{2m-1,2m-2}, \\ \mathfrak{W}_{2m-1,2m-2} = z^2\,\mathfrak{W}_{2m-3,2m-4} + (\nu+2m-2)\, g^2_{2m-3,\nu}(z), \end{cases}$$

so that

$$\mathfrak{W}_{2m,2m-2} = g^2_{2m-2,\nu}(z) + (\nu+2m)\sum_{r=1}^{m-1}(\nu+2r)\, z^{2m-2r-2} g^2_{2r-1,\nu}(z),$$

and therefore, if $m \geqslant 1$, $\mathfrak{W}_{2m,2m-2}$ is expressible as a sum of positive terms when $\nu > -2$.

* *Math. Ann.* XXXIII. (1889), pp. 254—256.

The monotonic property is therefore established, and it is obvious from a graph that the real zeros of $g_{2m-2,\nu}(z)$ separate those of $g_{2m,\nu}(z)$.

It follows from Sturm's theorem that the number of zeros of $g_{2m,\nu}(z)$ on any interval of the real axis is the *excess* of the number of alternations of sign in the set of expressions $g_{2m,\nu}(z), g_{2m-2,\nu}(z), \ldots, g_{0,\nu}(z)$ at the right-hand end of the interval over the number of alternations at the left-hand end.

The reason why the number of zeros is the *excess* and not the *deficiency* is that the quotient $g_{2m,\nu}(z)/g_{2m-2,\nu}(z)$ is a decreasing function, and not an increasing function of z, as in the usual version of Sturm's theorem. See Burnside and Panton, *Theory of Equations*, I. (1918), § 96.

The arrangements of signs for the set of functions when z has the values $-\infty$, 0, ∞ are as follows:

	$2m$	$2m-2$	$2m-4$		2	0
$-\infty$	$+$	$+$	$+$	...	$+$	$+$
0	$\pm$	$\pm$	$\pm$	...	$\pm$	$+$
∞	$(-)^m$	$(-)^{m-1}$	$(-)^{m-2}$	...	$-$	$+$

The upper or lower signs are to be taken according as $\nu+1$ is positive or negative; and the truth of Hurwitz' theorem is obvious from an inspection of this Table.

9·72. *Negative zeros of $g_{2m,\nu}(z)$ when $\nu < -2$.*

Let ν be less than -2, and let the positive integer s be defined by the inequalities

$$-2s > \nu > -2s-2.$$

It will now be shewn that*, *when ν lies between $-2s$ and $-2s-1$, $g_{2m,\nu}(z)$ has no negative zero; but that, when ν lies between $-2s-1$ and $-2s-2$, $g_{2m,\nu}(z)$ has one negative zero. Provided that, in each case, m is taken to be so large that $\nu+2m$ is positive.*

It will first be shewn that the negative zeros (if any) of $g_{2m,\nu}(z)$ alternate with those of $g_{2m-2,\nu}(z)$.

* This proof differs from the proof given by Hurwitz; see *Proc. London Math. Soc.* (2) XIX. (1921), pp. 266—272.

By means of the formulae quoted in § 9·7, it is clear that

$$z^{-\nu-2m+2} g^2_{2m-2,\nu}(z) \frac{d}{dz}\left\{\frac{z^\nu g_{2m,\nu}(z)}{z^{-2m+1} g_{2m-2,\nu}(z)}\right\}$$
$$= g_{2m-2,\nu}(z)\{z g_{2m-1,\nu}(z) + g_{2m+1,\nu-1}(z)\} - g_{2m,\nu}(z)\{g_{2m-1,\nu-1}(z) - g_{2m-1,\nu}(z)\}$$
$$= (\nu+2m) g^2_{2m-1,\nu}(z) + g_{2m-2,\nu}(z) g_{2m+1,\nu-1}(z) - g_{2m-1,\nu-1}(z) g_{2m,\nu}(z)$$
$$= (\nu+2m) g^2_{2m-1,\nu}(z) - z^{2m-1} g_{0,\nu}(z) g_{1,\nu+2m-1}(z)$$
$$= (\nu+2m)\{g^2_{2m-1,\nu}(z) - z^{2m-1}\}$$
$$> 0,$$

provided that $\nu + 2m$ is positive and z is negative. Therefore, in the circumstances postulated, the quotient

$$(-z)^{\nu+2m-1} g_{2m,\nu}(z)/g_{2m-2,\nu}(z)$$

is a decreasing function, and the alternation of the zeros is evident.

The existence of the system of equations § 9·7 (8) now shews that the set of functions

$$g_{2m,\nu}(z),\ g_{2m-2,\nu}(z),\ \ldots,\ g_{2s+2,\nu}(z),\ g_{2s,\nu}(z),$$
$$-g_{2s-2,\nu}(z),\ +g_{2s-4,\nu}(z),\ \ldots,\ (-)^s g_{0,\nu}(z)$$

form a set of Sturm's functions.

The signs of these functions when z is $-\infty$ are

$$+,\ +,\ \ldots,\ +,\ +,\ -,\ +,\ \ldots,\ (-)^s,$$

and there are s alternations of sign. When z is zero, the signs of the functions are

$$\pm,\ \pm,\ \ldots,\ \pm,\ +,\ -,\ +,\ \ldots,\ (-)^s,$$

the upper signs being taken when $-2s > \nu > -2s-1$, and the lower signs being taken when $-2s-1 > \nu > -2s-2$; there are s and $s+1$ alternations of sign in the respective cases. Hence, when $-2s > \nu > -2s-1$, $g_{2m,\nu}(z)$ has no negative zero; but when $-2s-1 > \nu > -2s-2$, $g_{2m,\nu}(z)$ has one negative zero. The theorem stated is therefore proved.

9·73. *Positive and complex zeros of $g_{2m,\nu}(z)$ when $\nu < -2$.*

As in § 9·72, define the positive integer s by the inequalities

$$-2s > \nu > -2s-2.$$

It will now be shewn* that *when ν lies between $-2s$ and $-2s-1$, $g_{2m,\nu}(z)$ has $m-2s$ positive zeros; but that, when ν lies between $-2s-1$ and $-2s-2$, $g_{2m,\nu}(z)$ has $m-2s-1$ positive zeros. Provided that, in each case, m is so large that $m+\nu$ is positive.*

* This proof is of a more elementary character than the proof given by Hurwitz; see the paper cited in § 9·72.

In the first place, it follows from Descartes' rule of signs that, in each case, $g_{2m,\nu}(z)$ cannot have more than the specified number of positive zeros. For, when ν lies between $-2s$ and $-2s-1$, the signs of the coefficients of

$$1, z, z^2, \ldots, z^{2s}, z^{2s+1}, z^{2s+2}, z^{2s+3}, \ldots, z^m \text{ in } g_{2m,\nu}(z)$$

are
$$+, +, +, \ldots, +, -, +, -, \ldots, (-)^m;$$

and since there are $m-2s$ alternations of sign, there cannot be more than $m-2s$ positive zeros. When ν lies between $-2s-1$ and $-2s-2$ the corresponding set of signs is

$$-, -, -, \ldots, -, -, +, -, \ldots, (-)^m;$$

and since there are $m-2s-1$ alternations of sign there cannot be more than $m-2s-1$ positive zeros.

Next, we shall prove by induction from the system of equations § 9·7 (8) that there are as many as the specified number of positive zeros.

When ν lies between $-2s$ and $-2s-1$, the coefficients in $g_{4s,\nu}(z)$ have no alternations of sign (being all +) and so this function has no positive zeros. On the other hand

$$g_{4s+2,\nu}(0) > 0, \quad g_{4s+2,\nu}(+\infty) = -\infty,$$

and so $g_{4s+2,\nu}(z)$ has one positive zero, $\alpha_{1,1}$ say; and, by reasoning already given, it has no other positive zeros. Next, take $g_{4s+4,\nu}(z)$; from § 9·7 (8) it follows that its signs at $0, \alpha_{1,1}, +\infty$ are $+, -, +$; hence it has two positive zeros, and by the reasoning already given it has no others. The process of induction (whereby we prove that the zeros of each function separate those of the succeeding function) is now evident, and we infer that $g_{2m,\nu}(z)$ has $m-2s$ positive zeros, and no more.

Again, when ν lies between $-2s-1$ and $-2s-2$, the coefficients in $g_{4s+2,\nu}(z)$ have no alternations in sign (being all −), and so this function has no positive zeros. On the other hand

$$g_{4s+4,\nu}(0) < 0, \quad g_{4s+4,\nu}(+\infty) = +\infty,$$

and so $g_{4s+4,\nu}(z)$ has one positive zero, and by the reasoning already given it has no other positive zero.

By appropriate modifications of the preceding reasoning we prove in succession that $g_{4s+6,\nu}(z)$, $g_{4s+8,\nu}(z)$, ... have 2, 3, ... positive zeros, and in general that $g_{2m,\nu}(z)$ has $m-2s-1$ positive zeros.

By combining these results with the result of § 9·72, we obtain Hurwitz' theorem, that, *when $\nu < -2$, and m is so large that $m+\nu$ is positive, $g_{2m,\nu}(z)$ has $2s$ complex zeros, where s is the integer such that*

$$-2s > \nu > -2s-2.$$

CHAPTER X

FUNCTIONS ASSOCIATED WITH BESSEL FUNCTIONS

10·1. *The functions* $\mathbf{J}_\nu(z)$ *and* $\mathbf{E}_\nu(z)$ *investigated by Anger and H. F. Weber.*

In this chapter we shall examine the properties of various functions whose definitions are suggested by certain representations of Bessel functions. We shall first investigate functions defined by integrals resembling Bessel's integral and Poisson's integral, and, after discussing the properties of several functions connected with $Y_n(z)$ we shall study a class of functions, first defined by Lommel, of which Bessel functions are a particular case.

The first function to be examined, $\mathbf{J}_\nu(z)$, is suggested by Bessel's integral. It is defined by the equation

(1) $$\mathbf{J}_\nu(z) = \frac{1}{\pi}\int_0^\pi \cos(\nu\theta - z\sin\theta)\,d\theta.$$

This function obviously reduces to $J_n(z)$ when ν has the integral value n. It follows from § 6·2 (4) that, when ν is not an integer, the two functions are distinct. A function of the same type as $\mathbf{J}_\nu(z)$ was studied by Anger*, but he took the upper limit of the integral to be 2π; and the function $\mathbf{J}_\nu(z)$ is conveniently described as Anger's function of argument z and order ν.

A similar function was discussed later by H. F. Weber†, and he also investigated the function $\mathbf{E}_\nu(z)$ defined by the equation

(2) $$\mathbf{E}_\nu(z) = \frac{1}{\pi}\int_0^\pi \sin(\nu\theta - z\sin\theta)\,d\theta.$$

In connexion with this function reference should also be made to researches by Lommel, *Math. Ann.* XVI. (1880), pp. 183–208.

It may be noted that the function $\frac{1}{2\pi}\int_0^{2\pi}\cos(\nu\theta - z\sin\theta)\,d\theta$ which was actually discussed by Anger is easily expressible in terms of $\mathbf{J}_\nu(z)$ and $\mathbf{E}_\nu(z)$; for, if we replace θ by $2\pi-\theta$ in the right-hand half of the range of integration, we get

$$\frac{1}{2\pi}\int_0^{2\pi}\cos(\nu\theta - z\sin\theta)\,d\theta = \frac{1}{2\pi}\int_0^{\pi}\cos(\nu\theta - z\sin\theta)\,d\theta + \frac{1}{2\pi}\int_0^{\pi}\cos(2\nu\pi - \nu\theta + z\sin\theta)\,d\theta$$
$$= \cos^2\nu\pi\,.\,\mathbf{J}_\nu(z) + \sin\nu\pi\cos\nu\pi\,.\,\mathbf{E}_\nu(z).$$

* *Neueste Schriften der Naturf. Ges. in Danzig*, v. (1855), pp. 1—29. It was shewn by Poisson that

$$\nabla_\nu\int_0^\pi \cos(\nu\theta - z\sin\theta)\,d\theta = (z-\nu)\sin\nu\pi,$$

Additions à la Conn. des Temps, 1836, p. 15 (cf. § 10·12), but as he did no more it seems reasonable to give Anger's name to the function.

† *Zürich Vierteljahrsschrift*, XXIV. (1879), pp. 33—76. Weber omits the factor $1/\pi$ in his definition of $\mathbf{E}_\nu(z)$.

To expand $\mathbf{J}_\nu(z)$ and $\mathbf{E}_\nu(z)$ in ascending powers of z, write $\frac{1}{2}\pi+\phi$ for θ in the integrals and proceed thus:

$$\int_0^\pi \sin^m\theta\sin\nu\theta\,d\theta=\int_{-\frac{1}{2}\pi}^{\frac{1}{2}\pi}\cos^m\phi\sin(\tfrac{1}{2}\nu\pi+\nu\phi)\,d\phi$$

$$=2\sin\tfrac{1}{2}\nu\pi\int_0^{\frac{1}{2}\pi}\cos^m\phi\cos\nu\phi\,d\phi$$

$$=\frac{\pi\,.\,m!\sin\frac{1}{2}\nu\pi}{2^m\,\Gamma(\frac{1}{2}m-\frac{1}{2}\nu+1)\,\Gamma(\frac{1}{2}m+\frac{1}{2}\nu+1)},$$

by a formula due to Cauchy*.

In like manner,

$$\int_0^\pi \sin^m\theta\cos\nu\theta\,d\theta=\frac{\pi\,.\,m!\cos\frac{1}{2}\nu\pi}{2^m\,\Gamma(\frac{1}{2}m-\frac{1}{2}\nu+1)\,\Gamma(\frac{1}{2}m+\frac{1}{2}\nu+1)}.$$

But, evidently,

$$\mathbf{J}_\nu(z)=\frac{1}{\pi}\sum_{m=0}^\infty\frac{(-)^m z^{2m}}{(2m)!}\int_0^\pi\sin^{2m}\theta\cos\nu\theta\,d\theta+\frac{1}{\pi}\sum_{m=0}^\infty\frac{(-)^m z^{2m+1}}{(2m+1)!}\int_0^\pi\sin^{2m+1}\theta\sin\nu\theta\,d\theta,$$

so that

$$(3)\qquad \mathbf{J}_\nu(z)=\cos\tfrac{1}{2}\nu\pi\sum_{m=0}^\infty\frac{(-)^m(\frac{1}{2}z)^{2m}}{\Gamma(m-\frac{1}{2}\nu+1)\,\Gamma(m+\frac{1}{2}\nu+1)}$$
$$+\sin\tfrac{1}{2}\nu\pi\sum_{m=0}^\infty\frac{(-)^m(\frac{1}{2}z)^{2m+1}}{\Gamma(m-\frac{1}{2}\nu+\frac{3}{2})\,\Gamma(m+\frac{1}{2}\nu+\frac{3}{2})},$$

and similarly

$$(4)\qquad \mathbf{E}_\nu(z)=\sin\tfrac{1}{2}\nu\pi\sum_{m=0}^\infty\frac{(-)^m(\frac{1}{2}z)^{2m}}{\Gamma(m-\frac{1}{2}\nu+1)\,\Gamma(m+\frac{1}{2}\nu+1)}$$
$$-\cos\tfrac{1}{2}\nu\pi\sum_{m=0}^\infty\frac{(-)^m(\frac{1}{2}z)^{2m+1}}{\Gamma(m-\frac{1}{2}\nu+\frac{3}{2})\,\Gamma(m+\frac{1}{2}\nu+\frac{3}{2})}.$$

These results may be written in the alternative forms

$$(5)\qquad \mathbf{J}_\nu(z)=\frac{\sin\nu\pi}{\nu\pi}\left[1-\frac{z^2}{2^2-\nu^2}+\frac{z^4}{(2^2-\nu^2)(4^2-\nu^2)}-\frac{z^6}{(2^2-\nu^2)(4^2-\nu^2)(6^2-\nu^2)}+\dots\right]$$
$$+\frac{\sin\nu\pi}{\pi}\left[\frac{z}{1^2-\nu^2}-\frac{z^3}{(1^2-\nu^2)(3^2-\nu^2)}+\frac{z^5}{(1^2-\nu^2)(3^2-\nu^2)(5^2-\nu^2)}-\dots\right],$$

$$(6)\qquad \mathbf{E}_\nu(z)=\frac{1-\cos\nu\pi}{\nu\pi}\left[1-\frac{z^2}{2^2-\nu^2}+\frac{z^4}{(2^2-\nu^2)(4^2-\nu^2)}-\dots\right]$$
$$-\frac{1+\cos\nu\pi}{\pi}\left[\frac{z}{1^2-\nu^2}-\frac{z^3}{(1^2-\nu^2)(3^2-\nu^2)}+\frac{z^5}{(1^2-\nu^2)(3^2-\nu^2)(5^2-\nu^2)}-\dots\right].$$

Results equivalent to these were given by Anger and Weber.

The formula corresponding to (5) was given by Anger (before the publication of his memoir) in a letter to Cauchy which was communicated to the French Academy on July 17, 1854; see *Comptes Rendus*, XXXIX. (1854), pp. 128—135.

* *Mém. sur les intégrales définies* (Paris, 1825), p. 40. Cf. *Modern Analysis*, p. 263.

For a reason which will be apparent subsequently (§ 10·7), it is convenient to write

(7) $$s_{0,\nu}(z) \equiv \frac{z}{1^2-\nu^2} - \frac{z^3}{(1^2-\nu^2)(3^2-\nu^2)} + \frac{z^5}{(1^2-\nu^2)(3^2-\nu^2)(5^2-\nu^2)} - \ldots,$$

(8) $$s_{-1,\nu}(z) \equiv -\frac{1}{\nu^2} + \frac{z^2}{\nu^2(2^2-\nu^2)} - \frac{z^4}{\nu^2(2^2-\nu^2)(4^2-\nu^2)} + \ldots,$$

and, with this notation, we have

(9) $$\mathbf{J}_\nu(z) = \frac{\sin \nu\pi}{\pi} s_{0,\nu}(z) - \frac{\nu \sin \nu\pi}{\pi} s_{-1,\nu}(z),$$

(10) $$\mathbf{E}_\nu(z) = -\frac{1+\cos \nu\pi}{\pi} s_{0,\nu}(z) - \frac{\nu(1-\cos \nu\pi)}{\pi} s_{-1,\nu}(z).$$

It is easy to deduce the following formulae from these results:

(11) $$\int_0^\pi \cos \nu\theta \,.\, \cos(z \sin\theta)\, d\theta = -\nu \sin \nu\pi \,.\, s_{-1,\nu}(z),$$

(12) $$\int_0^\pi \sin \nu\theta \,.\, \cos(z \sin\theta)\, d\theta = -\nu(1-\cos \nu\pi) \,.\, s_{-1,\nu}(z),$$

(13) $$\int_0^\pi \sin \nu\theta \,.\, \sin(z \sin\theta)\, d\theta = \sin \nu\pi \,.\, s_{0,\nu}(z),$$

(14) $$\int_0^\pi \cos \nu\theta \,.\, \sin(z \sin\theta)\, d\theta = (1+\cos \nu\pi) \,.\, s_{0,\nu}(z),$$

(15) $$\int_0^{\frac{1}{2}\pi} \cos \nu\phi \,.\, \cos(z \cos\phi)\, d\phi = -\nu \sin \tfrac{1}{2}\nu\pi \,.\, s_{-1,\nu}(z),$$

(16) $$\int_0^{\frac{1}{2}\pi} \cos \nu\phi \,.\, \sin(z \cos\phi)\, d\phi = \cos \tfrac{1}{2}\nu\pi \,.\, s_{0,\nu}(z).$$

Integrals somewhat resembling the integrals discussed in this section, namely

$$\int e^{\sin\theta} {\cos \atop \sin} (n\theta - \cos\theta)\, d\theta,$$

have been examined by Unferdinger, *Wiener Sitzungsberichte*, LVII. (2), (1868), pp. 611—620.

Also, Hardy, *Messenger*, XXXV. (1906), pp. 158—166, has investigated the integral

$$\int_0^\infty \sin(\nu\theta - z\sin\theta) \frac{d\theta}{\theta},$$

and has proved that, when ν is real, it is equal to $\frac{1}{2}\pi \sum_{n=-\infty}^{\infty} \eta_n J_n(z)$, where η_n is 1, 0 or −1 according as $\nu - n$ is positive, zero, or negative.

10·11. *Weber's formulae connecting his functions with Anger's functions.*

It is evident from the formulae § 10·1 (9), (10), (15) and (16) that

(1) $$\mathbf{J}_\nu(z) + \mathbf{J}_{-\nu}(z) = \frac{4\cos\frac{1}{2}\nu\pi}{\pi} \int_0^{\frac{1}{2}\pi} \cos \nu\phi \cos(z\cos\phi)\, d\phi,$$

(2) $$\mathbf{J}_\nu(z) - \mathbf{J}_{-\nu}(z) = \frac{4\sin\frac{1}{2}\nu\pi}{\pi} \int_0^{\frac{1}{2}\pi} \cos \nu\phi \sin(z\cos\phi)\, d\phi,$$

(3) $$\mathbf{E}_\nu(z)+\mathbf{E}_{-\nu}(z)=-\frac{4\cos\frac{1}{2}\nu\pi}{\pi}\int_0^{\frac{1}{2}\pi}\cos\nu\phi\sin(z\cos\phi)\,d\phi,$$

(4) $$\mathbf{E}_\nu(z)-\mathbf{E}_{-\nu}(z)=\frac{4\sin\frac{1}{2}\nu\pi}{\pi}\int_0^{\frac{1}{2}\pi}\cos\nu\phi\cos(z\cos\phi)\,d\phi.$$

It follows on addition that

$$\mathbf{J}_\nu(z)=\tfrac{1}{2}\cot\tfrac{1}{2}\nu\pi\,\{\mathbf{E}_\nu(z)-\mathbf{E}_{-\nu}(z)\}-\tfrac{1}{2}\tan\tfrac{1}{2}\nu\pi\,\{\mathbf{E}_\nu(z)+\mathbf{E}_{-\nu}(z)\},$$

so that

(5) $$\sin\nu\pi\,.\,\mathbf{J}_\nu(z)=\cos\nu\pi\,.\,\mathbf{E}_\nu(z)-\mathbf{E}_{-\nu}(z),$$

and similarly

(6) $$\sin\nu\pi\,.\,\mathbf{E}_\nu(z)=\mathbf{J}_{-\nu}(z)-\cos\nu\pi\,.\,\mathbf{J}_\nu(z).$$

The formulae (5) and (6) are due to Weber.

10·12. *Recurrence formulae for* $\mathbf{J}_\nu(z)$ *and* $\mathbf{E}_\nu(z)$.

The recurrence formulae which are satisfied by the functions of Anger and Weber have been determined by Weber.

It is evident from the definite integrals that

$$\begin{aligned}\mathbf{J}_{\nu-1}(z)+\mathbf{J}_{\nu+1}(z)-\frac{2\nu}{z}\mathbf{J}_\nu(z)&=\frac{2}{\pi}\int_0^\pi\left(\cos\theta-\frac{\nu}{z}\right)\cos(\nu\theta-z\sin\theta)\,d\theta\\&=-\frac{2}{\pi z}\int_0^\pi\frac{d}{d\theta}\{\sin(\nu\theta-z\sin\theta)\}\,d\theta\\&=-\frac{2\sin\nu\pi}{\pi z},\end{aligned}$$

and

$$\begin{aligned}\mathbf{E}_{\nu-1}(z)+\mathbf{E}_{\nu+1}(z)-\frac{2\nu}{z}\mathbf{E}_\nu(z)&=\frac{2}{\pi}\int_0^\pi\left(\cos\theta-\frac{\nu}{z}\right)\sin(\nu\theta-z\sin\theta)\,d\theta\\&=\frac{2}{\pi z}\int_0^\pi\frac{d}{d\theta}\{\cos(\nu\theta-z\sin\theta)\}\,d\theta\\&=-\frac{2(1-\cos\nu\pi)}{\pi z}.\end{aligned}$$

It is also very easy to prove that

$$\begin{cases}\mathbf{J}_{\nu-1}(z)-\mathbf{J}_{\nu+1}(z)-2\,\mathbf{J}_\nu'(z)=0,\\ \mathbf{E}_{\nu-1}(z)-\mathbf{E}_{\nu+1}(z)-2\,\mathbf{E}_\nu'(z)=0.\end{cases}$$

From these results we deduce the eight formulae

(1) $$\mathbf{J}_{\nu-1}(z)+\mathbf{J}_{\nu+1}(z)=\frac{2\nu}{z}\mathbf{J}_\nu(z)-\frac{2\sin\nu\pi}{\pi z},$$

(2) $$\mathbf{J}_{\nu-1}(z)-\mathbf{J}_{\nu+1}(z)=2\mathbf{J}_\nu'(z),$$

(3) $$(\vartheta+\nu)\,\mathbf{J}_\nu(z)=z\mathbf{J}_{\nu-1}(z)+(\sin\nu\pi)/\pi,$$

(4) $$(\vartheta-\nu)\,\mathbf{J}_\nu(z)=-z\mathbf{J}_{\nu+1}(z)-(\sin\nu\pi)/\pi,$$

(5) $$\mathbf{E}_{\nu-1}(z) + \mathbf{E}_{\nu+1}(z) = \frac{2\nu}{z}\mathbf{E}_\nu(z) - \frac{2(1-\cos\nu\pi)}{\pi z},$$

(6) $$\mathbf{E}_{\nu-1}(z) - \mathbf{E}_{\nu+1}(z) = 2\mathbf{E}_\nu'(z),$$

(7) $$(\vartheta+\nu)\mathbf{E}_\nu(z) = z\mathbf{E}_{\nu-1}(z) + (1-\cos\nu\pi)/\pi,$$

(8) $$(\vartheta-\nu)\mathbf{E}_\nu(z) = -z\mathbf{E}_{\nu+1}(z) - (1-\cos\nu\pi)/\pi,$$

where ϑ, as usual, stands for $z(d/dz)$.

Next we construct the differential equations; it is evident that

$$\begin{aligned}(\vartheta^2-\nu^2)\mathbf{J}_\nu(z) &= (\vartheta-\nu)\{z\mathbf{J}_{\nu-1}(z) + (\sin\nu\pi)/\pi\}\\ &= z(\vartheta+1-\nu)\mathbf{J}_{\nu-1}(z) - (\nu\sin\nu\pi)/\pi\\ &= -z^2\mathbf{J}_\nu(z) + (z\sin\nu\pi)/\pi - (\nu\sin\nu\pi)/\pi,\end{aligned}$$

so that

(9) $$\nabla_\nu\mathbf{J}_\nu(z) = \frac{(z-\nu)\sin\nu\pi}{\pi}.$$

We also have

$$\begin{aligned}(\vartheta^2-\nu^2)\mathbf{E}_\nu(z) &= (\vartheta-\nu)\{z\mathbf{E}_{\nu-1}(z) + (1-\cos\nu\pi)/\pi\}\\ &= z(\vartheta+1-\nu)\mathbf{E}_{\nu-1}(z) - \nu(1-\cos\nu\pi)/\pi\\ &= -z^2\mathbf{E}_\nu(z) - z(1+\cos\nu\pi)/\pi - \nu(1-\cos\nu\pi)/\pi,\end{aligned}$$

so that

(10) $$\nabla_\nu\mathbf{E}_\nu(z) = -\frac{z+\nu}{\pi} - \frac{(z-\nu)\cos\nu\pi}{\pi}.$$

Formulae equivalent to (9) and (10) were obtained by Anger, *Neueste Schriften der Naturf. Ges. in Danzig*, v. (1855), p. 17 and by Weber, *Zürich Vierteljahrsschrift*, XXIV. (1879), p. 47, respectively; formula (9) had been discovered earlier by Poisson (cf. § 10·1).

10·13. *Integrals expressible in terms of the functions of Anger and H. F. Weber.*

It is evident from the definitions that

(1) $$\mathbf{J}_\nu(z) \pm i\,\mathbf{E}_\nu(z) = \frac{1}{\pi}\int_0^\pi \exp\{\pm i(\nu\theta - z\sin\theta)\}\,d\theta.$$

By means of this result, combined with formulae obtained in §§ 6·2—6·22, it is possible to express numerous definite integrals in terms of the functions of Bessel, Anger and Weber. Thus, from § 6·2 (4) we have

(2) $$\int_0^\infty e^{-\nu t - z\sinh t}\,dt = \frac{\pi}{\sin\nu\pi}\{\mathbf{J}_\nu(z) - J_\nu(z)\},$$

when $|\arg z| < \frac{1}{2}\pi$; the result is valid when $|\arg z| = \frac{1}{2}\pi$, provided that $R(\nu) > 0$.

Again, we have

(3) $$\int_0^\infty e^{\nu t - z\sinh t}\,dt = \frac{\pi}{\sin\nu\pi}\{J_{-\nu}(z) - \mathbf{J}_{-\nu}(z)\},$$

so that, when we combine (2) and (3),

(4) $$\int_0^\infty e^{-z\sinh t}\cosh\nu t\,dt = \tfrac{1}{2}\pi\tan\tfrac{1}{2}\nu\pi\,\{\mathbf{J}_\nu(z) - J_\nu(z)\} - \tfrac{1}{2}\pi\,\{\mathbf{E}_\nu(z) + Y_\nu(z)\},$$

(5) $$\int_0^\infty e^{-z\sinh t}\sinh\nu t\,dt = \tfrac{1}{2}\pi\cot\tfrac{1}{2}\nu\pi\,\{J_\nu(z) - \mathbf{J}_\nu(z)\} - \tfrac{1}{2}\pi\,\{\mathbf{E}_\nu(z) + Y_\nu(z)\}.$$

The integral $\int_0^\infty e^{-z\cosh t}\cosh\nu t\,dt$ has already been evaluated (§ 6·3); but

$$\int_0^\infty e^{-z\cosh t}\sinh\nu t\,dt$$

does not appear to be expressible in a simple form; its expansion in ascending powers of z can be obtained from the formula of § 6·22 (4),

$$I_{-\nu}(z) + I_\nu(z) = \frac{2}{\pi}\int_0^\pi e^{z\cos\theta}\cos\nu\theta\,d\theta + \frac{2\sin\nu\pi}{\pi}\int_0^\infty e^{-z\cosh t}\sinh\nu t\,dt,$$

but, since

$$\int_0^\pi \cos^m\theta\cos\nu\theta\,d\theta = \frac{(-)^m\sin\nu\pi}{2^m(\nu+m)}\cdot{}_2F_1\left(-m,\ \frac{\nu+m}{2};\ 1-\frac{\nu\mp m}{2};\ -1\right),$$

the integral under consideration cannot be evaluated in any simple form*.

The formulae (2)—(5) are nugatory when ν is an integer, but from §§ 6·21, 9·33 we have

(6) $$\int_0^\infty e^{nt-z\sinh t}\,dt = \tfrac{1}{2}\{S_n(z) - \pi\mathbf{E}_n(z) - \pi Y_n(z)\},$$

(7) $$\int_0^\infty e^{-nt-z\sinh t}\,dt = \tfrac{1}{2}(-)^{n+1}\{S_n(z) + \pi\mathbf{E}_n(z) + \pi Y_n(z)\}.$$

The associated integrals

$$\int_0^\infty e^{-\nu t}\,{\cos \atop \sin}\,(x\sinh t)\,dt,\quad \int_0^\infty e^{-\nu t}\,{\cos \atop \sin}\,(x\cosh t)\,dt$$

have been noticed by Coates, *Quarterly Journal*, xx. (1885), p. 260.

Various integrals of these types occur in researches on diffraction by a prism; see, e.g. Whipple, *Proc. London Math. Soc.* (2) xvi. (1917), p. 106.

10·14. *Asymptotic expansions of Anger-Weber functions of large argument.*

It follows from § 10·13 (2) that, in order to obtain the asymptotic expansion of $\mathbf{J}_{\pm\nu}(z)$ when $|z|$ is large and $|\arg z| < \frac{1}{2}\pi$, it is sufficient to obtain the asymptotic expansion of the integrals

$$\int_0^\infty e^{\mp\nu t - z\sinh t}\,dt.$$

To carry out this investigation we shall first expand $\cosh\nu t/\cosh t$ and $\sinh\nu t/\cosh t$ in a series of ascending powers of $\sinh t$.

* See Anding, *Sechsstellige Tafeln der Besselschen Funktionen imaginären Arguments* (Leipzig, 1911) [*Jahrbuch über die Fortschritte der Math.* 1911, pp. 493—494], and Takeuchi, *Tôhoku Math. Journal*, xviii. (1920), pp. 295—296.

If $e^{2t} = u$, we have, after the manner of § 7·4,

$$u^{\frac{1}{2}\nu} + u^{-\frac{1}{2}\nu} = \frac{1}{2\pi i}\int^{(u+,\,1/u+)} \zeta^{\frac{1}{2}\nu-\frac{1}{2}} \left\{\frac{u^{\frac{1}{2}}}{\zeta - u} + \frac{1/u^{\frac{1}{2}}}{\zeta - 1/u}\right\} d\zeta,$$

so that

$$\frac{\cosh \nu t}{\cosh t} = \frac{1}{2\pi i}\int^{(u+,\,1/u+,\,1+)} \frac{\zeta^{\frac{1}{2}\nu-\frac{1}{2}}(\zeta-1)\,d\zeta}{(\zeta-1)^2 - 4\zeta\sinh^2 t}$$

$$= \frac{1}{2\pi i}\int^{(u+,\,1/u+,\,1+)} \zeta^{\frac{1}{2}\nu-\frac{1}{2}} \left[\sum_{m=0}^{p-1} \frac{2^{2m}\,\zeta^m \sinh^{2m} t}{(\zeta-1)^{2m+1}} + \frac{2^{2p}\,\zeta^p \sinh^{2p} t}{(\zeta-1)^{2p-1}\{(\zeta-1)^2 - 4\zeta\sinh^2 t\}}\right] d\zeta.$$

Now

$$\frac{1}{2\pi i}\int^{(1+)} \frac{\zeta^{\frac{1}{2}\nu+m-\frac{1}{2}}\,d\zeta}{(\zeta-1)^{2m+1}} = \frac{\Gamma(\frac{1}{2}\nu+m+\frac{1}{2})}{\Gamma(\frac{1}{2}\nu-m+\frac{1}{2}).(2m)!}$$

$$= \frac{(-)^m \cos\frac{1}{2}\nu\pi}{\pi}.\frac{\Gamma(m+\frac{1}{2}+\frac{1}{2}\nu)\,\Gamma(m+\frac{1}{2}-\frac{1}{2}\nu)}{(2m)!},$$

and, if we take p so large that $R(p+\frac{1}{2}\pm\frac{1}{2}\nu) > 0$, and then take the contour to be that shewn in Fig. 15 of § 7·4, we find that

$$\frac{1}{2\pi i}\int^{(u+,\,1/u+,\,1+)} \frac{\zeta^{p+\frac{1}{2}\nu-\frac{1}{2}}\,d\zeta}{(\zeta-1)^{2p-1}\{(\zeta-1)^2 - 4\zeta\sinh^2 t\}}$$

$$= \frac{(-)^p \cos\frac{1}{2}\nu\pi}{\pi}\int_0^1 \frac{x^{p-\frac{1}{2}\nu-\frac{1}{2}}(1-x)^{p+\frac{1}{2}\nu-\frac{1}{2}}\,dx}{1+4x(1-x)\sinh^2 t}.$$

If ν and t are real, the last expression may be written in the form

$$\theta_1 \frac{(-)^p\cos\frac{1}{2}\nu\pi}{\pi}\frac{\Gamma(p+\frac{1}{2}+\frac{1}{2}\nu)\,\Gamma(p+\frac{1}{2}-\frac{1}{2}\nu)}{(2p)!},$$

where $0 \leqslant \theta_1 \leqslant 1$, since $1 + 4x(1-x)\sinh^2 t \geqslant 1$.

It follows that, when $R(p+\frac{1}{2}\pm\frac{1}{2}\nu) \geqslant 0$, we have

$$\frac{\cosh\nu t}{\cosh t} = \frac{\cos\frac{1}{2}\nu\pi}{\pi}\left[\sum_{m=0}^{p-1} \frac{(-)^m\,\Gamma(m+\frac{1}{2}+\frac{1}{2}\nu)\,\Gamma(m+\frac{1}{2}-\frac{1}{2}\nu)}{(2m)!}(2\sinh t)^{2m} + \theta_1\frac{(-)^p\,\Gamma(p+\frac{1}{2}+\frac{1}{2}\nu)\,\Gamma(p+\frac{1}{2}-\frac{1}{2}\nu)}{(2p)!}(2\sinh t)^{2p}\right].$$

For complex values of ν and t this equation has to be modified by replacing the condition $0 \leqslant \theta_1 \leqslant 1$ by a less stringent condition, in a way with which the reader will be familiar in view of the similar analysis occurring in various sections of Chapter VII.

Similarly we have

$$u^{\frac{1}{2}\nu} - u^{-\frac{1}{2}\nu} = \frac{1}{2\pi i}\int^{(u+,\,1/u+)} \zeta^{\frac{1}{2}\nu}\left\{\frac{1}{\zeta-u} - \frac{1}{\zeta-1/u}\right\} d\zeta,$$

so that

$$\frac{\sinh \nu t}{\sinh 2t} = \frac{1}{2\pi i}\int^{(u+,1/u+,1+)} \frac{\zeta^{\frac{1}{2}\nu}\, d\zeta}{(\zeta-1)^2 - 4\zeta\sinh^2 t}$$

$$= \frac{1}{2\pi i}\int^{(u+,1/u+,1+)} \zeta^{\frac{1}{2}\nu}\left[\sum_{m=0}^{p-1}\frac{2^{2m}\zeta^m \sinh^{2m} t}{(\zeta-1)^{2m+2}} + \frac{2^{2p}\zeta^p\sinh^{2p} t}{(\zeta-1)^{2p}\{(\zeta-1)^2 - 4\zeta\sinh^2 t\}}\right] d\zeta,$$

whence it follows that, if we take p so large that $R(p+1\pm\frac{1}{2}\nu)>0$, then

$$\frac{\sinh \nu t}{\cosh t} = \frac{\sin\frac{1}{2}\nu\pi}{\pi}\left[\sum_{m=0}^{p-1}\frac{(-)^m\Gamma(m+1+\frac{1}{2}\nu)\Gamma(m+1-\frac{1}{2}\nu)}{(2m+1)!}(2\sinh t)^{2m+1}\right.$$

$$\left. + \theta_2\frac{(-)^p\Gamma(p+1+\frac{1}{2}\nu)\Gamma(p+1-\frac{1}{2}\nu)}{(2p+1)!}(2\sinh t)^{2p+1}\right].$$

On integrating these results, it follows that

$$\int_0^\infty \cosh \nu t\,.\,e^{-z\sinh t}\,dt \sim \frac{\cos\frac{1}{2}\nu\pi}{2\pi}\sum_{m=0}^{\infty}\frac{(-)^m\Gamma(m+\frac{1}{2}+\frac{1}{2}\nu)\Gamma(m+\frac{1}{2}-\frac{1}{2}\nu)}{(\frac{1}{2}z)^{2m+1}},$$

$$\int_0^\infty \sinh \nu t\,.\,e^{-z\sinh t}\,dt \sim \frac{\sin\frac{1}{2}\nu\pi}{2\pi}\sum_{m=0}^{\infty}\frac{(-)^m\Gamma(m+1+\frac{1}{2}\nu)\Gamma(m+1-\frac{1}{2}\nu)}{(\frac{1}{2}z)^{2m+2}}.$$

If ν is real and z is positive, these asymptotic expansions possess the property that the remainder after p terms is of the same sign as, and is numerically less than, the $(p+1)$th term when p is so large that $R(p+1\pm\frac{1}{2}\nu)\geqslant 0$.

It follows from §§ 10·13 (2) and (3) combined with § 10·11 (6) that

$$(1)\qquad \mathbf{J}_\nu(z) \sim J_\nu(z) + \frac{\sin\nu\pi}{\pi z}\left[1 - \frac{1^2-\nu^2}{z^2} + \frac{(1^2-\nu^2)(3^2-\nu^2)}{z^4} - \dots\right]$$

$$- \frac{\sin\nu\pi}{\pi z}\left[\frac{\nu}{z} - \frac{\nu(2^2-\nu^2)}{z^3} + \frac{\nu(2^2-\nu^2)(4^2-\nu^2)}{z^5} - \dots\right],$$

$$(2)\qquad \mathbf{E}_\nu(z) \sim -Y_\nu(z) - \frac{1+\cos\nu\pi}{\pi z}\left[1 - \frac{1^2-\nu^2}{z^2} + \frac{(1^2-\nu^2)(3^2-\nu^2)}{z^4} - \dots\right]$$

$$- \frac{1-\cos\nu\pi}{\pi z}\left[\frac{\nu}{z} - \frac{\nu(2^2-\nu^2)}{z^3} + \frac{\nu(2^2-\nu^2)(4^2-\nu^2)}{z^5} - \dots\right].$$

These results were stated without proof by Weber, *Zürich Vierteljahrsschrift*, XXIV. (1879), p. 48 and by Lommel, *Math. Ann.* XVI. (1880), pp. 186—188. They were proved as special cases of much more general formulae by Nielsen, *Handbuch der Theorie der Cylinderfunktionen* (Leipzig, 1904), p. 228. The proof of this section does not seem to have been given previously.

Since the only singularities of $\cosh \nu t/\cosh t$ and $\sinh \nu t/\cosh t$, *qua* functions of $\sinh t$, are at $\sinh t = \pm i$, it is possible to change the contours of integration into curves in the t-plane on which $\arg(\sinh t)$ is a positive or negative acute angle; and then we deduce in the usual manner (cf. § 6·1) that the formulae (1) and (2) are valid over the sector $|\arg z| < \pi$.

10·15. *Asymptotic expansions of Anger-Weber functions of large order and argument.*

We shall now obtain asymptotic expansions, of a type similar to the expansions investigated in Chapter VIII, which represent $\mathbf{J}_\nu(z)$ and $\mathbf{E}_\nu(z)$ when $|\nu|$ and $|z|$ are both large.

In view of the results obtained in § 10·13, it will be adequate to obtain asymptotic expansions of the two integrals

$$\frac{1}{\pi}\int_0^\infty e^{\mp\nu t - z\sinh t}\,dt.$$

As in Chapter VIII, we write

$$\nu = z\cosh(\alpha + i\beta) = z\cosh\gamma,$$

where $0 \leqslant \beta \leqslant \pi$ and γ is not nearly equal* to πi.

(I) We first consider the integral

$$\frac{1}{\pi}\int_0^\infty e^{-\nu t - z\sinh t}\,dt = \frac{1}{\pi}\int_0^\infty e^{-z(t\cosh\gamma + \sinh t)}\,dt,$$

in which it is supposed temporarily that ν/z is positive. When $\cosh\gamma$ is positive, $t\cosh\gamma + \sinh t$ steadily increases from 0 to ∞ as t increases from 0 to ∞; we shall take this function of t as a new variable τ.

It is easy to shew that t is a monogenic function of τ, except possibly when

$$\tau = (2n+1)\pi i\cosh\gamma \pm \sinh\gamma \mp \gamma\cosh\gamma,$$

where n is an integer; and, when $\cosh\gamma$ is positive, none of these values of τ is a real positive number; for, when γ is real, $(2n+1)\pi i\cosh\gamma$ does not vanish, and, when γ is a pure imaginary $(=i\beta)$, the singularities are on the imaginary axis and the origin is not one of them since γ is not equal to πi.

The expansion of $dt/d\tau$ in ascending powers of τ is

$$\frac{dt}{d\tau} = \sum_{m=0}^{\infty} \mathbf{a}_m \tau^{2m},$$

where
$$\mathbf{a}_m = \frac{1}{2\pi i}\int^{(0+)} \frac{1}{\tau^{2m+1}}\cdot\frac{dt}{d\tau}\,d\tau = \frac{1}{2\pi i}\int^{(0+)} \frac{dt}{\tau^{2m+1}},$$

and so $\mathbf{a}_m$ is the coefficient of $1/t$ in the expansion of τ^{-2m-1} in ascending powers of t. In particular we have

$$\mathbf{a}_0 = \frac{1}{1+\cosh\gamma},\quad \mathbf{a}_1 = -\frac{1}{2(1+\cosh\gamma)^4},\quad \mathbf{a}_2 = \frac{9-\cosh\gamma}{24(1+\cosh\gamma)^7},$$

$$\mathbf{a}_3 = -\frac{225 - 54\cosh\gamma + \cosh^2\gamma}{720(1+\cosh\gamma)^{10}}.$$

From the general theorem of § 8·3, we are now in a position to write down the expansion

$$\frac{1}{\pi}\int_0^\infty e^{-\nu t - z\sinh t}\,dt \sim \frac{1}{\pi}\sum_{m=0}^{\infty}\frac{(2m)!\,\mathbf{a}_m}{z^{2m+1}}. \tag{1}$$

* Expansions valid near $\gamma = \pi i$ are obtained at the end of this section.

This expansion is valid when ν/z is positive; it has, so far, been established on the hypothesis that $|\arg z| < \frac{1}{2}\pi$, but, by a process of swinging round the contour in the τ-plane, the range of validity may be extended to cover the domain in which $|\arg z| < \pi$.

Next, we consider the modifications caused by abandoning the hypothesis that $\cosh\gamma$ is real. If we write $t = u + iv$, the curve on which τ is real has for its equation

$$u \sinh\alpha \sin\beta + v \cosh\alpha \cos\beta + \cosh u \sin v = 0.$$

The shape of this curve has to be examined by methods resembling those of § 8·61. For brevity we write

$$u \sinh\alpha \sin\beta + v \cosh\alpha \cos\beta + \cosh u \sin v \equiv \Phi(u, v).$$

Since $\Phi(u, v)$ is unaffected by a change of sign of both u and α, we first study the curve in which $\alpha \geqslant 0$. It is evident that the curve has the origin as its centre.

Since $$\partial\Phi(u, v)/\partial u = \sinh\alpha \sin\beta + \sinh u \sin v,$$
it follows that, when v has any assigned value, $\partial\Phi/\partial u$ vanishes for only one value of u, and so the equation in u

$$\Phi(u, v) = 0$$

has, at most, two real roots; and one of these is infinite whenever v is a multiple of π.

When $0 > v > -\pi$, we have

$$\Phi(-\infty, v) = -\infty, \quad \Phi(+\infty, v) = -\infty;$$

and, when $v = \beta - \pi$, the maximum value of $\Phi(u, v)$, *qua* function of u, is at $u = \alpha$, the value of $\Phi(u, v)$ then being

$$-\cosh\alpha \sin\beta \{1 - \alpha \tanh\alpha + (\pi - \beta) \cot\beta\}.$$

If this is negative, the equation $\Phi(u, \beta - \pi) = 0$ has no real root, and so the contour does not meet the line $v = \beta - \pi$ or (by symmetry) the line $v = \pi - \beta$.

Hence provided that the point (α, β) lies in one of the domains numbered 1, 2, 3 in Fig. 21 of § 8·61, the contour $\Phi(u, v) = 0$ lies as in Fig. 25, the continuous curve indicating the shape of the contour when α is positive

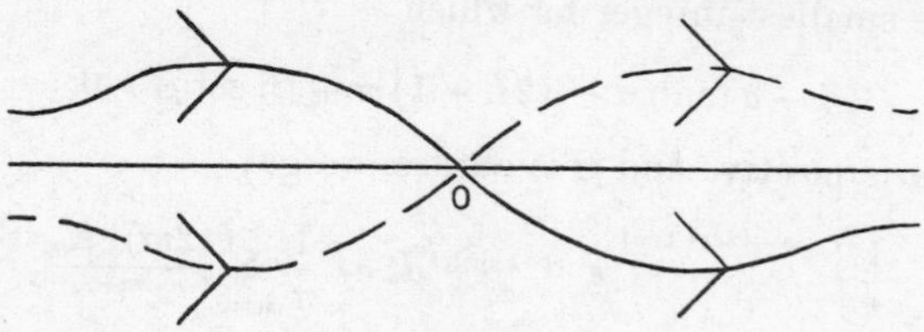

Fig. 25.

and the broken curve the shape when α is negative; the direction in which τ increases is marked by an arrow.

It follows that the expansion (1) is valid when (α, β) lies in any of the domains 1, 2, 3.

Next, we have to consider the asymptotic expansion when (α, β) does not lie in any of these domains. To effect our purpose we have to determine the destinations of the branch of the curve $\Phi(u, v) = 0$ which passes through the origin.

Consider first the case in which α is positive and β is acute. The function $\Phi(\alpha, v)$ has maxima at $v = (2n+1)\pi - \beta$ and minima at $v = (2n+1)\pi + \beta$, each minimum being greater than the preceding; and since $\phi(\alpha, \beta - \pi)$ is now positive, it follows that $\phi(\alpha, v)$ is positive when v is greater than $-\pi$.

Hence the curve cannot cross the line $u = \alpha$ above the point at which $v = -\pi$, and similarly it cannot cross the line $u = -\alpha$ below the point at which $v = \pi$. The branch which goes downwards at the origin is therefore confined to the strip $-\alpha < u < \alpha$ until it gets below the line $v = -2K\pi + \pi - \beta$, where K is the smallest integer for which

$$1 - \alpha \tanh \alpha + \{(2K+1)\pi - \beta\} \cot \beta > 0.$$

The curve cannot cross the line $v = -(2K+1)\pi + \beta$, and so it crosses the line $u = \alpha$ and goes off to infinity in the direction of the line $v = -2K\pi$.

Hence, if α is positive and β is acute, we get

$$\frac{1}{\pi}\int_0^{\infty - 2K\pi i} e^{-\nu t - z \sinh t}\, dt \sim \frac{1}{\pi}\sum_{m=0}^{\infty} \frac{(2m)!\, \mathbf{a}_m}{z^{2m+1}}, \tag{2}$$

while, if α is negative and β is acute, we get

$$\frac{1}{\pi}\int_0^{\infty + 2K\pi i} e^{-\nu t - z \sinh t}\, dt \sim \frac{1}{\pi}\sum_{m=0}^{\infty} \frac{(2m)!\, \mathbf{a}_m}{z^{2m+1}}. \tag{3}$$

By combining these results with those obtained in § 8·61, we obtain the asymptotic expansions for the domains $6a$ and $7a$.

If, however, β is obtuse and α is positive, the branch which goes below the axis of u at the origin cannot cross the line $u = \alpha$ below $(\alpha, \pi - \beta)$ and it does not cross the u-axis again, so it must go to $-\infty$ along the line $v = -(2L+1)\pi$, where L is the smallest integer for which

$$1 - \alpha \tanh \alpha - \{(2L+1)\pi + \beta\} \cot \beta > 0.$$

Hence, if α is positive and β is obtuse, we get

$$\frac{1}{\pi}\int_0^{-\infty - (2L+1)\pi i} e^{-\nu t - z \sinh t}\, dt \sim \frac{1}{\pi}\sum_{m=0}^{\infty} \frac{(2m)!\, \mathbf{a}_m}{z^{2m+1}}, \tag{4}$$

while, if α is negative and β is obtuse, we get

$$\frac{1}{\pi}\int_0^{-\infty + (2L+1)\pi i} e^{-\nu t - z \sinh t}\, dt \sim \frac{1}{\pi}\sum_{m=0}^{\infty} \frac{(2m)!\, \mathbf{a}_m}{z^{2m+1}}. \tag{5}$$

By combining these results with those obtained in § 8·61, we obtain the asymptotic expansions for the domains 4, 5, 6*b* and 7*b*.

Since formula (1) is the only one which is of practical importance, we shall not give the other expansions in greater detail.

An approximate formula for $\mathbf{a}_m$ when m is large and γ is zero, namely

$$\mathbf{a}_m \sim \frac{(-)^m \Gamma(\frac{1}{3})}{3^{1/6}\pi^{2m+5/3}m^{1/3}},$$

was obtained by Cauchy, *Comptes Rendus*, XXXVIII. (1854), p. 1106.

(II) Next consider the integral

$$\frac{1}{\pi}\int_0^\infty e^{\nu t - z\sinh t}\,dt = \frac{1}{\pi}\int_0^\infty e^{-z(-t\cosh\gamma + \sinh t)}\,dt.$$

The only difference between this and the previous integral is the change in the sign of $\cosh\gamma$; and so, when γ lies in any of the regions numbered 1, 4, 5 in Fig. 21 of § 8·61, we have

$$(6) \qquad \frac{1}{\pi}\int_0^\infty e^{\nu t - z\sinh t}\,dt \sim \frac{1}{\pi}\sum_{m=0}^\infty \frac{(2m)!\,\mathbf{a}_m'}{z^{2m+1}},$$

where $\mathbf{a}_m'$ is derived from $\mathbf{a}_m$ by changing the sign of $\cosh\gamma$, so that

$$\mathbf{a}_0' = \frac{1}{1-\cosh\gamma}, \quad \mathbf{a}_1' = -\frac{2}{(1-\cosh\gamma)^4}, \quad \mathbf{a}_2' = \frac{9+\cosh\gamma}{24(1-\cosh\gamma)^7},$$

....

This expansion fails to be significant when γ is small, just as the previous expansion (1) failed when γ was nearly equal to πi.

To deal with this case we write

$$\nu = z(1-\epsilon), \quad \tau = t - \sinh t,$$

after the method of § 8·42. It is thus found that

$$\frac{1}{\pi}\int_0^\infty e^{\nu t - z\sinh t}\,dt = \frac{1}{\pi}\int_0^{-\infty} e^{z\tau}e^{-\epsilon z t}\frac{dt}{d\tau}\,d\tau$$

$$= -\frac{1}{3\pi}\int_0^{-\infty} e^{z\tau}\sum_{m=0}^\infty 6^{\frac{1}{3}(m+1)} B_m(-\epsilon z)\,.\,(-\tau)^{\frac{1}{3}m-\frac{2}{3}}\,d\tau$$

and hence

$$(7) \qquad \frac{1}{\pi}\int_0^\infty e^{\nu t - z\sinh t}\,dt \sim \frac{1}{3\pi}\sum_{m=0}^\infty \frac{(-)^m\,\Gamma(\frac{1}{3}m+\frac{1}{3})\,B_m(\epsilon z)}{(\frac{1}{6}z)^{\frac{1}{3}(m+1)}}.$$

A result equivalent to this has been given by Airey, *Proc. Royal Soc.* XCIV. A, (1918), p. 313.

10·2. *Hardy's generalisations of Airy's integral.*

The integral considered by Airy and Stokes (§ 6·3) has been generalised by Hardy* in the following manner:

If $s = \sinh \phi$, then

$$\begin{cases} 2\cosh 2\phi = 4s^2 + 2 \\ 2\sinh 3\phi = 8s^3 + 6s \\ 2\cosh 4\phi = 16s^4 + 16s^2 + 2 \\ 2\sinh 5\phi = 32s^5 + 40s^3 + 10s, \end{cases}$$

and generally

$$2 \begin{matrix}\cosh \\ \sinh\end{matrix} n\phi = (2s)^n {}_2F_1(-\tfrac{1}{2}n, \tfrac{1}{2} - \tfrac{1}{2}n; 1-n; -1/s^2),$$

the cosh or sinh being taken according as n is even or odd.

Now write

$$T_n(t, \alpha) = t^n . {}_2F_1(-\tfrac{1}{2}n, \tfrac{1}{2} - \tfrac{1}{2}n; 1-n; -4\alpha/t^2),$$

so that

$$\begin{cases} T_2(t, \alpha) = t^2 + 2\alpha \\ T_3(t, \alpha) = t^3 + 3\alpha t \\ T_4(t, \alpha) = t^4 + 4\alpha t^2 + 2\alpha^2 \\ T_5(t, \alpha) = t^5 + 5\alpha t^3 + 5\alpha^2 t \\ \dots\dots\dots\dots\dots\dots \end{cases}$$

Then the following three integrals are generalisations† of Airy's integral:

$$(1) \qquad Ci_n(\alpha) = \int_0^\infty \cos T_n(t, \alpha)\, dt,$$

$$(2) \qquad Si_n(\alpha) = \int_0^\infty \sin T_n(t, \alpha)\, dt,$$

$$(3) \qquad Ei_n(\alpha) = \int_0^\infty \exp\{-T_n(t, \alpha)\}\, dt.$$

It may be shewn‡ that the first two integrals are convergent when α is real (whether positive or negative) if $n = 2, 3, 4, \dots$. But the third integral converges when α is complex; and it is indeed fairly obvious that $Ei_n(\alpha)$ is an integral function of α.

When n is an even integer, the three functions are expressible in terms of Bessel functions; but when n is odd, the first only is so expressible, the other two involving the function of H. F. Weber.

Before evaluating the integrals, we observe that integral functions exist which reduce to $Ci_n(\alpha)$ and $Si_n(\alpha)$ when α is real; for take the combination

$$Ci_n(\alpha) + iSi_n(\alpha) = \int_0^\infty \exp\{iT_n(t, \alpha)\}\, dt.$$

* *Quarterly Journal*, XLI. (1910), pp. 226—240.

† The sine-integral in the case $n=3$ was examined by Stokes, *Camb. Phil. Trans.* IX. (1856), pp. 168—182. [*Math. and Phys. Papers*, II. (1883), pp. 332—349.]

‡ Hardy, *loc. cit.*, p. 228.

By Jordan's lemma, the integral, when taken round an arc of a circle of radius R with centre at the origin (the arc being terminated by the points with complex coordinates R, $Re^{\frac{1}{2}\pi i/n}$), tends to zero as $R\to\infty$.

And therefore

$$Ci_n(\alpha)+iSi_n(\alpha)=\int_0^{\infty\exp(\frac{1}{2}\pi i/n)}\exp\{iT_n(t,\alpha)\}\,dt$$

$$=e^{\frac{1}{2}\pi i/n}\int_0^\infty \exp\{-T_n(\tau,\alpha e^{-\pi i/n})\}\,d\tau,$$

where $\tau=te^{-\frac{1}{2}\pi i/n}$; and the last integral is an integral function of α. The combination $Ci_n(\alpha)-iSi_n(\alpha)$ may be treated in a similar manner, and the result is then evident.

10·21. *The evaluation of Airy-Hardy integrals of even order.*

To evaluate the three integrals $Ci_n(\alpha)$, $Si_n(\alpha)$, $Ei_n(\alpha)$ when n is even, we suppose temporarily that α is positive, and then, making the substitution

$$t=2\alpha^{\frac{1}{2}}\sinh(u/n)$$

in the integrals, we find that, by § 6·21 (10),

$$Ci_n(\alpha)+iSi_n(\alpha)=\frac{2\alpha^{\frac{1}{2}}}{n}\int_0^\infty \exp(2\alpha^{\frac{1}{2}n}i\cosh u)\cosh(u/n)\,du$$

$$=\pi i\alpha^{\frac{1}{2}}n^{-1}e^{\frac{1}{2}\pi i/n}H_{1/n}{}^{(1)}(2\alpha^{\frac{1}{2}n}),$$

that is to say

$$Ci_n(\alpha)+iSi_n(\alpha)=\frac{\pi\alpha^{\frac{1}{2}}}{n\sin(\pi/n)}\{e^{\frac{1}{2}\pi i/n}J_{-1/n}(2\alpha^{\frac{1}{2}n})-e^{-\frac{1}{2}\pi i/n}J_{1/n}(2\alpha^{\frac{1}{2}n})\}.$$

If we equate real and imaginary parts, we have

(1) $$Ci_n(\alpha)=\frac{\pi\alpha^{\frac{1}{2}}}{2n\sin(\frac{1}{2}\pi/n)}\{J_{-1/n}(2\alpha^{\frac{1}{2}n})-J_{1/n}(2\alpha^{\frac{1}{2}n})\},$$

(2) $$Si_n(\alpha)=\frac{\pi\alpha^{\frac{1}{2}}}{2n\cos(\frac{1}{2}\pi/n)}\{J_{-1/n}(2\alpha^{\frac{1}{2}n})+J_{1/n}(2\alpha^{\frac{1}{2}n})\}.$$

In a similar manner,

$$Ei_n(\alpha)=\frac{2\alpha^{\frac{1}{2}}}{n}\int_0^\infty \exp(-2\alpha^{\frac{1}{2}n}\cosh u)\cosh(u/n)\,du,$$

so that, by § 6·22 (5),

(3) $$Ei_n(\alpha)=(2\alpha^{\frac{1}{2}}/n)\,K_{1/n}(2\alpha^{\frac{1}{2}n}).$$

These results have been obtained on the hypothesis that α is positive; and the expressions on the right are the integral functions of α which reduce to $Ci_n(\alpha)$, $Si_n(\alpha)$ and $Ei_n(\alpha)$ when α is real, whether positive or negative. Hence, when α is negative the equations (1), (2), (3) are still valid, so that, for example, we have

$$Ci_n(\alpha)=\frac{\pi}{2n\sin(\frac{1}{2}\pi/n)}\left\{\sum_{m=0}^\infty\frac{(-)^m\alpha^{mn}}{m!\,\Gamma(m+1-1/n)}-\alpha\sum_{m=0}^\infty\frac{(-)^m\alpha^{mn}}{m!\,\Gamma(m+1+1/n)}\right\},$$

whether α be positive or negative.

Hence, replacing α by $-\beta$, we see that, when β is positive and n is even, then

(4) $$Ci_n(-\beta) = \frac{\pi\beta^{\frac{1}{2}}}{2n\sin(\frac{1}{2}\pi/n)}\{J_{-1/n}(2\beta^{\frac{1}{2}n}) + J_{1/n}(2\beta^{\frac{1}{2}n})\},$$

(5) $$Si_n(-\beta) = \frac{\pi\beta^{\frac{1}{2}}}{2n\cos(\frac{1}{2}\pi/n)}\{J_{-1/n}(2\beta^{\frac{1}{2}n}) - J_{1/n}(2\beta^{\frac{1}{2}n})\},$$

(6) $$Ei_n(-\beta) = \frac{\pi\beta^{\frac{1}{2}}}{n\sin(\pi/n)}\{I_{-1/n}(2\beta^{\frac{1}{2}n}) + I_{1/n}(2\beta^{\frac{1}{2}n})\}.$$

It follows from § 4·31 (9) that, when n is even, the functions $Ci_n(\alpha)$ and $Si_n(\alpha)$ are annihilated by the operator

$$\frac{d^2}{d\alpha^2} + n^2\alpha^{n-2},$$

and that $Ei_n(\alpha)$ is annihilated by the operator

$$\frac{d^2}{d\alpha^2} - n^2\alpha^{n-2}.$$

In the case of the first two functions it is difficult to obtain this result* directly from the definitions, because the integrals obtained by differentiating twice under the integral sign are not convergent.

10·22. *The evaluation of Airy-Hardy integrals of odd order.*

To evaluate $Ci_n(\alpha)$ when n is odd, we suppose temporarily that α is positive, and then, by § 6·22 (13),

$$Ci_n(\alpha) = \frac{2\alpha^{\frac{1}{2}}}{n}\int_0^\infty \cos(2\alpha^{\frac{1}{2}n}\sinh u)\cosh(u/n)\,du$$
$$= \frac{2\alpha^{\frac{1}{2}}\cos(\frac{1}{2}\pi/n)}{n}K_{1/n}(2\alpha^{\frac{1}{2}n}).$$

That is to say,

(1) $$Ci_n(\alpha) = \frac{2\alpha^{\frac{1}{2}}\cos(\frac{1}{2}\pi/n)}{n}K_{1/n}(2\alpha^{\frac{1}{2}n})$$
$$= \frac{\pi\alpha^{\frac{1}{2}}}{2n\sin(\frac{1}{2}\pi/n)}\{I_{-1/n}(2\alpha^{\frac{1}{2}n}) - I_{1/n}(2\alpha^{\frac{1}{2}n})\}.$$

Using the device explained in § 10·21, we see that, when β is positive,

(2) $$Ci_n(-\beta) = \frac{\pi\beta^{\frac{1}{2}}}{2n\sin(\frac{1}{2}\pi/n)}\{J_{-1/n}(2\beta^{\frac{1}{2}n}) + J_{1/n}(2\beta^{\frac{1}{2}n})\}.$$

It follows that the equation § 10·21 (4) is true whether n be even or odd; and, whether n be even or odd, $Ci_n(\alpha)$ is annihilated by the operator

$$\frac{d^2}{d\alpha^2} + (-)^n n^2\alpha^{n-2},$$

for all real values of α.

* It has been proved by Hardy, *loc. cit.*, p. 229, with the aid of the theory of "generalised integrals."

Next we evaluate $Ei_n(\alpha)$ when α is positive; making the usual substitution, we find that, by § 10·13 (4),

$$Ei_n(\alpha) = \frac{2\alpha^{\frac{1}{2}}}{n}\int_0^\infty \exp(-2\alpha^{\frac{1}{2}n}\sinh u)\cosh(u/n)\,du$$

$$= \frac{\pi\alpha^{\frac{1}{2}}}{n}\{\tan(\tfrac{1}{2}\pi/n)\,\mathbf{J}_{1/n}(2\alpha^{\frac{1}{2}n}) - \mathbf{E}_{1/n}(2\alpha^{\frac{1}{2}n})\}$$

$$+ \frac{\pi\alpha^{\frac{1}{2}}}{n\sin(\pi/n)}\{J_{-1/n}(2\alpha^{\frac{1}{2}n}) - J_{1/n}(2\alpha^{\frac{1}{2}n})\}.$$

Hence the series which represents $Ei_n(\alpha)$ when n is odd and α may have any value is

(3) $$Ei_n(\alpha) = \frac{\pi\alpha^{\frac{1}{2}(n+1)}}{n\cos(\frac{1}{2}\pi/n)}\sum_{m=0}^\infty \frac{(-)^m\alpha^{mn}}{\Gamma(m+\frac{3}{2}-\frac{1}{2}/n)\,\Gamma(m+\frac{3}{2}+\frac{1}{2}/n)}$$

$$+ \frac{\pi}{n\sin(\pi/n)}\left\{\sum_{m=0}^\infty \frac{(-)^m\alpha^{mn}}{m!\,\Gamma(m+1-1/n)} - \alpha\sum_{m=0}^\infty \frac{(-)^m\alpha^{mn}}{m!\,\Gamma(m+1+1/n)}\right\},$$

and hence it follows that

(4) $$\left\{\frac{d^2}{d\alpha^2} + n^2\alpha^{n-2}\right\} Ei_n(\alpha) = n\alpha^{\frac{1}{2}(n-3)}.$$

Next consider $Ci_n(\alpha) + i\,Si_n(\alpha)$, where α is temporarily assumed to be positive. From § 10·13 (4) we deduce that

$$Ci_n(\alpha) + i\,Si_n(\alpha) = \frac{2\alpha^{\frac{1}{2}}}{n}\int_0^\infty \exp(2\alpha^{\frac{1}{2}n}\,i\sinh u)\cosh(u/n)\,du$$

$$= \frac{\pi\alpha^{\frac{1}{2}}}{n}\{\tan(\tfrac{1}{2}\pi/n)\,\mathbf{J}_{1/n}(-2\alpha^{\frac{1}{2}n}i) - \mathbf{E}_{1/n}(-2\alpha^{\frac{1}{2}n}i)\}$$

$$+ \frac{\pi\alpha^{\frac{1}{2}}}{n\sin(\pi/n)}\{J_{-1/n}(-2\alpha^{\frac{1}{2}n}i) - J_{1/n}(-2\alpha^{\frac{1}{2}n}i)\}$$

$$= -\frac{\pi\alpha^{\frac{1}{2}(n+1)}i}{n\cos(\frac{1}{2}\pi/n)}\sum_{m=0}^\infty \frac{\alpha^{mn}}{\Gamma(m+\frac{3}{2}-\frac{1}{2}/n)\,\Gamma(m+\frac{3}{2}+\frac{1}{2}/n)}$$

$$+ \frac{\pi\alpha^{\frac{1}{2}}}{n\sin(\pi/n)}\{e^{\frac{1}{2}\pi i/n}I_{-1/n}(2\alpha^{\frac{1}{2}n}) - e^{-\frac{1}{2}\pi i/n}I_{1/n}(2\alpha^{\frac{1}{2}n})\},$$

and therefore

(5) $$Si_n(\alpha) = -\frac{\pi\alpha^{\frac{1}{2}(n+1)}}{n\cos(\frac{1}{2}\pi/n)}\sum_{m=0}^\infty \frac{\alpha^{mn}}{\Gamma(m+\frac{3}{2}-\frac{1}{2}/n)\,\Gamma(m+\frac{3}{2}+\frac{1}{2}/n)}$$

$$+ \frac{\pi\alpha^{\frac{1}{2}}}{2n\cos(\frac{1}{2}\pi/n)}\{I_{-1/n}(2\alpha^{\frac{1}{2}n}) + I_{1/n}(2\alpha^{\frac{1}{2}n})\},$$

whence it follows that, when $\beta > 0$,

(6) $$Si_n(-\beta) = \frac{(-)^{\frac{1}{2}(n-1)}\pi\beta^{\frac{1}{2}(n+1)}}{n\cos(\frac{1}{2}\pi/n)}\sum_{m=0}^\infty \frac{(-)^m\beta^{mn}}{\Gamma(m+\frac{3}{2}-\frac{1}{2}/n)\,\Gamma(m+\frac{3}{2}+\frac{1}{2}/n)}$$

$$+ \frac{\pi\beta^{\frac{1}{2}}}{2n\cos(\frac{1}{2}\pi/n)}\{J_{-1/n}(2\beta^{\frac{1}{2}n}) - J_{1/n}(2\beta^{\frac{1}{2}n})\},$$

and hence, for all real values of α,

$$\left\{\frac{d^2}{d\alpha^2} - n^2\alpha^{n-2}\right\} Si_n(\alpha) = -n\alpha^{\frac{1}{2}(n-3)}. \tag{7}$$

This equation was given by Stokes in the case $n = 3$.

It should be noticed that

$$\begin{aligned} Si_n(\alpha) + (-)^{\frac{1}{2}(n+1)} Ei_n(\alpha) &= \frac{\pi\alpha^{\frac{1}{2}}}{n\sin(\pi/n)}\{\sin(\tfrac{1}{2}\pi/n) + (-1)^{\frac{1}{2}(n+1)}\} \\ &\qquad \times \{I_{-1/n}(2\alpha^{\frac{1}{2}n}) + I_{1/n}(2\alpha^{\frac{1}{2}n})\} \\ &= \frac{\pi\beta^{\frac{1}{2}}}{n\sin(\pi/n)}\{\sin(\tfrac{1}{2}\pi/n) + (-1)^{\frac{1}{2}(n+1)}\} \\ &\qquad \times \{J_{-1/n}(2\beta^{\frac{1}{2}n}) - J_{1/n}(2\beta^{\frac{1}{2}n})\}, \end{aligned} \tag{8}$$

where $\beta = -\alpha$, and α and β are real.

The formulae of the preceding three sections are due to Hardy, though his methods of obtaining them were different and he gave some of them only in the special case $n = 3$.

10·3. *Cauchy's numbers.*

In connexion with a generalisation of Bessel's integral which was defined by Bourget, and subsequently studied by Giuliani (see § 10·31), it is convenient to investigate a class of functions known as *Cauchy's numbers.*

The typical number, $N_{-n,k,m}$, is defined by Cauchy* as the coefficient of the term independent of t in the expansion of

$$t^{-n}\left(t + \frac{1}{t}\right)^k \left(t - \frac{1}{t}\right)^m$$

in ascending powers of t. It is supposed that n, k, and m are integers of which the last two are not negative.

It follows from Cauchy's theorem that

$$\begin{aligned} N_{-n,k,m} &= \frac{1}{2\pi i}\int^{(0+)} t^{-n-1}\left(t + \frac{1}{t}\right)^k \left(t - \frac{1}{t}\right)^m dt \\ &= \frac{2^{m+k}\, i^m}{2\pi}\int_{-\pi}^{\pi} e^{-ni\theta}\cos^k\theta\sin^m\theta\, d\theta \\ &= \frac{2^{m+k}\, i^m}{\pi}\int_0^{\pi} \tfrac{1}{2}\{e^{-ni\theta} + (-)^m e^{ni\theta}\}\cos^k\theta\sin^m\theta\, d\theta \\ &= \frac{2^{m+k}}{\pi}\int_0^{\pi}\cos(\tfrac{1}{2}m\pi - n\theta)\cos^k\theta\sin^m\theta\, d\theta. \end{aligned} \tag{1}$$

It is evident from the definition that $N_{-n,k,m}$ is zero if $-n + k + m$ is odd or if it is a negative integer.

* *Comptes Rendus*, XI. (1840), pp. 473—475, 510—511; XII. (1841), pp. 92—93; XIII. (1841), pp. 682—687, 850—854.

From (1) it is seen that

(2) $$N_{-n,k,m} = (-)^m N_{n,k,m} = (-)^{n-k} N_{n,k,m}.$$

These results, together with recurrence formulae from which successive numbers may be calculated, were given by Bourget*.

The recurrence formulae are

(3) $$N_{-n,k,m} = N_{-n+1,k-1,m} + N_{-n-1,k-1,m},$$

(4) $$N_{-n,k,m} = N_{-n+1,k,m-1} - N_{-n-1,k,m-1},$$

and they are immediate consequences of the identities

$$t^{-n}(t+1/t)^k(t-1/t)^m = t^{1-n}(t+1/t)^{k-1}(t-1/t)^m + t^{-n-1}(t+1/t)^{k-1}(t-1/t)^m,$$

$$t^{-n}(t+1/t)^k(t-1/t)^m = t^{1-n}(t+1/t)^k(t-1/t)^{m-1} - t^{-n-1}(t+1/t)^k(t-1/t)^{m-1}.$$

By means of these formulae any Cauchy's number is ultimately expressible in terms of numbers of the types $N_{-n,k,0}$, $N_{-n,0,m}$.

A different class of recurrence formulae, also due to Bourget, owes its existence to the equation

$$t\frac{d}{dt}\left(t \pm \frac{1}{t}\right) = t \mp \frac{1}{t}.$$

It follows that

$$N_{-n,k,m} = \frac{1}{2\pi i(m+1)}\int^{(0+)} t^{-n}\left(t+\frac{1}{t}\right)^{k-1}\frac{d}{dt}\left(t-\frac{1}{t}\right)^{m+1} dt$$

$$= -\frac{1}{2\pi i(m+1)}\int^{(0+)} \left(t-\frac{1}{t}\right)^{m+1}\frac{d}{dt}\left\{t^{-n}\left(t+\frac{1}{t}\right)^{k-1}\right\} dt,$$

by a partial integration. On performing the differentiation we see that

(5) $$(m+1)N_{-n,k,m} = nN_{-n,k-1,m+1} - (k-1)N_{-n,k-2,m+2},$$

and similarly

(6) $$(k+1)N_{-n,k,m} = nN_{-n,k+1,m-1} - (m-1)N_{-n,k+2,m-2}.$$

Developments due to Chessin, *Annals of Math.* x. (1895—6), pp. 1—2, are

(7) $$N_{-n,k,m} = \sum_{r=0}^{s} {}_sC_r \cdot N_{-n+s-2r,\,k-s,\,m},$$

(8) $$N_{-n,k,m} = \sum_{r=0}^{s} (-)^r {}_sC_r \cdot N_{-n+s-2r,\,k,\,m-s}.$$

These may be deduced by induction from (3) and (4).

Another formula due to Chessin is

(9) $$N_{-n,k,m} = \sum_{r=0}^{p} (-)^r {}_kC_{p-r} \cdot {}_mC_r,$$

where $p = \frac{1}{2}(k+m-n)$. This is proved by selecting the coefficient of t^n in the product

$$(t+1/t)^k \times (t-1/t)^m.$$

* *Journal de Math.* (2) vi. (1861), pp. 33—54.

10·31. *The functions of Bourget and Giuliani.*

The function $J_{n,k}(z)$ is defined by the generalisation of Bessel's integral

(1) $$J_{n,k}(z) = \frac{1}{2\pi i}\int^{(0+)} t^{-n-1}\left(t+\frac{1}{t}\right)^k \exp\left\{\tfrac{1}{2}z\left(t-\frac{1}{t}\right)\right\} dt,$$

where n is an integer, and k is a positive integer.

It follows that

$$J_{n,k}(z) = \frac{1}{2\pi}\int_{-\pi}^{\pi} \exp\{-i(n\theta - z\sin\theta)\} . (2\cos\theta)^k \, d\theta,$$

and therefore

(2) $$J_{n,k}(z) = \frac{1}{\pi}\int_0^{\pi} (2\cos\theta)^k \cos(n\theta - z\sin\theta)\, d\theta.$$

The function $J_{n,k}(z)$ has been studied by Bourget, *Journal de Math.* (2) VI. (1861), pp. 42—55, for the sake of various astronomical applications; while Giuliani, *Giornale di Mat.* XXVI. (1888), pp. 151—171, has constructed a linear differential equation of the fourth order satisfied by the function.

[NOTE. An earlier paper by Giuliani, *Giornale di Mat.* XXV. (1887), pp. 198—202, contains properties of another generalisation of Bessel's integral, namely

$$\frac{1}{\pi}\int_0^{\pi} \cos(n\theta - z^p \sin^p\theta)\, d\theta,$$

but parts of the analysis in this paper seem to be incorrect.]

If we expand the integrand of (1) in powers of z, we deduce from § 10·3 that

(3) $$J_{n,k}(z) = \sum_{m=0}^{\infty} \frac{(\frac{1}{2}z)^m}{m!} N_{-n,k,m};$$

and it is evident from (1) that

(4) $$J_{n,0}(z) \equiv J_n(z).$$

Again from § 10·3 (2) and (3) it is evident that

(5) $$J_{-n,k}(z) = (-)^{n-k} J_{n,k}(z),$$

(6) $$J_{n,k}(z) = J_{n-1,k-1}(z) + J_{n+1,k-1}(z);$$

and, if we take $k = 1$ in this formula,

(7) $$J_{n,1}(z) = \frac{2n}{z} J_n(z).$$

These results were obtained by Bourget; and the reader should have no difficulty in proving that

(8) $$2J'_{n,k}(z) = J_{n-1,k}(z) - J_{n+1,k}(z).$$

Other recurrence formulae (due to Bourget and Giuliani respectively) are

(9) $$J_{n,k+2}(z) = \frac{2n}{z} J_{n,k+1}(z) - \frac{2(k+1)}{z}\{J_{n-1,k}(z) - J_{n+1,k}(z)\},$$

(10) $$4J''_{n,k-2}(z) = J_{n,k}(z) - 4J_{n,k-2}(z).$$

The differential equation is most simply constructed by the method used by Giuliani; thus

$$\begin{aligned}\nabla_n J_{n,k}(z) &= \frac{1}{\pi}\int_0^\pi \frac{d}{d\theta}\{-(n+z\cos\theta)\sin(n\theta - z\sin\theta)\}(2\cos\theta)^k\, d\theta\\ &= -\frac{2k}{\pi}\int_0^\pi (n+z\cos\theta)\sin(n\theta-z\sin\theta)(2\cos\theta)^{k-1}\sin\theta\, d\theta\\ &= -2kzJ'_{n,k}(z) + \frac{2k}{\pi}\int_0^\pi\left\{\frac{d}{d\theta}\cos(n\theta - z\sin\theta)\right\}(2\cos\theta)^{k-1}\sin\theta\, d\theta\\ &= -2kzJ'_{n,k}(z) - \frac{2k}{\pi}\int_0^\pi \cos(n\theta-z\sin\theta)\frac{d}{d\theta}\{(2\cos\theta)^{k-1}\sin\theta\}\, d\theta,\end{aligned}$$

and so

$$\nabla_n J_{n,k}(z) = -2kzJ'_{n,k}(z) - k^2 J_{n,k}(z) + 4k(k-1)J_{n,k-2}(z).$$

Operating on this equation by $\dfrac{d^2}{dz^2}+1$, and using (10), it follows that

$$\left(\frac{d^2}{dz^2}+1\right)\{\nabla_n J_{n,k}(z) + 2kzJ'_{n,k}(z) + k^2 J_{n,k}(z)\} = k(k-1)J_{n,k}(z),$$

and hence we have Giuliani's equation

$$\text{(11)}\quad \begin{aligned} z^2 J^{\text{iv}}_{n,k}(z) + (2k+5)zJ'''_{n,k}(z) + \{2z^2 + (k+2)^2 - n^2\}J''_{n,k}(z)&\\ + (2k+5)zJ'_{n,k}(z) + (z^2+k+2-n^2)J_{n,k}(z) &= 0.\end{aligned}$$

It was also observed by Giuliani that

$$\text{(12)}\quad \begin{aligned} e^{iz\sin\theta}(2\cos\theta)^k = \sum_{n=0}^\infty \epsilon_{2n} J_{2n,k}(z)\cos 2n\theta&\\ + i\sum_{n=0}^\infty \epsilon_{2n+1}J_{2n+1,k}(z)\sin(2n+1)\theta&;\end{aligned}$$

this is verified by applying Fourier's rule (cf. § 2·2) to the function on the right.

A somewhat similar function $J(z;\,\nu,\,k)$ has been studied by Bruns, *Astr. Nach.* CIV. (1883), col. 1—8. This function is defined by the series

$$\text{(13)}\quad J(z;\,\nu,\,k) = \sum_{m=0}^\infty \frac{(-)^m(\tfrac{1}{2}z)^{\nu+2k+2m}}{m!\,\Gamma(\nu+2k+m)}\,\frac{2\nu}{(\nu+2k-2)(\nu+2k)(\nu+2k+2m+2)}.$$

The most important property of this function is that

$$\text{(14)}\quad J(z;\,\nu,\,k) - J(z;\,\nu,\,k+1) = \frac{2\nu J_{\nu+2k}(z)}{(\nu+2k-2)(\nu+2k+2)},$$

whence it follows that

$$\text{(15)}\quad J(z;\,\nu,\,k) = \sum_{m=k}^\infty \frac{2\nu J_{\nu+2m}(z)}{(\nu+2m-2)(\nu+2m+2)}.$$

10·4. *The definition of Struve's function* $\mathbf{H}_\nu(z)$.

Now that we have completely examined the functions defined by integrals resembling Bessel's integral, it is natural to investigate a function defined by an integral resembling Poisson's integral. This function is called Struve's function, although Struve investigated* only the special functions of this type of orders zero and unity. The properties of the general function have been examined at some length by Siemon† and by J. Walker‡.

Struve's function $\mathbf{H}_\nu(z)$, of order ν, is defined by the equations

$$\mathbf{H}_\nu(z) = \frac{2(\tfrac{1}{2}z)^\nu}{\Gamma(\nu+\tfrac{1}{2})\Gamma(\tfrac{1}{2})}\int_0^1 (1-t^2)^{\nu-\frac{1}{2}} \sin zt\,.\,dt \tag{1}$$

$$= \frac{2(\tfrac{1}{2}z)^\nu}{\Gamma(\nu+\tfrac{1}{2})\Gamma(\tfrac{1}{2})}\int_0^{\frac{1}{2}\pi} \sin(z\cos\theta)\sin^{2\nu}\theta\, d\theta,$$

provided that $R(\nu) > -\frac{1}{2}$.

By analysis similar to that of § 3·3, we have

$$\mathbf{H}_\nu(z) = \frac{2(\tfrac{1}{2}z)^\nu}{\Gamma(\nu+\tfrac{1}{2})\Gamma(\tfrac{1}{2})}\sum_{m=0}^\infty \frac{(-)^m z^{2m+1}}{(2m+1)!}\int_0^1 t^{2m+1}(1-t^2)^{\nu-\frac{1}{2}}\, dt$$

$$= \frac{(\tfrac{1}{2}z)^\nu}{\Gamma(\tfrac{1}{2})}\sum_{m=0}^\infty \frac{(-)^m z^{2m+1}\,.\, m!}{(2m+1)!\,\Gamma(\nu+m+\tfrac{3}{2})},$$

so that

$$\mathbf{H}_\nu(z) = \sum_{m=0}^\infty \frac{(-)^m (\tfrac{1}{2}z)^{\nu+2m+1}}{\Gamma(m+\tfrac{3}{2})\Gamma(\nu+m+\tfrac{3}{2})}. \tag{2}$$

The function $\mathbf{H}_\nu(z)$ is defined by this equation for all values of ν, whether $R(\nu)$ exceeds $-\frac{1}{2}$ or not. It is evident that $\mathbf{H}_\nu(z)$ is an integral function of ν and, if the factor $(\frac{1}{2}z)^\nu$ be suppressed, the resulting expression is also an integral function of z.

It is easy to see [cf. §§ 2·11 (5), 3·121 (1)] that

$$\mathbf{H}_\nu(z) = \frac{(\tfrac{1}{2}z)^{\nu+1}}{\Gamma(\tfrac{3}{2})\Gamma(\nu+\tfrac{3}{2})}(1+\theta), \tag{3}$$

where

$$|\theta| < \tfrac{2}{3}\exp\left\{\frac{\tfrac{1}{4}|z|^2}{|\nu_0+\tfrac{3}{2}|} - 1\right\}, \tag{4}$$

and $|\nu_0+\frac{3}{2}|$ is the smallest of the numbers $|\nu+\frac{3}{2}|$, $|\nu+\frac{5}{2}|$, $|\nu+\frac{7}{2}|$, ...

* *Mém. de l'Acad. Imp. des Sci. de St Pétersbourg*, (7) xxx. (1882), no. 8; *Ann. der Physik*, (3) xvii. (1882), pp. 1008—1016. See also Lommel, *Archiv der Math. und Phys.* xxxvi. (1861), p. 399.

† *Programm, Luisenschule*, Berlin, 1890. [*Jahrbuch über die Fortschritte der Math.* 1890, pp. 340—342.]

‡ *The Analytical Theory of Light* (Cambridge, 1904), pp. 392—395. The results contained in this section, with the exception of (3), (4), (10) and (11), are there given.

We can obtain recurrence formulae thus:

$$\frac{d}{dz}\{z^\nu \mathbf{H}_\nu(z)\} = \sum_{m=0}^{\infty} \frac{(-)^m (2\nu+2m+1)\, z^{2\nu+2m}}{2^{\nu+2m+1}\,\Gamma(m+\frac{3}{2})\,\Gamma(\nu+m+\frac{3}{2})}$$
$$= z^\nu \mathbf{H}_{\nu-1}(z),$$

and similarly

$$\frac{d}{dz}\{z^{-\nu} \mathbf{H}_\nu(z)\} = \sum_{m=0}^{\infty} \frac{(-)^m (2m+1)\, z^{2m}}{2^{\nu+2m+1}\,\Gamma(m+\frac{3}{2})\,\Gamma(\nu+m+\frac{3}{2})}$$
$$= \sum_{m=-1}^{\infty} \frac{(-)^{m+1}\, z^{2m+2}}{2^{\nu+2m+2}\,\Gamma(m+\frac{3}{2})\,\Gamma(\nu+m+\frac{5}{2})}$$
$$= \frac{1}{2^\nu\,\Gamma(\nu+\frac{3}{2})\,\Gamma(\frac{1}{2})} - z^{-\nu}\mathbf{H}_{\nu+1}(z).$$

On comparing these results, we find that

(5) $$\mathbf{H}_{\nu-1}(z) + \mathbf{H}_{\nu+1}(z) = \frac{2\nu}{z}\mathbf{H}_\nu(z) + \frac{(\frac{1}{2}z)^\nu}{\Gamma(\nu+\frac{3}{2})\,\Gamma(\frac{1}{2})},$$

(6) $$\mathbf{H}_{\nu-1}(z) - \mathbf{H}_{\nu+1}(z) = 2\mathbf{H}_\nu'(z) - \frac{(\frac{1}{2}z)^\nu}{\Gamma(\nu+\frac{3}{2})\,\Gamma(\frac{1}{2})},$$

(7) $$(\vartheta+\nu)\,\mathbf{H}_\nu(z) = z\mathbf{H}_{\nu-1}(z),$$

(8) $$(\vartheta-\nu)\,\mathbf{H}_\nu(z) = \frac{(\frac{1}{2}z)^{\nu+1}}{\Gamma(\nu+\frac{3}{2})\,\Gamma(\frac{3}{2})} - z\mathbf{H}_{\nu+1}(z).$$

In particular we have

(9) $$\frac{d}{dz}\{z\mathbf{H}_1(z)\} = z\mathbf{H}_0(z), \quad \frac{d}{dz}\{\mathbf{H}_0(z)\} = \frac{2}{\pi} - \mathbf{H}_1(z).$$

Again, from (7) and (8), we have

$$(\vartheta^2-\nu^2)\,\mathbf{H}_\nu(z) = (\vartheta-\nu)\,\{z\mathbf{H}_{\nu-1}(z)\}$$
$$= z(\vartheta-\nu+1)\,\mathbf{H}_{\nu-1}(z)$$
$$= \frac{4(\frac{1}{2}z)^{\nu+1}}{\Gamma(\nu+\frac{1}{2})\,\Gamma(\frac{1}{2})} - z^2\mathbf{H}_\nu(z),$$

so that $\mathbf{H}_\nu(z)$ satisfies the differential equation

(10) $$\nabla_\nu \mathbf{H}_\nu(z) = \frac{4(\frac{1}{2}z)^{\nu+1}}{\Gamma(\nu+\frac{1}{2})\,\Gamma(\frac{1}{2})}.$$

The function $\mathbf{L}_\nu(z)$ which bears the same relation to Struve's function as $I_\nu(z)$ bears to $J_\nu(z)$ has been studied (in the case $\nu=0$) by* Nicholson, *Quarterly Journal*, XLII. (1911), p. 218. This function is defined by the equation

(11) $$\mathbf{L}_\nu(z) = \sum_{m=0}^{\infty} \frac{(\frac{1}{2}z)^{\nu+2m+1}}{\Gamma(m+\frac{3}{2})\,\Gamma(\nu+m+\frac{3}{2})}$$
$$= \frac{2(\frac{1}{2}z)^\nu}{\Gamma(\nu+\frac{1}{2})\,\Gamma(\frac{1}{2})}\int_0^{\frac{1}{2}\pi} \sinh(z\cos\theta)\sin^{2\nu}\theta\, d\theta,$$

the integral formula being valid only when $R(\nu) > -\frac{1}{2}$.

The reader should have no difficulty in obtaining the fundamental properties of this function.

* See also Gubler, *Zürich Vierteljahrsschrift*, XLVII. (1902), p. 424.

10·41. *The loop-integral for* $\mathbf{H}_\nu(z)$.

It was noticed in § 10·4 that the integral definition of $\mathbf{H}_\nu(z)$ fails when $R(\nu) \leqslant -\frac{1}{2}$, because the integral does not converge at the upper limit. We can avoid this disability by considering a loop-integral in place of the definite integral.

Let us take

$$\int_0^{(1+)} (t^2-1)^{\nu-\frac{1}{2}} \sin zt \,.\, dt,$$

where the phase of $t^2 - 1$ vanishes at the point on the right of $t = 1$ at which the contour crosses the real axis, and the contour does not enclose the point $t = -1$.

If we suppose that $R(\nu) > -\frac{1}{2}$, we may deform the contour into the segment (0, 1) of the real axis, taken twice, and we find that

$$\int_0^{(1+)} (t^2-1)^{\nu-\frac{1}{2}} \sin zt \,.\, dt = 2i \cos \nu\pi \int_0^1 (1-t^2)^{\nu-\frac{1}{2}} \sin zt \,.\, dt,$$

where the phase of $1 - t^2$ is zero.

Hence, when $R(\nu) > -\frac{1}{2}$, we have

(1) $$\mathbf{H}_\nu(z) = \frac{\Gamma(\frac{1}{2}-\nu)\,.\,(\frac{1}{2}z)^\nu}{\pi i\,\Gamma(\frac{1}{2})} \int_0^{(1+)} (t^2-1)^{\nu-\frac{1}{2}} \sin zt \,.\, dt.$$

Both sides of this equation are analytic functions of ν for all* values of ν; and so, by the general theory of analytic continuation, equation (1) holds for all values of ν.

From this result, combined with § 6·1 (6), we deduce that

(2) $$J_\nu(z) + i\mathbf{H}_\nu(z) = \frac{\Gamma(\frac{1}{2}-\nu)\,.\,(\frac{1}{2}z)^\nu}{\pi i\,\Gamma(\frac{1}{2})} \int_0^{(1+)} e^{izt} (t^2-1)^{\nu-\frac{1}{2}} dt.$$

To transform this result, let ω be any acute angle (positive or negative), and let the phase of z lie between $-\frac{1}{2}\pi + \omega$ and $\frac{1}{2}\pi + \omega$. We then deform the contour into that shewn in Fig. 26, in which the four parallel lines make an angle $-\omega$ with the imaginary axis. It is evident that, as the lines parallel to the real axis move off to infinity, the integrals along them tend to zero. The integral along the path which starts from and returns to $1 + \infty\, ie^{-i\omega}$ is equal to $H_\nu^{(1)}(z)$; and on the lines through the origin we write $t = iu$, so that on them

$$(t^2-1)^{\nu-\frac{1}{2}} = e^{\mp(\nu-\frac{1}{2})\pi i}(1+u^2)^{\nu-\frac{1}{2}}.$$

It follows that

$$J_\nu(z) + i\mathbf{H}_\nu(z) = H_\nu^{(1)}(z) + \frac{2i(\frac{1}{2}z)^\nu}{\Gamma(\nu+\frac{1}{2})\,\Gamma(\frac{1}{2})} \int_0^{\infty \exp(-i\omega)} e^{-zu}(1+u^2)^{\nu-\frac{1}{2}} du,$$

* The isolated values $\frac{1}{2}, \frac{3}{2}, \frac{5}{2}, \ldots$ are excepted, because the expression on the right is then an undetermined form.

where the phase of $1 + u^2$ has its principal value; and hence

$$(3) \qquad \mathbf{H}_\nu(z) = Y_\nu(z) + \frac{2(\frac{1}{2}z)^\nu}{\Gamma(\nu+\frac{1}{2})\,\Gamma(\frac{1}{2})} \int_0^{\infty \exp(-i\omega)} e^{-zu}(1+u^2)^{\nu-\frac{1}{2}}\,du.$$

This result, which is true for unrestricted values of ν, and for any value of z for which $-\pi < \arg z < \pi$, will be applied immediately to obtain the asymptotic expansion of $\mathbf{H}_\nu(z)$ when $|z|$ is large.

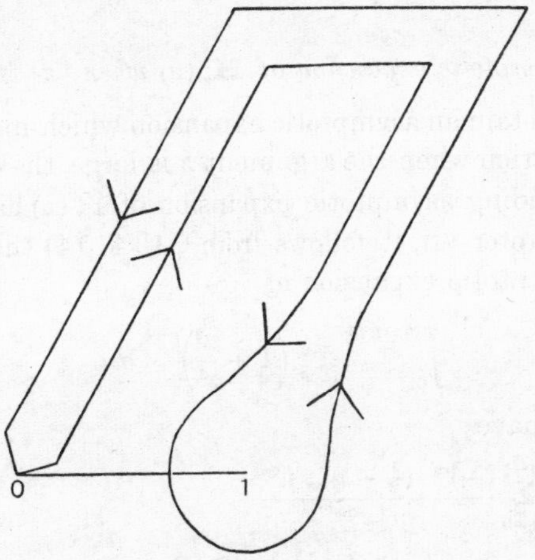

Fig. 26.

A result equivalent to (2) was obtained by J. Walker*, who assumed that $R(\nu) > -\frac{1}{2}$, $R(z) > 0$, so that ω might be taken to be zero. In the case $\nu = 0$, the result had previously been obtained by Rayleigh† with the aid of the method of Lipschitz (§ 7·21).

If, as in § 6·12, we replace ω by $\arg z - \beta$, it is evident that (3) may be written in the form

$$(4) \qquad \mathbf{H}_\nu(z) = Y_\nu(z) + \frac{(\frac{1}{2}z)^{\nu-1}}{\Gamma(\nu+\frac{1}{2})\,\Gamma(\frac{1}{2})} \int_0^{\infty \exp i\beta} e^{-u}\left(1+\frac{u^2}{z^2}\right)^{\nu-\frac{1}{2}} du,$$

where $-\frac{1}{2}\pi < \beta < \frac{1}{2}\pi$ and $-\frac{1}{2}\pi + \beta < \arg z < \frac{1}{2}\pi + \beta$.

This equation gives a representation of $\mathbf{H}_\nu(z)$ when $|\arg z| < \pi$. To obtain a representation valid near the negative half of the real axis, we define $\mathbf{H}_\nu(z)$ for unrestricted values of $\arg z$ by the equation

$$(5) \qquad \mathbf{H}_\nu(ze^{m\pi i}) = e^{m(\nu+1)\pi i}\,\mathbf{H}_\nu(z),$$

and use (4) with z replaced by $ze^{\mp\pi i}$.

* *The Analytical Theory of Light* (Cambridge, 1904), pp. 394—395.

† *Proc. London Math. Soc.* XIX. (1889), pp. 504—507. [*Scientific Papers*, III. (1902), pp. 44—46.]

If we write $z=ix$ in (3), where x is positive, we see that, when $R(\nu)<\frac{1}{2}$,

$$i\mathbf{L}_\nu(x)=e^{-\frac{1}{2}\nu\pi i}\,Y_\nu(ix)+\frac{2(\frac{1}{2}x)^\nu}{\Gamma(\nu+\frac{1}{2})\,\Gamma(\frac{1}{2})}\int_0^\infty e^{-ixu}(1+u^2)^{\nu-\frac{1}{2}}\,du,$$

and, by considering imaginary parts, we deduce that

$$\mathbf{L}_\nu(x)=I_{-\nu}(x)-\frac{2(\frac{1}{2}x)^\nu}{\Gamma(\nu+\frac{1}{2})\,\Gamma(\frac{1}{2})}\int_0^\infty \sin(xu).(1+u^2)^{\nu-\frac{1}{2}}\,du, \tag{6}$$

a result given by Nicholson, *Quarterly Journal*, XLII. (1911), p. 219, in the special case in which $\nu=0$.

10·42. *The asymptotic expansion of* $\mathbf{H}_\nu(z)$ *when* $|z|$ *is large.*

We shall now obtain an asymptotic expansion which may be used for tabulating Struve's function when the argument z is large, the order ν being fixed. Since the corresponding asymptotic expansion of $Y_\nu(z)$ has been completely investigated in Chapter VII, it follows from § 10·41 (4) that it is sufficient to determine the asymptotic expansion of

$$\int_0^{\infty\exp i\beta} e^{-u}\left(1+\frac{u^2}{z^2}\right)^{\nu-\frac{1}{2}} du.$$

As in § 7·2, we have

$$\left(1+\frac{u^2}{z^2}\right)^{\nu-\frac{1}{2}}=\sum_{m=0}^{p-1}\frac{(-)^m.(\frac{1}{2}-\nu)_m\,u^{2m}}{m!\,z^{2m}}$$
$$+\frac{(-)^p.(\frac{1}{2}-\nu)_p\,u^{2p}}{(p-1)!\,z^{2p}}\int_0^1(1-t)^{p-1}\left(1+\frac{u^2t}{z^2}\right)^{\nu-p-\frac{1}{2}}dt.$$

We take p so large that $R(\nu-p-\frac{1}{2})\leqslant 0$, and take δ to be any positive angle for which

$$|\beta|\leqslant\tfrac{1}{2}\pi-\delta,\quad |\arg z-\beta|\leqslant\tfrac{1}{2}\pi-\delta,$$

so that z is confined to the sector of the plane for which

$$-\pi+2\delta\leqslant\arg z\leqslant\pi-2\delta.$$

We then have

$$\left|\left(1\pm\frac{iu\sqrt{t}}{z}\right)\right|\geqslant\sin\delta,\quad \left|\arg\left(1\pm\frac{iu\sqrt{t}}{z}\right)\right|<\pi,$$

so that

$$\left|\left(1+\frac{u^2t}{z^2}\right)^{\nu-p-\frac{1}{2}}\right|\leqslant e^{2\pi|I(\nu)|}(\sin\delta)^{2R(\nu)-2p-1}=\mathbf{A}_p,$$

say, where $\mathbf{A}_p$ is independent of z.

It follows on integration that

$$\int_0^{\infty\exp i\beta} e^{-u}\left(1+\frac{u^2}{z^2}\right)^{\nu-\frac{1}{2}}du=\sum_{m=0}^{p-1}\frac{(-)^m(\frac{1}{2}-\nu)_m.(2m)!}{m!\,z^{2m}}+\mathbf{R}_p,$$

where

$$|\mathbf{R}_p|\leqslant\frac{\mathbf{A}_p}{p!}\left|\frac{(\frac{1}{2}-\nu)_p}{z^{2p}}\right|.\left|\int_0^{\infty\exp i\beta}e^{-u}\,u^{2p}\,du\right|$$
$$=O(z^{-2p}).$$

We deduce that, when $|\arg z| < \pi$ and $|z|$ is large,

$$(1)\quad \mathbf{H}_\nu(z) = Y_\nu(z) + \frac{(\frac{1}{2}z)^{\nu-1}}{\Gamma(\nu+\frac{1}{2})\,\Gamma(\frac{1}{2})}\left[\sum_{m=0}^{p-1}\frac{(-)^m(\frac{1}{2}-\nu)_m\,.\,(2m)!}{m!\,z^{2m}} + O(z^{-2p})\right],$$

provided that $R(p-\nu+\frac{1}{2}) \geqslant 0$; but, as in § 7·2, this last restriction may be removed.

This asymptotic expansion may also be written in the form

$$(2)\quad \mathbf{H}_\nu(z) = Y_\nu(z) + \frac{1}{\pi}\sum_{m=0}^{p-1}\frac{\Gamma(m+\frac{1}{2})}{\Gamma(\nu+\frac{1}{2}-m)(\frac{1}{2}z)^{2m-\nu+1}} + O(z^{\nu-2p-1}).$$

It may be proved without difficulty that, if ν is real and z is positive, the remainder after p terms in the asymptotic expansion is of the same sign as, and numerically less than the first term neglected, provided that $R(p+\frac{1}{2}-\nu) \geqslant 0$. This may be established by the method used in § 7·32.

The asymptotic expansion* was given by Rayleigh, *Proc. London Math. Soc.* XIX. (1888), p. 504 in the case $\nu=0$, by Struve, *Mém. de l'Acad. Imp. des Sci. de St Pétersbourg*, (7) XXX. (1882), no. 8, p. 101, and *Ann. der Phys. und Chemie*, (3) XVII. (1882), p. 1012 in the case $\nu=1$; the result for general values of ν was given by J. Walker, *The Analytical Theory of Light* (Cambridge, 1904), pp. 394—395.

If ν has any of the values $\frac{1}{2}, \frac{3}{2}, \ldots$, then $(1+u^2/z^2)^{\nu-\frac{1}{2}}$ is expressible as a *terminating* series and $Y_\nu(z)$ is also expressible in a finite form. It follows that, when ν is half of an odd positive integer, $\mathbf{H}_\nu(z)$ is expressible in terms of elementary functions. In particular

$$(3)\quad \begin{cases} \mathbf{H}_{\frac{1}{2}}(z) = \left(\dfrac{2}{\pi z}\right)^{\frac{1}{2}}(1-\cos z), \\ \mathbf{H}_{\frac{3}{2}}(z) = \left(\dfrac{z}{2\pi}\right)^{\frac{1}{2}}\left(1+\dfrac{2}{z^2}\right) - \left(\dfrac{2}{\pi z}\right)^{\frac{1}{2}}\left(\sin z + \dfrac{\cos z}{z}\right). \end{cases}$$

10·43. *The asymptotic expansion of Struve's functions of large order.*

We shall now obtain asymptotic expansions, of a type similar to the expansions investigated in Chapter VIII, which represent Struve's function $\mathbf{H}_\nu(z)$ when $|\nu|$ and $|z|$ are both large.

As usual, we shall write

$$\nu = z\cosh(\alpha+i\beta) = z\cosh\gamma$$

and, for simplicity, we shall confine the investigation to the special case in which $\cosh\gamma$ is real and positive. The more general case in which $\cosh\gamma$ is complex may be investigated by the methods used in § 8·6 and § 10·15, but it is of no great practical importance and it involves some rather intricate analysis.

* For an asymptotic expansion of the associated integral

$$\int_1^\infty \frac{e^{-u}}{u^2}\left(1+\frac{u^2}{z^2}\right)^{-\frac{1}{2}} du,$$

see Rayleigh, *Phil. Mag.* (6) VIII. (1904), pp. 481—487. [*Scientific Papers*, v. (1912), pp. 206—211.]

The method of steepest descents has to be applied to an integral of Poisson's type, and not, as in the previous investigations, to one of Bessel's type.

In view of the formula of § 10·41 (3), we consider the integral

$$\int e^{-wz}(1+w^2)^{\nu}\frac{dw}{\sqrt{(1+w^2)}},$$

which we write in the form

$$\int e^{-z\tau}\frac{dw}{\sqrt{(1+w^2)}},$$

where $\tau \equiv w - \cosh\gamma . \log(1+w^2)$.

It is evident that τ, *qua* function of w, has stationary points where $w = e^{\pm\gamma}$, so that, since γ is equal either to α or to $i\beta$, two cases have to be considered, which give rise to the stationary points

$$\text{(I)}\ e^{\pm\alpha}, \quad \text{(II)}\ e^{\pm i\beta}.$$

Accordingly we consider separately the cases (I) in which z/ν is less than 1, and (II) in which z/ν is greater than 1.

(I) When γ is a real positive number α, τ is real when w is real, and, as w increases from 0 to ∞, τ first increases from 0 to $e^{-\alpha} - \cosh\alpha . \log(1+e^{-2\alpha})$, then decreases to $e^{\alpha} - \cosh\alpha . \log(1+e^{2\alpha})$ and finally increases to $+\infty$.

In order to obtain a contour along which τ continually increases, we suppose that w first moves along the real axis from the origin to the point $e^{-\alpha}$, and then starts moving along a certain curve, which leaves the real axis at right angles, on which τ is positive and increasing.

To find the ultimate destination of this curve, it is convenient to make a change of variables by writing

$$w = \sinh\zeta, \quad \zeta = \xi + i\eta, \quad e^{-\alpha} = \sinh\xi_0,$$

where ξ, η and ξ_0 are real.

The curve in the ζ-plane, on which τ is real, has for its equation

$$\cosh\xi\sin\eta = 2\cosh\alpha \operatorname{arc\,tan}(\tanh\xi\tan\eta),$$

and it has a double point* at ξ_0.

We now write

$$F(\xi, \eta) \equiv \frac{2\operatorname{arc\,tan}(\tanh\xi\tan\eta)}{\cosh\xi\sin\eta}$$

and examine the values of $F(\xi, \eta)$ as ζ traces out the rectangle whose corners O, A, B, C have complex coordinates

$$0, \quad \operatorname{arc\,sinh} 1, \quad \operatorname{arc\,sinh} 1 + \tfrac{1}{2}\pi i, \quad \tfrac{1}{2}\pi i.$$

As ζ goes from O to A, $F(\xi, \eta)$ is equal to $2\sinh\xi/\cosh^2\xi$, and this steadily increases from 0 to 1.

* Except when $\alpha = 0$, in which case it has a triple point.

When ζ is on AB, $F(\xi, \eta)$ is equal to

$$\sqrt{2}\,.\,\text{arc tan}\left(\frac{\tan\eta}{\sqrt{2}}\right).\,\text{cosec}\,\eta,$$

and this steadily increases from 1 to $\pi/\sqrt{2}$ as η increases from 0 to $\frac{1}{2}\pi$.

NOTE. To establish this result, write $\tan\eta = t\sqrt{2}$ and observe that

$$\frac{d}{dt}\left\{\frac{\sqrt{(1+2t^2)}}{t}\,\text{arc tan}\,t\right\} = \frac{1}{t^2\sqrt{(1+2t^2)}}\left\{\frac{t+2t^3}{1+t^2} - \text{arc tan}\,t\right\} \geqslant 0,$$

because $\dfrac{t+2t^3}{1+t^2} - \text{arc tan}\,t$, which vanishes with t, has the positive derivate $\dfrac{2t^2(2+t^2)}{(1+t^2)^2}$.

When ζ is on BC, $F(\xi, \eta)$ is equal to $\pi\,\text{sech}\,\xi$, and this increases steadily from $\pi/\sqrt{2}$ to π as ζ goes from B to C; and finally when ζ is on CO, $F(\xi, \eta)$ is zero.

Hence the curve, on which $F(\xi, \eta)$ is equal to $\text{sech}\,\alpha$, cannot emerge from the rectangle $OABC$, except at the double point on the side OA; and so the part of the curve inside the rectangle must pass from this double point to the singular point C.

The contours in the w-plane for which a has the values 0, $\frac{1}{2}$ are shewn in Fig. 27 by broken and continuous curves respectively.

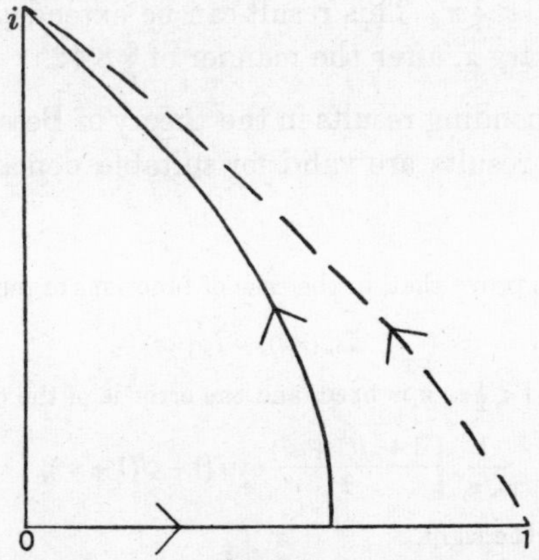

Fig. 27.

Consequently a contour in the w-plane, on which τ is real, consists of the part of the real axis joining the origin to e^{-a} and a curve from this point to the singular point i; and, as w traces out this contour, τ increases from 0 to $+\infty$.

It follows that, if the expansion of $d\zeta/d\tau$ in powers of τ is

$$\frac{d\zeta}{d\tau} = \sum_{m=0}^{\infty} b_m \tau^m,$$

then

$$\int_0^i e^{-zw}(1+w^2)^{\nu-\frac{1}{2}}\,dw = \int_0^\infty e^{-z\tau}\left\{\frac{1}{\sqrt{(1+w^2)}}\frac{dw}{d\tau}\right\}d\tau$$

$$\sim \sum_{m=0}^{\infty}\frac{m!\,b_m}{z^{m+1}},$$

and hence, by 10·4 (1), we have

(1) $$\mathbf{H}_\nu(z) \sim -i J_\nu(z) + \frac{2(\frac{1}{2}z)^\nu}{\Gamma(\nu+\frac{1}{2})\Gamma(\frac{1}{2})}\sum_{m=0}\frac{m!\,b_m}{z^{m+1}}.$$

It is easy to prove that

$b_0=1,\quad b_1=2\cosh\gamma,\quad b_2=6\cosh^2\gamma-\frac{1}{2},\quad b_3=20\cosh^3\gamma-4\cosh\gamma,\quad \ldots.$

(II) When γ is a pure imaginary $(=i\beta)$, τ is real and increases steadily from 0 to ∞ as w travels along the real axis from 0 to ∞; and so

$$\int_0^\infty e^{-zw}(1+w^2)^{\nu-\frac{1}{2}}\,dw = \int_0^\infty e^{-z\tau}\left\{\frac{1}{\sqrt{(1+w^2)}}\frac{dw}{d\tau}\right\}d\tau.$$

Hence, from § 10·41 (3) it follows that

(2) $$\mathbf{H}_\nu(z) \sim Y_\nu(z) + \frac{2(\frac{1}{2}z)^\nu}{\Gamma(\nu+\frac{1}{2})\Gamma(\frac{1}{2})}\sum_{m=0}^{\infty}\frac{m!\,b_m}{z^{m+1}},$$

provided that $|\arg z| < \frac{1}{2}\pi$. This result can be extended to a somewhat wider domain of values of $\arg z$, after the manner of § 8·42.

From the corresponding results in the theory of Bessel functions, it is to be expected that these results are valid for suitable domains of complex values of the arguments.

In particular, we can prove that, in the case of functions of purely imaginary argument,

(3) $$\mathbf{L}_\nu(\nu x) \sim I_\nu(\nu x)$$

when $|\nu|$ is large, $|\arg\nu| < \frac{1}{2}\pi$, x is fixed, and the error is of the order of magnitude of

$$\frac{1}{\sqrt{\nu}}\cdot\left[\frac{1+\sqrt{(1+x^2)}}{2}\exp\{1-\sqrt{(1+x^2)}\}\right]^\nu$$

times the expression on the right.

[NOTE. If in (I) we had taken the contour from $w=0$ to $w=e^{-\alpha}$ and thence to $w=-i$, we should have obtained the formula containing $iJ_\nu(z)$ in place of $-iJ_\nu(z)$. This indicates that we get a case of Stokes' phenomenon as γ crosses the line $\beta=0$.]

10·44. *The relation between* $\mathbf{H}_n(z)$ *and* $\mathbf{E}_n(z)$.

When the order n is a positive integer (or zero), we can deduce from § 10·1 (4) that $\mathbf{E}_n(z)$ differs from $-\mathbf{H}_n(z)$ by a polynomial in z; and when n is a negative integer, the two functions differ by a polynomial in $1/z$.

For, when n is a positive integer or zero, we have

$$\mathbf{J}_n(z) + i\,\mathbf{E}_n(z) = \sum_{m=0}^{\infty} \frac{e^{\frac{1}{2}(n-m)\pi i}(\frac{1}{2}z)^m}{\Gamma(\frac{1}{2}m - \frac{1}{2}n + 1)\,\Gamma(\frac{1}{2}m + \frac{1}{2}n + 1)}$$

$$= \sum_{m=-n}^{\infty} \frac{e^{-\frac{1}{2}m\pi i}(\frac{1}{2}z)^{n+m}}{\Gamma(\frac{1}{2}m+1)\,\Gamma(\frac{1}{2}m+n+1)},$$

and

$$J_n(z) - i\,\mathbf{H}_n(z) = \sum_{m=0}^{\infty} \frac{e^{-\frac{1}{2}m\pi i}(\frac{1}{2}z)^{n+m}}{\Gamma(\frac{1}{2}m+1)\,\Gamma(\frac{1}{2}m+n+1)},$$

and therefore, since $\mathbf{J}_n(z) = J_n(z)$, we have

$$\mathbf{E}_n(z) = \sum_{m=1}^{n} \frac{e^{\frac{1}{2}(m-1)\pi i}(\frac{1}{2}z)^{n-m}}{\Gamma(1-\frac{1}{2}m)\,\Gamma(n+1-\frac{1}{2}m)} - \mathbf{H}_n(z),$$

that is to say

(1) $$\mathbf{E}_n(z) = \frac{1}{\pi}\sum_{m=0}^{<\frac{1}{2}(n-1)} \frac{\Gamma(m+\frac{1}{2})\,.\,(\frac{1}{2}z)^{n-2m-1}}{\Gamma(n+\frac{1}{2}-m)} - \mathbf{H}_n(z).$$

In like manner, when $-n$ is a negative integer,

(2) $$\mathbf{E}_{-n}(z) = \frac{(-)^{n+1}}{\pi}\sum_{m=0}^{<\frac{1}{2}(n-1)} \frac{\Gamma(n-m-\frac{1}{2})\,.\,(\frac{1}{2}z)^{-n+2m+1}}{\Gamma(m+\frac{3}{2})} - \mathbf{H}_{-n}(z).$$

10·45. *The sign of Struve's function.*

We shall now prove the interesting result that $\mathbf{H}_\nu(x)$ is *positive* when x is positive and ν has any positive value greater than or equal to $\frac{1}{2}$. This result, which was pointed out by Struve* in the case $\nu = 1$, is derivable from a definite integral (which will be established in § 13·47) which is of considerable importance in the Theory of Diffraction.

To obtain the result by an elementary method, we integrate § 10·4 (1) by parts and then we see that, for values of ν exceeding $\frac{1}{2}$,

$$\mathbf{H}_\nu(x) = \frac{(\frac{1}{2}x)^{\nu-1}}{\Gamma(\nu+\frac{1}{2})\,\Gamma(\frac{1}{2})}\int_0^{\frac{1}{2}\pi} \frac{d\cos(x\cos\theta)}{d\theta}\sin^{2\nu-1}\theta\, d\theta$$

$$= \frac{(\frac{1}{2}x)^{\nu-1}}{\Gamma(\nu+\frac{1}{2})\,\Gamma(\frac{1}{2})}\left\{\Big[\cos(x\cos\theta)\sin^{2\nu-1}\theta\Big]_0^{\frac{1}{2}\pi} - (2\nu-1)\int_0^{\frac{1}{2}\pi}\cos(x\cos\theta)\sin^{2\nu-2}\theta\cos\theta\, d\theta\right\}$$

$$= \frac{(\frac{1}{2}x)^{\nu-1}}{\Gamma(\nu+\frac{1}{2})\,\Gamma(\frac{1}{2})}\left\{1 - (2\nu-1)\int_0^{\frac{1}{2}\pi}\cos(x\cos\theta)\sin^{2\nu-2}\theta\cos\theta\, d\theta\right\}$$

$$= \frac{(\frac{1}{2}x)^{\nu-1}(2\nu-1)}{\Gamma(\nu+\frac{1}{2})\,\Gamma(\frac{1}{2})}\int_0^{\frac{1}{2}\pi}\sin^{2\nu-2}\theta\cos\theta\,\{1-\cos(x\cos\theta)\}\, d\theta$$

$$\geqslant 0,$$

since the integrand is positive.

* *Mém. de l'Acad. Imp. des Sci. de St Pétersbourg*, (7) xxx. (1882), no. 8, pp. 100—101. The proof given here is the natural extension of Struve's proof.

When ν is less than $\frac{1}{2}$, the partial integration cannot be performed; and, when $\nu=\frac{1}{2}$, we have

$$\mathbf{H}_{\frac{1}{2}}(x)=\left(\frac{2}{\pi x}\right)^{\frac{1}{2}}(1-\cos x)\geqslant 0,$$

and the theorem is completely established.

A comparison of the asymptotic expansion which was proved in § 10·42 with that of $Y_\nu(x)$ given in § 7·21 shews that, *when x is sufficiently large and positive*, $\mathbf{H}_\nu(x)$ is positive if $\nu>\frac{1}{2}$ and that $\mathbf{H}_\nu(x)$ is not one-signed when $\nu<\frac{1}{2}$; for the dominant term of the asymptotic expansion of $\mathbf{H}_\nu(x)$ is

$$\frac{(\frac{1}{2}x)^{\nu-1}}{\Gamma(\nu+\frac{1}{2})\,\Gamma(\frac{1}{2})}\quad\text{or}\quad\left(\frac{2}{\pi x}\right)^{\frac{1}{2}}\sin(x-\tfrac{1}{2}\nu\pi-\tfrac{1}{4}\pi)$$

according as $\nu>\frac{1}{2}$ or $\nu<\frac{1}{2}$. The theorem of this section proves the more extended result that Struve's function is positive for *all* positive values of x when $\nu>\frac{1}{2}$ and not merely for sufficiently large values.

The theorem indicates an essential difference between Struve's function and Bessel functions; for the asymptotic expansions of Chapter VII shew that, for sufficiently large values of x, $J_\nu(x)$ and $Y_\nu(x)$ are not of constant sign.

10·46. *Theisinger's integral.*

If we take the equation

$$\begin{aligned}\frac{\pi^2}{2}\{I_0(x)-\mathbf{L}_0(x)\}&=\frac{\pi}{2}\int_{-\frac{1}{2}\pi}^{\frac{1}{2}\pi}e^{-x\cos\theta}\,d\theta\\&=I\int_{-\frac{1}{2}\pi}^{\frac{1}{2}\pi}e^{-x\cos\theta}\log\frac{1+ie^{i\theta}}{1-ie^{i\theta}}\,d\theta\\&=-R\int_{-i}^{i}\exp\left\{-\frac{x}{2}\left(z+\frac{1}{z}\right)\right\}\log\frac{1+iz}{1-iz}\frac{dz}{z},\end{aligned}$$

and choose the contour to be the imaginary axis, indented at the origin*, and then write $z=\pm i\tan\frac{1}{2}\phi$, we find that

$$\frac{\pi^2}{4}\{I_0(x)-\mathbf{L}_0(x)\}=\int_0^{\frac{1}{2}\pi}\cos(x\cot\phi)\log\tan(\tfrac{1}{4}\pi+\tfrac{1}{2}\phi)\frac{d\phi}{\sin\phi},$$

and so

(1) $$I_0(x)-\mathbf{L}_0(x)=\frac{4}{\pi^2}\int_0^{\frac{1}{2}\pi}\cos(x\tan\phi)\log\cot(\tfrac{1}{2}\phi)\frac{d\phi}{\cos\phi},$$

a formula given by Theisinger, *Monatshefte für Math. und Phys.* XXIV. (1913), p. 341.

If we replace x by $x\sin\theta$, multiply by $\sin\theta$, and integrate, we find, on changing the order of the integrations in the absolutely convergent integral on the right,

$$\int_0^{\frac{1}{2}\pi}\mathbf{E}_1(x\tan\phi)\log\cot(\tfrac{1}{2}\phi)\frac{d\phi}{\cos\phi}=\frac{\pi}{2}\int_0^{\frac{1}{2}\pi}\{I_0(x\sin\theta)-\mathbf{L}_0(x\sin\theta)\}\sin\theta\,d\theta$$

so that

(2) $$\int_0^{\frac{1}{2}\pi}\mathbf{E}_1(x\tan\phi)\log\cot(\tfrac{1}{2}\phi)\frac{d\phi}{\cos\phi}=\frac{\pi}{2}\cdot\frac{1-e^{-x}}{x},$$

on expanding the integrand on the right in powers of x. This curious result is also due to Theisinger.

* The presence of the logarithmic factor ensures the convergence of the integral round the indentation.

10·5. *Whittaker's integral.*

The integral

$$z^{\frac{1}{2}}\int_{-1}^{1} e^{izt}P_{\nu-\frac{1}{2}}(t)\,dt,$$

which is a solution of Bessel's equation only when 2ν is an odd integer, has been studied by Whittaker*.

It follows from § 6·17 that, for all values of ν,

$$(1)\qquad \nabla_\nu\left\{z^{\frac{1}{2}}\int_{-1}^{1} e^{izt}P_{\nu-\frac{1}{2}}(t)\,dt\right\} = -\lim_{t\to-1+0}\left[z^{\frac{1}{2}}e^{izt}(1-t^2)P'_{\nu-\frac{1}{2}}(t)\right]$$
$$= \frac{2}{\pi}\cos\nu\pi\,.\,z^{\frac{1}{2}}e^{-iz}.$$

If we expand the integrand (multiplied by e^{iz}) in ascending powers of z and integrate term-by-term† it is found that

$$(2)\qquad z^{\frac{1}{2}}\int_{-1}^{1} e^{izt}P_{\nu-\frac{1}{2}}(t)\,dt = 2z^{\frac{1}{2}}e^{-iz}\sum_{m=0}^{\infty}\frac{(2iz)^m\,.\,m!}{\Gamma(m+\frac{3}{2}-\nu)\,\Gamma(m+\frac{3}{2}+\nu)}.$$

The formula of § 3·32 suggests that we write

$$\frac{i^{\frac{1}{2}-\nu}z^{\frac{1}{2}}}{\sqrt{(2\pi)}}\int_{-1}^{1} e^{izt}P_{\nu-\frac{1}{2}}(t)\,dt = \mathbf{W}_\nu(z),$$

and then it is easy to verify the following recurrence formulae, either by using the series (2), or by using recurrence formulae for Legendre functions:

$$(3)\qquad \mathbf{W}_{\nu-1}(z)+\mathbf{W}_{\nu+1}(z) = \frac{2\nu}{z}\left[\mathbf{W}_\nu(z) - \frac{2i^{\frac{1}{2}-\nu}z^{\frac{1}{2}}e^{-iz}}{\sqrt{(2\pi)}\,\Gamma(\frac{3}{2}-\nu)\,\Gamma(\frac{3}{2}+\nu)}\right],$$

$$(4)\qquad \mathbf{W}_{\nu-1}(z)-\mathbf{W}_{\nu+1}(z) = 2\mathbf{W}_\nu'(z) - \frac{2i^{\frac{1}{2}-\nu}z^{-\frac{1}{2}}e^{-iz}}{\sqrt{(2\pi)}\,\Gamma(\frac{3}{2}-\nu)\,\Gamma(\frac{3}{2}+\nu)},$$

$$(5)\qquad (\vartheta+\nu)\,\mathbf{W}_\nu(z) = z\mathbf{W}_{\nu-1}(z) + \frac{2i^{\frac{1}{2}-\nu}z^{\frac{1}{2}}e^{-iz}}{\sqrt{(2\pi)}\,.\,\Gamma(\frac{3}{2}-\nu)\,\Gamma(\frac{1}{2}+\nu)},$$

$$(6)\qquad (\vartheta-\nu)\,\mathbf{W}_\nu(z) = -z\mathbf{W}_{\nu+1}(z) + \frac{2i^{\frac{1}{2}-\nu}z^{\frac{1}{2}}e^{-iz}}{\sqrt{(2\pi)}\,.\,\Gamma(\frac{1}{2}-\nu)\,\Gamma(\frac{3}{2}+\nu)}.$$

An asymptotic expansion of $\mathbf{W}_\nu(z)$ for large values of $|z|$ may be obtained by deforming the path of integration after the manner of Lipschitz (§ 7·21).

* *Proc. London Math. Soc.* XXXV. (1903), pp. 198—206.

† By a use of Legendre's equation the recurrence formula

$$\int_{-1}^{1}(1+t)^m P_{\nu-\frac{1}{2}}(t)\,dt = \frac{2m^2}{(m+\frac{1}{2})^2-\nu^2}\int_{-1}^{1}(1+t)^{m-1}P_{\nu-\frac{1}{2}}(t)\,dt$$

may be verified; and we can prove that $\int_{-1}^{1}P_{\nu-\frac{1}{2}}(t)\,dt = \frac{2}{\Gamma(\frac{3}{2}+\nu)\,\Gamma(\frac{3}{2}-\nu)}$ by expanding $F(\frac{1}{2}-\nu, \frac{1}{2}+\nu;\ 1;\ \frac{1}{2}-\frac{1}{2}t)$ in ascending powers of $1-t$, and integrating term-by-term.

The function is thus seen to be equal to

$$\frac{i^{\frac{1}{2}-\nu}z^{\frac{1}{2}}}{\sqrt{(2\pi)}}\int_{-1}^{-1+\infty} e^{izt}P_{\nu-\frac{1}{2}}(t)\,dt - \frac{i^{\frac{1}{2}-\nu}z^{\frac{1}{2}}}{\sqrt{(2\pi)}}\int_{1}^{1+\infty i} e^{izt}P_{\nu-\frac{1}{2}}(t)\,dt$$

$$= -\frac{i^{\frac{1}{2}-\nu}z^{\frac{1}{2}}e^{-iz}}{\sqrt{(2\pi)}}\int_{1}^{1-\infty i} e^{iz(1-t)}P_{\nu-\frac{1}{2}}(-t)\,dt - \frac{i^{\frac{1}{2}-\nu}z^{\frac{1}{2}}e^{iz}}{\sqrt{(2\pi)}}\int^{1+\infty i} e^{-iz(1-t)}P_{\nu-\frac{1}{2}}(t)\,dt.$$

Now it is known that*, near $t=1$,

$$P_{\nu-\frac{1}{2}}(t) = {}_2F_1(\tfrac{1}{2}-\nu, \tfrac{1}{2}+\nu;\, 1;\, \tfrac{1}{2}-\tfrac{1}{2}t),$$

$$P_{\nu-\frac{1}{2}}(-t) = -\left(\frac{\cos\nu\pi}{\pi}\right)^2 \sum_{m=0}^{\infty} \frac{\Gamma(m-\nu+\frac{1}{2})\,\Gamma(m+\nu+\frac{1}{2})}{(m!)^2}\left(\frac{1-t}{2}\right)^m$$
$$\times\left\{\log\left(\frac{1-t}{2}\right) - 2\psi(m+1) + \psi(m-\nu+\tfrac{1}{2}) + \psi(m+\nu+\tfrac{1}{2})\right\},$$

and since

$$\int_1^{1-\infty i} e^{iz(1-t)}\left(\frac{1-t}{2}\right)^\mu dt = \frac{i^{\mu-1}\,\Gamma(\mu+1)}{2^\mu z^{\mu+1}},$$

$$\int_1^{1+\infty i} e^{-iz(1-t)}\left(\frac{1-t}{2}\right)^\mu dt = \frac{i^{1-\mu}\,\Gamma(\mu+1)}{2^\mu z^{\mu+1}},$$

we obtain the asymptotic expansion

$$(7)\quad \mathbf{W}_\nu(z) \sim \tfrac{1}{2}H_\nu^{(1)}(z)$$
$$+\frac{e^{-i(z+\frac{1}{2}\nu\pi+\frac{1}{4}\pi)}\cos\nu\pi}{\pi\sqrt{(2\pi z)}}\left[\sum_{m=0}^{\infty}\frac{(\nu, m)}{(2iz)^m}\{\psi(m+\tfrac{1}{2}-\nu)\right.$$
$$\left.+\psi(m+\tfrac{1}{2}+\nu) - \psi(m+1) - \log 2z - \tfrac{1}{2}\pi i\}\right].$$

Some functions which satisfy equations of the same general type as (1) have been noticed by Nagaoka, *Journal of the Coll. of Sci. Imp. Univ. Japan*, IV. (1891), p. 310.

10·6. *The functions composing* $Y_n(z)$.

The reader will remember that the Bessel function of the second kind, of integral order, may be written in the form (§ 3·52)

$$\pi Y_n(z) = -\sum_{m=0}^{n-1}\frac{(n-m-1)!}{m!}(\tfrac{1}{2}z)^{-n+2m}$$
$$+\sum_{m=0}^{\infty}\frac{(-)^m(\tfrac{1}{2}z)^{n+2m}}{m!\,(n+m)!}\{2\log(\tfrac{1}{2}z) - \psi(m+1) - \psi(n+m+1)\}.$$

The series on the right may be expressed as the sum of four functions, each of which has fairly simple recurrence properties, thus

$$(1)\quad \pi Y_n(z) = 2\{\log(\tfrac{1}{2}z) - \psi(1)\}J_n(z) - S_n(z) + T_n(z) - 2U_n(z),$$

* Cf. Barnes, *Quarterly Journal*, XXXIX. (1908), p. 111.

where

$$(2)\qquad T_n(z) = -\sum_{m \geqslant \frac{1}{2}n}^{n-1} \frac{(n-m-1)!}{m!} (\tfrac{1}{2}z)^{-n+2m} + \sum_{m=0}^{\infty} \frac{(-)^m (\frac{1}{2}z)^{n+2m}}{m!\,(n+m)!} \{\psi(n+m+1) - \psi(m+1)\},$$

and (cf. § 3·582)

$$(3)\qquad U_n(z) = \sum_{m=0}^{\infty} \frac{(-)^m (\frac{1}{2}z)^{n+2m}}{m!\,(n+m)!} \{\psi(n+m+1) - \psi(1)\}.$$

The functions $T_n(z)$ and $U_n(z)$ have been studied by Schläfli, *Math. Ann.* III. (1871), pp. 142—147, though he used the slightly different notation indicated by the equations

$$S_n(z) = -2G_n(z), \quad T_n(z) = 2H_n(z), \quad U_n(z) = -E_n(z);$$

more recent investigations are due to Otti* and to Graf and Gubler†.

The function $T_n(z)$ is most simply represented by the definite integral

$$(4)\qquad T_n(z) = \frac{2}{\pi}\int_0^{\pi} (\tfrac{1}{2}\pi - \theta) \sin(z\sin\theta - n\theta)\, d\theta.$$

To establish this result, observe that

$$T_n(z) = \left[\frac{\partial}{\partial\epsilon} \sum_{m>-\frac{1}{2}n-\frac{1}{2}}^{\infty} \frac{(-)^m (\frac{1}{2}z)^{n+2m}}{\Gamma(m+1+\epsilon)\,\Gamma(n+m+1-\epsilon)}\right]_{\epsilon=0}$$

$$= \frac{1}{2\pi i}\left[\frac{\partial}{\partial\epsilon} \sum_{m>-\frac{1}{2}n-\frac{1}{2}}^{\infty} \frac{(-)^m (\frac{1}{2}z)^{n+2m}}{(n+2m)!} \int_{-1}^{(0+)} \frac{(1+t)^{n+2m}}{t^{m+1+\epsilon}}\, dt\right]_{\epsilon=0}$$

$$= -\frac{1}{2\pi i} \sum_{m>-\frac{1}{2}n-\frac{1}{2}}^{\infty} \frac{(-)^m (\frac{1}{2}z)^{n+2m}}{(n+2m)!} \int_{-1}^{(0+)} \frac{(1+t)^{n+2m}\log t}{t^{m+1}}\, dt$$

$$= \frac{2}{\pi i}\int_0^{\pi} \sum_{m>-\frac{1}{2}n-\frac{1}{2}}^{\infty} \frac{e^{ni\theta}(-iz\sin\theta)^{n+2m} . (\theta - \frac{1}{2}\pi)}{(n+2m)!}\, d\theta,$$

where t has been replaced by $e^{(2\theta-\pi)i}$.

It follows that

$$T_n(z) = \frac{2}{\pi i}\int_0^{\pi} (\theta - \tfrac{1}{2}\pi)\, e^{ni\theta} \left\{\sum_{m>-\frac{1}{2}n-\frac{1}{2}}^{\infty} \frac{(-iz\sin\theta)^{n+2m}}{(n+2m)!}\right\} d\theta.$$

Now

$$\sum_{m>-\frac{1}{2}n-\frac{1}{2}}^{\infty} \frac{(-iz\sin\theta)^{n+2m}}{(n+2m)!} = \begin{cases} \cosh(-iz\sin\theta) & (n \text{ even}) \\ \sinh(-iz\sin\theta) & (n \text{ odd}) \end{cases}$$

$$= \tfrac{1}{2}\{e^{-iz\sin\theta} + (-)^n e^{iz\sin\theta}\},$$

and so

$$T_n(z) = \frac{1}{\pi i}\int_0^{\pi} (\theta - \tfrac{1}{2}\pi)\{e^{ni\theta - iz\sin\theta} + e^{ni(\theta-\pi)+iz\sin\theta}\}\, d\theta.$$

* *Bern Mittheilungen*, 1898, pp. 1—56.

† *Einleitung in die Theorie der Bessel'schen Funktionen*, II. (Bern, 1900), pp. 42—69. Lommel's treatise, pp. 77—87, should also be consulted.

If θ is replaced by $\pi - \theta$ in the integral obtained by considering only the second of the two exponentials, the formula (4), which is due to Schläfli, is obtained at once.

The corresponding integral for $U_n(z)$ is obtained by observing that

$$U_n(z) = -\left[\frac{\partial}{\partial \epsilon} \sum_{m=0}^{\infty} \frac{(-)^m (\tfrac{1}{2}z)^{n+2m}\, \Gamma(1+\epsilon)}{m!\, \Gamma(n+m+1+\epsilon)}\right]_{\epsilon=0}$$

$$= -\left[\frac{\partial}{\partial \epsilon} \{(\tfrac{1}{2}z)^{-\epsilon}\, \Gamma(1+\epsilon)\, J_{n+\epsilon}(z)\}\right]_{\epsilon=0},$$

and so, from § 6·2 (4), we deduce that

(5) $$U_n(z) = \{\log(\tfrac{1}{2}z) - \psi(1)\} J_n(z) + \frac{1}{\pi}\int_0^{\pi} \theta \sin(n\theta - z\sin\theta)\, d\theta + (-)^n \int_0^{\infty} e^{-nt - z\sinh t}\, dt.$$

10·61. *Recurrence formulae for $T_n(z)$ and $U_n(z)$.*

From § 10·6 (4) we see that

$$T_{n-1}(z) + T_{n+1}(z) - (2n/z)\, T_n(z)$$
$$= \frac{2}{\pi}\int_0^{\pi} (\tfrac{1}{2}\pi - \theta) \sin(z\sin\theta - n\theta) . \{2\cos\theta - 2n/z\}\, d\theta$$
$$= -\frac{4}{\pi z}\int_0^{\pi} (\tfrac{1}{2}\pi - \theta) \frac{d}{d\theta}\{\cos(z\sin\theta - n\theta)\}\, d\theta$$
$$= \frac{4}{z}\cos^2 \tfrac{1}{2}n\pi - \frac{4}{z} J_n(z),$$

on integrating by parts and using Bessel's integral.

Thus

(1) $$T_{n-1}(z) + T_{n+1}(z) = (2n/z)\, T_n(z) + 4\{\cos^2 \tfrac{1}{2}n\pi - J_n(z)\}/z.$$

Again $$T_n'(z) = \frac{2}{\pi}\int_0^{\pi} (\tfrac{1}{2}\pi - \theta)\sin\theta \cos(z\sin\theta - n\theta)\, d\theta,$$

and so

(2) $$T_{n-1}(z) - T_{n+1}(z) = 2T_n'(z).$$

From these formulae it follows that

(3) $$(\vartheta + n)\, T_n(z) = zT_{n-1}(z) - 2\cos^2 \tfrac{1}{2}n\pi + 2J_n(z),$$

(4) $$(\vartheta - n)\, T_n(z) = -z\, T_{n+1}(z) + 2\cos^2 \tfrac{1}{2}n\pi - 2J_n(z),$$

and hence (cf. § 10·12) we find that

(5) $$\nabla_n T_n(z) = 2\{z\sin^2 \tfrac{1}{2}n\pi + n\cos^2 \tfrac{1}{2}n\pi\} - 4n\, J_n(z).$$

With the aid of these formulae combined with the corresponding formulae for $J_n(z)$, $\mathbf{Y}_n(z)$ and $S_n(z)$, we deduce from § 10·6 (1) that

(6) $$U_{n-1}(z) + U_{n+1}(z) = (2n/z)\, U_n(z) - (2/z)\, J_n(z),$$

(7) $$U_{n-1}(z) - U_{n+1}(z) = 2U_n'(z) - (2/z)\, J_n(z),$$

(8) $$(\vartheta + n)\, U_n(z) = zU_{n-1}(z) + 2J_n(z),$$

(9) $$(\vartheta - n)\, U_n(z) = -zU_{n+1}(z), \qquad \text{[cf. §§ 3·58 (1), 3·58 (2)]}$$

(10) $$\nabla_n U_n(z) = -2zJ_{n+1}(z).$$

The reader may verify these directly from the definition, § 10·6 (3).

It is convenient to define the function $T_{-n}(z)$, of negative order, by the equivalent of § 10·6 (4). If we replace θ by $\pi - \theta$ in the integral we find that

$$T_{-n}(z) = \frac{2}{\pi}\int_0^\pi (\tfrac{1}{2}\pi - \theta)\sin(z\sin\theta + n\theta)\, d\theta$$

$$= -\frac{2}{\pi}\int_0^\pi (\tfrac{1}{2}\pi - \theta)\sin(z\sin\theta - n\theta + n\pi)\, d\theta,$$

and so

(11) $$T_{-n}(z) = (-)^{n+1}\, T_n(z).$$

We now define $U_{-n}(z)$ by supposing § 10·6 (1) to hold for all values of n; it is then found that

(12) $$U_{-n}(z) = (-)^n \{U_n(z) - T_n(z) + S_n(z)\}.$$

10·62. *Series for $T_n(z)$ and $U_n(z)$.*

We shall now shew how to derive the expansion

(1) $$T_n(z) = \sum_{m=1}^{\infty} \frac{1}{m}\{J_{n+2m}(z) - J_{n-2m}(z)\}$$

from § 10·6 (4). The method which we shall use is to substitute

$$\tfrac{1}{2}\pi - \theta = \sum_{m=1}^{\infty} \frac{\sin 2m\theta}{m}$$

in the integral for $T_n(z)$, and then integrate term-by-term. This procedure needs justification, since the Fourier series does not converge uniformly near $\theta = 0$ and $\theta = \pi$, and, in fact, the equation just quoted is untrue for these two values of θ.

To justify the process*, let δ and ϵ be arbitrarily small positive numbers. Since the series converges uniformly when $\delta \leqslant \theta \leqslant \pi - \delta$, we can find an integer m_0 such that

$$\left| (\tfrac{1}{2}\pi - \theta) - \sum_{m=1}^{M} \frac{\sin 2m\theta}{m} \right| < \epsilon,$$

* The analysis immediately following is due to D. Jackson, *Palermo Rendiconti*, XXXII. (1911), pp. 257—262. The value of the constant A is 1·8519....

throughout the range $\delta \leqslant \theta \leqslant \pi - \delta$, for all values of M exceeding m_0. Again, for *all* values of θ between 0 and π, we have

$$\tfrac{1}{2}\pi - \theta - \sum_{m=1}^{M} \frac{\sin 2m\theta}{m} = \int_\theta^{\frac{1}{2}\pi} \{1 + 2\cos 2t + 2\cos 4t + \ldots + 2\cos 2Mt\}\, dt$$

$$= \int_\theta^{\frac{1}{2}\pi} \frac{\sin(2M+1)t}{t} \cdot \frac{t}{\sin t}\, dt$$

$$= \tfrac{1}{2}\pi \int_\phi^{\frac{1}{2}\pi} \frac{\sin(2M+1)t}{t}\, dt = \tfrac{1}{2}\pi \int_{(2M+1)\phi}^{(M+\frac{1}{2})\pi} \frac{\sin x}{x}\, dx$$

for some value of ϕ between θ and $\frac{1}{2}\pi$, by the second mean-value theorem, since $t/\sin t$ is a monotonic (increasing) function.

By drawing the graph of $x^{-1}\sin x$ it is easy to see that the last expression cannot exceed $\frac{1}{2}\pi \int_0^\pi \frac{\sin x}{x}\, dx$ in absolute value; if this be called $\frac{1}{2}\pi A$, we have

$$\left| T_n(z) - \frac{2}{\pi} \sum_{m=1}^{M} \int_0^\pi \frac{\sin 2m\theta}{m} \sin(z \sin\theta - n\theta)\, d\theta \right|$$

$$= \frac{2}{\pi} \left| \int_0^\pi \left\{ (\tfrac{1}{2}\pi - \theta) - \sum_{m=1}^{M} \frac{\sin 2m\theta}{m} \right\} \sin(z\sin\theta - n\theta)\, d\theta \right|$$

$$\leqslant \frac{2}{\pi} \left\{ \int_0^\delta + \int_\delta^{\pi-\delta} + \int_{\pi-\delta}^{\pi} \right\} \left| \left\{ (\tfrac{1}{2}\pi - \theta) - \sum_{m=1}^{M} \frac{\sin 2m\theta}{m} \right\} \right| . \left| \sin(z\sin\theta - n\theta) \right| d\theta$$

$$< \frac{2}{\pi} \{\pi A \delta + (\pi - 2\delta)\epsilon\} B,$$

where B is the upper bound of $|\sin(z\sin\theta - n\theta)|$.

Since $2(A\delta + \epsilon)B$ is arbitrarily small, it follows from the definition of an infinite series that*

$$T_n(z) = \frac{2}{\pi} \sum_{m=1}^{\infty} \int_0^\pi \frac{\sin 2m\theta}{m} \sin(z\sin\theta - n\theta)\, d\theta$$

$$= \sum_{m=1}^{\infty} \frac{1}{m} \{J_{n+2m}(z) - J_{n-2m}(z)\},$$

and the result is established.

It will be remembered that $U_n(z)$ has already been defined (§ 3·581) as a series of Bessel coefficients by the equation

$$U_n(z) = s_n J_n(z) + \sum_{m=1}^{\infty} \frac{(-)^m (n+2m)}{m(n+m)} J_{n+2m}(z),$$

and that, in § 3·582, this definition was identified with the definition of $U_n(z)$ as a power series given in § 10·6 (3).

10·63. *Graf's expansion of $T_n(z+t)$ as a series of Bessel coefficients.*

It is easy to obtain the expansion

(1) $$T_n(z+t) = \sum_{m=-\infty}^{\infty} T_{n-m}(t) J_m(z),$$

* This expansion was discovered by Schläfli, *Math. Ann.* III. (1871), p. 146.

for, from § 10·6 (4), it is evident that

$$T_n(z+t) = \frac{2}{\pi}\int_0^\pi (\tfrac{1}{2}\pi - \theta)\sin(t\sin\theta - n\theta + z\sin\theta)\,d\theta$$

$$= \frac{1}{\pi i}\int_0^\pi (\tfrac{1}{2}\pi - \theta)\sum_{m=-\infty}^{\infty} J_m(z)\{e^{i(t\sin\theta - n\theta + m\theta)} - e^{-i(t\sin\theta - n\theta + m\theta)}\}\,d\theta$$

$$= \frac{2}{\pi}\int_0^\pi (\tfrac{1}{2}\pi - \theta)\sum_{m=-\infty}^{\infty} J_m(z)\sin\{t\sin\theta - (n-m)\theta\}\,d\theta,$$

by using § 2·1; since the series under the integral sign is uniformly convergent, the order of summation and integration may be changed, and the result is evident.

The proof of the formula given by Graf, *Math. Ann.* XLIII. (1893), p. 141, is more complicated; it depends on the use of the series of § 10·62 combined with § 2·4.

There seems to be no equally simple expression for $U_n(z+t)$.

10·7. *The genesis of Lommel's functions $S_{\mu,\nu}(z)$ and $s_{\mu,\nu}(z)$.*

A function, which includes as special cases the polynomials $zO_n(z)$ and $S_n(z)$ of Neumann and Schläfli, was derived by Lommel, *Math. Ann.* IX. (1876), pp. 425—444, as a particular integral of the equation

(1) $$\nabla_\nu y = kz^{\mu+1},$$

where k and μ are constants. It is easy to shew that a particular integral of this equation, proceeding in ascending powers of z beginning with $z^{\mu+1}$, is

(2) $$y = k\left[\frac{z^{\mu+1}}{(\mu+1)^2 - \nu^2} - \frac{z^{\mu+3}}{\{(\mu+1)^2-\nu^2\}\{(\mu+3)^2-\nu^2\}} + \ldots\right]$$

$$= kz^{\mu-1}\sum_{m=0}^{\infty}\frac{(-)^m(\tfrac{1}{2}z)^{2m+2}}{(\tfrac{1}{2}\mu - \tfrac{1}{2}\nu + \tfrac{1}{2})_{m+1}\cdot(\tfrac{1}{2}\mu + \tfrac{1}{2}\nu + \tfrac{1}{2})_{m+1}}$$

$$= kz^{\mu-1}\sum_{m=0}^{\infty}\frac{(-)^m(\tfrac{1}{2}z)^{2m+2}\,\Gamma(\tfrac{1}{2}\mu - \tfrac{1}{2}\nu + \tfrac{1}{2})\,\Gamma(\tfrac{1}{2}\mu + \tfrac{1}{2}\nu + \tfrac{1}{2})}{\Gamma(\tfrac{1}{2}\mu - \tfrac{1}{2}\nu + m + \tfrac{3}{2})\,\Gamma(\tfrac{1}{2}\mu + \tfrac{1}{2}\nu + m + \tfrac{3}{2})}.$$

For brevity the expressions on the right are written in the form

$$ks_{\mu,\nu}(z).$$

The function $s_{\mu,\nu}(z)$ is evidently undefined when either of the numbers $\mu \pm \nu$ is an odd negative integer*. Apart from this restriction the general solution of (1) is evidently

(3) $$y = \mathscr{C}_\nu(z) + ks_{\mu,\nu}(z).$$

In like manner the general solution of

(4) $$\frac{d^2y}{dz^2} + \frac{a}{z}\frac{dy}{dz} + \left\{1 - \frac{\nu^2 - (\tfrac{1}{2}a - \tfrac{1}{2})^2}{z^2}\right\}y = kz^{\mu - \frac{1}{2}(a+1)}$$

is

(5) $$y = z^{-\frac{1}{2}(a-1)}\{\mathscr{C}_\nu(z) + ks_{\mu,\nu}(z)\}.$$

* The solution of the equation for such values of μ and ν is discussed in § 10·71.

Next let us consider the solution of (1) by the method of "variation of parameters." We assume as a solution*

$$y = A(z) J_\nu(z) + B(z) J_{-\nu}(z),$$

where $A(z)$ and $B(z)$ are functions of z determined by the equations

$$J_\nu(z) A'(z) + J_{-\nu}(z) B'(z) = 0,$$
$$J'_\nu(z) A'(z) + J'_{-\nu}(z) B'(z) = kz^{\mu-1}.$$

On using § 3·12 (2), we see that

$$A(z) = \frac{k\pi}{2 \sin \nu\pi} \int^z z^\mu J_{-\nu}(z)\, dz, \quad B(z) = -\frac{k\pi}{2 \sin \nu\pi} \int^z z^\mu J_\nu(z)\, dz.$$

Hence a solution† of (1) is

(6) $$y = \frac{k\pi}{2 \sin \nu\pi} \left[J_\nu(z) \int^z z^\mu J_{-\nu}(z)\, dz - J_{-\nu}(z) \int^z z^\mu J_\nu(z)\, dz \right],$$

where the lower limits of the integrals are arbitrary.

Similarly a solution of (1) which is valid for all values of ν, whether integers or not, is

(7) $$y = \tfrac{1}{2} k\pi \left[Y_\nu(z) \int^z z^\mu J_\nu(z)\, dz - J_\nu(z) \int^z z^\mu Y_\nu(z)\, dz \right].$$

It is easy to see that, if both of the numbers $\mu \pm \nu + 1$ have positive real parts, the lower limits in (6) and (7) may be taken to be zero. If we expand the integrands in ascending powers of z, we see that the expression on the right in (6) is expressible as a power series containing no powers of z other than $z^{\mu+1}, z^{\mu+3}, z^{\mu+5}, \ldots$. Hence, from (3), it follows that, *since neither of the numbers $\mu \pm \nu$ is an odd negative integer*, we must have

(8) $$s_{\mu,\nu}(z) = \frac{\pi}{2 \sin \nu\pi} \left[J_\nu(z) \int_0^z z^\mu J_{-\nu}(z)\, dz - J_{-\nu}(z) \int_0^z z^\mu J_\nu(z)\, dz \right].$$

In obtaining this result it was supposed that ν is not an integer; but if we introduce functions of the second kind, we find that

(9) $$s_{\mu,\nu}(z) = \tfrac{1}{2}\pi \left[Y_\nu(z) \int_0^z z^\mu J_\nu(z)\, dz - J_\nu(z) \int_0^z z^\mu Y_\nu(z)\, dz \right],$$

and in this formula we may proceed to the limit in making ν an integer.

It should be observed that, in Pochhammer's notation (§ 4·4),

(10) $$s_{\mu,\nu}(z) = \frac{z^{\mu+1}}{(\mu - \nu + 1)(\mu + \nu + 1)} \times {}_1F_2(1;\ \tfrac{1}{2}\mu - \tfrac{1}{2}\nu + \tfrac{3}{2},\ \tfrac{1}{2}\mu + \tfrac{1}{2}\nu + \tfrac{3}{2};\ -\tfrac{1}{4}z^2).$$

* Cf. Forsyth, *Treatise on Differential Equations* (1914), § 66; it is supposed temporarily that ν is not an integer.

† The generalisation of this result, obtained by replacing $z^{\mu+1}$ in (1) by an arbitrary function of z, was given by Chessin, *Comptes Rendus*, CXXXV. (1902), pp. 678—679; and it was applied by him, *Comptes Rendus*, CXXXVI. (1903), pp. 1124—1126, to solve a sequence of equations resembling Bessel's equation.

The associated function $S_{\mu,\nu}(z)$ is derived from a consideration of a solution of (1) in the form of a descending series. We now proceed to construct this solution and investigate its properties.

10·71. *The construction of the function* $S_{\mu,\nu}(z)$.

A particular integral of the equation § 10·7 (1), proceeding in descending powers of z, beginning with $z^{\mu-1}$, is

$$(1)\quad y = kz^{\mu-1}\left[1 - \frac{(\mu-1)^2-\nu^2}{z^2} + \frac{\{(\mu-1)^2-\nu^2\}\{(\mu-3)^2-\nu^2\}}{z^4} - \dots\right].$$

This series, however, does not converge unless it terminates; but if it terminates, it is a solution of § 10·7 (1), and it will be called $kS_{\mu,\nu}(z)$.

The series terminates if $\mu-\nu$ is an odd positive integer, or if $\mu+\nu$ is an odd positive integer, and in no other case.

In the former case we write $\mu = \nu + 2p + 1$, and then we have

$$\begin{aligned} S_{\mu,\nu}(z) &= z^{\mu-1}\sum_{m=0}^{p} \frac{(-)^m\,\Gamma(\tfrac12\mu-\tfrac12\nu+\tfrac12)\,\Gamma(\tfrac12\mu+\tfrac12\nu+\tfrac12)}{(\tfrac12 z)^{2m}\,\Gamma(\tfrac12\mu-\tfrac12\nu+\tfrac12-m)\,\Gamma(\tfrac12\mu+\tfrac12\nu+\tfrac12-m)} \\ &= (-)^p z^{\mu-1}\sum_{m=0}^{p}\frac{(-)^m(\tfrac12 z)^{2m-2p}\,\Gamma(p+1)\,\Gamma(\nu+p+1)}{\Gamma(m+1)\,\Gamma(\nu+m+1)} \\ &= (-)^p\, 2^{\mu-1}\,\Gamma(\tfrac12\mu-\tfrac12\nu+\tfrac12)\,\Gamma(\tfrac12\mu+\tfrac12\nu+\tfrac12)\,J_\nu(z) + s_{\mu,\nu}(z) \\ &= -\,2^{\mu-1}\,\Gamma(\tfrac12\mu-\tfrac12\nu+\tfrac12)\,\Gamma(\tfrac12\mu+\tfrac12\nu+\tfrac12)\,\frac{\cos\frac12(\mu+\nu)\pi}{\sin\nu\pi}\,J_\nu(z) + s_{\mu,\nu}(z). \end{aligned}$$

When $\mu-\nu = 2p+1$, the function

$$2^{\mu-1}\,\Gamma(\tfrac12\mu-\tfrac12\nu+\tfrac12)\,\Gamma(\tfrac12\mu+\tfrac12\nu+\tfrac12)\,\frac{\cos\frac12(\mu-\nu)\pi}{\sin\nu\pi}\,J_{-\nu}(z)$$

vanishes, and so, when $\mu-\nu$ is an odd positive integer, we have

$$(2)\quad S_{\mu,\nu}(z) = s_{\mu,\nu}(z) + \frac{2^{\mu-1}\,\Gamma(\tfrac12\mu-\tfrac12\nu+\tfrac12)\,\Gamma(\tfrac12\mu+\tfrac12\nu+\tfrac12)}{\sin\nu\pi} \times\left[\cos\tfrac12(\mu-\nu)\pi\,.\,J_{-\nu}(z) - \cos\tfrac12(\mu+\nu)\pi\,.\,J_\nu(z)\right].$$

Since both sides of this equation are even functions of ν, the equation is true also when $\mu+\nu$ is an odd positive integer, so that it holds in all cases in which $S_{\mu,\nu}(z)$ has, as yet, been defined. We adopt it as the general definition of $S_{\mu,\nu}(z)$, except that, when ν is an integer, we have to use the equivalent form

$$(3)\quad S_{\mu,\nu}(z) = s_{\mu,\nu}(z) + 2^{\mu-1}\,\Gamma(\tfrac12\mu-\tfrac12\nu+\tfrac12)\,\Gamma(\tfrac12\mu+\tfrac12\nu+\tfrac12) \times\left[\sin\tfrac12(\mu-\nu)\pi\,.\,J_\nu(z) - \cos\tfrac12(\mu-\nu)\pi\,.\,Y_\nu(z)\right].$$

It will be shewn in § 10·73 that $S_{\mu,\nu}(z)$ has a limit when $\mu+\nu$ or $\mu-\nu$ is an odd negative integer, i.e. when $s_{\mu,\nu}(z)$ is undefined; and so, of Lommel's two functions $s_{\mu,\nu}(z)$ and $S_{\mu,\nu}(z)$, it is frequently more convenient to use the latter.

It will appear in § 10·75 that the series (1), by means of which $S_{\mu,\nu}(z)$ is defined when either of the numbers $\mu \pm \nu$ is an odd positive integer, is still of significance when the numbers $\mu \pm \nu$ are not odd positive integers. It yields, in fact, an asymptotic expansion of $S_{\mu,\nu}(z)$ valid for large values of the variable z.

10·72. *Recurrence formulae satisfied by Lommel's functions.*

It is evident from § 10·7 (2) that

$$s_{\mu,\nu}(z) = \frac{z^{\mu+1} - s_{\mu+2,\nu}(z)}{(\mu+1)^2 - \nu^2},$$

that is to say

(1) $$s_{\mu+2,\nu}(z) = z^{\mu+1} - \{(\mu+1)^2 - \nu^2\}\, s_{\mu,\nu}(z).$$

Again, it is easy to verify that

$$\frac{d}{dz}\{z^\nu s_{\mu,\nu}(z)\} = (\mu+\nu-1)\, z^\nu s_{\mu-1,\nu-1}(z),$$

so that

(2) $$s'_{\mu,\nu}(z) + (\nu/z)\, s_{\mu,\nu}(z) = (\mu+\nu-1)\, s_{\mu-1,\nu-1}(z),$$

and similarly

(3) $$s'_{\mu,\nu}(z) - (\nu/z)\, s_{\mu,\nu}(z) = (\mu-\nu-1)\, s_{\mu-1,\nu+1}(z).$$

On subtracting and adding these results we obtain the formulae

(4) $$(2\nu/z)\, s_{\mu,\nu}(z) = (\mu+\nu-1)\, s_{\mu-1,\nu-1}(z) - (\mu-\nu-1)\, s_{\mu-1,\nu+1}(z),$$

(5) $$2s'_{\mu,\nu}(z) = (\mu+\nu-1)\, s_{\mu-1,\nu-1}(z) + (\mu-\nu-1)\, s_{\mu-1,\nu+1}(z).$$

The reader will find it easy to deduce from § 10·71 (2) that the functions of the type $s_{\mu,\nu}(z)$ may be replaced throughout these formulae by functions of the type $S_{\mu,\nu}(z)$; so that

(6) $$S_{\mu+2,\nu}(z) = z^{\mu+1} - \{(\mu+1)^2 - \nu^2\}\, S_{\mu,\nu}(z),$$

(7) $$S'_{\mu,\nu}(z) + (\nu/z)\, S_{\mu,\nu}(z) = (\mu+\nu-1)\, S_{\mu-1,\nu-1}(z),$$

(8) $$S'_{\mu,\nu}(z) - (\nu/z)\, S_{\mu,\nu}(z) = (\mu-\nu-1)\, S_{\mu-1,\nu+1}(z),$$

(9) $$(2\nu/z)\, S_{\mu,\nu}(z) = (\mu+\nu-1)\, S_{\mu-1,\nu-1}(z) - (\mu-\nu-1)\, S_{\mu-1,\nu+1}(z),$$

(10) $$2S'_{\mu,\nu}(z) = (\mu+\nu-1)\, S_{\mu-1,\nu-1}(z) + (\mu-\nu-1)\, S_{\mu-1,\nu+1}(z).$$

These formulae may be transformed in various ways by using (1) and (6). They are due to Lommel, *Math. Ann.* IX. (1876), pp. 429—432, but his methods of proving them were not in all cases completely satisfactory.

10·73. *Lommel's functions $S_{\mu,\nu}(z)$ when $\mu \pm \nu$ is an odd negative integer.*

The formula § 10·71 (2) assumes an undetermined form when $\mu - \nu$ or $\mu + \nu$ is an odd negative integer*. We can easily define $S_{\nu-2p-1,\nu}(z)$ in terms of $S_{\nu-1,\nu}(z)$ by a repeated use of § 10·72 (6) which gives

(1) $$S_{\nu-2p-1,\nu}(z) = \sum_{m=0}^{p-1} \frac{(-)^m z^{\nu-2p+2m}}{2^{2m+2}(-p)_{m+1}(\nu-p)_{m+1}} + \frac{(-)^p S_{\nu-1,\nu}(z)}{2^{2p}\, p!\, (1-\nu)_p}.$$

* Since $S_{\mu,\nu}(z)$ is an even function of ν, it is sufficient to consider the case in which $\mu - \nu$ is an odd negative integer.

We next define $S_{\nu-1,\nu}(z)$ by the limiting form of § 10·72 (6), namely

(2) $$S_{\nu-1,\nu}(z) = \lim_{\mu \to \nu-1} \left[\frac{z^{\mu+1} - S_{\mu+2,\nu}(z)}{(\mu-\nu+1)(\mu+\nu+1)} \right].$$

The numerator (which is an analytic function of μ near $\mu = \nu - 1$) vanishes when $\mu = \nu - 1$, and so, by L'Hospital's theorem*

$$S_{\nu-1,\nu}(z) = \frac{1}{2\nu} \left[z^\nu \log z - \frac{\partial S_{\mu+2,\nu}(z)}{\partial \mu} \right]_{\mu=\nu-1}.$$

Now it is easy to verify that

$$\left[\frac{\partial s_{\mu+2,\nu}(z)}{\partial \mu}\right]_{\mu=\nu-1} = \tfrac{1}{2} z^\nu \Gamma(\nu+1) \sum_{m=0}^{\infty} \frac{(-)^m (\frac{1}{2}z)^{2m+2}}{(m+1)!\, \Gamma(\nu+m+2)}$$
$$\times \{2 \log z + \psi(1) + \psi(\nu+1) - \psi(m+2) - \psi(\nu+m+2)\}.$$

Also

$$\left[\frac{\partial}{\partial \mu} \{2^{\mu+1} \Gamma(\tfrac{1}{2}\mu - \tfrac{1}{2}\nu + \tfrac{3}{2}) \Gamma(\tfrac{1}{2}\mu + \tfrac{1}{2}\nu + \tfrac{3}{2}) \cos \tfrac{1}{2}(\mu+\nu)\pi\}\right]_{\mu=\nu-1}$$
$$= 2^\nu \Gamma(\nu+1) \sin \nu\pi \{\log 2 + \tfrac{1}{2}\psi(1) + \tfrac{1}{2}\psi(\nu+1) + \tfrac{1}{2}\pi \cot \nu\pi\},$$

and

$$\left[\frac{\partial}{\partial \mu} \{2^{\mu+1} \Gamma(\tfrac{1}{2}\mu - \tfrac{1}{2}\nu + \tfrac{3}{2}) \Gamma(\tfrac{1}{2}\mu + \tfrac{1}{2}\nu + \tfrac{3}{2}) \cos \tfrac{1}{2}(\mu-\nu)\pi\}\right]_{\mu=\nu-1} = 2^{\nu-1} \pi \Gamma(\nu+1),$$

and hence it follows that

(3) $$S_{\nu-1,\nu}(z) = \tfrac{1}{4} z^\nu \Gamma(\nu) \sum_{m=0}^{\infty} \frac{(-)^m (\frac{1}{2}z)^{2m}}{m!\, \Gamma(\nu+m+1)}$$
$$\times \{2 \log \tfrac{1}{2}z - \psi(\nu+m+1) - \psi(m+1)\} - 2^{\nu-2} \pi \Gamma(\nu) Y_\nu(z),$$

and this formula, which appears to be nugatory whenever ν is a negative integer, is, in effect, nugatory only when $\nu = 0$; for when $\nu = -n$ (where n is a positive integer) we define the function by the formula

$$S_{-n-1,-n}(z) = S_{-n-1,n}(z),$$

in which the function on the right is defined by equation § 10·73 (1).

To discuss the case in which $\nu = 0$, we take the formula

$$S_{\mu,0}(z) = \frac{z^{\mu+1} - S_{\mu+2,0}(z)}{(\mu+1)^2}$$

which gives $$S_{-1,0}(z) = \frac{1}{2} \left[\frac{\partial^2}{\partial \mu^2} \{z^{\mu+1} - S_{\mu+2,0}(z)\}\right]_{\mu=-1}.$$

Since $$S_{\mu+2,0}(z) = z^{\mu+1} \{\Gamma(\tfrac{1}{2}\mu + \tfrac{3}{2})\}^2 \sum_{m=0}^{\infty} \frac{(-)^m (\frac{1}{2}z)^{2m+2}}{\{\Gamma(\frac{1}{2}\mu + m + \frac{5}{2})\}^2}$$
$$+ 2^{\mu+1} \{\Gamma(\tfrac{1}{2}\mu + \tfrac{3}{2})\}^2 \{\cos \tfrac{1}{2}\mu\pi \,.\, Y_0(z) - \sin \tfrac{1}{2}\mu\pi \,.\, J_0(z)\},$$

it follows, on reduction, that

(4) $$S_{-1,0}(z) = \frac{1}{2} \sum_{m=0}^{\infty} \frac{(-)^m (\frac{1}{2}z)^{2m}}{(m!)^2} [\{\log(\tfrac{1}{2}z) - \psi(m+1)\}^2 - \tfrac{1}{2}\psi'(m+1) + \tfrac{1}{4}\pi^2].$$

* Cf. Bromwich, *Theory of Infinite Series*, § 152.

10·74. *Functions expressible in terms of Lommel's functions.*

From the descending series given in § 10·71 (1) it is evident that Neumann's polynomial $O_n(z)$ is expressible in terms of Lommel's functions by the equations

(1) $$O_{2m}(z) = (1/z)\, S_{1, 2m}(z), \quad O_{2m+1}(z) = \{(2m+1)/z\}\, S_{0, 2m+1}(z),$$

and Schläfli's polynomial $S_n(z)$ is similarly expressible by the equations

(2) $$S_{2m}(z) = 4m\, S_{-1, 2m}(z), \quad S_{2m+1}(z) = 2S_{0, 2m+1}(z).$$

It is also possible to express the important integrals

$$\int^z z^\mu J_\nu(z)\, dz, \quad \int^z z^\mu Y_\nu(z)\, dz$$

in terms of Lommel's functions; thus we have

$$\frac{d}{dz}\{z^\nu J_\nu(z) \,.\, z^{1-\nu} S_{\mu-1,\nu-1}(z)\} = zJ_{\nu-1}(z)\, S_{\mu-1,\nu-1}(z) + (\mu-\nu-1)\, zJ_\nu(z)\, S_{\mu-2,\nu}(z),$$

$$\frac{d}{dz}\{z^{1-\nu} J_{\nu-1}(z) \,.\, z^{-\nu} S_{\mu,\nu}(z)\} = -zJ_\nu(z)\, S_{\mu,\nu}(z) + (\mu+\nu-1)\, zJ_{\nu-1}(z)\, S_{\mu-1,\nu-1}(z).$$

On eliminating $S_{\mu-1,\nu-1}(z)$ from the right of these equations, and using § 10·72 (6), we find by integrating that

(3) $$\int^z z^\mu J_\nu(z)\, dz = (\mu+\nu-1)\, zJ_\nu(z)\, S_{\mu-1,\nu-1}(z) - zJ_{\nu-1}(z)\, S_{\mu,\nu}(z),$$

and proofs of the same nature shew that

(4) $$\int^z z^\mu Y_\nu(z)\, dz = (\mu+\nu-1)\, zY_\nu(z)\, S_{\mu-1,\nu-1}(z) - zY_{\nu-1}(z)\, S_{\mu,\nu}(z),$$

and, more generally,

(5) $$\int^z z^\mu \mathscr{C}_\nu(z)\, dz = (\mu+\nu-1)\, z\mathscr{C}_\nu(z)\, S_{\mu-1,\nu-1}(z) - z\mathscr{C}_{\nu-1}(z)\, S_{\mu,\nu}(z).$$

Special cases of these formulae are obtained by choosing μ and ν so that the functions on the right reduce to Neumann's or Schläfli's polynomials, thus

(6) $$\int^z z\mathscr{C}_{2m}(z)\, dz = z^2 \left\{\frac{2m}{2m-1}\, \mathscr{C}_{2m}(z)\, O_{2m-1}(z) - \mathscr{C}_{2m-1}(z)\, O_{2m}(z)\right\},$$

(7) $$\int^z \mathscr{C}_{2m+1}(z)\, dz = \tfrac{1}{2} z\, \{\mathscr{C}_{2m+1}(z)\, S_{2m}(z) - \mathscr{C}_{2m}(z)\, S_{2m+1}(z)\}.$$

Of these results, (1), (3), (4) and (6) are contained in Lommel's paper, *Math. Ann.* IX. (1876), pp. 425—444; (6) and (7) were given by Nielsen, *Handbuch der Theorie der Cylinderfunktionen* (Leipzig, 1904), p. 100, but his formulae contain some misprints.

It should be noticed that Lommel's function, in those cases when it is expressible in finite terms, is equivalent to Gegenbauer's polynomial of § 9·2. The formulae connecting the functions are*

$$(8)\quad \begin{cases} A_{2m,\nu}(z) = \dfrac{2^\nu\,\Gamma(\nu+m)}{m!}\cdot\dfrac{\nu+2m}{z^{1-\nu}} S_{1-\nu,\nu+2m}(z), \\ A_{2m+1,\nu}(z) = \dfrac{2^{\nu+1}\,\Gamma(\nu+m+1)}{m!}\cdot\dfrac{\nu+2m+1}{z^{1-\nu}} S_{-\nu,\nu+2m+1}(z). \end{cases}$$

It follows that the most general case in which the integral (5) is expressible in terms of elementary functions and cylinder functions is given by the formula

$$(9)\quad \int^z z^{1-\nu}\mathscr{C}_{\nu+2m}(z)\,dz = \frac{m!\,z^{2-\nu}}{2^\nu\,\Gamma(\nu+m)}\left[\frac{\mathscr{C}_{\nu+2m}(z)\,A_{2m-1,\nu}(z)}{\nu+2m-1} - \frac{\mathscr{C}_{\nu+2m-1}(z)\,A_{2m,\nu}(z)}{\nu+2m}\right].$$

The function defined by the series

$$\sum_{m=0}^{\infty}\frac{(-)^m z^{\nu+2m}}{\Gamma(\nu+2m+1)} = \frac{z^{\frac{1}{2}}\,s_{\nu-\frac{3}{2},\frac{1}{2}}(z)}{\Gamma(\nu-1)}\,s_{\nu+\frac{1}{2},\frac{1}{2}}(z)$$

has been studied in great detail by W. H. Young†; this function possesses many properties analogous to those of Bessel functions, but the increase of simplicity over Lommel's more general function seems insufficient to justify an account of them here.

The integral $\nu\displaystyle\int_0^\infty \frac{J_\nu(t)}{t\,(z-t)}\,dt$ has been studied (when ν is an integer) by H. A. Webb, *Messenger*, XXXIII. (1904), p. 58; and he stated that, when $\nu=n$, its value is $O_n(z)$. This is incorrect (as was pointed out by Kapteyn); and the value for general values of ν is‡

$$\{S_{1,\nu}(-z) - \nu S_{0,\nu}(-z)\}/z,$$

when $R(\nu)>0$ and $|\arg(-z)|<\pi$.

10·75. *The asymptotic expansion of* $S_{\mu,\nu}(z)$.

We shall now shew by Barnes' method§ that, when $\mu\pm\nu$ are not odd positive integers, then $S_{\mu,\nu}(z)$ admits of the asymptotic expansion

$$(1)\quad S_{\mu,\nu}(z) \sim z^{\mu-1}\left[1 - \frac{(\mu-1)^2-\nu^2}{z^2} + \frac{\{(\mu-1)^2-\nu^2\}\{(\mu-3)^2-\nu^2\}}{z^4} - \ldots\right],$$

when $|z|$ is large and $|\arg z|<\pi$.

Let us take the integral

$$-\frac{z^{\mu-1}}{2\pi i}\int_{-\infty i-p+\frac{1}{2}}^{\infty i-p+\frac{1}{2}} \frac{\Gamma(\frac{1}{2}-\frac{1}{2}\mu+\frac{1}{2}\nu-s)\,\Gamma(\frac{1}{2}-\frac{1}{2}\mu-\frac{1}{2}\nu-s)}{\Gamma(\frac{1}{2}-\frac{1}{2}\mu+\frac{1}{2}\nu)\,\Gamma(\frac{1}{2}-\frac{1}{2}\mu-\frac{1}{2}\nu)}\cdot\frac{\pi(\frac{1}{2}z)^{2s}\,ds}{\sin s\pi}.$$

The contour is to be drawn by taking p to be an integer so large that the only poles of the integrand on the left of the contour are poles of cosec $s\pi$, the poles of the Gamma functions being on the right of the contour.

* Gegenbauer, *Wiener Sitzungsberichte*, LXXIV. (2), (1877), p. 126.

† *Quarterly Journal*, XLIII. (1911), pp. 161—177.

‡ Cf. Gubler, *Zürich Vierteljahrsschrift*, XLVII. (1902), pp. 422—428.

§ *Proc. London Math. Soc.* (2) V. (1907), pp. 59—118; cf. §§ 6·5, 7·5, 7·51.

The integral is convergent when $|\arg z| < \pi$, and it may be seen without difficulty that it is $O(z^{\mu-2p})$.

It may be shewn from the asymptotic expansion of the Gamma function that the same integrand, when integrated round a semicircle, of radius R with centre at $-p-\frac{1}{2}$, on the right of the contour, tends to zero as $R \to \infty$, provided that R tends to infinity in such a manner that the semicircle never passes through any of the poles of the integrand.

It follows that the expression given above is equal to the sum of the residues of

$$z^{\mu-1}\frac{\Gamma(\frac{1}{2}-\frac{1}{2}\mu+\frac{1}{2}\nu-s)\,\Gamma(\frac{1}{2}-\frac{1}{2}\mu-\frac{1}{2}\nu-s)}{\Gamma(\frac{1}{2}-\frac{1}{2}\mu+\frac{1}{2}\nu)\,\Gamma(\frac{1}{2}-\frac{1}{2}\mu-\frac{1}{2}\nu)}\cdot\frac{\pi(\frac{1}{2}z)^{2s}}{\sin s\pi}$$

at the points

$$\begin{array}{l} 0, -1, -2, \ldots, -(p-1), \\ 1, 2, 3, \ldots, \\ \frac{1}{2}-\frac{1}{2}\mu-\frac{1}{2}\nu,\quad \frac{3}{2}-\frac{1}{2}\mu-\frac{1}{2}\nu,\quad \frac{5}{2}-\frac{1}{2}\mu-\frac{1}{2}\nu, \ldots, \\ \frac{1}{2}-\frac{1}{2}\mu+\frac{1}{2}\nu,\quad \frac{3}{2}-\frac{1}{2}\mu+\frac{1}{2}\nu,\quad \frac{5}{2}-\frac{1}{2}\mu+\frac{1}{2}\nu, \ldots. \end{array}$$

When we calculate these residues we find that

$$\begin{aligned} & z^{\mu-1}\sum_{m=0}^{p-1}\frac{(-)^m\,\Gamma(\frac{1}{2}-\frac{1}{2}\mu+\frac{1}{2}\nu+m)\,\Gamma(\frac{1}{2}-\frac{1}{2}\mu-\frac{1}{2}\nu+m)}{(\frac{1}{2}z)^{2m}\,\Gamma(\frac{1}{2}-\frac{1}{2}\mu+\frac{1}{2}\nu)\,\Gamma(\frac{1}{2}-\frac{1}{2}\mu-\frac{1}{2}\nu)} \\ & \quad + z^{\mu-1}\sum_{m=1}^{\infty}\frac{(-)^m\,\Gamma(\frac{1}{2}+\frac{1}{2}\mu-\frac{1}{2}\nu)\,\Gamma(\frac{1}{2}+\frac{1}{2}\mu+\frac{1}{2}\nu)\,(\frac{1}{2}z)^{2m}}{\Gamma(\frac{1}{2}+\frac{1}{2}\mu-\frac{1}{2}\nu+m)\,\Gamma(\frac{1}{2}+\frac{1}{2}\mu+\frac{1}{2}\nu+m)} \\ & \quad - \frac{2^{\mu-1}\pi\Gamma(\frac{1}{2}+\frac{1}{2}\mu+\frac{1}{2}\nu)}{\Gamma(\frac{1}{2}-\frac{1}{2}\mu+\frac{1}{2}\nu)\sin\nu\pi}\sum_{m=0}^{\infty}\frac{(-)^m.(\frac{1}{2}z)^{-\nu+2m}}{m!\,\Gamma(1-\nu+m)} \\ & \quad + \frac{2^{\mu-1}\pi\,\Gamma(\frac{1}{2}+\frac{1}{2}\mu-\frac{1}{2}\nu)}{\Gamma(\frac{1}{2}-\frac{1}{2}\mu-\frac{1}{2}\nu)\sin\nu\pi}\sum_{m=0}^{\infty}\frac{(-)^m(\frac{1}{2}z)^{\nu+2m}}{m!\,\Gamma(\nu+m+1)} = O(z^{\mu-2p}), \end{aligned}$$

so that

$$\begin{aligned} & z^{\mu-1}\sum_{m=0}^{p-1}\frac{(-)^m\,\Gamma(\frac{1}{2}-\frac{1}{2}\mu+\frac{1}{2}\nu+m)\,\Gamma(\frac{1}{2}-\frac{1}{2}\mu-\frac{1}{2}\nu+m)}{(\frac{1}{2}z)^{2m}\,\Gamma(\frac{1}{2}-\frac{1}{2}\mu+\frac{1}{2}\nu)\,\Gamma(\frac{1}{2}-\frac{1}{2}\mu-\frac{1}{2}\nu)} - s_{\mu,\nu}(z) \\ & - \frac{2^{\mu-1}\Gamma(\frac{1}{2}+\frac{1}{2}\mu-\frac{1}{2}\nu)\,\Gamma(\frac{1}{2}+\frac{1}{2}\mu+\frac{1}{2}\nu)}{\sin\nu\pi} \\ & \qquad \times[\cos\tfrac{1}{2}(\mu-\nu)\pi.J_{-\nu}(z)-\cos\tfrac{1}{2}(\mu+\nu)\pi.J_\nu(z)] = O(z^{\mu-2p}), \end{aligned}$$

and so, by § 10·71 (2), we have the formula

$$S_{\mu,\nu}(z) = z^{\mu-1}\sum_{m=0}^{p-1}\frac{(-)^m\,\Gamma(\frac{1}{2}-\frac{1}{2}\mu+\frac{1}{2}\nu+m)\,\Gamma(\frac{1}{2}-\frac{1}{2}\mu-\frac{1}{2}\nu+m)}{(\frac{1}{2}z)^{2m}\,\Gamma(\frac{1}{2}-\frac{1}{2}\mu+\frac{1}{2}\nu)\,\Gamma(\frac{1}{2}-\frac{1}{2}\mu-\frac{1}{2}\nu)} + O(z^{\mu-2p}),$$

and this is equivalent to the asymptotic expansion stated in (1).

10·8. *Hemi-cylindrical functions.*

Functions $\mathbf{S}_n(z)$ which satisfy the single recurrence formula

(1) $$\mathbf{S}_{n-1}(z) - \mathbf{S}_{n+1}(z) = 2\mathbf{S}_n'(z)$$

combined with

(2) $$\mathbf{S}_1(z) = -\mathbf{S}_0'(z)$$

have been studied in great detail by Sonine*. They will be called hemi-cylindrical functions.

It is evident that $\mathbf{S}_n(z)$ is expressible in the form

$$\mathbf{S}_n(z) = f_n(D) \, . \, \mathbf{S}_0(z),$$

where $D \equiv d/dz$ and $f_n(D)$ is a polynomial in D of degree n; and the polynomial $f_n(\xi)$ satisfies the recurrence formula

$$f_{n-1}(\xi) - f_{n+1}(\xi) = 2\xi f_n(\xi)$$

combined with

$$f_0(\xi) = 1, \quad f_1(\xi) = -\xi.$$

It follows by induction (cf. § 9·14) that

$$f_n(\xi) = \tfrac{1}{2}[\{-\xi + \sqrt{(\xi^2+1)}\}^n + \{-\xi - \sqrt{(\xi^2+1)}\}^n],$$

and therefore

(3) $$\mathbf{S}_n(z) = \tfrac{1}{2}[\{-D + \sqrt{(D^2+1)}\}^n + \{-D - \sqrt{(D^2+1)}\}^n] \, . \, \mathbf{S}_0(z).$$

If it is supposed that (1) holds for negative values of n, it is easy to see that

(4) $$\mathbf{S}_{-n}(z) = (-)^n \mathbf{S}_n(z).$$

To obtain an alternative expression to (3), put $\xi = \sinh t$, and then†

$$f_n(\xi) = \begin{cases} \cosh nt & (n \text{ even}) \\ -\sinh nt & (n \text{ odd}) \end{cases}$$

$$= \begin{cases} 1 + \dfrac{n^2}{2!}\xi^2 + \dfrac{n^2(n^2-2^2)}{4!}\xi^4 + \dots & (n \text{ even}) \\ -\dfrac{n}{1!}\xi - \dfrac{n(n^2-1^2)}{3!}\xi^3 - \dots. & (n \text{ odd}) \end{cases}$$

Hence

(5) $$\mathbf{S}_n(z) = \begin{cases} \mathbf{S}_0(z) + \dfrac{n^2}{2!}\mathbf{S}_0''(z) + \dfrac{n^2(n^2-2^2)}{4!}\mathbf{S}_0^{\text{iv}}(z) + \dots & (n \text{ even}) \\ -\dfrac{n}{1!}\mathbf{S}_0'(z) - \dfrac{n(n^2-1^2)}{3!}\mathbf{S}_0'''(z) - \dots. & (n \text{ odd}) \end{cases}$$

It is to be noticed that $O_n(z)$, $T_n(z)$ and $\mathbf{E}_n(z)$ are hemi-cylindrical functions, but $S_n(z)$, $U_n(z)$ and $\mathbf{H}_n(z)$ are not hemi-cylindrical functions.

It should be remarked that the single recurrence formula

$$\Sigma_{\nu-1}(z) + \Sigma_{\nu+1}(z) = \frac{2\nu}{z}\Sigma_\nu(z)$$

gives rise to functions of no greater intrinsic interest than Lommel's polynomials.

* *Math. Ann.* XVI. (1880), pp. 1—9, 71—80.

† See e.g. Hobson, *Plane Trigonometry* (1918), § 264.

10·81. *The addition theorem for hemi-cylindrical functions.*

We shall now establish Sonine's important expansion*

(1) $$\mathbf{S}_m(z+t) = \sum_{n=-\infty}^{\infty} J_n(t)\,\mathbf{S}_{m-n}(z);$$

the expansion is valid when $z+t$ lies inside the largest circle, whose centre is at the point z, which does not contain any singularity of the hemi-cylindrical function under consideration.

Take as contour a circle C with centre z such that $\mathbf{S}_0(\zeta)$ has no singularity inside or on the circle. Then

$$\begin{aligned}\mathbf{S}_m(z+t) &= \frac{1}{2\pi i}\int_C \frac{\mathbf{S}_m(\zeta)\,d\zeta}{\zeta - z - t} \\ &= \frac{1}{2\pi i}\int_C \mathbf{S}_m(\zeta)\left\{\sum_{n=0}^{\infty} \epsilon_n O_n(\zeta - z)\,J_n(t)\right\} d\zeta.\end{aligned}$$

The series converges uniformly on the contour, and so we have

$$\begin{aligned}\mathbf{S}_m(z+t) &= \frac{1}{2\pi i}\sum_{n=0}^{\infty} \epsilon_n J_n(t)\int_C \mathbf{S}_m(\zeta)\,O_n(\zeta - z)\,d\zeta \\ &= \frac{1}{2\pi i}\sum_{n=0}^{\infty} \epsilon_n J_n(t)\,f_n\left(-\frac{d}{dz}\right)\int_C \frac{\mathbf{S}_m(\zeta)}{\zeta - z}\,d\zeta \\ &= \sum_{n=0}^{\infty} \epsilon_n J_n(t)\,f_n\left(-\frac{d}{dz}\right)\mathbf{S}_m(z)\,dz \\ &= \sum_{n=0}^{\infty} \epsilon_n J_n(t)\,f_n\left(-\frac{d}{dz}\right) f_m\left(\frac{d}{dz}\right)\mathbf{S}_0(z)\,dz.\end{aligned}$$

But it is easy to verify that

$$2f_n(-\xi)\,f_m(\xi) = f_{m-n}(\xi) + (-)^n f_{m+n}(\xi),$$

so that

$$\mathbf{S}_m(z+t) = J_0(t)\,\mathbf{S}_m(z) + \sum_{n=1}^{\infty} J_n(t)\,\{\mathbf{S}_{m-n}(z) + (-)^n\,\mathbf{S}_{m+n}(z)\},$$

whence Sonine's formula is obvious.

It should be noticed that, if $\mathbf{S}_n(z)$ denotes a function of a more general type than a hemi-cylindrical function, namely one which merely satisfies the equation

$$\mathbf{S}_{n-1}(z) - \mathbf{S}_{n+1}(z) = 2\mathbf{S}_n'(z),$$

without satisfying the equation $\mathbf{S}_1(z) = -\mathbf{S}_0'(z)$, we still have

$$f_n\left(-\frac{d}{dz}\right)\mathbf{S}_m(z) = \mathbf{S}_{m-n}(z) + (-)^n\,\mathbf{S}_{m+n}(z),$$

and so the formula (1) is still valid. We thus have an alternative proof of the formulae of §§ 5·3, 9·1, 9·34 and 10·63.

* *Math. Ann.* XVI. (1880), pp. 4—8. See also König, *Math. Ann.* V. (1872), pp. 310—340; *ibid.* XVII. (1880), pp. 85—86.

10·82. *Nielsen's functional equations.*

The pair of simultaneous equations

$$\begin{cases} (1) \quad F_{\nu-1}(z) - F_{\nu+1}(z) - 2F_\nu'(z) = 2f_\nu(z)/z, \\ (2) \quad F_{\nu-1}(z) + F_{\nu+1}(z) - (2\nu/z)\, F_\nu(z) = 2g_\nu(z)/z, \end{cases}$$

where $f_\nu(z)$ and $g_\nu(z)$ are given arbitrary functions of the variables ν and z, form an obvious generalisation of the pair of functional equations whereby cylinder functions are defined. It has been shewn by Nielsen* that the functions $f_\nu(z)$ and $g_\nu(z)$ must satisfy the relation

$$f_{\nu-1}(z) + f_{\nu+1}(z) - (2\nu/z)\, f_\nu(z) = g_{\nu-1}(z) - g_{\nu+1}(z) - 2g_\nu'(z);$$

and it has been proved† that, if this relation is satisfied, the system can be reduced to a pair of soluble difference equations of the first order.

For brevity write

$$f_\nu(z) + g_\nu(z) \equiv \alpha_\nu(z), \quad f_\nu(z) - g_\nu(z) \equiv \beta_\nu(z),$$

and the given system of equations is equivalent to

$$\begin{cases} (3) \quad (\vartheta + \nu)\, F_\nu(z) = zF_{\nu-1}(z) - \alpha_\nu(z), \\ (4) \quad (\vartheta - \nu)\, F_\nu(z) = -\, zF_{\nu+1}(z) - \beta_\nu(z). \end{cases}$$

It is now evident that

$$\begin{aligned} (\vartheta^2 - \nu^2)\, F_\nu(z) &= (\vartheta - \nu)\, [zF_{\nu-1}(z) - \alpha_\nu(z)] \\ &= -\, z^2 F_\nu(z) - z\beta_{\nu-1}(z) - (\vartheta - \nu)\, \alpha_\nu(z), \end{aligned}$$

so that $$\nabla_\nu F_\nu(z) = -\, z\beta_{\nu-1}(z) - (\vartheta - \nu)\, \alpha_\nu(z).$$

Again

$$\begin{aligned} (\vartheta^2 - \nu^2)\, F_\nu(z) &= (\vartheta + \nu)\, [-\, zF_{\nu+1}(z) - \beta_\nu(z)] \\ &= -\, z^2 F_\nu(z) + z\alpha_{\nu+1}(z) - (\vartheta + \nu)\, \beta_\nu(z). \end{aligned}$$

We are thus led to the equation

$$(5) \quad \nabla_\nu F_\nu(z) = z\varpi_\nu(z),$$

where

$$\begin{cases} (6) \quad z\varpi_\nu(z) = -\, z\beta_{\nu-1}(z) - (\vartheta - \nu)\, \alpha_\nu(z), \\ (7) \quad z\varpi_\nu(z) = +\, z\alpha_{\nu+1}(z) - (\vartheta + \nu)\, \beta_\nu(z). \end{cases}$$

On comparing these values of $\varpi_\nu(z)$, we are at once led to Nielsen's condition

$$(8) \quad f_{\nu-1}(z) + f_{\nu+1}(z) - (2\nu/z)\, f_\nu(z) = g_{\nu-1}(z) - g_{\nu+1}(z) - 2g_\nu'(z).$$

It now has to be shewn that Nielsen's condition is sufficient for the existence of a solution of the given system. To prove this, we assume (8) to be

* *Ann. di Mat.* (3) VI. (1901), pp. 51—59.

† Watson, *Messenger*, XLVIII. (1919), pp. 49—53.

given, and, after defining $\varpi_\nu(z)$ by (6) and (7), we solve (5) by the method of variation of parameters. The solution is

$$(9) \qquad F_\nu(z) = J_\nu(z)\left\{c_\nu - \tfrac{1}{2}\pi\int_a^z Y_\nu(t)\,\varpi_\nu(t)\,dt\right\} + Y_\nu(z)\left\{d_\nu + \tfrac{1}{2}\pi\int_b^z J_\nu(t)\,\varpi_\nu(t)\,dt\right\},$$

where a and b are arbitrary constants; and c_ν and d_ν may be taken to be independent of z, though they will, in general, depend on ν.

It remains to be shewn that c_ν and d_ν can be chosen so that the value of $F_\nu(z)$ given by (9) satisfies (1) and (2), or (what comes to the same thing) that it satisfies (3) and (4). If (3) is satisfied, then

$$\begin{aligned} zJ_{\nu-1}(z)&\left\{c_\nu - \tfrac{1}{2}\pi\int_a^z Y_\nu(t)\,\varpi_\nu(t)\,dt\right\} - \tfrac{1}{2}\pi z\,J_\nu(z)\,Y_\nu(z)\,\varpi_\nu(z) \\ &+ zY_{\nu-1}(z)\left\{d_\nu + \tfrac{1}{2}\pi\int_b^z J_\nu(t)\,\varpi_\nu(t)\,dt\right\} + \tfrac{1}{2}\pi z\,Y_\nu(z)\,J_\nu(z)\,\varpi_\nu(z) \\ = zJ_{\nu-1}(z)&\left\{c_{\nu-1} - \tfrac{1}{2}\pi\int_a^z Y_{\nu-1}(t)\,\varpi_{\nu-1}(t)\,dt\right\} \\ &+ zY_{\nu-1}(z)\left\{d_{\nu-1} + \tfrac{1}{2}\pi\int_b^z J_{\nu-1}(t)\,\varpi_{\nu-1}(t)\,dt\right\} - \alpha_\nu(z), \end{aligned}$$

that is to say,

$$\begin{aligned} zJ_{\nu-1}(z)&\left[c_\nu - c_{\nu-1} - \tfrac{1}{2}\pi\int_a^z \{Y_\nu(t)\,\varpi_\nu(t) - Y_{\nu-1}(t)\,\varpi_{\nu-1}(t)\}\,dt\right] \\ &+ zY_{\nu-1}(z)\left[d_\nu - d_{\nu-1} + \tfrac{1}{2}\pi\int_b^z \{J_\nu(t)\,\varpi_\nu(t) - J_{\nu-1}(t)\,\varpi_{\nu-1}(t)\}\,dt\right] + \alpha_\nu(z) = 0. \end{aligned}$$

But it is easy to verify that

$$\frac{d}{dz}\{\mathscr{C}_{\nu-1}(z)\,\beta_{\nu-1}(z) - \mathscr{C}_\nu(z)\,\alpha_\nu(z)\} = \varpi_\nu(z)\,\mathscr{C}_\nu(z) - \varpi_{\nu-1}(z)\,\mathscr{C}_{\nu-1}(z),$$

since (6) and (7) are satisfied; and so (3) is satisfied if

$$\begin{aligned} zJ_{\nu-1}(z)&\left\{c_\nu - c_{\nu-1} - \tfrac{1}{2}\pi\Big[Y_{\nu-1}(z)\,\beta_{\nu-1}(z) - Y_\nu(z)\,\alpha_\nu(z)\Big]_a^z\right\} \\ &+ zY_{\nu-1}(z)\left\{d_\nu - d_{\nu-1} + \tfrac{1}{2}\pi\Big[J_{\nu-1}(z)\,\beta_{\nu-1}(z) - J_\nu(z)\,\alpha_\nu(z)\Big]_b^z\right\} + \alpha_\nu(z) = 0, \end{aligned}$$

and this condition, by § 3·63 (12), reduces to

$$\begin{aligned} zJ_{\nu-1}(z)&\{c_\nu - c_{\nu-1} + \tfrac{1}{2}\pi[Y_{\nu-1}(a)\,\beta_{\nu-1}(a) - Y_\nu(a)\,\alpha_\nu(a)]\} \\ &+ zY_{\nu-1}(z)\{d_\nu - d_{\nu-1} - \tfrac{1}{2}\pi[J_{\nu-1}(b)\,\beta_{\nu-1}(b) - J_\nu(b)\,\alpha_\nu(b)]\} = 0. \end{aligned}$$

Consequently, so far as (3) is concerned, it is sufficient to choose c_ν and d_ν to satisfy the difference equations

$$\begin{cases} (10) \qquad c_\nu - c_{\nu-1} = -\tfrac{1}{2}\pi\{Y_{\nu-1}(a)\,\beta_{\nu-1}(a) - Y_\nu(a)\,\alpha_\nu(a)\}, \\ (11) \qquad d_\nu - d_{\nu-1} = \tfrac{1}{2}\pi\{J_{\nu-1}(b)\,\beta_{\nu-1}(b) - J_\nu(b)\,\alpha_\nu(b)\}; \end{cases}$$

and the reader will have no difficulty in verifying that, if these same two difference equations (with ν replaced by $\nu+1$ throughout) are satisfied, then the value of $F_\nu(z)$ given by (9) is a solution of (4).

These difference equations are of a type whose solutions may be regarded as known*; and so the condition (8) is a sufficient, as well as a necessary, condition for the existence of a solution of the given pair of functional equations (1) and (2).

If, as $z \to \infty$,

$$f_\nu(z) = O(z^{\frac{1}{2}-\delta}), \quad g_\nu(z) = O(z^{\frac{1}{2}-\delta}),$$

where $\delta > 0$, then we may make $a \to \infty$, $b \to \infty$, and we have

$$c_\nu = c_{\nu-1}, \quad d_\nu = d_{\nu-1},$$

so that the general solution may be written

$$(12) \qquad F_\nu(z) = J_\nu(z)\left\{\pi_1(\nu) + \tfrac{1}{2}\pi \int_z^\infty Y_\nu(t)\,\varpi_\nu(t)\,dt\right\} + Y_\nu(z)\left\{\pi_2(\nu) - \tfrac{1}{2}\pi \int_z^\infty J_\nu(t)\,\varpi_\nu(t)\,dt\right\},$$

where $\pi_1(\nu)$ and $\pi_2(\nu)$ are arbitrary periodic functions of ν with period unity.

NOTE. Some interesting properties of functions *which satisfy equation* (2) *only* are to be found in Nielsen's earlier paper, *Ann. di Mat.* (3) v. (1901), pp. 17—31. Thus, from a set of formulae of the type

$$F_{\nu-1}(z) + F_{\nu+1}(z) - (2\nu/z)\,F(z) = 2g_\nu(z)/z,$$

it is easy to deduce that

$$(13) \qquad F_{\nu+n}(z) = F_\nu(z)\,R_{n,\nu}(z) - F_{\nu-1}(z)\,R_{n-1,\nu+1}(z) + (2/z)\sum_{m=0}^{n-1} g_{\nu+m}(z)\,R_{n-m-1,\nu+m+1}(z);$$

the first two terms on the right are the complementary function of the difference equation, and the series is the particular integral.

* An account of various memoirs dealing with such equations is given by Barnes, *Proc. London Math. Soc.* (2) II. (1904), pp. 438—469.

CHAPTER XI

ADDITION THEOREMS

11·1. *The general nature of addition theorems.*

It has been proved (§ 4·73) that Bessel functions are not algebraic functions, and it is fairly obvious from the asymptotic expansions obtained in Chapter VII that they are not simply periodic functions, and, *a fortiori*, that they are not doubly periodic functions. Consequently, in accordance with a theorem due to Weierstrass*, it is not possible to express $J_\nu(Z+z)$ as an algebraic function of $J_\nu(Z)$ and $J_\nu(z)$. That is to say, that Bessel functions do not possess addition theorems in the strict sense of the term.

There are, however, two classes of formulae which are commonly described as addition theorems. In the case of functions of order zero the two classes coincide; and the formula for functions of the first kind is

$$J_0\{\sqrt{(Z^2+z^2-2Zz\cos\phi)}\} = \sum_{m=0}^{\infty} \epsilon_m J_m(Z)\, J_m(z) \cos m\phi,$$

which has already been indicated in § 4·82.

The simplest rigorous proof of this formula, which is due to Neumann†, depends on a transformation of Parseval's integral; another proof is due to Heine‡, who obtained the formula as a confluent form of the addition theorem for Legendre functions.

11·2. *Neumann's addition theorem*§.

We shall now establish the result

$$(1) \qquad J_0(\varpi) = \sum_{m=0}^{\infty} \epsilon_m J_m(Z)\, J_m(z) \cos m\phi,$$

where, for brevity, we write

$$\varpi = \sqrt{(Z^2+z^2-2Zz\cos\phi)},$$

and all the variables are supposed to have general complex values.

* The theorem was stated in §§ 1—3 of Schwarz' edition of Weierstrass' lectures (Berlin, 1893); see Phragmén, *Acta Math.* VII. (1885), pp. 33—42, and Forsyth, *Theory of Functions* (1918), Ch. XIII for proofs of the theorem.

† *Theorie der Bessel'schen Functionen* (Leipzig, 1867), pp. 59—70.

‡ *Handbuch der Kugelfunctionen*, I. (Berlin, 1878), pp. 340—343; cf. § 5·71 and *Modern Analysis*, § 15·7.

§ In addition to Neumann's treatise cited in § 11·1, see Beltrami, *Atti della R. Accad. di Torino*, XVI. (1880—1881), pp. 201—202.

We take the formula (Parseval's integral)

$$J_0(\varpi) = \frac{1}{2\pi}\int_{-\pi}^{\pi} e^{i\varpi\cos\theta}\,d\theta = \frac{1}{2\pi}\int_{-\pi}^{\pi} e^{i\varpi\cos(\theta-\alpha)}\,d\theta,$$

which is valid for all (complex) values of ϖ and α, the integrand being a periodic analytic function of θ with period 2π. We next suppose that α is defined by the equations

$$\varpi\sin\alpha = Z - z\cos\phi, \quad \varpi\cos\alpha = z\sin\phi,$$

and it is then apparent that

$$\begin{aligned} J_0(\varpi) &= \frac{1}{2\pi}\int_{-\pi}^{\pi} \exp\{i(Z - z\cos\phi)\sin\theta + iz\sin\phi\cos\theta\}\,d\theta \\ &= \frac{1}{2\pi}\int_{-\pi}^{\pi}\left\{\sum_{m=-\infty}^{\infty} J_m(Z)\,e^{mi\theta}\right\} e^{iz\sin(\phi-\theta)}\,d\theta \\ &= \frac{1}{2\pi}\sum_{m=-\infty}^{\infty} J_m(Z)\int_{-\pi}^{\pi} e^{mi\theta+iz\sin(\phi-\theta)}\,d\theta \\ &= \frac{1}{2\pi}\sum_{m=-\infty}^{\infty} J_m(Z)\int_{-\pi}^{\pi} e^{mi(\theta+\phi)-iz\sin\theta}\,d\theta \\ &= \sum_{m=-\infty}^{\infty} J_m(Z)\,J_m(z)\,e^{mi\phi}, \end{aligned}$$

the interchange of the order of summation and integration following from the uniformity of convergence of the series, and the next step following from the periodicity of the integrand.

If we group the terms for which the values of m differ only in sign, we immediately obtain Neumann's formula.

The corresponding formulae for Bessel functions of order $\pm\frac{1}{2}$ were obtained by Clebsch, *Journal für Math.* LXI. (1863), pp. 224—227, four years before the publication of Neumann's formula; see § 11·4.

11·3. *Graf's generalisation of Neumann's formula.*

Neumann's addition theorem has been extended to functions of arbitrary order ν in two different ways. The extension which seems to be of more immediate importance in physical applications is due to Graf*, whose formula is

$$(1) \qquad J_\nu(\varpi)\cdot\left\{\frac{Z - ze^{-i\phi}}{Z - ze^{i\phi}}\right\}^{\frac{1}{2}\nu} = \sum_{m=-\infty}^{\infty} J_{\nu+m}(Z)\,J_m(z)\,e^{mi\phi},$$

and this formula is valid provided that both of the numbers $|ze^{\pm i\phi}|$ are less than $|Z|$.

* *Math. Ann.* XLIII. (1893), pp. 142—144 and *Verhandlungen der Schweiz. Naturf. Ges.* 1896, pp. 59—61. A special case of the result has also been obtained by Nielsen, *Math. Ann.* LII. (1899), p. 241.

Graf's proof is based on the theory of contour integration, but, two years after it was published, an independent proof was given by G. T. Walker, *Messenger*, xxv. (1896), pp. 76—80; this proof is applicable to functions of integral order only, and it may be obtained from Graf's proof by replacing the contour integrals by definite integrals.

To prove the general formula, observe that the series on the right in (1) is convergent in the circumstances postulated, and so, if $\arg Z = \alpha$, we have

$$\sum_{m=-\infty}^{\infty} J_{\nu+m}(Z)\, J_m(z)\, e^{mi\phi}$$
$$= \frac{1}{2\pi i} \sum_{m=-\infty}^{\infty} \int_{-\infty \exp(-i\alpha)}^{(0+)} \exp\left\{\tfrac{1}{2}Z\left(t - \frac{1}{t}\right)\right\} t^{-\nu-m-1} J_m(z)\, e^{mi\phi}\, dt$$
$$= \frac{1}{2\pi i} \int_{-\infty \exp(-i\alpha)}^{(0+)} \exp\left\{\tfrac{1}{2}Z\left(t - \frac{1}{t}\right) - \tfrac{1}{2}z\left(\frac{t}{e^{i\phi}} - \frac{e^{i\phi}}{t}\right)\right\} \frac{dt}{t^{\nu+1}};$$

there is no special difficulty in interchanging the order of summation and integration*.

Now write

$$(Z - ze^{-i\phi})\, t = \varpi u, \quad (Z - ze^{i\phi})/t = \varpi/u,$$

where, as usual, $\varpi = \sqrt{(Z^2 + z^2 - 2Zz\cos\phi)}$,

and it is supposed now that *that value of the square root is taken which makes* $\varpi \to +Z$ when $z \to 0$.

For all admissible values of z, the phase of ϖ/Z is now an *acute* angle, positive or negative. This determination of ϖ renders it possible to take the u-contour to start from and end at $-\infty \exp(-i\beta)$, where $\beta = \arg \varpi$.

We then have

$$\sum_{m=-\infty}^{\infty} J_{\nu+m}(Z)\, J_m(z)\, e^{mi\phi} = \frac{1}{2\pi i}\left(\frac{Z - ze^{-i\phi}}{\varpi}\right)^{\nu} \int_{-\infty \exp(-i\beta)}^{(0+)} \exp\left\{\tfrac{1}{2}\varpi\left(u - \frac{1}{u}\right)\right\} \frac{du}{u^{\nu+1}}$$
$$= \left(\frac{Z - ze^{-i\phi}}{Z - ze^{i\phi}}\right)^{\frac{1}{2}\nu} J_\nu(\varpi),$$

by § 6·2 (2); and this is Graf's result.

If we define the angle ψ by the equations

$$Z - z\cos\phi = \varpi\cos\psi, \quad z\sin\phi = \varpi\sin\psi,$$

where $\psi \to 0$ as $z \to 0$ (so that, for real values of the variables, we obtain the relation indicated by Fig. 28), then Graf's formula may be written

$$e^{\nu i\psi} J_\nu(\varpi) = \sum_{m=-\infty}^{\infty} J_{\nu+m}(Z)\, J_m(z)\, e^{mi\phi}, \tag{2}$$

and, on changing the signs of ϕ and ψ, we have

$$e^{-\nu i\psi} J_\nu(\varpi) = \sum_{m=-\infty}^{\infty} J_{\nu+m}(Z)\, J_m(z)\, e^{-mi\phi}, \tag{3}$$

* Cf. Bromwich, *Theory of Infinite Series*, § 176.

whence it follows that

$$(4)\qquad J_\nu(\varpi)\,{\overset{\cos}{\underset{\sin}{\cdot}}}\,\nu\psi = \sum_{m=-\infty}^{\infty} J_{\nu+m}(Z)\,J_m(z)\,{\overset{\cos}{\underset{\sin}{\cdot}}}\,m\phi.$$

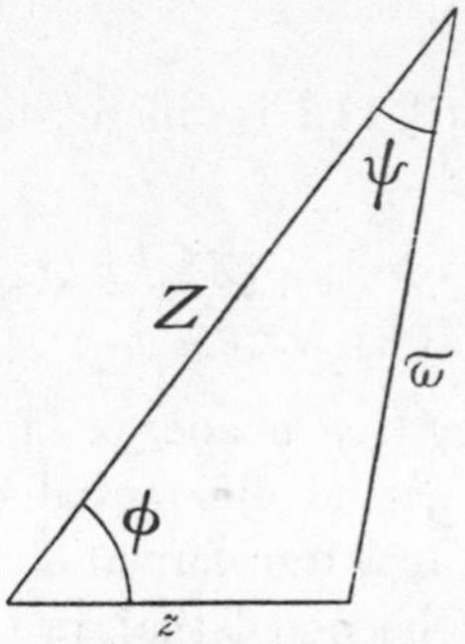

Fig. 28.

If, in this formula, we change the signs of ν and m, we readily deduce from § 3·54 that

$$(5)\qquad Y_\nu(\varpi)\,{\overset{\cos}{\underset{\sin}{\cdot}}}\,\nu\psi = \sum_{m=-\infty}^{\infty} Y_{\nu+m}(Z)\,J_m(z)\,{\overset{\cos}{\underset{\sin}{\cdot}}}\,m\phi,$$

and so

$$(6)\qquad \mathscr{C}_\nu(\varpi)\,{\overset{\cos}{\underset{\sin}{\cdot}}}\,\nu\psi = \sum_{m=-\infty}^{\infty} \mathscr{C}_{\nu+m}(Z)\,J_m(z)\,{\overset{\cos}{\underset{\sin}{\cdot}}}\,m\phi.$$

The formula (5) was given by Neumann in his treatise in the special case $\nu=0$; see also Sommerfeld, *Math. Ann.* XLV. (1894), p. 276; *ibid.* XLVII. (1896), p. 356. Some physical applications of the formulae are due to Schwarzschild, *Math. Ann.* LV. (1902), pp. 177—247.

If we replace Z, z and ϖ in these equations by iZ, iz and $i\varpi$ respectively, it is apparent that

$$(7)\qquad I_\nu(\varpi)\,{\overset{\cos}{\underset{\sin}{\cdot}}}\,\nu\psi = \sum_{m=-\infty}^{\infty} (-)^m I_{\nu+m}(Z)\,I_m(z)\,{\overset{\cos}{\underset{\sin}{\cdot}}}\,m\phi,$$

$$(8)\qquad K_\nu(\varpi)\,{\overset{\cos}{\underset{\sin}{\cdot}}}\,\nu\psi = \sum_{m=-\infty}^{\infty} K_{\nu+m}(Z)\,I_m(z)\,{\overset{\cos}{\underset{\sin}{\cdot}}}\,m\phi.$$

Of these results, (7) was stated by Beltrami, *Atti della R. Accad. di Torino*, XVI. (1880—1881), pp. 201—202.

The following special results, obtained by taking $\phi=\frac{1}{2}\pi$, should be noticed:

$$(9)\qquad \mathscr{C}_\nu(\varpi)\cos\nu\psi = \sum_{m=-\infty}^{\infty} (-)^m \mathscr{C}_{\nu+2m}(Z)\,J_{2m}(z),$$

$$(10)\qquad \mathscr{C}_\nu(\varpi)\sin\nu\psi = \sum_{m=-\infty}^{\infty} (-)^m \mathscr{C}_{\nu+2m+1}(Z)\,J_{2m+1}(z),$$

where $$Z=\varpi\cos\psi,\quad z=\varpi\sin\psi\ \text{ and }\ |z|<|Z|.$$

For the physical interpretation of these formulae the reader is referred to the papers by G. T. Walker and Schwarzschild; it should be observed that, in the special case in which ν is an integer and the only functions involved are of the first kind, the inequalities $|ze^{\pm i\phi}|<|Z|$ need not be in force.

11·4. *Gegenbauer's addition theorem.*

The second type of generalisation of Neumann's addition theorem was obtained by Gegenbauer* nearly twenty years before the publication of Graf's paper.

If Neumann's formula of § 11·1 is differentiated n times with respect to $\cos\phi$, we find that

$$(1)\qquad \frac{J_n(\varpi)}{\varpi^n} = \sum_{m=0}^{\infty} \epsilon_{m+n} \frac{J_{m+n}(Z)}{Z^n} \frac{J_{m+n}(z)}{z^n} \frac{d^n \cos(m+n)\phi}{d(\cos\phi)^n}.$$

This formula was extended by Gegenbauer to functions of non-integral order by means of the theory of partial differential equations (see § 11·42); but Sonine† gave a proof by a direct transformation of series, and this proof we shall now reproduce; it is to be noted that, in (1), z is not restricted (as in § 11·3) with reference to Z.

We take Lommel's expansion of § 5·22, namely

$$\frac{J_\nu\{\sqrt{(\zeta+h)}\}}{(\zeta+h)^{\frac{1}{2}\nu}} = \sum_{p=0}^{\infty} \frac{(-\frac{1}{2}h)^p}{p!} \frac{J_{\nu+p}(\sqrt{\zeta})}{\zeta^{\frac{1}{2}(\nu+p)}}$$

and replace ζ and h by Z^2+z^2 and $-2Zz\cos\phi$ respectively; if we write Ω in place of $J_\nu(\varpi)/\varpi^\nu$ for brevity, it is found that

$$\begin{aligned}\Omega &= \sum_{p=0}^{\infty} \frac{(Zz\cos\phi)^p}{p!} \frac{J_{\nu+p}\{\sqrt{(Z^2+z^2)}\}}{(Z^2+z^2)^{\frac{1}{2}(\nu+p)}} \\ &= \sum_{p=0}^{\infty}\sum_{q=0}^{\infty} \frac{(-)^q z^{p+2q}\cos^p\phi}{2^q \cdot p!\, q!} \frac{J_{\nu+p+q}(Z)}{Z^{\nu+q}},\end{aligned}$$

by a further application of Lommel's expansion with ζ and h replaced by Z^2 and z^2.

But, by § 5·21,

$$\frac{J_{\nu+p+q}(Z)}{Z^q} = \sum_{k=0}^{q} \frac{q!}{k!\,(q-k)!} \frac{\nu+p+2k}{2^q} \frac{\Gamma(\nu+p+k)}{\Gamma(\nu+p+q+k+1)} J_{\nu+p+2k}(Z),$$

and so

$$\Omega = \sum_{p=0}^{\infty}\sum_{q=0}^{\infty}\sum_{k=0}^{q} \frac{(-)^q(\nu+p+2k)\,\Gamma(\nu+p+k)\, z^{p+2q}\cos^p\phi}{2^{2q}p!\,k!\,(q-k)!\,\Gamma(\nu+p+q+k+1)} \frac{J_{\nu+p+2k}(Z)}{Z^\nu},$$

the triple series on the right being absolutely convergent, by comparison with

$$\sum_{p=0}^{\infty}\sum_{q=0}^{\infty}\sum_{k=0}^{q} \left|\frac{\Gamma(\nu+p+k)\, z^{p+2q}\, Z^{\nu+p+2k}}{2^{p+2q+2k}p!\,k!\,(q-k)!\,\Gamma(\nu+p+2k)\,\Gamma(\nu+p+q+k+1)}\right|.$$

* *Wiener Sitzungsberichte*, LXX. (2), (1875), pp. 6—16.

† *Math. Ann.* XVI. (1880), pp. 22—23.

But, for an absolutely convergent series,

$$\sum_{q=0}^{\infty}\sum_{k=0}^{q} u_{k,q} = \sum_{k=0}^{\infty}\sum_{n=0}^{\infty} u_{k,k+n},$$

and so

$$\begin{aligned}\Omega &= \sum_{p=0}^{\infty}\sum_{k=0}^{\infty}\sum_{n=0}^{\infty} \frac{(-)^{k+n}(\nu+p+2k)\,\Gamma(\nu+p+k)\,z^{p+2k+2n}\cos^p\phi}{2^{2k+2n}\,p!\,k!\,n!\,\Gamma(\nu+p+2k+n+1)}\,\frac{J_{\nu+p+2k}(Z)}{Z^\nu}\\
&= \sum_{p=0}^{\infty}\sum_{k=0}^{\infty} \frac{(-)^k 2^{\nu+p}(\nu+p+2k)\,\Gamma(\nu+p+k)\cos^p\phi}{p!\,k!}\,\frac{J_{\nu+p+2k}(Z)}{Z^\nu}\,\frac{J_{\nu+p+2k}(z)}{z^\nu}\\
&= \sum_{k=0}^{\infty}\sum_{m=2k}^{\infty} \frac{(-)^k 2^{\nu+m-2k}(\nu+m)\,\Gamma(\nu+m-k)\cos^{m-2k}\phi}{(m-2k)!\,k!}\,\frac{J_{\nu+m}(Z)}{Z^\nu}\,\frac{J_{\nu+m}(z)}{z^\nu}\\
&= \sum_{m=0}^{\infty}\sum_{k=0}^{\leqslant\frac{1}{2}m} \frac{(-)^k 2^{\nu+m-2k}(\nu+m)\,\Gamma(\nu+m-k)\cos^{m-2k}\phi}{(m-2k)!\,k!}\,\frac{J_{\nu+m}(Z)}{Z^\nu}\,\frac{J_{\nu+m}(z)}{z^\nu}.\end{aligned}$$

Now
$$\sum_{k=0}^{\leqslant\frac{1}{2}m} \frac{(-)^k 2^{m-2k}\,\Gamma(\nu+m-k)\cos^{m-2k}\phi}{(m-2k)!\,k!\,\Gamma(\nu)} = C_m^\nu(\cos\phi),$$

where, as in § 3·32, $C_m^\nu(\cos\phi)$ denotes the coefficient of α^m in the expansion of $(1-2\alpha\cos\phi+\alpha^2)^{-\nu}$ in ascending powers of α. We have therefore obtained the expansion

$$\frac{J_\nu(\varpi)}{\varpi^\nu} = 2^\nu\,\Gamma(\nu)\sum_{m=0}^{\infty}(\nu+m)\,\frac{J_{\nu+m}(Z)}{Z^\nu}\,\frac{J_{\nu+m}(z)}{z^\nu}\,C_m^\nu(\cos\phi), \tag{2}$$

which is valid for all values of Z, z, and ϕ, and for all values of ν with the exception of $0, -1, -2, \ldots$.

In the special case in which $\nu=\frac{1}{2}$, we have

$$\frac{\sin\varpi}{\varpi} = \pi\sum_{m=0}^{\infty}(m+\tfrac{1}{2})\,\frac{J_{m+\frac{1}{2}}(Z)}{\sqrt{Z}}\,.\,\frac{J_{m+\frac{1}{2}}(z)}{\sqrt{z}}\,P_m(\cos\phi). \tag{3}$$

This formula is due to Clebsch, *Journal für Math.* LXI. (1863), p. 227; it is also given by Heine, *Journal für Math.* LXIX. (1868), p. 133, and Neumann, *Leipziger Berichte*, 1886, pp. 75—82. The formula in which 2ν is a positive integer has been obtained by Hobson, *Proc. London Math. Soc.* XXV. (1894), pp. 60—61, from a consideration of solutions of Laplace's equation for space of $2\nu+2$ dimensions.

An extension of the expansion (2) has been given by Wendt, *Monatshefte für Math. und Phys.* XI. (1900), pp. 125—131; the effect of her generalisation is to express

$$\varpi^{-\nu-\rho}\sin^{2\rho}\phi\,J_{\nu+\rho}(\varpi)$$

as a series of Bessel functions in which the coefficients are somewhat complicated determinants.

11·41. *The modified form of Gegenbauer's addition theorem.*

The formula

$$\frac{J_{-\nu}(\varpi)}{\varpi^\nu} = 2^\nu\,\Gamma(\nu)\sum_{m=0}^{\infty}(-)^m(\nu+m)\,\frac{J_{-\nu-m}(Z)}{Z^\nu}\,\frac{J_{\nu+m}(z)}{z^\nu}\,C_m^\nu(\cos\phi) \tag{1}$$

may be established in the same manner as the Gegenbauer-Sonine formula of § 11·4. This formula does not seem to have been given previously explicitly, though it is used implicitly in obtaining some of the results given subsequently in this section.

Unlike the formulae of § 11·4, the formula is true only when $|z|$ is so small that both the inequalities $|ze^{\pm i\phi}| < |Z|$ are satisfied; but, in proving the formula, it is convenient first to suppose that the further inequalities

$$|2Zz\cos\phi| < |Z^2+z^2|, \quad |z| < |Z|$$

are satisfied.

We then use Lommel's expansion of § 5·22 (2) in the form

$$(\zeta+h)^{-\frac{1}{2}\nu} J_{-\nu}\{\sqrt{(\zeta+h)}\} = \sum_{p=0}^{\infty} \frac{(\frac{1}{2}h)^p}{p!} \zeta^{-\frac{1}{2}(\nu+p)} J_{-\nu-p}(\sqrt{\zeta}),$$

which is valid when $|h| < |\zeta|$.

It is then found by making slight alterations in the analysis of § 11·4 that

$$\frac{J_{-\nu}(\varpi)}{\varpi^\nu} = \sum_{p=0}^{\infty} \frac{(-Zz\cos\phi)^p}{p!} (Z^2+z^2)^{-\frac{1}{2}(\nu+p)} J_{-\nu-p}\{\sqrt{(Z^2+z^2)}\}$$

$$= \sum_{p=0}^{\infty}\sum_{q=0}^{\infty} \frac{(-)^p z^{p+2q}\cos^p\phi}{2^q p!\,q!} Z^{-\nu-q} J_{-\nu-p-q}(Z)$$

$$= \sum_{p=0}^{\infty}\sum_{q=0}^{\infty}\sum_{k=0}^{q} \frac{(-)^{p+1}(\nu+p+2k)\,\Gamma(-\nu-p-q-k)\, z^{p+2q}\cos^p\phi}{2^{2q} p!\,k!\,(q-k)!\,\Gamma(1-\nu-p-k)} \frac{J_{-\nu-p-2k}(Z)}{Z^\nu}$$

$$= \sum_{p=0}^{\infty}\sum_{k=0}^{\infty}\sum_{n=0}^{\infty} \frac{(-)^{p+1}(\nu+p+2k)\,\Gamma(-\nu-p-2k-n)\, z^{p+2k+2n}\cos^p\phi}{2^{2k+2n} p!\,k!\,n!\,\Gamma(1-\nu-p-k)} \frac{J_{-\nu-p-2k}(Z)}{Z^\nu}$$

$$= \sum_{p=0}^{\infty}\sum_{k=0}^{\infty} \frac{(-)^{p+k} 2^{\nu+p}(\nu+p+2k)\,\Gamma(\nu+p+k)\cos^p\phi}{p!\,k!} \frac{J_{-\nu-p-2k}(Z)}{Z^\nu} \frac{J_{\nu+p+2k}(z)}{z^\nu}$$

$$= \sum_{m=0}^{\infty}\sum_{k=0}^{\leq\frac{1}{2}m} \frac{(-)^{m-k} 2^{\nu+m-2k}(\nu+m)\,\Gamma(\nu+m-k)\cos^{m-2k}\phi}{(m-2k)!\,k!} \frac{J_{-\nu-m}(Z)}{Z^\nu} \frac{J_{\nu+m}(z)}{z^\nu}$$

$$= 2^\nu\,\Gamma(\nu) \sum_{m=0}^{\infty} (-)^m (\nu+m) \frac{J_{-\nu-m}(Z)}{Z^\nu} \frac{J_{\nu+m}(z)}{z^\nu} C_m^\nu(\cos\phi),$$

so the required result is established under the conditions

$$|2Zz\cos\phi| < |Z^2+z^2|, \quad |z| < |Z|.$$

Now the last expression is an analytic function of z when z lies inside the circle of convergence of the series*

$$\sum_{m=0}^{\infty} \frac{(\nu+m)\, Z^{-2\nu-m} z^m C_m^\nu(\cos\phi)}{\Gamma(1-\nu-m)\,\Gamma(1+\nu+m)},$$

and this circle is the circle of convergence of the series

$$\sum_{m=0}^{\infty} \left(\frac{z}{Z}\right)^m C_m^\nu(\cos\phi).$$

Hence the given series converges and represents an analytic function of z provided only that $|ze^{\pm i\phi}| < |Z|$; and, when this pair of inequalities is satisfied, $J_{-\nu}(\varpi)/\varpi^\nu$ is also an analytic function of z.

* Cf. § 5·22.

Hence, by the theory of analytic continuation, (1) is valid through the whole of the domain of values of z for which

$$|ze^{\pm i\phi}| < |Z|.$$

If in (1) we replace ν by $-\nu$ we find that

$$(2)\quad \frac{J_\nu(\varpi)}{\varpi^{-\nu}} = -2^{-\nu}\Gamma(-\nu)\sum_{m=0}^{\infty}(-)^m(\nu-m)Z^\nu J_{\nu-m}(Z)z^\nu J_{-\nu+m}(z)C_m^{-\nu}(\cos\phi).$$

Again, if we combine (1) with § 11·4 (2), we see that, for the domain of values of z now under consideration,

$$(3)\quad \frac{Y_\nu(\varpi)}{\varpi^\nu} = 2^\nu\Gamma(\nu)\sum_{m=0}^{\infty}(\nu+m)\frac{Y_{\nu+m}(Z)}{Z^\nu}\frac{J_{\nu+m}(z)}{z^\nu}C_m^\nu(\cos\phi),$$

and so, generally,

$$(4)\quad \frac{\mathscr{C}_\nu(\varpi)}{\varpi^\nu} = 2^\nu\Gamma(\nu)\sum_{m=0}^{\infty}(\nu+m)\frac{\mathscr{C}_{\nu+m}(Z)}{Z^\nu}\frac{J_{\nu+m}(z)}{z^\nu}C_m^\nu(\cos\phi).$$

If in (3) we make $\nu\to 0$ and use the formulae

$$C_0^0(\cos\phi)=1,\quad \lim_{\nu\to 0}\{\Gamma(\nu)(\nu+m)C_m^\nu(\cos\phi)\} = 2\cos m\phi,\quad (m\neq 0)$$

we find that

$$(5)\quad Y_0(\varpi) = \sum_{m=0}^{\infty}\epsilon_m Y_m(Z)J_m(z)\cos m\phi.$$

The formulae (1) and (2) have not been given previously; but (3) is due to Gegenbauer, and (5) was given by Neumann in his treatise (save that the functions Y_m were replaced by the functions $Y^{(m)}$). The formula (3) with ν equal to an integer has also been examined by Heine, *Handbuch der Kugelfunctionen*, I. (Berlin, 1878), pp. 463—464. Some developments of (4) are due to Ignatowsky, *Archiv der Math. und Phys.* (3) XVIII. (1911), pp. 322—327.

If we replace Z, z and ϖ by iZ, iz and $i\varpi$ in the formulae of § 11·4 and this section we find that

$$(6)\quad \frac{I_\nu(\varpi)}{\varpi^\nu} = 2^\nu\Gamma(\nu)\sum_{m=0}^{\infty}(-)^m(\nu+m)\frac{I_{\nu+m}(Z)}{Z^\nu}\frac{I_{\nu+m}(z)}{z^\nu}C_m^\nu(\cos\phi),$$

$$(7)\quad \frac{I_{-\nu}(\varpi)}{\varpi^\nu} = 2^\nu\Gamma(\nu)\sum_{m=0}^{\infty}(-)^m(\nu+m)\frac{I_{-\nu-m}(Z)}{Z^\nu}\frac{I_{\nu+m}(z)}{z^\nu}C_m^\nu(\cos\phi),$$

$$(8)\quad \frac{K_\nu(\varpi)}{\varpi^\nu} = 2^\nu\Gamma(\nu)\sum_{m=0}^{\infty}(\nu+m)\frac{K_{\nu+m}(Z)}{Z^\nu}\frac{I_{\nu+m}(z)}{z^\nu}C_m^\nu(\cos\phi).$$

Of these formulae, (8) is due to Macdonald, *Proc. London Math. Soc.* XXXII. (1900), pp. 156—157; while (6) and (7) were given by Neumann in the special case $\nu=\frac{1}{2}$.

The formulae of § 11·4 and of this section are of special physical importance in the case $\nu = \frac{1}{2}$. If we change the notation by writing ka, kr and θ for Z, z and ϕ we see that the formulae become

$$(9)\quad \frac{\sin k\sqrt{(r^2+a^2-2ar\cos\theta)}}{\sqrt{(r^2+a^2-2ar\cos\theta)}} = \pi \sum_{m=0}^{\infty} (m+\tfrac{1}{2}) \frac{J_{m+\frac{1}{2}}(ka)}{\sqrt{a}} \frac{J_{m+\frac{1}{2}}(kr)}{\sqrt{r}} P_m(\cos\theta),$$

$$(10)\quad \frac{\cos k\sqrt{(r^2+a^2-2ar\cos\theta)}}{\sqrt{(r^2+a^2-2ar\cos\theta)}} = \pi \sum_{m=0}^{\infty} (-)^m (m+\tfrac{1}{2}) \frac{J_{-m-\frac{1}{2}}(ka)}{\sqrt{a}} \frac{J_{m+\frac{1}{2}}(kr)}{\sqrt{r}} P_m(\cos\theta),$$

$$(11)\quad \frac{\exp\{-k\sqrt{(r^2+a^2-2ar\cos\theta)}\}}{\sqrt{(r^2+a^2-2ar\cos\theta)}} = \sum_{m=0}^{\infty} (2m+1) \frac{K_{m+\frac{1}{2}}(ka)}{\sqrt{a}} \frac{I_{m+\frac{1}{2}}(kr)}{\sqrt{r}} P_m(\cos\theta).$$

These formulae are of importance in problems in which pulsations emanate from a point on the axis of harmonics at distance a from the origin, in presence of a sphere whose centre is at the origin. Cf. Carslaw, *Math. Ann.* LXXV. (1914), p. 141 *et seq.*

The following special cases of (4) were pointed out by Gegenbauer, and are worth recording:

If $\phi = \pi$, we have

$$(12)\quad \frac{\mathscr{C}_\nu(Z+z)}{(Z+z)^\nu} = 2^\nu \Gamma(\nu) \sum_{m=0}^{\infty} (-)^m (\nu+m) \frac{\mathscr{C}_{\nu+m}(Z)}{Z^\nu} \frac{J_{\nu+m}(z)}{z^\nu} \cdot \frac{\Gamma(2\nu+m)}{m!\,\Gamma(2\nu)}.$$

If $\phi = \frac{1}{2}\pi$, we have

$$(13)\quad \frac{\mathscr{C}_\nu\{\sqrt{(Z^2+z^2)}\}}{(Z^2+z^2)^{\frac{1}{2}\nu}} = 2^\nu \sum_{m=0}^{\infty} (-)^m (\nu+2m) \frac{\mathscr{C}_{\nu+2m}(Z)}{Z^\nu} \frac{J_{\nu+2m}(z)}{z^\nu} \cdot \frac{\Gamma(\nu+m)}{m!}.$$

If $Z = z$, $\phi = 0$, and $\mathscr{C}_\nu$ is taken to be J_ν,

$$(14)\quad z^{2\nu} = \frac{2^{2\nu}\Gamma(\nu)\Gamma(\nu+1)}{\Gamma(2\nu)} \sum_{m=0}^{\infty} (\nu+m) \frac{\Gamma(2\nu+m)}{m!} J^2_{\nu+m}(z),$$

a formula already obtained (§ 5·5) by a different method; in this connexion the reader should consult Gegenbauer, *Wiener Sitzungsberichte*, LXXV. (2), (1877), p. 221.

More generally, taking $Z = z$, $\phi \neq 0$, $\mathscr{C}_\nu = J_\nu$, we have

$$(15)\quad \frac{J_\nu(2z\sin\frac{1}{2}\phi)}{(2z\sin\frac{1}{2}\phi)^\nu} = 2^\nu \Gamma(\nu) \sum_{m=0}^{\infty} (\nu+m) \left\{\frac{J_{\nu+m}(z)}{z^\nu}\right\}^2 C_m^\nu(\cos\phi).$$

Gegenbauer, *loc. cit.* gives also special cases of this formula, obtained by taking $\phi = \frac{1}{2}\pi$, $\phi = \pi$.

Again, it can be shewn that*, if $R(\nu) > -\frac{1}{2}$,

$$\int_0^\pi \sin^{2\nu}\phi\, C_m^\nu(\cos\phi)\, C_p^\nu(\cos\phi)\, d\phi \begin{cases} = 0 & (m \neq p) \\ = \dfrac{\pi\,\Gamma(2\nu+m)}{2^{2\nu-1}(\nu+m)\,.\,m!\,\{\Gamma(\nu)\}^2} & (m = p) \end{cases}$$

and so, provided that $R(\nu) > -\frac{1}{2}$,

$$(16) \qquad \int_0^\pi \frac{\mathscr{C}_\nu\{\sqrt{(Z^2+z^2-2Zz\cos\phi)}\}}{(Z^2+z^2-2Zz\cos\phi)^{\frac{1}{2}\nu}} \sin^{2\nu}\phi\, d\phi = 2^\nu\,\Gamma(\nu+\tfrac{1}{2})\,\Gamma(\tfrac{1}{2})\,\frac{\mathscr{C}_\nu(Z)}{Z^\nu}\,\frac{J_\nu(z)}{z^\nu}$$

and, more generally,

$$(17) \qquad \int_0^\pi \frac{\mathscr{C}_\nu\{\sqrt{(Z^2+z^2-2Zz\cos\phi)}\}}{(Z^2+z^2-2Zz\cos\phi)^{\frac{1}{2}\nu}} C_m^\nu(\cos\phi)\sin^{2\nu}\phi\, d\phi$$
$$= \frac{\pi\,\Gamma(2\nu+m)}{2^{\nu-1}\,.\,m!\,\Gamma(\nu)}\,\frac{\mathscr{C}_{\nu+m}(Z)}{Z^\nu}\,\frac{J_{\nu+m}(z)}{z^\nu}.$$

A simple proof of this formula†, in the special case in which $m=0$ and the cylinder functions are functions of the first kind, was given by Sonine, *Math. Ann.* XVI. (1880), p. 37. Another direct proof for functions of the first kind is due to Kluyver, *Proc. Section of Sci., K. Acad. van Wet. te Amsterdam*, XI. (1909), pp. 749—755. An indirect proof, depending on § 12·13 (1), is due to Gegenbauer, *Wiener Sitzungsberichte*, LXXXV. (2), (1882), pp. 491—502.

[NOTE. An interesting consequence of (4), which was noticed by Gegenbauer, *Wiener Sitzungsberichte*, LXXIV. (2), (1877), p. 127, is that, if $|ze^{\pm i\phi}| < |Z|$ throughout the contour of integration, then (cf. § 9·2)

$$(18) \qquad \frac{1}{2\pi i}\int^{(0+)} \frac{\mathscr{C}_\nu(\varpi)}{\varpi^\nu} A_{m,\nu}(z)\, dz = 2^\nu\,\Gamma(\nu)\,.\,(\nu+m)\,\frac{\mathscr{C}_{\nu+m}(Z)}{Z^\nu}\, C_m^\nu(\cos\phi).$$

Special cases of this formula, resembling the results of § 9·2, are obtainable by taking ϕ equal to 0 or π.]

11·42. *Gegenbauer's investigation of the addition theorem.*

The method used by Gegenbauer, *Wiener Sitzungsberichte*, LXX. (2), (1875), pp. 6—16, to obtain the addition theorem of § 11·4 is not quite so easy to justify as Sonine's transformation. It consists in proving that Ω is a solution of the partial differential equation

$$\frac{\partial^2\Omega}{\partial z^2} + \frac{2\nu+1}{z}\frac{\partial\Omega}{\partial z} + \frac{1}{z^2}\frac{\partial^2\Omega}{\partial\phi^2} + \frac{2\nu\cot\phi}{z^2}\frac{\partial\Omega}{\partial\phi} + \Omega = 0,$$

and assuming that Ω can be expanded in the form

$$\Omega = \sum_{m=0}^{\infty} B_m\,.\,C_m^\nu(\cos\phi),$$

where B_m is independent of ϕ, and $C_m^\nu(\cos\phi)$ is a polynomial of degree m in $\cos\phi$; it follows that

$$\left\{\frac{\partial^2}{\partial\phi^2} + 2\nu\cot\phi\,\frac{\partial}{\partial\phi}\right\} C_m^\nu(\cos\phi)$$

* Gegenbauer, *Wiener Sitzungsberichte*, LXX. (2), (1875), pp. 433—443, and Bateman, *Proc. London Math. Soc.* (2) IV. (1906), p. 472; cf. also Barnes, *Quarterly Journal*, XXXIX. (1908), p. 189; *Modern Analysis*, § 15·51 and *Proc. London. Math. Soc.* (2) XVII. (1919), pp. 241—246.

† Formula (16) has been given in the special case $\nu=0$ by Heaviside, *Electromagnetic Theory*, III. (London, 1912), p. 267, in a somewhat disguised form.

is a constant multiple of $C_m^\nu(\cos\phi)$, and so $C_m^\nu(\cos\phi)$ may be taken to be the coefficient of a^m in the expansion of $(1-2a\cos\phi+a^2)^{-\nu}$. And then B_m, *qua* function of z, satisfies the differential equation

$$\frac{\partial^2 B_m}{\partial z^2}+\frac{2\nu+1}{z}\frac{\partial B_m}{\partial z}+\left\{1-\frac{m(2\nu+m)}{z^2}\right\}B_m=0,$$

so that B_m is a multiple of $z^{-\nu}J_{\nu+m}(z)$, the other solution of this differential equation not being analytic near the origin.

From considerations of symmetry Gegenbauer inferred that B_m, *qua* function of Z, is a multiple of $Z^{-\nu}J_{\nu+m}(Z)$, so that

$$\Omega=\sum_{m=0}^{\infty} b_m\frac{J_{\nu+m}(Z)}{Z^\nu}\frac{J_{\nu+m}(z)}{z^\nu}C_m^\nu(\cos\phi),$$

where b_m is a function of ν and m only; and b_m is determined by comparing coefficients of $z^m Z^m\cos^m\phi$ in Ω and in the expression on the right.

A similar process was used by Gegenbauer to establish § 11·41 (3), but the analysis seems less convincing than in the case of functions of the first kind.

11·5. *The degenerate form of the addition theorem.*

The formula

$$\text{(1)}\qquad e^{iz\cos\phi}=\left(\frac{\pi}{2z}\right)^{\frac{1}{2}}\sum_{n=0}^{\infty}(2n+1)\,i^n J_{n+\frac{1}{2}}(z)\,P_n(\cos\phi)$$

was discovered by Bauer* as early as 1859; it was generalised by Gegenbauer†, who obtained the expansion

$$\text{(2)}\qquad e^{iz\cos\phi}=2^\nu\,\Gamma(\nu)\sum_{m=0}^{\infty}(\nu+m)\,i^m\frac{J_{\nu+m}(z)}{z^\nu}C_m^\nu(\cos\phi);$$

Bauer's result is obviously the special case of this expansion in which $\nu=\frac{1}{2}$. In the limit when $\nu\to 0$, the expansion becomes the fundamental expansion of § 2·1.

Gegenbauer's expansion is deducible from the expansion of § 11·41 (4) by multiplying by $Z^{\nu+\frac{1}{2}}$ and making $Z\to\infty$; it is then apparent from § 11·41 (9) and (10) that the physical interpretation of the expansion is that it gives the effect due to a train of plane waves coming from infinity on the axis of harmonics in a form suitable for the discussion of the disturbance produced by the introduction of a sphere with centre at the origin.

A simple analytical proof of the expansion consists in expanding $z^\nu e^{iz\cos\phi}$ in powers of z and substituting for each power the series of Bessel functions supplied by the formula of § 5·2; we thus find that

$$\begin{aligned}z^\nu e^{iz\cos\phi}&=\sum_{n=0}^{\infty}\frac{i^n\cos^n\phi}{n!}z^{\nu+n}\\&=\sum_{n=0}^{\infty}\frac{i^n\cos^n\phi}{n!}\sum_{k=0}^{\infty}\frac{2^{\nu+n}(\nu+n+2k)\,.\,\Gamma(\nu+n+k)}{k!}J_{\nu+n+2k}(z).\end{aligned}$$

* *Journal für Math.* LVI. (1859), pp. 104, 106.

† *Wiener Sitzungsberichte*, LXVIII. (2), (1874), pp. 355—367; LXXIV. (2), (1877), p. 128; and LXXV. (2), (1877), pp. 904—905.

If we rearrange the repeated series by writing $n = m - 2k$, we deduce that

$$z^\nu e^{iz\cos\phi} = \sum_{m=0}^{\infty} \sum_{k=0}^{\leqslant \frac{1}{2}m} \frac{i^{m-2k} \cos^{m-2k}\phi}{k!\,(m-2k)!} 2^{\nu+m-2k} (\nu+m)\,\Gamma(\nu+m-k)\, J_{\nu+m}(z)$$

$$= 2^\nu \Gamma(\nu) \sum_{m=0}^{\infty} (\nu+m)\, i^m J_{\nu+m}(z)\, C_m^\nu(\cos\phi),$$

and this is Gegenbauer's result.

Modified forms of this expansion, also due to Gegenbauer, are

(3) $$e^{z\cos\phi} = 2^\nu \Gamma(\nu) \sum_{m=0}^{\infty} (\nu+m) \frac{I_{\nu+m}(z)}{z^\nu} C_m^\nu(\cos\phi),$$

(4) $$e^{-z\cos\phi} = 2^\nu \Gamma(\nu) \sum_{m=0}^{\infty} (-)^m . (\nu+m) \frac{I_{\nu+m}(z)}{z^\nu} C_m^\nu(\cos\phi),$$

(5) $$\cos(z\cos\phi) = 2^\nu \Gamma(\nu) \sum_{m=0}^{\infty} (-)^m . (\nu+2m) \frac{J_{\nu+2m}(z)}{z^\nu} C^\nu_{2m}(\cos\phi),$$

(6) $$\sin(z\cos\phi) = 2^\nu \Gamma(\nu) \sum_{m=0}^{\infty} (-)^m . (\nu+2m+1) \frac{J_{\nu+2m+1}(z)}{z^\nu} C^\nu_{2m+1}(\cos\phi),$$

(7) $$1 = 2^\nu \sum_{m=0}^{\infty} (\nu+2m) \frac{J_{\nu+2m}(z)}{z^\nu} . \frac{\Gamma(\nu+m)}{m!},$$

(8) $$\int_0^\pi e^{iz\cos\phi} C_m^\nu(\cos\phi) \sin^{2\nu}\phi\, d\phi = \frac{2^\nu\, \Gamma(\nu+\frac{1}{2})\, \Gamma(\frac{1}{2})\, \Gamma(2\nu+m)}{m!\, \Gamma(2\nu)} i^m \frac{J_{\nu+m}(z)}{z^\nu}.$$

The last is a generalisation of Poisson's integral, which was obtained by a different method in § 3·32. It is valid only when $R(\nu) > -\frac{1}{2}$.

These formulae are to be found on pp. 363—365 of the first of Gegenbauer's memoirs to which reference has just been made.

Equation (1) was obtained by Hobson, *Proc. London Math. Soc.* xxv. (1894), p. 59, by a consideration of solutions of Laplace's equation in space of $2\nu+2$ dimensions, $2\nu+2$ being an integer.

A more general set of formulae may be derived from (2) by replacing $\cos\phi$ by $\cos\phi\cos\phi' + \sin\phi\sin\phi'\cos\psi$, multiplying by $\sin^{2\nu-1}\psi$, and integrating with respect to ψ. The integral*

$$\int_0^\pi C_m^\nu(\cos\phi\cos\phi' + \sin\phi\sin\phi'\cos\psi) \sin^{2\nu-1}\psi\, d\psi$$
$$= \frac{2^{2\nu-1} . m!\, \{\Gamma(\nu)\}^2}{\Gamma(2\nu+m)} C_m^\nu(\cos\phi)\, C_m^\nu(\cos\phi'),$$

which is valid when $R(\nu) > 0$, shews that

$$\int_0^\pi \exp[iz(\cos\phi\cos\phi' + \sin\phi\sin\phi'\cos\psi)] \sin^{2\nu-1}\psi\, d\psi$$
$$= 2^{3\nu-1} \{\Gamma(\nu)\}^3 \sum_{m=0}^{\infty} \frac{i^m . m!\, (\nu+m)}{\Gamma(2\nu+m)} \frac{J_{\nu+m}(z)}{z^\nu} C_m^\nu(\cos\phi)\, C_m^\nu(\cos\phi'),$$

* Cf. Gegenbauer, *Wiener Sitzungsberichte*, LXX. (2), (1874), p. 433; CII. (2*a*), (1893), p. 942.

and so

$$(9)\quad \frac{J_{\nu-\frac{1}{2}}(z\sin\phi\sin\phi')}{(z\sin\phi\sin\phi')^{\nu-\frac{1}{2}}}\exp[iz\cos\phi\cos\phi']$$

$$=\frac{2^{2\nu}\{\Gamma(\nu)\}^2}{\sqrt{(2\pi)}}\sum_{m=0}^{\infty}\frac{i^m\, m!\,(\nu+m)}{\Gamma(2\nu+m)}\frac{J_{\nu+m}(z)}{z^\nu}C_m^\nu(\cos\phi)\,C_m^\nu(\cos\phi').$$

The integral used in the proof converges only when $R(\nu)>0$, but the final result is true for all values of ν, by analytic continuation.

This result was given by Bauer, *Münchener Sitzungsberichte*, v. (1875), p. 263 in the case $\nu=\frac{1}{2}$; the general formula is due to Gegenbauer, *Monatshefte für Math. und Phys.* x. (1899), pp. 189—192; see also Bateman, *Messenger*, XXXIII. (1904), p. 182 and a letter from Gegenbauer to Kapteyn, *Proc. Section of Sci., K. Acad. van Wet. te Amsterdam*, IV. (1902) pp. 584—588.

Interesting special cases of the formula are obtained by taking ϕ' equal to ϕ or to $\frac{1}{2}\pi$; and, if we put ϕ' equal to $\frac{1}{2}\pi$, multiply by $e^{iZ\cos\phi}\sin^{2\nu}\phi$ and integrate, we find that

$$(10)\quad \frac{1}{z^{\nu-\frac{1}{2}}}\int_0^\pi J_{\nu-\frac{1}{2}}(z\sin\phi)\,e^{iZ\cos\phi}\sin^{2\nu}\phi\,d\phi$$

$$=2^\nu\sqrt{(2\pi)}\sum_{m=0}^{\infty}(-)^m\frac{\Gamma(\nu+m)\,.\,(\nu+2m)}{m!}\frac{J_{\nu+2m}(z)}{z^\nu}\frac{J_{\nu+2m}(Z)}{Z^\nu},$$

so that the expression on the left is a symmetric function of z and Z; this formula also was given by Bauer in the case $\nu=\frac{1}{2}$.

11·6. *Bateman's expansion.*

We shall now establish the general expansion

$$(1)\quad \tfrac{1}{2}z\,J_\mu(z\cos\phi\cos\Phi)\,J_\nu(z\sin\phi\sin\Phi)$$

$$=\cos^\mu\phi\cos^\mu\Phi\sin^\nu\phi\sin^\nu\Phi\sum_{n=0}^{\infty}(-)^n(\mu+\nu+2n+1)\,J_{\mu+\nu+2n+1}(z)$$

$$\times\frac{\Gamma(\mu+\nu+n+1)\,\Gamma(\nu+n+1)}{n!\,\Gamma(\mu+n+1)\,\{\Gamma(\nu+1)\}^2}\,.\,{}_2F_1(-n,\mu+\nu+n+1;\nu+1;\sin^2\phi)$$

$$\times{}_2F_1(-n,\mu+\nu+n+1;\nu+1;\sin^2\Phi),$$

which is valid for all values of μ and ν with the exception of negative integral values.

Some of the results of § 11·5 are special cases of this expansion, which was discovered by Bateman* from a consideration of the two types of normal solutions of the generalised equation of wave motions examined in § 4·84. We proceed to give a proof of the expansion by a direct transformation.

* *Messenger*, XXXIII. (1904), pp. 182—188; *Proc. London Math. Soc.* (2) III. (1905), pp. 111—123.

It is easy to deduce from the expansion (§ 5·21) of a Bessel function as a series of Bessel functions that

$$\tfrac{1}{2}z\,J_\mu(z\cos\phi\cos\Phi)\,J_\nu(z\sin\phi\sin\Phi)$$
$$=\sum_{m=0}^{\infty}\frac{(-)^m(\tfrac{1}{2}z)^{\mu+2m+1}(\cos\phi\cos\Phi)^{\mu+2m}}{m!\,\Gamma(\mu+m+1)}J_\nu(z\sin\phi\sin\Phi)$$
$$=\cos^\mu\phi\cos^\mu\Phi\sin^\nu\phi\sin^\nu\Phi\sum_{m=0}^{\infty}\left[\frac{(-)^m\cos^{2m}\phi\cos^{2m}\Phi}{m!\,\Gamma(\mu+m+1)}\right.$$
$$\times\sum_{n=0}^{\infty}\left\{(\mu+\nu+2m+2n+1)\frac{\Gamma(\mu+\nu+2m+n+1)}{n!\,\Gamma(\nu+1)}J_{\mu+\nu+2m+2n+1}(z)\right.$$
$$\left.\left.\times\,{}_2F_1(-n,\mu+\nu+2m+n+1;\nu+1;\sin^2\phi\sin^2\Phi)\right\}\right]$$
$$=\cos^\mu\phi\cos^\mu\Phi\sin^\nu\phi\sin^\nu\Phi\sum_{n=0}^{\infty}\left[(\mu+\nu+2n+1)J_{\mu+\nu+2n+1}(z)\right.$$
$$\times\sum_{m=0}^{n}\left\{\frac{(-)^m\cos^{2m}\phi\cos^{2m}\Phi\,.\,\Gamma(\mu+\nu+n+m+1)}{m!\,(n-m)!\,\Gamma(\nu+1)\,\Gamma(\mu+m+1)}\right.$$
$$\left.\left.\times\,{}_2F_1(m-n,\mu+\nu+m+n+1;\ \nu+1;\ \sin^2\phi\sin^2\Phi)\right\}\right]$$
$$=\cos^\mu\phi\cos^\mu\Phi\sin^\nu\phi\sin^\nu\Phi\sum_{n=0}^{\infty}\left[\frac{(\mu+\nu+2n+1)\,\Gamma(\mu+\nu+n+1)}{n!\,\Gamma(\mu+1)\,\Gamma(\nu+1)}J_{\mu+\nu+2n+1}(z)\right.$$
$$\left.\times\,\mathfrak{F}_4(-n,\mu+\nu+n+1;\ \mu+1,\nu+1;\ \cos^2\phi\cos^2\Phi,\sin^2\phi\sin^2\Phi)\right],$$

where $\mathfrak{F}_4$ denotes the fourth type of Appell's* hypergeometric functions of two variables, defined by the equation

$$\mathfrak{F}_4(\alpha,\beta;\gamma,\gamma';\xi,\eta)=\sum_{r=0}^{\infty}\sum_{s=0}^{\infty}\frac{(\alpha)_{r+s}(\beta)_{r+s}}{r!\,s!\,(\gamma)_r(\gamma')_s}\xi^r\eta^s.$$

We now have to transform† Appell's function into a product of hypergeometric functions in order to obtain equation (1); in effecting the transformation we assume that $R(\mu)>0$, though obviously this restriction may ultimately be removed by using the theory of analytic continuation.

The transformation is a consequence of the following analysis, in which series are rearranged, and a free use is made of Vandermonde's theorem:

$$\cos^{2\mu}\Phi\,.\,\mathfrak{F}_4(-n,\mu+\nu+n+1;\ \mu+1,\nu+1;\ \cos^2\phi\cos^2\Phi,\sin^2\phi\sin^2\Phi)$$
$$=\sum_{r=0}^{n}\sum_{s=0}^{n-r}\frac{(-n)_{r+s}(\mu+\nu+n+1)_{r+s}}{s!\,r!\,(\mu+1)_s(\nu+1)_r}\cos^{2s}\phi\cos^{2\mu+2s}\Phi\sin^{2r}\phi\sin^{2r}\Phi$$
$$=\sum_{r=0}^{n}\sum_{s=0}^{n-r}\frac{(-n)_{r+s}(\mu+\nu+n+1)_{r+s}}{r!\,(\nu+1)_r}\sum_{t=0}^{s}\frac{(-)^t\sin^{2r+2t}\phi}{t!\,(s-t)!}\sum_{u=0}^{\infty}\frac{(-)^u\sin^{2r+2u}\Phi}{u!\,(\mu+1)_{s-u}}$$

* *Comptes Rendus*, xc. (1880), pp. 296, 731.

† This transformation has not been previously noticed to exist except in the special case in which $\Phi=\phi$, see Appell, *Journal de Math.* (3) x. (1884), pp. 407—428; some associated researches are due to Tisserand, *Annales (Mémoires) de l'Observatoire* (Paris), XVIII. (1885), mém. C.

$$= \sum_{r=0}^{n} \sum_{s=0}^{n-r} \sum_{t=r}^{r+s} \sum_{u=r}^{\infty} \frac{(-n)_{r+s}(\mu+\nu+n+1)_{r+s}(-)^{t+u}\sin^{2t}\phi\sin^{2u}\Phi}{r!\,(\nu+1)_r\,(t-r)!\,(r+s-t)!\,(u-r)!\,(\mu+1)_{r+s-u}}$$

$$= \sum_{t=0}^{n} \sum_{u=0}^{\infty} \sum_{r=0}^{u} \sum_{s=t-r}^{n-r} \frac{(-n)_{r+s}(\mu+\nu+n+1)_{r+s}(-)^{t+u}\sin^{2t}\phi\sin^{2u}\Phi}{r!\,(\nu+1)_r\,(t-r)!\,(r+s-t)!\,(u-r)!\,(\mu+1)_{r+s-u}}$$

$$= \sum_{t=0}^{n} \sum_{u=0}^{\infty} \sum_{r=0}^{u} \frac{(-n)_t(\mu+\nu+n+1)_t(\nu+t+u+1)_{n-t}(-)^{n+u}\sin^{2t}\phi\sin^{2u}\Phi}{r!\,(\nu+1)_r\,(t-r)!\,(u-r)!\,(\mu+1)_{n-u}}$$

$$= \sum_{t=0}^{n} \sum_{u=0}^{\infty} \frac{(-n)_t\,(\mu+\nu+n+1)_t}{t!\,(\nu+1)_t}\sin^{2t}\phi\,.\,\frac{(\nu+u+1)_n\,(-\mu)_{u-n}}{u!}\sin^{2u}\Phi$$

$$= (-)^n\frac{(\nu+1)_n}{(\mu+1)_n}\,.\,{}_2F_1(-n,\,\mu+\nu+n+1\,;\;\nu+1\,;\;\sin^2\phi)$$

$$\times\,{}_2F_1(-\mu-n,\,\nu+n+1\,;\;\nu+1\,;\;\sin^2\Phi)$$

$$= (-)^n\frac{(\nu+1)_n}{(\mu+1)_n}\cos^{2\mu}\Phi\,.\,{}_2F_1(-n,\,\mu+\nu+n+1\,;\;\nu+1\,;\;\sin^2\phi)$$

$$\times\,{}_2F_1(-n,\,\mu+\nu+n+1\,;\;\nu+1\,;\;\sin^2\Phi).$$

Hence we at once obtain the result

$$\tfrac{1}{2}z\,J_\mu(z\cos\phi\cos\Phi)\,J_\nu(z\sin\phi\sin\Phi)$$

$$= \cos^\mu\phi\cos^\mu\Phi\sin^\nu\phi\sin^\nu\Phi\sum_{n=0}^{\infty}\frac{(\mu+\nu+2n+1)\,\Gamma(\mu+\nu+n+1)}{n!\,\Gamma(\mu+1)\,\Gamma(\nu+1)}J_{\mu+\nu+2n+1}(z)$$

$$\times(-)^n\frac{(\nu+1)_n}{(\mu+1)_n}\,.\,{}_2F_1(-n,\,\mu+\nu+n+1\,;\;\nu+1\,;\;\sin^2\phi)$$

$$\times\,{}_2F_1(-n,\,\mu+\nu+n+1\,;\;\nu+1\,;\;\sin^2\Phi),$$

from which Bateman's form of the expansion is evident.

CHAPTER XII

DEFINITE INTEGRALS

12·1. *Various types of definite integrals.*

In this chapter we shall investigate various definite integrals which contain either Bessel functions or functions of a similar character under the integral sign, and which have finite limits. The methods by which the integrals are evaluated are, for the most part, of an obvious character; the only novel feature is the fairly systematic use of a method by which a double integral is regarded as a surface integral over a portion of a sphere referred to one or other of two systems of polar coordinates. The most interesting integrals are those discussed in §§ 12·2—12·21, which are due to Kapteyn and Bateman. These integrals, for no very obvious reason, seem to be of a much more recondite character than the other integrals discussed in this chapter; their real significance has become apparent from the recent work by Hardy described in § 12·22. The numerous and important types of integrals, in which the upper limit of integration is infinite, are deferred to Chapter XIII.

The reader may here be reminded of the very important integral, due to Sonine and Gegenbauer, which has already been established in § 11·41, namely

$$\int_0^\pi \frac{\mathscr{C}_\nu\{\sqrt{(Z^2+z^2-2Zz\cos\phi)}\}}{(Z^2+z^2-2Zz\cos\phi)^{\frac{1}{2}\nu}} C_m^\nu(\cos\phi)\sin^{2\nu}\phi\,d\phi$$

$$= \frac{\pi\Gamma(2\nu+m)}{2^{\nu-1}.m!\,\Gamma(\nu)} \frac{\mathscr{C}_{\nu+m}(Z)}{Z^\nu} \frac{J_{\nu+m}(z)}{z^\nu}.$$

12·11. *Sonine's first finite integral.*

The formula

$$(1) \qquad J_{\mu+\nu+1}(z) = \frac{z^{\nu+1}}{2^\nu\Gamma(\nu+1)}\int_0^{\frac{1}{2}\pi} J_\mu(z\sin\theta)\sin^{\mu+1}\theta\cos^{2\nu+1}\theta\,d\theta,$$

which is valid when both $R(\mu)$ and $R(\nu)$ exceed -1, expresses any Bessel function in terms of an integral involving a Bessel function of lower order.

The formula was stated in a slightly different form by Sonine*, Rutgers† and Schafheitlin‡, and it may be proved quite simply by expanding the inte-

* *Math. Ann.* XVI. (1880), p. 36; see also Gegenbauer, *Wiener Sitzungsberichte*, LXXXVIII. (2), (1884), p. 979.

† *Nieuw Archief voor Wiskunde*, (2) VI. (1905), p. 370.

‡ *Die Theorie der Bessel'schen Funktionen* (Leipzig, 1908), p. 31. Schafheitlin seems to have been unaware of previous researches on what he describes as a new integral.

grand in powers of z and integrating term-by-term, thus

$$\frac{z^{\nu+1}}{2^\nu \Gamma(\nu+1)} \int_0^{\frac{1}{2}\pi} J_\mu(z \sin\theta) \sin^{\mu+1}\theta \cos^{2\nu+1}\theta\, d\theta$$

$$= \sum_{m=0}^{\infty} \frac{(-)^m z^{\mu+\nu+2m+1}}{2^{\mu+\nu+2m}\, m!\, \Gamma(\mu+m+1)\, \Gamma(\nu+1)} \int_0^{\frac{1}{2}\pi} \sin^{2\mu+2m+1}\theta \cos^{2\nu+1}\theta\, d\theta$$

$$= \sum_{m=0}^{\infty} \frac{(-)^m (\frac{1}{2}z)^{\mu+\nu+2m+1}}{m!\, \Gamma(\mu+\nu+m+2)},$$

and the truth of the formula is obvious.

It will be observed that the effect of the factor $\sin^{\mu+1}\theta$ in the integrand is to eliminate the factors $\Gamma(\mu+m+1)$ in the denominators. If we had taken $\sin^{1-\mu}\theta$ as the factor, we should have removed the factors $m!$. Hence, when $R(\nu) > -1$ and μ is unrestricted, we have

(2) $$\int_0^{\frac{1}{2}\pi} J_\mu(z\sin\theta) \sin^{1-\mu}\theta \cos^{2\nu+1}\theta\, d\theta = \frac{s_{\mu+\nu,\, \nu-\mu+1}(z)}{2^{\mu-1} z^{\nu+1} \Gamma(\mu)}.$$

In particular, by taking $\nu = -\frac{1}{2}$, we have

(3) $$\left(\frac{2z}{\pi}\right)^{\frac{1}{2}} \int_0^{\frac{1}{2}\pi} J_\mu(z\sin\theta) \sin^{1-\mu}\theta\, d\theta = \mathbf{H}_{\mu-\frac{1}{2}}(z).$$

A formula* which is easily obtained from (1) is

(4) $$\int_0^{\frac{1}{2}\pi} J_\mu(z\sin\theta)\, I_\nu(z\cos\theta) \tan^{\mu+1}\theta\, d\theta = \frac{\Gamma(\frac{1}{2}\nu - \frac{1}{2}\mu)(\frac{1}{2}z)^\mu}{\Gamma(\frac{1}{2}\nu + \frac{1}{2}\mu + 1)} J_\nu(z),$$

when $R(\nu) > R(\mu) > -1$. This may be proved by expanding $I_\nu(z\cos\theta)$ and integrating term-by-term, and finally making use of Lommel's expansion given in § 5·21.

The functional equation, obtained from (1) by substituting functions to be determined, F_μ and $F_{\mu+\nu+1}$, in place of the Bessel functions, has been examined by Sonine, *Math. Ann.* LIX. (1904), pp. 529—552.

Some special cases of the formulae of this section have been given by Beltrami, *Istituto Lombardo Rendiconti*, (2) XIII. (1880), p. 331, and Rayleigh, *Phil. Mag.* (5) XII. (1881), p. 92. [*Scientific Papers*, I. (1899), p. 528.]

It will be obvious to the reader that Poisson's integral is the special case of (1) obtained by taking $\mu = -\frac{1}{2}$.

For some developments of the formulae of this section, the reader should consult two papers by Rutgers, *Nieuw Archief voor Wiskunde*, (2) VI. (1905), pp. 368—373; (2) VII. (1907), pp. 88—90.

12·12. *The geometrical proof of Sonine's first integral.*

An instructive proof of the formula of the preceding section depends on the device (explained in § 3·33) of integrating over a portion of the surface of a unit sphere with various axes of polar coordinates.

If (l, m, n) are the direction cosines of the line joining the centre of the

* Due to Rutgers, *Nieuw Archief voor Wiskunde*, (2) VII. (1907), p. 175.

sphere to an element of surface $d\omega$ whose longitude and co-latitude are ϕ and θ, it is evident from an application of Poisson's integral that

$$\begin{aligned}\Gamma(\mu+\tfrac{1}{2})\,\Gamma(\tfrac{1}{2})\,(\tfrac{1}{2}z)^{\nu+1}\int_0^{\frac{1}{2}\pi} J_\mu(z\sin\theta)\sin^{\mu+1}\theta\cos^{2\nu+1}\theta\,d\theta\\
&=(\tfrac{1}{2}z)^{\mu+\nu+1}\int_0^{\frac{1}{2}\pi}\int_0^{\pi} e^{iz\sin\theta\cos\phi}\sin^{2\mu+1}\theta\cos^{2\nu+1}\theta\sin^{2\mu}\phi\,d\phi\,d\theta\\
&=(\tfrac{1}{2}z)^{\mu+\nu+1}\iint_{m\geqslant 0,\,n\geqslant 0} e^{izl}\,m^{2\mu}\,n^{2\nu+1}\,d\omega\\
&=(\tfrac{1}{2}z)^{\mu+\nu+1}\iint_{l\geqslant 0,\,m\geqslant 0} e^{izn}\,l^{2\mu}\,m^{2\nu+1}\,d\omega\\
&=(\tfrac{1}{2}z)^{\mu+\nu+1}\int_0^{\pi}\int_0^{\frac{1}{2}\pi} e^{iz\cos\theta}\sin^{2\mu+2\nu+2}\theta\cos^{2\mu}\phi\sin^{2\nu+1}\phi\,d\phi\,d\theta\\
&=\frac{\Gamma(\mu+\frac{1}{2})\,\Gamma(\nu+1)\,(\frac{1}{2}z)^{\mu+\nu+1}}{2\Gamma(\mu+\nu+\frac{3}{2})}\int_0^{\pi} e^{iz\cos\theta}\sin^{2\mu+2\nu+2}\theta\,d\theta\\
&=\tfrac{1}{2}\Gamma(\mu+\tfrac{1}{2})\,\Gamma(\nu+1)\,\Gamma(\tfrac{1}{2})\,J_{\mu+\nu+1}(z),\end{aligned}$$

and the truth of Sonine's formula is obvious.

An integral involving two Bessel functions which can be evaluated by the same device* is

$$\int_0^{\frac{1}{2}\pi} J_\nu(z\sin^2\theta)\,J_\nu(z\cos^2\theta)\sin^{2\nu+1}\theta\cos^{2\nu+1}\theta\,d\theta,$$

in which, to secure convergence, $R(\nu)>-\frac{1}{2}$.

If we write

$$\varpi^2\equiv\sin^4\theta+\cos^4\theta-2\sin^2\theta\cos^2\theta\cos\phi=1-\sin^2 2\theta\cos^2\tfrac{1}{2}\phi,$$

and use § 11·41 (16), we see that the integral is equal to

$$\begin{aligned}\frac{(\frac{1}{2}z)^\nu}{\Gamma(\nu+\frac{1}{2})\,\Gamma(\frac{1}{2})}\int_0^{\frac{1}{2}\pi}\int_0^{\pi}\frac{J_\nu(z\varpi)}{\varpi^\nu}\sin^{4\nu+1}\theta\cos^{4\nu+1}\theta\sin^{2\nu}\phi\,d\phi\,d\theta\\
&=\frac{(\frac{1}{2}z)^\nu}{2^{4\nu+1}\,\Gamma(\nu+\frac{1}{2})\,\Gamma(\frac{1}{2})}\int_0^{\pi}\int_0^{\frac{1}{2}\pi}\frac{J_\nu\{z\sqrt{(1-\sin^2\theta\cos^2\phi)}\}}{(1-\sin^2\theta\cos^2\phi)^{\frac{1}{2}\nu}}\sin^{4\nu+1}\theta\sin^{2\nu}2\phi\,d\phi\,d\theta\\
&=\frac{(\frac{1}{2}z)^\nu}{2^{2\nu+1}\,\Gamma(\nu+\frac{1}{2})\,\Gamma(\frac{1}{2})}\iint_{l\geqslant 0,\,m\geqslant 0}\frac{J_\nu\{z\sqrt{(1-l^2)}\}}{(1-l^2)^{\frac{1}{2}\nu}}\,l^{2\nu}\,m^{2\nu}\,d\omega\\
&=\frac{(\frac{1}{2}z)^\nu}{2^{2\nu+1}\,\Gamma(\nu+\frac{1}{2})\,\Gamma(\frac{1}{2})}\iint_{l\geqslant 0,\,n\geqslant 0}\frac{J_\nu\{z\sqrt{(1-n^2)}\}}{(1-n^2)^{\frac{1}{2}\nu}}\,l^{2\nu}\,n^{2\nu}\,d\omega\\
&=\frac{(\frac{1}{2}z)^\nu}{2^{2\nu+1}\,\Gamma(\nu+\frac{1}{2})\,\Gamma(\frac{1}{2})}\int_{-\frac{1}{2}\pi}^{\frac{1}{2}\pi}\int_0^{\frac{1}{2}\pi} J_\nu(z\sin\theta)\sin^{\nu+1}\theta\cos^{2\nu}\phi\cos^{2\nu}\theta\,d\theta\,d\phi\\
&=\frac{(\frac{1}{2}z)^\nu}{2^{2\nu+1}\,\Gamma(\nu+1)}\int_0^{\frac{1}{2}\pi} J_\nu(z\sin\theta)\sin^{\nu+1}\theta\cos^{2\nu}\theta\,d\theta,\end{aligned}$$

so that finally, by § 12·11 (1),

$$(1)\qquad \int_0^{\frac{1}{2}\pi} J_\nu(z\sin^2\theta)\,J_\nu(z\cos^2\theta)\sin^{2\nu+1}\theta\cos^{2\nu+1}\theta\,d\theta=\frac{\Gamma(\nu+\frac{1}{2})\,J_{2\nu+\frac{1}{2}}(z)}{2^{2\nu+\frac{3}{2}}\,\Gamma(\nu+1)\,z^{\frac{1}{2}}}.$$

* This integral has been evaluated by a different method by Rutgers, *Nieuw Archief voor Wiskunde*, (2) VII. (1907), p. 400; cf. also § 12·22.

Some integrals which resemble this, but which are much more difficult to evaluate, have been the subject of researches by Bateman, Kapteyn and Rutgers; see § 12·2.

As a simple example of an integral which may be evaluated by the same device, the reader may prove that, when $R(\nu) > -\frac{1}{2}$,

(2) $$\int_0^x (x^2-t^2)^{\frac{1}{2}\nu} \cos t \,.\, I_\nu\{\sqrt{(x^2-t^2)}\}\, dt = \frac{\Gamma(\frac{1}{2})\, x^{2\nu+1}}{2^{\nu+1}\,\Gamma(\nu+\frac{3}{2})},$$

by writing the integral on the left in the form

$$\frac{x^{2\nu+1}}{2^{\nu+1}\,\Gamma(\nu+\frac{1}{2})\,\Gamma(\frac{1}{2})} \int_0^\pi \int_0^\pi e^{ix\cos\theta + x\sin\theta\cos\phi} \sin^{2\nu+1}\theta \sin^{2\nu}\phi \, d\theta\, d\phi.$$

This formula was given (with $\nu=0$) by Bôcher, *Annals of Math.* VIII. (1894), p. 136.

12·13. *Sonine's second finite integral.*

The formula

(1) $$\int_0^{\frac{1}{2}\pi} J_\mu(z\sin\theta)\, J_\nu(Z\cos\theta) \sin^{\mu+1}\theta \cos^{\nu+1}\theta\, d\theta = \frac{z^\mu Z^\nu J_{\mu+\nu+1}\{\sqrt{(Z^2+z^2)}\}}{(Z^2+z^2)^{\frac{1}{2}(\mu+\nu+1)}},$$

which is valid when both $R(\mu)$ and $R(\nu)$ exceed -1, is also due to Sonine*; and, in fact, he obtained the formula of § 12·11 from it by dividing both sides of the equation by Z^ν and then making $Z \to 0$.

A simple method of proving the formula is to expand the integral in powers of z and Z and to verify that the terms of degree $\mu+\nu+2m$ on the left combine to form

$$\frac{(-)^r z^\mu Z^\nu (Z^2+z^2)^m}{m!\,\Gamma(\mu+\nu+m+2)}.$$

The proof by this method is left to the reader.

We proceed to establish Sonine's formula by integrating over portions of the surface of a unit sphere. Under the hypothesis that $R(\mu)$ and $R(\nu)$ exceed $-\frac{1}{2}$, we see that, with the notation of § 12·12, we have

$$\frac{\pi\Gamma(\mu+\frac{1}{2})\,\Gamma(\nu+\frac{1}{2})}{(\frac{1}{2}z)^\mu(\frac{1}{2}Z)^\nu} \int_0^{\frac{1}{2}\pi} J_\mu(z\sin\theta)\, J_\nu(Z\cos\theta)\sin^{\mu+1}\theta\cos^{\nu+1}\theta\, d\theta$$

$$= \int_0^{\frac{1}{2}\pi}\int_0^\pi\int_0^\pi e^{iz\sin\theta\cos\phi + iZ\cos\theta\cos\psi} \sin^{2\mu+1}\theta\cos^{2\nu+1}\theta \sin^{2\mu}\phi \sin^{2\nu}\psi\, d\phi\, d\psi\, d\theta$$

$$= \int_0^\pi \iint_{m\geqslant 0,\, n\geqslant 0} e^{izl+iZn\cos\psi}\, m^{2\mu}\, n^{2\nu+1} \sin^{2\nu}\psi\, d\omega\, d\psi$$

$$= \int_0^\pi \iint_{n\geqslant 0,\, l\geqslant 0} e^{izm+iZl\cos\psi}\, n^{2\mu}\, l^{2\nu+1} \sin^{2\nu}\psi\, d\omega\, d\psi$$

$$= \int_0^\pi \int_0^{\frac{1}{2}\pi}\int_{-\frac{1}{2}\pi}^{\frac{1}{2}\pi} e^{i\sin\theta(z\cos\phi + Z\sin\phi\cos\psi)} \cos^{2\mu}\theta\sin^{2\nu+1}\phi\sin^{2\nu+2}\theta\sin^{2\nu}\psi\, d\phi\, d\theta\, d\psi$$

$$= \int_0^{\frac{1}{2}\pi}\iint_{m\geqslant 0} e^{i\sin\theta\,(zn+Zl)}\, m^{2\nu} \cos^{2\mu}\theta \sin^{2\nu+2}\theta\, d\omega\, d\theta$$

$$= \int_0^{\frac{1}{2}\pi}\iint_{n\geqslant 0} e^{i\sin\theta\,(zl+Zm)}\, n^{2\nu} \cos^{2\mu}\theta \sin^{2\nu+2}\theta\, d\omega\, d\theta$$

$$= \int_0^{\frac{1}{2}\pi}\int_0^{\frac{1}{2}\pi}\int_0^{2\pi} e^{i\sin\theta\sin\phi\,(z\cos\psi + Z\sin\psi)} \cos^{2\nu}\phi \sin\phi \cos^{2\mu}\theta\sin^{2\nu+2}\theta\, d\psi\, d\phi\, d\theta.$$

* *Math. Ann.* XVI. (1880), pp. 35—36.

Now the exponential function involved here is a periodic analytic function of ψ with period 2π, and so, by Cauchy's theorem, the limits of integration with respect to ψ may be taken to be α and $2\pi+\alpha$, where α is defined by the equations

$$\varpi \cos \alpha = z, \quad \varpi \sin \alpha = Z,$$

and $\varpi = \sqrt{(Z^2+z^2)}$. If we adopt these limits of integration, and then write $\psi+\alpha$ for ψ, the triple integral becomes

$$\int_0^{\frac{1}{2}\pi}\int_0^{\frac{1}{2}\pi}\int_0^{2\pi} e^{i\varpi \sin\theta \sin\phi \cos\psi} \cos^{2\nu}\phi \sin\phi \cos^{2\mu}\theta \sin^{2\nu+2}\theta \, d\psi \, d\phi \, d\theta,$$

and this integral may also be obtained from its preceding form by replacing z by ϖ and Z by zero. On retracing the steps of the analysis with these substitutions we reduce the triple integral to

$$\int_0^{\frac{1}{2}\pi}\int_0^{\pi}\int_0^{\pi} e^{i\varpi \sin\theta\cos\phi} \sin^{2\mu+1}\theta \cos^{2\nu+1}\theta \sin^{2\mu}\phi \sin^{2\nu}\psi \, d\phi \, d\psi \, d\theta,$$

$$= \frac{\Gamma(\nu+\frac{1}{2})\,\Gamma(\frac{1}{2})}{\Gamma(\nu+1)} \iint_{m\geqslant 0,\, n\geqslant 0} e^{i\varpi l} m^{2\mu} n^{2\nu+1} \, d\omega$$

$$= \frac{\Gamma(\nu+\frac{1}{2})\,\Gamma(\frac{1}{2})}{\Gamma(\nu+1)} \iint_{l\geqslant 0,\, m\geqslant 0} e^{i\varpi n} l^{2\mu} m^{2\nu+1} \, d\omega$$

$$= \frac{\Gamma(\nu+\frac{1}{2})\,\Gamma(\frac{1}{2})}{\Gamma(\nu+1)} \int_0^{\pi}\int_0^{\frac{1}{2}\pi} e^{i\varpi\cos\theta} \sin^{2\mu+2\nu+2}\theta \cos^{2\mu}\phi \sin^{2\nu+1}\phi \, d\phi \, d\theta$$

$$= \frac{\Gamma(\mu+\frac{1}{2})\,\Gamma(\nu+\frac{1}{2})\,\Gamma(\frac{1}{2})}{2\Gamma(\mu+\nu+\frac{3}{2})} \int_0^{\pi} e^{i\varpi\cos\theta} \sin^{2\mu+2\nu+2}\theta \, d\theta$$

$$= \tfrac{1}{2}\pi\,\Gamma(\mu+\tfrac{1}{2})\,\Gamma(\nu+\tfrac{1}{2}) \frac{J_{\mu+\nu+1}(\varpi)}{(\frac{1}{2}\varpi)^{\mu+\nu+1}},$$

and we obtain Sonine's formula by a comparison of the initial and final expressions.

Sonine's own proof of this formula was based on the use of infinite discontinuous integrals, and the process of making it rigorous would be long and tedious.

The formula may be extended to the domains in which $-\frac{1}{2} \geqslant R(\mu) > -1$, and $-\frac{1}{2} \geqslant R(\nu) > -1$, by analytic continuation.

In Sonine's formula, replace Z by $\sqrt{(Z^2+\zeta^2-2Z\zeta\cos\phi)}$, multiply by $\sin^{2\nu}\phi/(Z^2+\zeta^2-2Z\zeta\cos\phi)^{\frac{1}{2}\nu}$, and integrate. It follows from § 11·41 (16) that

$$(2) \quad \int_0^{\frac{1}{2}\pi} J_\mu(z\sin\theta)\, J_\nu(Z\cos\theta)\, J_\nu(\zeta\cos\theta) \sin^{\mu+1}\theta \cos\theta \, d\theta$$

$$= \frac{z^\mu Z^\nu \zeta^\nu}{2^\nu\,\Gamma(\nu+\frac{1}{2})\,\Gamma(\frac{1}{2})} \int_0^{\pi} \frac{J_{\mu+\nu+1}\{\sqrt{(z^2+Z^2+\zeta^2-2Z\zeta\cos\phi)}\}}{(z^2+Z^2+\zeta^2-2Z\zeta\cos\phi)^{\frac{1}{2}(\mu+\nu+1)}} \sin^{2\nu}\phi \, d\phi,$$

provided that

$$R(\mu) > -1, \quad R(\nu) > -\tfrac{1}{2}.$$

This result is also due to Sonine, *ibid.* p. 45. In connexion with the formulae of this section the reader should consult Macdonald's memoir, *Proc. London Math. Soc.* XXXV (1903), pp. 442, 443

12·14. *Gegenbauer's finite integral.*

An integral which somewhat resembles the first of Sonine's integrals, namely

$$\int_0^\pi \begin{matrix}\cos\\ \sin\end{matrix}(z\cos\theta\cos\psi)\, J_{\nu-\frac{1}{2}}(z\sin\theta\sin\psi)\, C_r^\nu(\cos\theta)\sin^{\nu+\frac{1}{2}}\theta\, d\theta,$$

has been evaluated by Gegenbauer*; we shall adopt our normal procedure of using the method of integration over a unit sphere.

It is thus seen that

$$\int_0^\pi e^{iz\cos\theta\cos\psi} J_{\nu-\frac{1}{2}}(z\sin\theta\sin\psi)\, C_r^\nu(\cos\theta)\sin^{\nu+\frac{1}{2}}\theta\, d\theta$$

$$= \frac{(\frac{1}{2}z\sin\psi)^{\nu-\frac{1}{2}}}{\Gamma(\nu)\Gamma(\frac{1}{2})}\int_0^\pi\int_0^\pi e^{iz(\cos\theta\cos\psi+\sin\theta\sin\psi\cos\phi)} C_r^\nu(\cos\theta)\sin^{2\nu}\theta\sin^{2\nu-1}\phi\, d\phi\, d\theta$$

$$= \frac{(\frac{1}{2}z\sin\psi)^{\nu-\frac{1}{2}}}{\Gamma(\nu)\Gamma(\frac{1}{2})}\iint_{m\geqslant 0} e^{iz(n\cos\psi+l\sin\psi)} C_r^\nu(n)\, m^{2\nu-1}\, d\omega$$

$$= \frac{(\frac{1}{2}z\sin\psi)^{\nu-\frac{1}{2}}}{\Gamma(\nu)\Gamma(\frac{1}{2})}\iint_{n\geqslant 0} e^{iz(l\cos\psi+m\sin\psi)} C_r^\nu(l)\, n^{2\nu-1}\, d\omega$$

$$= \frac{(\frac{1}{2}z\sin\psi)^{\nu-\frac{1}{2}}}{\Gamma(\nu)\Gamma(\frac{1}{2})}\int_0^{\frac{1}{2}\pi}\int_0^{2\pi} e^{iz\sin\theta\cos(\phi-\psi)} C_r^\nu(\sin\theta\cos\phi)\cos^{2\nu-1}\theta\sin\theta\, d\phi\, d\theta$$

$$= \frac{(\frac{1}{2}z\sin\psi)^{\nu-\frac{1}{2}}}{\Gamma(\nu)\Gamma(\frac{1}{2})}\int_0^{\frac{1}{2}\pi}\int_0^{2\pi} e^{iz\sin\theta\cos\phi} C_r^\nu\{\sin\theta\cos(\phi+\psi)\}\cos^{2\nu-1}\theta\sin\theta\, d\phi\, d\theta,$$

since the penultimate integrand is a periodic analytic function of ϕ with period 2π.

If we retrace the steps of the analysis, using the last integral instead of its immediate predecessor, we find that the original integral is equal to

$$\frac{(\frac{1}{2}z\sin\psi)^{\nu-\frac{1}{2}}}{\Gamma(\nu)\Gamma(\frac{1}{2})}\iint_{n\geqslant 0} e^{izl} C_r^\nu(l\cos\psi - m\sin\psi)\, n^{2\nu-1}\, d\omega$$

$$= \frac{(\frac{1}{2}z\sin\psi)^{\nu-\frac{1}{2}}}{\Gamma(\nu)\Gamma(\frac{1}{2})}\iint_{m\geqslant 0} e^{izn} C_r^\nu(n\cos\psi - l\sin\psi)\, m^{2\nu-1}\, d\omega$$

$$= \frac{(\frac{1}{2}z\sin\psi)^{\nu-\frac{1}{2}}}{\Gamma(\nu)\Gamma(\frac{1}{2})}\int_0^\pi\int_0^\pi e^{iz\cos\theta} C_r^\nu(\cos\psi\cos\theta - \sin\psi\sin\theta\cos\phi)\sin^{2\nu}\theta\sin^{2\nu-1}\phi\, d\phi\, d\theta.$$

Now, by the addition theorem† for Gegenbauer's function,

$$C_r^\nu(\cos\psi\cos\theta - \sin\psi\sin\theta\cos\phi)$$

$$= \frac{\Gamma(2\nu-1)}{\{\Gamma(\nu)\}^2}\sum_{p=0}^{r}\frac{2^{2p}.(r-p)!\,\{\Gamma(\nu+p)\}^2}{\Gamma(2\nu+p+r)}(2\nu+2p-1)\sin^p\theta\sin^p\psi$$
$$\times C_{r-p}^{\nu+p}(\cos\theta)\, C_{r-p}^{\nu+p}(\cos\psi)\, C_p^{\nu-\frac{1}{2}}(\cos\phi).$$

* *Wiener Sitzungsberichte*, LXXV: (2), (1877), p. 221 and LXXXV. (2), (1882), pp. 491—502.

† This was proved by Gegenbauer, *Wiener Sitzungsberichte*, LXX. (2), (1874), p. 433; CII. (2*a*), (1893), p. 942.

When this is multiplied by $\sin^{2\nu-1}\phi$ and integrated, all the terms of the integral of the sum vanish except the first which is

$$\frac{r!\,\Gamma(2\nu)}{\Gamma(2\nu+r)}\,C_r^\nu(\cos\theta)\,C_r^\nu(\cos\psi)\int_0^\pi \sin^{2\nu-1}\phi\,d\phi.$$

We thus find that

$$\int_0^\pi e^{iz\cos\theta\cos\psi}\,J_{\nu-\frac{1}{2}}(z\sin\theta\sin\psi)\,C_r^\nu(\cos\theta)\sin^{\nu+\frac{1}{2}}\theta\,d\theta$$
$$=\frac{r!\,\Gamma(2\nu)\,.\,(\tfrac{1}{2}z\sin\psi)^{\nu-\frac{1}{2}}}{\Gamma(\nu+\frac{1}{2})\,\Gamma(2\nu+r)}\,C_r^\nu(\cos\psi)\int_0^\pi e^{iz\cos\theta}\,C_r^\nu(\cos\theta)\sin^{2\nu}\theta\,d\theta,$$

and hence, by § 3·32,

(1) $$\int_0^\pi e^{iz\cos\theta\cos\psi}\,J_{\nu-\frac{1}{2}}(z\sin\theta\sin\psi)\,C_r^\nu(\cos\theta)\sin^{\nu+\frac{1}{2}}\theta\,d\theta$$
$$=\left(\frac{2\pi}{z}\right)^{\frac{1}{2}} i^r\sin^{\nu-\frac{1}{2}}\psi\,C_r^\nu(\cos\psi)\,J_{\nu+r}(z).$$

If we equate real and imaginary parts, we obtain Gegenbauer's formulae

(2) $$\int_0^\pi \cos(z\cos\theta\cos\psi)\,J_{\nu-\frac{1}{2}}(z\sin\theta\sin\psi)\,C_r^\nu(\cos\theta)\sin^{\nu+\frac{1}{2}}\theta\,d\theta$$
$$=\begin{cases}(-)^{\frac{1}{2}r}\left(\dfrac{2\pi}{z}\right)^{\frac{1}{2}}\sin^{\nu-\frac{1}{2}}\psi\,C_r^\nu(\cos\psi)\,J_{\nu+r}(z), & (r \text{ even})\\ 0 & (r \text{ odd})\end{cases}$$

and

(3) $$\int_0^\pi \sin(z\cos\theta\cos\psi)\,J_{\nu-\frac{1}{2}}(z\sin\theta\sin\psi)\,C_r^\nu(\cos\theta)\sin^{\nu+\frac{1}{2}}\theta\,d\theta$$
$$=\begin{cases}0, & (r \text{ even})\\ (-)^{\frac{1}{2}(r-1)}\left(\dfrac{2\pi}{z}\right)^{\frac{1}{2}}\sin^{\nu-\frac{1}{2}}\psi\,C_r^\nu(\cos\psi)\,J_{\nu+r}(z). & (r \text{ odd})\end{cases}$$

12·2. *Integrals deduced from Bateman's expansion.*

In Bateman's expansion of § 11·6, write $\Phi=\phi$; and then, noting Jacobi's formula*

$$2\int_0^{\frac{1}{2}\pi}\{{}_2F_1(-n,\mu+\nu+n+1;\nu+1;\sin^2\phi)\}^2\cos^{2\mu+1}\phi\sin^{2\nu+1}\phi\,d\phi$$
$$=\frac{n!\,\Gamma(\mu+n+1)\,\{\Gamma(\nu+1)\}^2}{(\mu+\nu+2n+1)\,\Gamma(\mu+\nu+n+1)\,\Gamma(\nu+n+1)},$$

we deduce that, when $R(\mu)$ and $R(\nu)$ both exceed -1,

(1) $$z\int_0^{\frac{1}{2}\pi} J_\mu(z\cos^2\phi)\,J_\nu(z\sin^2\phi)\sin\phi\cos\phi\,d\phi=\sum_{n=0}^\infty(-)^n J_{\mu+\nu+2n+1}(z),$$

* *Journal für Math.* LVI. (1859), pp. 149—175 [*Werke*, VI. (1891), pp. 184—202].

that is to say

$$\int_0^z J_\mu(t) J_\nu(z-t)\, dt = 2 \sum_{n=0}^{\infty} (-)^n J_{\mu+\nu+2n+1}(z). \tag{2}$$

An important deduction from this result is that, when $R(\mu) > 0$ and $R(\nu) > -1$,

$$2\mu \int_0^z J_\mu(t) J_\nu(z-t) \frac{dt}{t} = \int_0^z \{J_{\mu-1}(t) + J_{\mu+1}(t)\} J_\nu(z-t)\, dt = 2J_{\mu+\nu}(z),$$

so that

$$\int_0^z J_\mu(t) J_\nu(z-t) \frac{dt}{t} = \frac{J_{\mu+\nu}(z)}{\mu}. \tag{3}$$

This formula is due to Bateman*; some special cases had been obtained independently by Kapteyn†, who considered integral values of μ and ν only.

It will be observed that we can deduce from (2), combined with § 2·22 (2), that

$$\int_0^z J_\mu(t) J_{-\mu}(z-t)\, dt = \sin z, \quad \int_0^z J_\mu(t) J_{1-\mu}(z-t)\, dt = J_0(z) - \cos z, \tag{4}$$

when $-1 < R(\mu) < 1$, and when $-1 < R(\mu) < 2$ respectively.

By interchanging μ with ν and t with $z-t$ in (3), we see that, if $R(\mu)$ and $R(\nu)$ are both positive, then

$$\int_0^z \frac{J_\mu(t)}{t} \cdot \frac{J_\nu(z-t)}{z-t}\, dt = \left(\frac{1}{\mu} + \frac{1}{\nu}\right) \frac{J_{\mu+\nu}(z)}{z}. \tag{5}$$

It seems unnecessary to give the somewhat complicated inductions by which Kapteyn deduced (3) from the special case in which $\mu = \nu = 1$, or to describe the disquisition by Rutgers‡ on the subject of the formulae generally.

12·21. *Kapteyn's trigonometrical integrals* §.

A simpler formula than those just considered is

$$\int_0^z \cos(z-t) J_0(t)\, dt = zJ_0(z). \tag{1}$$

To prove this, we put the left-hand side equal to u, and then it is easily verified that

$$\frac{d^2u}{dz^2} + u = -J_1(z),$$

and therefore

$$u = zJ_0(z) + A\cos z + B\sin z,$$

where A and B are constants of integration.

* *Proc. London Math. Soc.* (2) III. (1905), p. 120. Some similar integrals occurring in the theory of integral equations are examined by the same writer, *ibid.* (2) IV. (1906), p. 484.

† *Proc. Section of Sci., K. Akad. van Wet. te Amsterdam*, VII. (1905), p. 499; *Nieuw Archief voor Wiskunde*, (2) VII. (1907), pp. 20—25; *Mém. de la Soc. R. des Sci. de Liége*, (3) VI. (1906), no. 5.

‡ *Nieuw Archief voor Wiskunde*, (2) VII. (1907), pp. 385—405.

§ *Mém. de la Soc. R. des Sci. de Liége*, (3) VI. (1906), no. 5.

Now, when z is small,

$$u = z + O(z^3),$$

and so $A = B = 0$, and the result is established.

It follows from (1) by differentiation that

$$\int_0^z \sin(z-t) \,.\, J_0(t)\, dt = zJ_1(z), \tag{2}$$

and, by a partial integration,

$$\int_0^z \sin(z-t) \,.\, J_1(t)\, dt = \sin z - zJ_0(z). \tag{3}$$

The formula

$$\int_0^z \sin(z-t) \frac{J_\mu(t)}{t}\, dt = \frac{2}{\mu} \sum_{n=0}^{\infty} (-)^n J_{\mu+2n+1}(z), \tag{4}$$

which is valid when $R(\mu) > 0$, is of a more elaborate character, and the result of the preceding section is required to prove it.

We write
$$v = \int_0^z J_0(z-t) J_\mu(t)\, dt,$$

and then we have

$$\begin{aligned}
\frac{d^2v}{dz^2} + v &= \int_0^z \{J_0''(z-t) + J_0(z-t)\} J_\mu(t)\, dt + J_\mu'(z) \\
&= \int_0^z \frac{J_1(z-t)}{z-t} J_\mu(t)\, dt + J_\mu'(z) \\
&= \int_0^z J_\mu(z-t) \frac{J_1(t)}{t}\, dt + J_\mu'(z) \\
&= \mu J_\mu(z)/z,
\end{aligned}$$

by § 12·2.

By the method of variation of parameters (cf. § 7·33), we deduce that

$$v = A\cos z + B\sin z + \mu \int_0^z \sin(z-t) \frac{J_\mu(t)}{t}\, dt,$$

and, since
$$v = \frac{z^{\mu+1}}{2^\mu\, \Gamma(\mu+2)} + O(z^{\mu+3}),$$

when z is small, it follows that, when $R(\mu) > 0$,

$$A = B = 0.$$

Hence we obtain the required result.

By differentiating (4) with respect to z we find that

$$\int_0^z \cos(z-t) \frac{J_\mu(t)}{t}\, dt = \frac{1}{\mu} \sum_{n=0}^{\infty} (-)^n \epsilon_n J_{\mu+2n}(z). \tag{5}$$

12·22. *Hardy's method of evaluating finite integrals.*

As a typical example of a very powerful method of evaluating finite integrals*, we shall now give a proof of the formula (cf. § 12·12)

$$(1)\qquad \int_0^{\frac{1}{2}\pi} J_\mu(z\sin^2\theta)\, J_\nu(z\cos^2\theta)\sin^{2\mu+1}\theta\cos^{2\nu+1}\theta\, d\theta = \frac{\Gamma(\mu+\frac{1}{2})\,\Gamma(\nu+\frac{1}{2})}{2\Gamma(\frac{1}{2})\,\Gamma(\mu+\nu+1)}\cdot\frac{J_{\mu+\nu+\frac{1}{2}}(z)}{\sqrt{(2z)}},$$

which is valid when $R(\mu) > -\frac{1}{2}$ and $R(\nu) > -\frac{1}{2}$.

The method is more elaborate than any other method described in this chapter, because it involves the use of infinite integrals combined with an application of Lerch's theorem† on null-functions.

Let
$$\int_0^{\frac{1}{2}\pi} J_\mu(zr^2\sin^2\theta)\, J_\nu(zr^2\cos^2\theta)\, r^{2\mu+2\nu+3}\sin^{2\mu+1}\theta\cos^{2\nu+1}\theta\, d\theta \equiv f_1(r),$$

$$\frac{\Gamma(\mu+\frac{1}{2})\,\Gamma(\nu+\frac{1}{2})}{2\Gamma(\mu+\nu+1)\sqrt{(2\pi z)}}\, J_{\mu+\nu+\frac{1}{2}}(zr^2)\, r^{2\mu+2\nu+2} \equiv f_2(r).$$

By changing from polar coordinates (r, θ) to Cartesian coordinates (x, y) and using § 13·2 (5) we see that, whenever $t > |I(z)|$, then

$$\int_0^\infty \exp(-r^2t)\,.\,f_1(r)\,dr = \int_0^\infty \exp(-x^2t)\, J_\nu(zx^2)\, x^{2\nu+1}\,dx \int_0^\infty \exp(-y^2t)\, J_\mu(zy^2)\, y^{2\mu+1}\,dy$$
$$= \frac{(2z)^{\mu+\nu}\,\Gamma(\mu+\frac{1}{2})\,\Gamma(\nu+\frac{1}{2})}{4\pi\,(t^2+z^2)^{\mu+\nu+1}}$$
$$= \int_0^\infty \exp(-r^2t)\,.\,f_2(r)\,dr,$$

and hence, by an obvious modification of Lerch's theorem, $f_1(r)$ is identically equal to $f_2(r)$; and this establishes the truth of the formula.

12·3. *Chessin's integral for* $\mathbf{Y}_n(z)$.

A curious integral for $\mathbf{Y}_n(z)$ has been obtained by Chessin, *American Journal*, XVI. (1894), pp. 186—187, from the formula

$$\frac{1}{1}+\frac{1}{2}+\ldots+\frac{1}{n+m} = \int_0^1 \frac{1-t^{n+m}}{1-t}\,dt;$$

if we substitute this result in the coefficients of the ascending series for $\mathbf{Y}_n(z)$, we obtain the formula in question, namely

$$(1)\qquad \mathbf{Y}_n(z) = 2(\gamma+\log\tfrac{1}{2}z)\, J_n(z) - \sum_{m=0}^{n-1}\frac{(\frac{1}{2}z)^{-n+2m}(n-m-1)!}{m!}$$
$$-\int_0^1 \frac{2J_n(z)-(t^{-\frac{1}{2}n}+t^{\frac{1}{2}n})\, J_n(z\sqrt{t})}{1-t}\,dt.$$

* I must express my thanks to Professor Hardy for communicating the method to me before the publication of his own developments of it. The method was used by Ramanujan to evaluate many curious integrals; and the reader may use it to evaluate the integrals examined earlier in this chapter.

† *Acta Mathematica*, XXVII. (1903), pp. 339—352. The form of the theorem required here is that, if $f(r)$ is a continuous function of r when $r > 0$, such that

$$\int_0^\infty \exp(-r^2t)\,.\,f(r)\,dr = 0$$

for all sufficiently large positive values of t, then $f(r)$ is identically zero.

CHAPTER XIII

INFINITE INTEGRALS

13·1. *Various types of infinite integrals.*

The subject of this chapter is the investigation of various classes of infinite integrals which contain either Bessel functions or functions of a similar character under the integral sign. The methods of evaluating such integrals are not very numerous; they consist, for the most part, of the following devices:

(I) Expanding the Bessel function in powers of its argument and integrating term-by-term.

(II) Replacing the Bessel function by Poisson's integral, changing the order of the integrations, and then carrying out the integrations.

(III) Replacing the Bessel function by one of the generalisations of Bessel's integral, changing the order of the integrations, and then carrying out the integrations; this procedure has been carried out systematically by Sonine* in his weighty memoir.

(IV) When two Bessel functions of the same order occur as a product under the integral sign, they may be replaced by the integral of a single Bessel function by Gegenbauer's formula (cf. § 12·1), and the order of the integrations is then changed†.

(V) When two functions of different orders but of the same argument occur as a product under the integral sign, the product may be replaced by the integral of a single Bessel function by Neumann's formula (§ 5·43), and the order of the integrations is then changed.

(VI) The Bessel function under the integral sign may be replaced by the contour integral of Barnes' type (§ 6·5) involving Gamma functions, and the order of the integrations is then changed; this very powerful method has not previously been investigated in a systematic manner.

Infinite integrals involving Bessel functions under the integral sign are not only of great interest to the Pure Mathematician, but they are of extreme importance in many branches of Mathematical Physics. And the various types are so numerous that it is not possible to give more than a selection of the most important integrals, whose values will be worked out by the most suitable methods; care has been taken to evaluate several examples by each method. In spite of the incompleteness of this chapter, its length must be contrasted unfavourably with the length of the chapter on finite integrals.

* *Math. Ann.* XVI. (1880), pp. 33—60.

† This procedure has been carried out by Gegenbauer in a number of papers published in the *Wiener Sitzungsberichte.*

13·2. *The integral of Lipschitz, with Hankel's generalisations.*

It was shewn by Lipschitz* that

$$(1) \qquad \int_0^\infty e^{-at} J_0(bt)\, dt = \frac{1}{\sqrt{(a^2+b^2)}},$$

where $R(a) > 0$, and, in order to secure convergence at the upper limit of integration, both the numbers $R(a \pm ib)$ are positive. That value of the square root is taken which makes $|a + \sqrt{(a^2+b^2)}| > |b|$.

The simplest method of establishing this result is to replace the Bessel coefficient by Parseval's integral (§ 2·2) and then change the order of the integrations—a procedure which may be justified without difficulty. It is thus found that

$$\begin{aligned}\int_0^\infty e^{-at} J_0(bt)\, dt &= \frac{1}{\pi}\int_0^\infty e^{-at}\int_0^\pi e^{ibt\cos\theta}\, d\theta\, dt \\ &= \frac{1}{\pi}\int_0^\pi \frac{d\theta}{a - ib\cos\theta} \\ &= 1/\sqrt{(a^2+b^2)},\end{aligned}$$

and the formula is proved.

Now consider the more general integral

$$\int_0^\infty e^{-at} J_\nu(bt)\, t^{\mu-1}\, dt.$$

This integral was first investigated in all its generality by Hankel†, in a memoir published posthumously at about the same time as the appearance of two papers by Gegenbauer‡. These writers proved that, if $R(\mu+\nu) > 0$, to secure convergence at the origin, and the previous conditions concerning a and b are satisfied, to secure convergence at infinity, then the integral is equal to

$$\frac{(\tfrac{1}{2}b/a)^\nu\, \Gamma(\mu+\nu)}{a^\mu\, \Gamma(\nu+1)}\, {}_2F_1\left(\frac{\mu+\nu}{2}, \frac{\mu+\nu+1}{2}; \nu+1; -\frac{b^2}{a^2}\right).$$

To establish this result, first suppose that b is further restricted so that $|b| < |a|$. If we expand the integrand in powers of b and integrate term-by-term, we find that

$$\begin{aligned}\int_0^\infty e^{-at} J_\nu(bt)\, t^{\mu-1}\, dt &= \sum_{m=0}^\infty \frac{(-)^m (\tfrac{1}{2}b)^{\nu+2m}}{m!\,\Gamma(\nu+m+1)}\int_0^\infty t^{\mu+\nu+2m-1} e^{-at}\, dt \\ &= \sum_{m=0}^\infty \frac{(-)^m (\tfrac{1}{2}b)^{\nu+2m}}{m!\,\Gamma(\nu+m+1)}\, \frac{\Gamma(\mu+\nu+2m)}{a^{\mu+\nu+2m}}.\end{aligned}$$

* *Journal für Math.* LVI. (1859), pp. 191—192.

† *Math. Ann.* VIII. (1875), pp. 467—468.

‡ *Wiener Sitzungsberichte*, LXX. (2), (1875), pp. 433—443; *ibid.* LXXII. (2), (1876), pp. 343—344. In the former, the special case $\mu=\nu+1$ was investigated by the integral given in § 3·32; in the latter, Gegenbauer obtained the general result by substituting Poisson's integral for $J_\nu(bt)$.

The final series converges absolutely, since $|b| < |a|$, and so the process of term-by-term integration is justified*. Hence

$$(2) \quad \int_0^\infty e^{-at} J_\nu(bt)\, t^{\mu-1}\, dt$$

$$= \frac{(\tfrac{1}{2}b/a)^\nu \, \Gamma(\mu+\nu)}{a^\mu \, \Gamma(\nu+1)} \, {}_2F_1\left(\frac{\mu+\nu}{2}, \frac{\mu+\nu+1}{2}; \nu+1; -\frac{b^2}{a^2}\right).$$

The result has, as yet, been proved only when $R(a) > 0$ and $|b| < |a|$; but, so long as merely

$$R(a+ib) > 0 \text{ and } R(a-ib) > 0,$$

then both sides of (2) are analytic functions of b; and so, by the principle of analytic continuation, (2) is true for this more extensive range of values of b.

Again, by using transformations of the hypergeometric functions, (2) may be written in the following forms:

$$(3) \quad \int_0^\infty e^{-at} J_\nu(bt)\, t^{\mu-1}\, dt$$

$$= \frac{(\tfrac{1}{2}b/a)^\nu \, \Gamma(\mu+\nu)}{a^\mu \, \Gamma(\nu+1)} \left(1+\frac{b^2}{a^2}\right)^{\frac{1}{2}-\mu} {}_2F_1\left(\frac{\nu-\mu+1}{2}, \frac{\nu-\mu}{2}+1; \nu+1; -\frac{b^2}{a^2}\right)$$

$$= \frac{(\tfrac{1}{2}b)^\nu \, \Gamma(\mu+\nu)}{(a^2+b^2)^{\frac{1}{2}(\mu+\nu)} \, \Gamma(\nu+1)} \, {}_2F_1\left(\frac{\mu+\nu}{2}, \frac{1-\mu+\nu}{2}; \nu+1; \frac{b^2}{a^2+b^2}\right).$$

The formula (2) has been used by Gegenbauer† in expressing toroidal functions as infinite integrals; special cases of (2) are required in various physical researches, of which those by Lamb‡ may be regarded as typical.

By combining two Bessel functions, it is easy to deduce that

$$(4) \quad \int_0^\infty e^{-at} Y_\nu(bt)\, t^{\mu-1}\, dt$$

$$= \cot \nu\pi \, \frac{(\tfrac{1}{2}b)^\nu \, \Gamma(\mu+\nu)}{(a^2+b^2)^{\frac{1}{2}(\mu+\nu)} \, \Gamma(\nu+1)} \, {}_2F_1\left(\frac{\mu+\nu}{2}, \frac{1-\mu+\nu}{2}; \nu+1; \frac{b^2}{a^2+b^2}\right)$$

$$- \operatorname{cosec} \nu\pi \, \frac{(\tfrac{1}{2}b)^{-\nu} \, \Gamma(\mu-\nu)}{(a^2+b^2)^{\frac{1}{2}(\mu-\nu)} \, \Gamma(1-\nu)} \, {}_2F_1\left(\frac{\mu-\nu}{2}, \frac{1-\mu-\nu}{2}; 1-\nu; \frac{b^2}{a^2+b^2}\right),$$

provided $R(\mu) > |R(\nu)|$ and $R(a \pm ib) > 0$; special cases of this formula are due to Hobson, *Proc. London Math. Soc.* XXV. (1892), p. 75, and Heaviside, *Electromagnetic Theory*, III. (London, 1912), p. 85.

It is obvious that interesting special cases of the formulae so far discussed may be

* Cf. Bromwich, *Theory of Infinite Series*, § 176.

† *Wiener Sitzungsberichte*, c. (2), (1891), pp. 745—766; Gegenbauer also expressed series, whose general terms involve toroidal functions and Bessel functions, as integrals with Bessel functions under the integral sign.

‡ *Proc. London Math. Soc.* XXXIV. (1902), pp. 276—284; (2) VII. (1909), pp. 122—141. See also Macdonald, *Proc. London Math. Soc.* XXXV. (1903), pp. 428—443 and Basset, *Proc. Camb. Phil. Soc.* V. (1886), pp. 425—433.

obtained by choosing μ and ν so that the hypergeometric functions reduce to elementary functions. Thus, by taking μ equal to $\nu+1$ or $\nu+2$, we obtain the results

$$(5) \qquad \int_0^\infty e^{-at} J_\nu(bt)\, t^\nu\, dt = \frac{(2b)^\nu\, \Gamma(\nu+\frac{1}{2})}{(a^2+b^2)^{\nu+\frac{1}{2}}\sqrt{\pi}},$$

$$(6) \qquad \int_0^\infty e^{-at} J_\nu(bt)\, t^{\nu+1} dt = \frac{2a\,.\,(2b)^\nu\, \Gamma(\nu+\frac{3}{2})}{(a^2+b^2)^{\nu+\frac{3}{2}}\sqrt{\pi}},$$

provided that $R(\nu) > -\frac{1}{2}$, $R(\nu) > -1$ respectively.

These formulae were obtained by Gegenbauer, *Wiener Sitzungsberichte*, LXX. (2), (1875), pp. 433—443; they were also noticed by Sonine, *Math. Ann.* XVI. (1880), p. 45; and Hardy, *Trans. Camb. Phil. Soc.* XXI. (1912), p. 12; while Beltrami, *Atti della R. Accad. delle Sci. di Torino*, XVI. (1880—1881), p. 203, and *Bologna Memorie*, (4) II. (1880), pp. 461—505, has obtained various special formulae by taking $\mu=1$ and ν to be any integer.

Other special formulae are

$$(7) \qquad \int_0^\infty e^{-at} J_\nu(bt) \frac{dt}{t} = \frac{\{\sqrt{(a^2+b^2)}-a\}^\nu}{\nu b^\nu},$$

$$(8) \qquad \int_0^\infty e^{-at} J_\nu(bt)\, dt = \frac{\{\sqrt{(a^2+b^2)}-a\}^\nu}{b^\nu \sqrt{(a^2+b^2)}}.$$

[NOTE. It was observed by Pincherle, *Bologna Memorie*, (4) VIII. (1887), pp. 125—143, that these integrals are derivable from the generalised form of Bessel's integrals (§ 6·2) by Laplace's transformation (cf. § 9·15). This aspect of the subject has been studied by Macdonald, *Proc. London Math. Soc.* XXXV. (1903), pp. 428—443, and Cailler, *Mém. de la Soc. de Physique de Genève*, XXXIV. (1902—1905), pp. 295—368. The differential equations satisfied by (5) and (6), *qua* functions of a, have been examined by Kapteyn, *Archives Néerlandaises*, (2) VI. (1901), pp. 103—116.]

The integral $\int_0^\infty \frac{\sinh at}{\sinh \pi t} J_0(bt)\, t\, dt$ was obtained by Neumann, *Journal für Math.* LXXII. (1863), p. 46, as a limit of a series of Legendre functions (cf. § 14·64). The integral does not seem to be capable of being evaluated in finite terms, though it is easy to obtain a series for it by using the expansion

$$\operatorname{cosech} \pi t = 2 \sum_{n=0}^\infty e^{-(2n+1)\pi t}.$$

A series which converges more rapidly (when b is large) will be obtained in § 13·51.

Some integrals of the same general type are given by Weber, *Journal für Math.* LXXV. (1873), pp. 92—102; and more recently the formula

$$(9) \qquad \int_0^\infty \frac{J_\nu(bt)\, t^\nu\, dt}{e^{\pi t}-1} = \frac{(2b)^\nu\, \Gamma(\nu+\frac{1}{2})}{\sqrt{\pi}} \sum_{n=1}^\infty \frac{1}{(n^2\pi^2+b^2)^{\nu+\frac{1}{2}}},$$

which is valid when $R(\nu) > 0$ and $|I(b)| < \pi$, has been obtained by Kapteyn, *Mém. de la Soc. R. des Sci. de Liége*, (3) VI. (1906), no. 9.

13·21. *The Lipschitz-Hankel integrals expressed as Legendre functions.*

It was noticed by Hankel that the hypergeometric functions which occur in the integrals just discussed are of the special type associated with Legendre functions; subsequently Gegenbauer expressed the integrals in terms of toroidal functions (which are known to be expressible as Legendre functions), and a little later Hobson* gave the formulae in some detail.

* *Proc. London Math. Soc.* XXV. (1893), pp. 49—75.

To obtain the fundamental formulae* of this type, we shall change the notation by writing

$$a = \cosh \alpha, \quad b = i \sinh \alpha,$$

where α is a complex number such that

$$-\tfrac{1}{2}\pi \leqslant I(\alpha) \leqslant \tfrac{1}{2}\pi;$$

we thus obtain the formula

$$(1) \qquad \int_0^\infty e^{-t\cosh\alpha} I_\nu(t \sinh \alpha)\, t^\mu\, dt = \Gamma(\mu+\nu+1)\, P_\mu^{-\nu}(\cosh \alpha),$$

provided that $R(\mu+\nu) > -1$.

The special case of this formula in which $\nu=0$ had been given by Callandreau, *Bull. des Sci. Math.* (2) XV. (1891), pp. 121—124, two years before Hobson published the general formula.

It follows at once from (1) that

$$(2) \quad \int_0^\infty e^{-t\cosh\alpha} K_\nu(t\sinh\alpha)\, t^\mu\, dt = \frac{\sin \mu\pi}{\sin(\mu+\nu)\pi}\Gamma(\mu-\nu+1)\, Q_\mu^\nu(\cosh\alpha),$$

provided that $R(\mu+1) > |R(\nu)|$.

The modification of (1) which has to be used when the argument of the Legendre function† is positive and less than 1 is

$$(3) \qquad \int_0^\infty e^{-t\cos\beta} J_\nu(t\sin\beta)\, t^\mu\, dt = \Gamma(\mu+\nu+1)\, P_\mu^{-\nu}(\cos\beta),$$

and hence we find that

$$(4) \quad \int_0^\infty e^{-t\cos\beta} Y_\nu(t\sin\beta)\, t^\mu\, dt = -\frac{\sin\mu\pi}{\sin(\mu+\nu)\pi}\cdot\frac{\Gamma(\mu-\nu+1)}{\pi}$$
$$\times [Q_\mu^\nu(\cos\beta+0i)\, e^{\frac{1}{2}\nu\pi i} + Q_\mu^\nu(\cos\beta-0i)\, e^{-\frac{1}{2}\nu\pi i}].$$

Some special cases of this formula have been given by Hobson, *loc. cit.* p. 75, and by Heaviside, *Electromagnetic Theory*, III. (London, 1912), p. 85.

An apparently different formula, namely

$$(5) \qquad \int_0^\infty e^{-t\cosh\alpha} I_\nu(t)\frac{dt}{\sqrt{t}} = \frac{Q_{\nu-\frac{1}{2}}(\cosh\alpha)}{\sqrt{(\frac{1}{2}\pi)}},$$

has been studied by Steinthal‡. This formula is connected with formulae of the previous type by Whipple's§ transformation of Legendre functions, which

* Since, by a change of variable, the integrals are expressible in terms of the ratio of b to a, no generality is lost. The various expressions for Legendre functions as hypergeometric series which are required in this analysis are given by Barnes, *Quarterly Journal*, XXXIX. (1908), pp. 97—204.

† The reader will remember that it is customary to give a different definition for the Legendre function in such circumstances; cf. Hobson, *Phil. Trans. of the Royal Soc.* CLXXXVII. A, (1896), p. 471; and *Modern Analysis*, §§ 15·5, 15·6.

‡ *Quarterly Journal*, XVIII. (1882), pp. 337—340.

§ *Proc. London Math. Soc.* (2) XVI. (1917), pp. 301—314.

expresses a function of $\cosh\alpha$ in terms of a function of $\coth\alpha$. The more general formula of the same type is

$$(6)\qquad \int_0^\infty e^{-t\cosh\alpha} I_\nu(t)\, t^{\mu-1}\, dt = \frac{\cos\nu\pi}{\sin(\mu+\nu)\pi} \frac{Q_{\nu-\frac{1}{2}}^{\mu-\frac{1}{2}}(\cosh\alpha)}{\sqrt{(\frac{1}{2}\pi)} \,.\, \sinh^{\mu-\frac{1}{2}}\alpha}.$$

In these formulae, $R(\mu+\nu) > 0$ and $R(\cosh\alpha) > 1$.

On replacing ν by $-\nu$ in (6), we find that

$$(7)\qquad \int_0^\infty e^{-t\cosh\alpha} K_\nu(t)\, t^{\mu-1}\, dt = \sqrt{(\tfrac{1}{2}\pi)} \,.\, \Gamma(\mu-\nu)\,\Gamma(\mu+\nu) \frac{P_{\nu-\frac{1}{2}}^{\frac{1}{2}-\mu}(\cosh\alpha)}{\sinh^{\mu-\frac{1}{2}}\alpha},$$

and this formula is valid when $R(\mu) > |R(\nu)|$ and $R(\cosh\alpha) > -1$.

If we take $\cosh\alpha = 0$, we deduce that

$$(8)\qquad \int_0^\infty K_\nu(t)\, t^{\mu-1}\, dt = 2^{\mu-2}\,\Gamma\left(\frac{\mu-\nu}{2}\right)\Gamma\left(\frac{\mu+\nu}{2}\right),$$

a result given by Heaviside* in the case $\nu = 0$.

When $\mu = 1$, (7) becomes

$$(9)\qquad \int_0^\infty e^{-t\cosh\alpha} K_\nu(t)\, dt = \frac{\pi}{\sin\nu\pi}\frac{\sinh\nu\alpha}{\sinh\alpha},$$

and hence, if $\nu = 0$,

$$\int_0^\infty e^{-at} K_0(t)\, dt = \frac{\operatorname{arc\,sinh}\sqrt{(a^2-1)}}{\sqrt{(a^2-1)}} = \frac{\operatorname{arc\,sin}\sqrt{(1-a^2)}}{\sqrt{(1-a^2)}} = \frac{\operatorname{arc\,cos} a}{\sqrt{(1-a^2)}}.$$

If we replace a by $\pm ib$, we find that

$$\int_0^\infty e^{\mp ibt} K_0(t)\, dt = \frac{\frac{1}{2}\pi \mp i\operatorname{arc\,sinh} b}{\sqrt{(1+b^2)}},$$

and so, when $|I(b)| < 1$,

$$(10)\qquad \int_0^\infty \cos(bt) \,.\, K_0(t)\, dt = \frac{\frac{1}{2}\pi}{\sqrt{(1+b^2)}},$$

$$(11)\qquad \int_0^\infty \sin(bt) \,.\, K_0(t)\, dt = \frac{\operatorname{arc\,sinh} b}{\sqrt{(1+b^2)}}.$$

The former of these is due to Basset, *Hydrodynamics*, II. (Cambridge, 1889), p. 32.

[NOTE. Various writers have studied the Lipschitz-Hankel integrals from the aspect of potential theory; to take the simplest case, if (ρ, ϕ, z) are cylindrical coordinates, we have

$$\int_0^\infty e^{-\rho t} J_0(zt)\, dt = \frac{1}{\sqrt{(\rho^2+z^2)}}.$$

It is suggested that, since $e^{-\rho t} J_0(zt)$ is a potential function, the integral on the left is a potential function finite at all points of real space except the origin and that on† the plane $z=0$ it is equal to $1/\rho$, and so it is inferred that it must be the potential of a unit charge at the origin. But such an argument does not seem to preclude the possibility of the integral being a potential function with a complicated essential singularity at the origin, and so this reasoning must be regarded as suggestive rather than convincing.

* *Electromagnetic Theory*, III. (London, 1912), p. 269.

† On the axis of z, the integral is equal to a constant divided by $|z|$.

For various researches on potential theory with the aid of the integrals of this section, the reader may consult Hafen, *Math. Ann.* LXIX. (1910), pp. 517—537. For some developments based on the potential function

$$e^{-kz}\int_{-\infty}^{\infty} J_0[k\sqrt{\{(x-t)^2+y^2\}}]\,f(t)\,dt,$$

see Bateman, *Messenger*, XLI. (1912), p. 94.]

13·22. *Applications of the addition formula to the Lipschitz-Hankel integrals.*

It is easy to deduce from the results of the preceding sections combined with § 11·41 (16) that, if all four of the numbers $R(a \pm ib \pm ic)$ are positive and $R(\mu + 2\nu) > 0$, while ϖ is written in place of $\sqrt{(b^2+c^2-2bc\cos\phi)}$, then

$$(1)\qquad \int_0^\infty e^{-at} J_\nu(bt) J_\nu(ct)\, t^{\mu-1}\,dt$$
$$= \frac{(\tfrac12 bc)^\nu}{\Gamma(\nu+\tfrac12)\,\Gamma(\tfrac12)}\int_0^\infty\int_0^\pi e^{-at}\frac{J_\nu(\varpi t)}{\varpi^\nu}\, t^{\mu+\nu-1}\sin^{2\nu}\phi\, d\phi\, dt$$
$$= \frac{(bc)^\nu\,\Gamma(\mu+2\nu)}{\pi a^{\mu+2\nu}\,\Gamma(2\nu+1)}\int_0^\pi {}_2F_1\left(\frac{\mu+2\nu}{2}, \frac{\mu+2\nu+1}{2}; \nu+1; -\frac{\varpi^2}{a^2}\right)\sin^{2\nu}\phi\, d\phi.$$

The hypergeometric function reduces to an elementary function if $\mu = 1$ or 2; and so we have

$$(2)\qquad \int_0^\infty e^{-at} J_\nu(bt) J_\nu(ct)\,dt = \frac{1}{\pi\sqrt{(bc)}}\, Q_{\nu-\frac12}\left(\frac{a^2+b^2+c^2}{2bc}\right).$$

The case $\mu = 2$ may be derived from this by differentiation with respect to a.

These formulae, or special cases of them, have been examined by the following writers: Beltrami, *Bologna Memorie*, (4) II. (1880), pp. 461—505; *Atti della R. Accad. delle Sci. di Torino*, XVI. (1880—1881), pp. 201—205; Sommerfeld, *Königsberg Dissertation*, 1891; Gegenbauer, *Monatshefte für Math. und Phys.* V. (1894), p. 55; and Macdonald, *Proc. London Math. Soc.* XXVI. (1895), pp. 257—260.

By taking $\mu = -1$, $\nu = 1$ in (1), we find that

$$\int_0^\infty e^{-at}\frac{J_1^2(t)\,dt}{t^2} = \frac{1}{2\pi}\int_0^\pi \{\sqrt{(a^2+2-2\cos\phi)} - a\}(1+\cos\phi)\,d\phi,$$

so that the integral on the left, which was encountered by Rayleigh, *Phil. Mag.* (5) XLII. (1896), p. 195 [*Scientific Papers*, IV. (1904), p. 260], is expressible as an elliptic integral.

An integral which may be associated with (1) is

$$(3)\qquad \int_0^\infty \cos at\, I_0(bt)\, K_0(ct)\,dt = \int_0^{\frac12\pi}\frac{d\theta}{\sqrt{\{a^2+(b+c)^2-4bc\sin^2\theta\}}}.$$

This was discovered by Kirchhoff* as early as 1853; the reader should have no difficulty in deducing it from § 13·21 (10) combined with § 11·41 (16); it is valid if all the numbers

$$R(c \pm b \pm ia)$$

are positive.

* *Journal für Math.* XLVIII. (1854), p. 364.

A somewhat similar result, namely

$$(4)\qquad \int_0^\infty e^{-at}\, t^{\mu-\nu} J_\mu(bt)\, J_\nu(ct)\, dt$$
$$= \frac{(\tfrac{1}{2}b)^\mu (\tfrac{1}{2}c)^\nu \Gamma(2\mu+1)}{\Gamma(\nu+\tfrac{1}{2})\Gamma(\tfrac{1}{2})} \int_0^\pi \frac{\sin^{2\nu}\phi\, d\phi}{(a^2+2iac\cos\phi - c^2\cos^2\phi + b^2)^{\mu+\frac{1}{2}}},$$

which is valid when $R(a \pm ib \pm ic) > 0$ and $R(\mu) > -\tfrac{1}{2}$, is due to Gegenbauer, *Wiener Sitzungsberichte*, LXXXVIII. (2), (1884), p. 995. It is most easily proved by substituting integrals of Poisson's type for the Bessel functions. In the memoir cited Gegenbauer has also given a list of cases in which the integral on the right is expressible by elementary functions (cf. § 13·23).

13·23. *Gegenbauer's deductions from the integrals of Lipschitz and Hankel.*

A formula due to Gegenbauer, *Monatshefte für Math. und Phys.* IV. (1893), pp. 397—401, is obtained by combining the results of § 13·2 with the integral formula of § 5·43 for the product of two Bessel functions; it is thus possible to express certain exponential integrals which involve two Bessel functions by means of integrals of trigonometrical functions*. The general result obtained by Gegenbauer is deduced by taking the formula

$$J_\mu(bt)\, J_\nu(bt) = \frac{2}{\pi}\int_0^{\frac{1}{2}\pi} J_{\mu+\nu}(2bt\cos\phi)\cos(\mu-\nu)\phi\, d\phi,$$

multiplying it by $e^{-2at}\, t^{\mu+\nu}$ and integrating from 0 to ∞; it is thus found that, if $R(a) > |I(b)|$ and $R(\mu+\nu) > -\tfrac{1}{2}$, then

$$\int_0^\infty e^{-2at} J_\mu(bt)\, J_\nu(bt)\, t^{\mu+\nu}\, dt = \frac{2}{\pi}\int_0^\infty \int_0^{\frac{1}{2}\pi} e^{-2at} J_{\mu+\nu}(2bt\cos\phi)\, t^{\mu+\nu}\cos(\mu-\nu)\phi \,.\, d\phi\, dt$$
$$= \frac{2}{\pi}\int_0^{\frac{1}{2}\pi} \int_0^\infty e^{-2at} J_{\mu+\nu}(2bt\cos\phi)\, t^{\mu+\nu}\cos(\mu-\nu)\phi \,.\, dt\, d\phi$$
$$= \frac{2}{\pi}\int_0^{\frac{1}{2}\pi} \frac{(4b\cos\phi)^{\mu+\nu}\,\Gamma(\mu+\nu+\tfrac{1}{2})}{(4a^2+4b^2\cos^2\phi)^{\mu+\nu+\frac{1}{2}}\sqrt{\pi}} \cos(\mu-\nu)\phi \,.\, d\phi.$$

The inversion of the order of the integrations presents no great theoretical difficulties; hence

$$(1)\qquad \int_0^\infty e^{-2at} J_\mu(bt)\, J_\nu(bt)\, t^{\mu+\nu}\, dt$$
$$= \frac{\Gamma(\mu+\nu+\tfrac{1}{2})\, b^{\mu+\nu}}{\pi^{\frac{3}{2}}} \int_0^{\frac{1}{2}\pi} \frac{\cos^{\mu+\nu}\phi\cos(\mu-\nu)\phi}{(a^2+b^2\cos^2\phi)^{\mu+\nu+\frac{1}{2}}}\, d\phi.$$

This result, in the special case in which $\mu=\nu=0$, had been obtained previously by Beltrami, *Atti della R. Accad. delle Sci. di Torino*, XVI. (1880—1881), p. 204.

As particular cases of (1) take $\mu=1$ and ν equal to 0 and to -1. It is found that

$$(2)\qquad \int_0^\infty e^{-2at} J_1(bt)\, J_0(bt)\, t\, dt = \frac{K-E}{2\pi b\sqrt{(a^2+b^2)}},$$

$$(3)\qquad \int_0^\infty e^{-2at} J_1^{\,2}(bt)\, dt = \frac{(2a^2+b^2)\,K - 2\,(a^2+b^2)\,E}{\pi b^2\sqrt{(a^2+b^2)}},$$

* See also an earlier note by Gegenbauer, *ibid.* pp. 379—380.

where the modulus of the complete elliptic integrals K and E is $b/\sqrt{(a^2+b^2)}$. Beltrami's corresponding formula is

$$(4) \qquad \int_0^\infty e^{-2at} J_0^2(bt)\, dt = \frac{K}{\pi\sqrt{(a^2+b^2)}}.$$

Replacing b by ib, we deduce from (2) that

$$(5) \qquad \int_0^\infty e^{-2at} I_1(bt)\, I_0(bt)\, t\, dt = \frac{a\,(E_1 - k_1'^2 K_1)}{2\pi b\,(a^2-b^2)},$$

where $R(a) > |R(b)|$, and the modulus k_1 of the elliptic integrals is b/a. The formulae (3) and (4) may be modified in a similar manner.

It was stated by Gegenbauer that the integrals in (2), (3) and (5) are expressible by means of elliptic integrals, but he did not give the results in detail; some formulae deducible from the results of this section were given by Meissel, *Kiel Programm*, 1890. [*Jahrbuch über die Fortschritte der Math.* 1890, pp. 521—522.]

13·24. *Weber's infinite integral, after Schafheitlin.*

The formula

$$(1) \qquad \int_0^\infty \frac{J_\nu(t)\, dt}{t^{\nu-\mu+1}} = \frac{\Gamma(\tfrac{1}{2}\mu)}{2^{\nu-\mu+1}\,\Gamma(\nu - \tfrac{1}{2}\mu + 1)},$$

in which $0 < R(\mu) < R(\nu) + \tfrac{1}{2}$, was obtained by Weber* for integral values of ν. The result was extended to general values of ν by Sonine†; and the completely general result was also proved by Schafheitlin‡.

The formula is of a more recondite type than the exponential integral formulae given in § 13·2; it may be established as a limiting case of these formulae, for, since the conditions§ of convergence are satisfied, we have by § 13·2 (3)

$$\int_0^\infty \frac{J_\nu(t)\, dt}{t^{\nu-\mu+1}} = \lim_{a\to+0} \int_0^\infty e^{-at} \frac{J_\nu(t)\, dt}{t^{\nu-\mu+1}}$$

$$= \frac{\Gamma(\mu)}{2^\nu\,\Gamma(\nu+1)} \cdot {}_2F_1\left(\frac{\mu}{2}, \frac{1-\mu+2\nu}{2}; \nu+1; 1\right),$$

whence the formula is at once obtained.

A direct method of evaluating the integral is to substitute Poisson's integral for the Bessel function, and then change the order of the integrations; this is the method used by Schafheitlin, but the analysis is intricate because the result is established first for a limited range of values of μ and ν and then extended by the use of recurrence formulae and partial integrations.

Analytical difficulties are, to a large extent, avoided by using contour integrals instead of the definite integrals of Schafheitlin. If we suppose that

* *Journal für Math.* LXIX. (1868), p. 230. The special case in which $\nu = 0$ was set by Stokes as a Smith's Prize question, Jan. 29, 1867. [*Math. and Phys. Papers*, v. (1905), p. 347.]

† *Math. Ann.* XVI. (1880), p. 39.

‡ *Math. Ann.* XXX. (1887), pp. 157—161.

§ Cf. Bromwich, *Theory of Infinite Series*, § 172.

$R(\mu) < 0$ and $R(\nu) > -\frac{1}{2}$, we then have (the integrals being absolutely convergent)

$$\int_{+\infty}^{(0+)} \frac{J_\nu(-t)\,dt}{(-t)^{\nu-\mu+1}} = \frac{1}{2^{\nu-1}\,\Gamma(\nu+\frac{1}{2})\,\Gamma(\frac{1}{2})}\int_{+\infty}^{(0+)} (-t)^{\mu-1}\int_0^{\frac{1}{2}\pi} \cos(t\cos\theta)\sin^{2\nu}\theta\,d\theta\,dt$$

$$= \frac{-2i\sin\mu\pi\,.\,\Gamma(\mu)\cos\frac{1}{2}\mu\pi}{2^{\nu-1}\,\Gamma(\nu+\frac{1}{2})\,\Gamma(\frac{1}{2})}\int_0^{\frac{1}{2}\pi}\cos^{-\mu}\theta\sin^{2\nu}\theta\,d\theta$$

$$= \frac{-2i\sin\mu\pi\,.\,\Gamma(\frac{1}{2}\mu)}{2^{\nu-\mu+1}\,\Gamma(\nu-\frac{1}{2}\mu+1)}.$$

By the theory of analytic continuation, this result is valid when μ and ν are subjected to the single restriction $R(\mu) < R(\nu+\frac{3}{2})$.

When $R(\mu) > 0$, we deform the contour into the positive half of the real axis taken twice, and we at once obtain the Weber-Schafheitlin formula.

The integral*

$$\int_{+\infty}^{(0+)} \frac{\mathbf{H}_\nu(-t)\,dt}{(-t)^{\nu-\mu+1}}$$

may be treated in exactly the same manner; the only difference in the analysis is that $\cos(t\cos\theta)$ has to be replaced by $-\sin(t\cos\theta)$, and so, by Euler's formula (adapted for contour integrals), the factor $\cos\frac{1}{2}\mu\pi$ has to be replaced by $-\sin\frac{1}{2}\mu\pi$.

It is thus found that

$$\int_{+\infty}^{(0+)} \frac{\mathbf{H}_\nu(-t)\,dt}{(-t)^{\nu-\mu+1}} = \frac{2i\sin\mu\pi\,.\,\Gamma(\frac{1}{2}\mu)\tan\frac{1}{2}\mu\pi}{2^{\nu-\mu+1}\,\Gamma(\nu-\frac{1}{2}\mu+1)},$$

provided that $R(\mu) < R(\nu+\frac{3}{2})$ and $R(\mu) \leqslant 0$.

When $R(\mu) > -1$, the contour may be deformed into the positive half of the real axis taken twice, so that

$$(2) \qquad \int_0^\infty \frac{\mathbf{H}_\nu(t)\,dt}{t^{\nu-\mu+1}} = \frac{\Gamma(\frac{1}{2}\mu)\tan(\frac{1}{2}\mu\pi)}{2^{\nu-\mu+1}\,\Gamma(\nu-\frac{1}{2}\mu+1)},$$

provided that $-1 < R(\mu) \leqslant 0$ and $R(\mu) < R(\nu)+\frac{3}{2}$.

If we take $\mu=0$, $\nu=1$, we see that

$$(3) \qquad \int_0^\infty \frac{\mathbf{H}_1(t)\,dt}{t^2} = \tfrac{1}{4}\pi.$$

This result, combined with the asymptotic formula

$$\frac{1}{\pi}\int_x^\infty \frac{\mathbf{H}_1(2t)\,dt}{t^2} \sim \frac{2}{\pi^2}\left(\frac{1}{x^2}+\frac{1}{12x^3}\right) - \frac{\cos(2x+\frac{1}{4}\pi)}{2\pi^{3/2}\,x^{5/2}},$$

was used by Struve, *Ann. der Physik und Chemie*, (3) XVII. (1882), p. 1014, to tabulate

$$\frac{1}{\pi}\int_x^\infty \frac{\mathbf{H}_1(2t)\,dt}{t^2}$$

for both small and large values of x. The last integral is of importance in the Theory of Diffraction.

* Generalisations obtained by replacing Bessel functions by Lommel's functions (§ 10·7) in the integrals of this section and in many other integrals are discussed by Nielsen, *K. Danske Videnskabernes Selskabs Skrifter*, (7) v. (1910), pp. 1—37.

[NOTE. By differentiating (1) under the integral sign we obtain Weber's result

$$\int_0^\infty J_0(t)\log t\,dt = -\gamma - \log 2; \tag{4}$$

this formula has also been investigated by Lerch, *Monatshefte für Math. und Phys.* I. (1890), pp. 105—112.

The formula for functions of the second kind, corresponding to (1), is

$$\int_0^\infty \frac{Y_\nu(t)\,dt}{t^{\nu-\mu+1}} = -\frac{\Gamma(\tfrac{1}{2}\mu)\,\Gamma(\tfrac{1}{2}\mu-\nu)\cos(\tfrac{1}{2}\mu-\nu)\pi}{2^{\nu-\mu+1}\,\pi}, \tag{5}$$

provided that $|R(\nu)| < R(\mu-\nu) < \tfrac{3}{2}$. This result has been given by Heaviside, *Electromagnetic Theory*, III. (London, 1912), p. 273, when $\nu = 0$.]

13·3. *Weber's first exponential integral and its generalisations.*

The integral formula

$$\int_0^\infty J_0(at)\exp(-p^2t^2)\,.\,t\,dt = \frac{1}{2p^2}\exp\left(-\frac{a^2}{4p^2}\right) \tag{1}$$

was deduced by Weber* from his double integral formula which will be discussed in § 14·2. This integral differs from those considered earlier in the chapter by containing the square of the variable in the exponential function. It is supposed that $|\arg p| < \tfrac{1}{4}\pi$ to secure convergence, but a is an unrestricted complex number.

It is equally easy to prove Hankel's† more general formula,

$$\int_0^\infty J_\nu(at)\exp(-p^2t^2)\,.\,t^{\mu-1}\,dt \tag{2}$$
$$= \frac{\Gamma(\tfrac{1}{2}\nu+\tfrac{1}{2}\mu)\,.\,(\tfrac{1}{2}a/p)^\nu}{2p^\mu\,\Gamma(\nu+1)}\,{}_1F_1\left(\tfrac{1}{2}\nu+\tfrac{1}{2}\mu;\,\nu+1;\,-\frac{a^2}{4p^2}\right),$$

by a direct method. To secure convergence at the origin, it must now be supposed that‡

$$R(\mu+\nu) > 0.$$

To obtain the result, we observe that, since (by § 7·23)

$$\int_0^\infty I_{|\nu|}(|a|t)\,.\,|\exp(-p^2t^2)|\,.\,|t^{\mu-1}|\,dt$$

is *convergent*, it is permissible§ to evaluate the given integral by expanding $J_\nu(at)$ in powers of t and integrating term-by-term.

* *Journal für Math.* LXIX. (1868), p. 227. Weber also evaluated (2) in the case $\mu = \nu + 2$, ν being an integer.

† *Math. Ann.* VIII. (1875), p. 469. See also Gegenbauer, *Wiener Sitzungsberichte*, LXXII. (2), (1876), p. 346.

‡ This restriction may be disregarded if we replace the definite integral $\int_0^\infty$ by the contour integral $\int_\infty^{(0+)}$.

§ Cf. Bromwich, *Theory of Infinite Series*, § 176.

It is thus found that

$$\int_0^\infty J_\nu(at)\exp(-p^2t^2)\,.\,t^{\mu-1}dt = \sum_{m=0}^\infty \frac{(-)^m(\tfrac{1}{2}a)^{\nu+2m}}{m!\,\Gamma(\nu+m+1)}\int_0^\infty t^{\nu+\mu+2m-1}\exp(-p^2t^2)\,dt$$

$$= \sum_{m=0}^\infty \frac{(-)^m(\tfrac{1}{2}a)^{\nu+2m}}{m!\,\Gamma(\nu+m+1)}\cdot\frac{\Gamma(\tfrac{1}{2}\nu+\tfrac{1}{2}\mu+m)}{2p^{\nu+\mu+2m}},$$

and this is equivalent to the result stated.

If we apply Kummer's first transformation (§ 4·42) to the function on the right in (2), we find that

(3) $$\int_0^\infty J_\nu(at)\exp(-p^2t^2)\,.\,t^{\mu-1}\,dt$$

$$= \frac{\Gamma(\tfrac{1}{2}\nu+\tfrac{1}{2}\mu)\,.\,(\tfrac{1}{2}a/p)^\nu}{2p^\mu\,\Gamma(\nu+1)}\exp\left(-\frac{a^2}{4p^2}\right).\,{}_1F_1\left(\tfrac{1}{2}\nu-\tfrac{1}{2}\mu+1;\,\nu+1;\,\frac{a^2}{4p^2}\right),$$

and so the integral is expressible in finite terms whenever $\mu-\nu$ is an even positive integer.

In particular, we have

(4) $$\int_0^\infty J_\nu(at)\exp(-p^2t^2)\,.\,t^{\nu+1}\,dt = \frac{a^\nu}{(2p^2)^{\nu+1}}\exp\left(-\frac{a^2}{4p^2}\right),$$

provided that $R(\nu) > -1$. This integral is the basis of several investigations by Sonine, *Math. Ann.* XVI. (1880), pp. 35—38; some of these applications are discussed in § 13·47.

In order that the hypergeometric function on the right in (2) may be susceptible to Kummer's second transformation (§ 4·42), we take $\mu = 1$; and, if we replace ν by 2ν, we then find that

(5) $$\int_0^\infty J_{2\nu}(at)\exp(-p^2t^2)\,.\,dt = \frac{\sqrt{\pi}}{2p}\exp\left(-\frac{a^2}{8p^2}\right).\,I_\nu\left(\frac{a^2}{8p^2}\right),$$

a result given by Weber in the case $\nu = \frac{1}{2}$.

If we replace ν by $-\nu$, it is easy to see that

(6) $$\int_0^\infty Y_{2\nu}(at)\exp(-p^2t^2)\,dt$$

$$= -\frac{\sqrt{\pi}}{2p}\exp\left(-\frac{a^2}{8p^2}\right)\left[I_\nu\left(\frac{a^2}{8p^2}\right)\tan\nu\pi + \frac{1}{\pi}K_\nu\left(\frac{a^2}{8p^2}\right)\sec\nu\pi\right],$$

when $|R(\nu)| < \frac{1}{2}$; and, if we make $p \to 0$, (a being now positive), we find that

(7) $$\int_0^\infty Y_{2\nu}(at)\,dt = -\frac{\tan\nu\pi}{a},$$

when $|R(\nu)| < \frac{1}{2}$, by using § 7·23; and, in particular,

(8) $$\int_0^\infty Y_0(t)\,dt = 0.$$

Formulae (5) and (6) were given (when $\nu=0$) by Heaviside, *Electromagnetic Theory*, III. (London, 1912), p. 271.

Another method of evaluating the integral on the left of (3) is suggested by Basset, *Proc. Camb. Phil. Soc.* VIII. (1895), pp. 122—128; the integrals have also been evaluated with the help of Laplace's transformation by Macdonald, *Proc. London Math. Soc.* XXXV. (1903), pp. 428—443; see also Curzon, *Proc. London Math. Soc.* (2) XIII. (1914), pp. 417—440; and Hardy, *Trans. Camb. Phil. Soc.* XXI. (1912), pp. 10, 27, for formulae obtained by making p^2 a pure imaginary.

For some applications of the integrals of this section to the Theory of Conduction of Heat, see Rayleigh, *Phil. Mag.* (6) XXII. (1911), pp. 381—396 [*Scientific Papers*, VI. (1920), pp. 51—64].

13·31. *Weber's second exponential integral.*

The result of applying the formula § 11·41 (16) to the integral just discussed is to modify it by replacing the Bessel function under the integral sign by a product of two Bessel functions of the same order.

If $\varpi = \sqrt{(a^2 + b^2 - 2ab\cos\phi)}$ and if $R(\nu) > -\frac{1}{2}$, $R(2\nu+\mu) > 0$, $|\arg p| < \frac{1}{4}\pi$, we thus deduce that

$$\int_0^\infty \exp(-p^2t^2)\, J_\nu(at)\, J_\nu(bt)\, t^{\mu-1}\, dt$$

$$= \frac{(\frac{1}{2}ab)^\nu}{\Gamma(\nu+\frac{1}{2})\,\Gamma(\frac{1}{2})} \int_0^\infty \int_0^\pi \exp(-p^2t^2)\, \frac{J_\nu(\varpi t)}{\varpi^\nu}\, t^{\nu+\mu-1} \sin^{2\nu}\phi\, d\phi\, dt$$

$$= \frac{\Gamma(\nu+\frac{1}{2}\mu)}{2\pi p^\mu\, \Gamma(2\nu+1)} \left(\frac{ab}{p^2}\right)^\nu \int_0^\pi \exp\left(-\frac{\varpi^2}{4p^2}\right) {}_1F_1\left(1-\frac{\mu}{2};\ \nu+1;\ \frac{\varpi^2}{4p^2}\right) \sin^{2\nu}\phi\, d\phi.$$

The hypergeometric function reduces to unity when $\mu = 2$; so that

$$\int_0^\infty \exp(-p^2t^2)\, J_\nu(at)\, J_\nu(bt)\, t\, dt = \frac{(\frac{1}{4}ab/p^2)^\nu}{2p^2\Gamma(\nu+\frac{1}{2})\,\Gamma(\frac{1}{2})} \int_0^\pi \exp\left(-\frac{\varpi^2}{4p^2}\right) \sin^{2\nu}\phi\, d\phi,$$

$$= \frac{(\frac{1}{4}ab/p^2)^\nu}{2p^2\Gamma(\nu+\frac{1}{2})\,\Gamma(\frac{1}{2})} \exp\left(-\frac{a^2+b^2}{4p^2}\right) \int_0^\pi \exp\left(\frac{ab\cos\phi}{2p^2}\right) \sin^{2\nu}\phi\, d\phi.$$

If we expand the exponential under the integral sign, we find that

$$(1) \qquad \int_0^\infty \exp(-p^2t^2)\, J_\nu(at)\, J_\nu(bt)\, t\, dt = \frac{1}{2p^2} \exp\left(-\frac{a^2+b^2}{4p^2}\right) I_\nu\left(\frac{ab}{2p^2}\right).$$

This formula is valid if $R(\nu) > -1$ and $|\arg p| < \frac{1}{4}\pi$.

Like the result of § 13·3, this equation is due to Weber, *Journal für Math.* LXIX. (1868), p. 228; Weber gave a different proof of it, as also did Hankel, *Math. Ann.* VIII. (1875), pp. 469—470. The proof given here is due to Gegenbauer, *Wiener Sitzungsberichte*, LXXII. (2), (1876), p. 347. Other investigations are due to Sonine, *Math. Ann.* XVI. (1880), p. 40; Sommerfeld, *Königsberg Dissertation*, 1891; Macdonald, *Proc. London Math. Soc.* XXXV. (1903), p. 438; and Cailler, *Mém. de la Soc. Phys. de Genève*, XXXIV. (1902—1905), p. 331. Some physical applications are due to Carslaw, *Proc. London Math. Soc.* (2), VIII. (1910), pp. 365—374.

13·32. *Generalisations of Weber's second exponential integral.*

When the Bessel functions in integrals of the type just considered are not of the same order, it is usually impossible to express the result in any simple form. The only method of dealing with the most general integral

$$\int_0^\infty J_\mu(at)\, J_\nu(bt) \exp(-p^2t^2)\, t^{\lambda-1}\, dt$$

is to substitute the series of § 11·6 for the product of Bessel functions and integrate term-by-term, but it seems unnecessary to give the result here. In the special case in which $\lambda = \nu - \mu$, Macdonald* has shewn that the integral is equal to

$$\frac{(\frac{1}{2}a)^{\mu-\nu}}{p^2\Gamma(\mu-\nu)} \int_0^{\frac{1}{2}\pi} \cos^{2\mu-2\nu-1}\theta \sin^{\nu+1}\theta\, I_\nu\left(\frac{ab\sin\theta}{2p^2}\right) \exp\left(-\frac{b^2+a^2\sin^2\theta}{4p^2}\right) d\theta,$$

by a transformation based on the results of §§ 12·11, 13·7.

An exceptional case occurs when $a = b$; if $R(\lambda+\mu+\nu) > 0$, we then have

$$\int_0^\infty J_\mu(at)\, J_\nu(at) \exp(-p^2t^2)\, t^{\lambda-1}\, dt = \frac{a^{\mu+\nu}}{2^{\mu+\nu} p^{\lambda+\mu+\nu}} \frac{\Gamma\left(\frac{\lambda+\mu+\nu}{2}\right)}{\Gamma(\mu+1)\Gamma(\nu+1)}$$

$$\times {}_3F_3\left(\frac{\mu+\nu+1}{2}, \frac{\mu+\nu+2}{2}, \frac{\lambda+\mu+\nu}{2}; \mu+1, \nu+1, \mu+\nu+1; -\frac{a^2}{p^2}\right),$$

by using the expansion of § 5·41. Some special cases of this formula have been investigated by Gegenbauer†.

13·33. *Struve's integral involving products of Bessel functions.*

It will now be shewn that, when $R(\mu+\nu) > 0$, then

$$(1) \qquad \int_0^\infty \frac{J_\mu(t)\, J_\nu(t)}{t^{\mu+\nu}}\, dt = \frac{\Gamma(\mu+\nu)\,\Gamma(\frac{1}{2})}{2^{\mu+\nu}\,\Gamma(\mu+\nu+\frac{1}{2})\,\Gamma(\mu+\frac{1}{2})\,\Gamma(\nu+\frac{1}{2})}.$$

This result was obtained by Struve, *Mém. de l'Acad. Imp. des Sci. de St Pétersbourg*, (7) xxx. (1882), p. 91, in the special case $\mu=\nu=1$; the expression on the right is then equal to $4/(3\pi)$.

In evaluating the integral it is first convenient to suppose that $R(\mu)$ and $R(\nu)$ both exceed $\frac{1}{2}$. It then follows from § 3·3 (7) that

$$\int_0^\infty \frac{J_\mu(t)\, J_\nu(t)}{t^{\mu+\nu}}\, dt = \frac{(2\mu-1)(2\nu-1)}{2^{\mu+\nu-2}\pi\,\Gamma(\mu+\frac{1}{2})\,\Gamma(\nu+\frac{1}{2})}$$

$$\times \int_0^\infty \int_0^{\frac{1}{2}\pi} \int_0^{\frac{1}{2}\pi} \frac{\sin(t\sin\theta)\sin(t\sin\phi)}{t^2} \cos^{2\mu-2}\theta \cos^{2\nu-2}\phi \sin\theta \sin\phi\, d\theta\, d\phi\, dt.$$

* *Proc. London Math. Soc.* xxxv. (1903), p. 440.

† *Wiener Sitzungsberichte*, LXXXVIII. (1884), pp. 999—1000.

In view of the fact that $t^{-2}\sin(t\sin\theta)\sin(t\sin\phi)$ does not exceed numerically the smaller of $1/t^2$ and $\sin\theta\sin\phi$, the repeated integral converges absolutely, and the order of the integrations may be changed.

Since *

$$\int_0^\infty \frac{\sin(t\sin\theta)\sin(t\sin\phi)}{t^2}\,dt = \begin{cases}\tfrac{1}{2}\pi\sin\theta, & (\theta\leqslant\phi)\\ \tfrac{1}{2}\pi\sin\phi, & (\theta\geqslant\phi)\end{cases}$$

we find that the triple integral is equal to

$$\tfrac{1}{2}\pi\int_0^{\frac{1}{2}\pi}\int_0^{\phi}\cos^{2\mu-2}\theta\cos^{2\nu-2}\phi\sin^2\theta\sin\phi\,d\theta\,d\phi$$

$$+\tfrac{1}{2}\pi\int_0^{\frac{1}{2}\pi}\int_0^{\theta}\cos^{2\mu-2}\theta\cos^{2\nu-2}\phi\sin\theta\sin^2\phi\,d\phi\,d\theta.$$

But, by a partial integration, we have

$$(2\nu-1)\int_0^{\frac{1}{2}\pi}\cos^{2\nu-2}\phi\sin\phi\left\{\int_0^{\phi}\cos^{2\mu-2}\theta\sin^2\theta\,d\theta\right\}d\phi$$

$$=\left[-\cos^{2\nu-1}\phi\int_0^{\phi}\cos^{2\mu-2}\theta\sin^2\theta\,d\theta\right]_0^{\frac{1}{2}\pi}+\int_0^{\frac{1}{2}\pi}\cos^{2\nu-1}\phi\,.\cos^{2\mu-2}\phi\sin^2\phi\,d\phi$$

$$=\frac{\Gamma(\mu+\nu-1)\,\Gamma(\frac{3}{2})}{2\Gamma(\mu+\nu+\frac{1}{2})}.$$

The other integral is evaluated in the same manner, and so we have

$$\int_0^\infty \frac{J_\mu(t)\,J_\nu(t)}{t^{\mu+\nu}}\,dt=\frac{\Gamma(\mu+\nu-1)\,\Gamma(\frac{3}{2})\,\{(2\mu-1)+(2\nu-1)\}}{2^{\mu+\nu}\,\Gamma(\mu+\nu+\frac{1}{2})\,\Gamma(\mu+\frac{1}{2})\,\Gamma(\nu+\frac{1}{2})},$$

whence the result stated is evident. The extension over the range of values of μ and ν for which merely $R(\mu+\nu)>0$ is obtained by the theory of analytic continuation.

It may be shewn in a similar manner that, when $R(\mu+\nu)$ is positive, then also

(2) $$\int_0^\infty \frac{\mathbf{H}_\mu(t)\,\mathbf{H}_\nu(t)}{t^{\mu+\nu}}\,dt=\frac{\Gamma(\mu+\nu)\,\Gamma(\frac{1}{2})}{2^{\mu+\nu}\,\Gamma(\mu+\nu+\frac{1}{2})\,\Gamma(\mu+\frac{1}{2})\,\Gamma(\nu+\frac{1}{2})}.$$

This result was also obtained by Struve (*ibid.* p. 104) in the case $\mu=\nu=1$.

By using § 10·45 we find that, when $R(\mu)$ and $R(\nu)$ exceed $\frac{1}{2}$,

$$\int_0^\infty \frac{\mathbf{H}_\mu(t)\,\mathbf{H}_\nu(t)}{t^{\mu+\nu}}\,dt$$

$$=\frac{(2\mu-1)(2\nu-1)}{2^{\mu+\nu-2}\,\pi\Gamma(\mu+\frac{1}{2})\,\Gamma(\nu+\frac{1}{2})}\int_0^\infty\int_0^{\frac{1}{2}\pi}\int_0^{\frac{1}{2}\pi}\frac{\{1-\cos(t\sin\theta)\}\,\{1-\cos(t\sin\phi)\}}{t^2}$$

$$\times\cos^{2\mu-2}\theta\cos^{2\nu-2}\phi\sin\theta\sin\phi\,d\theta\,d\phi\,dt.$$

Now, if a and β are positive, it appears from a consideration of

$$\int\frac{(1-e^{aiz})\,(1-e^{\beta iz})}{z^2}\,dz$$

* This result is easily proved by contour integration.

round a contour consisting of the real axis and a large semicircle above it, that

$$\int_0^\infty \frac{\{1-\cos(at)\}\{1-\cos(\beta t)\}}{t^2}dt = \int_0^\infty \frac{\sin(at)\sin(\beta t)}{t^2}dt.$$

Hence the triple integral under consideration is equal to the triple integral evaluated in proving (1), and consequently (2) is established in the same way as (1).

The reader will prove in like manner that, if $R(\mu)$ and $R(\nu)$ both exceed $\frac{1}{2}$, then

$$\int_0^\infty \frac{\mathbf{H}_\mu(t)\,J_\nu(t)}{t^{\mu+\nu-1}}dt = \frac{2\nu-1}{2^{\mu+\nu}(\mu+\nu-1)\,\Gamma(\mu+\frac{1}{2})\,\Gamma(\nu+\frac{1}{2})} \tag{3}$$

and this may be extended over the range of values of μ and ν for which $R(\nu)>\frac{1}{2}$ and $R(\mu+\nu)>1$.

The integrals
$$\int_0^\infty \frac{\mathbf{H}_\mu(t)\,\mathbf{H}_\nu(t)}{t^{\mu+\nu+2}}dt,\quad \int_0^\infty \frac{\mathbf{H}_\mu(t)\,J_\nu(t)}{t^{\mu+\nu+1}}dt$$

may be evaluated in a similar manner, but the results are of no great interest*.

13·4. *The discontinuous integral of Weber and Schafheitlin.*

The integral

$$\int_0^\infty \frac{J_\mu(at)\,J_\nu(bt)}{t^\lambda}dt,$$

in which a and b are supposed to be positive to secure convergence at the upper limit, was investigated by Weber, *Journal für Math.* LXXV. (1873), pp. 75—80, in several special cases, namely,

$$\text{(i)}\ \lambda=\mu=0,\ \nu=1, \qquad \text{(ii)}\ \lambda=-\tfrac{1}{2},\ \mu=0,\ \nu=\pm\tfrac{1}{2}.$$

The integral was evaluated, for all values of λ, μ and ν for which it is convergent, by Sonine†, *Math. Ann.* XVI. (1880), pp. 51—52; but he did not examine the integral in very great detail, nor did he lay any stress on the discontinuities which occur when a and b become equal. Some years later the integral was investigated very thoroughly by Schafheitlin‡, but his preliminary analysis rests to a somewhat undue extent on the theory of linear differential equations.

The special case in which $\lambda=0$ was discussed in 1895 by Gubler§ who used a very elegant transformation of contour integrals; unfortunately, however, it seems impossible to adapt Gubler's analysis to the more general case in which $\lambda\neq0$. The analysis in the special case will be given subsequently (§ 13·44).

* Some related integrals have been evaluated by Siemon, *Programm, Luisenschule*, Berlin, 1890 [*Jahrbuch über die Fortschritte der Math.* 1890, p. 341].

† See also § 13·43 in connexion with the researches of Gegenbauer, *Wiener Sitzungsberichte*, LXXXVIII. (2), (1884), pp. 990—991.

‡ *Math. Ann.* XXX. (1887), pp. 161—178. The question of priority is discussed by Sonine, *Math. Ann.* XXX. (1887), pp. 582—583, and by Schafheitlin, *Math. Ann.* XXXI. (1888), p. 156.

§ *Math. Ann.* XLVIII. (1897), pp. 37—48. See also Graf and Gubler, *Einleitung in die Theorie der Bessel'schen Funktionen*, II. (Bern, 1900), pp. 136—148.

The first investigation which we shall give is based on the results of § 13·2. The conditions for convergence are*

$$\begin{cases} R(\mu+\nu+1) > R(\lambda) > -1, & (a \neq b) \\ R(\mu+\nu+1) > R(\lambda) > 0, & (a = b) \end{cases}$$

it being supposed, as already stated, that a and b are *positive.*

We shall first suppose that the former conditions are satisfied, and we shall also take $b < a$. The analysis is greatly shortened by choosing new constants α, β, γ defined by the equations

$$\begin{cases} 2\alpha = \mu + \nu - \lambda + 1, \\ 2\beta = \nu - \lambda - \mu + 1, \\ \gamma = \nu + 1, \end{cases} \qquad \begin{cases} \lambda = \gamma - \alpha - \beta, \\ \mu = \alpha - \beta, \\ \nu = \gamma - 1. \end{cases}$$

It will be supposed that these relations hold down to the end of § 13·41.

It is known that

$$\int_0^\infty \frac{J_\mu(at)\, J_\nu(bt)}{t^\lambda}\, dt = \lim_{c \to +0} \int_0^\infty e^{-ct}\, \frac{J_\mu(at)\, J_\nu(bt)}{t^\lambda}\, dt,$$

since the integral on the left is convergent; now, when c has any assigned positive value, the integral on the right is convergent for complex values of b; we replace b by z and the resulting integral is an analytic function of z when $R(z) > 0$ and $|I(z)| < c$.

Now $\displaystyle\int_0^\infty e^{-ct}\, \frac{J_{\alpha-\beta}(at)\, J_{\gamma-1}(zt)}{t^{\gamma-\alpha-\beta}}\, dt$

$$= \int_0^\infty e^{-ct}\, J_{\alpha-\beta}(at) \left\{ \sum_{m=0}^\infty \frac{(-)^m (\tfrac{1}{2}z)^{\gamma+2m-1}\, t^{\alpha+\beta+2m-1}}{m!\, \Gamma(\gamma+m)} \right\} dt$$

$$= \sum_{m=0}^\infty \frac{(-)^m (\tfrac{1}{2}z)^{\gamma+2m-1}}{m!\, \Gamma(\gamma+m)} \int_0^\infty e^{-ct}\, J_{\alpha-\beta}(at)\, t^{\alpha+\beta+2m-1}\, dt,$$

provided that†

$$\sum_{m=0} \frac{(-)^m (\tfrac{1}{2}z)^{\gamma+2m-1}}{m!\, \Gamma(\gamma+m)} \int_0^\infty e^{-ct}\, |J_{\alpha-\beta}(at)\, t^{\alpha+\beta+2m-1}|\, dt$$

is absolutely convergent; and it is easy to shew that this is the case when $|z| < c$.

Hence, when $|z| < c$,

$$\int_0^\infty e^{-ct}\, \frac{J_{\alpha-\beta}(at)\, J_{\gamma-1}(zt)}{t^{\gamma-\alpha-\beta}}\, dt = \sum_{m=0}^\infty \frac{(-)^m (\tfrac{1}{2}z)^{\gamma+2m-1}}{m!\, \Gamma(\gamma+m)} \cdot \frac{(\tfrac{1}{2}a)^{\alpha-\beta}\, \Gamma(2\alpha+2m)}{(a^2+c^2)^{\alpha+m}\, \Gamma(\alpha-\beta+1)}$$
$$\times {}_2F_1\left(\alpha+m, \tfrac{1}{2}-\beta-m;\ \alpha-\beta+1;\ \frac{a^2}{a^2+c^2}\right),$$

* It follows from the asymptotic expansions of the Bessel functions that the conditions

$$R(\mu+\nu+1) > R(\lambda) > -1$$

are sufficient to secure convergence when $a = b$, provided that $\mu - \nu$ is an *odd integer.*

† Cf. Bromwich, *Theory of Infinite Series*, § 176.

and the hypergeometric function on the right may be replaced by*

$$\frac{\Gamma(\alpha-\beta+1)\,\Gamma(\frac{1}{2})}{\Gamma(1-\beta-m)\,\Gamma(\alpha+m+\frac{1}{2})}\cdot{}_2F_1\left(\alpha+m,\ \tfrac{1}{2}-\beta-m;\ \tfrac{1}{2};\ \frac{c^2}{a^2+c^2}\right)$$
$$+\frac{\Gamma(\alpha-\beta+1)\,\Gamma(-\frac{1}{2})}{\Gamma(\frac{1}{2}-\beta-m)\,\Gamma(\alpha+m)}\frac{c}{\sqrt{(a^2+c^2)}}\cdot{}_2F_1\left(\alpha+m+\tfrac{1}{2},\ 1-\beta-m;\ \tfrac{3}{2};\ \frac{c^2}{a^2+c^2}\right).$$

Now the moduli of the terms in the expansion of

$${}_2F_1(\alpha+m,\ \tfrac{1}{2}-\beta-m;\ \tfrac{1}{2};\ x)$$

do not exceed in absolute value the alternate terms in the expansion of $(1-\sqrt{x})^{-A-2m}$, where A is the greater of $|2\alpha|$ and $|2\beta-1|$; and, similarly, the moduli of the terms in the expansion of

$${}_2F_1(\alpha+m+\tfrac{1}{2},\ 1-\beta-m;\ \tfrac{3}{2};\ x)$$

do not exceed in absolute value the alternate terms in the expansion of

$$(1-\sqrt{x})^{-A-2m-1}/\sqrt{x}.$$

Hence the terms in the infinite series which has been obtained do not exceed in absolute value the terms of the series

$$\sum_{m=0}^{\infty}\frac{(-)^m(\frac{1}{2}z)^{\gamma+2m-1}}{m!\,\Gamma(\gamma+m)}\frac{(\frac{1}{2}a)^{\alpha-\beta}\,\Gamma(2\alpha+2m)}{(a^2+c^2)^{\alpha+m}}\left[\frac{\Gamma(\frac{1}{2})\,(1-\sqrt{x})^{-A-2m}}{|\Gamma(1-\beta-m)\,\Gamma(\alpha+m+\frac{1}{2})|}+\frac{|\Gamma(-\frac{1}{2})|\,(1-\sqrt{x})^{-A-2m-1}}{|\Gamma(\frac{1}{2}-\beta-m)\,\Gamma(\alpha+m)|}\right]$$

where $x=c^2/(a^2+c^2)$. But this last series is absolutely convergent when $|z|<\sqrt{(a^2+c^2)}-c$, and it represents an analytic function of z in this domain.

Hence, by the general theory of analytic continuation,

$$\int_0^\infty e^{-ct}\frac{J_{\alpha-\beta}(at)\,J_{\gamma-1}(zt)}{t^{\gamma-\alpha-\beta}}\,dt$$
$$=\sum_{m=0}^{\infty}\frac{(-)^m(\frac{1}{2}z)^{\gamma+2m-1}}{m!\,\Gamma(\gamma+m)}\frac{(\frac{1}{2}a)^{\alpha-\beta}\,\Gamma(2\alpha+2m)}{(a^2+c^2)^{\alpha+m}\,\Gamma(\alpha-\beta+1)}$$
$$\times{}_2F_1\left(\alpha+m,\ \tfrac{1}{2}-\beta-m;\ \alpha-\beta+1;\ \frac{a^2}{a^2+c^2}\right),$$

provided that z satisfies the three conditions

$$R(z)>0,\quad |I(z)|<c,\quad |z|<\sqrt{(a^2+c^2)}-c.$$

Now take C to be a positive number so small that

$$b<\sqrt{(a^2+C^2)}-C,$$

and take $0<c\leqslant C$, so that also

$$b<\sqrt{(a^2+c^2)}-c.$$

* Cf. Forsyth, *Treatise on Differential Equations*, (1914), § 127.

Then in the last integral formula we may take $z = b$, and when this has been done, if we use *fonctions majorantes* just as before, we find that the resulting series has its terms less than the terms of an absolutely convergent series

$$\sum_{m=0}^{\infty} \frac{(-)^m (\frac{1}{2}b)^{\gamma+2m-1}}{m!\,\Gamma(\gamma+m)} \frac{(\frac{1}{2}a)^{\alpha-\beta}\,\Gamma(2\alpha+2m)}{(a^2+C^2)^{\alpha+m}} \left[\frac{\Gamma(\frac{1}{2})(1-\sqrt{X})^{-A-2m}}{|\Gamma(1-\beta-m)\,\Gamma(\alpha+m+\frac{1}{2})|} + \frac{|\Gamma(-\frac{1}{2})|\,(1-\sqrt{X})^{-A-2m-1}}{|\Gamma(\frac{1}{2}-\beta-m)\,\Gamma(\alpha+m)|}\right],$$

where $X = C^2/(a^2+C^2)$.

Hence, by the test of Weierstrass, the original series converges uniformly with respect to c when $0 \leqslant c \leqslant C$, and therefore the limit of the series when $c \to 0$ is the same as the value of the series when $c = 0$.

We have therefore proved that

$$\lim_{c\to 0} \int_0^\infty e^{-ct} \frac{J_{\alpha-\beta}(at)\, J_{\gamma-1}(bt)}{t^{\gamma-\alpha-\beta}}\, dt$$
$$= \sum_{m=0}^{\infty} \frac{(-)^m (\frac{1}{2}b)^{\gamma+2m-1}}{m!\,\Gamma(\gamma+m)} \frac{(\frac{1}{2}a)^{\alpha-\beta}\,\Gamma(2\alpha+2m)}{a^{2\alpha+2m}\,\Gamma(\alpha-\beta+1)}\, {}_2F_1(\alpha+m, \tfrac{1}{2}-\beta-m; \alpha-\beta+1; 1),$$

and therefore

$$\int_0^\infty \frac{J_{\alpha-\beta}(at)\, J_{\gamma-1}(bt)}{t^{\gamma-\alpha-\beta}}\, dt$$
$$= \sum_{m=0}^{\infty} \frac{(-)^m\, b^{\gamma+2m-1}\,\Gamma(2\alpha+2m)\,\Gamma(\frac{1}{2})}{m!\,\Gamma(\gamma+m)\, 2^{\alpha-\beta+\gamma+2m-1}\, a^{\alpha+\beta+2m}\,\Gamma(1-\beta-m)\,\Gamma(\alpha+m+\frac{1}{2})}.$$

It has therefore been shewn that

$$(1) \quad \int_0^\infty \frac{J_{\alpha-\beta}(at)\, J_{\gamma-1}(bt)}{t^{\gamma-\alpha-\beta}}\, dt = \frac{b^{\gamma-1}\,\Gamma(\alpha)}{2^{\gamma-\alpha-\beta}\, a^{\alpha+\beta}\,\Gamma(\gamma)\,\Gamma(1-\beta)} \cdot {}_2F_1\left(\alpha, \beta; \gamma; \frac{b^2}{a^2}\right),$$

that is to say

$$(2) \quad \int_0^\infty \frac{J_\mu(at)\, J_\nu(bt)}{t^\lambda}\, dt = \frac{b^\nu\,\Gamma(\frac{1}{2}\mu+\frac{1}{2}\nu-\frac{1}{2}\lambda+\frac{1}{2})}{2^\lambda\, a^{\nu-\lambda+1}\,\Gamma(\nu+1)\,\Gamma(\frac{1}{2}\lambda+\frac{1}{2}\mu-\frac{1}{2}\nu+\frac{1}{2})}$$
$$\times {}_2F_1\left(\frac{\mu+\nu-\lambda+1}{2}, \frac{\nu-\lambda-\mu+1}{2}; \nu+1; \frac{b^2}{a^2}\right),$$

provided that $0 < b < a$, and that the integral is convergent. This is the result obtained by Sonine and Schafheitlin.

If we interchange a and b, and also μ and ν, throughout the work, we find that, when $0 < a < b$ and the integral is convergent, then

$$(3) \quad \int_0^\infty \frac{J_{\alpha-\beta}(at)\, J_{\gamma-1}(bt)}{t^{\gamma-\alpha-\beta}}\, dt = \frac{a^{\alpha-\beta}\,\Gamma(\alpha)}{2^{\gamma-\alpha-\beta}\, b^{2\alpha-\gamma+1}\,\Gamma(\gamma-\alpha)\,\Gamma(\alpha-\beta+1)}$$
$$\times {}_2F_1\left(\alpha, \alpha-\gamma+1; \alpha-\beta+1; \frac{a^2}{b^2}\right).$$

Now it so happens that the expressions on the right in (1) and (3) *are not the analytic continuations of the same function.* There is consequently a discontinuity in the formula when $a = b$; and it will be necessary to examine this phenomenon in some detail.

13·41. *The critical case of the Weber-Schafheitlin integral.*

In the case of the integral now under consideration, when $a = b$, we have, as before,

$$\int_0^\infty \frac{J_\mu(at)\,J_\nu(at)}{t^\lambda}\,dt = \lim_{c\to+0}\int_0^\infty e^{-ct}\,\frac{J_\mu(at)\,J_\nu(at)}{t^\lambda}\,dt,$$

assuming that $R(\mu+\nu+1) > R(\lambda) > 0$, to secure convergence.

Now consider

$$\int_0^\infty e^{-zt}\,\frac{J_{\alpha-\beta}(at)\,J_{\gamma-1}(at)}{t^{\gamma-\alpha-\beta}}\,dt,$$

where z is a complex variable with $R(z)$ positive.

When $R(z) > 2a$ we may expand the integrand in ascending powers of a and integrate term-by-term, this procedure being justified by the fact that the resulting series is convergent.

We thus get, by using § 5·41,

$$\int_0^\infty e^{-zt}\,\frac{J_{\alpha-\beta}(at)\,J_{\gamma-1}(at)}{t^{\gamma-\alpha-\beta}}\,dt$$

$$= \sum_{m=0}^\infty \int_0^\infty \frac{e^{-zt}(-)^m(\tfrac{1}{2}a)^{\alpha-\beta+\gamma+2m-1}\,t^{2\alpha+2m-1}\,\Gamma(\alpha-\beta+\gamma+2m)}{m!\,\Gamma(\alpha-\beta+m+1)\,\Gamma(\gamma+m)\,\Gamma(\alpha-\beta+\gamma+m)}\,dt$$

$$= \sum_{m=0}^\infty \frac{(-)^m(\tfrac{1}{2}a)^{\alpha-\beta+\gamma+2m-1}\,\Gamma(2\alpha+2m)\,\Gamma(\alpha-\beta+\gamma+2m)}{z^{2\alpha+2m}\,m!\,\Gamma(\alpha-\beta+m+1)\,\Gamma(\gamma+m)\,\Gamma(\alpha-\beta+\gamma+m)}.$$

Now the integral on the left is an analytic function of z when $R(z) > 0$, and so its value, when z has the small positive value c, is the analytic continuation of the series on the right.

But, by Barnes' theory*, the series on the right and its analytic continuations may be represented by the integral

$$\frac{1}{2\pi i}\int_{-\infty i}^{\infty i} \frac{(\tfrac{1}{2}a)^{\alpha-\beta+\gamma+2s-1}\,\Gamma(2\alpha+2s)\,\Gamma(\alpha-\beta+\gamma+2s)}{z^{2\alpha+2s}\,\Gamma(\alpha-\beta+s+1)\,\Gamma(\gamma+s)\,\Gamma(\alpha-\beta+\gamma+s)}\,\Gamma(-s)\,ds;$$

and this integral represents a function of z which is analytic when $|\arg z| < \pi$. It is supposed that the contour consists of the imaginary axis with loops to ensure that the poles of $\Gamma(-s)$ lie on the right of the contour, while the poles of $\Gamma(2\alpha+2s)$ and of $\Gamma(\alpha-\beta+\gamma+2s)$ lie on the left of the contour.

When $|z| < 2a$ we may evaluate the integral by modifying the contour so as to enclose the poles on the left of the contour and evaluating the residues

* *Proc. London Math. Soc.* (2) v. (1907), pp. 59—118. See also §§ 6·5, 6·51 *supra.*

at them. The sum of these residues forms two convergent series proceeding in ascending powers of z; hence, when $R(z) > 0$ and $|z| < 2a$,

$$\int_0^\infty e^{-zt} \frac{J_{\alpha-\beta}(at)\, J_{\gamma-1}(at)}{t^{\gamma-\alpha-\beta}}\, dt$$

$$= \frac{1}{2} \sum_{m=0}^{\infty} \frac{(-)^m (\frac{1}{2}a)^{\gamma-\alpha-\beta-m-1} z^m\, \Gamma(\gamma-\alpha-\beta-m)\, \Gamma(\alpha+\frac{1}{2}m)}{m!\, \Gamma(1-\beta-\frac{1}{2}m)\, \Gamma(\gamma-\alpha-\frac{1}{2}m)\, \Gamma(\gamma-\beta-\frac{1}{2}m)}$$

$$+ \frac{1}{2} \sum_{m=0}^{\infty} \frac{(-)^m (\frac{1}{2}a)^{-m-1} z^{\gamma-\alpha-\beta+m}\, \Gamma(\alpha+\beta-\gamma-m)\, \Gamma(\frac{1}{2}\alpha+\frac{1}{2}\gamma-\frac{1}{2}\beta+\frac{1}{2}m)}{m!\, \Gamma(\frac{1}{2}\alpha-\frac{1}{2}\beta-\frac{1}{2}\gamma-\frac{1}{2}m+1)\, \Gamma(\frac{1}{2}\beta+\frac{1}{2}\gamma-\frac{1}{2}\alpha-\frac{1}{2}m)\, \Gamma(\frac{1}{2}\alpha-\frac{1}{2}\beta+\frac{1}{2}\gamma-\frac{1}{2}m)}.$$

Now $R(\gamma-\alpha-\beta) > 0$, and so, when we make z assume the positive value c and then make $c \to 0$, we deduce that

$$(1) \qquad \int_0^\infty \frac{J_{\alpha-\beta}(at)\, J_{\gamma-1}(at)}{t^{\gamma-\alpha-\beta}}\, dt = \frac{(\frac{1}{2}a)^{\gamma-\alpha-\beta-1}\, \Gamma(\gamma-\alpha-\beta)\, \Gamma(\alpha)}{2\Gamma(1-\beta)\, \Gamma(\gamma-\alpha)\, \Gamma(\gamma-\beta)}$$

provided that $\qquad R(\alpha) > 0, \quad R(\gamma-\alpha-\beta) > 0.$

From the Gaussian formula for ${}_2F_1(\alpha, \beta; \gamma; 1)$, there is therefore no discontinuity in the value of the integral, though there is a discontinuity in the formula which expresses that value as b increases through the value a.

The result may be written in the alternative form

$$(2) \qquad \int_0^\infty \frac{J_\mu(at)\, J_\nu(at)}{t^\lambda}\, dt$$

$$= \frac{(\frac{1}{2}a)^{\lambda-1}\, \Gamma(\lambda)\, \Gamma(\frac{1}{2}\mu+\frac{1}{2}\nu-\frac{1}{2}\lambda+\frac{1}{2})}{2\Gamma(\frac{1}{2}\lambda+\frac{1}{2}\nu-\frac{1}{2}\mu+\frac{1}{2})\, \Gamma(\frac{1}{2}\lambda+\frac{1}{2}\mu+\frac{1}{2}\nu+\frac{1}{2})\, \Gamma(\frac{1}{2}\lambda+\frac{1}{2}\mu-\frac{1}{2}\nu+\frac{1}{2})},$$

provided that $\qquad R(\mu+\nu+1) > R(\lambda) > 0.$

If $\mu-\nu$ is an odd integer the integral converges when $0 \geqslant R(\lambda) > -1$; this case next demands attention.

We shall make a change in notation by writing $\alpha+p$ and $\alpha-p-1$ in place of μ and ν in the preceding analysis; if $R(\lambda) > 0$, we then find that

$$\int_0^\infty e^{-zt} \frac{J_{\alpha+p}(at)\, J_{\alpha-p-1}(at)}{t^\lambda}\, dt$$

$$= \frac{1}{2\pi i} \int_{-\infty i}^{\infty i} \frac{(\frac{1}{2}a)^{2\alpha+2s-1}\, \Gamma(2\alpha+2s)\, \Gamma(2\alpha+2s-\lambda)}{z^{2\alpha+2s-\lambda}\, \Gamma(\alpha+p+s+1)\, \Gamma(\alpha-p+s)\, \Gamma(2\alpha+s)}\, \Gamma(-s)\, ds$$

$$= \frac{1}{2} \sum_{m=0}^{\infty} \frac{(-)^m (\frac{1}{2}a)^{\lambda-m-1} z^m\, \Gamma(\lambda-m)\, \Gamma(\alpha-\frac{1}{2}\lambda+\frac{1}{2}m)}{m!\, \Gamma(p-\frac{1}{2}m+\frac{1}{2}\lambda+1)\, \Gamma(-p-\frac{1}{2}m+\frac{1}{2}\lambda)\, \Gamma(\alpha+\frac{1}{2}\lambda-\frac{1}{2}m)}$$

$$+ \frac{1}{2} \sum_{m=1}^{\infty} \frac{(-)^m (\frac{1}{2}a)^{-m-1} z^{\lambda+m}\, \Gamma(-\lambda-m)\, \Gamma(\alpha+\frac{1}{2}m)}{m!\, \Gamma(p-\frac{1}{2}m+1)\, \Gamma(-p-\frac{1}{2}m)\, \Gamma(\alpha-\frac{1}{2}m)},$$

and hence

$$(3) \quad \int_0^\infty \frac{J_{\alpha+p}(at)\, J_{\alpha-p-1}(at)}{t^\lambda}\, dt = \frac{a^{\lambda-1}\,\Gamma(\lambda)\,\Gamma(\alpha-\tfrac{1}{2}\lambda)}{2^\lambda\,\Gamma(p+\tfrac{1}{2}\lambda+1)\,\Gamma(\alpha+\tfrac{1}{2}\lambda)\,\Gamma(\tfrac{1}{2}\lambda-p)},$$

unless $\lambda = 0$. This should be compared with the more general formulae obtained from § 13·4, namely that, when $b < a$,

$$(4) \quad \int_0^\infty \frac{J_{\alpha+p}(at)\, J_{\alpha-p-1}(bt)}{t^\lambda}\, dt$$
$$= \frac{b^{\alpha-p-1}\,\Gamma(\alpha-\tfrac{1}{2}\lambda)}{2^\lambda\, a^{\alpha-p-\lambda}\,\Gamma(\alpha-p)\,\Gamma(p+\tfrac{1}{2}\lambda+1)}\, {}_2F_1\left(\alpha-\tfrac{1}{2}\lambda, -p-\tfrac{1}{2}\lambda;\ \alpha-p;\ \frac{b^2}{a^2}\right),$$

and, when $b > a$,

$$(5) \quad \int_0^\infty \frac{J_{\alpha+p}(at)\, J_{\alpha-p-1}(bt)}{t^\lambda}\, dt$$
$$= \frac{a^{\alpha+p}\,\Gamma(\alpha-\tfrac{1}{2}\lambda)}{2^\lambda\, b^{\alpha+p-\lambda+1}\,\Gamma(\alpha+p+1)\,\Gamma(\tfrac{1}{2}\lambda-p)}\, {}_2F_1\left(\alpha-\tfrac{1}{2}\lambda, p+1-\tfrac{1}{2}\lambda;\ \alpha+p+1;\ \frac{a^2}{b^2}\right).$$

Since $\lambda \neq 0$, the functions on the right in (4) and (5) do not tend to limits when $a \to b$.

On the other hand, when λ *is* zero, the contour integral becomes

$$\frac{1}{2\pi i}\int_{-\infty i}^{\infty i} \frac{(\tfrac{1}{2}a)^{2\alpha+2s-1}\,\{\Gamma(2\alpha+2s)\}^2\,\Gamma(-s)}{z^{2\alpha+2s}\,\Gamma(\alpha+p+s+1)\,\Gamma(\alpha-p+s)\,\Gamma(2\alpha+s)}\, ds,$$

and the residue at $s = -\alpha$ is $(-)^p/(2a)$.

It follows that

$$(6) \quad \int_0^\infty J_{\alpha+p}(at)\, J_{\alpha-p-1}(bt)\, dt = \begin{cases} \dfrac{b^{\alpha-p-1}\,\Gamma(\alpha)}{a^{\alpha-p}\,\Gamma(\alpha-p)\,.\,p!}\, {}_2F_1\left(\alpha, -p;\ \alpha-p;\ \dfrac{b^2}{a^2}\right), \\ (-)^p/(2a), \\ 0, \end{cases}$$

according as $b < a$, $b = a$, $b > a$. Since

$${}_2F_1(\alpha, -p;\ \alpha-p;\ 1) = (-)^p\, p!\, \Gamma(\alpha-p)/\Gamma(\alpha),$$

it is evident that *the value of the integral when $b = a$ is the mean of its limits when $b = a - 0$ and $b = a + 0$.*

The result of taking $\lambda = 1$ in (2) is

$$(7) \qquad \int_0^\infty J_\mu(at)\, J_\nu(at)\, \frac{dt}{t} = \frac{2}{\pi}\, \frac{\sin \tfrac{1}{2}(\nu-\mu)\pi}{\nu^2-\mu^2},$$

which is also easily obtained by inserting limits in § 5·11 (13); this formula has been discussed in great detail by Kapteyn, *Proc. Section of Sci., K. Akad. van Wet. te Amsterdam*, IV. (1902), pp. 102—103; *Archives Néerlandaises*, (2) VI. (1901), pp. 103—116.

13·42. *Special cases of the discontinuous integral.*

Numerous special cases of interest are obtained by giving special values to the constants λ, μ, ν in the preceding analysis. To save repetition, when three values are given for an integral, the first is its value for $b < a$, the second for $b = a$, and the third for $b > a$; when two values only are given, the first is the value for $b \leqslant a$, the second for $b \geqslant a$; and the values are correct for all values of the constants which make the integrals convergent.

The following are the most important special cases*:

(1) $$\int_0^\infty \frac{J_\mu(at)\, J_\mu(bt)}{t}\, dt = \begin{cases} \tfrac{1}{2}(b/a)^\mu/\mu, \\ \tfrac{1}{2}(a/b)^\mu/\mu. \end{cases} \qquad [R(\mu) > 0]$$

(2) $$\int_0^\infty \frac{J_\mu(at)\sin bt}{t}\, dt = \begin{cases} \mu^{-1}\sin\{\mu \arcsin(b/a)\}, \\ \dfrac{a^\mu \sin\frac{1}{2}\mu\pi}{\mu\{b + \sqrt{(b^2 - a^2)}\}^\mu}. \end{cases} \qquad [R(\mu) > -1]$$

(3) $$\int_0^\infty \frac{J_\mu(at)\cos bt}{t}\, dt = \begin{cases} \mu^{-1}\cos\{\mu \arcsin(b/a)\}, \\ \dfrac{a^\mu \cos\frac{1}{2}\mu\pi}{\mu\{b + \sqrt{(b^2 - a^2)}\}^\mu}. \end{cases} \qquad [R(\mu) > 0]$$

(4) $$\int_0^\infty J_\mu(at)\sin bt\,.\,dt = \begin{cases} \dfrac{\sin\{\mu \arcsin(b/a)\}}{\sqrt{(a^2 - b^2)}}, \\ \infty \text{ or } 0, \\ \dfrac{a^\mu \cos\frac{1}{2}\mu\pi}{\sqrt{(b^2 - a^2)}\,.\,\{b + \sqrt{(b^2 - a^2)}\}^\mu}. \end{cases} \qquad [R(\mu) > -2]$$

(5) $$\int_0^\infty J_\mu(at)\cos bt\,.\,dt = \begin{cases} \dfrac{\cos\{\mu \arcsin(b/a)\}}{\sqrt{(a^2 - b^2)}}, \\ \infty \text{ or } 0, \\ -\dfrac{a^\mu \sin\frac{1}{2}\mu\pi}{\sqrt{(b^2 - a^2)}\,.\,\{b + \sqrt{(b^2 - a^2)}\}^\mu}. \end{cases} \qquad [R(\mu) > -1]$$

Special cases of preceding results are

(6) $$\int_0^\infty J_0(at)\sin bt\,.\,dt = \begin{cases} 0, \\ \infty, \\ 1/\sqrt{(b^2 - a^2)}. \end{cases}$$

(7) $$\int_0^\infty J_0(at)\cos bt\,.\,dt = \begin{cases} 1/\sqrt{(a^2 - b^2)}, \\ \infty, \\ 0. \end{cases}$$

These two formulae, which were given by Weber†, *Journal für Math.* LXXV. (1873), p. 77, are known as *Weber's discontinuous factors*; they are associated with the problem of determining the potential of an electrified circular disc‡.

* Numerous other special cases are given by Nielsen, *Ann. di Mat.* (3) XIV. (1908), pp. 82—90. The integrals in (4) and (5) diverge for certain values of μ when $a = b$.

† The former was known to Stokes many years earlier, and was, in fact, set by him as a Smith's prize examination question in Feb. 1853. [*Math. and Phys. Papers*, V. (1905), p. 319.]

‡ Cf. Gallop, *Quarterly Journal*, XXI. (1886), pp. 230—231.

Another special formula is

$$(8)\qquad \int_0^\infty J_\mu(at)\,J_{\mu-1}(bt)\,dt = \begin{cases} b^{\mu-1}/a^\mu, \\ 1/(2b), \\ 0; \end{cases} \qquad [R(\mu)>0]$$

and if we put $\mu=1$, we obtain Weber's result (*ibid.*, p. 80),

$$(9)\qquad \int_0^\infty J_0(at)\,J_1(bt)\,dt = \begin{cases} 0, \\ 1/(2b), \\ 1/b. \end{cases}$$

The result of putting $\mu=\frac{1}{2}$ in (8) is known as *Dirichlet's discontinuous factor*; see the article by Voss, *Encyclopädie der Math. Wiss.* II. (1), (1916), p. 109.

Some other special formulae have been found useful in the theory of Fourier series by W. H. Young, *Leipziger Berichte*, LXIII. (1911), pp. 369—387.

Another method of evaluating (5) has been given by Hopf and Sommerfeld, *Archiv der Math. und Phys.* (3) XVIII. (1911), pp. 1—16.

A consequence of formula (1) must be noted. When $\nu>0$, we have, by § 5·51 (5),

$$\begin{aligned}\sum_{n=0}^\infty \epsilon_n J^2{}_{\nu+n}(x) &= 2\nu\int_0^x J_\nu{}^2(t)\,\frac{dt}{t} \\ &\leqslant 2\nu\int_0^\infty J_\nu{}^2(t)\,\frac{dt}{t} \\ &= 1,\end{aligned}$$

and so

$$(10)\qquad |J_\nu(x)|\leqslant 1,\quad |J_{\nu+1}(x)|\leqslant 1/\sqrt{2},$$

provided only that ν be positive; this is an interesting generalisation of Hansen's inequality (§ 2·5) which was discovered by Lommel, *Münchener Abh.* XV. (1886), pp. 548—549.

The reader may find it interesting to deduce Bateman's integral*,

$$(11)\qquad \int_0^\infty J_0(at)\,\{1-J_0(bt)\}\,\frac{dt}{t} = \begin{cases} 0, \\ \log(b/a), \end{cases}$$

from the Weber-Schafheitlin theorem.

13·43. *Gegenbauer's investigation of the Weber-Schafheitlin integral.*

In the special case in which the Bessel functions are of the same order, Gegenbauer† found that by his method Weber's integral could be evaluated in a simple manner.

If $R(2\nu+1)>R(\lambda)>R(\nu+\frac{1}{2})$ we have

$$\begin{aligned}\int_0^\infty J_\nu(at)\,J_\nu(bt)\,\frac{dt}{t^\lambda} &= \frac{(\frac{1}{2}ab)^\nu}{\Gamma(\nu+\frac{1}{2})\,\Gamma(\frac{1}{2})}\int_0^\infty\int_0^\pi \frac{J_\nu\{t\sqrt{(a^2+b^2-2ab\cos\phi)}\}}{t^{\lambda-\nu}(a^2+b^2-2ab\cos\phi)^{\frac{1}{2}\nu}}\sin^{2\nu}\phi\,d\phi\,dt \\ &= \frac{(ab)^\nu\,\Gamma(\nu-\frac{1}{2}\lambda+\frac{1}{2})}{2^\lambda\,\Gamma(\nu+\frac{1}{2})\,\Gamma(\frac{1}{2})\,\Gamma(\frac{1}{2}\lambda+\frac{1}{2})}\int_0^\pi \frac{\sin^{2\nu}\phi\,d\phi}{(a^2+b^2-2ab\cos\phi)^{\nu-\frac{1}{2}\lambda+\frac{1}{2}}}\end{aligned}$$

* *Messenger*, XLI. (1912), p. 101; for a proof of the formula by another method, see Hardy, *Messenger*, XLII. (1913), pp. 92—93.

† *Wiener Sitzungsberichte*, LXXXVIII. (2), (1884), p. 991.

by § 13·22. When $b < a$, the expression on the right is

$$\frac{b^\nu \Gamma(\nu - \frac{1}{2}\lambda + \frac{1}{2})}{2^\lambda a^{\nu-\lambda+1}\Gamma(\nu + \frac{1}{2})\Gamma(\frac{1}{2})\Gamma(\frac{1}{2}\lambda + \frac{1}{2})} \sum_{n=0}^{\infty} \left(\frac{b}{a}\right)^n \int_0^\pi C_n^{\nu-\frac{1}{2}\lambda+\frac{1}{2}}(\cos\phi)\sin^{2\nu}\phi\, d\phi.$$

Now from the recurrence formulae

$$\frac{d}{dz}\{(1-z^2)^{-\frac{1}{2}n} C_n^\mu(z)\} = (n+2\mu-1)(1-z^2)^{-\frac{1}{2}n-1} C^\mu_{n-1}(z),$$

$$\frac{d}{dz}\{(1-z^2)^{\frac{1}{2}n+\mu} C_n^\mu(z)\} = -(n+1)(1-z^2)^{\frac{1}{2}n+\mu-1} C^\mu_{n+1}(z),$$

we see that

$$\begin{aligned}(n+1)&\int_{-1}^1 (1-z^2)^{\nu-\frac{1}{2}} C^\mu_{n+1}(z)\,dz \\ &= -\int_{-1}^1 (1-z^2)^{\nu-\mu-\frac{1}{2}n+\frac{1}{2}} \frac{d}{dz}\left\{(1-z^2)^{\frac{1}{2}n+\mu} C_n^\mu(z)\right\} dz \\ &= (n+2\mu-2\nu-1)\int_{-1}^1 z(1-z^2)^{\nu-\frac{1}{2}} C_n^\mu(z)\,dz \\ &= \frac{n+2\mu-2\nu-1}{2\nu+n+1}\int_{-1}^1 (1-z^2)^{\nu+\frac{1}{2}n+\frac{1}{2}} \frac{d}{dz}\left\{(1-z^2)^{-\frac{1}{2}n} C_n^\mu(z)\right\} dz \\ &= \frac{(n+2\mu-2\nu-1)(n+2\mu-1)}{2\nu+n+1}\int_{-1}^1 (1-z^2)^{\nu-\frac{1}{2}} C^\mu_{n-1}(z)\,dz,\end{aligned}$$

so that

$$\begin{aligned}\int_0^\pi & C^\mu_{n+1}(\cos\phi)\sin^{2\nu}\phi\,d\phi \\ &= \frac{(n+2\mu-2\nu-1)(n+2\mu-1)}{(2\nu+n+1)(n+1)}\int_0^\pi C^\mu_{n-1}(\cos\phi)\sin^{2\nu}\phi\,d\phi.\end{aligned}$$

Hence it follows that

$$\begin{aligned}\int_0^\infty J_\nu(at)J_\nu(bt)\frac{dt}{t^\lambda} &= \frac{b^\nu\Gamma(\nu-\frac{1}{2}\lambda+\frac{1}{2})}{2^\lambda a^{\nu-\lambda+1}\Gamma(\nu+\frac{1}{2})\Gamma(\frac{1}{2})\Gamma(\frac{1}{2}\lambda+\frac{1}{2})}\int_0^\pi \sin^{2\nu}\phi\,d\phi \\ &\times {}_2F_1\left(\nu-\tfrac{1}{2}\lambda+\tfrac{1}{2}, \tfrac{1}{2}-\tfrac{1}{2}\lambda; \nu+1; \frac{b^2}{a^2}\right),\end{aligned}$$

and this agrees with the result of § 13·4.

The method given here is substantially the same as Gegenbauer's; but he used slightly more complicated analysis in order to avoid the necessity of appealing to the theory of analytic continuation to establish the result over the more extended range $R(2\nu+1) > R(\lambda) > -1$.

By expanding the finite integral in powers of $\cos\phi$, we obtain the formula

$$\begin{aligned}(1)\quad \int_0^\infty J_\nu(at)J_\nu(bt)\frac{dt}{t^\lambda} &= \frac{(ab)^\nu\Gamma(\nu-\frac{1}{2}\lambda+\frac{1}{2})}{2^\lambda(a^2+b^2)^{\nu-\frac{1}{2}\lambda+\frac{1}{2}}\Gamma(\nu+1)} \\ &\times {}_2F_1\left(\frac{2\nu+1-\lambda}{4}, \frac{2\nu+3-\lambda}{4}; \nu+1; \frac{4a^2b^2}{(a^2+b^2)^2}\right),\end{aligned}$$

which is valid whether $a > b$ or $a < b$. This result was given by Gegenbauer, and with this form of the result the discontinuity is masked.

The reader will find it interesting to examine the critical case obtained by putting $b = a$ in the finite integral.

13·44. *Gubler's investigation of the Weber-Schafheitlin integral.*

The integral

$$\int_0^\infty J_\mu(at) J_\nu(bt)\, dt$$

will now be investigated by the method due to Gubler*. It is convenient first to consider the more general integral

$$\int_0^\infty \frac{J_\mu(at) J_\nu(bt)}{t^\lambda}\, dt$$

even though this integral cannot be evaluated in a simple manner by Gubler's methods. It is first supposed that $R(\nu) > 0$, $R(\lambda) > \frac{1}{2}$, $R(\mu - \lambda) > -1$; and, as usual, a and b are positive, and $a > b$.

From the generalisation of Bessel's integral, given by § 6·2 (2), it is evident that the integral is equal to

$$\frac{1}{2\pi i}\int_0^\infty \frac{J_\mu(at)}{t^\lambda}\int_{-\infty}^{(0+)} z^{-\nu-1} \exp\left\{\tfrac{1}{2}bt\left(z - \frac{1}{z}\right)\right\} dz\, dt.$$

We take the contour as shewn in Fig. 29 to meet the circle $|z| = 1$ and the

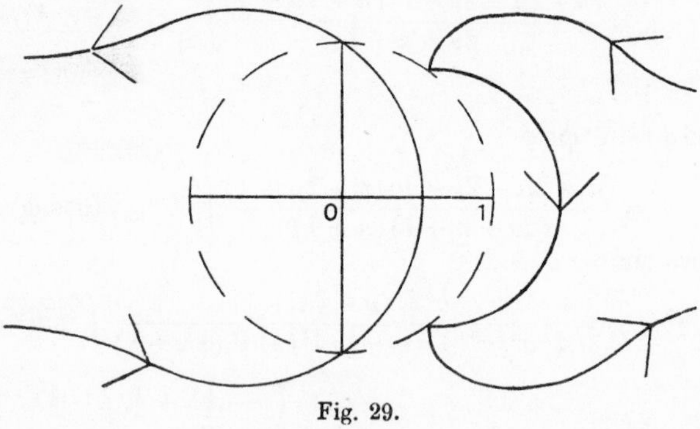

Fig. 29.

line $R(z) = 0$ only at $z = \pm i$; and then, for all the values of z and t under consideration,

$$R\{\tfrac{1}{2}bt(z - 1/z)\} \leqslant 0;$$

and the repeated integral converges absolutely, since

$$\int_0^\infty \left|\frac{J_\mu(at)}{t^\lambda}\right| dt \,.\int_{-\infty}^{0+} |z^{-\nu-1}\, dz|$$

is convergent. The order of the integrations may therefore be changed, and we have

$$\int_0^\infty \frac{J_\mu(at) J_\nu(bt)}{t^\lambda}\, dt = \frac{1}{2\pi i}\int_{-\infty}^{(0+)}\int_0^\infty \frac{J_\mu(at)}{t^\lambda} \exp\left\{\tfrac{1}{2}bt\left(z - \frac{1}{z}\right)\right\} z^{-\nu-1}\, dt dz.$$

If we write

$$b(z - 1/z) = -a(\zeta - 1/\zeta),$$

* *Math. Ann.* XLVIII. (1897), pp. 37—48.

and suppose that that value of ζ is taken for which $|\zeta| \geqslant 1$, we have, by § 13·2,

$$\int_0^\infty \frac{J_\mu(at)\,J_\nu(bt)}{t^\lambda}\,dt = \frac{1}{2\pi i}\int_{-\infty}^{(0+)} z^{-\nu-1}\frac{(\tfrac{1}{2}a)^\mu\,\Gamma(\mu-\lambda+1)}{\{\tfrac{1}{2}a(\zeta+1/\zeta)\}^{\mu-\lambda+1}\,\Gamma(\mu+1)}$$

$$\times {}_2F_1\left(\tfrac{1}{2}\mu-\tfrac{1}{2}\lambda+\tfrac{1}{2}, \tfrac{1}{2}\mu+\tfrac{1}{2}\lambda;\, \mu+1;\, \frac{4\zeta^2}{(\zeta^2+1)^2}\right)dz$$

$$= \frac{1}{2\pi i}\cdot\frac{(\tfrac{1}{2}a)^{\lambda-1}\,\Gamma(\mu-\lambda+1)}{\Gamma(\mu+1)}\int_{-\infty}^{(0+)}\frac{z^{-\nu-1}}{(\zeta+1/\zeta)^{\mu-\lambda+1}}$$

$$\times {}_2F_1\left(\mu-\lambda+1, \mu+\lambda;\, \mu+1;\, \frac{1}{1+\zeta^2}\right)dz,$$

by Kummer's transformation*.

Next consider the path described by ζ, when z describes its contour. Since the value of ζ with the greater modulus is chosen, the path is the curve on the right of the circle in Fig. 29; and the curve is irreducible because different branches of z, *qua* function of ζ, are taken on the different parts of it. The curve meets the unit circle only at $e^{\pm i\omega}$, where ω is the acute angle for which $b = a \sin \omega$.

Now both the original integral and the final contour integral are analytic functions of λ when $R(\lambda) > -1$, so long as $a \neq b$. Hence we may take† $\lambda = 0$, provided that $R(\mu) > -1$; and then we have

$$\int_0^\infty J_\mu(at)\,J_\nu(bt)\,dt = \frac{(2/a)}{2\pi i}\int_{-\infty}^{(0+)}\frac{z^{-\nu-1}}{\zeta^{\mu+1}}\,\frac{\zeta^2}{1+\zeta^2}\,dz.$$

Next write $\zeta = z\tau$ and then

$$z^2 = \frac{b\tau + a}{\tau(a\tau + b)}, \quad \zeta^2 = \frac{\tau(b\tau + a)}{a\tau + b},$$

and the τ contour is that shewn in Fig. 30; it starts from $-b/a$, encircles the origin clockwise, and returns to $-b/a$; where the contour crosses the positive half of the real axis, we have $\arg\tau = 0$.

Since $$\frac{dz}{z(1+\zeta^2)} = -\frac{1}{2}\,\frac{a\,d\tau}{\tau(b\tau + a)},$$

we find (on reversing the direction of the contour) that

$$\int_0^\infty J_\mu(at)\,J_\nu(bt)\,dt = \frac{1}{2\pi i}\int_{-b/a}^{(0+)} \tau^{\frac{1}{2}(\nu-\mu-1)}\,(b\tau + a)^{-\frac{1}{2}(\nu+\mu+1)}\,(a\tau + b)^{\frac{1}{2}(\nu+\mu-1)}\,d\tau$$

$$= \frac{b^\nu}{2\pi i a^{\nu+1}}\int_{-1}^{(0+)} w^{\frac{1}{2}(\nu-\mu-1)}\left(1 + \frac{b^2}{a^2}w\right)^{-\frac{1}{2}(\nu+\mu+1)}(1+w)^{\frac{1}{2}(\nu+\mu-1)}\,dw.$$

* *Journal für Math.* xv. (1836), p. 78, formula (57). See also Barnes, *Quarterly Journal*, xxxix. (1908), pp. 115—119.

† If $\lambda \neq 0$, the hypergeometric function does not in general reduce to an elementary function, and the analysis becomes intractable.

If we expand in ascending powers of b^2/a^2 and substitute the values of the Euler-Pochhammer integrals, then Gubler's result

$$\int_0^\infty J_\mu(at) J_\nu(bt)\, dt$$
$$= \frac{b^\nu \Gamma(\tfrac{1}{2}\mu + \tfrac{1}{2}\nu + \tfrac{1}{2})}{a^{\nu+1}\Gamma(\nu+1)\Gamma(\tfrac{1}{2}\mu - \tfrac{1}{2}\nu + \tfrac{1}{2})} \cdot {}_2F_1\left(\frac{\mu+\nu+1}{2}, \frac{\nu-\mu+1}{2}; \nu+1; \frac{b^2}{a^2}\right)$$

is manifest.

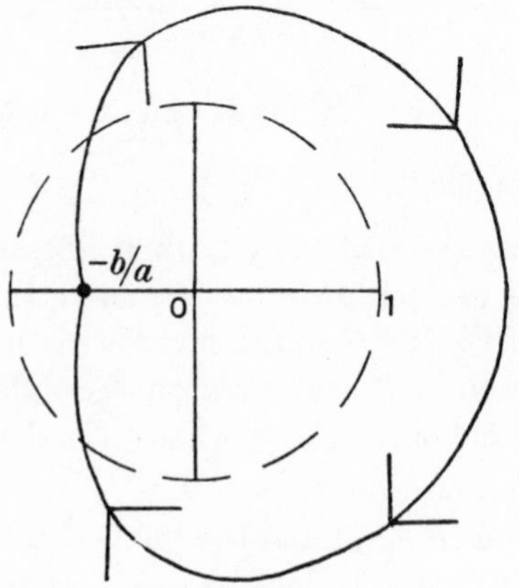

Fig. 30.

13·45. *A modification of the Weber-Schafheitlin integral.*

The integral
$$\int_0^\infty \frac{K_\mu(at) J_\nu(bt)}{t^\lambda} dt,$$
which converges if $R(a) > |I(b)|$ and $R(\nu + 1 - \lambda) > |R(\mu)|$, is expressible in terms of hypergeometric functions, like the Weber-Schafheitlin integral, but unlike that integral it has no discontinuity when $a = b$.

To evaluate it, expand $J_\nu(bt)$ in powers of b, assuming temporarily that $|b| < |a|$ in order that the result of term-by-term integration may be a convergent series. By using § 13·21 (8) it is found that

$$(1)\quad \int_0^\infty \frac{K_\mu(at) J_\nu(bt)}{t^\lambda} dt = \sum_{n=0}^\infty \frac{(-)^n (\tfrac{1}{2}b)^{\nu+2n}}{n!\,\Gamma(\nu+n+1)} \int_0^\infty t^{\nu-\lambda+2n} K_\mu(at)\, dt$$
$$= \frac{b^\nu \Gamma(\tfrac{1}{2}\nu - \tfrac{1}{2}\lambda + \tfrac{1}{2}\mu + \tfrac{1}{2}) \Gamma(\tfrac{1}{2}\nu - \tfrac{1}{2}\lambda - \tfrac{1}{2}\mu + \tfrac{1}{2})}{2^{\lambda+1} a^{\nu-\lambda+1} \Gamma(\nu+1)}$$
$$\times {}_2F_1\left(\frac{\nu-\lambda+\mu+1}{2}, \frac{\nu-\lambda-\mu+1}{2}; \nu+1; -\frac{b^2}{a^2}\right)$$

and, in particular,

$$(2)\quad \int_0^\infty K_\mu(at) J_\nu(bt)\, t^{\mu+\nu+1} dt = \frac{(2a)^\mu (2b)^\nu \Gamma(\mu+\nu+1)}{(a^2+b^2)^{\mu+\nu+1}},$$

provided that $R(\nu+1) > |R(\mu)|$ and $R(a) > |I(b)|$.

Formula (1) was given by Heaviside* when $\mu = \nu = 0$ and λ is 0 and -1.

* *Electromagnetic Theory*, III. (London, 1912), pp. 249, 268, 275.

13·46. *Generalisations of the Weber-Schafheitlin integral.*

To obtain the values of integrals containing three Bessel functions under the integral sign, take the integral

$$\int_0^\infty \frac{J_\mu(at)\,J_\nu(bt)}{t^\lambda}\,dt,$$

replace b by $\sqrt{(b^2+c^2-2bc\cos\phi)}$, where b and c are positive, multiply by $\sin^{2\nu}\phi/(b^2+c^2-2bc\cos\phi)^{\frac{1}{2}\nu}$ and integrate. It is thus found that

$$\int_0^\infty \frac{J_\mu(at)\,J_\nu(bt)\,J_\nu(ct)}{t^{\lambda+\nu}}\,dt = \frac{(\tfrac{1}{2}bc)^\nu}{\Gamma(\nu+\tfrac{1}{2})\,\Gamma(\tfrac{1}{2})}\int_0^\infty\int_0^\pi \frac{J_\mu(at)\,J_\nu(\varpi t)}{\varpi^\nu\, t^\lambda}\sin^{2\nu}\phi\,d\phi\,dt,$$

where $\varpi=\sqrt{(b^2+c^2-2bc\cos\phi)}$; and the integral on the right is absolutely convergent if

$$R(\nu)>-\tfrac{1}{2},\ R(\mu+\nu+2)>R(\lambda+1)>0.$$

Change the order of the integrations on the right; then the result of the integration with respect to t is an elementary function of ϖ if $\lambda+\nu+1=\pm\mu$, by the formulae

$$\int_0^\infty \frac{J_\mu(at)\,J_\nu(\varpi t)}{\varpi^\nu\,t^{\mu-\nu-1}}\,dt = \begin{cases} 0, & (a<\varpi) \\ \dfrac{(a^2-\varpi^2)^{\mu-\nu-1}}{2^{\mu-\nu-1}\,a^\mu\,\Gamma(\mu-\nu)}. & (a>\varpi)\end{cases}$$

It follows that

$$(1)\quad \int_0^\infty J_\mu(at)\,J_\nu(bt)\,J_\nu(ct)\,t^{1-\mu}\,dt$$

$$=\frac{(\tfrac{1}{2}bc)^\nu}{2^{\mu-\nu-1}\,a^\mu\,\Gamma(\mu-\nu)\,\Gamma(\nu+\tfrac{1}{2})\,\Gamma(\tfrac{1}{2})}\int_0^A (a^2-b^2-c^2+2bc\cos\phi)^{\mu-\nu-1}\sin^{2\nu}\phi\,d\phi,$$

in which the value of A is

$$0,\quad \operatorname{arc\,cos}\frac{b^2+c^2-a^2}{2bc},\quad \pi$$

according as a^2 is less than, between, or greater than the two numbers

$$(b-c)^2,\ (b+c)^2,$$

provided that both $R(\mu)$ and $R(\nu)$ exceed $-\tfrac{1}{2}$.

In particular

$$(2)\qquad \int_0^\infty J_{\nu+1}(at)\,J_\nu(bt)\,J_\nu(ct)\,\frac{dt}{t^\nu} = \frac{(\tfrac{1}{2}bc)^\nu}{a^{\nu+1}\,\Gamma(\nu+\tfrac{1}{2})\,\Gamma(\tfrac{1}{2})}\int_0^A \sin^{2\nu}\phi\,d\phi.$$

Multiply by $a^{\nu+1}$ and differentiate under the integral sign with respect to a; and we then obtain the interesting result that, if $R(\nu)>-\tfrac{1}{2}$,

$$(3)\qquad \int_0^\infty J_\nu(at)\,J_\nu(bt)\,J_\nu(ct)\,\frac{dt}{t^{\nu-1}} = \frac{2^{\nu-1}\,\Delta^{2\nu-1}}{(abc)^\nu\,\Gamma(\nu+\tfrac{1}{2})\,\Gamma(\tfrac{1}{2})},$$

when a, b, c are the sides of a triangle of area Δ; but if a, b, c are *not* sides of a triangle, the integral is zero.

This formula is due to Sonine, *Math. Ann.* XVI. (1880), p. 46; other aspects of it have been investigated by Dougall, *Proc. Edinburgh Math. Soc.* XXXVII. (1919), pp. 33—47.

It has been observed by Macdonald* that the integral on the left in (1) is always expressible in terms of Legendre functions. The expression may be derived from the integral on the right in the following manner:

When a, b, c are the sides of a triangle, by the substitution

$$\sin \tfrac{1}{2}\phi = \sin \tfrac{1}{2}A \sin \theta$$

we have

$$\int_0^A (a^2 - b^2 - c^2 + 2bc \cos \phi)^{\mu-\nu-1} \sin^{2\nu} \phi \, d\phi$$

$$= (2bc)^{\mu-\nu-1} \int_0^A (\cos \phi - \cos A)^{\mu-\nu-1} \sin^{2\nu} \phi \, d\phi$$

$$= 2^{2\mu-2\nu-1} (bc)^{\mu-\nu-1} \sin^{2\mu-1} \tfrac{1}{2}A \, . \int_0^{\frac{1}{2}\pi} (1 - \sin^2 \tfrac{1}{2}A \sin^2 \theta)^{\nu-\frac{1}{2}} \sin^{2\nu} \theta \cos^{2\mu-2\nu-1} \theta \, d\theta$$

$$= 4^{\mu-1} (bc)^{\mu-\nu-1} \sin^{2\mu-1} \tfrac{1}{2}A \, \frac{\Gamma(\nu+\frac{1}{2})\,\Gamma(\mu-\nu)}{\Gamma(\mu+\frac{1}{2})} \, . \, {}_2F_1(\tfrac{1}{2}+\nu, \tfrac{1}{2}-\nu; \mu+\tfrac{1}{2}; \sin^2 \tfrac{1}{2}A),$$

and therefore, if $R(\mu)$ and $R(\nu)$ exceed $-\frac{1}{2}$, and a, b, c are the sides of a triangle, we have

$$(4) \qquad \int_0^\infty J_\mu(at)\, J_\nu(bt)\, J_\nu(ct)\, t^{1-\mu}\, dt = \frac{(bc)^{\mu-1} \sin^{\mu-\frac{1}{2}} A}{(2\pi)^{\frac{1}{2}} a^\mu} P_{\nu-\frac{1}{2}}^{\frac{1}{2}-\mu}(\cos A).$$

If, however, $a^2 > (b+c)^2$, and we write

$$a^2 - b^2 - c^2 = 2bc \cosh \mathscr{A},$$

we have

$$\int_0^\pi (a^2 - b^2 - c^2 + 2bc \cos \phi)^{\mu-\nu-1} \sin^{2\nu} \phi \, d\phi$$

$$= (2bc)^{\mu-\nu-1} \int_0^\pi (\cosh \mathscr{A} + \cos \phi)^{\mu-\nu-1} \sin^{2\nu} \phi \, d\phi$$

$$= (2bc \cosh \mathscr{A})^{\mu-\nu-1} \frac{\Gamma(\nu+\frac{1}{2})\,\Gamma(\frac{1}{2})}{\Gamma(\nu+1)}$$

$$\times {}_2F_1\left(\frac{\nu-\mu}{2}+1, \frac{\nu-\mu+1}{2}; \nu+1; \operatorname{sech}^2 \mathscr{A}\right),$$

so that, when $a^2 > (b+c)^2$, we have

$$(5) \quad \int_0^\infty J_\mu(at)\, J_\nu(bt)\, J_\nu(ct)\, t^{1-\mu}\, dt = \frac{(bc)^{\mu-1} \cos \nu\pi \, . \sinh^{\mu-\frac{1}{2}} \mathscr{A}}{(\frac{1}{2}\pi^3)^{\frac{1}{2}} a^\mu} Q_{\nu-\frac{1}{2}}^{\frac{1}{2}-\mu}(\cosh \mathscr{A}).$$

In like manner, we deduce from § 13·45 (2) that

$$(6) \quad \int_0^\infty K_\mu(at)\, J_\nu(bt)\, J_\nu(ct)\, t^{\mu+1}\, dt = -\frac{a^\mu \cos \nu\pi \, . (X^2-1)^{-\frac{1}{2}\mu-\frac{1}{4}}}{(2\pi)^{\frac{1}{2}} (bc)^{\mu+1} \sin(\mu+\nu)\pi} Q_{\nu-\frac{1}{2}}^{\mu+\frac{1}{2}}(X),$$

where $2bc\,X = a^2 + b^2 + c^2$; and in this formula a, b, c may be complex, provided only that the four numbers

$$R(a \pm ib \pm ic)$$

are positive; this result is also due to Macdonald.

* *Proc. London Math. Soc.* (2) VII. (1909), pp. 142—149.

[NOTE. The apparent discrepancy between these formulae and the formulae of Macdonald's paper is a consequence of the different definitions adopted for the function Q_n^m; see § 5·71.]

Other formulae involving three Bessel functions may be obtained by taking formula § 11·6 (1), replacing z by x, multiplying by

$$2J_\rho(x\cos\theta)/x^\lambda$$

and integrating.

It is thus found that

(7) $$\int_0^\infty J_\mu(x\cos\phi\cos\Phi)\,J_\nu(x\sin\phi\sin\Phi)\,J_\rho(x\cos\theta)\frac{dx}{x^{\lambda-1}}$$

$$=\frac{\cos^\mu\phi\cos^\mu\Phi\sin^\nu\phi\sin^\nu\Phi\cos^\rho\theta}{2^{\lambda-1}\Gamma(\rho+1)\{\Gamma(\nu+1)\}^2}$$

$$\times\sum_{n=0}^\infty\left[(-)^n(\mu+\nu+2n+1)\frac{\Gamma(\mu+\nu+n+1)\,\Gamma(\nu+n+1)}{n!\,\Gamma(\mu+n+1)}\right.$$

$$\times\frac{\Gamma(\frac{1}{2}\mu+\frac{1}{2}\nu+\frac{1}{2}\rho-\frac{1}{2}\lambda+n+1)}{\Gamma(\frac{1}{2}\mu+\frac{1}{2}\nu-\frac{1}{2}\rho+\frac{1}{2}\lambda+n+1)}$$

$$\times{}_2F_1\left(\frac{\mu+\nu+\rho-\lambda}{2}+n+1,\ \frac{\rho-\lambda-\mu-\nu}{2}-n;\ \rho+1;\ \cos^2\theta\right)$$

$$\times{}_2F_1(-n,\ \mu+\nu+n+1;\ \nu+1;\ \sin^2\phi)$$

$$\left.\times{}_2F_1(-n,\ \mu+\nu+n+1;\ \nu+1;\ \sin^2\Phi)\right],$$

when

$$R(\mu+\nu+\rho+2)>R(\lambda)>-\tfrac{1}{2}$$

and $\cos\theta$ is not equal to $\pm\cos(\Phi\pm\phi)$.

Some special cases of this result have been given by Gegenbauer in a letter to Kapteyn, *Proc. Section of Sci., K. Acad. van Wet. te Amsterdam*, IV. (1902), pp. 584—588.

Some extensions of formula (3) have been given recently by Nicholson*. If $a_1, a_2, \ldots a_m$ are positive numbers arranged in descending order of magnitude it is easy to shew that, if

$$a_1>\sum_{n=2}^m a_m,\quad R(\nu)>-1,$$

then

(8) $$\int_0^\infty \prod_{n=1}^m J_\nu(a_n t)\frac{dt}{t^{\nu m-2\nu-1}}=0;$$

the simplest method of establishing this result is by induction, by substituting Gegenbauer's formula of § 11·41 [on the assumption that $R(\nu)>-\frac{1}{2}$] for $J_\nu(a_{m-1}t)\,J_\nu(a_m t)$, and then changing the order of the integrations.

When $a_1, a_2, \ldots a_m$ are such that they can be the lengths of the sides of a polygon, the integral is intractable unless $m=3$ (the case already considered), or $m=4$.

* *Quarterly Journal*, XLVIII. (1920), pp. 321—329. Some associated integrals will be discussed in § 13·48.

When a_1, a_2, a_3, a_4 can form the sides of a quadrilateral, we write

$$16\Delta^2 = \prod_{n=1}^{4} (a_1 + a_2 + a_3 + a_4 - 2a_n),$$

so that Δ is the area of the cyclic quadrilateral with sides a_1, a_2, a_3, a_4.

The integral can be evaluated in a simple form only* when $\nu = 0$; but to deduce its value, it is simplest first to obtain an expression for the integral when $R(\nu) > \frac{1}{2}$, and deduce the value for $\nu = 0$ by analytic continuation; the value of the integral assumes different forms according as†

$$a_1 + a_4 \lesseqgtr a_2 + a_3,$$

i.e. according as

$$\Delta^2 \gtrless a_1 a_2 a_3 a_4.$$

We write $\varpi^2 = a_2^2 + a_3^2 - 2a_2 a_3 \cos\phi$, and replace $J_\nu(a_2 t) J_\nu(a_3 t)$ by Gegenbauer's formula, so that

$$\int_0^\infty \prod_{n=1}^{4} J_\nu(a_n t) \frac{dt}{t^{2\nu-1}} = \frac{(\frac{1}{2}a_2 a_3)^\nu}{\Gamma(\nu+\frac{1}{2})\Gamma(\frac{1}{2})} \int_0^\infty \int_0^\pi \frac{J_\nu(a_1 t) J_\nu(a_4 t) J_\nu(\varpi t)}{\varpi^\nu t^{\nu-1}} \sin^{2\nu}\phi \, d\phi \, dt$$

$$= \frac{(a_2 a_3)^\nu (a_1 a_4)^{-\nu}}{2^{4\nu-1}\{\Gamma(\nu+\frac{1}{2})\Gamma(\frac{1}{2})\}^2} \int \{(a_1+a_4)^2 - \varpi^2\}^{\nu-\frac{1}{2}} \{\varpi^2 - (a_1-a_4)^2\}^{\nu-\frac{1}{2}} \frac{\sin^{2\nu}\phi \, d\phi}{\varpi^{2\nu}},$$

where the lower limit is given by $\varpi = a_1 - a_4$ and the upper limit by $\varpi = a_1 + a_4$ or $a_2 + a_3$, whichever is the smaller.

We write

$$\frac{\varpi^2 - (a_1 - a_4)^2}{\varpi^2 - (a_2 - a_3)^2} = \frac{(a_1+a_4)^2 - (a_1-a_4)^2}{(a_1+a_4)^2 - (a_2-a_3)^2} x^2,$$

so that the upper limit for x is 1 or $\Delta/\sqrt{(a_1 a_2 a_3 a_4)}$: this expression will be called $1/k$.

We now carry out the process of analytic continuation (unless $a_1 + a_4 = a_2 + a_3$, when the integrals diverge at the upper limit if $\nu = 0$), and we get

$$\int_0^\infty \prod_{n=1}^{4} J_0(a_n t)\, t\, dt$$

$$= \frac{4}{\pi^2} \int [\{(a_1+a_4)^2 - \varpi^2\}\{\varpi^2 - (a_1-a_4)^2\}\{\varpi^2 - (a_2-a_3)^2\}\{(a_2+a_3)^2 - \varpi^2\}]^{-\frac{1}{2}} \varpi \, d\varpi$$

$$= \frac{1}{\pi^2 \Delta} \int_0^{1 \text{ or } 1/k} \frac{dx}{\sqrt{\{(1-x^2)(1-k^2x^2)\}}}.$$

Hence

$$(9) \qquad \int_0^\infty \prod_{n=1}^{4} J_0(a_n t)\, t\, dt = \begin{cases} \dfrac{1}{\pi^2 \Delta} K\left(\dfrac{\sqrt{(a_1 a_2 a_3 a_4)}}{\Delta}\right), \\[2ex] \dfrac{1}{\pi^2 \sqrt{(a_1 a_2 a_3 a_4)}} K\left(\dfrac{\Delta}{\sqrt{(a_1 a_2 a_3 a_4)}}\right), \end{cases}$$

where K denotes the complete elliptic integral of the first kind, and that one whose modulus is less than unity is to be taken.

* For other values of ν it is expressible as a hypergeometric function of three variables.

† We still suppose that $a_1 \geqslant a_2 \geqslant a_3 \geqslant a_4$.

Nicholson has also evaluated

$$\int_0^\infty \{J_\nu(at)\}^4 \frac{dt}{t^{2\nu-1}}$$

when $R(\nu) > 0$ and $a > 0$. The simplest procedure is to regard the integral as a special case of the last, so that it is equal to

$$\frac{1}{2^{4\nu-1}\{\Gamma(\nu+\frac{1}{2})\,\Gamma(\frac{1}{2})\}^2}\int_0^\pi \{2a^2(1+\cos\phi)\}^{\nu-\frac{1}{2}} \frac{\sin^{2\nu}\phi\, d\phi}{\{2a^2(1-\cos\phi)\}^{\frac{1}{2}}},$$

and hence*

$$(10) \qquad \int_0^\infty \{J_\nu(at)\}^4 \frac{dt}{t^{2\nu-1}} = \frac{a^{2\nu-2}\,\Gamma(2\nu)\,\Gamma(\nu)}{2\pi\,\Gamma(3\nu)\,\{\Gamma(\nu+\frac{1}{2})\}^2}.$$

13·47. *The discontinuous integrals of Sonine and Gegenbauer.*

Several discontinuous integrals, of a more general character than the Weber-Schafheitlin type, have been investigated by Sonine† and Gegenbauer‡; some modifications of these integrals are of importance in physical problems.

The first example§ which we shall take is due to Sonine, namely

$$(1) \quad \int_0^\infty J_\mu(bt)\,\frac{J_\nu\{a\sqrt{(t^2+z^2)}\}}{(t^2+z^2)^{\frac{1}{2}\nu}}\, t^{\mu+1}\,dt$$

$$= \begin{cases} 0, & (a<b) \\ \dfrac{b^\mu}{a^\nu}\left\{\dfrac{\sqrt{(a^2-b^2)}}{z}\right\}^{\nu-\mu-1} J_{\nu-\mu-1}\{z\sqrt{(a^2-b^2)}\}. & (a>b) \end{cases}$$

To secure convergence, a and b are taken to be positive and $R(\nu) > R(\mu) > -1$; if $a = b$, then we take $R(\nu) > R(\mu+1) > 0$. The number z is an unrestricted complex number, and the integral reduces to a case of the Weber-Schafheitlin integral when z is zero.

The integrals involved being absolutely convergent||, we see from § 6·2 (8) that, if $c > 0$, then

$$\int_0^\infty J_\mu(bt)\,\frac{J_\nu\{a\sqrt{(t^2+z^2)}\}}{(t^2+z^2)^{\frac{1}{2}\nu}}\, t^{\mu+1}\,dt$$

$$= \frac{1}{2\pi i}\int_0^\infty \int_{c-\infty i}^{c+\infty i} J_\mu(bt)\, t^{\mu+1}\, u^{-\nu-1} \exp\left[\tfrac{1}{2}a\left(u - \frac{t^2+z^2}{u}\right)\right] du\,dt$$

$$= \frac{b^\mu}{2\pi i a^{\mu+1}} \int_{c-\infty i}^{c+\infty i} u^{\mu-\nu} \exp\left[\frac{(a^2-b^2)\,u}{2a} - \frac{az^2}{2u}\right] du.$$

* An arithmetical error in Nicholson's work has been corrected. The result for values of $R(\nu)$ between 0 and $\frac{1}{2}$ is obtained by analytic continuation.

† *Math. Ann.* XVI. (1880), p. 38 *et seq.*

‡ *Wiener Sitzungsberichte*, LXXXVIII. (1884), pp. 990—1003.

§ This formula is also investigated by Cailler, *Mém. de la Soc. de phys. de Genève*, XXXIV. (1902—1905), pp. 348—349.

|| The convergence is absolute only when $R(\nu) > R(\mu+1) > 0$; for values of ν not covered by this condition, the formula is to be established by analytic continuation.

When $a<b$ the contour involved in the last integral may be deformed into an indefinitely great semicircle on the right of the imaginary axis, and the integral along this is zero; but, when $a \geqslant b$, we have to apply § 6·2 (8), and then we obtain the formula stated*.

A related integral

$$(2) \quad \int_0^\infty J_\mu(bt) \frac{K_\nu\{a\sqrt{(t^2+z^2)}\}}{(t^2+z^2)^{\frac{1}{2}\nu}} t^{\mu+1}\,dt = \frac{b^\mu}{a^\nu}\left\{\frac{\sqrt{(a^2+b^2)}}{z}\right\}^{\nu-\mu-1} K_{\nu-\mu-1}\{z\sqrt{(a^2+b^2)}\}$$

may be evaluated in a similar manner.

We suppose that a and b are positive†, and that $R(\mu) > -1$; in evaluating the integral it is convenient to suppose that $|\arg z| < \frac{1}{4}\pi$, though we may subsequently extend the range of values of z to $|\arg z| < \frac{1}{2}\pi$ by analytic continuation.

From § 6·22 (8) it follows that the integral on the left of (2) is equal to

$$\frac{1}{2}\int_0^\infty \int_0^\infty J_\mu(bt)\, t^{\mu+1} u^{-\nu-1} \exp\left[-\tfrac{1}{2}a\left(u+\frac{t^2+z^2}{u}\right)\right] du\,dt$$

$$= \frac{b^\mu}{2a^{\mu+1}} \int_0^\infty u^{\mu-\nu} \exp\left[-\frac{(a^2+b^2)u}{2a} - \frac{az^2}{2u}\right] du$$

$$= \frac{b^\mu}{a^\nu}\left\{\frac{\sqrt{(a^2+b^2)}}{z}\right\}^{\nu-\mu-1} K_{\nu-\mu-1}\{z\sqrt{(a^2+b^2)}\},$$

by § 6·22 (8); and this is the result stated.

Now make $\arg z \to \pm\frac{1}{2}\pi$. If we put $z=iy$, where $y>0$, we find that

$$(3) \quad \int_0^\infty J_\mu(bt) \frac{K_\nu\{a\sqrt{(t^2-y^2)}\}}{(t^2-y^2)^{\frac{1}{2}\nu}} t^{\mu+1}\,dt$$

$$= \tfrac{1}{2}\pi e^{-\frac{1}{2}\pi(\nu-\mu-\frac{1}{2})} \frac{b^\mu}{a^\nu}\left\{\frac{\sqrt{(a^2+b^2)}}{y}\right\}^{\nu-\mu-1} [J_{\nu-\mu-1}\{y\sqrt{(a^2+b^2)}\} - iY_{\nu-\mu-1}\{y\sqrt{(a^2+b^2)}\}],$$

provided that $R(\nu) < 1$; and it is supposed that the path of integration avoids the singularity $t=y$ by an indentation *above* the singular point, and that interpretation is given to $\sqrt{(t^2-y^2)}$ which makes the expression positive when $t>y$.

If we had put $z=-iy$, we should have had the indentation *below* the real axis and the sign of i would have been changed throughout (3).

In particular

$$(4) \qquad \int_0^\infty J_0(bt) \frac{\exp\{-a\sqrt{(t^2-y^2)}\}}{(t^2-y^2)^{\frac{1}{2}}} t\,dt = \frac{\exp\{\mp iy\sqrt{(a^2+b^2)}\}}{\sqrt{(a^2+b^2)}}$$

where the upper or lower sign is taken according as the indentation passes above or below the axis of y.

* For physical applications of this integral, see Lamb, *Proc. London Math. Soc.* (2) VII. (1909), pp. 122—141.

† With certain limitations, a and b may be complex.

The last formula (with the lower sign*) has been used in physical investigations by Sommerfeld, *Ann. der Physik und Chemie*, (4) XXVIII. (1909), pp. 682—683; see also Bateman, *Electrical and Optical Wave-Motion* (Cambridge, 1915), p. 72.

If in (1) we divide by b^μ and make $b \to 0$, we obtain Sonine's formula

$$\int_0^\infty \frac{J_\nu\{a\sqrt{(t^2+z^2)}\}}{(t^2+z^2)^{\frac{1}{2}\nu}} t^{2\mu+1}\,dt = \frac{2^\mu\,\Gamma(\mu+1)}{a^{\mu+1}\,z^{\nu-\mu-1}} J_{\nu-\mu-1}(az), \tag{5}$$

provided that $a \geqslant 0$ and $R(\frac{1}{2}\nu - \frac{1}{4}) > R(\mu) > -1$; this might have been established independently by the same method.

Similarly, from (2) we have

$$\int_0^\infty \frac{K_\nu\{a\sqrt{(t^2+z^2)}\}}{(t^2+z^2)^{\frac{1}{2}\nu}} t^{2\mu+1}\,dt = \frac{2^\mu\,\Gamma(\mu+1)}{a^{\mu+1}\,z^{\nu-\mu-1}} K_{\nu-\mu-1}(az), \tag{6}$$

if $a > 0$ and $R(\mu) > -1$.

In (5) replace ν by 2ν, a by $2\sin\theta$ and integrate from $\theta = 0$ to $\theta = \frac{1}{2}\pi$. It follows that

$$\int_0^\infty \frac{J_\nu^2\{\sqrt{(t^2+z^2)}\}}{(t^2+z^2)^\nu} t^{2\mu+1}\,dt = \frac{\Gamma(\mu+1)}{\pi z^{2\nu-\mu-1}} \int_0^{\frac{1}{2}\pi} \frac{J_{2\nu-\mu-1}(2z\sin\theta)\,d\theta}{\sin^{\mu+1}\theta}; \tag{7}$$

this is valid when $R(\nu - \frac{1}{2}) > R(\mu) > -1$.

The integral on the right is easily expansible in powers of z; but the only case of interest is when $2\nu = 2\mu + 3$, and we then have

$$\int_0^\infty \frac{J_\nu^2\{\sqrt{(t^2+z^2)}\}}{(t^2+z^2)^\nu} t^{2\nu-2}\,dt = \frac{\Gamma(\nu-\frac{1}{2})}{2z^{\nu+1}\sqrt{\pi}} \mathbf{H}_\nu(2z), \tag{8}$$

so that

$$\int_z^\infty \frac{J_\nu^2(u)}{u^{2\nu-1}} (u^2 - z^2)^{\nu-\frac{3}{2}}\,du = \frac{\Gamma(\nu-\frac{1}{2})}{2z^{\nu+1}\sqrt{\pi}} \mathbf{H}_\nu(2z); \tag{9}$$

and these are valid if $R(\nu) > \frac{1}{2}$. The last formula was established in a different manner (when $\nu = 1$) by Struve†; and from it we deduce the important theorem that‡, *when $\nu > \frac{1}{2}$ and $x > 0$, $\mathbf{H}_\nu(x)$ is positive.* Struve's integral is of considerable value in the Theory of Diffraction.

Some variations of Sonine's discontinuous integral are obtainable by multiplying by $b^{\mu+1}$ and then integrating with respect to b from 0 to b.

It is thus found that

$$\int_0^\infty J_{\mu+1}(bt) \frac{J_\nu\{a\sqrt{(t^2+z^2)}\}}{(t^2+z^2)^{\frac{1}{2}\nu}} t^\mu\,dt$$
$$= \frac{1}{a^\nu b^{\mu+1}} \int_0 u^{2\mu+1} \left\{\frac{\sqrt{(a^2-u^2)}}{z}\right\}^{\nu-\mu-1} J_{\nu-\mu-1}\{z\sqrt{(a^2-u^2)}\}\,du,$$

the upper limit in the last integral being b or a, whichever is the smaller.

* My thanks are due to Professor Love for pointing out to me the desirability of emphasizing the ambiguity of sign.

† *Ann. der Physik und Chemie*, (3) XVII. (1882), pp. 1010—1011.

‡ Cf. § 10·45.

If $b < a$, the integral on the right seems intractable, but, when $b > a$, we put $u = a \sin \theta$ and deduce that

$$(10) \qquad \int_0^\infty J_{\mu+1}(bt) \frac{J_\nu\{a\sqrt{(t^2+z^2)}\}}{(t^2+z^2)^{\frac{1}{2}\nu}} t^\mu \, dt = \frac{2^\mu \Gamma(\mu+1)}{b^{\mu+1}} \frac{J_\nu(az)}{z^\nu},$$

provided that $R(\nu+1) > R(\mu) > -1$; this is one of Sonine's integrals.

If we replace a by u in (1) and then take $a \leqslant b$ and integrate with respect to u from a to ∞ after dividing by $u^{\nu-1}$, we find that, *when z is restricted to be positive,*

$$\int_0^\infty J_\mu(bt) \frac{J_{\nu-1}\{a\sqrt{(t^2+z^2)}\}}{(t^2+z^2)^{\frac{1}{2}\nu+1}} t^{\mu+1} \, dt$$

$$= a^{\nu-1} b^\mu \int_b^\infty \left\{\frac{\sqrt{(u^2-b^2)}}{z}\right\}^{\nu-\mu-1} J_{\nu-\mu-1}\{z\sqrt{(u^2-b^2)}\} \frac{du}{u^{2\nu-1}}$$

$$= \frac{a^{\nu-1} b^\mu}{z^{\nu-\mu-1}} \int_0^\infty \frac{v^{\nu-\mu} J_{\nu-\mu-1}(vz) \, dv}{(v^2+b^2)^\nu}$$

$$= \frac{2a^{\nu-1} b^\mu}{\Gamma(\nu) z^{\nu-\mu-1}} \int_0^\infty \int_0^\infty t^{2\nu-1} v^{\nu-\mu} J_{\nu-\mu-1}(vz) \exp\{-t^2(v^2+b^2)\} \, dv dt$$

$$= \frac{2^{\mu-\nu+1} a^{\nu-1} b^\mu}{\Gamma(\nu)} \int_0^\infty t^{2\mu-1} \exp\left(-t^2 b^2 - \frac{z^2}{4t^2}\right) dt,$$

by § 13·3 (4), and thence we see that

$$(11) \qquad \int_0^\infty J_\mu(bt) \frac{J_{\nu-1}\{a\sqrt{(t^2+z^2)}\}}{(t^2+z^2)^{\frac{1}{2}\nu+1}} t^{\mu+1} \, dt = \frac{a^{\nu-1} z^\mu}{2^{\nu-1}\Gamma(\nu)} K_\mu(bz),$$

provided that $a < b$ and $R(\nu+2) > R(\mu) > -1$; the restriction that z is positive may now be removed.

Formula (10), which may be written in the form

$$(12) \qquad \int_0^\infty J_\mu(bt) \frac{J_\nu\{a\sqrt{(t^2+z^2)}\}}{(t^2+z^2)^{\frac{1}{2}\nu}} t^{\mu-1} \, dt = \frac{2^{\mu-1}\Gamma(\mu)}{b^\mu} \frac{J_\nu(az)}{z^\nu},$$

where $R(\nu+2) > R(\mu) > 0$ and $b > a$, has been generalised in two ways by Gegenbauer*, by the usual methods of substituting Neumann's integral and Gegenbauer's integral (cf. § 13·1) for the second Bessel function.

The first method gives

$$(13) \qquad \int_0^\infty J_\mu(bt) \frac{J_\nu\{a\sqrt{(t^2+z^2)}\} J_\lambda\{a\sqrt{(t^2+z^2)}\}}{(t^2+z^2)^{\frac{1}{2}(\lambda+\nu)}} t^{\mu-1} \, dt$$

$$= \frac{2}{\pi} \int_0^\infty \int_0^{\frac{1}{2}\pi} J_\mu(bt) \frac{J_{\lambda+\nu}\{2a\cos\phi \,.\, \sqrt{(t^2+z^2)}\}}{(t^2+z^2)^{\frac{1}{2}(\lambda+\nu)}} t^{\mu-1} \cos(\lambda-\nu)\phi \, d\phi \, dt$$

$$= \frac{2^{\mu-1}\Gamma(\mu)}{b^\mu} \frac{J_\nu(az) J_\lambda(az)}{z^{\lambda+\nu}},$$

provided that $b > 2a$ and $R(\nu+\lambda+\frac{5}{2}) > R(\mu) > 0$.

* *Wiener Sitzungsberichte*, LXXXVIII. (2), (1884), pp. 1002—1003.

If $\varpi = \sqrt{(a^2 + c^2 - 2ac\cos\phi)}$, the second method gives

$$(14)\quad \int_0^\infty J_\mu(bt) \frac{J_\nu\{a\sqrt{(t^2+z^2)}\}\, J_\nu\{c\sqrt{(t^2+z^2)}\}}{(t^2+z^2)^\nu} t^{\mu-1}\,dt$$

$$= \frac{(\frac{1}{2}ac)^\nu}{\Gamma(\nu+\frac{1}{2})\,\Gamma(\frac{1}{2})} \int_0^\infty \int_0^\pi J_\mu(bt) \frac{J_\nu\{\varpi\sqrt{(t^2+z^2)}\}}{\varpi^\nu (t^2+z^2)^{\frac{1}{2}\nu}} t^{\mu-1} \sin^{2\nu}\phi\, d\phi\, dt$$

$$= \frac{2^{\mu-1}\Gamma(\mu)}{b^\mu} \frac{J_\nu(az)}{z^\nu} \frac{J_\nu(cz)}{z^\nu},$$

if $b > a + c$ and $R(2\nu + \frac{5}{2}) > R(\mu) > 0$.

By induction it follows that, if $b > \Sigma a$,

$$(15)\qquad \int_0^\infty J_\mu(bt) \frac{\prod\limits_a [J_\nu\{a\sqrt{(t^2+z^2)}\}]}{(t^2+z^2)^{\frac{1}{2}n\nu}} t^{\mu-1}\,dt = \frac{2^{\mu-1}\Gamma(\mu)}{b^\mu} \prod_a \left[\frac{J_\nu(az)}{z^\nu}\right],$$

where the product applies to n values of a, and

$$R(n\nu + \tfrac{1}{2}n + \tfrac{1}{2}) > R(\mu) > 0.$$

If the induction of the second method is used after applying the first method once, we find still further generalisations.

The special case of (15) when $z \to 0$ is

$$(16)\qquad \int_0^\infty J_\mu(bt) \prod_a [J_\nu(at)]\, t^{\mu-n\nu-1}\,dt = \frac{2^{\mu-1}\Gamma(\mu)}{b^\mu} \prod_a \left[\frac{(\frac{1}{2}a)^\nu}{\Gamma(\nu+1)}\right];$$

this has been pointed out by Kluyver, *Proc. Section of Sci., K. Akad. van Wet. te Amsterdam*, XI. (1909), pp. 749—755.

13·48. *The problem of random flights.*

A problem which was propounded by Pearson* (in the case of two-dimensional displacements) is as follows:

"A man starts from a point O and walks a distance a in a straight line; he then turns through any angle whatever and walks a distance a in a second straight line. He repeats this process n times.

"I require the probability that after these n stretches he is at a distance between r and $r + \delta r$ from his starting point, O."

The generalised form of the problem, in which the stretches may be taken to be unequal, say $a_1, a_2, \ldots, a_n$, has been solved by Kluyver† with the help of the discontinuous integrals which were discussed in § 13·42; and subsequently Rayleigh‡ gave the full details of the analysis of the problem (which had been examined somewhat briefly by Kluyver), and then obtained the solution of the corresponding problem for flights in three dimensions.

If s_m is the resultant of $a_1, a_2, \ldots, a_m$ $(m = 1, 2, \ldots, n-1)$, and if θ_m is the

* *Nature*, LXXII. (1905), pp. 294, 342 (see also p. 318); *Drapers' Company Research Memoirs*, Biometric Series, III. (1906).

† *Proc. Section of Sci., K. Akad. van Wet. te Amsterdam*, VIII. (1906), pp. 341—350.

‡ *Phil. Mag.* (6) XXXVII. (1919), pp. 321—347. [*Scientific Papers*, VI. (1920), pp. 604—626.]

angle between s_m and a_{m+1}, then, in the two-dimensional problem, all values of the angle θ_m between $-\pi$ and π are equally probable.

Now let $P_n(r; a_1, a_2, \ldots, a_n)$ denote the probability that after n stretches the distance from the starting point shall be *less* than r, so that the probability that the distance lies between r and $r + \delta r$ is

$$\frac{dP_n(r; a_1, a_2, \ldots, a_n)}{dr}\,\delta r.$$

It is then evident that

$$P_n(r; a_1, a_2, \ldots, a_n) = \frac{1}{(2\pi)^{n-1}} \int_{-\pi}^{\pi} \int_{-\pi}^{\pi} \ldots \int_{-\pi}^{\pi} \int d\theta_{n-1}\, d\theta_{n-2} \ldots d\theta_2\, d\theta_1,$$

where $\theta_1, \theta_2, \ldots, \theta_{n-2}$ assume all values between $-\pi$ and π, while θ_{n-1} is to assume only such values as make*

$$s_n \leqslant r,$$

for each set of values of $\theta_1, \theta_2, \ldots, \theta_{n-2}$.

Now (§ 13·42)

$$r\int_0^\infty J_1(rt)\, J_0(s_n t)\, dt = \begin{cases} 1, & (s_n < r) \\ 0, & (s_n > r) \end{cases}$$

and so, *if this discontinuous factor is inserted in the* $(n-1)$-*tuple integral*, the range of values of θ_{n-1} may be taken to be $(-\pi, \pi)$.

We change the order of the integrations with respect to θ_{n-1} and t, and, remembering that

$$s^2{}_n = s^2{}_{n-1} + a^2{}_n - 2s_{n-1}\, a_n \cos\theta_{n-1},$$

we get

$$r\int_{-\pi}^{\pi}\int_0^\infty J_1(rt)\, J_0(s_n t)\, dt\, d\theta_{n-1} = 2\pi r \int_0^\infty J_1(rt)\, J_0(s_{n-1} t)\, J_0(a_n t)\, dt$$

by § 11·41 (16). We next make the substitution

$$s^2{}_{n-1} = s^2{}_{n-2} + a^2{}_{n-1} - 2s_{n-2}\, a_{n-1} \cos\theta_{n-2},$$

and perform the integration with respect to θ_{n-2}. By repetitions of this process we deduce ultimately that

$$P_n(r; a_1, a_2, \ldots, a_n) = r\int_0^\infty J_1(rt) \prod_{m=1}^{n} J_0(a_m t)\, dt,$$

and this is Kluyver's result.

We shall now consider the corresponding problem for space of p dimensions. In this problem it is no longer the case that all values of θ_m are equally likely. If generalised polar coordinates (in which θ_m is regarded as a co-latitude) are used, the element of generalised solid angle contains θ_m only by the factor $\sin^{p-2}\theta_m\, d\theta_m$, and θ_m varies from 0 to π. The symmetry with respect to the polar axis enables us to disregard the factor depending on the longitudes.

* It is to be remembered that s_m is a function of the variables $\theta_1, \theta_2, \ldots, \theta_{m-1}$.

If $P_n(r; a_1, a_2, \ldots, a_n | p)$ denotes the probability that the final distance is less than r, we deduce, as before, that

$$P_n(r; a_1, a_2, \ldots, a_n | p)$$
$$= \left\{\frac{\Gamma(\frac{1}{2}p)}{\Gamma(\frac{1}{2}p-\frac{1}{2})\Gamma(\frac{1}{2})}\right\}^{n-1} \int_0^\pi \int_0^\pi \ldots \int_0^\pi \int \prod_{m=1}^{n-1} \sin^{p-2}\theta_m \, . \, d\theta_{n-1}\, d\theta_{n-2} \ldots d\theta_2 d\theta_1,$$

where the integration with respect to θ_{n-1} extends over the values of θ_{n-1} which make $s_n < r$.

The discontinuous factor which we now introduce is

$$r^{\frac{1}{2}p} \int_0^\infty J_{\frac{1}{2}p}(rt) \frac{J_{\frac{1}{2}p-1}(s_n t)}{s_n^{\frac{1}{2}p-1}} dt = \begin{cases} 1, & (s_n < r) \\ 0, & (s_n > r) \end{cases}$$

and then, since by § 11·41 (16),

$$\int_0^\pi \frac{J_{\frac{1}{2}p-1}(s_m t)}{s_m^{\frac{1}{2}p-1}} \sin^{p-2}\theta_{m-1} d\theta_{m-1} = \Gamma(\tfrac{1}{2}p-\tfrac{1}{2})\,\Gamma(\tfrac{1}{2}) \frac{J_{\frac{1}{2}p-1}(s_{m-1}t)}{(s_{m-1})^{\frac{1}{2}p-1}} \cdot \frac{J_{\frac{1}{2}p-1}(a_m t)}{(\frac{1}{2}a_m t)^{\frac{1}{2}p-1}},$$

we infer that

$$P_n(r; a_1, a_2, \ldots, a_n | p) = r\{\Gamma(\tfrac{1}{2}p)\}^{n-1} \int_0^\infty (\tfrac{1}{2}rt)^{\frac{1}{2}p-1} . J_{\frac{1}{2}p}(rt) \prod_{m=1}^{n} \left\{\frac{J_{\frac{1}{2}p-1}(a_m t)}{(\frac{1}{2}a_m t)^{\frac{1}{2}p-1}}\right\} dt.$$

When the displacements $a_1, a_2, \ldots, a_n$ are all equal to a, and n is large, we may approximate to the value of the integral by Laplace's* process. The important part of the integrand is the part for which t is small, and, for such values of t,

$$J_{\frac{1}{2}p-1}(at) \sim \frac{(\frac{1}{2}at)^{\frac{1}{2}p-1}}{\Gamma(\frac{1}{2}p)} \exp\left(-\frac{a^2t^2}{2p}\right),$$

so that (§ 13·3)

$$P_n(r; a, a, \ldots, a | p) \sim \frac{r}{\Gamma(\frac{1}{2}p)} \int_0^\infty (\tfrac{1}{2}rt)^{\frac{1}{2}p-1} J_{\frac{1}{2}p}(rt) \exp\left(-\frac{na^2t^2}{2p}\right) dt$$
$$= \frac{1}{\Gamma(\frac{1}{2}p+1)} \left(\frac{r^2p}{2na^2}\right)^{\frac{1}{2}p} . {}_1F_1\left(\tfrac{1}{2}p; \tfrac{1}{2}p+1; -\frac{r^2p}{2na^2}\right).$$

This process of approximation has been carried much further by Rayleigh in the cases $p=2$, $p=3$, while Pearson has published various arithmetical tables connected with the problem.

13·49. *The discontinuous integrals of Gallop and Hardy.*

The integral

$$\int_{-\infty}^\infty \frac{J_\mu\{a(z+t)\}}{(z+t)^\mu} \frac{J_\nu\{b(\zeta+t)\}}{(\zeta+t)^\nu} dt$$

is convergent if a and b are positive and $R(\mu+\nu) > -1$; when $a=b$ the last condition must be replaced by $R(\mu+\nu) > 0$.

The special case of the integral in which $\mu=0$, $\nu=\frac{1}{2}$ has been investigated by Gallop, *Quarterly Journal*, XXI. (1886), pp. 232—234; and the case in which $a=b$ has been investigated by Hardy, *Proc. London Math. Soc.* (2) VII. (1909), pp. 469. The integral is obviously to be associated with the discontinuous integrals of Weber and Schafheitlin.

* *La théorie analytique des Probabilités* (Paris, 1812), chapter III. The process may be recognised as a somewhat disguised form of the method of steepest descents.

To evaluate the integral in the general case, the method discovered by Hardy is effective; suppose that $a \geqslant b$, and at first let us take $R(\nu) > -\frac{1}{2}$, $R(\mu) > \frac{1}{2}$, so that Poisson's integral may be substituted for the second Bessel function and all the integrals which will be used are absolutely convergent. Write t in place of $t+\zeta$, and let $z-\zeta = Z$, so that the integral to be evaluated becomes

$$\int_{-\infty}^{\infty} \frac{J_\mu\{a(Z+t)\}}{(Z+t)^\mu} \frac{J_\nu(bt)}{t^\nu}\, dt$$

$$= \frac{(\frac{1}{2}b)^\nu}{\Gamma(\nu+\frac{1}{2})\Gamma(\frac{1}{2})} \int_{-\infty}^{\infty}\int_0^\pi \frac{J_\mu\{a(Z+t)\}}{(Z+t)^\mu} \cos(bt\cos\phi)\sin^{2\nu}\phi\, d\phi\, dt$$

$$= \frac{(\frac{1}{2}b)^\nu}{\Gamma(\nu+\frac{1}{2})\Gamma(\frac{1}{2})} \int_{-\infty}^{\infty}\int_0^\pi \frac{J_\mu(at)}{t^\mu} \cos\{b(t-Z)\cos\phi\}\sin^{2\nu}\phi\, d\phi\, dt$$

$$= \frac{2\,.\,(\frac{1}{2}b)^\nu}{\Gamma(\nu+\frac{1}{2})\Gamma(\frac{1}{2})} \int_0^{\infty}\int_0^\pi \frac{J_\mu(at)}{t^\mu} \cos(bt\cos\phi)\cos(bZ\cos\phi)\sin^{2\nu}\phi\, d\phi\, dt$$

$$= \frac{2\,.\,(\frac{1}{2}b)^\nu}{(2a)^\mu\,\Gamma(\mu+\frac{1}{2})\Gamma(\nu+\frac{1}{2})} \int_0^\pi (a^2-b^2\cos^2\phi)^{\mu-\frac{1}{2}}\cos(bZ\cos\phi)\sin^{2\nu}\phi\, d\phi,$$

by a special case of § 13·4 (2).

This integral is expressible in a simple manner only when $\mu = \frac{1}{2}$, a case considered by Gallop, or when $a = b$, the case considered by Hardy.

We easily obtain Gallop's two results

$$\text{(1)} \qquad \int_{-\infty}^{\infty} \frac{\sin a(z+t)}{z+t} J_0(bt)\, dt = \pi J_0(bz), \qquad (b \leqslant a)$$

$$\text{(2)} \qquad \int_{-\infty}^{\infty} \frac{\sin a(z+t)}{z+t} J_0(bt)\, dt = 2\int_0^a \frac{\cos uz\,.\,du}{\sqrt{(b^2-u^2)}}, \qquad (b \geqslant a)$$

and Hardy's formula

$$\text{(3)} \quad \int_{-\infty}^{\infty} \frac{J_\mu\{a(z+t)\}}{(z+t)^\mu} \frac{J_\nu\{a(\zeta+t)\}}{(\zeta+t)^\nu}\, dt = \frac{\Gamma(\mu+\nu)\Gamma(\frac{1}{2})}{\Gamma(\mu+\frac{1}{2})\Gamma(\nu+\frac{1}{2})} \times \left(\frac{2}{a}\right)^{\frac{1}{2}} \frac{J_{\mu+\nu-\frac{1}{2}}\{a(z-\zeta)\}}{(z-\zeta)^{\mu+\nu-\frac{1}{2}}}$$

The reader will find it interesting to obtain (1) by integrating

$$\int \frac{e^{ai(z+t)}}{z+t} J_0(bt)\, dt$$

round the contour formed by the real axis and an indefinitely great semicircle above it; it has to be supposed that there is an indentation at $-z$ when z is real.

The integral

$$\int_{-\infty}^{\infty} \frac{|t|\sin a(z+t)}{z+t} J_0(bt)\, dt$$

has also been considered by Gallop. To evaluate it, we observe that

$$\frac{t}{z+t} = 1 - \frac{z}{z+t},$$

and so the integral may be written in the form

$$\int_{-\infty}^{0} \{-\sin a(z+t)\} J_0(bt)\,dt + \int_0^\infty \sin a(z+t) J_0(bt)\,dt$$

$$+ z\int_{-\infty}^{\infty} \frac{\sin a(z+t)}{z+t} J_0(bt)\,dt - 2z\int_0^\infty \frac{\sin a(z+t)}{z+t} J_0(bt)\,dt$$

$$= 2\cos az \int_0^\infty \sin at\, J_0(bt)\,dt + z\int_0^a \int_{-\infty}^{\infty} \cos u(z+t) J_0(bt)\,dt\,du$$

$$- 2z\int_0^a \int_0^\infty \cos u(z+t) J_0(bt)\,dt\,du$$

$$= 2\cos az \int_0^\infty \sin at\, J_0(bt)\,dt + 2z\int_0^a \int_0^\infty \sin uz \sin ut\, J_0(bt)\,dt\,du.$$

Hence, when $a > b$,

$$(4)\quad \int_{-\infty}^{\infty} \frac{|t| \sin a(z+t)}{z+t} J_0(bt)\,dt = \frac{2\cos az}{\sqrt{(a^2-b^2)}} + 2z\int_b^a \frac{\sin uz}{\sqrt{(u^2-b^2)}}\,du$$

$$= \frac{2\cos az}{\sqrt{(a^2-b^2)}} + 2z\int_0^{\operatorname{arc\,cosh} a/b} \sin(zb\cosh\theta)\,d\theta,$$

but, when $a < b$,

$$(5)\quad \int_{-\infty}^{\infty} \frac{|t| \sin a(z+t)}{z+t} J_0(bt)\,dt = 0.$$

13·5. *Definite integrals evaluated by contour integration.*

A large number of definite integrals can be evaluated by considering integrals of the forms

$$\frac{1}{2\pi i}\int \phi(z) H_\nu^{(1)}(az)\,dz, \quad \frac{1}{2\pi i}\int \phi(z)\,\mathscr{C}_\mu(bz) H_\nu^{(1)}(az)\,dz,$$

taken round suitable contours; it is supposed that $\phi(z)$ is an algebraic function, and that a is positive.

The appropriate contours are of two types. We take the first type when $\phi(z)$ has no singularities except poles in the upper half-plane; the contour is taken to be a large semicircle above the real axis with its centre at the origin, together with that part of the real axis (indented at the origin) which joins the ends of the semicircle.

We take the second type when $\phi(z)$ has branch points in the upper half-plane; the contour is derived from the first type by inserting loops starting from and ending at the indentation, one loop passing round each branch point, so that the integrand has no singularity inside the contour.

A more powerful method (cf. § 13·1) which is effective in evaluating integrals with Bessel functions under the integral sign is to substitute for the Bessel function one of the integrals discussed in § 6·5, and change the order of the integrations; since the integrand in § 6·5 (7) is $O(x^{\nu-\delta})$, *qua* function of x, where δ is an arbitrarily small positive number, the double integral usually converges absolutely when the original integral does so, and the interchange produces no theoretical difficulties.

13·51. *Hankel's integrals involving one Bessel function.*

Before Hankel investigated the more abstruse integrals which will be discussed in Chapter XIV, he evaluated a large class of definite integrals* by considering

$$\frac{1}{2\pi i}\int \frac{z^{\rho-1} H_\nu^{(1)}(az)}{(z^2-r^2)^{m+1}}\,dz$$

taken round the first type of contour described in § 13·5. In this integral, a is positive, m is a positive integer (zero included), r is a complex number with positive imaginary part, and

$$|R(\nu)| < R(\rho) < 2m + \frac{7}{2}.$$

The first inequality secures the convergence of the integral when the radius of the indentation tends to zero; and (as a consequence of Jordan's lemma) the second inequality ensures that the integral round the large semicircle tends to zero as the radius tends to infinity.

The only singularity of the integrand inside the contour is the point r. It follows that

$$\frac{1}{2\pi i}\int_0^\infty \frac{x^{\rho-1}\{H_\nu^{(1)}(ax) - e^{\rho\pi i} H_\nu^{(1)}(axe^{\pi i})\}}{(x^2-r^2)^{m+1}}\,dx = \frac{1}{2\pi i}\int^{(r+)} \frac{z^{\rho-1} H_\nu^{(1)}(az)\,dz}{(z^2-r^2)^{m+1}}$$
$$= \frac{1}{4\pi i}\int^{(r^2+)} \frac{\zeta^{\frac{1}{2}\rho-1} H_\nu^{(1)}(a\sqrt{\zeta})\,d\zeta}{(\zeta-r^2)^{m+1}} = \frac{1}{2\,.\,m!}\left(\frac{d}{dr^2}\right)^m \{r^{\rho-2} H_\nu^{(1)}(ar)\}.$$

It follows from § 3·62 (5) that

$$(1)\quad \int_0^\infty [(1+e^{(\rho-\nu)\pi i}) J_\nu(ax) + i(1-e^{(\rho-\nu)\pi i}) Y_\nu(ax)]\frac{x^{\rho-1}\,dx}{(x^2-r^2)^{m+1}}$$
$$= \frac{\pi i}{2^m\,.\,m!}\left(\frac{d}{r\,dr}\right)^m \{r^{\rho-2} H_\nu^{(1)}(ar)\}.$$

This result can be expressed in a neater form by writing $r = ik$, so that $R(k) > 0$. It is thus found that†

$$(2)\quad \int_0^\infty [\cos\tfrac{1}{2}(\rho-\nu)\pi\,.\,J_\nu(ax) + \sin\tfrac{1}{2}(\rho-\nu)\pi\,.\,Y_\nu(ax)]\frac{x^{\rho-1}\,dx}{(x^2+k^2)^{m+1}}$$
$$= \frac{(-)^{m+1}}{2^m\,.\,m!}\left(\frac{d}{k\,dk}\right)^m \{k^{\rho-2} K_\nu(ak)\}.$$

* Hankel's work was published posthumously, *Math. Ann.* VIII. (1875), pp. 458—461. A partial investigation of the integral with $\nu = n$, $\rho = 2n+2$, $m = 2n$ was given by Neumann, *Theorie der Bessel'schen Functionen* (Leipzig, 1867), p. 58.

† The evaluation of integrals of this character which contain only one of the two Bessel functions is effected in § 13·6.

The reader should notice the following special cases of this formula:

(3) $$\int_0^\infty \{\cos\nu\pi \,.\, J_\nu(ax) - \sin\nu\pi \,.\, Y_\nu(ax)\}\frac{x^{1-\nu}\,dx}{(x^2+k^2)^{m+1}} = \frac{a^m K_{\nu+m}(ak)}{2^m \,.\, m!\, k^{\nu+m}},$$

(4) $$\int_0^\infty \frac{x^{\nu+1} J_\nu(ax)\,dx}{(x^2+k^2)^{m+1}} = \frac{a^m k^{\nu-m} K_{\nu-m}(ak)}{2^m \,.\, m!}.$$

The former is valid when $-2m-\frac{3}{2} < R(\nu) < 1$, and the latter when $-1 < R(\nu) < 2m+\frac{3}{2}$. For an extension of (4) to the case when m is not an integer, see § 13·6 (2).

The special formula

(5) $$\int_0^\infty \frac{xJ_0(ax)\,dx}{x^2+k^2} = K_0(ak),$$

has been pointed out by Mehler, *Math. Ann.* XVIII. (1881), p. 194, and Basset, *Hydrodynamics*, II. (Cambridge, 1889), p. 19; while Nicholson, *Quarterly Journal*, XLII. (1911), p. 220, has obtained another special formula

(6) $$\int_0^\infty \frac{Y_0(ax)\,dx}{x^2+k^2} = -\frac{K_0(ak)}{k},$$

by a complicated transformation of repeated integrals.

Some integrals resembling those just given may be established here, though it is most convenient to prove them without using Cauchy's theorem.

Thus, Nicholson has observed that

$$\begin{aligned}\int_0^\infty \frac{J_0(ax)\,dx}{x^2+k^2} &= \frac{2}{\pi}\int_0^\infty\int_0^{\frac{1}{2}\pi} \frac{\cos(ax\cos\theta)}{x^2+k^2}\,d\theta\,dx \\ &= \frac{1}{k}\int_0^{\frac{1}{2}\pi} e^{-ak\cos\theta}\,d\theta \\ &= \frac{\pi}{2k}\{I_0(ak) - \mathbf{L}_0(ak)\},\end{aligned}$$

by § 10·4 (11), provided that a and $R(k)$ are both positive; so that

(7) $$\int_0^\infty \frac{J_0(ax)\,dx}{x^2+k^2} = \frac{\pi}{2k}\{I_0(ak) - \mathbf{L}_0(ak)\}.$$

More generally, if $R(\nu) > -\frac{1}{2}$, we have

$$I_\nu(ak) - \mathbf{L}_\nu(ak) = \frac{2(\frac{1}{2}ak)^\nu}{\Gamma(\nu+\frac{1}{2})\Gamma(\frac{1}{2})}\int_0^{\frac{1}{2}\pi} e^{-ak\cos\theta}\sin^{2\nu}\theta\,d\theta,$$

and since, by a special form of (2),

$$\int_{-\infty}^\infty \frac{x^\nu e^{aix\cos\theta}\,dx}{x^2+k^2} = \frac{\pi}{k}(ik)^\nu e^{-ak\cos\theta},$$

provided that $R(\nu) < 2$ and a is positive, it follows that

$$\begin{aligned}I_\nu(ak) - \mathbf{L}_\nu(ak) &= \frac{2ki^{-\nu}(\frac{1}{2}a)^\nu}{\pi\Gamma(\nu+\frac{1}{2})\Gamma(\frac{1}{2})}\int_0^{\frac{1}{2}\pi}\int_{-\infty}^\infty \frac{x^\nu\sin^{2\nu}\theta \,.\, e^{aix\cos\theta}}{x^2+k^2}\,dx\,d\theta \\ &= \frac{ki^{-\nu}}{\pi}\int_{-\infty}^\infty \frac{J_\nu(ax) + i\mathbf{H}_\nu(ax)}{x^2+k^2}\,dx \\ &= \frac{ki^{-\nu}}{\pi}\int_0^\infty [(1+e^{\nu\pi i})J_\nu(ax) + i(1-e^{\nu\pi i})\mathbf{H}_\nu(ax)]\frac{dx}{x^2+k^2}\end{aligned}$$

and so we have the formula

$$(8)\quad \int_0^\infty [\cos \tfrac{1}{2}\nu\pi \,.\, J_\nu(ax) + \sin \tfrac{1}{2}\nu\pi \,.\, \mathbf{H}_\nu(ax)] \frac{dx}{x^2+k^2} = \frac{\pi}{2k}\{I_\nu(ak) - \mathbf{L}_\nu(ak)\},$$

where $a > 0$, $R(k) > 0$ and $-\frac{1}{2} < R(\nu) < 2$. The change in the order of the integrations presents no great theoretical difficulties.

A somewhat similar integral is

$$\int_0^\infty \frac{x^\nu K_\nu(ax)\,dx}{x^2+k^2},$$

which converges if $R(\nu) > -\frac{1}{2}$ and $R(a) > 0$.

If we choose k so that $R(k) > 0$, we have, by § 6·16 (1),

$$\begin{aligned}\int_0^\infty \frac{x^\nu K_\nu(ax)\,dx}{x^2+k^2} &= \frac{\Gamma(\nu+\frac{1}{2})}{\Gamma(\frac{1}{2})}\int_0^\infty\int_0^\infty \frac{(2a)^\nu \cos xu \,.\, du\,dx}{(x^2+k^2)(u^2+a^2)^{\nu+\frac{1}{2}}} \\ &= \frac{\pi\,\Gamma(\nu+\frac{1}{2})}{2k\,\Gamma(\frac{1}{2})}\int_0^\infty \frac{(2a)^\nu e^{-uk}\,du}{(u^2+a^2)^{\nu+\frac{1}{2}}} \\ &= \frac{\pi^2 k^{\nu-1}}{4\cos\nu\pi}\{\mathbf{H}_{-\nu}(ak) - Y_{-\nu}(ak)\},\end{aligned}$$

when we use § 10·41 (3). Hence, when $R(\nu) > -\frac{1}{2}$,

$$(9)\quad \int_0^\infty \frac{x^\nu K_\nu(ax)\,dx}{x^2+k^2} = \frac{\pi^2 k^{\nu-1}}{4\cos\nu\pi}\{\mathbf{H}_{-\nu}(ak) - Y_{-\nu}(ak)\},$$

and therefore, when $R(\nu) < \frac{1}{2}$,

$$(10)\quad \int_0^\infty \frac{K_\nu(ax)}{x^2+k^2}\frac{dx}{x^\nu} = \frac{\pi^2}{4k^{\nu+1}\cos\nu\pi}\{\mathbf{H}_\nu(ak) - Y_\nu(ak)\}.$$

These formulae (when $\nu = 0$) are due to Nicholson, and the last has also been given by Heaviside.

The integral
$$\int_0^\infty \frac{J_\nu(ax)}{x^2+k^2}\frac{dx}{x^\nu}$$
has been investigated by Gegenbauer*. To evaluate it, we suppose that $R(\nu) > -\frac{1}{2}$ and that a and $R(k)$ are both positive; we then have

$$\begin{aligned}\int_0^\infty \frac{J_\nu(ax)}{x^2+k^2}\frac{dx}{x^\nu} &= \frac{2(\frac{1}{2}a)^\nu}{\Gamma(\nu+\frac{1}{2})\Gamma(\frac{1}{2})}\int_0^\infty\int_0^{\frac{1}{2}\pi} \frac{\cos(ax\cos\theta)}{x^2+k^2}\sin^{2\nu}\theta\,d\theta\,dx \\ &= \frac{(\frac{1}{2}a)^\nu\pi}{\Gamma(\nu+\frac{1}{2})\Gamma(\frac{1}{2})k}\int_0^{\frac{1}{2}\pi} e^{-ak\cos\theta}\sin^{2\nu}\theta\,d\theta,\end{aligned}$$

and so

$$(11)\quad \int_0^\infty \frac{J_\nu(ax)}{x^2+k^2}\frac{dx}{x^\nu} = \frac{\pi}{2k^{\nu+1}}\{I_\nu(ak) - \mathbf{L}_\nu(ak)\}.$$

* *Wiener Sitzungsberichte*, LXXII. (2), (1876), p. 349. Gegenbauer's result is incorrect because he omitted to insert the term $-\mathbf{L}_\nu(ak)$; and consequently the results which he deduced from his formula are also incorrect. A similar error was made by Basset, *Proc. Camb. Phil. Soc.* VI. (1889), p. 11. The correct result was given by Gubler, *Zürich Vierteljahrsschrift*, XLVII. (1902), pp. 422—424.

The condition $R(\nu) > -\frac{1}{2}$ may now be replaced by the less stringent condition $R(\nu) > -\frac{5}{2}$, by analytic continuation.

An integral which may be evaluated in the form of a series by this method is

$$\int_0^\infty \frac{\sinh ax}{\sinh \pi x} J_\nu(bx)\, x^{\nu+1}\, dx,$$

which is a generalisation of Neumann's integral described in § 13·2; it is supposed that

$$|R(a)| + |I(b)| < \pi \text{ and } R(\nu) > -1.$$

By taking
$$\frac{1}{2\pi i}\int \frac{\sinh az}{\sinh \pi z} H_\nu^{(1)}(bz)\, z^{\nu+1}\, dz$$

round the contour used in this section, we find that the definite integral is πi times the sum of the residues of

$$\frac{\sinh az}{\sinh \pi z} H_\nu^{(1)}(bz)\, z^{\nu+1}$$

at the points i, $2i$, $3i$, ... It follows that

$$\int_0^\infty \frac{\sinh ax}{\sinh \pi x} J_\nu(bx)\, x^{\nu+1}\, dx = \frac{2}{\pi}\sum_{n=1}^\infty (-)^{n-1} n^{\nu+1} \sin na \,.\, K_\nu(nb). \tag{12}$$

The series converges rapidly if b is at all large.

An integral expressible as a similar series was investigated by Riemann, *Ann. der Physik und Chemie*, (2) xcv. (1855), pp. 132—135.

13·52. *The generalisation of Hankel's integral.*

Let us next consider the integral

$$\frac{1}{2\pi i}\int \frac{z^{\rho-1} H_\nu^{(1)}(az)\, dz}{(z^2+k^2)^{\mu+1}}.$$

This differs from Hankel's integral in containing the (complex) number μ in place of the integer m. The conditions for convergence (with the second type of contour specified in § 13·5) are*

$$a > 0, \quad |R(\nu)| < R(\rho) < 2R(\mu) + \tfrac{7}{2}.$$

The contour is chosen with a loop to exclude the point ik, as shewn in Fig. 31, and then there are no poles inside the contour; and the integral round the large semicircle tends to zero as the radius tends to infinity. Hence

$$\frac{1}{2\pi i}\int_0^\infty \left[H_\nu^{(1)}(ax) - e^{(\rho-2\mu)\pi i} H_\nu^{(1)}(axe^{\pi i})\right] \frac{x^{\rho-1}\, dx}{(x^2+k^2)^{\mu+1}}$$
$$= \frac{1}{2\pi i}\int_0^{(ik+)} \frac{z^{\rho-1} H_\nu^{(1)}(az)\, dz}{(z^2+k^2)^{\mu+1}}.$$

Now
$$\frac{1}{2\pi i}\int_0^{(ik+)} \frac{z^{\rho-1}\, dz}{(z^2+k^2)^{\mu+1}} = -\tfrac{1}{2} e^{(\frac{1}{2}\rho-\mu)\pi i} \frac{k^{\rho-2\mu-2}\,\Gamma(\frac{1}{2}\rho)}{\Gamma(\frac{1}{2}\rho-\mu)\,\Gamma(\mu+1)}.$$

* As in § 13·51, we take $R(k) > 0$.

Hence, when we expand $H_\nu^{(1)}(az)$ in ascending powers of z, we find that

$$\int_0^\infty [(1+e^{(\rho-\nu-2\mu)\pi i}) J_\nu(ax) + i(1-e^{(\rho-\nu-2\mu)\pi i}) Y_\nu(ax)] \frac{x^{\rho-1}\,dx}{(x^2+k^2)^{\mu+1}}$$

$$= \frac{\pi e^{\frac{1}{2}(\rho-\nu-2\mu)\pi i}}{\sin\nu\pi\,.\,\Gamma(\mu+1)} \left[(\tfrac{1}{2}a)^\nu k^{\rho+\nu-2\mu-2} \sum_{m=0}^\infty \frac{\Gamma(\frac{1}{2}\rho+\frac{1}{2}\nu+m)\,.\,(\frac{1}{2}ak)^{2m}}{m!\,\Gamma(\nu+m+1)\,\Gamma(\frac{1}{2}\rho+\frac{1}{2}\nu-\mu+m)} \right.$$

$$\left. - (\tfrac{1}{2}a)^{-\nu} k^{\rho-\nu-2\mu-2} \sum_{m=0}^\infty \frac{\Gamma(\frac{1}{2}\rho-\frac{1}{2}\nu+m)\,.\,(\frac{1}{2}ak)^{2m}}{m!\,\Gamma(-\nu+m+1)\,\Gamma(\frac{1}{2}\rho-\frac{1}{2}\nu-\mu+m)} \right],$$

and therefore

$$(1)\quad \int_0^\infty [\cos(\tfrac{1}{2}\rho-\tfrac{1}{2}\nu-\mu)\pi\,.\,J_\nu(ax) + \sin(\tfrac{1}{2}\rho-\tfrac{1}{2}\nu-\mu)\pi\,.\,Y_\nu(ax)] \frac{x^{\rho-1}\,dx}{(x^2+k^2)^{\mu+1}}$$

$$= \frac{\pi k^{\rho-2\mu-2}}{2\sin\nu\pi\,.\,\Gamma(\mu+1)}$$

$$\times \left[\frac{(\frac{1}{2}ak)^\nu\,\Gamma(\frac{1}{2}\rho+\frac{1}{2}\nu)}{\Gamma(\nu+1)\,\Gamma(\frac{1}{2}\rho+\frac{1}{2}\nu-\mu)}\,{}_1F_2\left(\frac{\rho+\nu}{2};\ \frac{\rho+\nu}{2}-\mu,\ \nu+1;\ \frac{a^2k^2}{4}\right) \right.$$

$$\left. - \frac{(\frac{1}{2}ak)^{-\nu}\,\Gamma(\frac{1}{2}\rho-\frac{1}{2}\nu)}{\Gamma(1-\nu)\,\Gamma(\frac{1}{2}\rho-\frac{1}{2}\nu-\mu)}\,{}_1F_2\left(\frac{\rho-\nu}{2};\ \frac{\rho-\nu}{2}-\mu,\ 1-\nu;\ \frac{a^2k^2}{4}\right) \right].$$

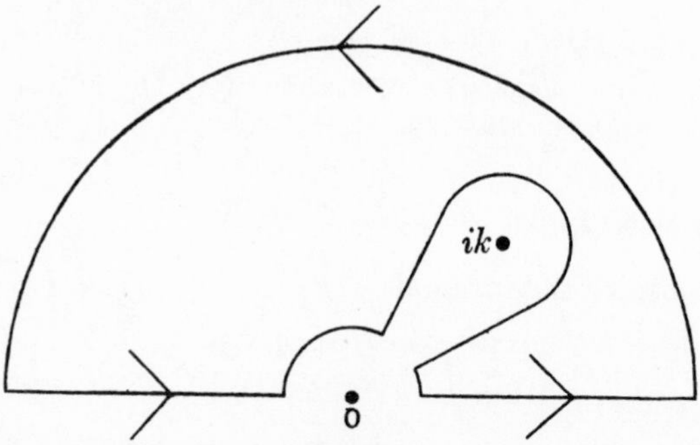

Fig. 31.

It is natural to enquire whether the integral of this type which contains a single Bessel function cannot be evaluated; it seems that the only effective method of evaluating it is the method which will be explained in § 13·6.

13·53. *Hankel's integrals involving a product of Bessel functions.*

Integrals resembling those of § 13·51, except that they contain a product of Bessel functions instead of a single Bessel function, have been investigated by Hankel* by applying Cauchy's theorem to the integral

$$\frac{1}{2\pi i}\int z^{\rho-1}\,\mathscr{C}_\mu(bz)\,\frac{H_\nu^{(1)}(az)\,dz}{(z^2-r^2)^{m+1}},$$

in which $a \geqslant b > 0$, m is a positive integer, r is a complex number with a positive imaginary part, $\mathscr{C}_\mu$ denotes any cylinder function of order μ, and

$$|R(\nu)| + |R(\mu)| < R(\rho) < 2m+4.$$

* *Math. Ann.* VIII. (1875), pp. 461—467.

[When $\mathscr{C}_\mu$ is a Bessel function of the first kind, $|R(\mu)|$ may be replaced by $-R(\mu)$ in this inequality.]

When $a=b$, the presence of a non-oscillatory term* in the asymptotic expansion of the integrand shews that we must replace $2m+4$ by $2m+3$ in the inequality in order to make the integral, when taken round a large semicircle above the real axis, tend to zero as the radius tends to infinity.

The contour to be taken is that of § 13·52; and if we proceed in the manner of that section, we find that

$$(1)\quad \frac{1}{2\pi i}\int_0^\infty [\mathscr{C}_\mu(bx)\,H_\nu^{(1)}(ax)-e^{\rho\pi i}\,\mathscr{C}_\mu(bxe^{\pi i})\,H_\nu^{(1)}(axe^{\pi i})]\frac{x^{\rho-1}\,dx}{(x^2-r^2)^{m+1}}$$
$$=\frac{1}{2^{m+1}.\,m!}\left(\frac{d}{rdr}\right)^m[r^{\rho-2}\,\mathscr{C}_\mu(br)\,H_\nu^{(1)}(ar)].$$

Numerous special cases of this result are given by Hankel.

It must be pointed out that, when $\rho=2m+3$ and $a=b$, the integral round the large semicircle tends to a non-zero limit as the radius tends to infinity; and, if

$$\mathscr{C}_\mu(az)\equiv c_1 H_\mu^{(1)}(az)+c_2 H_\mu^{(2)}(az),$$

we then obtain the new formula

$$(2)\quad \frac{1}{2\pi i}\int_0^\infty [\mathscr{C}_\mu(ax)\,H_\nu^{(1)}(ax)+\mathscr{C}_\mu(axe^{\pi i})\,H_\nu^{(1)}(axe^{\pi i})]\frac{x^{2m+2}\,dx}{(x^2-r^2)^{m+1}}$$
$$=\frac{1}{2^{m+1}.\,m!}\left(\frac{d}{rdr}\right)^m[r^{2m+1}\,\mathscr{C}_\mu(ar)\,H_\nu^{(1)}(ar)]-\frac{c_2 e^{\frac{1}{2}(\mu-\nu)\pi i}}{\pi a}.$$

The particular case of (1) in which $\rho=2$, $m=0$ and $\mathscr{C}_\mu$ is a Bessel function of the first kind deserves special mention; it is

$$(3)\quad \int_0^\infty J_\mu(bx)[\cos\tfrac{1}{2}(\mu-\nu)\pi\,.\,J_\nu(ax)+\sin\tfrac{1}{2}(\mu-\nu)\pi\,.\,Y_\nu(ax)]\frac{x\,dx}{x^2-r^2}$$
$$=\tfrac{1}{2}\pi i\,e^{\frac{1}{2}(\nu-\mu)\pi i}\,J_\mu(br)\,H_\nu^{(1)}(ar),$$

provided that $a\geqslant b>0$ and $R(\mu)>|R(\nu)|-2$.

If we take $\mu=\nu$ and $R(\nu)>-1$, we see that

$$(4)\quad \int_0^\infty J_\nu(ax)\,J_\nu(bx)\frac{x\,dx}{x^2-r^2}=\begin{cases}\tfrac{1}{2}\pi i\,J_\nu(br)\,H_\nu^{(1)}(ar),\\ \tfrac{1}{2}\pi i\,J_\nu(ar)\,H_\nu^{(1)}(br),\end{cases}$$

according as $a\gtrless b$.

The existence of the discontinuity in the expression for this integral was pointed out by Hankel.

If we modify formula (3) we see that, if $a\geqslant b>0$ and $R(k)>0$, then

$$(5)\quad \int_0^\infty \frac{x}{x^2+k^2}J_\mu(bx)[\cos\tfrac{1}{2}(\mu-\nu)\pi\,.\,J_\nu(ax)+\sin\tfrac{1}{2}(\mu-\nu)\pi\,.\,Y_\nu(ax)]\,dx$$
$$=I_\mu(bk)\,K_\nu(ak).$$

* Since $H_\mu^{(2)}(az)\,H_\nu^{(1)}(az)\sim\dfrac{2}{\pi az}e^{\frac{1}{2}(\mu-\nu)\pi i}$ when $|z|$ is large.

More generally, taking equation (1) with $m=0$ and $\mathscr{C}_\mu \equiv J_\mu$, we have

$$(6)\quad \int_0^\infty \frac{x^{\rho-1} J_\mu(bx)}{x^2+k^2}[\cos\tfrac{1}{2}(\rho+\mu-\nu)\pi . J_\nu(ax) + \sin\tfrac{1}{2}(\rho+\mu-\nu)\pi . Y_\nu(ax)]\,dx = -I_\mu(bk)\,K_\nu(ak)\,k^{\rho-2}.$$

In this result replace ρ by $\rho+\nu$, a by $\sqrt{(a^2+c^2-2ac\cos\theta)}$, where $a-c>b$, multiply by $\sin^{2\nu}\theta/(a^2+c^2-2ac\cos\theta)^{\frac{1}{2}\nu}$, and integrate with respect to θ from 0 to π; we find from Gegenbauer's formula, § 11·41 (16),

$$(7)\quad \int_0^\infty \frac{x^{\rho-1} J_\mu(bx)\,J_\nu(cx)}{x^2+k^2}[\cos\tfrac{1}{2}(\rho+\mu)\pi\, J_\nu(ax) + \sin\tfrac{1}{2}(\rho+\mu)\pi\, Y_\nu(ax)]\,dx = -I_\mu(bk)\,I_\nu(ck)\,K_\nu(ak)\,k^{\rho-2}.$$

This process may be repeated as often as we please; and we find that, if $a>b+\Sigma c$, then

$$(8)\quad \int_0^\infty \frac{x^{\rho-1} J_\mu(bx)}{x^2+k^2}\prod_{n=1}^{N} J_\nu(c_n x)\,[\cos\tfrac{1}{2}\{\rho+\mu+(N-1)\nu\}\pi . J_\nu(ax) + \sin\tfrac{1}{2}\{\rho+\mu+(N-1)\nu\}\pi . Y_\nu(ax)]\,dx = -I_\mu(bk)\prod_{n=1}^{N} I_\nu(c_n k) . K_\nu(ak) . k^{\rho-2}.$$

Again, by considering

$$\frac{1}{2\pi i}\int \frac{z^{\rho-1}}{z^2+k^2}[\Pi\, J_\mu(bz)]\,H_\nu^{(1)}(az)\,dz$$

round the contour previously used, where both b and μ differ in the different factors of the product, we obtain the slightly more general result

$$(9)\quad \int_0^\infty \frac{x^{\rho-1}}{x^2+k^2}[\Pi\, J_\mu(bx)] . [\cos\tfrac{1}{2}(\rho+\Sigma\mu-\nu)\pi . J_\nu(ax) + \sin\tfrac{1}{2}(\rho+\Sigma\mu-\nu)\pi . Y_\nu(ax)]\,dx = -[\Pi\, I_\mu(bk)]\,K_\nu(ak)\,k^{\rho-2}$$

provided that $a>\Sigma\,|R(b)|$ and $R\{\rho+\Sigma(\mu)\}>|R(\nu)|$.

If $\rho+\Sigma\mu-\nu$ is an even integer, the integral on the left involves functions of the first kind only; a result involving the integrals of products of functions of the first kind of this type was given by Gegenbauer, who overlooked the necessity for this restriction (cf. § 13·51).

An extension of Hankel's results is obtained by considering

$$\frac{1}{2\pi i}\int z^{\rho-1}\frac{J_\mu\{b\sqrt{(z^2+\zeta^2)}\}}{(z^2+\zeta^2)^{\frac{1}{2}\mu}}\,\frac{H_\nu^{(1)}(az)}{(z^2-r^2)^{m+1}}\,dz$$

round the contour, where $a \geqslant b > 0$, m is a positive integer, and

$$|R(\nu)| < R(\rho) < 2m+4+R(\mu).$$

It follows that

$$\frac{1}{2\pi i}\int_0^\infty \frac{x^{\rho-1}}{(x^2-r^2)^{m+1}}\,\frac{J_\mu\{b\sqrt{(x^2+\zeta^2)}\}}{(x^2+\zeta^2)^{\frac{1}{2}\mu}}\{H_\nu^{(1)}(ax) - e^{\rho\pi i}H_\nu^{(1)}(axe^{\pi i})\}\,dx = \frac{1}{2^{m+1}\,m!}\left(\frac{d}{r\,dr}\right)^m\left[r^{\rho-2}\frac{J_\mu\{b\sqrt{(r^2+\zeta^2)}\}}{(r^2+\zeta^2)^{\frac{1}{2}\mu}}H_\nu^{(1)}(ar)\right],$$

and, in particular,

$$(10)\quad \int_0^\infty \frac{x^{\nu+1}}{x^2+k^2}\frac{J_\mu\{b\sqrt{(x^2+\zeta^2)}\}}{(x^2+\zeta^2)^{\frac{1}{2}\mu}} J_\nu(ax)\,dx = \frac{J_\mu\{b\sqrt{(\zeta^2-k^2)}\}}{(\zeta^2-k^2)^{\frac{1}{2}\mu}} k^\nu K_\nu(ak),$$

a result obtained in a much more elaborate manner by Sonine, *Math. Ann.* XVI. (1880), pp. 56—60.

13·54. *Generalisations of Nicholson's integral.*

An interesting consequence of Mehler's integral of § 13·51 (5) is due to Nicholson*, namely that, when a and k are positive,

$$(1)\qquad \int_0^\infty \frac{J_0(ax)\,x\,dx}{\sqrt{(x^4+4k^4)}} = K_0(ak)\,J_0(ak)$$

The method by which this result is obtained is as follows:

$$K_0(ak)\,J_0(ak) = \int_0^\infty \frac{\rho\,J_0(a\rho)\,J_0(ak)}{\rho^2+k^2}\,d\rho$$

$$= \frac{1}{\pi}\int_0^\infty\int_0^\pi \frac{\rho}{\rho^2+k^2} J_0\{a\sqrt{(\rho^2+k^2-2\rho k\cos\phi)}\}\,d\phi\,d\rho.$$

This repeated integral may be regarded as an absolutely convergent double integral, since the integrand is $O(\rho^{-\frac{3}{2}})$ when ρ is large. Now make a change of origin of the polar coordinates by writing

$$\rho\cos\phi = k + r\cos\theta,\quad \rho\sin\phi = r\sin\theta,$$

and we have

$$K_0(ak)\,J_0(ak) = \frac{1}{\pi}\int_0^\infty\int_0^\pi \frac{J_0(ar)\,r\,d\theta\,dr}{r^2+2kr\cos\theta+2k^2} = \int_0^\infty \frac{J_0(ar)\,r\,dr}{\sqrt{(r^4+4k^4)}},$$

and this is the result to be established.

To generalise the result consider

$$\int \frac{z^{\rho-1}H_\nu^{(1)}(az)\,dz}{(z^4+4k^4)^{\mu+1}}$$

taken round the contour shewn in Fig. 32.

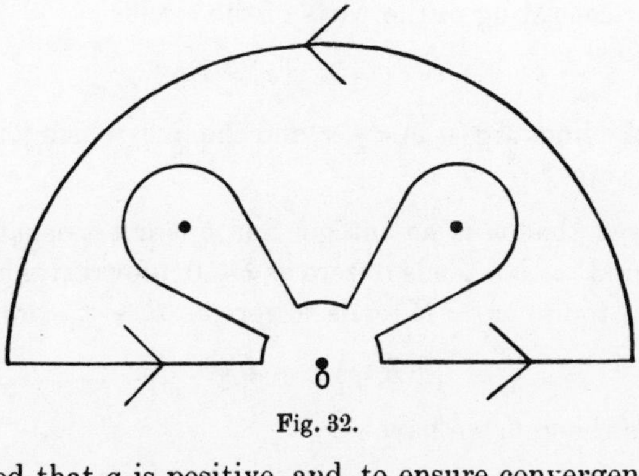

Fig. 32.

It is supposed that a is positive, and, to ensure convergence,

$$|R(\nu)| < R(\rho) < 4R(\mu) + \tfrac{11}{2}.$$

* *Quarterly Journal*, XLII. (1911), p. 224.

It is also supposed that $|\arg k| < \frac{1}{4}\pi$, and the loops in the contour surround the points

$$e^{\frac{1}{4}\pi i} k\sqrt{2}, \quad e^{\frac{3}{4}\pi i} k\sqrt{2}.$$

By analysis resembling that of § 13·52, the reader will find that

$$(2) \quad \int_0^\infty [\cos(\tfrac{1}{2}\rho - \tfrac{1}{2}\nu - 2\mu)\pi \,.\, J_\nu(ax) + \sin(\tfrac{1}{2}\rho - \tfrac{1}{2}\nu - 2\mu)\pi \,.\, Y_\nu(ax)] \frac{x^{\rho-1}dx}{(x^4+4k^4)^{\mu+1}}$$

$$= \frac{\pi(k\sqrt{2})^{\rho-4\mu-4}}{2\sin\nu\pi \,.\, \Gamma(\mu+1)} \left[\sum_{m=0}^\infty \frac{(ak/\sqrt{2})^{\nu+2m}\,\Gamma(\frac{1}{4}\rho+\frac{1}{4}\nu+\frac{1}{2}m)}{m!\,\Gamma(\nu+m+1)\,\Gamma(\frac{1}{4}\rho+\frac{1}{4}\nu-\mu+\frac{1}{2}m)} \cos\frac{\rho+\nu-4\mu+2m}{4}\pi \right.$$

$$\left. - \sum_{m=0}^\infty \frac{(ak/\sqrt{2})^{-\nu+2m}\,\Gamma(\frac{1}{4}\rho-\frac{1}{4}\nu+\frac{1}{2}m)}{m!\,\Gamma(-\nu+m+1)\,\Gamma(\frac{1}{4}\rho-\frac{1}{4}\nu-\mu+\frac{1}{2}m)} \cos\frac{\rho-\nu-4\mu+2m}{4}\pi \right].$$

If the series on the right are compared with those given in § 5·41, it is seen that the former is expressible as a product of Bessel functions if $\rho - 2 = \nu = \mu + \frac{1}{2}$ or if $\rho - 4 = \nu = \mu + \frac{1}{2}$, while the latter is so expressible if $\rho - 2 = -\nu = \mu + \frac{1}{2}$ or if $\rho - 4 = -\nu = \mu + \frac{1}{2}$.

The corresponding integral which contains a single Bessel function will be considered in § 13·6.

13·55. *Sonine's integrals.*

A number of definite integrals, of which special forms were given by Sonine, *Math. Ann.* XVI. (1880), pp. 63—66, can be evaluated by the method of contour integration.

The most general contour integral to be taken is

$$\frac{1}{2\pi i}\int \frac{e^{i(z+k)}}{(z+k)^{m+1}} (-z)^{\rho-1} H_\nu^{(1)}(-z)\,dz$$

round a contour consisting of the parts of the circles

$$|z| = \delta, \quad |z| = R,$$

terminated by the lines $\arg(-z) = \pm\pi$, and the lines which join the extremities of these circular arcs*.

It is supposed that m is an integer and k is not a negative real number. The integral round $|z| = \delta$ tends to zero as $\delta \to 0$, provided that $R(\rho) > |R(\nu)|$, and the integral round $|z| = R$ tends to zero as $R \to \infty$, provided that

$$R(\rho) < m + \tfrac{3}{2}.$$

By Cauchy's theorem we have

$$\frac{1}{2\pi i}\int \frac{e^{i(z+k)}}{(z+k)^{m+1}} (-z)^{\rho-1} H_\nu^{(1)}(-z)\,dz = \frac{(-)^m e^{ik}}{m!} \frac{d^m}{dk^m}\{e^{-ik} k^{\rho-1} H_\nu^{(1)}(k)\}$$

* Cf. *Modern Analysis*, § 6·2; or § 7·4 *supra*.

and thus we have

$$(1)\quad \frac{(-)^m e^{ik}}{m!}\frac{d^m}{dk^m}\{e^{-ik}k^{\rho-1}H_\nu^{(1)}(k)\}$$

$$=\frac{1}{2\pi i}\int_0^\infty \frac{e^{i(x+k)}x^{\rho-1}}{(x+k)^{m+1}}\{e^{\rho\pi i}H_\nu^{(1)}(xe^{\pi i})-e^{-\rho\pi i}H_\nu^{(1)}(xe^{-\pi i})\}\,dx$$

$$=\frac{1}{\pi}\int_0^\infty \frac{e^{i(x+k)}x^{\rho-1}}{(x+k)^{m+1}}[J_\nu(x)\{\sin(\rho+\nu)\pi+2i\cos\nu\pi\cos\rho\pi\}$$
$$+iY_\nu(x)\sin(\rho-\nu)\pi]\,dx.$$

In particular, taking $m=0$, we get

$$(2)\quad k^{\rho-1}H_\nu^{(1)}(k)=\frac{1}{\pi}\int_0^\infty \frac{e^{i(x+k)}x^{\rho-1}}{x+k}[J_\nu(x)\{\sin(\rho+\nu)\pi+2i\cos\nu\pi\cos\rho\pi\}$$
$$+iY_\nu(x)\sin(\rho-\nu)\pi]\,dx.$$

If we consider the integral

$$\frac{1}{2\pi i}\int\frac{e^{-i(z+k)}}{z+k}(-z)^{\rho-1}H_\nu^{(2)}(-z)\,dz,$$

we find that

$$(3)\quad k^{\rho-1}H_\nu^{(2)}(k)=\frac{1}{\pi}\int_0^\infty \frac{e^{-i(x+k)}x^{\rho-1}}{x+k}[J_\nu(x)\{\sin(\rho+\nu)\pi-2i\cos\nu\pi\cos\rho\pi\}$$
$$-iY_\nu(x)\sin(\rho-\nu)\pi]\,dx.$$

If we take $\rho=1$, $\nu=0$, we get

$$(4)\qquad J_0(k)=\frac{2}{\pi}\int_0^\infty\frac{\sin(x+k)}{x+k}J_0(x)\,dx,$$

$$(5)\qquad Y_0(k)=-\frac{2}{\pi}\int_0^\infty\frac{\cos(x+k)}{x+k}J_0(x)\,dx.$$

The last two results are due to Sonine*.

More generally, taking $\rho=\nu+1$ and $-\tfrac{1}{2}<R(\nu)<\tfrac{1}{2}$, we get

$$(6)\qquad \frac{1}{\pi}\int_0^\infty\frac{e^{\pm i(x+k)}x^\nu}{x+k}J_\nu(x)\,dx=\mp\frac{e^{\pm\nu\pi i}}{2i\cos\nu\pi}k^\nu\{J_\nu(k)\pm iY_\nu(k)\},$$

a result also due to Sonine.

By writing $$\frac{\cos(x+k)}{x+k}=\frac{1}{x+k}-\int_0^1\sin t(x+k)\,dt,$$

and using the formulae (6) and (7) of § 13·42, Sonine deduced from (5) that

$$(7)\qquad Y_0(k)=-\frac{2}{\pi}\int_0^\infty\frac{J_0(x)}{x+k}dx+\frac{2}{\pi}\int_0^{\frac{1}{2}\pi}\sin(k\cos\theta)\,d\theta,$$

and hence from §§ 3·56 (2), 9·11 (2),

$$(8)\quad Y_n(k)=-\frac{2}{\pi}\int_0^\infty O_n(x+k)J_0(x)\,dx+\frac{2}{\pi}\int_0^{\frac{1}{2}\pi}\sin(k\cos\theta-\tfrac{1}{2}n\pi)\cos n\theta\,d\theta.$$

* See also Lerch, *Monatshefte für Math. und Phys.* I. (1890), pp. 105—112.

13·6. *A new method* of evaluating definite integrals.*

We shall now evaluate various definite integrals by substituting for the Bessel function, under the integral sign, the definite integral of § 6·5, and reversing the order of the integrations.

As a first example consider the integral of Hankel's type

$$\int_0^\infty \frac{x^{\rho-1} J_\nu(ax)}{(x^2+k^2)^{\mu+1}}\,dx,$$

in which it is at first supposed that

$$R(\nu) > 0, \quad R(2\mu+2) > R(\rho+\nu) > 0;$$

and a is a *real* (positive) number, in order that the integral may converge.

The integral is equal to†

$$\frac{1}{2\pi i}\int_0^\infty \int_{-\infty i}^{\infty i} \frac{\Gamma(-s)}{\Gamma(\nu+s+1)} \frac{x^{\rho-1}(\frac{1}{2}ax)^{\nu+2s}}{(x^2+k^2)^{\mu+1}}\,ds\,dx = \frac{1}{2\pi i}\int_{-\infty i}^{\infty i} \frac{\Gamma(-s)}{\Gamma(\nu+s+1)}$$

$$\times (\tfrac{1}{2}a)^{\nu+2s} \cdot \tfrac{1}{2} k^{\rho+\nu+2s-2\mu-2} \frac{\Gamma(\frac{1}{2}\rho+\frac{1}{2}\nu+s)\,\Gamma(\mu+1-\frac{1}{2}\rho-\frac{1}{2}\nu-s)}{\Gamma(\mu+1)}\,ds.$$

When this is evaluated (by swinging round the contour so as to enclose the poles on the right of the contour) we find that

$$(1) \quad \int_0^\infty \frac{x^{\rho-1} J_\nu(ax)\,dx}{(x^2+k^2)^{\mu+1}}$$

$$= \frac{a^\nu k^{\rho+\nu-2\mu-2}\,\Gamma(\frac{1}{2}\rho+\frac{1}{2}\nu)\,\Gamma(\mu+1-\frac{1}{2}\rho-\frac{1}{2}\nu)}{2^{\nu+1}\,\Gamma(\mu+1)\,\Gamma(\nu+1)} \cdot {}_1F_2\left(\frac{\rho+\nu}{2};\ \frac{\rho+\nu}{2}-\mu,\ \nu+1;\ \frac{a^2k^2}{4}\right)$$

$$+ \frac{a^{2\mu+2-\rho}\,\Gamma(\frac{1}{2}\nu+\frac{1}{2}\rho-\mu-1)}{2^{2\mu+3-\rho}\,\Gamma(\mu+2+\frac{1}{2}\nu-\frac{1}{2}\rho)} \cdot {}_1F_2\left(\mu+1;\ \mu+2+\frac{\nu-\rho}{2},\ \mu+2-\frac{\nu+\rho}{2};\ \frac{a^2k^2}{4}\right).$$

The hypergeometric functions on the right are reducible to Bessel functions in certain circumstances; the former if $\rho=\nu+2$ or $\rho=\nu+2\mu+2$, the latter if $\rho = 2 \pm \nu$.

By the principle of analytic continuation (1) is valid when

$$-R(\nu) < R(\rho) < 2R(\mu) + \tfrac{7}{2}.$$

In particular, taking $\rho=\nu+2$, we find that

$$(2) \quad \int_0^\infty \frac{x^{\nu+1} J_\nu(ax)\,dx}{(x^2+k^2)^{\mu+1}} = \frac{a^\mu k^{\nu-\mu}}{2^\mu\,\Gamma(\mu+1)} K_{\nu-\mu}(ak),$$

a formula obtained by another method by Sonine, *Math. Ann.* XVI. (1880), p. 50; it is valid when

$$-1 < R(\nu) < 2R(\mu) + \tfrac{3}{2}.$$

* This method is due to Lerch, *Rozpravy*, v. (1896), no. 23 [*Jahrbuch über die Fortschritte der Math.* (1896), p. 233]; he shewed that

$$\int_{2x}^\infty J_0(t)\frac{dt}{t} = -\frac{1}{4\pi i}\int_{-\infty i}^{\infty i} \frac{\Gamma(-s)\,x^{2s}\,ds}{s\Gamma(s+1)},$$

but no other use has been made of the method.

† The change of the order of the integrations may be justified without difficulty.

A formula of some interest is obtained by making $\rho = 1$, $\mu = -\frac{1}{2}$, the hypergeometric functions then reducing to squares or products of Bessel functions; and another such formula is found by making $\rho = 1 - \nu$, $\mu = \nu - \frac{1}{2}$. It is thus deduced that (cf. § 5·41)

$$\text{(3)}\qquad \int_0^\infty \frac{J_\nu(ax)\,dx}{(x^2+k^2)^{\frac{1}{2}}} = I_{\frac{1}{2}\nu}(\tfrac{1}{2}ak)\,K_{\frac{1}{2}\nu}(\tfrac{1}{2}ak),$$

provided that $R(\nu) > -1$; and that

$$\text{(4)}\qquad \int_0^\infty \frac{x^{-\nu}J_\nu(ax)\,dx}{(x^2+k^2)^{\nu+\frac{1}{2}}} = \frac{(2a/k^2)^\nu\,\Gamma(\nu+1)}{\Gamma(2\nu+1)} I_\nu(\tfrac{1}{2}ak)\,K_\nu(\tfrac{1}{2}ak),$$

provided that $R(\nu) > -\frac{1}{2}$.

Next consider

$$\int_0^\infty \frac{x^{\rho-1}J_\nu(ax)\,dx}{(x^4+4k^4)^{\mu+1}}$$

in which $a > 0$ and $|\arg k| < \frac{1}{4}\pi$. It is first to be supposed that

$$R(\nu) > 0,\quad R(4\mu+4) > R(\rho+\nu) > 0.$$

The integral is equal to

$$\frac{1}{2\pi i}\int_0^\infty \int_{-\infty i}^{\infty i} \frac{\Gamma(-s)}{\Gamma(\nu+s+1)} \frac{x^{\rho-1}(\frac{1}{2}ax)^{\nu+2s}}{(x^4+4k^4)^{\mu+1}}\,ds\,dx$$

$$= \frac{1}{8\pi i}\int_{-\infty i}^{\infty i} \frac{\Gamma(-s)\,\Gamma(\frac{1}{4}\rho+\frac{1}{4}\nu+\frac{1}{2}s)\,\Gamma(\mu+1-\frac{1}{4}\rho-\frac{1}{4}\nu-\frac{1}{2}s)}{\Gamma(\nu+s+1)\,\Gamma(\mu+1)} \times (\tfrac{1}{2}a)^{\nu+2s}(k\sqrt{2})^{\rho+\nu+2s-4\mu-4}\,ds$$

$$= \frac{\frac{1}{2}\pi(k\sqrt{2})^{\rho-4\mu-4}}{\sin(\frac{1}{2}\rho+\frac{1}{2}\nu-2\mu)\pi\,\Gamma(\mu+1)} \Bigg[\sum_{m=0}^\infty \frac{(ak/\sqrt{2})^{\nu+2m}\,\Gamma(\frac{1}{4}\rho+\frac{1}{4}\nu+\frac{1}{2}m)}{m!\,\Gamma(\nu+m+1)\,\Gamma(\frac{1}{4}\rho+\frac{1}{4}\nu-\mu+\frac{1}{2}m)} \times\cos(\tfrac{1}{4}\rho+\tfrac{1}{4}\nu-\mu+\tfrac{1}{2}m)\pi$$

$$+\sum_{m=0}^\infty \frac{(-)^m(ak/\sqrt{2})^{4\mu+4-\rho+4m}\,\Gamma(\mu+m+1)}{m!\,\Gamma(2\mu-\frac{1}{2}\rho+\frac{1}{2}\nu+2m+3)\,\Gamma(2\mu-\frac{1}{2}\rho-\frac{1}{2}\nu+2m+3)}\Bigg].$$

This expansion is a representation of the integral when the conditions to be laid on μ, ν and ρ are

$$4R(\mu) + \tfrac{11}{2} > R(\rho) > -R(\nu).$$

Now take the cases in which the first series reduces to a product of Bessel functions, namely

$$\rho - 2 = \nu = \mu + \tfrac{1}{2} \quad\text{or}\quad \rho - 4 = \nu = \mu + \tfrac{1}{2}.$$

By § 5·41 we then obtain the formulae

$$\text{(5)}\qquad \int_0^\infty \frac{x^{\nu+1}J_\nu(ax)\,dx}{(x^4+4k^4)^{\nu+\frac{1}{2}}} = \frac{(\frac{1}{2}a)^\nu\sqrt{\pi}}{(2k)^{2\nu}\,\Gamma(\nu+\frac{1}{2})} J_\nu(ak)\,K_\nu(ak),$$

$$\text{(6)}\qquad \int_0^\infty \frac{x^{\nu+3}J_\nu(ax)\,dx}{(x^4+4k^4)^{\nu+\frac{1}{2}}} = \frac{(\frac{1}{2}a)^\nu\sqrt{\pi}}{2\,.\,(2k)^{2\nu-2}\,\Gamma(\nu+\frac{1}{2})} J_{\nu-1}(ak)\,K_{\nu-1}(ak).$$

The former of these is valid when $R(\nu) > -\frac{1}{2}$, the latter when $R(\nu) > \frac{1}{6}$, and, in both, $a > 0$ and $|\arg k| < \frac{1}{4}\pi$.

Finally, as an example suggested by § 13·55, we shall consider

$$\int_0^\infty \frac{x^{\rho-1} J_\nu(ax)\,dx}{(x+k)^{\mu+1}},$$

in which $a > 0$ and $|\arg k| < \pi$. It is first to be supposed that

$$R(\nu) > 0, \quad R(\mu+1) > R(\rho+\nu) > 0.$$

The integral is equal to

$$\frac{1}{2\pi i}\int_0^\infty \int_{-\infty i}^{\infty i} \frac{\Gamma(-s)}{\Gamma(\nu+s+1)} \frac{x^{\rho-1}(\frac{1}{2}ax)^{\nu+2s}}{(x+k)^{\mu+1}}\,ds\,dx$$

$$= \frac{1}{2\pi i}\int_{-\infty i}^{\infty i} \frac{\Gamma(-s)\,\Gamma(\rho+\nu+2s)\,\Gamma(\mu+1-\rho-\nu-2s)}{\Gamma(\nu+s+1)\,\Gamma(\mu+1)} (\tfrac{1}{2}a)^{\nu+2s} k^{\rho+\nu+2s-\mu-1}\,ds$$

$$= \frac{\pi k^{\rho-\mu-1}}{\sin(\rho+\nu-\mu)\pi \,.\, \Gamma(\mu+1)} \left[\sum_{m=0}^\infty \frac{(-)^m (\frac{1}{2}ak)^{\nu+2m}\,\Gamma(\rho+\nu+2m)}{m!\,\Gamma(\nu+m+1)\,\Gamma(\rho+\nu-\mu+2m)} \right.$$
$$\left. - \sum_{m=0}^\infty \frac{(\frac{1}{2}ak)^{\mu+1-\rho+m}\,\Gamma(\mu+m+1)\sin\frac{1}{2}(\rho+\nu-\mu-m)\pi}{m!\,\Gamma(\frac{1}{2}\mu+\frac{1}{2}\nu-\frac{1}{2}\rho+\frac{1}{2}m+\frac{3}{2})\,\Gamma(\frac{1}{2}\mu-\frac{1}{2}\nu-\frac{1}{2}\rho+\frac{1}{2}m+\frac{3}{2})}\right].$$

The first series reduces to $J_\nu(ak)$ when $\mu = 0$, and the second series is then expressible by Lommel's functions (cf. § 10·7). In particular we have

(7) $$\int_0^\infty \frac{x^\nu J_\nu(ax)\,dx}{x+k} = \frac{\pi k^\nu}{2\cos\nu\pi}[\mathbf{H}_{-\nu}(ak) - Y_{-\nu}(ak)]$$

provided that $-\frac{1}{2} < R(\nu) < \frac{3}{2}$.

The reader will find that a large number of the integrals discussed in this chapter may be evaluated by this method.

13·61. *Integrals involving products of Bessel functions.*

If an integral involves the product of two Bessel functions of the same argument (but not necessarily of the same order), it is likely that the integral is capable of being evaluated either by replacing the product by Neumann's integral (§ 5·43) and using the method just described, or else by replacing the product $J_\mu(x)\,J_\nu(x)$ by

$$\frac{1}{2\pi i}\int_{-\infty i}^{\infty i} \frac{\Gamma(-s)\,\Gamma(\mu+\nu+2s+1)(\frac{1}{2}x)^{\mu+\nu+2s}}{\Gamma(\mu+s+1)\,\Gamma(\nu+s+1)\,\Gamma(\mu+\nu+s+1)}\,ds,$$

in which the poles of $\Gamma(-s)$ are on the right of the contour while those of $\Gamma(\mu+\nu+2s+1)$ are on the left; this expression is easily derived from § 5·41 by using the method of obtaining § 6·5.

The reader may find it interesting to evaluate

$$\int_0^\infty \frac{x^{\rho-1} J_\mu(ax)\,J_\nu(ax)\,dx}{(x^2+k^2)^{\lambda+1}}$$

by these methods. The result is a combination of two functions of the type ${}_3F_4$, and the final element in each function is a^2k^2.

Another integral formula, obtainable by replacing $J_\nu(b/x)$ by an integral, is

$$(1)\quad \int_0^\infty x^{\rho-1} J_\mu(ax)\, J_\nu(b/x)\, dx$$

$$= \frac{a^{\nu-\rho}\, b^\nu\, \Gamma(\tfrac{1}{2}\mu+\tfrac{1}{2}\rho-\tfrac{1}{2}\nu)}{2^{2\nu-\rho+1}\,\Gamma(\nu+1)\,\Gamma(\tfrac{1}{2}\mu+\tfrac{1}{2}\nu-\tfrac{1}{2}\rho+1)} \cdot {}_0F_3\left(\nu+1, \frac{\nu-\mu-\rho}{2}+1, \frac{\nu+\mu-\rho}{2}+1;\ \frac{a^2b^2}{16}\right)$$

$$+ \frac{a^\mu\, b^{\mu+\rho}\, \Gamma(\tfrac{1}{2}\nu-\tfrac{1}{2}\mu-\tfrac{1}{2}\rho)}{2^{2\mu+\rho+1}\,\Gamma(\mu+1)\,\Gamma(\tfrac{1}{2}\mu+\tfrac{1}{2}\nu+\tfrac{1}{2}\rho+1)} \cdot {}_0F_3\left(\mu+1, \frac{\mu-\nu+\rho}{2}+1, \frac{\nu+\mu+\rho}{2}+1;\ \frac{a^2b^2}{16}\right)$$

This is valid when a and b are positive and

$$-R(\mu+\tfrac{3}{2}) < R(\rho) < R(\nu+\tfrac{3}{2}).$$

The general formula was given by Hanumanta Rao, *Messenger*, XLVII. (1918), pp. 134—137; special cases had been given previously by Cailler, *Mém. de la Soc. de Phys. de Genève*, XXXIV. (1902—1905), p. 352; Bateman, *Trans. Camb. Phil. Soc.* XXI. (1912), pp. 185, 186; and Hardy, who discussed the case of functions of orders $\pm\tfrac{1}{2}$, (see § 6·23).

An interesting example of an integral* which contains a product is

$$\lim_{\delta\to+0}\left[\int_\delta^\infty \exp(-p^2x^2)\, J_\nu(x)\, J_{1-\nu}(x)\, \frac{dx}{x^2} + \frac{\sin\nu\pi}{4\pi\nu(1-\nu)}\{1+2\log\tfrac{1}{2}\delta - \psi(\nu+1)-\psi(2-\nu)\}\right],$$

which may be written in the form

$$\int_0^\infty \exp(-p^2x^2)\left\{J_\nu(x)\, J_{1-\nu}(x) - \frac{\tfrac{1}{2}x\sin\nu\pi}{\pi\nu(1-\nu)}\right\}\frac{dx}{x^2} + \frac{\sin\nu\pi}{4\pi\nu(1-\nu)}\{\psi(2)-\psi(\nu+1)-\psi(2-\nu)-2\log 2p\}.$$

It is easily proved that

$$\int_0^\infty \exp(-p^2x^2)\left\{J_\nu(x)\, J_{1-\nu}(x) - \frac{\tfrac{1}{2}x\sin\nu\pi}{\pi\nu(1-\nu)}\right\}\frac{dx}{x^2}$$

$$= \frac{1}{8\pi i}\int_0^\infty\int_{\frac{1}{2}-\infty i}^{\frac{1}{2}+\infty i} \exp(-p^2x^2)\, \frac{\Gamma(-s)\,\Gamma(2s+2)\,.\,(\tfrac{1}{2}x)^{2s-1}}{\Gamma(\nu+s+1)\,\Gamma(2-\nu+s)\,\Gamma(s+2)}\, ds\, dx$$

$$= \frac{1}{8\pi i}\int_{\frac{1}{2}-\infty i}^{\frac{1}{2}+\infty i} \frac{\Gamma(-s)\,\Gamma(2s+2)}{s(s+1)\,\Gamma(\nu+s+1)\,\Gamma(2-\nu+s)}\,\frac{ds}{(2p)^{2s}}$$

$$= \frac{1}{4}\sum_{n=1}^\infty \frac{(2n+1)!\,(2p)^{-2n}}{n\,.\,(n+1)!\,\Gamma(\nu+n+1)\,\Gamma(2-\nu+n)},$$

and this series is an integral function of $1/p$.

To obtain an asymptotic representation of the integral, valid when $|p|$ is small and $|\arg p| < \tfrac{1}{4}\pi$, we observe that the last integrand has double poles at 0 and -1, and simple poles at $-\tfrac{3}{2}, -2, -\tfrac{5}{2}, -3, \ldots$.

* This integral was brought to my notice by Mr C. G. Darwin, who encountered it in a problem of Diffusion of Salts in a circular cylinder of liquid.

Hence we find

$$\frac{1}{8\pi i}\int_{\frac{1}{2}-\infty i}^{\frac{1}{2}+\infty i} \frac{\Gamma(-s)\,\Gamma(2s+2)}{s(s+1)\,\Gamma(\nu+s+1)\,\Gamma(2-\nu+s)}\frac{ds}{(2p)^{2s}}$$

$$\sim -\frac{\sin\nu\pi}{4\pi\nu(1-\nu)}\{\psi(2)-\psi(\nu+1)-\psi(2-\nu)-2\log 2p\}\{1+2\nu(1-\nu)p^2\}$$

$$-\frac{(1-2\nu)\sin\nu\pi}{2\pi\nu(1-\nu)}p^2+\sum_{n=3}^{\infty}\frac{(-)^n(2p)^n\,\Gamma(\frac{1}{2}n)}{2n(n-2)\,\Gamma(\nu+1-\frac{1}{2}n)\,\Gamma(2-\nu-\frac{1}{2}n).(n-2)!}$$

and so

$$(2)\quad \lim_{\delta\to+0}\Bigg[\int_\delta^\infty \exp(-p^2x^2)\,J_\nu(x)\,J_{1-\nu}(x)\frac{dx}{x^2} + \frac{\sin\nu\pi}{4\pi\nu(1-\nu)}\{1+2\log\tfrac{1}{2}\delta-\psi(\nu+1)-\psi(2-\nu)\}\Bigg]$$

$$\sim -\frac{\sin\nu\pi}{2\pi}\{\psi(2)-\psi(\nu+1)-\psi(2-\nu)-2\log 2p\}\,p^2-\frac{(1-2\nu)\sin\nu\pi}{2\pi\nu(1-\nu)}p^2$$

$$+\sum_{n=3}^{\infty}\frac{(-)^n(2p)^n\,\Gamma(\frac{1}{2}n)}{2n(n-2)\,\Gamma(\nu+1-\frac{1}{2}n)\,\Gamma(2-\nu-\frac{1}{2}n).(n-2)!}.$$

In the special case $\nu=0$, we find that

$$(3)\quad \lim_{\delta\to0}\left[\frac{\gamma+\log(\frac{1}{2}\delta)}{2}+\int_\delta^\infty \exp(-p^2x^2)\,J_0(x)\,J_1(x)\frac{dx}{x^2}\right]$$

$$\sim -\tfrac{1}{2}p^2+\sum_{m=0}^{\infty}\frac{\{\Gamma(m+\frac{3}{2})\}^3(2p)^{2m+3}}{\pi^2(2m+1)^2(2m+3).(2m+1)!}.$$

13·7. *Integral representations of products of Bessel functions.*

From Gegenbauer's formula of § 11·41 (16) an interesting result is obtainable by taking the cylinder function to be of the first kind and substituting the result of § 6·2 (8) for the function under the integral sign.

This procedure gives

$$2^\nu\Gamma(\nu+\tfrac{1}{2})\,\Gamma(\tfrac{1}{2})\frac{J_\nu(Z)}{Z^\nu}\frac{J_\nu(z)}{z^\nu}$$

$$=\frac{1}{2\pi i}\int_0^\pi\int_{c-\infty i}^{c+\infty i}\sin^{2\nu}\phi\,.\,t^{-\nu-1}\exp\left\{\tfrac{1}{2}t-\frac{Z^2+z^2-2zZ\cos\phi}{2t}\right\}dt\,d\phi,$$

and if we change the order of the integrations, we find that

$$(1)\qquad J_\nu(Z)\,J_\nu(z)=\frac{1}{2\pi i}\int_{c-\infty i}^{c+\infty i}\exp\left\{\tfrac{1}{2}t-\frac{Z^2+z^2}{2t}\right\}.\,I_\nu\left(\frac{zZ}{t}\right)\frac{dt}{t}.$$

This result is proved when $R(\nu)>-\frac{1}{2}$ and $|z|<|Z|$; but the former restriction may obviously be replaced by $R(\nu)>-1$, and the latter may be removed on account of the symmetry in z and Z. It is also permissible to proceed to the limit by making $|z|\to|Z|$.

By using the results of § 6·21 (4) and (5), we find in the same way that

$$(2) \qquad H_\nu^{(1)}(Z)\, J_\nu(z) = \frac{1}{\pi i}\int_0^{c+\infty i} \exp\left\{\tfrac{1}{2}t - \frac{Z^2+z^2}{2t}\right\}. I_\nu\left(\frac{zZ}{t}\right)\frac{dt}{t},$$

$$(3) \qquad H_\nu^{(2)}(Z)\, J_\nu(z) = -\frac{1}{\pi i}\int_0^{c-\infty i} \exp\left\{\tfrac{1}{2}t - \frac{Z^2+z^2}{2t}\right\}. I_\nu\left(\frac{zZ}{t}\right)\frac{dt}{t},$$

provided that $R(\nu) > -1$ and $|z| < |Z|$.

The formula (1) was obtained by Macdonald, *Proc. London Math. Soc.* XXXII. (1900), pp. 152—155, from the theory of linear differential equations, and he deduced Gegenbauer's integral by reversing the steps of the analysis which we have given. The formulae (2) and (3) were given by Macdonald, though they are also to be found in a modified form in Sonine's memoir, *Math. Ann.* XVI. (1880), p. 61.

A further modification of the integrals on the right in (2) and (3) was given by Sonine, the object of the change being to remove the exponential functions.

For physical applications of these integrals, see Macdonald, *Proc. London Math. Soc.* (2) XIV. (1915), pp. 410—427.

13·71. *The expression of $K_\nu(Z)\, K_\nu(z)$ as an integral.*

We shall next obtain a formula, due to Macdonald*, which represents the product $K_\nu(Z)\, K_\nu(z)$ as an integral involving a single function of the type K_ν, namely

$$(1) \qquad K_\nu(Z)\, K_\nu(z) = \frac{1}{2}\int_0^\infty \exp\left[-\frac{v}{2} - \frac{Z^2+z^2}{2v}\right] K_\nu\left(\frac{Zz}{v}\right)\frac{dv}{v}.$$

This formula is valid for all values of ν when

$$|\arg Z| < \pi, \quad |\arg z| < \pi \text{ and } |\arg(Z+z)| < \tfrac{1}{4}\pi;$$

but it is convenient to prove it when Z, z have positive values X, x, and to extend it by the theory of analytic continuation; the formula, which is obviously to be associated with § 11·41 (16), is of some importance in dealing with the zeros of functions of the type $K_\nu(z)$. It is possible to prove the formula without the rather elaborate transformations used in proving § 11·41 (16); the following proof, which differs from Macdonald's, is on the lines of § 2·6.

By § 6·22 (7) we have

$$K_\nu(X)\, K_\nu(x) = \frac{1}{4}\int_{-\infty}^\infty\int_{-\infty}^\infty e^{-\nu(t+u) - X\cosh t - x\cosh u}\, dt\, du$$
$$= \frac{1}{2}\int_{-\infty}^\infty\int_{-\infty}^\infty e^{-2\nu T - X\cosh(T+U) - x\cosh(T-U)}\, dU\, dT.$$

If $(Xe^T + xe^{-T})\, e^U$ be taken as a new variable v, in the integral

$$\int_{-\infty}^\infty e^{-X\cosh(T+U) - x\cosh(T-U)}\, dU,$$

* *Proc. London Math. Soc.* XXX. (1899), pp. 169—171.

it becomes $$\int_0^\infty \exp\left[-\frac{1}{2}\left\{v + \frac{X^2 + x^2 + 2Xx\cosh 2T}{v}\right\}\right]\frac{dv}{v},$$

and so we have

$$K_\nu(X)\,K_\nu(x) = \frac{1}{2}\int_{-\infty}^\infty \int_0^\infty e^{-2\nu T - (Xx/v)\cosh 2T} \exp\left[-\left\{\frac{v}{2} - \frac{X^2 + x^2}{v}\right\}\right]\frac{dv}{v}\,dT,$$

and, on performing the integration with respect to T, we at once obtain Macdonald's theorem when the variables X and x are positive.

13·72. *Nicholson's integral representations of products.*

We shall now discuss a series of integral representations of Bessel functions which are to be associated with Neumann's integral of § 5·43.

The formulae of this type have been developed by Nicholson*, and the two which are most easily proved are

(1) $$K_\mu(z)K_\nu(z) = 2\int_0^\infty K_{\mu+\nu}(2z\cosh t)\cosh(\mu-\nu)t\,dt$$
$$= 2\int_0^\infty K_{\mu-\nu}(2z\cosh t)\cosh(\mu+\nu)t\,dt,$$

when $|\arg z| < \frac{1}{2}\pi$, while μ and ν are unrestricted.

To obtain these formulae we use § 6·22 (5) which shews that

$$K_\mu(z)\,K_\nu(z) = \frac{1}{4}\int_{-\infty}^\infty \int_{-\infty}^\infty e^{-z(\cosh t + \cosh u)}\cosh\mu t\cosh\nu u\,dt\,du.$$

The repeated integral is absolutely convergent, and it may be regarded as a double integral. In the double integral make the transformation

$$t + u = 2T,\quad t - u = 2U,$$

and it is apparent that

$$K_\mu(z)\,K_\nu(z) = \frac{1}{2}\int_{-\infty}^\infty \int_{-\infty}^\infty e^{-2z\cosh T\cosh U}\cosh\mu(T+U)\cosh\nu(T-U)\,dT\,dU.$$

But $2\cosh\mu(T+U)\cosh\nu(T-U)$
$$= \cosh(\mu+\nu)T\cosh(\mu-\nu)U + \cosh(\mu-\nu)T\cosh(\mu+\nu)U$$
$$+ \sinh(\mu+\nu)T\sinh(\mu-\nu)U + \sinh(\mu-\nu)T\sinh(\mu+\nu)U.$$

The integrals corresponding to the last two of these four terms obviously vanish; and, if we interchange the parametric variables T and U in the integral corresponding to the second of the four terms, we obtain the formula

$$K_\mu(z)\,K_\nu(z) = \frac{1}{2}\int_{-\infty}^\infty \int_{-\infty}^\infty e^{-2z\cosh T\cosh U}\cosh(\mu+\nu)T\cosh(\mu-\nu)U\,dT\,dU.$$

If we integrate with respect to U we obtain the first form of (1), and if we integrate with respect to T we obtain the second form of (1).

* *Quarterly Journal*, XLII. (1911), pp. 220—223.

The formula

(2) $$I_\mu(z)\, I_\nu(z) = \frac{2}{\pi}\int_0^{\frac{1}{2}\pi} I_{\mu+\nu}(2z\cos\theta)\cos(\mu-\nu)\,\theta\, d\theta,$$

which is valid when $R(\mu+\nu)$ exceeds -1, is at once deducible from Neumann's formula.

If we take $\mu = 0$ and change the sign of ν, we find that

(3) $$I_0(z)\, K_\nu(z) = \frac{2}{\pi}\int_0^{\frac{1}{2}\pi} K_\nu(2z\cos\theta)\cos\nu\theta\, d\theta.$$

More generally, if we take $\mu = -m$ and then replace μ and ν by m and $-\nu$, we find that, if $|R(\nu - m)| < 1$, then

(4) $$I_m(z)\, K_\nu(z) = \frac{2(-)^m}{\pi}\int_0^{\frac{1}{2}\pi} K_{\nu-m}(2z\cos\theta)\cos(m+\nu)\,\theta\, d\theta.$$

If we combine (3) with § 6·16 (1) we find that

$$I_0(z)\, K_\nu(z) = \frac{2}{\pi}\frac{\Gamma(\nu+\frac{1}{2})}{\Gamma(\frac{1}{2})}\int_0^{\frac{1}{2}\pi}\int_0^\infty \frac{(4z)^\nu}{\cos^\nu\theta}\,\frac{\cos(u\cos\theta)\cos\nu\theta}{(u^2+4z^2)^{\nu+\frac{1}{2}}}\, du\, d\theta,$$

provided that $-\frac{1}{2} < R(\nu) < 1$; and in particular

(5) $$I_0(z)\, K_0(z) = \int_0^\infty \frac{J_0(u)\, du}{\sqrt{(u^2+4z^2)}},$$

a result of which a more general form has been given in § 13·6, formula (3).

13·73. *Nicholson's integral* for* $J_\nu^2(z) + Y_\nu^2(z)$.

The integral, corresponding to those just discussed, which represents $J_\nu^2(z) + Y_\nu^2(z)$ is difficult to establish rigorously. It is first necessary to assume that the argument is positive $(= x)$, and it is also necessary to appeal to Hardy's theory of generalised integrals, or some such principle, in the course of the proof.

Take the formula (§ 6·21)

$$H_\nu^{(1)}(x) = \frac{1}{\pi i}\int_{-\infty}^{\infty+\pi i} e^{x\sinh w - \nu w}\, dw.$$

From the manner in which the integrand tends to zero as $|w| \to \infty$ on the contour, it is clear that when an exponential factor $\exp\{-\lambda w^2\}$ is inserted, the resulting integral converges uniformly with regard to λ, and so it is a continuous function of λ. Hence†

$$H_\nu^{(1)}(x) = \lim_{\lambda\to+0}\frac{1}{\pi i}\int_{-\infty}^{\infty+\pi i}\exp\{-\lambda w^2\}\, e^{x\sinh w - \nu w}\, dw.$$

* *Phil. Mag.* (6) XIX. (1910), p. 234; *Quarterly Journal*, XLII. (1911), p. 221.

† Hardy, *Quarterly Journal*, XXXV. (1904), pp. 22—66; *Trans. Camb. Phil. Soc.* XXI. (1912), pp. 1—48. In this integral (as distinguished from those which follow) the sign lim is commutative with the integral sign.

By Cauchy's theorem, the contour may be deformed into the line $I(w) = \frac{1}{2}\pi$, so long as λ has an assigned positive value; writing $t + \frac{1}{2}\pi i$ for w, we get

$$H_\nu^{(1)}(x) = \lim_{\lambda \to +0} \frac{e^{-\frac{1}{2}\nu\pi i}}{\pi i} \int_{-\infty}^{\infty} \exp\{-\lambda(t + \tfrac{1}{2}\pi i)^2\} e^{ix\cosh t - \nu t}\, dt$$

$$= \frac{e^{-\frac{1}{2}\nu\pi i}}{\pi i} G \int_{-\infty}^{\infty} e^{ix\cosh t - \nu t}\, dt,$$

in Hardy's notation. In like manner

$$H_\nu^{(2)}(x) = -\frac{e^{\frac{1}{2}\nu\pi i}}{\pi i} G \int_{-\infty}^{\infty} e^{-ix\cosh u - \nu u}\, du,$$

with an implied exponential factor $\exp\{-\lambda(u - \frac{1}{2}\pi i)^2\}$.

Since the requisite convergence conditions are fulfilled when $\lambda > 0$, we may regard the product of the two integrals

$$\int_{-\infty}^{\infty} e_1(t)\, e^{ix\cosh t - \nu t}\, dt \times \int_{-\infty}^{\infty} e_2(u)\, e^{-ix\cosh u - \nu u}\, du$$

(in which $e_1(t)$ and $e_2(u)$ stand for the exponential factors) as a double integral

$$\int_{-\infty}^{\infty}\int_{-\infty}^{\infty} e_1(t)\, e_2(u)\, e^{ix(\cosh t - \cosh u) - \nu(t+u)}\, (dt\, du).$$

We thus find that

$$\pi^2 H_\nu^{(1)}(x)\, H_\nu^{(2)}(x) = G \int_{-\infty}^{\infty}\int_{-\infty}^{\infty} e^{ix(\cosh t - \cosh u) - \nu(t+u)}\, (dt\, du),$$

with the implied exponential factor

$$\exp\{-\lambda(t + \tfrac{1}{2}\pi i)^2 - \lambda(u - \tfrac{1}{2}\pi i)^2\}.$$

Make the substitution $t + u = 2T$, $t - u = 2U$ and then

$$\tfrac{1}{2}\pi^2 \{J_\nu^2(x) + Y_\nu^2(x)\} = G \int_{-\infty}^{\infty}\int_{-\infty}^{\infty} e^{2ix\sinh T \sinh U - 2\nu T}\, (dT\, dU),$$

with an implied exponential factor

$$\exp\{-2\lambda T^2 - 2\lambda(U + \tfrac{1}{2}\pi i)^2\}.$$

In view of the absolute convergence of the integral, it may be replaced by the repeated integral in which the integration with respect to U is performed first, so that

$$\tfrac{1}{2}\pi^2 \{J_\nu^2(x) + Y_\nu^2(x)\} = G \left[\int_0^{\infty}\int_{-\infty}^{\infty} e^{2ix\sinh T \sinh U - 2\nu T}\, dU\, dT + \int_0^{\infty}\int_{-\infty}^{\infty} e^{-2ix\sinh T \sinh U + 2\nu T}\, dU\, dT\right],$$

with an implied exponential factor in each case equal to

$$\exp\{-2\lambda T^2 - 2\lambda(U + \tfrac{1}{2}\pi i)^2\}.$$

We first consider the integral

$$\int_{-\infty}^{\infty} \exp\{-2\lambda(U+\tfrac{1}{2}\pi i)^2\}\, e^{2ix\sinh T \sinh U}\, dU,$$

in which T is positive.

When T is positive, the U-path of integration may be deformed into the contour $I(U)=\frac{1}{2}\pi$; if we then write $U=v+\frac{1}{2}\pi i$, where v is real, the integral becomes

$$\begin{aligned}\int_{-\infty}^{\infty} &\exp\{-2\lambda(v+\pi i)^2\}\, e^{-2x\sinh T\cosh v}\, dv \\ &= 2\exp(2\lambda\pi^2)\int_0^\infty \exp(-2\lambda v^2)\cos 4\pi\lambda v\,.\, e^{-2x\sinh T\cosh v}\, dv \\ &= 2\exp(2\lambda\pi^2)\int_0^\infty e^{-2x\sinh T\cosh v}\, dv \\ &\quad - 2\exp(2\lambda\pi^2)\int_0^\infty \{1-\exp(-2\lambda v^2)\cos 4\pi\lambda v\}\, e^{-2x\sinh T\cosh v}\, dv.\end{aligned}$$

To approximate to the latter integral when λ is small, we use the inequalities

$$\begin{aligned}0 \leqslant 1 &- \exp(-2\lambda v^2)\cos 4\pi\lambda v \\ &= 1-\exp(-2\lambda v^2) + 2\exp(-2\lambda v^2)\sin^2 2\pi\lambda v \leqslant 2\lambda v^2 + 8\pi^2\lambda^2 v^2,\end{aligned}$$

so that, for some value of θ between 0 and 1,

$$\begin{aligned}\int_{-\infty}^{\infty} &\exp\{-2\lambda(U+\tfrac{1}{2}\pi i)^2\}\, e^{2ix\sinh T\sinh U}\, dU \\ &= 2\exp(2\lambda\pi^2)\left[\int_0^\infty e^{-2x\sinh T\cosh v}\, dv - (2\lambda+8\pi^2\lambda^2)\,\theta\int_0^\infty v^2 e^{-2x\sinh T\cosh v}\, dv\right] \\ &= 2\exp(2\lambda\pi^2)\left[K_0(2x\sinh T) - (2\lambda+8\pi^2\lambda^2)\,\theta\left\{\frac{\partial^2}{\partial\mu^2}K_\mu(2x\sinh T)\right\}_{\mu=0}\right].\end{aligned}$$

If we treat the integral

$$\int_{-\infty}^{\infty} \exp\{-2\lambda(U+\tfrac{1}{2}\pi i)^2\}\, e^{-2ix\sinh T\sinh U}\, dU$$

in a similar manner, we find it equal to

$$2K_0(2x\sinh T) - 2\lambda\theta_1\left\{\frac{\partial^2}{\partial\mu^2}K_\mu(2x\sinh T)\right\}_{\mu=0},$$

where $0\leqslant\theta_1\leqslant 1$, provided that T is positive.

On collecting the results and remembering that $\int_0^\infty$ means the same thing as $\lim\limits_{\delta\to+0}\int_\delta^\infty$, we find that

$$\begin{aligned}\tfrac{1}{4}\pi^2&\{J_\nu^2(x)+Y_\nu^2(x)\} \\ &= \lim_{\lambda\to+0}\lim_{\delta\to+0}\int_\delta^\infty \Big[\exp(2\lambda\pi^2)K_0(2x\sinh T) \\ &\qquad - (2\lambda+8\pi^2\lambda^2)\,\theta\exp(2\lambda\pi^2)\left\{\frac{\partial^2}{\partial\mu^2}K_\mu(2x\sinh T)\right\}_{\mu=0}\Big]\, e^{-2\nu T}\, dT \\ &\quad + \lim_{\lambda\to+0}\lim_{\delta\to+0}\int_\delta^\infty\left[K_0(2x\sinh T) - 2\lambda\theta_1\left\{\frac{\partial^2}{\partial\mu^2}K_\mu(2x\sinh T)\right\}_{\mu=0}\right] e^{2\nu T}\, dT.\end{aligned}$$

Now, *qua* function of T,

$$\left\{\frac{\partial^2}{\partial\mu^2} K_\mu(2x\sinh T)\right\}_{\mu=0} = O\{(\log\sinh T)^2\},$$

when T is small, and so we may proceed at once to the limit by making $\delta \to 0$, since the integral is convergent; and, since the integrals

$$\int_0^\infty \left\{\frac{\partial^2}{\partial\mu^2} K_\mu(2x\sinh T)\right\}_{\mu=0} e^{\pm 2\nu T}\,dT$$

are convergent, the result of making $\lambda \to 0$ is

$$\tfrac{1}{4}\pi^2\{J_\nu^2(x) + Y_\nu^2(x)\} = \int_0^\infty K_0(2x\sinh T)(e^{-2\nu T} + e^{2\nu T})\,dT.$$

It is therefore proved that, when $x > 0$,

$$J_\nu^2(x) + Y_\nu^2(x) = \frac{8}{\pi^2}\int_0^\infty K_0(2x\sinh T)\cosh 2\nu T\,dT.$$

If we replace x by z, both sides of this equation become analytic functions of z, provided that $R(z) > 0$. Hence, by the theory of analytic continuation, we have the result

(1) $$J_\nu^2(z) + Y_\nu^2(z) = \frac{8}{\pi^2}\int_0^\infty K_0(2z\sinh t)\cosh 2\nu t\,dt,$$

provided that $R(z) > 0$.

Another integral formula which can be established by the same method* is

(2) $$J_\nu(z)\frac{\partial Y_\nu(z)}{\partial\nu} - Y_\nu(z)\frac{\partial J_\nu(z)}{\partial\nu} = -\frac{4}{\pi}\int_0^\infty K_0(2z\sinh t)\,e^{-2\nu t}\,dt.$$

To prove this formula, we first suppose that z is a positive variable (which we replace by x), and then

$$\begin{aligned}
J_\nu(x)\frac{\partial Y_\nu(x)}{\partial\nu} &- Y_\nu(x)\frac{\partial J_\nu(x)}{\partial\nu} \\
&= \frac{1}{2i}\left\{H_\nu^{(2)}(x)\frac{\partial H_\nu^{(1)}(x)}{\partial\nu} - H_\nu^{(1)}(x)\frac{\partial H_\nu^{(2)}(x)}{\partial\nu}\right\} \\
&= \frac{1}{2\pi^2 i}\,G\int_{-\infty}^\infty\int_{-\infty}^\infty (u - t - \pi i)\,e^{ix(\cosh t - \cosh u)}\,e^{-\nu(t+u)}\,.\,(dt\,du) \\
&= -\frac{1}{\pi^2 i}\,G\int_{-\infty}^\infty\int_{-\infty}^\infty (2U + \pi i)\,e^{2ix\sinh T\sinh U}\,e^{-2\nu T}\,.\,(dT\,dU) \\
&= -\frac{1}{\pi^2 i}\,G\int_0^\infty\int_{-\infty}^\infty (2U + \pi i)\,e^{2ix\sinh T\sinh U}\,e^{-2\nu T}\,dU\,dT \\
&\quad - \frac{1}{\pi^2 i}\,G\int_0^\infty\int_{-\infty}^\infty (2U + \pi i)\,e^{-2ix\sinh T\sinh U}\,e^{2\nu T}\,dU\,dT.
\end{aligned}$$

* For the full details of the analysis, see Watson, *Proc. Royal Soc.* XCIV. A, (1918), pp. 197—202.

Now, T being positive, we have

$$\int_{-\infty}^{\infty} (2U + \pi i) \exp\{-2\lambda(U + \tfrac{1}{2}\pi i)^2\} e^{2ix\sinh T \sinh U} dU$$

$$= 2\int_{-\infty}^{\infty} (v + \pi i) \exp\{-2\lambda(v + \pi i)^2\} e^{-2x\sinh T\cosh v} dv$$

and, since $ve^{-2x\sinh T\cosh v}$ is an odd function of v, it may be proved that the last integral is

$$2\pi i \int_{-\infty}^{\infty} e^{-2x\sinh T\cosh v} dv + O(\lambda),$$

where the constant implied in the symbol $O(\lambda)$ is a function of T such that its integral with respect to T from 0 to ∞ is convergent.

In like manner,

$$\int_{-\infty}^{\infty} (2U + \pi i) \exp\{-2\lambda(U + \tfrac{1}{2}\pi i)^2\} e^{-2ix\sinh T\sinh U} dU$$

$$= \int_{-\infty}^{\infty} 2v \exp(-2\lambda v^2) e^{-2x\sinh T\cosh v} dv = 0.$$

Hence it follows that

$$J_\nu(x)\frac{\partial Y_\nu(x)}{\partial\nu} - Y_\nu(x)\frac{\partial J_\nu(x)}{\partial\nu} = -\frac{2}{\pi}\int_0^\infty\int_{-\infty}^{\infty} e^{-2x\sinh T\cosh v - 2\nu T} dv\,dT$$

$$= -\frac{4}{\pi}\int_0^\infty K_0(2x\sinh T)\, e^{-2\nu T} dT.$$

The extension to the case in which the argument of the Bessel functions is complex with a positive real part is made as in (1).

It should be mentioned that formula (2) is of importance in the discussion of descriptive properties of zeros of Bessel functions.

The reader may find it interesting to prove that

$$\nabla_\nu\left[J_\nu(z)\frac{\partial Y_\nu(z)}{\partial\nu} - Y_\nu(z)\frac{\partial J_\nu(z)}{\partial\nu}\right]$$

$$= 2z^2\left[J_\nu'(z)\frac{\partial Y_\nu'(z)}{\partial\nu} - Y_\nu'(z)\frac{\partial J_\nu'(z)}{\partial\nu}\right] - (z^2-\nu^2)\left[J_\nu(z)\frac{\partial Y_\nu(z)}{\partial\nu} - Y_\nu(z)\frac{\partial J_\nu(z)}{\partial\nu}\right],$$

and hence that

(3) $$J_\nu'(z)\frac{\partial Y_\nu'(z)}{\partial\nu} - Y_\nu'(z)\frac{\partial J_\nu'(z)}{\partial\nu} = -\frac{4}{\pi z^2}\int_0^\infty (z^2\cosh 2T - \nu^2)\, K_0(2z\sinh T)\, e^{-2\nu T} dT.$$

Other formulae which may be established by the methods of this section are

(4) $$J_\mu(z)\, J_\nu(z) + Y_\mu(z)\, Y_\nu(z)$$

$$= \frac{4}{\pi^2}\int_0^\infty K_{\nu-\mu}(2z\sinh t)\,.\,\{e^{(\mu+\nu)t} + e^{-(\mu+\nu)t}\cos(\mu-\nu)\pi\}\, dt,$$

(5) $$J_\mu(z)\, Y_\nu(z) - J_\nu(z)\, Y_\mu(z) = \frac{4\sin(\mu-\nu)\pi}{\pi^2}\int_0^\infty K_{\nu-\mu}(2z\sinh t)\, e^{-(\mu+\nu)t} dt;$$

these are valid when $R(z) > 0$ and $|R(\mu-\nu)| < 1$; they do not appear to have been previously published.

13·74. *Deductions from Nicholson's integrals.*

Since $K_0(\xi)$ is a decreasing* function of ξ, it is clear from § 13·73 (1) that

$$J_\nu^2(x) + Y_\nu^2(x)$$

is a decreasing function of x for any real fixed value of ν, when x is positive.

Since this function is approximately equal to $2/(\pi x)$, when x is large, we shall investigate

$$x\{J_\nu^2(x) + Y_\nu^2(x)\}$$

and prove that it is a *decreasing* function of x when $\nu > \frac{1}{2}$, and that it is an *increasing* function of x when $\nu < \frac{1}{2}$.

It is clear that

$$\begin{aligned}\frac{d}{dx}[x\{J_\nu^2(x) + Y_\nu^2(x)\}]&\\ &= \frac{8}{\pi^2}\int_0^\infty \{K_0(2x\sinh T) + 2x\sinh T\, K_0'(2x\sinh T)\}\cosh 2\nu T\,dT\\ &= \frac{8}{\pi^2}\left[K_0(2x\sinh T)\tanh T\cosh 2\nu T\right]_0^\infty\\ &\quad + \frac{8}{\pi^2}\int_0^\infty K_0(2x\sinh T)\left[\cosh 2\nu T - \frac{d}{dT}\{\tanh T\cosh 2\nu T\}\right]dT,\end{aligned}$$

on integrating the second term in the integral by parts. Hence

$$\begin{aligned}\frac{d}{dx}[x\{J_\nu^2(x) + Y_\nu^2(x)\}]&\\ &= -\frac{8}{\pi^2}\int_0^\infty K_0(2x\sinh T)\tanh T\cosh 2\nu T\{\tanh T - 2\nu\tanh 2\nu T\}\,dT.\end{aligned}$$

Now $\lambda \tanh \lambda T$ is an increasing function of λ when $\lambda > 0$, and so the last integrand is negative or positive according as $2\nu > 1$ or $0 < 2\nu < 1$; and this establishes the result.

Next we prove that, when $x \geqslant \nu \geqslant 0$,

$$(x^2 - \nu^2)^{\frac{1}{2}}\{J_\nu^2(x) + Y_\nu^2(x)\}$$

is an *increasing* function of x.

If we omit the positive factor $8(x^2 - \nu^2)^{-\frac{1}{2}}/\pi^2$ from the derivate of the expression under consideration we get

$$\int_0^\infty \{xK_0(2x\sinh t) + 2(x^2 - \nu^2)\sinh t\,.\,K_0'(2x\sinh t)\}\cosh 2\nu t\,dt,$$

and to establish the theorem stated it is sufficient to prove that this integral is positive.

* This is obvious from the formula

$$K_0(\xi) = \int_0^\infty e^{-\xi\cosh t}\,dt.$$

We twice integrate by parts the last portion of the second term in the integral thus

$$2\nu^2 \int_0^\infty \sinh t\, K_0'(2x\sinh t)\cosh 2\nu t\, dt$$

$$= \Big[\nu \sinh t \sinh 2\nu t\, K_0'(2x \sinh t)\Big]_0^\infty - \nu \int_0^\infty \frac{d}{dt}\{\sinh t\, K_0'(2x\sinh t)\}\sinh 2\nu t\, dt$$

$$= -\nu \int_0^\infty \frac{d}{dt}\{\sinh t\, K_0'(2x\sinh t)\}\sinh 2\nu t\, dt$$

$$= -\nu \int_0^\infty 2x \sinh t \cosh t\, K_0(2x\sinh t)\sinh 2\nu t\, dt$$

$$= \Big[-x\sinh t\cosh t\, K_0(2x\sinh t)\cosh 2\nu t\Big]_0^\infty$$

$$+ x\int_0^\infty \frac{d}{dt}[\sinh t \cosh t\, K_0(2x\sinh t)]\cosh 2\nu t\, dt$$

$$= \int_0^\infty [x\cosh 2t\, K_0(2x\sinh t) + 2x^2 \sinh t\cosh^2 t\, K_0'(2x\sinh t)]\cosh 2\nu t\, dt;$$

the simplification after the second step is produced by using the differential equation

$$zK_0''(z) + K_0'(z) - zK_0(z) = 0.$$

The integral under discussion consequently reduces to

$$\int_0^\infty [-2x\sinh^2 t\, K_0(2x\sinh t) - 2x^2\sinh^3 t\, K_0'(2x\sinh t)]\cosh 2\nu t\, dt$$

$$= \left[-\frac{x\sinh^3 t}{\cosh t} K_0(2x\sinh t)\cosh 2\nu t\right]_0^\infty$$

$$+ x\int_0^\infty K_0(2x\sinh t)\left[-2\sinh^2 t\cosh 2\nu t + \frac{d}{dt}\left\{\frac{\sinh^3 t}{\cosh t}\cosh 2\nu t\right\}\right]dt$$

$$= x\int_0^\infty K_0(2x\sinh t)[\tanh^2 t\cosh 2\nu t + 2\nu\sinh^3 t\,\mathrm{sech}\, t\sinh 2\nu t]\, dt,$$

and this is positive because the integrand is positive; hence the differential coefficient of

$$(x^2-\nu^2)^{\frac{1}{2}}\{J_\nu^2(x) + Y_\nu^2(x)\}$$

is positive, and the result is established.

Since the limits of both the functions

$$x\{J_\nu^2(x) + Y_\nu^2(x)\}, \quad (x^2-\nu^2)^{\frac{1}{2}}\{J_\nu^2(x) + Y_\nu^2(x)\}$$

are $2/\pi$, it follows from the last two results that when $x \geqslant \nu \geqslant \frac{1}{2}$,

$$(1) \qquad \frac{2/\pi}{(x^2-\nu^2)^{\frac{1}{2}}} > J_\nu^2(x) + Y_\nu^2(x) > \frac{2/\pi}{x}.$$

An elementary proof of the last inequality (with various related inequalities) was deduced by Schafheitlin, *Berliner Sitzungsberichte*, v. (1906), p. 86, from the formula (cf. § 5·14)

$$(4\nu^2-1)\int_x^\infty \mathscr{C}_\nu^2(t)\frac{dt}{t^2}$$

$$= \frac{4}{\pi}(a^2+b^2) - x\left[\left\{\frac{\mathscr{C}_\nu(x)}{x} + \mathscr{C}_\nu'(x)\right\}^2 + 2\left(1-\frac{\nu^2}{x^2}\right)\mathscr{C}_\nu^2(x) + \mathscr{C}_\nu'^2(x)\right],$$

where $\mathscr{C}_\nu(x) \equiv aJ_\nu(x) + bY_\nu(x)$.

The next consequence which we shall deduce from the integrals of § 13·73 is that, when ν is positive,

$$J_\nu(\nu)\frac{dY_\nu(\nu)}{d\nu} - Y_\nu(\nu)\frac{dJ_\nu(\nu)}{d\nu} > 0.$$

To obtain this result, we observe that the expression on the left may be written in the form

$$\left[J_\nu(x)\frac{\partial Y_\nu(x)}{\partial\nu} - Y_\nu(x)\frac{\partial J_\nu(x)}{\partial\nu}\right]_{x=\nu}$$
$$+\left[J_\nu(x)\frac{\partial Y_\nu(x)}{\partial x} - Y_\nu(x)\frac{\partial J_\nu(x)}{\partial x}\right]_{x=\nu}$$
$$=\frac{2}{\pi\nu} - \frac{4}{\pi}\int_0^\infty K_0(2\nu\sinh T)\,e^{-2\nu T}\,dT$$
$$=\frac{2}{\pi\nu}\left[1 - \int_0^\infty e^{-t}K_0\left(2\nu\sinh\frac{t}{2\nu}\right)dt\right].$$

But, for each positive value of t, $2\nu\sinh(\tfrac{1}{2}t/\nu)$ is a decreasing function of ν, and so, since $K_0(x)$ is a positive decreasing function of its argument, we see that

$$1 - \int_0^\infty e^{-t}K_0\left(2\nu\sinh\frac{t}{2\nu}\right)dt$$

is a decreasing function of ν, and therefore

$$\nu\left\{J_\nu(\nu)\frac{dY_\nu(\nu)}{d\nu} - Y_\nu(\nu)\frac{dJ_\nu(\nu)}{d\nu}\right\}$$
$$\geqslant \lim_{\nu\to\infty}\left[\nu\left\{J_\nu(\nu)\frac{dY_\nu(\nu)}{d\nu} - Y_\nu(\nu)\frac{dJ_\nu(\nu)}{d\nu}\right\}\right]$$
$$=\lim_{\nu\to\infty}\left[\frac{4}{105\pi\nu^2} - \ldots\right] = 0,$$

by using the asymptotic expansions of § 8·42; and this establishes the result stated.

13·75. *The asymptotic expansion of* $J_\nu^2(z) + Y_\nu^2(z)$.

It is easy to deduce the asymptotic expansion of $J_\nu^2(z) + Y_\nu^2(z)$ from Nicholson's formula obtained in § 13·73, namely

$$J_\nu^2(z) + Y_\nu^2(z) = \frac{8}{\pi^2}\int_0^\infty K_0(2z\sinh t)\cosh 2\nu t\,dt;$$

for we have, by § 7·4 (4),

$$\frac{\cosh 2\nu t}{\cosh t} = \sum_{m=0}^{p-1}\frac{m!\,(\nu, m)}{(2m)!}2^{2m}\sinh^{2m}t + R_p,$$

where
$$|R_p| < \left|\frac{\cos\nu\pi}{\cos R(\nu\pi)}\right|\frac{p!\,|(R(\nu), p)|}{(2p)!}2^{2p}\sinh^{2p}t;$$

and, when ν is real and p is so large that $p + \tfrac{1}{2} > \nu$, R_p lies between 0 and

$$\frac{p!\,(\nu, p)}{(2p)!}2^{2p}\sinh^{2p}t.$$

We at once deduce the asymptotic expansion

$$J_\nu^2(z)+Y_\nu^2(z)\sim\frac{4}{\pi^2}\sum_{m=0}^{\infty}\frac{m!(\nu,m)}{(2m)!}2^{2m}\int_0^\infty K_0(2zu)\,u^{2m}\,du,$$

that is to say, by § 13·21 (8),

(1) $$J_\nu^2(z)+Y_\nu^2(z)\sim\frac{2}{\pi z}\sum_{m=0}^{\infty}\{1.3\ldots(2m-1)\}\frac{(\nu,m)}{2^m z^{2m}};$$

this is proved when $R(z)>0$, but it may be extended over the wider range $|\arg z|<\pi$; and, if ν is real and z is positive, and p exceeds $\nu-\frac{1}{2}$, the remainder after p terms is of the same sign as, and numerically less than, the $(p+1)$th term.

13·8. *Ramanujan's integrals.*

Some extraordinary integrals have been obtained by Ramanujan* from an application of Fourier's integral theorem† to Cauchy's well-known formula

$$\int_{-\frac{1}{2}\pi}^{\frac{1}{2}\pi}\cos^{\mu+\nu-2}\theta\,.\,e^{i\theta(\mu-\nu+2\xi)}\,d\theta=\frac{\pi\Gamma(\mu+\nu-1)}{2^{\mu+\nu-2}\Gamma(\mu+\xi)\Gamma(\nu-\xi)},$$

which is valid if $R(\mu+\nu)>1$. The application shews that

$$\int_{-\infty}^{\infty}\frac{e^{it\xi}\,d\xi}{\Gamma(\mu+\xi)\Gamma(\nu-\xi)}=\begin{cases}\dfrac{(2\cos\frac{1}{2}t)^{\mu+\nu-2}}{\Gamma(\mu+\nu-1)}e^{\frac{1}{2}it(\nu-\mu)}, & (|t|<\pi),\\ 0, & (|t|>\pi),\end{cases}$$

where t is any real number.

By expanding in ascending powers of x and y, and then applying this formula, it is seen that

(1) $$\int_{-\infty}^{\infty}\frac{J_{\mu+\xi}(x)}{x^{\mu+\xi}}\frac{J_{\nu-\xi}(y)}{y^{\nu-\xi}}e^{it\xi}\,d\xi$$
$$=\left(\frac{2\cos\frac{1}{2}t}{x^2e^{-\frac{1}{2}it}+y^2e^{\frac{1}{2}it}}\right)^{\frac{1}{2}(\mu+\nu)}e^{\frac{1}{2}it(\nu-\mu)}J_{\mu+\nu}[\sqrt{\{2\cos\tfrac{1}{2}t\,(x^2e^{-\frac{1}{2}it}+y^2e^{\frac{1}{2}it})\}}],$$

if $-\pi<t<\pi$; for other real values of t, the integral is zero.

In particular

(2) $$\int_{-\infty}^{\infty}J_{\mu+\xi}(x)\,J_{\nu-\xi}(x)\,d\xi=J_{\mu+\nu}(2x).$$

In view of the researches of March, *Ann. der Physik und Chemie*, (4) XXXVII. (1912), pp. 29—50 and Rybczyński, *Ann. der Physik und Chemie*, (4) XLI. (1913), pp. 191—208, it seems quite likely that, in spite of the erroneous character of the analysis of these writers‡, these integrals evaluated by Ramanujan may prove to be of the highest importance in the theory of the transmission of Electric Waves.

* *Quarterly Journal*, XLVIII. (1920), pp. 294—310.

† Cf. *Modern Analysis*, §§ 9·7, 11·1.

‡ Cf. Love, *Phil. Trans. of the Royal Soc.* CCXV. A, (1915), pp. 123—124.

CHAPTER XIV

MULTIPLE INTEGRALS

14·1. *Problems connected with multiple integrals.*

The difference between the subjects of this chapter and the last is more than one of mere degree produced by the insertion of an additional integral sign. In Chapter XIII we were concerned with the discussion of integrals of perfectly definite functions of the variable and of a number of auxiliary parameters; in the integrals which are now to be discussed the functions under the integral sign are to a greater or less extent arbitrary. Thus, in the first problem which will be discussed, the integral involves a function which has merely to satisfy the conditions of being a solution of a partial differential equation, and of having continuous differential coefficients at all points of real three-dimensional space.

In subsequent problems, which are generalisations of Fourier's integral formula, the arbitrary element has to satisfy even more general restrictions such as having an absolutely convergent integral, and having limited total fluctuation.

14·2. *Weber's infinite integrals.*

The integrals which will now be considered involve Bessel functions only incidentally; but it seems desirable to investigate them somewhat fully because many of the formulae of Chapter XIII may easily be derived from them, and were, in fact, discovered by Weber as special cases of the results of this section.

Weber's researches* are based upon a result discovered by Fourier† to the effect that a solution of the equation of Conduction of Heat

$$\frac{\partial u}{\partial t} = \frac{\partial^2 u}{\partial x^2} + \frac{\partial^2 u}{\partial y^2} + \frac{\partial^2 u}{\partial z^2}$$

is

$$u = \frac{1}{\pi^{\frac{3}{2}}} \int_{-\infty}^{\infty} \int_{-\infty}^{\infty} \int_{-\infty}^{\infty} \Phi(x + 2X\sqrt{t}, y + 2Y\sqrt{t}, z + 2Z\sqrt{t}) \times \exp\{-(X^2 + Y^2 + Z^2)\}\, dX\, dY\, dZ,$$

where Φ is an arbitrary function of its three variables.

Weber first proved that, if $\Phi(x, y, z)$ is restricted to be a solution of the equation

$$(1) \qquad \frac{\partial^2 \Phi}{\partial x^2} + \frac{\partial^2 \Phi}{\partial y^2} + \frac{\partial^2 \Phi}{\partial z^2} = -k^2 \Phi,$$

* *Journal für Math.* LXIX. (1868), pp. 222—237.

† *La Théorie Analytique de la Chaleur* (Paris, 1822), § 372. The simpler equation with only one term on the right had previously been solved by Laplace, *Journal de l'École polytechnique*, VIII. (1809), pp. 235—244.

then

(2) $$u = \exp(-k^2 t)\,\Phi(x, y, z),$$

provided that Φ has continuous first and second differential coefficients, and the integral converges in such a way* that transformations to polar coordinates are permissible.

The method by which this result is established is successful in expressing a more general triple integral as a single integral [cf. equation (4) below].

If we change to polar coordinates by writing

$$2X\sqrt{t} = r\sin\theta\cos\phi, \quad 2Y\sqrt{t} = r\sin\theta\sin\phi, \quad 2Z\sqrt{t} = r\cos\theta,$$

we get

$$u = \frac{1}{(4\pi t)^{\frac{3}{2}}}\int_0^\infty\int_0^\pi\int_{-\pi}^\pi \Phi(x + r\sin\theta\cos\phi,\ y + r\sin\theta\sin\phi,\ z + r\cos\theta) \times \exp\left(-\frac{r^2}{4t}\right) r^2 \sin\theta\, d\phi\, d\theta\, dr.$$

Now consider the function of r, $\varpi(r)$, defined by the equation

$$\varpi(r) = \int_0^\pi\int_{-\pi}^\pi \Phi(x + r\sin\theta\cos\phi,\ y + r\sin\theta\sin\phi,\ z + r\cos\theta)\sin\theta\, d\phi\, d\theta.$$

It is a continuous function of r, with continuous first and second differential coefficients when r has any positive value; and the result of applying the operator

$$\frac{1}{r^2}\frac{d}{dr}\left(r^2\frac{d}{dr}\right) + k^2$$

to $\varpi(r)$ is

$$\int_0^\pi\int_{-\pi}^\pi \left\{\frac{1}{r^2}\frac{\partial}{\partial r}\left(r^2\frac{\partial\Phi}{\partial r}\right) + k^2\Phi\right\}\sin\theta\, d\phi\, d\theta.$$

We proceed to shew that the last integral is zero. If we make use of the differential equation (1), which Φ satisfies, we find that

$$\iint\left\{\frac{1}{r^2}\frac{\partial}{\partial r}\left(r^2\frac{\partial\Phi}{\partial r}\right) + k^2\Phi\right\}\sin\theta\, d\phi\, d\theta = -\frac{1}{r^2}\iint\left\{\frac{1}{\sin\theta}\frac{\partial}{\partial\theta}\left(\sin\theta\frac{\partial\Phi}{\partial\theta}\right) + \frac{1}{\sin^2\theta}\frac{\partial^2\Phi}{\partial\phi^2}\right\}\sin\theta\, d\phi\, d\theta.$$

To avoid the difficulty† caused by the apparent singularity of the last integrand on the polar axis, we consider the integral taken over the surface of a sphere with the exception of a small cap of angular radius δ at each pole; since the integrand on the left is bounded at the poles, the integrals over the caps can be made arbitrarily small by taking δ sufficiently small.

If we perform the integration of the second term on the right with respect

* A sufficient condition is that Φ should be bounded when the variables assume all real values, infinite values of the variables being included. Cf. the corresponding two-dimensional investigation, *Modern Analysis*, § 12·41.

† This difficulty was overlooked by Weber.

to ϕ, we see that its integral vanishes because $\partial\Phi/\partial\phi$ is supposed to be a one-valued function of position. The first term on the right gives

$$-\frac{1}{r^2}\int_{-\pi}^{\pi}\left[\sin\theta\frac{\partial\Phi}{\partial\theta}\right]_{\delta}^{\pi-\delta} d\phi,$$

and this can be made arbitrarily small by taking δ sufficiently small since

$$\partial\Phi/(\sin\theta\,\partial\theta)$$

is continuous and therefore bounded.

Hence $$\frac{1}{r^2}\frac{d}{dr}\left(r^2\frac{d\varpi(r)}{dr}\right)+k^2\varpi(r)$$

can be made arbitrarily small by taking δ sufficiently small, and therefore it is zero.

Consequently

(3) $$\frac{d^2\{r\varpi(r)\}}{dr^2}+k^2 r\varpi(r)=0,$$

so that $$\varpi(r)=\frac{A\sin kr+B\cos kr}{r},$$

where A and B are constants; since $\varpi(r)$ and its derivate are continuous for all values of r, A and B must have the same constant values for all values of r.

If we make $r\to 0$, we see that

$$B=0,\quad A=4\pi\Phi(x, y, z)/k.$$

Hence*

$$u=\frac{\Phi(x, y, z)}{2k\sqrt{(\pi t^3)}}\int_0^\infty \exp\left(-\frac{r^2}{4t}\right)\sin kr\,.\,r\,dr=\exp(-k^2t)\,\Phi(x, y, z),$$

and this establishes Weber's result.

A similar change to polar coordinates shews that, if $\Phi(x, y, z)$ is a solution of (1) of the type already considered, and if $f(r)$ is an arbitrary continuous function of r, then

(4) $$\int_{-\infty}^{\infty}\int_{-\infty}^{\infty}\int_{-\infty}^{\infty}\Phi(X, Y, Z)\,f[\sqrt{\{(X-x)^2+(Y-y)^2+(Z-z)^2\}}]\,dX\,dY\,dZ$$
$$=\frac{4\pi\Phi(x, y, z)}{k}\int_0^\infty f(r)\sin kr\,.\,r\,dr.$$

The reader will have no difficulty in enunciating sufficient conditions concerning absoluteness of convergence to make the various changes in the

* This integral is most easily evaluated by differentiating the well-known formula

$$\int_0^\infty \exp\left(-\frac{r^2}{4t}\right)\cos kr\,.\,dr=\sqrt{(\pi t)}\exp(-k^2t)$$

with respect to k.

integrations permissible. One such set of conditions is that Φ should be bounded as the variables tend to infinity, and that

$$f(r) = O(r^{-p}), \quad (r \to 0); \quad f(r) = O(r^{-q}), \quad (r \to \infty),$$

where $p < 3$, $q > 1$.

A somewhat simpler formula established at about the same time by Weber* is that, if $u(r, \theta)$ is a function of the polar coordinates (r, θ) which has continuous first and second differential coefficients at all points such that $0 \leqslant r \leqslant a$, whose value at the origin is u_0, and which is a solution of the equation

$$\frac{\partial^2 u}{\partial x^2} + \frac{\partial^2 u}{\partial y^2} + k^2 u = 0,$$

then

$$\int_{-\pi}^{\pi} u(r, \theta)\, d\theta = 2\pi u_0 J_0(kr),$$

when $0 \leqslant r \leqslant a$. The proof of this is left to the reader.

14·3. *General discussion of Neumann's integral.*

The formula

$$(1) \quad \int_0^\infty u\,du \int_0^\infty \int_{-\pi}^{\pi} F(R, \Phi) \,.\, J_0[u\sqrt{\{R^2 + r^2 - 2Rr\cos(\Phi - \phi)\}}]\, R\,(d\Phi\, dR) = 2\pi F(r, \phi)$$

was given by Neumann in his treatise† published in 1862. In this formula, $F(R, \Phi)$ is an arbitrary function of the two variables (R, Φ), and the integration over the plane of the polar coordinates (R, Φ) is a double integration.

In the special case in which the arbitrary function is independent of Φ, we replace the double integral by a repeated integral, and then perform the integration with respect to Φ; the formula reduces to

$$(2) \quad \int_0^\infty u\,du \int_0^\infty F(R)\, J_0(uR)\, J_0(ur)\, R\,dR = F(r),$$

a result which presents a closer resemblance to Fourier's integral‡ than (1).

The extension of (2) to functions of any order, namely

$$(3) \quad \int_0^\infty u\,du \int_0^\infty F(R)\, J_\nu(uR)\, J_\nu(ur)\, R\,dR = F(r),$$

was effected by Hankel§. In this result it is apparently necessary that $\nu \geqslant -\frac{1}{2}$, though a modified form of the theorem (§§ 14·5—14·52) is valid for all real values of ν; when $\nu = \pm\frac{1}{2}$, (3) is actually a case of Fourier's formula.

The formulae (2) and (3) are, naturally, much more easy to prove than (1); and the proof of (3) is of precisely the same character as that of (2), the

* *Math. Ann.* I. (1869), pp. 8—11.

† *Allgemeine Lösung des Problemes über den stationären Temperaturzustand eines homogenen Körpers, welcher von zwei nichtconcentrischen Kugelflächen begrenzt wird* (Halle, 1862), pp. 147—151. Cf. Gegenbauer, *Wiener Sitzungsberichte*, xcv. (2), (1887), pp. 409—410.

‡ Cf. *Modern Analysis*, § 9·7.

§ *Math. Ann.* VIII. (1875), pp. 476—483.

arbitrariness of the order of the Bessel functions not introducing any additional complications.

Following Hankel, many writers* describe the integrals (2) and (3) as "Fourier integrals" or "Fourier-Bessel integrals."

On account of its greater simplicity, we shall give a proof of (3) before proving (1); and at this stage it is convenient to give a brief account of the researches of the various writers who have investigated the formulae.

As has already been stated, Hankel was the first writer† to give the general formula (3). He transformed the integral into

$$\lim_{\lambda\to\infty}\int_0^\infty RF(R)\,dR\int_0^\lambda J_\nu(uR)\,J_\nu(ur)\,u\,du$$
$$= \lim_{\lambda\to\infty}\int_0^\infty RF(R)\,[RJ_{\nu+1}(\lambda R)\,J_\nu(\lambda r) - rJ_{\nu+1}(\lambda r)\,J_\nu(\lambda R)\,\frac{\lambda\,dR}{R^2-r^2}$$

and then applied the second mean-value theorem to the integrand just as in the evaluation of Dirichlet's integrals. Substantially the same proof was given by Sheppard‡ who laid stress on the important fact that the value of the integral depends only on that part of the R-range of integration which is in the immediate neighbourhood of r, so that the value of the integral is independent of the values which $F(R)$ assumes when R is not nearly equal to r.

A different mode of proof, based on the theory of discontinuous integrals has been given by Sonine§, who integrated the formula (§ 13·42)

$$r^{\nu+1}\int_0^\infty J_{\nu+1}(ur)\,J_\nu(uR)\,du = \begin{cases} R^\nu, & (R<r) \\ 0, & (R>r)\end{cases}$$

after multiplication by $F(R)\,R\,dR$, from 0 to ∞, so as to get

$$r^{\nu+1}\int_0^\infty\int_0^\infty J_{\nu+1}(ur)\,J_\nu(uR)\,F(R)\,R\,dR\,du = \int_0^r R^{\nu+1}F(R)\,dR;$$

and then, by differentiating both sides with respect to r, formula (3) is at once obtained; but the whole of this procedure is difficult to justify.

A proof of a more directly physical character has been given by Basset‖ but, according to Gray and Mathews, it is open to various objections.

A proof depending on the theory of integral equations has been constructed by Weyl¶.

The extension of Hankel's formula, which is effected by replacing the

* See e.g. Orr's paper cited later in this section.

† A statement of a mode of deducing (3) from (1) when ν is an integer was made by Weber *Math. Ann.* VI. (1873), p. 149, but this was probably later than Hankel's researches, since it is dated 1872, while Hankel's memoir is dated 1869.

‡ *Quarterly Journal*, XXIII. (1889), pp. 223—244. § *Math. Ann.* XVI. (1880), p. 47.

‖ *Proc. Camb. Phil. Soc.* V. (1886), pp. 425—433. See Gray and Mathews, *A Treatise on Bessel Functions* (London, 1895), pp. 80—82.

¶ *Math. Ann.* LXVI. (1909), p. 324.

Bessel functions by arbitrary cylinder functions, was obtained by Weber*, and it will be discussed in §§ 14·5—14·52.

An attempt has been made by Orr† to replace the Bessel functions by any cylinder functions, the u-path of integration being a contour which avoids the origin; but some of the integrals used by him appear to be divergent, so it is difficult to say to what extent his results are correct. The same criticism applies to the discussion of Weber's problem in Nielsen's treatise. It will be shewn (§ 14·5) that if, as Nielsen assumes, the two cylinder functions under the integral sign are not necessarily of the same type, the repeated integral is not, of necessity, convergent.

It should be stated that, if r be a point of discontinuity of $F(R)$, the expressions on the right in (2) and (3) must be replaced by‡

$$\tfrac{1}{2}\{F(r-0)+F(r+0)\},$$

just as in Fourier's theorem.

For the more recent researches by Neumann, the reader should consult his treatise *Ueber die nach Kreis-, Kugel- und Cylinder-functionen fortschreitenden Entwickelungen* (Leipzig, 1881).

Neumann's formula (1) was obtained by Mehler§ as a limiting case of a formula involving Legendre functions; in fact, it was apparently with this object in view that he obtained the formula of § 5·71,

$$\lim_{n\to\infty} P_n\{\cos(z/n)\}=J_0(z),$$

but it does not seem easy to construct a rigorous proof on these lines (cf. § 14·64). A more direct method of proof is given in a difficult memoir by Du Bois Reymond‖ on the general theory of integrals resembling Fourier's integral. The proof which we shall give subsequently (§§ 14·6 *et seq.*) is based on these researches.

Subsequently Ermakoff¶ pointed out that the formula is also derivable from a result obtained by Du Bois Reymond which is the direct extension to two variables of Fourier's theorem for one variable, namely

$$\Psi(x,y)$$
$$=\frac{1}{4\pi^2}\int_{-\infty}^{\infty}\int_{-\infty}^{\infty}\int_{-\infty}^{\infty}\int_{-\infty}^{\infty}\Psi(X,Y)\cos\{\alpha(X-x)+\beta(Y-y)\}\,.\,(dX\,dY)\,d\alpha\,d\beta.$$

Ermakoff deduced the formula by changing to polar coordinates by means of the substitution

$$\alpha=u\cos\omega,\quad \beta=u\sin\omega,$$

and effecting the integration with respect to ω.

* *Math. Ann.* VI. (1873), pp. 146—161.

† *Proc. Royal Irish Acad.* XXVII. A, (1909), pp. 205—248.

‡ The value of the integral at a point of discontinuity has been examined with some care by Cailler, *Archives des Sci.* (*Soc. Helvétique*), (4) XIV. (1902), pp. 347—350.

§ *Math. Ann.* V. (1872), pp. 135—137.

‖ *Math. Ann.* IV. (1871), pp. 362—390.

¶ *Math. Ann.* V. (1872), pp. 639—640.

If (r, ϕ) and (R, Φ) be the polar coordinates corresponding to the Cartesian coordinates (x, y) and (X, Y) respectively, the formal result is fairly obvious when we replace $\Psi(X, Y)$ by $F(R, \Phi)$; but the investigation by this method is not without difficulties, since it seems to be by no means easy to prove that the repeated integral taken over an infinite rectangle in the (α, β) plane may be replaced by a repeated integral taken over the area of an indefinitely great circle.

If the arbitrary function $F(R, \Phi)$ is not continuous, the factor $F(r, \phi)$ which occurs on the right in (1) must be replaced by the limit of the mean value of $F(R, \Phi)$ on a circle of radius δ with centre at (r, ϕ) when $\delta \to 0$. This was, in effect, proved by Neumann in his treatise of 1881, and the proof will be given in §§ 14·6—14·63. The reader might anticipate this result from what he knows of the theory of Fourier series.

A formula which is more recondite than (3), namely

$$\frac{1}{2}\int_{-\infty}^{\infty}\int_{r}^{\infty} J_0(u-r)\frac{J_1(u-R)}{u-R}F(R)\,du\,dR = F(r), \tag{4}$$

has been examined by Bateman, *Proc. London Math. Soc.* (2) IV. (1906), p. 484; cf. § 12·2.

14·4. *Hankel's repeated integral.*

The generalisation of Neumann's integral formula which was effected by Hankel (cf. § 14·3) in the case of functions of a single variable, may be formally stated as follows:

Let $F(R)$ be an arbitrary function of the real variable R subject to the condition that

$$\int_0^\infty F(R)\sqrt{R}\,.\,dR$$

exists and is absolutely convergent; and let the order ν of the Bessel functions be not less than $-\frac{1}{2}$. Then*

$$\int_0^\infty u\,du\int_0^\infty F(R)\,J_\nu(uR)\,J_\nu(ur)\,R\,dR = \tfrac{1}{2}\{F(r+0)+F(r-0)\}, \tag{1}$$

provided that the positive number r lies inside *an interval in which $F(R)$ has limited total fluctuation.*

The proof which we shall now give is substantially Hankel's proof, and it is of the same general character as the proof of Fourier's theorem; it will be set out in the same manner as the proof of Fourier's theorem given in *Modern Analysis*, Chapter IX. It is first convenient to prove a number of lemmas.

* It seems not unlikely that it is sufficient for ν to be greater than -1; but the proof for the more extended range of values of ν would be more difficult.

14·41. *The analogue of the Riemann-Lebesgue lemma.*

A result, which resembles the lemma of Riemann-Lebesgue* in the theory of Fourier series, and which is required in the proof of Hankel's integral theorem is as follows:

Let† $\int_a^b F(R)\sqrt{R}\,.\,dR$ *exist, and (if it is an improper integral) let it be absolutely convergent; and let* $\nu \geqslant -\frac{1}{2}$. *Then, as* $\lambda \to \infty$,

$$\int_a^b F(R)\, J_\nu(\lambda R)\, R\, dR = o(1/\sqrt{\lambda}).$$

It is convenient to divide the proof into three parts; in the first part it is assumed that $F(R)\sqrt{R}$ is bounded, and that b is finite; in the second part the restriction that b is finite is removed; and in the third part the restriction that $F(R)\sqrt{R}$ is bounded is also removed.

(I) Let the upper bound of $|F(R)\sqrt{R}|$ be K. Divide the range of integration (a, b) into n equal intervals by the points $x_1, x_2, \ldots x_{n-1}$ ($x_0 = a$, $x_n = b$), and choose n so large that

$$\sum_{m=1}^{n} (U_m - L_m)(x_m - x_{m-1}) < \epsilon,$$

where ϵ is an arbitrarily small positive number and U_m and L_m are the upper and lower bounds of $F(R)\sqrt{R}$ in the mth interval.

Write $F(R)\sqrt{R} = F(R_{m-1})\sqrt{R_{m-1}} + \omega_m(R)$, so that, when R lies in the mth interval, $|\omega_m(R)| \leqslant U_m - L_m$. Now, when $\nu \geqslant -\frac{1}{2}$, both of the functions of x,

$$x^{\frac{1}{2}} J_\nu(x), \qquad \int_0^x t^{\frac{1}{2}} J_\nu(t)\, dt,$$

are bounded when $x \geqslant 0$, even though the integral is not convergent as $x \to \infty$. Let A and B be the upper bounds of the moduli of these functions. It is then clear that

$$\begin{aligned}
&\left| \int_a^b F(R)\, J_\nu(\lambda R)\, R\, dR \right| \\
&\quad = \left| \sum_{m=1}^{n} F(R_{m-1})\sqrt{R_{m-1}} \int_{x_{m-1}}^{x_m} J_\nu(\lambda R)\sqrt{R}\,.\,dR \right. \\
&\qquad\qquad \left. + \sum_{m=1}^{n} \int_{x_{m-1}}^{x_m} \omega_m(R)\, J_\nu(\lambda R)\sqrt{R}\,.\,dR \right| \\
&\quad \leqslant \sum_{m=1}^{n} \frac{|F(R_{m-1})\sqrt{R_{m-1}}|}{\lambda^{\frac{3}{2}}} \left| \int_{\lambda x_{m-1}}^{\lambda x_m} J_\nu(x)\sqrt{x}\,.\,dx \right| + \sum_{m=1}^{n} \int_{x_{m-1}}^{x_m} |\omega_m(R)| \frac{A\, dR}{\sqrt{\lambda}} \\
&\quad \leqslant \frac{2B}{\lambda^{\frac{3}{2}}} \sum_{m=1}^{n} F(R_{m-1})\sqrt{R_{m-1}} + \frac{A\epsilon}{\sqrt{\lambda}} \\
&\quad \leqslant \frac{2BnK}{\lambda^{\frac{3}{2}}} + \frac{A\epsilon}{\sqrt{\lambda}}.
\end{aligned}$$

* Cf. *Modern Analysis*, § 9·41.

† The upper limit of the integral may be infinite; and $a \geqslant 0$. The apparently irrelevant factor R preserves the analogy with § 14·3 (3).

By taking λ sufficiently large (n remaining fixed after ϵ has been chosen) the last expression can be made less than $2A\epsilon/\sqrt{\lambda}$, and so the original integral is $o(1/\sqrt{\lambda})$.

(II) If the upper limit is infinite, choose c so that

$$\int_c^\infty |F(R)| \sqrt{R} . dR < \epsilon,$$

and use the inequality

$$\left|\int_a^\infty F(R) J_\nu(\lambda R) R dR\right| \leqslant \left|\int_a^c F(R) J_\nu(\lambda R) R dR\right| + \frac{A}{\sqrt{\lambda}} \int_c^\infty |F(R)| \sqrt{R} . dR$$

then, proceeding as in case (I), we get

$$\left|\int_a^\infty F(R) J_\nu(\lambda R) R dR\right| < \frac{2BnK}{\lambda^{\frac{3}{2}}} + \frac{2A\epsilon}{\sqrt{\lambda}}.$$

The choice of n now depends on ϵ through the choice of c as well as by the mode of subdivision of the range of integration (a, c); but the choice of n is still independent of λ, and so we can infer that the integral (with upper limit infinite) is still $o(1/\sqrt{\lambda})$.

(III) If $F(R)\sqrt{R}$ is unbounded*, we may enclose the points at which it is unbounded in a number p of intervals δ such that

$$\sum_\delta \int_\delta |F(R)| \sqrt{R} . dR < \epsilon.$$

By applying the arguments of (I) and (II) to the parts of (a, b) outside these intervals, we get

$$\left|\int_a^b F(R) J_\nu(\lambda R) R dR\right| < \frac{2Bn(p+1)K}{\lambda^{\frac{3}{2}}} + \frac{3A\epsilon}{\sqrt{\lambda}},$$

where K is now the upper bound of $|F(R)|\sqrt{R}$ outside the intervals δ. The choices of both K and n now depend on ϵ, but are still independent of λ, so that we can still infer that the integral is $o(1/\sqrt{\lambda})$.

14·42. *The inversion of Hankel's repeated integral.*

We shall next prove that, *when* $\nu \geqslant -\frac{1}{2}$, *and* $\int_0^\infty F(R)\sqrt{R} . dR$ *exists and is absolutely convergent, then*

$$\int_0^\infty u\,du \int_0^\infty F(R) J_\nu(uR) J_\nu(ur) R dR$$
$$= \lim_{\lambda \to \infty} \int_0^\infty F(R) \left\{\int_0^\lambda J_\nu(uR) J_\nu(ur) u\,du\right\} R dR,$$

provided that the limit on the right exists.

* Cf. *Modern Analysis*, § 9·41.

For any assigned value of λ, and any arbitrary positive number ϵ, *ex hypothesi* there exists a number β such that

$$\int_\beta^\infty |F(R)| \sqrt{R} \,.\, dR < \frac{\epsilon \sqrt{r}}{2A^2\lambda},$$

where A is the constant defined in § **14·41**.

If we write $$F(R) J_\nu(uR) J_\nu(ur) uR = \phi(R, u),$$

it is clear that*

$$\left| \int_0^\infty \left\{ \int_0^\lambda \phi(R, u)\, du \right\} dR - \int_0^\lambda \left\{ \int_0^\infty \phi(R, u)\, dR \right\} du \right|$$

$$= \left| \int_\beta^\infty \left\{ \int_0^\lambda \phi(R, u)\, du \right\} dR - \int_0^\lambda \left\{ \int_\beta^\infty \phi(R, u)\, dR \right\} du \right|$$

$$\leqslant \int_\beta^\infty \left\{ \int_0^\lambda |\phi(R, u)|\, du \right\} dR + \int_0^\lambda \left\{ \int_\beta^\infty |\phi(R, u)|\, dR \right\} du$$

$$\leqslant \int_\beta^\infty \int_0^\lambda \frac{A^2}{\sqrt{r}} |F(R)| \sqrt{R} \,.\, du dR + \int_0^\lambda \int_\beta^\infty \frac{A^2}{\sqrt{r}} |F(R)| \sqrt{R} \,.\, dR du$$

$$< \epsilon.$$

Since this result is true for arbitrarily small values of ϵ, we infer that

$$\int_0^\infty \int_0^\lambda \phi(R, u)\, du dR = \int_0^\lambda \int_0^\infty \phi(R, u)\, dR du,$$

the integral on the left existing because the integral on the right is assumed to exist. If the integral on the left has a limit as $\lambda \to \infty$, it is evident from the definition of an infinite integral that

$$\int_0^\infty u du \int_0^\infty F(R) J_\nu(uR) J_\nu(ur) R dR$$

$$= \lim_{\lambda \to \infty} \int_0^\lambda u du \int_0^\infty F(R) J_\nu(uR) J_\nu(ur) R dR$$

$$= \lim_{\lambda \to \infty} \int_0^\infty F(R) \left\{ \int_0^\lambda J_\nu(uR) J_\nu(ur)\, u du \right\} R dR,$$

and this is the inversion formula which had to be proved.

14·43. *The relevant part of the range of integration in Hankel's repeated integral.*

Next we shall prove that, in Hankel's integral, the only part of the R-range of integration which contributes anything to the value of the integral is the part of the path *in the immediate vicinity of* r, provided merely that $F(R)\sqrt{R}$ has an absolutely convergent integral.

* The justification of the inversion of the order of integration for a *finite* rectangle whose sides are λ and β presents no great theoretical difficulties.

To effect this, it is sufficient to prove that, if r is *not* a point of the interval* (a, b), then

$$\int_0^\infty u\,du \int_a^b F(R)\,J_\nu(uR)\,J_\nu(ur)\,R\,dR = 0.$$

We invert the order of the integrations, as in § 14·42, and we find that, if the limits on the right exist,

$$\int_0^\infty u\,du \int_a^b F(R)\,J_\nu(uR)\,J_\nu(ur)\,R\,dR$$

$$= \lim_{\lambda\to\infty} \int_a^b F(R) \left\{ \int_0^\lambda J_\nu(uR)\,J_\nu(ur)\,u\,du \right\} R\,dR$$

$$= \lim_{\lambda\to\infty} \int_a^b F(R)\,[R J_{\nu+1}(\lambda R)\,J_\nu(\lambda r) - r J_{\nu+1}(\lambda r)\,J_\nu(\lambda R)]\,\frac{\lambda\,.\,R\,dR}{R^2 - r^2}$$

$$= \lim_{\lambda\to\infty} \lambda J_\nu(\lambda r) \int_a^b \frac{F(R)\,R^2}{R^2 - r^2}\,J_{\nu+1}(\lambda R)\,dR$$

$$- \lim_{\lambda\to\infty} \lambda r J_{\nu+1}(\lambda r) \int_a^b \frac{F(R)\,R}{R^2 - r^2}\,J_\nu(\lambda R)\,dR.$$

Since both the integrals

$$\int_a^b \frac{F(R)\,R^{\frac{3}{2}}}{R^2 - r^2}\,dR, \quad \int_a^b \frac{F(R)\,R^{\frac{1}{2}}}{R^2 - r^2}\,dR$$

are *ex hypothesi* absolutely convergent, it follows from the generalised Riemann-Lebesgue lemma (§ 14·41) that the last two limits are zero; and so

$$\int_0^\infty u\,du \int_a^b F(R)\,J_\nu(uR)\,J_\nu(ur)\,R\,dR = 0$$

provided that r is *not* such that $a \leqslant r \leqslant b$.

14·44. *The boundedness of* $\int_a^b \int_0^\lambda J_\nu(uR)\,J_\nu(ur)\,uR^{\frac{1}{2}}\,du\,dR$.

It will now be shewn that, *as* $\lambda \to \infty$, *the repeated integral*

$$\int_a^b \int_0^\lambda J_\nu(uR)\,J_\nu(ur)\,uR^{\frac{1}{2}}\,du\,dR$$

remains bounded, provided that a and b have any (bounded) positive values. It is permissible for a and b to be functions of λ *of which one (or both) may tend to r as* $\lambda \to \infty$.

Let us first consider the integral obtained by taking the dominant terms of the asymptotic expansions, namely

$$\frac{2}{\pi\sqrt{r}} \int_a^b \int_0^\lambda \cos(uR - \tfrac{1}{2}\nu\pi - \tfrac{1}{4}\pi)\cos(ur - \tfrac{1}{2}\nu\pi - \tfrac{1}{4}\pi)\,du\,dR$$

$$= \frac{1}{\pi\sqrt{r}} \int_a^b \left[\frac{\sin\lambda(R-r)}{R-r} - \frac{\cos\{\lambda(R+r) - \nu\pi\} - \cos\nu\pi}{R+r} \right] dR$$

$$= \frac{1}{\pi\sqrt{r}} \left[\int_{\lambda(a-r)}^{\lambda(b-r)} \frac{\sin x}{x}\,dx - \int_{\lambda(a+r)}^{\lambda(b+r)} \frac{\cos(x - \nu\pi)}{x}\,dx + \cos\nu\pi \log\frac{b+r}{a+r} \right].$$

* It is permissible for b to be infinite.

The first integral is bounded because $\int_{-\infty}^{\infty} \frac{\sin x}{x}\, dx$ is convergent; and the second integral is bounded because $\int^{\infty} \frac{\cos(x - \nu\pi)}{x}\, dx$ is convergent; and so the integral now under consideration is bounded, and its limit, as $\lambda \to \infty$, is the limit of

$$\frac{1}{\pi \sqrt{r}} \left[\int_{\lambda(a-r)}^{\lambda(b-r)} \frac{\sin x}{x}\, dx + \cos \nu\pi \log \frac{b+r}{a+r} \right],$$

provided that this limit exists.

But we may write

$$\int_a^b \int_0^\lambda J_\nu(uR)\, J_\nu(ur)\, uR^{\frac{1}{2}}\, du\, dR$$

$$= \frac{2}{\pi\sqrt{r}} \int_a^b \int_0^\infty [\tfrac{1}{2}\pi u J_\nu(uR)\, J_\nu(ur) \sqrt{(Rr)}$$

$$- \cos(uR - \tfrac{1}{2}\nu\pi - \tfrac{1}{4}\pi) \cos(ur - \tfrac{1}{2}\nu\pi - \tfrac{1}{4}\pi)]\, du\, dR$$

$$- \frac{2}{\pi\sqrt{r}} \int_a^b \int_\lambda^\infty [\tfrac{1}{2}\pi u J_\nu(uR)\, J_\nu(ur) \sqrt{(Rr)}$$

$$- \cos(uR - \tfrac{1}{2}\nu\pi - \tfrac{1}{4}\pi) \cos(ur - \tfrac{1}{2}\nu\pi - \tfrac{1}{4}\pi)]\, du\, dR$$

$$+ \frac{2}{\pi\sqrt{r}} \int_a^b \int_0^\lambda \cos(uR - \tfrac{1}{2}\nu\pi - \tfrac{1}{4}\pi) \cos(ur - \tfrac{1}{2}\nu\pi - \tfrac{1}{4}\pi)\, du\, dR.$$

Now, of the integrals on the right, the first is the integral with respect to R of an integral (with respect to u) which converges uniformly in any positive domain of values of R and r, and so it is a continuous (and therefore bounded) function of r when r is positive and bounded.

The third integral has been shewn to be bounded, and it converges to a limit whenever

$$\int_{\lambda(a-r)}^{\lambda(b-r)} \frac{\sin x}{x}\, dx$$

does so.

The second integral may be written in the form

$$- \frac{4\nu^2 - 1}{4\pi\sqrt{r}} \int_a^b \int_\lambda^\infty \left[\frac{1}{uR} \sin(uR - \tfrac{1}{2}\nu\pi - \tfrac{1}{4}\pi) \cos(ur - \tfrac{1}{2}\nu\pi - \tfrac{1}{4}\pi) \right.$$

$$\left. + \frac{1}{ur} \cos(uR - \tfrac{1}{2}\nu\pi - \tfrac{1}{4}\pi) \sin(ur - \tfrac{1}{2}\nu\pi - \tfrac{1}{4}\pi) + O(1/u^2) \right] du\, dR$$

$$= - \frac{4\nu^2 - 1}{4\pi\sqrt{r}} \int_a^b \left[\frac{1}{R} \phi_1(\lambda) + \frac{1}{r} \phi_2(\lambda) + \phi_3(\lambda) \right] dR,$$

where $\phi_1(\lambda)$, $\phi_2(\lambda)$ and $\phi_3(\lambda)$ are functions of λ and R which tend uniformly to zero as $\lambda \to \infty$.

Hence, for *all* bounded positive values of a, b, r, the integral

$$\int_a^b \int_0^\lambda J_\nu(uR)\, J_\nu(ur)\, uR^{\frac{1}{2}}\, du dR$$

is bounded as $\lambda \to \infty$; and it converges to a limit whenever

$$\int_{\lambda(a-r)}^{\lambda(b-r)} \frac{\sin x}{x}\, dx$$

does so.

14·45. *Proof of Hankel's integral theorem.*

Now that all the preliminary lemmas have been proved, the actual proof of Hankel's theorem is quite simple.

Since $F(R)$ has limited fluctuation in an interval of which r is an internal point, so also has $F(R)\sqrt{R}$; and therefore we may write

$$F(R)\sqrt{R} = \chi_1(R) - \chi_2(R),$$

where $\chi_1(R)$ and $\chi_2(R)$ are monotonic (positive) increasing functions.

After choosing a positive number ϵ arbitrarily, we choose a positive number δ so small that $F(R)$ has limited total fluctuation in the interval $(r-\delta, r+\delta)$ and also

$$\left.\begin{array}{l} \chi_1(r+\delta) - \chi_1(r+0) < \epsilon \\ \chi_2(r+\delta) - \chi_2(r+0) < \epsilon \end{array}\right\}, \quad \left.\begin{array}{l} \chi_1(r-0) - \chi_1(r-\delta) < \epsilon \\ \chi_2(r-0) - \chi_2(r-\delta) < \epsilon \end{array}\right\}.$$

If we apply the second mean-value theorem, we find that there exists a number ξ intermediate in value between 0 and δ such that

$$\int_r^{r+\delta} \int_0^\lambda \chi_1(R)\, J_\nu(uR)\, J_\nu(ur)\, u\sqrt{R}\,.\, du dR$$
$$= \chi_1(r+0) \int_r^{r+\delta} \int_0^\lambda J_\nu(uR)\, J_\nu(ur)\, u\sqrt{R}\,.\, du dR$$
$$+ \{\chi_1(r+\delta) - \chi_1(r+0)\} \int_{r+\xi}^{r+\delta} \int_0^\lambda J_\nu(uR)\, J_\nu(ur)\, u\sqrt{R}\,.\, du dR.$$

Since

$$\int_0^{\lambda\delta} \frac{\sin x}{x}\, dx \to \tfrac{1}{2}\pi,$$

as $\lambda \to \infty$, δ remaining fixed, it follows from § 14·44 that the first term on the right tends to a limit as $\lambda \to \infty$ while δ remains fixed. And the second term on the right does not exceed $C\epsilon$ in absolute value, where C is the upper bound of the modulus of the repeated integral (cf. § 14·44).

Hence, if

$$\lim_{\lambda\to\infty} \int_r^{r+\delta} \int_0^\lambda J_\nu(uR)\, J_\nu(ur)\, uR^{\frac{1}{2}}\, du dR = C_1/\sqrt{r},$$

it follows that

$$\lim_{\lambda\to\infty} \int_r^{r+\delta} \int_0^\lambda \chi_1(R)\, J_\nu(uR)\, J_\nu(ur)\, uR^{\frac{1}{2}}\, du dR$$

exists and is equal to $C_1\chi_1(r+0)/\sqrt{r}$.

We treat $\chi_2(R)$ in a similar manner, and also apply similar reasoning to the interval $(r-\delta, r)$; and we infer that, if

$$\lim_{\lambda\to\infty}\int_{r-\delta}^{r}\int_0^{\lambda} J_\nu(uR)\, J_\nu(ur)\, uR^{\frac{1}{2}}\, du\, dR = C_2/\sqrt{r},$$

then
$$\lim_{\lambda\to\infty}\int_{r-\delta}^{r+\delta}\int_0^{\lambda} F(R)\, J_\nu(uR)\, J_\nu(ur)\, uR\, du\, dR$$

exists and is equal to

$$C_1 F(r+0) + C_2 F(r-0).$$

We now have to evaluate C_1 and C_2. By the theory of generalised integrals*, we have

$$\begin{aligned}\frac{C_1}{\sqrt{r}} &= \int_0^\infty \int_r^{r+\delta} J_\nu(uR)\, J_\nu(ur)\, uR^{\frac{1}{2}}\, dR\, du\\
&= \lim_{p\to 0}\int_0^\infty \exp(-p^2u^2)\int_r^{r+\delta} J_\nu(uR)\, J_\nu(ur)\, uR^{\frac{1}{2}}\, dR\, du\\
&= \lim_{p\to 0}\int_r^{r+\delta}\int_0^\infty \exp(-p^2u^2)\, J_\nu(uR)\, J_\nu(ur)\, uR^{\frac{1}{2}}\, du\, dR\\
&= \lim_{p\to 0}\int_r^{r+\delta}\frac{1}{2p^2}\exp\left\{-\frac{R^2+r^2}{4p^2}\right\}.\, I_\nu\left(\frac{Rr}{2p^2}\right) R^{\frac{1}{2}}\, dR,\end{aligned}$$

by § 13·31 (1).

Now, throughout the range of integration,

$$I_\nu\left(\frac{Rr}{2p^2}\right) = \frac{p}{\sqrt{(\pi Rr)}}\{1 + O(p^2)\}\exp\left(\frac{Rr}{2p^2}\right),$$

and
$$\int_r^{r+\delta}\frac{O(p^2)}{2p\sqrt{(Rr)}}\exp\left\{-\frac{(R-r)^2}{4p^2}\right\} R^{\frac{1}{2}}\, dR \to 0,$$

as $p\to 0$.

Hence
$$\begin{aligned}C_1 &= \lim_{p\to 0}\frac{1}{2p\sqrt{\pi}}\int_r^{r+\delta}\exp\left\{-\frac{(R-r)^2}{4p^2}\right\} dR\\
&= \lim_{p\to 0}\frac{1}{\sqrt{\pi}}\int_0^{\frac{1}{2}\delta/p}\exp(-x^2)\, dx = \tfrac{1}{2},\end{aligned}$$

and similarly

$$C_2 = \lim_{p\to 0}\frac{1}{\sqrt{\pi}}\int_{-\frac{1}{2}\delta/p}^{0}\exp(-x^2)\, dx = \tfrac{1}{2}.$$

We have therefore shewn that

$$\lim_{\lambda\to\infty}\int_{r-\delta}^{r+\delta}\int_0^{\lambda} F(R)\, J_\nu(uR)\, J_\nu(ur)\, uR\, du\, dR$$

exists and is equal to

$$\tfrac{1}{2}\{F(r+0) + F(r-0)\}.$$

* Hardy, *Quarterly Journal*, XXXV. (1904), pp. 22—66. For a different method of calculating C_1 and C_2, see § 14·52.

But, if this limit exists, then, by § 14·42,

$$\int_0^\infty udu \int_0^\infty F(R)\, J_\nu(uR)\, J_\nu(ur)\, RdR$$

also exists and is equal to it; and so we have proved Hankel's theorem, as stated in § 14·4.

The use of generalised integrals in the proof of the theorem seems to be due to Sommerfeld, in his Königsberg Dissertation, 1891. For some applications of such methods combined with the general results of this chapter to the *problème des moments* of Stieltjes, see a recent paper by Hardy, *Messenger*, XLVII. (1918), pp. 81—88.

14·46. *Note on Hankel's proof of his theorem.*

The proof given by Hankel of his formula seems to discuss two points somewhat inadequately. The first is in the discussion of

$$\lim_{\lambda\to\infty} \int_0^\infty \int_0^\lambda F(R)\, J_\nu(uR)\, J_\nu(ur)\, uRdudR,$$

which he replaces by

$$\lim_{\lambda\to\infty} \int_0^\infty F(R)\left[RJ_{\nu+1}(\lambda R)\, J_\nu(\lambda r) - rJ_{\nu+1}(\lambda r)\, J_\nu(\lambda R)\right] \frac{\lambda RdR}{R^2-r^2}.$$

In order to approximate to this integral, he substitutes the first terms of the asymptotic expansions of the Bessel functions without considering whether the integrals arising from the second and following terms are negligible (which seems a fatal objection to the proof), and without considering the consequences of λR vanishing at the lower limit of the path of integration.

The second point, which is of a similar character, is in the discussion of

$$\lim_{\lambda\to\infty} \int_{r+\xi}^{r+\delta} \int_0^\lambda J_\nu(uR)\, J_\nu(ur)\, uRdudR;$$

after proving by the method just explained that this is zero if ξ tends to a positive limit and is $\frac{1}{2}$ if $\xi=0$, he takes it for granted that it must be bounded if $\xi\to 0$ as $\lambda\to\infty$; and this does not seem *prima facie* obvious.

14·5. *Extensions of Hankel's theorem to any cylinder functions.*

We shall now discuss integrals of the type

$$\int^\infty udu \int^\infty F(R)\, \mathscr{C}_\nu(uR)\, \mathscr{C}_\nu(ur)\, RdR,$$

in which the order ν of the unrestricted cylinder function $\mathscr{C}_\nu(z)$ is any real* number. The lower limits of the integrals will be specified subsequently, since it is convenient to give them values which depend on the value of ν.

For definiteness we shall suppose that

$$\mathscr{C}_\nu(z) \equiv \sigma\left\{\cos\alpha \,.\, J_\nu(z) + \sin\alpha \,.\, Y_\nu(z)\right\},$$

where σ and α are constants.

* The subsequent discussion is simplified and no generality is lost by assuming that $\nu \geqslant 0$.

The analogue of the Riemann-Lebesgue lemma (§ 14·41), namely *that*

$$\int_a^b F(R)\,\mathscr{C}_\nu(\lambda R)\,R\,dR = o\,(1/\sqrt{\lambda}),$$

provided that
$$\int_a^b F(R)\sqrt{R}\,.\,dR$$

exists and is absolutely convergent, may obviously be proved by precisely the methods of § 14·41, provided that $a < b \leqslant \infty$, and

$$\begin{cases} a \geqslant 0 \text{ if } 0 \leqslant \nu \leqslant \tfrac{1}{2}, \\ a > 0 \text{ if } \nu > \tfrac{1}{2}. \end{cases}$$

The theorem of § 14·44 has to be modified slightly in form. The modified theorem is that the repeated integral

$$\int_a^b \int_\tau^\lambda \mathscr{C}_\nu(uR)\,\mathscr{C}_\nu(ur)\,u\sqrt{R}\,.\,du\,dR$$

is bounded as $\lambda \to \infty$ while τ remains fixed; as in § 14·44, a and b may be functions of λ which have finite limits as $\lambda \to \infty$. The number τ is positive, though it is permissible for it to be zero when $0 \leqslant \nu \leqslant \frac{1}{2}$.

Also the repeated integral and the integral

$$\int_{\lambda(a-r)}^{\lambda(b-r)} \frac{\sin x}{x}\,dx$$

both converge or both oscillate as $\lambda \to \infty$.

[NOTE. If the two cylinder functions in the repeated integral were not of the same type, i.e. if we considered the integral

$$\int_a^b \int_\tau^\lambda \mathscr{C}_\nu(uR)\,\overline{\mathscr{C}}_\nu(ur)\,u\sqrt{R}\,.\,du\,dR,$$

it would be found that the convergence of this integral necessitates the convergence of the integral

$$\int_{\lambda(a-r)}^{\lambda(b-r)} \frac{1-\cos x}{x}\,dx\,;$$

and so, if $\lambda(b-r) \to \infty$ as $\lambda \to \infty$, the repeated integral is divergent*.]

14·51. *The extension of Hankel's theorem when $0 \leqslant \nu \leqslant \frac{1}{2}$.*

Retaining the notation of §§ 14·4—14·5, we shall now prove the following theorem.

Let $\int_0^\infty F(R)\sqrt{R}\,.\,dR$ *exist and be an absolutely convergent integral, and let* $0 \leqslant \nu \leqslant \frac{1}{2}$. *Then*

$$(1)\quad \int_0^\infty u\,du \int_0^\infty F(R)\,\mathscr{C}_\nu(uR)\,\mathscr{C}_\nu(ur)\,R\,dR$$
$$= \tfrac{1}{2}\sigma^2\{F(r+0)+F(r-0)\} - \frac{2\sigma^2 \sin\alpha \sin(\alpha+\nu\pi)}{\pi \sin\nu\pi}\int_0^\infty \frac{R^{2\nu}-r^{2\nu}}{R^{\nu-1}r^\nu(R^2-r^2)}\,F(R)\,dR,$$

* This point was overlooked by Nielsen, *Handbuch der Theorie der Cylinderfunktionen* (Leipzig, 1904), p. 365, in his exposition of Hankel's theorem.

provided that the positive number r lies inside *an interval in which $F(R)$ has limited total fluctuation.*

As in § 14·42, we may shew that

$$\int_0^\infty u\,du \int_0^\infty F(R)\,\mathscr{C}_\nu(uR)\,\mathscr{C}_\nu(ur)\,R\,dR$$
$$= \lim_{\lambda\to\infty} \int_0^\infty F(R) \int_0^\lambda \mathscr{C}_\nu(uR)\,\mathscr{C}_\nu(ur)\,uR\,du\,dR,$$

provided that the limit on the right exists.

But now we observe that

$$\int_0^\lambda \mathscr{C}_\nu(uR)\,\mathscr{C}_\nu(ur)\,u\,du$$
$$= \frac{1}{R^2-r^2}\left[uR\,\mathscr{C}_{\nu+1}(uR)\,\mathscr{C}_\nu(ur) - ur\,\mathscr{C}_{\nu+1}(ur)\,\mathscr{C}_\nu(uR)\right]_0^\lambda$$
$$= \frac{\lambda}{R^2-r^2}\left[R\,\mathscr{C}_{\nu+1}(\lambda R)\,\mathscr{C}_\nu(\lambda r) - r\,\mathscr{C}_{\nu+1}(\lambda r)\,\mathscr{C}_\nu(\lambda R)\right]$$
$$- \frac{2\sigma^2 \sin\alpha \sin(\alpha+\nu\pi)}{\pi \sin\nu\pi}\,\frac{R^{2\nu}-r^{2\nu}}{R^\nu r^\nu (R^2-r^2)}.$$

Hence we infer that, if r is *not* a point of the interval (a, b), then

$$\int_a^b F(R) \int_0^\lambda \mathscr{C}_\nu(uR)\,\mathscr{C}_\nu(ur)\,uR\,du\,dR$$
$$= o(1) - \frac{2\sigma^2 \sin\alpha \sin(\alpha+\nu\pi)}{\pi \sin\nu\pi} \int_a^b \frac{R^{2\nu}-r^{2\nu}}{R^\nu r^\nu (R^2-r^2)}\,F(R)\,R\,dR,$$

as $\lambda \to \infty$; and so the last repeated integral has a limit when $\lambda \to \infty$.

Now choose an arbitrary positive number ϵ, and then choose δ so small that $F(R)$ has limited total fluctuation in the interval $(r-\delta, r+\delta)$ and so that

$$\begin{cases} |F(R) - F(r+0)| < \epsilon \text{ if } r < R \leqslant r+\delta, \\ |F(R) - F(r-0)| < \epsilon \text{ if } r-\delta \leqslant R < r. \end{cases}$$

Now take
$$\int_0^\infty F(R) \int_0^\lambda \mathscr{C}_\nu(uR)\,\mathscr{C}_\nu(ur)\,u\,du\,dR,$$

and divide the R-path of integration into four parts, namely

$$(0, r-\delta),\ (r-\delta, r),\ (r, r+\delta),\ (r+\delta, \infty).$$

Apply the second mean-value theorem as in § 14·45, and we find that

$$\int_0^\infty F(R) \int_0^\lambda \mathscr{C}_\nu(uR)\,\mathscr{C}_\nu(ur)\,uR\,du\,dR$$
$$= -\frac{2\sigma^2 \sin\alpha \sin(\alpha+\nu\pi)}{\pi \sin\nu\pi}\left\{\int_0^{r-\delta} + \int_{r+\delta}^\infty\right\} \frac{R^{2\nu}-r^{2\nu}}{R^{\nu-1} r^\nu (R^2-r^2)}\,F(R)\,dR$$
$$+ F(r+0)\sqrt{r}\,.\int_r^{r+\delta}\int_0^\lambda \mathscr{C}_\nu(uR)\,\mathscr{C}_\nu(ur)\,uR^{\frac{1}{2}}\,du\,dR$$
$$+ F(r-0)\sqrt{r}\,.\int_{r-\delta}^{r}\int_0^\lambda \mathscr{C}_\nu(uR)\,\mathscr{C}_\nu(ur)\,uR^{\frac{1}{2}}\,du\,dR$$
$$+ \eta,$$

where $|\eta|$ has an upper bound which is independent of λ and which is arbitrarily small when ϵ is arbitrarily small.

The integrals on the right converge to limits when $\lambda \to \infty$, and so, by making $\epsilon \to 0$ after $\lambda \to \infty$, we infer that

$$\int_0^\infty u\,du \int_0^\infty F(R)\,\mathscr{C}_\nu(uR)\,\mathscr{C}_\nu(ur)\,R\,dR$$

is convergent and equal to

$$-\frac{2\sigma^2 \sin\alpha \sin(\alpha+\nu\pi)}{\pi \sin \nu\pi}\int_0^\infty \frac{R^{2\nu}-r^{2\nu}}{R^{\nu-1}r^\nu(R^2-r^2)}F(R)\,dR$$

$$+F(r+0)\sqrt{r}\,.\lim_{\delta\to 0}\int_0^\infty\int_r^{r+\delta}\mathscr{C}_\nu(uR)\,\mathscr{C}_\nu(ur)\,uR^{\frac{1}{2}}\,dR\,du$$

$$+F(r-0)\sqrt{r}\,.\lim_{\delta\to 0}\int_0^\infty\int_{r-\delta}^{r}\mathscr{C}_\nu(uR)\,\mathscr{C}_\nu(ur)\,uR^{\frac{1}{2}}\,dR\,du,$$

provided that the limits on the right exist.

To prove that the limits exist and to evaluate them simultaneously, take $F(R)=R^\nu$ when $r<R<r+\delta$ and $F(R)=0$ for all other values of R.

We thus find that

$$r^{\nu+\frac{1}{2}}\lim_{\delta\to 0}\int_0^\infty\int_r^{r+\delta}\mathscr{C}_\nu(uR)\,\mathscr{C}_\nu(ur)\,uR^{\frac{1}{2}}\,dR\,du$$
$$=\lim_{\delta\to 0}\int_0^\infty\int_r^{r+\delta}\mathscr{C}_\nu(uR)\,\mathscr{C}_\nu(ur)\,uR^{\nu+1}\,dR\,du,$$

provided that this repeated limit exists; and similarly

$$r^{\nu+\frac{1}{2}}\lim_{\delta\to 0}\int_0^\infty\int_{r-\delta}^{r}\mathscr{C}_\nu(uR)\,\mathscr{C}_\nu(ur)\,uR^{\frac{1}{2}}\,dR\,du$$
$$=\lim_{\delta\to 0}\int_0^\infty\int_{r-\delta}^{r}\mathscr{C}_\nu(uR)\,\mathscr{C}_\nu(ur)\,uR^{\nu+1}\,dR\,du.$$

For brevity we write b in place of $r+\delta$. We then have

$$\int_0^\infty\int_r^b\mathscr{C}_\nu(uR)\,\mathscr{C}_\nu(ur)\,uR^{\nu+1}\,dR\,du$$
$$=\int_0^\infty\{b^{\nu+1}\mathscr{C}_{\nu+1}(ub)-r^{\nu+1}\mathscr{C}_{\nu+1}(ur)\}\,\mathscr{C}_\nu(ur)\,du$$
$$=\lim_{\rho\to+0}\int_0^\infty\{b^{\nu+1}\mathscr{C}_{\nu+1}(ub)-r^{\nu+1}\mathscr{C}_{\nu+1}(ur)\}\,\mathscr{C}_\nu(ur)\frac{du}{u^{2\rho}},$$

since the second of these three integrals is convergent, and the third is absolutely convergent when $0<\rho<1-\nu$.

Now the last expression can be replaced by a combination of the four integrals of the types

$$\int_0^\infty\{b^{\nu+1}J_{\pm(\nu+1)}(ub)-r^{\nu+1}J_{\pm(\nu+1)}(ur)\}\,\frac{J_\nu(ur)}{J_{-\nu}(ur)}\cdot\frac{du}{u^{2\rho}},$$

and these are all absolutely convergent. They may be evaluated as cases of Weber's discontinuous integral of § 13·4, and hence we find that

$$\int_0^\infty \{b^{\nu+1}\mathscr{C}_{\nu+1}(ub) - r^{\nu+1}\mathscr{C}_{\nu+1}(ur)\}\,\mathscr{C}_\nu(ur)\frac{du}{u^{2\rho}}$$

$$= \frac{\sigma^2 r^\nu \sin(\alpha+\rho\pi)\sin(\alpha+\nu\pi)\,.\,\Gamma(\nu+1-\rho)}{2^{2\rho}\sin\rho\pi\sin\nu\pi\,.\,\Gamma(\nu+1)\Gamma(\rho+1)}$$

$$\times\left[b^{2\rho}\,.\,{}_2F_1\left(\nu+1-\rho, -\rho;\ \nu+1;\ \frac{r^2}{b^2}\right) - r^{2\rho}\frac{\Gamma(\nu+1)\Gamma(2\rho)}{\Gamma(\rho)\Gamma(\nu+1+\rho)}\right]$$

$$-\frac{\sigma^2 r^{-\nu}\sin\alpha\sin(\alpha+\rho\pi+\nu\pi)\,.\,\Gamma(1-\rho)}{2^{2\rho}\sin(\rho\pi+\nu\pi)\sin\nu\pi\,.\,\Gamma(1-\nu)\Gamma(\nu+\rho+1)}$$

$$\times\left[b^{2\nu+2\rho}\,.\,{}_2F_1\left(1-\rho, -\nu-\rho;\ 1-\nu;\ \frac{r^2}{b^2}\right) - r^{2\nu+2\rho}\frac{\Gamma(1-\nu)\Gamma(2\rho)}{\Gamma(\rho-\nu)\Gamma(\rho+1)}\right].$$

The limit of this expression, when $\rho\to 0$, is reducible to

$$\frac{\sigma^2 r^\nu \sin\alpha\sin(\alpha+\nu\pi)}{\pi\sin\nu\pi}\left[\log\left(1-\frac{r^2}{b^2}\right) + 2\log\frac{b}{r} - \frac{b^{2\nu}}{\nu r^{2\nu}}\,{}_2F_1\left(1, -\nu;\ 1-\nu;\ \frac{r^2}{b^2}\right)\right.$$

$$\left. + \tfrac{1}{2}\pi\cot\alpha - \tfrac{1}{2}\pi\cot(\alpha+\nu\pi) - \psi(1) + \psi(-\nu)\right],$$

after some algebra; and the limit of the last expression, when $b\to r+0$, is simply $\frac{1}{2}\sigma^2 r^\nu$.

In like manner it may be shewn that

$$\lim_{\delta\to+0}\int_0^\infty\int_{r-\delta}^r \mathscr{C}_\nu(uR)\,\mathscr{C}_\nu(ur)\,uR^{\nu+1}\,dR\,du = \tfrac{1}{2}\sigma^2 r^\nu,$$

and so we have proved that

$$\int_0^\infty u\,du\int_0^\infty F(R)\,\mathscr{C}_\nu(uR)\,\mathscr{C}_\nu(ur)\,R\,dR$$

$$= \tfrac{1}{2}\sigma^2\{F(r+0)+F(r-0)\} - \frac{2\sigma^2\sin\alpha\sin(\alpha+\nu\pi)}{\pi\sin\nu\pi}\int_0^\infty \frac{R^{2\nu}-r^{2\nu}}{R^{\nu-1}r^\nu(R^2-r^2)}F(R)\,dR,$$

provided that $0\leqslant\nu\leqslant\frac{1}{2}$, $F(R)$ is subject to the conditions stated in § 14·4, and

$$\mathscr{C}_\nu(z)\equiv\sigma\{\cos\alpha J_\nu(z)+\sin\alpha Y_\nu(z)\};$$

and this is the general theorem stated at the beginning of the section.

14·52. *Weber's integral theorem.*

It is evident from § 14·51 that, if $\int_a^\infty F(R)\sqrt{R}\,dR$ exists and is absolutely convergent, where $a>0$, then

$$(1)\quad \lim_{\lambda\to\infty}\int_a^\infty F(R)\left[R\mathscr{C}_{\nu+1}(\lambda R)\,\mathscr{C}_\nu(\lambda r) - r\mathscr{C}_{\nu+1}(\lambda r)\,\mathscr{C}_\nu(\lambda R)\right]\frac{\lambda R\,dR}{R^2-r^2}$$

$$= \tfrac{1}{2}\sigma^2\{F(r+0)+F(r-0)\},$$

provided that r lies inside an interval in which $F(R)$ has limited total fluctuation and $F(R)$ is defined to be zero when $0\leqslant R<a$, if the order of the cylinder functions lies between $-\frac{1}{2}$ and $\frac{1}{2}$.

We shall now establish the truth of this formula for cylinder functions of unrestricted order.

Let $[R\mathscr{C}_{\nu+1}(\lambda R)\mathscr{C}_\nu(\lambda r) - r\mathscr{C}_{\nu+1}(\lambda r)\mathscr{C}_\nu(\lambda R)]\dfrac{\lambda}{R^2-r^2} \equiv \Phi_\nu(R, r;\, \lambda).$

It is an easy deduction from the recurrence formulae that

$$\Phi_\nu(R, r;\, \lambda) - \Phi_{\nu-1}(R, r;\, \lambda) = \frac{\lambda}{R+r}[\mathscr{C}_{\nu-1}(\lambda R)\mathscr{C}_\nu(\lambda r) + \mathscr{C}_{\nu-1}(\lambda r)\mathscr{C}_\nu(\lambda R)],$$

and so, by the analogue of the Riemann-Lebesgue lemma (§ 14·41), we have

$$(2) \qquad \lim_{\lambda\to\infty}\int_a^\infty [\Phi_\nu(R, r;\, \lambda) - \Phi_{\nu-1}(R, r;\, \lambda)]\, RF(R)\, dR = 0.$$

Hence, by adding up repetitions of this result,

$$(3) \qquad \lim_{\lambda\to\infty}\int_a^\infty [\Phi_\nu(R, r;\, \lambda) - \Phi_{\nu\pm n}(R, r;\, \lambda)]\, RF(R)\, dR = 0,$$

where n is any positive integer.

Choose n so that one of the integers $\nu \pm n$ lies between $\pm\frac{1}{2}$, and then from (1)

$$\lim_{\lambda\to\infty}\int_a^\infty \Phi_{\nu\pm n}(R, r;\, \lambda)\, RF(R)\, dR = \tfrac{1}{2}\sigma^2\{F(r+0) + F(r-0)\},$$

and so, for all real values of ν, we deduce from (3) that

$$(4) \qquad \lim_{\lambda\to\infty}\int_a^\infty \Phi_\nu(R, r;\, \lambda)\, RF(R)\, dR = \tfrac{1}{2}\sigma^2\{F(r+0) + F(r-0)\}.$$

This result is practically due to Weber*, and it was obtained by the method indicated in § 14·46.

To obtain the result in Weber's form, let

$$(5) \qquad \begin{cases} \mathscr{C}_\nu(z) \equiv Y_\nu(r)J_\nu(z) - J_\nu(r)Y_\nu(z), \\ \overline{\mathscr{C}}_\nu(z) \equiv Y_\nu(R)J_\nu(z) - J_\nu(R)Y_\nu(z). \end{cases}$$

Then

$$u\left[\overline{\mathscr{C}}_\nu(uR)\frac{d\mathscr{C}_\nu(ur)}{du} - \mathscr{C}_\nu(ur)\frac{d\overline{\mathscr{C}}_\nu(uR)}{du}\right] = (R^2 - r^2)\int_1^u \mathscr{C}_\nu(ur)\overline{\mathscr{C}}_\nu(uR)\, u\, du,$$

and the expression on the left is also equal to

$$\begin{aligned}
&u[R\overline{\mathscr{C}}_{\nu+1}(uR)\mathscr{C}_\nu(ur) - r\mathscr{C}_{\nu+1}(ur)\overline{\mathscr{C}}_\nu(uR)] \\
&\quad = uR[Y_\nu(R)J_{\nu+1}(uR) - J_\nu(R)Y_{\nu+1}(uR)][Y_\nu(r)J_\nu(ur) - J_\nu(r)Y_\nu(ur)] \\
&\qquad - ur[Y_\nu(r)J_{\nu+1}(ur) - J_\nu(r)Y_{\nu+1}(ur)][Y_\nu(R)J_\nu(uR) - J_\nu(R)Y_\nu(uR)] \\
&\quad = uY_\nu(R)Y_\nu(r)[RJ_{\nu+1}(uR)J_\nu(ur) - rJ_{\nu+1}(ur)J_\nu(uR)] \\
&\qquad + \tfrac{1}{2}u\{J_\nu(R)Y_\nu(r) - J_\nu(r)Y_\nu(R)\}[RJ_{\nu+1}(uR)Y_\nu(ur) - RY_{\nu+1}(uR)J_\nu(ur) \\
&\qquad\qquad - rY_{\nu+1}(ur)J_\nu(uR) + rJ_{\nu+1}(ur)Y_\nu(uR)] \\
&\qquad - \tfrac{1}{4}u\{J_\nu(R)Y_\nu(r) + J_\nu(r)Y_\nu(R)\}[RD_{\nu+1}(uR)D_\nu(ur) - rD_{\nu+1}(ur)D_\nu(uR) \\
&\qquad\qquad - R\overline{D}_{\nu+1}(uR)\overline{D}_\nu(ur) + r\overline{D}_{\nu+1}(ur)\overline{D}_\nu(uR)] \\
&\qquad - uJ_\nu(R)J_\nu(r)[RY_{\nu+1}(uR)Y_\nu(ur) - rY_{\nu+1}(ur)Y_\nu(uR)],
\end{aligned}$$

where $\begin{cases} D_\nu(z) = J_\nu(z) + Y_\nu(z), \\ \overline{D}_\nu(z) = J_\nu(z) - Y_\nu(z). \end{cases}$

* *Math. Ann.* VI. (1873), pp. 146—161.

Now suppose that

$$\int_a^\infty f(R)\, R dR$$

exists and is absolutely convergent; and consider

$$\lim_{\lambda\to\infty} \int_a^\infty f(R) \int_1^\lambda \mathscr{C}_\nu(ur)\overline{\mathscr{C}}_\nu(uR)\, uR du dR.$$

Carry out the integration with respect to u, and replace the integrated part by the sum of the four terms written above, divided by $R^2 - r^2$.

Since

$$\frac{J_\nu(R) Y_\nu(r) - J_\nu(r) Y_\nu(R)}{R^2 - r^2}$$

is bounded near r, and has limited total fluctuation in any bounded interval containing r, it follows that the integrals corresponding to the second group of terms tend to zero as $\lambda \to \infty$, by the generalised Riemann-Lebesgue lemma.

Corresponding to the third group of terms we get a pair of integrals which happen to cancel.

When we use (1), we are therefore left with the result that

$$\lim_{\lambda\to\infty} \int_a^\infty f(R) \int_1^\lambda \mathscr{C}_\nu(ur)\overline{\mathscr{C}}_\nu(uR)\, uR\,.\, du dR$$
$$= \tfrac{1}{2}\{J_\nu^2(r) + Y_\nu^2(r)\}\,.\,\{f(r+0) + f(r-0)\},$$

that is to say

$$(6) \quad \int_1^\infty u du \int_a^\infty f(R)\, \mathscr{C}_\nu(ur)\overline{\mathscr{C}}_\nu(uR)\, R dR$$
$$= \tfrac{1}{2}\{J_\nu^2(r) + Y_\nu^2(r)\}\,.\,\{f(r+0) + f(r-0)\},$$

in which the cylinder functions are defined by (5), and r lies inside an interval in which $f(R)$ has limited total fluctuation.

Apart from details of notation, this is the result obtained by Weber in the case of functions of integral order.

14·6. *Formal statement of Neumann's integral theorem.*

We shall now state precisely the theorem which will be the subject of discussion in the sections immediately following. It is convenient to enunciate the theorem with Du Bois Reymond's* generalisation, obtained by replacing the Bessel function by any function which satisfies certain general conditions. The generalised theorem is as follows:

(I) *Let $\Psi(X, Y)$ be a bounded arbitrary function of the pair of real variables (X, Y), which is such that the double integral*

$$\int_{-\infty}^\infty \int_{-\infty}^\infty \Psi(X, Y)\,.\,(X^2 + Y^2)^{-\frac{1}{4}}\,.\,(dX dY)$$

exists and is absolutely convergent.

* *Math. Ann.* IV. (1871), pp. 383—390. Neumann's formula (cf. § 14·3) is obtained by writing $g(t) \equiv J_0(t)$, and the conditions (I)—(III) are substantially those given in Neumann's treatise published in 1881.

(II) *When $\Psi(X, Y)$ is expressed in terms of polar coordinates, let it be denoted by $F(R, \Phi)$, and let $F(R, \Phi)$ have the property that (for all values of Φ between $\pm\pi$), $F(R, \Phi)$, qua function of R, has limited total fluctuation in the interval $(0, \infty)$; and let this fluctuation and also $F(+0, \Phi)$ be integrable functions of Φ.*

(III) *If $\Omega(R, \Phi)$ denote the total fluctuation of $F(R, \Phi)$ in the interval $(\pm 0, R)$, let $\Omega(R, \Phi)$ tend to zero uniformly with respect to Φ as $R \to 0$, throughout the whole of the interval $(-\pi, \pi)$, with the exception* of values of Φ in a number of sectors the sum of whose angles may be assumed arbitrarily small.*

Since $|F(R, \Phi) - F(+0, \Phi)| \leqslant \Omega(R, \Phi)$, this condition necessitates that $F(R; \Phi) \to F(+0, \Phi)$ uniformly except in the exceptional sectors.

(IV) *Let $g(R)$ be a continuous function of the positive variable R, such that $g(R)\sqrt{R}$ is bounded both when $R \to 0$ and when $R \to \infty$.*

Let $\int_0^R g(t)\, t\, dt = G(R)$, and let $\int_0^\infty G(t)\frac{dt}{t}$ be convergent.

Then
$$\int_0^\infty u\, du \int_{-\infty}^\infty \int_{-\infty}^\infty \Psi(X, Y) \,.\, g\{u\sqrt{(X^2+Y^2)}\} \,.\, (dX\, dY)$$
is convergent, and is equal to
$$2\pi \,.\, [\mathfrak{M} F(+0, \Phi)] \int_0^\infty G(t)\frac{dt}{t},$$
where $\mathfrak{M} F(+0, \Phi)$ means†
$$\frac{1}{2\pi}\int_{-\pi}^{\pi} F(+0, \Phi)\, d\Phi.$$

Before proving the main theorem, we shall prove a number of Lemmas, just as in the case of Hankel's integral.

14·61. *The analogue of the Riemann-Lebesgue lemma.*

Corresponding to the result of § 14·41, we have the theorem that *if T is an unbounded domain‡ surrounding the origin, of which the origin is not an interior point or a boundary point, then, as $\lambda \to \infty$,*
$$\iint_T F(R, \Phi)\, G(\lambda R)\frac{(dR\, d\Phi)}{R} = o(1).$$

* The object of the exception is to ensure that the reasoning is applicable to the case (which is of considerable physical importance) in which $\Psi(X, Y)$ is zero outside a region bounded by one or more analytic curves and is, say, a positive constant inside the region, the origin being on the boundary of the region.

† The discovery that the repeated integral is equal to an expression involving the mean value of $F(R, \Phi)$ when the origin is a point of discontinuity of $F(R, \Phi)$ was made by Neumann, *Ueber die nach Kreis-, Kugel- und Cylinder-functionen fortschreitenden Entwickelungen* (Leipzig, 1881), pp. 130—131.

‡ For instance T might be the whole of the plane outside a circle of radius δ with centre at the origin.

It will be observed that this is a theorem of a much weaker character than the theorem of § 14·41, in view of hypothesis (II) of § 14·6. The reason of this is the fact that $G(\lambda R)$ may be* $O(\sqrt{\lambda})$ for certain values of R, and this seems to make arguments of the type used in § 14·41 inapplicable.

To prove the lemma, suppose first that T is bounded. Then, for any value of Φ, $F(R, \Phi)$ may be expressed as the difference† of two (increasing) monotonic functions $\chi_1(R, \Phi)$, $\chi_2(R, \Phi)$, whose sum is the total fluctuation of $F(R, \Phi)$ in the interval $(0, R)$.

If R_0 and R_1 are the extreme values of R for any particular value of Φ, it follows from the second mean-value theorem that, for some value of R_2 between R_0 and R_1,

$$\int_{R_0}^{R_1} \chi_1(R, \Phi)\, G(\lambda R)\frac{dR}{R} = \chi_1(R_0, \Phi)\int_{R_0}^{R_2} G(\lambda R)\frac{dR}{R} + \chi_1(R_1, \Phi)\int_{R_2}^{R_1} G(\lambda R)\frac{dR}{R}$$

$$= \chi_1(R_0, \Phi)\int_{\lambda R_0}^{\lambda R_2} G(t)\frac{dt}{t} + \chi_1(R_1, \Phi)\int_{\lambda R_2}^{\lambda R_1} G(t)\frac{dt}{t}.$$

Since $\int^{\infty} G(t)\frac{dt}{t}$ is convergent, if ϵ is an arbitrary positive number, we can choose λ so large that

$$\left|\int_{\lambda\xi}^{\infty} G(t)\frac{dt}{t}\right| < \epsilon,$$

for *all* values of ξ not less than the smallest value of R_0. Also

$$|\chi_1(R, \Phi)| \leqslant \{\chi_1(R, \Phi) - \tfrac{1}{2}F(+0, \Phi)\} + \tfrac{1}{2}|F(+0, \Phi)|$$

$$\leqslant \chi_1(\infty, \Phi) - \tfrac{1}{2}F(+0, \Phi) + \tfrac{1}{2}|F(+0, \Phi)|,$$

and similarly

$$|\chi_2(R, \Phi)| \leqslant \chi_2(\infty, \Phi) + \tfrac{1}{2}F(+0, \Phi) + \tfrac{1}{2}|F(+0, \Phi)|,$$

whence it follows that

$$\left|\iint_T F(R, \Phi)\, G(\lambda R)\frac{(dR\, d\Phi)}{R}\right|$$

$$< 2\epsilon\int_{-\pi}^{\pi}\{\chi_1(\infty, \Phi) + \chi_2(\infty, \Phi) + |F(+0, \Phi)|\}\, d\Phi$$

$$= 2\epsilon\int_{-\pi}^{\pi}\{\Omega(\infty, \Phi) + |F(+0, \Phi)|\}\, d\Phi;$$

and, since $F(+0, \Phi)$ is bounded, this can be made arbitrarily small by taking ϵ sufficiently small, and it is independent of the outer boundary of T. Hence we may proceed to the limit when the outer boundary tends to infinity.

* This is the case when $g(R) \equiv J_0(R)$; then $G(R) = RJ_1(R)$. It is by no means impossible that some of the conditions imposed on $F(R, \Phi)$ are superfluous.

† Cf. *Modern Analysis*, § 3·64.

We infer that, if T has no outer boundary, the modulus of

$$\iint_T F(R, \Phi)\, G(\lambda R)\, \frac{(dRd\Phi)}{R}$$

can be made arbitrarily small by taking λ sufficiently large; and this is the theorem to be proved.

14·62. *The inversion of Neumann's repeated integral.*

We shall next prove that *the existence and absolute convergence of the integral*

$$\int_{-\infty}^{\infty}\int_{-\infty}^{\infty} \Psi(X, Y)\frac{(dX\,dY)}{(X^2+Y^2)^{\frac{1}{4}}}$$

are sufficient conditions that

$$\int_0^\infty u\,du \int_{-\infty}^{\infty}\int_{-\infty}^{\infty} \Psi(X, Y)\,.\,g\{u\sqrt{(X^2+Y^2)}\}\,.\,(dX\,dY)$$

$$= \lim_{\lambda\to\infty}\int_{-\infty}^{\infty}\int_{-\infty}^{\infty} \Psi(X, Y)\int_0^\lambda g\{u\sqrt{(X^2+Y^2)}\}\,u\,du\,(dX\,dY),$$

provided that the limit on the right exists.

For any given value of λ and any arbitrary positive value of ϵ, there exists a number β such that

$$\int_{-\pi}^{\pi}\int_{\beta}^{\infty} |F(R, \Phi)|\,R^{\frac{1}{2}}\,(dRd\Phi) < \frac{3\epsilon}{4A\lambda^{\frac{3}{2}}},$$

where A is the upper bound of $|g(u)|\sqrt{u}$.

We then have

$$\Bigg|\int_{-\pi}^{\pi}\int_0^\infty\int_0^\lambda F(R, \Phi)\, g(uR)\, u\,du\,.\,R\,(dRd\Phi)$$
$$-\int_0^\lambda\int_{-\pi}^{\pi}\int_0^\infty F(R, \Phi)\, g(uR)\, R\,(dRd\Phi)\,u\,du\,\Bigg|$$
$$=\Bigg|\int_{-\pi}^{\pi}\int_\beta^\infty\int_0^\lambda F(R, \Phi)\, g(uR)\, u\,du\,.\,R\,(dRd\Phi)$$
$$-\int_0^\lambda\int_{-\pi}^{\pi}\int_\beta^\infty F(R, \Phi)\, g(uR)\, R\,(dRd\Phi)\,u\,du\,\Bigg|$$
$$\leqslant A\int_{-\pi}^{\pi}\int_\beta^\infty\int_0^\lambda |F(R, \Phi)|\, u^{\frac{1}{2}}\,du\, R^{\frac{1}{2}}\,(dRd\Phi)$$
$$+ A\int_0^\lambda\int_{-\pi}^{\pi}\int_\beta^\infty |F(R, \Phi)|\, R^{\frac{1}{2}}\,(dRd\Phi)\,u^{\frac{1}{2}}\,du$$
$$< \epsilon.$$

Since this is true for arbitrarily small values of ϵ, we infer that

$$\int_0^\infty\int_{-\pi}^{\pi}\int_0^\infty F(R, \Phi)\, g(uR)\, R\,(dRd\Phi)\,u\,du$$
$$= \lim_{\lambda\to\infty}\int_{-\pi}^{\pi}\int_0^\infty\int_0^\lambda F(R, \Phi)\, g(uR)\, u\,du\,.\,R\,(dRd\Phi),$$

the integral on the left existing because the limit on the right is assumed to exist.

Hence it follows that, if the limit on the right exists, then

$$\int_0^\infty u\,du \int_{-\pi}^{\pi} \int_0^\infty F(R, \Phi)\, g(uR)\, R\,(dR\,d\Phi)$$
$$= \lim_{\lambda \to \infty} \int_{-\pi}^{\pi} \int_0^\infty F(R, \Phi)\, G(\lambda R) \frac{(dR\,d\Phi)}{R}.$$

14·63. *The proof of Neumann's integral theorem.*

We are now in a position to prove without difficulty the theorem due to Neumann stated in § 14·6. We first take an arbitrarily small positive number ϵ and then choose the sectors in which the convergence of $\Omega(R, \Phi)$ to zero is uniform, in such a way that the sum of their angles exceeds $2\pi - \epsilon$. We then choose δ so small that $\Omega(R, \Phi) < \epsilon$ in these sectors whenever $R \leqslant \delta$; and we take the upper bounds of

$$\Omega(R, \Phi) + |F(R, \Phi)| \text{ and } \left| \int_\xi^\infty G(u) \frac{du}{u} \right|$$

to be B and C.

We then apply the second mean-value theorem. We have

$$\int_0^\delta \chi_1(R, \Phi)\, G(\lambda R) \frac{dR}{R}$$
$$= \chi_1(+0, \Phi) \int_0^\delta G(\lambda R) \frac{dR}{R} + \{\chi_1(\delta, \Phi) - \chi_1(+0, \Phi)\} \int_\xi^\delta G(\lambda R) \frac{dR}{R},$$

where $0 \leqslant \xi \leqslant \delta$.

Now
$$\left| \int_\xi^\delta G(\lambda R) \frac{dR}{R} \right| = \left| \int_{\lambda\xi}^{\lambda\delta} G(u) \frac{du}{u} \right| < 2C.$$

Hence
$$\int_0^\delta F(R, \Phi)\, G(\lambda R) \frac{dR}{R} = F(+0, \Phi) \int_0^{\lambda\delta} G(u) \frac{du}{u} + \eta,$$

where $|\eta|$ is less than $2\epsilon C$ inside the sectors in which convergence is uniform, and is less than $2BC$ in the exceptional sectors.

Hence it follows that

$$\left| \int_{-\pi}^{\pi} \int_0^\delta F(R, \Phi)\, G(\lambda R) \frac{(dR\,d\Phi)}{R} - 2\pi \mathfrak{M} F(+0, \Phi) \int_0^{\lambda\delta} G(u) \frac{du}{u} \right|$$
$$< 2\pi \,.\, 2\epsilon C + \epsilon \,.\, 2BC$$
$$= 2\epsilon C\,\{2\pi + B\}.$$

Hence, for large values of λ,

$$\left| \int_{-\pi}^{\pi} \int_0^\infty F(R, \Phi)\, G(\lambda R) \frac{(dR\,d\Phi)}{R} - 2\pi \mathfrak{M} F(+0, \Phi) \int_0^{\lambda\delta} G(u) \frac{du}{u} \right|$$
$$< 2\epsilon C\,(2\pi + B) + o(1),$$

that is to say

$$\varlimsup_{\lambda \to \infty} \left| \int_{-\pi}^{\pi} \int_0^\infty F(R, \Phi)\, G(\lambda R) \frac{(dR\,d\Phi)}{R} - 2\pi \mathfrak{M} F(+0, \Phi) \int_0^\infty G(u) \frac{du}{u} \right|$$
$$\leqslant 2\epsilon C\,(2\pi + B).$$

Now the expression on the left is independent of ϵ; and so since ϵ is arbitrarily small, we infer that the limit is zero. That is to say,

$$\lim_{\lambda\to\infty}\int_{-\pi}^{\pi}\int_0^\infty F(R,\Phi)\,G(\lambda R)\,\frac{(dRd\Phi)}{R}$$

exists and is equal to

$$2\pi\,\mathfrak{M}F(+0,\Phi)\,.\int_0^\infty G(u)\,\frac{du}{u}.$$

Applying the result of § 14·62, we see that Neumann's theorem has now been proved.

In the special case in which $g(u)=J_0(u)$, we have

$$\int_0^u t\,g(t)\,dt = uJ_1(u),$$

so that

$$G(u)=uJ_1(u),$$

and

$$\int_0^\infty \frac{G(u)\,du}{u}=\int_0^\infty \{-J_0'(u)\}\,du=1.$$

Hence we have

$$(1)\quad \int_0^\infty udu\int_{-\infty}^\infty\int_{-\infty}^\infty \Psi(X,Y)\,.\,J_0\{u\sqrt{(X^2+Y^2)}\}\,.\,(dX\,dY)$$
$$=2\pi\,\mathfrak{M}\Psi(+0\,.\cos\Phi,\ +0\,.\sin\Phi).$$

If we change the origin, we deduce that

$$(2)\quad \int_0^\infty udu\int_{-\infty}^\infty\int_{-\infty}^\infty \Psi(X,Y)\,.\,J_0[u\sqrt{\{(X-x)^2+(Y-y)^2\}}]\,.\,(dX\,dY)$$
$$=2\pi\,\mathfrak{M}\Psi(x+0\cos\Phi,\ y+0\sin\Phi),$$

and finally, changing to polar coordinates,

$$(3)\quad \int_0^\infty udu\int_{-\pi}^{\pi}\int_{-\infty}^\infty F(R,\Phi)\,J_0[u\sqrt{(R^2+r^2-2Rr\cos(\Phi-\phi)}]\,RdRd\Phi$$
$$=2\pi\,\mathfrak{M}F(r,\phi),$$

where $\mathfrak{M}F(r,\phi)$ now means the mean of the values of $F(R,\Phi)$ when (R,Φ) traverses the circumference of an indefinitely small circle with centre (r,ϕ).

14·64. *Mehler's investigation of Neumann's integral.*

Neumann's integral has been deduced by Mehler* from the formula

$$f(\theta,\phi)=\sum_{n=0}^\infty\frac{2n+1}{4\pi}\int_0^\pi\int_{-\pi}^{\pi} f(\Theta,\Phi)\,P_n(\cos\gamma)\sin\Theta\,d\Phi\,d\Theta$$

by a limiting process; in this formula

$$\cos\gamma=\cos\theta\cos\Theta+\sin\theta\sin\Theta\cos(\Phi-\phi).$$

The formula is obtained† by constructing a solution of Laplace's equation, valid inside a sphere of radius κ, which has an assigned value $f(\theta,\phi)$ on the surface of the sphere.

* *Math. Ann.* v. (1872), pp. 135—137; cf. Lamb, *Proc. London Math. Soc.* (2) II. (1905), p. 384.

† Cf. *Modern Analysis*, § 18·4.

The limiting process used by Mehler is that suggested by the result of § 5·71; the radius of the sphere is made indefinitely large, and new variables R, r are defined by the equations

$$R = \kappa\Theta, \quad r = \kappa\theta,$$

so that R, r are substantially cylindrical coordinates of the points with polar coordinates (κ, Θ, Φ), (κ, θ, ϕ); the function of position $f(\Theta, \Phi)$ is then denoted by $F(R, \Phi)$, and $P_n(\cos\gamma)$ becomes approximately equal to $J_0(n\varpi/\kappa)$, where

$$\varpi^2 = R^2 + r^2 - 2Rr\cos(\Phi - \phi).$$

We are thus led to the equation

$$F(r, \phi) = \lim_{\kappa \to \infty} \sum_{n=0}^{\infty} \frac{2n+1}{4\pi} \int_0^{\kappa\pi} \int_{-\pi}^{\pi} F(R, \Phi) J_0(n\varpi/\kappa) \frac{RdR}{\kappa^2} d\Phi.$$

If now we write $n/\kappa = u$, and replace the summation by an integration (taking $1/\kappa$ as the differential element), we get

$$F(r, \phi) = \frac{1}{2\pi} \int_0^\infty u\,du \int_0^\infty \int_{-\pi}^{\pi} F(R, \Phi) J_0(u\varpi) RdRd\Phi,$$

which is Neumann's result.

But this procedure can hardly be made the basis of a rigorous proof, because there are so many steps which require justification.

Thus, although we know that

$$\sum_{n=0}^{\infty} \left(\frac{\rho}{\kappa}\right)^n \frac{2n+1}{4\pi} \int_0^\pi \int_{-\pi}^{\pi} f(\Theta, \Phi) P_n(\cos\gamma) \sin\Theta\, d\Phi\, d\Theta$$

is a potential function (when $r < \kappa$), which assumes the value $f(\theta, \phi)$ on the surface of the sphere, the theorem that we may put $\rho = \kappa$ in the series necessitates a discussion of the convergence of the series on the surface of the sphere; and the transition from the surface of a sphere to a plane, by making $\kappa \to \infty$, with the corresponding transition from a series to an integral, is one of considerable theoretical difficulty.

It is possible that the method which has just been described is the method by which Neumann discovered his integral formula in 1862. Concerning his method he stated that "Die Methode, durch welche ich diese Formel so eben abgeleitet habe, ist nicht vollständig strenge."

CHAPTER XV

THE ZEROS OF BESSEL FUNCTIONS

15·1. *Problems connected with the zeros of Bessel functions.*

There are various classes of problems, connected with the zeros of Bessel functions, which will be investigated in this chapter. We shall begin by proving quite general theorems mainly concerned with the fact that Bessel functions have an infinity of zeros, and with the relative situations of the zeros of different functions. Next, we shall examine the reality of the zeros of Bessel functions (and cylinder functions) whose order is real, and discuss the intervals in which the real zeros lie, either by elementary methods or by the use of Poisson-Schafheitlin integrals. Next, we shall consider the zeros of $J_\nu(z)$ when ν is not necessarily real, and proceed to represent this function as a Weierstrassian product. We then proceed to the numerical calculation of zeros of functions of assigned order, and finally consider the rates of growth of the zeros with the increase of the order, and the situation of the zeros of cylinder functions of unrestrictedly large order. A full discussion of the applications of the results contained in this chapter to problems of Mathematical Physics is beyond the scope of this book, though references to such applications will be made in the course of the chapter.

Except in §§ 15·4—15·54, it is supposed that *the order ν, of the functions under consideration, is real.*

The zeros of functions whose order is half an odd integer obviously lend themselves to discussion more readily than the zeros of other functions. In particular the zeros of $\dfrac{d\,\{x^{-\frac{1}{2}} J_{n+\frac{1}{2}}(x)\}}{dx}$ have been investigated by Schwerd and by Rayleigh*; and more recently Hermite† has examined the zeros of $J_{n+\frac{1}{2}}(x)$. The zeros of this function have also been the subject of papers by Rudski‡ who used the methods of Sturm; but it has been pointed out by Porter and by Schafheitlin§ that some of Rudski's results are not correct, and, in particular, his theorem that the smallest positive zero of $J_{n+\frac{1}{2}}(x)$ lies between $\frac{1}{2}(n+1)\pi$ and $\frac{1}{2}(n+2)\pi$ is untrue. Such a theorem is incompatible with the inequality given in § 15·3 (5) and the formulae of §§ 15·81, 15·83.

* Schwerd, *Die Beugungserscheinungen* (Mannheim, 1835); cf. Verdet, *Leçons d'Optique Physique*, I. (Paris, 1869), p. 266; Rayleigh, *Proc. London Math. Soc.* IV. (1873), pp. 95—103.

† *Archiv der Math. und Phys.* (3) I. (1901), pp. 20—21.

‡ *Mém. de la Soc. R. des Sci. de Liége*, (2) XVIII. (1895), no. 3. See also *Prace Matematyczno-Fizyczne*, III. (1892), pp. 69—81. [*Jahrbuch über die Fortschritte der Math.* 1892, pp. 107—108.]

§ Porter, *American Journal of Math.* XX. (1898), p. 198; Schafheitlin, *Journal für Math.* CXXII. (1900), p. 304.

15·2. *The Bessel-Lommel theorem on the zeros of $J_\nu(z)$.*

It was stated by Daniel Bernoulli* and Fourier† that $J_0(z)$ has an infinity of real zeros; and a formal proof of this result by an analysis of Parseval's integral is due to Bessel‡. It was subsequently observed by Lommel§ that Bessel's arguments are immediately applicable to Poisson's integral for $J_\nu(z)$, provided that $-\frac{1}{2} < \nu \leqslant \frac{1}{2}$. A straightforward application of Rolle's theorem to $x^{\pm\nu} J_\nu(x)$ is then adeqnate to prove Lommel's theorem that *$J_\nu(z)$ has an infinity of real zeros, for any given real value of ν.*

The Bessel-Lommel investigation consists in proving that when $-\frac{1}{2} < \nu \leqslant \frac{1}{2}$, and x lies between $m\pi$ and $(m+\frac{1}{2})\pi$, then $J_\nu(x)$ is positive for even values of m, (0, 2, 4, ...), and is negative for odd values of m, (1, 3, 5, ...). Since $J_\nu(x)$ is a continuous function of x when $x \geqslant 0$, it is obvious that $J_\nu(x)$ has an odd number of zeros in each of the intervals $(\frac{1}{2}\pi, \pi)$, $(\frac{3}{2}\pi, 2\pi)$, $(\frac{5}{2}\pi, 3\pi)$, ...

Some more precise results of a similar character will be given in §§ 15·32—15·36.

To prove Lommel's theorem, let $x = (m + \frac{1}{2}\theta)\pi$ where $0 \leqslant \theta \leqslant 1$; then, by obvious transformations of Poisson's integral, we have

$$J_\nu(x) = \frac{2(\frac{1}{4}\pi)^\nu}{\Gamma(\nu+\frac{1}{2})\,\Gamma(\frac{1}{2})\,.\,(2m+\theta)^\nu} \int_0^{2m+\theta} \frac{\cos\frac{1}{2}\pi u}{\{(2m+\theta)^2 - u^2\}^{\frac{1}{2}-\nu}}\,du,$$

and so

$$\operatorname{sgn} J_\nu(x) = \operatorname{sgn} \int_0^{2m+\theta} \frac{\cos\frac{1}{2}\pi u}{\{(2m+\theta)^2 - u^2\}^{\frac{1}{2}-\nu}}\,du.$$

Now the last integral may be written in the form

$$\sum_{r=1}^{m} (-)^r v_r + (-)^m v_m',$$

where

$$(-)^r v_r = \int_{2r-2}^{2r} \frac{\cos\frac{1}{2}\pi u}{\{(2m+\theta)^2 - u^2\}^{\frac{1}{2}-\nu}}\,du,$$

$$(-)^m v_m' = \int_{2m}^{2m+\theta} \frac{\cos\frac{1}{2}\pi u}{\{(2m+\theta)^2 - u^2\}^{\frac{1}{2}-\nu}}\,du.$$

If now we write $u = 2r - 1 \pm U$, and then put

$$\{(2m+\theta)^2 - (2r-1+U)^2\}^{\nu-\frac{1}{2}} - \{(2m+\theta)^2 - (2r-1-U)^2\}^{\nu-\frac{1}{2}} \equiv f_r(U),$$

it is clear that

$$v_r = \int_0^1 f_r(U) \sin\tfrac{1}{2}\pi U\,.\,dU,$$

and, since|| $\nu \leqslant \frac{1}{2}$, $f_r(U)$ is a positive increasing¶ function of r.

* *Comm. Acad. Sci. Imp. Petrop.* VI. (1732—3) [1738], p. 116.

† *La Théorie Analytique de la Chaleur* (Paris, 1822), § 308.

‡ *Berliner Abh.*, 1824, p. 39.

§ *Studien über die Bessel'schen Functionen* (Leipzig, 1868), pp. 65—67.

|| This is the point at which the condition $\nu \leqslant \frac{1}{2}$ is required; the condition $\nu > -\frac{1}{2}$ ensures the convergence of the integral.

¶ The reader will prove this without any difficulty by regarding r as a continuous variable and then differentiating $f_r(U)$ with respect to r.

It follows that

$$0 \leqslant v_1 \leqslant v_2 \leqslant v_3 \leqslant \ldots \leqslant v_m,$$

and so

$$\begin{aligned}\operatorname{sgn} J_\nu(m\pi + \tfrac{1}{2}\theta\pi) &= \operatorname{sgn}\left[(-)^m\{v_m' + (v_m - v_{m-1}) + (v_{m-2} - v_{m-3}) + \ldots\}\right] \\ &= \operatorname{sgn}(-1)^m,\end{aligned}$$

since v_m' is obviously not negative.

That is to say, when $-\frac{1}{2} < \nu \leqslant \frac{1}{2}$,

$$\operatorname{sgn} J_\nu(m\pi + \tfrac{1}{2}\theta\pi) = \begin{cases} +, & (m = 0, 2, 4, \ldots) \\ -, & (m = 1, 3, 5, \ldots) \end{cases}$$

and from this result Lommel's theorem follows in the manner already stated.

The zeros of $J_1'(x)$, as well as those of $J_0(x)$, have been investigated by Baehr, *Archives Néerlandaises*, VII. (1872), pp. 351—358, with the help of a method which resembles the Bessel-Lommel method. Baehr's result for $J_1(x)$ is that the function is positive when x lies in the intervals $(0, \pi)$, $(\frac{5}{2}\pi, 3\pi)$, $(\frac{9}{2}\pi, 5\pi)$, ..., and that it is negative when x lies in the intervals $(\frac{3}{2}\pi, 2\pi)$, $(\frac{7}{2}\pi, 4\pi)$, $(\frac{11}{2}\pi, 6\pi)$, ... The function $J_1(x)$ has also been investigated in this way by C. N. Moore, *Annals of Math.* (2) IX. (1908), pp. 156—162.

The results just stated are of a less exact nature than the results obtained with the aid of slightly more refined analysis by Schafheitlin (§§ 15·33—15·35).

It was noted by Whewell, *Trans. Camb. Phil. Soc.* IX. (1856), p. 156, that $J_0(x)$ has a zero between 2 and $2\sqrt{2}$, and that the function $\mathbf{H}_0(z)$ has some real zeros.

15·21. *The non-repetition of zeros of cylinder functions.*

It is easy to prove that $\mathscr{C}_\nu(z)$ has no repeated zeros, with the possible exception of the origin*. For, if $\mathscr{C}_\nu(z)$ and $\mathscr{C}_\nu'(z)$ vanished simultaneously, it would follow, by repeated differentiations of the differential equation $\nabla_\nu \mathscr{C}_\nu(z) = 0$, that *all* the differential coefficients of $\mathscr{C}_\nu(z)$ would vanish at the common zero of $\mathscr{C}_\nu(z)$ and $\mathscr{C}_\nu'(z)$, and then, by Taylor's theorem, $\mathscr{C}_\nu(z)$ would be identically zero.

15·22. *The interlacing of zeros of Bessel functions.*

It will now be shewn that if $j_{\nu,1}, j_{\nu,2}, \ldots$ are the positive zeros of $J_\nu(x)$, arranged in ascending order of magnitude, then, if $\nu > -1$,

$$0 < j_{\nu,1} < j_{\nu+1,1} < j_{\nu,2} < j_{\nu+1,2} < j_{\nu,3} < \ldots.$$

This result is sometimes expressed by saying that the positive zeros of $J_\nu(x)$ are interlaced with those of $J_{\nu+1}(x)$.

To prove the result we use the recurrence formulae

$$\frac{d}{dx}\{x^{-\nu} J_\nu(x)\} = -x^{-\nu} J_{\nu+1}(x), \quad \frac{d}{dx}\{x^{\nu+1} J_{\nu+1}(x)\} = x^{\nu+1} J_\nu(x);$$

the first of these shews that between each consecutive pair of zeros of $x^{-\nu} J_\nu(x)$ there is at least one zero of $x^{-\nu} J_{\nu+1}(x)$, and the second shews that between each consecutive pair of zeros of $x^{\nu+1} J_{\nu+1}(x)$ there is at least one zero of $x^{\nu+1} J_\nu(x)$; and the result is now obvious.

* This is a special case of a theorem proved by Sturm, *Journal de Math.* I. (1836), p. 109.

If $\nu \leqslant -1$, the zeros are obviously still interlaced but the smallest zero of $J_{\nu+1}(x)$ is nearer the origin than the smallest zero of $J_\nu(x)$.

The result concerning interlacing of positive zeros is obviously true for any real cylinder function* $\mathscr{C}_\nu(x)$ and the contiguous function $\mathscr{C}_{\nu+1}(x)$.

This fundamental and simple property of Bessel functions appears never to have been proved until about a quarter of a century ago†, when four mathematicians published proofs almost simultaneously; the proof which has just been given is due to Gegenbauer‡ and Porter§; the other proofs, which are of a slightly more elaborate character, were given by Hobson|| and van Vleck¶.

It has been pointed out by Porter that, since

$$J_\nu(x) + J_{\nu+2}(x) = \frac{2(\nu+1)}{x} J_{\nu+1}(x),$$

at any positive zero of $J_\nu(x)$ the functions $J_{\nu+1}(x)$ and $J_{\nu+2}(x)$ have the same sign; but at successive zeros of $J_\nu(x)$ the function $J_{\nu+1}(x)$ alternates in sign, and so there are an odd number of zeros of $J_{\nu+2}(x)$ between each consecutive pair of positive zeros of $J_\nu(x)$; interchanging the functions $J_{\nu+2}(x)$ and $J_\nu(x)$ throughout this argument, we obtain Porter's theorem that the positive zeros of $J_{\nu+2}(x)$ are interlaced with those of $J_\nu(x)$.

15·23. *Dixon's theorem on the interlacing of zeros.*

A result of a slightly more general character than the theorem of § 15·22 is due to A. C. Dixon**, namely that, when $\nu > -1$, and A, B, C, D are constants such that $AD \neq BC$, then the positive zeros of $AJ_\nu(x) + BxJ_\nu'(x)$ are interlaced with those of $CJ_\nu(x) + DxJ_\nu'(x)$, and that no function of this type can have a repeated zero other than $x = 0$.

The latter part of the theorem is an immediate consequence of the formula, deducible from § 5·11 (11),

$$\int_0^x J_\nu^2(t)\,t\,dt = -\tfrac{1}{2}x \begin{vmatrix} J_\nu(x), & xJ_\nu'(x) \\ \dfrac{d\{J_\nu(x)\}}{dx}, & \dfrac{d\{xJ_\nu'(x)\}}{dx} \end{vmatrix},$$

for the integral is positive when x is positive and the expression on the right would vanish at a repeated zero of $AJ_\nu(x) + BxJ_\nu'(x)$.

* A real cylinder function is an expression of the form

$$\alpha J_\nu(x) + \beta Y_\nu(x)$$

in which α, β and ν are real, and x is positive.

† Cf. Gray and Mathews, *A Treatise on Bessel Functions* (London, 1895), p. 50.

‡ *Monatshefte für Math. und Phys.* VIII. (1897), pp. 383—384.

§ *Bulletin American Math. Soc.* IV. (1898), pp. 274—275.

|| *Proc. London Math. Soc.* XXVIII. (1897), pp. 372—373.

¶ *American Journal of Math.* XIX. (1897), pp. 75—85.

** *Messenger*, XXXII. (1903), p. 7; see also Bryan, *Proc. Camb. Phil. Soc.* VI. (1889), pp. 248—264.

To prove the former part of the theorem, we observe that, if

$$\phi(x) \equiv \frac{C J_\nu(x) + Dx J_\nu'(x)}{A J_\nu(x) + Bx J_\nu'(x)},$$

then

$$\phi'(x) = -\frac{2}{x\{AJ_\nu(x) + BxJ_\nu'(x)\}^2} \begin{vmatrix} A, & B \\ C, & D \end{vmatrix} \int_0^x J_\nu^2(t)\, t\, dt,$$

and so $\phi(x)$ is monotonic. The positive zeros of $\phi(x)$ are therefore interlaced with the positive poles, and from this result the former part of the theorem is obvious.

If the function $J_\nu(x)$ is replaced by a real cylinder function $\alpha J_\nu(x) + \beta Y_\nu(x)$, we have

$$-\int_0^x \mathscr{C}_\nu^2(t)\, t\, dt = \tfrac{1}{2}x \begin{vmatrix} \mathscr{C}_\nu(x), & x\mathscr{C}_\nu'(x) \\ \dfrac{d\mathscr{C}_\nu(x)}{dx}, & \dfrac{d\{x\mathscr{C}_\nu'(x)\}}{dx} \end{vmatrix} + \frac{2\nu\beta(\alpha \sin \nu\pi + \beta \cos \nu\pi)}{\pi \sin \nu\pi},$$

provided that $-1 < \nu < 1$; and so the theorems concerning non-repetition and interlacing of zeros are true for $A\mathscr{C}_\nu(x) + Bx\mathscr{C}_\nu'(x)$ and $C\mathscr{C}_\nu(x) + Dx\mathscr{C}_\nu'(x)$ provided that $\beta(\alpha \sin \nu\pi + \beta \cos \nu\pi)$ is positive.

Again, since

$$\tfrac{1}{2}x \begin{vmatrix} \mathscr{C}_\nu(x), & x\mathscr{C}_\nu'(x) \\ \dfrac{d\mathscr{C}_\nu(x)}{dx}, & \dfrac{d\{x\mathscr{C}_\nu'(x)\}}{dx} \end{vmatrix} = -\tfrac{1}{2}\{(x^2 - \nu^2)\mathscr{C}_\nu^2(x) + x^2\mathscr{C}_\nu'^2(x)\},$$

the theorem is true for zeros exceeding $+\sqrt{\nu^2}$, whether ν lies between -1 and 1 or not.

The result of § 15·22 is the special case of Dixon's theorem in which $A = 1$, $B = 0$, $C = \nu$, $D = -1$.

15·24. *The interlacing of zeros of cylinder functions of order ν.*

Let $\mathscr{C}_\nu(x)$ and $\overline{\mathscr{C}}_\nu(x)$ be any distinct cylinder functions of the same order; we shall prove that their positive zeros are interlaced*.

If $\mathscr{C}_\nu(x) = \alpha J_\nu(x) + \beta Y_\nu(x)$, $\overline{\mathscr{C}}_\nu(x) = \gamma J_\nu(x) + \delta Y_\nu(x)$,

then

$$\mathscr{C}_\nu(x)\overline{\mathscr{C}}_\nu'(x) - \overline{\mathscr{C}}_\nu(x)\mathscr{C}_\nu'(x) = \frac{2(\alpha\delta - \beta\gamma)}{\pi x}.$$

Now it is known that, at consecutive positive zeros of $\mathscr{C}_\nu(x)$, $\mathscr{C}_\nu'(x)$ has opposite signs, and therefore, from the last equation, $\overline{\mathscr{C}}_\nu(x)$ has opposite signs; that is to say $\overline{\mathscr{C}}_\nu(x)$ has an odd number of zeros between each consecutive pair of positive zeros of $\mathscr{C}_\nu(x)$; similarly $\mathscr{C}_\nu(x)$ has an odd number of zeros between each consecutive pair of positive zeros of $\overline{\mathscr{C}}_\nu(x)$; and so the zeros must be interlaced.

If we take one of the cylinder functions to be a function of the first kind, we deduce that *all* real cylinder functions have an infinity of positive zeros.

* Olbricht, *Nova Acta Caes.-Leop.-Acad.* (*Halle*), 1888, pp. 43—48, has given an elaborate discussion of this result with some instructive diagrams.

15·25. *Lommel's theorem on the reality of the zeros of $J_\nu(z)$.*

An extension of a theorem due to Fourier*, that the function $J_0(z)$ has no zeros which are not real, has been effected by Lommel†. The extended theorem is that, *if the order ν exceeds -1, then the function $J_\nu(z)$ has no zeros which are not real.*

To prove Lommel's theorem, suppose, if possible, that α is a zero of $J_\nu(z)$ which is not real. It follows from the series for $J_\nu(z)$ that α is not a pure imaginary, because then

$$\sum_{m=0}^{\infty} \frac{(-)^m (\frac{1}{2}\alpha)^{2m}}{m!\,\Gamma(\nu+m+1)}$$

would be a series of positive terms.

Let α_0 be the complex number conjugate to α, so that α_0 is also a zero of $J_\nu(z)$, because $J_\nu(z)$ is a real function of z.

Since $\nu > -1$, it follows from § 5·11 (8) that

$$\int_0^x tJ_\nu(\alpha t)\,J_\nu(\alpha_0 t)\,dt = \frac{x}{\alpha^2-\alpha_0^2}\left[J_\nu(\alpha x)\frac{dJ_\nu(\alpha_0 x)}{dx} - J_\nu(\alpha_0 x)\frac{dJ_\nu(\alpha x)}{dx}\right],$$

and so, since $\alpha^2 \neq \alpha_0^2$,

$$\int_0^1 tJ_\nu(\alpha t)\,J_\nu(\alpha_0 t)\,dt = 0.$$

The integrand on the left is positive, and so we have obtained a contradiction. Hence the number α cannot exist, and the theorem is proved.

Similar arguments‡ may be used to shew that, if A and B are real and $\nu > -1$, the function $AJ_\nu(z) + BzJ_\nu'(z)$ has all its zeros real, except that it has two purely imaginary zeros when $(A/B)+\nu<0$.

These results follow from the series for $\frac{d}{dz}\{z^{A/B}J_\nu(z)\}$ combined with the formula

$$\int_0^1 tJ_\nu(\beta t)\,J_\nu(\beta_0 t)\,dt = 0,$$

which is satisfied if β and β_0 are any zeros of $AJ_\nu(z)+BzJ_\nu'(z)$ such that $\beta^2 \neq \beta_0^2$.

15·26. *The analogue of Lommel's theorem for functions of the second kind.*

It is not possible to prove by the methods of § 15·25 that§ $Y_\nu(z)$ has no complex zeros in the region‖ in which $|\arg z| < \pi$. But it has been proved by Schafheitlin¶ that $Y_0(z)$ has no zeros with a *positive* real part, other than the real zeros.

* *La Théorie Analytique de la Chaleur* (Paris, 1822), § 308; see also Stearn, *Quarterly Journal*, XVII. (1880), p. 93.

† *Studien über die Bessel'schen Functionen* (Leipzig, 1868), p. 69.

‡ See A. C. Dixon, *Messenger*, XXXII. (1903), p. 7.

§ Or, more generally, $\mathscr{C}_\nu(z)$.

‖ When $\arg z = \pm\pi$, $Y_\nu(z) = e^{\mp\nu\pi i}\,Y_\nu(-z) \pm 2i\cos\nu\pi\,J_\nu(-z)$, and hence, by § 3·63 (1), $Y_\nu(z)$ cannot vanish unless ν is half of an odd integer. This type of reasoning is due to Macdonald, *Proc. London Math. Soc.* XXX. (1899), pp. 165—179.

¶ *Archiv der Math. und Phys.* (3) I. (1901), pp. 133—137. In this paper Schafheitlin also subjects the complex zeros of $Y_1(z)$ to a similar treatment.

For let β be a complex zero of $Y_0(z)$, and let β_0 be the conjugate complex, so that β_0 is also a zero of $Y_0(z)$. Then, by §§ 5·11 (8) and 3·51 (1),

$$\int_0^x tY_0(\beta t)\,Y_0(\beta_0 t)\,dt$$
$$= \frac{x}{\beta^2-\beta_0^2}\left[Y_0(\beta x)\frac{dY_0(\beta_0 x)}{dx} - Y_0(\beta_0 x)\frac{dY_0(\beta x)}{dx}\right] - \frac{4}{\pi^2(\beta^2-\beta_0^2)}\log\frac{\beta}{\beta_0},$$

and so, if $\beta = \rho e^{i\omega}$, we have

$$\int_0^1 tY_0(\beta t)\,Y_0(\beta_0 t)\,dt = -\frac{4\omega}{\pi^2\rho^2\sin 2\omega},$$

and the expression on the left is positive while the expression on the right is negative when ω is an acute angle.

15·27. *The theorems of Hurwitz on the zeros of $J_\nu(z)$.*

The proof which was given by Fourier that the zeros of $J_0(z)$ are all real was made more rigorous and extensive by Hurwitz*, who proved (i) that when $\nu > -1$, the zeros of $J_\nu(z)$ are all real, (ii) that, if s is a positive integer or zero and ν lies between $-(2s+1)$ and $-(2s+2)$, $J_\nu(z)$ has $4s+2$ complex zeros, of which 2 are purely imaginary, (iii) that, if s is a positive integer and ν lies between $-2s$ and $-(2s+1)$, $J_\nu(z)$ has $4s$ complex zeros, of which none are purely imaginary. To establish these results, we use the notation of § 9·7.

We take the function $g_{2m,\nu}(\zeta)$ which has, in the respective cases (i) m positive zeros, (ii) $m-2s-1$ positive zeros, 1 negative zero and $2s$ complex zeros, (iii) $m-2s$ positive zeros and $2s$ complex zeros.

We now prove that, if $f_\nu(\zeta) \equiv \sum_{n=0}^{\infty} \frac{(-)^n \zeta^n}{n!\,\Gamma(\nu+n+1)}$, then the function $f_\nu(\zeta)$ has at least as many complex zeros as $g_{2m,\nu}(\zeta)$. After Hurwitz, we write

$$\phi_m(\xi, \eta) = \frac{g_{2m+1,\nu}(\zeta)\,g_{2m,\nu}(\zeta') - g_{2m+1,\nu}(\zeta')\,g_{2m,\nu}(\zeta)}{\zeta'-\zeta},$$

where ξ, η are real and $\zeta = \xi + i\eta$, $\zeta' = \xi - i\eta$. The terms of highest degree in $\phi_m(\xi, \eta)$ are easily shewn to be

$$\tfrac{1}{6}m(m+1)(\nu+m)(\nu+m+1)\{(\nu+m)(2m+1)+m-1\}(\xi^2+\eta^2)^{m-1};$$

and since $g_{2m,\nu}$ is a real function, it follows that if ζ is a complex zero of $g_{2m,\nu}(\zeta)$, so also is ζ'; and therefore the complex zeros satisfy the equation

$$\phi_m(\xi, \eta) = 0.$$

Again, it is not difficult to deduce from the recurrence formulae (§ 9·7) that

$$\phi_{m+1}(\xi, \eta) = (\nu+2m+2)\,g_{2m+1,\nu}(\zeta)\,g_{2m+1,\nu}(\zeta') + (\xi^2+\eta^2)\,\phi_m(\xi, \eta).$$

* *Math. Ann.* XXXIII. (1889), pp. 246—266; cf. also Segar, *Messenger*, XXII. (1893), pp. 171—181, for a discussion of the Bessel coefficients. The analysis of this section differs in some respects from that of Hurwitz; see Watson, *Proc. London Math. Soc.* (2) XIX. (1921), pp. 266—272.

Hence, for sufficiently large values of m (i.e. those for which $\nu + 2m$ is positive), the curve $\phi_m(\xi, \eta) = 0$ lies in the finite part of the plane, and $\phi_m(\xi, \eta)$ is negative when $\phi_{m+1}(\xi, \eta)$ is zero so the curve $\phi_{m+1}(\xi, \eta) = 0$ lies wholly *inside* one or other of the closed branches which compose the curve $\phi_m(\xi, \eta) = 0$.

Hence as $m \to \infty$, the complex zeros of $g_{2m,\nu}(\zeta)$ lie in bounded regions of the ζ-plane, and consequently have limit-points.

Now, since, by §§ 9·65, 9·7,

$$\left| f_\nu(\zeta) - \frac{g_{2m,\nu}(\zeta)}{\Gamma(\nu + 2m + 1)} \right|$$

can be made arbitrarily small in any bounded domain of the ζ-plane, by taking m sufficiently large, it follows from Lagrange's expansion* that the number of zeros of $f_\nu(\zeta)$ in any small area is at least equal to the number of zeros of $g_{2m,\nu}(\zeta)$ in that area when m is sufficiently large; and so $f_\nu(\zeta)$ has $2s$ complex zeros. None of these zeros is real, for if one of them were real it would be a limit point of two conjugate complex zeros of $g_{2m,\nu}(\zeta)$, and so it would count as a double zero of $f_\nu(\zeta)$; and $f_\nu(\zeta)$ has no double zeros.

Again, from the series for $f_\nu(\zeta)$ it is seen that, when ν lies between $-(2s+1)$ and $-(2s+2)$, $f_\nu(\zeta)$ must have one negative zero, and it cannot have more than one negative zero, for then $g_{m,\nu}(\zeta)$ could be made to change sign more than once as ζ varied from 0 to $-\infty$ [since $g_{2m,\nu}(\zeta)$ can be made to differ from $f_\nu(\zeta)$ by an arbitrarily small number], and this is impossible.

For similar reasons $f_\nu(\zeta)$ cannot have *more* than $2s$ complex zeros.

If we replace ζ by $\frac{1}{4}z^2$, so that negative values of ζ correspond to purely imaginary values of z, we obtain the results stated in the case of $J_\nu(z)$.

For a discussion of zeros of Bessel functions in association with zeros of polynomials based on rather different ideas, the reader should consult Lindner, *Sitz. der Berliner Math. Ges.* XI. (1911), pp. 3—5. It may be mentioned that Hurwitz has extended his results to generalised Bessel functions in a brief paper, *Hamburger Mittheilungen*, II. (1890), pp. 25—31.

15·28. *Bourget's hypothesis.*

It has been conjectured by Bourget† that, *when ν is a positive integer (zero included)*, the functions $J_\nu(z)$, $J_{\nu+m}(z)$ have no common zeros, other than the origin, for all positive integral values of m.

It seems that this theorem was not proved before 1929, except in some simple special cases, such as $m = 1$ and $m = 2$ (cf. § 15·22).

The formula

$$J_{\nu+m}(z) = J_\nu(z)\, R_{m,\nu}(z) - J_{\nu-1}(z)\, R_{m-1,\nu+1}(z)$$

* Cf. *Modern Analysis*, § 7·32.

† *Ann. Sci. de l'École norm. sup.* III. (1866), pp. 55—95.

shews that, since $J_\nu(z)$ and $J_{\nu-1}(z)$ have no common zeros, the common zeros of $J_\nu(z)$ and $J_{\nu+m}(z)$ must satisfy the equation

$$R_{m-1,\nu+1}(z)=0,$$

i.e. they must be algebraic numbers.

It has, however, been proved by Siegel* that $J_\nu(z)$ is not an algebraic number when ν is a rational number and z is an algebraic number other than zero; hence follow theorems which include Bourget's hypothesis as a special case.

When ν is half of an odd integer, it is easy to shew that $J_\nu(z)$ and $J_{\nu+m}(z)$ have no common zeros†; for such zeros are algebraic numbers and it is known that no algebraic number‡ can satisfy the equation

$$\cot(z-\tfrac{1}{2}\nu\pi-\tfrac{1}{4}\pi)=\frac{Q(z,\nu)}{P(z,\nu)},$$

since the right-hand side is algebraic in z when ν is half of an odd integer.

The proof‡ given by Lambert and Legendre that π^2 is irrational may be applied to § 5·6 (6) to prove that $J_\nu(z)$ has no zero whose square is rational when ν is rational; an inspection of the polynomial $R_{m-1,\nu+1}(z)$ now immediately yields an elementary proof of Bourget's hypothesis in the cases $m=3$ and $m=4$.

15·3. *Elementary properties of the zeros of $J_\nu(x)$.*

It is possible to acquire a considerable amount of interesting information concerning the smallest zeros of $J_\nu(x)$ and related functions, when ν is positive, by a discussion of the differential equation satisfied by $J_\nu(x)$ together with the recurrence formulae; we shall now establish the truth of a selection of theorems concerning such zeros.

The reader will find a more systematic investigation§ of these theorems in various papers by Schafheitlin, notably *Journal für Math.* CXXII. (1900), pp. 299—321; *Archiv der Math. und Phys.* (3) I. (1901), pp. 133—137; *Berliner Sitzungsberichte*, III. (1904), pp. 83—85.

For brevity, the smallest positive zeros of $J_\nu(x)$, $J_\nu'(x)$, $J_\nu''(x)$, ... will be called j_ν, j_ν', j_ν'', ... The smallest positive zeros of $Y_\nu(x)$, $Y_\nu'(x)$, $Y_\nu''(x)$, ... will similarly be called y_ν, y_ν', y_ν'', ...

We first prove that

(1) $$j_\nu>\nu,\quad j_\nu'>\nu.$$

It is obvious from the power series for $J_\nu(x)$ and $J_\nu'(x)$ that these functions

* *Abhandlungen Akad. Berlin*, 1929, pp. 1—70. This abstruse and important memoir contains numerous applications of Siegel's fundamental theorem.

† This was noticed by Porter, *American Journal of Math.* XX. (1898), p. 203.

‡ Cf. Hobson, *Squaring the Circle* (Cambridge, 1913), pp. 44, 51—53.

§ Some related results are due to Watson, *Proc. London Math. Soc.* (2) XVI. (1917), pp. 165—171.

are positive for sufficiently small positive values of x; and, from the differential equation

$$x\frac{d}{dx}\left\{x\frac{dJ_\nu(x)}{dx}\right\} = (\nu^2 - x^2)\, J_\nu(x),$$

it is evident that, so long as $x < \nu$ and $J_\nu(x)$ is positive, $xJ_\nu'(x)$ is positive and increasing, and so $J_\nu(x)$ increases with x.

Therefore, so long as $0 < x < \nu$, both $J_\nu(x)$ and $xJ_\nu'(x)$ are positive increasing functions so that j_ν and j_ν' cannot* be less than ν.

Again, from the differential equation

$$\nu J_\nu''(\nu) = - J_\nu'(\nu) < 0,$$

and so $J_\nu''(x)$ has become negative before x has increased to the value ν from zero. Hence, when† $\nu > 1$,

(2) $$j_\nu'' < \nu.$$

Next, since

$$J'_{\nu+2}(x) = \frac{2(\nu+1)}{x}\left\{1 - \frac{\nu(\nu+2)}{x^2}\right\} J_\nu(x) - \left\{1 - \frac{2(\nu+1)(\nu+2)}{x^2}\right\} J_\nu'(x),$$

the expression on the right is positive so long as $x < \nu + 2$. Now, if j_ν' were less than $\sqrt{\{\nu(\nu+2)\}}$, the expression on the right would be negative when x is equal to j_ν' (which, from a graph, is obviously less than j_ν), and this is not the case.

Therefore

(3) $$j_\nu' > \sqrt{\{\nu(\nu+2)\}}.$$

Now, from § 15·22 it follows that

$$j_\nu < j_{\nu+1} < j_{\nu+2},$$

and, as has just been stated,

$$j_\nu' < j_\nu,$$

so that $J_\nu(j_\nu')$ and $J_{\nu+2}(j_\nu')$ are both positive. If now we put $x = j_\nu'$ in the formula

$$J_{\nu+2}(x) = -\left\{1 - \frac{2\nu(\nu+1)}{x^2}\right\} J_\nu(x) - \frac{2(\nu+1)}{x} J_\nu'(x),$$

it is obvious that

(4) $$j_\nu' < \sqrt{\{2\nu(\nu+1)\}}.$$

Similarly, by putting $x = j_\nu$ in the formula

$$(\nu+3)\, J_\nu(x) + 2(\nu+2)\left\{1 - \frac{2(\nu+1)(\nu+3)}{x^2}\right\} J_{\nu+2}(x) + (\nu+1)\, J_{\nu+4}(x) = 0,$$

we deduce that

$$j_\nu < \sqrt{\{2(\nu+1)(\nu+3)\}},$$

and therefore

(5) $$\sqrt{\{\nu(\nu+2)\}} < j_\nu < \sqrt{\{2(\nu+1)(\nu+3)\}}.$$

* Cf. Riemann, *Partielle Differentialgleichungen* (Brunswick, 1876), p. 269.

† When $0 < \nu < 1$, $J_\nu''(x)$ is negative for sufficiently small values of x.

In like manner, we can deduce from the formulae

$$\frac{J_{\nu+1}(x)}{x} = \left\{1 - \frac{\nu(\nu-1)}{x^2}\right\} J_\nu(x) + J_\nu''(x)$$

and
$$J'_{\nu+1}(x) = -\nu\left\{1 - \frac{\nu^2-1}{x^2}\right\} J_\nu(x) - (\nu+1) J_\nu''(x)$$

that

(6) $$\sqrt{\{\nu(\nu-1)\}} < j_\nu'' < \sqrt{(\nu^2-1)}.$$

Some rather better inequalities than these are obtainable by taking more complicated formulae; thus, from the equation

$$J_{\nu+6}(x) = J_{\nu+1}(x) R_{5,\nu+1}(x) - J_\nu(x) R_{4,\nu+2}(x),$$

Schafheitlin* deduced that

$$R_{5,\nu+1}(j_\nu) > 0,$$

i.e. $$3j_\nu^4 - 16(\nu+2)(\nu+4) j_\nu^2 + 16(\nu+1)(\nu+2)(\nu+4)(\nu+5) > 0.$$

Since j_ν^2 is certainly less than $\frac{8}{3}(\nu+2)(\nu+4)$, by results already proved, j_ν must be less than the smaller positive root of the equation

$$3x^4 - 16(\nu+2)(\nu+4)x^2 + 16(\nu+1)(\nu+2)(\nu+4)(\nu+5) = 0,$$

and hence, *a fortiori*,

(7) $$j_\nu < \sqrt{\{\tfrac{4}{3}(\nu+1)(\nu+5)\}}.$$

Similarly, from the equation

$$4J'_{\nu+4}(x) = J_\nu(x)\{R_{3,\nu}(x) + R_{3,\nu+2}(x) - R_{5,\nu}(x) - R_{1,\nu+2}(x)\} - 2J_\nu'(x)\{R_{2,\nu+1}(x) - R_{4,\nu+1}(x)\},$$

Schafheitlin deduced that

(8) $$j_\nu' > \sqrt{\{\nu(\nu+2)\}},$$

and, when $\nu > 4$,

(9) $$j_\nu' > \sqrt{\{\nu(\nu+3)\}},$$

these inequalities being derived from the consideration that j_ν' lies between the positive roots of the equation

$$x^4 - 3(\nu+2)^2 x^2 + 2\nu(\nu+1)(\nu+3)(\nu+4) = 0.$$

The discussion of y_ν requires slightly more abstruse reasoning. We use the result that

$$J_\nu^2(x) + Y_\nu^2(x)$$

is a *decreasing* function of x; this is obvious from § 13·73. Hence it follows that $Y_\nu^2(x)$ decreases through the interval $(0, j_\nu')$, and so y_ν exceeds j_ν'; again, in this interval $Y_\nu(x)$ is negative, and it follows from § 3·63 (1) that $Y_\nu(j_\nu)$ is positive, since $J_\nu'(j_\nu)$ is obviously negative.

Hence

(10) $$j_\nu' < y_\nu < j_\nu.$$

This inequality (with j_ν' replaced by $\nu + \frac{1}{2}$) was established by Schafheitlin† with the aid of rather elaborate analysis.

* *Berliner Sitzungsberichte*, III. (1904), p. 83.

† *Journal für Math.* CXXII. (1900), pp. 317—321.

15·31. *Stationary values of cylinder functions**.

It has already been seen that the cylinder function $J_\nu(x)\cos\alpha - Y_\nu(x)\sin\alpha$, or $\mathscr{C}_\nu(x)$, has an infinite number of positive zeros, and so there are an infinite number of positive values of x for which it is stationary. Such values of x which exceed the order ν (supposed positive) will be called $\mu_1, \mu_2, \mu_3, \ldots$, where

$$\mu_1 < \mu_2 < \mu_3 < \ldots.$$

We shall now study some of the simpler properties of the sequence

$$\mathscr{C}_\nu(\mu_1),\ \mathscr{C}_\nu(\mu_2),\ \mathscr{C}_\nu(\mu_3),\ \ldots.$$

The first theorem which we shall establish is that

$$|\mathscr{C}_\nu(\mu_1)| > |\mathscr{C}_\nu(\mu_2)| > |\mathscr{C}_\nu(\mu_3)| > \ldots.$$

To prove this, observe that the function $\Lambda(x)$ defined as

$$\mathscr{C}_\nu^2(x) + \frac{x^2\mathscr{C}_\nu'^2(x)}{x^2-\nu^2}$$

has the negative derivate

$$-2x^3\mathscr{C}_\nu'^2(x)/(x^2-\nu^2)^2,$$

and so
$$\Lambda(\mu_1) > \Lambda(\mu_2) > \Lambda(\mu_3) > \ldots.$$

Since $\Lambda(\mu_n) = \mathscr{C}_\nu^2(\mu_n)$, the truth of the theorem is now evident.

A more interesting result is suggested by Hankel's asymptotic formula (§ 7·21)

$$\mathscr{C}_\nu(x) = \left(\frac{2}{\pi x}\right)^{\frac{1}{2}}\cos(x+\alpha-\tfrac{1}{2}\nu\pi-\tfrac{1}{4}\pi) + O\left(\frac{1}{x^{\frac{3}{2}}}\right).$$

This indicates the possibility of proving inequalities consistent with

$$|\mathscr{C}_\nu(\mu_n)| = O(1/\sqrt{\mu_n})$$

when μ_n is large.

It can in fact be shewn that

(I) *The values assumed by* $(x^2-\nu^2)^{\frac{1}{4}}|\mathscr{C}_\nu(x)|$ *when* x *takes the values* $\mu_1, \mu_2, \mu_3, \ldots$ *form an increasing sequence whose members are less than* $\sqrt{(2/\pi)}$.

(II) *The values assumed by* $x^{\frac{1}{2}}|\mathscr{C}_\nu(x)|$ *when* x *takes the values* $\mu_r, \mu_{r+1}, \mu_{r+2}, \ldots$ *form a decreasing sequence whose members are greater than* $\sqrt{(2/\pi)}$ *provided that*

$$\text{(i)}\ \nu > \tfrac{1}{2}\sqrt{3}, \qquad \text{(ii)}\ \mu_r^2 > \nu^2\{4\nu^2+4+\sqrt{(48\nu^2+13)}\}/(4\nu^2-3).$$

Consider the function

$$A(x)\mathscr{C}_\nu^2(x) + 2B(x)\mathscr{C}_\nu(x)\mathscr{C}_\nu'(x) + C(x)\mathscr{C}_\nu'^2(x) \equiv \Theta(x),$$

where $A(x)$, $B(x)$, $C(x)$ are to be suitably chosen. We have

$$\begin{aligned}\Theta'(x) = &\{A'(x) - 2(x^2-\nu^2)B(x)/x^2\}\mathscr{C}_\nu^2(x)\\ &+ 2\{B'(x) + A(x) - B(x)/x - (x^2-\nu^2)C(x)/x^2\}\mathscr{C}_\nu(x)\mathscr{C}_\nu'(x)\\ &+ \{C'(x) + 2B(x) - 2C(x)/x\}\mathscr{C}_\nu'^2(x)\\ = &D(x)\mathscr{C}_\nu'^2(x),\end{aligned}$$

where
$$D(x) \equiv C'(x) + 2B(x) - 2C(x)/x,$$

* Cf. *Proc. London Math. Soc.* (2) XVI. (1917), pp. 170—171.

provided that $A(x)$ is chosen arbitrarily and that $B(x)$ and $C(x)$ are then defined by the equations

$$2B(x) = x^2 A'(x)/(x^2 - \nu^2),$$
$$C(x) = x^2 \{B'(x) + A(x) - B(x)/x\}/(x^2 - \nu^2).$$

(I) If $A(x) \equiv (x^2 - \nu^2)^{\frac{1}{2}}$, then

$$2D(x)(x^2 - \nu^2)^{\frac{5}{2}} \equiv x^3 (3x^4 + 14x^2\nu^2 + 4\nu^4) > 0,$$

and so $\Theta(x)$ is an increasing function of x which is therefore less than

$$\lim_{x \to \infty} \Theta(x) = 2/\pi.$$

Since $\Theta(\mu_n) = (\mu_n{}^2 - \nu^2)^{\frac{1}{2}} \mathscr{C}_\nu{}^2(\mu_n)$ we see that when n assumes the values 1, 2, ..., then the numbers $(\mu_n{}^2 - \nu^2)^{\frac{1}{4}} |\mathscr{C}_\nu(\mu_n)|$ form an increasing sequence less than $\sqrt{(2/\pi)}$.

(II) If $A(x) \equiv x$, then

$$2D(x)(x^2 - \nu^2)^4 = -x^2 \{(4\nu^2 - 3)x^4 - 8\nu^2(\nu^2 + 1)x^2 + \nu^4(4\nu^2 - 1)\}$$
$$< 0,$$

provided that $4\nu^2 > 3$ and x exceeds the greatest root of the equation

$$(4\nu^2 - 3)x^4 - 8\nu^2(\nu^2 + 1)x^2 + \nu^4(4\nu^2 - 1) = 0.$$

In this case $\Theta(x)$ is a *decreasing* function and we can apply arguments, similar to those used in theorem (I), to deduce the truth of theorem (II).

15·32. *Schafheitlin's investigation of the zeros of* $J_0(x)$.

By means of the integrals which have been given in § 6·12, it has been shewn by Schafheitlin* that the only positive zeros of $J_0(x)$ lie in the intervals $(m\pi + \frac{3}{4}\pi, m\pi + \frac{7}{8}\pi)$ and the only positive zeros of $Y_0(x)$ lie in the intervals $(m\pi + \frac{1}{4}\pi, m\pi + \frac{3}{8}\pi)$, where $m = 0, 1, 2, \ldots$.

We shall first give Schafheitlin's investigation for $J_0(x)$, with slight modifications, and then we shall prove similar results for cylinder functions of the type

$$J_\nu(x)\cos\alpha - Y_\nu(x)\sin\alpha$$

(where ν lies between $-\frac{1}{2}$ and $\frac{5}{2}$), by the methods used by Schafheitlin. Schafheitlin's investigations were confined to the values 0 and $\frac{1}{2}\pi$ of α.

From an inspection of the formula of § 6·12 (7),

$$J_0(x) = \frac{2}{\pi}\int_0^{\frac{1}{2}\pi} \frac{\sin(x + \frac{1}{2}\theta)}{\sin\theta\sqrt{\cos\theta}} e^{-2x\cot\theta}\, d\theta,$$

it is obvious that, when $m\pi < x < m\pi + \frac{3}{4}\pi$,

$$\operatorname{sgn}\{\sin(x + \tfrac{1}{2}\theta)\} = \operatorname{sgn}(-1)^m,$$

and so

$$\operatorname{sgn} J_0(x) = \operatorname{sgn}(-1)^m.$$

Consequently $J_0(x)$ has no zeros in the intervals $(m\pi, m\pi + \frac{3}{4}\pi)$.

* *Journal für Math.* CXIV. (1894), pp. 31—44.

To prove that $J_0(x)$ has no zeros in the intervals $(m\pi + \frac{7}{8}\pi, m\pi + \pi)$, write

$$x = (m+1)\pi - \phi,$$

and then
$$J_0(x) = \frac{2(-)^{m+1}}{\pi} \int_0^{\frac{1}{2}\pi} \frac{\sin(\frac{1}{2}\theta - \phi)}{\sin\theta\sqrt{\cos\theta}} e^{-2x\cot\theta}\, d\theta.$$

The last integrand is negative or positive according as

$$0 < \theta < 2\phi \text{ or } 2\phi < \theta < \tfrac{1}{2}\pi.$$

Since $\phi < \frac{1}{8}\pi$, the second of these intervals is the longer; and the function

$$\frac{e^{-2x\cot\theta}}{\sin\theta\sqrt{\cos\theta}}$$

is an increasing function* of θ when $x > \frac{7}{8}\pi$ and θ is an acute angle.

Hence to each value of θ between 0 and 2ϕ there corresponds a value between 2ϕ and $\frac{1}{2}\pi$ for which $\sin(\frac{1}{2}\theta - \phi)$ has the same numerical value, but has the positive sign, and the cofactor of $\sin(\frac{1}{2}\theta - \phi)$ is greater for the second set of values of θ than for the first set. The integral under consideration is consequently positive, and so $J_0(x)$ cannot have a zero in any of the intervals $(m\pi + \frac{7}{8}\pi, m\pi + \pi)$. Therefore the only positive zeros of $J_0(x)$ are in the intervals $(m\pi + \frac{3}{4}\pi, m\pi + \frac{7}{8}\pi)$.

15·33. *Theorems of Schafheitlin's type, when* $-\frac{1}{2} < \nu \leqslant \frac{1}{2}$.

We shall now extend Schafheitlin's results to functions of the type

$$\mathscr{C}_\nu(x) \equiv J_\nu(x)\cos\alpha - Y_\nu(x)\sin\alpha,$$

where $0 \leqslant \alpha < \pi$ and $-\frac{1}{2} < \nu \leqslant \frac{1}{2}$.

We shall first prove the crude result that the only positive zeros of $\mathscr{C}_\nu(x)$ lie in the intervals

$$(m\pi + \tfrac{3}{4}\pi + \tfrac{1}{2}\nu\pi - \alpha,\ m\pi + \pi - \alpha)$$

where $m = 0, 1, 2, \ldots$. This result follows at once from the formulae of § 6·12, which shew that

$$\mathscr{C}_\nu(x) = \frac{2^{\nu+1} x^\nu}{\Gamma(\nu+\frac{1}{2})\Gamma(\frac{1}{2})} \int_0^{\frac{1}{2}\pi} \frac{\cos^{\nu-\frac{1}{2}}\theta \sin(x + \alpha - \nu\theta + \frac{1}{2}\theta)}{\sin^{2\nu+1}\theta} e^{-2x\cot\theta}\, d\theta,$$

for, when
$$m\pi - \alpha < x < m\pi + \tfrac{3}{4}\pi + \tfrac{1}{2}\nu\pi - \alpha,$$
we have
$$\operatorname{sgn}[\sin(x + \alpha - \nu\theta + \tfrac{1}{2}\theta)] = \operatorname{sgn}(-1)^m,$$

and so, for such values of x, $\mathscr{C}_\nu(x)$ is not zero. Consequently the only zeros of $\mathscr{C}_\nu(x)$ lie in the specified intervals, and there are an odd number of zeros in each interval, with the possible exception of the first if $\alpha > \frac{3}{4}\pi + \frac{1}{2}\nu\pi$.

Next we obtain the more precise result that *the only positive zeros of* $\mathscr{C}_\nu(x)$ *lie in the intervals*

$$(m\pi + \tfrac{3}{4}\pi + \tfrac{1}{2}\nu\pi - \alpha,\ m\pi + \tfrac{7}{8}\pi + \tfrac{1}{4}\nu\pi - \alpha)$$

* Its logarithmic derivate is

$$(2x - \sin\theta\cos\theta)\operatorname{cosec}^2\theta + \tfrac{1}{2}\tan\theta.$$

where $m = 0, 1, 2, \ldots$, except that, if α is sufficiently near to π, there may be zeros* in the interval

$$(\tfrac{7}{8}\pi + \tfrac{1}{4}\nu\pi - \alpha,\ \pi - \alpha).$$

We shall prove this result by proving that $\mathscr{C}_\nu(x)$ has a fixed sign throughout each of the intervals†

$$(m\pi + \tfrac{7}{8}\pi + \tfrac{1}{4}\nu\pi - \alpha,\ m\pi + \pi - \alpha).$$

Write $$x = (m+1)\pi - \alpha - (1-2\nu)\phi,$$

where ϕ is an angle between 0 and $\frac{1}{8}\pi$.

With this value of x,

$$\mathscr{C}_\nu(x) = \frac{(-)^{m+1}\, 2^{\nu+1}\, x^\nu}{\Gamma(\nu+\frac{1}{2})\,\Gamma(\frac{1}{2})} \int_0^{\frac{1}{2}\pi} \frac{\cos^{\nu-\frac{1}{2}}\theta \sin\{(1-2\nu)(\frac{1}{2}\theta - \phi)\}}{\sin^{2\nu+1}\theta}\, e^{-2x\cot\theta}\, d\theta.$$

To each value of θ between 0 and 2ϕ there corresponds a value between 2ϕ and $\frac{1}{2}\pi$ for which $\sin\{(1-2\nu)(\frac{1}{2}\theta - \phi)\}$ has the same numerical value, but has the positive sign.

Again $$e^{-2x\cot\theta}\cos^{\nu-\frac{1}{2}}\theta/\sin^{2\nu+1}\theta$$

is an increasing function of θ provided that

$$2x > \max\left[(2\nu+1)\sin\theta\cos\theta + (\nu - \tfrac{1}{2})\frac{\sin^3\theta}{\cos\theta}\right],$$

and this condition is satisfied when $x > \frac{1}{2}$ since $\nu \leqslant \frac{1}{2}$.

Hence, if $x > \frac{1}{2}$ and

$$m\pi + \tfrac{7}{8}\pi + \tfrac{1}{4}\nu\pi - \alpha < x < m\pi + \pi - \alpha,$$

we have $$\operatorname{sgn}\mathscr{C}_\nu(x) = \operatorname{sgn}(-1)^{m+1},$$

and this proves the more precise theorem.

15·34. *Theorems of Schafheitlin's type, when* $\frac{1}{2} < \nu < \frac{5}{2}$.

We next consider the function

$$\mathscr{C}_\nu(x) \equiv J_\nu(x)\cos\alpha - Y_\nu(x)\sin\alpha,$$

where $0 \leqslant \alpha < \pi$, as before, in which it is now supposed that $\frac{1}{2} < \nu < \frac{5}{2}$.

We shall first prove the crude result that the only positive zeros of $\mathscr{C}_\nu(x)$ lie in the intervals

$$(m\pi - \alpha,\ m\pi - \tfrac{1}{4}\pi + \tfrac{1}{2}\nu\pi - \alpha)$$

where‡ $m = 0, 1, 2, \ldots$. This result follows at once from the formulae of § 6·12, which shew that

$$\mathscr{C}_\nu(x) = \frac{2^{\nu+1}\, x^\nu}{\Gamma(\nu+\frac{1}{2})\,\Gamma(\frac{1}{2})} \int_0^{\frac{1}{2}\pi} \frac{\cos^{\nu-\frac{1}{2}}\theta \sin(x + \alpha - \nu\theta + \frac{1}{2}\theta)}{\sin^{2\nu+1}\theta}\, e^{-2x\cot\theta}\, d\theta;$$

for when $$m\pi - \tfrac{1}{4}\pi + \tfrac{1}{2}\nu\pi - \alpha < x < (m+1)\pi - \alpha,$$

* By taking as an alternative function $J_{|\nu|}(x)$ and applying the theorem of § 15·24, we see that there cannot be more than one such zero.

† If $x < \frac{1}{2}$ and $m = 0$, the reasoning fails when $\frac{7}{8}\pi + \frac{1}{4}\nu\pi - \alpha < \frac{1}{2}$.

‡ If $\alpha > (\frac{1}{2}\nu - \frac{1}{4})\pi$, the interval for which $m = 0$ is, of course, to be omitted.

we have $$\operatorname{sgn}[\sin(x+\alpha-\nu\theta+\tfrac{1}{2}\theta)] = \operatorname{sgn}(-1)^m,$$
whence the theorem stated is obvious.

Next we obtain the more precise result that *the only positive zeros of* $\mathscr{C}_\nu(x)$ *lie in the intervals*
$$(m\pi - \tfrac{1}{8}\pi + \tfrac{1}{4}\nu\pi - \alpha,\ m\pi - \tfrac{1}{4}\pi + \tfrac{1}{2}\nu\pi - \alpha),$$
where $m = 0, 1, 2, \ldots$, except that, if α is sufficiently near to π, there may be a zero in the interval $(0, \tfrac{1}{4}\nu\pi - \tfrac{1}{8}\pi - \alpha)$, and there may be one in the interval $(\pi - \alpha, \tfrac{7}{8}\pi + \tfrac{1}{4}\nu\pi - \alpha)$.

We use the same notation and reasoning as in § 15·33; only now, if
$$e^{-2x\cot\theta}\cos^{\nu-\frac{1}{2}}\theta/\sin^{2\nu+1}\theta \equiv f(\theta),$$
$f(\theta)$ is not necessarily an increasing function of θ; *but it is sufficient to prove that, when* $0 < \psi < 2\phi$, *then*
$$f(2\phi - \psi) < f(2\phi + \psi).$$

To obtain this result, observe that
$$\frac{d}{d\psi}\log\frac{f(2\phi+\psi)}{f(2\phi-\psi)} = 2x\{\operatorname{cosec}^2(2\phi+\psi) + \operatorname{cosec}^2(2\phi-\psi)\}$$
$$-\sin 4\phi\left[\frac{\nu-\frac{1}{2}}{\cos(2\phi+\psi)\cos(2\phi-\psi)} + \frac{2\nu+1}{\sin(2\phi+\psi)\sin(2\phi-\psi)}\right].$$

But $$[\operatorname{cosec}^2(2\phi+\psi) + \operatorname{cosec}^2(2\phi-\psi)]\sin(2\phi+\psi)\sin(2\phi-\psi)$$
is an increasing function of ψ, and therefore, *a fortiori*,
$$[\operatorname{cosec}^2(2\phi+\psi) + \operatorname{cosec}^2(2\phi-\psi)]\cos(2\phi+\psi)\cos(2\phi-\psi)$$
is an increasing function, since this function exceeds the former by an increasing function because 4ϕ is an acute angle; and so $\dfrac{d}{d\psi}\log\dfrac{f(2\phi+\psi)}{f(2\phi-\psi)}$ is *always* positive if it is positive when $\psi = 0$, i.e. if
$$4x > \{(\nu - \tfrac{1}{2})\tan^2 2\phi + (2\nu+1)\}\sin 4\phi,$$
and this is the case when $x > \tfrac{3}{4}\nu + \tfrac{1}{8}$.

Hence, when $\tfrac{1}{2} < \nu < \tfrac{5}{2}$, the only zeros of $\mathscr{C}_\nu(x)$, which exceed $\tfrac{3}{4}\nu + \tfrac{1}{8}$, lie in the intervals
$$(m\pi - \tfrac{1}{8}\pi + \tfrac{1}{4}\nu\pi - \alpha,\ m\pi - \tfrac{1}{4}\pi + \tfrac{1}{2}\nu\pi - \alpha).$$
The method seems inapplicable for larger values of ν on account of the oscillatory character of $\sin(x + \alpha - \nu\theta + \tfrac{1}{2}\theta)$ as θ increases from 0 to $\tfrac{1}{2}\pi$; a method which is effective for these larger values will now be explained.

15·35. *Schafheitlin's investigations of the zeros of cylinder functions of unrestrictedly large order.*

We shall now prove that, if $\nu > \tfrac{1}{2}$, *those zeros of the cylinder function*
$$J_\nu(x)\cos\alpha - Y_\nu(x)\sin\alpha$$
which exceed $(2\nu+1)(2\nu+3)/\pi$ *lie in the intervals*
$$(m\pi - \alpha + \tfrac{1}{2}\nu\pi + \tfrac{1}{2}\pi,\ m\pi - \alpha + \tfrac{1}{2}\nu\pi + \tfrac{3}{4}\pi)$$
where m assumes integer values.

The method used to obtain this result is due to Schafheitlin*; but he considered the case of functions of the first kind and of integral order only, and his reasoning is made lengthy and obscure by the use of arguments equivalent to the use of the second mean-value theorem when the explicit use of that theorem is obviously desirable.

As in the preceding analysis, write

$$\mathscr{C}_\nu(x) \equiv J_\nu(x)\cos\alpha - Y_\nu(x)\sin\alpha,$$

so that

$$\mathscr{C}_\nu(x) = \frac{2^{\nu+1}x^\nu}{\Gamma(\nu+\frac{1}{2})\,\Gamma(\frac{1}{2})}\int_0^{\frac{1}{2}\pi}\frac{\cos^{\nu-\frac{1}{2}}\theta\sin(x+\alpha-\nu\theta+\frac{1}{2}\theta)}{\sin^{2\nu+1}\theta}e^{-2x\cot\theta}\,d\theta$$

$$= \frac{2^{\nu+1}x^\nu}{\Gamma(\nu+\frac{3}{2})\,\Gamma(\frac{1}{2})}\int_0^{\frac{1}{2}\pi}\cot^{2\nu+1}\theta\,.\,e^{-2x\cot\theta}\frac{d}{d\theta}\left\{\frac{\cos(x+\alpha-\nu\theta-\frac{1}{2}\theta)}{\cos^{\nu+\frac{1}{2}}\theta}\right\}d\theta.$$

Now $\cot^{2\nu+1}\theta\,.\,e^{-2x\cot\theta}$ increases as θ increases from 0 to θ_2 and then decreases as θ increases from θ_2 to $\frac{1}{2}\pi$, where $\theta_2 = \text{arc tan}\dfrac{x}{\nu+\frac{1}{2}}$.

It will be observed that θ_2 is nearly equal to $\frac{1}{2}\pi$ when x is large compared with ν.

Now suppose that x lies between

$$m\pi - \alpha + \tfrac{1}{2}\pi(\nu-\tfrac{1}{2}) \text{ and } m\pi - \alpha + \tfrac{1}{2}\pi(\nu-\tfrac{1}{2}) + \tfrac{3}{4}\pi,$$

and then choose θ_1 so that

$$x + \alpha - (\nu+\tfrac{3}{2})\,\theta_1 = m\pi.$$

It is easy to verify that

$$\frac{2\nu-1}{2\nu+3}\cdot\frac{\pi}{2} < \theta_1 < \frac{2\nu+2}{2\nu+3}\cdot\frac{\pi}{2},$$

so that θ_1 is a positive angle less than θ_2, provided that

$$\text{arc tan}\frac{\nu+\frac{1}{2}}{x} < \frac{\pi}{4\nu+6}.$$

We suppose now that

$$x > (2\nu+1)(2\nu+3)/\pi,$$

so that θ_1 is certainly less than θ_2.

Then, by the second mean-value theorem, there exists a number θ_0, between 0 and θ_1, such that

$$(\nu+\tfrac{1}{2})\int_0^{\theta_1}\frac{\cos^{\nu-\frac{1}{2}}\theta\sin(x+\alpha-\nu\theta+\frac{1}{2}\theta)}{\sin^{2\nu+1}\theta}e^{-2x\cot\theta}\,d\theta$$

$$= \left\{\cot^{2\nu+1}\theta_1\,.\,e^{-2x\cot\theta_1}\right\}.\int_{\theta_0}^{\theta_1}\frac{d}{d\theta}\left\{\frac{\cos(x+\alpha-\nu\theta-\frac{1}{2}\theta)}{\cos^{\nu+\frac{1}{2}}\theta}\right\}d\theta$$

$$= \left\{\cot^{2\nu+1}\theta_1\,.\,e^{-2x\cot\theta_1}\right\}\left\{\frac{\cos(m\pi+\theta_1)}{\cos^{\nu+\frac{1}{2}}\theta_1} - \frac{\cos(x+\alpha-\nu\theta_0-\frac{1}{2}\theta_0)}{\cos^{\nu+\frac{1}{2}}\theta_0}\right\}.$$

* *Journal für Math.* CXXII. (1900), pp. 299—321.

Now *qua* function of θ,

$$\cos(x+\alpha-\nu\theta-\tfrac{1}{2}\theta)/\cos^{\nu+\frac{1}{2}}\theta$$

is stationary when $\sin(x+\alpha-\nu\theta+\frac{1}{2}\theta)=0$, and for such values of θ the fraction is equal to $\pm 1/\cos^{\nu-\frac{1}{2}}\theta$.

Hence
$$\frac{\cos(x+\alpha-\nu\theta_0-\frac{1}{2}\theta_0)}{\cos^{\nu+\frac{1}{2}}\theta_0}$$

cannot exceed numerically the greatest value of $1/\cos^{\nu-\frac{1}{2}}\theta$ in the interval $(0, \theta_1)$, and therefore

$$\operatorname{sgn}\left\{\frac{\cos(m\pi+\theta_1)}{\cos^{\nu+\frac{1}{2}}\theta_1}-\frac{\cos(x+\alpha-\nu\theta_0-\frac{1}{2}\theta_0)}{\cos^{\nu+\frac{1}{2}}\theta_0}\right\}=\operatorname{sgn}(-1)^m.$$

Therefore, since the sign of $\sin(x+\alpha-\nu\theta+\frac{1}{2}\theta)$ is the sign of $(-1)^m$ when θ lies between θ_1 and $\frac{1}{2}\pi$, we see that, for the values of x under consideration,

$$\operatorname{sgn}\mathscr{C}_\nu(x)=\operatorname{sgn}(-1)^m.$$

Hence, when x exceeds $(2\nu+1)(2\nu+3)/\pi$, $\mathscr{C}_\nu(x)$ has no zeros in intervals of the type

$$(m\pi-\alpha+\tfrac{1}{2}\nu\pi-\tfrac{1}{4}\pi,\ m\pi-\alpha+\tfrac{1}{2}\nu\pi+\tfrac{1}{2}\pi),$$

and so the only zeros of $\mathscr{C}_\nu(x)$ which exceed $(2\nu+1)(2\nu+3)/\pi$ lie in intervals of the type

$$(m\pi-\alpha+\tfrac{1}{2}\nu\pi+\tfrac{1}{2}\pi,\ m\pi-\alpha+\tfrac{1}{2}\nu\pi+\tfrac{3}{4}\pi),$$

and this reduces to Schafheitlin's result* when $\alpha=0$ and ν is an integer.

The reader will observe that this theorem gives no information concerning the smaller zeros of $\mathscr{C}_\nu(x)$ when ν is large; it will be apparent in § 15·8 that there are a large number of zeros less than $(2\nu+1)(2\nu+3)/\pi$, and that interesting information can be obtained concerning them by using Debye's integrals.

15·36. *Bôcher's theorem*† *on the zeros of* $\mathscr{C}_0(x)$.

A result of a slightly different character from those just established was discovered by Bôcher from a consideration of the integral formula § 11·41 (16). The theorem in question is that $\mathscr{C}_0(x)$ has an infinite number of positive zeros, and the distance between consecutive zeros does not exceed $2j_0$ where j_0 is the smallest positive zero of $J_0(x)$.

To establish this result, write $\nu=0$, $z=j_0$ in § 11·41 (16), and then

$$\int_0^\pi \mathscr{C}_0(\varpi)\,d\phi=\pi\mathscr{C}_0(Z)\,J_0(j_0)=0.$$

Hence $\mathscr{C}_0(\varpi)$ cannot be one-signed as ϕ increases from 0 to π, i.e. as ϖ increases from $Z-j_0$ to $Z+j_0$; and so $\mathscr{C}_0(\varpi)$ must vanish for at least one value‡ of ϖ in the interval $(Z-j_0, Z+j_0)$. Since Z is an arbitrary positive number (greater than j_0), Bôcher's theorem is now evident.

* Schafheitlin gives $(2\nu+3)(2\nu+5)/\pi$ as the lower limit of the values of x for which the zeros lie in the specified intervals.

† *Bulletin American Math. Soc.* v. (1899), pp. 385—388.

‡ Cf. *Modern Analysis*, § 3·63.

[NOTE. By a form of Green's theorem,

$$\int u \frac{\partial v}{\partial \nu} ds = \int v \frac{\partial u}{\partial \nu} ds,$$

where u, v are two solutions of $\dfrac{\partial^2 u}{\partial x^2} + \dfrac{\partial^2 u}{\partial y^2} + u = 0$ with continuous second differential coefficients inside the closed curve s, and $\partial/\partial\nu$ indicates differentiation along the normal.

By taking $v = J_0\{\sqrt{(x^2+y^2)}\}$ and the curve to be $x^2+y^2=j_0^2$, Weber* deduced that u must vanish at least twice on any circle of radius j_0.

Bôcher inferred from this result that since $\mathscr{C}_n(r)\cos n\theta$ satisfies the requisite conditions except at the origin, if a circle of radius j_0 is drawn with centre on the axis of x and subtending an angle less than π/n at the origin, $\mathscr{C}_n(r)$ must vanish somewhere on the circle. Hence the positive zeros of $\mathscr{C}_n(r)$ are such that consecutive zeros are at a distance apart less than $2j_0$, and the distance from the origin of the smallest of them does not exceed $j_0\left\{1+\operatorname{cosec}\dfrac{\pi}{2n}\right\}$.

These results are of interest on account of the extreme simplicity of the methods used to prove them.]

15·4. *On the number of zeros of $J_\nu(z)$ in an assigned strip of the z-plane.*

We shall next give the expression for $J_\nu(z)$ as a Weierstrassian product, and then develop expressions involving quotients of Bessel functions in the form of partial fractions; but as a preliminary it is convenient to prove the following theorem, which gives some indication as to the situation of those zeros of $J_\nu(z)$ which are of large modulus. In this investigation it is not supposed that ν is restricted to be a real number, though it is convenient to suppose that ν is not a negative integer. When ν is real the results of § 15·2 to some extent take the place of the theorem which will now be proved.

Let C be the rectangular contour whose vertices are

$$\pm iB + \tfrac{1}{2}\pi i I(\nu),\ \pm iB + m\pi + \tfrac{1}{2}\nu\pi + \tfrac{1}{4}\pi,$$

where B is a (large) positive number.

We shall shew that when m is a sufficiently large integer the number of zeros of $z^{-\nu}J_\nu(z)$ inside C is precisely equal† to m.

Since $z^{-\nu}J_\nu(z)$ is an integral function of z, the number of its zeros inside C is

$$\frac{1}{2\pi i}\int_C \frac{d\log\{w^{-\nu}J_\nu(w)\}}{dw}\,dw = -\frac{1}{2\pi i}\int_C \frac{J_{\nu+1}(w)}{J_\nu(w)}\,dw.$$

* *Math. Ann.* I. (1869), p. 10.

† When ν is a real negative number (and for certain complex values of ν) there may be pairs of zeros on the imaginary axis; in such circumstances the contour C has to be indented, and each pair of zeros is to be reckoned as a single zero.

We now consider the four sides of C in turn. It is first to be observed that on all the sides of C,

$$H_\nu^{(1)}(w) = \left(\frac{2}{\pi w}\right)^{\frac{1}{2}} e^{i(w-\frac{1}{2}\nu\pi-\frac{1}{4}\pi)}\{1+\eta_{1,\nu}(w)\},$$

$$H_\nu^{(2)}(w) = \left(\frac{2}{\pi w}\right)^{\frac{1}{2}} e^{-i(w-\frac{1}{2}\nu\pi-\frac{1}{4}\pi)}\{1+\eta_{2,\nu}(w)\},$$

where $\eta_{1,\nu}(w)$ and $\eta_{2,\nu}(w)$ are $O(1/w)$ when $|w|$ is large.

Now, since the integrand is an odd function*, we have, as $B \to \infty$,

$$-\frac{1}{2\pi i}\int_{iB+\frac{1}{2}\pi iI(\nu)}^{-iB+\frac{1}{2}\pi iI(\nu)} \frac{J_{\nu+1}(w)}{J_\nu(w)}\,dw = \frac{1}{2\pi i}\int_{iB-\frac{1}{2}\pi iI(\nu)}^{iB+\frac{1}{2}\pi iI(\nu)} \frac{J_{\nu+1}(w)}{J_\nu(w)}\,dw$$

$$\to \frac{1}{2\pi i}\int_{iB-\frac{1}{2}\pi iI(\nu)}^{iB+\frac{1}{2}\pi iI(\nu)} i\,dw = \tfrac{1}{2} iI(\nu).$$

Next take the integral along the upper horizontal side of C; this is equal to

$$-\frac{1}{2\pi i}\int_{iB+m\pi+\frac{1}{2}\nu\pi+\frac{1}{4}\pi}^{iB+\frac{1}{2}\pi iI(\nu)} \frac{J_{\nu+1}(w)}{J_\nu(w)}\,dw$$

$$= \frac{1}{2\pi}\int_{iB+\frac{1}{2}\pi iI(\nu)}^{iB+m\pi+\frac{1}{2}\nu\pi+\frac{1}{4}\pi} \left\{\frac{1+\eta_{2,\nu+1}(w)}{1+\eta_{2,\nu}(w)}\right\}[1+O(e^{2iw})]\,dw$$

$$= \frac{1}{2\pi}\left[m\pi + \tfrac{1}{2}\pi R(\nu) + \tfrac{1}{4}\pi + \frac{2\nu+1}{2i}\log\frac{iB+m\pi+\frac{1}{2}\nu\pi+\frac{1}{4}\pi}{iB+\frac{1}{2}\pi iI(\nu)} + O(1/B)\right]$$

$$\to \tfrac{1}{2}m + \tfrac{1}{4}R(\nu) + \tfrac{1}{8},$$

as $B \to \infty$.

Similarly the integral along the lower side tends to the same value, and so the limit of the integral along the three sides now considered is $m + \frac{1}{2}\nu + \frac{1}{4}$.

Lastly we have to consider the integral along the fourth side, and to do this we first investigate the difference

$$\frac{J_{\nu+1}(w)}{J_\nu(w)} - \tan(w - \tfrac{1}{2}\nu\pi - \tfrac{1}{4}\pi),$$

which, when $|w|$ is large, is equal to

$$\frac{2\nu+1}{2w} + O\left(\frac{1}{w^2}\right).$$

Now
$$\int_{-iB+m\pi+\frac{1}{2}\nu\pi+\frac{1}{4}\pi}^{iB+m\pi+\frac{1}{2}\nu\pi+\frac{1}{4}\pi} \tan(w - \tfrac{1}{2}\nu\pi - \tfrac{1}{4}\pi)\,dw = 0,$$

and so

$$-\frac{1}{2\pi i}\int_{-iB+m\pi+\frac{1}{2}\nu\pi+\frac{1}{4}\pi}^{iB+m\pi+\frac{1}{2}\nu\pi+\frac{1}{4}\pi} \frac{J_{\nu+1}(w)}{J_\nu(w)}\,dw$$

$$= -\frac{1}{2\pi i}\int_{-iB+m\pi+\frac{1}{2}\nu\pi+\frac{1}{4}\pi}^{iB+m\pi+\frac{1}{2}\nu\pi+\frac{1}{4}\pi} \left\{\frac{2\nu+1}{2w} + O\left(\frac{1}{w^2}\right)\right\}dw$$

$$\sim -\tfrac{1}{4}(2\nu+1) + O(1/m).$$

Hence the limit of the integral round the whole rectangle is $m + O(1/m)$.

* Allowance is made for the indentations, just specified, in the first step of the following analysis.

If we take m sufficiently large, we can ensure that the expression which is $O(1/m)$ is numerically less than 1; and since the integral round the rectangle must be an integer, it is equal to m.

That is to say, the number of zeros of $z^{-\nu} J_\nu(z)$ between the imaginary axis and the line on which

$$R(z) = m\pi + \{\tfrac{1}{2} R(\nu) + \tfrac{1}{4}\}\pi$$

is exactly m.

NOTE. The approximate formulae quoted for the functions of the third kind shew that the large zeros cannot have a large imaginary part; and so all the zeros of $J_\nu(z)$ lie inside a strip whose sides are parallel to the real axis and at distances from it which are bounded when $|\nu|$ is bounded.

15·41. *The expression of $J_\nu(z)$ as an infinite product.*

It is possible to express $J_\nu(z)$ as a product of 'simple factors' of Weierstrassian type, each factor vanishing at one of the zeros of $J_\nu(z)$. In order to express $J_\nu(z)$ in this form, it is convenient first to express the logarithmic derivate of $z^{-\nu} J_\nu(z)$ as a series of rational fractions by Mittag-Leffler's theorem*.

The zeros of $z^{-\nu} J_\nu(z)$ are taken to be $\pm j_{\nu,1}, \pm j_{\nu,2}, \pm j_{\nu,3}, \ldots$ where† $R(j_{\nu,n}) > 0$ and $|R(j_{\nu,1})| \leqslant |R(j_{\nu,2})| \leqslant |R(j_{\nu,3})| \leqslant \ldots$, the values of $j_{\nu,1}, j_{\nu,2}, j_{\nu,3}, \ldots$ being all unequal (§ 15·21). We draw a (large) rectangle D, whose vertices are $\pm A \pm iB$, where A and B are positive, and we suppose that $\pm j_{\nu,m}$ are the zeros of highest rank which are inside the rectangle.

We now consider

$$\frac{1}{2\pi i} \int_D \frac{z}{w(w-z)} \cdot \frac{J_{\nu+1}(w)}{J_\nu(w)}\, dw,$$

where z is any point inside the rectangle, other than a zero of $J_\nu(w)$, and ν is not a negative integer.

The only poles of the integrand inside the rectangle are $z, \pm j_{\nu,1}, \pm j_{\nu,2}, \ldots, \pm j_{\nu,m}$.

The residue at z is $J_{\nu+1}(z)/J_\nu(z)$ and the residues at $\pm j_{\nu,n}$ are

$$\left\{\frac{1}{z \mp j_{\nu,n}} \pm \frac{1}{j_{\nu,n}}\right\},$$

since $J_\nu'(z) = -J_{\nu+1}(z)$ when $z = \pm j_{\nu,n}$, by § 3·2.

It follows that

$$\frac{J_{\nu+1}(z)}{J_\nu(z)} + \sum_{n=1}^{m} \left\{\frac{1}{z - j_{\nu,n}} + \frac{1}{j_{\nu,n}}\right\} + \sum_{n=1}^{m} \left\{\frac{1}{z + j_{\nu,n}} - \frac{1}{j_{\nu,n}}\right\}$$
$$= \frac{1}{2\pi i} \int_D \frac{z}{w(w-z)} \cdot \frac{J_{\nu+1}(w)}{J_\nu(w)}\, dw.$$

* *Acta Soc. Scient. Fennicae*, XI. (1880), pp. 273—293. Cf. *Modern Analysis*, § 7·4.

† If $R(\pm j_{\nu,n}) = 0$ for any value of n, we choose $j_{\nu,n}$ to have its imaginary part positive.

We next shew that, by giving A and B suitable sequences of values which increase without limit, $J_{\nu+1}(w)/J_\nu(w)$ can be taken to be bounded on D. Since this function is an odd function of w, it is sufficient to consider the right-hand half of D.

We take $A = M\pi + R(\frac{1}{2}\nu + \frac{1}{4})\pi$, where M is a positive integer; and then we take M to be at least so large that $M = m$, which is possible by § 15·4, and also to be so large that we can take the functions $\eta_{\nu,1}(w)$, $\eta_{\nu,2}(w)$, defined in § 15·4, to be less than, say, $\frac{1}{4}$ in absolute value.

Then $J_{\nu+1}(w)/J_\nu(w)$ is bounded whenever

$$e^{2i(w-\frac{1}{2}\nu\pi-\frac{1}{4}\pi)}$$

is* less than $\frac{1}{2}$ or greater than 2; and when the expression does not lie within these limits, $I(w)$ is bounded and w is not arbitrarily near a zero of $J_\nu(w)$; so that, from the asymptotic expansion of § 7·21, $J_{\nu+1}(w)/J_\nu(w)$ is bounded on the part of the rectangle within this strip.

That is to say $J_{\nu+1}(w)/J_\nu(w)$ is bounded on the whole of the perimeter of the rectangle D as B and M tend to infinity.

Hence

$$\frac{1}{2\pi i}\int_D \frac{z}{w(w-z)}\frac{J_{\nu+1}(w)}{J_\nu(w)}dw \to 0,$$

and therefore†

$$-\frac{J_{\nu+1}(z)}{J_\nu(z)} = \sum_{n=1}^{\infty}\left\{\frac{1}{z-j_{\nu,n}} + \frac{1}{j_{\nu,n}}\right\} + \sum_{n=1}^{\infty}\left\{\frac{1}{z+j_{\nu,n}} - \frac{1}{j_{\nu,n}}\right\}. \tag{1}$$

When we integrate, we find that

$$\exp\left\{-\int_0^z \frac{J_{\nu+1}(t)}{J_\nu(t)}dt\right\} = \prod_{n=1}^{\infty}\left\{\left(1-\frac{z}{j_{\nu,n}}\right)\exp\left(\frac{z}{j_{\nu,n}}\right)\right\}\prod_{n=1}^{\infty}\left\{\left(1+\frac{z}{j_{\nu,n}}\right)\exp\left(-\frac{z}{j_{\nu,n}}\right)\right\},$$

and hence

$$J_\nu(z) = \frac{(\frac{1}{2}z)^\nu}{\Gamma(\nu+1)}\prod_{n=1}^{\infty}\left\{\left(1-\frac{z}{j_{\nu,n}}\right)\exp\left(\frac{z}{j_{\nu,n}}\right)\right\}\prod_{n=1}^{\infty}\left\{\left(1+\frac{z}{j_{\nu,n}}\right)\exp\left(-\frac{z}{j_{\nu,n}}\right)\right\} \tag{2}$$

This is the expression of $J_\nu(z)$ in the specified form.

The formula may also be written in the modified form

$$J_\nu(z) = \frac{(\frac{1}{2}z)^\nu}{\Gamma(\nu+1)}\prod_{n=1}^{\infty}\left\{1-\frac{z^2}{j^2_{\nu,n}}\right\}. \tag{3}$$

This formula was assumed by Euler, *Acta Acad. Petrop.* v. pars 1, (1781) [1784], p. 170, when $\nu=0$, and subsequently by various writers for other values of ν; cf. §§ 15·5, 15·51.

The analysis of this section is due in substance to Graf and Gubler, *Einleitung in die Theorie der Bessel'schen Funktionen*, I. (Bern, 1898), pp. 123—130, and it was given explicitly by Kapteyn, *Monatshefte für Math. und Phys.* XIV. (1903), pp. 281—282.

* Because $\frac{1-\frac{1}{4}}{1+\frac{1}{4}} > \frac{1}{2}$.

† If we take the rectangle to have its vertices at $A \pm iB$, $-A' \pm iB$, we see that the two series on the right converge separately.

The expansion on the right of (1) is evidently expansible in a power series; the coefficients in such a series have been expressed as determinants by Kapteyn, *Proc. Section of Sci., K. Acad. van Wet. te Amsterdam*, VIII. (1905), pp. 547—549, 640—642; *Archives Néerlandaises*, (2) XI. (1906), pp. 149—168. Some associated formulae have just been published by Forsyth, *Messenger*, L. (1921), pp. 129—149.

15·42. *The Kneser-Sommerfeld expansion.*

An expansion which, in some respects, resembles the partial fraction formula obtained in § 15·41 is as follows:

$$\sum_{n=1}^{\infty} \frac{J_\nu(j_{\nu,n} x)\, J_\nu(j_{\nu,n} X)}{(z^2 - j^2_{\nu,n})\, j_{\nu,n}\, J_\nu'^2(j_{\nu,n})} = \frac{\pi J_\nu(xz)}{4z J_\nu(z)} \{J_\nu(z)\, Y_\nu(Xz) - Y_\nu(z)\, J_\nu(Xz)\},$$

in which x and X are positive numbers such that

$$0 \leqslant x \leqslant X \leqslant 1,$$

while z and ν are unrestricted (complex) *numbers, except that it is convenient to take $R(z) > 0$.*

The expansion was discovered in the case $\nu = 0$, as a special form of an expansion occurring in the theory of integral equations, by Kneser, *Math. Ann.* LXIII. (1907), pp. 511—517. Proofs of this and of related expansions for integral values of ν were published later by Sommerfeld, *Jahresbericht der Deutschen Math. Vereinigung*, XXI. (1913), pp. 309—353, but Sommerfeld's method of proof has been criticised adversely by Carslaw, *Proc. London Math. Soc.* (2) XIII. (1914), p. 239.

It may be noticed that the expansion has some connexion with the 'Fourier-Bessel' expansions which will be discussed in Chapter XVIII.

To obtain a proof of the expansion, consider the integral

$$\frac{1}{2\pi i} \int \frac{H_\nu^{(1)}(Xw)\, H_\nu^{(2)}(w) - H_\nu^{(2)}(Xw)\, H_\nu^{(1)}(w)}{z^2 - w^2} \frac{J_\nu(xw)}{J_\nu(w)}\, dw,$$

in which the path of integration is a rectangle with vertices $\pm Bi$, $A \pm Bi$, and it is supposed that the left side of the rectangle is indented at the origin.

The integral round the indentation tends to zero with the radius of the indentation, whether ν be an integer or not; and the integrals along the two parts of the imaginary axis cancel.

Also, when x and X satisfy the specified inequalities, the function

$$\{H_\nu^{(1)}(Xw)\, H_\nu^{(2)}(w) - H_\nu^{(2)}(Xw)\, H_\nu^{(1)}(w)\}\, J_\nu(xw)/J_\nu(w)$$

remains bounded on the other three sides of the rectangle when $B \to \infty$ and when $A \to \infty$ through the values specified in § 15·41.

Hence the limit of the integral round the rectangle is zero, and so the limit of the sum of the residues of the integrand at the poles on the right of the imaginary axis is zero.

Now the residue at z is

$$\frac{i J_\nu(xz)}{z J_\nu(z)} \{J_\nu(Xz)\, Y_\nu(z) - J_\nu(z)\, Y_\nu(Xz)\},$$

while the residue at $j_{\nu,n}$ is

$$-2i\,J_\nu(j_{\nu,n}X)\,Y_\nu(j_{\nu,n})\,J_\nu(j_{\nu,n}x)/\{J_\nu'(j_{\nu,n})(z^2-j^2_{\nu,n})\}$$

$$=\frac{-2i\,J_\nu(j_{\nu,n}X)\,J_\nu(j_{\nu,n}x)}{J_\nu'^2(j_{\nu,n})(z^2-j^2_{\nu,n})}\{Y_\nu(j_{\nu,n})\,J_\nu'(j_{\nu,n})-J_\nu(j_{\nu,n})\,Y_\nu'(j_{\nu,n})\}$$

$$=\frac{4i\,J_\nu(j_{\nu,n}X)\,J_\nu(j_{\nu,n}x)}{\pi j_{\nu,n}\,J_\nu'^2(j_{\nu,n})(z^2-j^2_{\nu,n})},$$

and on summing the residues we at once obtain the stated expansion.

For a generalisation of this expansion, obtained by replacing $J_\nu(xw)/J_\nu(w)$ by $\mathscr{C}_\nu(xw)/\mathscr{C}_\nu(w)$ in the contour integral, see Carslaw, *Proc. London Math. Soc.* (2) XVI. (1917), pp. 84—93; Carslaw has also constructed some similar series which contain Legendre functions as well as Bessel functions, and these series represent the Green's functions appropriate to certain physical problems. See also Beltrami, *Lombardo Rendiconti*, (2) XIII. (1880), p. 336; and Lorenz, *Oeuvres Scientifiques*, II. (1899), p. 506.

15·5. *Euler's investigation of the zeros of $J_0(2\sqrt{z})$.*

An ingenious method of calculating the smallest zeros of a function was devised by Euler*, and applied by him to determine the three smallest zeros of $J_0(2\sqrt{z})$.

If the zeros arranged in ascending order† of magnitude be $\alpha_1, \alpha_2, \alpha_3, \ldots$, then by § 15·41,

$$J_0(2\sqrt{z})=\prod_{n=1}^{\infty}\left(1-\frac{z}{\alpha_n}\right).$$

As has already been stated (§ 15·41), this formula was assumed by Euler; if it is differentiated logarithmically, then

$$-\frac{d}{dz}\log J_0(2\sqrt{z})=\sum_{n=1}^{\infty}\frac{1}{\alpha_n-z}$$

$$=\sum_{n=1}^{\infty}\sum_{m=0}^{\infty}\frac{z^m}{\alpha_n^{m+1}}$$

provided that $|z|<\alpha_1$; and the last series is then absolutely convergent.

Put $\sum_{n=1}^{\infty}1/\alpha_n^{m+1}\equiv\sigma_{m+1}$ and change the order of the summations; then

$$-\frac{d}{dz}J_0(2\sqrt{z})\equiv J_0(2\sqrt{z})\sum_{m=0}^{\infty}\sigma_{m+1}z^m.$$

Replace $J_0(2\sqrt{z})$ on each side by

$$1-\frac{z}{1^2}+\frac{z^2}{1^2.2^2}-\frac{z^3}{1^2.2^2.3^2}+\ldots,$$

* *Acta Acad. Petrop.* v. pars 1, (1781) [1784], pp. 170 *et seq.* A paper by Stern, *Journal für Math.* XXXIII. (1846), pp. 363—365 should also be consulted.

† From § 15·25 it follows that the zeros are positive and unequal.

multiply out the product on the right, and equate coefficients of the various powers of z in the identity; we thus obtain the system* of equations

$$1 = \sigma_1, \; -\tfrac{1}{2} = \sigma_2 - \sigma_1,$$
$$\tfrac{1}{12} = \sigma_3 - \sigma_2 + \tfrac{1}{4}\sigma_1,$$
$$-\tfrac{1}{144} = \sigma_4 - \sigma_3 + \tfrac{1}{4}\sigma_2 - \tfrac{1}{36}\sigma_1,$$
$$\tfrac{1}{2880} = \sigma_5 - \sigma_4 + \tfrac{1}{4}\sigma_3 - \tfrac{1}{36}\sigma_2 + \tfrac{1}{576}\sigma_1,$$
$$-\tfrac{1}{86400} = \sigma_6 - \sigma_5 + \tfrac{1}{4}\sigma_4 - \tfrac{1}{36}\sigma_3 + \tfrac{1}{576}\sigma_2 - \tfrac{1}{14400}\sigma_1,$$
$$\dots\dots\dots\dots\dots\dots\dots\dots\dots\dots$$

whence

$$\sigma_1 = 1, \sigma_2 = \tfrac{1}{2}, \sigma_3 = \tfrac{1}{3}, \sigma_4 = \tfrac{11}{48}, \sigma_5 = \tfrac{19}{120}, \sigma_6 = \tfrac{473}{4320}, \dots .$$

Since $0 < \alpha_1 < \alpha_2 < \alpha_3 < \dots$, it is evident that

$$1/\alpha_1{}^m < \sigma_m, \quad \sigma_{m+1} < \sigma_m/\alpha_1,$$

and so

$$\sigma_m{}^{-1/m} < \alpha_1 < \sigma_m/\sigma_{m+1}.$$

By extrapolating from the following Table:

m	$\sigma_m{}^{-1/m}$	σ_m/σ_{m+1}
1	1·000 000	2·000 000
2	1·414 213	1·500 000
3	1·442 250	1·454 545
4	1·445 314	1·447 368
5	1·445 724	1·446 089
6	1·445 785	—

Euler inferred that $\alpha_1 = 1{\cdot}445795$, whence

$$1/\alpha_1 = 0{\cdot}691661, \quad 2\sqrt{\alpha_1} = 2{\cdot}404824.$$

By adopting this value for α_1, writing

$$\sum_{n=2}^{\infty} 1/\alpha_n{}^m = \sigma_m{}',$$

and then using the inequalities

$$1/\alpha_2{}^m < \sigma'_m, \quad \sigma'_{m+1} < \sigma'_m/\alpha_2,$$

Euler deduced that $\alpha_2 = 7{\cdot}6658$, and hence that $\alpha_3 = 18{\cdot}63$, by carrying the process a stage further.

These results should be compared with the values

$$\alpha_1 = 1{\cdot}445796, \quad \alpha_2 = 7{\cdot}6178, \quad \alpha_3 = 18{\cdot}72,$$

derived from the Tables of Willson and Peirce, *Bulletin American Math. Soc.* III. (1898), pp. 153—155.

The value of α_1 is given by Poisson† as 1·446796491 (misprinted as 1·46796491); according to Freeman‡ this result was calculated by Largeteau for Poisson by solving the quartic obtained by equating to zero the first five terms of the series for $J_0(2\sqrt{z})$; the magnitude of the sixth term is quite sufficient to account for the error.

* This system is an obvious extension of Newton's system for an algebraic equation.

† *Mém. de l'Acad. R. des Sci.* XII. (1833), p. 330.

‡ *Proc. Camb. Phil. Soc.* III. (1880), pp. 375—377. Cf. Freeman's translation of Fourier's *La Théorie Analytique de la Chaleur*, p. 310, footnote.

15·51. *Rayleigh's extension of Euler's formula.*

The method just described was used independently by Rayleigh* to calculate the smallest positive zero of $J_\nu(z)$.

Taking the formula (§ 15·41)

$$J_\nu(z) = \frac{(\frac{1}{2}z)^\nu}{\Gamma(\nu+1)} \prod_{n=1}^{\infty} \left\{1 - \frac{z^2}{j^2_{\nu,n}}\right\},$$

and writing

$$\sum_{n=1}^{\infty} \frac{1}{j^{2r}_{\nu,n}} \equiv \sigma_\nu^{(r)},$$

we find, after Rayleigh, that

$$\sigma_\nu^{(1)} = \frac{1}{2^2(\nu+1)},\quad \sigma_\nu^{(2)} = \frac{1}{2^4(\nu+1)^2(\nu+2)},\quad \sigma_\nu^{(3)} = \frac{1}{2^5(\nu+1)^3(\nu+2)(\nu+3)},$$

$$\sigma_\nu^{(4)} = \frac{5\nu+11}{2^8(\nu+1)^4(\nu+2)^2(\nu+3)(\nu+4)},$$

$$\sigma_\nu^{(5)} = \frac{7\nu+19}{2^9(\nu+1)^5(\nu+2)^2(\nu+3)(\nu+4)(\nu+5)}.$$

The smallest positive zeros of $J_0(z)$ and $J_1(z)$ are deduced to be 2·404826 and 3·831706.

Immediately afterwards Cayley † noticed that $\sigma_\nu^{(r)}$ can be calculated rapidly when r is a power of 2 by a process which he attributed to Encke ‡, but which is more usually known as Graeffe's § method of solving an equation.

The method consists in calculating $\sigma_\nu^{(r)}$ when r is a power of 2 by starting with the given equation and forming from it a sequence of equations each of which has for its roots the squares of the roots of its predecessor; and $\sigma_\nu^{(r)}$ then rapidly tends to a ratio of equality with $1/j^{2r}_{\nu,1}$.

Cayley thus found $\sigma_\nu^{(8)}$ to be

$$\frac{429\nu^5+7640\nu^4+53752\nu^3+185430\nu^2+311387\nu+202738}{2^{16}(\nu+1)^8(\nu+2)^4(\nu+3)^2(\nu+4)^2(\nu+5)(\nu+6)(\nu+7)(\nu+8)}.$$

It was observed by Graf and Gubler|| that the value of $\sigma_\nu^{(r)}$ can easily be checked by the formula

$$\sigma_{\frac{1}{2}}^{(r)} = 2^{2r-1} B_r/(2r)!,$$

where B_r is the rth Bernoullian number; this formula is an evident consequence of the equation

$$J_{\frac{1}{2}}(z) = \left(\frac{2}{\pi z}\right)^{\frac{1}{2}} \sin z.$$

Extensions of some of these results to the zeros of $zJ_\nu'(z)+hJ_\nu(z)$, where h is a constant, have been made by Lamb, *Proc. London Math. Soc.* XV. (1884), p. 273.

The smallest zero of $J_\nu(z)$, for various values of ν between 0 and 1, has recently been tabulated by Airey, *Phil. Mag.* (6) XLI. (1921), pp. 200—205, with the aid of the Rayleigh-Cayley formulae.

* *Proc. London Math. Soc.* v. (1874), pp. 119—124. [*Scientific Papers*, I. (1899), pp. 190—195.]

† *Proc. London Math. Soc.* v. (1874), pp. 123—124. [*Collected Papers*, IX. (1896), pp. 19—20.]

‡ *Journal für Math.* XXII. (1841), pp. 193—248.

§ *Die Auflösung der höheren numerischen Gleichungen* (Zürich, 1837).

|| *Einleitung in die Theorie der Bessel'schen Funktionen*, I. (Bern, 1898), pp. 130—131.

[NOTE. The procedure of calculating the sum of the rth powers of the roots of an equation in order to obtain the numerical value of its largest root seems to be due to Waring, *Meditationes Analyticae* (Cambridge, 1776), p. 311; other writers who were acquainted with such a method before Graeffe are Euler (cf. § 15·5); Dandelin*, *Mém. de l'Acad. R. des Sci. de Bruxelles*, III. (1826), p. 48; Lobatschevsky*, *Algebra, or Calculus of Finites* (Kazan, 1834), § 257.]

15·52. *The large zeros of* $J_0(x)$.

The most effective method of calculating the large zeros of cylinder functions (when the order ν is not too large) is, in substance, due to Stokes†, though subsequent writers have, to some extent, improved on his analysis.

Stokes' method will be sufficiently illustrated by his own example‡ $J_0(x)$, whose zeros are the roots of the equation

$$\cot(x - \tfrac{1}{4}\pi) = \frac{Q(x, 0)}{P(x, 0)},$$

with the notation of § 7·3. It will be remembered that the asymptotic expansions of $P(x, 0)$ and $Q(x, 0)$ are

$$P(x, 0) \sim 1 - \frac{1 . 9}{2!(8x)^2} + \frac{1 . 9 . 25 . 49}{4!(8x)^4} - \ldots,$$

$$Q(x, 0) \sim -\frac{1}{1!8x} + \frac{1 . 9 . 25}{3!(8x)^3} - \ldots.$$

For sufficiently large values of x, $P(x, 0)$ is positive, $Q(x, 0)$ is negative and the quotient $Q(x, 0)/P(x, 0)$ is a negative increasing§ function of x. The function $\cot(x - \frac{1}{4}\pi)$ is a decreasing function which vanishes when $x = n\pi - \frac{1}{4}\pi$, and so it is obvious from a graph of $\cot(x - \frac{1}{4}\pi)$ that there exists a positive integer N such that when $n > N$, $J_0(x)$ has precisely one zero in each of the intervals $(n\pi - \frac{1}{4}\pi, n\pi + \frac{1}{4}\pi)$, and that the distance of the zero from the left-hand end of the interval tends to zero as $n \to \infty$.

Again, if u_r, v_r denote the $(r+1)$th terms of $P(x, 0)$ and $Q(x, 0)$ we may write

$$P(x, 0) = \sum_{r=0}^{m-1} u_r + \theta u_m, \quad Q(x, 0) = \sum_{r=0}^{m-1} v_r + \theta_1 v_m,$$

where θ and θ_1 are certain functions of x and m which lie between 0 and 1.

* I owe these two references to Professor Whittaker.

† *Camb. Phil. Trans.* IX. (1856), pp. 182—184. [*Math. and Phys. Papers*, II. (1883), pp. 350—353.]

‡ Stokes also considered Airy's integral (§ 6·4) and $J_1(x)$, for the purpose of investigating the position of the dark bands seen in artificial rainbows.

§ The reader may verify, by § 3·63, that its derivate is

$$\{1 - P^2 - Q^2\}/P^2,$$

where P, Q stand for $P(x, 0)$, $Q(x, 0)$; and, by the asymptotic expansions, this is ultimately positive.

Now consider the equation

$$\cot(x - \tfrac{1}{4}\pi) = \frac{\sum_{r=0}^{m-1} v_r + \theta_1 v_m}{\sum_{r=0}^{m-1} u_r + \theta u_m},$$

in which it is temporarily supposed that θ and θ_1, instead of having their actual values, are *any* numbers which lie between 0 and 1.

The equation now under consideration involves no functions more complicated than trigonometrical functions. If x were supposed complex, there would be a number of contours in the x-plane each of which enclosed one of the points $n\pi - \frac{1}{4}\pi$ and on which $|\cot(x - \frac{1}{4}\pi)|$ exceeded the modulus of the quotient on the right.

By Bürmann's theorem* the modified equation would have one root inside the part of the contour which surrounds $n\pi - \frac{1}{4}\pi$, and this root can be expanded in descending powers of $n\pi - \frac{1}{4}\pi$.

We thus obtain an expansion for the root of the equation in the form

$$x = (n\pi - \tfrac{1}{4}\pi) + \sum_{r=0}^{\infty} \frac{f_r(\theta, \theta_1)}{(n\pi - \frac{1}{4}\pi)^{2r+1}},$$

in which the coefficients $f_r(\theta, \theta_1)$ are independent of n but depend on θ and θ_1; and it is readily perceived that the first m of the coefficients are actually independent of θ and θ_1, so that, when $r < m$, we may write

$$f_r(\theta, \theta_1) \equiv f_r.$$

Now the sum of the terms after the mth is a bounded function of θ and θ_1 as θ and θ_1 vary between 0 and 1; and it is clear that the upper bound of the modulus of the function in question is $O(n^{-2m-1})$ as $n \to \infty$. Hence, when θ and θ_1 are given their actual values which they have at the zero under consideration, the sum of the terms after the mth is still $O(n^{-2m-1})$.

That is to say, it has been proved that there exists one zero (nearly equal to $n\pi - \frac{1}{4}\pi$), and its value may be written

$$x = (n\pi - \tfrac{1}{4}\pi) + \sum_{r=0}^{m-1} \frac{f_r}{(n\pi - \frac{1}{4}\pi)^{2r+1}} + O(n^{-2m-1}).$$

Hence the *asymptotic expansion* of the zero is

$$x \sim n\pi - \tfrac{1}{4}\pi + \sum_{r=0}^{\infty} \frac{f_r}{(n\pi - \frac{1}{4}\pi)^{2r+1}}.$$

It remains to calculate the first few of the coefficients f_r. If

$$\tan\psi \equiv -\frac{Q(x, 0)}{P(x, 0)},$$

where $\psi \to 0$ as $x \to \infty$, then

$$\tan\psi \sim \frac{1}{8x} - \frac{33}{512x^3} + \frac{3417}{16384x^5} - \dots,$$

* Cf. *Modern Analysis*, § 7·31.

so that
$$\psi \sim \frac{1}{8x} - \frac{25}{384x^3} + \frac{1073}{5120x^5} - \ldots,$$

and therefore the equation to be solved assumes the form

$$x - (n\pi - \tfrac{1}{4}\pi) \sim \frac{1}{8x} - \frac{25}{384x^3} + \frac{1073}{5120x^5} - \ldots.$$

The result of reverting the series is

$$x = (n\pi - \tfrac{1}{4}\pi) + \frac{1}{8(n\pi - \frac{1}{4}\pi)} - \frac{31}{384(n\pi - \frac{1}{4}\pi)^3} + \frac{3779}{15360(n\pi - \frac{1}{4}\pi)^5} - \ldots.$$

This series is adequate for calculating all the zeros of $J_0(x)$, to at least five places of decimals, except the smallest zero, for which $n = 1$.

15·53. *The large zeros of cylinder functions.*

It is easy to see that the large zeros of any cylinder function,

$$J_\nu(z)\cos\alpha - Y_\nu(z)\sin\alpha,$$

where ν and α are not necessarily real, may be calculated by Stokes' method from a consideration of the equation

$$\cot(z - \tfrac{1}{2}\nu\pi - \tfrac{1}{4}\pi + \alpha) = \frac{Q(z, \nu)}{P(z, \nu)}.$$

It seems unnecessary to prove the existence of such zeros (with large positive real parts) or the fact that they may be calculated as though the series for $P(z, \nu)$ and $Q(z, \nu)$ were convergent, because the proof differs from the investigation of the preceding section only in tedious details.

The expression for the large zeros of a cylinder function of any given order was calculated after the manner of Stokes by McMahon*; but the subsequent memoirs of Kalähne† and Marshall‡ have made the investigation more simple and have carried the approximation a stage further with no greater expenditure of work in the calculation.

Following Marshall we define§ two functions of z, called M and ψ, by the equations

$$M\cos\psi = P(z, \nu), \quad M\sin\psi = -Q(z, \nu),$$

on the understanding that $M \to +1$ and $\psi \to 0$ as $z \to +\infty$.

It is then clear that

$$J_\nu(z)\cos\alpha - Y_\nu(z)\sin\alpha = \left(\frac{2}{\pi z}\right)^{\frac{1}{2}} M\cos(z - \tfrac{1}{2}\nu\pi - \tfrac{1}{4}\pi + \alpha - \psi).$$

* *Annals of Math.* IX. (1895), pp. 23—25; see also Airey, *Proc. Phys. Soc.* 1911, pp. 219—224, 225—232.

† *Zeitschrift für Math. und Phys.* LIV. (1907), pp. 55—86.

‡ *Annals of Math.* (2) XI. (1910), pp. 153—160.

§ Cf. Nicholson, *Phil. Mag.* (6) XIX. (1910), pp. 228—249.

Again

$$\operatorname{arc\,tan}\left\{\frac{Y_\nu(z)}{J_\nu(z)}\right\} = z - \tfrac{1}{2}\nu\pi - \tfrac{1}{4}\pi - \psi,$$

and, when we differentiate this equation, and use § 3·63 (3), we find that

$$1 - \frac{d\psi}{dz} = \frac{2/(\pi z)}{J_\nu^2(z) + Y_\nu^2(z)},$$

so that, by § 7·51,

$$1 - \frac{d\psi}{dz} \sim \left[\sum_{m=0}^{\infty} \{1 . 3 \ldots (2m-1)\} \frac{(\nu, m)}{2^m z^{2m}}\right]^{-1}.$$

When the expression on the right is expanded as far as the term involving $1/z^8$, we find that

$$1 - \frac{d\psi}{dz} \sim 1 - \frac{\mu - 1}{2^3 z^2} - \frac{(\mu-1)(\mu-25)}{2^7 z^4} - \frac{(\mu-1)(\mu^2 - 114\mu + 1073)}{2^{10} z^6}$$
$$- \frac{(\mu-1)(5\mu^3 - 1535\mu^2 + 54703\mu - 375733)}{2^{15} z^8} - \ldots;$$

in this equation μ has been written in place of $4\nu^2$ for brevity. It follows, on integration, that

$$-\psi \sim \frac{\mu-1}{2^3 z} + \frac{(\mu-1)(\mu-25)}{3 . 2^7 z^3} + \frac{(\mu-1)(\mu^2 - 114\mu + 1073)}{5 . 2^{10} z^5}$$
$$+ \frac{(\mu-1)(5\mu^3 - 1535\mu^2 + 54703\mu - 375733)}{7 . 2^{15} z^7} + \ldots,$$

and so the equation to be solved is[*]

$$z - n\pi - \tfrac{1}{2}\nu\pi + \tfrac{1}{4}\pi + \alpha \sim -\frac{\mu-1}{2^3 z} - \frac{(\mu-1)(\mu-25)}{3 . 2^7 z^3} - \ldots.$$

If $\beta = (n + \frac{1}{2}\nu - \frac{1}{4})\pi - \alpha$, the result of reversion is

$$z \sim \beta - \frac{\mu-1}{2^3\beta} - \frac{(\mu-1)(7\mu-31)}{3 . 2^7 \beta^3} - \frac{(\mu-1)(83\mu^2 - 982\mu + 3779)}{15 . 2^{10}\beta^5}$$
$$- \frac{(\mu-1)(6949\mu^3 - 153855\mu^2 + 1585743\mu - 6277237)}{105 . 2^{15}\beta^7} - \ldots.$$

Therefore the large zeros of $J_\nu(z)\cos\alpha - Y_\nu(z)\sin\alpha$ are given by the asymptotic expansion

$$(n + \tfrac{1}{2}\nu - \tfrac{1}{4})\pi - \alpha - \frac{4\nu^2 - 1}{8\{(n + \frac{1}{2}\nu - \frac{1}{4})\pi - \alpha\}} - \frac{(4\nu^2-1)(28\nu^2-31)}{384\{(n + \frac{1}{2}\nu - \frac{1}{4})\pi - \alpha\}^3} - \ldots.$$

* This equation (in the case $\nu = 1$) was given by Gauss in his notebook with the date Oct. 16, 1797, but no clue is given concerning the method by which he obtained it. [Cf. *Math. Ann.* LVII. (1902), p. 19.]

[NOTE. The fact that $J_\nu^2(z)+Y_\nu^2(z)$ has a simple asymptotic expansion shortens the analysis in a manner which was not noticed by Marshall; he used the equations

$$1+\frac{d\psi}{dz}=\frac{1}{M^2},\quad \left[\frac{d^2}{dz^2}+1-\frac{\nu^2-\frac{1}{4}}{z^2}\right]M=\frac{1}{M^3},$$

and he solved the latter by assuming a descending series for M.]

15·54. *Zeros of functions related to cylinder functions.*

The method of Stokes is, of course, applicable to functions other than those just investigated. Thus McMahon* has calculated the large zeros of $\mathscr{C}_\nu'(z)$ and of $\dfrac{d\{z^{-\frac{1}{2}}\mathscr{C}_\nu(z)\}}{dz}$ when the cylinder function is a Bessel function of the first or second kind.

The general formula for the large zeros of $\mathscr{C}_\nu'(z)$ is

$$\beta_1-\frac{\mu+3}{8\beta_1}-\frac{7\mu^2+82\mu-9}{384\beta_1^3}-\dots,$$

where $\beta_1=(n+\frac{1}{2}\nu+\frac{1}{4})\pi-\alpha$, while the corresponding formula for the large zeros of $\dfrac{d\{z^{-\frac{1}{2}}\mathscr{C}_\nu(z)\}}{dz}$ is

$$\beta_1-\frac{\mu+7}{8\beta_1}-\frac{7\mu^2+154\mu+95}{384\beta_1^3}-\dots.$$

The zeros of

$$J_\nu(z)\,Y_\nu(kz)-Y_\nu(z)\,J_\nu(kz),$$

where k is constant, and of

$$J_\nu'(z)\,Y_\nu'(kz)-Y_\nu'(z)\,J_\nu'(kz)$$

have been treated in a similar manner by McMahon. Kalähne† has constructed tables of the zeros of the former function when k has the values 1·2, 1·5, 2·0 and ν is 0, $\frac{1}{2}$, 1, $\frac{3}{2}$, 2, $\frac{5}{2}$; while it has been proved by Carslaw, *Conduction of Heat* (London, 1922), p. 128, that these zeros are all real when ν and k are real. The zeros of $J_\nu'(z)\,Y_\nu(kz)-Y_\nu'(z)\,J_\nu(kz)$ have been examined by Sasaki, *Tôhoku Math. Journal*, v. (1914), pp. 45—47.

15·6. *The mode of variation of the zeros of a cylinder function when its order is varied.*

The equation in z

$$J_\nu(z)=0$$

has an infinite number of roots, the values of which depend on ν; since $J_\nu(z)$ is an analytic function of both z and ν, so long as $z\neq 0$, it follows that each root of the equation is (within certain limits) an analytic function of ν. A similar statement holds good when the function of the first kind is replaced by any cylinder function of the type $J_\nu(z)\cos\alpha-Y_\nu(z)\sin\alpha$, where α is any constant.

If j denotes any particular zero of $J_\nu(z)$, the rate of change of j, as ν varies, is given by the ordinary formula of partial differentiation

(1) $$J_\nu'(j)\frac{dj}{d\nu}+\left[\frac{\partial J_\nu(z)}{\partial\nu}\right]_{z=j}=0.$$

* *Annals of Math.* IX. (1895), pp. 25—29.

† *Zeitschrift für Math. und Phys.* LIV. (1907), pp. 55—86.

Since $J_\nu(j)=0$, it follows that $J_\nu'(j)=-J_{\nu+1}(j)\neq 0$, so long as j is not zero, and hence, from § 5·11 (15), when $R(\nu)>0$,

$$\frac{dj}{d\nu}=\frac{2\nu}{jJ^2_{\nu+1}(j)}\int_0^j J_\nu^2(t)\frac{dt}{t}. \tag{2}$$

This formula shews that *when ν is positive, the positive zeros of $J_\nu(x)$ increase as ν is increased.*

Equation (2) was stated without proof by Schläfli, *Math. Ann.* x. (1876), p. 137; and the deduction from it was established in a different manner by Gegenbauer*, *Mém. de la Soc. R. des Sci. de Liége*, (3) II. (1900), no. 3, in the case of the smallest zero of $J_\nu(x)$.

We proceed to extend the results already obtained to the positive zeros of

$$\mathscr{C}_\nu(z)\equiv J_\nu(z)\cos\alpha - Y_\nu(z)\sin\alpha,$$

where ν is an unrestricted real variable, and α is constant (i.e. independent of ν).

The extended theorem is as follows:

Any positive zero, c, of $\mathscr{C}_\nu(z)$ is definable as a continuous increasing function of the real variable ν.

To prove this theorem we observe that c is a function of ν such that

$$\arctan\left\{\frac{Y_\nu(c)}{J_\nu(c)}\right\}$$

is constant, so that

$$\frac{dc}{d\nu}\left[\frac{\partial}{\partial z}\arctan\left\{\frac{Y_\nu(z)}{J_\nu(z)}\right\}\right]_{z=c}+\left[\frac{\partial}{\partial\nu}\arctan\left\{\frac{Y_\nu(z)}{J_\nu(z)}\right\}\right]_{z=c}=0,$$

and therefore

$$\frac{2}{\pi c}\frac{dc}{d\nu}+\left[J_\nu(z)\frac{\partial Y_\nu(z)}{\partial\nu}-Y_\nu(z)\frac{\partial J_\nu(z)}{\partial\nu}\right]_{z=c}=0.$$

Hence, by § 13·73 (2), we have

$$\frac{dc}{d\nu}=2c\int_0^\infty K_0(2c\sinh t)\,e^{-2\nu t}\,dt. \tag{3}$$

Since the integrand is positive, this formula shews that c is an increasing function of ν.

A less general theorem, namely that, if c is a zero *which is greater than the order ν (supposed positive)*, then c is an increasing function of ν, has been proved by Schafheitlin† with the aid of very elaborate analysis.

It will be observed from the definition of $Y_\nu(z)$ that c tends to zero only when ν tends to *any negative value* which satisfies the equation

$$\sin(\alpha-\nu\pi)=0.$$

* The reader should note that the analysis in the latter part of Gegenbauer's memoir is vitiated by his use of Rudski's erroneous results (§ 15·1).

† *Berliner Sitzungsberichte*, v. (1906), pp. 82—93; *Jahresbericht der Deutschen Math. Vereinigung*, XVI. (1907), pp. 272—279.

It should be noticed that (3) shews that, when ν is taken to be a complex number and c is a (complex) number, with a *positive* real part, then c is an analytic function of ν; and so, as ν varies, the zeros of $\mathscr{C}_\nu(z)$ vary continuously, and they can only come into existence or disappear when c fails to be an analytic function of ν, i.e. when $c = 0$.

It follows that the positive zeros of $\mathscr{C}_\nu(z)$ are derived from those of $\mathscr{C}_{\frac{1}{2}}(z)$ by a process of continuous variation as ν varies, except that one positive zero disappears whenever ν passes through one of the specified negative values.

If we now choose α so that* $0 \leqslant \alpha < \pi$, we see that, as ν varies from $\frac{1}{2}$ to *any* value exceeding $(\alpha/\pi) - 1$, no zeros disappear during the process of variation of ν, and so in the case of zeros which are so large that the formula of Stokes' type (§ 15·53) is available, the formula

$$n\pi + \tfrac{1}{2}\nu\pi - \tfrac{1}{4}\pi - \alpha - \frac{4\nu^2 - 1}{8(n\pi + \frac{1}{2}\nu\pi - \frac{1}{4}\pi - \alpha)} - \dots$$

gives the *nth* positive zero, when the positive zeros are regarded as arranged in order of magnitude.

If, however, ν has varied so that it finally lies between $(\alpha/\pi) - k$ and $(\alpha/\pi) - k - 1$, where k is a positive integer, k zeros have disappeared, and so the formula just quoted gives the $(n-k)$th positive zero.

This type of argument is due to Macdonald, *Proc. London Math. Soc.* XXIX. (1898), pp. 575—584; it was applied by him to the discussion of the zeros of Bessel functions of the first kind of order exceeding -1.

If we draw the curve $\mathscr{C}_x(y) = 0$, it evidently consists of a number of branches starting from points on the negative half of the x-axis and moving upwards towards the right, both x and y increasing without limit on each branch.

If we take any point with *positive* coordinates (ν_0, y_0) and draw from it a line to the right and a line downwards terminated by the x-axis, it is evident that the curve $\mathscr{C}_x(y) = 0$ meets each of the lines in the same number of points. It follows that the number of zeros of $\mathscr{C}_\nu(y_0)$, *qua* function of ν, which exceed ν_0 is equal to the number of positive zeros of $\mathscr{C}_{\nu_0}(y)$ *qua* function of y which are less than y_0. This is a generalisation of a theorem due to Macdonald†, who took $\nu_0 = 0$ and the cylinder function to be a function of the first kind.

Fig. 33 illustrates the general shape of the curves $J_x(y) = 0$, the length of the sides of the squares being 2 units. A much larger and more elaborate diagram of the same character has been constructed by Gasser‡, who has also constructed the corresponding diagram for $Y_x(y) = 0$. The diagram for $\mathscr{C}_x(y) = 0$ is of the same general character as that for $J_x(y) = 0$, except that

* This does not lead to any real loss of generality.

† See a letter from Macdonald to Carslaw, *Proc. London Math. Soc.* (2) XIII. (1914), p. 239.

‡ *Bern Mittheilungen*, 1904, p. 135.

the portions of the curves below the axis of x consist merely of a number of isolated points on the lines on which $2x$ is an odd integer.

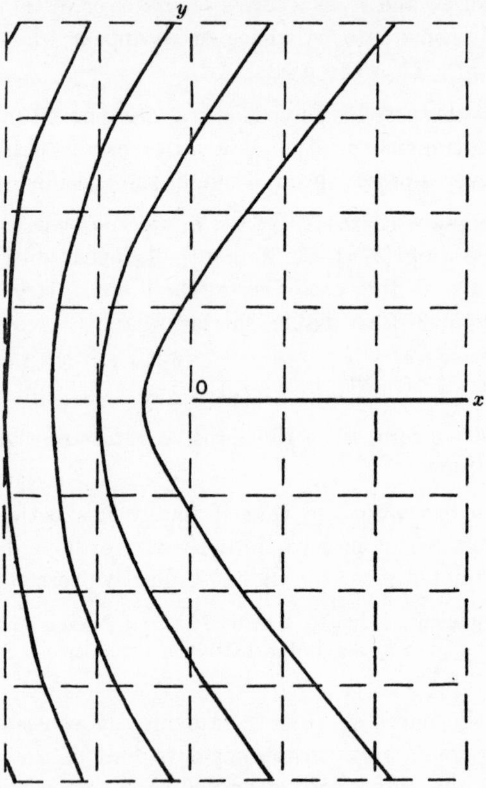

Fig. 33.

[NOTE. The reader will find it interesting to deduce from § 13·73 (3) that, if c' is a zero of $\mathscr{C}_\nu'(z)$, then

(4) $$\frac{dc'}{d\nu} = \frac{2c'}{c'^2 - \nu^2} \int_0^\infty (c'^2 \cosh 2t - \nu^2)\, K_0(2c' \sinh t)\, e^{-2\nu t}\, dt,$$

and hence, if the variables are real and $c' > |\nu| > 0$, then c' increases with ν.

The sign of $dc'/d\nu$ has also been discussed (by more elementary methods) by Schafheitlin, *Jahresbericht der Deutschen Math. Vereinigung*, XVI. (1907), pp. 272—279; but the analysis used by Schafheitlin is extremely complicated.]

15·61. *The problem of the vibrating membrane.*

The mode of increase of the zeros of $J_\nu(x)$ when ν is increased has been examined by Rayleigh* with the aid of arguments depending on properties of transverse vibrations of a membrane in the form of a circular sector. If the membrane is bounded by the lines $\theta = 0$ and $\theta = \pi/\nu$ (where $\nu > \frac{1}{2}$), and by

* *Phil. Mag.* (6) XXI. (1911), pp. 53—58 [*Scientific Papers*, VI. (1920), pp. 1—5]. Cf. *Phil. Mag.* (6) XXXII. (1916), pp. 544—546 [*Scientific Papers*, VI. (1920), pp. 444—446].

the circle $r = a$, and if the straight edges of the membrane are fixed, the displacement in a normal vibration is proportional to

$$J_\nu (rp/c) \sin \nu\theta \cos (pt + \epsilon),$$

where c is the velocity of propagation of vibrations. If the circular boundary of the membrane is fixed, the values of ap/c are the zeros of $J_\nu (x)$, while if the boundary is free to move transversely they are the zeros of $J_\nu' (x)$.

The effect of introducing constraints in the form of clamps which gradually diminish the effective angle of the sector is to increase ν and to shorten the periods of vibration, so that p is an increasing function of ν, and therefore (since a and c are unaltered) ap/c is an increasing function of ν. That is to say, the zeros of $J_\nu (x)$ and $J_\nu' (x)$ increase with ν.

By using arguments of this character, Rayleigh has given proofs of a number of theorems which are proved elsewhere in this chapter by analytical methods.

15·7. *The zeros of* $K_\nu (z)$.

The zeros of the function $K_\nu (z)$, where ν is a given positive number (zero included), and z lies in the domain in which $|\arg z| < \frac{3}{2}\pi$, have been studied qualitatively by Macdonald*.

From the generalisation of Bessel's integral, given in § 6·22, it is obvious that $K_\nu (z)$ has no positive zeros; and it has been shewn further by Macdonald that $K_\nu (z)$ has no zeros for which $|\arg z| \leqslant \frac{1}{2}\pi$. This may be proved at once from a consideration of the integral given in § 13·71; for, if $z = re^{i\alpha}$ were such a zero $(r > 0, -\frac{1}{2}\pi < \alpha < \frac{1}{2}\pi)$, then $z = re^{-i\alpha}$ would be another zero; but the integral shews that

$$K_\nu (re^{i\alpha}) K_\nu (re^{-i\alpha}) = \frac{1}{2} \int_0^\infty \exp \left\{ -\frac{v}{2} - \frac{r^2 \cos 2\alpha}{v} \right\} K_\nu \left(\frac{r^2}{v} \right) \frac{dv}{v}$$
$$> 0,$$

which is contrary to hypothesis.

If α is equal to $\pm \frac{1}{2}\pi$, we have

$$|K_\nu (re^{\pm \frac{1}{2}\pi i})| = \tfrac{1}{2}\pi \sqrt{\{J_\nu^2 (r) + Y_\nu^2 (r)\}},$$

and so $K_\nu (z)$ has no purely imaginary zeros.

Next we study the zeros for which $R(z)$ is negative, the phase of z lying either between $\frac{1}{2}\pi$ and π or between $-\frac{1}{2}\pi$ and $-\pi$.

It may be shewn that the total number of zeros in this pair of quadrants is the even integer† nearest to $\nu - \frac{1}{2}$, unless $\nu - \frac{1}{2}$ is an integer, in which case the number is $\nu - \frac{1}{2}$.

In the first place, there are no zeros on the lines $\arg z = \pm \pi$, unless $\nu - \frac{1}{2}$

* *Proc. London Math. Soc.* xxx. (1899), pp. 165—179.

† This is not the number given by Macdonald.

is an integer; for $K_\nu(re^{\pm\pi i}) = e^{\mp\nu\pi i} K_\nu(r) \mp \pi i I_\nu(r)$, and, if both the real and the imaginary parts of this expression are to vanish, we must have

$$\cos\nu\pi . K_\nu(r) = 0, \quad \sin\nu\pi . K_\nu(r) + \pi I_\nu(r) = 0.$$

Since the Wronskian of the pair of functions on the left of the equations is $(\pi/r)\cos\nu\pi$, they cannot vanish simultaneously unless $\cos\nu\pi = 0$.

Now consider the change in phase of $z^\nu K_\nu(z)$ as z describes a contour consisting of arcs of large and small circles terminated by the lines $\arg z = \pm\pi$, together with the parts of these lines terminated by the circular arcs. (Cf. Fig. 15 of § 7·4.)

If the circles be called Γ and γ, their equations being $|z| = R$ and $|z| = \delta$, it is evident that the number of zeros of $K_\nu(z)$ in the pair of quadrants under consideration is equal to the number of zeros of $z^\nu K_\nu(z)$ inside the contour, and this is equal to $1/(2\pi)$ times the change in phase of $z^\nu K_\nu(z)$ as z traverses the contour.

Now the change in phase is

$$\Big[\arg\{z^\nu K_\nu(z)\}\Big]_\Gamma - \Big[\arg\{z^\nu K_\nu(z)\}\Big]_\gamma + \Big[\arg\{z^\nu K_\nu(z)\}\Big]_{R\exp\pi i}^{\delta\exp\pi i} + \Big[\arg\{z^\nu K_\nu(z)\}\Big]_{\delta\exp(-\pi i)}^{R\exp(-\pi i)}$$

As $R \to \infty$ and $\delta \to 0$, the first two terms* tend to $2\pi(\nu - \tfrac{1}{2})$ and 0 respectively, because when $|z|$ is large or small on the contour,

$$z^\nu K_\nu(z) \sim z^{\nu-\frac{1}{2}} e^{-z}\sqrt{(\tfrac{1}{2}\pi)}, \quad z^\nu K_\nu(z) \sim 2^{\nu-1}\Gamma(\nu)$$

respectively†.

The last two terms become

$$\lim_{\left(\substack{\delta\to 0\\ R\to\infty}\right)} 2\left[\arc\tan\frac{\pi\cos\nu\pi . I_\nu(r)}{K_\nu(r) + \pi\sin\nu\pi . I_\nu(r)}\right]_\delta^R .$$

Now $K_\nu(r)$ is a positive decreasing function of r while $I_\nu(r)$ is a positive increasing function, and so the last denominator has one zero if $\sin\nu\pi$ is negative, and no zero if $\sin\nu\pi$ is positive.

If therefore we take the inverse function to vanish when $r \to 0$, its limit when $r \to \infty$ is $\arc\tan(\cot\nu\pi)$, the value assigned to the inverse function being numerically less than two right angles and having the same sign as the sign of $\cos\nu\pi$.

Hence the total number of zeros of $K_\nu(z)$ in the pair of quadrants‡ in which $R(z)$ is negative and $|\arg z| < \pi$ is

$$\nu - \tfrac{1}{2} + \frac{1}{\pi}\arc\tan(\cot\nu\pi),$$

* This is evident from the consideration that the asymptotic expansion of § 7·23 is valid when $|\arg z| \leqslant \pi$.

† The second of these approximate formulae requires modification when $\nu = 0$.

‡ The two zeros of $K_2(z)$ are not very far from the points $-1·29 \pm 0·44i$.

and the reader will find it easy to verify that this number is the even integer which is nearest to $\nu - \frac{1}{2}$.

When $\nu - \frac{1}{2}$ is an integer, $K_\nu(z)$ is a polynomial in z multiplied by a function with no zeros in the finite part of the plane, and so the number of zeros for which $R(z) < 0$ is exactly $\nu - \frac{1}{2}$.

Next consider the portion of the plane for which $\pi < \arg z \leqslant 2\pi$.

If we write $z = \zeta e^{\frac{3}{2}\pi i}$, we have

$$K_\nu(z) = -\tfrac{1}{2}\pi e^{-\frac{3}{2}\nu\pi i}[Y_\nu(\zeta) + i(1 + 2e^{2\nu\pi i}) J_\nu(\zeta)],$$

and so $K_\nu(z)$ has a sequence of zeros lying near the negative part of the imaginary axis. The zeros of large modulus which belong to this sequence are given approximately by the roots of the equation

$$\tan(\zeta - \tfrac{1}{2}\nu\pi - \tfrac{1}{4}\pi) = -i(1 + 2e^{2\nu\pi i});$$

it may be verified that they are ultimately on the right or left of the imaginary axis in the z-plane according as $\cos^2 \nu\pi$ is less than or greater than $\frac{1}{4}$; i.e. according as ν differs from the nearest integer by more or less than $\frac{1}{3}$. The sequence does not exist when $e^{2\nu\pi i} = -1$, i.e. when ν is half of an odd integer.

There is a corresponding sequence of zeros near the line $\arg z = -\frac{3}{2}\pi$.

15·8. *Zeros of Bessel functions of unrestrictedly large order.*

The previous investigations, based mainly on integrals of Poisson's type, have resulted in the determination of properties of zeros of Bessel functions, when the order ν is not unduly large. This is, of course, consistent with the fact that Hankel's asymptotic expansions, discussed in Chapter VII, are significant only when ν^2 is fairly small in comparison with the argument of the Bessel function.

The fact that Debye's integrals of § 8·31 afford representations of functions of large order suggests that these integrals may form an effective means of discussing the zeros of Bessel functions of large order; and this, in fact, proves to be the case*. Moreover, the majority of the results which will be obtained are valid for functions of *any* positive order, though they gain in importance with the increase of the order.

We shall adopt the notation of § 8·31, so that†

$$H_\nu^{(1)}(\nu \sec\beta) = \frac{e^{\nu i(\tan\beta - \beta)}}{\pi i}\int_{-\infty - i\beta}^{\infty + \pi i - i\beta} e^{-\nu\tau}\, dw,$$

where $-\tau = \sinh w - w + i\tan\beta(\cosh w - 1)$, and the contour in the plane of the complex variable w is chosen so that τ is positive on it.

* Watson, *Proc. Royal Soc.* XCIV. A, (1918), pp. 190—206.

† We shall use the symbols x and $\nu\sec\beta$ indifferently when $x \geqslant \nu$.

If $w = u + iv$, where u and v are real, u and v both increase steadily as w describes the contour, so that

$$\int_{-\infty}^{\infty} e^{-\nu\tau}\, du, \qquad \int_{-\beta}^{\pi-\beta} e^{-\nu\tau}\, dv$$

are both positive.

Hence, if we regard β as variable, and *define*

$$\arg \int_{-\infty - i\beta}^{\infty + \pi i - i\beta} e^{-\nu\tau}\, dw$$

to be a positive acute angle when $\beta = 0$, and to vary continuously with β, it will remain a positive acute angle for *all* values of β between 0 and $\frac{1}{2}\pi$; and, moreover, by § 8·32, it cannot exceed $\frac{1}{3}\pi$, since $dv/du \leqslant \sqrt{3}$.

This positive acute angle will be called χ, and then Ψ will be defined by the equation

$$\Psi = \nu(\tan\beta - \beta) + \chi - \tfrac{1}{2}\pi.$$

It is then evident that

$$H_\nu^{(1)}(\nu \sec\beta) = \mathfrak{M} e^{i\Psi},$$

where $\mathfrak{M}$ is positive (not zero); and

$$J_\nu(x) = \mathfrak{M} \cos\Psi, \quad Y_\nu(x) = \mathfrak{M} \sin\Psi.$$

If
$$\mathscr{C}_\nu(x) \equiv J_\nu(x)\cos\alpha - Y_\nu(x)\sin\alpha,$$

it is clear that the *only* zeros of $\mathscr{C}_\nu(x)$, greater than ν, are derived from the values of Ψ which make $\Psi + \alpha$ equal to an odd number of right angles.

It is easy to shew that Ψ increases with x, when ν remains constant. For we have

$$\Psi = \arc\tan \frac{Y_\nu(x)}{J_\nu(x)}, \quad \frac{d\Psi}{dx} = \frac{2/(\pi x)}{J_\nu^2(x) + Y_\nu^2(x)} > 0.$$

Hence, as x increases, Ψ increases steadily, and so, to each of the values of Ψ for which

$$\Psi = (m + \tfrac{1}{2})\pi - \alpha,$$

corresponds *one and only one* positive zero of $\mathscr{C}_\nu(x)$.

Next we shall prove that χ is also an increasing function of x. This is a theorem of a much deeper character, since the result of § 13·74 is required to prove it; we thence have

$$\frac{d\chi}{dx} = \frac{d\Psi}{dx} - \nu \frac{d(\tan\beta - \beta)}{dx} = \frac{2/(\pi x)}{J_\nu^2(x) + Y_\nu^2(x)} - \frac{\sqrt{(x^2 - \nu^2)}}{x} > 0.$$

From Hankel's asymptotic expansion it is clear that

$$\lim_{x\to\infty} \chi = \lim_{x\to\infty} [x - \tfrac{1}{2}\nu\pi - \tfrac{1}{4}\pi - \sqrt{(x^2 - \nu^2)} + \nu \arc\cos(\nu/x) + \tfrac{1}{2}\pi + O(1/x)]$$
$$= \tfrac{1}{4}\pi,$$

and so
$$\arc\tan\left\{\frac{J_\nu(\nu)}{-Y_\nu(\nu)}\right\} < \chi < \tfrac{1}{4}\pi,$$

in which the expression on the left is a positive acute angle.

To form an estimate of the value of χ when ν is large, we write

$$J_\nu(\nu) = -Y_\nu(\nu)\tan\gamma_\nu;$$

hence, from § 8·42, we have

$$\lim_{\nu\to\infty}\gamma_\nu = \tfrac{1}{6}\pi,$$

and, when ν is large,

$$\tan\gamma_\nu = \frac{1}{\sqrt{3}}\left\{1 - \frac{\Gamma(\frac{2}{3})}{210\,\Gamma(\frac{1}{3}).(\frac{1}{6}\nu)^{\frac{4}{3}}} + \dots\right\},$$

so that, when ν is large, γ_ν is an increasing function of ν.

The following Table gives the sexagesimal measure (to the nearest half minute) of the angle whose circular measure is γ_ν; it exhibits the closeness of γ_ν to its limit, even when ν is quite small:

ν	0	$\frac{1}{2}$	1	$\frac{3}{2}$	2	3	4	6	8	10	12	∞
γ_ν	0	28° 39′	29° 23½′	29° 38′	29° 45′	29° 51′	29° 54′	29° 56½′	29° 57½′	29° 58′	29° 58½′	30°

The table suggests that γ_ν is an increasing function of ν for all positive values of ν; to prove that this is the case we need the theorem that

$$J_\nu(\nu)\frac{dY_\nu(\nu)}{d\nu} - Y_\nu(\nu)\frac{dJ_\nu(\nu)}{d\nu} > 0;$$

and this inequality has already been established in § 13·74.

We are now in a position to infer that $\mathscr{C}_\nu(\nu\sec\beta)$ has a zero for which

$$\nu(\tan\beta - \beta) + \chi = m\pi - \alpha,$$

where m is any positive integer (unity possibly excepted); and therefore at the positive zeros of $\mathscr{C}_\nu(\nu\sec\beta)$, with the possible exception of the first, $\nu(\tan\beta - \beta)$ has a value between

$$(m - \tfrac{1}{4})\pi - \alpha \text{ and } m\pi - \gamma_\nu - \alpha,$$

and the difference of these expressions (when ν is not small) slightly exceeds $\frac{1}{12}\pi$.

From the result of § 15·3 (10), it follows that the phase of $H_\nu^{(1)}(X)\,e^{i\alpha}$ increases from $\alpha - \frac{1}{2}\pi$ to $\Psi + \alpha$ as X increases from 0 to any value exceeding ν, and so, if $0 < \alpha < \pi$, the number of zeros of $\mathscr{C}_\nu(X)$ in the interval $(0, x)$ is the greatest integer contained in $(\Psi + \alpha)/\pi + \frac{1}{2}$.

A simple theorem which may be noted here is that, when $\nu^2 > \frac{1}{4}$, it follows from § 13·74 that, if δ is any positive number, $\int_\delta^{\delta+\pi}\frac{dx}{d\Psi}\,d\Psi$ decreases as δ is increased. This result shews that the interval between consecutive zeros of $\mathscr{C}_\nu(x)$ decreases when the rank of the zeros is increased.

A different proof of this theorem is due to Porter; cf. § 15·82.

15·81. *The smallest zeros of $J_\nu(x)$ and $Y_\nu(x)$.*

It has been seen (§ 15·3) that $J_\nu(x)$ and $Y_\nu(x)$ have no zeros in the interval $(0, \nu)$, when ν is positive, and it is fairly obvious from the asymptotic formulae obtained in § 8·42 that they have no zeros of the form $\nu + o(\nu^{\frac{1}{3}})$ when ν is large.

The asymptotic formulae which were quoted in § 8·43 shew that, with the notation of § 15·8,

$$\Psi = \nu(\tan\beta - \beta) - \tfrac{1}{4}\pi + \text{arc tan}\,\frac{Q(\frac{1}{3}\nu\tan^3\beta, \frac{1}{3})}{P(\frac{1}{3}\nu\tan^3\beta, \frac{1}{3})} + O\left(\frac{1}{\sqrt{\nu}}\right),$$

where the inverse tangent denotes a negative acute angle.

Hence, at the smallest zero of $J_\nu(x)$,

$$\tan\{\nu(\tan\beta - \beta) - \tfrac{3}{4}\pi + O(1/\sqrt{\nu})\} = -\frac{Q(\frac{1}{3}\nu\tan^3\beta, \frac{1}{3})}{P(\frac{1}{3}\nu\tan^3\beta, \frac{1}{3})}.$$

As β increases from 0 to $\frac{1}{2}\pi$, the expression on the left increases from $1 + O(1/\sqrt{\nu})$, while the expression on the right decreases from 0·2679 to 0; the smallest root occurs for a value of β for which $\nu(\tan\beta - \beta)$ lies between $\frac{3}{4}\pi$ and $\frac{5}{4}\pi$, so that

$$\nu(\tan\beta - \beta) = \tfrac{1}{3}\nu\tan^3\beta + O(\nu^{-\frac{2}{3}}).$$

Hence, if we solve the equation

$$\tan(\xi - \tfrac{3}{4}\pi) = -Q(\xi, \tfrac{1}{3})/P(\xi, \tfrac{1}{3}),$$

the value of ξ so obtained is the value of $\frac{1}{3}\nu\tan^3\beta$ at the zero with an error which is $O(\nu^{-\frac{2}{3}})$. The value of ξ is approximately 2·383447, and hence the smallest positive zero of $J_\nu(x)$ is

$$\nu + \nu^{\frac{1}{3}} \times 1{\cdot}855757 + O(1).$$

In like manner, by solving the equation

$$\tan(\xi - \tfrac{1}{4}\pi) = -Q(\xi, \tfrac{1}{3})/P(\xi, \tfrac{1}{3}),$$

of which the smallest root is approximately $\xi = 0{\cdot}847719$, we find that the smallest zero of $Y_\nu(x)$ is

$$\nu + \nu^{\frac{1}{3}} \times 0{\cdot}931577 + O(1).$$

The formula for the smallest zero of $J_\nu(x)$ has been given by Airey, *Phil. Mag.* (6) XXXIV. (1917), p. 193. Airey's formula was derived by using Debye's asymptotic expansion of § 8·42 for $J_\nu(x)$ when x has a value such that

$$x - \nu = O(\nu^{\frac{1}{3}}) \neq o(\nu^{\frac{1}{3}}).$$

For such values of the variables, it has not been proved that Debye's expansion is valid, and although Airey's method gives the two dominant terms of the smallest zero of $J_\nu(x)$ correctly, the numerical result which Airey gives for the smallest zero of $J_\nu'(x)$ is not the same as that of § 15·83. The reason why Airey's method gives correct results is that $J_\nu(\nu + \zeta)$ is expansible *in powers of* ζ so long as ζ is $o(\nu)$, and in this expansion it is permissible to substitute Debye's formulae for $J_\nu(\nu)$, $J_\nu'(\nu)$, $J_\nu''(\nu)$, ...

A formula for the smallest zero of $J_{-\nu}(x)$ was given by Airey. This zero may lie anywhere between 0 and the smallest zero of $J_\nu(x)$, according to the value of ν.

It does not seem to be possible to make further progress by the methods used in this section. We shall now make a digression to explain the methods of Sturm (which have been applied to Bessel's equation by various mathematicians), and we shall then give an investigation which leads to the fascinating result that the two expressions which, in this section, were proved to be $O(1)$ are in reality $O(\nu^{-\frac{1}{3}})$, so that approximations are obtained for the smallest zeros of $J_\nu(x)$ and $Y_\nu(x)$ in which the errors are $O(\nu^{-\frac{1}{3}})$, i.e. the errors become negligible when ν is large.

[NOTE. An elementary result concerning the smallest zero of $J_\nu(x)$ has been obtained from the formula of § 5·43 by Gegenbauer, *Wiener Sitzungsberichte*, CXI. (2a), (1902), p. 571; if $\mu = \nu + \epsilon$ where $0 < \epsilon < 1$, then the smallest zero of $J_{2\nu+\epsilon}(x)$ is less than twice the smallest zero of $J_\nu(x)$, because, for the latter value of x the integrand cannot be one-signed.]

15·82. *Applications of Sturm's methods.*

Various writers have discussed properties of Bessel functions by means of the general methods invented by Sturm* for the investigation of any linear differential equation of the second order. The results hitherto obtained in this manner are of some interest, though they are not of a particularly deep character, and most of them have already been proved in this chapter by other methods.

The theorem which is at the base of the investigations in question is that, given a differential equation of the second order in its normal form

$$\frac{d^2u}{dx^2} + Iu = 0,$$

in which the invariant I is positive, then the greater the value of I, the more rapidly do the solutions of the equation oscillate as x increases.

As an example of an application of this result, we may take a theorem due to Sturm (*ibid.* pp. 174—175) and Bourget, *Ann. sci. de l'École norm. sup.* III. (1866), p. 72, that, if $\nu^2 - \frac{1}{4}$ be positive and c be any zero of $\mathscr{C}_\nu(x)$ which exceeds $\sqrt{(\nu^2 - \frac{1}{4})}$, then the zero of $\mathscr{C}_\nu(x)$ which is next greater than c does not exceed $c + \dfrac{\pi c}{\sqrt{(c^2 - \nu^2 + \frac{1}{4})}}$.

This result follows at once from the consideration of the facts that the function $x^{\frac{1}{2}}\mathscr{C}_\nu(x)$ is annihilated by the operator (§ 4·3)

$$\frac{d^2}{dx^2} + \left(1 - \frac{\nu^2 - \frac{1}{4}}{x^2}\right),$$

and that, when $x \geqslant c$,

$$1 - \frac{\nu^2 - \frac{1}{4}}{x^2} \geqslant \frac{c^2 - \nu^2 + \frac{1}{4}}{c^2}.$$

A slightly more abstruse result is due to Porter, *American Journal of Math.* XX. (1898), pp. 196—198, to the effect that, if $\nu^2 > \frac{1}{4}$ and if the zeros of $\mathscr{C}_\nu(x)$, greater than $\sqrt{(\nu^2 - \frac{1}{4})}$, in ascending order of magnitude are $c_1, c_2, c_3, \ldots$ then $c_{n+1} - c_n$ decreases as n increases. This has already been proved in § 15·8 by another method.

Other theorems of like nature are due to Bôcher, *Bulletin American Math. Soc.* III. (1897), pp. 205—213; VII. (1901), pp. 333—340; and to Gasser, *Bern Mittheilungen*, 1904, pp. 92—135.

* *Journal de Math.* I. (1836), pp. 106—186; an account of recent researches on differential equations by Sturm's methods is given in a lecture by Bôcher, *Proc. Int. Congress of Math.* I. (Cambridge, 1912), pp. 163—195.

15·83. *Applications of Sturm's methods to functions of large order.*

We proceed to establish a number of results concerning cylinder functions of large order which are based on the following theorem of Sturm's type:

Let $u_1(x)$ and $u_2(x)$ be solutions of the equations

$$\frac{d^2u_1}{dx^2}+I_1u_1=0, \quad \frac{d^2u_2}{dx^2}+I_2u_2=0$$

such that, when $x=a$,

$$u_1(a)=u_2(a), \quad u_1'(a)=u_2'(a),$$

and let I_1 and I_2 be continuous in the interval $a \leqslant x \leqslant b$, and also let $u_1'(x)$ and $u_2'(x)$ be continuous in the same interval.

Then, if $I_1 \geqslant I_2$ throughout the interval, $|u_2(x)|$ exceeds $|u_1(x)|$ so long as x lies between a and the first zero of $u_1(x)$ in the interval, so that the first zero of $u_1(x)$ in the interval is on the left of the first zero of $u_2(x)$.*

Further, if $u_1'(a)$ has the same sign as $u_1(a)$, the first maximum point of $|u_1(x)|$ in the interval is on the left of the first maximum point of $|u_2(x)|$, and, moreover

$$\max|u_1(x)|<\max|u_2(x)|.$$

To prove the theorem†, observe that, so long as $u_1(x)$ and $u_2(x)$ are both positive,

$$u_1\frac{d^2u_2}{dx^2}-u_2\frac{d^2u_1}{dx^2}=(I_1-I_2)\,u_1u_2 \geqslant 0,$$

and so, when we integrate,

$$\left[u_1\frac{du_2}{dx}-u_2\frac{du_1}{dx}\right]_a^x \geqslant 0.$$

Since the expression now under consideration vanishes at the lower limit, we have

$$u_1\frac{du_2}{dx}-u_2\frac{du_1}{dx} \geqslant 0. \tag{1}$$

Hence we have

$$\frac{d\,(u_2/u_1)}{dx} \geqslant 0,$$

and therefore

$$\left[\frac{u_2}{u_1}\right]_a^x \geqslant 0,$$

that is to say,

$$\frac{u_2(x)}{u_1(x)} \geqslant \frac{u_2(a)}{u_1(a)}=1. \tag{2}$$

* To simplify the presentation of the proof of the theorem, it is convenient to change the signs of $u_1(x)$ and $u_2(x)$, if necessary, so that $u_1(x)$ is positive immediately on the right of $x=a$; the signs indicating moduli may then be omitted throughout the enunciation.

† The theorem is practically due to Sturm, *Journal de Math.* I. (1836), pp. 125—127, 145—147.

It follows that just before $u_1(x)$ vanishes for the first time $u_2(x)$ is still positive, and it has remained positive while x has increased from the value a.

The first part of the theorem is therefore proved.

Again, if $u_1'(a)$ is positive, as well as $u_1(a)$, then $u_1(x)$ must have a maximum before it vanishes, and at this point, μ_1, we have from (1)

$$u_1(\mu_1)\, u_2'(\mu_1) \geqslant 0,$$

so that $u_2'(\mu_1)$ is positive and $u_2'(x)$ must be positive in the interval $(0, \mu_1)$. Therefore the first maximum point μ_2 of $u_2(x)$ must be on the right of μ_1.

Finally we have

$$\max u_1(x) = u_1(\mu_1) \leqslant u_2(\mu_1) \leqslant u_2(\mu_2) = \max u_2(x),$$

and the theorem is completely proved.

When two functions, $u_1(x)$ and $u_2(x)$, are related in the manner postulated in this theorem, it is convenient to say that $u_1(x)$ is *more oscillatory** than $u_2(x)$ and that $u_2(x)$ is *less oscillatory* than $u_1(x)$.

We shall now apply the theorem just proved to obtain results† concerning $J_\nu(x)$ and $Y_\nu(x)$ when ν is large and $x - \nu$ is $O(\nu^{\frac{1}{3}})$. Our procedure will be to construct pairs of functions which are respectively slightly less and slightly more oscillatory than the functions in question.

In the first place we reduce Bessel's equation to its normal form by writing $x = \nu e^\theta$; we then have

$$\left[\frac{d^2}{d\theta^2} + \nu^2 (e^{2\theta} - 1)\right] \mathscr{C}_\nu(\nu e^\theta) = 0. \tag{3}$$

A function which is obviously slightly less oscillatory than $\mathscr{C}_\nu(\nu e^\theta)$ for small positive values of θ is obtainable by solving the equation

$$\left[\frac{d^2}{d\theta^2} + 2\nu^2\theta\right] u = 0, \tag{4}$$

since $e^{2\theta} - 1 \geqslant 2\theta$ when $\theta \geqslant 0$.

The general solution of (4) is

$$u = \theta^{\frac{1}{2}} \overline{\mathscr{C}}_{\frac{1}{3}}\left(\frac{2^{\frac{3}{2}}\nu\theta^{\frac{3}{2}}}{3}\right);$$

and the constants implied in this cylinder function have to be adjusted so that u and its differential coefficient are equal to $\mathscr{C}_\nu(\nu e^\theta)$ and its differential coefficient at $\theta = 0$.

* The reason for the use of these terms is obvious from a consideration of the special case in which I_1 and I_2 are positive constants.

† These results supersede the inequalities obtained by Watson, *Proc. London Math. Soc.* (2) XVI. (1917), pp. 166—169.

It follows that a function which is (slightly) less oscillatory than $\mathscr{C}_\nu(x)$, when $x \geqslant \nu$, is

$$(2\theta)^{\frac{1}{2}}\left[\Gamma(\tfrac{2}{3})(\tfrac{1}{6}\nu)^{\frac{1}{3}}\mathscr{C}_\nu(\nu)J_{-\frac{1}{3}}\left(\frac{2^{\frac{3}{2}}\nu\theta^{\frac{3}{2}}}{3}\right)+\Gamma(\tfrac{1}{3})(\tfrac{1}{6}\nu)^{\frac{2}{3}}\mathscr{C}_\nu'(\nu)J_{\frac{1}{3}}\left(\frac{2^{\frac{3}{2}}\nu\theta^{\frac{3}{2}}}{3}\right)\right].$$

We now endeavour to construct a function which is (slightly) more oscillatory than $\mathscr{C}_\nu(x)$, in order that we may have $\mathscr{C}_\nu(x)$ trapped between two functions which are more easily investigated than $\mathscr{C}_\nu(x)$.

The formula for the less oscillatory function, combined with the result stated in § 8·43, suggests that we should construct a function of the type*

$$\sqrt{\{\psi(\theta)/\psi'(\theta)\}}\,.\,\overline{\mathscr{C}}_{\frac{1}{3}}\{\nu\psi(\theta)\},$$

where $\psi(\theta)$ is a function of θ to be determined. It might be anticipated from § 8·43 that the suitable form for $\psi(\theta)$ would be $\frac{1}{3}\tan^3\beta$, where $\sec\beta = e^\theta$; but it appears that this function leads to a differential equation whose solution is such that its degree of oscillation depends on the relative values of ν and θ, and we are not able to obtain any information thereby.

The invariant of the equation determined by

$$\sqrt{\{\psi(\theta)/\psi'(\theta)\}}\,.\,\overline{\mathscr{C}}_{\frac{1}{3}}\{\nu\psi(\theta)\}$$

is known to be (§ 4·31)

$$\frac{1}{2}\frac{\psi'''(\theta)}{\psi'(\theta)}-\frac{3}{4}\left\{\frac{\psi''(\theta)}{\psi'(\theta)}\right\}^2+\frac{5}{36}\left\{\frac{\psi'(\theta)}{\psi(\theta)}\right\}^2+\nu^2\{\psi'(\theta)\}^2,$$

and it is requisite that this should slightly exceed $\nu^2(e^{2\theta}-1)$. It is consequently natural to test the value of $\psi(\theta)$ which is given by the equations

$$\psi'(\theta)=\sqrt{(e^{2\theta}-1)},\quad \psi(0)=0,$$

by determining whether, for this value of $\psi(\theta)$,

$$\frac{1}{2}\frac{\psi'''(\theta)}{\psi'(\theta)}-\frac{3}{4}\left\{\frac{\psi''(\theta)}{\psi'(\theta)}\right\}^2+\frac{5}{36}\left\{\frac{\psi'(\theta)}{\psi(\theta)}\right\}^2\geqslant 0.$$

When we replace e^θ by $\sec\beta$, we find that

$$\psi'(\theta)=\tan\beta,\qquad \psi(\theta)=\tan\beta-\beta,$$

$$\psi''(\theta)=\sec\beta\operatorname{cosec}\beta,\quad \psi'''(\theta)=-\frac{4\cos 2\beta}{\tan\beta\sin^2 2\beta},$$

and hence we have to test the truth of the inequality

$$(5)\qquad 3(\tan\beta-\beta)\leqslant\frac{\sin^3\beta}{\cos^2\beta\sqrt{(1+\frac{1}{5}\tan^2\beta)}}.$$

Now $$\frac{d}{d\beta}\left[3(\tan\beta-\beta)-\frac{\sin^3\beta}{\cos^2\beta\sqrt{(1+\frac{1}{5}\tan^2\beta)}}\right]$$

is negative when $\tan^2\beta<\sqrt{24}-3$, and it is positive for greater values of $\tan^2\beta$.

* The multiple of the cylinder function is taken so that the product satisfies a differential equation in its normal form; cf. § 4·31 (17).

Hence, since (5) is true when $\beta = 0$, it is true when $0 \leqslant \beta \leqslant \beta_0$, where β_0 is a certain angle between $\arctan \sqrt{\{\sqrt{24} - 3\}}$ and $\frac{1}{2}\pi$ The sexagesimal measure of β_0 is 59° 39′ 24″·27.

Proceeding as in the former case, we find that the function

$$\sqrt{\{3(1-\beta\cot\beta)\}}\,.\,[\Gamma(\tfrac{2}{3})(\tfrac{1}{6}\nu)^{\frac{1}{3}}\mathscr{C}_\nu(\nu)\,J_{-\frac{1}{3}}\{\nu(\tan\beta-\beta)\}$$
$$+\,\Gamma(\tfrac{1}{3})(\tfrac{1}{6}\nu)^{\frac{2}{3}}\mathscr{C}_\nu'(\nu)\,J_{\frac{1}{3}}\{\nu(\tan\beta-\beta)\}]$$

is slightly more oscillatory than $J_\nu(\nu\sec\beta)$, so long as* $0 \leqslant \beta \leqslant \beta_0$.

We can now obtain an extremely important result concerning the smallest zero of $\mathscr{C}_\nu(x)$ which is greater than ν; for let $\frac{1}{3}\lambda_\nu^3$ be the smallest value of ξ which makes

$$\Gamma(\tfrac{2}{3})(\tfrac{1}{6}\nu)^{\frac{1}{3}}\mathscr{C}_\nu(\nu)\,J_{-\frac{1}{3}}(\xi) + \Gamma(\tfrac{1}{3})(\tfrac{1}{6}\nu)^{\frac{2}{3}}\mathscr{C}_\nu'(\nu)\,J_{\frac{1}{3}}(\xi)$$

vanish. Then *both* of the equations

$$2\theta = \lambda_\nu^2/\nu^{\frac{2}{3}}, \quad \nu(\tan\beta-\beta) = \tfrac{1}{3}\lambda_\nu^3$$

give

$$x = \nu + \tfrac{1}{2}\lambda_\nu^2\nu^{\frac{1}{3}} + O(\nu^{-\frac{1}{3}}).$$

Since, by Sturm's theorem, the zero of $\mathscr{C}_\nu(x)$ lies between two expressions of this form, we see that *the value of the zero of $\mathscr{C}_\nu(x)$ which is next greater than ν is expressible in the form*

$$\nu + \tfrac{1}{2}\lambda_\nu^2\nu^{\frac{1}{3}} + O(\nu^{-\frac{1}{3}}).$$

When $\mathscr{C}_\nu(x)$ is equal to $J_\nu(x)$ it is easy to verify from a Table of Bessel functions of orders $\pm\frac{1}{3}$ that

$$\lambda_\nu = 1{\cdot}926529 + O(\nu^{-\frac{2}{3}}),$$

and so the smallest zero of $J_\nu(x)$, when ν is large, is

$$\nu + \nu^{\frac{1}{3}} \times 1{\cdot}855757 + O(\nu^{-\frac{1}{3}}).$$

In like manner, the smallest zero of $Y_\nu(x)$ is

$$\nu + \nu^{\frac{1}{3}} \times 0{\cdot}931577 + O(\nu^{-\frac{1}{3}}).$$

The first maximum of $J_\nu(x)$ may be obtained in a similar manner, by differentiating† the two expressions constructed as approximations.

The result is that if $\frac{1}{3}\mu_\nu^3$ is the smallest value of ξ which makes

$$\Gamma(\tfrac{2}{3})(\tfrac{1}{6}\nu)^{\frac{1}{3}}J_\nu(\nu)\,J_{\frac{2}{3}}(\xi) - \Gamma(\tfrac{1}{3})(\tfrac{1}{6}\nu)^{\frac{2}{3}}J_\nu'(\nu)\,J_{-\frac{2}{3}}(\xi)$$

vanish‡, then the first maximum of $J_\nu(x)$ is at the point

$$\nu + \tfrac{1}{2}\mu_\nu^2\nu^{\frac{1}{3}} + O(\nu^{-\frac{1}{3}}),$$

i.e. at the point

$$\nu + \nu^{\frac{1}{3}} \times 0{\cdot}808618 + O(\nu^{-\frac{1}{3}}).$$

The first maximum of the function $Y_\nu(x)$ cannot be treated in this manner because its first maximum is on the right of its first zero; this follows at once from § 15·3, because $Y_\nu(x)$ increases from $-\infty$ to 0 as x increases from 0 to the first zero.

For an investigation of the maximum value of $J_\nu(x)$ *qua* function of ν the reader should consult a paper by Meissel, *Astr. Nach.* CXXVIII. (1891), cols. 435—438.

* This restriction is trivial because we have to consider values of β for which $\nu\beta^3$ is bounded; i.e. *small* values of β.

† The permissibility of this follows from the second part of Sturm's theorem just given.

‡ This value of ξ is approximately 0·685548.

CHAPTER XVI

NEUMANN SERIES AND LOMMEL'S FUNCTIONS OF TWO VARIABLES

16·1. *The definition of Neumann series.*

The object of this chapter and of Chapter XVII is the investigation of various types of expansions of *analytic functions of complex variables* in series whose general terms contain one or more Bessel functions or related functions. These expansions are to some extent analogous to the well known expansions of an analytic function by the theorems of Taylor and Laurent. The expansions analogous to Fourier's expansion of a function of a real variable are of a much more recondite character, and they will be discussed in Chapters XVIII and XIX.

Any series of the type

$$\sum_{n=0}^{\infty} a_n J_{\nu+n}(z)$$

is called a ***Neumann series***, although in fact Neumann considered* only the special type of series for which ν is an integer; the investigation of the more general series is due to Gegenbauer†.

To distinguish these series from the types discussed in § 16·14, the description 'Neumann series of the first kind' has been suggested by Nielsen, *Math. Ann.* LV. (1902), p. 493.

The reader will remember that various expansions of functions as Neumann series have already been discussed in Chapter V. It will be sufficient to quote here the following formulae:

$$(\tfrac{1}{2}z)^{\mu} = \sum_{n=0}^{\infty} \frac{(\mu+2n)\,.\,\Gamma(\mu+n)}{n!} J_{\mu+2n}(z),$$

$$J_\nu(z+t) = \sum_{m=-\infty}^{\infty} J_{\nu-m}(t)\, J_m(z),$$

$$\frac{J_\nu(\varpi)}{\varpi^\nu} = 2^\nu\, \Gamma(\nu) \sum_{m=0}^{\infty} (\nu+m) \frac{J_{\nu+m}(Z)}{Z^\nu} \frac{J_{\nu+m}(z)}{z^\nu} C_m^{\nu}(\cos\phi),$$

where $\varpi^2 = Z^2 + z^2 - 2Zz\cos\phi$.

We shall first discuss the possibility of expanding an arbitrary function into a Neumann series; then we shall investigate the singularities of the analytic function defined by a Neumann series with given coefficients; and finally we shall discuss the expansions of various particular functions.

For a very general discussion of generalisations of all kinds of series of Bessel functions, the reader may consult memoirs by Nielsen, *Journal für Math.* CXXXII. (1907), pp. 138—146; *Leipziger Berichte*, LXI. (1909), pp. 33—61.

* *Theorie der Bessel'schen Functionen* (Leipzig, 1867), pp. 33—35.

† *Wiener Sitzungsberichte*, LXXIV. (2), (1877), pp. 125—127.

Various expansions of types which resemble Neumann's (other than those given in this chapter) are due to H. A. Webb, *Phil. Trans. of the Royal Soc.* CCIV. (1905), p. 487 and Nielsen, *Atti della R. Accad. dei Lincei*, (5) XV. (1906), pp. 490—497.

16·11. *Neumann's expansion* of an arbitrary function in a series of Bessel coefficients.*

Let $f(z)$ be a function of z which is analytic inside and on a circle of radius R with centre at the origin. If C denotes the contour formed by this circle and if z is any point inside it, it follows from Cauchy's theorem that

$$f(z)=\frac{1}{2\pi i}\int_C \frac{f(t)}{t-z}\,dt.$$

Now, by § 9·1,

$$\frac{1}{t-z}=\sum_{n=0}^{\infty}\epsilon_n O_n(t)\,.\,J_n(z);$$

and this expansion converges uniformly on the contour. It follows at once that

(1) $$f(z)=\sum_{n=0}^{\infty} a_n J_n(z),$$

where

(2) $$a_n=\frac{\epsilon_n}{2\pi i}\int_C f(t)\,O_n(t)\,dt;$$

and this is Neumann's expansion.

If the Maclaurin expansion of $f(z)$ is

$$f(z)=\sum_{n=0}^{\infty} b_n z^n,$$

we see that

$$\begin{aligned} a_n &= \frac{\epsilon_n}{2\pi i}\int_C O_n(t)\left\{\sum_{m=0}^{\infty} b_m t^m\right\}dt \\ &= \frac{\epsilon_n}{2\pi i}\sum_{m=0}^{\infty} b_m \int_C t^m O_n(t)\,dt \\ &= \sum_{m=0}^{\leqslant \frac{1}{2}n} b_{n-2m}\,.\,2^{n-2m}\frac{n\,.\,(n-m-1)!}{m!} \qquad (n\geqslant 1) \end{aligned}$$

and so

(3) $$\begin{cases} a_0=b_0, \\ a_n=n\displaystyle\sum_{m=0}^{\leqslant\frac{1}{2}n} 2^{n-2m}\frac{(n-m-1)!}{m!}b_{n-2m}. \qquad (n\geqslant 1) \end{cases}$$

This result shews that the Neumann series corresponding to a given function assumes a simple form whenever a simple expression can be found for the sum

$$\sum_{m=0}^{\leqslant\frac{1}{2}n} 2^{n-2m}\frac{(n-m-1)!}{m!}b_{n-2m}.$$

The construction of the Neumann series when the Maclaurin expansion is given is consequently now merely a matter of analytical ingenuity.

* *Theorie der Bessel'schen Functionen* (Leipzig, 1867), pp. 33—35. See also König, *Math. Ann.* v. (1872), pp. 338—340.

16·12. *Neumann's* analogue of Laurent's theorem.*

Let $f(z)$ be a function of z which is analytic and one-valued in the ring-shaped region defined by the inequalities

$$r \leqslant |z| \leqslant R.$$

Let C and c be the contours formed by the circles

$$|z| = R, \quad |z| = r,$$

both contours being taken counter-clockwise; then, if z be a point of the region between the circles, we have

$$\begin{aligned} f(z) &= \frac{1}{2\pi i}\int_C \frac{f(t)\,dt}{t-z} + \frac{1}{2\pi i}\int_c \frac{f(t)\,dt}{z-t} \\ &= \sum_{n=0}^{\infty} \frac{\epsilon_n}{2\pi i} J_n(z)\int_C f(t)\,O_n(t)\,dt + \sum_{n=0}^{\infty} \frac{\epsilon_n}{2\pi i} O_n(z)\int_c f(t)\,J_n(t)\,dt. \end{aligned}$$

Consequently $f(z)$ is expansible in the form

(1) $$f(z) = \sum_{n=0}^{\infty} a_n J_n(z) + \sum_{n=0}^{\infty} a_n' O_n(z),$$

where

(2) $$a_n = \frac{\epsilon_n}{2\pi i}\int_C f(t)\,O_n(t)\,dt, \quad a_n' = \frac{\epsilon_n}{2\pi i}\int_c f(t)\,J_n(t)\,dt.$$

If the Laurent expansion of $f(z)$ in the annulus is

$$f(z) = \sum_{n=0}^{\infty} b_n z^n + \sum_{n=1}^{\infty} \frac{b_n'}{z^n},$$

we have, as in § 16·11,

(3) $$\begin{cases} a_0 = b_0, \\ a_n = n \sum\limits_{m=0}^{\leqslant \frac{1}{2}n} 2^{n-2m} \dfrac{(n-m-1)!}{m!} b_{n-2m}, & (n \geqslant 1). \end{cases}$$

(4) $$a_n' = \sum_{m=0}^{\infty} \frac{(-)^m}{2^{n+2m}\, m!\,(n+m)!} b'_{n+2m+1}.$$

16·13. *Gegenbauer's generalisation of Neumann's expansion.*

By using the polynomial $A_{n,\nu}(t)$ defined in § 9·2, Gegenbauer† has generalised the formula given in § 16·11.

If $f(z)$ is analytic inside and on the circle $|z| = R$, and if C denotes the contour formed by this circle, we have

$$\begin{aligned} z^\nu f(z) &= \frac{1}{2\pi i}\int_C \frac{z^\nu f(t)\,dt}{t-z} \\ &= \frac{1}{2\pi i}\int_C \left\{\sum_{n=0}^{\infty} J_{\nu+n}(z)\,A_{n,\nu}(t)\right\} f(t)\,dt, \end{aligned}$$

and so

(1) $$z^\nu f(z) = \sum_{n=0}^{\infty} a_n J_{\nu+n}(z),$$

* *Theorie der Bessel'schen Functionen* (Leipzig, 1867), pp. 36—39.

† *Wiener Sitzungsberichte*, LXXIV. (2), (1877), pp. 124—130. See *Wiener Denkschriften*, XLVIII. (1884), pp. 293—316 for some special cases of the expansion.

where

(2) $$a_n = \frac{1}{2\pi i}\int_C f(t)\, A_{n,\nu}(t)\, dt,$$

provided only that ν is not a negative integer.

If, as in § 16·11, the Maclaurin expansion of $f(z)$ is

$$f(z) = \sum_{n=0}^{\infty} b_n z^n,$$

then

(3) $$a_n = (\nu + n) \sum_{m=0}^{\leqslant \frac{1}{2}n} 2^{\nu+n-2m} \frac{\Gamma(\nu + n - m)}{m!} b_{n-2m}.$$

Neumann's expansion of § 16·12 may be generalised in a similar manner.

16·14. *The Neumann-Gegenbauer expansion of a function as a series of squares or products.*

From the expansion of § 9·5, namely

$$\frac{z^{\mu+\nu}}{t-z} = \sum_{n=0}^{\infty} B_{n;\mu,\nu}(t)\, J_{\mu+\frac{1}{2}n}(z)\, J_{\nu+\frac{1}{2}n}(z),$$

which is valid when $|z| < |t|$, we can at once infer that, if $f(z)$ is analytic when $|z| \leqslant r$, then the expansion

(1) $$z^{\mu+\nu} f(z) = \sum_{n=0}^{\infty} a_n J_{\mu+\frac{1}{2}n}(z)\, J_{\nu+\frac{1}{2}n}(z)$$

is valid when $|z| < r$, and the coefficients are given by the formula

(2) $$a_n = \frac{1}{2\pi i}\int_C f(t)\, B_{n;\mu,\nu}(t)\, dt,$$

C being the contour formed by the circle $|z| = r$. This expansion is due to Gegenbauer*; an expansion closely connected with this, namely that

(3) $$f(z) = \sum_{n=0}^{\infty} a_n' J_n^2(z),$$

where

(4) $$a_n' = \frac{\epsilon_n}{2\pi i}\int_C t f(t)\, \Omega_n(t)\, dt,$$

and $\Omega_n(t)$ is Neumann's second polynomial (§ 9·4), is valid *provided that* $f(z)$ *is an even analytic function*; this expansion was obtained by Neumann†.

Gegenbauer's formula has been investigated more recently by Nielsen, *Nouv. Ann. de Math.* (4) II. (1902), pp. 407—410.

A type of series slightly different from those previously considered is derived from the formula of § 5·22 (7) in the form

$$z^\nu = 2^\nu\, \Gamma(\tfrac{1}{2}\nu + 1) \sum_{n=0}^{\infty} \frac{(\frac{1}{2}z)^{\frac{1}{2}\nu+n}}{n!} J_{\frac{1}{2}\nu+n}(z),$$

* *Wiener Sitzungsberichte*, LXXV. (2), (1877), pp. 218—222.

† *Math. Ann.* III. (1871), p. 599.

which shews that

$$\sum_{n=0}^{\infty} b_n z^{\nu+n} = \sum_{n=0}^{\infty} a_n (\tfrac{1}{2}z)^{\frac{1}{2}(\nu+n)} J_{\frac{1}{2}(\nu+n)}(z), \tag{5}$$

where

$$a_n = \sum_{m=0}^{\leq \frac{1}{2}n} \frac{2^{\nu+n-2m}\,\Gamma(\tfrac{1}{2}\nu + \tfrac{1}{2}n - m + 1)}{m!} b_{n-2m}. \tag{6}$$

Expansions of this type have been the topic of a detailed investigation by Nielsen*.

16·2. *Pincherle's theorem and its generalisations.*

Let $\sum_{n=0}^{\infty} a_n J_{\nu+n}(z)$ be any Neumann series, and let the function defined by this series and its analytic continuations be called $f(z)$.

Let also

$$\sum_{n=0}^{\infty} \frac{a_n (\tfrac{1}{2}z)^{\nu+n}}{\Gamma(\nu+n+1)} \equiv f(z)_N.$$

The function defined by $f(z)_N$ and its analytic continuations will be called the *associated power series* of $f(z)$.

The Neumann series converges throughout the domain in which

$$\overline{\lim_{n\to\infty}} \sqrt[n]{|\{a_n J_{\nu+n}(z)\}|} < 1,$$

and this domain is identical with the domain in which

$$\overline{\lim_{n\to\infty}} \sqrt[n]{\left|\frac{a_n (\tfrac{1}{2}z)^{\nu+n}}{\Gamma(\nu+n+1)}\right|} < 1,$$

by Horn's asymptotic formula (§ 8·1).

It follows that a Neumann series has a circle of convergence, just like a power series, and the circles of convergence of a Neumann series and of the associated power series are identical.

The theorem that the convergence of a Neumann series resembles that of a power series is due to Pincherle†; but it is possible to go much further, and, in fact, it can be proved that $f(z)$ has no singularities which are not also singularities of $f(z)_N$.

To prove this theorem, we write

$$2\sum_{n=0}^{\infty} \frac{a_n (\tfrac{1}{2}z)^{\nu+n}}{\Gamma(\nu+n+\tfrac{1}{2})\,\Gamma(\tfrac{1}{2})} \equiv \phi(z),$$

* *Nyt Tidsskrift*, IX. (B), (1898), pp. 77—79.

† *Bologna Memorie*, (4) III. (1881), pp. 151—180; see also Nielsen, *Math. Ann.* LV. (1902), pp. 493—496.

and then, inside the circle of convergence*,

$$f(z)=\int_0^1 \cos\{z\sqrt{(1-t^2)}\}\,.\,\phi(zt^2)\frac{dt}{\sqrt{(1-t^2)}}.$$

From the theory of analytic continuation it follows that, if $\phi(z)$ is analytic for any value of z, so also is $f(z)$, provided that the path of integration is suitably chosen; and so all the singularities of $f(z)$ must be singularities of $\phi(z)$.

Now the series defining $\phi(z)$ may be written in the form

$$\frac{2z^\nu}{\sqrt{\pi}}\sum_{n=0}^{\infty}\frac{a_n}{2^{\nu+n}\,\Gamma(\nu+n+1)}\frac{\Gamma(\nu+n+1)}{\Gamma(\nu+n+\frac{1}{2})}z^n,$$

and a theorem due to Hadamard† states that, if

$$F_1(z)=\sum_{n=0}^{\infty} b_n z^n,\quad F_2(z)=\sum_{n=0}^{\infty} c_n z^n,\quad F_3(z)=\sum_{n=0}^{\infty} b_n c_n z^n,$$

then all the singularities of $F_3(z)$ are expressible in the form $\beta\gamma$, where β is some singularity of $F_1(z)$ and γ is some singularity of $F_2(z)$.

Since the only finite singularity of the hypergeometric function

$$\sum_{n=0}^{\infty}\frac{\Gamma(\nu+n+1)}{\Gamma(\nu+n+\frac{1}{2})}z^n$$

is at the point $z=1$, it follows that all the singularities of $\phi(z)$ are singularities of $f(z)_N$; and therefore all the singularities of $f(z)$ are singularities of $f(z)_N$; and this is the theorem which was to be proved.

The reader should have no difficulty in enunciating and proving similar theorems‡ connected with the other types of expansions which are dealt with in this chapter.

16·3. *Various special Neumann series.*

The number of Neumann series, in which the coefficients are of simple forms, whose sums represent functions with important analytical properties is not large; we shall now give investigations of some such series which are of special interest.

By using the expansion

$$(t^4-2t^2\cos 2\theta+1)^{-\frac{1}{2}}=\sum_{n=0}^{\infty} t^{-2n-2}P_n(\cos 2\theta),$$

* It is assumed that $R(\nu+\frac{1}{2})$ is positive; if not, the several series under discussion have to be truncated by the omission of the terms for which $R(\nu+n+\frac{1}{2})$ is negative, but the general argument is unaffected.

† *Acta Math.* XXII. (1899), pp. 55—64; Hadamard, *La Série de Taylor* (Paris, 1901), p. 69.

‡ For such theorems concerning the expansion of § 16·14, see Nielsen, *Math. Ann.* LII. (1899), p. 230 *et seq.*

Pincherle* has observed that

$$\sum_{n=0}^{\infty} J_{2n+1}(z) P_n(\cos 2\theta) = \frac{1}{2\pi i}\int^{(0+)} \frac{\exp\{\frac{1}{2}z(t-1/t)\}\,dt}{\sqrt{(t^4-2t^2\cos 2\theta+1)}},$$

where the contour lies wholly outside the circle $|t|=1$.

If now we write $\frac{1}{2}(t-1/t)=w$, so that the contour in the w-plane is a (large) closed curve surrounding the origin, we find that

$$\sum_{n=0}^{\infty} J_{2n+1}(z) P_n(\cos 2\theta) = \frac{1}{4\pi i}\int^{(0+)} \frac{e^{zw}\,dw}{\sqrt{\{(w^2+1)(w^2+\sin^2\theta)\}}},$$

and so we obtain the formula

$$(1)\quad \sum_{n=0}^{\infty} J_{2n+1}(z) P_n(\cos 2\theta) = \frac{1}{4\pi}\int^{(0+)} e^{-iz\,\mathrm{ns}\,u}\,du = \frac{1}{4\pi}\int^{(iK'+)} e^{-iz\sin\theta\,\mathrm{sn}\,u}\,du,$$

where the modulus of the elliptic function is $\sin\theta$.

The interesting expansion

$$(2)\quad J_\nu^2(\tfrac{1}{2}z) = \sum_{n=0}^{\infty} \frac{1.3\ldots(2n-1)}{2.4\ldots(2n)}\,\frac{\Gamma(2\nu+2n+1)}{\{2^{\nu+n}\Gamma(\nu+n+1)\}^2} \times \{J_{2\nu+4n}(z)+J_{2\nu+4n+2}(z)\}$$

has been given by Jolliffe†, who proved that the series on the right satisfied the same differential equation as $J_\nu^2(\frac{1}{2}z)$. This expansion is easily derived as a special case of § 11·6 (1), but the following direct proof is not without interest:

By Neumann's formula (§ 5·43) we have‡

$$J_\nu^2(\tfrac{1}{2}z) = \frac{1}{\pi}\int_0^1 \frac{J_{2\nu}(z\sqrt{t})\,dt}{\sqrt{\{t(1-t)\}}},$$

and, if we expand $J_{2\nu}(z\sqrt{t})$ into the series

$$J_{2\nu}(z\sqrt{t}) = \frac{2t^\nu}{z}\sum_{m=0}^{\infty}\frac{(2\nu+2m+1)\Gamma(2\nu+m+1)}{m!\,\Gamma(2\nu+1)} \times {}_2F_1(-m,\,2\nu+m+1;\;2\nu+1;\;t)\,J_{2\nu+2m+1}(z),$$

we find on integration that

$$J_\nu^2(\tfrac{1}{2}z) = \sum_{m=0}^{\infty}\frac{2(2\nu+2m+1)}{z}\,a_m J_{2\nu+2m+1}(z),$$

* *Bologna Memorie*, (4) VIII. (1887), pp. 125—143. Pincherle used elliptic functions of modulus cosec θ in his result.

† *Messenger*, XLV. (1916), p. 16. The corresponding expansion of $z^{\frac{1}{2}} J_{\nu-\frac{1}{2}}(\frac{1}{2}z) J_\nu(\frac{1}{2}z)$ was obtained by Nielsen, *Nyt Tidsskrift*, IX. B, (1898), p. 80.

‡ If $R(\nu+\frac{1}{2})<0$, we use loop integrals instead of definite integrals.

where

$$a_m = \frac{\Gamma(2\nu+m+1)}{\pi . m! \, \Gamma(2\nu+1)} \int_0^1 t^{\nu-\frac{1}{2}} (1-t)^{-\frac{1}{2}} {}_2F_1(-m, 2\nu+m+1; 2\nu+1; t)\, dt$$

$$= \frac{1}{\pi . m!} \int_0^1 t^{-\nu-\frac{1}{2}} (1-t)^{-\frac{1}{2}} \frac{d^m \{t^{2\nu+m}(1-t)^m\}}{dt^m} dt$$

$$= \frac{(-)^m}{\pi . m!} \int_0^1 t^{2\nu+m}(1-t)^m \frac{d^m \{t^{-\nu-\frac{1}{2}}(1-t)^{-\frac{1}{2}}\}}{dt^m} dt,$$

by m partial integrations.

It follows that a_m is the coefficient of h^m in the expansion of

$$\frac{1}{\pi} \int_0^1 t^{2\nu} \{t - ht(1-t)\}^{-\nu-\frac{1}{2}} \{1 - t + ht(1-t)\}^{-\frac{1}{2}} dt$$

in ascending powers of h; and this expansion is absolutely convergent when $|h| < 1$.

If we write

$$t \equiv \frac{u(1-h)}{1-hu},$$

we find that

$$\sum_{m=0}^{\infty} a_m h^m = \frac{1}{\pi} \int_0^1 t^{\nu-\frac{1}{2}} \{1 - h(1-t)\}^{-\nu-\frac{1}{2}} (1-t)^{-\frac{1}{2}} (1+ht)^{-\frac{1}{2}} dt$$

$$= \frac{1}{\pi} \int_0^1 u^{\nu-\frac{1}{2}} (1-u)^{-\frac{1}{2}} (1-h^2 u)^{-\frac{1}{2}} du.$$

It is now evident that $a_{2n+1} = 0$ and that

$$a_{2n} = \frac{1}{\pi} \cdot \frac{1.3 \ldots (2n-1)}{2.4 \ldots (2n)} \int_0^1 u^{\nu+n-\frac{1}{2}} (1-u)^{-\frac{1}{2}} du$$

$$= \frac{1}{\pi} \cdot \frac{1.3 \ldots (2n-1)}{2.4 \ldots (2n)} \frac{\Gamma(\nu+n+\frac{1}{2})\,\Gamma(\frac{1}{2})}{\Gamma(\nu+n+1)},$$

and this formula at once gives Jolliffe's form of the expansion.

16·31. *The Neumann series summed by Lommel.*

The effects of transforming Neumann series by means of recurrence formulae have been studied systematically by Lommel*; and he has succeeded by this means in obtaining the sums of various series of the type

$$\Sigma a_n J_{\nu+n}(z)$$

in which a_n is a polynomial in n.

Take the functions

$$\begin{cases} \mathscr{A}_{\nu,m}(z) = \sum_{n=0}^{\infty} (\nu+2n+1) f_{2m}(\nu+2n+1) J_{\nu+2n+1}(z), \\ \mathscr{B}_{\nu,m}(z) = \sum_{n=0}^{\infty} (\nu+2n+2) f_{2m-1}(\nu+2n+2) J_{\nu+2n+2}(z), \end{cases}$$

where $f_m(\nu)$ is a function to be determined presently.

* *Studien über die Bessel'schen Functionen* (Leipzig, 1868), pp. 46—49.

By the recurrence formula we have

$$(1)\qquad \frac{2}{z}\mathscr{A}_{\nu,m}(z) = \sum_{n=0}^{\infty} f_{2m}(\nu+2n+1)\{J_{\nu+2n}(z)+J_{\nu+2n+2}(z)\}$$
$$= f_{2m}(\nu+1)J_\nu(z) + 2\mathscr{B}_{\nu,m}(z),$$

provided that $f_m(\nu)$ *satisfies the equation of mixed differences**

$$f_{2m}(\nu+2n+3) + f_{2m}(\nu+2n+1) = 2(\nu+2n+2) f_{2m-1}(\nu+2n+2).$$

A solution of this equation is

$$f_m(\nu) = 2^{m+1}\frac{\Gamma(\tfrac{1}{2}\nu+\tfrac{1}{2}m+1)}{\Gamma(\tfrac{1}{2}\nu-\tfrac{1}{2}m)}.$$

We adopt this value of $f_m(\nu)$ and then it is found by the same method that

$$(2)\qquad \frac{2}{z}\mathscr{B}_{\nu,m}(z) = -f_{2m-1}(\nu)J_{\nu+1}(z) + 2\mathscr{A}_{\nu,m-1}(z).$$

Hence it follows that

$$\mathscr{A}_{\nu,m}(z) = \tfrac{1}{2}z f_{2m}(\nu+1)J_\nu(z) - \tfrac{1}{2}z^2 f_{2m-1}(\nu)J_{\nu+1}(z) + z^2\mathscr{A}_{\nu,m-1}(z),$$

and so
$$\mathscr{A}_{\nu,m}(z) = z^{2m}\sum_{n=0}^{m}\{\tfrac{1}{2}z^{1-2n} f_{2n}(\nu+1)J_\nu(z)$$
$$- \tfrac{1}{2}z^{2-2n} f_{2n-1}(\nu)J_{\nu+1}(z)\} + z^{2m+2}\mathscr{A}_{\nu,-1}(z).$$

Therefore, since

$$\mathscr{A}_{\nu,-1}(z) = \sum_{n=0}^{\infty} J_{\nu+2n+1}(z) = \frac{1}{2}\int_0^z J_\nu(t)\,dt,$$

we have

$$(3)\qquad \sum_{n=0}^{\infty}(\nu+2n+1)\frac{\Gamma(\tfrac{1}{2}\nu+n+m+\tfrac{3}{2})}{\Gamma(\tfrac{1}{2}\nu+n-m+\tfrac{1}{2})}J_{\nu+2n+1}(z)$$
$$= (\tfrac{1}{2}z)^{2m+2}\int_0^z J_\nu(t)\,dt + J_\nu(z)\left\{\sum_{n=0}^{m}\frac{\Gamma(\tfrac{1}{2}\nu+n+\tfrac{3}{2})}{\Gamma(\tfrac{1}{2}\nu-n+\tfrac{1}{2})}(\tfrac{1}{2}z)^{2m-2n+1}\right\}$$
$$- J_{\nu+1}(z)\left\{\sum_{n=0}^{m}\frac{\Gamma(\tfrac{1}{2}\nu+n+\tfrac{1}{2})}{\Gamma(\tfrac{1}{2}\nu-n+\tfrac{1}{2})}(\tfrac{1}{2}z)^{2m-2n+2}\right\},$$

and similarly, from the expression for $\mathscr{B}_{\nu,m}(z)$,

$$(4)\qquad \sum_{n=0}^{\infty}(\nu+2n+2)\frac{\Gamma(\tfrac{1}{2}\nu+n+m+\tfrac{3}{2})}{\Gamma(\tfrac{1}{2}\nu+n-m+\tfrac{3}{2})}J_{\nu+2n+2}(z)$$
$$= (\tfrac{1}{2}z)^{2m+1}\int_0^z J_\nu(t)\,dt + J_\nu(z)\left\{\sum_{n=0}^{m-1}\frac{\Gamma(\tfrac{1}{2}\nu+n+\tfrac{3}{2})}{\Gamma(\tfrac{1}{2}\nu-n+\tfrac{1}{2})}(\tfrac{1}{2}z)^{2m-2n}\right\}$$
$$- J_{\nu+1}(z)\left\{\sum_{n=0}^{m}\frac{\Gamma(\tfrac{1}{2}\nu+n+\tfrac{1}{2})}{\Gamma(\tfrac{1}{2}\nu-n+\tfrac{1}{2})}(\tfrac{1}{2}z)^{2m-2n+1}\right\}.$$

* Recent applications of Neumann series to the solution of equations of mixed differences are due to Bateman, *Proc. Int. Congress of Math.* I. (Cambridge, 1912), pp. 291—294.

The potentialities of the other recurrence formula

$$2J_\nu'(z) = J_{\nu-1}(z) - J_{\nu+1}(z)$$

were also investigated by Lommel, but the results are not so interesting. As examples of his expansions the reader may notice that

$$2^{2m} \sum_{n=m}^{\infty} \frac{(n+m-1)!}{(n-m)!} J_{\nu+2n}^{(2m)}(z) = (2m-1)!\, J_\nu(z),$$

$$2^{2m+1} \sum_{n=m}^{\infty} \frac{(n+m)!}{(n-m)!} J_{\nu+2n+1}^{(2m+1)}(z) = (2m)!\, J_\nu(z).$$

These results were given by Lommel, though his formulae contain numerical errors.

16·32. *The Neumann series summed by Kapteyn.*

The sum of the series

$$\sum_{n=1}^{\infty} nJ_n(z)\, J_n(\alpha)$$

is expressible as the integral

$$\tfrac{1}{2}z \int_0^\alpha \frac{J_1(z-v)}{z-v} J_0(\alpha - v)\, dv;$$

the sums of the alternate terms of the series have been expressed by Kapteyn* in the form of integrals from which this integral may be deduced, and conversely Kapteyn's formulae may be deduced from this integral.

We proceed to establish this result by a simplified form of Kapteyn's methods.

The series may be written in the form†

$$\tfrac{1}{2}z \sum_{n=0}^{\infty} \{J_{n-1}(z) + J_{n+1}(z)\}\, J_n(\alpha) = \frac{\tfrac{1}{2}z}{(2\pi i)^2} \int^{(0+)} \int^{(0+)} \sum_{n=0}^{\infty} (t^{-n} + t^{-n-2})\, u^{-n-1}$$

$$\times \exp\left\{\tfrac{1}{2}z\left(t - \frac{1}{t}\right) + \tfrac{1}{2}\alpha\left(u - \frac{1}{u}\right)\right\} du\,dt$$

$$= \frac{\tfrac{1}{2}z}{(2\pi i)^2} \int^{(0+)} \int^{(0+)} \frac{t^2+1}{t(tu-1)} \exp\left\{\tfrac{1}{2}z\left(t - \frac{1}{t}\right) + \tfrac{1}{2}\alpha\left(u - \frac{1}{u}\right)\right\} du\,dt,$$

where the contours may be taken to be the circles $|u| = 1$, $|t| = A > 1$.

Now, let

$$I = \frac{1}{2\pi i} \int^{(0+)} \frac{1}{tu-1} \exp\left\{\tfrac{1}{2}\alpha\left(u - \frac{1}{u}\right)\right\} du.$$

Then, if $m = \tfrac{1}{2}(t - 1/t)$, we have

$$\frac{dI}{d\alpha} + mI = \frac{1}{2\pi i} \int^{(0+)} \frac{1}{2}\left(\frac{1}{t} + \frac{1}{u}\right) \exp\left\{\tfrac{1}{2}\alpha\left(u - \frac{1}{u}\right)\right\} du = \tfrac{1}{2}\{J_0(\alpha) - J_1(\alpha)/t\}.$$

* *Nieuw Archief voor Wiskunde*, (2) VII. (1907), pp. 20—25; *Proc. Section of Sci., K. Akad. van Wet. te Amsterdam*, VII. (1905), pp. 494—500; Kapteyn has subsequently summed other series, *ibid.* XIV. (1912), pp. 962—969.

† The interchange of summation and integration is permissible so long as $|tu| > 1$, where t, u are any points on the contours.

Therefore, on integration,

$$I = Ce^{-m\alpha} + \frac{1}{2}\int_0^\alpha e^{-m(\alpha-v)}\{J_0(v) - J_1(v)/t\}\,dv,$$

where C is independent of α. By taking $\alpha = 0$, we see that $C = 1/t$.

Hence we have

$$\sum_{n=1}^{\infty} nJ_n(z)\,J_n(\alpha) = \frac{\frac{1}{2}z}{2\pi i}\int^{(0+)}\frac{t^2+1}{t^2}\left[\exp\left\{\tfrac{1}{2}(z-\alpha)\left(t-\frac{1}{t}\right)\right\}\right.$$
$$\left. + \frac{1}{2}\int_0^\alpha \exp\left\{\tfrac{1}{2}(z-\alpha+v)\left(t-\frac{1}{t}\right)\right\}\{tJ_0(v) - J_1(v)\}\,dv\right]dt$$
$$= \frac{z}{4}\int_0^\alpha [J_0(z-\alpha+v) + J_2(z-\alpha+v)]\,J_0(v)\,dv,$$

the majority of the terms having a zero residue at $t = 0$.

Consequently

$$\sum_{n=1}^{\infty} nJ_n(z)\,J_n(\alpha) = \frac{z}{2}\int_0^\alpha \frac{J_1(z-\alpha+v)}{z-\alpha+v}\,J_0(v)\,dv,$$

that is to say

$$(1) \qquad \sum_{n=1}^{\infty} nJ_n(z)\,J_n(\alpha) = \frac{z}{2}\int_0^\alpha \frac{J_1(z-v)}{z-v}\,J_0(\alpha-v)\,dv.$$

If we select the odd and even parts of the functions of z on each side of this equation, we find that

$$(2) \qquad \sum_{n=0}^{\infty} (2n+1)\,J_{2n+1}(z)\,J_{2n+1}(\alpha)$$
$$= \frac{z}{4}\int_0^\alpha \left\{\frac{J_1(z-v)}{z-v} + \frac{J_1(z+v)}{z+v}\right\} J_0(\alpha-v)\,dv,$$

which is one of Kapteyn's formulae; and

$$\frac{d}{d\alpha}\sum_{n=1}^{\infty} 2n\,J_{2n}(z)\,J_{2n}(\alpha) = \frac{z}{4}\left\{\frac{J_1(z-\alpha)}{z-\alpha} - \frac{J_1(z+\alpha)}{z+\alpha}\right\}$$
$$- \frac{z}{4}\int_0^\alpha \left\{\frac{J_1(z-v)}{z-v} - \frac{J_1(z+v)}{z+v}\right\} J_1(\alpha-v)\,dv$$
$$= \frac{z}{4}\int_0^\alpha \left\{\frac{J_2(z+v)}{z+v} + \frac{J_2(z-v)}{z-v}\right\} J_0(\alpha-v)\,dv,$$

when we integrate by parts.

Hence it follows that

$$(3) \qquad \sum_{n=1}^{\infty} 2n\,J_{2n}(z)\,J_{2n}(\alpha) = \frac{z}{4}\int_0^\alpha dt\int_0^t \left\{\frac{J_2(z+v)}{z+v} + \frac{J_2(z-v)}{z-v}\right\} J_0(t-v)\,dv,$$

which is the other of Kapteyn's results.

The reader should have no difficulty in proving by similar methods that, when $R(\nu) > 0$,

$$(4)\qquad \sum_{n=0}^{\infty} (\nu+n) J_{\nu+n}(z) J_{\nu+n}(\alpha) = \frac{(\nu-1)z}{2} \int_0^\alpha \frac{J_{\nu-1}(z-v)}{z-v} J_\nu(\alpha-v)\,dv + \frac{\nu z}{2} \int_0^\alpha \frac{J_\nu(z-v)}{z-v} J_{\nu-1}(\alpha-v)\,dv.$$

16·4. *The Webb-Kapteyn theory of Neumann series.*

Neumann series have been studied from the standpoint of the theory of functions of real variables by H. A. Webb*. His theory has been developed by Kapteyn† and subsequently by Bateman‡. The theory is not so important as it appears to be at first sight, because, as the reader will presently realise, it has to deal with functions which must not only behave in a prescribed manner as the variable tends to $\pm\infty$, but must also satisfy an intricate integral equation. In fact, the functions which are amenable to the theory seem to be included in the functions to which the complex theory is applicable, and simple functions have been constructed to which the real variable theory is inapplicable.

The result on which the theory is based is that (§ 13·42)

$$\int_0^\infty J_{2m+1}(t) J_{2n+1}(t) \frac{dt}{t} = \begin{cases} 0 & (m \neq n), \\ 1/(4n+2) & (m = n), \end{cases}$$

so that, if an *odd* function $f(x)$ admits of an expansion of the type

$$f(x) = \sum_{n=0}^{\infty} a_{2n+1} J_{2n+1}(x),$$

and if term-by-term integration is permissible, we have

$$\int_0^\infty f(t) J_{2n+1}(t) \frac{dt}{t} = \frac{a_{2n+1}}{4n+2}.$$

We are therefore led to consider the possibility that

$$(1)\qquad f(x) = \sum_{n=0}^{\infty} (4n+2) J_{2n+1}(x) \int_0^\infty \frac{J_{2n+1}(t)}{t} f(t)\,dt;$$

and we shall establish the truth of this expansion under the following conditions:

(I) *The integral*

$$\int_0^\infty f(t)\,dt$$

exists and is absolutely convergent.

* *Messenger*, XXXIII. (1904), p. 55.

† *Messenger*, XXXV. (1906), pp. 122—125.

‡ *Messenger*, XXXVI. (1907), pp. 31—37.

(II) *The function $f(t)$ has a continuous differential coefficient for all positive values of the variable which do not exceed x.*

(III) *The function $f(t)$ satisfies the equation*

$$2f'(t) = \int_0^\infty \frac{J_1(v)}{v} \{f(v+t) + f(v-t)\}\, dv \tag{2}$$

when t does not exceed x.

We now proceed to sum the series

$$S \equiv \sum_{n=0}^{\infty} (4n+2) J_{2n+1}(x) \int_0^\infty \frac{J_{2n+1}(t)}{t} f(t)\, dt,$$

and we first interchange the order of summation and integration. It is evident that

$$\sum_{n=0}^{\infty} J_{2n+1}(x) \{J_{2n}(t) + J_{2n+2}(t)\}$$

converges uniformly with respect to t for positive (unbounded) values of t, since $|J_{2n}(t)| \leqslant 1$ and $\Sigma |J_{2n+1}(x)|$ is convergent. Hence, since $f(t)$ possesses an absolutely convergent integral, we may effect the interchange, and then, by § 16·32,

$$\begin{aligned}
S &= \int_0^\infty f(t) \left\{ \sum_{n=0}^{\infty} (4n+2) J_{2n+1}(x) J_{2n+1}(t) \right\} \frac{dt}{t} \\
&= \frac{1}{2} \int_0^\infty f(t) \int_0^x \left\{ \frac{J_1(t-v)}{t-v} + \frac{J_1(t+v)}{t+v} \right\} J_0(x-v)\, dv\, dt \\
&= \frac{1}{2} \int_0^x \int_0^\infty \left\{ \frac{J_1(t-v)}{t-v} + \frac{J_1(t+v)}{t+v} \right\} J_0(x-v) f(t)\, dt\, dv \\
&= \frac{1}{2} \int_0^x J_0(x-v) \int_0^\infty \frac{J_1(t)}{t} \{f(t+v) + f(t-v)\}\, dt\, dv \\
&\quad + \int_0^x J_0(x-v) \int_0^v f(v-t) \frac{J_1(t)}{t}\, dt\, dv.
\end{aligned}$$

We now transform the last integral by using § 12·2, and then we have*

$$\begin{aligned}
&\int_0^x \int_0^v J_0(x-v) f(v-t) \frac{J_1(t)}{t}\, dt\, dv \\
&\qquad = \int_0^x \int_0^u J_0(u-t) f(x-u) \frac{J_1(t)}{t}\, dt\, du \\
&\qquad = \int_0^x J_1(u) f(x-u)\, du \\
&\qquad = f(x) - \int_0^x J_0(u) f'(x-u)\, du.
\end{aligned}$$

* The first transformation is effected by writing

$$v = x + t - u.$$

Hence

$$(3)\quad \sum_{n=0}^{\infty}(4n+2)\,J_{2n+1}(x)\int_0^{\infty}\frac{J_{2n+1}(t)}{t}f(t)\,dt$$
$$=f(x)-\int_0^x J_0(x-v)\left[f'(v)-\frac{1}{2}\int_0^{\infty}\frac{J_1(t)}{t}\{f(t+v)+f(t-v)\}\,dt\right]dv.$$

Now write

$$f'(v)-\frac{1}{2}\int_0^{\infty}\frac{J_1(t)}{t}\{f(t+v)+f(t-v)\}\,dt\equiv F(v),$$

so that $F(v)$ is a continuous function of v, since $J_1(t)/t$ has an absolutely convergent integral.

If then we are to have $S=f(x)$ when x has any value in such an interval as $(0, X)$, we must have

$$\int_0^x J_0(x-v)\,F(v)\,dv\equiv 0,$$

throughout this interval; and, differentiating with respect to x,

$$F(x)=\int_0^x J_1(x-v)\,F(v)\,dv.$$

Since $|J_1(x-v)|\leqslant 1/\sqrt{2}$, it follows by induction from this equation, since

$$|F(x)|\leqslant\frac{1}{\sqrt{2}}\int_0^x|F(v)|\,dv,$$

that

$$|F(x)|\leqslant\frac{A\,.\,x^n}{2^{\frac{1}{2}n}\,.\,n!},$$

where A is the upper bound of $|F(x)|$ in the interval and n is any positive integer.

If we make $n\to\infty$, it is clear that $F(x)\equiv 0$, and so the *necessity* of equation (2) is established.

The *sufficiency* of equation (2) for the truth of the expansion* is evident from (3).

It has been pointed out by Kapteyn that the function $\sin(x\operatorname{cosec}\alpha)$ is one for which equation (2) is *not* satisfied; and Bateman has consequently endeavoured to determine general criteria for functions which satisfy equation (2); but no simple criteria have, as yet, been discovered.

[NOTE. If $f(x)$ is not an odd function, we expand the two odd functions

$$\tfrac{1}{2}\{f(x)-f(-x)\},\quad \tfrac{1}{2}x\,\{f(x)+f(-x)\}$$

separately; and then it is easy to prove, by rearranging the second expansion, that

$$f(x)=\sum_{n=0}^{\infty}a_n'J_n(x),$$

where

$$a_0=\frac{1}{2}\int_{-\infty}^{\infty}f(x)\,J_1(|x|)\,dx,$$
$$a_n=n\int_{-\infty}^{\infty}f(x)\,J_n(x)\,\frac{dx}{|x|},\qquad (n>0),$$

provided that the appropriate integral equations are satisfied.]

* The sufficiency (but not the necessity) of the equation was proved by Kapteyn.

16·41. *Cailler's theory of reduced functions.*

The Webb-Kapteyn theory of Neumann series which has just been expounded has several points of contact with a theory due to Cailler*. This theory is based on Borel's integral connecting a pair of functions. Thus, if

$$f(z) = \sum_{n=0}^{\infty} c_n z^n,$$

then the function $f(z)_R$ defined by the series

$$f(z)_R = \sum_{n=0}^{\infty} c_n \,.\, n!\, z^n,$$

supposed convergent for sufficiently small values of $|z|$, may be represented by the integral

$$f(z)_R = \int_0^\infty e^{-t} f(tz)\, dt.$$

The function $f(z)_R$ may be termed the *reduced function* (*la réduite*) of $f(z)$.

If the Neumann series which represents $f(z)$ is

$$f(z) = \sum_{n=0}^{\infty} a_n J_n(z),$$

then we have, formally,

$$\begin{aligned} f(z)_R &= \sum_{n=0}^{\infty} a_n \int_0^\infty e^{-t} J_n(tz)\, dt \\ &= \frac{1}{\sqrt{(1+z^2)}} \sum_{n=0}^{\infty} a_n \left\{\frac{\sqrt{(1+z^2)}-1}{z}\right\}^n \end{aligned}$$

Now put

$$\frac{\sqrt{(1+z^2)}-1}{z} = \zeta$$

and we see that

$$\frac{1+\zeta^2}{1-\zeta^2} f\left(\frac{2\zeta}{1-\zeta^2}\right)_R = \sum_{n=0}^{\infty} a_n \zeta^n.$$

Hence, if the Neumann series for $f(z)$ is $\sum_{n=0}^{\infty} a_n J_n(z)$, then the generating function of $\sum_{n=0}^{\infty} a_n \zeta^n$ is

$$\frac{1+\zeta^2}{1-\zeta^2} f\left(\frac{2\zeta}{1-\zeta^2}\right)_R,$$

provided that this function is analytic near the origin.

More generally, if $f(z)$ has a branch-point near the origin of such a nature that

$$f(z) = \sum_{n=0}^{\infty} a_n J_{\nu+n}(z), \tag{1}$$

then

$$\sum_{n=0}^{\infty} a_n \zeta^{\nu+n} = \frac{1+\zeta^2}{1-\zeta^2} f\left(\frac{2\zeta}{1-\zeta^2}\right)_R. \tag{2}$$

* *Mém. de la Soc. de Phys. de Genève*, XXXIV. (1902—1905), pp. 295—368.

In like manner, if

(3) $$f(z) = \sum_{n=0}^{\infty} a_n z^{\nu+n} J_{\nu+n}(z),$$

then

(4) $$\sum_{n=0}^{\infty} \frac{\Gamma(2\nu+2n+1)}{2^{\nu+n}\,\Gamma(\nu+n+1)} a_n \zeta^{2\nu+2n} = \frac{1}{\sqrt{(1-\zeta^2)}}\, f\left(\frac{\zeta}{\sqrt{(1-\zeta^2)}}\right)_R.$$

[NOTE. If $$e^{az}\sin bz = \sum_{n=1}^{\infty} a_n J_n(z),$$

then $$\sum_{n=1}^{\infty} a_n z^n = \frac{2bz(1+z^2)}{(1-2az-z^2)^2+4b^2z^2}.$$

This result, which is immediately deducible from Cailler's theory, was set as a problem in the Mathematical Tripos, 1896.]

16·5. *Lommel's functions of two variables.*

Two functions, which are of considerable importance in the theory of Diffraction and which are defined by simple series of Neumann's type, have been discussed exhaustively by Lommel* in his great memoirs on Diffraction at a Circular Aperture and Diffraction at a Straight Edge.

The functions of integral order n, denoted by the symbols $U_n(w, z)$ and $V_n(w, z)$, are defined by the equations

(1) $$U_n(w, z) = \sum_{m=0}^{\infty} (-)^m \left(\frac{w}{z}\right)^{n+2m} J_{n+2m}(z),$$

(2) $$V_n(w, z) = \sum_{m=0}^{\infty} (-)^m \left(\frac{w}{z}\right)^{-n-2m} J_{-n-2m}(z).$$

It is easy to see from § 2·22 (3) that

(3) $$U_n(w, z) - V_{-n+2}(w, z) = \sum_{m=-\infty}^{\infty} (-)^m \left(\frac{w}{z}\right)^{n+2m} J_{n+2m}(z)$$
$$= \cos\left(\frac{w}{2} + \frac{z^2}{2w} - \frac{n\pi}{2}\right),$$

(4) $$U_{n+1}(w, z) - V_{-n+1}(w, z) = \sin\left(\frac{w}{2} + \frac{z^2}{2w} - \frac{n\pi}{2}\right).$$

The last equation may be derived from the preceding equation by replacing n by $n+1$.

There is no difficulty in extending (1) to define functions of non-integral order; for unrestricted values of ν we write

(5) $$U_\nu(w, z) = \sum_{m=0}^{\infty} (-)^m \left(\frac{w}{z}\right)^{\nu+2m} J_{\nu+2m}(z).$$

* *Abh. der math. phys. Classe der k. b. Akad. der Wiss.* (*München*), xv. (1886), pp. 229—328, 529—664. The first memoir deals with functions of integral order; and the definition of $V_n(w, z)$ in it differs from that adopted subsequently by the factor $(-1)^n$. Much of Lommel's work is reproduced by J. Walker, *The Analytical Theory of Light* (Cambridge, 1904). The occurrence of Lommel's functions in a different physical problem has been noticed by Pocklington, *Nature*, LXXI. (1905), pp. 607—608.

The expression on the right is an integral function of z, and (when the factor w^ν is removed) an integral function of w.

The corresponding generalisation of (2) gives a series which converges only when ν is an integer. And consequently it is convenient to define $V_\nu(w, z)$ for unrestricted values of ν by means of the natural generalisation of (3), namely

$$V_\nu(w, z) = \cos\left(\frac{w}{2} + \frac{z^2}{2w} + \frac{\nu\pi}{2}\right) + U_{-\nu+2}(w, z). \tag{6}$$

It is evident that

$$U_\nu(w, z) + U_{\nu+2}(w, z) = \left(\frac{w}{z}\right)^\nu J_\nu(z), \tag{7}$$

$$V_\nu(w, z) + V_{\nu+2}(w, z) = \left(\frac{w}{z}\right)^{-\nu} J_{-\nu}(z). \tag{8}$$

As special formulae, we deduce from § 2·22 that

$$U_0(z, z) = V_0(z, z) = \tfrac{1}{2}\{J_0(z) + \cos z\}, \tag{9}$$

$$U_1(z, z) = -V_1(z, z) = \tfrac{1}{2}\sin z; \tag{10}$$

and hence, by (7) and (8),

$$U_{2n}(z, z) = V_{2n}(z, z) = \tfrac{1}{2}(-)^n\left\{\cos z - \sum_{m=0}^{n-1}(-)^m \epsilon_{2m} J_{2m}(z)\right\}, \tag{11}$$

$$U_{2n+1}(z, z) = -V_{2n+1}(z, z) = \tfrac{1}{2}(-)^n\left\{\sin z - \sum_{m=0}^{n-1}(-)^m \epsilon_{2m+1} J_{2m+1}(z)\right\}; \tag{12}$$

provided that $n \geqslant 1$ in (11), and $n \geqslant 0$ in (12).

It is also to be observed that, as a generalisation of these formulae,

$$V_n(w, z) = (-)^n U_n(z^2/w, z). \tag{13}$$

The functions $\sum\limits_{m=0}^{\infty} \begin{matrix}\sin\\ \cos\end{matrix} m\theta \,.\, J_m(z), \quad \sum\limits_{m=0}^{\infty} \sin(m+\tfrac{1}{2})\theta \,.\, J_{m+\frac{1}{2}}(z),$

which are closely associated with Lommel's functions, have been studied by Kapteyn, *Proc. Section of Sci., K. Akad. van Wet. te Amsterdam*, VII. (1905), pp. 375—376, and by Hargreaves, *Phil. Mag.* (6) XXXVI. (1918), pp. 191—199, respectively.

16·51. *The differential equations for Lommel's functions of two variables.*

It is evident by differentiating § 16·5 (1) that

$$\frac{\partial}{\partial z} U_\nu(w, z) = -\frac{z}{w} U_{\nu+1}(w, z), \tag{1}$$

and hence

$$\frac{\partial^2}{\partial z^2} U_\nu(w, z) = \frac{z^2}{w^2} U_{\nu+2}(w, z) - \frac{1}{w} U_{\nu+1}(w, z),$$

and consequently

$$\left\{\frac{\partial^2}{\partial z^2} - \frac{1}{z}\frac{\partial}{\partial z} + \frac{z^2}{w^2}\right\} U_\nu(w, z) = \frac{z^2}{w^2}\{U_{\nu+2}(w, z) + U_\nu(w, z)\}.$$

It is now evident that $U_\nu(w, z)$ is a particular integral of the equation

$$\frac{\partial^2 y}{\partial z^2} - \frac{1}{z}\frac{\partial y}{\partial z} + \frac{z^2 y}{w^2} = \left(\frac{w}{z}\right)^{\nu-2} J_\nu(z). \tag{2}$$

Since the complementary function of this equation is

$$A\cos\frac{z^2}{2w} + B\sin\frac{z^2}{2w},$$

where A and B are independent of z, it is clear from § 16·5 (6) that $V_{-\nu+2}(w, z)$ is also a particular integral. Therefore $V_\nu(w, z)$ is a particular integral of

$$\frac{\partial^2 y}{\partial z^2} - \frac{1}{z}\frac{\partial y}{\partial z} + \frac{z^2 y}{w^2} = \left(\frac{z}{w}\right)^{\nu} J_{-\nu+2}(z). \tag{3}$$

These equations are due to Lommel, *Münchener Abh.* xv. (1886), pp. 561—563.

16·52. *Recurrence formulae for Lommel's functions of two variables.*

We have just obtained one recurrence formula for $U_\nu(w, z)$, namely

$$\frac{\partial}{\partial z} U_\nu(w, z) = -\frac{z}{w} U_{\nu+1}(w, z). \tag{1}$$

To obtain other formulae, we observe that

$$\begin{aligned}\frac{\partial}{\partial w} U_\nu(w, z) &= \sum_{m=0}^{\infty} (-)^m (\nu+2m)(w/z)^{\nu+2m-1} J_{\nu+2m}(z)/z \\ &= \frac{1}{2}\sum_{m=0}^{\infty} (-)^m (w/z)^{\nu+2m-1} \{J_{\nu+2m-1}(z) + J_{\nu+2m+1}(z)\},\end{aligned}$$

and so

$$2\frac{\partial}{\partial w} U_\nu(w, z) = U_{\nu-1}(w, z) + (z/w)^2 U_{\nu+1}(w, z). \tag{2}$$

Again, by differentiating § 16·5 (6) we deduce that

$$\frac{\partial}{\partial z} V_\nu(w, z) = -\frac{z}{w} V_{\nu-1}(w, z), \tag{3}$$

$$2\frac{\partial}{\partial w} V_\nu(w, z) = V_{\nu+1}(w, z) + (z/w)^2 V_{\nu-1}(w, z). \tag{4}$$

If now we take $w = cz$, where c is constant, we deduce that

$$2\frac{d}{dz} U_\nu(cz, z) = cU_{\nu-1}(cz, z) - (1/c)\, U_{\nu+1}(cz, z), \tag{5}$$

$$2\frac{d}{dz} V_\nu(cz, z) = cV_{\nu+1}(cz, z) - (1/c)\, V_{\nu-1}(cz, z). \tag{6}$$

Hence we get

$$\begin{aligned}4\frac{d^2}{dz^2} U_\nu(cz, z) &= c^2 U_{\nu-2}(cz, z) - 2U_\nu(cz, z) + (1/c^2)\, U_{\nu+2}(cz, z) \\ &= c^\nu J_{\nu-2}(z) + c^{\nu-2} J_\nu(z) - (c + 1/c)^2 U_\nu(cz, z).\end{aligned}$$

Hence it follows that $U_\nu(cz, z)$, and similarly $V_{-\nu+2}(cz, z)$, are particular integrals of the equation

(7) $$4\frac{d^2y}{dz^2}+\left(c+\frac{1}{c}\right)^2 y = c^\nu J_{\nu-2}(z)+c^{\nu-2}J_\nu(z).$$

The particular case in which $z=0$ is of some interest; we have

(8) $$U_\nu(w, 0)=\sum_{m=0}^{\infty}\frac{(-)^m(\frac{1}{2}w)^{\nu+2m}}{\Gamma(\nu+2m+1)},$$

and so $U_\nu(w, 0)$ and $V_{-\nu+2}(w, 0)$ are expressible in terms of Lommel's functions of one variable by the equations

(9) $$U_\nu(w, 0)=\frac{(\frac{1}{2}w)^{\frac{1}{2}}s_{\nu-\frac{3}{2},\frac{1}{2}}(\frac{1}{2}w)}{\Gamma(\nu-1)},$$

(10) $$V_{-\nu+2}(w, 0)=\frac{(\frac{1}{2}w)^{\frac{1}{2}}S_{\nu-\frac{3}{2},\frac{1}{2}}(\frac{1}{2}w)}{\Gamma(\nu-1)}.$$

Of these results, (1)—(8) were given in Lommel's memoir.

The following formulae, valid when n is a positive integer (zero included), should be noticed:

(11) $$U_{2n}(w, 0)=(-)^n\left[\cos\tfrac{1}{2}w-\sum_{m=0}^{n-1}\frac{(-)^m(\frac{1}{2}w)^{2m}}{(2m)!}\right],$$

(12) $$U_{2n+1}(w, 0)=(-)^n\left[\sin\tfrac{1}{2}w-\sum_{m=0}^{n-1}\frac{(-)^m(\frac{1}{2}w)^{2m+1}}{(2m+1)!}\right],$$

(13) $$U_{-n}(w, 0)=\cos(\tfrac{1}{2}w+\tfrac{1}{2}n\pi).$$

Hence it follows that

(14) $$V_0(w, 0)=1,\quad V_{n+1}(w, 0)=0,$$

(15) $$V_{-2n}(w, 0)=(-)^n\sum_{m=0}^{n}\frac{(-)^m(\frac{1}{2}w)^{2m}}{(2m)!},$$

(16) $$V_{-2n-1}(w, 0)=(-)^n\sum_{m=0}^{n}\frac{(-)^m(\frac{1}{2}w)^{2m+1}}{(2m+1)!}.$$

16·53. *Integral representations of Lommel's functions.*

The formulae

(1) $$U_\nu(w, z)=\frac{w^\nu}{z^{\nu-1}}\int_0^1 J_{\nu-1}(zt)\,.\cos\{\tfrac{1}{2}w(1-t^2)\}\,.\,t^\nu\,dt,$$

(2) $$U_{\nu+1}(w, z)=\frac{w^\nu}{z^{\nu-1}}\int_0^1 J_{\nu-1}(zt)\,.\sin\{\tfrac{1}{2}w(1-t^2)\}\,.\,t^\nu\,dt,$$

which are valid when $R(\nu)>0$, may be verified immediately by expanding the integrands in powers of w and then using the result of § 12·11 (1) in performing term-by-term integrations. For other values of ν, they may be replaced by the equations

$$(3)\qquad U_\nu(w, z) = -\frac{w^\nu}{2iz^{\nu-1}\sin 2\nu\pi}\int_1^{(0+)} J_{\nu-1}(-zt)\,.\cos\{\tfrac{1}{2}w(1-t^2)\}\,.(-t)^\nu\,dt,$$

$$(4)\qquad U_{\nu+1}(w, z) = -\frac{w^\nu}{2iz^{\nu-1}\sin 2\nu\pi}\int_1^{(0+)} J_{\nu-1}(-zt)\,.\sin\{\tfrac{1}{2}w(1-t^2)\}\,.(-t)^\nu\,dt,$$

in which the phase of $-t$ increases from $-\pi$ to π as t describes the contour.

It is clear that, when $R(\nu) > 0$,

$$(5)\qquad U_\nu(w, z) \pm iU_{\nu+1}(w, z) = \frac{w^\nu}{z^{\nu-1}}\int_0^1 J_{\nu-1}(zt)\exp\{\pm\tfrac{1}{2}iw(1-t^2)\}\,.t^\nu\,dt.$$

By modifying this formula we can obtain integral representations of $V_\nu(w, z)$ valid for *positive* values of w and z. Let us consider

$$\frac{w^\nu}{z^{\nu-1}}\int_0^\infty J_{\nu-1}(zt)\exp\{\pm\tfrac{1}{2}iw(1-t^2)\}\,.t^\nu\,dt.$$

The integral converges at the lower limit when $R(\nu) > 0$ and at the upper limit when $R(\nu) < \frac{3}{2}$, if w and z are restricted to be positive.

To evaluate the last integral, swing the contour round until it coincides with the ray $\arg t = \mp\frac{1}{4}\pi$, this ambiguity in sign being determined by the ambiguity in sign in the integral; such a modification in the contour is permissible by Jordan's lemma.

When we expand the new integral in ascending powers of z, as in § 13·3, we find that

$$\frac{w^\nu}{z^{\nu-1}}\int_0^\infty J_{\nu-1}(zt)\exp\{\pm\tfrac{1}{2}iw(1-t^2)\}\,.t^\nu\,dt$$

$$= \frac{w^\nu e^{\pm\frac{1}{2}iw}}{2^{\nu-1}}\sum_{m=0}^\infty \frac{(-)^m(\frac{1}{2}z)^{2m}}{m!\,\Gamma(\nu+m)}\int_0^{\infty\exp(\mp\frac{1}{4}\pi i)} t^{2\nu+2m-1}\exp(\mp\tfrac{1}{2}iwt^2)\,dt$$

$$= \frac{w^\nu e^{\pm\frac{1}{2}iw}}{2^\nu}\sum_{m=0}^\infty \frac{(-)^m(\frac{1}{2}z)^{2m}}{m!}\left(\mp\frac{2i}{w}\right)^{\nu+m},$$

that is to say

$$(6)\qquad \frac{w^\nu}{z^{\nu-1}}\int_0^\infty J_{\nu-1}(zt)\exp\{\pm\tfrac{1}{2}iw(1-t^2)\}\,.t^\nu\,dt = \exp\left(\pm\frac{iw}{2}\pm\frac{iz^2}{2w}\mp\frac{\nu\pi i}{2}\right).$$

When we combine the results contained in this formula, we see that, if $w > 0$, $z > 0$, and $0 < R(\nu) < \frac{3}{2}$, then

$$(7)\qquad \frac{w^\nu}{z^{\nu-1}}\int_0^\infty J_{\nu-1}(zt)\cos\{\tfrac{1}{2}w(1-t^2)\}\,.t^\nu\,dt = \cos\left(\frac{w}{2}+\frac{z^2}{2w}-\frac{\nu\pi}{2}\right),$$

$$(8)\qquad \frac{w^\nu}{z^{\nu-1}}\int_0^\infty J_{\nu-1}(zt)\sin\{\tfrac{1}{2}w(1-t^2)\}\,.t^\nu\,dt = \sin\left(\frac{w}{2}+\frac{z^2}{2w}-\frac{\nu\pi}{2}\right).$$

It follows at once from (1) and (2) combined with § 16·5 (6) that

$$(9) \qquad V_{2-\nu}(w, z) = -\frac{w^\nu}{z^{\nu-1}} \int_1^\infty J_{\nu-1}(zt) \cos\{\tfrac{1}{2}w(1-t^2)\} . t^\nu \, dt,$$

$$(10) \qquad V_{1-\nu}(w, z) = -\frac{w^\nu}{z^{\nu-1}} \int_1^\infty J_{\nu-1}(zt) \sin\{\tfrac{1}{2}w(1-t^2)\} . t^\nu \, dt.$$

Since convergence at the origin is now unnecessary, the theory of analytic continuation enables us to remove the restriction $R(\nu) > 0$.

Changing the notation, we see that

$$(11) \qquad V_\nu(w, z) = -\frac{z^{\nu-1}}{w^{\nu-2}} \int_1^\infty J_{1-\nu}(zt) \cos\{\tfrac{1}{2}w(1-t^2)\} \frac{dt}{t^{\nu-2}},$$

$$(12) \qquad V_{\nu-1}(w, z) = -\frac{z^{\nu-1}}{w^{\nu-2}} \int_1^\infty J_{1-\nu}(zt) \sin\{\tfrac{1}{2}w(1-t^2)\} \frac{dt}{t^{\nu-2}},$$

provided that w and z are positive and $R(\nu) > \tfrac{1}{2}$.

The following special formulae are worth mention:

$$(13) \qquad \frac{U_{2n}(z, z)}{z} = \int_0^1 J_{2n-1}(zt) \cos\{\tfrac{1}{2}z(1-t^2)\} . t^{2n} \, dt$$

$$= \int_0^1 J_{2n-2}(zt) \sin\{\tfrac{1}{2}z(1-t^2)\} . t^{2n-1} \, dt,$$

$$(14) \qquad \frac{U_{2n+1}(z, z)}{z} = \int_0^1 J_{2n}(zt) \cos\{\tfrac{1}{2}z(1-t^2)\} . t^{2n+1} \, dt$$

$$= \int_0^1 J_{2n-1}(zt) \sin\{\tfrac{1}{2}z(1-t^2)\} . t^{2n} \, dt.$$

Again, from (6), we see that

$$(15) \qquad \int_0^\infty J_{\nu-1}(zt) \exp\left(\mp \frac{iwt^2}{2}\right) . t^\nu \, dt = \frac{z^{\nu-1}}{w^\nu} \exp\left(\pm \frac{iz^2}{2w} \mp \frac{\nu\pi i}{2}\right),$$

and, in particular,

$$(16) \qquad \int_0^\infty J_0(zt) \begin{matrix}\cos\\ \sin\end{matrix} \left(\frac{wt^2}{2}\right) t\,dt = \frac{1}{w} \begin{matrix}\sin\\ \cos\end{matrix} \left(\frac{z^2}{2w}\right).$$

The last results should be compared with § 13·3; see also Hardy, *Trans. Camb. Phil. Soc.* XXI. (1912), pp. 10, 11.

The formulae of this section (with the exception of the contour integrals) are all to be found in one or other of Lommel's two memoirs.

16·54. *Lommel's reciprocation formulae.*

It is evident from § 16·5 (13) that functions of the type $U_\nu(z^2/w, z)$ are closely connected with functions of the type $U_\nu(w, z)$ *provided that* ν *is an integer.*

To appreciate the significance of such relations observe that

$$\frac{d}{dt}\left[\cos(\tfrac{1}{2}wt^2) . U_\nu\left(\frac{z^2}{w}, zt\right) + \sin(\tfrac{1}{2}wt^2) . U_{\nu+1}\left(\frac{z^2}{w}, zt\right)\right]$$

$$= \sin(\tfrac{1}{2}wt^2) . \left[-wtU_\nu\left(\frac{z^2}{w}, zt\right) - wtU_{\nu+2}\left(\frac{z^2}{w}, zt\right)\right]$$

$$= -zJ_\nu(zt)\sin(\tfrac{1}{2}wt^2) . (wt/z)^{1-\nu}.$$

On integration we find that

(1) $$\frac{z^\nu}{w^{\nu-1}}\int_0^1 J_\nu(zt)\sin(\tfrac{1}{2}wt^2)\, t^{1-\nu}\,dt$$

$$= -\cos\tfrac{1}{2}w . U_\nu\left(\frac{z^2}{w}, z\right) - \sin\tfrac{1}{2}w . U_{\nu+1}\left(\frac{z^2}{w}, z\right) + U_\nu\left(\frac{z^2}{w}, 0\right),$$

and, similarly,

(2) $$\frac{z^\nu}{w^{\nu-1}}\int_0^1 J_\nu(zt)\cos(\tfrac{1}{2}wt^2)\, t^{1-\nu}\,dt$$

$$= \sin\tfrac{1}{2}w . U_\nu\left(\frac{z^2}{w}, z\right) - \cos\tfrac{1}{2}w . U_{\nu+1}\left(\frac{z^2}{w}, z\right) + U_{\nu+1}\left(\frac{z^2}{w}, 0\right).$$

Hence it follows that

(3) $$\frac{w^\nu}{z^{\nu-1}}\int_0^1 J_{1-\nu}(zt)\cos\{\tfrac{1}{2}w(1-t^2)\} . t^\nu\,dt$$

$$= -U_{2-\nu}\left(\frac{z^2}{w}, z\right) + \sin\tfrac{1}{2}w . U_{1-\nu}\left(\frac{z^2}{w}, 0\right) + \cos\tfrac{1}{2}w . U_{2-\nu}\left(\frac{z^2}{w}, 0\right),$$

and

(4) $$\frac{w^\nu}{z^{\nu-1}}\int_0^1 J_{1-\nu}(zt)\sin\{\tfrac{1}{2}w(1-t^2)\} . t^\nu\,dt$$

$$= U_{1-\nu}\left(\frac{z^2}{w}, z\right) - \cos\tfrac{1}{2}w . U_{1-\nu}\left(\frac{z^2}{w}, 0\right) + \sin\tfrac{1}{2}w . U_{2-\nu}\left(\frac{z^2}{w}, 0\right),$$

and these integrals differ from the corresponding integrals of the preceding section only in the sign of the order of the Bessel function.

The reader will find some additional formulae concerning Lommel's functions in a paper by Schafheitlin, *Berliner Sitzungsberichte*, VIII. (1909), pp. 62—67.

16·55. *Pseudo-addition formulae for functions of orders $\frac{1}{2}$ and $\frac{3}{2}$.*

Some very curious formulae have been obtained by Lommel, which connect functions of the type $U_\nu(w, z)$ with functions of the same type in which the second variable is zero, provided that ν is equal to $\frac{1}{2}$ or $\frac{3}{2}$.

When we write $\nu = \frac{1}{2}$ in § 16·53 (5), we get

$$U_{\frac{1}{2}}(w, z) \pm iU_{\frac{3}{2}}(w, z) = \left(\frac{w}{2\pi}\right)^{\frac{1}{2}}\int_0^1 \exp\{\pm i(\tfrac{1}{2}w - zt - \tfrac{1}{2}wt^2)\}\,dt$$

$$+ \left(\frac{w}{2\pi}\right)^{\frac{1}{2}}\int_0^1 \exp\{\pm i(\tfrac{1}{2}w + zt - \tfrac{1}{2}wt^2)\}\,dt.$$

Now write

$$\sigma = \frac{(w+z)^2}{2w}, \quad \delta = \frac{(w-z)^2}{2w},$$

and we find that

$$U_{\frac{1}{2}}(w, z) \pm i U_{\frac{3}{2}}(w, z) = e^{\mp iz} \left(\frac{\sigma}{\pi}\right)^{\frac{1}{2}} \int_{\sqrt{\{z^2/(2w\sigma)\}}}^{1} \exp\{\pm \sigma i (1-\xi^2)\}\, d\xi$$
$$+ e^{\pm iz} \left(\frac{\delta}{\pi}\right)^{\frac{1}{2}} \int_{-\sqrt{\{z^2/(2w\delta)\}}}^{1} \exp\{\pm \delta i (1-\xi^2)\}\, d\xi,$$

where

$$\xi = \left(\frac{w}{2\sigma}\right)^{\frac{1}{2}} \left(t + \frac{z}{w}\right), \quad \left(\frac{w}{2\delta}\right)^{\frac{1}{2}} \left(t - \frac{z}{w}\right)$$

in the respective integrals, *and* $\sqrt{\sigma}$, $\sqrt{\delta}$ *are to be interpreted by the conventions**

$$\sqrt{\sigma} = \frac{w+z}{\sqrt{(2w)}}, \quad \sqrt{\delta} = \frac{w-z}{\sqrt{(2w)}}.$$

Hence we have

$$U\ (w, z) \pm i U_{\frac{3}{2}}(w, z) = \tfrac{1}{2} e^{\mp iz} \{U_{\frac{1}{2}}(2\sigma, 0) \pm i U_{\frac{3}{2}}(2\sigma, 0)\}$$
$$+ \tfrac{1}{2} e^{\pm iz} \{U_{\frac{1}{2}}(2\delta, 0) \pm i U_{\frac{3}{2}}(2\delta, 0)\}$$
$$- e^{\mp iz} \left(\frac{\sigma}{\pi}\right)^{\frac{1}{2}} \int_0^{\sqrt{\{z^2/(2w\sigma)\}}} \exp\{\pm \sigma i (1-\xi^2)\}\, d\xi$$
$$+ e^{\pm iz} \left(\frac{\delta}{\pi}\right)^{\frac{1}{2}} \int_0^{\sqrt{\{z^2/(2w\delta)\}}} \exp\{\pm \delta i (1-\xi^2)\}\, d\xi.$$

When we take $\sigma\xi^2$ and $\delta\xi^2$ as new variables in the last two integrals respectively, these integrals are seen to cancel; and so we have the two results combined in the formula

(1) $$U_{\frac{1}{2}}(w, z) \pm i U_{\frac{3}{2}}(w, z) = \tfrac{1}{2} e^{\mp iz} \{U_{\frac{1}{2}}(2\sigma, 0) \pm i U_{\frac{3}{2}}(2\sigma, 0)\} + \tfrac{1}{2} e^{\pm iz} \{U_{\frac{1}{2}}(2\delta, 0) \pm i U_{\frac{3}{2}}(2\delta, 0)\},$$

and, as a corollary,

(2) $$U_{\frac{1}{2}}(z, z) \pm i U_{\frac{3}{2}}(z, z) = \tfrac{1}{2} e^{\mp iz} \{U_{\frac{1}{2}}(4z, 0) \pm i U_{\frac{3}{2}}(4z, 0)\}.$$

These formulae are due to Lommel, *Münchener Abh.* xv. (1886), pp. 601–605; they are reproduced by Walker, *The Analytical Theory of Light* (Cambridge, 1904), pp. 401–402.

16·56. *Fresnel's integrals.*

It is easy to see from § 16·53 (1) and (2) that, when $R(\nu) > 0$,

$$U_\nu(w, 0) = \frac{w^\nu}{2^{\nu-1}\Gamma(\nu)} \int_0^1 t^{2\nu-1} \cos\{\tfrac{1}{2} w (1-t^2)\}\, dt,$$

$$U_{\nu+1}(w, 0) = \frac{w^\nu}{2^{\nu-1}\Gamma(\nu)} \int_0^1 t^{2\nu-1} \sin\{\tfrac{1}{2} w (1-t^2)\}\, dt,$$

so that

(1) $$\frac{w^\nu}{2^{\nu-1}\Gamma(\nu)} \int_0^1 t^{2\nu-1} \cos(\tfrac{1}{2} w t^2)\, dt = U_\nu(w, 0) \cos\tfrac{1}{2} w + U_{\nu+1}(w, 0) \sin\tfrac{1}{2} w,$$

(2) $$\frac{w^\nu}{2^{\nu-1}\Gamma(\nu)} \int_0^1 t^{2\nu-1} \sin(\tfrac{1}{2} w t^2)\, dt = U_\nu(w, 0) \sin\tfrac{1}{2} w - U_{\nu+1}(w, 0) \cos\tfrac{1}{2} w.$$

* These are not the same as the conventions used by Lommel.

If we take $\nu = \frac{1}{2}$ and modify the notation by writing $\frac{1}{2}w = z = \frac{1}{2}\pi u^2$, we see that

$$(3)\quad \int_0^u \cos(\tfrac{1}{2}\pi t^2)\,dt = \frac{1}{2}\int_0^z \left(\frac{2}{\pi t}\right)^{\frac{1}{2}} \cos t\,dt = \frac{1}{2}\int_0^z J_{-\frac{1}{2}}(t)\,dt$$
$$= [U_{\frac{1}{2}}(2z, 0)\cos z + U_{\frac{3}{2}}(2z, 0)\sin z]/\sqrt{2}$$
$$= \tfrac{1}{2} + [V_{\frac{1}{2}}(2z, 0)\sin z + V_{\frac{3}{2}}(2z, 0)\cos z]/\sqrt{2},$$

and

$$(4)\quad \int_0^u \sin(\tfrac{1}{2}\pi t^2)\,dt = \frac{1}{2}\int_0^z \left(\frac{2}{\pi t}\right)^{\frac{1}{2}} \sin t\,dt = \frac{1}{2}\int_0^z J_{\frac{1}{2}}(t)\,dt$$
$$= [U_{\frac{1}{2}}(2z, 0)\sin z - U_{\frac{3}{2}}(2z, 0)\cos z]/\sqrt{2}$$
$$= \tfrac{1}{2} - [V_{\frac{1}{2}}(2z, 0)\cos z - V_{\frac{3}{2}}(2z, 0)\sin z]/\sqrt{2}.$$

We thus obtain ascending series and asymptotic expansions for *Fresnel's integrals**

$$\int_0^u \cos(\tfrac{1}{2}\pi t^2)\,dt, \quad \int_0^u \sin(\tfrac{1}{2}\pi t^2)\,dt.$$

The ascending series, originally given by Knockenhauer, *Ann. der Physik und Chemie*, (2) XLI. (1837), p. 104, are readily derived from the U-series, namely

$$(5)\quad U_{\frac{1}{2}}(2z, 0) = \left(\frac{4z}{\pi}\right)^{\frac{1}{2}} \left\{1 - \frac{(2z)^2}{1.3.5} + \frac{(2z)^4}{1.3.5.7.9} - \dots\right\},$$

$$(6)\quad U_{\frac{3}{2}}(2z, 0) = \left(\frac{4z}{\pi}\right)^{\frac{1}{2}} \left\{\frac{2z}{1.3} - \frac{(2z)^3}{1.3.5.7} + \dots\right\},$$

while the asymptotic expansions, due to Cauchy, *Comptes Rendus*, XV. (1842), pp. 554, 573, are derived with equal ease from the V-series, namely

$$(7)\quad V_{\frac{1}{2}}(2z, 0) \sim \left(\frac{4z}{\pi}\right)^{\frac{1}{2}} \left\{\frac{1}{2z} - \frac{1.3}{(2z)^3} + \frac{1.3.5.7}{(2z)^5} - \dots\right\},$$

$$(8)\quad V_{\frac{3}{2}}(2z, 0) \sim -\left(\frac{4z}{\pi}\right)^{\frac{1}{2}} \left\{\frac{1}{(2z)^2} - \frac{1.3.5}{(2z)^4} + \dots\right\}.$$

Tables of Fresnel's integrals were constructed by Gilbert, *Mém. couronnées de l'Acad. R. des Sci. de Bruxelles*, XXXI. (1863), pp. 1—52, and Lindstedt, *Ann. der Physik und Chemie*, (3) XVII. (1882), p. 720; and by Lommel in his second memoir.

Lommel has given various representations of Fresnel's integrals by series which are special cases of the formulae†

$$(9)\quad \int_0^z J_\nu(t)\,dt = 2\sum_{n=0}^{\infty} J_{\nu+2n+1}(z)$$
$$= \sum_{n=0}^{\infty} \frac{z^{n+1} J_{\nu+n}(z)}{(\nu+1)(\nu+3)\dots(\nu+2n+1)},$$

$$(10)\quad \int_z^\infty J_\nu(t)\,dt \sim \sum_{n=0}^{\infty} \frac{(\nu+1)(\nu+3)\dots(\nu+2n-1)}{z^n} J_{\nu+n}(z).$$

* *Mém. de l'Acad. des Sci.* v. (1818), p. 339. [*Oeuvres*, I. (1866), p. 176.]

† It is supposed in (9) that $R(\nu) > -1$.

These are readily verified by differentiation. Other formulae also due to Lommel are

$$(11)\quad \int_0^z J_{-\frac{1}{2}}(t)\,dt = 2^{\frac{3}{2}}\cos\tfrac{1}{2}z\left[\sum_{n=0}^{\infty}(-)^n J_{2n+\frac{1}{2}}(\tfrac{1}{2}z)\right] + 2^{\frac{3}{2}}\sin\tfrac{1}{2}z\left[\sum_{n=0}^{\infty}(-)^n J_{2n+\frac{3}{2}}(\tfrac{1}{2}z)\right],$$

$$(12)\quad \int_0^z J_{\frac{1}{2}}(t)\,dt = 2^{\frac{3}{2}}\sin\tfrac{1}{2}z\left[\sum_{n=0}^{\infty}(-)^n J_{2n+\frac{1}{2}}(\tfrac{1}{2}z)\right] - 2^{\frac{3}{2}}\cos\tfrac{1}{2}z\left[\sum_{n=0}^{\infty}(-)^n J_{2n+\frac{3}{2}}(\tfrac{1}{2}z)\right].$$

These may also be verified by differentiation.

16·57. *Hardy's integrals for Lommel's functions.*

The fact that the integrals

$$\int_0^\infty \cos\left(at-\frac{b}{t}\right)\frac{dt}{1\pm t^2},\quad \int_0^\infty \sin\left(at-\frac{b}{t}\right)\frac{t\,dt}{1\pm t^2}$$

are expressible in terms of elementary functions* when a and b are positive suggested to Hardy† the consideration of the integrals

$$\int_0^\infty \cos\left(at+\frac{b}{t}\right)\frac{dt}{1\pm t^2},\quad \int_0^\infty \sin\left(at+\frac{b}{t}\right)\frac{t\,dt}{1\pm t^2},$$

and he found them to be expressible in terms of Lommel's functions of two variables of orders zero and unity respectively. This discovery is important because the majority of the integrals representing such functions contain Bessel functions under the integral sign.

If $1/t$ be written in place of t, it is seen that

$$(1)\quad \int_0^\infty \cos\left(at+\frac{b}{t}\right)\frac{dt}{1\pm t^2} = \pm\int_0^\infty \cos\left(bt+\frac{a}{t}\right)\frac{dt}{1\pm t^2},$$

$$(2)\quad \int_0^\infty \sin\left(at+\frac{b}{t}\right)\frac{t\,dt}{1\pm t^2} = \pm\int_0^\infty \sin\left(at+\frac{b}{t}\right)\frac{dt}{t} - \int_0^\infty \sin\left(bt+\frac{a}{t}\right)\frac{t\,dt}{1\pm t^2},$$

and since, by § 6·13 (3),

$$\frac{2}{\pi}\int_0^\infty \sin\left(at+\frac{b}{t}\right)\frac{dt}{t} = J_0\{2\sqrt{(ab)}\},$$

it is sufficient to confine our attention to the case in which $b < a$.

We now write

$$c = \sqrt{(b/a)},\quad x = 2\sqrt{(ab)},\quad \theta = \tfrac{1}{2}(1-c^2)/c,$$

* Hardy, *Quarterly Journal*, XXXII. (1901), p. 374. When the lower sign is taken it is supposed that the integrals have their principal values.

† *Messenger*, XXXVIII. (1909), pp. 129—132.

and then the substitutions $t = ce^u$ and $\cosh u = \tau$ shew that

$$\int_0^\infty \cos\left(at + \frac{b}{t}\right)\frac{dt}{1+t^2} = \int_{-\infty}^\infty \frac{\cos(x\cosh u)\,du}{ce^u + 1/(ce^u)}$$

$$= \int_0^\infty \cos(x\cosh u)\left\{\frac{1}{ce^u + 1/(ce^u)} + \frac{1}{ce^{-u} + 1/(ce^{-u})}\right\}du$$

$$= \frac{a+b}{x}\int_1^\infty \frac{\cos(x\tau)\,.\,\tau\,d\tau}{(\theta^2+\tau^2)\sqrt{(\tau^2-1)}}.$$

Now consider

$$\frac{1}{2\pi i}\int_\Gamma \frac{e^{ix\tau}\,\tau\,d\tau}{(\theta^2+\tau^2)\sqrt{(\tau^2-1)}},$$

where Γ is a contour consisting of the real axis and a large semicircle above it, the real axis being indented at $\tau = \pm 1$.

The only pole of the integrand inside the contour is at $i\theta$, and so

$$\frac{1}{2\pi i}\int_\Gamma \frac{e^{ix\tau}\,\tau\,d\tau}{(\theta^2+\tau^2)\sqrt{(\tau^2-1)}} = \frac{e^{-x\theta}}{2i\sqrt{(\theta^2+1)}}.$$

As the radius of the large semicircle tends to infinity, the integral round it tends to zero by Jordan's lemma, and hence

$$\int_1^\infty \frac{\cos(x\tau)\,.\,\tau\,d\tau}{(\theta^2+\tau^2)\sqrt{(\tau^2-1)}} + \frac{1}{2}\int_{-1}^1 \frac{\sin(x\tau)\,.\,\tau\,d\tau}{(\theta^2+\tau^2)\sqrt{(1-\tau^2)}} = \frac{\pi e^{-x\theta}}{2\sqrt{(\theta^2+1)}}.$$

Thus we have

$$\int_0^\infty \cos\left(at+\frac{b}{t}\right)\frac{dt}{1+t^2} = \frac{\pi e^{-(a-b)}}{2} - \frac{a+b}{2x}\int_0^\pi \frac{\sin(x\cos\phi)\,.\,\cos\phi\,d\phi}{\theta^2+\cos^2\phi}.$$

But $$\frac{\cos\phi}{\theta^2+\cos^2\phi} = \frac{4c}{1+c^2}\{c\cos\phi - c^3\cos 3\phi + c^5\cos 5\phi - \ldots\},$$

and so we find that

(3) $$\int_0^\infty \cos\left(at+\frac{b}{t}\right)\frac{dt}{1+t^2} = \frac{\pi e^{-(a-b)}}{2} - \pi\sum_{m=1}^\infty c^{2m-1}J_{2m-1}(x).$$

Similarly it is found that*

(4) $$\int_0^\infty \sin\left(at+\frac{b}{t}\right)\frac{t\,dt}{1+t^2} = \frac{\pi e^{-(a-b)}}{2} - \pi\sum_{m=1}^\infty c^{2m}J_{2m}(x),$$

and

(5) $$P\int_0^\infty \cos\left(at+\frac{b}{t}\right)\frac{dt}{1-t^2} = \tfrac{1}{2}\pi\sin(a+b) - \pi\sum_{m=1}^\infty (-)^{m-1}c^{2m-1}J_{2m-1}(x),$$

(6) $$P\int_0^\infty \sin\left(at+\frac{b}{t}\right)\frac{t\,dt}{1-t^2} = -\tfrac{1}{2}\pi\cos(a+b) - \pi\sum_{m=1}^\infty (-)^{m-1}c^{2m}J_{2m}(x).$$

* The details of the analysis will be found in Hardy's paper.

The last two results may be written in the form

(7) $$U_1(w, x) + V_1(w, x) = -\frac{2}{\pi} P \int_0^\infty \cos\left(\frac{x^2 t}{2w} + \frac{w}{2t}\right) \frac{dt}{1-t^2},$$

(8) $$U_2(w, x) + V_0(w, x) = -\frac{2}{\pi} P \int_0^\infty \sin\left(\frac{x^2 t}{2w} + \frac{w}{2t}\right) \frac{t\,dt}{1-t^2},$$

provided that $0 < w < x$.

16·58. *Integrals of Gilbert's type for Lommel's functions.*

An obvious method of representing $U_\nu(w, z)$ and $V_\nu(w, z)$ by integrals is to substitute the Bessel-Schläfli integral of § 6·2 for each Bessel function in the appropriate series. We thus get

$$U_\nu(w, z) = \frac{1}{2\pi i} \sum_{m=0}^{\infty} (-)^m (\tfrac{1}{2}w)^{\nu+2m} \int_{-\infty}^{(0+)} \exp\left(t - \frac{z^2}{4t}\right) \frac{dt}{t^{\nu+2m+1}}.$$

When the contour is so chosen that it lies wholly outside the circle on which $|t| = \frac{1}{2}|w|$, we may change the order of summation and integration and get

(1) $$U_\nu(w, z) = \frac{1}{2\pi i} \int_{-\infty}^{(0+,\, \frac{1}{2}iw+,\, -\frac{1}{2}iw+)} \frac{(\frac{1}{2}w/t)^\nu}{1 + \frac{1}{4}w^2/t^2} \exp\left(t - \frac{z^2}{4t}\right) \frac{dt}{t}.$$

Now the residues of the integrand at $\pm \frac{1}{2}iw$ are

$$\tfrac{1}{2} \exp\left\{\pm \frac{iw}{2} \pm \frac{iz^2}{2w} \mp \frac{\nu\pi i}{2}\right\},$$

and so

$$V_{2-\nu}(w, z) = \frac{1}{2\pi i} \int_{-\infty}^{(0+)} \frac{(\frac{1}{2}w/t)^\nu}{1 + \frac{1}{4}w^2/t^2} \exp\left(t - \frac{z^2}{4t}\right) \frac{dt}{t}.$$

Making a slight change in the notation, we deduce that

(2) $$V_\nu(w, z) = \frac{1}{2\pi i} \int_{-\infty}^{(0+)} \frac{(t/w)^\nu}{1 + t^2/w^2} \exp\left(\frac{t}{2} - \frac{z^2}{2t}\right) \frac{dt}{t},$$

and, in this integral, the points $\pm iw$ lie outside the contour.

In general it is impossible to modify the contour in (2) into the negative half of the real axis taken twice, in consequence of the essential singularity of the integrand at the origin. The exception occurs when $z = 0$, because then the essential singularity disappears, and

(3) $$V_\nu(w, 0) = \frac{1}{2\pi i} \int_{-\infty}^{(0+)} \frac{(t/w)^\nu e^{\frac{1}{2}t}}{1 + t^2/w^2} \frac{dt}{t},$$

and hence

(4) $$V_\nu(w, 0) = \frac{\sin \nu\pi}{\pi} \int_0^{\infty \exp i\alpha} \frac{u^{\nu-1} e^{-\frac{1}{2}uw}}{1 + u^2} du,$$

provided that $R(\nu) > 0$ and α is an acute angle (positive or negative) such that

$$|\alpha + \arg w| < \tfrac{1}{2}\pi.$$

If ν is equal to $\frac{1}{2}$ or $\frac{3}{2}$, the integral on the right in (4) is called *Gilbert's integral**.

Formula (4) was obtained by Lommel† from the formula of § 16·53 (11) by a transformation of infinite integrals.

From (4) it is clear that, when ν and w are positive, $V_\nu(w, 0)$ has the same sign as, and is numerically less than

$$\frac{\sin \nu\pi}{\pi}\int_0^\infty u^{\nu-1}\, e^{-\frac{1}{2}uw}\, du = \frac{1}{\Gamma(1-\nu)\,.\,(\frac{1}{2}w)^\nu}.$$

A similar but less exact inequality was obtained by Lommel.

The reader will also observe that $V_\nu(w, 0)/\sin\nu\pi$ is a positive decreasing function of w when w is positive.

16·59. *Asymptotic expansions of Lommel's functions of two variables.*

From Gilbert's integrals it is easy to deduce asymptotic expansions of $V_\nu(w, 0)$ and $U_\nu(w, 0)$ for large values of $|w|$; thus, from § 16·5 (8), we have

$$V_\nu(w, 0) = \sum_{m=0}^{p-1} \frac{(-)^m}{\Gamma(1-\nu-2m)\,.\,(\frac{1}{2}w)^{\nu+2m}} + (-)^p V_{\nu+2p}(w, 0),$$

where p is any positive integer. We choose p to be so large that $R(\nu+2p) > 0$ and then, by § 16·58 (4), we have

$$\begin{aligned}(-)^p V_{\nu+2p}(w, 0) &= \frac{\sin\nu\pi}{\pi}\int_0^{\infty \exp i\alpha} \frac{u^{\nu+2p-1}\, e^{-\frac{1}{2}uw}\, du}{1+u^2}\\ &= O\left\{\int_0^{\infty\exp i\alpha} u^{\nu+2p-1}\, e^{-\frac{1}{2}uw}\, du\right\}\\ &= O(w^{-\nu-2p}),\end{aligned}$$

when $|w|$ is large and, as in the similar analysis of § 7·2, $|\arg w| < \pi$.

Hence

(1) $$V_\nu(w, 0) \sim \sum_{m=0}^{\infty} \frac{(-)^m}{\Gamma(1-\nu-2m)\,.\,(\frac{1}{2}w)^{\nu+2m}}$$

for the values of w under consideration.

When $\nu+2p$ and w are both positive, $(-)^p V_{\nu+2p}(w, 0)$ has the same sign as, and is numerically less than

$$\frac{\sin\nu\pi}{\pi}\int_0^\infty u^{\nu+2p-1}\, e^{-\frac{1}{2}uw}\, du = \frac{(-)^p}{\Gamma(1-\nu-2p)\,.\,(\frac{1}{2}w)^{\nu+2p}},$$

so that the remainder after p terms in (1) has the same sign as, and is numerically less than the $(p+1)$th term.

It may be proved in like manner from § 16·53 (11) that

(2) $$V_\nu(w, z) \sim \sum_{m=0}^{\infty} (-)^m (z/w)^{\nu+2m} J_{-\nu-2m}(z)$$

when $|w|$ is large while ν and z are fixed; but it is not easy to obtain a simple expression which gives the magnitude and sign of the remainder.

* *Mém. couronnées de l'Acad. R. des Sci. de Bruxelles*, XXXI. (1863), pp. 1—52.

† *Münchener Abh.* XV. (1886), pp. 582—585.

It is evident from § 16·5 (6) that the corresponding formulae for $U_\nu(w, z)$ are

$$(3) \qquad U_\nu(w, 0) \sim \cos(\tfrac{1}{2}w - \tfrac{1}{2}\nu\pi) + \sum_{m=0}^{\infty} \frac{(-)^m}{\Gamma(\nu - 1 - 2m)(\frac{1}{2}w)^{2m-\nu+2}},$$

$$(4) \qquad U_\nu(w, z) \sim \cos(\tfrac{1}{2}w + \tfrac{1}{2}z^2/w - \tfrac{1}{2}\nu\pi) + \sum_{m=0}^{\infty} (-)^m (z/w)^{2m-\nu+2} J_{\nu-2-2m}(z).$$

These results were given by Lommel*, but he did not investigate them in any detail.

The asymptotic expansion of $V_\nu(cx, x)$, when ν is 0 or 1 and c is fixed, while x is large and positive, has been investigated by Mayall†.

The dominant term for general (real) values of ν greater than $-\frac{1}{2}$ is readily derived from § 16·53 (12) which shews that

$$V_\nu(cx, x) \sim -\frac{x}{c^{\nu-1}} \int_1^\infty \left(\frac{2}{\pi x t}\right)^{\frac{1}{2}} \cos(xt + \tfrac{1}{2}\nu\pi - \tfrac{1}{4}\pi) \sin\{\tfrac{1}{2}cx(1 - t^2)\} \frac{dt}{t^{\nu-1}}.$$

Now, if $c > 1$, the functions $\frac{1}{2}cx(1-t^2) \pm (xt + \frac{1}{2}\nu\pi - \frac{1}{4}\pi)$ vary monotonically as t increases from 1 to ∞, and hence it may be verified by partial integrations that

$$(5) \qquad V_\nu(cx, x) \sim \left(\frac{2}{\pi x}\right)^{\frac{1}{2}} \frac{c^{2-\nu}}{c^2 - 1} \cos(x + \tfrac{1}{2}\nu\pi - \tfrac{1}{4}\pi),$$

the next term in the asymptotic expansion being $O(x^{-\frac{3}{2}})$.

If, however, $c < 1$, then $\frac{1}{2}cx(1 - t^2) + (xt + \frac{1}{2}\nu\pi - \frac{1}{4}\pi)$, *qua* function of t, has a maximum at $1/c$; and hence, by the principle of stationary phase (§ 8·2), it follows that

$$(6) \qquad V_\nu(cx, x) \sim \frac{1}{c^{\nu-1}} \cos\left\{\tfrac{1}{2}x\left(c + \frac{1}{c}\right) + \tfrac{1}{2}\nu\pi\right\}.$$

Finally, when $c = 1$, the maximum-point is at one end of the range of integration, and so the expression on the right in (6) must be halved. We consequently have

$$(7) \qquad V_\nu(x, x) \sim \tfrac{1}{2}\cos(x + \tfrac{1}{2}\nu\pi).$$

This equation, like (5) and (6), has been established on the hypothesis that $\nu > -\frac{1}{2}$; the three equations may now be proved for all real values of ν by using the recurrence formula § 16·5 (8).

* *Münchener Abh.* xv. (1886), pp. 540, 572—573.

† *Proc. Camb. Phil. Soc.* ix. (1898), pp. 259—269.

CHAPTER XVII

KAPTEYN SERIES

17·1. *Definition of Kapteyn series.*

Any series of the type

$$\sum_{n=0}^{\infty} \alpha_n J_{\nu+n}\{(\nu+n)z\},$$

in which* ν and the coefficients α_n are constants, is called a *Kapteyn series.*

Such series owe their name to the fact that they were first systematically investigated, *qua* functions of the complex variable z, by Kapteyn† in an important memoir published in 1893. In this memoir Kapteyn examined the question of the possibility of expanding an arbitrary analytic function into such a series, and generally he endeavoured to put the theory of such series into a position similar to that which was then occupied by Neumann series.

Although the properties of Kapteyn series are, in general, of a more recondite character than properties of Neumann series, yet Kapteyn series are of more practical importance; they first made their appearance in the solution of Kepler's problem which was discovered by Lagrange‡ and rediscovered half a century later by Bessel§; and related series are of general occurrence in a class of problems concerning elliptic motion under the inverse square law, of which Kepler's problem may be taken as typical. More recently, in the hands of Schott|| they have proved to be of frequent occurrence in the modern theory of Electromagnetic Radiation.

The astronomical problems, in which all the variables concerned are real, are of a much more simple analytical character than the problems investigated by Kapteyn; and in order to develop the theory of Kapteyn series in a simple manner, it seems advisable to begin with a description of the series which occur in connexion with elliptic motion.

17·2. *Kepler's problem and allied problems discussed by Bessel.*

The notation which will be used in this section in the discussion of the motion in an ellipse of a particle under the action of a centre of force at the focus, attracting the particle according to the inverse square law, is as follows:

The semi-major axis, semi-minor axis, and the eccentricity of the ellipse are denoted by a, b, and ϵ. The axes of the ellipse are taken as coordinate

* It will, for the most part, be assumed that ν is zero.

† *Ann. sci. de l'École norm. sup.* (3) x. (1893), pp. 91—120.

‡ *Hist. de l'Acad. R. des Sci. de Berlin*, xxv. (1769) [1770], pp. 204—233. [*Oeuvres*, III. (1869), pp. 113—138.]

§ *Berliner Abh.* 1816—7 [1819], pp. 49—55.

|| *Electromagnetic Radiation* (Cambridge, 1912).

axes, the direction of the axis of x being from the centre of the ellipse to the centre of force. The centre of force is taken as origin of polar coordinates, the radius vector to the particle being r, and the *true anomaly*, namely the angle between the radius vector and the axis of x, being w. The *eccentric anomaly*, namely the eccentric angle of the particle on the ellipse, is denoted by E. The time which has elapsed from an instant when the particle was at the positive end of the major axis is called t.

The *mean anomaly* M is defined as the angle through which the radius vector would turn in time t if the radius vector rotated uniformly in such a way as to perform complete revolutions in the time it actually takes to perform complete revolutions.

The geometrical properties of the ellipse supply the equations*

$$(1) \qquad r = \frac{a(1-\epsilon^2)}{1+\epsilon\cos w} = a(1-\epsilon\cos E),$$

from which the equations

$$(2) \qquad \tan\tfrac{1}{2}w = \sqrt{\left(\frac{1+\epsilon}{1-\epsilon}\right)}\tan\tfrac{1}{2}E,$$

$$(3) \qquad \sin w = \frac{\sqrt{(1-\epsilon^2)}\,.\sin E}{1-\epsilon\cos E}, \quad \sin E = \frac{\sqrt{(1-\epsilon^2)}\,.\sin w}{1+\epsilon\cos w}$$

are deducible; and an integrated form of the equations of motion (the analytical expression of Kepler's Second Law) supplies the equation

$$(4) \qquad M = E - \epsilon\sin E.$$

Kepler's problem is that of expressing the various coordinates r, w, E, which determine the position of the particle†, in terms of the time t, that is, effectively, in terms of M. It is of course supposed that the variables are real and, since the motion is elliptic (or parabolic, as a limiting case), $0 < \epsilon \leqslant 1$.

The solution of the problem which was effected by Lagrange was of an approximate character, because he calculated only the first few terms in the expansions of E and r.

The more complete solution given by Bessel depends on the fact that (4) defines E as a continuous increasing function of M such that the effect of increasing M by 2π is to increase E by 2π.

It follows that any function of E with limited total fluctuation is a function of M with limited total fluctuation, and so such functions of E are expansible in Fourier series, *qua* functions of M.

* The construction of these equations will be found in any text-book on Astronomy or Dynamics of a Particle. See, e.g. Plummer, *Dynamical Astronomy* (Cambridge, 1918), Ch. III.

† Kepler himself was concerned with the expression of E in terms of M.

In particular $\epsilon \sin E$ is an odd periodic function of M, and so, for all real values of E, it is expansible into the Fourier sine-series

$$\epsilon \sin E = \sum_{n=1}^{\infty} A_n \sin nM,$$

where
$$\begin{aligned} A_n &= \frac{2}{\pi}\int_0^{\pi} \epsilon \sin E \sin nM\,dM \\ &= \left[-\frac{2\epsilon \sin E \cos nM}{n\pi}\right]_0^{\pi} + \frac{2}{n\pi}\int_0^{\pi} \cos nM \frac{d(\epsilon \sin E)}{dM}\,dM \\ &= \frac{2}{n\pi}\int_0^{\pi} \cos nM \frac{dE - dM}{dM}\,dM \\ &= \frac{2}{n\pi}\int_0^{\pi} \cos nM\,.\,dE \\ &= \frac{2}{n} J_n(n\epsilon). \end{aligned}$$

Hence it follows that

$$E = M + \sum_{n=1}^{\infty} \frac{2}{n} J_n(n\epsilon) \sin nM, \tag{5}$$

and this result gives the complete analytical solution of Kepler's problem concerning the eccentric anomaly. The series on the right is a Kapteyn series which converges rapidly when $\epsilon < 1$, and it is still convergent when $\epsilon = 1$; cf. §§ 8·4, 8·42.

The radius vector is similarly expansible as a cosine series, thus

$$\frac{r}{a} = B_0 + \sum_{n=1}^{\infty} B_n \cos nM,$$

where
$$\begin{aligned} B_0 &= \frac{1}{\pi}\int_0^{\pi} (1 - \epsilon \cos E)\,dM \\ &= \frac{1}{\pi}\int_0^{\pi} (1 - \epsilon \cos E)^2\,dE \\ &= 1 + \tfrac{1}{2}\epsilon^2, \end{aligned}$$

while, when $n \neq 0$,

$$\begin{aligned} B_n &= \frac{2}{\pi}\int_0^{\pi} (1 - \epsilon \cos E) \cos nM\,dM \\ &= \left[\frac{2(1 - \epsilon \cos E) \sin nM}{n\pi}\right]_0^{\pi} + \frac{2}{n\pi}\int_0^{\pi} \sin nM \frac{d(\epsilon \cos E)}{dM}\,dM \\ &= -\frac{2\epsilon}{n\pi}\int_0^{\pi} \sin E \sin(nE - n\epsilon \sin E)\,dE \\ &= -\frac{2\epsilon}{n} J_n'(n\epsilon), \end{aligned}$$

so that

$$\frac{r}{a} = 1 + \tfrac{1}{2}\epsilon^2 - \sum_{n=1}^{\infty} \frac{2\epsilon}{n} J_n'(n\epsilon) \cos nM. \tag{6}$$

The expansion of the true anomaly is derived from the consideration that $w - M$ is an odd periodic function of M, and so

$$w - M = \sum_{n=1}^{\infty} C_n \sin nM,$$

where
$$C_n = \frac{2}{\pi}\int_0^\pi (w - M)\sin nM\, dM$$
$$= \left[-\frac{2(w-M)\cos nM}{n\pi}\right]_0^\pi + \frac{2}{n\pi}\int_0^\pi \cos nM \,.\left(\frac{dw}{dM} - 1\right) dM$$
$$= \frac{2}{n\pi}\int_0^\pi \cos nM \,.\, \frac{dw}{dE}\, dE$$
$$= \frac{2\sqrt{(1-\epsilon^2)}}{n\pi}\int_0^\pi \frac{\cos(nE - n\epsilon\sin E)}{1 - \epsilon\cos E}\, dE.$$

This expression is not such a simple transcendent as the coefficients A_n and B_n. The most effective method of evaluating it is due to Bessel*, who used the expansion

$$\frac{\sqrt{(1-\epsilon^2)}}{1-\epsilon\cos E} = 1 + 2f\cos E + 2f^2\cos 2E + 2f^3\cos 3E + \ldots,$$

where
$$f = \frac{\epsilon}{1 + \sqrt{(1-\epsilon^2)}}.$$

On making the substitution, we find at once that

$$C_n = \frac{2}{n}\left[J_n(n\epsilon) + \sum_{m=1}^{\infty} f^m\{J_{n-m}(n\epsilon) + J_{n+m}(n\epsilon)\}\right].$$

17·21. *Expansions associated with the Kepler-Bessel expansions.*

A large class of expressions associated with the radius vector, true anomaly and eccentric anomaly, are expansible in series of much the same type as those just discussed. Such series have been investigated in a systematic manner by Herz†, and we shall now state a few of the more important of them; they are all obtainable by Fourier's rule, and it seems unnecessary to write out in detail the analysis, which the reader will easily construct for himself.

First, we have

$$r\cos w = x - a\epsilon = \frac{a(1-\epsilon^2) - r}{\epsilon},$$

so that

$$(1)\qquad \frac{r\cos w}{a} = -\tfrac{3}{2}\epsilon + \sum_{n=1}^{\infty}\frac{2}{n}J_n'(n\epsilon)\cos nM,$$

and next

$$(2)\qquad \frac{r\sin w}{a} = \frac{b}{a}\sin E = \frac{\sqrt{(1-\epsilon^2)}}{\epsilon}\sum_{n=1}^{\infty}\frac{2}{n}J_n(n\epsilon)\sin nM,$$

* *Berliner Abh.* 1824 [1826], p. 42.

† *Astr. Nach.* CVII. (1884), col. 17—28. Various expansions had also been given by Plana, *Mem. della R. Accad. delle Sci. di Torino*, (2) x. (1849), pp. 249—332. In connexion with their convergence, see Cauchy, *Comptes Rendus*, XVIII. (1844), pp. 625—643. [*Oeuvres*, (1) VIII. (1893), pp. 168—188.]

while

(3) $$\cos E = \frac{a-r}{a\epsilon} = -\tfrac{1}{2}\epsilon + \sum_{n=1}^{\infty} \frac{2}{n} J_n'(n\epsilon)\cos nM.$$

Next, if m is any positive integer*,

(4) $$\cos mE = m \sum_{n=1}^{\infty} \frac{1}{n} \{J_{n-m}(n\epsilon) - J_{n+m}(n\epsilon)\} \cos nM,$$

(5) $$\sin mE = m \sum_{n=1}^{\infty} \frac{1}{n} \{J_{n-m}(n\epsilon) + J_{n+m}(n\epsilon)\} \sin nM.$$

The expansion of a/r is particularly simple, namely,

(6) $$\frac{a}{r} = 1 + 2 \sum_{n=1}^{\infty} J_n(n\epsilon)\cos nM.$$

The expansions of $\cos w$ and $\sin w$ are

(7) $$\cos w = -\epsilon + \frac{1-\epsilon^2}{\epsilon} \sum_{n=}^{\infty} 2J_n(n\epsilon)\cos nM,$$

(8) $$\sin w = \sqrt{(1-\epsilon^2)} \sum_{n=1}^{\infty} 2J_n'(n\epsilon)\sin nM.$$

The expansions of $\cos w/r^2$, $\sin w/r^2$ are of a simple form, namely

(9) $$\frac{a^2}{r^2}\cos w = \sum_{n=1}^{\infty} 2nJ_n'(n\epsilon)\cos nM,$$

(10) $$\frac{a^2}{r^2}\sin w = \frac{\sqrt{(1-\epsilon^2)}}{\epsilon} \sum_{n=1}^{\infty} 2n\, J_n(n\epsilon)\sin nM.$$

[NOTE. It is pointed out by Plummer, *Dynamical Astronomy* (Cambridge, 1918), p. 39, that these are readily derived from the Cartesian equations of motion in the form

$$\frac{d^2x}{dM^2} + \frac{a^3\cos w}{r^2} = 0, \quad \frac{d^2y}{dM^2} + \frac{a^3 \sin w}{r^2} = 0,$$

combined with (1) and (2).]

17·22. *Sums of special Kapteyn series.*

The reader will observe that, in the case of the expansions of even functions of M, the results simplify when we take the particle to be at one of the ends of the major axis, because then the three anomalies are all equal to 0 or to π, while the radius vector is equal to $a(1-\epsilon)$ or to $a(1+\epsilon)$. From the results of the last section we thus obtain the following formulae, which were given by Herz in the paper already quoted:

(1) $$\tfrac{1}{2} + \tfrac{1}{4}\epsilon = \sum_{n=1}^{\infty} \frac{J_n'(n\epsilon)}{n}, \qquad \tfrac{1}{2} - \tfrac{1}{4}\epsilon = \sum_{n=1}^{\infty} (-)^{n-1} \frac{J_n'(n\epsilon)}{n},$$

(2) $$\frac{\tfrac{1}{2}\epsilon}{1-\epsilon} = \sum_{n=1}^{\infty} J_n(n\epsilon), \qquad \frac{\tfrac{1}{2}\epsilon}{1+\epsilon} = \sum_{n=1}^{\infty} (-)^{n-1} J_n(n\epsilon),$$

(3) $$\frac{\tfrac{1}{2}}{(1-\epsilon)^2} = \sum_{n=1}^{\infty} nJ_n'(n\epsilon), \qquad \frac{\tfrac{1}{2}}{(1+\epsilon)^2} = \sum_{n=}^{\infty} (-)^{n-1} nJ_n'(n\epsilon).$$

* It is seen from (3) that, when m is equal to 1, the expansion (4) has to be modified by the insertion of a constant term. These two formulae were given by Jacobi, *Astr. Nach.* XXVIII. (1849), col. 69. [*Ges. Math. Werke*, VII. (1891), p. 149.]

More generally we find by differentiating § 17·21 (6) that

(4) $$\frac{(-)^m}{2}\left[\frac{d^{2m}}{dM^{2m}}\frac{1}{1-\epsilon\cos E}\right]_{M=0}=\sum_{n=1}^{\infty} n^{2m} J_n(n\epsilon),$$

(5) $$\frac{(-)^{m-1}}{2}\left[\frac{d^{2m}}{dM^{2m}}\frac{1}{1-\epsilon\cos E}\right]_{M=\pi}=\sum_{n=1}^{\infty}(-)^{n-1} n^{2m} J_n(n\epsilon).$$

Since $\dfrac{d}{dM}=\dfrac{1}{1-\epsilon\cos E}\dfrac{d}{dE}$, the expressions on the left in (4) and (5) can be calculated for any positive integral value of m, with sufficient labour.

Again, if we regard ϵ and M as the independent variables, it is easily seen that

$$\begin{aligned}\frac{\partial}{\partial\epsilon}\left\{\frac{1}{\sin E(1-\epsilon\cos E)}\right\}&=-\frac{\cos E}{\sin^2 E(1-\epsilon\cos E)}\frac{\partial E}{\partial\epsilon}\\&\quad-\frac{1}{\sin E(1-\epsilon\cos E)^2}\left\{-\cos E+\epsilon\sin E\frac{\partial E}{\partial\epsilon}\right\}\\&=-\frac{\epsilon\sin E}{(1-\epsilon\cos E)^3}\\&=\frac{\partial}{\partial M}\frac{1}{1-\epsilon\cos E},\end{aligned}$$

so that, by § 17·21 (6)

$$\frac{\partial}{\partial\epsilon}\left\{\frac{1}{\sin E(1-\epsilon\cos E)}\right\}=-2\sum_{n=1}^{\infty} nJ_n(n\epsilon)\sin nM,$$

and therefore, if we integrate with $\epsilon=0$ as the lower limit,

(6) $$\frac{1}{\sin E(1-\epsilon\cos E)}-\frac{1}{\sin M}=-2\sum_{n=1}^{\infty} n\sin nM.\int_0^{\epsilon} J_n(nx)\,dx.$$

If we differentiate with respect to M, we find that

(7) $$\frac{\cos E}{\sin^2 E(1-\epsilon\cos E)^2}-\frac{\cos M}{\sin^2 M}+\frac{\epsilon}{(1-\epsilon\cos E)^3}=2\sum_{n=1}^{\infty} n^2\cos nM.\int_0^{\epsilon} J_n(nx)\,dx.$$

The last two expansions do not appear to have been published previously.

Expressions resembling those on the right of (6) and (7) have occurred in the researches of Schott, *Electromagnetic Radiation* (Cambridge, 1912) *passim.*

Thus, as cases of (4) and (5), Schott proved (*ibid.* p. 110) that

(8) $$\sum_{n=1}^{\infty} n^2 J_{2n}(2n\epsilon)=\frac{\epsilon^2(1+\epsilon^2)}{2(1-\epsilon^2)^4},\quad \sum_{n=1}^{\infty} n^2\int_0^{\epsilon} J_{2n}(2nx)\,dx=\frac{\epsilon^3}{6(1-\epsilon^2)^3}.$$

The last of these may be obtained by taking M equal to 0 and π in (7).

17·23. *Meissel's expansions of Kapteyn's type.*

Two extremely interesting series, namely

(1) $$2\sum_{n=1}^{\infty}\frac{J_{2n}(2n\epsilon)}{n^2+\xi^2}=\frac{\epsilon^2}{1^2+\xi^2}+\frac{\epsilon^4\xi^2}{(1^2+\xi^2)(2^2+\xi^2)}+\frac{\epsilon^6\xi^4}{(1^2+\xi^2)(2^2+\xi^2)(3^2+\xi^2)}+\dots,$$

(2) $$2\sum_{n=1}^{\infty}\frac{J_{2n-1}\{(2n-1)\epsilon\}}{(2n-1)^2+\xi^2}=\frac{\epsilon}{1^2+\xi^2}+\frac{\epsilon^3\xi^2}{(1^2+\xi^2)(3^2+\xi^2)}+\frac{\epsilon^5\xi^4}{(1^2+\xi^2)(3^2+\xi^2)(5^2+\xi^2)}+\dots,$$

have been stated by Meissel* who deduced various consequences from them; it is to be supposed at present† that $0<\epsilon\leqslant 1$, and ξ is real.

The simplest method of procedure to adopt in establishing these expansions is to take the Fourier series‡

$$\sum_{n=1}^{\infty}\frac{\cos 2nM}{n^2+\xi^2}=\frac{\pi\cosh(\pi-2M)\xi}{2\xi\sinh\pi\xi}-\frac{1}{2\xi^2},$$

(which is valid when $0\leqslant M\leqslant\pi$), replace M by $E-\epsilon\sin E$, and integrate from 0 to π. It is thus found that

$$\begin{aligned}2\sum_{n=1}^{\infty}\frac{J_{2n}(2n\epsilon)}{n^2+\xi^2}&=\frac{1}{\pi}\int_0^{\pi}\left\{\frac{\pi\cosh(\pi-2E+2\epsilon\sin E)\xi}{\xi\sinh\pi\xi}-\frac{1}{\xi^2}\right\}dE\\&=\frac{1}{\pi}\int_{-\frac{1}{2}\pi}^{\frac{1}{2}\pi}\left\{\frac{\pi\cosh(2\theta+2\epsilon\cos\theta)\xi}{\xi\sinh\pi\xi}-\frac{1}{\xi^2}\right\}d\theta\\&=\frac{2}{\pi}\int_0^{\frac{1}{2}\pi}\left\{\frac{\pi\cosh 2\xi\theta\,.\cosh(2\epsilon\xi\cos\theta)}{\xi\sinh\pi\xi}-\frac{1}{\xi^2}\right\}d\theta.\end{aligned}$$

Now the last expression is an even integral function of ϵ, and hence it is expansible in the form§

$$\begin{aligned}\sum_{m=1}^{\infty}\frac{2^{2m+1}\xi^{2m-1}\epsilon^{2m}}{(2m)!}\int_0^{\frac{1}{2}\pi}\frac{\cos^{2m}\theta\cosh 2\xi\theta}{\sinh\pi\xi}d\theta\\=\sum_{m=1}^{\infty}\frac{\Gamma(1+i\xi)\,\Gamma(1-i\xi)}{\Gamma(m+1+i\xi)\,\Gamma(m+1-i\xi)}\cdot\xi^{2m-2}\epsilon^{2m},\end{aligned}$$

by a formula due to Cauchy‖; and the truth of Meissel's first formula is now evident.

The second formula follows in like manner from the Fourier series

$$\sum_{n=1}^{\infty}\frac{\cos(2n-1)M}{(2n-1)^2+\xi^2}=\frac{\pi\sinh(\frac{1}{2}\pi-M)\xi}{4\xi\cosh\frac{1}{2}\pi\xi}.$$

* *Astr. Nach.* CXXX. (1892), col. 363—368.

† The extension to complex variables is made in § 17·31.

‡ See Legendre, *Exercices de Calc. Int.* II. (Paris, 1817), p. 166.

§ It is easy to see that the term independent of ϵ vanishes.

‖ *Mém. sur les intégrales définies* (Paris, 1825), p. 40. Cf. *Modern Analysis*, p. 263.

Now, since the series obtained from (1) and (2) by differentiations with respect to ξ^2 are uniformly convergent throughout any bounded domain of real values of ξ, we may differentiate any number of times and then make $\xi \to 0$.

We thus deduce that

$$\sum_{n=1}^{\infty} \frac{J_{2m}(2n\epsilon)}{n^{2m}}, \quad \sum_{n=1}^{\infty} \frac{J_{2m-1}\{(2n-1)\epsilon\}}{(2n-1)^{2m}}$$

are polynomials in* ϵ; the former is an even polynomial of degree $2m$, and the latter is an odd polynomial of degree $2m-1$.

The values of the former polynomial were given by Meissel in the cases $m=1, 2, 3, 4, 5$; the values for $m=1, 2, 3$ are

$$\frac{\epsilon^2}{2}, \quad \frac{\epsilon^2}{2} - \frac{\epsilon^4}{8}, \quad \frac{\epsilon^2}{2} - \frac{5\epsilon^4}{32} + \frac{\epsilon^6}{72}.$$

The values of the latter polynomial for $m=1, 2, 3$ are

$$\frac{\epsilon}{2}, \quad \frac{\epsilon}{2} - \frac{\epsilon^3}{18}, \quad \frac{\epsilon}{2} - \frac{5\epsilon^3}{81} + \frac{\epsilon^5}{450}.$$

Meissel also gave the values of the latter polynomial for $m=4, 5$.

Conversely, it is evident that every even polynomial of degree $2m$ is expressible in the form

$$\sum_{n=0}^{\infty} a_n J_{2n}(2n\epsilon),$$

and that every odd polynomial, of degree $2m-1$, is expressible in the form

$$\sum_{n=1}^{\infty} b_n J_{2n-1}\{(2n-1)\epsilon\},$$

where a_n and b_n are even polynomials in $1/n$ and $1/(2n-1)$ respectively, of degree $2m$.

17·3. *Simple Kapteyn series with complex variables.*

It was stated in § 17·1 that, in general, Kapteyn series are of a more recondite character than Neumann series, and we shall now explain one of the characteristic differences between the two types of series.

In the case of Neumann series it is, in general, possible to expand each of the Bessel functions in the form of a power series in the variable, and then to rearrange the resulting double series as a power series whose domain of convergence is that of the original Neumann series.

* It is to be noted that the coefficients of ϵ^{2m} and ϵ^{2m-1} in the respective polynomials are not zero; they are

$$\frac{(-)^{m-1}}{2.(m!)^2} \text{ and } \frac{(-)^{m-1}}{2.1^2.3^2\ldots(2m-1)^2}.$$

The corresponding property of Kapteyn series is quite different; for the Kapteyn series

$$\Sigma a_n J_{\nu+n}\{(\nu+n)z\}$$

is convergent and represents an analytic function (cf. § 8·7) throughout the domain in which

$$\left|\frac{z\exp\sqrt{(1-z^2)}}{1+\sqrt{(1-z^2)}}\right| < \lim_{n\to\infty}\frac{1}{|\sqrt[\nu+n]{a_n}|},$$

while the double series obtained by expanding each Bessel function in powers of z is absolutely convergent only throughout the domain in which

$$\frac{|z|\exp\sqrt{(1-|z|^2)}}{1+\sqrt{(1-|z|^2)}} < \lim_{n\to\infty}\frac{1}{|\sqrt[\nu+n]{a_n}|},$$

and the latter domain is smaller than the former; thus, when the limit is 1, the first domain is the interior of the curve shewn in Fig. 24 of § 8·7, in which the longest diameter joins the points ± 1, while the shortest joins the points $\pm i \times 0{\cdot}6627434$; while the second domain* is only the interior of the circle $|z| = 0{\cdot}6627434$.

Hence, when we are dealing with Kapteyn series, if we use the method of expansion into double series we succeed, at best, in proving theorems only for a portion of the domain of their validity; and the proof for the remainder of the domain either has to take the form of an appeal to the theory of analytic continuation or else it has to be effected by a completely different method.

As an example of the methods which have to be employed, we shall give Kapteyn's† proof of the theorem that

$$\frac{1}{1-z} = 1 + 2\sum_{n=1}^{\infty} J_n(nz), \tag{1}$$

provided that z lies in the open domain in which

$$\left|\frac{z\exp\sqrt{(1-z^2)}}{1+\sqrt{(1-z^2)}}\right| < 1.$$

This domain occurs so frequently in the following analysis that it is convenient to describe it as the domain K; it is the interior of the curve shewn in Fig. 24 of § 8·7.

Formula (1) is, of course, suggested by formula (2) of § 17·22.

To establish the truth of the expansion, we write

$$1 + 2\sum_{n=1}^{\infty} J_n(nz) \equiv S(z),$$

and then it has to be proved that $S(z) = 1/(1-z)$.

Since
$$J_n(nz) = \frac{1}{2\pi i}\int^{(0+)}\left[\frac{\exp\{\tfrac{1}{2}z(t-1/t)\}}{t}\right]^n \frac{dt}{t},$$

* For an investigation of the magnitude of this domain, see Puiseux, *Journal de Math.* XIV. (1849), pp. 33—39, 242—246.

† *Nieuw Arcnief voor Wiskunde*, XX. (1893), pp. 123—126; *Ann. sci. de l'École norm. sup.* (3) X. (1893), pp. 96—102.

we see that, if we can find a circle Γ with centre at the origin of such a radius that on it the inequality

(2) $$\left|\frac{\exp\{\frac{1}{2}z(t-1/t)\}}{t}\right|<1$$

is true, then

(3) $$S(z)=\frac{1}{2\pi i}\int_{(\Gamma+)}\frac{1+t^{-1}\exp\{\frac{1}{2}z(t-1/t)\}}{1-t^{-1}\exp\{\frac{1}{2}z(t-1/t)\}}\frac{dt}{t}.$$

To investigate (2), we recall the analysis of § 8·7. If $z=\rho e^{i\alpha}$, $t=e^{u+i\theta}$, where ρ, u, α, θ are all real (ρ and u being positive), then (2) is satisfied for all values of θ if

$$\rho\sqrt{(\sinh^2 u+\sin^2\alpha)}-u<0;$$

and when u is chosen so that the last expression on the left has its least value, this value is (§ 8·7)

$$\log\left|\frac{z\exp\sqrt{(1-z^2)}}{1+\sqrt{(1-z^2)}}\right|,$$

which is negative when z lies in the domain K. Hence, when z lies in the domain K, we can find a positive value of u such that the inequality (2) is satisfied when $|t|=e^u$.

Again, if we write $1/t$ in place of t in (3) we find that

(4) $$S(z)=\frac{1}{2\pi i}\int_{(\gamma+)}\frac{1+t\exp\{-\frac{1}{2}z(t-1/t)\}}{1-t\exp\{-\frac{1}{2}z(t-1/t)\}}\frac{dt}{t},$$

where γ is the circle $|t|=e^{-u}$.

When we combine (3) and (4) we find that

$$2S(z)=\frac{1}{2\pi i}\int_{(\Gamma+,\gamma-)}\frac{t+\exp\{\frac{1}{2}z(t-1/t)\}}{t-\exp\{\frac{1}{2}z(t-1/t)\}}\frac{dt}{t},$$

and so $2S(z)$ is the sum of the residues of the integrand at its poles which lie inside the annulus bounded by Γ and γ.

We next prove that *there is only one pole inside the annulus**, and, having proved this, we notice that this pole is obviously $t=1$.

For the number of poles is equal to

$$\frac{1}{2\pi i}\int_{(\Gamma+,\gamma-)}\frac{d\log[1-t^{-1}\exp\{\frac{1}{2}z(t-1/t)\}]}{dt}dt$$

$$=\frac{1}{2\pi i}\int_{(\Gamma+)}\frac{d\log[1-t^{-1}\exp\{\frac{1}{2}z(t-1/t)\}]}{dt}dt$$
$$+\frac{1}{2\pi i}\int_{(\Gamma+)}\frac{d\log[1-t\exp\{-\frac{1}{2}z(t-1/t)\}]}{dt}dt$$
$$=\frac{1}{\pi i}\int_{(\Gamma+)}\frac{d\log[1-t^{-1}\exp\{\frac{1}{2}z(t-1/t)\}]}{dt}dt$$
$$+\frac{1}{2\pi i}\int_{(\Gamma+)}\frac{d\log[-t\exp\{-\frac{1}{2}z(t-1/t)\}]}{dt}dt.$$

* The corresponding part of Kapteyn's investigation does not seem to be quite so convincing as the investigation given in the text.

Now the first of these integrals vanishes; for, if we write

$$t^{-1}\exp\{\tfrac{1}{2}z(t-1/t)\}\equiv U,$$

then $|U|<1$ on Γ, and so the expression under consideration may be written in the form

$$-\frac{1}{\pi i}\int_{(\Gamma+)}\left\{\sum_{n=0}^{\infty}U^n\right\}\frac{dU}{dt}\,dt,$$

and the integral of each term of the uniformly convergent series involved is zero.

Hence the number of zeros of $1-t^{-1}\exp\{\tfrac{1}{2}z(t-1/t)\}$ in the annulus is equal to

$$\frac{1}{2\pi i}\int_{\Gamma+}\left[\frac{1}{t}-\frac{z}{2}\left(1+\frac{1}{t^2}\right)\right]dt=1.$$

It follows that $2S(z)$ is equal to the residue of

$$\frac{t+\exp\{\tfrac{1}{2}z(t-1/t)\}}{t-\exp\{\tfrac{1}{2}z(t-1/t)\}}$$

at $t=1$; and this residue is easily calculated to be $2/(1-z)$.

It has therefore been shewn that $S(z)$ is equal to $1/(1-z)$ throughout the domain K, *i.e. throughout the whole of the open domain in which the series defining* $S(z)$ *is convergent.*

[NOTE. It is possible to prove that $S(z)$ converges to the sum $1/(1-z)$ on the boundary of K, except at $z=1$, but the proof requires an appeal to be made to theorems of an Abelian type; cf. § 17·8.]

17·31. *The extension of Meissel's expansions to the case of complex variables.*

We shall now shew how to obtain the expansions

(1) $$2\sum_{n=1}^{\infty}\frac{J_{2n}(2nz)}{n^2+\zeta^2}=\frac{z^2}{1^2+\zeta^2}+\frac{z^4\zeta^2}{(1^2+\zeta^2)(2^2+\zeta^2)}+\frac{z^6\zeta^4}{(1^2+\zeta^2)(2^2+\zeta^2)(3^2+\zeta^2)}+\dots,$$

(2) $$2\sum_{n=1}^{\infty}\frac{J_{2n-1}\{(2n-1)z\}}{(2n-1)^2+\zeta^2}=\frac{z}{1^2+\zeta^2}+\frac{z^3\zeta^2}{(1^2+\zeta^2)(3^2+\zeta^2)}+\frac{z^5\zeta^4}{(1^2+\zeta^2)(3^2+\zeta^2)(5^2+\zeta^2)}+\dots,$$

which are valid when z lies in the domain K and ζ is a complex variable which is unrestricted apart from the obvious condition that ζi must not be an integer in (1) nor an odd integer in (2). These results are the obvious extensions of Meissel's formulae of § 17·23.

[NOTE. The expansions when ζ is a pure imaginary have to be established by a limiting process by making ζ approach the imaginary axis; since the functions involved in (1) and (2) are all even functions of ζ, no generality is lost by assuming that $R(\zeta)$ is positive.]

In order to establish these formulae, it is first convenient to effect the generalisation to complex variables of the expansion of the reciprocal of the radius vector given by § 17·21 (6). That is to say, we take the expansion

$$1 + 2 \sum_{n=1}^{\infty} J_n(nz) \cos n\phi,$$

which we denote by the symbol $S(z, \phi)$, and proceed to sum it by Kapteyn's method (explained in § 17·3), on the hypotheses that ϕ is a real variable and that z lies in the domain K. We define a complex variable ψ by the equation

$$\phi = \psi - z \sin \psi.$$

The singularities of ψ, *qua* function of ϕ, are given by $\cos \psi = 1/z$, that is

$$\phi = \operatorname{arc\,sec} z - \sqrt{(z^2 - 1)}.$$

None of these values of ϕ is real* if z lies in the domain K; and, as ϕ increases from 0 to ∞ through real values, ψ describes an undulating curve which can be reconciled with the real axis in the ψ-plane without passing over any singular points.

It follows that if, for brevity, we write

$$U \equiv t^{-1} \exp \{\tfrac{1}{2} z (t - 1/t)\},$$

then

$$S(z, \phi) = \frac{1}{2\pi i} \int_{(\Gamma+)} \frac{1 - U^2}{1 - 2U \cos \phi + U^2} \frac{dt}{t},$$

with the notation of § 17·3. By the methods of that section we have

$$2S(z, \phi) = \frac{1}{2\pi i} \int_{(\Gamma+, \gamma-)} \frac{1 - U^2}{1 - 2U \cos \phi + U^2} \frac{dt}{t},$$

and so $2S(z, \phi)$ *is equal to the sum of the residues of the integrand at those of its poles which lie inside the annulus bounded by* Γ *and* γ.

We shall now shew that *there are only two poles inside the annulus*, and, having proved this, we then notice that these poles are obviously $t = e^{\pm i\psi}$.

By Cauchy's theorem, the number of poles is equal to

$$\begin{aligned}
&\frac{1}{2\pi i} \int_{(\Gamma+, \gamma-)} \frac{d \log (1 - 2U \cos \phi + U^2)}{dt} dt \\
&= \frac{1}{\pi i} \int_{(\Gamma+)} \frac{d \log (1 - 2U \cos \phi + U^2)}{dt} dt \\
&\quad + \frac{1}{2\pi i} \int_{(\Gamma+)} \frac{d \log [t^2 \exp \{-z(t - 1/t)\}]}{dt} dt \\
&= -\frac{2}{\pi i} \int_{(\Gamma+)} \left[\sum_{n=0}^{\infty} U^n \cos (n+1) \phi \right] \frac{dU}{dt} dt + 2 \\
&= 2,
\end{aligned}$$

* It is easy to shew that such values of ϕ satisfy the equation

$$e^{\pm i\phi} = \frac{z \exp \sqrt{(1 - z^2)}}{1 + \sqrt{(1 - z^2)}},$$

so that $|e^{\pm i\phi}| < 1$.

the integral of each term of the uniformly convergent series vanishing, just as in § 17·3.

Now the residues of

$$\frac{1-U^2}{1-2U\cos\phi+U^2}\cdot\frac{1}{t}$$

at $t=e^{\pm i\psi}$ are both equal to $1/(1-z\cos\psi)$; and therefore we have proved that

$$(3)\qquad \frac{1}{1-z\cos\psi}=1+2\sum_{n=1}^{\infty}J_n(nz)\cos n\phi,$$

in the circumstances postulated; and the series on the right is a periodic function of ϕ which converges uniformly in the unbounded range of real values of ϕ.

Hence, when $R(\zeta)>0$, we may multiply by $e^{-\zeta\phi}$ and integrate thus:

$$\int_0^\infty e^{-\zeta\phi}\,d\phi+2\sum_{n=1}^{\infty}J_n(nz)\int_0^\infty e^{-\zeta\phi}\cos n\phi\,d\phi=\int_0^\infty\frac{e^{-\zeta\phi}}{1-z\cos\psi}\,d\phi.$$

That is to say,

$$(4)\qquad \int_0^\infty e^{-\zeta(\psi-z\sin\psi)}\,d\psi=\frac{1}{\zeta}+2\sum_{n=1}^{\infty}\frac{\zeta J_n(nz)}{n^2+\zeta^2},$$

where the path of integration is the undulatory curve in the ψ-plane which corresponds to the real axis in the ϕ-plane; and, by Cauchy's theorem, this undulatory curve may be reconciled with the real axis.

Now, when the path of integration is the real axis, *the integral on the left in* (4) *is an integral function of* z; and this function may be expanded in the form

$$\sum_{m=0}^{\infty}\frac{z^m\zeta^m}{m!}\int_0^\infty e^{-\zeta\psi}\sin^m\psi\,d\psi.$$

By changing the sign of z throughout the work we infer the two formulae

$$(5)\qquad 2\sum_{n=1}^{\infty}\frac{J_{2n}(2nz)}{4n^2+\zeta^2}=\sum_{m=1}^{\infty}\frac{z^{2m}\zeta^{2m-1}}{(2m)!}\int_0^\infty e^{-\zeta\psi}\sin^{2m}\psi\,d\psi,$$

$$(6)\qquad 2\sum_{n=1}^{\infty}\frac{J_{2n-1}\{(2n-1)z\}}{(2n-1)^2+\zeta^2}=\sum_{m=1}^{\infty}\frac{z^{2m-1}\zeta^{2m-2}}{(2m-1)!}\int_0^\infty e^{-\zeta\psi}\sin^{2m-1}\psi\,d\psi,$$

which are now established on the hypotheses that z lies in the domain K and that $R(\zeta)>0$.

By dividing the paths of integration into the intervals $(0,\pi)$, $(\pi,2\pi)$, ... and writing $\frac{1}{2}\pi+\theta$, $\frac{3}{2}\pi+\theta$, ... for ψ in the respective intervals, we infer that

$$\int_0^\infty e^{-\zeta\psi}\sin^{2m}\psi\,d\psi=\frac{1}{\sinh\frac{1}{2}\pi\zeta}\int_0^{\frac{1}{2}\pi}\cosh\zeta\theta\,.\cos^{2m}\theta\,d\theta$$

$$=\frac{1}{\zeta\{\zeta^2+2^2\}\{\zeta^2+4^2\}\dots\{\zeta^2+4m^2\}},$$

and that

$$\int_0^\infty e^{-\zeta\psi} \sin^{2m-1}\psi\, d\psi = \frac{1}{\cosh\frac{1}{2}\pi\zeta}\int_0^{\frac{1}{2}\pi} \cosh\zeta\theta\,.\cos^{2m-1}\theta\, d\theta$$

$$= \frac{1}{\zeta\{\zeta^2+1^2\}\{\zeta^2+3^2\}\ldots\{\zeta^2+(2m-1)^2\}}.$$

By substitution in (5) and (6) and writing 2ζ for ζ in (5) we at once infer the truth of (1) and (2) when $R(\zeta) > 0$; and the mode of extending the results to all other values of ζ has already been explained. The required generalisations of Meissel's expansions are therefore completely established.

17·32. *The expansion of z^n into a Kapteyn series.*

With the aid of Meissel's generalised formula it is easy to obtain the expansion of any integral power of z in the form of a Kapteyn series. It is convenient to consider even powers and odd powers separately.

In the case of an even power, z^{2n}, we take the equation given by § 17·31 (1) in the form

$$(1)\qquad \frac{1}{2\pi i}\int \frac{2\Gamma(n+1+i\zeta)\,\Gamma(n+1-i\zeta)}{\zeta^{2n-1}\,\Gamma(1+i\zeta)\,\Gamma(1-i\zeta)} \sum_{m=1}^{\infty} \frac{J_{2m}(2mz)}{m^2+\zeta^2}\, d\zeta$$

$$= \frac{1}{2\pi i}\int \sum_{m=1}^{\infty} \frac{\Gamma(n+1+i\zeta)\,\Gamma(n+1-i\zeta)}{\Gamma(m+1+i\zeta)\,\Gamma(m+1-i\zeta)} z^{2m}\,\zeta^{2m-2n-1}\, d\zeta,$$

where the contour of integration is the circle $|\zeta| = n+\frac{1}{2}$. Since both series converge uniformly on the circle, when z lies in the domain K, term-by-term integrations are permissible.

Consider now the value of

$$\frac{1}{2\pi i}\int_{|\zeta|=n+\frac{1}{2}} \frac{(1^2+\zeta^2)(2^2+\zeta^2)\ldots(n^2+\zeta^2)}{\zeta^{2n-1}(m^2+\zeta^2)}\, d\zeta.$$

When $m \leqslant n$, there are no poles outside the contour, and so the contour may be deformed into an infinitely great circle, and the expression is seen to be equal to unity; but when $m > n$, the poles $\pm im$ are outside the circle and the expression is equal to unity minus the sum of the residues of the integrand at these two poles, i.e. to

$$1 - \frac{(m+n)!}{m^{2n+1}\,.\,(m-n-1)!}.$$

The expression on the left of (1) is therefore equal to

$$2\sum_{m=1}^{\infty} J_{2m}(2mz) - 2\sum_{m=n+1}^{\infty} \frac{(m+n)!\,J_{2m}(2mz)}{m^{2n+1}\,.\,(m-n-1)!}.$$

Next we evaluate

$$\frac{1}{2\pi i}\int_{|\zeta|=n+\frac{1}{2}} \frac{\Gamma(n+1+i\zeta)\,\Gamma(n+1-i\zeta)}{\Gamma(m+1+i\zeta)\,\Gamma(m+1-i\zeta)}\,\frac{d\zeta}{\zeta^{2n-2m+1}}.$$

When $m \leqslant n$, the origin is the only pole of the integrand, and, if we take the contour to be an infinitely great circle, the expression is seen to be equal to 1.

But, when $m > n$, there are no poles inside the circle $|\zeta| = n + \frac{1}{2}$, and the expression is zero.

Hence we have

$$(2)\qquad 2\sum_{m=1}^{\infty} J_{2m}(2mz) - 2\sum_{m=n+1}^{\infty} \frac{(m+n)!\,J_{2m}(2mz)}{m^{2n+1}.\,(m-n-1)!} = z^2 + z^4 + \ldots + z^{2n}.$$

If we replace n by $n-1$ and subtract the result so obtained from (2), we find that

$$z^{2n} = 2\sum_{m=n}^{\infty} \frac{(m+n-1)!\,J_{2m}(2mz)}{m^{2n-1}.\,(m-n)!} - 2\sum_{m=n+1}^{\infty} \frac{(m+n)!\,J_{2m}(2mz)}{m^{2n+1}.\,(m-n-1)!},$$

and so

$$(3)\qquad z^{2n} = 2n^2 \sum_{m=n}^{\infty} \frac{(m+n-1)!\,J_{2m}(2mz)}{m^{2n+1}.\,(m-n)!}.$$

If $n = 1$, equation (3) is at once deducible from equation (2), without the intervening analysis.

When we have to deal with an odd power, z^{2n-1}, we take the equation given by § 17·31 (2) in the form

$$(4)\qquad \frac{1}{2\pi i}\int_{|\zeta|=2n} \frac{2.\{1^2+\zeta^2\}\{3^2+\zeta^2\}\ldots\{(2n-1)^2+\zeta^2\}}{\zeta^{2n-1}} \times \sum_{m=1}^{\infty} \frac{J_{2m-1}\{(2m-1)z\}}{\zeta^2+(2m-1)^2}\,d\zeta$$

$$= \frac{1}{2\pi i}\int_{|\zeta|=2n} \{1^2+\zeta^2\}\{3^2+\zeta^2\}\ldots\{(2n-1)^2+\zeta^2\} \times \sum_{m=1}^{\infty} \frac{z^{2m-1}\,\zeta^{2m-2n-1}}{\{1^2+\zeta^2\}\{3^2+\zeta^2\}\ldots\{(2m-1)^2+\zeta^2\}}\,d\zeta,$$

and we deduce in a similar manner that

$$(5)\qquad 2\sum_{m=1}^{\infty} J_{2m-1}\{(2m-1)z\} - 2\sum_{m=n+1}^{\infty} \frac{(m+n-1)!\,J_{2m-1}\{(2m-1)z\}}{(m-\frac{1}{2})^{2n}.\,(m-n-1)!} = z + z^3 + \ldots + z^{2n-1}.$$

Hence

$$(6)\qquad z^{2n-1} = 2(n-\tfrac{1}{2})^2 \sum_{m=n}^{\infty} \frac{(m+n-2)!\,J_{2m-1}\{(2m-1)z\}}{(m-\frac{1}{2})^{2n}.\,(m-n)!}.$$

The formulae (3) and (6) may be combined into the single formula

$$(7)\qquad (\tfrac{1}{2}z)^n = n^2 \sum_{m=0}^{\infty} \frac{(n+m-1)!\,J_{n+2m}\{(n+2m)z\}}{(n+2m)^{n+1}.\,m!},$$

which is obviously valid throughout the domain K when n has any of the values 1, 2, 3, ...

This formula was discovered by Kapteyn*; the proof of it which has just been given, though somewhat artificial, seems rather less so than Kapteyn's proof.

* *Ann. sci. de l'École norm. sup.* (3) x. (1893), p. 103.

17·33. *The investigation of the Kapteyn series for z^n by the method of induction.*

We shall now give an alternative method* of investigating the expansion of z^n as a Kapteyn series, which has the advantage of using no result more abstruse than the equations

$$(1) \qquad \frac{1}{1-z} = 1 + 2\sum_{m=1}^{\infty} J_m(mz), \qquad \frac{1}{1+z} = 1 + 2\sum_{m=1}^{\infty} (-)^m J_m(mz),$$

which were proved for real variables in § 17·22 and for complex variables in § 17·3; it is, of course, supposed that, if z is real, then $-1 < z < 1$, and, if z is complex, then z lies in the domain K.

The induction which will be used depends on the fact that when the sum, $f(z)$, of the Kapteyn series $\sum_{m=1}^{\infty} a_m J_m(mz)$ is known, then the sum $F(z)$ of the series $\sum_{m=1}^{\infty} \frac{a_m J_m(mz)}{m^2}$ can be obtained by two quadratures, if the former series converges uniformly. To establish this result, observe that, by term-by-term differentiations,

$$z^2 \frac{d^2 F(z)}{dz^2} + z\frac{dF(z)}{dz} = \sum_{m=1}^{\infty} a_m \{z^2 J_m''(mz) + (z/m) J_m'(mz)\}$$
$$= (1-z^2)\sum_{m=1}^{\infty} a_m J_m(mz),$$

so that

$$\left(z\frac{d}{dz}\right)^2 F(z) = (1-z^2) f(z);$$

it follows at once that $F(z)$ can be determined in terms of $f(z)$ by quadratures.

Now, from (1), we have

$$\sum_{m=1}^{\infty} J_{2m}(2mz) = \frac{\tfrac{1}{2}z^2}{1-z^2},$$

and so, if

$$F(z) \equiv \sum_{m=1}^{\infty} \frac{J_{2m}(2mz)}{4m^2},$$

then

$$\left(z\frac{d}{dz}\right)^2 F(z) = \tfrac{1}{2}z^2.$$

Therefore, in the domain K,

$$\sum_{m=1}^{\infty} \frac{J_{2m}(2mz)}{4m^2} = \tfrac{1}{8}z^2 + A\log z + B,$$

where A and B are constants of integration. If we make $z \to 0$, we see that

$$A = B = 0.$$

Consequently

$$(2) \qquad z^2 = 2\sum_{m=1}^{\infty} \frac{J_{2m}(2mz)}{m^2}.$$

* Watson, *Messenger*, XLVI. (1917), pp. 150—157.

In like manner, we deduce from (1) that

$$\sum_{m=0}^{\infty} J_{2m+1}\{(2m+1)z\} = \frac{\frac{1}{2}z}{1-z^2},$$

and hence that

(3) $$z = 2\sum_{m=0}^{\infty} \frac{J_{2m+1}\{(2m+1)z\}}{(2m+1)^2}.$$

The expansions of z^n when n is 1 or 2 are therefore constructed.

Now assume that, for some particular value of n, z^n is expansible in the form

$$z^n = n^2 \sum_{m=1}^{\infty} b_{m.n} J_m(mz),$$

and consider the function $\phi(z)$ defined by the equation

$$\phi(z) = (n+2)^2 \sum_{m=1}^{\infty} \frac{m^2-n^2}{m^2} b_{m,n} J_m(mz).$$

By the process of differentiation already used, we have

$$\begin{aligned} z^2\frac{d^2\phi(z)}{dz^2} + z\frac{d\phi(z)}{dz} &= (n+2)^2(1-z^2)\sum_{m=1}^{\infty}(m^2-n^2)\,b_{m,n}J_m(mz) \\ &= (n+2)^2\left\{z^2\frac{d^2}{dz^2}+z\frac{d}{dz}\right\}\frac{z^n}{n^2} - (n+2)^2(1-z^2)\,z^n \\ &= (n+2)^2 z^{n+2}. \end{aligned}$$

On integration we deduce that

$$\phi(z) = z^{n+2} + A'\log z + B'.$$

It is obvious that $A' = B' = 0$ from a consideration of the behaviour of $\phi(z)$ near the origin.

Hence the expansion of z^{n+2} is

$$z^{n+2} = (n+2)^2 \sum_{m=1}^{\infty} b_{m,n+2} J_m(mz),$$

where $$b_{m,n+2} = \frac{m^2-n^2}{m^2} b_{m,n}.$$

It follows at once by induction that

$$b_{m,2n} = \frac{2^{2n-1}\,\Gamma(\frac{1}{2}m+n)}{m^{2n-1}\,\Gamma(\frac{1}{2}m-n+1)} b_{m,2},$$

and so $$z^{2n} = n^2 \sum_{m=1}^{\infty} \frac{2^{2n}\,\Gamma(m+n)\,J_{2m}(2mz)}{(2m)^{2n-1}.m^2\,\Gamma(m-n+1)}.$$

That is to say

$$z^{2n} = 2n^2 \sum_{m=n}^{\infty} \frac{\Gamma(m+n)\,J_{2m}(2mz)}{m^{2n+1}\,\Gamma(m-n+1)},$$

and this is equation (3) of § 17·32. The expansion of z^{2n-1} is obtained in the same way from the expansion of z; the analysis in this case is left to the reader.

We therefore obtain the expansion

$$(4)\qquad (\tfrac{1}{2}z)^n = n^2 \sum_{m=0}^{\infty} \frac{\Gamma(n+m) \,.\, J_{n+2m}\{(n+2m)z\}}{(n+2m)^{n+1} \,.\, m!},$$

which is the expansion obtained by other methods in § 17·32; and the expansion is valid throughout the domain K.

Since the series

$$\sum_{m=0}^{\infty} \frac{\Gamma(n+m)}{(n+2m)^{n+1} \,.\, m!}$$

is absolutely convergent (being comparable with $\Sigma\, 1/m^2$), the expansion (4) converges uniformly throughout K and its boundary. The expansion is therefore valid (from considerations of continuity) on the boundary of K, and in particular at the points $z = \pm 1$, as well as throughout the domain K.

17·34. *The expansion of* $1/(t-z)$ *in a Kapteyn series.*

From the expansion of z^n, obtained in the two preceding sections, we can deduce, after Kapteyn*, the expansion of $1/(t-z)$ when z lies in the domain K and t lies outside a certain domain whose extent will be defined later in this section.

Assuming that $|t| > |z|$, we have

$$\frac{1}{t-z} = \frac{1}{t} + \sum_{n=1}^{\infty} \frac{z^n}{t^{n+1}} = \frac{1}{t} + \sum_{n=1}^{\infty} \frac{2^n n^2}{t^{n+1}} \sum_{m=0}^{\infty} \frac{\Gamma(n+m)\, J_{n+2m}\{(n+2m)z\}}{(n+2m)^{n+1} \,.\, m!}.$$

Now, if

$$\left| \frac{z \exp \sqrt{(1-z^2)}}{1+\sqrt{(1-z^2)}} \right| = V,$$

the repeated series is expressible as an absolutely convergent double series if the double series

$$\sum_{n=1}^{\infty} \sum_{m=0}^{\infty} \frac{2^n n^2\, \Gamma(n+m)\, V^{n+2m}}{(n+2m)^{n+1} . m!\, |t|^{n+1}}$$

is convergent. But the terms in this series are less than the terms of the double series

$$\sum_{n=1}^{\infty} \sum_{m=0}^{\infty} \frac{2^n V^{n+2m}}{m!\, |t|^{n+1}} = \frac{2V \exp V^2}{|t|\,(|t| - 2V)},$$

provided that $|t| > 2V$.

Hence, when

$$|t| > 2 \left| \frac{z \exp \sqrt{(1-z^2)}}{1+\sqrt{(1-z^2)}} \right|,$$

rearrangement of the repeated series for $1/(t-z)$ is permissible, and, when we arrange it as a Kapteyn series, we obtain the formula

$$(1)\qquad \frac{1}{t-z} = \Theta_0(t) + 2 \sum_{n=1}^{\infty} \Theta_n(t)\, J_n(nz),$$

* *Ann. sci. de l'École norm. sup.* (3) x. (1893), pp. 113—120.

where*

$$(2)\qquad O_0(t) = 1/t,$$

$$(3)\qquad O_n(t) = \frac{1}{4}\sum_{m=0}^{<\frac{1}{2}n} \frac{(n-2m)^2.(n-m-1)!}{m!\,(\frac{1}{2}nt)^{n-2m+1}}.$$

From the last formula we may deduce a very remarkable theorem discovered by Kapteyn; we have

$$\left\{t\frac{d}{dt}\right\}^2 . \frac{1}{4}\sum_{m=0}^{<\frac{1}{2}n} \frac{(n-m-1)!}{m!\,(\frac{1}{2}nt)^{n-2m}} = \frac{1}{4}\sum_{m=0}^{<\frac{1}{2}n} \frac{(n-2m)^2.(n-m-1)!}{m!\,(\frac{1}{2}nt)^{n-2m}},$$

and therefore, by § 9·1 (2),

$$\tfrac{1}{2}nt\,O_n(t) = \frac{1}{n}\left(t\frac{d}{dt}\right)^2 \{\tfrac{1}{2}nt\,O_n(nt)\},$$

so that, by § 9·12 (1),

$$(4)\qquad O_n(t) = n(1-t^2)\,O_n(nt) + \sin^2 \tfrac{1}{2}n\pi + t\cos^2 \tfrac{1}{2}n\pi$$

when $n = 1, 2, 3, \ldots$.

Kapteyn's polynomial $O_n(t)$ is therefore expressible in terms of Neumann's polynomial $O_n(nt)$.

It is now possible to extend the domain of validity of the expansion (1); for, by § 8·7 combined with § 9·17, it follows that the series on the right of (1) is a uniformly convergent series of analytic functions of z and t when z and t lie in domains such that

$$(5)\qquad \Omega(z) < \Omega(t), \quad \Omega(z) < \Omega(1),$$

where

$$\Omega(z) \equiv \left|\frac{z\exp\sqrt{(1-z^2)}}{1+\sqrt{(1-z^2)}}\right|.$$

The expansion (1) is therefore valid throughout the domains in which both of the inequalities (5) are satisfied.

[NOTE. This result gives a somewhat more extensive domain of values of t than was contemplated by Kapteyn; he ignored the theorem proved in § 9·17, and observed that (since the coefficients in the series for $O_n(t)$ are positive) when $|t| \geqslant 1$,

$$|O_n(t)| \leqslant O_n(|t|) \leqslant O_n(1) = 1,$$

by (4); so that Kapteyn proved that (1) is valid when

$$\Omega(z) < \Omega(1), \quad |t| \geqslant 1.]$$

17·35. *Alternative proofs of the expansion of* $1/(t-z)$ *into a Kapteyn series.*

Now that explicit expressions have been obtained for the coefficients in the expansion

$$\frac{1}{t-z} = O_0(t) + 2\sum_{n=1}^{\infty} O_n(t)\,J_n(nz),$$

it is possible to verify this expansion in various ways. Thus, if $O_n(t)$ be *defined* as

$$n(1-t^2)\,O_n(nt) + \sin^2\tfrac{1}{2}n\pi + t\cos^2\tfrac{1}{2}n\pi,$$

the reader will find it an interesting analysis to take the series

$$\frac{1}{t} + \frac{z+tz^2}{1-z^2} + 2(1-t^2)\sum_{n=1}^{\infty} n\,O_n(nt)\,J_n(nz),$$

* Cf. Kapteyn, *Nieuw Archief voor Wiskunde*, xx. (1893), p. 122.

substitute suitable integrals for the Bessel coefficients and Neumann polynomials, and reduce the result to $1/(t-z)$ after the manner of § 9·14.

Or again, if we differentiate the expansion twice with respect to z we find that

$$\left\{\frac{2z^2}{(t-z)^3}+\frac{z}{(t-z)^2}\right\}=(1-z^2)\sum_{n=0}^{\infty}\epsilon_n n^2 \mathfrak{O}_n(t)\, J_n(nz),$$

and then, dividing by $1-z^2$, and making use of § 17·3 (1), we find that

$$-\frac{2t^2}{(t^2-1)(t-z)^3}-\frac{t(t^2+3)}{(t^2-1)^2(t-z)^2}-\frac{t^4+6t^2+1}{(t^2-1)^3(t-z)^3}$$
$$=\sum_{n=0}^{\infty}\epsilon_n n^2 J_n(nz)\left\{\mathfrak{O}_n(t)-\frac{(t^4+6t^2+1)\sin^2\frac{1}{2}n\pi}{(t^2-1)^3}-\frac{4t(t^2+1)\cos^2\frac{1}{2}n\pi}{(t^2-1)^3}\right\},$$

whence the differential equation for $\mathfrak{O}_n(t)$ is easily constructed in the form

$$-\frac{t^2}{t^2-1}\mathfrak{O}_n''(t)+\frac{t(t^2+3)}{(t^2-1)^2}\mathfrak{O}_n'(t)-\frac{t^4+6t^2+1}{(t^2-1)^3}\mathfrak{O}_n(t)$$
$$=n^2\left\{\mathfrak{O}_n(t)-\frac{(t^4+\frac{1}{2}t^2+1)\sin^2\frac{1}{2}n\pi}{(t^2-1)^3}-\frac{4t(t^2+1)\cos^2\frac{1}{2}n\pi}{(t^2-1)^3}\right\},$$

and hence it follows that

$$\mathfrak{O}_n(t)=n(1-t^2)\,O_n(nt)+\sin^2\tfrac{1}{2}n\pi+t\cos^2\tfrac{1}{2}n\pi+t^{-1}\{A_n J_n(nt)+B_n Y_n(nt)\},$$

where A_n and B_n are independent of t; but it does not seem easy to prove that $A_n=B_n=0$.

17·4. *The expansion of an arbitrary analytic function into a Kapteyn series.*

We shall now prove the following expansion-theorem:

Let $f(z)$ be a function which is analytic throughout the region in which $\Omega(z)\leqslant a$, where $a\leqslant 1$.

Then, at all points z inside *the region,*

(1) $$f(z)=\alpha_0+2\sum_{n=1}^{\infty}\alpha_n J_n(nz),$$

where

(2) $$\alpha_n=\frac{1}{2\pi i}\int \mathfrak{O}_n(t)\, f(t)\, dt,$$

and the path of integration is the curve on which $\Omega(t)=a$.

This result is obvious when we substitute the uniformly convergent expansion

$$\mathfrak{O}_0(t)+2\sum_{n=1}^{\infty}\mathfrak{O}_n(t)\, J_n(nz)$$

for $1/(t-z)$ in the equation

$$f(z)=\frac{1}{2\pi i}\int\frac{f(t)\,dt}{t-z};$$

since $\Omega(t)=a$ on the contour, while both $\Omega(z)<1$ and $\Omega(z)<\Omega(t)$ when z is inside the contour.

This theorem is due to Kapteyn.

It is easy to deduce that, if the Maclaurin series for $f(z)$ is

$$f(z) = \sum_{n=0} a_n z^n,$$

then

(3) $$\alpha_0 = a_0,$$

(4) $$\alpha_n = \frac{1}{4} \sum_{m=0}^{<\frac{1}{2}n} \frac{(n-2m)^2 . (n-m-1)!\, a_{n-2m}}{m!\,(\frac{1}{2}n)^{n-2m+1}}.$$

17·5. *Kapteyn series in which ν is not zero.*

The theory of Kapteyn series of the type

$$\sum_{m=0}^{\infty} a_m J_{\nu+m} \{(\nu+m) z\},$$

in which ν is not zero or an integer, can be made to depend on the expansion of z^ν. The result of § 17·33 suggests that it may be possible to prove that

(1) $$(\tfrac{1}{2}z)^\nu = \nu^2 \sum_{m=0}^{\infty} \frac{\Gamma(\nu+m)}{(\nu+2m)^{\nu+1} . m!} J_{\nu+2m}\{(\nu+2m) z\},$$

throughout the domain K.

It is easy enough to establish this expansion* when $|z| < 0{\cdot}6627434$; but no direct proof of the validity of the expansion throughout the remainder of the domain K is known, and the expansion has to be inferred by the theory of analytic continuation.

To obtain the expansion throughout the interior of the specified circle, expand the series on the right in powers of z. The coefficient of $z^{\nu+2r}$ is

$$\sum_{m=0}^{r} \frac{\Gamma(\nu+m)}{(\nu+2m)^{\nu+1} . m!} \cdot \frac{(-)^{r-m} (\nu+2m)^{\nu+2r}}{2^{\nu+2r} (r-m)!\, \Gamma(\nu+r+m+1)}$$
$$= \frac{\Gamma(\nu)}{2^{\nu+2r}\, \Gamma(\nu+2r+1)} \sum_{m=0}^{r} \frac{(-)^{r-m} (\nu+2m)^{2r-1}}{m!\,(r-m)!} \frac{\Gamma(\nu+m)}{\Gamma(\nu)} \frac{\Gamma(\nu+2r+1)}{\Gamma(\nu+r+m+1)}.$$

When $r \geqslant 1$, the last series is a polynomial in ν of degree $3r-1$ which is known to vanish identically whenever ν is an integer. It therefore vanishes identically for all values of ν. The expansion (1) is therefore established (inside the circle) by a comparison of the coefficient of z^ν on each side of the equation.

From this result, we can prove that, under the conditions specified in § 17·4,

(2) $$\frac{z^\nu}{t-z} = \sum_{n=0}^{\infty} \mathscr{A}_{n,\nu}(t)\, J_{\nu+n}\{(\nu+n) z\},$$

where

(3) $$\mathscr{A}_{n,\nu}(t) = \frac{1}{2} \sum_{m=0}^{<\frac{1}{2}n} \frac{(\nu+n-2m)^2\, \Gamma(\nu+n-m)}{(\frac{1}{2}\nu+n)^{\nu+n-2m+1} . m!\, t^{n-2m+1}}.$$

* This was done when $|z| < 0{\cdot}659$ by Nielsen, *Ann. sci. de l'École norm. sup.* (3) XVIII. (1901), pp. 42—46.

It is not difficult* to express $\mathscr{A}_{n,\nu}(t)$ in terms of Gegenbauer's polynomial $A_{n,\nu}(nt+\frac{1}{2}\nu t)$, defined in § 9·2.

And the reader will easily prove that if $f(z)$ satisfies the conditions specified in § 17·4, then

(4) $$z^\nu f(z) = \sum_{n=0}^{\infty} a_{n,\nu} J_{\nu+n}\{(\nu+n)z\},$$

where

(5) $$a_{n,\nu} = \frac{1}{2\pi i}\int f(t)\,\mathscr{A}_{n,\nu}(t)\,dt,$$

in which the contour of integration surrounds the origin; and hence

(6) $$a_{n,\nu} = \frac{1}{2}\sum_{m=0}^{\leqslant\frac{1}{2}n} \frac{(\nu+n-2m)^2\,\Gamma(\nu+n-m)\,a_{n-2m}}{(\frac{1}{2}\nu+n)^{\nu+n-2m+1}\,.\,m!},$$

where $a_0, a_1, \ldots$ are the coefficients in the Maclaurin series for $f(z)$.

[NOTE. Jacobi in one of his later papers, *Astr. Nach.* XXVIII. (1849), col. 257—270 [*Ges. Math. Werke*, VII. (1891), pp. 175—188] has criticised Carlini for stating that certain expansions are valid only when $|z| < 0{\cdot}663\ldots$. But Carlini had some excuse for his statement because the expansions are obtained by rearrangements of repeated series which are permissible only in this domain, although the expansions are actually valid throughout the domain K.]

17·6. *Kapteyn series of the second kind.*

Series of the type

$$\Sigma\,\beta_n J_{\mu+n}\left\{\left(\frac{\mu+\nu}{2}+n\right)z\right\} J_{\nu+n}\left\{\left(\frac{\mu+\nu}{2}+n\right)z\right\}$$

have been studied in some detail by Nielsen†. But the only series of this type which have, as yet, proved to be of practical importance‡, are some special series with $\mu=\nu$, and with simple coefficients. The results required in the applications just specified are obtainable by integrating Meissel's expansion of § 17·31 (1) after replacing z by $z\sin\theta$. It is thus found that, throughout the domain K,

$$2\sum_{n=1}^{\infty}\frac{J_n^2(nz)}{n^2+\zeta^2} = \sum_{n=1}^{\infty}\frac{z^{2n}\zeta^{2n-2}}{(1^2+\zeta^2)(2^2+\zeta^2)\ldots(n^2+\zeta^2)}\cdot\frac{2}{\pi}\int_0^{\frac{1}{2}\pi}\sin^{2n}\theta\,d\theta,$$

so that

(1) $$2\sum_{n=1}^{\infty}\frac{J_n^2(nz)}{n^2+\zeta^2} = \frac{1}{2}\,\frac{z^2}{1^2+\zeta^2} + \frac{1.3}{2.4}\cdot\frac{z^4\zeta^2}{(1^2+\zeta^2)(2^2+\zeta^2)}+\ldots,$$

and hence we deduce that $\Sigma J_n^2(nz)/n^{2m}$ is a polynomial in z^2 of degree m; while the sum of series of the type $\Sigma n^{2m} J_n^2(nz)$ may be found in a similar manner from the corresponding expansion $\Sigma n^{2m} J_{2n}(2nz)$.

* Cf. Nielsen, *Ann. sci. de l'École norm. sup.* (3) XVIII. (1901), p. 60.

† *Ann. sci. de l'École norm. sup.* (3) XVIII. (1901), pp. 39—75.

‡ Cf. Schott, *Electromagnetic Radiation* (Cambridge, 1912), Chapter VIII.

Thus Schott* has shewn that

$$\sum_{n=1}^{\infty} J_n^2(nz) = \frac{1}{2\sqrt{(1-z^2)}} - \frac{1}{2}, \tag{2}$$

$$\sum_{n=1}^{\infty} n^2 J_n^2(nz) = \frac{z^2(4+z^2)}{16(1-z^2)^{\frac{7}{2}}}. \tag{3}$$

A general theory resembling that of § 16·14 is deducible from the expansion

$$(\tfrac{1}{2}z)^{2\nu} = \frac{2\nu\{\Gamma(\nu+1)\}^2}{\Gamma(2\nu+1)} \sum_{m=0}^{\infty} \frac{\Gamma(2\nu+m)}{(\nu+m)^{2\nu+1}\,.\,m!} J^2_{\nu+m}\{(\nu+m)z\} \tag{4}$$

which is easily derived from § 17·5 (1) and is valid throughout K; but it seems unnecessary to go into details which the reader should have no difficulty in constructing, in the unlikely event of his requiring them.

17·7. *Kapteyn series which converge outside the domain K.*

If

$$\overline{\lim_{n\to\infty}} \left| \sqrt[n]{\alpha_n} \right| = 1,$$

we have seen that the Kapteyn series $\Sigma \alpha_n J_n(nz)$ represents an analytic function throughout the domain K. But since, when x is real, $|J_n(nx)| < 1$, the series may converge along the whole of the real axis, although when $|z| > 1$, the series does not converge at points which are not on the real axis.

The behaviour of such a Kapteyn series may be summed up† by saying that it resembles a power-series throughout the domain K and that it resembles a Fourier series on the real axis outside K.

As an example, let us consider the series

$$S \equiv \sum_{n=1}^{\infty} \frac{J_n(nx)}{n^2}.$$

It is evident that, if $\phi = \psi - x \sin \psi$, then

$$S = \frac{1}{\pi} \int_0^{\pi} \sum_{n=1}^{\infty} \frac{\cos n\phi}{n^2}\, d\psi,$$

since the Fourier series is uniformly convergent.

Now, when $x > 1$, ϕ decreases as ψ increases from 0 to $\operatorname{arc}\cos(1/x)$ and then increases to π as ψ increases from $\operatorname{arc}\cos(1/x)$ to π. If m be the integer such that the minimum value of ϕ lies between $-2m\pi$ and $-2(m+1)\pi$, let the values of ψ corresponding to the values

$$0,\ -2\pi,\ -4\pi,\ \ldots,\ -2m\pi,\ -2m\pi,\ \ldots,\ -2\pi,\ 0$$

of ϕ be $\gamma_0, \gamma_1, \ldots \gamma_m, \delta_m, \delta_{m-1}, \ldots \delta_1, \delta_0$, and then

$$S = \frac{1}{\pi}\left\{ \sum_{r=0}^{m-1} \int_{\gamma_r}^{\gamma_{r+1}} + \int_{\gamma_m}^{\delta_m} + \sum_{r=0}^{m-1} \int_{\delta_{r+1}}^{\delta_r} + \int_{\delta_0}^{\pi} \right\} \sum_{n=1}^{\infty} \frac{\cos n\phi}{n^2}\, d\psi.$$

* *Electromagnetic Radiation* (Cambridge, 1912), p. 120.

† The suggestion of these analogies was made by Professor Hardy.

Now when ψ lies in the intervals (γ_r, γ_{r+1}) and (δ_r, δ_{r+1}) the sum of the series under the integral sign is

$$\tfrac{1}{4}\phi^2 - \tfrac{1}{2}\pi\phi + \tfrac{1}{6}\pi^2 + r(r+1)\pi^2 + (r+1)\pi\phi,$$

and, since

$$\int(\psi - x\sin\psi)\,d\psi = \tfrac{1}{2}\psi^2 + x\cos\psi,$$

$$\int(\psi - x\sin\psi)^2\,d\psi = \tfrac{1}{3}\psi^3 + 2x(\psi\cos\psi - \sin\psi) + \tfrac{1}{2}x^2(\psi - \sin\psi\cos\psi),$$

it may be shewn without much difficulty that

$$S = \tfrac{1}{8}x^2 + \tfrac{1}{2}x + \sum_{r=0}^{m}\{\tfrac{1}{2}(\delta_r^2 - \gamma_r^2) + x(\cos\delta_r - \cos\gamma_r)\} + 2\pi\sum_{r=1}^{m} r(\delta_r - \gamma_r).$$

The reader will see that a large class of Kapteyn series may be summed by this method*.

17·8. *The convergence of Kapteyn series on the boundary of the domain K.*

With the exception of the points ± 1, the boundary of K presents no features of special interest; because, by means of Debye's asymptotic expansion the consideration of the convergence of the Kapteyn series $\Sigma a_n J_{\nu+n}\{(\nu+n)z\}$ is reducible to that of the power series

$$\Sigma\frac{a_n}{\sqrt{n}}\left\{\frac{z\exp\sqrt{(1-z^2)}}{1+\sqrt{(1-z^2)}}\right\}^n,$$

and that of two similar series† with $\sqrt{n^3}$, $\sqrt{n^5}$ written for $\sqrt{n}$.

The points ± 1 present more interest, because the ordinary asymptotic expansions fail. But the lacuna thereby produced is filled, for real values of ν, by the following theorem of an Abelian type:

The convergence of

$$\sum_{n=1}^{\infty}\frac{a_n}{n^{\frac{1}{3}}}$$

is sufficient to ensure both the convergence of $\Sigma a_n J_{\nu+n}(\nu+n)$ *and the continuity of* $\Sigma a_n J_{\nu+n}\{(\nu+n)x\}$ *throughout the interval*‡ $0 \leqslant x \leqslant 1$.

Since $\Sigma a_n/n^{\frac{1}{3}}$ converges and $\{n/(\nu+n)\}^{\frac{1}{3}}$ is monotonic, with a limit as $n\to\infty$, it follows§ that $\Sigma a_n/(\nu+n)^{\frac{1}{3}}$ converges; and since, by § 8·54, $(\nu+n)^{\frac{1}{3}}J_{\nu+n}(\nu+n)$ is monotonic, with a limit as $n\to\infty$, it follows that $\Sigma a_n J_{\nu+n}(\nu+n)$ converges.

* In this connexion the researches by Nielsen, *Oversigt K. Danske Videnskabernes Selskabs*, 1901, pp. 127—146, should be consulted.

† If $a_n/\sqrt{n}$ does not tend to zero the series cannot converge; and if it does tend to zero $\Sigma a_n/\sqrt{n^5}$ is absolutely convergent, and so, if we replace each Bessel function by the first two terms of the asymptotic expansion with a remainder term, the series of remainder terms is absolutely convergent.

‡ Due allowance has to be made for the origin if $\nu < 0$.

§ Bromwich, *Theory of Infinite Series*, § 19.

Again, since
$$\frac{J_{\nu+n}\{(\nu+n)x\}}{J_{\nu+n}(\nu+n)}$$
is a function of n which *does not increase* as n increases, for all values of x in the interval $0 \leqslant x \leqslant 1$, it follows from the test of Abel's type for uniformity of convergence* that
$$\Sigma\, a_n J_{\nu+n}\{(\nu+n)x\}$$
is uniformly convergent (and therefore continuous) throughout the interval $0 \leqslant x \leqslant 1$; and this proves the theorem.

By reversing the reasoning, it may be shewn that if $\Sigma\, a_n J_{\nu+n}(\nu+n)$ converges, so does $\Sigma a_n/n^{\frac{1}{3}}$, so that the convergence of $\Sigma a_n/n^{\frac{1}{3}}$ is both necessary and sufficient for the theorem to be true; the theorem is therefore the best theorem of its kind†.

* Bromwich, *Theory of Infinite Series*, § 44.

† This was pointed out by Professor Hardy. Cf. Watson, *Proc. London Math. Soc.* (2) XVI. (1917), pp. 171—174.

CHAPTER XVIII

SERIES OF FOURIER-BESSEL AND DINI

18·1. *Fourier's formal expansion of an arbitrary function.*

In his researches on the Theory of Conduction of Heat, Fourier* was led to consider the expansion of an arbitrary function $f(x)$ of a real variable of x in the form

(1) $$f(x) = \sum_{m=1}^{\infty} a_m J_0(j_m x),$$

where $j_1, j_2, j_3, \ldots$ denote the positive zeros of $J_0(z)$ arranged in ascending order of magnitude.

The necessity of expanding an arbitrary function in this manner arises also in Daniel Bernoulli's problem of a chain oscillating under gravity and in Euler's problem of the vibrations of a circular membrane with an initial arbitrary symmetrical displacement (§§ 1·3, 1·5).

In order to determine the coefficients a_m in the expansion, Fourier multiplied both sides of (1) by $x J_0(j_m x)$ and integrated between the limits 0 and 1. It follows from § 5·11 that

$$\int_0^1 x J_0(j_m x) J_0(j_n x)\,dx = \begin{cases} 0, & m \neq n, \\ \tfrac{1}{2} J_1^2(j_m), & m = n, \end{cases}$$

and hence Fourier inferred that

(2) $$a_m = \frac{2}{J_1^2(j_m)} \int_0^1 t f(t) J_0(j_m t)\,dt.$$

If we now change the significance of the symbols j_m, so that† $j_1, j_2, j_3, \ldots$ denote the positive zeros of the function $J_\nu(z)$, arranged in ascending order of magnitude, then

(3) $$f(x) = \sum_{m=1}^{\infty} a_m J_\nu(j_m x),$$

where

(4) $$a_m = \frac{2}{J_{\nu+1}^2(j_m)} \int_0^1 t f(t) J_\nu(j_m t)\,dt.$$

This more general result was stated by Lommel‡; but, of course, neither in the general case nor in the special case $\nu = 0$ does the procedure which has been indicated establish the validity of the expansion; it merely indicates how the coefficients are to be determined on the hypothesis that the expansion exists and is uniformly convergent.

* *La Théorie Analytique de la Chaleur* (Paris, 1822), §§ 316—319.

† The omission of the suffix ν, associated with $j_1, j_2, j_3, \ldots$, should cause no confusion, and it considerably improves the appearance of the formulae.

‡ *Studien über die Bessel'schen Functionen* (Leipzig, 1868), pp. 69—73.

In fact the simplicity of the procedure is somewhat deceptive; for the reader might anticipate that, if the function $f(x)$ is subjected to appropriate restrictions, the expansion would be valid for all values of ν for which the integral

$$\int_0^1 tJ_\nu(j_m t)\, J_\nu(j_n t)\, dt$$

is convergent, i.e. when $\nu \geqslant -1$.

Dini, however, remarked that he was unable to deal with the range $-1<\nu<-\frac{1}{2}$, and limited himself to the range $\nu \geqslant -\frac{1}{2}$. Several subsequent writers, while proving theorems for the latter range, asserted that the extension to the former range was merely a matter of detail; but it was not until after 1922 that anyone took the trouble to supply the detail which is tedious and of no great interest. In the exposition given here, it will be supposed that $\nu \geqslant -\frac{1}{2}$.

The first attempt at a rigorous proof of the expansions (1) and (3) is contained in some notes compiled by Hankel* in 1869 and published posthumously. A more complete investigation was given by Schläfli† a year after the publication of Hankel's work; and an important paper by Harnack‡ contains an investigation of the expansion (3) by methods which differed appreciably from those of earlier writers.

A few years after the appearance of the researches of Hankel and Schläfli, the more general expansion

(5) $$f(x) = \sum_{m=1}^{\infty} b_m J_\nu(\lambda_m x),$$

where $\lambda_1, \lambda_2, \lambda_3, \ldots$ denote the positive zeros (in ascending order of magnitude) of the function

$$z^{-\nu}\{zJ_\nu'(z) + HJ_\nu(z)\},$$

when $\nu \geqslant -\frac{1}{2}$ and H is any given constant, was investigated by Dini§.

The coefficients in the expansion are given by the formula

(6) $$\{(\lambda_m^2 - \nu^2) J_\nu^2(\lambda_m) + \lambda_m^2 J_\nu'^2(\lambda_m)\}\, b_m = 2\lambda_m^2 \int_0^1 tf(t)\, J_\nu(\lambda_m t)\, dt.$$

The mode of determination of the numbers λ_m subjects $f(x)$ to what is known as a 'mixed boundary condition,' namely that $f'(x) + Hf(x)$ should formally vanish at $x = 1$.

The expansion (5) was examined by Fourier (when $\nu = 0$) in the problem of the propagation of heat in a circular cylinder when heat is radiated from the cylinder; in this problem the physical significance of H is the ratio of the external conductivity of the cylinder to the internal conductivity.

* *Math. Ann.* VIII. (1875), pp. 471—494. In the course of this paper, Hankel obtained the integral formula of § 14·4 as a limiting case of (3).

† *Math. Ann.* X. (1876), pp. 137—142.

‡ *Leipziger Berichte*, XXXIX. (1887), pp. 191—214; *Math. Ann.* XXXV. (1889), pp. 41—62.

§ *Serie di Fourier* (Pisa, 1880), pp. 190—277.

It was pointed out by Dini that the expansion (5) must be modified* by the insertion of an initial term when $H+\nu=0$; and, although Dini's analysis contains a numerical error, this discovery seems to make it advisable to associate Dini's name rather than Fourier's with the expansion.

The researches which have now been described depend ultimately on a set of lemmas which are proved by Cauchy's theory of residues. The use of complex variables has, however, been abandoned, so far as possible, by Kneser † and Hobson‡, who have constructed the expansion by using the theory of integral equations as a basis.

On aesthetic grounds there is a great deal to be said for this procedure, because it seems somewhat unnatural to use complex variables in proving theorems which are essentially theorems concerning functions of real variables. On the other hand, researches based on the theory of integral equations are liable to give rise to uneasy feelings of suspicion in the mind of the ultra-orthodox mathematician.

The theory has recently been made distinctly more complete by the important memoir of W. H. Young§, who has thrown new light on many parts of the subject by using modern knowledge of the theory of functions of real variables in conjunction with the calculus of residues. An earlier paper by Filon|| which makes some parts of the analysis appreciably less synthetic must also be mentioned here.

The question of the permissibility of term-by-term differentiation of the expansion which represents a function as a series of Bessel functions has been discussed by Ford¶, who has obtained important results with the help of quite simple analysis (cf. § 18·4).

More recondite investigations are due to C. N. Moore**, who, after studying the summability of the expansion by Cesàro's means, has investigated the uniformity of the convergence of the expansion in the neighbourhood of the origin, and also the uniformity of the summability of the expansion (when not necessarily convergent) in this neighbourhood.

The reason why the uniformity of the convergence (or summability) of the expansion in the neighbourhood of the origin needs rather special consideration is that it is necessary to use asymptotic formulae for $J_\nu(\lambda_m x)$ which are valid when $\lambda_m x$ is large; and, as x approaches zero, the smallest value of m, for which the asymptotic formulae are significant, is continually increasing.

* Details of necessary modifications when $H+\nu \leqslant 0$ will be given in § 18·3. The modification was also noticed by Kirchhoff, *Berliner Sitzungsberichte*, 1883, pp. 519—524.

† *Archiv der Math. und Phys.* (3) VII. (1903), pp. 123—133; *Math. Ann.* LXIII. (1907), pp. 477—524.

‡ *Proc. London Math. Soc.* (2) VII. (1909), pp. 359—388.

§ *Ibid.* (2) XVIII. (1920), pp. 163—200.

|| *Ibid.* (2) IV. (1906), pp. 396—430. Cf. §§ 19·21—19·24.

¶ *Trans. American Math. Soc.* IV. (1903), pp. 178—184.

** *Ibid.* X. (1909), pp. 391—435; XII. (1911), pp. 181—206; XXI. (1920), pp. 107—156.

In the exposition which will be given in this chapter, the methods of the calculus of residues will be used to a far greater extent than has been usual in recent researches; this is a reversion to the practice of Hankel and Schläfli and (in the special case of Fourier series) of Cauchy. The advantage of this procedure is that it results in a great simplification in the general appearance of the analysis throughout the whole theory. And, although it seems impracticable to prove certain theorems (notably those* relating to fractional orders of summability) with the help of complex variables, the gain in simplicity is so marked that it has been possible to include in this chapter very many more theorems than would have been possible if the methods of the theory of functions of real variables had been used more exclusively.

As an example of the simplicity produced by using complex variables, it may be mentioned that comparatively crude inequalities, such as

$$|J_\nu(z)| < \frac{c_1 \exp |I(z)|}{\sqrt{|z|}},$$

where c_1 is a constant, independent of z, when ν is given and exceeds $-\frac{1}{2}$, are sufficient to prove all the requisite theorems concerning convergence at a point (or summability at a point) and they are also sufficient to prove theorems concerning uniformity of *summability* throughout an interval of which the origin may be an end point. Direct proofs of theorems concerning uniformity of *convergence* throughout such an interval require more elaborate inequalities, but in this work the use of such inequalities is evaded by deducing uniformity of convergence from uniformity of summability by an application of Hardy's convergence theorem†.

It may be stated here that the theorems of this chapter correspond exactly to the theorems concerning Fourier series which are given in *Modern Analysis*.

In addition to the memoirs which have already been cited, the following may be mentioned: Beltrami, *R. Ist. Lombardo Rendiconti*, (2) XIII. (1880), pp. 327—337; Gegenbauer, *Wiener Sitzungsberichte*, LXXXVIII. (2) (1884), pp. 975—1003; Alexander, *Trans. Edinburgh Royal Soc.* XXXIII. (1888), pp. 313—320; Sheppard, *Quarterly Journal*, XXIII. (1889), pp. 223—260; Volterra, *Ann. di Mat.* (2) XXV. (1897), p. 145; Stephenson, *Phil. Mag.* (6) XIV. (1907), pp. 547—549; *Messenger*, XXXIII. (1904), pp. 70—77, 178—182; Rutgers, *Nieuw Archief*, (2) VIII. (1909), pp. 375—380; Orr, *Proc. R. Irish Acad.* XXVII. A, (1910), pp. 233—248; and Dinnik, *Kief Polyt. Inst.* (*Engineering Section*), 1911, no. 1, pp. 83—85. [*Jahrbuch über die Fortschritte der Math.* 1911, p. 492.]

The investigations by Alexander are mainly based on operational methods, while Orr dealt with expansions in which functions of the second kind are involved.

18·11. *The various types of series.*

In the special case of series of circular functions, it is necessary, as the reader will remember, to make a distinction‡ between any *trigonometrical series*

$$\tfrac{1}{2}a_0 + \sum_{m=1}^{\infty} (a_m \cos mx + b_m \sin mx),$$

* Such theorems have been investigated by Moore and Young.

† *Modern Analysis*, § 8·5.

‡ Cf. *Modern Analysis*, § 9·1.

and a *Fourier series* in which the coefficients are expressed as integrals,

$$a_m = \frac{1}{\pi}\int_{-\pi}^{\pi} f(t)\cos mt\,dt,\quad b_m = \frac{1}{\pi}\int_{-\pi}^{\pi} f(t)\sin mt\,dt.$$

It is necessary to make a similar distinction* between the types of series which will be dealt with in this chapter; any series of the type

$$\sum_{m=1}^{\infty} a_m J_\nu(j_m x),$$

in which the coefficients a_m merely form a given sequence of constants, will be called *a series of Bessel functions.*

If, however, the coefficients in this series are expressible by the formula†

$$a_m = \frac{2}{J^2_{\nu+1}(j_m)}\int_0^1 tf(t)\,J_\nu(j_m t)\,dt,$$

the series will be called *the Fourier-Bessel series associated with* $f(x)$.

And if, further, the series converges to the sum $f(x)$ for any point x of the interval (0, 1), the series will be described as *the Fourier-Bessel expansion of* $f(x)$.

In like manner, the series

$$\sum_{m=1}^{\infty} b_m J_\nu(\lambda_m x),$$

where $\lambda_1, \lambda_2, \lambda_3, \ldots$ are the positive zeros of

$$zJ_\nu'(z) + HJ_\nu(z),$$

will be called *Dini's series of Bessel functions.*

If the coefficients b_m are determined by the formula‡

$$\{(\lambda_m^2 - \nu^2)\,J_\nu^2(\lambda_m) + \lambda_m^2 J_\nu'^2(\lambda_m)\}\,b_m = 2\lambda_m^2\int_0^1 tf(t)\,J_\nu(\lambda_m t)\,dt,$$

the series will be called the *Dini series associated with* $f(x)$.

And if, further, the series converges to the sum $f(x)$ for any point x of the interval (0, 1), the series will be described as *the Dini expansion of* $f(x)$.

Some writers have been inclined to regard Fourier-Bessel expansions as merely a special case of Dini expansions, obtainable by making $H \to \infty$; but there are certain distinctions between the two expansions which make this view somewhat misleading (cf. §§ 18·26, 18·34, 18·35).

18·12. *Special cases of Fourier-Bessel and Dini expansions.*

There are very few expansions of simple functions in which the coefficients assume a simple form.

One function whose expansion has simple coefficients has already been

* The greater part of the terminology is due to Young, *Proc. London. Math. Soc.* (2) XVIII. (1920), pp. 167—168.

† It is supposed that the integral is convergent for all positive integral values of m.

‡ It is supposed that the series is modified, as in § 18·34, when $H + \nu \leqslant 0$.

investigated in § 15·42. Another is x^ν, which gives rise to the formal expansions

(1) $$x^\nu = \sum_{m=1}^{\infty} \frac{2J_\nu(j_m x)}{j_m J_{\nu+1}(j_m)},$$

(2) $$x^\nu = \sum_{m=1}^{\infty} \frac{2\lambda_m J_\nu(\lambda_m x) J_{\nu+1}(\lambda_m)}{(\lambda_m^2 - \nu^2) J_\nu^2(\lambda_m) + \lambda_m^2 J_\nu'^2(\lambda_m)}.$$

It will be seen subsequently that (1) is valid when $0 \leqslant x < 1$, and (2) when $0 \leqslant x \leqslant 1$, if $H + \nu > 0$. Cf. §§ 18·22, 18·35.

The reduction formula

$$\lambda_m^2 \int_0^1 t^{\nu+2n+1} J_\nu(\lambda_m t)\, dt = (\nu+2n) J_\nu(\lambda_m) - \lambda_m J_\nu'(\lambda_m) - 4n(\nu+n) \int_0^1 t^{\nu+2n-1} J_\nu(\lambda_m t)\, dt$$

is easily established, so that the Dini expansion of $x^{\nu+2n}$ may be determined when ν is any positive integer. The Dini expansion of $x^{\nu+2n+1}$ may similarly be determined; in this case the general coefficient is expressible in terms of known functions and

$$\int_0^1 t^\nu J_\nu(\lambda_m t)\, dt.$$

In order to calculate this when ν is an integer, McMahon* has proposed to tabulate the function

$$\frac{1}{2}\int_0^x J_0(t)\, dt = J_1(x) + J_3(x) + J_5(x) + \ldots,$$

which is a special form of one of Lommel's functions of two variables (§§ 16·5, 16·56).

18·2. *The methods of Hankel and Schläfli.*

The earlier investigations which were described in § 18·1 are based on the analysis used by Dirichlet† in his researches on trigonometrical series of Fourier's type; this method of proceeding is obviously suggested by the fact that the trigonometrical series are special cases of the Fourier-Bessel expansion, obtained by giving ν the values $\pm\frac{1}{2}$.

In the case of Fourier's theorem, to prove that

$$f(x) = \tfrac{1}{2}a_0 + \sum_{m=1}^{\infty} (a_m \cos mx + b_m \sin mx),$$

where $$a_m = \frac{1}{\pi}\int_{-\pi}^{\pi} f(t) \cos mt\, dt, \qquad b_m = \frac{1}{\pi}\int_{-\pi}^{\pi} f(t) \sin mt\, dt,$$

it is sufficient to prove that

$$f(x) = \lim_{n\to\infty} \frac{1}{\pi}\int_{-\pi}^{\pi} \{\tfrac{1}{2} + \cos(x-t) + \cos 2(x-t) + \ldots + \cos n(x-t)\} f(t)\, dt,$$

i.e. that $$f(x) = \lim_{n\to\infty} \frac{1}{2\pi}\int_{-\pi}^{\pi} \frac{\sin(n+\frac{1}{2})(x-t)}{\sin\frac{1}{2}(x-t)} f(t)\, dt.$$

* *Proc. American Assoc.* 1900, pp. 42—43. The tabulation is most simply effected by using § 10·74(3) in conjunction with Table I. (pp. 666—697); see Table VIII.

† *Journal für Math.* IV. (1829), pp. 157—169.

In the case of the general Fourier-Bessel expansion, the corresponding limit to be evaluated is

$$\lim_{n\to\infty} \sum_{m=1}^{n} \frac{2J_\nu(j_m x)}{J^2_{\nu+1}(j_m)} \int_0^1 tf(t) J_\nu(j_m t)\,dt,$$

and so it is necessary to investigate the behaviour of the sum

$$\sum_{m=1}^{n} \frac{2J_\nu(j_m x) J_\nu(j_m t)}{J^2_{\nu+1}(j_m)},$$

when n is large; and it is in this investigation that the use of the calculus of residues is more than desirable.

In the case of Dini's expansion, the corresponding sum which needs examination is

$$\sum_{m=1}^{n} \frac{2\lambda_m^2 J_\nu(\lambda_m x) J_\nu(\lambda_m t)}{(\lambda_m^2 - \nu^2) J_\nu^2(\lambda_m) + \lambda_m^2 J_\nu'^2(\lambda_m)}.$$

An application of the calculus of residues which will be described in §§ 18·3—18·33 shews that the difference of the two sums is readily amenable to discussion, and so we are spared the necessity of repeating the whole of the analysis of the Fourier-Bessel expansion with the modifications appropriate to the more general case of the Dini expansion.

18·21. *The Hankel-Schläfli contour integral.*

We shall now begin the attack on the problem of Fourier-Bessel expansions by discussing properties of the function $T_n(t, x)$, defined by the equation

$$(1)\qquad T_n(t, x) = \sum_{m=1}^{n} \frac{2J_\nu(j_m x) J_\nu(j_m t)}{J^2_{\nu+1}(j_m)},$$

where $0 < x \leqslant 1$, $0 \leqslant t \leqslant 1$, and the order ν is real and is subject to the condition

$$\nu + \tfrac{1}{2} \geqslant 0.$$

The method which will be used is due to Hankel* and Schläfli†, though many of the details of the analysis are suggested by Young's‡ recent memoir.

The function $T_n(t, x)$ is obviously as fundamental in the theory of Fourier-Bessel expansions as is the function

$$\frac{\sin(n+\frac{1}{2})(x-t)}{\sin\frac{1}{2}(x-t)},$$

in the special theory of Fourier series.

In order to obtain the formulae connected with $T_n(t, x)$ which are subsequently required it is necessary to express the mth term of the sum for $T_n(t, x)$ as the residue at j_m of a function, of the complex variable w, which has poles at $j_1, j_2, j_3, \dots j_n$. When this has been done, we express

* *Math. Ann.* VIII. (1875), pp. 471—494.

† *Ibid.* X. (1876), pp. 137—142.

‡ *Proc. London Math. Soc.* (2) XVIII. (1920), pp. 163—200.

$T_n(t, x)$ as the integral of this function round a rectangle of which one of the sides lies along the imaginary axis while the opposite side passes between j_n and j_{n+1}. The sides parallel to the real axis are then moved off to infinity in opposite directions, so that, in order to secure the convergence of the integral, it is necessary to prescribe the behaviour of the integrand as $|I(w)| \to \infty$.

There are three integrands which we shall study, namely

$$(2) \qquad 2\{tJ_\nu(xw)J_{\nu+1}(tw) - xJ_\nu(tw)J_{\nu+1}(xw)\}/\{(t^2-x^2)J_\nu^2(w)\},$$

$$(3) \qquad \pi w\{J_\nu(w)Y_\nu(xw) - J_\nu(xw)Y_\nu(w)\}J_\nu(tw)/J_\nu(w),$$

$$(4) \qquad \pi w\{J_\nu(w)Y_\nu(tw) - J_\nu(tw)Y_\nu(w)\}J_\nu(xw)/J_\nu(w).$$

The first of these was the integrand studied by Schläfli; the other two are suggested by the work of Kneser and Carslaw which was described in § 15·42.

A study of the asymptotic values of these integrands indicates that (2) is suitable for discussions in which $x \neq t$ and $0 < x + t < 2$; (3) when* $0 \leqslant t < x < 1$; and (4) when $0 \leqslant x < t < 1$.

We proceed to verify that the integrands all have the same residue, namely

$$2J_\nu(j_m x)J_\nu(j_m t)/J^2_{\nu+1}(j_m),$$

at $w = j_m$. In the case of (2), we define the function† $g(w)$ by the formula

$$(5) \qquad g(w) \equiv \frac{2w}{t^2 - x^2}\{tJ_\nu(xw)J_{\nu+1}(tw) - xJ_\nu(tw)J_{\nu+1}(xw)\},$$

and then, if $w = j_m + \theta$, where θ is small, we have

$$J_\nu(w) = \theta J_\nu'(j_m) + \tfrac{1}{2}\theta^2 J_\nu''(j_m) + \ldots,$$

so that

$$wJ_\nu^2(w) = \theta^2 j_m J_\nu'^2(j_m) + \theta^3 J_\nu'(j_m)\{j_m J_\nu''(j_m) + J_\nu'(j_m)\} + \ldots.$$

It is easy to verify, by using Bessel's differential equation, that the coefficient of θ^3 on the right vanishes; and hence the residue of $g(w)/\{wJ_\nu^2(w)\}$ at j_m is

$$g'(j_m)/\{j_m J_\nu'^2(j_m)\},$$

and this is easily reduced to

$$2J_\nu(j_m x)J_\nu(j_m t)/J^2_{\nu+1}(j_m)$$

by using recurrence formulae.

In the case of (3), the residue at j_m is

$$\begin{aligned}\pi j_m\{J_\nu(j_m)Y_\nu(j_m x) - J_\nu(j_m x)Y_\nu(j_m)\}J_\nu(j_m t)/J_\nu'(j_m) \\ = -\pi j_m J_\nu(j_m x)J_\nu(j_m t)Y_\nu(j_m)/J_\nu'(j_m) \\ = 2J_\nu(j_m x)J_\nu(j_m t)/J_\nu'^2(j_m),\end{aligned}$$

by § 3·63, and this is the expression required; the integrand (4) is dealt with in the same way.

* This is most easily seen by writing the integrand in the form

$$\tfrac{1}{2}\pi i w\{H_\nu^{(1)}(w)H_\nu^{(2)}(xw) - H_\nu^{(1)}(xw)H_\nu^{(2)}(w)\}J_\nu(tw)/J_\nu(w).$$

† The results obtainable by using the integrand (2) are discussed in great detail by Graf and Gubler, *Einleitung in die Theorie der Bessel'schen Funktionen*, I. (Bern, 1898), pp. 131—139.

We next take the contour of integration to be a rectangle with vertices at $\pm Bi$, $A_n \pm Bi$, where B will be made to tend to ∞, and A_n is chosen so that $j_n < A_n < j_{n+1}$. When it is desired to assign a definite value to A_n, we shall take it to be equal to $(n + \frac{1}{2}\nu + \frac{1}{4})\pi$, which lies between j_n and j_{n+1} when n is sufficiently large (§ 15·53).

Now it is easy to verify that the three integrands are odd functions of w, and so the three integrals along the left sides of the rectangles vanish*.

Again, if $w = u + iv$, it may be verified that when v is large, and either positive or negative, while $u \geqslant 0$, then the three integrands are respectively

$$O(e^{-(2-x-t)|v|}), \quad O(e^{-(x-t)|v|}), \quad O(e^{-(t-x)|v|}),$$

and so, for any assigned value of A_n, the integrals along the upper and lower sides of the rectangle tend to zero as $B \to \infty$ when x and t have the relative values which have already been specified.

We thus obtain the three formulae

$$(6) \qquad T_n(t, x) = \frac{1}{2\pi i}\int_{A_n-\infty i}^{A_n+\infty i} \frac{g(w)\, dw}{w J_\nu^2(w)}$$

$$= \frac{1}{2\pi i}\int_{A_n-\infty i}^{A_n+\infty i}\int_0^1 \frac{2\theta w J_\nu(x\theta w) J_\nu(t\theta w)\, d\theta\, dw}{J_\nu^2(w)},$$

$$(0 < x + t < 2;\ x \neq t)$$

$$(7) \qquad T_n(t, x) = \frac{1}{2i}\int_{A_n-\infty i}^{A_n+\infty i} w\{J_\nu(w) Y_\nu(xw) - J_\nu(xw) Y_\nu(w)\}\frac{J_\nu(tw)\, dw}{J_\nu(w)},$$

$$(0 < t < x < 1)$$

$$(8) \qquad T_n(t, x) = \frac{1}{2i}\int_{A_n-\infty i}^{A_n+\infty i} w\{J_\nu(w) Y_\nu(tw) - J_\nu(tw) Y_\nu(w)\}\frac{J_\nu(xw)\, dw}{J_\nu(w)}.$$

$$(0 < x < t < 1)$$

From equation (6) it is easy to obtain an upper bound for $|T_n(t, x)|$; for it is evident from the asymptotic expansion of § 7·21 that, when $\nu + \frac{1}{2}$ is positive (or zero) and bounded, there exist positive constants c_1 and c_2 such that

$$(9) \qquad |J_\nu(tw)| \leqslant \frac{c_1 \exp\{|I(tw)|\}}{\sqrt{|tw|}}, \qquad |J_\nu(w)| \geqslant \frac{c_2 \exp\{|I(w)|\}}{\sqrt{|w|}}$$

when w is on the line joining $A_n - \infty i$ to $A_n + \infty i$ and $t \geqslant 0$, provided that n exceeds a value which depends on ν. Hence

$$|(t^2 - x^2) T_n(t, x)| \leqslant \frac{2c_1^2}{\pi c_2^2 \sqrt{(xt)}}\int_{-\infty}^{\infty} \exp\{-(2 - x - t)|v|\}\, dv,$$

so that

$$(10) \qquad |T_n(t, x)| \leqslant \frac{4c_1^2}{\pi c_2^2 |t^2 - x^2| (2 - x - t)\sqrt{(xt)}}.$$

This inequality gives the upper bound in question.

* It is necessary to make an indentation at the origin, but the integral round the indentation tends to zero with the radius of the indentation.

It is also easy to see that

$$\int_0^t t^{\nu+1} T_n(t,x)(t^2-x^2)\,dt$$
$$= \frac{t^{\nu+1}}{\pi i}\int_{A_n-\infty i}^{A_n+\infty i} \{t\,J_\nu(xw)\,J_{\nu+2}(tw) - xJ_{\nu+1}(tw)\,J_{\nu+1}(xw)\}\frac{dw}{wJ_\nu^2(w)},$$

and hence

$$\left|\int_0^t t^{\nu+1} T_n(t,x)(t^2-x^2)\,dt\right| \leqslant \frac{4c_1^2 t^{\nu+1}}{\pi c_2^2 A_n(2-x-t)\sqrt{(xt)}}. \tag{11}$$

[NOTE. Theorems obtained by a consideration of integrals involving Bessel functions of the first kind only can usually be made to cover the origin, in view of the fact that the constant c_1 in equation (9) is independent of t in the interval $0 \leqslant t \leqslant 1$. Thus (11) may be written

$$\left|x^{\frac{1}{2}}\int_0^t t^{\nu+1} T_n(t,x)(t^2-x^2)\,dt\right| \leqslant \frac{4c_1^2 t^{\nu+\frac{1}{2}}}{\pi c_2^2 A_n(2-x-t)},$$

valid when $0 \leqslant x \leqslant 1$, $0 \leqslant t \leqslant 1$. This extension is not so easily effected when integrals involving functions of the second kind have been used because the simplest inequality corresponding to (9) is

$$|Y_\nu(tw)| \leqslant c_1'\{|tw|^{-\nu}\log|tw| + |tw|^{-\frac{1}{2}}\}\exp\{|I(tw)|\},$$

and it is a somewhat tedious matter to obtain a simple upper bound to the integrand in (8) from this inequality.]

Equation (6) was used by Schläfli to prove that, when n is large, then

$$T_n(t,x) \sim \frac{1}{2\sqrt{(xt)}}\left[\frac{\sin A_n(t-x)}{\sin\frac{1}{2}\pi(t-x)} - \frac{\sin A_n(t+x)}{\sin\frac{1}{2}\pi(t+x)}\right],$$

but, since the order of magnitude of the error in this approximation is not evident, we shall next evaluate some integrals involving $T_n(t,x)$ by means of which difficulties caused by the unknown error may be evaded.

18·22. *Integrals involving $T_n(t,x)$.*

The two fundamental formulae which we shall now obtain are as follows:

$$\lim_{n\to\infty}\int_0^1 t^{\nu+1} T_n(t,x)\,dt = x^\nu, \qquad (0<x<1) \tag{1}$$

$$\lim_{n\to\infty}\int_0^x t^{\nu+1} T_n(t,x)\,dt = \tfrac{1}{2}x^\nu. \qquad (0<x<1) \tag{2}$$

From these it is obvious that

$$\lim_{n\to\infty}\int_x^1 t^{\nu+1} T_n(t,x)\,dt = \tfrac{1}{2}x^\nu. \qquad (0<x<1) \tag{3}$$

In the course of proving (1) it will be apparent that

$$x^{\frac{1}{2}}\int_0^1 t^{\nu+1} T_n(t,x)\,dt \to x^{\nu+\frac{1}{2}}$$

uniformly as $n \to \infty$ when x lies in the interval

$$0 \leqslant x \leqslant 1-\Delta,$$

where Δ is any positive number.

We shall also investigate the boundedness of

$$\int_0^t t^{\nu+1} T_n(t, x)\, dt$$

in the interval in which $0 < t \leqslant 1$.

Of these results, (1) was given by Young, *Proc. London Math. Soc.* (2) XVIII. (1920), pp. 173—174, and the proof of it, which will now be given, is his. Formula (2) seems to be new, though it is contained implicitly in Hobson's memoir.

It is evident that

$$\int_0^1 t^{\nu+1} T_n(t, x)\, dt = \sum_{m=1}^{n} \frac{2J_\nu(j_m x)}{j_m J_{\nu+1}(j_m)}.$$

When we transform the sum on the right into a contour integral after the manner of § 18·21, we find that it is equal to

$$\frac{1}{2\pi i}\int_{-\infty i}^{\infty i} \frac{2J_\nu(xw)\, dw}{wJ_\nu(w)} - \frac{1}{2\pi i}\int_{A_n-\infty i}^{A_n+\infty i} \frac{2J_\nu(xw)\, dw}{wJ_\nu(w)}.$$

In the former of these two integrals, the origin has to be avoided by an indentation on the right of the imaginary axis.

Since the integrand is an odd function of w, the value of the first integral reduces to πi times the residue of the integrand at the origin, so that

$$\int_0^1 t^{\nu+1} T_n(t, x)\, dt = x^\nu - \frac{1}{2\pi i}\int_{A_n-\infty i}^{A_n+\infty i} \frac{2J_\nu(xw)\, dw}{wJ_\nu(w)}.$$

Now

$$\left|\int_{A_n-\infty i}^{A_n+\infty i} \frac{2J_\nu(xw)\, dw}{wJ_\nu(w)}\right| < \frac{2c_1}{c_2 A_n \sqrt{x}}\int_{-\infty}^{\infty} \exp\{-(1-x)\,|\,v\,|\}\, dv$$

$$= \frac{4c_1}{c_2 A_n (1-x)\sqrt{x}},$$

and, from this result, (1) is evident; it is also evident that

$$x^{\frac{1}{2}}\int_0^1 t^{\nu+1} T_n(t, x)\, dt - x^{\nu+\frac{1}{2}}$$

tends uniformly to zero as $n \to \infty$ so long as $0 \leqslant x \leqslant 1 - \Delta$.

It will be observed that the important expansion

$$x^{\nu+\frac{1}{2}} = x^{\frac{1}{2}} \sum_{m=1}^{\infty} \frac{2J_\nu(j_m x)}{j_m J_{\nu+1}(j_m)}, \qquad (0 \leqslant x < 1) \tag{4}$$

which was formally obtained in § 18·12, is an immediate consequence of (1).

Formula (2) can be proved in a somewhat similar manner (though the details of the proof are rather more elaborate) by using an integrand involving functions of the second kind. It is easy to see that

$$\int_0^x t^{\nu+1} T_n(t, x)\, dt = \sum_{m=1}^{n} \frac{2x^{\nu+1} J_\nu(j_m x) J_{\nu+1}(j_m x)}{j_m J^2_{\nu+1}(j_m)}$$

$$= \lim_{B\to\infty} \frac{x^{\nu+1}}{2i}\int_{A_n-Bi}^{A_n+Bi} \{J_\nu(w)\, Y_\nu(xw) - J_\nu(xw)\, Y_\nu(w)\} \frac{J_{\nu+1}(xw)\, dw}{J_\nu(w)}.$$

Now take $0 < x \leqslant 1$ in the last integral and substitute for the Bessel functions the *dominant* terms of their respective asymptotic expansions, valid when $|w|$ is large (§ 7·21). The error produced thereby in the integrand is, at most, $O(1/w^2)$ when $0 < x \leqslant 1$; and, as $n \to \infty$, we have

$$\int_{A_n - \infty i}^{A_n + \infty i} \left| \frac{dw}{w^2} \right| = \frac{\pi}{A_n} = O\left(\frac{1}{n}\right).$$

Now the result of substituting these dominant terms is

$$- \lim_{B\to\infty} \frac{x^\nu}{\pi i} \int_{A_n - Bi}^{A_n + Bi} \frac{\sin w(1-x)\sin(xw - \frac{1}{2}\nu\pi - \frac{1}{4}\pi)}{w \cos(w - \frac{1}{2}\nu\pi - \frac{1}{4}\pi)} dw$$

$$= \lim_{B\to\infty} \frac{x^\nu}{2\pi i} \int_{A_n - Bi}^{A_n + Bi} \frac{\cos(w - \frac{1}{2}\nu\pi - \frac{1}{4}\pi) - \cos(2xw - w - \frac{1}{2}\nu\pi - \frac{1}{4}\pi)}{w\cos(w - \frac{1}{2}\nu\pi - \frac{1}{4}\pi)} dw$$

$$= \tfrac{1}{2}x^\nu - \lim_{B\to\infty} \int_{A_n - Bi}^{A_n + Bi} \frac{\cos(2xw - w - \frac{1}{2}\nu\pi - \frac{1}{4}\pi)}{w\cos(w - \frac{1}{2}\nu\pi - \frac{1}{4}\pi)} dw.$$

We shall have to discuss, almost immediately, several integrals of this general type; so it is convenient at this stage to prove a lemma concerning their boundedness as $n \to \infty$.

LEMMA. *The integral*

$$\lim_{B\to\infty} \int_{A_n - Bi}^{A_n + Bi} \frac{\cos(\lambda w - \frac{1}{2}\nu\pi - \frac{1}{4}\pi)}{w\cos(w - \frac{1}{2}\nu\pi - \frac{1}{4}\pi)} dw$$

is $O(1/n)$, *as* $n \to \infty$, *if* $-1 < \lambda < 1$; *and the integral is bounded if* $0 \leqslant \lambda \leqslant 1$.

If we put $w = A_n \pm iv$, where A_n, as usual, stands for $(n + \frac{1}{2}\nu + \frac{1}{4})\pi$, the expression under consideration may be written in the form

$$2i\left[A_n \cos(\lambda - 1)A_n \, . \int_0^\infty \frac{\cosh \lambda v \, . \, dv}{(A_n^2 + v^2)\cosh v} - \sin(\lambda - 1)A_n \, . \int_0^\infty \frac{v \sinh \lambda v \, . \, dv}{(A_n^2 + v^2)\cosh v}\right].$$

When $-1 < \lambda < 1$, the modulus of this does not exceed

$$\frac{2}{A_n}\int_0^\infty \frac{\cosh \lambda v \, . \, dv}{\cosh v} + \frac{2}{A_n^2}\int_0^\infty \frac{v\,|\sinh \lambda v|\,dv}{\cosh v},$$

and the first part of the Lemma is obvious.

Again, if $0 \leqslant \lambda \leqslant 1$ and $v(1-\lambda) = \xi$, we have*

$$0 \leqslant v(1-\lambda)\sinh \lambda v = \xi \sinh(v - \xi) \leqslant \cosh v,$$

so the integral to be considered does not exceed (in absolute value)

$$2A_n \int_0^\infty \frac{dv}{A_n^2 + v^2} + 2A_n \int_0^\infty \frac{dv}{A_n^2 + v^2} = 2\pi,$$

and the second part of the Lemma is proved.

It follows immediately from the Lemma that

$$\int_0^x t^{\nu+1} T_n(t, x)\, dt = \tfrac{1}{2}x^\nu + O(1/n),$$

when $0 < x < 1$; and this is equivalent to (2).

* The function $\xi \sinh(v - \xi)$ has one maximum, at ξ_0 say, and its value there is equal to $\sinh^2(v - \xi_0)/\cosh(v - \xi_0)$ which is less than $\sinh(v - \xi_0)$.

Moreover, if we close the range of values of x on the *right* so that $0 < x \leqslant 1$, we infer from the Lemma that the integrals

$$\int_0^x t^{\nu+1} T_n(t, x)\, dt, \quad \int_0^1 t^{\nu+1} T_n(t, x)\, dt$$

are bounded as $n \to \infty$ when $0 < x \leqslant 1$.

Lastly we shall consider

$$\int_0^t t^{\nu+1} T_n(t, x)\, dt,$$

and we shall prove that, *when $0 < t \leqslant 1$ and $0 < x \leqslant 1$, this integral is a bounded function of n, x and t, as $n \to \infty$.*

It is easy to shew by the methods which have just been used that, when $1 - x + t \leqslant 1$, i.e. when $t \leqslant x$, then

$$\begin{aligned}
&\int_0^t t^{\nu+1} T_n(t, x)\, dt \\
&\quad = \sum_{m=1}^{n} \frac{2t^{\nu+1} J_\nu(j_m x) J_{\nu+1}(j_m t)}{j_m J^2_{\nu+1}(j_m)} \\
&\quad = \lim_{B\to\infty} \frac{t^{\nu+1}}{2i} \int_{A_n - Bi}^{A_n+Bi} \{J_\nu(w) Y_\nu(xw) - J_\nu(xw) Y_\nu(w)\} \frac{J_{\nu+1}(tw)}{J_\nu(w)}\, dw \\
&\quad = O\left(\frac{1}{A_n}\right) - \lim_{B\to\infty} \frac{t^{\nu+\frac{1}{2}}}{\pi i x^{\frac{1}{2}}} \int_{A_n - Bi}^{A_n+Bi} \frac{\sin w(1-x) \,.\, \sin(tw - \frac{1}{2}\nu\pi - \frac{1}{4}\pi)}{w \cos(w - \frac{1}{2}\nu\pi - \frac{1}{4}\pi)}\, dw \\
&\quad = O\left(\frac{1}{A_n}\right) - \lim_{B\to\infty} \frac{t^{\nu+\frac{1}{2}}}{2\pi i x^{\frac{1}{2}}} \int_{A_n - Bi}^{A_n+Bi} \frac{\cos(xw + tw - w - \frac{1}{2}\nu\pi - \frac{1}{4}\pi)}{w \cos(w - \frac{1}{2}\nu\pi - \frac{1}{4}\pi)}\, dw \\
&\qquad + \lim_{B\to\infty} \frac{t^{\nu+\frac{1}{2}}}{2\pi i x^{\frac{1}{2}}} \int_{A_n - Bi}^{A_n+Bi} \frac{\cos(w + tw - xw - \frac{1}{2}\nu\pi - \frac{1}{4}\pi)}{w \cos(w - \frac{1}{2}\nu\pi - \frac{1}{4}\pi)}\, dw.
\end{aligned}$$

These integrals are of the type examined in the Lemma given earlier in this section; and so the original integral is bounded when $-1 < x + t - 1 \leqslant 1$ and $-1 < 1 + t - x \leqslant 1$, i.e. when $0 < t \leqslant x \leqslant 1$.

To prove that the integral is bounded when $0 < x \leqslant t \leqslant 1$, we first shew that

$$\begin{aligned}
&\int_0^t t^{\nu+1} T_n(t, x)\, dt \\
&\quad = \lim_{B\to\infty} \frac{t^{\nu+1}}{2i} \int_{A_n - Bi}^{A_n+Bi} \{J_\nu(w) Y_{\nu+1}(tw) - J_{\nu+1}(tw) Y_\nu(w)\} \frac{J_\nu(xw)\, dw}{J_\nu(w)},
\end{aligned}$$

and then apply the arguments just used in order to approximate to the integral on the right; the details of the analysis are left to the reader.

It has therefore been proved that, if Δ be an arbitrary positive number, then

$$\left| \int_0^t t^{\nu+1} T_n(t, x)\, dt \right| < U,$$

where U is independent of n, x and t when $\Delta \leqslant x \leqslant 1$, $\Delta \leqslant t \leqslant 1$.

These results constitute the necessary preliminary theorems concerning $T_n(t, x)$, and we are now in a position to discuss integrals, involving $T_n(t, x)$, which occur in the investigation of the Fourier-Bessel expansion associated with an arbitrary function $f(x)$.

18·23. *The analogue of the Riemann-Lebesgue Lemma**.

We shall now prove that, *if* (a, b) *is any part of the closed interval* $(0, 1)$, *such that* x *is not an internal point or an end point of* (a, b), *then the existence and the absolute convergence of*

$$\int_a^b t^{\frac{1}{2}} f(t)\, dt$$

are sufficient to ensure that, as $n \to \infty$,

$$\int_a^b t f(t)\, T_n(t, x)\, dt = o(1),$$

where† $0 < x \leqslant 1$.

The reader will observe that this theorem asserts that the only part of the path of integration in

$$\int_0^1 t f(t)\, T_n(t, x)\, dt$$

which is of any significance, as $n \to \infty$, is the part *in the immediate vicinity of the point* x.

It is convenient to prove the theorem in three stages. It is first supposed that $t^{\frac{1}{2}} f(t)$ is bounded and that the origin is not an end point of (a, b). In the second stage we remove the restriction of boundedness, and in the third stage we remove the restriction concerning the origin.

(I) Let
$$t^{-\nu} f(t) \equiv F(t)(t^2 - x^2),$$
and let the upper bound of $|F(t)|$ in (a, b) be K. Divide (a, b) into p equal parts by the points $t_1, t_2, \ldots t_{p-1}$, $(t_0 = a,\ t_p = b)$; and, after choosing an arbitrary positive number ϵ, take p to be so large that

$$\sum_{m=1}^{p} (U_m - L_m)(t_m - t_{m-1}) < \epsilon,$$

where U_m and L_m are the upper and lower bounds of $F(t)$ in (t_{m-1}, t_m).

Let
$$F(t) \equiv F(t_{m-1}) + \omega_m(t),$$
so that $|\omega_m(t)| \leqslant U_m - L_m$ in (t_{m-1}, t_m).

It is then evident that

$$\int_a^b t f(t)\, T_n(t, x)\, dt = \sum_{m=1}^{p} F(t_{m-1}) \int_{t_{m-1}}^{t_m} t^{\nu+1}\, T_n(t, x)(t^2 - x^2)\, dt$$
$$+ \sum_{m=1}^{p} \int_{t_{m-1}}^{t_m} t^{\nu+1}\, T_n(t, x)(t^2 - x^2)\, \omega_m(t)\, dt,$$

* Cf. *Modern Analysis*, § 9·41.

† If $x = 1$, it is, of course, supposed that $b < 1$.

and hence, by the inequalities (10) and (11) of § 18·21,

$$\left|\int_a^b tf(t)\,T_n(t,x)\,dt\right| \leqslant \frac{8c_1^2Kp}{\pi c_2^2 A_n(2-x-b)\sqrt{x}} + \frac{4c_1^2}{\pi c_2^2(2-x-b)\sqrt{x}}\sum_{m=1}^{p}(U_m-L_m)\int_{t_{m-1}}^{t_m} dt,$$

that is to say

$$\left|\int_a^b tf(t)\,T_n(t,x)\,dt\right| \leqslant \frac{4c_1^2}{\pi c_2^2(2-x-b)\sqrt{x}}\left[\frac{2Kp}{A_n}+\epsilon\right].$$

Now the choice of ϵ fixes p; when ϵ (and therefore p) has been chosen, we are at liberty to choose A_n so large that $A_n > 2Kp/\epsilon$. That is to say, by a suitable choice of A_n, we may make the integral on the left less than

$$\frac{8c_1^2\epsilon}{\pi c_2^2(2-x-b)\sqrt{x}},$$

which is arbitrarily small. Consequently the integral is $o(1)$ as $A_n \to \infty$, and this is the theorem to be proved.

(II) When $F(t)$ is not bounded throughout (a, b), let it be possible to choose r intervals μ, such that $F(t)$ is bounded outside these intervals and such that

$$\sum_\mu \int_\mu |F(t)|\,dt < \epsilon.$$

When t lies in one of the intervals μ we use the inequality

$$|t^{\nu+1}(t^2-x^2)\,T_n(t,x)| \leqslant \frac{4c_1^2}{\pi c_2^2(2-x-b)\sqrt{x}},$$

and hence, if K is the upper bound of $|F(t)|$ in the parts of (a, b) outside the intervals μ, by applying (I) to each of these parts, we have

$$\left|\int_a^b tf(t)\,T_n(t,x)\,dt\right| \leqslant \frac{8c_1^2}{\pi c_2^2(2-x-b)\sqrt{x}}\left[\frac{(r+1)Kp}{A_n}+\epsilon\right].$$

If we take ϵ sufficiently small (thus fixing K) and then take A_n to be sufficiently large, we can make the expression on the right (and therefore also the expression on the left) arbitrarily small, and this is the result which had to be proved.

(III) If $\int_0^b t^{\frac{1}{2}} f(t)\,dt$ exists and is absolutely convergent, we can choose η so small that

$$\int_0^\eta \left|\frac{t^{\frac{1}{2}} f(t)}{t^2-x^2}\right| dt < \epsilon,$$

and then, since we have

$$\left|\int_0^\eta tf(t)\,T_n(t,x)\,dt\right| < \frac{4c_1^2}{\pi c_2^2(2-x-b)\sqrt{x}}\int^\eta \left|\frac{t^{\frac{1}{2}} f(t)}{t^2-x^2}\right| dt,$$

it follows from (II) that

$$\left|\int_0^b tf(t)\,T_n(t,x)\,dt\right| < \frac{8c_1^2}{\pi c_2^2(2-x-b)\sqrt{x}}\left[\frac{(r+1)Kp}{A_n}+\frac{3\epsilon}{2}\right],$$

where K is the upper bound of $|F(t)|$ in (η, b) when the intervals μ are omitted.

Hence it follows that the expression on the left can be made arbitrarily small by taking n sufficiently large, and so the analogue of the Riemann-Lebesgue Lemma is completely proved.

18·24. *The Fourier-Bessel expansion.*

We shall now prove the following theorem*, by means of which the sum of the Fourier-Bessel expansion associated with a given function is determined:

Let $f(t)$ be a function defined arbitrarily in the interval $(0, 1)$; *and let* $\int_0^1 t^{\frac{1}{2}} f(t)\,dt$ *exist and (if it is an improper integral) let it be absolutely convergent.*

Let
$$a_m = \frac{2}{J^2_{\nu+1}(j_m)}\int_0^1 tf(t)\,J_\nu(j_m t)\,dt,$$
where $\nu+\frac{1}{2}\geqslant 0$.

Let x be any internal point of an interval (a, b) such that $0<a<b<1$ and such that $f(t)$ has limited total fluctuation in (a, b).

Then the series
$$\sum_{m=1}^{\infty} a_m J_\nu(j_m x)$$
is convergent and its sum is $\frac{1}{2}\{f(x+0)+f(x-0)\}$.

We first observe that, by §§ 18·21, 18·22,

$$\sum_{m=1}^{n} a_m J_\nu(j_m x) = \int_0^1 tf(t)\,T_n(t,x)\,dt,$$

$$\tfrac{1}{2}\{f(x-0)+f(x+0)\} = \lim_{n\to\infty} x^{-\nu} f(x-0)\int_0^x t^{\nu+1}\,T_n(t,x)\,dt$$
$$+\lim_{n\to\infty} x^{-\nu} f(x+0)\int_x^1 t^{\nu+1}\,T_n(t,x)\,dt.$$

Hence, if

$$S_n(x) \equiv \int_0^x t^{\nu+1}\{t^{-\nu}f(t)-x^{-\nu}f(x-0)\}\,T_n(t,x)\,dt$$
$$+\int_x^1 t^{\nu+1}\{t^{-\nu}f(t)-x^{-\nu}f(x+0)\}\,T_n(t,x)\,dt,$$

it is sufficient to prove that $S_n(x)\to 0$ as $n\to\infty$ in order to establish the convergence of

$$\sum_{m=1}^{\infty} a_m J_\nu(j_m x)$$

to the sum $\frac{1}{2}\{f(x+0)+f(x-0)\}$.

* Hobson, *Proc. London Math. Soc.* (2) VII. (1909), pp. 387—388.

We now discuss

$$\int_x^1 t^{\nu+1} \{t^{-\nu} f(t) - x^{-\nu} f(x+0)\} T_n(t, x)\, dt$$

in detail, and the reader can then investigate the other integral involved in $S_n(x)$ in precisely the same manner.

The function $t^{-\nu} f(t) - x^{-\nu} f(x+0)$ has limited total fluctuation in (x, b), and so we may write*

$$t^{-\nu} f(t) - x^{-\nu} f(x+0) \equiv \chi_1(t) - \chi_2(t),$$

where $\chi_1(t)$ and $\chi_2(t)$ are bounded positive increasing functions of t in (x, b), such that

$$\chi_1(x+0) = \chi_2(x+0) = 0.$$

Hence, when an arbitrary positive number ϵ is chosen, there exists a positive number δ not exceeding $b - x$, such that

$$0 \leqslant \chi_1(t) < \epsilon, \quad 0 \leqslant \chi_2(t) < \epsilon,$$

whenever $x \leqslant t \leqslant x + \delta$.

We then have

$$\int_x^1 t^{\nu+1} \{t^{-\nu} f(t) - x^{-\nu} f(x+0)\} T_n(t, x)\, dt$$
$$= \int_{x+\delta}^1 t^{\nu+1} \{t^{-\nu} f(t) - x^{-\nu} f(x+0)\} T_n(t, x)\, dt$$
$$+ \int_x^{x+\delta} t^{\nu+1} \chi_1(t)\, T_n(t, x)\, dt - \int_x^{x+\delta} t^{\nu+1} \chi_2(t)\, T_n(t, x)\, dt.$$

We now obtain inequalities satisfied by the three integrals on the right.

It follows from the analogue of the Riemann-Lebesgue lemma that the modulus of the first can be made less than ϵ by taking n sufficiently large.

Next, from the second mean-value theorem it follows that there is a number ξ between 0 and δ such that

$$\int_x^{x+\delta} t^{\nu+1} \chi_1(t)\, T_n(t, x)\, dt = \chi_1(x+\delta) \int_{x+\xi}^{x+\delta} t^{\nu+1} T_n(t, x)\, dt,$$

and, by § 18·22, the modulus of this does not exceed $2U\epsilon$; and similarly the modulus of the third integral does not exceed $2U\epsilon$. By treating the integral between the limits 0 and x in a similar manner, we deduce that, by taking n sufficiently large, we can make the difference between

$$\sum_{m=1}^{n} a_m J_\nu(j_m x) \text{ and } \tfrac{1}{2}\{f(x+0) + f(x-0)\}$$

numerically less than $(8U + 2)\,\epsilon$; and this is arbitrarily small.

Hence, by the definition of an infinite series, we have proved that, in the circumstances postulated, $\sum_{m=1}^{\infty} a_m J_\nu(j_m x)$ is convergent and its sum is

$$\tfrac{1}{2}\{f(x+0) + f(x-0)\};$$

and this is the theorem to be proved.

* Cf. *Modern Analysis*, § 3·64.

18·25. *The uniformity of the convergence of the Fourier-Bessel expansion.*

Let $f(t)$ satisfy the conditions enunciated in § 18·24, and also let $f(t)$ be *continuous* (in addition to having limited total fluctuation) in the interval (a, b).

Then the Fourier-Bessel expansion associated with $f(t)$ converges uniformly to the sum $f(x)$ throughout the interval $(a+\Delta, b-\Delta)$ where Δ is any positive number.

This theorem is analogous to the usual theorem concerning uniformity of convergence of Fourier series *; the discussion of the uniformity of the convergence of the Fourier-Bessel expansion near $x=1$ and near $x=0$ requires rather more careful consideration, in the first place because formula § 18·22 (1) is untrue when $x=1$, and in the second place because it is not practicable to examine the bounds of

$$\int_0^t t^{\nu+1} T_n(t, x)\, dt,$$

when x and t are small, without using approximations for Bessel functions of the second kind.

The difficulties in the case of the neighbourhood of $x=1$ are easy to overcome (cf. § 18·26); but the difficulties in the case of the neighbourhood of the origin are of a graver character; and the discussion of them is deferred to § 18·55.

We shall prove the theorem concerning uniformity of convergence throughout $(a+\Delta, b-\Delta)$ by a recapitulation of the arguments of the preceding section.

In the first place, since continuity involves uniformity of continuity†, the choice of δ which was made in § 18·24 is independent of x when x lies in $(a+\Delta, b-\Delta)$.

Next we discuss such an integral as

$$\int_{x+\delta}^1 t^{\nu+1} \{t^{-\nu} f(t) - x^{-\nu} f(x)\}\, T_n(t, x)\, dt.$$

Since δ is independent of x, it follows from the proof of the Riemann-Lebesgue lemma (§ 18·23) that this integral tends to zero uniformly as $n \to \infty$, provided that

$$\int_{x+\delta}^1 t^{\nu+\frac{1}{2}} \{t^{-\nu} f(t) - x^{-\nu} f(x)\}\, dt$$

is a bounded function of x.

Now

$$\left|\int_{x+\delta}^1 t^{\nu+\frac{1}{2}} \{t^{-\nu} f(t) - x^{-\nu} f(x)\}\, dt\right| \leqslant \int_0^1 |t^{\frac{1}{2}} f(t)|\, dt + |x^{-\nu} f(x)| \int_0^1 t^{\nu+\frac{1}{2}}\, dt,$$

and this is bounded in $(a+\Delta, b-\Delta)$ since $f(x)$ is continuous and therefore bounded in this interval.

* Cf. *Modern Analysis*, § 9·44.

† Cf. *Modern Analysis*, § 3·61. It is now convenient to place an additional (trivial) restriction on δ, namely that it should be less than Δ, in order that the interval $(x-\delta, x+\delta)$ may lie inside the interval (a, b).

Similarly the other integrals introduced in § 18·24 tend to zero uniformly, and so

$$\sum_{m=1}^{n} a_m J_\nu(j_m x) - f(x)$$

tends to zero uniformly as $n \to \infty$, and this proves the theorem stated.

18·26. *The uniformity of the convergence of the Fourier-Bessel expansion near $x = 1$.*

It is evident that all the terms of the Fourier-Bessel expansion vanish at the point $x = 1$, so that, at that point, the sum of the terms of the expansion is zero.

Since uniformity of convergence of a series of continuous functions involves the continuity of the sum, it is evident that the condition

$$f(1-0) = 0$$

is *necessary* in order that the convergence of the Fourier-Bessel expansion associated with $f(t)$ may be uniform near $x = 1$.

We shall now prove that the conditions that $f(x)$ is to be continuous in $(a, 1)$ and that $f(1)$ is zero, combined with the conditions stated* in § 18·24, are *sufficient* for the convergence to be uniform throughout $(a + \Delta, 1)$.

The analysis is almost identical with that of the preceding section; we take

$$\int_0^1 t^{\nu+1} \{t^{-\nu} f(t) - x^{-\nu} f(x)\} \, T_n(t, x) \, dt,$$

just as before, and we then divide the interval $(0, 1)$ either into three parts $(0, x-\delta)$, $(x-\delta, x+\delta)$, $(x+\delta, 1)$, if $x \leqslant 1-\delta$, or into two parts $(0, x-\delta)$, $(x-\delta, 1)$, if $x \geqslant 1-\delta$. And we then prove that the three integrals (or the two integrals, as the case may be) tend uniformly to zero.

Again, when $f(1) = 0$, we can choose δ_1 so that

$$|x^{-\nu} f(x)| < \epsilon, \quad |f(x)| < \epsilon$$

when $1 - \delta_1 \leqslant x \leqslant 1$.

Then the expression

$$\left| f(x) - x^{-\nu} f(x) \int_0^1 t^{\nu+1} T_n(t, x) \, dt \right|$$

tends uniformly to zero† as $n \to \infty$ when x lies in $(a + \Delta, 1 - \delta_1)$, and the expression does not exceed $(U + 1)\epsilon$ for *any* value of n when x lies in $(1 - \delta_1, 1)$.

* The interval (a, b) is, of course, to be replaced by the interval $(a, 1)$.

† Because the integral involved tends to x^ν uniformly throughout $(\Delta, 1-\delta_1)$, by § 18·21.

Hence we can make

$$\left| f(x) - x^{-\nu} f(x) \int_0^1 t^{\nu+1} T_n(t, x)\, dt \right|$$

arbitrarily small for all values of x in $(a+\Delta, 1)$ by a choice of n which is independent of x; and this establishes the uniformity of the convergence of

$$\int_0^1 t f(t) T_n(t, x)\, dt$$

to the sum $f(x)$ in $(a+\Delta, 1)$ in the postulated circumstances.

18·27. *The order of magnitude of the terms in the Fourier-Bessel series.*

It is easy to prove that, *if $t^{\frac{1}{2}} f(t)$ has limited total fluctuation in (a, b), where (a, b) is any part (or the whole) of the interval $(0, 1)$, then*

$$\int_a^b t f(t) J_\nu(\lambda t)\, dt = O\left(\frac{1}{\lambda^{\frac{3}{2}}}\right)$$

as $\lambda \to \infty$.

From this theorem we at once obtain Sheppard's result* that

$$\frac{2J_\nu(j_m x)}{J^2_{\nu+1}(j_m)} \int_0^1 t f(t) J_\nu(j_m t)\, dt = O\left(\frac{1}{j_m}\right)$$

when $0 < x \leqslant 1$; this equation, of course, has a well-known parallel in the theory of Fourier series.

We first observe that, as a consequence of the asymptotic expansion of § 7·21,

$$\left| \int_0^t t^{\frac{1}{2}} J_\nu(t)\, dt \right| < c,$$

where c is a constant, independent of t when t lies in the interval $(0, \infty)$.

Now write $t^{\frac{1}{2}} f(t) = \psi_1(t) - \psi_2(t)$, where $\psi_1(t)$ and $\psi_2(t)$ are monotonic in (a, b); and then a number ξ exists such that

$$\begin{aligned} \left| \int_a^b \psi_1(t)\, t^{\frac{1}{2}} J_\nu(\lambda t)\, dt \right| &= \left| \psi_1(a) \int_a^\xi t^{\frac{1}{2}} J_\nu(\lambda t)\, dt + \psi_1(b) \int_\xi^b t^{\frac{1}{2}} J_\nu(\lambda t)\, dt \right| \\ &< 2c \{ |\psi_1(a)| + |\psi_1(b)| \} \lambda^{-\frac{3}{2}} \\ &= O(\lambda^{-\frac{3}{2}}). \end{aligned}$$

A similar result holds for $\psi_2(t)$, and hence the theorem stated is evident.

If it is known merely that

$$\int_a^b t^{\frac{1}{2}} f(t)\, dt$$

exists and is absolutely convergent, then all that can be proved is the theorem that

$$\int_a^b t f(t) J_\nu(\lambda t)\, dt = o(1/\sqrt{\lambda}).$$

* *Quarterly Journal*, XXIII. (1889), p. 247.

This theorem is due to W. H. Young*, and it may be proved in precisely the same manner as the theorem of § 18·23. We shall write out the proof when $|t^{\frac{1}{2}}f(t)|$ is bounded, with upper bound K, and leave the reader to construct the proof, when the function is unbounded, on the lines of § 18·23.

Divide (a, b) into p equal parts by the points $t_1, t_2, \ldots, t_{p-1}$ $(t_0 = a,\ t_p = b)$, and let the parts be so numerous that

$$\sum_{m=1}^{p} (t_m - t_{m-1})(U_m - L_m) < \epsilon,$$

where U_m and L_m are the upper and lower bounds of $t^{\frac{1}{2}}f(t)$ in (t_{m-1}, t_m).

Next let $\quad t^{\frac{1}{2}}f(t) \equiv F(t), \quad F(t) \equiv F(t_{m-1}) + \omega_m(t),$

and then

$$\left|\int_a^b tf(t)\,J_\nu(\lambda t)\,dt\right| < K\sum_{m=1}^{p}\left|\int_{t_{m-1}}^{t_m} t^{\frac{1}{2}}J_\nu(\lambda t)\,dt\right| + \sum_{m=1}^{p}\int_{t_{m-1}}^{t_m} |t^{\frac{1}{2}}J_\nu(\lambda t)\,\omega_m(t)|\,dt$$
$$\leqslant 2Kcp\,\lambda^{-\frac{3}{2}} + c'\epsilon\lambda^{-\frac{1}{2}},$$

where c' is the upper bound of $|t^{\frac{1}{2}}J_\nu(t)|$ in the interval $(0, \infty)$. Hence, by reasoning resembling that used in § 18·23, the integral on the left is $o(\lambda^{-\frac{1}{2}})$, and this is the theorem to be proved.

The theorems of this section can be made to cover the *closed* interval $(0 \leqslant x \leqslant 1)$ in the forms

$$\frac{2x^{\frac{1}{2}}J_\nu(j_m x)}{J^2_{\nu+1}(j_m)}\int_0^1 tf(t)\,J_\nu(j_m t)\,dt = \begin{cases} O(1/j_m), \\ o(1). \end{cases}$$

This is evident when it is remembered that

$$|(j_m x)^{\frac{1}{2}}J_\nu(j_m x)| \leqslant c'.$$

Hence *the general term in the Fourier-Bessel series associated with $f(x)$ tends to zero (after multiplication by $\sqrt{x}$) throughout the interval $(0 \leqslant x \leqslant 1)$ if $x^{\frac{1}{2}}f(x)$ has an integral which is absolutely convergent; and, if this function has limited total fluctuation, the general term tends to zero as rapidly as $1/j_m$.*

18·3. *The application of the Hankel-Schläfli methods to Dini's expansion.*

We shall now consider a class of contour integrals by means of which we can obtain theorems concerning Dini's expansion, analogous to those which have been proved for Fourier-Bessel expansions, either in a direct manner or by means of the corresponding theorems for Fourier-Bessel expansions.

The Dini expansion associated with $f(x)$ is

$$\sum_{m=1}^{\infty} b_m J_\nu(\lambda_m x),$$

where $\lambda_1, \lambda_2, \lambda_3, \ldots$ are the positive zeros (arranged in ascending order of magnitude) of the function

$$zJ_\nu'(z) + HJ_\nu(z),$$

* *Proc. London Math. Soc.* (2) XVIII. (1920), pp. 169—171.

where H and ν are real constants, and

$$\nu + \tfrac{1}{2} \geqslant 0.$$

The coefficients b_m are to be determined by the formula

$$b_m \int_0^1 t J_\nu^2 (\lambda_m t)\, dt = \int_0^1 t f(t)\, J_\nu (\lambda_m t)\, dt$$

so that

$$b_m = \frac{2\lambda_m^2 \int_0^1 t f(t)\, J_\nu (\lambda_m t)\, dt}{(\lambda_m^2 - \nu^2)\, J_\nu^2 (\lambda_m) + \lambda_m^2 J_\nu'^2 (\lambda_m)}.$$

Before proceeding further, we shall explain a phenomenon, peculiar to certain Dini expansions, which has no analogue in the theory of Fourier-Bessel expansions.

The investigation of Dini expansions is based on properties of a function which has poles at the zeros of

$$z^{-\nu} \{z J_\nu' (z) + H J_\nu (z)\};$$

and, when $H + \nu = 0$, *this last function has a zero at the origin.*

Further, if $H + \nu$ is negative, *the function has two purely imaginary zeros.*

It is only to be expected that these zeros should contribute to the terms of the series, and such a contribution in fact is made.

If $H + \nu = 0$, an initial term

$$2(\nu + 1)\, x^\nu \int_0^1 t^{\nu+1} f(t)\, dt \tag{1}$$

has to be inserted on account of the zero at the origin.

If $H + \nu$ is negative and the purely imaginary zeros are $\pm i\lambda_0$, then an initial term

$$\frac{2\lambda_0^2 I_\nu (\lambda_0 x)}{(\lambda_0^2 + \nu^2)\, I_\nu^2 (\lambda_0) - \lambda_0^2 I_\nu'^2 (\lambda_0)} \int_0^1 t f(t)\, I_\nu (\lambda_0 t)\, dt \tag{2}$$

must be inserted on account of the zeros $\pm i\lambda_0$.

These initial terms in the respective cases will be denoted by the common symbol $\mathscr{B}_0 (x)$, so that the series which will actually be considered is

$$\mathscr{B}_0 (x) + \sum_{m=1}^{\infty} b_m J_\nu (\lambda_m x),$$

where $\mathscr{B}_0 (x)$ is zero when $H + \nu$ is positive and is defined as the expression (1) or (2) in the respective cases $H + \nu = 0$, $H + \nu < 0$.

[NOTE. The fact that an initial term must be inserted when $H + \nu = 0$ was noticed by Dini, *Serie di Fourier* (Pisa, 1880), p. 268, but Dini gave its value incorrectly, the factor x^ν being omitted. Dini's formula was misquoted by Nielsen, *Handbuch der Theorie der Cylinderfunktionen* (Leipzig, 1904), p. 354. For corrections of these errors, see Bridgeman, *Phil. Mag.* (6) XVI. (1908), pp. 947—948; Chree, *Phil. Mag.* (6) XVII. (1909), pp. 329—331; and C. N. Moore, *Trans. American Math. Soc.* X. (1909), pp. 419—420.]

We now consider the function

$$\frac{2w J_\nu(xw) J_\nu(tw)}{J_\nu(w)\{wJ_\nu'(w)+HJ_\nu(w)\}}.$$

This function has poles at $j_1, j_2, j_3, \ldots, \lambda_1, \lambda_2, \lambda_3, \ldots,$ (0 or $\pm i\lambda_0$).

The residue of the function at j_m is

$$\frac{2J_\nu(j_m x) J_\nu(j_m t)}{J_\nu'^2(j_m)}.$$

The residue at λ_m is

$$\frac{2\lambda_m J_\nu(\lambda_m x) J_\nu(\lambda_m t)}{J_\nu(\lambda_m)\{\lambda_m J_\nu''(\lambda_m)+J_\nu'(\lambda_m)+HJ_\nu'(\lambda_m)\}}$$
$$= -\frac{2\lambda_m^2 J_\nu(\lambda_m x) J_\nu(\lambda_m t)}{(\lambda_m^2-\nu^2) J_\nu^2(\lambda_m)+\lambda_m^2 J_\nu'^2(\lambda_m)}.$$

The residue at the origin when $H+\nu=0$ is

$$-4(\nu+1)x^\nu t^\nu.$$

The residues at $\pm i\lambda_0$ when $H+\nu$ is negative are both equal to

$$-\frac{2\lambda_0^2 I_\nu(\lambda_0 x) I_\nu(\lambda_0 t)}{(\lambda_0^2+\nu^2) I_\nu^2(\lambda_0)-\lambda_0^2 I_\nu'^2(\lambda_0)}.$$

Now let D_n be a number, which lies between λ_n and λ_{n+1}, so chosen that it is not equal to any of the numbers j_m; and let j_N be the greatest of the numbers j_m which does not exceed D_n.

$$\text{Let } S_n(t, x; H) = \sum_{m=1}^{N} \frac{2J_\nu(j_m x) J_\nu(j_m t)}{J^2_{\nu+1}(j_m)} - \mathscr{A}_0(x,t)$$
$$- \sum_{m=1}^{n} \frac{2\lambda_m^2 J_\nu(\lambda_m x) J_\nu(\lambda_m t)}{(\lambda_m^2-\nu^2) J_\nu^2(\lambda_m)+\lambda_m^2 J_\nu'^2(\lambda_m)},$$

where $\mathscr{A}_0(x,t)$ is defined to be 0, $2(\nu+1)x^\nu t^\nu$ or

$$\frac{2\lambda_0^2 I_\nu(\lambda_0 x) I_\nu(\lambda_0 t)}{(\lambda_0^2+\nu^2) I_\nu^2(\lambda_0)-\lambda_0^2 I_\nu'^2(\lambda_0)},$$

according as $H+\nu$ is positive, zero or negative.

Then, evidently,

$$\sum_{m=1}^{N} a_m J_\nu(j_m x) - \mathscr{B}_0(x) - \sum_{m=1}^{n} b_m J_\nu(\lambda_m x) = \int_0^1 t f(t) S_n(t, x; H)\, dt.$$

We shall now prove a number of theorems leading up to the result that, when $0 < x < 1$, the existence and absolute convergence of

$$\int_0^1 t^{\frac{1}{2}} f(t)\, dt$$

are sufficient to ensure that, as $n \to \infty$,

$$\int_0^1 t f(t) S_n(t, x; H)\, dt = o(1).$$

This equation enables us to deduce the properties of Dini's series in respect of convergence* from the corresponding properties of the Fourier-Bessel series.

18·31. *The contour integral for $S_n(t, x; H)$.*

It is evident from Cauchy's theory of residues that

$$S_n(t, x; H) = \frac{1}{2\pi i}\int_{D_n-\infty i}^{D_n+\infty i} \frac{2w\, J_\nu(xw)\, J_\nu(tw)\, dw}{J_\nu(w)\{wJ_\nu'(w)+HJ_\nu(w)\}} - \frac{1}{2\pi i}\cdot P\int_{-\infty i}^{\infty i} \frac{2w\, J_\nu(xw)\, J_\nu(tw)\, dw}{J_\nu(w)\{wJ_\nu'(w)+HJ_\nu(w)\}},$$

where the symbol P denotes Cauchy's 'principal value.' The integrand being an odd function of w, the second integral vanishes, and so we have

(1) $$S_n(t, x; H) = \frac{1}{2\pi i}\int_{D_n-\infty i}^{D_n+\infty i} \frac{2w\, J_\nu(xw)\, J_\nu(tw)\, dw}{J_\nu(w)\{wJ_\nu'(w)+HJ_\nu(w)\}}.$$

An immediate consequence of this formula (cf. § 18·21) is that

(2) $$|S_n(t, x; H)| \leqslant \frac{c_3}{(2-x-t)\sqrt{(xt)}},$$

where c_3 is independent of n, x and t.

Also

$$\int_0^t t^{\nu+1} S_n(t, x; H)\, dt = \frac{t^{\nu+1}}{2\pi i}\int_{D_n-\infty i}^{D_n+\infty i} \frac{2J_\nu(xw)\, J_{\nu+1}(tw)\, dw}{J_\nu(w)\{wJ_\nu'(w)+HJ_\nu(w)\}},$$

and hence

$$\left|\int_0^t t^{\nu+1} S_n(t, x; H)\, dt\right| \leqslant \frac{c_4\, t^{\nu+\frac{1}{2}}}{(2-x-t)\, D_n \sqrt{x}},$$

where c_4 is independent of n, x and t.

18·32. *The analogue for $S_n(t, x; H)$ of the Riemann-Lebesgue lemma.*

We shall now prove the theorem that, *if (a, b) is any part (or the whole) of the interval $(0, 1)$, then the existence and absolute convergence of*

$$\int_a^b t^{\frac{1}{2}} f(t)\, dt$$

are sufficient to ensure that, as $n \to \infty$,

$$\int_a^b t f(t)\, S_n(t, x; H)\, dt = o(1),$$

provided that $0 < x < 1$. And, if $b < 1$, the theorem is valid when $0 < x \leqslant 1$.

The proof has to be divided into three stages just as in the corresponding theorem (§ 18·23) for $T_n(t, x)$. We shall now give the proof of the first stage, when it is supposed that $t^{\frac{1}{2}} f(t)$ is bounded and $a > 0$. The proofs of the remaining stages should be constructed by the reader without difficulty.

* Except at the point $x = 1$.

Let
$$t^{-\nu} f(t) \equiv F(t),$$
and let the upper bound of $|F(t)|$ in (a, b) be K.

Divide (a, b) into p equal parts by the points $t_1, t_2, \ldots, t_{p-1}$ $(t_0 = a,\ t_p = b)$, and, after choosing an arbitrary positive number ϵ, take p to be so large that
$$\sum_{m=1}^{p} (U_m - L_m)(t_m - t_{m-1}) < \epsilon,$$
where U_m and L_m are the upper and lower bounds of $F(t)$ in (t_{m-1}, t_m).

Let
$$F(t) \equiv F(t_{m-1}) + \omega_m(t),$$
so that $|\omega_m(t)| \leqslant U_m - L_m$ in (t_{m-1}, t_m).

Then
$$\int_a^b tf(t)\, S_n(t, x; H)\, dt$$
$$= \sum_{m=1}^{p} F(t_{m-1}) \int_{t_{m-1}}^{t_m} t^{\nu+1} S_n(t, x; H)\, dt + \sum_{m=1}^{p} \int_{t_{m-1}}^{t_m} t^{\nu+1} \omega_m(t)\, S_n(t, x; H)\, dt.$$
Hence, by § 18·31,
$$\left|\int_a^b tf(t)\, S_n(t, x; H)\, dt\right| < \frac{1}{(2 - x - b)\sqrt{x}} \left[\frac{2Kpc_4}{D_n} + \epsilon c_3\right],$$
and if we now take n so large that $D_n \epsilon c_3 > 2Kp\, c_4$, we have
$$\left|\int_a^b tf(t)\, S_n(t, x; H)\, dt\right| < \frac{2\epsilon c_3}{(2 - x - b)\sqrt{x}},$$
and the expression on the right is arbitrarily small. Hence the integral on the left tends to zero as $n \to \infty$.

When the reader has removed the restrictions concerning boundedness and the magnitude of a by the method of § 18·23, the theorem is completely proved.

As a corollary, it should be observed that
$$x^{\frac{1}{2}} \int_a^b tf(t)\, S_n(t, x; H)\, dt$$
tends *uniformly* to zero as $n \to \infty$ when $0 \leqslant x \leqslant 1$ if $b < 1$, and when $0 \leqslant x \leqslant 1 - \Delta$ if $b \leqslant 1$, where Δ is an arbitrary positive number.

18·33. *Dini's expansion of an arbitrary function.*

An immediate consequence of the result of the preceding section is that the existence and absolute convergence of the integral
$$\int_0^1 t^{\frac{1}{2}} f(t)\, dt$$
are sufficient to ensure that the Dini expansion associated with $f(x)$ behaves in the same manner, as regards convergence (or summability), as the Fourier-Bessel expansion throughout the interval $(0 < x < 1)$.

For it is evident that

$$\mathscr{B}_0(x) + \sum_{m=1}^{n} b_m J_\nu(\lambda_m x) - \sum_{m=1}^{N} a_m J_\nu(j_m x)$$

tends to zero as $n \to \infty$ when $0 < x < 1$; and this sum (multiplied by $\sqrt{x}$) tends *uniformly* to zero when $0 \leqslant x \leqslant 1 - \Delta$.

Now, since the numbers λ_m and j_m which exceed $|\nu|$ are interlaced (§ 15·23), it follows that D_n may be chosen so that $n - N$ has the same value for all values of n after a certain stage.

Therefore, since

$$x^{\frac{1}{2}} \sum_{m=N+1}^{n} a_m J_\nu(j_m x) \to 0,$$

uniformly throughout $(0, 1)$, we have proved that

$$x\ \mathscr{B}_0(x) + \sum_{m=1}^{n} x^{\frac{1}{2}} \{b_m J_\nu(\lambda_m x) - a_m J_\nu(j_m x)\}$$

tends to zero, as $n \to \infty$, uniformly throughout $(0, 1 - \Delta)$.

That is to say, the series

$$x^{\frac{1}{2}} \mathscr{B}_0(x) + \sum_{m=1}^{\infty} x^{\frac{1}{2}} \{b_m J_\nu(\lambda_m x) - a_m J_\nu(j_m x)\}$$

is uniformly convergent throughout $(0, 1 - \Delta)$ and its sum is zero.

It follows from the 'consistency theorems' concerning convergent series* that, when the series is 'summed' by Cesàro's means, or any similar method, it is (uniformly) summable and its 'sum' is zero.

Hence, if, for any particular value of x in the interval $(0, 1 - \Delta)$, the series

$$\sum_{m=1}^{\infty} x^{\frac{1}{2}} a_m J_\nu(j_m x),$$

associated with $f(x)$, is convergent (or is summable by some method), then the series

$$x^{\frac{1}{2}} \mathscr{B}_0(x) + \sum_{m=1}^{\infty} x^{\frac{1}{2}} b_m J_\nu(\lambda_m x)$$

is convergent (or is summable by the same method) and the two series have the same 'sum.'

And if, further, the Fourier-Bessel series (multiplied by $\sqrt{x}$) is uniformly convergent (or uniformly summable) throughout an interval (a, b), where

$$0 \leqslant a < b < 1,$$

then also the Dini series (multiplied by $\sqrt{x}$) is uniformly convergent (or uniformly summable) throughout (a, b).

In particular, if $f(x)$ has limited total fluctuation in (a, b) where

$$0 \leqslant a < b \leqslant 1,$$

* Cf. Bromwich, *Theory of Infinite Series*, § 100.

then the series

$$\mathscr{B}_0(x) + \sum_{m=1}^{\infty} b_m J_\nu(\lambda_m x)$$

converges to the sum

$$\tfrac{1}{2}\{f(x+0) + f(x-0)\}$$

at all points x such that $a + \Delta \leqslant x \leqslant b - \Delta$, where Δ is arbitrarily small; and the convergence is uniform if $f(x)$ is continuous in (a, b).

18·34. *The value of Dini's series at $x = 1$.*

We shall now complete the investigation of the value of the sum of Dini's series by considering the point $x = 1$; and we shall prove the theorem, due to Hobson*, that, if $f(x)$ has limited total fluctuation in the interval $(a, 1)$, the sum of the Dini expansion at $x = 1$ is $f(1-0)$.

We first write

$$T_n(t, x;\ H) \equiv T_n(t, x) - S_n(t, x;\ H)$$
$$= \mathscr{A}_0(x, t) + \sum_{m=1}^{n} \frac{2\lambda_m^2 J_\nu(\lambda_m x) J_\nu(\lambda_m t)}{(\lambda_m^2 - \nu^2) J_\nu^2(\lambda_m) + \lambda_m^2 J_\nu'^2(\lambda_m)},$$

and then we have

$$T_n(t, x;\ H) = \begin{cases} \dfrac{1}{2\pi i}\displaystyle\int_{D_n - \infty i}^{D_n + \infty i} \frac{w\phi(w, x) J_\nu(tw)\, dw}{wJ_\nu'(w) + HJ_\nu(w)}, \\[2ex] \dfrac{1}{2\pi i}\displaystyle\int_{D_n - \infty i}^{D_n + \infty i} \frac{w\phi(w, t) J_\nu(xw)\, dw}{wJ_\nu'(w) + HJ_\nu(w)}, \end{cases}$$

where

$$\phi(w, x) \equiv \pi[\{wJ_\nu'(w) + HJ_\nu(w)\} Y_\nu(xw) - \{wY_\nu'(w) + HY_\nu(w)\} J_\nu(xw)].$$

The former representation of $T_n(t, x;\ H)$ is valid when $0 < t < x \leqslant 1$, the latter when $0 < x < t \leqslant 1$.

[NOTE. These representations of $T_n(t, x;\ H)$ are strictly analogous to the representations of $T_n(t, x)$ given by § 18·21 (7) and § 18·21 (8); the fact that there is no formula for $T_n(t, x;\ H)$ analogous to § 18·21 (6) is the reason why Dini series were discussed in § 18·33 with the help of the theory of Fourier-Bessel series.]

Now consider the value of

$$\int_0^t t^{\nu+1} T_n(t, 1;\ H)\, dt$$

when $0 < t \leqslant 1$. We have

$$\int_0^t t^{\nu+1} T_n(t, 1;\ H)\, dt = \lim_{B\to\infty} \frac{t^{\nu+1}}{2\pi i}\int_{D_n - Bi}^{D_n + Bi} \frac{\phi(w, 1) J_{\nu+1}(tw)\, dw}{wJ_\nu'(w) + HJ_\nu(w)}$$
$$= -\lim_{B\to\infty} \frac{t^{\nu+1}}{\pi i}\int_{D_n - Bi}^{D_n + Bi} \frac{J_{\nu+1}(tw)\, dw}{wJ_\nu'(w) + HJ_\nu(w)}$$
$$= \lim_{B\to\infty} \frac{t^{\nu+1}}{\pi i}\int_{D_n - Bi}^{D_n + Bi} \left\{\frac{J_{\nu+1}(tw)}{wJ_{\nu+1}(w)} + O\left(\frac{1}{w^2}\right)\right\} dw.$$

* *Proc. London Math. Soc.* (2) VII. (1909), p. 388.

For any given positive value of δ, it follows from § 18·21 that this is a bounded function of t in the interval $(\delta, 1)$. When $\delta \leqslant t \leqslant 1-\delta$, it is $O(1/D_n)$. And when $t=1$, it has the limit 1 when $n \to \infty$.

It follows that

$$\mathscr{B}_0(1) + \sum_{m=1}^{n} b_m J_\nu(\lambda_m) - f(1-0) = \int_0^1 t^{\nu+1} \{t^{-\nu} f(t) - f(1-0)\} T_n(t, 1; H)\, dt.$$

Since $t^{-\nu} f(t) - f(1-0)$ has limited total fluctuation in $(a, 1)$ we may write it in the form $\chi_1(t) - \chi_2(t)$, where $\chi_1(t)$ and $\chi_2(t)$ are bounded positive decreasing functions of t such that

$$\chi_1(1-0) = \chi_2(1-0) = 0.$$

Hence, given an arbitrary positive number ϵ, we can choose a positive number δ, not exceeding $1-a$, such that

$$0 \leqslant \chi_1(t) < \epsilon, \quad 0 \leqslant \chi_2(t) < \epsilon$$

whenever $1-\delta \leqslant t \leqslant 1$.

We then have

$$\begin{aligned}\int_0^1 & t^{\nu+1} \{t^{-\nu} f(t) - f(1-0)\} T_n(t, 1; H)\, dt \\ &= \int_0^{1-\delta} t^{\nu+1} \{t^{-\nu} f(t) - f(1-0)\} T_n(t, 1; H)\, dt \\ &\quad + \int_{1-\delta}^1 t^{\nu+1} \chi_1(t) T_n(t, 1; H)\, dt - \int_{1-\delta}^1 t^{\nu+1} \chi_2(t) T_n(t, 1; H)\, dt.\end{aligned}$$

By arguments similar to those used in § 18·24, the first integral on the right is $o(1)$ as $n \to \infty$; and neither the second nor the third exceeds

$$2\epsilon \overline{\lim} \left| \int_0^t t^{\nu+1} T_n(t, 1; H)\, dt \right|$$

in absolute value (cf. § 18·24), and this expression is arbitrarily small.

It follows that

$$\lim_{n \to \infty} \int_0^1 t^{\nu+1} \{t^{-\nu} f(t) - f(1-0)\} T_n(t, 1; H)\, dt = 0,$$

and so we have proved that, in the circumstances postulated at the beginning of this section,

$$\mathscr{B}_0(1) + \sum_{m=1}^{\infty} b_m J_\nu(\lambda_m)$$

converges to the sum $f(1-0)$.

This discrepancy between the behaviours of Dini series and of Fourier-Bessel series (§ 18·26) is somewhat remarkable.

18·35. *The uniformity of the convergence of Dini's expansion in an interval extending to $x = 1$.*

Because Dini series do not vanish identically at $x = 1$, it seems not unlikely that the condition that $f(x)$ is *continuous** in $(a, 1)$, combined with the existence and absolute convergence of

$$\int_0^1 t^{\frac{1}{2}} f(t)\,dt,$$

and the condition that $f(x)$ has limited total fluctuation in $(a, 1)$, may be sufficient to ensure the uniformity of the convergence of the Dini expansion in $(a + \Delta, 1)$.

We shall prove that this is, in fact, the case.

The reason for the failure in the uniformity of the convergence of the Fourier-Bessel expansion (§ 18·26) near $x = 1$ was the fact that

$$\int_0^1 t^{\nu+1} T_n(t, x)\,dt$$

does not converge uniformly to x^ν in $(\Delta, 1)$, as was seen in § 18·22. We shall prove that, on the contrary,

$$\int_0^1 t^{\nu+1} T_n(t, x;\, H)\,dt$$

does converge uniformly to x^ν in $(\Delta, 1)$, and the cause of the failure is removed.

A consideration of § 18·26 should then enable the reader to see without difficulty that the Dini expansion converges uniformly in $(a + \Delta, 1)$.

It is easy to see, from § 18·34, that

$$\int_0^t t^{\nu+1} T_n(t, x;\, H)\,dt$$

is the sum of the residues of

$$\pi t^{\nu+1} [\{wJ_\nu'(w) + HJ_\nu(w)\}\, Y_{\nu+1}(tw) - \{wY_\nu'(w) + HY_\nu(w)\}\, J_{\nu+1}(tw)] \\ \times J_\nu(xw)/\{wJ_\nu'(w) + HJ_\nu(w)\},$$

at $\lambda_1, \lambda_2, \ldots, \lambda_n$, plus half the residues at 0 or $\pm i\lambda_0$ if $H + \nu \leqslant 0$.

Hence
$$\int_0^1 t^{\nu+1}\, T_n(t, x;\, H)\,dt$$

is the sum of the residues of

$$-(2/w)(H + \nu)\, J_\nu(xw)/\{wJ_\nu'(w) + HJ_\nu(w)\},$$

and hence, when $0 < x \leqslant 1$,

$$\int_0^1 t^{\nu+1} T_n(t, x;\, H)\,dt = x^\nu - \frac{H + \nu}{\pi i} \int_{D_n - \infty i}^{D_n + \infty i} \frac{J_\nu(xw)\,dw}{w\,\{wJ_\nu'(w) + HJ_\nu(w)\}},$$

* Without restriction on the value of $f(1 - 0)$.

and the integrand on the right* is of the order of magnitude of

$$\frac{\exp\{-(1-x)\,|\,I(w)\,|\}}{w^2\sqrt{x}},$$

and so the integral on the right converges uniformly to zero like $1/(D_n\sqrt{x})$ when $\Delta \leqslant x \leqslant 1$. That is to say

$$\int_0^1 t^{\nu+1}\,T_n(t, x;\, H)\,dt$$

converges uniformly to x^ν in $(\Delta, 1)$; and we have just seen that this is a sufficient condition for the uniformity of the convergence of the Dini series associated with $f(t)$ to the sum $f(x)$ in $(a+\Delta, 1)$ under the conditions postulated concerning $f(t)$.

18·4. *The differentiability of Fourier-Bessel expansions.*

In the earlier part of this chapter we obtained an expansion which, when written in full, assumes the form

$$(1) \qquad f(x) = \sum_{m=1}^{\infty} a_m J_\nu(j_{m,\nu}\,x).$$

We shall now study the circumstances in which, given this expansion, it is permissible to deduce that

$$(2) \qquad f'(x) = \sum_{m=1}^{\infty} a_m j_{m,\nu} J_\nu'(j_{m,\nu}\,x).$$

This problem was examined by Ford†, and his investigation is analogous to Stokes' researches on the differentiability of Fourier series‡.

Ford also investigated the differentiability of Dini's expansion when $H = -\nu$, but his method is not applicable to other values of H.

It is evident that we can prove the truth of (2) if we can succeed in proving that

$$(3) \qquad f'(x) - \frac{\nu}{x} f(x) = -\sum_{m=1}^{\infty} a_m j_{m,\nu} J_{\nu+1}(j_{m,\nu}\,x);$$

and the numbers $j_{m,\nu}$ are the positive zeros of

$$z^{-\nu-1}\{zJ'_{\nu+1}(z) + (\nu+1)\,J_{\nu+1}(z)\}.$$

Now we know that $f'(x) - (\nu/x) f(x)$ admits of the Dini expansion

$$\sum_{m=1}^{\infty} b_m J_{\nu+1}(j_{m,\nu}\,x)$$

inside any interval in which the function has limited fluctuation, provided that

$$\int_0^1 t^{\frac{1}{2}}\left\{f'(t) - \frac{\nu}{t} f(t)\right\} dt$$

exists and is absolutely convergent.

* The term in $wJ_\nu'(w)$ is more important than the term in $J_\nu(w)$ *except in the limit when H is infinite*; this shews clearly the reason for the difference in the behaviour of the Dini expansion from that of the Fourier-Bessel expansion (cf. § 18·26).

† *Trans. American Math. Soc.* IV. (1903), pp. 178—184.

‡ Cf. *Modern Analysis*, § 9·31.

The coefficients b_m are given by the formula

$$b_m = \frac{2j^2_{m,\nu}\int_0^1 \{tf'(t) - \nu f(t)\} J_{\nu+1}(j_{m,\nu} t)\,dt}{\{j^2_{m,\nu} - (\nu+1)^2\} J^2_{\nu+1}(j_{m,\nu}) + j^2_{m,\nu} J'^2_{\nu+1}(j_{m,\nu})}$$

$$= \frac{2}{J^2_{\nu+1}(j_{m,\nu})}\int_0^1 \frac{d}{dt}\{t^{-\nu} f(t)\}\, t^{\nu+1} J_{\nu+1}(j_{m,\nu} t)\,dt$$

$$= \frac{2}{J^2_{\nu+1}(j_{m,\nu})}\left\{\left[tf(t)\,J_{\nu+1}(j_{m,\nu} t)\right]_0^1 - j_{m,\nu}\int_0^1 tf(t)\,J_\nu(j_{m,\nu} t)\,dt\right\}$$

$$= -j_{m,\nu}\, a_m,$$

provided that

$$\left[tf(t)\,J_{\nu+1}(j_{m,\nu} t)\right]_0^1 = 0.$$

Sufficient conditions that this may be the case are

(i) $t^{\nu+2} f(t) \to 0$ as $t \to 0$,

(ii) $f(1-0) = 0$,

(iii) $f(t)$ is continuous in the open interval in which $0 < t < 1$.

These conditions combined with the existence and absolute convergence of

$$\int_0^1 t^{\nu+\frac{1}{2}} \frac{d}{dt}\{t^{-\nu} f(t)\}\,dt$$

are sufficient to ensure the truth of (2) in any interval in which

$$f'(x) - (\nu/x) f(x)$$

has limited total fluctuation.

18·5. *The summability of Fourier-Bessel series.*

A consideration of the values of the coefficients in the Fourier-Bessel series associated with $f(x)$, combined with the expression of $T_n(t, x)$ as a contour integral, suggests that it is no easy matter to discuss by direct methods the question of the summability, by Cesàro's means, of the Fourier-Bessel expansion.

It is, however, very easy to investigate the summability when the method of Riesz* is used to 'sum' the series, and then the summability $(C\,1)$ can be inferred with the help of quite elementary analysis.

The expression which will be taken as the 'sum' of the series by the method of Riesz is

$$\lim_{n\to\infty} \sum_{m=1}^{n} \left(1 - \frac{j_m}{A_n}\right) a_m J_\nu(j_m x);$$

and when this limit exists, the Fourier-Bessel series will be said to be *summable* (R).

It is evident that

$$\sum_{m=1}^{n} \left(1 - \frac{j_m}{A_n}\right) a_m J_\nu(j_m x) = \int_0^1 tf(t)\, T_n(t, x \mid R)\,dt, \tag{1}$$

* Cf. Hardy, *Proc. London Math. Soc.* (2) VIII. (1910), p. 309.

where

$$(2) \qquad T_n(t, x \mid R) = \sum_{m=1}^{n} \left(1 - \frac{j_m}{A_n}\right) \frac{2J_\nu(j_m x) J_\nu(j_m t)}{J^2_{\nu+1}(j_m)},$$

and so it will be convenient to discuss the properties of $T_n(t, x \mid R)$ after the manner of § 18·22 before we make further progress with the main problem.

18·51. *Theorems concerning* $T_n(t, x \mid R)$.

When $T_n(t, x \mid R)$ is defined by equation (2) of § 18·5, it is a symmetric function of t and x, and so we shall proceed to establish the properties of the function on the hypothesis that $0 \leqslant t \leqslant x \leqslant 1$, and we can then write down the corresponding properties when $0 \leqslant x \leqslant t \leqslant 1$ by interchanging t and x in the results already obtained.

We first observe that $T_n(t, x \mid R)$ is the sum of the residues of

$$\pi w \left(1 - \frac{w}{A_n}\right) \{J_\nu(w) Y_\nu(xw) - J_\nu(xw) Y_\nu(w)\} \frac{J_\nu(tw)}{J_\nu(w)}$$

at $j_1, j_2, j_3, \ldots, j_n$.

For brevity we write

$$w \{J_\nu(w) Y_\nu(xw) - J_\nu(xw) Y_\nu(w)\} \equiv \Phi(w, x),$$

and then it is obvious that, when* $t < x$,

$$\begin{aligned} T_n(t, x \mid R) &= \frac{1}{2i} \left[\int_{A_n - \infty i}^{A_n + \infty i} - \int_{-\infty i}^{\infty i}\right] \left(1 - \frac{w}{A_n}\right) \Phi(w, x) \frac{J_\nu(tw)\, dw}{J_\nu(w)} \\ &= \frac{1}{2i} \int_{A_n - \infty i}^{A_n + \infty i} \left(1 - \frac{w}{A_n}\right) \Phi(w, x) \frac{J_\nu(tw)\, dw}{J_\nu(w)} \\ &\quad + \frac{1}{2iA_n} \int_{-\infty i}^{\infty i} w\, \Phi(w, x) \frac{J_\nu(tw)\, dw}{J_\nu(w)}, \end{aligned}$$

since $\Phi(w, x) J_\nu(tw)/J_\nu(w)$ is an odd function of w.

We shall now obtain some upper bounds for

$$|\Phi(w, x) J_\nu(tw)/J_\nu(w)|$$

both when w is on the line joining $A_n - \infty i$ to $A_n + \infty i$, and when w is on the imaginary axis; the formulae which will be discussed are valid when $0 \leqslant x \leqslant 1$ and $0 \leqslant t \leqslant 1$, the sign of $x - t$ being immaterial.

To obtain these inequalities, we shall use series of ascending powers of w when $|w|$ is not large, and inequalities derived from the formulae of Chapter VII when $|w|$ is not small.

* When $t \geqslant x$, the integrals taken along the lines joining $\pm iB$ to $A_n \pm iB$ do not tend to zero as $B \to \infty$. There is no need to make an indentation at the origin, because $\Phi(w, x)$ is analytic at the origin.

We first deal with the factor $J_\nu(tw)/J_\nu(w)$. We observe that*

$$\left|\frac{J_\nu(tw)}{J_\nu(w)}\right| < \frac{k_1}{\sqrt{t}} \exp\{-(1-t)\,|\,I(w)\,|\} \tag{1}$$

when w is on either contour; this follows from inequalities of the type § 18·21 (9) when $|w|$ is not small, and from the ascending series when $|w|$ is not large (i.e. less than j_m).

We next consider $\Phi(w, x)$, which is equal to

$$\tfrac{1}{2}iw\,\{H_\nu^{(1)}(w)\,H_\nu^{(2)}(xw) - H_\nu^{(1)}(xw)\,H_\nu^{(2)}(w)\};$$

it is convenient to make two investigations concerning this function, the former being valid when $-\frac{1}{2} \leqslant \nu \leqslant \frac{1}{2}$, the second when $\nu \geqslant \frac{1}{2}$.

(I) The first investigation is quite simple. It follows from § 3·6 and § 7·33 that

$$|\,H_\nu^{(1)}(xw)\,| < \frac{k_2\,|\,e^{ixw}\,|}{|\,xw\,|^{\frac{1}{2}}}, \quad |\,H_\nu^{(2)}(xw)\,| < \frac{k_2\,|\,e^{-ixw}\,|}{|\,xw\,|^{\frac{1}{2}}} \tag{2}$$

for *all* the values of w and x under consideration when $-\frac{1}{2} \leqslant \nu \leqslant \frac{1}{2}$. Hence

$$|\,\Phi(w, x)\,| < \frac{k_2^2}{\sqrt{x}} \exp\{(1-x)\,|\,I(w)\,|\}. \tag{3}$$

(II) When $\nu \geqslant \frac{1}{2}$ and $|w|$ is not large, it is easy to deduce from the ascending series for $J_\nu(w)$, $Y_\nu(w)$, $J_\nu(xw)$ and $Y_\nu(xw)$ that

$$|\,\Phi(w, x)\,| < k_3\,|\,w\,|\,x^{-\nu}. \tag{4}$$

If $|w|$ is not small, we use the inequalities (deduced from § 7·33)

$$|\,H_\nu^{(1)}(w)\,| < \frac{k_2\,|\,e^{iw}\,|}{|\,w\,|^{\frac{1}{2}}}, \quad |\,H_\nu^{(2)}(w)\,| < \frac{k_2\,|\,e^{-iw}\,|}{|\,w\,|^{\frac{1}{2}}}, \tag{5}$$

together with the inequalities

$$\begin{cases} |\,H_\nu^{(1)}(xw)\,| < k_4\,\{|\,xw\,|^{-\frac{1}{2}} + |\,xw\,|^{-\nu}\}\,|\,e^{ixw}\,|, \\ |\,H_\nu^{(2)}(xw)\,| < k_4\,\{|\,xw\,|^{-\frac{1}{2}} + |\,xw\,|^{-\nu}\}\,|\,e^{-ixw}\,|. \end{cases} \tag{6}$$

It follows from § 3·6 and § 7·33 that the inequalities (6) are true whether $|xw|$ is large or not. Hence,

$$|\,\Phi(w, x)\,| < k_2 k_4\,\{x^{-\frac{1}{2}} + x^{-\nu}\,|\,w\,|^{\frac{1}{2}-\nu}\}\exp\{(1-x)\,|\,I(w)\,|\}, \tag{7}$$

when $\nu \geqslant \frac{1}{2}$ and $|w|$ is large, whatever be the magnitude† of $|xw|$.

If we now combine the results contained in formulae (3), (4) and (7) we deduce that, whether $-\frac{1}{2} \leqslant \nu \leqslant \frac{1}{2}$ or $\nu \geqslant \frac{1}{2}$,

$$|\,\Phi(w, x)\,| < k_5\,(x^{-\frac{1}{2}} + x^{-\nu})\exp\{(1-x)\,|\,I(w)\,|\}, \tag{8}$$

* It is supposed that the numbers $k_1, k_2, k_3, \ldots$ are positive and independent of w, x and t; their values may, however, depend on the value of ν.

† Provided of course that $0 \leqslant x \leqslant 1$.

when w is any point of either contour and $0 \leqslant x \leqslant 1$. Hence, by (1), it follows that

(9) $$\left| \Phi(w, x) \frac{J_\nu(tw)}{J_\nu(w)} \right| < k_6 t^{-\frac{1}{2}} (x^{-\frac{1}{2}} + x^{-\nu}) \exp\{-(x-t) \mid I(w) \mid\},$$

when $0 \leqslant x \leqslant 1$ and $0 \leqslant t \leqslant 1$.

We now return to the integral formula for $T_n(t, x \mid R)$. If we replace w by $A_n \pm iv$ and $\pm iv$ in the first and second contour integrals respectively, we deduce that, when $0 \leqslant t < x \leqslant 1$,

$$| T_n(t, x \mid R) | < \frac{2k_6}{A_n \sqrt{t}} (x^{-\frac{1}{2}} + x^{-\nu}) \int_0^\infty v e^{-(x-t)v} dv = \frac{2k_6 (x^{-\frac{1}{2}} + x^{-\nu})}{A_n (x-t)^2 \sqrt{t}}.$$

We have consequently proved the two inequalities

(10) $$| T_n(t, x \mid R) | < \frac{2k_6 (x^{-\frac{1}{2}} + x^{-\nu})}{A_n (x-t)^2 \sqrt{t}} \qquad (0 \leqslant t < x \leqslant 1),$$

(11) $$| T_n(t, x \mid R) | < \frac{2k_6 (t^{-\frac{1}{2}} + t^{-\nu})}{A_n (x-t)^2 \sqrt{x}} \qquad (0 \leqslant x < t \leqslant 1).$$

It is to be remembered that k_6 is independent of x and t, so that we may make $| x - t |$ tend to zero, if we desire to do so.

One other pair of inequalities is required in order to discuss the behaviour of $T_n(t, x \mid R)$ when x and t are nearly equal. To obtain them, we write

$$T_n(t, x \mid R) = \frac{1}{2i} \int \left(1 - \frac{w}{A_n}\right) \Phi(w, x) \frac{J_\nu(tw)\, dw}{J_\nu(w)},$$

when $0 \leqslant t \leqslant x \leqslant 1$; in this integral the contour is taken to be a rectangle with vertices $\pm iA_n$, $A_n \pm iA_n$.

It is easy to see that (9) is satisfied whether w be on the horizontal sides or on the vertical sides of this rectangle; and the factor $1 - (w/A_n)$ does not exceed $\sqrt{2}$ in absolute value at any point of the contour.

Consequently the modulus of the integrand does not exceed

$$k_6 t^{-\frac{1}{2}} (x^{-\frac{1}{2}} + x^{-\nu}) \sqrt{2};$$

and since the length of the contour is $6A_n$, we infer that, when $0 \leqslant t \leqslant x \leqslant 1$,

(12) $$| T_n(t, x \mid R) | < \frac{3A_n k_6 (1 + x^{\frac{1}{2}-\nu})}{\sqrt{(\frac{1}{2} tx)}},$$

and similarly, when $0 \leqslant x \leqslant t \leqslant 1$,

(13) $$| T_n(t, x \mid R) | < \frac{3A_n k_6 (1 + t^{\frac{1}{2}-\nu})}{\sqrt{(\frac{1}{2} tx)}}.$$

The last four inequalities are sufficient to enable us to discuss adequately the summability (R) of Fourier-Bessel series. The reader will observe that the consideration of small values of x has increased the length of the analysis to an appreciable but not to an undue extent.

18·52. *The analogue of Fejér's theorem.*

We can now prove that *the existence and the absolute convergence of*

$$\int_0^1 t^{\frac{1}{2}} f(t)\,dt$$

are sufficient to ensure that the Fourier-Bessel series associated with $f(t)$ *is summable* (R) *at all points* x *of the open interval* $(0, 1)$ *at which the two limits* $f(x \pm 0)$ *exist. And the sum* (R) *of the series is*

$$\tfrac{1}{2}\{f(x+0)+f(x-0)\}.$$

This theorem is obviously the analogue of Fejér's theorem* concerning Fourier series.

Since† a series which is *convergent* is summable (R), it follows from § 18·35 that, when $0 < x < 1$,

$$\lim_{n\to\infty}\int_0^x t^{\nu+1}\, T_n(t, x \mid R)\,dt = \lim_{n\to\infty}\int_x^1 t^{\nu+1}\, T_n(t, x \mid R)\,dt$$
$$= \tfrac{1}{2}x^{\nu}.$$

Hence it follows that, when the limits $f(x \pm 0)$ exist, then

$$\lim_{n\to\infty}\int_0^x t^{\nu+1}\, T_n(t, x \mid R)\,x^{-\nu} f(x-0)\,dt + \lim_{n\to\infty}\int_x^1 t^{\nu+1}\, T_n(t, x \mid R)\,x^{-\nu} f(x+0)\,dt$$
$$= \tfrac{1}{2}\{f(x+0)+f(x-0)\}.$$

We are now in a position to consider the sum $S_n(x \mid R)$, defined as

$$\sum_{m=1}^{n}\left(1-\frac{j_m}{A_n}\right) a_m J_\nu(j_m x) - \int_0^x t^{\nu+1}\, T_n(t, x \mid R)\,x^{-\nu} f(x-0)\,dt$$
$$-\int_x^1 t^{\nu+1}\, T_n(t, x \mid R)\,x^{-\nu} f(x+0)\,dt,$$

and we shall prove that it can be made arbitrarily small by taking n sufficiently large.

The sum $S_n(x \mid R)$ is equal to

$$\int_0^x t^{\nu+1}\{t^{-\nu} f(t) - x^{-\nu} f(x-0)\}\, T_n(t, x \mid R)\,dt$$
$$+\int_x^1 t^{\nu+1}\{t^{-\nu} f(t) - x^{-\nu} f(x+0)\}\, T_n(t, x \mid R)\,dt.$$

Now, on the hypothesis that the limits $f(x \pm 0)$ exist, if we choose an arbitrary positive number ϵ, there exists a positive number‡ δ such that

$$\begin{cases} |t^{-\nu} f(t) - x^{-\nu} f(x+0)| < \epsilon, & (x \leqslant t \leqslant x+\delta), \\ |t^{-\nu} f(t) - x^{-\nu} f(x-0)| < \epsilon, & (x \geqslant t \geqslant x-\delta). \end{cases}$$

We now choose a positive function of n, say $\sigma(n)$, which is less than δ for sufficiently large values of n, and divide the interval $(0, 1)$ into six parts by the points $x \pm \delta$, $x \pm \sigma(n)$, x.

* Cf. *Modern Analysis*, § 9·4.

† Cf. *Modern Analysis*, § 8·43.

‡ It is convenient to take δ less than x and $1 - x$.

In the intervals $(0, x-\delta)$, $(x-\delta,\ x-\sigma(n))$ and also in the intervals $(x+\sigma(n), x+\delta)$, $(x+\delta, 1)$ we use inequalities of the form given in § 18·51 (10) and (11); and in the intervals $(x-\sigma(n), x)$, $(x,\ x+\sigma(n))$ we use inequalities of the form given in § 18·51 (12) and (13).

It is thus found that $|S_n(x \mid R)|$ does not exceed

$$\frac{2k_6(x^{-\frac{1}{2}}+x^{-\nu})}{A_n\delta^2}\int_0^{x-\delta}|\{t^{\frac{1}{2}}f(t)-t^{\nu+\frac{1}{2}}x^{-\nu}f(x-0)\}|\,dt$$
$$+\frac{2k_6\epsilon(x^{-\frac{1}{2}}+x^{-\nu})}{A_n}\left[\int_{x-\delta}^{x-\sigma(n)}\frac{t^{\nu+\frac{1}{2}}dt}{(x-t)^2}+\frac{3A_n^2}{\sqrt{2}}\int_{x-\sigma(n)}^{x}t^{\nu+\frac{1}{2}}\,dt\right]$$
$$+\frac{4k_6\epsilon}{A_n\sqrt{x}}\left[\frac{3A_n^2}{\sqrt{2}}\int_x^{x+\sigma(n)}dt+\int_{x+\sigma(n)}^{x+\delta}\frac{dt}{(x-t)^2}\right]$$
$$+\frac{2k_6}{A_n\delta^2\sqrt{x}}\int_{x+\delta}^{1}(t^{\nu+\frac{1}{2}}+t)\,|\{t^{-\nu}f(t)-x^{-\nu}f(x+0)\}|\,dt.$$

For any given value of ϵ (and therefore of δ), the first and last terms in this expression can be made arbitrarily small by taking n sufficiently large, on account of the convergence of

$$\int_0^1|t^{\frac{1}{2}}f(t)|\,dt.$$

The remaining terms do not exceed

$$\frac{2k_6\epsilon(3x^{-\frac{1}{2}}+x^{-\nu})}{A_n}\left\{\frac{1}{\sigma(n)}+\frac{3A_n^2\sigma(n)}{\sqrt{2}}\right\},$$

and, if we take $\sigma(n)\equiv 1/A_n$, this is independent of n, and it can be made as small as we please by taking ϵ sufficiently small initially.

We can therefore make the intermediate terms in the expression for $|S_n(x \mid R)|$ as small as we please by taking ϵ sufficiently small, and when this has been done, the first and last terms can be made as small as we please by taking n sufficiently large.

That is to say, $|S_n(x \mid R)|$ can be made arbitrarily small by taking n sufficiently large, so that

$$\lim_{n\to\infty} S_n(x \mid R)=0.$$

Hence

$$\lim_{n\to\infty}\sum_{m=1}^{n}\left(1-\frac{j_m}{A_n}\right)a_m J_\nu(j_m x)=x^{-\nu}f(x-0)\lim_{n\to\infty}\int_0^x t^{\nu+1}T_n(t, x \mid R)\,dt$$
$$+x^{-\nu}f(x+0)\lim_{n\to\infty}\int_x^1 t^{\nu+1}T_n(t, x \mid R)\,dt,$$

since the limits on the right exist.

Since each of the limits on the right is equal to $\frac{1}{2}x^\nu$, it has now been proved that

$$\sum_{m=1} a_m J_\nu (j_m x)$$

is summable (R) with sum $\frac{1}{2}\{f(x+0)+f(x-0)\}$ provided that the limits $f(x \pm 0)$ exist; and this is the theorem to be established.

As a corollary, the reader should be able to prove without difficulty that, if $f(t)$ is *continuous* in (a, b), the summability (R) is *uniform* throughout the interval in which $a+\Delta \leqslant x \leqslant b-\Delta$, where Δ is any positive number. Cf. § 18·25.

18·53. *Uniformity of summability of the Fourier-Bessel series near the origin.*

We shall now examine the *uniformity* of the summability (R) of the Fourier-Bessel expansion throughout an interval of which the origin is an end-point. It will be supposed that the expansion is modified by being multiplied throughout by $\sqrt{x}$, and it will then be proved that, *if $t^{-\nu} f(t)$ is continuous in the interval* $(0, b)$, *then the modified expansion is uniformly summable throughout* $(0, b-\Delta)$, *where Δ is any positive number.*

Given ϵ, we can now choose δ (less than Δ) so that

$$|\{t^{-\nu} f(t) - x^{-\nu} f(x)\}| < \epsilon$$

whenever $x-\delta \leqslant t \leqslant x+\delta$ and $t \geqslant 0$, provided that x lies in $(0, b-\Delta)$.

Since continuity involves uniformity of continuity, this choice of δ may be taken to be independent of x.

We now write

$$S_n(x \mid R) = \int_0^1 t^{\nu+1} \{t^{-\nu} f(t) - x^{-\nu} f(x)\}\, T_n(t, x \mid R)\, dt$$

and then examine $|x^{\frac{1}{2}} S_n(x \mid R)|$ after the manner of § 18·52.

We express $x^{\frac{1}{2}} S_n(x \mid R)$ as the sum of six integrals (some of which are to be omitted when $x < \delta$), and we see that $|x^{\frac{1}{2}} S_n(x \mid R)|$ does not exceed

$$\frac{2k_6(x^{\nu+\frac{1}{2}}+x)}{A_n\delta^2}\int_0^{x-\delta} |\{t^{-\nu} f(t) - x^{-\nu} f(x)\}|\, dt$$
$$+\frac{2k_6\epsilon(x^{\nu+\frac{1}{2}}+x)}{A_n}\left[\int_{x-\delta}^{x-\sigma(n)}\frac{dt}{(t-x)^2}+\frac{3A_n^2}{\sqrt{2}}\int_{x-\sigma(n)}^{x} dt\right]$$
$$+\frac{4k_6\epsilon}{A_n}\left[\frac{3A_n^2}{\sqrt{2}}\int_x^{x+\sigma(n)} dt + \int_{x+\sigma(n)}^{x+\delta}\frac{dt}{(t-x)^2}\right]$$
$$+\frac{4k_6}{A_n\delta^2}\int_{x+\delta}^1 |\{t^{-\nu} f(t) - x^{-\nu} f(x)\}|\, dt.$$

In this formula any of the limits of integration which are negative are supposed to be replaced by zero.

Now this upper bound for $|x^{\frac{1}{2}} S_n(x \mid R)|$ does not exceed

$$\frac{4k_6}{A_n \delta^2}\int_0^1 |\{t^{-\nu} f(t) - x^{-\nu} f(x)\}|\, dt + \frac{8k_6 \epsilon}{A_n}\left\{\frac{1}{\sigma(n)} + \frac{3A_n{}^2 \sigma(n)}{\sqrt{2}}\right\},$$

and, since $x^{-\nu} f(x)$ is bounded (because it is continuous), this can be made arbitrarily small by a choice of n which is independent of x.

Consequently $x^{\frac{1}{2}} S_n(x \mid R)$ tends to zero uniformly as $n \to \infty$.

Now it has already been shewn (§ 18·22) that

$$x^{\frac{1}{2}} \int_0^1 t^{\nu+1} T_n(t, x)\, dt$$

is uniformly convergent in $(0, 1 - \Delta)$, and so, since uniformity of convergence involves uniformity of summability,

$$x^{\frac{1}{2}-\nu} f(x) \int_0^1 t^{\nu+1} T_n(t, x \mid R)\, dt$$

tends uniformly to $x^{\frac{1}{2}} f(x)$ in $(0, b - \Delta)$.

Hence, since $x^{\frac{1}{2}} S_n(x \mid R)$ tends to zero uniformly,

$$x^{\frac{1}{2}} \int_0^1 t f(t)\, T_n(t, x \mid R)\, dt$$

tends uniformly to $x^{\frac{1}{2}-\nu} f(x) \int_0^1 t^{\nu+1} T_n(t, x \mid R)\, dt$, i.e. to $x^{\frac{1}{2}} f(x)$ in $(0, b - \Delta)$.

It has therefore been proved that

$$\sum_{m=1}^{\infty} a_m x^{\frac{1}{2}} J_\nu(j_m x)$$

is uniformly summable (R) in $(0, b - \Delta)$ with sum $x^{\frac{1}{2}} f(x)$, provided that

$$\int_0^1 t^{\frac{1}{2}} f(t)\, dt$$

exists and is absolutely convergent, and that $t^{-\nu} f(t)$ is continuous in $(0, b)$.

18·54. *Methods of 'summing' Fourier-Bessel series.*

We shall now investigate various methods of summing the Fourier-Bessel series*

$$\sum_{m=0}^{\infty} a_m x^{\frac{1}{2}} J_\nu(j_m x)$$

on the hypotheses (i) that the limits $f(x \pm 0)$ exist, (ii) that

$$\int_0^1 t^{\frac{1}{2}} f(t)\, dt$$

exists and is absolutely convergent, and (iii) that the series is summable (R).

It conduces to brevity to write $f_m(x)$ in place of $a_m x^{\frac{1}{2}} J_\nu(j_m x)$, so that $f_m(x)$ tends uniformly to zero (§ 18·27) as $m \to \infty$ when x lies in $(0, 1)$.

* The factor $x^{\frac{1}{2}}$ is inserted merely in order that the discussion may cover the investigation of uniformity of summability near the origin.

Consider first the limit

$$\lim_{n\to\infty} \sum_{m=1}^{n} \left(1 - \frac{j_m}{j_n}\right) f_m(x)$$

which gives the most natural method (of Riesz' type) for summing the series.

Since $(j_n/A_n) \to 1$, it is evident that

$$\lim_{n\to\infty} \sum_{m=1}^{n} \left(\frac{A_n - j_m}{j_n}\right) f_m(x)$$

exists and is equal to

$$\lim_{n\to\infty} \sum_{m=1}^{n} \left(1 - \frac{j_m}{A_n}\right) f_m(x).$$

Again, since $f_n(x) = o(1)$, it is easy to see that

$$\sum_{m=1}^{n} f_m(x) = o(n),$$

so that

$$\lim_{n\to\infty} \left(\frac{A_n - j_n}{j_n}\right) \sum_{m=1}^{n} f_m(x) = 0,$$

and therefore

$$\lim_{n\to\infty} \sum_{m=1}^{n} \left(1 - \frac{j_m}{j_n}\right) f_m(x) = \lim_{n\to\infty} \sum_{m=1}^{n} \left(1 - \frac{j_m}{A_n}\right) f_m(x);$$

the limit on the right exists in consequence of the hypotheses made at the beginning of the section.

Again, since

$$\frac{j_m}{j_n} - \frac{m}{n} = O\left(\frac{1}{n}\right),$$

whether m be $o(n)$ or $O(n)$, it follows that

$$\lim_{n\to\infty} \sum_{m=1}^{n} \left(\frac{j_m}{j_n} - \frac{m}{n}\right) f_m(x) = 0,$$

and so

$$\lim_{n\to\infty} \sum_{m=1}^{n} \left(1 - \frac{m}{n}\right) f_m(x) = \lim_{n\to\infty} \sum_{m=1}^{n} \left(1 - \frac{j_m}{j_n}\right) f_m(x).$$

Consequently the hypotheses that the limits $f(x \pm 0)$ exist $(0 < x < 1)$ and that the integral

$$\int_0^1 t^{\frac{1}{2}} f(t)$$

exists and is absolutely convergent *are sufficient to ensure that*

$$\sum_{m=1}^{\infty} a_m x^{\frac{1}{2}} J_\nu(j_m x)$$

is summable $(C\,1)$ *with sum* $\frac{1}{2} x^{\frac{1}{2}} \{f(x+0) + f(x-0)\}$.

By the same reasoning, if $f(x)$ is continuous in (a, b), the summability $(C\,1)$ is *uniform* in $(a + \Delta, b - \Delta)$; and, if $a = 0$ and $t^{-\nu} f(t)$ has a limit as $t \to 0$, the summability $(C\,1)$ is *uniform* in $(0, b - \Delta)$.

18·55. *Uniformity of convergence of the Fourier-Bessel expansion near the origin.*

We can now prove, by using Hardy's convergence theorem*, that, *if* $t^{\frac{1}{2}} f(t)$ *has limited total fluctuation in* $(0, b)$, while $f(t)$ is also subject to the conditions of § 18·53, then

$$\sum_{m=1}^{\infty} a_m x^{\frac{1}{2}} J_\nu(j_m x)$$

is *uniformly convergent* in $(0, b-\Delta)$ with sum $x^{\frac{1}{2}} f(x)$.

Let $h(t)$ be an auxiliary function defined to be equal to $f(t)$ in $(0, b)$ and equal to zero in $(b, 1)$; and let the Fourier-Bessel series associated with $h(t)$ be

$$\sum_{m=1}^{\infty} \alpha_m J_\nu(j_m x).$$

Then, by § 18·54, $\sum_{m=1}^{\infty} \alpha_m x^{\frac{1}{2}} J_\nu(j_m x)$ is *uniformly summable* $(C1)$ throughout $(0, b-\Delta)$ with sum $x^{\frac{1}{2}} f(x)$, and, by Sheppard's theorem (§ 18·27), $\alpha_m/\sqrt{j_m}$ is $O(1/m)$, while $(j_m x)^{\frac{1}{2}} J_\nu(j_m x)$ is a bounded function of x and m. Hence, by Hardy's convergence theorem,

$$\sum_{m=1} \alpha_m x^{\frac{1}{2}} J_\nu(j_m x)$$

is *uniformly convergent* throughout $(0, b-\Delta)$, with sum $x^{\frac{1}{2}} f(x)$.

Again

$$\sum_{m=1}^{n} (a_m - \alpha_m) x^{\frac{1}{2}} J_\nu(j_m x) = x^{\frac{1}{2}} \int_b^1 t f(t)\, T_n(t, x)\, dt,$$

and this tends uniformly to zero in $(0, b-\Delta)$ as $n \to \infty$ by an analogue of the Riemann-Lebesgue lemma (§ 18·23).

Hence $\sum_{m=1}^{n} a_m x^{\frac{1}{2}} J_\nu(j_m x)$ tends uniformly to the sum $x^{\frac{1}{2}} f(x)$ in $(0, b-\Delta)$ as $n \to \infty$; and this is the theorem to be established.

18·56. *Summability of Dini series.*

Except when $x = 1$, the summability $(C1)$ of the Dini series associated with $f(t)$ may be inferred by combining the results of § 18·33 and §§ 18·51—18·53.

The summability $(C1)$ may, however, be established independently† for all points x such that $0 < x \leqslant 1$ by replacing A_n and the functions $J_\nu(w)$ and $Y_\nu(w)$, which occur in § 18·5, by D_n and the functions $wJ_\nu'(w) + HJ_\nu(w)$ and $wY_\nu'(w) + HY_\nu(w)$ respectively; the details of the analysis may be left to the reader, and he will find that when $x = 1$ the expression $\frac{1}{2}\{f(x+0) + f(x-0)\}$ must be replaced by $f(1-0)$.

* Cf. *Modern Analysis*, § 8·5.

† Of course on the hypotheses concerning $f(t)$ which were assumed in § 18·53.

The uniformity of the summability in the interval $(a+\Delta, 1)$ when $f(x)$ is continuous in $(a, 1)$ may be dealt with in the same way as the uniformity of convergence was dealt with in §§ 18·33, 18·35.

The summability of Dini series (and of Fourier-Bessel series) by a modification of Abel's method is of some physical importance. Thus, in Fourier's* problem of the Conduction of Heat in an infinite solid cylinder of radius unity, the temperature v at distance r from the axis satisfies the equation

$$\frac{dv}{dt} = k\left\{\frac{d^2v}{dr^2} + \frac{1}{r}\frac{dv}{dr}\right\},$$

with the boundary condition

$$\left[\frac{dv}{dr} + Hv\right]_{r=1} = 0,$$

if the initial distribution of heat is symmetrical.

Normal solutions of the differential equation satisfying the boundary condition are

$$J_0(\lambda_m r)\exp(-k\lambda_m^2 t),$$

and so the temperature v is given by the series†

$$\sum_{m=1}^{\infty} b_m J_0(\lambda_m r)\exp(-k\lambda_m^2 t),$$

where the coefficients b_m are to be determined from the consideration that

$$\sum_{m=1}^{\infty} b_m J_0(\lambda_m r)$$

is the Dini series associated with the initial temperature $f(r)$. It is evident that the initial temperature is expressible as

$$\lim_{t\to+0}\sum_{m=1}^{\infty} b_m J_0(\lambda_m r)\exp(-k\lambda_m^2 t);$$

and this limit exists when the Dini series is summable (R).

18·6. *The uniqueness of Fourier-Bessel series and Dini series.*

It has been shewn by Young‡ that the existence and the absolute convergence of

$$\int_0^1 t^{\frac{1}{2}} f(t)\,dt$$

are sufficient to ensure that *if all the coefficients a_m of the Dini series (or the Fourier-Bessel series) associated with $f(t)$ are zero, then the function $f(t)$ must be a null-function.*

* *La Théorie Analytique de la Chaleur* (Paris, 1822), §§ 306—320. Cf. Rayleigh, *Phil. Mag.* (6) XII. (1906), pp. 106—107 [*Scientific Papers*, v. (1912), pp. 338—339]; and Kirchhoff, *Berliner Sitzungsberichte*, 1883, pp. 519—524.

† In this physical problem, $H > 0$, and so there is no initial term to be inserted.

‡ *Proc. London Math. Soc.* (2) XVIII. (1920), pp. 174—175.

To prove this theorem we observe that, when $p = 0, 1, 2, \ldots$, we may write

$$t^{\nu+2p+\frac{1}{2}} = \sum_{m=0}^{\infty} \bar{a}_m t^{\frac{1}{2}} J_\nu(j_m t),$$

where the coefficients $\bar{a}_m$ are determined by the formula

$$\bar{a}_m = \frac{2}{J^2_{\nu+1}(j_m)} \int_0^1 t^{\nu+2p+1} J_\nu(j_m t)\, dt;$$

and the series on the right converges uniformly in $(0, 1-\Delta)$ and oscillates boundedly in $(1-\Delta, 1)$. It is therefore permissible to multiply the expansion by $t^{\frac{1}{2}} f(t)$ and integrate term-by-term.

It follows that

$$\int_0^1 t^{\nu+2p+1} f(t)\, dt = \sum_{m=0}^{\infty} \bar{a}_m \int_0^1 t f(t) J_\nu(j_m t)\, dt$$
$$= 0.$$

Since all the integrals

$$\int_0^1 t^{\nu+2p+1} f(t)\, dt \qquad (p = 1, 2, 3, \ldots)$$

are zero, it follows that $t^\nu f(t)$ is a null-function, by Lerch's theorem*, and the theorem stated is proved for Fourier-Bessel series. The theorem for Dini series can be proved in precisely the same way, and it is theoretically simpler because the Dini series associated with $t^{\nu+2p}$ does not fail to converge uniformly in $(1-\Delta, 1)$.

It is possible to construct a theory of series of Bessel functions of the types

$$\sum_{m=1}^{\infty} a_m J_\nu(j_m x), \qquad \sum_{m=1}^{\infty} b_m J_\nu(\lambda_m x),$$

(where the coefficients a_m and b_m are any constants) which resembles Riemann's theory of trigonometrical series†.

Such a theory is, however, more directly associated with Schlömilch's series of Bessel functions, which will be discussed in Chapter XIX; and it seems convenient to defer the examination of the series

$$\sum_{m=1}^{\infty} a_m J_\nu(j_m x), \qquad \sum_{m=1}^{\infty} b_m J_\nu(\lambda_m x)$$

by Riemann's methods to § 19·7, when the discussion of the series forms a simple corollary to the discussion of Schlömilch series.

* Lerch, *Acta Mathematica*, XXVII. (1903), pp. 345—347; Young, *Messenger*, XL. (1910), pp. 37—43. Cf. § 12·22.

† Cf. *Modern Analysis*, §§ 9·6—9·632.

CHAPTER XIX

SCHLÖMILCH SERIES

19·1. *Schlömilch's expansion of a function of a real variable.*

In Chapter XVIII we dealt with the expansion of a function $f(x)$ of the real variable x in the form

$$f(x) = \sum_{m=1}^{\infty} a_m J_\nu(j_m x),$$

where j_m is the mth positive zero of $J_\nu(z)$, so that, for large values of m,

$$j_m = (m + \tfrac{1}{2}\nu - \tfrac{1}{4})\pi + O(1/m).$$

That is to say, the argument of the Bessel function in a term of high rank in the series is approximately proportional to the rank of the term.

In this chapter we shall discuss the series in which the argument of the Bessel function in each term is exactly proportional to the rank of the term. By choosing a suitable variable, such a series may be taken to be

$$\sum_{m=1}^{\infty} a_m J_\nu(mx).$$

It will appear subsequently that it is convenient to add an initial term (§ 19·11; cf. § 18·33); and the analysis is simplified by making a slight modification in the form of the coefficients in the series (§ 19·2).

Series of this type were first investigated by Schlömilch*. They are not of such great importance to the Physicist as Fourier-Bessel series, though Rayleigh† has pointed out that (when $\nu = 0$) they present themselves naturally in the investigation of a periodic transverse vibration of a two-dimensional membrane, if the vibration is composed of an unlimited number of equal one-dimensional transverse vibrations uniformly distributed in direction through the two dimensions of the membrane.

Apart from applications the series present various features of purely mathematical interest; and, in particular, it is remarkable that a null-function can be represented by such a series in which the coefficients are not all zero (§ 19·41).

In some respects the series are more amenable to analysis than Fourier-Bessel series, but the two types of series have many properties in common; and the reader will be right when he infers from a comparison of the arguments $j_m x$ and mx that the relevant range of values of x is $(0, \pi)$ for Schlömilch series, corresponding to the range (0, 1) for Fourier-Bessel series.

* *Zeitschrift für Math. und Phys.* II. (1857), pp. 155—158; Schlömilch considered only the special cases $\nu = 0$ and $\nu = 1$.

† *Phil. Mag.* (6) XXI. (1911), pp. 567—571 [*Scientific Papers*, VI. (1920), pp. 22—25].

19·11. *Schlömilch's expansion in a series of Bessel functions of order zero.*

We now state and prove the expansion theorem discovered by Schlömilch. The theorem is concerned with the expansion of an arbitrary function $f(x)$ of the real variable x, and, with modern terminology, it is to the following effect:

Let $f(x)$ be an arbitrary function, with a derivate $f'(x)$ which is continuous in the closed interval $(0, \pi)$ and which has limited total fluctuation in this interval.

Then $f(x)$ admits of the expansion

$$f(x) = \tfrac{1}{2}a_0 + \sum_{m=1}^{\infty} a_m J_0(mx) \tag{1}$$

where

$$\begin{cases} a_0 = 2f(0) + \dfrac{2}{\pi}\displaystyle\int_0^{\pi}\int_0^{\frac{1}{2}\pi} u f'(u\sin\phi)\, d\phi\, du, \\ a_m = \dfrac{2}{\pi}\displaystyle\int_0^{\pi}\int_0^{\frac{1}{2}\pi} u f'(u\sin\phi)\cos mu\, d\phi\, du; \qquad (m>0) \end{cases} \tag{2}$$

and this expansion is valid, and the series is convergent, throughout the closed interval $(0, \pi)$.

Schlömilch's investigation is based on a discussion of the integral equation

$$f(x) = \frac{2}{\pi}\int_0^{\frac{1}{2}\pi} g(x\sin\theta)\, d\theta, \tag{3}$$

of which he proved that a continuous solution is

$$g(x) = f(0) + x\int_0^{\frac{1}{2}\pi} f'(x\sin\phi)\, d\phi. \tag{4}$$

We proceed to verify that the function $g(x)$ defined by (4) actually is a solution of (3); we substitute the value given by (4) in the expression on the right of (3), and then we see that

$$\begin{aligned} \frac{2}{\pi}\int_0^{\frac{1}{2}\pi} g(x\sin\theta)\, d\theta &= \frac{2}{\pi}\int_0^{\frac{1}{2}\pi}\left[f(0) + x\sin\theta\int_0^{\frac{1}{2}\pi} f'(x\sin\theta\sin\phi)\, d\phi\right] d\theta \\ &= f(0) + \frac{2x}{\pi}\int_0^{\frac{1}{2}\pi}\int_0^{\frac{1}{2}\pi} f'(x\sin\theta\sin\phi)\sin\theta\, d\phi\, d\theta. \end{aligned}$$

Now replace θ by a new variable χ defined by the equation

$$\sin\chi = \sin\theta\sin\phi$$

and change the order of the integrations. We deduce that

$$\begin{aligned}
\frac{2}{\pi}\int_0^{\frac{1}{2}\pi} g(x\sin\theta)\,d\theta - f(0) &= \frac{2x}{\pi}\int_0^{\frac{1}{2}\pi}\int_0^{\frac{1}{2}\pi} f'(x\sin\theta\sin\phi)\sin\theta\, d\phi\, d\theta \\
&= \frac{2x}{\pi}\int_0^{\frac{1}{2}\pi}\int_0^{\theta} f'(x\sin\chi)\frac{\sin\theta\cos\chi\, d\chi\, d\theta}{\sqrt{(\sin^2\theta-\sin^2\chi)}} \\
&= \frac{2x}{\pi}\int_0^{\frac{1}{2}\pi}\int_{\chi}^{\frac{1}{2}\pi} f'(x\sin\chi)\frac{\sin\theta\cos\chi\, d\theta\, d\chi}{\sqrt{(\cos^2\chi-\cos^2\theta)}} \\
&= \frac{2x}{\pi}\int_0^{\frac{1}{2}\pi} f'(x\sin\chi)\left[-\operatorname{arc\,sin}\left(\frac{\cos\theta}{\cos\chi}\right)\right]_{\chi}^{\frac{1}{2}\pi}\cos\chi\, d\chi \\
&= x\int_0^{\frac{1}{2}\pi} f'(x\sin\chi)\cos\chi\, d\chi \\
&= f(x) - f(0),
\end{aligned}$$

and so, when $g(x)$ is defined by (4), $g(x)$ is a solution of (3).

Now it is easy to verify from (4) that, when $f'(x)$ is a continuous function with limited total fluctuation in the interval $(0, \pi)$, so also is $g(x)$; and therefore, by Fourier's theorem, $g(x)$ is expansible in the form

$$g(x) = \tfrac{1}{2}a_0 + \sum_{m=1}^{\infty} a_m \cos mx,$$

where

$$\begin{aligned}
a_m &= \frac{2}{\pi}\int_0^{\pi} g(u)\cos mu\, du \\
&= \frac{2}{\pi}\int_0^{\pi}\left[f(0) + u\int_0^{\frac{1}{2}\pi} f'(u\sin\phi)\, d\phi\right]\cos mu\, du,
\end{aligned}$$

and this series for $g(x)$ converges uniformly throughout the interval $(0, \pi)$.

Hence term-by-term integrations are permissible, and so we have

$$\begin{aligned}
f(x) &= \frac{2}{\pi}\int_0^{\frac{1}{2}\pi} g(x\sin\theta)\, d\theta \\
&= \frac{2}{\pi}\int_0^{\frac{1}{2}\pi}\left\{\tfrac{1}{2}a_0 + \sum_{m=1}^{\infty} a_m\cos(mx\sin\theta)\right\} d\theta \\
&= \tfrac{1}{2}a_0 + \sum_{m=1}^{\infty} a_m J_0(mx),
\end{aligned}$$

and this is the expansion to be established. It is easy to verify that the values obtained for the coefficients a_m are the same as those given by equation (2).

When the restriction concerning the limited total fluctuation of $f'(x)$ is removed, the Fourier series associated with $g(x)$ is no longer necessarily convergent, though the continuity of $f'(x)$ ensures that the Fourier series is uniformly summable $(C\,1)$ throughout $(0, \pi)$; and hence, by term-by-term integration, the series

$$\tfrac{1}{2}a_0 + \sum_{m=1}^{\infty} a_m J_0(mx)$$

is uniformly summable $(C\,1)$ throughout $(0, \pi)$, with sum $f(x)$; an application of Hardy's convergence theorem* then shews that the additional condition

$$a_m = O(1/\sqrt{m})$$

is sufficient to ensure the convergence of the Schlömilch series to the sum $f(x)$ when x lies in the *half-open* interval in which $0 < x \leqslant \pi$.

For further theorems concerning the summability of Schlömilch series, the reader should consult a memoir by Chapman†.

[NOTE. The integral equation connecting $f(x)$ and $g(x)$ is one which was solved in 1823 by Abel, *Journal für Math.* I. (1826), p. 153. It has subsequently been investigated‡ by Beltrami, *Ist. Lombardo Rendiconti*, (2) XIII. (1880), pp. 327, 402; Volterra, *Ann. di Mat.* (2) XXV. (1897), p. 104; C. E. Smith, *Trans. American Math. Soc.* VIII. (1907), pp. 92—106.

The equation

$$\frac{2x}{\pi}\int_0^{\frac{1}{2}\pi}\int_0^{\frac{1}{2}\pi} f'(x\sin\theta\sin\phi)\sin\theta\, d\phi\, d\theta = f(x) - f(0)$$

is most simply established by the method of changing axes of polar coordinates, explained in § 3·33; this method was used by Gwyther, *Messenger*, XXXIII. (1904), pp. 97—107, but in view of the arbitrary character of $f(x)$ the analytical proof given in the text seems preferable. In connexion with the changes in the order of the integrations, cf. *Modern Analysis*, § 4·51.

19·2. *The definition of Schlömilch series.*

We have now investigated Schlömilch's problem of expanding an arbitrary function into a series of Bessel functions of order zero, the argument of the function in the $(m+1)$th term being proportional to m; and the expansion is valid for the range of values $(0, \pi)$ of the variable.

Such series may be generalised by replacing the functions of order zero by functions of arbitrary order ν; and a further generalisation may be effected by taking the general term to contain not only the function $J_\nu(mx)$ but also a function which bears to the Bessel function the same kind of relation as the sine does to the cosine. The latter generalisation is, of course, suggested by the theory of Fourier series, and we are thus led to expect the existence of expansions valid for the range of values $(-\pi, \pi)$ of the variable.

The functions which naturally come under consideration for insertion are

* Cf. *Modern Analysis*, § 8·5.

† *Quarterly Journal*, XLIII. (1911), p. 34.

‡ Some interesting applications of Fourier's integral theorem to the integral equation have been made by Stearn, *Quarterly Journal*, XVII. (1880), pp. 90—104.

Bessel functions of the second kind and Struve's functions; and the types of series to be considered may be written in the forms*:

$$\frac{\frac{1}{2}a_0}{\Gamma(\nu+1)} + \sum_{m=1}^{\infty} \frac{a_m J_\nu(mx) + b_m Y_\nu(mx)}{(\frac{1}{2}mx)^\nu},$$

$$\frac{\frac{1}{2}a_0}{\Gamma(\nu+1)} + \sum_{m=1}^{\infty} \frac{a_m J_\nu(mx) + b_m \mathbf{H}_\nu(mx)}{(\frac{1}{2}mx)^\nu}.$$

Series of the former type (with $\nu = 0$) have been considered by Coates†; but his proof of the possibility of expanding an arbitrary function $f(x)$ into such a series seems to be invalid except in the trivial case in which $f(x)$ is defined to be periodic (with period 2π) and to tend to zero as $x \to \infty$.

Series of the latter type are of much greater interest, and they form a direct generalisation of trigonometrical series. They will be called *generalised Schlömilch series.*

Two types of investigation suggest themselves in connexion with generalised Schlömilch series. The first is the problem of expanding an arbitrary function into such a series; and the second is the problem of determining the properties of such a series with given coefficients and, in particular, the construction of analysis (resembling Riemann's analysis of trigonometrical series) with the object of determining whether a generalised Schlömilch series, in which the coefficients are not all zero, can represent a null-function.

Generalised Schlömilch series have been discussed in a series of memoirs by Nielsen, *Math. Ann.* LII. (1899), pp. 582—587; *Nyt Tidsskrift*, X. B (1899), pp. 73—81; *Oversigt K. Danske Videnskabernes Selskabs*, 1899, pp. 661—665; 1900, pp. 55—60; 1901, pp. 127—146; *Ann. di Mat.* (3) VI. (1901), pp. 301—329.

Nielsen‡ has given the forms for the coefficients in the generalised Schlömilch expansion of an arbitrary function and he has investigated with great detail the actual construction of Schlömilch series which represent null-functions, but his researches are of a distinctly different character from those which will be given in this chapter.

The investigation which we shall now give of the possibility of expanding an arbitrary function into a generalised Schlömilch series is based on the investigation given by Filon§ for the case $\nu = 0$ in his memoir on applications of the calculus of residues to the expansions of arbitrary functions in series of functions of given form. It seems to be of some importance to give such an investigation|| because there is no obvious method of modifying the set of

* The reason for inserting the factor x^ν in the denominators is to make the terms of the second series one-valued (cf. § 19·21).

† *Quarterly Journal*, XXI. (1886), pp. 189—190.

‡ See e.g. his *Handbuch der Theorie der Cylinderfunktionen* (Leipzig, 1904), p. 348.

§ *Proc. London Math. Soc.* (2) IV. (1906), pp. 396—430.

|| It has to be assumed that $-\frac{1}{2} < \nu < \frac{1}{2}$. The results which will be proved in §§ 19·41—19·62 suggest that it is only to be expected that difficulties should arise for other values of ν.

functions $J_\nu(mx)$, $\mathbf{H}_\nu(mx)$ so as to obtain a set which is a normal orthogonal set for the interval $(-\pi, \pi)$; and consequently there is no method of obtaining the coefficients in a Schlömilch expansion in so simple a manner as that in which the coefficients in a Fourier-Bessel expansion are obtained (§ 18·1).

The investigation, which forms the latter part of the chapter, concerning the representation of null-functions by generalised Schlömilch series, is of exactly the same character as the exposition of Riemann's researches on trigonometrical series given in *Modern Analysis*, §§ 9·6—9·632.

19·21. *The application of the calculus of residues to the generalised Schlömilch expansion.*

We shall now explain the method* by which it is possible to discover the values of the coefficients in the generalised Schlömilch expansion which represents an arbitrary function $f(x)$, when the order ν of the Bessel functions lies between $-\frac{1}{2}$ and $\frac{1}{2}$. When this has been done, we shall not consider the validity of the processes by which the discovery has been made, but we shall prove directly that the Schlömilch series in which the coefficients have the specified values actually does converge to the sum $f(x)$.

This is analogous to the procedure which is adopted in Dirichlet's proof of Fourier's theorem: in the expansion

$$f(x) = \tfrac{1}{2}a_0 + \sum_{m=1}^{\infty} (a_m \cos mx + \beta_m \sin mx)$$

the values of the coefficients are discovered by multiplying the expansion by $\cos mx$ and by $\sin mx$, and integrating, so that the values of a_m and β_m are taken to be given by the equations

$$a_m = \frac{1}{\pi}\int_{-\pi}^{\pi} f(t) \cos mt\, dt, \quad \beta_m = \frac{1}{\pi}\int_{-\pi}^{\pi} f(t) \sin mt\, dt.$$

We then take the series in which the coefficients have these values, namely

$$\frac{1}{2\pi}\int_{-\pi}^{\pi} f(t)\, dt + \frac{1}{\pi}\sum_{m=1}^{\infty}\int_{-\pi}^{\pi} f(t) \cos m(x-t)\, dt,$$

and prove that it actually converges to the sum $f(x)$.

It conduces to brevity to deal with the pair of functions

$$\frac{J_\nu(mx) \pm i\,\mathbf{H}_\nu(mx)}{(\frac{1}{2}mx)^\nu},$$

instead of with the pair of functions

$$J_\nu(mx)/(\tfrac{1}{2}mx)^\nu, \quad \mathbf{H}_\nu(mx)/(\tfrac{1}{2}mx)^\nu.$$

We shall write

$$\frac{J_\nu(z) + i\,\mathbf{H}_\nu(z)}{(\frac{1}{2}z)^\nu} \equiv \phi_\nu(z), \tag{1}$$

* Apart from details of notation, the following analysis is due to Filon; it was given by him, in the memoir just cited, for the special case $\nu=0$, but the extension to values of ν between $\pm\frac{1}{2}$ presents no difficulty.

so that* $\phi_\nu(z)$ is analytic and uniform for all finite values of the complex variable z; and evidently

$$\frac{J_\nu(mx) \pm i\mathbf{H}_\nu(mx)}{(\frac{1}{2}mx)^\nu} = \phi_\nu(\pm mx).$$

We now observe that $(-)^m \phi_\nu(mx)$ is the residue at $z = m$ of the function

$$\frac{\pi\phi_\nu(xz)}{\sin \pi z}$$

where $m = 0, \pm 1, \pm 2, \ldots$; and so we shall consider the integral

$$\frac{1}{2\pi i}\int_C F(z)\frac{\pi\phi_\nu(xz)}{\sin \pi z}dz,$$

in which the contour C is a circle, of radius $M + \frac{1}{2}$, with its centre at the origin, and M is an integer which will be made to tend to infinity.

The function $F(z)$ is assumed to be one-valued throughout the z-plane, and to be analytic at infinity (cf. § 19·24); its only singularity in the finite part of the plane is an essential singularity at the origin.

By Jordan's lemma, the integral tends to zero as M tends to infinity, provided that $\nu > -\frac{1}{2}$.

It is evident, by calculating residues, that

$$\sum_{m=1}^{\infty} (-)^m \{F(m)\,\phi_\nu(mx) + F(-m)\,\phi_\nu(-mx)\}$$

is equal to the residue at the origin of

$$-F(z)\frac{\pi\,\phi_\nu(xz)}{\sin \pi z},$$

that is to say

$$\sum_{m=1}^{\infty} (-)^m \{F(m)\,\phi_\nu(mx) + F(-m)\,\phi_\nu(-mx)\} = -\frac{1}{2\pi i}\int^{(0+)} F(z)\frac{\pi\phi_\nu(xz)}{\sin \pi z}dz. \tag{2}$$

The problem of expanding an arbitrary function $f(x)$ into a generalised Schlömilch series is consequently reduced to the determination of the form of $F(z)$ in such a way as to make

$$-\frac{1}{2\pi i}\int^{(0+)} F(z)\frac{\pi\phi_\nu(xz)}{\sin \pi z}dz$$

differ by a constant from $f(x)$.

* The insertion of the factor $(\frac{1}{2}z)^\nu$ in the denominator makes $\phi_\nu(z)$ amenable to Cauchy's theorem when the contour of integration completely surrounds the origin.

19·22. *The construction of the function $F(z)$.*

We now take the contour integral

$$-\frac{1}{2\pi i}\int^{(0+)} F(z)\,\frac{\pi\phi_\nu(xz)}{\sin\pi z}\,dz,$$

and, in order to calculate it in a simple manner, we shall suppose that $F(z)$ is expansible in a series of Filon's type*

(1) $$F(z)=\sum_{n=1}^{\infty}\frac{p_n\,\psi_n(z)}{z^{n+1}},$$

where $\psi_n(z)$ denotes the sum of those terms in the expansion of $\pi^{-1}\sin\pi z$ whose degree does not exceed n, and the coefficients p_n will be defined later.

The reader will observe that

$$\psi_1(z)=\psi_2(z)=z,$$
$$\psi_3(z)=\psi_4(z)=z-\tfrac{1}{6}\pi^2z^3,$$
$$\dots\dots\dots\dots\dots\dots\dots\dots$$

With this definition of $F(z)$, it is evident that, for small values of $|z|$,

$$\begin{aligned}F(z)\frac{\pi\phi_\nu(xz)}{\sin\pi z}&=\sum_{n=1}^{\infty}p_n\left\{\frac{\pi\psi_n(z)}{z^{n+1}\sin\pi z}\right\}\phi_\nu(xz)\\&=\sum_{n=1}^{\infty}p_n\left\{\frac{1}{z^{n+1}}-\frac{\pi^{n+1}\cos\frac{1}{2}n\pi+O(z)}{(n+1)!\,\sin\pi z}\right\}\phi_\nu(xz).\end{aligned}$$

It follows immediately that

(2) $$\begin{aligned}&-\frac{1}{2\pi i}\int^{(0+)}F(z)\,\frac{\pi\phi_\nu(xz)}{\sin\pi z}\,dz\\&=\phi_\nu(0)\sum_{n=1}^{\infty}\frac{p_n\,\pi^n\cos\frac{1}{2}n\pi}{(n+1)!}-\sum_{n=1}^{\infty}\frac{p_n\,(\frac{1}{2}ix)^n}{\Gamma(\frac{1}{2}n+1)\,\Gamma(\frac{1}{2}n+\nu+1)},\end{aligned}$$

and consequently we proceed to identify

$$-\sum_{n=1}^{\infty}\frac{p_n\,(\frac{1}{2}ix)^n}{\Gamma(\frac{1}{2}n+1)\,\Gamma(\frac{1}{2}n+\nu+1)}$$

with $f(x)-f(0)$. For this purpose we have to assume temporarily that $f(x)$ has differential coefficients of all orders at the origin, and then we define the coefficients p_n by the equation

(3) $$\frac{f^{(n)}(0)}{n!}=-\frac{p_n\,(\frac{1}{2}i)^n}{\Gamma(\frac{1}{2}n+1)\,\Gamma(\frac{1}{2}n+\nu+1)}.\qquad(n=1,2,3,\dots).$$

We next transform this equation defining p_n in such a way that the sum of the series, by which $F(z)$ is defined, is expressible in a compact symbolic form; the transformation of the series for $F(z)$ can be effected by expressing

* This type of series is fundamental in Filon's theory, and is not peculiar to Schlömilch expansions; thus, in his work on Fourier-Bessel series, $\sin\pi z$ is replaced by $z^{-\nu}J_\nu(z)$ and $\psi_n(z)$ denotes the sum of the terms whose degree does not exceed n in the expansion of that function.

the coefficients p_n in a form which involves n only as an exponent. For this purpose we make use of Eulerian integrals of the first kind, and, in order that they may be convergent, we shall find that it is necessary to suppose that $-\frac{1}{2} < \nu < \frac{1}{2}$. We then have

$$p_n = -\frac{\Gamma(\frac{1}{2})\,\Gamma(\frac{1}{2}n+\nu+1)}{i^n\,\Gamma(\frac{1}{2}n+\frac{1}{2})} f^{(n)}(0)$$

$$= -\frac{\Gamma(\frac{1}{2})\,(n+2\nu)}{i^n\,\Gamma(\frac{1}{2}-\nu)} f^{(n)}(0) \int_0^1 (1-t^2)^{-\frac{1}{2}-\nu}\, t^{n+2\nu-1}\,dt$$

$$= -\frac{\Gamma(\frac{1}{2})}{i^n\,\Gamma(\frac{1}{2}-\nu)} \int_0^1 (1-t^2)^{-\frac{1}{2}-\nu} \frac{d}{dt}\left[t^{2\nu} \frac{d^n f(tu)}{du^n}\right]_{u=0} dt,$$

and so we obtain the symbolic formula

$$(4) \qquad p_n = -\frac{\Gamma(\frac{1}{2})}{i^n\,\Gamma(\frac{1}{2}-\nu)} \int_0^1 (1-t^2)^{-\frac{1}{2}-\nu} \frac{d}{dt}\left[t^{2\nu} D^n f(tu)\right]_{u=0} dt,$$

where D stands for d/du.

Now, if we arrange the series

$$\sum_{n=}^{\infty} \frac{\psi_n(z)\,D^n}{i^n\,z^{n+1}}$$

in descending powers of z, it is easy to verify that

$$\sum_{n=1} \frac{\psi_n(z)\,D^n}{i^n\,z^{n+1}} = \frac{\sinh \pi D}{\pi(iz-D)},$$

and therefore

$$(5) \qquad F(z) = \frac{\Gamma(\frac{1}{2})}{\Gamma(\frac{1}{2}-\nu)} \int_0^1 (1-t^2)^{-\frac{1}{2}-\nu} \frac{d}{dt}\left[t^{2\nu} \frac{\sinh \pi D}{\pi(D-iz)} f(tu)\right]_{u=0} dt.$$

Again, a consideration of (2) shews that we need to sum the series

$$\sum_{n=1}^{\infty} \frac{p_n\,\pi^n \cos \frac{1}{2}n\pi}{(n+1)!},$$

and we are able to effect our purpose by making use of formula (4), whence we find that

$$(6) \quad \sum_{n=1}^{\infty} \frac{p_n\,\pi^n \cos \frac{1}{2}n\pi}{(n+1)!}$$

$$= -\frac{\Gamma(\frac{1}{2})}{\Gamma(\frac{1}{2}-\nu)} \int_0^1 (1-t^2)^{-\frac{1}{2}-\nu} \frac{d}{dt}\left[t^{2\nu} \left\{\frac{\sinh \pi D}{\pi D} - 1\right\} f(tu)\right]_{u=0} dt.$$

We have now obtained symbolic expressions for all the coefficients in the generalised Schlömilch expansion of $f(x)$, but it is necessary to transform these expressions into more useful forms, by finding the significance to be attached to the symbolic operator $\dfrac{\sinh \pi D}{\pi(D-iz)}$, both for general values of z and for the value zero of z.

19·23. *The transformation of the symbolic operators in the generalised Schlömilch expansion.*

We proceed to obtain an interpretation* of the symbolic expression

$$\left[\frac{\sinh \pi D}{\pi (D - iz)} f(tu)\right]_{u=0}.$$

The usual interpretation of $\dfrac{1}{D - iz} f(tu)$ is

$$e^{izu} \int_a^u e^{-izv} f(tv)\, dv,$$

where a is a constant of integration; and therefore

$$\begin{aligned}\frac{\sinh \pi D}{\pi (D - iz)} f(tu) &= \frac{\sinh \pi D}{\pi} \left[e^{izu} \int_a^u e^{-izv} f(tv)\, dv\right] \\ &= e^{izu} \frac{\sinh \pi (D + iz)}{\pi} \int_a^u e^{-izv} f(tv)\, dv.\end{aligned}$$

Now, by the symbolic form of Taylor's theorem, we have

$$e^{\pm \pi D} \chi(u) = \chi(u \pm \pi),$$

where $\chi(u)$ is an arbitrary function of u; and hence it follows that

$$\begin{aligned}\left[\frac{\sinh \pi D}{\pi (D - iz)} f(tu)\right]_{u=0} &= \left[\frac{\sinh \pi (D + iz)}{\pi} \int_a^u e^{-izv} f(tv)\, dv\right]_{u=0} \\ &= \frac{1}{2\pi}\left[e^{\pi iz} \int_a^{u+\pi} e^{-izv} f(tv)\, dv\right. \\ &\qquad \left. - e^{-\pi iz} \int_a^{u-\pi} e^{-izv} f(tv)\, dv\right]_{u=0},\end{aligned}$$

that is to say†

$$(1) \quad \left[\frac{\sinh \pi D}{\pi (D - iz)} f(tu)\right]_{u=0} = \frac{\cos \pi z}{2\pi} \int_{-\pi}^{\pi} e^{-izv} f(tv)\, dv + \frac{i \sin \pi z}{2\pi}\left[\int_a^{\pi} + \int_a^{-\pi}\right] e^{-izv} f(tv)\, dv.$$

The second term on the right has simple zeros at all the points at which $z = 0, \pm 1, \pm 2, \ldots$.

Therefore, so far as the calculation of residues of

$$F(z) \frac{\pi \phi_\nu (xz)}{\sin \pi z}$$

* The interpretations of numerous expressions involving symbolic operators of the types under consideration have been discussed by Gregory, *Cambridge Math. Journal*, I. (1839), pp. 22—32 and by Boole, *Differential Equations* (London, 1872), chapters XVI and XVII.

† The expression for $F(z)$ which is derived from this formula does not appear to have a singularity at the origin unless a is infinite or is a function of z; but it seems unreasonable to be perturbed by this when we consider the nature of some of the analysis which has already been used in the course of this investigation.

at $0, \pm 1, \pm 2, \ldots$ *is concerned, we may omit the second term on the right in* (1), *and calculate the residues of*

$$\bar{F}(z)\,\frac{\pi\phi_\nu(xz)}{\sin \pi z},$$

where $\bar{F}(z)$ *is defined by the formula*

$$(2)\qquad \bar{F}(z) = \frac{\cos \pi z}{2\Gamma(\frac{1}{2}-\nu)\Gamma(\frac{1}{2})}\int_0^1 (1-t^2)^{-\frac{1}{2}-\nu}\frac{d}{dt}\left[t^{2\nu}\int_{-\pi}^{\pi} e^{-izv} f(tv)\,dv\right]dt.$$

Again, from § 19·22 (6) and equation (1) of this section we have

$$(3)\qquad \sum_{n=1}^{\infty}\frac{p_n\pi^n \cos\frac{1}{2}n\pi}{(n+1)!} = \frac{\Gamma(\frac{1}{2})f(0)}{\Gamma(\frac{1}{2}-\nu)}\int_0^1 (1-t^2)^{-\frac{1}{2}-\nu}\frac{dt^{2\nu}}{dt}\,dt$$
$$-\frac{1}{2\Gamma(\frac{1}{2}-\nu)\Gamma(\frac{1}{2})}\int_0^1 (1-t^2)^{-\frac{1}{2}-\nu}\frac{d}{dt}\left[t^{2\nu}\int_{-\pi}^{\pi} f(tv)\,dv\right]dt.$$

The first term on the right in (3) is equal to $\Gamma(\nu+1)f(0)$, except when* $\nu=0$; when $\nu=0$, the value of the term in question is zero.

We thus obtain the expansion

$$(4)\quad f(x) = \phi_\nu(0)\bar{F}(0) + \sum_{m=1}^{\infty}(-)^m[\bar{F}(m)\phi_\nu(mx) + \bar{F}(-m)\phi_\nu(-mx)].$$

In the special case in which $\nu=0$, the modified form of (3) shews that an additional term $f(0)$ must be inserted on the right in (4).

When we change the notation to the notation normally used for Bessel functions and Struve's functions, the expansion becomes

$$(5)\qquad f(x) = \frac{\frac{1}{2}a_0}{\Gamma(\nu+1)} + \sum_{m=1}^{\infty}\frac{a_m J_\nu(mx) + b_m \mathbf{H}_\nu(mx)}{(\frac{1}{2}mx)^\nu},$$

where†

$$(6)\qquad \begin{cases} a_m = \dfrac{1}{\Gamma(\frac{1}{2}-\nu)\Gamma(\frac{1}{2})}\displaystyle\int_0^1 (1-t^2)^{-\frac{1}{2}-\nu}\frac{d}{dt}\left[t^{2\nu}\int_{-\pi}^{\pi} f(tv)\cos mv\,dv\right]dt, \\ b_m = \dfrac{1}{\Gamma(\frac{1}{2}-\nu)\Gamma(\frac{1}{2})}\displaystyle\int_0^1 (1-t^2)^{-\frac{1}{2}-\nu}\frac{d}{dt}\left[t^{2\nu}\int_{-\pi}^{\pi} f(tv)\sin mv\,dv\right]dt. \end{cases}$$

This is the generalised form of Schlömilch's expansion.

19·24. *The boundedness of* $F(z)$, *as* $|z|\to\infty$.

We shall now prove that, when the function $f(x)$ is restricted in a suitable manner, the function $F(z)$ is bounded when $|z|\to\infty$, whatever be the value of $\arg z$. The reader will remember that the assumption that $F(z)$ is bounded was made in § 19·21 to secure the convergence of the contour integral.

We take the series of § 19·22 (1), by which $F(z)$ was originally defined, namely

$$\sum_{n=1}^{\infty}\frac{p_n\psi_n(z)}{z^{n+1}},$$

* When ν is negative it is necessary to use a modified expression for the integrals; cf. § 19·3.

† When $\nu=0$, the expression for a_0 has to be modified by the insertion of the term $2f(0)$, in consequence of the discontinuity in value of the expression on the right of (3).

and divide it into two parts, namely the first N terms and the remainder of the terms, where N is the integer such that

$$N \leqslant \pi |z| < N+1.$$

When $n \leqslant N$, the terms of $\psi_n(z)$ do not exceed $\pi^{n-1}|z|^n/n!$, and therefore, when $n \leqslant N$,

$$\left|\frac{\psi_n(z)}{z^{n+1}}\right| < \frac{n\pi^{n-1}|z|^n/(n!)}{|z|^{n+1}} = \frac{\pi^{n-1}}{|z|.(n-1)!}.$$

When $n \geqslant N$, we have $|\psi_n(z)| < \pi^{-1}\sinh \pi |z|$, and therefore

$$|F(z)| < \frac{1}{|z|}\sum_{n=1}^{N} \frac{|p_n|\pi^{n-1}}{(n-1)!} + \frac{\sinh \pi |z|}{\pi |z|^{N+1}} \sum_{n=0}^{\infty} \frac{|p_{n+N+1}|}{|z|^n}.$$

Since

$$\frac{\sinh \pi |z|}{|z|^{N+1}}$$

tends to zero as $|z| \to \infty$, it is evident that a sufficient condition for $F(z)$ to be bounded as $|z| \to \infty$ is that the series

$$\sum_{n=1}^{\infty} |p_n|$$

should be convergent; and this is the case if $f(x)$ is such that

$$\sum_{n=1}^{\infty} n^{\nu+\frac{1}{2}} |f^{(n)}(0)|$$

is convergent.

19·3. *The expansion of an arbitrary function into a generalised Schlömilch series.*

Now that the forms of the coefficients in the generalised Schlömilch expansion have been ascertained by Filon's method, it is an easy matter to specify sufficient conditions for the validity of the expansion and then to establish it.

The theorem which we shall prove* is as follows:

Let ν be a number such that $-\frac{1}{2} < \nu < \frac{1}{2}$; and let $f(x)$ be defined arbitrarily in the interval $(-\pi, \pi)$, subject† to the following conditions:

(I) *The function $h(x)$, defined by the equation*

$$h(x) = 2\nu f(x) + xf'(x),$$

exists and is continuous in the closed interval $(-\pi, \pi)$.

(II) *The function $h(x)$ has limited total fluctuation in the interval $(-\pi, \pi)$.*

(III) *If ν is negative‡ the integral*

$$\int_0^{\Delta} \frac{d}{dx}[|x|^{2\nu}\{f(x) - f(0)\}]\,dx$$

is absolutely convergent when Δ is a (small) number either positive or negative.

* The expansion is stated by Nielsen, *Handbuch der Theorie der Cylinderfunktionen* (Leipzig, 1904), p. 348; but the formulae which he gives for the coefficients in the expansion seem to be quite inconsistent with those given by equation (2).

† The effect of conditions (I) and (II) is merely to ensure the uniformity of the convergence of a certain Fourier series connected with $h(x)$.

‡ If ν is positive, this Lipschitz condition is satisfied by reason of (II).

Then $f(x)$ admits of the expansion

(1) $$f(x) = \frac{\frac{1}{2}a_0}{\Gamma(\nu+1)} + \sum_{m=1}^{\infty} \frac{a_m J_\nu(mx) + b_m \mathbf{H}_\nu(mx)}{(\frac{1}{2}mx)^\nu},$$

where

(2) $$\begin{cases} a_m = \displaystyle\int_{-\pi}^{\pi}\int_0^{\frac{1}{2}\pi} \frac{\sec^{2\nu+1}\phi}{\Gamma(\frac{1}{2}-\nu)\,\Gamma(\frac{1}{2})}\frac{d}{d\phi}[\sin^{2\nu}\phi\,\{f(u\sin\phi)-f(0)\}]\cos mu\,d\phi\,du, \\ b_m = \displaystyle\int_{-\pi}^{\pi}\int_0^{\frac{1}{2}\pi} \frac{\sec^{2\nu+1}\phi}{\Gamma(\frac{1}{2}-\nu)\,\Gamma(\frac{1}{2})}\frac{d}{d\phi}[\sin^{2\nu}\phi\,\{f(u\sin\phi)-f(0)\}]\sin mu\,d\phi\,du, \end{cases}$$

when $m > 0$; the value of a_0 is obtained by inserting an additional term

$$2\Gamma(\nu+1)f(0)$$

on the right in the first equation of the system (2).

We shall base the investigation on a discussion of the integral equation

(3) $$f(x) = \frac{2}{\Gamma(\nu+\frac{1}{2})\,\Gamma(\frac{1}{2})}\int_0^{\frac{1}{2}\pi}\cos^{2\nu}\theta\, g(x\sin\theta)\,d\theta;$$

it will be proved that a continuous solution is given by the formula

(4) $$g(x) = \Gamma(\nu+1)f(0) + \frac{\Gamma(\frac{1}{2})}{\Gamma(\frac{1}{2}-\nu)}\int_0^{\frac{1}{2}\pi}\sec^{2\nu+1}\phi\,\frac{d}{d\phi}[\sin^{2\nu}\phi\,\{f(x\sin\phi)-f(0)\}]\,d\phi.$$

[NOTE. The (absolute) convergence of the integral contained in this formula is secured by condition (III). It should be observed that the aggregate of terms containing $f(0)$ in equation (4) may be omitted when ν is positive in view of the formula

$$\int_0^{\frac{1}{2}\pi}\sec^{2\nu+1}\phi\,\frac{d\sin^{2\nu}\phi}{d\phi}\,d\phi = \frac{\Gamma(\nu+1)\,\Gamma(\frac{1}{2}-\nu)}{\Gamma(\frac{1}{2})},$$

which is valid only when ν is positive.]

We proceed to verify that the function $g(x)$ defined by (4) actually is a solution of (3), by taking $g(x)$ to be defined by (4), substituting in the expression on the right of (3), and reducing the result to $f(x)$.

The result of substitution is

$$\frac{2\cos\nu\pi}{\pi}\int_0^{\frac{1}{2}\pi}\int_0^{\frac{1}{2}\pi}\cos^{2\nu}\theta\sec^{2\nu+1}\phi\,\frac{d}{d\phi}[\sin^{2\nu}\phi\,\{f(x\sin\theta\sin\phi)-f(0)\}]\,d\phi\,d\theta + f(0).$$

Hence we have to prove that

$$\frac{2\cos\nu\pi}{\pi}\int_0^{\frac{1}{2}\pi}\int_0^{\frac{1}{2}\pi}\cos^{2\nu}\theta\sec^{2\nu+1}\phi\,\frac{d}{d\phi}[\sin^{2\nu}\phi\,\{f(x\sin\theta\sin\phi)-f(0)\}]\,d\phi\,d\theta = f(x)-f(0).$$

Replace ϕ on the left by a new variable χ defined by the equation

$$\sin\chi = \sin\theta\sin\phi,$$

change the order of the integrations in the resulting absolutely convergent integral, and then replace θ by a new variable ψ defined by the equation

$$\cos\theta = \cos\chi\sin\psi.$$

We thus deduce that

$$\int_0^{\frac{1}{2}\pi}\int_0^{\frac{1}{2}\pi} \cos^{2\nu}\theta \sec^{2\nu+1}\phi \frac{d}{d\phi}[\sin^{2\nu}\phi\{f(x\sin\theta\sin\phi)-f(0)\}]\,d\phi d\theta$$

$$= \int_0^{\frac{1}{2}\pi}\int_0^{\theta} \frac{\sin\theta\cos^{2\nu}\theta}{(\sin^2\theta-\sin^2\chi)^{\nu+\frac{1}{2}}}\frac{d}{d\chi}[\sin^{2\nu}\chi\{f(x\sin\chi)-f(0)\}]\,d\chi d\theta$$

$$= \int_0^{\frac{1}{2}\pi}\int_{\chi}^{\frac{1}{2}\pi} \frac{\sin\theta\cos^{2\nu}\theta}{(\cos^2\chi-\cos^2\theta)^{\nu+\frac{1}{2}}}\frac{d}{d\chi}[\sin^{2\nu}\chi\{f(x\sin\chi)-f(0)\}]\,d\theta d\chi$$

$$= \int_0^{\frac{1}{2}\pi}\int_0^{\frac{1}{2}\pi} \tan^{2\nu}\psi\, d\psi \,.\, \frac{d}{d\chi}[\sin^{2\nu}\chi\{f(x\sin\chi)-f(0)\}]\,d\chi$$

$$= \tfrac{1}{2}\Gamma(\nu+\tfrac{1}{2})\,\Gamma(\tfrac{1}{2}-\nu)\,\{f(x)-f(0)\},$$

and hence the formula to be established is evident; and so, when $g(x)$ is defined by (4), then equation (3) is satisfied.

Now, by Fourier's theorem,

$$g(x) = \tfrac{1}{2}a_0 + \sum_{m=1}^{\infty}(a_m\cos mx + b_m\sin mx),$$

where

$$(5)\qquad \begin{cases} a_m = \dfrac{1}{\pi}\displaystyle\int_{-\pi}^{\pi} g(u)\cos mu\, du, \\ b_m = \dfrac{1}{\pi}\displaystyle\int_{-\pi}^{\pi} g(u)\sin mu\, du; \end{cases}$$

and it is easy to verify that when $f(x)$ is a continuous function with limited total fluctuation in the interval $(-\pi, \pi)$ so also is $g(x)$, and therefore the expansion for $g(x)$ is uniformly convergent when $-\pi+\delta \leqslant x \leqslant \pi-\delta$, where δ is an arbitrarily small positive number.

Replace x by $x\sin\theta$ in the expansion of $g(x)$, multiply by $\cos^{2\nu}\theta$, which has an absolutely convergent integral, and integrate term-by-term; we deduce at once that

$$f(x) = \frac{\frac{1}{2}a_0}{\Gamma(\nu+1)} + \sum_{m=1}^{\infty}\frac{a_m J_\nu(mx) + b_m \mathbf{H}_\nu(mx)}{(\frac{1}{2}mx)^\nu};$$

and this expansion converges uniformly when $-\pi+\delta \leqslant x \leqslant \pi-\delta$.

The values of a_m and b_m given by formula (5) are easily reconciled with those given by formula (2).

It should be noticed that, by the Riemann-Lebesgue lemma, a_m and b_m are both $O(1/m)$ when m is large. This seems to be connected with the fact that when we come to deal with *any* Schlömilch series (§ 19·62) we are unable to make any progress without assuming that $\Sigma b_m/m$ is convergent (or some equivalent hypothesis); this assumption will appear in § 19·62 to be necessary because the differential equation which Struve's function satisfies is not homogeneous, so that Struve's function is not of a type which occurs in solutions of Laplace's equation or the wave equation; there would consequently seem to be reasons of a physical character for the limitations which have been placed on $f(x)$ in order to ensure the existence of the Schlömilch expansion.

[NOTE. Just as in § 19·11, if condition (II) concerning the limited total fluctuation of $2\nu f(x)+xf'(x)$ is not satisfied, then all statements made in this section up to this point about convergence of series have to be replaced by statements about summability $(C\,1)$.]

There is one important consequence which follows from the fact that a_m and b_m are both $O(1/m)$ when $2\nu f(x)+xf'(x)$ has limited total fluctuation in $(-\pi, \pi)$, namely, that in the neighbourhoods of $-\pi$ and π, the general term of the Schlömilch expansion is $O(1/m^{\nu+\frac{3}{2}})$, and so the expansion represents a continuous function; hence the expansion converges (uniformly) to the sum $f(x)$ throughout the interval $(-\pi, \pi)$.

19·4. *Special functions represented by Schlömilch series.*

There are a few problems of Mathematical Physics (other than the problem mentioned in § 19·1) in which Schlömilch series occur in a natural manner, and we shall now give an account of various researches in which Schlömilch series are to be found.

A very simple series is

$$1+\sum_{m=1}^{\infty} e^{-mz} J_0(m\rho);$$

this series is convergent when ρ and z are positive, and, if ρ and z denote cylindrical-polar coordinates, it is a solution of Laplace's equation at all points of space above the plane $z=0$.

Various transformations of the series have been given by Whittaker*; thus, by changing to Cartesian coordinates (x, y, z) and using § 2·21, we have

$$(1)\quad 1+\sum_{m=1}^{\infty} e^{-mz} J_0(m\rho)=\frac{1}{2\pi}\int_{-\pi}^{\pi}\frac{du}{1-\exp\{-(z+ix\cos u+iy\sin u)\}}.$$

When $x^2+y^2+z^2<1$, the integrand may be expanded in ascending powers of $z+ix\cos u+iy\sin u$.

If this is done, we get†

$$(2)\quad 1+\sum_{m=1}^{\infty} e^{-mz} J_0(m\rho)=\frac{1}{2\pi}\int_{-\pi}^{\pi}\frac{du}{z+ix\cos u+iy\sin u}+\frac{1}{2}$$
$$+\frac{1}{2\pi}\sum_{m=1}^{\infty}\frac{(-)^{m-1}B_m}{(2m)!}\int_{-\pi}^{\pi}(z+ix\cos u+iy\sin u)^{2m-1}\,du$$
$$=\frac{1}{r}+\frac{1}{2}+\sum_{m=1}^{\infty}\frac{(-)^{m-1}B_m}{(2m)!}r^{2m-1}P_{2m-1}(\cos\theta),$$

where (r, θ) are the polar coordinates corresponding to the cylindrical-polar coordinates (ρ, z), and $B_1, B_2, B_3, \ldots$ are Bernoulli's numbers.

* *Math. Ann.* LVII. (1903), pp. 341—342.

† Cf. § 4·8 and *Modern Analysis*, §§ 7·2, 18·31.

Another transformation of the series, also given by Whittaker, is obtained from the expansion for $1/(1-e^{-t})$ in partial fractions; this expansion is

$$\frac{1}{1-e^{-t}}=\frac{1}{t}+\frac{1}{2}+\sum_{m=1}^{\infty}\left\{\frac{1}{t-2m\pi i}+\frac{1}{t+2m\pi i}\right\},$$

whence we deduce that

$$(3)\qquad 1+\sum_{m=1}^{\infty}e^{-mz}J_0(m\rho)=\frac{1}{r}+\frac{1}{2}$$
$$+\sum_{m=1}^{\infty}\left[\frac{1}{\sqrt{\{(2m\pi i+z)^2+x^2+y^2\}}}-\frac{1}{\sqrt{\{(2m\pi i-z)^2+x^2+y^2\}}}\right].$$

It follows that the series represents the electrostatic potential due to a set of unit charges (some positive and some negative) at the origin and at a set of imaginary points.

The reader may find it interesting to discuss the Lipschitz-Hankel integral of § 13·2 as a limiting form of a series of Whittaker's type.

Some other series have been examined by Nagaoka* in connexion with a problem of Diffraction. One such series is derived from the Fourier series for the function which is equal to $1/\sqrt{(1-x^2)}$ in the interval $(-1, 1)$.

The Fourier series in question is

$$(4)\qquad \frac{1}{\sqrt{(1-x^2)}}=\tfrac{1}{2}\pi+\pi\sum_{m=1}^{\infty}J_0(m\pi)\cos m\pi x,$$

and it converges uniformly throughout the interval $(-1+\Delta, 1-\Delta)$, where Δ is any positive number.

Multiply by e^{axi} and integrate, and we then obtain the formula (also due to Nagaoka)

$$(5)\qquad \frac{1}{\pi}\int^{x}\frac{e^{axi}\,dx}{\sqrt{(1-x^2)}}=-\frac{ie^{axi}}{2}\left[\frac{1}{a}+2\sum_{m=1}^{\infty}J_0(m\pi)\frac{a\cos m\pi x-m\pi i\sin m\pi x}{a^2-m^2\pi^2}\right].$$

The series on the right in (5) converges uniformly throughout the interval $(-1, 1)$ and so we may take -1 and 1 as limits of integration.

Hence, for all values (real and complex) of a,

$$(6)\qquad J_0(a)=\frac{\sin a}{a}\left[1+2a^2\sum_{m=1}^{\infty}\frac{(-)^m J_0(m\pi)}{a^2-m^2\pi^2}\right].$$

A more general result, valid when $R(\nu+\frac{3}{2})>0$, is

$$(7)\qquad J_\nu(a)=\frac{\sin a}{a}\left[\frac{(\frac{1}{2}a)^\nu}{\Gamma(\nu+1)}+2a^2\left(\frac{a}{\pi}\right)^\nu\sum_{m=1}^{\infty}\frac{(-)^m J_\nu(m\pi)}{m^\nu(a^2-m^2\pi^2)}\right].$$

* *Journal of the Coll. of Sci., Imp. Univ. of Japan*, IV. (1891), pp. 301—322. Some of Nagaoka's formulae are quoted by Cinelli, *Nuovo Cimento*, (4) I. (1895), p. 152.

This expansion is also obtainable by expressing $\dfrac{J_\nu(a)}{a^\nu \sin a}$ as a sum of partial fractions*.

Various representations of the integral on the left of (5) were obtained by Nagaoka; the formula quoted seems to be the most interesting of them.

Finally we shall give the formula†

$$\sum_{m=1}^{\infty} \frac{J_0\{(2m-1)x\}}{(2m-1)^2} = \frac{\pi^2}{8} - \frac{|x|}{2}. \qquad (-\pi < x < \pi) \tag{8}$$

This is deducible from the Fourier series

$$\sum_{m=1}^{\infty} \frac{\cos(2m-1)x}{(2m-1)^2} = \frac{\pi}{8}(\pi - 2|x|), \qquad (-\pi < x < \pi)$$

by replacing x by $x \sin\theta$ and integrating with respect to θ from 0 to $\frac{1}{2}\pi$.

As an example of the calculation of the sum of a Schlömilch series when the variable lies outside the interval $(-\pi, \pi)$, we shall take $\pi < x < 2\pi$, and then, if $f(x)$ denotes the sum of the Fourier series, we see that

$$\begin{aligned}\sum_{m=1}^{\infty} \frac{J_0\{(2m-1)x\}}{(2m-1)^2} &= \frac{2}{\pi}\int_0^{\frac{1}{2}\pi} f(x\sin\theta)\,d\theta \\ &= \frac{2}{\pi}\left\{\int_0^{\arcsin(\pi/x)} + \int_{\arcsin(\pi/x)}^{\frac{1}{2}\pi}\right\} f(x\sin\theta)\,d\theta \\ &= \frac{2}{\pi}\int_0^{\arcsin(\pi/x)} \frac{\pi}{8}(\pi - 2x\sin\theta)\,d\theta \\ &\quad + \frac{2}{\pi}\int_{\arcsin(\pi/x)}^{\frac{1}{2}\pi} \frac{\pi}{8}(2x\sin\theta - 3\pi)\,d\theta,\end{aligned}$$

so that, when $\pi < x < 2\pi$, we have

$$\sum_{m=1}^{\infty} \frac{J_0\{(2m-1)x\}}{(2m-1)^2} = \sqrt{(x^2-\pi^2)} - \tfrac{1}{2}x - \pi \operatorname{arc\,cos}\left(\frac{\pi}{x}\right) + \frac{\pi^2}{8}. \tag{9}$$

19·41. *Null-functions expressed as Schlömilch series.*

We shall now prove the remarkable theorem that

$$\frac{1}{2} + \sum_{m=1}^{\infty} (-)^m J_0(mx) = 0, \tag{1}$$

provided that $0 < x < \pi$; the series oscillates when $x = 0$ and diverges to $+\infty$ when $x = \pi$.

This theorem has no analogue in the theory of Fourier series, and, in fact, it is definitely known‡ that a Fourier cosine-series cannot represent a null-function throughout the interval $(0, \pi)$.

* Cf. *Modern Analysis*, § 7·4.

† This was set as a problem in the Mathematical Tripos, 1895.

‡ Cf. *Modern Analysis*, §§ 9·6—9·632.

It is easy to prove (1) by using Parseval's integral; when M is a large integer, we have

$$\frac{1}{2}+\sum_{m=1}^{M}(-)^m J_0(mx)=\frac{2}{\pi}\int_0^{\frac{1}{2}\pi}\left\{\frac{1}{2}+\sum_{m=1}^{M}(-)^m\cos(mx\sin t)\right\}dt$$
$$=\frac{(-)^M}{\pi}\int_0^{\frac{1}{2}\pi}\frac{\cos\{(M+\frac{1}{2})\,x\sin t\}}{\cos(\frac{1}{2}x\sin t)}\,dt$$
$$=\frac{(-)^M}{\pi}\int_0^{x}\frac{\cos(M+\frac{1}{2})\,u}{\cos\frac{1}{2}u}\cdot\frac{du}{\sqrt{(x^2-u^2)}}$$
$$=o(1),$$

as $M\to\infty$, by the Riemann-Lebesgue lemma*, which is applicable because the integral

$$\int_0^x\frac{du}{\cos\frac{1}{2}u\,.\sqrt{(x^2-u^2)}}$$

exists and is absolutely convergent when $0<x<\pi$.

Hence we have proved that

$$\lim_{M\to\infty}\left[\frac{1}{2}+\sum_{m=1}^{M}(-)^m J_0(mx)\right]=0$$

when $0<x<\pi$; and this is the theorem stated.

It is easy to prove in a similar manner that

(2) $$\frac{\frac{1}{2}}{\Gamma(\nu+1)}+\sum_{m=1}^{\infty}\frac{(-)^m J_\nu(mx)}{(\frac{1}{2}mx)^\nu}=0,$$

(i) when $-\frac{1}{2}<\nu\leqslant\frac{1}{2}$ and $0<x<\pi$, (ii) when $\nu>\frac{1}{2}$ and $0<x\leqslant\pi$.

By using Poisson's integral we have (since $\nu>-\frac{1}{2}$)

$$\frac{\frac{1}{2}}{\Gamma(\nu+1)}+\sum_{m=1}^{M}\frac{(-)^m J_\nu(mx)}{(\frac{1}{2}mx)^\nu}$$
$$=\frac{2}{\Gamma(\nu+\frac{1}{2})\,\Gamma(\frac{1}{2})}\int_0^{\frac{1}{2}\pi}\left\{\frac{1}{2}+\sum_{m=1}^{M}(-)^m\cos(mx\sin t)\right\}\cos^{2\nu}t\,dt$$
$$=\frac{(-)^M}{\Gamma(\nu+\frac{1}{2})\,\Gamma(\frac{1}{2})\,x^{2\nu}}\int_0^x\frac{\cos(M+\frac{1}{2})u}{\cos\frac{1}{2}u}\,.\,(x^2-u^2)^{\nu-\frac{1}{2}}\,du$$
$$=o(1),$$

as $M\to\infty$, provided that the integral

$$\int_0^x\frac{(x^2-u^2)^{\nu-\frac{1}{2}}}{\cos\frac{1}{2}u}\,du$$

exists and is absolutely convergent; and this is the case when x and ν satisfy the conditions stated.

The truth of (2) is now evident.

If n is a positive integer, and if ν is so large that $\nu-2n>-\frac{1}{2}$, the operator

$$\frac{d}{xdx}\left\{x\frac{d}{dx}+2\nu\right\}$$

* Cf. *Modern Analysis*, § 9·41.

may be applied n times to equation (2). The effect of applying the operator once to the function $J_\nu(mx)/(\frac{1}{2}mx)^\nu$ is to multiply the function by $-m^2$; and therefore, when $0 < x < \pi$,

$$\sum_{m=1}^{\infty} \frac{(-\frac{1}{4}m^2)^n (-)^m J_\nu(mx)}{(\frac{1}{2}mx)^\nu} = 0,$$

that is to say,

$$(3) \qquad \sum_{m=1}^{\infty} \frac{(-)^m J_\nu(mx)}{(\frac{1}{2}mx)^{\nu-2n}} = 0,$$

provided that either

(i) $-\frac{1}{2} < \nu - 2n \leqslant \frac{1}{2}$ and $0 \leqslant x < \pi$, or (ii) $\nu - 2n > \frac{1}{2}$ and $0 \leqslant x \leqslant \pi$.

The formulae given in this section are due to Nielsen*, *Math. Ann.* LII. (1899), pp. 582—587; two other papers by Nielsen on this subject were published at about the same time, *Nyt Tidsskrift*, x. B (1899), pp. 73—81; *Oversigt K. Danske Videnskabernes Selskabs*, 1899, pp. 661—665. In the first two of these three papers integral values of ν only were considered, the extension to general values of ν being made in the third paper.

Shortly afterwards† Nielsen gave a formula for the sum of the series in (2) when $x > \pi$; this formula is easily obtained from the integral of Dirichlet's type

$$(-)^M \int_0^x \frac{\cos(M+\frac{1}{2})u}{\cos\frac{1}{2}u} \cdot (x^2-u^2)^{\nu-\frac{1}{2}}\, du,$$

by considering the behaviour of the integrand at $u = \pi, 3\pi, 5\pi, \ldots$.

It is thus found that, when x is positive and q is the integer such that

$$(2q-1)\pi < x < (2q+1)\pi,$$

then

$$(4) \qquad \frac{\frac{1}{2}}{\Gamma(\nu+1)} + \sum_{m=1}^{\infty} \frac{(-)^m J_\nu(mx)}{(\frac{1}{2}mx)^\nu} = \frac{2\Gamma(\frac{1}{2})}{x\Gamma(\nu+\frac{1}{2})} \sum_{n=1}^{q} \left\{1 - \frac{(2n-1)^2\pi^2}{x^2}\right\}^{\nu-\frac{1}{2}}.$$

The importance of Nielsen's formulae lies in the fact that they make it evident that, when a function $f(x)$ is defined for the interval $(-\pi, \pi)$, if the function can be represented by a Schlömilch series throughout the interval (except possibly at a finite number of points) the representation is not unique and there are an unlimited number of Schlömilch series which are equal to the function $f(x)$ throughout the interval, except at a finite number of points, namely the points already specified together with the origin and (when $-\frac{1}{2} < \nu \leqslant \frac{1}{2}$) the end-points $\pm\pi$.

The converse theorem, that *the only Schlömilch series with non-vanishing coefficients which represent null-functions at all points of the interval* $-\pi < x < \pi$, *(when‡* $-\frac{1}{2} < \nu \leqslant \frac{1}{2}$*) except the origin are constant multiples of*

$$\frac{\frac{1}{2}}{\Gamma(\nu+1)} + \sum_{m=1}^{\infty} \frac{(-)^m J_\nu(mx)}{(\frac{1}{2}mx)^\nu},$$

* Formula (1) was rediscovered by Gwyther, *Messenger*, XXXIII. (1904), p. 101.

† *Oversigt K. Danske Videnskabernes Selskabs*, 1900, pp. 55—60; see also a later paper by Nielsen, *Ann. di Mat.* (3) VI. (1901), pp. 301—329 for more complicated results. Cf. § 19·4 (9).

‡ The theorem is untrue when $\nu > \frac{3}{2}$; cf. formula (3). It would be interesting to know whether any Schlömilch series other than the one given can represent a null-function when $\frac{1}{2} < \nu < \frac{3}{2}$.

is, of course, of a much deeper character, and it seems that no proof of it has yet been published. We shall now discuss a series of propositions which lead up to this theorem; the analysis which will be used resembles, in its main features, the analysis*, due to Riemann, which is applicable to trigonometrical series.

19·5. *Theorems concerning the convergence of Schlömilch series.*

We shall now discuss the special type of Schlömilch series in which $\nu = 0$, and in which Struve's functions do not appear; the object of taking this particular case is to avoid the loss of clearness due to the greater complication in the appearance of the formulae in the more general case. With a few exceptions, the complications in the general case are complications in detail only; those which are not matters of detail will be dealt with fully in §§ 19·6—19·62.

The series now to be considered is

$$\tfrac{1}{2}a_0 + \sum_{m=1}^{\infty} a_m J_0(mx), \tag{1}$$

in which the coefficients a_m are arbitrarily given functions of m.

We shall first prove the analogue of Cantor's lemma†, namely that *the condition that $a_m J_0(mx) \to 0$ as $m \to \infty$, at all points of* any *interval of values of x, is sufficient to ensure that*

$$a_m = o(\sqrt{m}).$$

[NOTE. If the origin is a point of the interval in question, then the theorem that

$$a_m = o(1)$$

is obviously true.]

Take any portion‡ of the interval which does not contain the origin, and let this portion be called I_1. Let the length of I_1 be L_1.

Throughout I_1 we have (cf. § 7·3)

$$a_m J_0(mx) = a_m \sqrt{\left(\frac{2}{m\pi x}\right)} \cdot [P(mx, 0)\cos(mx - \tfrac{1}{4}\pi) - Q(mx, 0)\sin(mx - \tfrac{1}{4}\pi)];$$

and, as $m \to \infty$,

$$P(mx, 0) \to 1, \quad Q(mx, 0) \to 0.$$

Hence, for all sufficiently large values of m, (say all values exceeding m_0)

$$P(mx, 0) \geqslant \tfrac{1}{2}, \quad |Q(mx, 0)| \leqslant \tfrac{1}{2}$$

at all points of I_1.

Now suppose that a_m is *not* $o(\sqrt{m})$; we have to shew that this hypothesis leads to a contradiction.

* Cf. *Modern Analysis*, §§ 9·6—9·632.

† *Ibid.* § 9·61.

‡ Since $J_0(mx)$ is an even function of x, the portion may be supposed to be on the right of the origin without loss of generality.

If a_m is not $o(\sqrt{m})$, a positive number ϵ must exist such that

$$|a_m| > \epsilon\sqrt{m}$$

whenever m is given any value belonging to a certain unending sequence* $m_1, m_2, m_3, \ldots$. Let the smallest member of this sequence which exceeds both m_0 and $2\pi/L_1$ be called m_1'.

Then $\cos(m_1' x - \frac{1}{4}\pi)$ goes through all its phases in I_1, and so there must be a portion† of I_1, say I_2, such that

$$|\cos(m_1' x - \tfrac{1}{4}\pi)| \geqslant \tfrac{1}{2}\sqrt{3}, \quad |\sin(m_1' x - \tfrac{1}{4}\pi)| < \tfrac{1}{2}$$

at all points of I_2. If L_2 is the length of I_2, then $L_2 = \frac{1}{3}\pi/m_1'$.

Next let the smallest member of the sequence m_r which exceeds both m_1' and $2\pi/L_2$ be called m_2'.

Then $\cos(m_2' x - \frac{1}{4}\pi)$ goes through all its phases in I_2, and so there must be a portion of I_2, say I_3, such that

$$|\cos(m_2' x - \tfrac{1}{4}\pi)| \geqslant \tfrac{1}{2}\sqrt{3}, \quad |\sin(m_2' x - \tfrac{1}{4}\pi)| \leqslant \tfrac{1}{2}$$

at all points of I_3. If L_3 is the length of I_3, then $L_3 = \frac{1}{3}\pi/m_2'$.

By continuing this process, we obtain a sequence of intervals $I_1, I_2, I_3, \ldots$ such that each is contained in its predecessor; there is therefore a point X which lies inside all these intervals, and at this point we have

$$|\cos(mX - \tfrac{1}{4}\pi)| \geqslant \tfrac{1}{2}\sqrt{3}, \quad |\sin(mX - \tfrac{1}{4}\pi)| \leqslant \tfrac{1}{2},$$

when m has any of the values $m_1', m_2', m_3', \ldots$.

For such values of m we consequently have

$$\begin{aligned} |a_m J_0(mX)| &\geqslant |a_m| \sqrt{\left(\frac{2}{m\pi X}\right)} \\ &\quad \times [P(mX, 0) . |\cos(mX - \tfrac{1}{4}\pi)| - |Q(mX, 0)| . |\sin(mX - \tfrac{1}{4}\pi)|] \\ &> \epsilon \frac{\sqrt{3} - 1}{4} \sqrt{\left(\frac{2}{\pi X}\right)}, \end{aligned}$$

and this is inconsistent with the hypothesis that $a_m J_0(mx)$ tends to zero at all points of I_1.

The contradiction which has now been obtained shews that a_m must be $o(\sqrt{m})$.

The next theorem which we shall prove is that, *if the Schlömilch series converges throughout any interval, then the necessary and sufficient condition*

* It is supposed that $m_1 < m_2 < m_3 < \ldots$.

† There are, in fact, at least two such portions of I_1; in order that I_2 may be uniquely determined, we take I_2 to be that portion which lies on the left of the others.

that the series should converge for any *positive value of x (whether a point of the interval or not) is that the series*

$$\sum_{m=1}^{\infty} a_m \sqrt{\left(\frac{2}{m\pi x}\right)} \cdot \left[\cos(mx - \tfrac{1}{4}\pi) + \frac{1}{8mx}\sin(mx - \tfrac{1}{4}\pi)\right]$$

should be convergent for that value of x.

This theorem is evident from the fact that the general term of the trigonometrical series differs from $a_m J_0(mx)$ by a function of m which is $O(a_m m^{-\frac{5}{2}}) = o(m^{-2})$; and $\Sigma o(m^{-2})$ is a convergent series.

19·51. *The associated function.*

Let the sum of the series

$$\tfrac{1}{2}a_0 + \sum_{m=1}^{\infty} a_m J_0(mx),$$

at any point at which the series is convergent, be called $f(x)$.

Let

$$\mathbf{F}(x) = \tfrac{1}{8}a_0 x^2 - \sum_{m=1}^{\infty} \frac{a_m J_0(mx)}{m^2}. \tag{1}$$

Then $\mathbf{F}(x)$ will be called the function associated with the Schlömilch series whose sum is $f(x)$.

It is easy to see that, *if the series defining $f(x)$ converges at all points of any interval, then the series defining $\mathbf{F}(x)$ converges for all real values of x.*

For $a_m J_0(mx) \to 0$ as $m \to \infty$ at all points of the interval, and therefore (§ 19·5)

$$a_m = o(\sqrt{m}).$$

Again, by § 2·5 (5), for all real values of x

$$|J_0(mx)| \leqslant 1;$$

and consequently

$$\frac{a_m J_0(mx)}{m^2} = o\left(\frac{1}{m^{\frac{3}{2}}}\right).$$

Since

$$\sum_{m=1}^{\infty} o\left(\frac{1}{m^{\frac{3}{2}}}\right)$$

is convergent, it is obvious that the series on the right in (1) must be convergent.

It is evident, moreover, not only that the convergence is absolute, but also that it is uniform throughout any domain of values of the real variable x.

19·52. *Lemma* I.

We shall now prove that, *if* $\mathbf{F}(x)$ *is the function associated with the Schlömilch series whose sum is* $f(x)$, *and if*

(1) $$\mathbf{G}(x,\alpha)=\frac{(x+\alpha)\mathbf{F}(x+2\alpha)+(x-\alpha)\mathbf{F}(x-2\alpha)-2x\mathbf{F}(x)}{4\alpha^2},$$

then

(2) $$\lim_{\alpha\to 0}\mathbf{G}(x,\alpha)=xf(x)$$

at any point x *at which the series defining* $f(x)$ *is convergent, provided that**

$$a_m=o(\sqrt{m}).$$

It is easy to deduce from (1) that

$$\mathbf{G}(x,\alpha)=\tfrac{1}{2}a_0x-\sum_{m=1}^{\infty}\frac{a_m}{4\alpha^2m^2}$$
$$\times[(x+\alpha)J_0(mx+2m\alpha)+(x-\alpha)J_0(mx-2m\alpha)-2xJ_0(mx)];$$

and, from l'Hospital's theorem, it follows that

$$\lim_{\alpha\to 0}\frac{1}{4m^2\alpha^2}[(x+\alpha)J_0(mx+2m\alpha)+(x-\alpha)J_0(mx-2m\alpha)-2xJ_0(mx)]$$
$$=\lim_{\alpha\to 0}\frac{1}{8m^2\alpha}[J_0(mx+2m\alpha)-J_0(mx-2m\alpha)$$
$$+2m(x+\alpha)J_0'(mx+2m\alpha)-2m(x-\alpha)J_0'(mx-2m\alpha)]$$
$$=xJ_0''(mx)+J_0'(mx)/m$$
$$=-xJ_0(mx).$$

Consequently the limits of the individual terms of the series defining $\mathbf{G}(x,\alpha)$ are the individual terms of the series defining $xf(x)$.

It is therefore sufficient to prove that the series for $\mathbf{G}(x,\alpha)$ converges uniformly with respect to α in an interval including the point $\alpha=0$ when x has any value such that the series for $f(x)$ is convergent.

It may be assumed, without loss of generality, that x is *positive*†, and we shall then take $|\alpha|$ so small that it does not exceed $\frac{1}{4}x$; we shall now prove that the series for $\mathbf{G}(x,\alpha)$ converges uniformly when $-\frac{1}{4}x\leqslant\alpha\leqslant\frac{1}{4}x$.

By observing that

$$x\pm\alpha-\sqrt{\{x(x\pm 2\alpha)\}}=\frac{\alpha^2}{x\pm\alpha+\sqrt{\{x(x\pm 2\alpha)\}}}$$
$$<\tfrac{3}{4}\alpha^2/x,$$

and that the series

$$\sum_{m=1}^{\infty}\frac{a_m}{m^2}\frac{J_0(mx\pm 2m\alpha)}{[x\pm\alpha+\sqrt{\{x(x\pm 2\alpha)\}}]}$$

* Since we are not assuming more than the convergence of $f(x)$ *at a single point*, it is not permissible to infer from § 19·5 that a_m must be $o(\sqrt{m})$.

† The functions under consideration are even functions of x; and since $\mathbf{G}(0,\alpha)=0$, the special case in which $x=0$ needs no further consideration.

is uniformly convergent (upper or lower signs throughout being taken), we see that $\mathbf{G}(x, \alpha)$ differs from

$$\tfrac{1}{2}a_0 x - \sum_{m=1}^{\infty} \frac{a_m \sqrt{x}}{4m^2\alpha^2}$$
$$[\sqrt{(x+2\alpha)}\,.\,J_0(mx+2m\alpha) + \sqrt{(x-2\alpha)}\,.\,J_0(mx-2m\alpha) - 2\sqrt{x}\,.\,J_0(mx)]$$

by the sum of two series, each of which is uniformly convergent.

It is therefore sufficient to establish the uniformity of the convergence of the last series which has been written down.

Now take the general term of this series, namely

$$-\frac{a_m\sqrt{x}}{4m^2\alpha^2}[\sqrt{(x+2\alpha)}\,.\,J_0(mx+2m\alpha) + \sqrt{(x-2\alpha)}\,.\,J_0(mx-2m\alpha) - 2\sqrt{x}\,.\,J_0(mx)],$$

and write it in the form

$$a_m\sqrt{\left(\frac{2x}{m\pi}\right)}\left[\cos(mx-\tfrac{1}{4}\pi) + \frac{\sin(mx-\tfrac{1}{4}\pi)}{8mx}\right]\left(\frac{\sin m\alpha}{m\alpha}\right)^2$$
$$+ a_m\sqrt{\left(\frac{2x}{m\pi}\right)}\,.\,\frac{\cos(mx-\tfrac{1}{4}\pi)}{8m(x^2-4\alpha^2)}\,.\,\frac{\sin 2m\alpha}{m^2\alpha}$$
$$- a_m\sqrt{\left(\frac{2x}{m\pi}\right)}\,.\,\frac{\sin(mx-\tfrac{1}{4}\pi)}{8mx(x^2-4\alpha^2)}\,.\,\frac{\cos 2m\alpha}{m^2}$$
$$+ a_m\sqrt{\left(\frac{2x}{m\pi}\right)}\,.\left[\frac{2\Phi(mx) - \Phi(mx+2m\alpha) - \Phi(mx-2m\alpha)}{4m^2\alpha^2}\right],$$

where $\Phi(y)$ is defined by the formula

$$\Phi(y) = \{P(y,0) - 1\}\cos(y-\tfrac{1}{4}\pi) - \left\{\frac{1}{8y} + Q(y,0)\right\}\sin(y-\tfrac{1}{4}\pi).$$

The general term is thus expressed as the sum of four terms, and we proceed to prove that each of the four series, of which these terms are the general terms, is uniformly convergent.

The first two series are proved to be uniformly convergent, in connexion with the theory of trigonometrical series*; and the third is obviously uniformly convergent from the test of Weierstrass.

To deal with the fourth series, we observe that, by the first mean-value theorem, numbers† θ and θ_1 exist such that

$$-1 < \theta < 1, \quad -1 < \theta_1 < 1,$$

* Cf. *Modern Analysis*, §§ 9·62, 9·621. It has been the general (but not invariable) custom to obtain various properties of the series without establishing the uniformity of their convergence. The *convergence* of the series for $f(x)$ is required to deal with the first series; the second series can be dealt with in consequence of the less stringent hypothesis that $a_m = o(\sqrt{m})$.

† The number θ is a function of a variable t which will be introduced immediately.

for which

$$\begin{aligned} &2\Phi(mx) - \Phi(mx + 2m\alpha) - \Phi(mx - 2m\alpha) \\ &= 2m\alpha \int_0^1 \{\Phi'(mx - 2m\alpha t) - \Phi'(mx + 2m\alpha t)\}\, dt \\ &= -2m\alpha \int_0^1 4m\alpha t\, \Phi''(mx - 2m\alpha\theta t)\, dt \\ &= -4m^2\alpha^2\, \Phi''(mx - 2m\alpha\theta_1). \end{aligned}$$

Since $\Phi''(y) = O(1/y^2)$ when y is large, it is evident that

$$\sum_{m=1}^{\infty} \Phi''(mx - 2m\alpha\theta_1)$$

is uniformly convergent with respect to α.

Hence $\mathbf{G}(x, \alpha)$ is expressed as the sum of six series each of which converges uniformly with respect to α when $-\frac{1}{4}x < \alpha < \frac{1}{4}x$; and therefore

$$\lim_{\alpha\to 0} \mathbf{G}(x, \alpha)$$

is equal to the sum of the limits of the terms of the series for $\mathbf{G}(x, \alpha)$, i.e. it is equal to $xf(x)$, provided that the series for $f(x)$ is convergent; and this is the lemma to be proved.

19·53. *Lemma II.*

We shall next prove that, *with the notation of* §§ 19·51, 19·52, *the condition that* $a_m = o(\sqrt{m})$ *is sufficient to ensure that*

$$\lim_{\alpha\to 0} \frac{(x+\alpha)\mathbf{F}(x+2\alpha) + (x-\alpha)\mathbf{F}(x-2\alpha) - 2x\mathbf{F}(x)}{\alpha} = 0,$$

for all values of x.

As in § 19·52, we need consider positive values of x only; and we express the series for $\alpha\,\mathbf{G}(x, \alpha)$ as the sum of six series each of which is easily seen to be uniformly convergent when $-\frac{1}{4}x < \alpha < \frac{1}{4}x$, by applying the theorems concerning trigonometrical series which were used in § 19·52.

Hence

$$\begin{aligned} &\lim_{\alpha\to 0} [\alpha\, \mathbf{G}(x, \alpha)] \\ &= \lim_{\alpha\to 0} (\tfrac{1}{2}a_0\alpha) - \sum_{m=1}^{\infty} \lim_{\alpha\to 0} \frac{a_m}{4m^2\alpha} [(x+\alpha) J_0(mx + 2m\alpha) \\ &\qquad + (x-\alpha) J_0(mx - 2m\alpha) - 2x J_0(mx)] \\ &= 0, \end{aligned}$$

and this is the lemma to be proved.

19·54. *The analogue of Riemann's theorem* on trigonometrical series.*

We can now prove that, if two Schlömilch series of the type now under consideration (i.e. with $\nu = 0$, and with Struve's function absent) converge

* Cf. *Modern Analysis*, § 9·63.

and have the same sum-function throughout the interval $(0, \pi)$, then corresponding coefficients in the two series are equal. The formal statement of the theorem is as follows:

Two Schlömilch series, of the special type, which converge and are equal at all points of the closed interval $(0, \pi)$, *with the possible exception of a finite number of points, must have corresponding coefficients equal, unless the end-points* 0 *and* π *are both exceptional points.*

If these points are exceptional points, the two series may differ by a constant multiple of the series

$$\tfrac{1}{2} + \sum_{m=1}^{\infty} (-)^m J_0(mx).$$

Let the difference of the two series be

$$\tfrac{1}{2}a_0 + \sum_{m=1}^{\infty} a_m J_0(mx),$$

and let the sum of this series be $f(x)$, so that $f(x)$ converges to zero for all values of x between 0 and π, except the exceptional values.

Let ξ_1, ξ_2 be any points (except the origin) of the interval $(0, \pi)$, such that there are no exceptional points inside* the interval (ξ_1, ξ_2).

We proceed to prove that, if $\mathbf{F}(x)$ is the function associated with the Schlömilch series for $f(x)$, then $\mathbf{F}(x)$ is a linear function of $\log x$ in the interval (ξ_1, ξ_2). This is the analogue of Schwarz' lemma†.

If $\theta = 1$, or if $\theta = -1$, and if

$$\phi(x) = \theta\left[\mathbf{F}(x) - \mathbf{F}(\xi_1) - \frac{\log(x/\xi_1)}{\log(\xi_2/\xi_1)}\{\mathbf{F}(\xi_2) - \mathbf{F}(\xi_1)\}\right] + h^2\left[x - \xi_1 - \frac{\log(x/\xi_1)}{\log(\xi_2/\xi_1)}(\xi_2 - \xi_1)\right],$$

then $\phi(x)$ is continuous when $\xi_1 \leqslant x \leqslant \xi_2$, and

$$\phi(\xi_1) = \phi(\xi_2) = 0.$$

If the first term of $\phi(x)$ is not zero‡ throughout the interval (ξ_1, ξ_2), there will be some point c at which it is not zero. Choose the sign of θ so that the first term of $\phi(c)$ is positive at c, and then choose h so small that $\phi(c)$ is still positive.

Since $\phi(x)$ is continuous in (ξ_1, ξ_2), it attains its upper bound which is positive since $\phi(c)$ is positive. Let it attain its upper bound at c_1, so that $\xi_1 < c_1 < \xi_2$.

Now by Lemma I (§ 19·52)

$$\lim_{\alpha \to 0} \frac{(c_1 + \alpha)\phi(c_1 + 2\alpha) + (c_1 - \alpha)\phi(c_1 - 2\alpha) - 2c_1\phi(c_1)}{4\alpha^2} = h^2.$$

* The points ξ_1, ξ_2 themselves may be exceptional points.

† Cf. *Modern Analysis*, § 9·631.

‡ If it *is* zero throughout (ξ_1, ξ_2), then $\mathbf{F}(x)$ is obviously a linear function of $\log x$.

But $\phi(c_1+2\alpha) \leqslant \phi(c_1)$, $\phi(c_1-2\alpha) \leqslant \phi(c_1)$, so the limit on the left must be negative or zero. This contradiction shews that the first term of $\phi(x)$ must be zero throughout (ξ_1, ξ_2), that is to say that $\mathbf{F}(x)$ must be a linear function of $\log x$; and this is the theorem to be proved.

Hence the curve whose equation is $y = \mathbf{F}(x)$ consists of a set of segments of logarithmic curves with equations of the type

$$y = A \log x + B.$$

Now, by § 19·51, $\mathbf{F}(x)$ is continuous in $(0, \pi)$, and so these logarithmic curves are connected at the exceptional points; and the curve $y = \mathbf{F}(x)$ cannot have an abrupt change of direction at an exceptional point, because, by Lemma II,

$$\lim_{\alpha \to 0} \left[(\xi+\alpha) \frac{\mathbf{F}(\xi+2\alpha) - \mathbf{F}(\xi)}{2\alpha} - (\xi-\alpha) \frac{\mathbf{F}(\xi) - \mathbf{F}(\xi-2\alpha)}{2\alpha} \right] = 0,$$

even when ξ is an exceptional point; that is to say

$$\xi \mathbf{F}'(\xi+0) = \xi \mathbf{F}'(\xi-0).$$

Hence the constants A and B cannot be discontinuous at the exceptional points, and so they have the same values for all values of x in the interval $(0, \pi)$.

Consequently, when $0 < x < \pi$,

$$\tfrac{1}{8} a_0 x^2 - \sum_{m=1}^{\infty} \frac{a_m J_0(mx)}{m^2} = A \log x + B.$$

Make $x \to 0$; the series on the left has a limit, namely

$$-\sum_{m=1}^{\infty} \frac{a_m}{m^2},$$

because it is uniformly convergent. Therefore $A \log x + B$ has a limit when $x \to 0$, and so A is zero.

Consequently, when $0 \leqslant x \leqslant \pi$,

$$\sum_{m=1}^{\infty} \frac{a_m J_0(mx)}{m^2} = \tfrac{1}{8} a_0 x^2 - B;$$

and the series on the left converges uniformly throughout $(0, \pi)$, so integrations term-by-term are permissible.

Replace x by $x \sin\theta$, multiply by $\sin\theta$, and integrate from 0 to $\tfrac{1}{2}\pi$. Then, by § 12·11,

$$\begin{aligned} \tfrac{1}{12} a_0 x^2 - B &= \sum_{m=1}^{\infty} \frac{a_m}{m^2} \int_0^{\frac{1}{2}\pi} J_0(mx \sin\theta) \sin\theta \, d\theta \\ &= \sum_{m=1}^{\infty} \frac{a_m \sin mx}{m^3 x}. \end{aligned}$$

Hence, when $0 \leqslant x \leqslant \pi$,

$$\sum_{m=1}^{\infty} \frac{a_m \sin mx}{m^3} = \tfrac{1}{12} a_0 x^3 - Bx.$$

Multiply by $\sin mx$ and integrate from 0 to π; it is then evident that

$$\frac{\pi a_m}{2m^3} = (-)^m \left[\frac{\pi a_0}{2m^3} + \frac{\pi B - \frac{1}{12}\pi^3 a_0}{m}\right].$$

Since a_m is given to be $o(\sqrt{m})$, this equation shews that

$$B = \tfrac{1}{12}\pi^2 a_0, \quad a_m = (-)^m a_0.$$

Hence we must have

$$f(x) = a_0 \left[\tfrac{1}{2} + \sum_{m=1}^{\infty} (-)^m J_0(mx)\right].$$

From the results contained in § 19·41 concerning the behaviour of the series on the right at $x=0$ and at $x=\pi$, it is evident that $f(x)$ cannot be a convergent Schlömilch series at either point unless a_0 is zero; and this proves the theorem stated at the beginning of this section.

19·6. *Theorems concerning the convergence of generalised Schlömilch series.*

We shall now study briefly the series

$$\frac{\frac{1}{2}a_0}{\Gamma(\nu+1)} + \sum_{m=1}^{\infty} \frac{a_m J_\nu(mx) + b_m \mathbf{H}_\nu(mx)}{(\frac{1}{2}mx)^\nu}. \tag{1}$$

We shall first prove that, *when* $\nu < \frac{1}{2}$, *the condition that the* $(m+1)$*th term of the series tends to zero as* $m \to \infty$ *at all points of* any *interval of values of* x is sufficient *to ensure that*

$$a_m = o(m^{\nu+\frac{1}{2}}), \quad b_m = o(m^{\nu+\frac{1}{2}}).$$

[NOTE. If the origin is a point of the interval in question, then the theorem that

$$a_m = o(1)$$

is obviously true.]

Since the series under consideration is unaffected by a change in the sign of x if the signs of all the coefficients b_m are also changed, no generality is lost by considering an interval on the right of the origin.

We call this interval I_1; and, at all points of I_1, we have, by § 10·41 (4),

$$\frac{a_m J_\nu(mx) + b_m \mathbf{H}_\nu(mx)}{(\frac{1}{2}mx)^\nu} = \frac{c_m}{(\frac{1}{2}mx)^{\nu+\frac{1}{2}}\sqrt{\pi}}[P(mx,\nu)\cos(mx - \tfrac{1}{2}\nu\pi - \tfrac{1}{4}\pi - \eta_m)$$
$$- Q(mx,\nu)\sin(mx - \tfrac{1}{2}\nu\pi - \tfrac{1}{4}\pi - \eta_m)] + b_m \,.\, O(m^{-1}),$$

where $a_m = c_m \cos \eta_m$, $b_m = c_m \sin \eta_m$.

We now suppose that a_m and b_m are *not* both $o(m^{\nu+\frac{1}{2}})$; we have to shew that this hypothesis leads to a contradiction.

If a_m and b_m are not both $o(m^{\nu+\frac{1}{2}})$, a positive number ϵ must exist such that

$$c_m > \epsilon m^{\nu+\frac{1}{2}}$$

whenever m is given any value belonging to a certain unending sequence $m_1, m_2, m_3, \ldots$.

We now prove, exactly as in § 19·5, that, at some point X of I_1, the inequalities

$$P(mX, \nu) \geqslant \tfrac{1}{2}, \quad |Q(mX, \nu)| \leqslant \tfrac{1}{2},$$

$$|\cos(mX - \tfrac{1}{2}\nu\pi - \tfrac{1}{4}\pi - \eta_m)| \geqslant \tfrac{1}{2}\sqrt{3}, \quad |\sin(mX - \tfrac{1}{2}\nu\pi - \tfrac{1}{4}\pi - \eta_m)| \leqslant \tfrac{1}{2}$$

are satisfied whenever m has any value belonging to a sequence (m_r') which is a sub-sequence of the sequence (m_r).

For values of m which belong to this sub-sequence we have

$$\left|\frac{a_m J_\nu(mX) + b_m \mathbf{H}_\nu(mX)}{(\frac{1}{2}mX)^\nu}\right| \geqslant \frac{\frac{1}{4}\epsilon(\sqrt{3}-1)}{(\frac{1}{2}X)^{\nu+\frac{1}{2}}\sqrt{\pi}} - \epsilon \,.\, O(m^{\nu-\frac{1}{2}})$$

and, since $\nu - \frac{1}{2}$ is negative, the expression on the right cannot be arbitrarily small. This is the contradiction which is sufficient to prove that a_m and b_m must both be $o(m^{\nu+\frac{1}{2}})$ if the $(m+1)$th term of the Schlömilch series tends to zero at all points of I_1.

The reader may now prove (as in § 19·5) that, *when* $\nu < \frac{1}{2}$, *if the generalised Schlömilch series converges throughout any interval, the necessary and sufficient condition that it may converge for any positive value of* x *(whether a point of the interval or not) is that the series*

$$\sum_{m=1}^{\infty}\left[\frac{c_m}{m^\nu}\sqrt{\left(\frac{2}{m\pi x}\right)}\left\{\cos(mx - \tfrac{1}{2}\nu\pi - \tfrac{1}{4}\pi - \eta_m) \right.\right.$$
$$\left.\left. - \frac{4\nu^2 - 1}{8mx}\sin(mx - \tfrac{1}{2}\nu\pi - \tfrac{1}{4}\pi - \eta_m)\right\} + \frac{b_m}{m^\nu} \cdot \frac{(\frac{1}{2}mx)^{\nu-1}}{\Gamma(\nu+\frac{1}{2})\Gamma(\frac{1}{2})}\right]$$

should be convergent for that value of x.

19·61. *The associated function.*

Let us take $-\frac{1}{2} < \nu < \frac{1}{2}$, and let the sum of the series

$$\frac{\frac{1}{2}a_0}{\Gamma(\nu+1)} + \sum_{m=1}^{\infty} \frac{a_m J_\nu(mx) + b_m \mathbf{H}_\nu(mx)}{(\frac{1}{2}mx)^\nu}$$

at any point at which the series is convergent be called $f_\nu(x)$.

Let

(1) $$\mathbf{F}_\nu(x) = \frac{a_0 x^2}{8\Gamma(\nu+2)} - \sum_{m=1}^{\infty} \frac{a_m J_\nu(mx) + b_m \mathbf{H}_\nu(mx)}{m^2 \,.\, (\frac{1}{2}mx)^\nu}.$$

Then $\mathbf{F}_\nu(x)$ will be called the function associated with the Schlömilch series whose sum is $f_\nu(x)$.

It is easy to prove that, *if the series defining* $f_\nu(x)$ *converges at all points of any interval, then the series defining* $\mathbf{F}_\nu(x)$ *converges for all real values of* x.

The only respect in which the proof differs from the analysis used in § 19·51 is that the additional theorem that $\mathbf{H}_\nu(x)/x^\nu$ is a bounded function of the real variable x has to be used.

Again, let

(2) $$\mathbf{G}_\nu(x, \alpha) = \tfrac{1}{4}[(x + 2\nu\alpha + \alpha)\mathbf{F}_\nu(x + 2\alpha) + (x - 2\nu\alpha - \alpha)\mathbf{F}_\nu(x - 2\alpha) - 2x\,\mathbf{F}_\nu(x)]/\alpha^2.$$

Then, just as in § 19·52, we may prove that*

$$\lim_{\alpha\to 0} \mathbf{G}_\nu(x, \alpha) = x f_\nu(x) - \frac{2}{\Gamma(\nu+\frac{1}{2})\,\Gamma(\frac{1}{2})} \sum_{m=1}^{\infty} \frac{b_m}{m}$$

at any point x at which the series defining $f_\nu(x)$ is convergent, provided that a_m and b_m are both $o(m^{\nu+\frac{1}{2}})$ and that the series $\sum_{m=1}^{\infty} b_m/m$ is convergent.

Further we may prove that

$$\lim_{\alpha\to 0} [\alpha\,\mathbf{G}_\nu(x, \alpha)] = 0,$$

provided only that a_m and b_m are both $o(m^{\nu+\frac{1}{2}})$, whether the series $\sum_{m=1}^{\infty} b_m/m$ is convergent or not.

19·62. *The analogue of Riemann's theorem.*

We can now prove that, *if two generalised Schlömilch series of the same order ν (where $-\frac{1}{2} < \nu < \frac{1}{2}$) converge and have the same sum-function at all points of the closed interval $(-\pi, \pi)$ with the possible exception of a finite number of points (it is supposed that the origin and the points $\pm\pi$ are not all exceptional points), and if the coefficients of the terms containing Struve's functions in the two series are sufficiently nearly equal† each to each, then all corresponding coefficients in the two series are equal.*

Let the difference of the two series be

$$\frac{\frac{1}{2}a_0}{\Gamma(\nu+1)} + \sum_{m=1}^{\infty} \frac{a_m J_\nu(mx) + b_m \mathbf{H}_\nu(mx)}{(\frac{1}{2}mx)^\nu}$$

and let the sum of this series be $f_\nu(x)$, so that the series for $f_\nu(x)$ converges to zero at all points of the interval $(-\pi, \pi)$ with a finite number of exceptions.

The convergence of the series for $f_\nu(x)$ nearly everywhere in the interval $(-\pi, \pi)$ necessitates the equations

$$a_m = o(m^{\nu+\frac{1}{2}}), \quad b_m = o(m^{\nu+\frac{1}{2}}).$$

The statement that the coefficients of the terms containing Struve's functions in the two series are to be sufficiently nearly equal is to be interpreted to mean that $b_m \to 0$ as $m \to \infty$ in such a way that $\sum_{m=1}^{\infty} \frac{b_m}{m}$ is convergent.

We now discuss the function $\mathbf{F}_\nu(x)$ associated with the Schlömilch series for $f_\nu(x)$. It can be proved‡ that if the interval (ξ_1, ξ_2) is such that the origin

* The presence of the series on the right is due to the lack of homogeneity in the differential equation satisfied by Struve's function.

† This statement will be made definite immediately.

‡ It seems unnecessary to repeat the arguments already used in § 19·54.

and the exceptional points (if any) are not internal points of the interval, then $\mathbf{F}_\nu(x)$ is a linear function* of $x^{-2\nu}$ in the interval. It may then be shewn that the exceptional points do not cause any discontinuity in the form of $\mathbf{F}_\nu(x)$, and hence we deduce that

$$\begin{cases} \mathbf{F}_\nu(x) = Ax^{-2\nu} + B, & (0 < x \leqslant \pi) \\ \mathbf{F}_\nu(x) = A' \,|\, x \,|^{-2\nu} + B', & (0 > x \geqslant -\pi) \end{cases}$$

where A, B, A', B' are constants.

Now take the equation

$$\mathbf{F}_\nu(x) = \frac{a_0 x^2}{8\Gamma(\nu+2)} - \sum_{m=1}^{\infty} \frac{a_m J_\nu(mx) + b_m \mathbf{H}_\nu(mx)}{m^2 \,.\, (\tfrac{1}{2}mx)^\nu},$$

replace x by $x \sin\theta$, multiply by $\sin^{2\nu+1}\theta/\cos^{2\nu}\theta$, (which has an absolutely convergent integral) and integrate from 0 to $\frac{1}{2}\pi$. The series for $\mathbf{F}_\nu(x\sin\theta)$ converges uniformly in this interval of values of θ, so term-by-term integrations are permissible.

It is thus found that

$$\int_0^{\frac{1}{2}\pi} \mathbf{F}_\nu(x\sin\theta) \frac{\sin^{2\nu+1}\theta}{\cos^{2\nu}\theta}\, d\theta = \frac{a_0 x^2}{8\Gamma(\nu+2)} \int_0^{\frac{1}{2}\pi} \frac{\sin^{2\nu+3}\theta}{\cos^{2\nu}\theta}\, d\theta$$

$$- \sum_{m=1}^{\infty} \int_0^{\frac{1}{2}\pi} \frac{a_m J_\nu(mx\sin\theta) + b_m \mathbf{H}_\nu(mx\sin\theta)}{m^2 \,.\, (\tfrac{1}{2}mx)^\nu} \cdot \frac{\sin^{\nu+1}\theta}{\cos^{2\nu}\theta}\, d\theta$$

$$= \frac{a_0 x^2 \,\Gamma(\tfrac{1}{2}-\nu)}{12\Gamma(\tfrac{1}{2})} - \frac{\Gamma(\tfrac{1}{2}-\nu)}{\Gamma(\tfrac{1}{2})} \sum_{m=1}^{\infty} \frac{a_m \sin mx + b_m(1-\cos mx)}{m^3 x}.$$

When we substitute for $\mathbf{F}_\nu(x\sin\theta)$ we deduce that

$$(1) \qquad \sum_{m=1}^{\infty} \frac{a_m \sin mx + b_m(1-\cos mx)}{m^3} = \frac{a_0 x^3}{12} - \frac{Ax^{1-2\nu}\,\Gamma(\tfrac{3}{2})}{\Gamma(\tfrac{3}{2}-\nu)} - Bx\,\Gamma(\nu+1),$$

when $0 \leqslant x \leqslant \pi$; and a similar equation may be obtained when $0 \geqslant x \geqslant -\pi$.

Since a_m and b_m are both $o(m^{\nu+\frac{1}{2}})$, it is permissible to differentiate (1) twice term-by-term when $0 > \nu > -\frac{1}{2}$; but it may only be differentiated once if $0 \leqslant \nu < \frac{1}{2}$.

If we differentiate, twice or once as the case may be, the resulting series on the left tends to a limit as $x \to 0$, but the resulting expression on the right fails to do so *unless A is zero*.

We infer that $A = 0$, and in like manner A' must be zero; the continuity of $\mathbf{F}_\nu(x)$ at the origin then shews that B and B' must be equal.

It now follows from (1) that

$$(2) \qquad \sum_{m=1}^{\infty} \frac{a_m \sin mx + b_m(1-\cos mx)}{m^3} = \frac{a_0 x^3}{12} - Bx\,\Gamma(\nu+1)$$

when $-\pi \leqslant x \leqslant \pi$.

* When ν is zero $x^{-2\nu}$ has to be replaced by $\log x$.

Multiply (2) by $\cos mx$ and integrate from $-\pi$ to π; and then

$$b_m = 0. \tag{3}$$

Again, multiply by $\sin mx$ and integrate; and then

$$(-)^m a_m = a_0 + m^2 \{2B\Gamma(\nu+1) - \tfrac{1}{6} a_0 \pi^2\}.$$

This equation is inconsistent with the fact that a_m is $o(m^{\nu+\frac{1}{2}})$ unless

$$2B\Gamma(\nu+1) = \tfrac{1}{6} a_0 \pi^2,$$

and then

$$a_m = (-)^m a_0.$$

Hence the series for $f_\nu(x)$ must reduce to

$$a_0 \left[\frac{\frac{1}{2}}{\Gamma(\nu+1)} + \sum_{m=1}^{\infty} \frac{(-)^m J_\nu(mx)}{(\frac{1}{2}mx)^\nu}\right].$$

Now at least one of the points 0, π, $-\pi$ is not an exceptional point; and the series for $f_\nu(x)$ cannot converge at that point unless a_0 is zero, so that a_m is also zero.

We have therefore proved that, if the series $\Sigma\, b_m/m$ is convergent *all the coefficients* a_m *and* b_m *must vanish*; that is to say, the two Schlömilch series with which we started must have corresponding coefficients equal. And this is the theorem to be proved.

We have therefore established for Schlömilch series in which $-\frac{1}{2} < \nu < \frac{1}{2}$ theorems analogous to the usual theorems concerning the representation of null-functions by trigonometrical series.

19·7. *Theorems of Riemann's type concerning series of Bessel functions and Dini's series of Bessel functions.*

We shall now give a very brief sketch of the method by which the series discussed in Chapter XVIII, namely

$$\sum_{m=1}^{\infty} a_m J_\nu(j_m x), \qquad \sum_{m=1}^{\infty} b_m J_\nu(\lambda_m x),$$

(in which $\nu > -\frac{1}{2}$) may be investigated after the manner of Riemann's investigation of trigonometrical series.

The method is identical with the method of investigation of Schlömilch series just given in §§ 19·6—19·62, though there are various points of detail*, which do not arise in the case of Schlömilch series, due to the fact that j_m and λ_m are not linear functions of m.

* These points of detail are very numerous and there is no special difficulty in discussing any of them; but it is a tedious and lengthy process to set them out in full, and they do not seem to be of sufficient importance to justify the use of the space which they would require. The reader who desires to appreciate the details necessary in such investigations may consult the papers by C. N. Moore, *Trans. American Math. Soc.* x. (1909), pp. 391—435; xii. (1911), pp. 181—206; xxi. (1920), pp. 107—156.

In the first place, it is easy to prove by the method used in § 19·5 that if the series

$$\sum_{m=1}^{\infty} a_m J_\nu(j_m x), \quad \sum_{m=1}^{\infty} b_m J_\nu(\lambda_m x)$$

converge throughout any interval of values of x, then

$$a_m = o(\sqrt{m}), \quad b_m = o(\sqrt{m}).$$

Next we consider the associated function; we write

$$f(x) = \sum_{m=1}^{\infty} a_m J_\nu(j_m x),$$

and then the function associated with $f(x)$ is defined by the equation

$$F(x) = \sum_{m=1}^{\infty} \frac{a_m}{j_m^2} \frac{J_\nu(j_m x)}{x^\nu}.$$

It may be proved that, when x has any positive value for which the series defining $f(x)$ is convergent, and if the expression

$$\frac{1}{4\alpha^2}[(x+2\nu\alpha+\alpha)F(x+2\alpha) - 2x\,F(x) + (x-2\nu\alpha-\alpha)\,F(x-2\alpha)]$$

is arranged as a series in which the mth term has a_m for a factor, then the latter series is uniformly convergent with respect to α in an interval containing the point $\alpha = 0$, and that its limit when $\alpha \to 0$ is $-x^{1-\nu} f(x)$.

It may also be proved that, whether the series for $f(x)$ converges or not, the condition that $a_m = o(\sqrt{m})$ is sufficient to ensure that

$$\frac{1}{4\alpha}[(x+2\nu\alpha+\alpha)F(x+2\alpha) - 2x\,F(x) + (x-2\nu\alpha-\alpha)\,F(x-2\alpha)]$$

tends to zero with α.

The proofs of these theorems depend on a number of lemmas such as the lemma* that

$$\sum_{m=1}^{\infty} \left| \frac{\sin^2 j_{m+1}\alpha}{j_{m+1}^2\alpha^2} - \frac{\sin^2 j_m\alpha}{j_m^2\alpha^2} \right|$$

is a bounded function of α; proofs of the lemmas can be constructed on the lines of the proofs in the special (trigonometrical) case in which $\nu = \frac{1}{2}$.

It now follows in the usual manner (cf. § 19·54) that, when $f(x)$ is a null-function throughout the interval $(0, 1)$, then $F(x)$ satisfies the differential equation

$$x\frac{d^2F(x)}{dx^2} + (2\nu+1)\frac{dF(x)}{dx} = 0,$$

and so†

$$F(x) = A + Bx^{-2\nu},$$

where A and B are constants. This equation is valid when $0 < x \leqslant 1$.

* Cf. *Modern Analysis*, § 9·62.

† When $\nu = 0$, $F(x) = A + B' \log x$.

Now since $\nu > -\frac{1}{2}$, $J_\nu(j_m x)/(j_m x)^\nu$ is bounded when $0 \leqslant x \leqslant 1$ whatever be the value of m; and so, when* $\nu < \frac{1}{2}$, the series

$$\sum_{m=1}^{\infty} \frac{a_m}{j_m^{2-\nu}} \frac{J_\nu(j_m x)}{(j_m x)^\nu}$$

converges uniformly when $0 \leqslant x \leqslant 1$, by the test of Weierstrass.

Hence $F(x)$ is a continuous function of x in the closed interval and so B is zero when ν is positive; and B' is zero in the case $\nu = 0$.

For any assigned value of n multiplying the series for $F(x)$ by $x^{\nu+1} J_\nu(j_n x)$ does not destroy the uniformity of its convergence; and, when we integrate, we find that

$$a_n \,.\, J_\nu'^2(j_n) = j_n^2 \int_0^1 (Ax^\nu + Bx^{-\nu})\, x\, J_\nu(j_n x)\, dx$$

$$= j_n \left[B . \frac{(\frac{1}{2} j_n)^{\nu-1}}{\Gamma(\nu)} - (A + B)\, J_\nu'(j_n) \right].$$

Now, when n is large,

$$|J_\nu'(j_n)| \sim \sqrt{\left(\frac{2}{\pi j_n}\right)},$$

and so the formula just obtained for a_n is inconsistent with the equation $a_n = o(\sqrt{n})$ when $\nu > -\frac{1}{2}$ unless both $A + B$ and B are zero†; and then a_n is zero.

Hence a series of Bessel functions (in which $\nu > -\frac{1}{2}$) cannot converge to the sum zero at all points of the interval $(0, 1)$, with the possible exception of a finite number of points (the origin not being an exceptional point‡ when $\nu > \frac{1}{2}$), unless all the coefficients in the series are zero.

We infer that two series of Bessel functions, in which $\nu > -\frac{1}{2}$, cannot converge and be equal at all points of the interval $(0, 1)$, with the possible exception of a finite number of points, unless corresponding coefficients in the two series are equal.

Dini's series§

$$f(x) \equiv \sum_{m=1}^{\infty} b_m J_\nu(\lambda_m x)$$

may be dealt with in the same manner. The associated function is defined by the equation

$$F(x) = \sum_{m=1}^{\infty} \frac{b_m}{\lambda_m^2} \frac{J_\nu(\lambda_m x)}{x^\nu},$$

* When $\nu \geqslant \frac{1}{2}$, the convergence of the series for $f(x)/x^\nu$ at $x = 0$ is sufficient to ensure the uniformity of the convergence.

† An exception might occur when $\nu - 1 = -\frac{1}{2}$; but this is the trigonometrical case.

‡ The series divided by x^ν then has to converge at the origin.

§ It is supposed for the present that $H + \nu > 0$, so that no initial term is inserted.

and it is inferred that when $f(x)$ is a null-function throughout (0, 1), then constants A and B exist such that

$$F(x) = A + Bx^{-2\nu},$$

$$b_n\left[\left(1-\frac{\nu^2}{\lambda_n^2}\right)J_\nu^2(\lambda_n)+J_\nu'^2(\lambda_n)\right] = \lambda_n^2\int_0^1 (Ax^\nu + Bx^{-\nu})\,xJ_\nu(\lambda_n x)\,dx$$

$$= \lambda_n\left[AJ_{\nu+1}(\lambda_n) + B\left\{\frac{(\frac{1}{2}\lambda_n)^{\nu-1}}{\Gamma(\nu)} - J_{\nu-1}(\lambda_n)\right\}\right];$$

and $B=0$ when $\nu \geqslant 0$.

Now, when n is large

$$\left(1-\frac{\nu^2}{\lambda_n^2}\right)J_\nu^2(\lambda_n)+J_\nu'^2(\lambda_n) \sim \frac{2}{\pi\lambda_n},$$

$$J_{\nu+1}(\lambda_n) = O(\lambda_n^{-\frac{1}{2}}),\quad J_{\nu-1}(\lambda_n) = O(\lambda_n^{-\frac{1}{2}}),$$

so that, if $B \neq 0$,

$$b_n \sim \frac{\pi B\lambda_n^{\nu+1}}{2^\nu\Gamma(\nu)},$$

and this is inconsistent with the equation

$$b_n = o(\sqrt{n}),$$

since $\nu > -\frac{1}{2}$.

Hence B is zero, and therefore

$$b_n\left[\left(1-\frac{\nu^2}{\lambda_n^2}\right)J_\nu^2(\lambda_n)+J_\nu'^2(\lambda_n)\right] = A\lambda_n J_{\nu+1}(\lambda_n).$$

This equation is inconsistent with the equation

$$b_n = o(\sqrt{n})$$

unless A is zero, since $J_{\nu+1}(\lambda_n)$ is not zero; and then b_n is zero.

We next consider what happens when $H+\nu$ is zero or negative; in these cases Dini's series assume the forms

$$b_0 x^\nu + \sum_{m=1}^\infty b_m J_\nu(\lambda_m x),$$

$$b_0 I_\nu(\lambda_0 x) + \sum_{m=1}^\infty b_m J_\nu(\lambda_m x),$$

respectively.

In the second of the two cases the previous arguments are unaffected by the insertion of an initial term; the first of the two cases needs more careful consideration because the initial term to be inserted in the associated function is

$$-\frac{b_0 x^2}{4(\nu+1)},$$

and hence, when $n \geqslant 1$,

$$b_n J_\nu^2(\lambda_n) = \lambda_n^2\int_0^1\left\{Ax^\nu + \frac{b_0 x^{\nu+2}}{4(\nu+1)} + Bx^{-\nu}\right\}xJ_\nu(\lambda_n x)\,dx$$

$$= 2B\nu\left\{\frac{(\frac{1}{2}\lambda_n)^\nu}{\Gamma(\nu+1)} - J_\nu(\lambda_n)\right\} + \frac{b_0 J_\nu(\lambda_n)}{2(\nu+1)}.$$

Since $b_n = o(\sqrt{n})$ we infer first that $B = 0$, by considering the term in $(\frac{1}{2}\lambda_n)^\nu$, and then that $b_0 = 0$; and so $b_n = 0$ for all values of n.

We infer also that, as in the limiting case of series of Bessel functions, Dini's series of Bessel functions cannot represent a null-function throughout the interval (0, 1), and that if two of Dini's series (with the same ν and H) converge and are equal at all points of the interval (0, 1), with the exception of a finite number of points, then corresponding coefficients in the two series are equal.

CHAPTER XX

THE TABULATION OF BESSEL FUNCTIONS

20·1. *Tables of Bessel Functions and associated functions.*

It is evident from a consideration of the analysis contained in Chapters VII, VIII and XV that a large part of the theory of Bessel Functions has been constructed expressly for the purpose of facilitating numerical computations connected with the functions. To the Mathematician such computations are of less interest and importance* than the construction of the theories which make them possible; but to the Physicist numerical results have a significance† which formulae may fail to convey.

As an application of various portions of the Theory of Bessel Functions, it has been considered desirable to insert this Chapter, which contains an historical account of Tables of Bessel Functions which have been previously published, together with a collection of those tables which seem to be of the greatest value for the present requirements of the Physicist.

The reader will not be concerned with the monotony and technical irrelevance of this Chapter when he realises that it can be read without the efforts required to master the previous chapters and to amplify arguments so ruthlessly condensed.

The first Tables of $J_0(x)$ and $J_1(x)$ were published by Bessel himself in his memoir on Planetary Perturbations, *Berliner Abhandlungen*, 1824 [1826], pp. 46—52. These tables give the values of $J_0(x)$ and $J_1(x)$ to ten places of decimals for a range of values of x from $x = 0$ to $x = 3{\cdot}20$ with interval 0·01.

A short Table of $J_0(x)$ and $J_0^2(x)$ to four places of decimals was constructed by Airy, *Phil. Mag.* (3) XVIII. (1841), p. 7; its range is from $x = 0$ to $x = 10{\cdot}0$ with interval 0·2. Airy‡ had previously constructed a Table of $2J_1(x)/x$, of the same scope.

The function $J_1(x)/x$ was subsequently tabulated to six places of decimals by Lommel, *Zeitschrift für Math. und Phys.* XV. (1870), pp. 164—167, with a range from $x = 0$ to $x = 20{\cdot}0$ with interval 0·1; this Table, with a Table of $J_1^2(x)/x^2$, was republished by Lommel, *Münchener Abhandlungen*, XV. (1886), pp. 312—315.

* Cf. Love, *Proc. London Math. Soc.* (2) XIV. (1915), p. 184.

† Cf. Lord Kelvin's statement "I have no satisfaction in formulas unless I feel their arithmetical magnitude—at all events when formulas are intended for definite dynamical or physical problems." *Baltimore Lectures* (Cambridge, 1904), p. 76.

‡ *Trans. Camb. Phil. Soc.* v. (1835), p 291. A Table of $2J_1(x)/x$ and its square, to four or five places of decimals, in which the range is from 0 to the circular measure of 1125° (with interval 15°), was given by Schwerd, *Die Beugungserscheinungen* (Mannheim, 1835), p. 146.

In consequence of the need of Tables of $J_n(x)$ with fairly large values of n and x for Astronomical purposes, Hansen constructed a Table of $J_0(x)$ and $J_1(x)$ to six places of decimals with a range from $x=0$ to $x=10{\cdot}0$ with interval 0·1; this was published in his *Ermittelung der absoluten Störungen in Ellipsen von beliebiger Excentricität und Neigung* (Gotha, 1843). Hansen's Table was reprinted by Schlömilch* and also by Lommel† who extended it to $x=20$.

These tables, however, are superseded by Meissel's great Table of $J_0(x)$ and $J_1(x)$ to twelve places of decimals‡, published in the *Berliner Abhandlungen*, 1888; its range is from $x=0$ to $x=15{\cdot}50$ with interval 0·01.

Meissel's Table was reprinted in full by Gray and Mathews, *A Treatise on Bessel Functions* (London, 1895), pp. 247—266, and an abridgement of it is given in Table I *infra*, pp. 666—697.

A Table of $J_0(x)$ and $J_1(x)$ to twenty-one places of decimals, from $x=0$ to $x=6{\cdot}0$ with interval 0·1, has been constructed by Aldis, *Proc. Royal Soc.* LXVI. (1900), p. 40.

A Table of $J_0(n\pi)$ to six places of decimals for $n=1, 2, 3, \ldots, 50$ has been computed by Nagaoka, *Journal of the Coll. of Sci. Imp. Univ. Japan*, IV. (1891), p. 313.

The value of $J_0(40)$ was computed by W. R. Hamilton from the ascending series, *Phil. Mag.* (4) XIV. (1857), p. 375.

A Table of $J_1(x)$ to six places of decimals from $x=20{\cdot}1$ to $x=41$ with interval 0·1 or 0·2 has been published by Steiner, *Math. und Naturwiss. Berichte aus Ungarn*, XI. (1894), pp. 372—373.

The earliest table of functions of the second kind was constructed by B. A. Smith, *Messenger*, XXVI. (1896), pp. 98—101; this is a Table to four places of decimals of Neumann's functions $Y^{(0)}(x)$ and $Y^{(1)}(x)$. Its range is from $x=0$ to $x=1{\cdot}00$ with interval 0·01 and from $x=1{\cdot}0$ to $x=10{\cdot}2$ with interval 0·1.

A more extensive table of these functions is given in the *British Association Report*, 1914, pp. 76—82; this is a Table to six places of decimals whose range is from $x=0$ to $x=15{\cdot}50$ with interval 0·02; a year later a table was published, *ibid.* 1915, p. 33, in which the values of $Y^{(0)}(x)$ and $Y^{(1)}(x)$ were given to ten places of decimals for a range from $x=0$ to $x=6{\cdot}0$ with interval 0·2 and from $x=6{\cdot}0$ to $x=16{\cdot}0$ with interval 0·5.

Shortly after the appearance of Smith's Table, an elaborate table was constructed by Aldis, *Proc. Royal Soc.* LXVI. (1900), p. 41, of Heine's functions§ $G_0(x)$ and $G_1(x)$ to twenty-one places of decimals; the reader should be

* *Zeitschrift für Math. und Phys.* II. (1857), pp. 158—165.

† *Studien über die Bessel'schen Functionen* (Leipzig, 1868), pp. 127—135.

‡ Meissel's Table contains a misprint, the correct value of $J_0(0{\cdot}62)$ being $+0{\cdot}90618\ldots$, not $+0{\cdot}90518\ldots$. An additional misprint was made in the reprint of the Table by Gray and Mathews.

§ These functions were also tabulated by B. A. Smith, *Phil. Mag.* (5), XLV. (1898), pp. 122—123; the scope of this table is the same as that of his Table of $Y^{(0)}(x)$ and $Y^{(1)}(x)$.

reminded that these functions are equal to $-\frac{1}{2}\pi Y_0(x)$ and $-\frac{1}{2}\pi Y_1(x)$ respectively. The range of Aldis' Table is from $x=0$ to $x=6{\cdot}0$ with interval 0·1.

Another table of these functions with a smaller interval was published in the *British Association Report*, 1913, pp. 116—130; this table gives the functions to seven places of decimals for a range from $x=0$ to $x=16{\cdot}00$ with interval 0·01. The *Report* for 1915, p. 33, contains a table to ten places of decimals from $x=6{\cdot}5$ to $x=15{\cdot}5$ with interval 0·5.

The functions $Y_0(x)$ and $Y_1(x)$ are tabulated to seven places of decimals in Table I *infra*; this table has an appreciable advantage over the British Association Tables*, in that the auxiliary tables make interpolation a trivial matter; in the British Association Tables interpolation is impracticable.

By means of the recurrence formulae combined with the use of the tables which have now been described, it is an easy matter to construct tables of functions whose order is any integer. Such tables of $J_n(x)$ were constructed by Hansen and reprinted by Schlömilch and Lommel after their Tables of $J_0(x)$ and $J_1(x)$. Subsequently Lommel, *Münchener Abhandlungen*, XV. (1886), pp. 315—316, published a Table of $J_n(x)$ to six places of decimals, in which $n=0, 1, 2, \dots, 20$, and $x=0, 1, 2, \dots, 12$; this Table is reprinted in Table IV *infra*, pp. 730—731. A Table of $J_n(x)$ of practically the same scope was also published by Meissel, *Astr. Nach.* CXXVIII. (1891), col. 154—155.

A much more extensive Table of $J_n(x)$ was computed by Meissel, but it seems that he never published it. He communicated it to Gray and Mathews for publication in their *Treatise*, pp. 267—279. This table gives $J_n(x)$ to eighteen places of decimals when $n=0, 1, 2, \dots, 60$, and $x=0, 1, 2, \dots, 24$.

Some graphs of $J_n(x)$ were constructed, with the help of the last-mentioned table, by Hague, *Proc. Phys. Soc.* XXIX. (1917), pp. 211—214.

The corresponding Tables of functions of the second kind are not so extensive.

The *British Association Report*, 1914, pp. 83—86 contains Tables of $G_n(x)$ to five places of decimals for† $n=0, 1, 2, \dots, 13$ for the range $x=0$ to 6·0 with interval 0·1 and $x=6{\cdot}0$ to 16·0 with interval 0·5.

Similar Tables† of $Y^{(n)}(x)$ to six places of decimals (with the intervals in the earlier part equal to 0·2) appeared in the *Report* for 1914, pp. 34—36.

Some values of Hankel's function $\mathbf{Y}_n(x)$ had been given previously by Nicholson, *Proc. London Math. Soc.* (2) XI. (1913), pp. 113—114.

A Table of $Y_n(x)$ to seven (or more) significant figures is contained in Table IV *infra*. This has been computed from Aldis' Table of $G_0(x)$ and $G_1(x)$.

* In the course of computing Table I, a small part of the British Association Table of $G_0(x)$ and $G_1(x)$ was checked, and the last digits in it were found to be unreliable in about 5 % of the entries checked.

† For the larger values of n the functions are not tabulated for small values of x.

Tables of $\log_{10}[\sqrt{(\frac{1}{2}\pi x)}\,.\,|H_\nu^{(1)}(x)|]$ to eight significant figures are given in the *British Association Report*, 1907, pp. 94—97. The values assigned to ν are $0, \frac{1}{2}, 1, 1\frac{1}{2}, \ldots, 6\frac{1}{2}$, and the range of values of x is from $x=10$ to 100 (interval 10) and 100 to 1000 (interval 100). For this range of values of x, the asymptotic expansion (§ 7·51) gives so rapid an approximation that the Table is of less value than a table in which the values of x and the intervals are considerably smaller.

Functions of the first kind with imaginary argument have been tabulated in the *British Association Reports*. The *Report* for 1896, pp. 99—149, contained a Table of $I_0(x)$ to nine places of decimals, its range being from $x=0$ to $x=5{\cdot}100$ with interval 0·001. A Table of $I_1(x)$ of the same scope had been published previously in the *Report* for 1893, pp. 229—279; an abridgement of this (with interval 0·01) was given by Gray and Mathews in their *Treatise*, pp. 282—284.

Tables of $I_0(x)$ and $I_1(x)$ to twenty-one places of decimals have been constructed by Aldis, *Proc. Royal Soc.* LXIV. (1899), p. 218. The range of these Tables is $x=0$ to $x=6{\cdot}0$ with interval 0·1; Aldis also gave (*ibid.* p. 221) the values of $I_0(x)$ and $I_1(x)$ for $x=7, 8, 9, 10, 11$.

Extensive tables connected with $I_0(x)$ and $I_1(x)$ have been published by Anding, *Sechsstellige Tafeln der Bessel'schen Funktionen imaginären Arguments* (Leipzig, 1911). These tables give $\log_{10} I_0(x)$ and $\log_{10}\{I_1(x)/x\}$ from $x=0$ to $x=10{\cdot}00$ with interval 0·01. They also give the values of the functions

$$\sqrt{(2\pi x)}\,.\,e^{-x} I_0(x), \quad \sqrt{(2\pi x)}\,.\,e^{-x} I_1(x), \quad \log_{10}\{\sqrt{x}\,.\,I_0(x)\} \text{ and } \log_{10}\{\sqrt{x}\,.\,I_1(x)\}$$

for values of x from $x=10{\cdot}0$ to $x=50{\cdot}0$ (interval 0·1), $x=50$ to $x=200$ (interval 1), $x=200$ to $x=1000$ (interval 10), and for various larger values of x.

Table II *infra*, pp. 698—713, gives the values of $e^{-x}I_0(x)$ and $e^{-x}I_1(x)$; these have been computed, for the most part, by interpolation in Aldis' Table.

The earliest tables of functions of the type $K_n(x)$ were constructed by Aldis, *Proc. Royal Soc.* LXIV. (1899), pp. 219—221. These give $K_0(x)$ and $K_1(x)$ to twenty-one places of decimals for values of x from $x=0$ to $x=6{\cdot}0$ with interval 0·1, and also to between seven and thirteen significant figures from $x=5{\cdot}0$ to $x=12{\cdot}0$ with interval 0·1.

The values of $e^x K_0(x)$ and $e^x K_1(x)$ in Table II *infra* were computed with the help of Aldis' Table, like the values of $e^{-x}I_0(x)$ and $e^{-x}I_1(x)$.

By means of recurrence formulae, $I_n(x)$ has been tabulated to twelve significant figures for $n=0, 1, 2, \ldots, 11$ over the range of values of x from $x=0$ to $x=6{\cdot}0$ with interval 0·2. These Tables of $I_n(x)$ were published in the *British Association Report*, 1889, pp. 29—32, and reprinted by Gray and Mathews in their *Treatise*, pp. 285—288. An abridgement (to five significant figures) of these Tables has been given by Isherwood, who added to them

Tables of $K_n(x)$ to five significant figures for $n = 0, 1, 2, \ldots, 10$ over the range of values of x from $x = 0$ to $x = 6{\cdot}0$ with interval 0·2. Isherwood's Tables were published in the *Mem. and Proc., Manchester Lit. and Phil. Soc.*, 1903—1904, no. 19.

Tables of $e^{-x} I_n(x)$ and $K_n(x)$ to seven places of decimals are given in Table IV *infra*, pp. 736—739.

The earliest Tables of Bessel functions of large order were constructed by Meissel, who has calculated $J_{2n}(n)$ to twelve significant figures for $n = 10, 11, \ldots, 21$, *Astr. Nach.* CXXIX. (1892), col. 284; Meissel also calculated $J_n(1000)$ to seven significant figures for $n = 1000, 999, \ldots, 981$, *ibid.* CXXVIII. (1891), col. 154—155. The values of $J_n(n)$, $J_{n-1}(n)$, $Y^{(n)}(n)$, $Y^{(n-1)}(n)$, $G_n(n)$, $G_{n-1}(n)$ to six places of decimals for values of n from $n = 1$ to $n = 50$ (interval 1), $n = 50$ to $n = 100$ (interval 5), $n = 100$ to $n = 200$ (interval 10), $n = 200$ to $n = 400$ (interval 20), $n = 400$ to $n = 1000$ (interval 50), $n = 1000$ to $n = 2000$ (interval 100) and for various larger values of n, are given in the *British Association Report*, 1916, pp. 93—96.

Tables of $J_n(n)$, $J_n'(n)$, $Y_n(n)$, $Y_n'(n)$ to seven places of decimals are given in Table VI *infra*, pp. 746—747.

The functions ber (x), bei (x), ker (x) and kei (x) have been extensively tabulated on account of their importance in the theory of alternating currents.

A brief Table of ber (x) and bei (x), computed by Maclean, was published by Kelvin, *Math. and Phys. Papers*, III. (1890), p. 493. Tables of $J_0(x\sqrt{i})$ and $\sqrt{2} . J_1(x\sqrt{i})$ to twenty-one places of decimals have been constructed by Aldis, *Proc. Royal Soc.* LXVI. (1900), pp. 42—43; their range is from $x = 0$ to $x = 6{\cdot}0$ with interval 0·1. These are extensions of the Table of $J_0(x\sqrt{i})$ to nine places of decimals for the range from $x = 0$ to $x = 6{\cdot}0$ with interval 0·2 published in the *British Association Report*, 1893, p. 228, and reprinted by Gray and Mathews in their *Treatise*, p. 281.

Tables of ber (x), bei (x), ker (x) and kei (x) to four significant figures for $x = 1, 2, 3, \ldots, 30$, have been published by Savidge, *Phil. Mag.* (6) XIX. (1910), p. 53.

The functions ber (x), bei (x), ber′ (x) and bei′ (x) are tabulated to nine places of decimals, from $x = 0$ to $x = 10{\cdot}0$ with interval 0·1 in the *British Association Report*, 1912, pp. 57—68; and a Table of ker (x), kei (x), ker′ (x) and kei′ (x) of the same scope (except that only six or seven significant figures were given) appeared in the *Report* for 1915, pp. 36—38. Tables of squares and products of the functions to six significant figures from $x = 0$ to $x = 10{\cdot}0$ with interval 0·2 were given in the *Report* for 1916, pp. 118—121.

The functions $J_{\pm(n+\frac{1}{2})}(x)$ have been tabulated to six places of decimals by Lommel, *Münchener Abh.* XV. (1886), pp. 644—647, for $n = 0, 1, 2, \ldots, 6$ with $x = 1, 2, \ldots, 50$, and (in the case of functions of positive order) $n = 7, 8, \ldots, 14$

with $x=1, 2, \ldots, 20$, and $n=15, \ldots, 34$ with smaller ranges of values of x; see Table V *infra*, pp. 740—743. A Table of the same functions to four places of decimals with $n=0, 1, 2$ and from $x=0$ to $x=8{\cdot}0$ with interval 0·2 is given by Dinnik, *Archiv der Math. und Phys.* (3) XX. (1913), pp. 238—240.

Functions related to $J_{\pm(n+\frac{1}{2})}(x)$ have recently been tabulated in the *British Association Reports.* The notation used is

$$\sqrt{(\tfrac{1}{2}\pi x)} \, . \, J_{n+\frac{1}{2}}(x) = S_n(x), \quad (-)^n \sqrt{(\tfrac{1}{2}\pi x)} \, . \, J_{-n-\frac{1}{2}}(x) = C_n(x),$$
$$E_n(x) = |C_n(x) + i\, S_n(x)|,$$

and the functions tabulated are $S_n(x)$, $C_n(x)$, $E_n^2(x)$, $S_n'(x)$, $C_n'(x)$, $E_n'^2(x)$, and their logarithms. In the *Report* for 1914, pp. 88—102, the functions are tabulated to seven significant figures for $n=0, 1, 2, \ldots, 17$ and $x=1, 2, 3, \ldots, 10$, and in the *Report* for 1916, pp. 97—107, for $n=0, 1, 2, \ldots, 10$ and $x=1{\cdot}1$, $1{\cdot}2, \ldots, 1{\cdot}9$.

Functions of order $\pm\frac{1}{3}$, $\pm\frac{2}{3}$, have been tabulated by Dinnik, *Archiv der Math. und Phys.* (3) XVIII. (1911), pp. 337—338, to four places of decimals; the functions tabulated are $\Gamma(1\pm\frac{1}{3})\, J_{\pm\frac{1}{3}}(x)$, $\Gamma(1\pm\frac{2}{3})\, J_{\pm\frac{2}{3}}(x)$ from $x=0$ to $x=8{\cdot}0$ with interval 0·2; and Dinnik has also tabulated $I_{\pm\frac{1}{3}}(x)$, $I_{\pm\frac{2}{3}}(x)$, *ibid.* (3) XXII. (1914), pp. 226—227 and $J_{\pm\frac{1}{4}}(x)$, $J_{\pm\frac{3}{4}}(x)$, *ibid.* (3) XXI. (1913), pp. 324—326. All these tables have the range $x=0$ to $x=8{\cdot}0$ with interval 0·2. The Tables of $\Gamma(1\pm\frac{1}{3})\, J_{\pm\frac{1}{3}}(x)$ are less extensive than Table III *infra*, pp. 714—729; but, with the exception of Dinnik's tables, there exist no tables of functions of orders $\frac{2}{3}$, $\frac{1}{4}$ and $\frac{3}{4}$.

In connexion with functions of order $\pm\frac{1}{3}$, Airy's Table of his integral

$$\int_0^\infty \cos \tfrac{1}{2}\pi\,(w^3 - mw)\, dw$$

must be mentioned; Airy calculated by quadratures and by ascending series the values of this integral for values of m from $-5{\cdot}6$ to $+5{\cdot}6$ with interval 0·2; a seven-figure Table from $m=-4$ to $m=4$ is given in the *Trans. Camb. Phil. Soc.* VI. (1838), p. 402, and a five-figure Table from $m=-5{\cdot}6$ to $m=5{\cdot}6$, *ibid.* VIII. (1849), p. 599.

Apart from the work of Euler described in § 15·5 the earliest computation of the zeros of $J_0(x)$ and $J_1(x)$ is due to Stokes, *Trans. Camb. Phil. Soc.* IX. (1856), p. 186 [*Math. and Phys. Papers*, II. (1883), p. 355]. Stokes gave the values of the first twelve zeros (divided by π) of $J_0(x)$ and $J_1(x)$ to four places of decimals. In the same memoir he gave the first fifty zeros of Airy's integral, and the first ten stationary points of this integral.

The first nine zeros of $J_0(x)$, $J_1(x)$, ..., $J_5(x)$ were computed by Bourget, *Ann. sci. de l'École norm. sup.* III. (1866), pp. 82—87. Bourget's results are given to three places of decimals; some corrections in his Tables have recently been made by Airey*.

* *Phil. Mag.* (6) XXXII. (1916), pp. 7—14.

Bourget's Tables have been reprinted so frequently that their authorship has been overlooked by the writers of the articles on Bessel Functions in the *Encyclopädie der Math. Wiss.* and the *Encyclopédie des Sci. Math.*

The first five zeros of $J_1(x)$ and $J_2(x)$ were given to six places of decimals by Lommel, *Zeitschrift für Math. und. Phys.* XV. (1870), p. 167 and *Münchener Abhandlungen*, XV. (1886), p. 315.

The first ten zeros of $J_0(x)$ were computed to ten places of decimals by Meissel, *Berliner Abhandlungen*, 1888.

The first fifty zeros (and their logarithms) of $J_0(x)$ were given to ten places of decimals by Willson and Peirce, *Bulletin American Math. Soc.* III. (1897), pp. 153—155; they also gave the values of $J_1(x)$ and $\log|J_1(x)|$ at these zeros to eight and seven places of decimals respectively.

The first fifty zeros of $J_1(x)$ and the corresponding values of $J_0(x)$ were computed to sixteen places of decimals by Meissel*, *Kiel Programm*, 1890; this Table is reprinted by Gray and Mathews in their *Treatise*, p. 280.

Tables of roots of the equation

$$J_n(x)\,Y_n(kx) - J_n(kx)\,Y_n(x) = 0$$

have been constructed by Kalähne, *Zeitschrift für Math. und Phys.* LIV. (1907), pp. 55—86; the values taken for k are 1·2, 1·5 and 2·0, while n is given the values 0, $\frac{1}{2}$, 1, $\frac{3}{2}$, 2, $\frac{5}{2}$.

Dinnik in his Tables of functions of fractional order mentions the values of a few of the zeros of each function, while Airey, *Phil. Mag.* (6) XLI. (1921), pp. 200—205, has computed the value of the smallest zero of $J_\nu(x)$ for small fractional values of ν by Euler's method.

Rayleigh, *Proc. London Math. Soc.* X. (1878), pp. 6—7 [*Scientific Papers*, I. (1899), pp. 363—364], has calculated that

$$(1 - x^2)\,xI_1(x)/I_0(x)$$

has a maximum when $x^2 = 0{\cdot}4858$.

Airey, *Archiv der Math. und Phys.* (3) XX. (1913), p. 291, has computed the first ten zeros of $3xJ_0(x) - 2J_1(x)$ and of $2xJ_0(x) - J_1(x)$ to four places of decimals.

In his memoirs on Diffraction, *Münchener Abhandlungen*, XV. (1886), Lommel has published tables connected with his functions of two variables, but these tables are so numerous that a detailed account of them will not be given here. His Table of Fresnel's integrals (p. 648) to six places of decimals from $x = 0$ to $x = 50{\cdot}0$ with interval 0·5 (with auxiliary tables for purposes of interpolation) must, however, be mentioned, and with it his Table of the first sixteen maxima and minima of these integrals.

* *Jahrbuch über die Fortschritte der Math.* 1890, p. 521. In consequence of the inaccessibility of Meissel's table, the zeros of $J_1(x)$ were recomputed (to ten places of decimals) for insertion in Table VII, p. 748.

Lommel's form for Fresnel's integrals was

$$\tfrac{1}{2}\int_0^x J_{\pm\frac{1}{2}}(t)\,dt\,;$$

a different form was tabulated earlier by Lindstedt, *Ann. der Physik und Chemie,* (3) XVII. (1882), p. 725.

Defining the functions $M(x)$ and $N(x)$ by the equations

$$\int_x^\infty \cos t^2 dt = M(x)\cos x^2 - N(x)\sin x^2,$$

$$\int_x^\infty \sin t^2 dt = M(x)\sin x^2 + N(x)\cos x^2,$$

and writing $x = \{(y+\tfrac{1}{2})\pi\}^{\frac{1}{2}}$, Lindstedt tabulated $M(x)$ and $N(x)$ to six places of decimals from $y = 0$ to $y = 9{\cdot}0$ with interval 0·1.

The function $I(x)$ defined as

$$\frac{1}{\pi}\int_x^\infty \frac{\mathbf{H}_1(2t)}{t^2}\,dt$$

has been tabulated to four places of decimals by Struve, *Ann. der Physik und Chemie,* (3) XVII. (1882), pp. 1008—1016, from $x = 0$ to 4·0 (interval 0·1), from $x = 4{\cdot}0$ to 7·0 (interval 0·2) and from $x = 7{\cdot}0$ to 11·0 (interval 0·4).

A table of values of the integral

$$\int \frac{J_1^{\,2}(x)}{x}\,dx,$$

in which the limits are consecutive zeros (up to the forty-ninth) of $J_1(x)$, has been published by Steiner, *Math. und Naturwiss. Berichte aus Ungarn* XI. (1894), pp. 366—367; this integral occurs in the problem of Diffraction by a Circular Aperture.

No Tables of Struve's functions seem to have been constructed before the Table of $\mathbf{H}_0(x)$ and $\mathbf{H}_1(x)$ which is given on pp. 666—697.

20·2. *Description of the Tables contained in this book.*

Preliminary considerations on the magnitude and character of the tables to be included in this book led to the following decisions:

(I) That space did not usually admit of the inclusion of more than seven places of decimals in the tables.

(II) That the tables should be so constructed as to minimise the difficulty of making interpolations. In particular, it was decided that a table with a moderately large interval (such as 0·02), together with an auxiliary table to facilitate interpolation, would be more useful than a table with a smaller interval (such as 0·01), occupying the same space as the first table and its auxiliary, in which interpolation was impracticable.

(III) That in computing tables, calculations should be carried to ten places of decimals in order to ensure that the number of cases of inaccuracy in the last figure of the published results should be trivial*. This does not apply to the auxiliary tables of angles which are entered in Tables I and III.

In order to obtain seven-figure accuracy, it is not sufficient to tabulate to tenths of a second of arc, because the differences per minute of arc in a seven-figure table of natural sines may be as large as 0·0002909; on the other hand, an error of a hundredth of a second does not affect the value of the sine by more than 0·00000005. Hence, for seven-figure accuracy, it was considered adequate to compute to nine places of decimals the sines (or cosines) of the angles tabulated and then to compute the angles from Gifford's *Natural Sines* (Manchester, 1914); these are eight-figure tables with an interval† of 1″.

The angles tabulated may consequently frequently be in error as to the last digit, but, in all probability, the error never exceeds a unit (i.e. a hundredth of a second of arc).

We now proceed to describe the tables in detail.

Table I consists primarily of Tables of $J_0(x)$, $Y_0(x)$, $J_1(x)$ and $Y_1(x)$ from $x=0$ to 16·00 with interval of 0·02. The values of $J_0(x)$ and $J_1(x)$ up to 15·50 are taken from Meissel's Table‡, while the values of $Y_0(x)$ and $Y_1(x)$ were computed partly by interpolation in Aldis' Table of $G_0(x)$ and $G_1(x)$ and partly from the asymptotic expansions of $J_0^2(x)+Y_0^2(x)$ and $J_1^2(x)+Y_1^2(x)$ given in § 7·51.

The auxiliary tables§ give the values of $|H_n^{(1)}(x)|$ and $\arg H_n^{(1)}(x)$ for $n=0$ and $n=1$. In these tables the first differences are sufficiently steady (except for quite small values of x) to enable interpolations to be effected with but little trouble on the part of the reader; thus, when x is about 10 the second differences of $|H_0^{(1)}(x)|$ do not exceed 0·0000009.

The values‖ of $|H_n^{(1)}(x)|$ and $\arg H_n^{(1)}(x)$ can consequently be computed by the reader for *any* value of x less than 16, with the exception of quite small values. The corresponding values of $J_n(x)$ and $Y_n(x)$ can then be calculated immediately by the use of seven-figure logarithm tables.

* The tables were differenced before removing the last three figures, and it was found that the ten-figure results were rarely in error by more than a unit in the tenth place; so it is hoped that the number of errors in the last figure retained does not exceed about one in every thousand entries.

† No tables with a smaller interval have been published; the use of any tables with a larger interval and a greater number of decimal places would have very greatly increased the labour of constructing the auxiliary tables of angles, and the increased accuracy so obtained would be of no advantage to anyone using the auxiliary tables for purposes of interpolation.

‡ I must here express my cordial thanks to the Preussische Akademie der Wissenschaften zu Berlin for permitting me to make use of this Table.

§ The idea of constructing the auxiliary tables grew out of a conversation with Professor Love, in the course of which he remarked that it was frequently not realised how closely Bessel functions of any given order resemble circular functions multiplied by a damping factor in which the rate of decay is slow.

‖ The remarks immediately following of course presuppose that n is 0 or 1.

The relation between the various functions tabulated may be expressed most briefly by regarding $|H_n^{(1)}(x)|$ and arg $H_n^{(1)}(x)$ as the polar coordinates of a point in a plane; then the Cartesian coordinates of this point are $J_n(x)$ and $Y_n(x)$. Thus, from the entry for $x = 8{\cdot}00$,

$$+0{\cdot}1716508 = 0{\cdot}2818259 \cos 412^\circ\, 28'\, 40''{\cdot}60,$$

$$+0{\cdot}2235215 = 0{\cdot}2818259 \sin 412^\circ\, 28'\, 40''{\cdot}60.$$

Table I also contains the values of Struve's functions $\mathbf{H}_0(x)$ and $\mathbf{H}_1(x)$. These functions are included in Table I, instead of being contained in a separate Table, to facilitate interpolation; by § 10·41 (4), the difference $\mathbf{H}_n(x) - Y_n(x)$ is a positive monotonic function, and it varies sufficiently steadily for interpolation to be easy when x is not small.

The Tables of Struve's functions were computed by calculating the values of $\mathbf{H}_0(x)$ and $\mathbf{H}_0'(x)$ directly from the ascending series when $x = 1, 2, 3, \ldots$, and then calculating $\mathbf{H}_0''(x)$, $\mathbf{H}_0'''(x), \ldots$ for these values of x from the differential equation § 10·4 (10) and the equations obtained by differentiating it.

A few differential coefficients are adequate to calculate $\mathbf{H}_0(x)$ and $\mathbf{H}_0'(x)$ by Taylor's theorem for the values 0·5, 0·6, 0·7, ... of x. Interpolation to fiftieths of the unit is then effected by using Taylor's theorem in the same manner. This process, though it seems at first sight to be complicated and lengthy, is, in reality, an extremely rapid one when a machine* is used. It is very much more effective than the use of asymptotic expansions or the process suggested in the *British Association Report*, 1913, p. 116. As an example of the rapidity of the process, it may be stated that the values of $e^{-x} I_0(x)$ and $e^{-x} I_1(x)$ in Table II took less than a fortnight to compute; of course the time taken over this tabulation was appreciably shortened by the use of Aldis' Table as a framework for interpolation.

Table II consists of Tables of $e^{-x} I_0(x)$, $e^{-x} I_1(x)$, $e^x K_0(x)$, and $e^x K_1(x)$, and a Table of e^x is inserted, in case the reader should require the values of the functions† $I_0(x)$, $I_1(x)$, $K_0(x)$ and $K_1(x)$; the functions are tabulated from 0 to 16·00 with interval 0·02.

Interpolation by differencing is easy in the case of the first four functions throughout the greater part of the range.

The Table of e^x was constructed with the help of Newman's Table of e^{-x}, *Trans. Camb. Phil. Soc.* XIII. (1883), pp. 145—241. Unlike the other Tables in this book, the Table of e^x is given to eight significant figures‡, and care has been taken that the last digit given is accurate in every entry. Interpolation in this Table is, of course, effected by multiplying or dividing entries by exponentials of numbers not exceeding 0·01; such exponentials can be calculated without difficulty.

* The machine on which the calculations were carried out is a Marchant Calculating Machine, 10 × 9 recording to 18 figures.

† These functions were not tabulated because tables of them are unsuited for interpolation.

‡ Nine figures are given in parts of the Table to avoid spoiling its appearance.

Newman's Table gives e^{-x} to a large number of places of decimals, but the actual number of significant figures in the latter part of the Table is small; and less than half of the Table of e^x was constructed by the process of calculating reciprocals; the rest was constructed from Newman's Table by using the values of e^{13} and e^{16} given by Glaisher*, and the value of $e^{-15\cdot 6}$ given by Newman in a short table of e^{-x} with interval 0·1. These exponentials were employed because the tenth significant figures in all three and the eleventh significant figures in the first and third are zero.

Table III consists of Tables of $J_{\frac{1}{3}}(x)$, $Y_{\frac{1}{3}}(x)$, $|H_{\frac{1}{3}}^{(1)}(x)|$ and $|\arg H_{\frac{1}{3}}^{(1)}(x)|$ of the same scope as Table I, and interpolations are effected in the manner already explained. A Table of $e^x K_{\frac{1}{3}}(x)$ is also included. These Tables are of importance in dealing with approximations to Bessel functions of large order (§ 8·43), and also in the theory of Airy's integral.

The reader can easily compute values of $J_{-\frac{1}{3}}(x)$ from this table by means of the formula

$$J_{-\frac{1}{3}}(x) = |H_{\frac{1}{3}}^{(1)}(x)| \cos\{60° + \arg H_{\frac{1}{3}}^{(1)}(x)\}.$$

Table IV gives the values of $J_n(x)$, $Y_n(x)$, $e^{-x}I_n(x)$ and $K_n(x)$ for various values of x and n. The values of $J_n(x)$ are taken from Lommel's Table†, with some corrections, but the remainder of Table IV, with the exception of some values of $K_n(x)$ taken from Isherwood's Table†, is new; they have been constructed in part by means of Aldis' Tables of functions of orders zero and unity.

Table V is Lommel's Table† of $J_{\pm(n+\frac{1}{2})}(x)$ and Fresnel's integrals with some modifications and corrections.

Table VI gives the values of $J_n(n)$, $Y_n(n)$, $J_n'(n)$, $Y_n'(n)$ and $n^{\frac{1}{3}}J_n(n)$, $n^{\frac{1}{3}}Y_n(n)$, $n^{\frac{2}{3}}J_n'(n)$, $n^{\frac{2}{3}}Y_n'(n)$ for $n = 1, 2, 3, \ldots 50$. Interpolation in the tables of the last four of the eight functions is easy.

Table VII gives the first forty zeros of $J_n(x)$ and $Y_n(x)$ for various values of n; part of this Table is taken from the Tables of Willson and Peirce†. Forty zeros of various cylinder functions of order one-third are also given.

Table VIII gives the values of

$$\tfrac{1}{2}\int_0^x J_0(t)\,dt, \quad \tfrac{1}{2}\int_0^x Y_0(t)\,dt,$$

from $x = 0$ to 50 with interval 1, together with the first sixteen maxima and minima of the integrals. The former table of maxima and minima can be used to compute the coefficients (cf. § 18·12) in certain Fourier-Bessel series for which $\nu = 0$.

* *Trans. Camb. Phil. Soc.* XIII. (1883), p. 245.

† I must here express my cordial thanks to the Bayerische Akademie der Wissenschaften zu München, to the Manchester Literary and Philosophical Society, and to the American Mathematical Society for permitting me to make use of these Tables. The non-existence of adequate trigonometrical tables of angles in radian measure has made it impracticable to check the last digits in the entries in the greater part of Table V.

TABLES OF BESSEL FUNCTIONS

Table I. Functions of order zero

x	$J_0(x)$	$Y_0(x)$	$\lvert H_0^{(1)}(x)\rvert$	$\arg H_0^{(1)}(x)$	$\mathbf{H}_0(x)$	x
0	+ 1·0000000	$-\infty$	∞	− 90°	0·0000000	0
0·02	+ 0·9999000	− 2·5639554	2·7520297	− 68° 41′ 42″·16	+ 0·0127318	0·02
0·04	+ 0·9996000	− 2·1219006	2·3455622	− 64° 46′ 31″·73	+ 0·0254603	0·04
0·06	+ 0·9991002	− 1·8626264	2·1136647	− 61° 47′ 28″·29	+ 0·0381819	0·06
0·08	+ 0·9984006	− 1·6780254	1·9525811	− 59° 14′ 52″·70	+ 0·0508934	0·08
0·10	+ 0·9975016	− 1·5342387	1·8299993	− 56° 58′ 10″·79	+ 0·0635913	0·10
0·12	+ 0·9964032	− 1·4161969	1·7315984	− 54° 52′ 14″·42	+ 0·0762722	0·12
0·14	+ 0·9951060	− 1·3158701	1·6497727	− 52° 54′ 7″·85	+ 0·0889328	0·14
0·16	+ 0·9936102	− 1·2284710	1·5800007	− 51° 2′ 0″·32	+ 0·1015697	0·16
0·18	+ 0·9919164	− 1·1509166	1·5193772	− 49° 14′ 37″·05	+ 0·1141796	0·18
0·20	+ 0·9900250	− 1·0811053	1·4659257	− 47° 31′ 4″·89	+ 0·1267590	0·20
0·22	+ 0·9879366	− 1·0175412	1·4182414	− 45° 50′ 44″·63	+ 0·1393046	0·22
0·24	+ 0·9856518	− 0·9591221	1·3752907	− 44° 13′ 6″·41	+ 0·1518131	0·24
0·26	+ 0·9831713	− 0·9050133	1·3362914	− 42° 37′ 46″·91	+ 0·1642813	0·26
0·28	+ 0·9804958	− 0·8545676	1·3006375	− 41° 4′ 27″·57	+ 0·1767056	0·28
0·30	+ 0·9776262	− 0·8072736	1·2678500	− 39° 32′ 53″·32	+ 0·1890829	0·30
0·32	+ 0·9745634	− 0·7627204	1·2375444	− 38° 2′ 51″·54	+ 0·2014099	0·32
0·34	+ 0·9713081	− 0·7205732	1·2094070	− 36° 34′ 12″·42	+ 0·2136834	0·34
0·36	+ 0·9678615	− 0·6805558	1·1831788	− 35° 6′ 47″·25	+ 0·2258999	0·36
0·38	+ 0·9642245	− 0·6424376	1·1586436	− 33° 40′ 28″·14	+ 0·2380565	0·38
0·40	+ 0·9603982	− 0·6060246	1·1356190	− 32° 15′ 8″·97	+ 0·2501497	0·40
0·42	+ 0·9563838	− 0·5711520	1·1139500	− 30° 50′ 44″·27	+ 0·2621765	0·42
0·44	+ 0·9521825	− 0·5376789	1·0935036	− 29° 27′ 9″·25	+ 0·2741336	0·44
0·46	+ 0·9477955	− 0·5054836	1·0741648	− 28° 4′ 19″·71	+ 0·2860180	0·46
0·48	+ 0·9432242	− 0·4744608	1·0558337	− 26° 42′ 11″·96	+ 0·2978265	0·48
0·50	+ 0·9384698	− 0·4445187	1·0384231	− 25° 20′ 42″·72	+ 0·3095559	0·50
0·52	+ 0·9335339	− 0·4155768	1·0218560	− 23° 59′ 49″·07	+ 0·3212033	0·52
0·54	+ 0·9284179	− 0·3875642	1·0060645	− 22° 39′ 28″·42	+ 0·3327655	0·54
0·56	+ 0·9231233	− 0·3604182	0·9909884	− 21° 19′ 38″·41	+ 0·3442396	0·56
0·58	+ 0·9176518	− 0·3340833	0·9765738	− 20° 0′ 16″·94	+ 0·3556226	0·58
0·60	+ 0·9120049	− 0·3085099	0·9627727	− 18° 41′ 22″·10	+ 0·3669114	0·60
0·62	+ 0·9061843	− 0·2836537	0·9495417	− 17° 22′ 52″·16	+ 0·3781032	0·62
0·64	+ 0·9001918	− 0·2594751	0·9368418	− 16° 4′ 45″·55	+ 0·3891950	0·64
0·66	+ 0·8940292	− 0·2359383	0·9246378	− 14° 47′ 0″·82	+ 0·4001841	0·66
0·68	+ 0·8876982	− 0·2130113	0·9128976	− 13° 29′ 36″·67	+ 0·4110675	0·68
0·70	+ 0·8812009	− 0·1906649	0·9015920	− 12° 12′ 31″·88	+ 0·4218424	0·70
0·72	+ 0·8745391	− 0·1688729	0·8906945	− 10° 55′ 45″·35	+ 0·4325061	0·72
0·74	+ 0·8677147	− 0·1476114	0·8801807	− 9° 39′ 16″·05	+ 0·4430558	0·74
0·76	+ 0·8607300	− 0·1268587	0·8700283	− 8° 23′ 3″·05	+ 0·4534888	0·76
0·78	+ 0·8535868	− 0·1065950	0·8602168	− 7° 7′ 5″·48	+ 0·4638026	0·78
0·80	+ 0·8462874	− 0·0868023	0·8507273	− 5° 51′ 22″·51	+ 0·4739944	0·80
0·82	+ 0·8388338	− 0·0674640	0·8415424	− 4° 35′ 53″·41	+ 0·4840616	0·82
0·84	+ 0·8312284	− 0·0485651	0·8326455	− 3° 20′ 37″·48	+ 0·4940018	0·84
0·86	+ 0·8234734	− 0·0300917	0·8240231	− 2° 5′ 34″·06	+ 0·5038124	0·86
0·88	+ 0·8155711	− 0·0120311	0·8156598	− 0° 50′ 42″·55	+ 0·5134909	0·88
0·90	+ 0·8075238	+ 0·0056283	0·8075434	0° 23′ 57″·61	+ 0·5230350	0·90
0·92	+ 0·7993339	+ 0·0228974	0·7996618	1° 38′ 26″·96	+ 0·5324422	0·92
0·94	+ 0·7910039	+ 0·0397860	0·7920038	2° 52′ 45″·99	+ 0·5417103	0·94
0·96	+ 0·7825361	+ 0·0563032	0·7845590	4° 6′ 55″·16	+ 0·5508368	0·96
0·98	+ 0·7739332	+ 0·0724576	0·7773177	5° 20′ 54″·92	+ 0·5598197	0·98
1·00	+ 0·7651977	+ 0·0882570	0·7702706	6° 34′ 45″·67	+ 0·5686566	1·00

Table I. Functions of order unity

x	$J_1(x)$	$Y_1(x)$	$\lvert H_1^{(1)}(x)\rvert$	arg $H_1^{(1)}(x)$	$\mathbf{H}_1(x)$	x
0	0·0000000	− ∞	∞	− 90°	0·0000000	0
0·02	+ 0·0099995	− 31·8598128	31·8598144	− 89° 58′ 55″·26	+ 0·0000849	0·02
0·04	+ 0·0199960	− 15·9643089	15·9643214	− 89° 55′ 41″·64	+ 0·0003395	0·04
0·06	+ 0·0299865	− 10·6757892	10·6758314	− 89° 50′ 20″·64	+ 0·0007638	0·06
0·08	+ 0·0399680	− 8·0376696	8·0377690	− 89° 42′ 54″·34	+ 0·0013575	0·08
0·10	+ 0·0499375	− 6·4589511	6·4591441	− 89° 33′ 25″·29	+ 0·0021207	0·10
0·12	+ 0·0598921	− 5·4094402	5·4097717	− 89° 21′ 56″·38	+ 0·0030528	0·12
0·14	+ 0·0698286	− 4·6619853	4·6625082	− 89° 8′ 30″·73	+ 0·0041539	0·14
0·16	+ 0·0797443	− 4·1030547	4·1038295	− 88° 53′ 11″·68	+ 0·0054232	0·16
0·18	+ 0·0896360	− 3·6696037	3·6706983	− 88° 36′ 2″·65	+ 0·0068607	0·18
0·20	+ 0·0995008	− 3·3238250	3·3253140	− 88° 17′ 7″·17	+ 0·0084657	0·20
0·22	+ 0·1093358	− 3·0416730	3·0436375	− 87° 56′ 28″·81	+ 0·0102377	0·22
0·24	+ 0·1191381	− 2·8071277	2·8096547	− 87° 34′ 11″·11	+ 0·0121762	0·24
0·26	+ 0·1289046	− 2·6091059	2·6122883	− 87° 10′ 17″·63	+ 0·0142806	0·26
0·28	+ 0·1386325	− 2·4396971	2·4436328	− 86° 44′ 51″·87	+ 0·0165502	0·28
0·30	+ 0·1483188	− 2·2931051	2·2978968	− 86° 17′ 57″·28	+ 0·0189843	0·30
0·32	+ 0·1579607	− 2·1649866	2·1707415	− 85° 49′ 37″·22	+ 0·0215820	0·32
0·34	+ 0·1675553	− 2·0520233	2·0588527	− 85° 19′ 55″·00	+ 0·0243427	0·34
0·36	+ 0·1770997	− 1·9516372	1·9596561	− 84° 48′ 53″·80	+ 0·0272652	0·36
0·38	+ 0·1865911	− 1·8617949	1·8711218	− 84° 16′ 36″·72	+ 0·0303489	0·38
0·40	+ 0·1960266	− 1·7808720	1·7916282	− 83° 43′ 6″·78	+ 0·0335925	0·40
0·42	+ 0·2054034	− 1·7075549	1·7198647	− 83° 8′ 26″·86	+ 0·0369952	0·42
0·44	+ 0·2147188	− 1·6407704	1·6547603	− 82° 32′ 39″·76	+ 0·0405559	0·44
0·46	+ 0·2239699	− 1·5796331	1·5954320	− 81° 55′ 48″·18	+ 0·0442733	0·46
0·48	+ 0·2331540	− 1·5234063	1·5411449	− 81° 17′ 54″·69	+ 0·0481463	0·48
0·50	+ 0·2422685	− 1·4714724	1·4912830	− 80° 39′ 1″·79	+ 0·0521737	0·50
0·52	+ 0·2513105	− 1·4233094	1·4453258	− 79° 59′ 11″·85	+ 0·0563542	0·52
0·54	+ 0·2602774	− 1·3784737	1·4028308	− 79° 18′ 27″·16	+ 0·0606865	0·54
0·56	+ 0·2691665	− 1·3365858	1·3634193	− 78° 36′ 49″·90	+ 0·0651691	0·56
0·58	+ 0·2779752	− 1·2973191	1·3267657	− 77° 54′ 22″·18	+ 0·0698006	0·58
0·60	+ 0·2867010	− 1·2603913	1·2925880	− 77° 11′ 5″·99	+ 0·0745797	0·60
0·62	+ 0·2953412	− 1·2255572	1·2606415	− 76° 27′ 3″·26	+ 0·0795046	0·62
0·64	+ 0·3038932	− 1·1926026	1·2307120	− 75° 42′ 15″·81	+ 0·0845739	0·64
0·66	+ 0·3123547	− 1·1613400	1·2026122	− 74° 56′ 45″·41	+ 0·0897860	0·66
0·68	+ 0·3207230	− 1·1316043	1·1761767	− 74° 10′ 33″·73	+ 0·0951392	0·68
0·70	+ 0·3289957	− 1·1032499	1·1512595	− 73° 23′ 42″·38	+ 0·1006317	0·70
0·72	+ 0·3371705	− 1·0761476	1·1277312	− 72° 36′ 12″·87	+ 0·1062619	0·72
0·74	+ 0·3452448	− 1·0501828	1·1054763	− 71° 48′ 6″·68	+ 0·1120279	0·74
0·76	+ 0·3532164	− 1·0252532	1·0843920	− 70° 59′ 25″·22	+ 0·1179279	0·76
0·78	+ 0·3610829	− 1·0012677	1·0643861	− 70° 10′ 9″·80	+ 0·1239601	0·78
0·80	+ 0·3688420	− 0·9781442	1·0453758	− 69° 20′ 21″·73	+ 0·1301225	0·80
0·82	+ 0·3764916	− 0·9558093	1·0272864	− 68° 30′ 2″·20	+ 0·1364131	0·82
0·84	+ 0·3840292	− 0·9341970	1·0100507	− 67° 39′ 12″·39	+ 0·1428299	0·84
0·86	+ 0·3914529	− 0·9132475	0·9936077	− 66° 47′ 53″·42	+ 0·1493710	0·86
0·88	+ 0·3987603	− 0·8929069	0·9779022	− 65° 56′ 6″·35	+ 0·1560343	0·88
0·90	+ 0·4059495	− 0·8731266	0·9628837	− 65° 3′ 52″·21	+ 0·1628175	0·90
0·92	+ 0·4130184	− 0·8538622	0·9485066	− 64° 11′ 11″·97	+ 0·1697186	0·92
0·94	+ 0·4199649	− 0·8350735	0·9347290	− 63° 18′ 6″·57	+ 0·1767354	0·94
0·96	+ 0·4267871	− 0·8167241	0·9215126	− 62° 24′ 36″·89	+ 0·1838656	0·96
0·98	+ 0·4334829	− 0·7987806	0·9088223	− 61° 30′ 43″·79	+ 0·1911070	0·98
1·00	+ 0·4400506	− 0·7812128	0·8966259	− 60° 36′ 28″·10	+ 0·1984573	1·00

Table I. Functions of order zero

x	$J_0(x)$	$Y_0(x)$	$\lvert H_0^{(1)}(x)\rvert$	$\arg H_0^{(1)}(x)$	$\mathbf{H}_0(x)$	x
1·02	+ 0·7563321	+ 0·1037085	0·7634092	7° 48′ 27″·80	+ 0·5773455	1·02
1·04	+ 0·7473390	+ 0·1188188	0·7567255	9° 2′ 1″·68	+ 0·5858842	1·04
1·06	+ 0·7382212	+ 0·1335943	0·7502120	10° 15′ 27″·64	+ 0·5942706	1·06
1·08	+ 0·7289813	+ 0·1480406	0·7438614	11° 28′ 46″·01	+ 0·6025028	1·08
1·10	+ 0·7196220	+ 0·1621632	0·7376671	12° 41′ 57″·10	+ 0·6105787	1·10
1·12	+ 0·7101461	+ 0·1759670	0·7316228	13° 55′ 1″·20	+ 0·6184965	1·12
1·14	+ 0·7005564	+ 0·1894567	0·7257225	15° 7′ 58″·58	+ 0·6262544	1·14
1·16	+ 0·6908557	+ 0·2026367	0·7199606	16° 20′ 49″·49	+ 0·6338504	1·16
1·18	+ 0·6810469	+ 0·2155111	0·7143317	17° 33′ 34″·20	+ 0·6412828	1·18
1·20	+ 0·6711327	+ 0·2280835	0·7088309	18° 46′ 12″·92	+ 0·6485500	1·20
1·22	+ 0·6611163	+ 0·2403577	0·7034533	19° 58′ 45″·88	+ 0·6556502	1·22
1·24	+ 0·6510004	+ 0·2523369	0·6981944	21° 11′ 13″·30	+ 0·6625819	1·24
1·26	+ 0·6407880	+ 0·2640243	0·6930499	22° 23′ 35″·37	+ 0·6693434	1·26
1·28	+ 0·6304822	+ 0·2754228	0·6880157	23° 35′ 52″·28	+ 0·6759334	1·28
1·30	+ 0·6200860	+ 0·2865354	0·6830879	24° 48′ 4″·22	+ 0·6823503	1·30
1·32	+ 0·6096023	+ 0·2973645	0·6782630	26° 0′ 11″·36	+ 0·6885928	1·32
1·34	+ 0·5990343	+ 0·3079127	0·6735372	27° 12′ 13″·86	+ 0·6946595	1·34
1·36	+ 0·5883850	+ 0·3181824	0·6689073	28° 24′ 11″·88	+ 0·7005492	1·36
1·38	+ 0·5776576	+ 0·3281758	0·6643701	29° 36′ 5″·57	+ 0·7062606	1·38
1·40	+ 0·5668551	+ 0·3378951	0·6599226	30° 47′ 55″·08	+ 0·7117925	1·40
1·42	+ 0·5559807	+ 0·3473424	0·6555618	31° 59′ 40″·54	+ 0·7171439	1·42
1·44	+ 0·5450376	+ 0·3565195	0·6512850	33° 11′ 22″·08	+ 0·7223136	1·44
1·46	+ 0·5340289	+ 0·3654285	0·6470895	34° 22′ 59″·82	+ 0·7273008	1·46
1·48	+ 0·5229579	+ 0·3740710	0·6429728	35° 34′ 33″·90	+ 0·7321043	1·48
1·50	+ 0·5118277	+ 0·3824489	0·6389325	36° 46′ 4″·43	+ 0·7367235	1·50
1·52	+ 0·5006415	+ 0·3905639	0·6349662	37° 57′ 31″·50	+ 0·7411573	1·52
1·54	+ 0·4894026	+ 0·3984176	0·6310717	39° 8′ 55″·24	+ 0·7454051	1·54
1·56	+ 0·4781143	+ 0·4060116	0·6272469	40° 20′ 15″·74	+ 0·7494662	1·56
1·58	+ 0·4667797	+ 0·4133476	0·6234898	41° 31′ 33″·10	+ 0·7533398	1·58
1·60	+ 0·4554022	+ 0·4204269	0·6197983	42° 42′ 47″·42	+ 0·7570255	1·60
1·62	+ 0·4439850	+ 0·4272512	0·6161706	43° 53′ 58″·77	+ 0·7605226	1·62
1·64	+ 0·4325313	+ 0·4338219	0·6126049	45° 5′ 7″·26	+ 0·7638306	1·64
1·66	+ 0·4210446	+ 0·4401404	0·6090994	46° 16′ 12″·95	+ 0·7669493	1·66
1·68	+ 0·4095280	+ 0·4462083	0·6056526	47° 27′ 15″·94	+ 0·7698781	1·68
1·70	+ 0·3979849	+ 0·4520270	0·6022627	48° 38′ 16″·30	+ 0·7726168	1·70
1·72	+ 0·3864185	+ 0·4575979	0·5989282	49° 49′ 14″·10	+ 0·7751652	1·72
1·74	+ 0·3748321	+ 0·4629223	0·5956477	51° 0′ 9″·41	+ 0·7775230	1·74
1·76	+ 0·3632292	+ 0·4680019	0·5924198	52° 11′ 2″·31	+ 0·7796902	1·76
1·78	+ 0·3516128	+ 0·4728378	0·5892429	53° 21′ 52″·85	+ 0·7816666	1·78
1·80	+ 0·3399864	+ 0·4774317	0·5861159	54° 32′ 41″·11	+ 0·7834523	1·80
1·82	+ 0·3283532	+ 0·4817849	0·5830374	55° 43′ 27″·14	+ 0·7850474	1·82
1·84	+ 0·3167166	+ 0·4858989	0·5800061	56° 54′ 11″·00	+ 0·7864518	1·84
1·86	+ 0·3050797	+ 0·4897751	0·5770210	58° 4′ 52″·75	+ 0·7876658	1·86
1·88	+ 0·2934460	+ 0·4934149	0·5740809	59° 15′ 32″·45	+ 0·7886897	1·88
1·90	+ 0·2818186	+ 0·4968200	0·5711845	60° 26′ 10″·14	+ 0·7895236	1·90
1·92	+ 0·2702008	+ 0·4999917	0·5683310	61° 36′ 45″·88	+ 0·7901680	1·92
1·94	+ 0·2585959	+ 0·5029315	0·5655192	62° 47′ 19″·73	+ 0·7906233	1·94
1·96	+ 0·2470071	+ 0·5056411	0·5627481	63° 57′ 51″·71	+ 0·7908898	1·96
1·98	+ 0·2354376	+ 0·5081220	0·5600168	65° 8′ 21″·89	+ 0·7909681	1·98
2·00	+ 0·2238908	+ 0·5103757	0·5573243	66° 18′ 50″·32	+ 0·7908588	2·00

Table I. Functions of order unity

x	$J_1(x)$	$Y_1(x)$	$\lvert H_1^{(1)}(x)\rvert$	$\arg H_1^{(1)}(x)$	$\mathbf{H}_1(x)$	x
1·02	+ 0·4464882	− 0·7639930	0·8848938	− 59° 41′ 50″·60	+ 0·2059142	1·02
1·04	+ 0·4527939	− 0·7470959	0·8735987	− 58° 46′ 52″·03	+ 0·2134753	1·04
1·06	+ 0·4589660	− 0·7304984	0·8627154	− 57° 51′ 33″·12	+ 0·2211382	1·06
1·08	+ 0·4650027	− 0·7141794	0·8522205	− 56° 55′ 54″·55	+ 0·2289005	1·08
1·10	+ 0·4709024	− 0·6981196	0·8420926	− 55° 59′ 56″·99	+ 0·2367597	1·10
1·12	+ 0·4766634	− 0·6823011	0·8323117	− 55° 3′ 41″·07	+ 0·2447133	1·12
1·14	+ 0·4822840	− 0·6667078	0·8228591	− 54° 7′ 7″·40	+ 0·2527589	1·14
1·16	+ 0·4877629	− 0·6513248	0·8137178	− 53° 10′ 16″·56	+ 0·2608939	1·16
1·18	+ 0·4930984	− 0·6361385	0·8048715	− 52° 13′ 9″·10	+ 0·2691157	1·18
1·20	+ 0·4982891	− 0·6211364	0·7963055	− 51° 15′ 45″·57	+ 0·2774218	1·20
1·22	+ 0·5033336	− 0·6063070	0·7880056	− 50° 18′ 6″·48	+ 0·2858095	1·22
1·24	+ 0·5082305	− 0·5916398	0·7799589	− 49° 20′ 12″·33	+ 0·2942761	1·24
1·26	+ 0·5129786	− 0·5771253	0·7721533	− 48° 22′ 3″·58	+ 0·3028191	1·26
1·28	+ 0·5175766	− 0·5627546	0·7645772	− 47° 23′ 40″·70	+ 0·3114357	1·28
1·30	+ 0·5220232	− 0·5485197	0·7572200	− 46° 25′ 4″·12	+ 0·3201231	1·30
1·32	+ 0·5263174	− 0·5344133	0·7500717	− 45° 26′ 14″·26	+ 0·3288788	1·32
1·34	+ 0·5304580	− 0·5204287	0·7431229	− 44° 27′ 11″·54	+ 0·3376999	1·34
1·36	+ 0·5344439	− 0·5065597	0·7363647	− 43° 27′ 56″·34	+ 0·3465837	1·36
1·38	+ 0·5382741	− 0·4928008	0·7297888	− 42° 28′ 29″·03	+ 0·3555273	1·38
1·40	+ 0·5419477	− 0·4791470	0·7233873	− 41° 28′ 49″·98	+ 0·3645280	1·40
1·42	+ 0·5454638	− 0·4655936	0·7171528	− 40° 28′ 59″·54	+ 0·3735830	1·42
1·44	+ 0·5488215	− 0·4521367	0·7110785	− 39° 28′ 58″·03	+ 0·3826894	1·44
1·46	+ 0·5520200	− 0·4387723	0·7051576	− 38° 28′ 45″·78	+ 0·3918443	1·46
1·48	+ 0·5550586	− 0·4254973	0·6993840	− 37° 28′ 23″·11	+ 0·4010450	1·48
1·50	+ 0·5579365	− 0·4123086	0·6937518	− 36° 27′ 50″·30	+ 0·4102885	1·50
1·52	+ 0·5606532	− 0·3992036	0·6882554	− 35° 27′ 7″·65	+ 0·4195719	1·52
1·54	+ 0·5632079	− 0·3861800	0·6828896	− 34° 26′ 15″·44	+ 0·4288924	1·54
1·56	+ 0·5656003	− 0·3732356	0·6776493	− 33° 25′ 13″·93	+ 0·4382471	1·56
1·58	+ 0·5678298	− 0·3603688	0·6725298	− 32° 24′ 3″·39	+ 0·4476330	1·58
1·60	+ 0·5698959	− 0·3475780	0·6675267	− 31° 22′ 44″·05	+ 0·4570472	1·60
1·62	+ 0·5717984	− 0·3348619	0·6626355	− 30° 21′ 16″·17	+ 0·4664869	1·62
1·64	+ 0·5735368	− 0·3222194	0·6578524	− 29° 19′ 39″·97	+ 0·4759490	1·64
1·66	+ 0·5751108	− 0·3096498	0·6531734	− 28° 17′ 55″·68	+ 0·4854306	1·66
1·68	+ 0·5765204	− 0·2971522	0·6485948	− 27° 16′ 3″·52	+ 0·4949288	1·68
1·70	+ 0·5777652	− 0·2847262	0·6441131	− 26° 14′ 3″·70	+ 0·5044407	1·70
1·72	+ 0·5788453	− 0·2723716	0·6397250	− 25° 11′ 56″·41	+ 0·5139633	1·72
1·74	+ 0·5797604	− 0·2600881	0·6354274	− 24° 9′ 41″·85	+ 0·5234937	1·74
1·76	+ 0·5805107	− 0·2478757	0·6312171	− 23° 7′ 20″·21	+ 0·5330289	1·76
1·78	+ 0·5810962	− 0·2357345	0·6270914	− 22° 4′ 51″·68	+ 0·5425661	1·78
1·80	+ 0·5815170	− 0·2236649	0·6230473	− 21° 2′ 16″·43	+ 0·5521021	1·80
1·82	+ 0·5817731	− 0·2116671	0·6190823	− 19° 59′ 34″·63	+ 0·5616342	1·82
1·84	+ 0·5818649	− 0·1997416	0·6151939	− 18° 56′ 46″·45	+ 0·5711594	1·84
1·86	+ 0·5817926	− 0·1878891	0·6113796	− 17° 53′ 52″·04	+ 0·5806748	1·86
1·88	+ 0·5815566	− 0·1761102	0·6076371	− 16° 50′ 51″·57	+ 0·5901775	1·88
1·90	+ 0·5811571	− 0·1644058	0·6039642	− 15° 47′ 45″·18	+ 0·5996645	1·90
1·92	+ 0·5805946	− 0·1527766	0·6003588	− 14° 44′ 33″·01	+ 0·6091329	1·92
1·94	+ 0·5798695	− 0·1412236	0·5968189	− 13° 41′ 15″·22	+ 0·6185800	1·94
1·96	+ 0·5789825	− 0·1297478	0·5933424	− 12° 37′ 51″·93	+ 0·6280027	1·96
1·98	+ 0·5779341	− 0·1183504	0·5899276	− 11° 34′ 23″·27	+ 0·6373982	1·98
2·00	+ 0·5767248	− 0·1070324	0·5865726	− 10° 30′ 49″·38	+ 0·6467637	2·00

Table I. Functions of order zero

x	$J_0(x)$	$Y_0(x)$	$\|H_0^{(1)}(x)\|$	arg $H_0^{(1)}(x)$	$\mathbf{H}_0(x)$	x
2·02	+ 0·2123697	+ 0·5124038	0·5546698	67° 29′ 17″·02	+ 0·7905626	2·02
2·04	+ 0·2008776	+ 0·5142080	0·5520523	68° 39′ 42″·05	+ 0·7900800	2·04
2·06	+ 0·1894177	+ 0·5157900	0·5494710	69° 50′ 5″·44	+ 0·7894119	2·06
2·08	+ 0·1779931	+ 0·5171513	0·5469250	71° 0′ 27″·23	+ 0·7885590	2·08
2·10	+ 0·1666070	+ 0·5182937	0·5444137	72° 10′ 47″·47	+ 0·7875222	2·10
2·12	+ 0·1552625	+ 0·5192190	0·5419362	73° 21′ 6″·18	+ 0·7863025	2·12
2·14	+ 0·1439626	+ 0·5199289	0·5394917	74° 31′ 23″·41	+ 0·7849006	2·14
2·16	+ 0·1327106	+ 0·5204252	0·5370796	75° 41′ 39″·19	+ 0·7833178	2·16
2·18	+ 0·1215095	+ 0·5207097	0·5346991	76° 51′ 53″·55	+ 0·7815550	2·18
2·20	+ 0·1103623	+ 0·5207843	0·5323496	78° 2′ 6″·53	+ 0·7796135	2·20
2·22	+ 0·0992720	+ 0·5206508	0·5300304	79° 12′ 18″·15	+ 0·7774943	2·22
2·24	+ 0·0882416	+ 0·5203112	0·5277408	80° 22′ 28″·45	+ 0·7751986	2·24
2·26	+ 0·0772742	+ 0·5197675	0·5254803	81° 32′ 37″·46	+ 0·7727279	2·26
2·28	+ 0·0663726	+ 0·5190215	0·5232482	82° 42′ 45″·20	+ 0·7700834	2·28
2·30	+ 0·0555398	+ 0·5180754	0·5210439	83° 52′ 51″·71	+ 0·7672665	2·30
2·32	+ 0·0447786	+ 0·5169311	0·5188670	85° 2′ 57″·01	+ 0·7642787	2·32
2·34	+ 0·0340921	+ 0·5155908	0·5167167	86° 13′ 1″·12	+ 0·7611214	2·34
2·36	+ 0·0234828	+ 0·5140565	0·5145926	87° 23′ 4″·08	+ 0·7577962	2·36
2·38	+ 0·0129538	+ 0·5123304	0·5124942	88° 33′ 5″·90	+ 0·7543047	2·38
2·40	+ 0·0025077	+ 0·5104147	0·5104209	89° 43′ 6″·62	+ 0·7506485	2·40
2·42	− 0·0078527	+ 0·5083116	0·5083723	90° 53′ 6″·25	+ 0·7468293	2·42
2·44	− 0·0181247	+ 0·5060233	0·5063478	92° 3′ 4″·83	+ 0·7428488	2·44
2·46	− 0·0283057	+ 0·5035522	0·5043471	93° 13′ 2″·36	+ 0·7387088	2·46
2·48	− 0·0383929	+ 0·5009004	0·5023696	94° 22′ 58″·88	+ 0·7344112	2·48
2·50	− 0·0483838	+ 0·4980704	0·5004149	95° 32′ 54″·40	+ 0·7299577	2·50
2·52	− 0·0582758	+ 0·4950645	0·4984826	96° 42′ 48″·94	+ 0·7253504	2·52
2·54	− 0·0680664	+ 0·4918851	0·4965722	97° 52′ 42″·52	+ 0·7205912	2·54
2·56	− 0·0777531	+ 0·4885347	0·4946834	99° 2′ 35″·17	+ 0·7156821	2·56
2·58	− 0·0873334	+ 0·4850157	0·4928157	100° 12′ 26″·90	+ 0·7106251	2·58
2·60	− 0·0968050	+ 0·4813306	0·4909688	101° 22′ 17″·74	+ 0·7054223	2·60
2·62	− 0·1061654	+ 0·4774820	0·4891422	102° 32′ 7″·69	+ 0·7000759	2·62
2·64	− 0·1154123	+ 0·4734724	0·4873357	103° 41′ 56″·77	+ 0·6945880	2·64
2·66	− 0·1245434	+ 0·4693043	0·4855488	104° 51′ 45″·01	+ 0·6889609	2·66
2·68	− 0·1335565	+ 0·4649805	0·4837812	106° 1′ 32″·41	+ 0·6831967	2·68
2·70	− 0·1424494	+ 0·4605035	0·4820325	107° 11′ 19″·00	+ 0·6772977	2·70
2·72	− 0·1512198	+ 0·4558761	0·4803025	108° 21′ 4″·79	+ 0·6712664	2·72
2·74	− 0·1598658	+ 0·4511009	0·4785907	109° 30′ 49″·80	+ 0·6651050	2·74
2·76	− 0·1683852	+ 0·4461806	0·4768970	110° 40′ 34″·04	+ 0·6588160	2·76
2·78	− 0·1767759	+ 0·4411181	0·4752209	111° 50′ 17″·53	+ 0·6524017	2·78
2·80	− 0·1850360	+ 0·4359160	0·4735621	113° 0′ 0″·27	+ 0·6458646	2·80
2·82	− 0·1931636	+ 0·4305772	0·4719204	114° 9′ 42″·29	+ 0·6392073	2·82
2·84	− 0·2011568	+ 0·4251045	0·4702955	115° 19′ 23″·59	+ 0·6324323	2·84
2·86	− 0·2090137	+ 0·4195008	0·4686871	116° 29′ 4″·20	+ 0·6255420	2·86
2·88	− 0·2167325	+ 0·4137689	0·4670950	117° 38′ 44″·12	+ 0·6185392	2·88
2·90	− 0·2243115	+ 0·4079118	0·4655187	118° 48′ 23″·36	+ 0·6114264	2·90
2·92	− 0·2317491	+ 0·4019323	0·4639582	119° 58′ 1″·94	+ 0·6042062	2·92
2·94	− 0·2390434	+ 0·3958334	0·4624131	121° 7′ 39″·86	+ 0·5968814	2·94
2·96	− 0·2461931	+ 0·3896181	0·4608831	122° 17′ 17″·15	+ 0·5894546	2·96
2·98	− 0·2531964	+ 0·3832893	0·4593681	123° 26′ 53″·82	+ 0·5819286	2·98
3·00	− 0·2600520	+ 0·3768500	0·4578678	124° 36′ 29″·87	+ 0·5743061	3·00

Table I. Functions of order unity

x	$J_1(x)$	$Y_1(x)$	$\lvert H_1^{(1)}(x)\rvert$	arg $H_1^{(1)}(x)$	$\mathbf{H}_1(x)$	x
2·02	+ 0·5753554	− 0·0957951	0·5832757	− 9° 27′ 10″·37	+ 0·6560964	2·02
2·04	+ 0·5738267	− 0·0846398	0·5800353	− 8° 23′ 26″·38	+ 0·6653933	2·04
2·06	+ 0·5721393	− 0·0735677	0·5768497	− 7° 19′ 37″·51	+ 0·6746517	2·06
2·08	+ 0·5702942	− 0·0625801	0·5737174	− 6° 15′ 43″·88	+ 0·6838688	2·08
2·10	+ 0·5682921	− 0·0516786	0·5706370	− 5° 11′ 45″·59	+ 0·6930418	2·10
2·12	+ 0·5661342	− 0·0408645	0·5676071	− 4° 7′ 42″·76	+ 0·7021680	2·12
2·14	+ 0·5638212	− 0·0301393	0·5646262	− 3° 3′ 35″·49	+ 0·7112445	2·14
2·16	+ 0·5613543	− 0·0195045	0·5616930	− 1° 59′ 23″·88	+ 0·7202688	2·16
2·18	+ 0·5587345	− 0·0089616	0·5588064	− 0° 55′ 8″·03	+ 0·7292381	2·18
2·20	+ 0·5559630	+ 0·0014878	0·5559650	0° 9′ 11″·98	+ 0·7381496	2·20
2·22	+ 0·5530410	+ 0·0118422	0·5531678	1° 13′ 36″·04	+ 0·7470008	2·22
2·24	+ 0·5499696	+ 0·0220999	0·5504135	2° 18′ 4″·07	+ 0·7557890	2·24
2·26	+ 0·5467502	+ 0·0322594	0·5477011	3° 22′ 35″·97	+ 0·7645117	2·26
2·28	+ 0·5433841	+ 0·0423191	0·5450295	4° 27′ 11″·68	+ 0·7731661	2·28
2·30	+ 0·5398725	+ 0·0522773	0·5423977	5° 31′ 51″·10	+ 0·7817498	2·30
2·32	+ 0·5362170	+ 0·0621324	0·5398047	6° 36′ 34″·16	+ 0·7902603	2·32
2·34	+ 0·5324190	+ 0·0718828	0·5372496	7° 41′ 20″·77	+ 0·7986950	2·34
2·36	+ 0·5284801	+ 0·0815267	0·5347315	8° 46′ 10″·86	+ 0·8070514	2·36
2·38	+ 0·5244016	+ 0·0910627	0·5322494	9° 51′ 4″·37	+ 0·8153272	2·38
2·40	+ 0·5201853	+ 0·1004889	0·5298025	10° 56′ 1″·21	+ 0·8235198	2·40
2·42	+ 0·5158327	+ 0·1098039	0·5273901	12° 1′ 1″·31	+ 0·8316270	2·42
2·44	+ 0·5113456	+ 0·1190059	0·5250111	13° 6′ 4″·62	+ 0·8396463	2·44
2·46	+ 0·5067256	+ 0·1280934	0·5226650	14° 11′ 11″·06	+ 0·8475755	2·46
2·48	+ 0·5019745	+ 0·1370647	0·5203509	15° 16′ 20″·57	+ 0·8554122	2·48
2·50	+ 0·4970941	+ 0·1459181	0·5180682	16° 21′ 33″·10	+ 0·8631542	2·50
2·52	+ 0·4920863	+ 0·1546522	0·5158160	17° 26′ 48″·57	+ 0·8707993	2·52
2·54	+ 0·4869528	+ 0·1632654	0·5135938	18° 32′ 6″·93	+ 0·8783453	2·54
2·56	+ 0·4816957	+ 0·1717560	0·5114009	19° 37′ 28″·12	+ 0·8857900	2·56
2·58	+ 0·4763168	+ 0·1801226	0·5092366	20° 42′ 52″·08	+ 0·8931314	2·58
2·60	+ 0·4708183	+ 0·1883635	0·5071003	21° 48′ 18″·76	+ 0·9003674	2·60
2·62	+ 0·4652020	+ 0·1964774	0·5049913	22° 53′ 48″·11	+ 0·9074958	2·62
2·64	+ 0·4594700	+ 0·2044627	0·5029092	23° 59′ 20″·07	+ 0·9145148	2·64
2·66	+ 0·4536245	+ 0·2123179	0·5008534	25° 4′ 54″·60	+ 0·9214224	2·66
2·68	+ 0·4476676	+ 0·2200416	0·4988232	26° 10′ 31″·65	+ 0·9282167	2·68
2·70	+ 0·4416014	+ 0·2276324	0·4968182	27° 16′ 11″·16	+ 0·9348957	2·70
2·72	+ 0·4354281	+ 0·2350890	0·4948378	28° 21′ 53″·09	+ 0·9414577	2·72
2·74	+ 0·4291500	+ 0·2424099	0·4928816	29° 27′ 37″·39	+ 0·9479008	2·74
2·76	+ 0·4227693	+ 0·2495937	0·4909490	30° 33′ 24″·03	+ 0·9542233	2·76
2·78	+ 0·4162882	+ 0·2566393	0·4890395	31° 39′ 12″·95	+ 0·9604235	2·78
2·80	+ 0·4097092	+ 0·2635454	0·4871528	32° 45′ 4″·11	+ 0·9664998	2·80
2·82	+ 0·4030346	+ 0·2703106	0·4852883	33° 50′ 57″·48	+ 0·9724504	2·82
2·84	+ 0·3962667	+ 0·2769339	0·4834456	34° 56′ 53″·01	+ 0·9782739	2·84
2·86	+ 0·3894079	+ 0·2834140	0·4816243	36° 2′ 50″·66	+ 0·9839687	2·86
2·88	+ 0·3824607	+ 0·2897497	0·4798240	37° 8′ 50″·39	+ 0·9895333	2·88
2·90	+ 0·3754275	+ 0·2959401	0·4780443	38° 14′ 52″·17	+ 0·9949663	2·90
2·92	+ 0·3683108	+ 0·3019839	0·4762847	39° 20′ 55″·96	+ 1·0002663	2·92
2·94	+ 0·3611130	+ 0·3078802	0·4745449	40° 27′ 1″·72	+ 1·0054318	2·94
2·96	+ 0·3538368	+ 0·3136281	0·4728245	41° 33′ 9″·42	+ 1·0104617	2·96
2·98	+ 0·3464846	+ 0·3192264	0·4711232	42° 39′ 19″·02	+ 1·0153547	2·98
3·00	+ 0·3390590	+ 0·3246744	0·4694406	43° 45′ 30″·50	+ 1·0201096	3·00

Table I. Functions of order zero

x	$J_0(x)$	$Y_0(x)$	$\lvert H_0^{(1)}(x)\rvert$	$\arg H_0^{(1)}(x)$	$\mathbf{H}_0(x)$	x
3·02	− 0·2667583	+ 0·3703033	0·4563820	125° 46′ 5″·31	+ 0·5665900	3·0
3·04	− 0·2733140	+ 0·3636522	0·4549104	126° 55′ 40″·16	+ 0·5587829	3·0
3·06	− 0·2797178	+ 0·3568997	0·4534528	128° 5′ 14″·42	+ 0·5508877	3·0
3·08	− 0·2859683	+ 0·3500489	0·4520090	129° 14′ 48″·11	+ 0·5429073	3·0
3·10	− 0·2920643	+ 0·3431029	0·4505787	130° 24′ 21″·23	+ 0·5348444	3·1
3·12	− 0·2980048	+ 0·3360648	0·4491618	131° 33′ 53″·80	+ 0·5267021	3·1
3·14	− 0·3037884	+ 0·3289376	0·4477581	132° 43′ 25″·82	+ 0·5184831	3·1
3·16	− 0·3094142	+ 0·3217245	0·4463673	133° 52′ 57″·31	+ 0·5101905	3·1
3·18	− 0·3148811	+ 0·3144287	0·4449893	135° 2′ 28″·27	+ 0·5018270	3·1
3·20	− 0·3201882	+ 0·3070533	0·4436239	136° 11′ 58″·71	+ 0·4933957	3·2
3·22	− 0·3253345	+ 0·2996013	0·4422708	137° 21′ 28″·65	+ 0·4848996	3·2
3·24	− 0·3303193	+ 0·2920760	0·4409300	138° 30′ 58″·08	+ 0·4763415	3·2
3·26	− 0·3351416	+ 0·2844806	0·4396011	139° 40′ 27″·02	+ 0·4677245	3·2
3·28	− 0·3398009	+ 0·2768182	0·4382841	140° 49′ 55″·47	+ 0·4590516	3·2
3·30	− 0·3442963	+ 0·2690920	0·4369787	141° 59′ 23″·45	+ 0·4503257	3·3
3·32	− 0·3486272	+ 0·2613052	0·4356849	143° 8′ 50″·95	+ 0·4415499	3·3
3·34	− 0·3527931	+ 0·2534609	0·4344023	144° 18′ 18″·00	+ 0·4327272	3·3
3·36	− 0·3567934	+ 0·2455624	0·4331310	145° 27′ 44″·59	+ 0·4238607	3·3
3·38	− 0·3606277	+ 0·2376128	0·4318706	146° 37′ 10″·74	+ 0·4149532	3·3
3·40	− 0·3642956	+ 0·2296153	0·4306210	147° 46′ 36″·44	+ 0·4060080	3·4
3·42	− 0·3677967	+ 0·2215732	0·4293822	148° 56′ 1″·71	+ 0·3970279	3·4
3·44	− 0·3711306	+ 0·2134896	0·4281539	150° 5′ 26″·56	+ 0·3880161	3·4
3·46	− 0·3742972	+ 0·2053678	0·4269360	151° 14′ 50″·98	+ 0·3789757	3·4
3·48	− 0·3772963	+ 0·1972108	0·4257283	152° 24′ 15″·00	+ 0·3699095	3·4
3·50	− 0·3801277	+ 0·1890219	0·4245308	153° 33′ 38″·61	+ 0·3608208	3·5
3·52	− 0·3827914	+ 0·1808043	0·4233432	154° 43′ 1″·82	+ 0·3517124	3·5
3·54	− 0·3852873	+ 0·1725612	0·4221655	155° 52′ 24″·63	+ 0·3425876	3·5
3·56	− 0·3876155	+ 0·1642956	0·4209974	157° 1′ 47″·05	+ 0·3334492	3·5
3·58	− 0·3897760	+ 0·1560109	0·4198389	158° 11′ 9″·10	+ 0·3243003	3·5
3·60	− 0·3917690	+ 0·1477100	0·4186898	159° 20′ 30″·77	+ 0·3151440	3·6
3·62	− 0·3935947	+ 0·1393962	0·4175501	160° 29′ 52″·06	+ 0·3059833	3·6
3·64	− 0·3952533	+ 0·1310727	0·4164195	161° 39′ 12″·99	+ 0·2968211	3·6
3·66	− 0·3967452	+ 0·1227424	0·4152980	162° 48′ 33″·56	+ 0·2876605	3·6
3·68	− 0·3980707	+ 0·1144086	0·4141855	163° 57′ 53″·78	+ 0·2785044	3·6
3·70	− 0·3992302	+ 0·1060743	0·4130817	165° 7′ 13″·65	+ 0·2693559	3·7
3·72	− 0·4002242	+ 0·0977426	0·4119867	166° 16′ 33″·17	+ 0·2602179	3·7
3·74	− 0·4010532	+ 0·0894167	0·4109003	167° 25′ 52″·36	+ 0·2510933	3·7
3·76	− 0·4017178	+ 0·0810994	0·4098223	168° 35′ 11″·21	+ 0·2419852	3·7
3·78	− 0·4022187	+ 0·0727939	0·4087528	169° 44′ 29″·74	+ 0·2328964	3·7
3·80	− 0·4025564	+ 0·0645032	0·4076915	170° 53′ 47″·93	+ 0·2238298	3·8
3·82	− 0·4027318	+ 0·0562303	0·4066383	172° 3′ 5″·81	+ 0·2147883	3·8
3·84	− 0·4027456	+ 0·0479782	0·4055933	173° 12′ 23″·37	+ 0·2057749	3·8
3·86	− 0·4025986	+ 0·0397498	0·4045561	174° 21′ 40″·63	+ 0·1967923	3·8
3·88	− 0·4022918	+ 0·0315481	0·4035269	175° 30′ 57″·57	+ 0·1878435	3·8
3·90	− 0·4018260	+ 0·0233759	0·4025054	176° 40′ 14″·22	+ 0·1789312	3·9
3·92	− 0·4012023	+ 0·0152362	0·4014915	177° 49′ 30″·56	+ 0·1700582	3·9
3·94	− 0·4004218	+ 0·0071319	0·4004853	178° 58′ 46″·60	+ 0·1612273	3·9
3·96	− 0·3994854	− 0·0009343	0·3994865	180° 8′ 2″·36	+ 0·1524412	3·9
3·98	− 0·3983943	− 0·0089594	0·3984951	181° 17′ 17″·86	+ 0·1437027	3·9
4·00	− 0·3971498	− 0·0169407	0·3975110	182° 26′ 33″·05	+ 0·1350146	4·0

Table I. Functions of order unity

x	$J_1(x)$	$Y_1(x)$	$\lvert H_1^{(1)}(x)\rvert$	arg $H_1^{(1)}(x)$	$\mathbf{H}_1(x)$	x
3·02	+ 0·3315626	+ 0·3299712	0·4677763	44° 51′ 43″·81	+ 1·0247251	3·02
3·04	+ 0·3239979	+ 0·3351158	0·4661301	45° 57′ 58″·93	+ 1·0292003	3·04
3·06	+ 0·3163677	+ 0·3401076	0·4645016	47° 4′ 15″·82	+ 1·0335340	3·06
3·08	+ 0·3086746	+ 0·3449457	0·4628904	48° 10′ 34″·45	+ 1·0377252	3·08
3·10	+ 0·3009211	+ 0·3496295	0·4612963	49° 16′ 54″·80	+ 1·0417730	3·10
3·12	+ 0·2931100	+ 0·3541583	0·4597190	50° 23′ 16″·83	+ 1·0456765	3·12
3·14	+ 0·2852440	+ 0·3585314	0·4581582	51° 29′ 40″·52	+ 1·0494347	3·14
3·16	+ 0·2773257	+ 0·3627483	0·4566135	52° 36′ 5″·84	+ 1·0530469	3·16
3·18	+ 0·2693579	+ 0·3668084	0·4550847	53° 42′ 32″·76	+ 1·0565124	3·18
3·20	+ 0·2613432	+ 0·3707113	0·4535716	54° 49′ 1″·25	+ 1·0598303	3·20
3·22	+ 0·2532845	+ 0·3744565	0·4520738	55° 55′ 31″·29	+ 1·0630001	3·22
3·24	+ 0·2451844	+ 0·3780436	0·4505911	57° 2′ 2″·86	+ 1·0660211	3·24
3·26	+ 0·2370457	+ 0·3814723	0·4491233	58° 8′ 35″·92	+ 1·0688928	3·26
3·28	+ 0·2288711	+ 0·3847421	0·4476701	59° 15′ 10″·45	+ 1·0716147	3·28
3·30	+ 0·2206635	+ 0·3878529	0·4462312	60° 21′ 46″·43	+ 1·0741863	3·30
3·32	+ 0·2124255	+ 0·3908045	0·4448064	61° 28′ 23″·84	+ 1·0766072	3·32
3·34	+ 0·2041599	+ 0·3935966	0·4433955	62° 35′ 2″·65	+ 1·0788770	3·34
3·36	+ 0·1958696	+ 0·3962292	0·4419983	63° 41′ 42″·85	+ 1·0809955	3·36
3·38	+ 0·1875574	+ 0·3987021	0·4406145	64° 48′ 24″·40	+ 1·0829624	3·38
3·40	+ 0·1792259	+ 0·4010153	0·4392439	65° 55′ 7″·28	+ 1·0847774	3·40
3·42	+ 0·1708779	+ 0·4031689	0·4378863	67° 1′ 51″·48	+ 1·0864406	3·42
3·44	+ 0·1625163	+ 0·4051628	0·4365415	68° 8′ 36″·98	+ 1·0879516	3·44
3·46	+ 0·1541439	+ 0·4069973	0·4352093	69° 15′ 23″·75	+ 1·0893106	3·46
3·48	+ 0·1457634	+ 0·4086724	0·4338895	70° 22′ 11″·77	+ 1·0905175	3·48
3·50	+ 0·1373775	+ 0·4101884	0·4325819	71° 29′ 1″·04	+ 1·0915723	3·50
3·52	+ 0·1289892	+ 0·4115455	0·4312864	72° 35′ 51″·51	+ 1·0924752	3·52
3·54	+ 0·1206010	+ 0·4127440	0·4300026	73° 42′ 43″·18	+ 1·0932264	3·54
3·56	+ 0·1122159	+ 0·4137843	0·4287305	74° 49′ 36″·04	+ 1·0938260	3·56
3·58	+ 0·1038365	+ 0·4146667	0·4274699	75° 56′ 30″·05	+ 1·0942743	3·58
3·60	+ 0·0954655	+ 0·4153918	0·4262206	77° 3′ 25″·21	+ 1·0945716	3·60
3·62	+ 0·0871059	+ 0·4159599	0·4249824	78° 10′ 21″·50	+ 1·0947183	3·62
3·64	+ 0·0787602	+ 0·4163716	0·4237552	79° 17′ 18″·89	+ 1·0947147	3·64
3·66	+ 0·0704312	+ 0·4166275	0·4225387	80° 24′ 17″·38	+ 1·0945614	3·66
3·68	+ 0·0621215	+ 0·4167282	0·4213329	81° 31′ 16″·94	+ 1·0942589	3·68
3·70	+ 0·0538340	+ 0·4166744	0·4201376	82° 38′ 17″·57	+ 1·0938077	3·70
3·72	+ 0·0455712	+ 0·4164668	0·4189527	83° 45′ 19″·23	+ 1·0932084	3·72
3·74	+ 0·0373359	+ 0·4161062	0·4177779	84° 52′ 21″·94	+ 1·0924617	3·74
3·76	+ 0·0291307	+ 0·4155934	0·4166131	85° 59′ 25″·66	+ 1·0915683	3·76
3·78	+ 0·0209582	+ 0·4149293	0·4154582	87° 6′ 30″·37	+ 1·0905289	3·78
3·80	+ 0·0128210	+ 0·4141147	0·4143131	88° 13′ 36″·07	+ 1·0893444	3·80
3·82	+ 0·0047218	+ 0·4131506	0·4131776	89° 20′ 42″·75	+ 1·0880156	3·82
3·84	− 0·0033369	+ 0·4120381	0·4120516	90° 27′ 50″·38	+ 1·0865434	3·84
3·86	− 0·0113524	+ 0·4107780	0·4109349	91° 34′ 58″·96	+ 1·0849288	3·86
3·88	− 0·0193223	+ 0·4093717	0·4098274	92° 42′ 8″·47	+ 1·0831727	3·88
3·90	− 0·0272440	+ 0·4078200	0·4087290	93° 49′ 18″·89	+ 1·0812762	3·90
3·92	− 0·0351151	+ 0·4061243	0·4076396	94° 56′ 30″·22	+ 1·0792403	3·92
3·94	− 0·0429330	+ 0·4042858	0·4065590	96° 3′ 42″·44	+ 1·0770662	3·94
3·96	− 0·0506953	+ 0·4023056	0·4054871	97° 10′ 55″·54	+ 1·0747551	3·96
3·98	− 0·0583995	+ 0·4001851	0·4044238	98° 18′ 9″·50	+ 1·0723082	3·98
4·00	− 0·0660433	+ 0·3979257	0·4033691	99° 25′ 24″·33	+ 1·0697267	4·00

Table I. Functions of order zero

x	$J_0(x)$	$Y_0(x)$	$\lvert H_0^{(1)}(x)\rvert$	arg $H_0^{(1)}(x)$	$\mathbf{H}_0(x)$	x
4·02	− 0·3957530	− 0·0248755	0·3965340	183° 35′ 47″·96	+ 0·1263794	4·02
4·04	− 0·3942053	− 0·0327610	0·3955643	184° 45′ 2″·62	+ 0·1177998	4·04
4·06	− 0·3925079	− 0·0405944	0·3946015	185° 54′ 16″·99	+ 0·1092784	4·06
4·08	− 0·3906622	− 0·0483732	0·3936457	187° 3′ 31″·11	+ 0·1008179	4·08
4·10	− 0·3886697	− 0·0560946	0·3926967	188° 12′ 44″·96	+ 0·0924208	4·10
4·12	− 0·3865318	− 0·0637561	0·3917546	189° 21′ 58″·54	+ 0·0840896	4·12
4·14	− 0·3842500	− 0·0713550	0·3908191	190° 31′ 11″·87	+ 0·0758269	4·14
4·16	− 0·3818259	− 0·0788889	0·3898903	191° 40′ 24″·95	+ 0·0676351	4·16
4·18	− 0·3792610	− 0·0863551	0·3889680	192° 49′ 37″·78	+ 0·0595166	4·18
4·20	− 0·3765571	− 0·0937512	0·3880522	193° 58′ 50″·36	+ 0·0514740	4·20
4·22	− 0·3737157	− 0·1010748	0·3871428	195° 8′ 2″·71	+ 0·0435095	4·22
4·24	− 0·3707386	− 0·1083234	0·3862397	196° 17′ 14″·81	+ 0·0356255	4·24
4·26	− 0·3676276	− 0·1154947	0·3853428	197° 26′ 26″·67	+ 0·0278243	4·26
4·28	− 0·3643845	− 0·1225863	0·3844522	198° 35′ 38″·31	+ 0·0201081	4·28
4·30	− 0·3610111	− 0·1295959	0·3835676	199° 44′ 49″·71	+ 0·0124793	4·30
4·32	− 0·3575093	− 0·1365213	0·3826891	200° 54′ 0″·88	+ 0·0049399	4·32
4·34	− 0·3538810	− 0·1433602	0·3818166	202° 3′ 11″·83	− 0·0025077	4·34
4·36	− 0·3501281	− 0·1501104	0·3809499	203° 12′ 22″·56	− 0·0098616	4·36
4·38	− 0·3462527	− 0·1567699	0·3800891	204° 21′ 33″·08	− 0·0171197	4·38
4·40	− 0·3422568	− 0·1633365	0·3792341	205° 30′ 43″·37	− 0·0242798	4·40
4·42	− 0·3381424	− 0·1698081	0·3783848	206° 39′ 53″·45	− 0·0313400	4·42
4·44	− 0·3339116	− 0·1761827	0·3775411	207° 49′ 3″·33	− 0·0382984	4·44
4·46	− 0·3295666	− 0·1824583	0·3767030	208° 58′ 12″·99	− 0·0451530	4·46
4·48	− 0·3251095	− 0·1886330	0·3758705	210° 7′ 22″·45	− 0·0519019	4·48
4·50	− 0·3205425	− 0·1947050	0·3750434	211° 16′ 31″·70	− 0·0585433	4·50
4·52	− 0·3158678	− 0·2006723	0·3742217	212° 25′ 40″·76	− 0·0650755	4·52
4·54	− 0·3110877	− 0·2065332	0·3734053	213° 34′ 49″·62	− 0·0714966	4·54
4·56	− 0·3062045	− 0·2122859	0·3725943	214° 43′ 58″·28	− 0·0778050	4·56
4·58	− 0·3012204	− 0·2179287	0·3717885	215° 53′ 6″·75	− 0·0839990	4·58
4·60	− 0·2961378	− 0·2234600	0·3709878	217° 2′ 15″·03	− 0·0900771	4·60
4·62	− 0·2909591	− 0·2288780	0·3701923	218° 11′ 23″·12	− 0·0960376	4·62
4·64	− 0·2856866	− 0·2341813	0·3694018	219° 20′ 31″·03	− 0·1018790	4·64
4·66	− 0·2803228	− 0·2393683	0·3686164	220° 29′ 38″·75	− 0·1075998	4·66
4·68	− 0·2748700	− 0·2444376	0·3678359	221° 38′ 46″·29	− 0·1131987	4·68
4·70	− 0·2693308	− 0·2493876	0·3670603	222° 47′ 53″·65	− 0·1186742	4·70
4·72	− 0·2637076	− 0·2542172	0·3662896	223° 57′ 0″·83	− 0·1240251	4·72
4·74	− 0·2580029	− 0·2589248	0·3655237	225° 6′ 7″·84	− 0·1292500	4·74
4·76	− 0·2522193	− 0·2635093	0·3647625	226° 15′ 14″·68	− 0·1343477	4·76
4·78	− 0·2463592	− 0·2679693	0·3640061	227° 24′ 21″·34	− 0·1393170	4·78
4·80	− 0·2404253	− 0·2723038	0·3632543	228° 33′ 27″·84	− 0·1441567	4·80
4·82	− 0·2344201	− 0·2765116	0·3625071	229° 42′ 34″·17	− 0·1488659	4·82
4·84	− 0·2283462	− 0·2805915	0·3617645	230° 51′ 40″·33	− 0·1534435	4·84
4·86	− 0·2222062	− 0·2845427	0·3610265	232° 0′ 46″·33	− 0·1578884	4·86
4·88	− 0·2160027	− 0·2883640	0·3602929	233° 9′ 52″·17	− 0·1621997	4·88
4·90	− 0·2097383	− 0·2920546	0·3595637	234° 18′ 57″·85	− 0·1663766	4·90
4·92	− 0·2034158	− 0·2956136	0·3588389	235° 28′ 3″·37	− 0·1704182	4·92
4·94	− 0·1970377	− 0·2990401	0·3581185	236° 37′ 8″·74	− 0·1743238	4·94
4·96	− 0·1906067	− 0·3023335	0·3574023	237° 46′ 13″·95	− 0·1780925	4·96
4·98	− 0·1841255	− 0·3054928	0·3566904	238° 55′ 19″·01	− 0·1817237	4·98
5·00	− 0·1775968	− 0·3085176	0·3559828	240° 4′ 23″·93	− 0·1852168	5·00

Table I. Functions of order unity

x	$J_1(x)$	$Y_1(x)$	$\lvert H_1^{(1)}(x)\rvert$	$\arg H_1^{(1)}(x)$	$\mathbf{H}_1(x)$	x
4·02	− 0·0736243	+ 0·3955287	0·4023226	100° 32′ 40″·00	+ 1·0670119	4·02
4·04	− 0·0811401	+ 0·3929956	0·4012845	101° 39′ 56″·49	+ 1·0641653	4·04
4·06	− 0·0885884	+ 0·3903277	0·4002545	102° 47′ 13″·81	+ 1·0611881	4·06
4·08	− 0·0959669	+ 0·3875267	0·3992325	103° 54′ 31″·94	+ 1·0580818	4·08
4·10	− 0·1032733	+ 0·3845940	0·3982185	105° 1′ 50″·87	+ 1·0548479	4·10
4·12	− 0·1105054	+ 0·3815313	0·3972123	106° 9′ 10″·59	+ 1·0514880	4·12
4·14	− 0·1176609	+ 0·3783401	0·3962138	107° 16′ 31″·08	+ 1·0480034	4·14
4·16	− 0·1247378	+ 0·3750222	0·3952229	108° 23′ 52″·34	+ 1·0443959	4·16
4·18	− 0·1317339	+ 0·3715792	0·3942396	109° 31′ 14″·36	+ 1·0406671	4·18
4·20	− 0·1386469	+ 0·3680128	0·3932638	110° 38′ 37″·12	+ 1·0368186	4·20
4·22	− 0·1454750	+ 0·3643248	0·3922953	111° 46′ 0″·62	+ 1·0328522	4·22
4·24	− 0·1522160	+ 0·3605171	0·3913340	112° 53′ 24″·85	+ 1·0287695	4·24
4·26	− 0·1588679	+ 0·3565914	0·3903799	114° 0′ 49″·79	+ 1·0245724	4·26
4·28	− 0·1654287	+ 0·3525497	0·3894328	115° 8′ 15″·45	+ 1·0202627	4·28
4·30	− 0·1718966	+ 0·3483938	0·3884928	116° 15′ 41″·80	+ 1·0158422	4·30
4·32	− 0·1782695	+ 0·3441256	0·3875596	117° 23′ 8″·84	+ 1·0113128	4·32
4·34	− 0·1845457	+ 0·3397472	0·3866333	118° 30′ 36″·56	+ 1·0066764	4·34
4·36	− 0·1907233	+ 0·3352606	0·3857136	119° 38′ 4″·96	+ 1·0019350	4·36
4·38	− 0·1968005	+ 0·3306677	0·3848007	120° 45′ 34″·02	+ 0·9970906	4·38
4·40	− 0·2027755	+ 0·3259707	0·3838942	121° 53′ 3″·74	+ 0·9921451	4·40
4·42	− 0·2086467	+ 0·3211716	0·3829943	123° 0′ 34″·10	+ 0·9871006	4·42
4·44	− 0·2144125	+ 0·3162725	0·3821008	124° 8′ 5″·10	+ 0·9819591	4·44
4·46	− 0·2200710	+ 0·3112757	0·3812136	125° 15′ 36″·73	+ 0·9767229	4·46
4·48	− 0·2256209	+ 0·3061832	0·3803327	126° 23′ 8″·99	+ 0·9713939	4·48
4·50	− 0·2310604	+ 0·3009973	0·3794579	127° 30′ 41″·86	+ 0·9659744	4·50
4·52	− 0·2363882	+ 0·2957202	0·3785893	128° 38′ 15″·34	+ 0·9604664	4·52
4·54	− 0·2416027	+ 0·2903542	0·3777267	129° 45′ 49″·42	+ 0·9548724	4·54
4·56	− 0·2467026	+ 0·2849015	0·3768700	130° 53′ 24″·09	+ 0·9491944	4·56
4·58	− 0·2516864	+ 0·2793644	0·3760193	132° 0′ 59″·35	+ 0·9434347	4·58
4·60	− 0·2565528	+ 0·2737452	0·3751744	133° 8′ 35″·18	+ 0·9375956	4·60
4·62	− 0·2613006	+ 0·2680464	0·3743352	134° 16′ 11″·59	+ 0·9316793	4·62
4·64	− 0·2659284	+ 0·2622702	0·3735018	135° 23′ 48″·56	+ 0·9256883	4·64
4·66	− 0·2704352	+ 0·2564190	0·3726740	136° 31′ 26″·09	+ 0·9196249	4·66
4·68	− 0·2748196	+ 0·2504952	0·3718517	137° 39′ 4″·17	+ 0·9134914	4·68
4·70	− 0·2790807	+ 0·2445013	0·3710350	138° 46′ 42″·80	+ 0·9072901	4·70
4·72	− 0·2832174	+ 0·2384397	0·3702237	139° 54′ 21″·96	+ 0·9010236	4·72
4·74	− 0·2872286	+ 0·2323128	0·3694177	141° 2′ 1″·65	+ 0·8946941	4·74
4·76	− 0·2911133	+ 0·2261230	0·3686171	142° 9′ 41″·87	+ 0·8883042	4·76
4·78	− 0·2948707	+ 0·2198730	0·3678218	143° 17′ 22″·60	+ 0·8818563	4·78
4·80	− 0·2984999	+ 0·2135652	0·3670317	144° 25′ 3″·85	+ 0·8753528	4·80
4·82	− 0·3019999	+ 0·2072020	0·3662467	145° 32′ 45″·61	+ 0·8687963	4·82
4·84	− 0·3053702	+ 0·2007860	0·3654668	146° 40′ 27″·87	+ 0·8621891	4·84
4·86	− 0·3086098	+ 0·1943198	0·3646919	147° 48′ 10″·62	+ 0·8555338	4·86
4·88	− 0·3117182	+ 0·1878058	0·3639221	148° 55′ 53″·86	+ 0·8488330	4·88
4·90	− 0·3146947	+ 0·1812467	0·3631571	150° 3′ 37″·59	+ 0·8420890	4·90
4·92	− 0·3175386	+ 0·1746449	0·3623971	151° 11′ 21″·79	+ 0·8353045	4·92
4·94	− 0·3202495	+ 0·1680031	0·3616418	152° 19′ 6″·47	+ 0·8284820	4·94
4·96	− 0·3228269	+ 0·1613238	0·3608913	153° 26′ 51″·61	+ 0·8216241	4·96
4·98	− 0·3252702	+ 0·1546097	0·3601456	154° 34′ 37″·22	+ 0·8147332	4·98
5·00	− 0·3275791	+ 0·1478631	0·3594045	155° 42′ 23″·28	+ 0·8078119	5·00

Table I. Functions of order zero

x	$J_0(x)$	$Y_0(x)$	$\lvert H_0^{(1)}(x)\rvert$	$\arg H_0^{(1)}(x)$	$\mathbf{H}_0(x)$	x
5·02	− 0·1710232	− 0·3114072	0·3552793	241° 13′ 28″·68	− 0·1885712	5·02
5·04	− 0·1644075	− 0·3141609	0·3545799	242° 22′ 33″·30	− 0·1917864	5·04
5·06	− 0·1577524	− 0·3167784	0·3538847	243° 31′ 37″·76	− 0·1948618	5·06
5·08	− 0·1510606	− 0·3192590	0·3531934	244° 40′ 42″·09	− 0·1977971	5·08
5·10	− 0·1443347	− 0·3216024	0·3525062	245° 49′ 46″·28	− 0·2005919	5·10
5·12	− 0·1375776	− 0·3238083	0·3518230	246° 58′ 50″·32	− 0·2032458	5·12
5·14	− 0·1307919	− 0·3258764	0·3511437	248° 7′ 54″·22	− 0·2057586	5·14
5·16	− 0·1239803	− 0·3278063	0·3504684	249° 16′ 57″·98	− 0·2081301	5·16
5·18	− 0·1171456	− 0·3295978	0·3497968	250° 26′ 1″·61	− 0·2103600	5·18
5·20	− 0·1102904	− 0·3312509	0·3491291	251° 35′ 5″·11	− 0·2124483	5·20
5·22	− 0·1034176	− 0·3327654	0·3484652	252° 44′ 8″·47	− 0·2143949	5·22
5·24	− 0·0965297	− 0·3341413	0·3478051	253° 53′ 11″·70	− 0·2161998	5·24
5·26	− 0·0896295	− 0·3353785	0·3471487	255° 2′ 14″·80	− 0·2178630	5·26
5·28	− 0·0827198	− 0·3364772	0·3464959	256° 11′ 17″·76	− 0·2193846	5·28
5·30	− 0·0758031	− 0·3374373	0·3458469	257° 20′ 20″·60	− 0·2207647	5·30
5·32	− 0·0688822	− 0·3382591	0·3452014	258° 29′ 23″·32	− 0·2220035	5·32
5·34	− 0·0619598	− 0·3389428	0·3445595	259° 38′ 25″·91	− 0·2231013	5·34
5·36	− 0·0550386	− 0·3394886	0·3439212	260° 47′ 28″·37	− 0·2240583	5·36
5·38	− 0·0481211	− 0·3398969	0·3432863	261° 56′ 30″·72	− 0·2248748	5·38
5·40	− 0·0412101	− 0·3401679	0·3426550	263° 5′ 32″·95	− 0·2255513	5·40
5·42	− 0·0343082	− 0·3403021	0·3420271	264° 14′ 35″·05	− 0·2260882	5·42
5·44	− 0·0274180	− 0·3402999	0·3414027	265° 23′ 37″·04	− 0·2264860	5·44
5·46	− 0·0205422	− 0·3401619	0·3407816	266° 32′ 38″·91	− 0·2267451	5·46
5·48	− 0·0136833	− 0·3398886	0·3401639	267° 41′ 40″·66	− 0·2268662	5·48
5·50	− 0·0068439	− 0·3394806	0·3395496	268° 50′ 42″·30	− 0·2268499	5·50
5·52	− 0·0000266	− 0·3389385	0·3389385	269° 59′ 43″·83	− 0·2266969	5·52
5·54	+ 0·0067661	− 0·3382631	0·3383307	271° 8′ 45″·24	− 0·2264079	5·54
5·56	+ 0·0135315	− 0·3374550	0·3377262	272° 17′ 46″·54	− 0·2259836	5·56
5·58	+ 0·0202673	− 0·3365151	0·3371249	273° 26′ 47″·73	− 0·2254249	5·58
5·60	+ 0·0269709	− 0·3354442	0·3365267	274° 35′ 48″·81	− 0·2247327	5·60
5·62	+ 0·0336398	− 0·3342432	0·3359317	275° 44′ 49″·79	− 0·2239078	5·62
5·64	+ 0·0402716	− 0·3329130	0·3353399	276° 53′ 50″·65	− 0·2229512	5·64
5·66	+ 0·0468638	− 0·3314545	0·3347511	278° 2′ 51″·41	− 0·2218639	5·66
5·68	+ 0·0534141	− 0·3298689	0·3341655	279° 11′ 52″·07	− 0·2206469	5·68
5·70	+ 0·0599200	− 0·3281571	0·3335828	280° 20′ 52″·62	− 0·2193014	5·70
5·72	+ 0·0663792	− 0·3263203	0·3330033	281° 29′ 53″·07	− 0·2178284	5·72
5·74	+ 0·0727894	− 0·3243597	0·3324267	282° 38′ 53″·42	− 0·2162291	5·74
5·76	+ 0·0791482	− 0·3222763	0·3318530	283° 47′ 53″·67	− 0·2145048	5·76
5·78	+ 0·0854533	− 0·3200715	0·3312824	284° 56′ 53″·82	− 0·2126567	5·78
5·80	+ 0·0917026	− 0·3177464	0·3307146	286° 5′ 53″·87	− 0·2106861	5·80
5·82	+ 0·0978937	− 0·3153025	0·3301498	287° 14′ 53″·83	− 0·2085942	5·82
5·84	+ 0·1040245	− 0·3127411	0·3295878	288° 23′ 53″·68	− 0·2063827	5·84
5·86	+ 0·1100928	− 0·3100636	0·3290286	289° 32′ 53″·44	− 0·2040527	5·86
5·88	+ 0·1160964	− 0·3072714	0·3284723	290° 41′ 53″·11	− 0·2016058	5·88
5·90	+ 0·1220334	− 0·3043659	0·3279188	291° 50′ 52″·68	− 0·1990435	5·90
5·92	+ 0·1279015	− 0·3013488	0·3273681	292° 59′ 52″·16	− 0·1963672	5·92
5·94	+ 0·1336987	− 0·2982215	0·3268201	294° 8′ 51″·54	− 0·1935787	5·94
5·96	+ 0·1394230	− 0·2949856	0·3262749	295° 17′ 50″·84	− 0·1906794	5·96
5·98	+ 0·1450725	− 0·2916428	0·3257324	296° 26′ 50″·05	− 0·1876711	5·98
6·00	+ 0·1506453	− 0·2881947	0·3251925	297° 35′ 49″·16	− 0·1845553	6·00

Table I. Functions of order unity

x	$J_1(x)$	$Y_1(x)$	$\lvert H_1^{(1)}(x)\rvert$	$\arg H_1^{(1)}(x)$	$\mathbf{H}_1(x)$	x
5·02	− 0·3297533	+ 0·1410869	0·3586680	156° 50′ 9″·80	+ 0·8008629	5·02
5·04	− 0·3317925	+ 0·1342835	0·3579362	157° 57′ 56″·75	+ 0·7938886	5·04
5·06	− 0·3336963	+ 0·1274556	0·3572088	159° 5′ 44″·15	+ 0·7868916	5·06
5·08	− 0·3354646	+ 0·1206057	0·3564859	160° 13′ 31″·99	+ 0·7798745	5·08
5·10	− 0·3370972	+ 0·1137364	0·3557675	161° 21′ 20″·26	+ 0·7728398	5·10
5·12	− 0·3385940	+ 0·1068504	0·3550534	162° 29′ 8″·96	+ 0·7657902	5·12
5·14	− 0·3399550	+ 0·0999502	0·3543437	163° 36′ 58″·08	+ 0·7587281	5·14
5·16	− 0·3411802	+ 0·0930384	0·3536383	164° 44′ 47″·61	+ 0·7516562	5·16
5·18	− 0·3422695	+ 0·0861176	0·3529371	165° 52′ 37″·56	+ 0·7445770	5·18
5·20	− 0·3432230	+ 0·0791903	0·3522402	167° 0′ 27″·92	+ 0·7374930	5·20
5·22	− 0·3440409	+ 0·0722592	0·3515474	168° 8′ 18″·68	+ 0·7304068	5·22
5·24	− 0·3447234	+ 0·0653269	0·3508587	169° 16′ 9″·84	+ 0·7233211	5·24
5·26	− 0·3452707	+ 0·0583958	0·3501742	170° 24′ 1″·40	+ 0·7162382	5·26
5·28	− 0·3456831	+ 0·0514685	0·3494936	171° 31′ 53″·34	+ 0·7091607	5·28
5·30	− 0·3459608	+ 0·0445476	0·3488171	172° 39′ 45″·68	+ 0·7020912	5·30
5·32	− 0·3461043	+ 0·0376356	0·3481446	173° 47′ 38″·39	+ 0·6950321	5·32
5·34	− 0·3461140	+ 0·0307351	0·3474759	174° 55′ 31″·49	+ 0·6879861	5·34
5·36	− 0·3459903	+ 0·0238485	0·3468112	176° 3′ 24″·96	+ 0·6809555	5·36
5·38	− 0·3457337	+ 0·0169784	0·3461503	177° 11′ 18″·80	+ 0·6739428	5·38
5·40	− 0·3453448	+ 0·0101273	0·3454933	178° 19′ 13″·00	+ 0·6669506	5·40
5·42	− 0·3448242	+ 0·0032975	0·3448400	179° 27′ 7″·56	+ 0·6599812	5·42
5·44	− 0·3441725	− 0·0035083	0·3441904	180° 35′ 2″·49	+ 0·6530372	5·44
5·46	− 0·3433905	− 0·0102879	0·3435445	181° 42′ 57″·78	+ 0·6461209	5·46
5·48	− 0·3424788	− 0·0170386	0·3429024	182° 50′ 53″·41	+ 0·6392347	5·48
5·50	− 0·3414382	− 0·0237582	0·3422638	183° 58′ 49″·39	+ 0·6323810	5·50
5·52	− 0·3402696	− 0·0304443	0·3416288	185° 6′ 45″·72	+ 0·6255623	5·52
5·54	− 0·3389739	− 0·0370944	0·3409975	186° 14′ 42″·38	+ 0·6187809	5·54
5·56	− 0·3375518	− 0·0437062	0·3403696	187° 22′ 39″·38	+ 0·6120390	5·56
5·58	− 0·3360045	− 0·0502774	0·3397452	188° 30′ 36″·72	+ 0·6053391	5·58
5·60	− 0·3343328	− 0·0568056	0·3391243	189° 38′ 34″·39	+ 0·5986835	5·60
5·62	− 0·3325379	− 0·0632886	0·3385069	190° 46′ 32″·39	+ 0·5920743	5·62
5·64	− 0·3306208	− 0·0697241	0·3378928	191° 54′ 30″·71	+ 0·5855138	5·64
5·66	− 0·3285826	− 0·0761099	0·3372821	193° 2′ 29″·35	+ 0·5790044	5·66
5·68	− 0·3264245	− 0·0824437	0·3366748	194° 10′ 28″·30	+ 0·5725481	5·68
5·70	− 0·3241477	− 0·0887233	0·3360708	195° 18′ 27″·57	+ 0·5661472	5·70
5·72	− 0·3217534	− 0·0949466	0·3354700	196° 26′ 27″·15	+ 0·5598038	5·72
5·74	− 0·3192429	− 0·1011115	0·3348725	197° 34′ 27″·04	+ 0·5535201	5·74
5·76	− 0·3166176	− 0·1072157	0·3342782	198° 42′ 27″·23	+ 0·5472981	5·76
5·78	− 0·3138787	− 0·1132573	0·3336871	199° 50′ 27″·72	+ 0·5411399	5·78
5·80	− 0·3110277	− 0·1192341	0·3330991	200° 58′ 28″·52	+ 0·5350476	5·80
5·82	− 0·3080661	− 0·1251442	0·3325143	202° 6′ 29″·60	+ 0·5290231	5·82
5·84	− 0·3049952	− 0·1309855	0·3319326	203° 14′ 30″·98	+ 0·5230685	5·84
5·86	− 0·3018166	− 0·1367560	0·3313540	204° 22′ 32″·65	+ 0·5171858	5·86
5·88	− 0·2985318	− 0·1424539	0·3307784	205° 30′ 34″·61	+ 0·5113768	5·88
5·90	− 0·2951424	− 0·1480772	0·3302059	206° 38′ 36″·85	+ 0·5056434	5·90
5·92	− 0·2916501	− 0·1536240	0·3296363	207° 46′ 39″·37	+ 0·4999876	5·92
5·94	− 0·2880563	− 0·1590925	0·3290697	208° 54′ 42″·17	+ 0·4944111	5·94
5·96	− 0·2843629	− 0·1644809	0·3285061	210° 2′ 45″·25	+ 0·4889157	5·96
5·98	− 0·2805715	− 0·1697874	0·3279453	211° 10′ 48″·60	+ 0·4835031	5·98
6·00	− 0·2766839	− 0·1750103	0·3273875	212° 18′ 52″·23	+ 0·4781753	6·00

Table I. Functions of order zero

x	$J_0(x)$	$Y_0(x)$	$\|H_0^{(1)}(x)\|$	$\arg H_0^{(1)}(x)$	$\mathbf{H}_0(x)$	x
6·02	+ 0·1561393	− 0·2846430	0·3246553	298° 44′ 48″·19	− 0·1813339	6·02
6·04	+ 0·1615527	− 0·2809893	0·3241208	299° 53′ 47″·13	− 0·1780085	6·04
6·06	+ 0·1668837	− 0·2772356	0·3235889	301° 2′ 45″·98	− 0·1745809	6·06
6·08	+ 0·1721306	− 0·2733835	0·3230596	302° 11′ 44″·74	− 0·1710529	6·08
6·10	+ 0·1772914	− 0·2694349	0·3225328	303° 20′ 43″·43	− 0·1674264	6·10
6·12	+ 0·1823646	− 0·2653917	0·3220087	304° 29′ 42″·03	− 0·1637033	6·12
6·14	+ 0·1873484	− 0·2612556	0·3214870	305° 38′ 40″·54	− 0·1598854	6·14
6·16	+ 0·1922411	− 0·2570287	0·3209679	306° 47′ 38″·97	− 0·1559746	6·16
6·18	+ 0·1970413	− 0·2527128	0·3204513	307° 56′ 37″·32	− 0·1519730	6·18
6·20	+ 0·2017472	− 0·2483100	0·3199371	309° 5′ 35″·59	− 0·1478824	6·20
6·22	+ 0·2063574	− 0·2438221	0·3194255	310° 14′ 33″·77	− 0·1437050	6·22
6·24	+ 0·2108705	− 0·2392513	0·3189162	311° 23′ 31″·88	− 0·1394427	6·24
6·26	+ 0·2152848	− 0·2345996	0·3184094	312° 32′ 29″·90	− 0·1350977	6·26
6·28	+ 0·2195991	− 0·2298691	0·3179050	313° 41′ 27″·85	− 0·1306719	6·28
6·30	+ 0·2238120	− 0·2250617	0·3174029	314° 50′ 25″·73	− 0·1261676	6·30
6·32	+ 0·2279222	− 0·2201798	0·3169032	315° 59′ 23″·52	− 0·1215867	6·32
6·34	+ 0·2319283	− 0·2152253	0·3164059	317° 8′ 21″·24	− 0·1169316	6·34
6·36	+ 0·2358292	− 0·2102005	0·3159109	318° 17′ 18″·88	− 0·1122043	6·36
6·38	+ 0·2396237	− 0·2051075	0·3154182	319° 26′ 16″·44	− 0·1074071	6·38
6·40	+ 0·2433106	− 0·1999486	0·3149278	320° 35′ 13″·94	− 0·1025422	6·40
6·42	+ 0·2468888	− 0·1947259	0·3144396	321° 44′ 11″·36	− 0·0976117	6·42
6·44	+ 0·2503573	− 0·1894417	0·3139537	322° 53′ 8″·71	− 0·0926181	6·44
6·46	+ 0·2537151	− 0·1840982	0·3134701	324° 2′ 5″·98	− 0·0875634	6·46
6·48	+ 0·2569612	− 0·1786977	0·3129887	325° 11′ 3″·18	− 0·0824500	6·48
6·50	+ 0·2600946	− 0·1732424	0·3125094	326° 20′ 0″·31	− 0·0772802	6·50
6·52	+ 0·2631145	− 0·1677348	0·3120324	327° 28′ 57″·38	− 0·0720564	6·52
6·54	+ 0·2660201	− 0·1621770	0·3115575	328° 37′ 54″·37	− 0·0667807	6·54
6·56	+ 0·2688106	− 0·1565714	0·3110848	329° 46′ 51″·29	− 0·0614556	6·56
6·58	+ 0·2714853	− 0·1509204	0·3106143	330° 55′ 48″·14	− 0·0560834	6·58
6·60	+ 0·2740434	− 0·1452262	0·3101458	332° 4′ 44″·93	− 0·0506665	6·60
6·62	+ 0·2764843	− 0·1394913	0·3096795	333° 13′ 41″·65	− 0·0452073	6·62
6·64	+ 0·2788074	− 0·1337179	0·3092152	334° 22′ 38″·30	− 0·0397080	6·64
6·66	+ 0·2810122	− 0·1279085	0·3087530	335° 31′ 34″·88	− 0·0341711	6·66
6·68	+ 0·2830981	− 0·1220655	0·3082929	336° 40′ 31″·40	− 0·0285990	6·68
6·70	+ 0·2850647	− 0·1161911	0·3078348	337° 49′ 27″·86	− 0·0229940	6·70
6·72	+ 0·2869117	− 0·1102879	0·3073788	338° 58′ 24″·25	− 0·0173587	6·72
6·74	+ 0·2886385	− 0·1043582	0·3069248	340° 7′ 20″·58	− 0·0116953	6·74
6·76	+ 0·2902449	− 0·0984043	0·3064727	341° 16′ 16″·84	− 0·0060063	6·76
6·78	+ 0·2917307	− 0·0924287	0·3060227	342° 25′ 13″·04	− 0·0002941	6·78
6·80	+ 0·2930956	− 0·0864339	0·3055746	343° 34′ 9″·18	+ 0·0054389	6·80
6·82	+ 0·2943394	− 0·0804221	0·3051285	344° 43′ 5″·26	+ 0·0111903	6·82
6·84	+ 0·2954620	− 0·0743958	0·3046843	345° 52′ 1″·27	+ 0·0169576	6·84
6·86	+ 0·2964633	− 0·0683573	0·3042421	347° 0′ 57″·22	+ 0·0227386	6·86
6·88	+ 0·2973434	− 0·0623092	0·3038017	348° 9′ 53″·12	+ 0·0285306	6·88
6·90	+ 0·2981020	− 0·0562537	0·3033633	349° 18′ 48″·95	+ 0·0343315	6·90
6·92	+ 0·2987395	− 0·0501933	0·3029268	350° 27′ 44″·73	+ 0·0401386	6·92
6·94	+ 0·2992557	− 0·0441303	0·3024921	351° 36′ 40″·45	+ 0·0459497	6·94
6·96	+ 0·2996510	− 0·0380671	0·3020593	352° 45′ 36″·10	+ 0·0517624	6·96
6·98	+ 0·2999254	− 0·0320062	0·3016283	353° 54′ 31″·70	+ 0·0575743	6·98
7·00	+ 0·3000793	− 0·0259497	0·3011992	355° 3′ 27″·25	+ 0·0633830	7·00

Table I. Functions of order unity

x	$J_1(x)$	$Y_1(x)$	$\lvert H_1^{(1)}(x)\rvert$	arg $H_1^{(1)}(x)$	$\mathbf{H}_1(x)$	x
6·02	− 0·2727017	− 0·1801479	0·3268325	213° 26′ 56″·12	+ 0·4729337	6·02
6·04	− 0·2686269	− 0·1851985	0·3262804	214° 35′ 0″·27	+ 0·4677800	6·04
6·06	− 0·2644612	− 0·1901605	0·3257311	215° 43′ 4″·69	+ 0·4627160	6·06
6·08	− 0·2602066	− 0·1950322	0·3251846	216° 51′ 9″·37	+ 0·4577431	6·08
6·10	− 0·2558648	− 0·1998122	0·3246409	217° 59′ 14″·31	+ 0·4528629	6·10
6·12	− 0·2514378	− 0·2044989	0·3240999	219° 7′ 19″·50	+ 0·4480769	6·12
6·14	− 0·2469275	− 0·2090908	0·3235616	220° 15′ 24″·94	+ 0·4433866	6·14
6·16	− 0·2423358	− 0·2135865	0·3230261	221° 23′ 30″·64	+ 0·4387935	6·16
6·18	− 0·2376649	− 0·2179846	0·3224932	222° 31′ 36″·59	+ 0·4342988	6·18
6·20	− 0·2329166	− 0·2222836	0·3219630	223° 39′ 42″·78	+ 0·4299040	6·20
6·22	− 0·2280930	− 0·2264824	0·3214354	224° 47′ 49″·21	+ 0·4256104	6·22
6·24	− 0·2231961	− 0·2305796	0·3209104	225° 55′ 55″·89	+ 0·4214192	6·24
6·26	− 0·2182281	− 0·2345740	0·3203880	227° 4′ 2″·81	+ 0·4173317	6·26
6·28	− 0·2131910	− 0·2384643	0·3198682	228° 12′ 9″·97	+ 0·4133490	6·28
6·30	− 0·2080869	− 0·2422495	0·3193509	229° 20′ 17″·35	+ 0·4094724	6·30
6·32	− 0·2029180	− 0·2459284	0·3188362	230° 28′ 24″·98	+ 0·4057028	6·32
6·34	− 0·1976865	− 0·2494998	0·3183239	231° 36′ 32″·83	+ 0·4020415	6·34
6·36	− 0·1923944	− 0·2529629	0·3178142	232° 44′ 40″·91	+ 0·3984894	6·36
6·38	− 0·1870440	− 0·2563166	0·3173069	233° 52′ 49″·22	+ 0·3950474	6·38
6·40	− 0·1816375	− 0·2595599	0·3168020	235° 0′ 57″·75	+ 0·3917166	6·40
6·42	− 0·1761771	− 0·2626920	0·3162996	236° 9′ 6″·51	+ 0·3884978	6·42
6·44	− 0·1706650	− 0·2657119	0·3157996	237° 17′ 15″·48	+ 0·3853919	6·44
6·46	− 0·1651035	− 0·2686190	0·3153020	238° 25′ 24″·68	+ 0·3823996	6·46
6·48	− 0·1594949	− 0·2714123	0·3148067	239° 33′ 34″·09	+ 0·3795218	6·48
6·50	− 0·1538413	− 0·2740913	0·3143138	240° 41′ 43″·72	+ 0·3767591	6·50
6·52	− 0·1481451	− 0·2766551	0·3138232	241° 49′ 53″·56	+ 0·3741123	6·52
6·54	− 0·1424086	− 0·2791032	0·3133350	242° 58′ 3″·61	+ 0·3715819	6·54
6·56	− 0·1366341	− 0·2814349	0·3128490	244° 6′ 13″·86	+ 0·3691685	6·56
6·58	− 0·1308238	− 0·2836498	0·3123653	245° 14′ 24″·33	+ 0·3668728	6·58
6·60	− 0·1249802	− 0·2857473	0·3118839	246° 22′ 35″·00	+ 0·3646951	6·60
6·62	− 0·1191054	− 0·2877269	0·3114047	247° 30′ 45″·88	+ 0·3626360	6·62
6·64	− 0·1132019	− 0·2895883	0·3109277	248° 38′ 56″·96	+ 0·3606958	6·64
6·66	− 0·1072720	− 0·2913310	0·3104529	249° 47′ 8″·24	+ 0·3588749	6·66
6·68	− 0·1013179	− 0·2929548	0·3099804	250° 55′ 19″·71	+ 0·3571737	6·68
6·70	− 0·0953421	− 0·2944593	0·3095099	252° 3′ 31″·38	+ 0·3555923	6·70
6·72	− 0·0893469	− 0·2958444	0·3090417	253° 11′ 43″·25	+ 0·3541310	6·72
6·74	− 0·0833346	− 0·2971098	0·3085756	254° 19′ 55″·31	+ 0·3527901	6·74
6·76	− 0·0773076	− 0·2982554	0·3081116	255° 28′ 7″·56	+ 0·3515696	6·76
6·78	− 0·0712681	− 0·2992811	0·3076497	256° 36′ 20″·01	+ 0·3504696	6·78
6·80	− 0·0652187	− 0·3001869	0·3071899	257° 44′ 32″·64	+ 0·3494901	6·80
6·82	− 0·0591615	− 0·3009727	0·3067322	258° 52′ 45″·45	+ 0·3486313	6·82
6·84	− 0·0530989	− 0·3016385	0·3062765	260° 0′ 58″·45	+ 0·3478930	6·84
6·86	− 0·0470332	− 0·3021846	0·3058229	261° 9′ 11″·64	+ 0·3472751	6·86
6·88	− 0·0409669	− 0·3026109	0·3053713	262° 17′ 25″·01	+ 0·3467775	6·88
6·90	− 0·0349021	− 0·3029176	0·3049217	263° 25′ 38″·56	+ 0·3464001	6·90
6·92	− 0·0288412	− 0·3031051	0·3044741	264° 33′ 52″·29	+ 0·3461426	6·92
6·94	− 0·0227866	− 0·3031734	0·3040285	265° 42′ 6″·19	+ 0·3460047	6·94
6·96	− 0·0167404	− 0·3031230	0·3035849	266° 50′ 20″·27	+ 0·3459862	6·96
6·98	− 0·0107051	− 0·3029541	0·3031432	267° 58′ 34″·53	+ 0·3460867	6·98
7·00	− 0·0046828	− 0·3026672	0·3027035	269° 6′ 48″·96	+ 0·3463057	7·00

Table I. Functions of order zero

x	$J_0(x)$	$Y_0(x)$	$\lvert H_0^{(1)}(x)\rvert$	arg $H_0^{(1)}(x)$	$\mathbf{H}_0(x)$	x
7·02	+ 0·3001128	− 0·0199002	0·3007719	356° 12′ 22″·74	+ 0·0691861	7·02
7·04	+ 0·3000264	− 0·0138600	0·3003464	357° 21′ 18″·17	+ 0·0749812	7·04
7·06	+ 0·2998204	− 0·0078314	0·2999227	358° 30′ 13″·54	+ 0·0807662	7·06
7·08	+ 0·2994953	− 0·0018167	0·2995008	359° 39′ 8″·86	+ 0·0865385	7·08
7·10	+ 0·2990514	+ 0·0041818	0·2990806	360° 48′ 4″·12	+ 0·0922958	7·10
7·12	+ 0·2984893	+ 0·0101617	0·2986622	361° 56′ 59″·33	+ 0·0980360	7·12
7·14	+ 0·2978096	+ 0·0161208	0·2982456	363° 5′ 54″·49	+ 0·1037565	7·14
7·16	+ 0·2970128	+ 0·0220568	0·2978307	364° 14′ 49″·59	+ 0·1094553	7·16
7·18	+ 0·2960996	+ 0·0279674	0·2974175	365° 23′ 44″·64	+ 0·1151299	7·18
7·20	+ 0·29[illegible]0707	+ 0·0338504	0·2970060	366° 32′ 39″·63	+ 0·1207782	7·20
7·22	+ 0·2939268	+ 0·0397036	0·2965962	367° 41′ 34″·57	+ 0·1263979	7·22
7·24	+ 0·2926686	+ 0·0455247	0·2961881	368° 50′ 29″·46	+ 0·1319868	7·24
7·26	+ 0·2912970	+ 0·0513115	0·2957817	369° 59′ 24″·30	+ 0·1375427	7·26
7·28	+ 0·2898128	+ 0·0570620	0·2953769	371° 8′ 19″·09	+ 0·1430634	7·28
7·30	+ 0·2882169	+ 0·0627739	0·2949738	372° 17′ 13″·83	+ 0·1485467	7·30
7·32	+ 0·2865103	+ 0·0684451	0·2945724	373° 26′ 8″·52	+ 0·1539905	7·32
7·34	+ 0·2846939	+ 0·0740734	0·2941726	374° 35′ 3″·16	+ 0·1593927	7·34
7·36	+ 0·2827687	+ 0·0796569	0·2937743	375° 43′ 57″·75	+ 0·1647511	7·36
7·38	+ 0·2807358	+ 0·0851934	0·2933777	376° 52′ 52″·29	+ 0·1700638	7·38
7·40	+ 0·2785962	+ 0·0906809	0·2929827	378° 1′ 46″·78	+ 0·1753286	7·40
7·42	+ 0·2763512	+ 0·0961173	0·2925893	379° 10′ 41″·22	+ 0·1805435	7·42
7·44	+ 0·2740018	+ 0·1015007	0·2921975	380° 19′ 35″·61	+ 0·1857066	7·44
7·46	+ 0·2715492	+ 0·1068292	0·2918072	381° 28′ 29″·96	+ 0·1908158	7·46
7·48	+ 0·2689947	+ 0·1121007	0·2914185	382° 37′ 24″·26	+ 0·1958692	7·48
7·50	+ 0·2663397	+ 0·1173133	0·2910313	383° 46′ 18″·52	+ 0·2008648	7·50
7·52	+ 0·2635853	+ 0·1224652	0·2906457	384° 55′ 12″·72	+ 0·2058008	7·52
7·54	+ 0·2607329	+ 0·1275545	0·2902616	386° 4′ 6″·88	+ 0·2106753	7·54
7·56	+ 0·2577839	+ 0·1325793	0·2898790	387° 13′ 1″·00	+ 0·2154865	7·56
7·58	+ 0·2547397	+ 0·1375379	0·2894979	388° 21′ 55″·07	+ 0·2202325	7·58
7·60	+ 0·2516018	+ 0·1424285	0·2891183	389° 30′ 49″·09	+ 0·2249115	7·60
7·62	+ 0·2483717	+ 0·1472494	0·2887401	390° 39′ 43″·07	+ 0·2295219	7·62
7·64	+ 0·2450508	+ 0·1519988	0·2883635	391° 48′ 37″·00	+ 0·2340620	7·64
7·66	+ 0·2416407	+ 0·1566751	0·2879883	392° 57′ 30″·89	+ 0·2385299	7·66
7·68	+ 0·2381429	+ 0·1612765	0·2876146	394° 6′ 24″·74	+ 0·2429241	7·68
7·70	+ 0·2345591	+ 0·1658016	0·2872424	395° 15′ 18″·54	+ 0·2472429	7·70
7·72	+ 0·2308910	+ 0·1702488	0·2868715	396° 24′ 12″·30	+ 0·2514848	7·72
7·74	+ 0·2271400	+ 0·1746164	0·2865021	397° 33′ 6″·02	+ 0·2556482	7·74
7·76	+ 0·2233081	+ 0·1789029	0·2861341	398° 41′ 59″·70	+ 0·2597315	7·76
7·78	+ 0·2193967	+ 0·1831070	0·2857676	399° 50′ 53″·33	+ 0·2637334	7·78
7·80	+ 0·2154078	+ 0·1872272	0·2854024	400° 59′ 46″·92	+ 0·2676524	7·80
7·82	+ 0·2113430	+ 0·1912620	0·2850386	402° 8′ 40″·47	+ 0·2714870	7·82
7·84	+ 0·2072042	+ 0·1952101	0·2846763	403° 17′ 33″·98	+ 0·2752358	7·84
7·86	+ 0·2029932	+ 0·1990701	0·2843152	404° 26′ 27″·45	+ 0·2788977	7·86
7·88	+ 0·1987118	+ 0·2028408	0·2839556	405° 35′ 20″·87	+ 0·2824711	7·88
7·90	+ 0·1943618	+ 0·2065209	0·2835973	406° 44′ 14″·26	+ 0·2859549	7·90
7·92	+ 0·1899452	+ 0·2101093	0·2832404	407° 53′ 7″·61	+ 0·2893479	7·92
7·94	+ 0·1854639	+ 0·2136046	0·2828848	409° 2′ 0″·92	+ 0·2926488	7·94
7·96	+ 0·1809198	+ 0·2170058	0·2825305	410° 10′ 54″·19	+ 0·2958566	7·96
7·98	+ 0·1763147	+ 0·2203118	0·2821776	411° 19′ 47″·41	+ 0·2989700	7·98
8·00	+ 0·1716508	+ 0·2235215	0·2818259	412° 28′ 40″·60	+ 0·3019881	8·00

Table I. Functions of order unity

x	$J_1(x)$	$Y_1(x)$	$\lvert H_1^{(1)}(x)\rvert$	$\arg H_1^{(1)}(x)$	$\mathbf{H}_1(x)$	x
7·02	+ 0·0013241	− 0·3022627	0·3022656	270° 15′ 3″·56	+ 0·3466429	7·02
7·04	+ 0·0073134	− 0·3017411	0·3018297	271° 23′ 18″·33	+ 0·3470978	7·04
7·06	+ 0·0132828	− 0·3011029	0·3013957	272° 31′ 33″·27	+ 0·3476699	7·06
7·08	+ 0·0192302	− 0·3003486	0·3009636	273° 39′ 48″·37	+ 0·3483585	7·08
7·10	+ 0·0251533	− 0·2994789	0·3005333	274° 48′ 3″·65	+ 0·3491631	7·10
7·12	+ 0·0310498	− 0·2984943	0·3001049	275° 56′ 19″·08	+ 0·3500830	7·12
7·14	+ 0·0369177	− 0·2973957	0·2996784	277° 4′ 34″·69	+ 0·3511175	7·14
7·16	+ 0·0427547	− 0·2961837	0·2992536	278° 12′ 50″·45	+ 0·3522659	7·16
7·18	+ 0·0485586	− 0·2948590	0·2988307	279° 21′ 6″·37	+ 0·3535275	7·18
7·20	+ 0·0543274	− 0·2934226	0·2984096	280° 29′ 22″·45	+ 0·3549013	7·20
7·22	+ 0·0600589	− 0·2918752	0·2979903	281° 37′ 38″·69	+ 0·3563867	7·22
7·24	+ 0·0657511	− 0·2902177	0·2975727	282° 45′ 55″·09	+ 0·3579825	7·24
7·26	+ 0·0714017	− 0·2884511	0·2971569	283° 54′ 11″·65	+ 0·3596880	7·26
7·28	+ 0·0770089	− 0·2865763	0·2967429	285° 2′ 28″·36	+ 0·3615021	7·28
7·30	+ 0·0825704	− 0·2845944	0·2963306	286° 10′ 45″·22	+ 0·3634239	7·30
7·32	+ 0·0880844	− 0·2825063	0·2959200	287° 19′ 2″·23	+ 0·3654523	7·32
7·34	+ 0·0935488	− 0·2803132	0·2955112	288° 27′ 19″·40	+ 0·3675862	7·34
7·36	+ 0·0989617	− 0·2780161	0·2951041	289° 35′ 36″·71	+ 0·3698244	7·36
7·38	+ 0·1043211	− 0·2756163	0·2946986	290° 43′ 54″·17	+ 0·3721659	7·38
7·40	+ 0·1096251	− 0·2731149	0·2942948	291° 52′ 11″·78	+ 0·3746094	7·40
7·42	+ 0·1148718	− 0·2705132	0·2938927	293° 0′ 29″·54	+ 0·3771537	7·42
7·44	+ 0·1200593	− 0·2678124	0·2934923	294° 8′ 47″·44	+ 0·3797976	7·44
7·46	+ 0·1251857	− 0·2650138	0·2930935	295° 17′ 5″·48	+ 0·3825396	7·46
7·48	+ 0·1302494	− 0·2621187	0·2926963	296° 25′ 23″·67	+ 0·3853786	7·48
7·50	+ 0·1352484	− 0·2591285	0·2923007	297° 33′ 42″·00	+ 0·3883131	7·50
7·52	+ 0·1401811	− 0·2560446	0·2919068	298° 42′ 0″·46	+ 0·3913417	7·52
7·54	+ 0·1450456	− 0·2528684	0·2915145	299° 50′ 19″·07	+ 0·3944630	7·54
7·56	+ 0·1498404	− 0·2496015	0·2911237	300° 58′ 37″·82	+ 0·3976756	7·56
7·58	+ 0·1545636	− 0·2462451	0·2907345	302° 6′ 56″·70	+ 0·4009779	7·58
7·60	+ 0·1592138	− 0·2428010	0·2903469	303° 15′ 15″·72	+ 0·4043684	7·60
7·62	+ 0·1637892	− 0·2392706	0·2899609	304° 23′ 34″·87	+ 0·4078456	7·62
7·64	+ 0·1682883	− 0·2356555	0·2895764	305° 31′ 54″·16	+ 0·4114078	7·64
7·66	+ 0·1727096	− 0·2319574	0·2891935	306° 40′ 13″·58	+ 0·4150535	7·66
7·68	+ 0·1770516	− 0·2281778	0·2888120	307° 48′ 33″·13	+ 0·4187811	7·68
7·70	+ 0·1813127	− 0·2243185	0·2884321	308° 56′ 52″·82	+ 0·4225888	7·70
7·72	+ 0·1854916	− 0·2203810	0·2880537	310° 5′ 12″·63	+ 0·4264750	7·72
7·74	+ 0·1895868	− 0·2163672	0·2876768	311° 13′ 32″·58	+ 0·4304379	7·74
7·76	+ 0·1935970	− 0·2122788	0·2873014	312° 21′ 52″·65	+ 0·4344758	7·76
7·78	+ 0·1975208	− 0·2081175	0·2869274	313° 30′ 12″·85	+ 0·4385870	7·78
7·80	+ 0·2013569	− 0·2038851	0·2865549	314° 38′ 33″·17	+ 0·4427696	7·80
7·82	+ 0·2051041	− 0·1995834	0·2861839	315° 46′ 53″·62	+ 0·4470219	7·82
7·84	+ 0·2087611	− 0·1952143	0·2858143	316° 55′ 14″·20	+ 0·4513419	7·84
7·86	+ 0·2123267	− 0·1907797	0·2854462	318° 3′ 34″·89	+ 0·4557279	7·86
7·88	+ 0·2157999	− 0·1862813	0·2850795	319° 11′ 55″·71	+ 0·4601780	7·88
7·90	+ 0·2191794	− 0·1817211	0·2847142	320° 20′ 16″·66	+ 0·4646902	7·90
7·92	+ 0·2224642	− 0·1771010	0·2843503	321° 28′ 37″·72	+ 0·4692627	7·92
7·94	+ 0·2256533	− 0·1724229	0·2839878	322° 36′ 58″·90	+ 0·4738934	7·94
7·96	+ 0·2287457	− 0·1676888	0·2836267	323° 45′ 20″·20	+ 0·4785806	7·96
7·98	+ 0·2317403	− 0·1629007	0·2832670	324° 53′ 41″·62	+ 0·4833221	7·98
8·00	+ 0·2346363	− 0·1580605	0·2829087	326° 2′ 3″·16	+ 0·4881160	8·00

Table I. Functions of order zero

x	$J_0(x)$	$Y_0(x)$	$\|H_0^{(1)}(x)\|$	$\arg H_0^{(1)}(x)$	$\mathbf{H}_0(x)$	x
8·02	+ 0·1669299	+ 0·2266339	0·2814756	413° 37′ 33″·75	+ 0·3049098	8·02
8·04	+ 0·1621542	+ 0·2296480	0·2811266	414° 46′ 26″·87	+ 0·3077342	8·04
8·06	+ 0·1573255	+ 0·2325628	0·2807789	415° 55′ 19″·94	+ 0·3104602	8·06
8·08	+ 0·1524459	+ 0·2353776	0·2804324	417° 4′ 12″·98	+ 0·3130870	8·08
8·10	+ 0·1475175	+ 0·2380913	0·2800873	418° 13′ 5″·98	+ 0·3156137	8·10
8·12	+ 0·1425423	+ 0·2407033	0·2797434	419° 21′ 58″·95	+ 0·3180394	8·12
8·14	+ 0·1375223	+ 0·2432126	0·2794007	420° 30′ 51″·87	+ 0·3203635	8·14
8·16	+ 0·1324598	+ 0·2456187	0·2790594	421° 39′ 44″·76	+ 0·3225852	8·16
8·18	+ 0·1273568	+ 0·2479207	0·2787192	422° 48′ 37″·62	+ 0·3247036	8·18
8·20	+ 0·1222153	+ 0·2501180	0·2783803	423° 57′ 30″·43	+ 0·3267183	8·20
8·22	+ 0·1170375	+ 0·2522101	0·2780427	425° 6′ 23″·21	+ 0·3286286	8·22
8·24	+ 0·1118256	+ 0·2541963	0·2777062	426° 15′ 15″·96	+ 0·3304339	8·24
8·26	+ 0·1065816	+ 0·2560762	0·2773710	427° 24′ 8″·67	+ 0·3321337	8·26
8·28	+ 0·1013077	+ 0·2578492	0·2770370	428° 33′ 1″·35	+ 0·3337274	8·28
8·30	+ 0·0960061	+ 0·2595150	0·2767042	429° 41′ 53″·99	+ 0·3352147	8·30
8·32	+ 0·0906789	+ 0·2610730	0·2763725	430° 50′ 46″·59	+ 0·3365952	8·32
8·34	+ 0·0853282	+ 0·2625230	0·2760421	431° 59′ 39″·16	+ 0·3378684	8·34
8·36	+ 0·0799563	+ 0·2638647	0·2757128	433° 8′ 31″·70	+ 0·3390341	8·36
8·38	+ 0·0745652	+ 0·2650977	0·2753847	434° 17′ 24″·21	+ 0·3400920	8·38
8·40	+ 0·0691573	+ 0·2662219	0·2750578	435° 26′ 16″·68	+ 0·3410418	8·40
8·42	+ 0·0637345	+ 0·2672370	0·2747321	436° 35′ 9″·11	+ 0·3418834	8·42
8·44	+ 0·0582992	+ 0·2681430	0·2744075	437° 44′ 1″·51	+ 0·3426166	8·44
8·46	+ 0·0528534	+ 0·2689397	0·2740840	438° 52′ 53″·88	+ 0·3432414	8·46
8·48	+ 0·0473994	+ 0·2696271	0·2737617	440° 1′ 46″·22	+ 0·3437576	8·48
8·50	+ 0·0419393	+ 0·2702051	0·2734405	441° 10′ 38″·53	+ 0·3441653	8·50
8·52	+ 0·0364752	+ 0·2706738	0·2731204	442° 19′ 30″·80	+ 0·3444644	8·52
8·54	+ 0·0310094	+ 0·2710333	0·2728015	443° 28′ 23″·04	+ 0·3446550	8·54
8·56	+ 0·0255440	+ 0·2712837	0·2724836	444° 37′ 15″·25	+ 0·3447373	8·56
8·58	+ 0·0200812	+ 0·2714251	0·2721669	445° 46′ 7″·43	+ 0·3447114	8·58
8·60	+ 0·0146230	+ 0·2714577	0·2718513	446° 54′ 59″·58	+ 0·3445775	8·60
8·62	+ 0·0091717	+ 0·2713818	0·2715368	448° 3′ 51″·69	+ 0·3443357	8·62
8·64	+ 0·0037293	+ 0·2711977	0·2712233	449° 12′ 43″·77	+ 0·3439865	8·64
8·66	− 0·0017019	+ 0·2709056	0·2709109	450° 21′ 35″·83	+ 0·3435301	8·66
8·68	− 0·0071200	+ 0·2705060	0·2705996	451° 30′ 27″·85	+ 0·3429669	8·68
8·70	− 0·0125227	+ 0·2699992	0·2702894	452° 39′ 19″·84	+ 0·3422972	8·70
8·72	− 0·0179081	+ 0·2693857	0·2699803	453° 48′ 11″·80	+ 0·3415216	8·72
8·74	− 0·0232739	+ 0·2686660	0·2696722	454° 57′ 3″·73	+ 0·3406404	8·74
8·76	− 0·0286182	+ 0·2678405	0·2693651	456° 5′ 55″·64	+ 0·3396543	8·76
8·78	− 0·0339388	+ 0·2669100	0·2690591	457° 14′ 47″·51	+ 0·3385638	8·78
8·80	− 0·0392338	+ 0·2658749	0·2687541	458° 23′ 39″·35	+ 0·3373694	8·80
8·82	− 0·0445011	+ 0·2647360	0·2684502	459° 32′ 31″·17	+ 0·3360719	8·82
8·84	− 0·0497387	+ 0·2634939	0·2681473	460° 41′ 22″·95	+ 0·3346718	8·84
8·86	− 0·0549445	+ 0·2621493	0·2678454	461° 50′ 14″·71	+ 0·3331700	8·86
8·88	− 0·0601167	+ 0·2607030	0·2675445	462° 59′ 6″·44	+ 0·3315672	8·88
8·90	− 0·0652532	+ 0·2591558	0·2672446	464° 7′ 58″·14	+ 0·3298642	8·90
8·92	− 0·0703522	+ 0·2575085	0·2669458	465° 16′ 49″·81	+ 0·3280617	8·92
8·94	− 0·0754116	+ 0·2557620	0·2666479	466° 25′ 41″·45	+ 0·3261608	8·94
8·96	− 0·0804295	+ 0·2539172	0·2663510	467° 34′ 33″·07	+ 0·3241622	8·96
8·98	− 0·0854042	+ 0·2519751	0·2660552	468° 43′ 24″·66	+ 0·3220669	8·98
9·00	− 0·0903336	+ 0·2499367	0·2657603	469° 52′ 16″·22	+ 0·3198760	9·00

Table I. Functions of order unity

x	$J_1(x)$	$Y_1(x)$	$\lvert H_1^{(1)}(x)\rvert$	$\arg H_1^{(1)}(x)$	$\mathbf{H}_1(x)$	x
8·02	+ 0·2374329	− 0·1531702	0·2825517	327° 10′ 24″·81	+ 0·4929604	8·02
8·04	+ 0·2401291	− 0·1482318	0·2821961	328° 18′ 46″·58	+ 0·4978531	8·04
8·06	+ 0·2427241	− 0·1432475	0·2818419	329° 27′ 8″·46	+ 0·5027922	8·06
8·08	+ 0·2452173	− 0·1382191	0·2814889	330° 35′ 30″·46	+ 0·5077756	8·08
8·10	+ 0·2476078	− 0·1331488	0·2811373	331° 43′ 52″·57	+ 0·5128012	8·10
8·12	+ 0·2498950	− 0·1280386	0·2807871	332° 52′ 14″·79	+ 0·5178671	8·12
8·14	+ 0·2520782	− 0·1228906	0·2804381	334° 0′ 37″·12	+ 0·5229711	8·14
8·16	+ 0·2541570	− 0·1177069	0·2800905	335° 8′ 59″·56	+ 0·5281111	8·16
8·18	+ 0·2561306	− 0·1124896	0·2797441	336° 17′ 22″·11	+ 0·5332850	8·18
8·20	+ 0·2579986	− 0·1072407	0·2793991	337° 25′ 44″·78	+ 0·5384907	8·20
8·22	+ 0·2597605	− 0·1019624	0·2790553	338° 34′ 7″·55	+ 0·5437262	8·22
8·24	+ 0·2614159	− 0·0966569	0·2787128	339° 42′ 30″·42	+ 0·5489893	8·24
8·26	+ 0·2629644	− 0·0913261	0·2783716	340° 50′ 53″·40	+ 0·5542779	8·26
8·28	+ 0·2644056	− 0·0859723	0·2780316	341° 59′ 16″·49	+ 0·5595898	8·28
8·30	+ 0·2657393	− 0·0805975	0·2776929	343° 7′ 39″·69	+ 0·5649229	8·30
8·32	+ 0·2669651	− 0·0752040	0·2773554	344° 16′ 2″·99	+ 0·5702752	8·32
8·34	+ 0·2680829	− 0·0697937	0·2770191	345° 24′ 26″·39	+ 0·5756443	8·34
8·36	+ 0·2690924	− 0·0643690	0·2766841	346° 32′ 49″·90	+ 0·5810283	8·36
8·38	+ 0·2699936	− 0·0589319	0·2763503	347° 41′ 13″·51	+ 0·5864249	8·38
8·40	+ 0·2707863	− 0·0534845	0·2760177	348° 49′ 37″·22	+ 0·5918321	8·40
8·42	+ 0·2714704	− 0·0480290	0·2756864	349° 58′ 1″·03	+ 0·5972476	8·42
8·44	+ 0·2720460	− 0·0425676	0·2753562	351° 6′ 24″·94	+ 0·6026694	8·44
8·46	+ 0·2725131	− 0·0371023	0·2750272	352° 14′ 48″·95	+ 0·6080953	8·46
8·48	+ 0·2728717	− 0·0316353	0·2746994	353° 23′ 13″·05	+ 0·6135232	8·48
8·50	+ 0·2731220	− 0·0261687	0·2743728	354° 31′ 37″·26	+ 0·6189510	8·50
8·52	+ 0·2732640	− 0·0207046	0·2740473	355° 40′ 1″·56	+ 0·6243764	8·52
8·54	+ 0·2732981	− 0·0152452	0·2737230	356° 48′ 25″·97	+ 0·6297975	8·54
8·56	+ 0·2732244	− 0·0097926	0·2733998	357° 56′ 50″·46	+ 0·6352120	8·56
8·58	+ 0·2730432	− 0·0043488	0·2730779	359° 5′ 15″·06	+ 0·6406180	8·58
8·60	+ 0·2727548	+ 0·0010840	0·2727570	360° 13′ 39″·74	+ 0·6460132	8·60
8·62	+ 0·2723596	+ 0·0065038	0·2724373	361° 22′ 4″·52	+ 0·6513957	8·62
8·64	+ 0·2718580	+ 0·0119084	0·2721187	362° 30′ 29″·39	+ 0·6567633	8·64
8·66	+ 0·2712504	+ 0·0172958	0·2718012	363° 38′ 54″·35	+ 0·6621139	8·66
8·68	+ 0·2705372	+ 0·0226640	0·2714849	364° 47′ 19″·41	+ 0·6674455	8·68
8·70	+ 0·2697190	+ 0·0280110	0·2711696	365° 55′ 44″·57	+ 0·6727561	8·70
8·72	+ 0·2687964	+ 0·0333346	0·2708555	367° 4′ 9″·81	+ 0·6780436	8·72
8·74	+ 0·2677699	+ 0·0386328	0·2705424	368° 12′ 35″·14	+ 0·6833060	8·74
8·76	+ 0·2666402	+ 0·0439037	0·2702305	369° 21′ 0″·57	+ 0·6885413	8·76
8·78	+ 0·2654079	+ 0·0491453	0·2699196	370° 29′ 26″·08	+ 0·6937475	8·78
8·80	+ 0·2640737	+ 0·0543556	0·2696098	371° 37′ 51″·68	+ 0·6989226	8·80
8·82	+ 0·2626384	+ 0·0595326	0·2693011	372° 46′ 17″·36	+ 0·7040647	8·82
8·84	+ 0·2611028	+ 0·0646744	0·2689934	373° 54′ 43″·14	+ 0·7091718	8·84
8·86	+ 0·2594677	+ 0·0697790	0·2686868	375° 3′ 9″·00	+ 0·7142420	8·86
8·88	+ 0·2577339	+ 0·0748447	0·2683813	376° 11′ 34″·95	+ 0·7192734	8·88
8·90	+ 0·2559024	+ 0·0798694	0·2680768	377° 20′ 0″·98	+ 0·7242641	8·90
8·92	+ 0·2539740	+ 0·0848513	0·2677733	378° 28′ 27″·10	+ 0·7292122	8·92
8·94	+ 0·2519497	+ 0·0897886	0·2674709	379° 36′ 53″·30	+ 0·7341159	8·94
8·96	+ 0·2498306	+ 0·0946795	0·2671694	380° 45′ 19″·59	+ 0·7389734	8·96
8·98	+ 0·2476176	+ 0·0995220	0·2668691	381° 53′ 45″·96	+ 0·7437828	8·98
9·00	+ 0·2453118	+ 0·1043146	0·2665697	383° 2′ 12″·41	+ 0·7485424	9·00

Table I. Functions of order zero

x	$J_0(x)$	$Y_0(x)$	$\lvert H_0^{(1)}(x)\rvert$	$\arg H_0^{(1)}(x)$	$\mathbf{H}_0(x)$	x
9·02	− 0·0952160	+ 0·2478029	0·2654663	471° 1′ 7″·75	+ 0·3175904	9·02
9·04	− 0·1000496	+ 0·2455748	0·2651734	472° 9′ 59″·26	+ 0·3152111	9·04
9·06	− 0·1048325	+ 0·2432536	0·2648814	473° 18′ 50″·73	+ 0·3127393	9·06
9·08	− 0·1095629	+ 0·2408402	0·2645904	474° 27′ 42″·18	+ 0·3101761	9·08
9·10	− 0·1142392	+ 0·2383360	0·2643003	475° 36′ 33″·60	+ 0·3075226	9·10
9·12	− 0·1188596	+ 0·2357420	0·2640112	476° 45′ 25″·01	+ 0·3047800	9·12
9·14	− 0·1234224	+ 0·2330595	0·2637230	477° 54′ 16″·39	+ 0·3019495	9·14
9·16	− 0·1279258	+ 0·2302898	0·2634358	479° 3′ 7″·73	+ 0·2990324	9·16
9·18	− 0·1323684	+ 0·2274341	0·2631495	480° 11′ 59″·05	+ 0·2960300	9·18
9·20	− 0·1367484	+ 0·2244937	0·2628641	481° 20′ 50″·35	+ 0·2929435	9·20
9·22	− 0·1410642	+ 0·2214700	0·2625796	482° 29′ 41″·62	+ 0·2897743	9·22
9·24	− 0·1453143	+ 0·2183644	0·2622961	483° 38′ 32″·86	+ 0·2865238	9·24
9·26	− 0·1494972	+ 0·2151782	0·2620135	484° 47′ 24″·08	+ 0·2831934	9·26
9·28	− 0·1536113	+ 0·2119130	0·2617318	485° 56′ 15″·28	+ 0·2797845	9·28
9·30	− 0·1576552	+ 0·2085701	0·2614510	487° 5′ 6″·44	+ 0·2762985	9·30
9·32	− 0·1616274	+ 0·2051510	0·2611711	488° 13′ 57″·59	+ 0·2727370	9·32
9·34	− 0·1655265	+ 0·2016573	0·2608921	489° 22′ 48″·70	+ 0·2691014	9·34
9·36	− 0·1693511	+ 0·1980905	0·2606140	490° 31′ 39″·80	+ 0·2653933	9·36
9·38	− 0·1730999	+ 0·1944522	0·2603368	491° 40′ 30″·87	+ 0·2616143	9·38
9·40	− 0·1767716	+ 0·1907439	0·2600604	492° 49′ 21″·92	+ 0·2577659	9·40
9·42	− 0·1803648	+ 0·1869673	0·2597850	493° 58′ 12″·95	+ 0·2538498	9·42
9·44	− 0·1838783	+ 0·1831240	0·2595104	495° 7′ 3″·94	+ 0·2498675	9·44
9·46	− 0·1873109	+ 0·1792157	0·2592367	496° 15′ 54″·92	+ 0·2458208	9·46
9·48	− 0·1906615	+ 0·1752440	0·2589638	497° 24′ 45″·87	+ 0·2417112	9·48
9·50	− 0·1939287	+ 0·1712106	0·2586918	498° 33′ 36″·80	+ 0·2375406	9·50
9·52	− 0·1971117	+ 0·1671174	0·2584206	499° 42′ 27″·70	+ 0·2333107	9·52
9·54	− 0·2002092	+ 0·1629659	0·2581503	500° 51′ 18″·58	+ 0·2290231	9·54
9·56	− 0·2032202	+ 0·1587580	0·2578809	502° 0′ 9″·44	+ 0·2246796	9·56
9·58	− 0·2061437	+ 0·1544955	0·2576123	503° 9′ 0″·28	+ 0·2202821	9·58
9·60	− 0·2089787	+ 0·1501801	0·2573445	504° 17′ 51″·08	+ 0·2158322	9·60
9·62	− 0·2117244	+ 0·1458137	0·2570775	505° 26′ 41″·87	+ 0·2113319	9·62
9·64	− 0·2143797	+ 0·1413982	0·2568114	506° 35′ 32″·65	+ 0·2067829	9·64
9·66	− 0·2169439	+ 0·1369352	0·2565461	507° 44′ 23″·40	+ 0·2021871	9·66
9·68	− 0·2194161	+ 0·1324268	0·2562816	508° 53′ 14″·12	+ 0·1975464	9·68
9·70	− 0·2217955	+ 0·1278748	0·2560180	510° 2′ 4″·82	+ 0·1928625	9·70
9·72	− 0·2240814	+ 0·1232810	0·2557551	511° 10′ 55″·49	+ 0·1881375	9·72
9·74	− 0·2262730	+ 0·1186475	0·2554931	512° 19′ 46″·15	+ 0·1833732	9·74
9·76	− 0·2283698	+ 0·1139760	0·2552318	513° 28′ 36″·79	+ 0·1785715	9·76
9·78	− 0·2303710	+ 0·1092686	0·2549714	514° 37′ 27″·40	+ 0·1737344	9·78
9·80	− 0·2322760	+ 0·1045271	0·2547117	515° 46′ 18″·00	+ 0·1688637	9·80
9·82	− 0·2340844	+ 0·0997535	0·2544529	516° 55′ 8″·57	+ 0·1639615	9·82
9·84	− 0·2357955	+ 0·0949498	0·2541948	518° 3′ 59″·12	+ 0·1590296	9·84
9·86	− 0·2374090	+ 0·0901178	0·2539375	519° 12′ 49″·65	+ 0·1540700	9·86
9·88	− 0·2389243	+ 0·0852597	0·2536810	520° 21′ 40″·15	+ 0·1490847	9·88
9·90	− 0·2403411	+ 0·0803773	0·2534253	521° 30′ 30″·64	+ 0·1440757	9·90
9·92	− 0·2416590	+ 0·0754727	0·2531703	522° 39′ 21″·11	+ 0·1390449	9·92
9·94	− 0·2428777	+ 0·0705477	0·2529161	523° 48′ 11″·56	+ 0·1339943	9·94
9·96	− 0·2439968	+ 0·0656045	0·2526626	524° 57′ 1″·99	+ 0·1289259	9·96
9·98	− 0·2450163	+ 0·0606450	0·2524100	526° 5′ 52″·39	+ 0·1238417	9·98
10·00	− 0·2459358	+ 0·0556712	0·2521580	527° 14′ 42″·78	+ 0·1187437	10·00

Table I. Functions of order unity

x	$J_1(x)$	$Y_1(x)$	$\lvert H_1^{(1)}(x)\rvert$	arg $H_1^{(1)}(x)$	$\mathbf{H}_1(x)$	x
9·02	+ 0·2429143	+ 0·1090553	0·2662713	384° 10′ 38″·94	+ 0·7532504	9·02
9·04	+ 0·2404263	+ 0·1137425	0·2659740	385° 19′ 5″·56	+ 0·7579051	9·04
9·06	+ 0·2378489	+ 0·1183744	0·2656776	386° 27′ 32″·26	+ 0·7625048	9·06
9·08	+ 0·2351833	+ 0·1229495	0·2653823	387° 35′ 59″·04	+ 0·7670477	9·08
9·10	+ 0·2324307	+ 0·1274659	0·2650879	388° 44′ 25″·90	+ 0·7715322	9·10
9·12	+ 0·2295925	+ 0·1319221	0·2647945	389° 52′ 52″·84	+ 0·7759567	9·12
9·14	+ 0·2266698	+ 0·1363164	0·2645021	391° 1′ 19″·86	+ 0·7803195	9·14
9·16	+ 0·2236640	+ 0·1406474	0·2642107	392° 9′ 46″·95	+ 0·7846192	9·16
9·18	+ 0·2205765	+ 0·1449133	0·2639202	393° 18′ 14″·12	+ 0·7888540	9·18
9·20	+ 0·2174087	+ 0·1491128	0·2636307	394° 26′ 41″·37	+ 0·7930226	9·20
9·22	+ 0·2141618	+ 0·1532443	0·2633421	395° 35′ 8″·70	+ 0·7971234	9·22
9·24	+ 0·2108375	+ 0·1573063	0·2630545	396° 43′ 36″·11	+ 0·8011549	9·24
9·26	+ 0·2074370	+ 0·1612974	0·2627679	397° 52′ 3″·59	+ 0·8051156	9·26
9·28	+ 0·2039620	+ 0·1652162	0·2624822	399° 0′ 31″·15	+ 0·8090043	9·28
9·30	+ 0·2004139	+ 0·1690613	0·2621974	400° 8′ 58″·78	+ 0·8128195	9·30
9·32	+ 0·1967943	+ 0·1728314	0·2619135	401° 17′ 26″·49	+ 0·8165598	9·32
9·34	+ 0·1931047	+ 0·1765251	0·2616306	402° 25′ 54″·27	+ 0·8202240	9·34
9·36	+ 0·1893468	+ 0·1801413	0·2613486	403° 34′ 22″·13	+ 0·8238107	9·36
9·38	+ 0·1855221	+ 0·1836785	0·2610675	404° 42′ 50″·06	+ 0·8273187	9·38
9·40	+ 0·1816322	+ 0·1871357	0·2607873	405° 51′ 18″·06	+ 0·8307469	9·40
9·42	+ 0·1776789	+ 0·1905116	0·2605080	406° 59′ 46″·14	+ 0·8340939	9·42
9·44	+ 0·1736637	+ 0·1938050	0·2602297	408° 8′ 14″·28	+ 0·8373587	9·44
9·46	+ 0·1695884	+ 0·1970150	0·2599522	409° 16′ 42″·50	+ 0·8405401	9·46
9·48	+ 0·1654548	+ 0·2001403	0·2596756	410° 25′ 10″·79	+ 0·8436371	9·48
9·50	+ 0·1612644	+ 0·2031799	0·2593999	411° 33′ 39″·15	+ 0·8466485	9·50
9·52	+ 0·1570192	+ 0·2061329	0·2591250	412° 42′ 7″·58	+ 0·8495735	9·52
9·54	+ 0·1527208	+ 0·2089982	0·2588511	413° 50′ 36″·08	+ 0·8524110	9·54
9·56	+ 0·1483711	+ 0·2117749	0·2585780	414° 59′ 4″·64	+ 0·8551601	9·56
9·58	+ 0·1439718	+ 0·2144621	0·2583058	416° 7′ 33″·28	+ 0·8578198	9·58
9·60	+ 0·1395248	+ 0·2170590	0·2580344	417° 16′ 1″·99	+ 0·8603894	9·60
9·62	+ 0·1350319	+ 0·2195646	0·2577639	418° 24′ 30″·76	+ 0·8628679	9·62
9·64	+ 0·1304950	+ 0·2219783	0·2574943	419° 32′ 59″·61	+ 0·8652546	9·64
9·66	+ 0·1259159	+ 0·2242992	0·2572255	420° 41′ 28″·52	+ 0·8675487	9·66
9·68	+ 0·1212965	+ 0·2265267	0·2569576	421° 49′ 57″·49	+ 0·8697495	9·68
9·70	+ 0·1166386	+ 0·2286600	0·2566904	422° 58′ 26″·53	+ 0·8718563	9·70
9·72	+ 0·1119443	+ 0·2306986	0·2564242	424° 6′ 55″·64	+ 0·8738685	9·72
9·74	+ 0·1072154	+ 0·2326417	0·2561587	425° 15′ 24″·82	+ 0·8757855	9·74
9·76	+ 0·1024537	+ 0·2344889	0·2558941	426° 23′ 54″·06	+ 0·8776066	9·76
9·78	+ 0·0976613	+ 0·2362395	0·2556303	427° 32′ 23″·36	+ 0·8793314	9·78
9·80	+ 0·0928401	+ 0·2378932	0·2553673	428° 40′ 52″·73	+ 0·8809594	9·80
9·82	+ 0·0879920	+ 0·2394495	0·2551052	429° 49′ 22″·16	+ 0·8824901	9·82
9·84	+ 0·0831189	+ 0·2409079	0·2548438	430° 57′ 52″·66	+ 0·8839230	9·84
9·86	+ 0·0782229	+ 0·2422681	0·2545833	432° 6′ 21″·22	+ 0·8852579	9·86
9·88	+ 0·0733059	+ 0·2435297	0·2543235	433° 14′ 50″·84	+ 0·8864944	9·88
9·90	+ 0·0683698	+ 0·2446924	0·2540646	434° 23′ 20″·52	+ 0·8876322	9·90
9·92	+ 0·0634167	+ 0·2457560	0·2538064	435° 31′ 50″·27	+ 0·8886710	9·92
9·94	+ 0·0584484	+ 0·2467203	0·2535491	436° 40′ 20″·08	+ 0·8896106	9·94
9·96	+ 0·0534670	+ 0·2475850	0·2532925	437° 48′ 49″·95	+ 0·8904508	9·96
9·98	+ 0·0484745	+ 0·2483501	0·2530367	438° 57′ 19″·88	+ 0·8911914	9·98
10·00	+ 0·0434727	+ 0·2490154	0·2527816	440° 5′ 49″·87	+ 0·8918325	10·00

Table I. Functions of order zero

x	$J_0(x)$	$Y_0(x)$	$\lvert H_0^{(1)}(x)\rvert$	$\arg H_0^{(1)}(x)$	$\mathbf{H}_0(x)$	x
10·02	− 0·2467551	+ 0·0506850	0·2519069	528° 23′ 33″·15	+ 0·1136338	10·02
10·04	− 0·2474743	+ 0·0456886	0·2516564	529° 32′ 23″·49	+ 0·1085142	10·04
10·06	− 0·2480931	+ 0·0406838	0·2514068	530° 41′ 13″·82	+ 0·1033867	10·06
10·08	− 0·2486116	+ 0·0356728	0·2511578	531° 50′ 4″·13	+ 0·0982533	10·08
10·10	− 0·2490297	+ 0·0306574	0·2509096	532° 58′ 54″·41	+ 0·0931162	10·10
10·12	− 0·2493474	+ 0·0256397	0·2506622	534° 7′ 44″·68	+ 0·0879771	10·12
10·14	− 0·2495649	+ 0·0206216	0·2504154	535° 16′ 34″·93	+ 0·0828382	10·14
10·16	− 0·2496822	+ 0·0156052	0·2501694	536° 25′ 25″·16	+ 0·0777014	10·16
10·18	− 0·2496996	+ 0·0105924	0·2499241	537° 34′ 15″·37	+ 0·0725687	10·18
10·20	− 0·2496171	+ 0·0055852	0·2496795	538° 43′ 5″·56	+ 0·0674420	10·20
10·22	− 0·2494350	+ 0·0005856	0·2494357	539° 51′ 55″·73	+ 0·0623234	10·22
10·24	− 0·2491536	− 0·0044044	0·2491925	541° 0′ 45″·88	+ 0·0572148	10·24
10·26	− 0·2487732	− 0·0093830	0·2489501	542° 9′ 36″·02	+ 0·0521181	10·26
10·28	− 0·2482942	− 0·0143481	0·2487084	543° 18′ 26″·13	+ 0·0470353	10·28
10·30	− 0·2477168	− 0·0192978	0·2484674	544° 27′ 16″·23	+ 0·0419684	10·30
10·32	− 0·2470416	− 0·0242303	0·2482270	545° 36′ 6″·31	+ 0·0369192	10·32
10·34	− 0·2462690	− 0·0291435	0·2479874	546° 44′ 56″·37	+ 0·0318896	10·34
10·36	− 0·2453994	− 0·0340355	0·2477484	547° 53′ 46″·41	+ 0·0268817	10·36
10·38	− 0·2444335	− 0·0389045	0·2475102	549° 2′ 36″·44	+ 0·0218972	10·38
10·40	− 0·2433718	− 0·0437486	0·2472726	550° 11′ 26″·44	+ 0·0169381	10·40
10·42	− 0·2422148	− 0·0485659	0·2470357	551° 20′ 16″·43	+ 0·0120062	10·42
10·44	− 0·2409633	− 0·0533546	0·2467995	552° 29′ 6″·41	+ 0·0071034	10·44
10·46	− 0·2396178	− 0·0581128	0·2465640	553° 37′ 56″·36	+ 0·0022315	10·46
10·48	− 0·2381792	− 0·0628386	0·2463291	554° 46′ 46″·30	− 0·0026077	10·48
10·50	− 0·2366482	− 0·0675304	0·2460949	555° 55′ 36″·22	− 0·0074123	10·50
10·52	− 0·2350255	− 0·0721862	0·2458614	557° 4′ 26″·12	− 0·0121806	10·52
10·54	− 0·2333120	− 0·0768043	0·2456285	558° 13′ 16″·00	− 0·0169108	10·54
10·56	− 0·2315085	− 0·0813830	0·2453963	559° 22′ 5″·87	− 0·0216012	10·56
10·58	− 0·2296158	− 0·0859205	0·2451648	560° 30′ 55″·72	− 0·0262499	10·58
10·60	− 0·2276350	− 0·0904152	0·2449339	561° 39′ 45″·55	− 0·0308553	10·60
10·62	− 0·2255670	− 0·0948652	0·2447037	562° 48′ 35″·37	− 0·0354157	10·62
10·64	− 0·2234127	− 0·0992689	0·2444741	563° 57′ 25″·17	− 0·0399295	10·64
10·66	− 0·2211732	− 0·1036247	0·2442451	565° 6′ 14″·95	− 0·0443948	10·66
10·68	− 0·2188495	− 0·1079309	0·2440168	566° 15′ 4″·71	− 0·0488103	10·68
10·70	− 0·2164427	− 0·1121859	0·2437891	567° 23′ 54″·46	− 0·0531741	10·70
10·72	− 0·2139539	− 0·1163881	0·2435621	568° 32′ 44″·20	− 0·0574847	10·72
10·74	− 0·2113843	− 0·1205360	0·2433357	569° 41′ 33″·91	− 0·0617406	10·74
10·76	− 0·2087349	− 0·1246281	0·2431099	570° 50′ 23″·61	− 0·0659402	10·76
10·78	− 0·2060071	− 0·1286627	0·2428847	571° 59′ 13″·30	− 0·0700821	10·78
10·80	− 0·2032020	− 0·1326384	0·2426602	573° 8′ 2″·97	− 0·0741646	10·80
10·82	− 0·2003208	− 0·1365537	0·2424363	574° 16′ 52″·62	− 0·0781864	10·82
10·84	− 0·1973650	− 0·1404073	0·2422130	575° 25′ 42″·26	− 0·0821461	10·84
10·86	− 0·1943357	− 0·1441977	0·2419903	576° 34′ 31″·88	− 0·0860421	10·86
10·88	− 0·1912343	− 0·1479234	0·2417683	577° 43′ 21″·48	− 0·0898731	10·88
10·90	− 0·1880622	− 0·1515832	0·2415468	578° 52′ 11″·07	− 0·0936378	10·90
10·92	− 0·1848208	− 0·1551757	0·2413260	580° 1′ 0″·64	− 0·0973349	10·92
10·94	− 0·1815115	− 0·1586996	0·2411057	581° 9′ 50″·20	− 0·1009630	10·94
10·96	− 0·1781356	− 0·1621537	0·2408861	582° 18′ 39″·75	− 0·1045209	10·96
10·98	− 0·1746947	− 0·1655367	0·2406671	583° 27′ 29″·28	− 0·1080073	10·98
11·00	− 0·1711903	− 0·1688473	0·2404486	584° 36′ 18″·79	− 0·1114210	11·00

Table I. Functions of order unity

x	$J_1(x)$	$Y_1(x)$	$\lvert H_1^{(1)}(x) \rvert$	arg $H_1^{(1)}(x)$	$\mathbf{H}_1(x)$	x
10·02	+ 0·0384638	+ 0·2495809	0·2525274	441° 14′ 19″·93	+ 0·8923738	10·02
10·04	+ 0·0334497	+ 0·2500465	0·2522739	442° 22′ 50″·04	+ 0·8928155	10·04
10·06	+ 0·0284322	+ 0·2504122	0·2520212	443° 31′ 20″·21	+ 0·8931574	10·06
10·08	+ 0·0234135	+ 0·2506782	0·2517692	444° 39′ 50″·44	+ 0·8933996	10·08
10·10	+ 0·0183955	+ 0·2508444	0·2515180	445° 48′ 20″·73	+ 0·8935423	10·10
10·12	+ 0·0133801	+ 0·2509111	0·2512676	446° 56′ 51″·08	+ 0·8935856	10·12
10·14	+ 0·0083694	+ 0·2508783	0·2510179	448° 5′ 21″·49	+ 0·8935295	10·14
10·16	+ 0·0033652	+ 0·2507464	0·2507690	449° 13′ 51″·95	+ 0·8933744	10·16
10·18	− 0·0016305	+ 0·2505154	0·2505208	450° 22′ 22″·48	+ 0·8931204	10·18
10·20	− 0·0066157	+ 0·2501858	0·2502733	451° 30′ 53″·06	+ 0·8927679	10·20
10·22	− 0·0115886	+ 0·2497578	0·2500266	452° 39′ 23″·69	+ 0·8923172	10·22
10·24	− 0·0165471	+ 0·2492319	0·2497806	453° 47′ 54″·39	+ 0·8917685	10·24
10·26	− 0·0214895	+ 0·2486082	0·2495353	454° 56′ 25″·14	+ 0·8911224	10·26
10·28	− 0·0264137	+ 0·2478874	0·2492907	456° 4′ 55″·94	+ 0·8903793	10·28
10·30	− 0·0313178	+ 0·2470699	0·2490469	457° 13′ 26″·80	+ 0·8895395	10·30
10·32	− 0·0362001	+ 0·2461562	0·2488038	458° 21′ 57″·72	+ 0·8886036	10·32
10·34	− 0·0410586	+ 0·2451468	0·2485614	459° 30′ 28″·69	+ 0·8875722	10·34
10·36	− 0·0458914	+ 0·2440423	0·2483197	460° 38′ 59″·71	+ 0·8864459	10·36
10·38	− 0·0506967	+ 0·2428434	0·2480787	461° 47′ 30″·79	+ 0·8852252	10·38
10·40	− 0·0554728	+ 0·2415506	0·2478385	462° 56′ 1″·93	+ 0·8839107	10·40
10·42	− 0·0602176	+ 0·2401646	0·2475989	464° 4′ 33″·12	+ 0·8825033	10·42
10·44	− 0·0649296	+ 0·2386862	0·2473600	465° 13′ 4″·36	+ 0·8810035	10·44
10·46	− 0·0696068	+ 0·2371162	0·2471218	466° 21′ 35″·65	+ 0·8794122	10·46
10·48	− 0·0742475	+ 0·2354552	0·2468843	467° 30′ 7″·00	+ 0·8777301	10·48
10·50	− 0·0788500	+ 0·2337042	0·2466475	468° 38′ 38″·40	+ 0·8759580	10·50
10·52	− 0·0834125	+ 0·2318640	0·2464114	469° 47′ 9″·85	+ 0·8740969	10·52
10·54	− 0·0879333	+ 0·2299355	0·2461760	470° 55′ 41″·35	+ 0·8721475	10·54
10·56	− 0·0924107	+ 0·2279195	0·2459412	472° 4′ 12″·91	+ 0·8701109	10·56
10·58	− 0·0968431	+ 0·2258171	0·2457071	473° 12′ 44″·52	+ 0·8679879	10·58
10·60	− 0·1012287	+ 0·2236293	0·2454736	474° 21′ 16″·18	+ 0·8657796	10·60
10·62	− 0·1055659	+ 0·2213570	0·2452409	475° 29′ 47″·89	+ 0·8634870	10·62
10·64	− 0·1098532	+ 0·2190013	0·2450088	476° 38′ 19″·65	+ 0·8611110	10·64
10·66	− 0·1140889	+ 0·2165633	0·2447773	477° 46′ 51″·46	+ 0·8586529	10·66
10·68	− 0·1182715	+ 0·2140441	0·2445465	478° 55′ 23″·32	+ 0·8561136	10·68
10·70	− 0·1223994	+ 0·2114448	0·2443164	480° 3′ 55″·23	+ 0·8534944	10·70
10·72	− 0·1264711	+ 0·2087666	0·2440869	481° 12′ 27″·19	+ 0·8507964	10·72
10·74	− 0·1304852	+ 0·2060107	0·2438581	482° 20′ 59″·20	+ 0·8480208	10·74
10·76	− 0·1344401	+ 0·2031783	0·2436299	483° 29′ 31″·25	+ 0·8451689	10·76
10·78	− 0·1383343	+ 0·2002707	0·2434024	484° 38′ 3″·36	+ 0·8422418	10·78
10·80	− 0·1421666	+ 0·1972891	0·2431755	485° 46′ 35″·51	+ 0·8392408	10·80
10·82	− 0·1459354	+ 0·1942349	0·2429492	486° 55′ 7″·71	+ 0·8361673	10·82
10·84	− 0·1496394	+ 0·1911093	0·2427236	488° 3′ 39″·96	+ 0·8330226	10·84
10·86	− 0·1532774	+ 0·1879138	0·2424986	489° 12′ 12″·26	+ 0·8298080	10·86
10·88	− 0·1568479	+ 0·1846498	0·2422742	490° 20′ 44″·61	+ 0·8265250	10·88
10·90	− 0·1603497	+ 0·1813185	0·2420505	491° 29′ 17″·00	+ 0·8231749	10·90
10·92	− 0·1637815	+ 0·1779215	0·2418273	492° 37′ 49″·43	+ 0·8197592	10·92
10·94	− 0·1671422	+ 0·1744602	0·2416048	493° 46′ 21″·92	+ 0·8162792	10·94
10·96	− 0·1704305	+ 0·1709362	0·2413829	494° 54′ 54″·45	+ 0·8127366	10·96
10·98	− 0·1736452	+ 0·1673507	0·2411616	496° 3′ 27″·03	+ 0·8091327	10·98
11·00	− 0·1767853	+ 0·1637055	0·2409410	497° 11′ 59″·65	+ 0·8054691	11·00

Table I. Functions of order zero

x	$J_0(x)$	$Y_0(x)$	$\lvert H_0^{(1)}(x)\rvert$	$\arg H_0^{(1)}(x)$	$\mathbf{H}_0(x)$	x
11·02	− 0·1676238	− 0·1720845	0·2402308	585° 45′ 8″·29	− 0·1147608	11·02
11·04	− 0·1639968	− 0·1752470	0·2400135	586° 53′ 57″·77	− 0·1180257	11·04
11·06	− 0·1603109	− 0·1783338	0·2397968	588° 2′ 47″·23	− 0·1212144	11·06
11·08	− 0·1565675	− 0·1813437	0·2395807	589° 11′ 36″·69	− 0·1243260	11·08
11·10	− 0·1527683	− 0·1842758	0·2393652	590° 20′ 26″·13	− 0·1273593	11·10
11·12	− 0·1489149	− 0·1871289	0·2391503	591° 29′ 15″·55	− 0·1303133	11·12
11·14	− 0·1450089	− 0·1899021	0·2389359	592° 38′ 4″·96	− 0·1331870	11·14
11·16	− 0·1410520	− 0·1925944	0·2387222	593° 46′ 54″·35	− 0·1359795	11·16
11·18	− 0·1370458	− 0·1952050	0·2385090	594° 55′ 43″·73	− 0·1386899	11·18
11·20	− 0·1329919	− 0·1977329	0·2382963	596° 4′ 33″·09	− 0·1413173	11·20
11·22	− 0·1288922	− 0·2001772	0·2380843	597° 13′ 22″·44	− 0·1438608	11·22
11·24	− 0·1247483	− 0·2025372	0·2378728	598° 22′ 11″·78	− 0·1463196	11·24
11·26	− 0·1205618	− 0·2048121	0·2376618	599° 31′ 1″·10	− 0·1486929	11·26
11·28	− 0·1163346	− 0·2070011	0·2374514	600° 39′ 50″·41	− 0·1509799	11·28
11·30	− 0·1120685	− 0·2091034	0·2372416	601° 48′ 39″·70	− 0·1531801	11·30
11·32	− 0·1077650	− 0·2111185	0·2370323	602° 57′ 28″·98	− 0·1552926	11·32
11·34	− 0·1034261	− 0·2130457	0·2368236	604° 6′ 18″·25	− 0·1573169	11·34
11·36	− 0·0990535	− 0·2148843	0·2366155	605° 15′ 7″·51	− 0·1592522	11·36
11·38	− 0·0946491	− 0·2166338	0·2364078	606° 23′ 56″·75	− 0·1610982	11·38
11·40	− 0·0902145	− 0·2182937	0·2362008	607° 32′ 45″·97	− 0·1628541	11·40
11·42	− 0·0857517	− 0·2198634	0·2359942	608° 41′ 35″·18	− 0·1645196	11·42
11·44	− 0·0812623	− 0·2213425	0·2357882	609° 50′ 24″·38	− 0·1660942	11·44
11·46	− 0·0767484	− 0·2227306	0·2355828	610° 59′ 13″·56	− 0·1675773	11·46
11·48	− 0·0722117	− 0·2240273	0·2353779	612° 8′ 2″·73	− 0·1689687	11·48
11·50	− 0·0676539	− 0·2252321	0·2351735	613° 16′ 51″·89	− 0·1702681	11·50
11·52	− 0·0630771	− 0·2263449	0·2349696	614° 25′ 41″·04	− 0·1714749	11·52
11·54	− 0·0584830	− 0·2273652	0·2347663	615° 34′ 30″·17	− 0·1725891	11·54
11·56	− 0·0538735	− 0·2282930	0·2345635	616° 43′ 19″·28	− 0·1736103	11·56
11·58	− 0·0492505	− 0·2291278	0·2343612	617° 52′ 8″·39	− 0·1745384	11·58
11·60	− 0·0446157	− 0·2298697	0·2341595	619° 0′ 57″·48	− 0·1753731	11·60
11·62	− 0·0399711	− 0·2305185	0·2339582	620° 9′ 46″·56	− 0·1761144	11·62
11·64	− 0·0353184	− 0·2310740	0·2337575	621° 18′ 35″·63	− 0·1767622	11·64
11·66	− 0·0306597	− 0·2315362	0·2335573	622° 27′ 24″·68	− 0·1773163	11·66
11·68	− 0·0259967	− 0·2319050	0·2333576	623° 36′ 13″·72	− 0·1777768	11·68
11·70	− 0·0213313	− 0·2321806	0·2331584	624° 45′ 2″·75	− 0·1781436	11·70
11·72	− 0·0166653	− 0·2323629	0·2329598	625° 53′ 51″·76	− 0·1784169	11·72
11·74	− 0·0120006	− 0·2324520	0·2327616	627° 2′ 40″·77	− 0·1785968	11·74
11·76	− 0·0073391	− 0·2324481	0·2325639	628° 11′ 29″·75	− 0·1786832	11·76
11·78	− 0·0026825	− 0·2323513	0·2323668	629° 20′ 18″·73	− 0·1786765	11·78
11·80	+ 0·0019672	− 0·2321618	0·2321701	630° 29′ 7″·70	− 0·1785768	11·80
11·82	+ 0·0066082	− 0·2318798	0·2319740	631° 37′ 56″·65	− 0·1783843	11·82
11·84	+ 0·0112388	− 0·2315056	0·2317783	632° 46′ 45″·59	− 0·1780994	11·84
11·86	+ 0·0158571	− 0·2310396	0·2315831	633° 55′ 34″·52	− 0·1777222	11·86
11·88	+ 0·0204612	− 0·2304820	0·2313884	635° 4′ 23″·44	− 0·1772532	11·88
11·90	+ 0·0250494	− 0·2298332	0·2311942	636° 13′ 12″·34	− 0·1766928	11·90
11·92	+ 0·0296200	− 0·2290937	0·2310005	637° 22′ 1″·24	− 0·1760413	11·92
11·94	+ 0·0341710	− 0·2282638	0·2308073	638° 30′ 50″·12	− 0·1752992	11·94
11·96	+ 0·0387007	− 0·2273441	0·2306146	639° 39′ 38″·99	− 0·1744669	11·96
11·98	+ 0·0432074	− 0·2263351	0·2304223	640° 48′ 27″·84	− 0·1735451	11·98
12·00	+ 0·0476893	− 0·2252373	0·2302306	641° 57′ 16″·69	− 0·1725341	12·00

Table I. Functions of order unity

x	$J_1(x)$	$Y_1(x)$	$\lvert H_1^{(1)}(x)\rvert$	arg $H_1^{(1)}(x)$	$\mathbf{H}_1(x)$	x
11·02	− 0·1798496	+ 0·1600021	0·2407209	498° 20′ 32″·32	+ 0·8017474	11·02
11·04	− 0·1828371	+ 0·1562419	0·2405014	499° 29′ 5″·03	+ 0·7979691	11·04
11·06	− 0·1857467	+ 0·1524266	0·2402826	500° 37′ 37″·79	+ 0·7941357	11·06
11·08	− 0·1885774	+ 0·1485578	0·2400643	501° 46′ 10″·59	+ 0·7902489	11·08
11·10	− 0·1913283	+ 0·1446371	0·2398466	502° 54′ 43″·44	+ 0·7863103	11·10
11·12	− 0·1939984	+ 0·1406661	0·2396296	504° 3′ 16″·34	+ 0·7823215	11·12
11·14	− 0·1965868	+ 0·1366465	0·2394131	505° 11′ 49″·27	+ 0·7782842	11·14
11·16	− 0·1990926	+ 0·1325799	0·2391972	506° 20′ 22″·25	+ 0·7742001	11·16
11·18	− 0·2015150	+ 0·1284681	0·2389819	507° 28′ 55″·28	+ 0·7700707	11·18
11·20	− 0·2038531	+ 0·1243127	0·2387671	508° 37′ 28″·35	+ 0·7658979	11·20
11·22	− 0·2061063	+ 0·1201154	0·2385530	509° 46′ 1″·46	+ 0·7616832	11·22
11·24	− 0·2082738	+ 0·1158779	0·2383394	510° 54′ 34″·61	+ 0·7574286	11·24
11·26	− 0·2103549	+ 0·1116021	0·2381264	512° 3′ 7″·81	+ 0·7531356	11·26
11·28	− 0·2123488	+ 0·1072896	0·2379140	513° 11′ 41″·05	+ 0·7488060	11·28
11·30	− 0·2142550	+ 0·1029422	0·2377021	514° 20′ 14″·33	+ 0·7444416	11·30
11·32	− 0·2160729	+ 0·0985617	0·2374909	515° 28′ 47″·65	+ 0·7400442	11·32
11·34	− 0·2178019	+ 0·0941498	0·2372801	516° 37′ 21″·01	+ 0·7356155	11·34
11·36	− 0·2194415	+ 0·0897083	0·2370700	517° 45′ 54″·42	+ 0·7311573	11·36
11·38	− 0·2209912	+ 0·0852391	0·2368604	518° 54′ 27″·87	+ 0·7266715	11·38
11·40	− 0·2224506	+ 0·0807440	0·2366513	520° 3′ 1″·36	+ 0·7221598	11·40
11·42	− 0·2238192	+ 0·0762247	0·2364428	521° 11′ 34″·89	+ 0·7176241	11·42
11·44	− 0·2250966	+ 0·0716830	0·2362349	522° 20′ 8″·46	+ 0·7130661	11·44
11·46	− 0·2262825	+ 0·0671209	0·2360275	523° 28′ 42″·07	+ 0·7084877	11·46
11·48	− 0·2273766	+ 0·0625402	0·2358207	524° 37′ 15″·72	+ 0·7038907	11·48
11·50	− 0·2283786	+ 0·0579425	0·2356144	525° 45′ 49″·41	+ 0·6992769	11·50
11·52	− 0·2292883	+ 0·0533299	0·2354086	526° 54′ 23″·14	+ 0·6946483	11·52
11·54	− 0·2301055	+ 0·0487042	0·2352034	528° 2′ 56″·91	+ 0·6900066	11·54
11·56	− 0·2308300	+ 0·0440671	0·2349987	529° 11′ 30″·72	+ 0·6853536	11·56
11·58	− 0·2314617	+ 0·0394206	0·2347946	530° 20′ 4″·58	+ 0·6806913	11·58
11·60	− 0·2320005	+ 0·0347665	0·2345910	531° 28′ 38″·47	+ 0·6760215	11·60
11·62	− 0·2324463	+ 0·0301065	0·2343879	532° 37′ 12″·39	+ 0·6713459	11·62
11·64	− 0·2327992	+ 0·0254427	0·2341854	533° 45′ 46″·36	+ 0·6666665	11·64
11·66	− 0·2330591	+ 0·0207768	0·2339833	534° 54′ 20″·37	+ 0·6619851	11·66
11·68	− 0·2332261	+ 0·0161106	0·2337818	536° 2′ 54″·41	+ 0·6573035	11·68
11·70	− 0·2333002	+ 0·0114460	0·2335809	537° 11′ 28″·49	+ 0·6526236	11·70
11·72	− 0·2332817	+ 0·0067849	0·2333804	538° 20′ 2″·61	+ 0·6479472	11·72
11·74	− 0·2331707	+ 0·0021289	0·2331804	539° 28′ 36″·77	+ 0·6432761	11·74
11·76	− 0·2329674	− 0·0025199	0·2329810	540° 37′ 10″·96	+ 0·6386122	11·76
11·78	− 0·2326720	− 0·0071598	0·2327821	541° 45′ 45″·19	+ 0·6339572	11·78
11·80	− 0·2322847	− 0·0117890	0·2325837	542° 54′ 19″·46	+ 0·6293131	11·80
11·82	− 0·2318060	− 0·0164057	0·2323858	544° 2′ 53″·77	+ 0·6246815	11·82
11·84	− 0·2312361	− 0·0210081	0·2321884	545° 11′ 28″·11	+ 0·6200643	11·84
11·86	− 0·2305754	− 0·0255944	0·2319915	546° 20′ 2″·49	+ 0·6154633	11·86
11·88	− 0·2298243	− 0·0301628	0·2317951	547° 28′ 36″·91	+ 0·6108802	11·88
11·90	− 0·2289832	− 0·0347115	0·2315993	548° 37′ 11″·36	+ 0·6063169	11·90
11·92	− 0·2280528	− 0·0392388	0·2314039	549° 45′ 45″·85	+ 0·6017751	11·92
11·94	− 0·2270334	− 0·0437430	0·2312090	550° 54′ 20″·37	+ 0·5972565	11·94
11·96	− 0·2259255	− 0·0482223	0·2310146	552° 2′ 54″·93	+ 0·5927628	11·96
11·98	− 0·2247299	− 0·0526749	0·2308207	553° 11′ 29″·53	+ 0·5882959	11·98
12·00	− 0·2234471	− 0·0570992	0·2306273	554° 20′ 4″·16	+ 0·5838573	12·00

Table I. Functions of order zero

x	$J_0(x)$	$Y_0(x)$	$\|H_0^{(1)}(x)\|$	$\arg H_0^{(1)}(x)$	$\mathbf{H}_0(x)$	x
12·02	+ 0·0521447	− 0·2240513	0·2300393	643° 6′ 5″·52	− 0·1714348	12·02
12·04	+ 0·0565718	− 0·2227778	0·2298485	644° 14′ 54″·35	− 0·1702475	12·04
12·06	+ 0·0609690	− 0·2214173	0·2296581	645° 23′ 43″·16	− 0·1689731	12·06
12·08	+ 0·0653346	− 0·2199706	0·2294682	646° 32′ 31″·96	− 0·1676121	12·08
12·10	+ 0·0696668	− 0·2184384	0·2292788	647° 41′ 20″·74	− 0·1661654	12·10
12·12	+ 0·0739640	− 0·2168214	0·2290899	648° 50′ 9″·52	− 0·1646336	12·12
12·14	+ 0·0782245	− 0·2151204	0·2289014	649° 58′ 58″·29	− 0·1630175	12·14
12·16	+ 0·0824468	− 0·2133362	0·2287134	651° 7′ 47″·04	− 0·1613180	12·16
12·18	+ 0·0866292	− 0·2114698	0·2285259	652° 16′ 35″·78	− 0·1595359	12·18
12·20	+ 0·0907701	− 0·2095218	0·2283388	653° 25′ 24″·51	− 0·1576720	12·20
12·22	+ 0·0948680	− 0·2074933	0·2281522	654° 34′ 13″·23	− 0·1557274	12·22
12·24	+ 0·0989212	− 0·2053852	0·2279660	655° 43′ 1″·94	− 0·1537028	12·24
12·26	+ 0·1029283	− 0·2031985	0·2277803	656° 51′ 50″·64	− 0·1515994	12·26
12·28	+ 0·1068877	− 0·2009341	0·2275950	658° 0′ 39″·32	− 0·1494180	12·28
12·30	+ 0·1107980	− 0·1985931	0·2274102	659° 9′ 28″·00	− 0·1471598	12·30
12·32	+ 0·1146576	− 0·1961765	0·2272258	660° 18′ 16″·67	− 0·1448257	12·32
12·34	+ 0·1184651	− 0·1936855	0·2270419	661° 27′ 5″·32	− 0·1424169	12·34
12·36	+ 0·1222191	− 0·1911210	0·2268585	662° 35′ 53″·96	− 0·1399344	12·36
12·38	+ 0·1259182	− 0·1884844	0·2266754	663° 44′ 42″·60	− 0·1373794	12·38
12·40	+ 0·1295610	− 0·1857766	0·2264928	664° 53′ 31″·22	− 0·1347532	12·40
12·42	+ 0·1331462	− 0·1829990	0·2263107	666′ 2′ 19″·83	− 0·1320567	12·42
12·44	+ 0·1366724	− 0·1801527	0·2261290	667° 11′ 8″·43	− 0·1292914	12·44
12·46	+ 0·1401382	− 0·1772390	0·2259477	668° 19′ 57″·02	− 0·1264583	12·46
12·48	+ 0·1435426	− 0·1742591	0·2257669	669° 28′ 45″·60	− 0·1235588	12·48
12·50	+ 0·1468841	− 0·1712143	0·2255865	670° 37′ 34″·17	− 0·1205943	12·50
12·52	+ 0·1501615	− 0·1681060	0·2254065	671° 46′ 22″·73	− 0·1175659	12·52
12·54	+ 0·1533737	− 0·1649354	0·2252270	672° 55′ 11″·28	− 0·1144750	12·54
12·56	+ 0·1565195	− 0·1617040	0·2250479	674° 3′ 59″·82	− 0·1113230	12·56
12·58	+ 0·1595977	− 0·1584131	0·2248692	675° 12′ 48″·35	− 0·1081112	12·58
12·60	+ 0·1626073	− 0·1550641	0·2246909	676° 21′ 36″·86	− 0·1048412	12·60
12·62	+ 0·1655471	− 0·1516585	0·2245131	677° 30′ 25″·37	− 0·1015142	12·62
12·64	+ 0·1684160	− 0·1481976	0·2243357	678° 39′ 13″·87	− 0·0981318	12·64
12·66	+ 0·1712131	− 0·1446830	0·2241587	679° 48′ 2″·36	− 0·0946953	12·66
12·68	+ 0·1739374	− 0·1411161	0·2239821	680° 56′ 50″·84	− 0·0912064	12·68
12·70	+ 0·1765879	− 0·1374984	0·2238059	682° 5′ 39″·30	− 0·0876664	12·70
12·72	+ 0·1791636	− 0·1338314	0·2236302	683° 14′ 27″·76	− 0·0840769	12·72
12·74	+ 0·1816637	− 0·1301168	0·2234548	684° 23′ 16″·21	− 0·0804395	12·74
12·76	+ 0·1840872	− 0·1263559	0·2232799	685° 32′ 4″·65	− 0·0767556	12·76
12·78	+ 0·1864334	− 0·1225504	0·2231054	686° 40′ 53″·08	− 0·0730269	12·78
12·80	+ 0·1887014	− 0·1187019	0·2229313	687° 49′ 41″·50	− 0·0692549	12·80
12·82	+ 0·1908904	− 0·1148120	0·2227576	688° 58′ 29″·91	− 0·0654413	12·82
12·84	+ 0·1929997	− 0·1108823	0·2225843	690° 7′ 18″·31	− 0·0615876	12·84
12·86	+ 0·1950286	− 0·1069143	0·2224114	691° 16′ 6″·70	− 0·0576954	12·86
12·88	+ 0·1969764	− 0·1029098	0·2222389	692° 24′ 55″·08	− 0·0537665	12·88
12·90	+ 0·1988424	− 0·0988704	0·2220668	693° 33′ 43″·45	− 0·0498024	12·90
12·92	+ 0·2006261	− 0·0947977	0·2218951	694° 42′ 31″·81	− 0·0458049	12·92
12·94	+ 0·2023269	− 0·0906934	0·2217238	695° 51′ 20″·16	− 0·0417755	12·94
12·96	+ 0·2039441	− 0·0865592	0·2215529	697° 0′ 8″·51	− 0·0377160	12·96
12·98	+ 0·2054773	− 0·0823968	0·2213824	698° 8′ 56″·85	− 0·0336280	12·98
13·00	+ 0·2069261	− 0·0782079	0·2212123	699° 17′ 45″·17	− 0·0295133	13·00

Table I. Functions of order unity

x	$J_1(x)$	$Y_1(x)$	$\lvert H_1^{(1)}(x)\rvert$	$\arg H_1^{(1)}(x)$	$\mathbf{H}_1(x)$	x
12·02	− 0·2220777	− 0·0614935	0·2304343	555° 28′ 38″·83	+ 0·5794489	12·02
12·04	− 0·2206225	− 0·0658561	0·2302419	556° 37′ 13″·53	+ 0·5750722	12·04
12·06	− 0·2190821	− 0·0701853	0·2300499	557° 45′ 48″·27	+ 0·5707290	12·06
12·08	− 0·2174574	− 0·0744794	0·2298584	558° 54′ 23″·04	+ 0·5664209	12·08
12·10	− 0·2157490	− 0·0787369	0·2296674	560° 2′ 57″·84	+ 0·5621495	12·10
12·12	− 0·2139578	− 0·0829561	0·2294769	561° 11′ 32″·68	+ 0·5579164	12·12
12·14	− 0·2120846	− 0·0871355	0·2292868	562° 20′ 7″·55	+ 0·5537234	12·14
12·16	− 0·2101303	− 0·0912733	0·2290973	563° 28′ 42″·46	+ 0·5495718	12·16
12·18	− 0·2080958	− 0·0953682	0·2289082	564° 37′ 17″·40	+ 0·5454634	12·18
12·20	− 0·2059820	− 0·0994184	0·2287195	565° 45′ 52″·38	+ 0·5413995	12·20
12·22	− 0·2037900	− 0·1034226	0·2285313	566° 54′ 27″·39	+ 0·5373819	12·22
12·24	− 0·2015206	− 0·1073791	0·2283436	568° 3′ 2″·43	+ 0·5334119	12·24
12·26	− 0·1991749	− 0·1112866	0·2281564	569° 11′ 37″·50	+ 0·5294911	12·26
12·28	− 0·1967540	− 0·1151435	0·2279696	570° 20′ 12″·61	+ 0·5256209	12·28
12·30	− 0·1942588	− 0·1189484	0·2277833	571° 28′ 47″·75	+ 0·5218027	12·30
12·32	− 0·1916907	− 0·1226999	0·2275974	572° 37′ 22″·93	+ 0·5180381	12·32
12·34	− 0·1890506	− 0·1263966	0·2274120	573° 45′ 58″·13	+ 0·5143282	12·34
12·36	− 0·1863397	− 0·1300372	0·2272270	574° 54′ 33″·37	+ 0·5106746	12·36
12·38	− 0·1835591	− 0·1336202	0·2270425	576° 3′ 8″·64	+ 0·5070786	12·38
12·40	− 0·1807102	− 0·1371444	0·2268585	577° 11′ 43″·95	+ 0·5035415	12·40
12·42	− 0·1777942	− 0·1406084	0·2266749	578° 20′ 19″·28	+ 0·5000646	12·42
12·44	− 0·1748122	− 0·1440111	0·2264917	579° 28′ 54″·65	+ 0·4966492	12·44
12·46	− 0·1717656	− 0·1473511	0·2263090	580° 37′ 30″·05	+ 0·4932964	12·46
12·48	− 0·1686557	− 0·1506272	0·2261267	581° 46′ 5″·48	+ 0·4900076	12·48
12·50	− 0·1654838	− 0·1538383	0·2259449	582° 54′ 40″·94	+ 0·4867839	12·50
12·52	− 0·1622513	− 0·1569831	0·2257635	584° 3′ 16″·43	+ 0·4836265	12·52
12·54	− 0·1589594	− 0·1600606	0·2255826	585° 11′ 51″·95	+ 0·4805365	12·54
12·56	− 0·1556097	− 0·1630696	0·2254020	586° 20′ 27″·51	+ 0·4775151	12·56
12·58	− 0·1522036	− 0·1660091	0·2252220	587° 29′ 3″·09	+ 0·4745632	12·58
12·60	− 0·1487423	− 0·1688779	0·2250423	588° 37′ 38″·71	+ 0·4716820	12·60
12·62	− 0·1452276	− 0·1716752	0·2248631	589° 46′ 14″·35	+ 0·4688725	12·62
12·64	− 0·1416606	− 0·1743998	0·2246843	590° 54′ 50″·03	+ 0·4661356	12·64
12·66	− 0·1380431	− 0·1770509	0·2245059	592° 3′ 25″·74	+ 0·4634724	12·66
12·68	− 0·1343765	− 0·1796274	0·2243280	593° 12′ 1″·47	+ 0·4608838	12·68
12·70	− 0·1306622	− 0·1821286	0·2241505	594° 20′ 37″·24	+ 0·4583706	12·70
12·72	− 0·1269019	− 0·1845534	0·2239734	595° 29′ 13″·04	+ 0·4559337	12·72
12·74	− 0·1230971	− 0·1869012	0·2237967	596° 37′ 48″·86	+ 0·4535740	12·74
12·76	− 0·1192494	− 0·1891710	0·2236204	597° 46′ 24″·72	+ 0·4512923	12·76
12·78	− 0·1153604	− 0·1913621	0·2234446	598° 55′ 0″·60	+ 0·4490893	12·78
12·80	− 0·1114316	− 0·1934738	0·2232692	600° 3′ 36″·51	+ 0·4469659	12·80
12·82	− 0·1074646	− 0·1955054	0·2230942	601° 12′ 12″·45	+ 0·4449226	12·82
12·84	− 0·1034612	− 0·1974561	0·2229196	602° 20′ 48″·43	+ 0·4429602	12·84
12·86	− 0·0994229	− 0·1993253	0·2227454	603° 29′ 24″·43	+ 0·4410794	12·86
12·88	− 0·0953513	− 0·2011125	0·2225716	604° 38′ 0″·46	+ 0·4392807	12·88
12·90	− 0·0912483	− 0·2028170	0·2223982	605° 46′ 36″·51	+ 0·4375647	12·90
12·92	− 0·0871153	− 0·2044382	0·2222253	606° 55′ 12″·60	+ 0·4359320	12·92
12·94	− 0·0829541	− 0·2059758	0·2220527	608° 3′ 48″·71	+ 0·4343830	12·94
12·96	− 0·0787663	− 0·2074291	0·2218805	609° 12′ 24″·85	+ 0·4329183	12·96
12·98	− 0·0745538	− 0·2087978	0·2217088	610° 21′ 1″·02	+ 0·4315383	12·98
13·00	− 0·0703181	− 0·2100814	0·2215374	611° 29′ 37″·22	+ 0·4302435	13·00

Table I. Functions of order zero

x	$J_0(x)$	$Y_0(x)$	$\|H_0^{(1)}(x)\|$	$\arg H_0^{(1)}(x)$	$\mathbf{H}_0(x)$	x
13·02	+ 0·2082899	− 0·0739941	0·2210426	700° 26′ 33″·48	− 0·0253735	13·02
13·04	+ 0·2095684	− 0·0697573	0·2208732	701° 35′ 21″·79	− 0·0212104	13·04
13·06	+ 0·2107612	− 0·0654990	0·2207043	702° 44′ 10″·08	− 0·0170257	13·06
13·08	+ 0·2118679	− 0·0612211	0·2205357	703° 52′ 58″·37	− 0·0128211	13·08
13·10	+ 0·2128882	− 0·0569253	0·2203676	705° 1′ 46″·65	− 0·0085984	13·10
13·12	+ 0·2138219	− 0·0526132	0·2201998	706° 10′ 34″·92	− 0·0043592	13·12
13·14	+ 0·2146687	− 0·0482867	0·2200324	707° 19′ 23″·18	− 0·0001054	13·14
13·16	+ 0·2154284	− 0·0439475	0·2198654	708° 28′ 11″·43	+ 0·0041614	13·16
13·18	+ 0·2161009	− 0·0395973	0·2196987	709° 36′ 59″·68	+ 0·0084393	13·18
13·20	+ 0·2166859	− 0·0352379	0·2195325	710° 45′ 47″·91	+ 0·0127268	13·20
13·22	+ 0·2171835	− 0·0308710	0·2193666	711° 54′ 36″·14	+ 0·0170219	13·22
13·24	+ 0·2175935	− 0·0264983	0·2192010	713° 3′ 24″·35	+ 0·0213229	13·24
13·26	+ 0·2179159	− 0·0221217	0·2190359	714° 12′ 12″·56	+ 0·0256282	13·26
13·28	+ 0·2181508	− 0·0177428	0·2188711	715° 21′ 0″·76	+ 0·0299359	13·28
13·30	+ 0·2182981	− 0·0133634	0·2187067	716° 29′ 48″·96	+ 0·0342443	13·30
13·32	+ 0·2183579	− 0·0089853	0·2185427	717° 38′ 37″·14	+ 0·0385517	13·32
13·34	+ 0·2183304	− 0·0046101	0·2183790	718° 47′ 25″·31	+ 0·0428564	13·34
13·36	+ 0·2182156	− 0·0002396	0·2182158	719° 56′ 13″·47	+ 0·0471565	13·36
13·38	+ 0·2180138	+ 0·0041244	0·2180528	721° 5′ 1″·63	+ 0·0514504	13·38
13·40	+ 0·2177252	+ 0·0084802	0·2178903	722° 13′ 49″·78	+ 0·0557363	13·40
13·42	+ 0·2173499	+ 0·0128262	0·2177281	723° 22′ 37″·92	+ 0·0600126	13·42
13·44	+ 0·2168884	+ 0·0171605	0·2175662	724° 31′ 26″·05	+ 0·0642775	13·44
13·46	+ 0·2163409	+ 0·0214816	0·2174047	725° 40′ 14″·17	+ 0·0685292	13·46
13·48	+ 0·2157076	+ 0·0257876	0·2172436	726° 49′ 2″·29	+ 0·0727662	13·48
13·50	+ 0·2149892	+ 0·0300770	0·2170829	727° 57′ 50″·39	+ 0·0769866	13·50
13·52	+ 0·2141858	+ 0·0343480	0·2169225	729° 6′ 38″·49	+ 0·0811889	13·52
13·54	+ 0·2132981	+ 0·0385990	0·2167624	730° 15′ 26″·58	+ 0·0853714	13·54
13·56	+ 0·2123263	+ 0·0428282	0·2166027	731° 24′ 14″·66	+ 0·0895324	13·56
13·58	+ 0·2112712	+ 0·0470341	0·2164434	732° 33′ 2″·73	+ 0·0936702	13·58
13·60	+ 0·2101332	+ 0·0512150	0·2162844	733° 41′ 50″·80	+ 0·0977832	13·60
13·62	+ 0·2089128	+ 0·0553693	0·2161257	734° 50′ 38″·86	+ 0·1018698	13·62
13·64	+ 0·2076107	+ 0·0594954	0·2159674	735° 59′ 26″·90	+ 0·1059283	13·64
13·66	+ 0·2062276	+ 0·0635916	0·2158095	737° 8′ 14″·94	+ 0·1099573	13·66
13·68	+ 0·2047641	+ 0·0676565	0·2156519	738° 17′ 2″·98	+ 0·1139550	13·68
13·70	+ 0·2032208	+ 0·0716883	0·2154946	739° 25′ 51″·00	+ 0·1179199	13·70
13·72	+ 0·2015986	+ 0·0756856	0·2153377	740° 34′ 39″·02	+ 0·1218505	13·72
13·74	+ 0·1998982	+ 0·0796469	0·2151811	741° 43′ 27″·03	+ 0·1257452	13·74
13·76	+ 0·1981203	+ 0·0835705	0·2150249	742° 52′ 15″·03	+ 0·1296025	13·76
13·78	+ 0·1962659	+ 0·0874551	0·2148690	744° 1′ 3″·03	+ 0·1334209	13·78
13·80	+ 0·1943356	+ 0·0912990	0·2147134	745° 9′ 51″·01	+ 0·1371989	13·80
13·82	+ 0·1923305	+ 0·0951009	0·2145582	746° 18′ 38″·99	+ 0·1409350	13·82
13·84	+ 0·1902515	+ 0·0988593	0·2144033	747° 27′ 26″·96	+ 0·1446278	13·84
13·86	+ 0·1880993	+ 0·1025727	0·2142488	748° 36′ 14″·92	+ 0·1482758	13·86
13·88	+ 0·1858751	+ 0·1062398	0·2140945	749° 45′ 2″·87	+ 0·1518777	13·88
13·90	+ 0·1835799	+ 0·1098592	0·2139407	750° 53′ 50″·82	+ 0·1554320	13·90
13·92	+ 0·1812145	+ 0·1134294	0·2137871	752° 2′ 38″·76	+ 0·1589374	13·92
13·94	+ 0·1787801	+ 0·1169492	0·2136339	753° 11′ 26″·70	+ 0·1623925	13·94
13·96	+ 0·1762777	+ 0·1204172	0·2134810	754° 20′ 14″·63	+ 0·1657960	13·96
13·98	+ 0·1737085	+ 0·1238321	0·2133284	755° 29′ 2″·54	+ 0·1691466	13·98
14·00	+ 0·1710735	+ 0·1271926	0·2131762	756° 37′ 50″·45	+ 0·1724429	14·00

Table I. Functions of order unity

x	$J_1(x)$	$Y_1(x)$	$\|H_1^{(1)}(x)\|$	arg $H_1^{(1)}(x)$	$\mathbf{H}_1(x)$	x
13·02	− 0·0660609	− 0·2112796	0·2213665	612° 38′ 13″·45	+ 0·4290341	13·02
13·04	− 0·0617841	− 0·2123920	0·2211959	613° 46′ 49″·71	+ 0·4279106	13·04
13·06	− 0·0574892	− 0·2134183	0·2210257	614° 55′ 25″·99	+ 0·4268732	13·06
13·08	− 0·0531781	− 0·2143582	0·2208559	616° 4′ 2″·30	+ 0·4259222	13·08
13·10	− 0·0488525	− 0·2152115	0·2206866	617° 12′ 38″·63	+ 0·4250579	13·10
13·12	− 0·0445140	− 0·2159780	0·2205176	618° 21′ 14″·99	+ 0·4242805	13·12
13·14	− 0·0401645	− 0·2166575	0·2203490	619° 29′ 51″·38	+ 0·4235900	13·14
13·16	− 0·0358056	− 0·2172499	0·2201807	620° 38′ 27″·80	+ 0·4229868	13·16
13·18	− 0·0314391	− 0·2177551	0·2200129	621° 47′ 4″·24	+ 0·4224709	13·18
13·20	− 0·0270667	− 0·2181729	0·2198455	622° 55′ 40″·71	+ 0·4220422	13·20
13·22	− 0·0226902	− 0·2185034	0·2196784	624° 4′ 17″·21	+ 0·4217010	13·22
13·24	− 0·0183113	− 0·2187466	0·2195117	625° 12′ 53″·73	+ 0·4214472	13·24
13·26	− 0·0139317	− 0·2189025	0·2193454	626° 21′ 30″·28	+ 0·4212807	13·26
13·28	− 0·0095532	− 0·2189712	0·2191795	627° 30′ 6″·85	+ 0·4212014	13·28
13·30	− 0·0051775	− 0·2189527	0·2190139	628° 38′ 43″·45	+ 0·4212094	13·30
13·32	− 0·0008063	− 0·2188473	0·2188487	629° 47′ 20″·08	+ 0·4213044	13·32
13·34	+ 0·0035587	− 0·2186550	0·2186839	630° 55′ 56″·73	+ 0·4214062	13·34
13·36	+ 0·0079157	− 0·2183761	0·2185195	632° 4′ 33″·41	+ 0·4217547	13·36
13·38	+ 0·0122630	− 0·2180108	0·2183555	633° 13′ 10″·12	+ 0·4221096	13·38
13·40	+ 0·0165990	− 0·2175595	0·2181918	634° 21′ 46″·85	+ 0·4225507	13·40
13·42	+ 0·0209219	− 0·2170223	0·2180285	635° 30′ 23″·60	+ 0·4230776	13·42
13·44	+ 0·0252301	− 0·2163997	0·2178655	636° 39′ 0″·38	+ 0·4236901	13·44
13·46	+ 0·0295218	− 0·2156920	0·2177029	637° 47′ 37″·19	+ 0·4243876	13·46
13·48	+ 0·0337954	− 0·2148996	0·2175407	638° 56′ 14″·02	+ 0·4251699	13·48
13·50	+ 0·0380493	− 0·2140229	0·2173788	640° 4′ 50″·87	+ 0·4260365	13·50
13·52	+ 0·0422817	− 0·2130625	0·2172174	641° 13′ 27″·75	+ 0·4269869	13·52
13·54	+ 0·0464911	− 0·2120188	0·2170562	642° 22′ 4″·66	+ 0·4280206	13·54
13·56	+ 0·0506758	− 0·2108924	0·2168954	643° 30′ 41″·59	+ 0·4291371	13·56
13·58	+ 0·0548341	− 0·2096838	0·2167350	644° 39′ 18″·54	+ 0·4303358	13·58
13·60	+ 0·0589646	− 0·2083936	0·2165750	645° 47′ 55″·52	+ 0·4316161	13·60
13·62	+ 0·0630655	− 0·2070225	0·2164153	646° 56′ 32″·52	+ 0·4329775	13·62
13·64	+ 0·0671353	− 0·2055711	0·2162559	648° 5′ 9″·55	+ 0·4344191	13·64
13·66	+ 0·0711725	− 0·2040400	0·2160969	649° 13′ 46″·60	+ 0·4359404	13·66
13·68	+ 0·0751755	− 0·2024302	0·2159382	650° 22′ 23″·67	+ 0·4375406	13·68
13·70	+ 0·0791428	− 0·2007421	0·2157799	651° 31′ 0″·77	+ 0·4392190	13·70
13·72	+ 0·0830728	− 0·1989768	0·2156220	652° 39′ 37″·89	+ 0·4409748	13·72
13·74	+ 0·0869640	− 0·1971349	0·2154644	653° 48′ 15″·04	+ 0·4428071	13·74
13·76	+ 0·0908150	− 0·1952173	0·2153071	654° 56′ 52″·21	+ 0·4447152	13·76
13·78	+ 0·0946243	− 0·1932249	0·2151502	656° 5′ 29″·41	+ 0·4466981	13·78
13·80	+ 0·0983905	− 0·1911585	0·2149936	657° 14′ 6″·62	+ 0·4487550	13·80
13·82	+ 0·1021121	− 0·1890191	0·2148374	658° 22′ 43″·86	+ 0·4508850	13·82
13·84	+ 0·1057877	− 0·1868077	0·2146815	659° 31′ 21″·12	+ 0·4530871	13·84
13·86	+ 0·1094160	− 0·1845252	0·2145260	660° 39′ 58″·41	+ 0·4553603	13·86
13·88	+ 0·1129955	− 0·1821726	0·2143708	661° 48′ 35″·72	+ 0·4577036	13·88
13·90	+ 0·1165249	− 0·1797510	0·2142159	662° 57′ 13″·05	+ 0·4601160	13·90
13·92	+ 0·1200029	− 0·1772613	0·2140614	664° 5′ 50″·40	+ 0·4625965	13·92
13·94	+ 0·1234282	− 0·1747048	0·2139072	665° 14′ 27″·78	+ 0·4651439	13·94
13·96	+ 0·1267995	− 0·1720824	0·2137533	666° 23′ 5″·18	+ 0·4677571	13·96
13·98	+ 0·1301156	− 0·1693954	0·2135998	667° 31′ 42″·60	+ 0·4704350	13·98
14·00	+ 0·1333752	− 0·1666448	0·2134466	668° 40′ 20″·04	+ 0·4731766	14·00

Table I. Functions of order zero

x	$J_0(x)$	$Y_0(x)$	$\|H_0^{(1)}(x)\|$	arg $H_0^{(1)}(x)$	$\mathbf{H}_0(x)$	x
14·02	+ 0·1683739	+ 0·1304974	0·2130243	757° 46′ 38″·36	+ 0·1756839	14·02
14·04	+ 0·1656108	+ 0·1337454	0·2128727	758° 55′ 26″·25	+ 0·1788681	14·04
14·06	+ 0·1627855	+ 0·1369354	0·2127214	760° 4′ 14″·14	+ 0·1819944	14·06
14·08	+ 0·1598991	+ 0·1400660	0·2125705	761° 13′ 2″·02	+ 0·1850617	14·08
14·10	+ 0·1569529	+ 0·1431362	0·2124198	762° 21′ 49″·89	+ 0·1880687	14·10
14·12	+ 0·1539481	+ 0·1461449	0·2122695	763° 30′ 37″·76	+ 0·1910143	14·12
14·14	+ 0·1508861	+ 0·1490909	0·2121195	764° 39′ 25″·62	+ 0·1938974	14·14
14·16	+ 0·1477681	+ 0·1519731	0·2119699	765° 48′ 13″·47	+ 0·1967169	14·16
14·18	+ 0·1445954	+ 0·1547905	0·2118205	766° 57′ 1″·32	+ 0·1994718	14·18
14·20	+ 0·1413694	+ 0·1575421	0·2116715	768° 5′ 49″·16	+ 0·2021610	14·20
14·22	+ 0·1380914	+ 0·1602268	0·2115227	769° 14′ 36″·99	+ 0·2047836	14·22
14·24	+ 0·1347629	+ 0·1628437	0·2113743	770° 23′ 24″·81	+ 0·2073385	14·24
14·26	+ 0·1313851	+ 0·1653919	0·2112262	771° 32′ 12″·63	+ 0·2098248	14·26
14·28	+ 0·1279596	+ 0·1678704	0·2110784	772° 41′ 0″·44	+ 0·2122416	14·28
14·30	+ 0·1244877	+ 0·1702783	0·2109310	773° 49′ 48″·24	+ 0·2145880	14·30
14·32	+ 0·1209709	+ 0·1726147	0·2107838	774° 58′ 36″·03	+ 0·2168632	14·32
14·34	+ 0·1174107	+ 0·1748790	0·2106369	776° 7′ 23″·82	+ 0·2190663	14·34
14·36	+ 0·1138085	+ 0·1770701	0·2104904	777° 16′ 11″·60	+ 0·2211964	14·36
14·38	+ 0·1101658	+ 0·1791875	0·2103441	778° 24′ 59″·38	+ 0·2232530	14·38
14·40	+ 0·1064841	+ 0·1812302	0·2101982	779° 33′ 47″·15	+ 0·2252351	14·40
14·42	+ 0·1027650	+ 0·1831977	0·2100525	780° 42′ 34″·91	+ 0·2271421	14·42
14·44	+ 0·0990100	+ 0·1850893	0·2099072	781° 51′ 22″·66	+ 0·2289733	14·44
14·46	+ 0·0952206	+ 0·1869042	0·2097621	783° 0′ 10″·41	+ 0·2307281	14·46
14·48	+ 0·0913984	+ 0·1886420	0·2096174	784° 8′ 58″·16	+ 0·2324058	14·48
14·50	+ 0·0875449	+ 0·1903019	0·2094729	785° 17′ 45″·89	+ 0·2340059	14·50
14·52	+ 0·0836617	+ 0·1918835	0·2093288	786° 26′ 33″·61	+ 0·2355279	14·52
14·54	+ 0·0797504	+ 0·1933862	0·2091849	787° 35′ 21″·33	+ 0·2369710	14·54
14·56	+ 0·0758127	+ 0·1948095	0·2090414	788° 44′ 9″·05	+ 0·2383350	14·56
14·58	+ 0·0718500	+ 0·1961530	0·2088981	789° 52′ 56″·76	+ 0·2396194	14·58
14·60	+ 0·0678641	+ 0·1974163	0·2087552	791° 1′ 44″·46	+ 0·2408236	14·60
14·62	+ 0·0638565	+ 0·1985989	0·2086125	792° 10′ 32″·15	+ 0·2419474	14·62
14·64	+ 0·0598288	+ 0·1997005	0·2084701	793° 19′ 19″·84	+ 0·2429903	14·64
14·66	+ 0·0557827	+ 0·2007208	0·2083280	794° 28′ 7″·52	+ 0·2439521	14·66
14·68	+ 0·0517198	+ 0·2016595	0·2081862	795° 36′ 55″·20	+ 0·2448324	14·68
14·70	+ 0·0476418	+ 0·2025163	0·2080447	796° 45′ 42″·87	+ 0·2456309	14·70
14·72	+ 0·0435503	+ 0·2032910	0·2079035	797° 54′ 30″·53	+ 0·2463475	14·72
14·74	+ 0·0394470	+ 0·2039834	0·2077626	799° 3′ 18″·19	+ 0·2469820	14·74
14·76	+ 0·0353334	+ 0·2045933	0·2076219	800° 12′ 5″·84	+ 0·2475341	14·76
14·78	+ 0·0312113	+ 0·2051206	0·2074816	801° 20′ 53″·48	+ 0·2480038	14·78
14·80	+ 0·0270823	+ 0·2055652	0·2073415	802° 29′ 41″·12	+ 0·2483909	14·80
14·82	+ 0·0229481	+ 0·2059270	0·2072017	803° 38′ 28″·75	+ 0·2486955	14·82
14·84	+ 0·0188102	+ 0·2062060	0·2070622	804° 47′ 16″·37	+ 0·2489173	14·84
14·86	+ 0·0146704	+ 0·2064022	0·2069229	805° 56′ 3″·98	+ 0·2490565	14·86
14·88	+ 0·0105303	+ 0·2065157	0·2067840	807° 4′ 51″·60	+ 0·2491132	14·88
14·90	+ 0·0063915	+ 0·2065464	0·2066453	808° 13′ 39″·21	+ 0·2490872	14·90
14·92	+ 0·0022558	+ 0·2064946	0·2065069	809° 22′ 26″·81	+ 0·2489788	14·92
14·94	− 0·0018753	+ 0·2063603	0·2063688	810° 31′ 14″·40	+ 0·2487881	14·94
14·96	− 0·0060002	+ 0·2061436	0·2062309	811° 40′ 1″·99	+ 0·2485152	14·96
14·98	− 0·0101171	+ 0·2058449	0·2060934	812° 48′ 49″·57	+ 0·2481604	14·98
15·00	− 0·0142245	+ 0·2054643	0·2059561	813° 57′ 37″·14	+ 0·2477238	15·00

Table I. Functions of order unity

x	$J_1(x)$	$Y_1(x)$	$\lvert H_1^{(1)}(x)\rvert$	$\arg H_1^{(1)}(x)$	$\mathbf{H}_1(x)$	x
14·02	+ 0·1365770	− 0·1638320	0·2132937	669° 48′ 57″·51	+ 0·4759804	14·02
14·04	+ 0·1397201	− 0·1609579	0·2131411	670° 57′ 35″·00	+ 0·4788455	14·04
14·06	+ 0·1428030	− 0·1580240	0·2129889	672° 6′ 12″·51	+ 0·4817705	14·06
14·08	+ 0·1458248	− 0·1550314	0·2128370	673° 14′ 50″·04	+ 0·4847542	14·08
14·10	+ 0·1487844	− 0·1519813	0·2126855	674° 23′ 27″·60	+ 0·4877954	14·10
14·12	+ 0·1516805	− 0·1488752	0·2125342	675° 32′ 5″·17	+ 0·4908927	14·12
14·14	+ 0·1545122	− 0·1457142	0·2123833	676° 40′ 42″·77	+ 0·4940449	14·14
14·16	+ 0·1572785	− 0·1424998	0·2122327	677° 49′ 20″·39	+ 0·4972506	14·16
14·18	+ 0·1599783	− 0·1392332	0·2120824	678° 57′ 58″·03	+ 0·5005085	14·18
14·20	+ 0·1626107	− 0·1359159	0·2119325	680° 6′ 35″·69	+ 0·5038172	14·20
14·22	+ 0·1651747	− 0·1325491	0·2117828	681° 15′ 13″·38	+ 0·5071753	14·22
14·24	+ 0·1676695	− 0·1291344	0·2116335	682° 23′ 51″·08	+ 0·5105814	14·24
14·26	+ 0·1700940	− 0·1256731	0·2114845	683° 32′ 28″·81	+ 0·5140341	14·26
14·28	+ 0·1724475	− 0·1221667	0·2113358	684° 41′ 6″·55	+ 0·5175320	14·28
14·30	+ 0·1747291	− 0·1186166	0·2111874	685° 49′ 44″·32	+ 0·5210736	14·30
14·32	+ 0·1769380	− 0·1150243	0·2110394	686° 58′ 22″·11	+ 0·5246575	14·32
14·34	+ 0·1790734	− 0·1113912	0·2108916	688° 6′ 59″·92	+ 0·5282821	14·34
14·36	+ 0·1811346	− 0·1077189	0·2107442	689° 15′ 37″·75	+ 0·5319460	14·36
14·38	+ 0·1831209	− 0·1040089	0·2105971	690° 24′ 15″·60	+ 0·5356477	14·38
14·40	+ 0·1850317	− 0·1002626	0·2104502	691° 32′ 53″·47	+ 0·5393857	14·40
14·42	+ 0·1868661	− 0·0964816	0·2103037	692° 41′ 31″·37	+ 0·5431583	14·42
14·44	+ 0·1886237	− 0·0926676	0·2101575	693° 50′ 9″·28	+ 0·5469642	14·44
14·46	+ 0·1903038	− 0·0888219	0·2100116	694° 58′ 47″·22	+ 0·5508016	14·46
14·48	+ 0·1919059	− 0·0849462	0·2098660	696° 7′ 25″·18	+ 0·5546691	14·48
14·50	+ 0·1934295	− 0·0810421	0·2097207	697° 16′ 3″·16	+ 0·5585651	14·50
14·52	+ 0·1948740	− 0·0771111	0·2095757	698° 24′ 41″·14	+ 0·5624879	14·52
14·54	+ 0·1962389	− 0·0731549	0·2094310	699° 33′ 19″·15	+ 0·5664361	14·54
14·56	+ 0·1975240	− 0·0691749	0·2092866	700° 41′ 57″·18	+ 0·5704080	14·56
14·58	+ 0·1987287	− 0·0651730	0·2091425	701° 50′ 35″·23	+ 0·5744019	14·58
14·60	+ 0·1998527	− 0·0611506	0·2089987	702° 59′ 13″·31	+ 0·5784163	14·60
14·62	+ 0·2008956	− 0·0571093	0·2088552	704° 7′ 51″·40	+ 0·5824496	14·62
14·64	+ 0·2018572	− 0·0530509	0·2087120	705° 16′ 29″·51	+ 0·5865001	14·64
14·66	+ 0·2027371	− 0·0489769	0·2085691	706° 25′ 7″·64	+ 0·5905662	14·66
14·68	+ 0·2035352	− 0·0448890	0·2084265	707° 33′ 45″·80	+ 0·5946463	14·68
14·70	+ 0·2042513	− 0·0407888	0·2082842	708° 42′ 23″·97	+ 0·5987387	14·70
14·72	+ 0·2048851	− 0·0366779	0·2081422	709° 51′ 2″·16	+ 0·6028418	14·72
14·74	+ 0·2054365	− 0·0325580	0·2080004	710° 59′ 40″·36	+ 0·6069539	14·74
14·76	+ 0·2059054	− 0·0284307	0·2078590	712° 8′ 18″·59	+ 0·6110734	14·76
14·78	+ 0·2062918	− 0·0242978	0·2077178	713° 16′ 56″·84	+ 0·6151987	14·78
14·80	+ 0·2065956	− 0·0201607	0·2075769	714° 25′ 35″·10	+ 0·6193280	14·80
14·82	+ 0·2068167	− 0·0160212	0·2074363	715° 34′ 13″·39	+ 0·6234599	14·82
14·84	+ 0·2069553	− 0·0118809	0·2072960	716° 42′ 51″·69	+ 0·6275926	14·84
14·86	+ 0·2070113	− 0·0077415	0·2071560	717° 51′ 30″·01	+ 0·6317244	14·86
14·88	+ 0·2069849	− 0·0036045	0·2070163	719° 0′ 8″·36	+ 0·6358538	14·88
14·90	+ 0·2068762	+ 0·0005283	0·2068768	720° 8′ 46″·72	+ 0·6399791	14·90
14·92	+ 0·2066853	+ 0·0046553	0·2067377	721° 17′ 25″·09	+ 0·6440987	14·92
14·94	+ 0·2064124	+ 0·0087750	0·2065988	722° 26′ 3″·49	+ 0·6482109	14·94
14·96	+ 0·2060577	+ 0·0128857	0·2064602	723° 34′ 41″·90	+ 0·6523142	14·96
14·98	+ 0·2056215	+ 0·0169858	0·2063219	724° 43′ 20″·33	+ 0·6564068	14·98
15·00	+ 0·2051040	+ 0·0210736	0·2061838	725° 51′ 58″·79	+ 0·6604873	15·00

Table I. Functions of order zero

x	$J_0(x)$	$Y_0(x)$	$\lvert H_0^{(1)}(x)\rvert$	$\arg H_0^{(1)}(x)$	$\mathbf{H}_0(x)$	x
15·02	− 0·0183207	+ 0·2050021	0·2058191	815° 6′ 24″·71	+ 0·2472058	15·02
15·04	− 0·0224042	+ 0·2044585	0·2056823	816° 15′ 12″·27	+ 0·2466066	15·04
15·06	− 0·0264732	+ 0·2038339	0·2055459	817° 23′ 59″·83	+ 0·2459265	15·06
15·08	− 0·0305263	+ 0·2031287	0·2054097	818° 32′ 47″·38	+ 0·2451659	15·08
15·10	− 0·0345619	+ 0·2023432	0·2052737	819° 41′ 34″·93	+ 0·2443252	15·10
15·12	− 0·0385782	+ 0·2014779	0·2051381	820° 50′ 22″·47	+ 0·2434048	15·12
15·14	− 0·0425738	+ 0·2005332	0·2050027	821° 59′ 10″·01	+ 0·2424052	15·14
15·16	− 0·0465472	+ 0·1995096	0·2048675	823° 7′ 57″·54	+ 0·2413268	15·16
15·18	− 0·0504967	+ 0·1984076	0·2047327	824° 16′ 45″·06	+ 0·2401701	15·18
15·20	− 0·0544208	+ 0·1972277	0·2045981	825° 25′ 32″·57	+ 0·2389357	15·20
15·22	− 0·0583180	+ 0·1959705	0·2044638	826° 34′ 20″·08	+ 0·2376242	15·22
15·24	− 0·0621868	+ 0·1946367	0·2043297	827° 43′ 7″·58	+ 0·2362361	15·24
15·26	− 0·0660256	+ 0·1932268	0·2041959	828° 51′ 55″·08	+ 0·2347721	15·26
15·28	− 0·0698330	+ 0·1917415	0·2040624	830° 0′ 42″·58	+ 0·2332329	15·28
15·30	− 0·0736075	+ 0·1901815	0·2039291	831° 9′ 30″·07	+ 0·2316191	15·30
15·32	− 0·0773477	+ 0·1885475	0·2037961	832° 18′ 17″·55	+ 0·2299315	15·32
15·34	− 0·0810521	+ 0·1868403	0·2036633	833° 27′ 5″·03	+ 0·2281707	15·34
15·36	− 0·0847192	+ 0·1850607	0·2035308	834° 35′ 52″·50	+ 0·2263377	15·36
15·38	− 0·0883477	+ 0·1832093	0·2033986	835° 44′ 39″·96	+ 0·2244331	15·38
15·40	− 0·0919362	+ 0·1812872	0·2032666	836° 53′ 27″·42	+ 0·2224578	15·40
15·42	− 0·0954833	+ 0·1792950	0·2031349	838° 2′ 14″·87	+ 0·2204127	15·42
15·44	− 0·0989876	+ 0·1772338	0·2030034	839° 11′ 2″·32	+ 0·2182987	15·44
15·46	− 0·1024478	+ 0·1751045	0·2028722	840° 19′ 49″·76	+ 0·2161166	15·46
15·48	− 0·1058626	+ 0·1729078	0·2027412	841° 28′ 37″·20	+ 0·2138674	15·48
15·50	− 0·1092307	+ 0·1706449	0·2026105	842° 37′ 24″·63	+ 0·2115520	15·50
15·52	− 0·1125507	+ 0·1683167	0·2024801	843° 46′ 12″·06	+ 0·2091715	15·52
15·54	− 0·1158215	+ 0·1659242	0·2023499	844° 54′ 59″·48	+ 0·2067269	15·54
15·56	− 0·1190418	+ 0·1634685	0·2022199	846° 3′ 46″·89	+ 0·2042191	15·56
15·58	− 0·1222103	+ 0·1609506	0·2020902	847° 12′ 34″·30	+ 0·2016493	15·58
15·60	− 0·1253260	+ 0·1583715	0·2019607	848° 21′ 21″·71	+ 0·1990185	15·60
15·62	− 0·1283875	+ 0·1557325	0·2018315	849° 30′ 9″·11	+ 0·1963278	15·62
15·64	− 0·1313938	+ 0·1530346	0·2017026	850° 38′ 56″·50	+ 0·1935784	15·64
15·66	− 0·1343438	+ 0·1502790	0·2015739	851° 47′ 43″·89	+ 0·1907714	15·66
15·68	− 0·1372363	+ 0·1474668	0·2014454	852° 56′ 31″·27	+ 0·1879079	15·68
15·70	− 0·1400702	+ 0·1445992	0·2013172	854° 5′ 18″·65	+ 0·1849893	15·70
15·72	− 0·1428446	+ 0·1416775	0·2011892	855° 14′ 6″·02	+ 0·1820166	15·72
15·74	− 0·1455583	+ 0·1387029	0·2010615	856° 22′ 53″·39	+ 0·1789911	15·74
15·76	− 0·1482104	+ 0·1356766	0·2009340	857° 31′ 40″·75	+ 0·1759141	15·76
15·78	− 0·1507998	+ 0·1326000	0·2008067	858° 40′ 28″·11	+ 0·1727868	15·78
15·80	− 0·1533257	+ 0·1294742	0·2006797	859° 49′ 15″·46	+ 0·1696105	15·80
15·82	− 0·1557872	+ 0·1263006	0·2005530	860° 58′ 2″·81	+ 0·1663866	15·82
15·84	− 0·1581832	+ 0·1230805	0·2004264	862° 6′ 50″·15	+ 0·1631163	15·84
15·86	− 0·1605130	+ 0·1198153	0·2003001	863° 15′ 37″·48	+ 0·1598009	15·86
15·88	− 0·1627757	+ 0·1165064	0·2001741	864° 24′ 24″·81	+ 0·1564420	15·88
15·90	− 0·1649705	+ 0·1131550	0·2000483	865° 33′ 12″·14	+ 0·1530407	15·90
15·92	− 0·1670966	+ 0·1097626	0·1999227	866° 41′ 59″·46	+ 0·1495986	15·92
15·94	− 0·1691532	+ 0·1063305	0·1997974	867° 50′ 46″·78	+ 0·1461170	15·94
15·96	− 0·1711396	+ 0·1028603	0·1996723	868° 59′ 34″·09	+ 0·1425972	15·96
15·98	− 0·1730551	+ 0·0993533	0·1995474	870° 8′ 21″·39	+ 0·1390409	15·98
16·00	− 0·1748991	+ 0·0958110	0·1994228	871° 17′ 8″·69	+ 0·1354493	16·00

Table I. Functions of order unity

x	$J_1(x)$	$Y_1(x)$	$\lvert H_1^{(1)}(x) \rvert$	$\arg H_1^{(1)}(x)$	$\mathbf{H}_1(x)$	x
15·02	+ 0·2045057	+ 0·0251476	0·2060460	727° 0′ 37″·25	+ 0·6645540	15·02
15·04	+ 0·2038267	+ 0·0292062	0·2059085	728° 9′ 15″·74	+ 0·6686052	15·04
15·06	+ 0·2030675	+ 0·0332478	0·2057713	729° 17′ 54″·25	+ 0·6726395	15·06
15·08	+ 0·2022286	+ 0·0372707	0·2056344	730° 26′ 32″·77	+ 0·6766552	15·08
15·10	+ 0·2013102	+ 0·0412735	0·2054977	731° 35′ 11″·31	+ 0·6806508	15·10
15·12	+ 0·2003130	+ 0·0452546	0·2053613	732° 43′ 49″·87	+ 0·6846246	15·12
15·14	+ 0·1992373	+ 0·0492124	0·2052252	733° 52′ 28″·45	+ 0·6885753	15·14
15·16	+ 0·1980838	+ 0·0531454	0·2050893	735° 1′ 7″·04	+ 0·6925011	15·16
15·18	+ 0·1968530	+ 0·0570521	0·2049537	736° 9′ 45″·65	+ 0·6964006	15·18
15·20	+ 0·1955454	+ 0·0609309	0·2048184	737° 18′ 24″·28	+ 0·7002724	15·20
15·22	+ 0·1941618	+ 0·0647803	0·2046834	738° 27′ 2″·93	+ 0·7041148	15·22
15·24	+ 0·1927027	+ 0·0685990	0·2045486	739° 35′ 41″·59	+ 0·7079263	15·24
15·26	+ 0·1911688	+ 0·0723853	0·2044141	740° 44′ 20″·27	+ 0·7117056	15·26
15·28	+ 0·1895608	+ 0·0761378	0·2042798	741° 52′ 58″·97	+ 0·7154512	15·28
15·30	+ 0·1878794	+ 0·0798551	0·2041459	743° 1′ 37″·69	+ 0·7191616	15·30
15·32	+ 0·1861255	+ 0·0835358	0·2040121	744° 10′ 16″·42	+ 0·7228353	15·32
15·34	+ 0·1842998	+ 0·0871785	0·2038787	745° 18′ 55″·17	+ 0·7264711	15·34
15·36	+ 0·1824032	+ 0·0907816	0·2037455	746° 27′ 33″·94	+ 0·7300674	15·36
15·38	+ 0·1804363	+ 0·0943440	0·2036125	747° 36′ 12″·72	+ 0·7336229	15·38
15·40	+ 0·1784003	+ 0·0978642	0·2034799	748° 44′ 51″·52	+ 0·7371363	15·40
15·42	+ 0·1762958	+ 0·1013409	0·2033475	749° 53′ 30″·34	+ 0·7406061	15·42
15·44	+ 0·1741239	+ 0·1047727	0·2032153	751° 2′ 9″·17	+ 0·7440312	15·44
15·46	+ 0·1718855	+ 0·1081584	0·2030834	752° 10′ 48″·02	+ 0·7474101	15·46
15·48	+ 0·1695816	+ 0·1114966	0·2029518	753° 19′ 26″·89	+ 0·7507416	15·48
15·50	+ 0·1672132	+ 0·1147861	0·2028204	754° 28′ 5″·77	+ 0·7540245	15·50
15·52	+ 0·1647812	+ 0·1180258	0·2026892	755° 36′ 44″·67	+ 0·7572574	15·52
15·54	+ 0·1622868	+ 0·1212142	0·2025584	756° 45′ 23″·59	+ 0·7604392	15·54
15·56	+ 0·1597310	+ 0·1243503	0·2024278	757° 54′ 2″·52	+ 0·7635687	15·56
15·58	+ 0·1571149	+ 0·1274329	0·2022974	759° 2′ 41″·47	+ 0·7666447	15·58
15·60	+ 0·1544396	+ 0·1304608	0·2021673	760° 11′ 20″·43	+ 0·7696660	15·60
15·62	+ 0·1517062	+ 0·1334329	0·2020374	761° 19′ 59″·41	+ 0·7726316	15·62
15·64	+ 0·1489160	+ 0·1363480	0·2019078	762° 28′ 38″·41	+ 0·7755402	15·64
15·66	+ 0·1460700	+ 0·1392052	0·2017784	763° 37′ 17″·43	+ 0·7783909	15·66
15·68	+ 0·1431695	+ 0·1420033	0·2016493	764° 45′ 56″·46	+ 0·7811825	15·68
15·70	+ 0·1402157	+ 0·1447413	0·2015204	765° 54′ 35″·50	+ 0·7839140	15·70
15·72	+ 0·1372099	+ 0·1474181	0·2013918	767° 3′ 14″·56	+ 0·7865845	15·72
15·74	+ 0·1341533	+ 0·1500329	0·2012634	768° 11′ 53″·64	+ 0·7891929	15·74
15·76	+ 0·1310471	+ 0·1525847	0·2011353	769° 20′ 32″·73	+ 0·7917383	15·76
15·78	+ 0·1278927	+ 0·1550724	0·2010074	770° 29′ 11″·84	+ 0·7942197	15·78
15·80	+ 0·1246913	+ 0·1574953	0·2008798	771° 37′ 50″·96	+ 0·7966362	15·80
15·82	+ 0·1214444	+ 0·1598524	0·2007524	772° 46′ 30″·10	+ 0·7989870	15·82
15·84	+ 0·1181532	+ 0·1621429	0·2006252	773° 55′ 9″·25	+ 0·8012712	15·84
15·86	+ 0·1148192	+ 0·1643659	0·2004983	775° 3′ 48″·42	+ 0·8034880	15·86
15·88	+ 0·1114436	+ 0·1665207	0·2003717	776° 12′ 27″·61	+ 0·8056365	15·88
15·90	+ 0·1080279	+ 0·1686064	0·2002452	777° 21′ 6″·81	+ 0·8077161	15·90
15·92	+ 0·1045735	+ 0·1706224	0·2001190	778° 29′ 46″·03	+ 0·8097259	15·92
15·94	+ 0·1010818	+ 0·1725680	0·1999931	779° 38′ 25″·26	+ 0·8116653	15·94
15·96	+ 0·0975542	+ 0·1744424	0·1998674	780° 47′ 4″·50	+ 0·8135336	15·96
15·98	+ 0·0939922	+ 0·1762450	0·1997419	781° 55′ 43″·76	+ 0·8153300	15·98
16·00	+ 0·0903972	+ 0·1779752	0·1996167	783° 4′ 23″·04	+ 0·8170541	16·00

Table II. Functions of imaginary argument, and e^x

x	$e^{-x} I_0(x)$	$e^{-x} I_1(x)$	$e^x K_0(x)$	$e^x K_1(x)$	e^x	x
0	1·0000000	0·0000000	∞	∞	1·0000000	0
0·02	0·9802967	0·0098025	4·1098376	50·9638701	1·0202013	0·02
0·04	0·9611738	0·0192196	3·4727083	25·9404241	1·0408108	0·04
0·06	0·9426123	0·0282657	3·1142387	17·5879738	1·0618365	0·06
0·08	0·9245939	0·0369542	2·8679911	13·4048206	1·0832871	0·08
0·10	0·9071009	0·0452984	2·6823261	10·8901827	1·1051709	0·10
0·12	0·8901162	0·0533111	2·5344522	9·2102792	1·1274969	0·12
0·14	0·8736233	0·0610043	2·4123173	8·0076794	1·1502738	0·14
0·16	0·8576062	0·0683899	2·3087874	7·1036124	1·1735109	0·16
0·18	0·8420496	0·0754792	2·2192980	6·3987260	1·1972174	0·18
0·20	0·8269386	0·0822831	2·1407573	5·8333860	1·2214028	0·20
0·22	0·8122587	0·0888122	2·0709767	5·3696274	1·2460767	0·22
0·24	0·7979961	0·0950766	2·0083522	4·9821285	1·2712492	0·24
0·26	0·7841375	0·1010861	1·9516748	4·6533504	1·2969301	0·26
0·28	0·7706698	0·1068501	1·9000114	4·3707591	1·3231298	0·28
0·30	0·7575806	0·1123776	1·8526273	4·1251578	1·3498588	0·30
0·32	0·7448578	0·1176774	1·8089345	3·9096449	1·3771278	0·32
0·34	0·7324896	0·1227579	1·7684552	3·7189398	1·4049476	0·34
0·36	0·7204648	0·1276272	1·7307961	3·5489328	1·4333294	0·36
0·38	0·7087725	0·1322931	1·6956301	3·3963772	1·4622846	0·38
0·40	0·6974022	0·1367632	1·6626821	3·2586739	1·4918247	0·40
0·42	0·6863436	0·1410447	1·6317188	3·1337176	1·5219616	0·42
0·44	0·6755870	0·1451446	1·6025406	3·0197845	1·5527072	0·44
0·46	0·6651228	0·1490697	1·5749758	2·9154495	1·5840740	0·46
0·48	0·6549419	0·1528263	1·5488754	2·8195242	1·6160744	0·48
0·50	0·6450353	0·1564208	1·5241094	2·7310097	1·6487213	0·50
0·52	0·6353945	0·1598592	1·5005638	2·6490599	1·6820276	0·52
0·54	0·6260111	0·1631473	1·4781381	2·5729537	1·7160069	0·54
0·56	0·6168773	0·1662906	1·4567432	2·5020724	1·7506725	0·56
0·58	0·6079851	0·1692946	1·4363000	2·4358821	1·7860384	0·58
0·60	0·5993272	0·1721644	1·4167376	2·3739200	1·8221188	0·60
0·62	0·5908962	0·1749051	1·3979927	2·3157825	1·8589280	0·62
0·64	0·5826853	0·1775214	1·3800080	2·2611160	1·8964809	0·64
0·66	0·5746875	0·1800181	1·3627320	2·2096099	1·9347923	0·66
0·68	0·5668963	0·1823995	1·3461180	2·1609894	1·9738777	0·68
0·70	0·5593055	0·1846700	1·3301237	2·1150113	2·0137527	0·70
0·72	0·5519089	0·1868337	1·3147102	2·0714590	2·0544332	0·72
0·74	0·5447006	0·1888946	1·2998425	2·0301393	2·0959355	0·74
0·76	0·5376748	0·1908567	1·2854881	1·9908791	2·1382762	0·76
0·78	0·5308260	0·1927235	1·2716174	1·9535227	2·1814723	0·78
0·80	0·5241489	0·1944987	1·2582031	1·9179303	2·2255409	0·80
0·82	0·5176383	0·1961857	1·2452202	1·8839752	2·2704998	0·82
0·84	0·5112892	0·1977879	1·2326455	1·8515429	2·3163670	0·84
0·86	0·5050967	0·1993083	1·2204575	1·8205294	2·3631607	0·86
0·88	0·4990561	0·2007502	1·2086362	1·7908399	2·4108997	0·88
0·90	0·4931630	0·2021165	1·1971634	1·7623882	2·4596031	0·90
0·92	0·4874128	0·2034101	1·1860217	1·7350954	2·5092904	0·92
0·94	0·4818015	0·2046336	1·1751953	1·7088892	2·5599814	0·94
0·96	0·4763248	0·2057898	1·1646692	1·6837033	2·6116965	0·96
0·98	0·4709788	0·2068813	1·1544294	1·6594768	2·6644562	0·98
1·00	0·4657596	0·2079104	1·1444631	1·6361535	2·7182818	1·00

Table II. Functions of imaginary argument, and e^x

x	$e^{-x} I_0(x)$	$e^{-x} I_1(x)$	$e^x K_0(x)$	$e^x K_1(x)$	e^x	x
1·02	0·4606636	0·2088796	1·1347579	1·6136817	2·7731948	1·02
1·04	0·4556871	0·2097912	1·1253024	1·5920135	2·8292170	1·04
1·06	0·4508266	0·2106473	1·1160860	1·5711046	2·8863710	1·06
1·08	0·4460788	0·2114501	1·1070984	1·5509141	2·9446796	1·08
1·10	0·4414404	0·2122016	1·0983303	1·5314038	3·0041660	1·10
1·12	0·4369082	0·2129039	1·0897726	1·5125382	3·0648542	1·12
1·14	0·4324792	0·2135587	1·0814169	1·4942846	3·1267684	1·14
1·16	0·4281504	0·2141680	1·0732553	1·4766120	3·1899333	1·16
1·18	0·4239190	0·2147335	1·0652802	1·4594919	3·2543742	1·18
1·20	0·4197821	0·2152569	1·0574845	1·4428976	3·3201169	1·20
1·22	0·4157371	0·2157398	1·0498615	1·4268038	3·3871877	1·22
1·24	0·4117813	0·2161838	1·0424048	1·4111872	3·4556135	1·24
1·26	0·4079123	0·2165905	1·0351082	1·3960258	3·5254215	1·26
1·28	0·4041277	0·2169613	1·0279662	1·3812990	3·5966397	1·28
1·30	0·4004249	0·2172976	1·0209732	1·3669873	3·6692967	1·30
1·32	0·3968018	0·2176008	1·0141239	1·3530725	3·7434214	1·32
1·34	0·3932561	0·2178721	1·0074136	1·3395375	3·8190435	1·34
1·36	0·3897857	0·2181129	1·0008375	1·3263660	3·8961933	1·36
1·38	0·3863885	0·2183243	0·9943910	1·3135429	3·9749016	1·38
1·40	0·3830625	0·2185076	0·9880700	1·3010537	4·0552000	1·40
1·42	0·3798057	0·2186638	0·9818703	1·2888849	4·1371204	1·42
1·44	0·3766162	0·2187941	0·9757881	1·2770235	4·2206958	1·44
1·46	0·3734922	0·2188994	0·9698197	1·2654575	4·3059595	1·46
1·48	0·3704319	0·2189809	0·9639615	1·2541752	4·3929457	1·48
1·50	0·3674336	0·2190394	0·9582101	1·2431659	4·4816891	1·50
1·52	0·3644956	0·2190759	0·9525622	1·2324190	4·5722252	1·52
1·54	0·3616163	0·2190913	0·9470148	1·2219249	4·6645903	1·54
1·56	0·3587941	0·2190865	0·9415648	1·2116741	4·7588212	1·56
1·58	0·3560275	0·2190623	0·9362095	1·2016578	4·8549558	1·58
1·60	0·3533150	0·2190195	0·9309460	1·1918676	4·9530324	1·60
1·62	0·3506552	0·2189589	0·9257717	1·1822954	5·0530903	1·62
1·64	0·3480467	0·2188812	0·9206842	1·1729335	5·1551695	1·64
1·66	0·3454881	0·2187872	0·9156810	1·1637748	5·2593108	1·66
1·68	0·3429782	0·2186775	0·9107597	1·1548123	5·3655560	1·68
1·70	0·3405157	0·2185528	0·9059181	1·1460392	5·4739474	1·70
1·72	0·3380993	0·2184138	0·9011541	1·1374494	5·5845285	1·72
1·74	0·3357279	0·2182610	0·8964656	1·1290368	5·6973434	1·74
1·76	0·3334003	0·2180952	0·8918506	1·1207955	5·8124374	1·76
1·78	0·3311153	0·2179167	0·8873072	1·1127201	5·9298564	1·78
1·80	0·3288719	0·2177263	0·8828335	1·1048054	6·0496475	1·80
1·82	0·3266691	0·2175244	0·8784278	1·0970461	6·1718584	1·82
1·84	0·3245058	0·2173115	0·8740882	1·0894376	6·2965383	1·84
1·86	0·3223810	0·2170881	0·8698132	1·0819752	6·4237368	1·86
1·88	0·3202938	0·2168548	0·8656012	1·0746544	6·5535049	1·88
1·90	0·3182432	0·2166119	0·8614506	1·0674709	6·6858944	1·90
1·92	0·3162282	0·2163599	0·8573599	1·0604208	6·8209585	1·92
1·94	0·3142481	0·2160993	0·8533277	1·0535000	6·9587510	1·94
1·96	0·3123020	0·2158304	0·8493526	1·0467048	7·0993271	1·96
1·98	0·3103890	0·2155536	0·8454332	1·0400316	7·2427430	1·98
2·00	0·3085083	0·2152693	0·8415682	1·0334768	7·3890561	2·00

Table II. Functions of imaginary argument, and e^x

x	$e^{-x} I_0(x)$	$e^{-x} I_1(x)$	$e^x K_0(x)$	$e^x K_1(x)$	e^x	x
2·02	0·3066592	0·2149779	0·8377564	1·0270373	7·5383249	2·02
2·04	0·3048408	0·2146797	0·8339966	1·0207097	7·6906092	2·04
2·06	0·3030525	0·2143750	0·8302875	1·0144909	7·8459698	2·06
2·08	0·3012935	0·2140643	0·8266281	1·0083780	8·0044689	2·08
2·10	0·2995631	0·2137477	0·8230172	1·0023681	8·1661699	2·10
2·12	0·2978606	0·2134256	0·8194537	0·9964584	8·3311375	2·12
2·14	0·2961855	0·2130983	0·8159366	0·9906463	8·4994376	2·14
2·16	0·2945369	0·2127660	0·8124650	0·9849292	8·6711377	2·16
2·18	0·2929144	0·2124291	0·8090377	0·9793046	8·8463063	2·18
2·20	0·2913173	0·2120877	0·8056540	0·9737702	9·0250135	2·20
2·22	0·2897451	0·2117422	0·8023128	0·9683236	9·2073308	2·22
2·24	0·2881970	0·2113927	0·7990133	0·9629626	9·3933313	2·24
2·26	0·2866727	0·2110396	0·7957545	0·9576851	9·5830892	2·26
2·28	0·2851715	0·2106829	0·7925358	0·9524890	9·7766804	2·28
2·30	0·2836930	0·2103230	0·7893561	0·9473722	9·9741825	2·30
2·32	0·2822366	0·2099600	0·7862149	0·9423329	10·1756743	2·32
2·34	0·2808018	0·2095941	0·7831112	0·9373692	10·3812366	2·34
2·36	0·2793881	0·2092256	0·7800443	0·9324793	10·5909515	2·36
2·38	0·2779951	0·2088545	0·7770135	0·9276613	10·8049029	2·38
2·40	0·2766223	0·2084811	0·7740181	0·9229137	11·0231764	2·40
2·42	0·2752693	0·2081055	0·7710575	0·9182347	11·2458593	2·42
2·44	0·2739356	0·2077279	0·7681308	0·9136228	11·4730407	2·44
2·46	0·2726209	0·2073485	0·7652376	0·9090764	11·7048115	2·46
2·48	0·2713246	0·2069674	0·7623771	0·9045941	11·9412644	2·48
2·50	0·2700464	0·2065846	0·7595487	0·9001744	12·1824940	2·50
2·52	0·2687860	0·2062005	0·7567518	0·8958159	12·4285967	2·52
2·54	0·2675429	0·2058151	0·7539859	0·8915172	12·6796710	2·54
2·56	0·2663168	0·2054285	0·7512504	0·8872771	12·9358173	2·56
2·58	0·2651072	0·2050408	0·7485447	0·8830942	13·1971382	2·58
2·60	0·2639140	0·2046523	0·7458682	0·8789673	13·4637380	2·60
2·62	0·2627367	0·2042628	0·7432205	0·8748952	13·7357236	2·62
2·64	0·2615749	0·2038727	0·7406011	0·8708767	14·0132036	2·64
2·66	0·2604285	0·2034820	0·7380094	0·8669107	14·2962891	2·66
2·68	0·2592970	0·2030907	0·7354449	0·8629961	14·5850933	2·68
2·70	0·2581801	0·2026991	0·7329072	0·8591319	14·8797317	2·70
2·72	0·2570776	0·2023071	0·7303957	0·8553169	15·1803222	2·72
2·74	0·2559892	0·2019148	0·7279102	0·8515502	15·4869851	2·74
2·76	0·2549146	0·2015224	0·7254500	0·8478308	15·7998429	2·76
2·78	0·2538534	0·2011299	0·7230148	0·8441577	16·1190209	2·78
2·80	0·2528055	0·2007374	0·7206041	0·8405301	16·4446468	2·80
2·82	0·2517706	0·2003450	0·7182176	0·8369469	16·7768507	2·82
2·84	0·2507484	0·1999527	0·7158548	0·8334074	17·1157655	2·84
2·86	0·2497387	0·1995606	0·7135154	0·8299106	17·4615269	2·86
2·88	0·2487412	0·1991688	0·7111989	0·8264557	17·8142732	2·88
2·90	0·2477557	0·1987773	0·7089050	0·8230420	18·1741454	2·90
2·92	0·2467820	0·1983862	0·7066333	0·8196687	18·5412875	2·92
2·94	0·2458198	0·1979955	0·7043834	0·8163349	18·9158463	2·94
2·96	0·2448690	0·1976053	0·7021551	0·8130399	19·2979718	2·96
2·98	0·2439292	0·1972157	0·6999479	0·8097830	19·6878167	2·98
3·00	0·2430004	0·1968267	0·6977616	0·8065635	20·0855369	3·00

Table II. Functions of imaginary argument, and e^x

x	$e^{-x} I_0(x)$	$e^{-x} I_1(x)$	$e^x K_0(x)$	$e^x K_1(x)$	e^x	x
3·02	0·2420822	0·1964383	0·6955958	0·8033807	20·491292	3·02
3·04	0·2411745	0·1960506	0·6934501	0·8002339	20·905243	3·04
3·06	0·2402772	0·1956637	0·6913243	0·7971224	21·327557	3·06
3·08	0·2393899	0·1952775	0·6892181	0·7940457	21·758402	3·08
3·10	0·2385126	0·1948921	0·6871311	0·7910030	22·197951	3·10
3·12	0·2376451	0·1945076	0·6850631	0·7879938	22·646380	3·12
3·14	0·2367871	0·1941240	0·6830138	0·7850176	23·103867	3·14
3·16	0·2359385	0·1937412	0·6809829	0·7820736	23·570596	3·16
3·18	0·2350991	0·1933594	0·6789701	0·7791613	24·046754	3·18
3·20	0·2342688	0·1929786	0·6769751	0·7762803	24·532530	3·20
3·22	0·2334475	0·1925988	0·6749978	0·7734299	25·028120	3·22
3·24	0·2326348	0·1922200	0·6730377	0·7706096	25·533722	3·24
3·26	0·2318308	0·1918423	0·6710948	0·7678189	26·049537	3·26
3·28	0·2310352	0·1914657	0·6691687	0·7650573	26·575773	3·28
3·30	0·2302480	0·1910902	0·6672592	0·7623243	27·112639	3·30
3·32	0·2294689	0·1907158	0·6653660	0·7596194	27·660351	3·32
3·34	0·2286978	0·1903425	0·6634890	0·7569422	28·219127	3·34
3·36	0·2279346	0·1899704	0·6616278	0·7542922	28·789191	3·36
3·38	0·2271792	0·1895995	0·6597823	0·7516690	29·370771	3·38
3·40	0·2264314	0·1892299	0·6579523	0·7490721	29·964100	3·40
3·42	0·2256911	0·1888614	0·6561375	0·7465010	30·569415	3·42
3·44	0·2249582	0·1884941	0·6543377	0·7439555	31·186958	3·44
3·46	0·2242325	0·1881282	0·6525527	0·7414350	31·816977	3·46
3·48	0·2235140	0·1877634	0·6507823	0·7389391	32·459722	3·48
3·50	0·2228024	0·1874000	0·6490263	0·7364675	33·115452	3·50
3·52	0·2220978	0·1870378	0·6472846	0·7340199	33·784428	3·52
3·54	0·2214000	0·1866770	0·6455569	0·7315957	34·466919	3·54
3·56	0·2207089	0·1863174	0·6438430	0·7291947	35·163197	3·56
3·58	0·2200243	0·1859592	0·6421427	0·7268165	35·873541	3·58
3·60	0·2193462	0·1856022	0·6404560	0·7244607	36·598234	3·60
3·62	0·2186745	0·1852467	0·6387825	0·7221270	37·337568	3·62
3·64	0·2180091	0·1848924	0·6371221	0·7198150	38·091837	3·64
3·66	0·2173498	0·1845396	0·6354747	0·7175245	38·861343	3·66
3·68	0·2166966	0·1841880	0·6338401	0·7152551	39·646394	3·68
3·70	0·2160494	0·1838379	0·6322181	0·7130065	40·447304	3·70
3·72	0·2154081	0·1834891	0·6306085	0·7107784	41·264394	3·72
3·74	0·2147726	0·1831416	0·6290112	0·7085704	42·097990	3·74
3·76	0·2141429	0·1827956	0·6274261	0·7063823	42·948426	3·76
3·78	0·2135187	0·1824509	0·6258529	0·7042139	43·816042	3·78
3·80	0·2129001	0·1821076	0·6242916	0·7020647	44·701184	3·80
3·82	0·2122870	0·1817657	0·6227419	0·6999345	45·604208	3·82
3·84	0·2116793	0·1814251	0·6212038	0·6978232	46·525474	3·84
3·86	0·2110768	0·1810860	0·6196771	0·6957302	47·465351	3·86
3·88	0·2104796	0·1807482	0·6181617	0·6936555	48·424215	3·88
3·90	0·2098875	0·1804119	0·6166573	0·6915988	49·402449	3·90
3·92	0·2093005	0·1800769	0·6151640	0·6895598	50·400445	3·92
3·94	0·2087186	0·1797433	0·6136814	0·6875382	51·418601	3·94
3·96	0·2081415	0·1794111	0·6122096	0·6855339	52·457326	3·96
3·98	0·2075693	0·1790803	0·6107484	0·6835466	53·517034	3·98
4·00	0·2070019	0·1787508	0·6092977	0·6815759	54·598150	4·00

Table II. Functions of imaginary argument, and e^x

x	$e^{-x} I_0(x)$	$e^{-x} I_1(x)$	$e^x K_0(x)$	$e^x K_1(x)$	e^x	x
4·02	0·2064393	0·1784228	0·6078573	0·6796219	55·701106	4·02
4·04	0·2058812	0·1780961	0·6064270	0·6776840	56·826343	4·04
4·06	0·2053278	0·1777709	0·6050069	0·6757623	57·974311	4·06
4·08	0·2047789	0·1774470	0·6035968	0·6738564	59·145470	4·08
4·10	0·2042345	0·1771245	0·6021965	0·6719662	60·340288	4·10
4·12	0·2036945	0·1768033	0·6008060	0·6700914	61·559242	4·12
4·14	0·2031589	0·1764836	0·5994251	0·6682318	62·802821	4·14
4·16	0·2026275	0·1761652	0·5980537	0·6663872	64·071523	4·16
4·18	0·2021003	0·1758482	0·5966917	0·6645575	65·365853	4·18
4·20	0·2015774	0·1755325	0·5953390	0·6627424	66·686331	4·20
4·22	0·2010585	0·1752182	0·5939955	0·6609418	68·033484	4·22
4·24	0·2005438	0·1749053	0·5926611	0·6591553	69·407852	4·24
4·26	0·2000330	0·1745937	0·5913357	0·6573830	70·809983	4·26
4·28	0·1995262	0·1742835	0·5900192	0·6556246	72·240440	4·28
4·30	0·1990233	0·1739746	0·5887114	0·6538798	73·699794	4·30
4·32	0·1985242	0·1736671	0·5874124	0·6521486	75·188628	4·32
4·34	0·1980290	0·1733609	0·5861220	0·6504308	76·707539	4·34
4·36	0·1975375	0·1730560	0·5848400	0·6487262	78·257134	4·36
4·38	0·1970497	0·1727525	0·5835665	0·6470346	79·838033	4·38
4·40	0·1965656	0·1724502	0·5823013	0·6453559	81·450869	4·40
4·42	0·1960851	0·1721493	0·5810443	0·6436899	83·096285	4·42
4·44	0·1956081	0·1718497	0·5797954	0·6420364	84·774942	4·44
4·46	0·1951347	0·1715515	0·5785546	0·6403953	86·487509	4·46
4·48	0·1946648	0·1712545	0·5773218	0·6387665	88·234673	4·48
4·50	0·1941983	0·1709588	0·5760968	0·6371498	90·017131	4·50
4·52	0·1937352	0·1706644	0·5748796	0·6355450	91·835598	4·52
4·54	0·1932754	0·1703713	0·5736701	0·6339521	93·690800	4·54
4·56	0·1928190	0·1700795	0·5724683	0·6323708	95·583480	4·56
4·58	0·1923658	0·1697890	0·5712740	0·6308010	97·514394	4·58
4·60	0·1919159	0·1694997	0·5700872	0·6292426	99·484316	4·60
4·62	0·1914692	0·1692117	0·5689078	0·6276955	101·494032	4·62
4·64	0·1910256	0·1689250	0·5677357	0·6261595	103·544348	4·64
4·66	0·1905851	0·1686395	0·5665708	0·6246345	105·636082	4·66
4·68	0·1901478	0·1683553	0·5654131	0·6231203	107·770073	4·68
4·70	0·1897134	0·1680723	0·5642625	0·6216169	109·947172	4·70
4·72	0·1892821	0·1677905	0·5631189	0·6201241	112·168253	4·72
4·74	0·1888538	0·1675100	0·5619823	0·6186418	114·434202	4·74
4·76	0·1884283	0·1672307	0·5608525	0·6171699	116·745926	4·76
4·78	0·1880058	0·1669526	0·5597295	0·6157082	119·104350	4·78
4·80	0·1875862	0·1666757	0·5586133	0·6142566	121·510418	4·80
4·82	0·1871694	0·1664000	0·5575038	0·6128151	123·965091	4·82
4·84	0·1867554	0·1661256	0·5564008	0·6113834	126·469352	4·84
4·86	0·1863442	0·1658523	0·5553045	0·6099616	129·024202	4·86
4·88	0·1859357	0·1655802	0·5542145	0·6085494	131·630664	4·88
4·90	0·1855300	0·1653093	0·5531310	0·6071468	134·289780	4·90
4·92	0·1851269	0·1650396	0·5520539	0·6057537	137·002613	4·92
4·94	0·1847265	0·1647710	0·5509830	0·6043699	139·770250	4·94
4·96	0·1843287	0·1645036	0·5499184	0·6029955	142·593796	4·96
4·98	0·1839335	0·1642374	0·5488599	0·6016301	145·474382	4·98
5·00	0·1835408	0·1639723	0·5478076	0·6002739	148·413159	5·00

Table II. Functions of imaginary argument, and e^x

x	$e^{-x} I_0(x)$	$e^{-x} I_1(x)$	$e^x K_0(x)$	$e^x K_1(x)$	e^x	x
5·02	0·1831507	0·1637083	0·5467613	0·5989266	151·41130	5·02
5·04	0·1827631	0·1634455	0·5457209	0·5975881	154·47002	5·04
5·06	0·1823780	0·1631838	0·5446865	0·5962584	157·59052	5·06
5·08	0·1819953	0·1629233	0·5436580	0·5949375	160·77406	5·08
5·10	0·1816151	0·1626639	0·5426354	0·5936250	164·02191	5·10
5·12	0·1812373	0·1624055	0·5416184	0·5923211	167·33537	5·12
5·14	0·1808618	0·1621483	0·5406072	0·5910256	170·71577	5·14
5·16	0·1804887	0·1618922	0·5396017	0·5897384	174·16446	5·16
5·18	0·1801180	0·1616372	0·5386017	0·5884594	177·68281	5·18
5·20	0·1797495	0·1613833	0·5376074	0·5871886	181·27224	5·20
5·22	0·1793833	0·1611304	0·5366185	0·5859258	184·93418	5·22
5·24	0·1790194	0·1608787	0·5356350	0·5846710	188·67010	5·24
5·26	0·1786577	0·1606280	0·5346570	0·5834241	192·48149	5·26
5·28	0·1782982	0·1603784	0·5336843	0·5821850	196·36988	5·28
5·30	0·1779409	0·1601298	0·5327170	0·5809536	200·33681	5·30
5·32	0·1775857	0·1598823	0·5317549	0·5797299	204·38388	5·32
5·34	0·1772327	0·1596358	0·5307980	0·5785137	208·51271	5·34
5·36	0·1768818	0·1593904	0·5298462	0·5773050	212·72495	5·36
5·38	0·1765331	0·1591460	0·5288996	0·5761038	217·02228	5·38
5·40	0·1761863	0·1589026	0·5279580	0·5749099	221·40642	5·40
5·42	0·1758417	0·1586603	0·5270215	0·5737233	225·87912	5·42
5·44	0·1754991	0·1584189	0·5260899	0·5725438	230·44218	5·44
5·46	0·1751585	0·1581786	0·5251633	0·5713715	235·09742	5·46
5·48	0·1748199	0·1579393	0·5242416	0·5702062	239·84671	5·48
5·50	0·1744833	0·1577010	0·5233247	0·5690480	244·69193	5·50
5·52	0·1741486	0·1574637	0·5224127	0·5678966	249·63504	5·52
5·54	0·1738159	0·1572274	0·5215054	0·5667521	254·67800	5·54
5·56	0·1734850	0·1569920	0·5206028	0·5656144	259·82284	5·56
5·58	0·1731561	0·1567576	0·5197049	0·5644834	265·07161	5·58
5·60	0·1728291	0·1565242	0·5188116	0·5633590	270·42641	5·60
5·62	0·1725039	0·1562918	0·5179230	0·5622413	275·88938	5·62
5·64	0·1721806	0·1560603	0·5170389	0·5611300	281·46272	5·64
5·66	0·1718591	0·1558298	0·5161593	0·5600253	287·14864	5·66
5·68	0·1715394	0·1556002	0·5152842	0·5589269	292·94943	5·68
5·70	0·1712215	0·1553716	0·5144136	0·5578348	298·86740	5·70
5·72	0·1709054	0·1551439	0·5135474	0·5567491	304·90492	5·72
5·74	0·1705911	0·1549171	0·5126855	0·5556695	311·06441	5·74
5·76	0·1702785	0·1546913	0·5118280	0·5545962	317·34833	5·76
5·78	0·1699676	0·1544664	0·5109748	0·5535289	323·75919	5·78
5·80	0·1696584	0·1542424	0·5101258	0·5524676	330·29956	5·80
5·82	0·1693509	0·1540193	0·5092811	0·5514124	336·97205	5·82
5·84	0·1690451	0·1537971	0·5084406	0·5503631	343·77934	5·84
5·86	0·1687410	0·1535758	0·5076042	0·5493197	350·72414	5·86
5·88	0·1684385	0·1533554	0·5067719	0·5482821	357·80924	5·88
5·90	0·1681377	0·1531359	0·5059438	0·5472503	365·03747	5·90
5·92	0·1678384	0·1529172	0·5051197	0·5462242	372·41171	5·92
5·94	0·1675408	0·1526995	0·5042996	0·5452037	379·93493	5·94
5·96	0·1672448	0·1524826	0·5034835	0·5441889	387·61012	5·96
5·98	0·1669503	0·1522666	0·5026713	0·5431796	395·44037	5·98
6·00	0·1666574	0·1520515	0·5018631	0·5421759	403·42879	6·00

Table II. Functions of imaginary argument, and e^x

x	$e^{-x} I_0(x)$	$e^{-x} I_1(x)$	$e^x K_0(x)$	$e^x K_1(x)$	e^x	x
6·02	0·1663661	0·1518372	0·5010588	0·5411776	411·57860	6·02
6·04	0·1660763	0·1516237	0·5002584	0·5401848	419·89303	6·04
6·06	0·1657880	0·1514111	0·4994618	0·5391973	428·37544	6·06
6·08	0·1655012	0·1511994	0·4986689	0·5382151	437·02919	6·08
6·10	0·1652159	0·1509885	0·4978799	0·5372382	445·85777	6·10
6·12	0·1649321	0·1507784	0·4970946	0·5362666	454·86469	6·12
6·14	0·1646498	0·1505691	0·4963130	0·5353001	464·05357	6·14
6·16	0·1643689	0·1503607	0·4955351	0·5343387	473·42807	6·16
6·18	0·1640894	0·1501531	0·4947608	0·5333825	482·99196	6·18
6·20	0·1638114	0·1499463	0·4939902	0·5324313	492·74904	6·20
6·22	0·1635348	0·1497403	0·4932232	0·5314851	502·70323	6·22
6·24	0·1632596	0·1495351	0·4924597	0·5305438	512·85851	6·24
6·26	0·1629858	0·1493307	0·4916998	0·5296075	523·21894	6·26
6·28	0·1627134	0·1491271	0·4909434	0·5286761	533·78866	6·28
6·30	0·1624424	0·1489243	0·4901905	0·5277494	544·57191	6·30
6·32	0·1621727	0·1487223	0·4894411	0·5268276	555·57299	6·32
6·34	0·1619044	0·1485211	0·4886950	0·5259105	566·79631	6·34
6·36	0·1616374	0·1483206	0·4879524	0·5249982	578·24636	6·36
6·38	0·1613717	0·1481209	0·4872132	0·5240905	589·92771	6·38
6·40	0·1611073	0·1479220	0·4864773	0·5231874	601·84504	6·40
6·42	0·1608443	0·1477238	0·4857448	0·5222889	614·00311	6·42
6·44	0·1605825	0·1475264	0·4850156	0·5213950	626·40680	6·44
6·46	0·1603220	0·1473297	0·4842896	0·5205056	639·06106	6·46
6·48	0·1600628	0·1471338	0·4835669	0·5196207	651·97095	6·48
6·50	0·1598048	0·1469386	0·4828474	0·5187402	665·14163	6·50
6·52	0·1595481	0·1467442	0·4821312	0·5178642	678·57839	6·52
6·54	0·1592927	0·1465505	0·4814181	0·5169925	692·28658	6·54
6·56	0·1590385	0·1463576	0·4807082	0·5161251	706·27169	6·56
6·58	0·1587855	0·1461653	0·4800014	0·5152620	720·53933	6·58
6·60	0·1585337	0·1459738	0·4792978	0·5144032	735·09519	6·60
6·62	0·1582831	0·1457830	0·4785972	0·5135486	749·94510	6·62
6·64	0·1580336	0·1455930	0·4778997	0·5126982	765·09499	6·64
6·66	0·1577854	0·1454036	0·4772053	0·5118520	780·55094	6·66
6·68	0·1575384	0·1452149	0·4765138	0·5110099	796·31911	6·68
6·70	0·1572925	0·1450270	0·4758254	0·5101719	812·40583	6·70
6·72	0·1570477	0·1448397	0·4751400	0·5093380	828·81751	6·72
6·74	0·1568042	0·1446532	0·4744575	0·5085080	845·56074	6·74
6·76	0·1565617	0·1444673	0·4737779	0·5076821	862·64220	6·76
6·78	0·1563204	0·1442821	0·4731013	0·5068602	880·06872	6·78
6·80	0·1560802	0·1440976	0·4724276	0·5060421	897·84729	6·80
6·82	0·1558411	0·1439138	0·4717567	0·5052280	915·98501	6·82
6·84	0·1556031	0·1437306	0·4710887	0·5044178	934·48913	6·84
6·86	0·1553662	0·1435481	0·4704235	0·5036114	953·36707	6·86
6·88	0·1551304	0·1433663	0·4697612	0·5028088	972·62636	6·88
6·90	0·1548956	0·1431852	0·4691016	0·5020099	992·27472	6·90
6·92	0·1546619	0·1430047	0·4684449	0·5012149	1012·31999	6·92
6·94	0·1544293	0·1428248	0·4677908	0·5004235	1032·77021	6·94
6·96	0·1541978	0·1426457	0·4671395	0·4996359	1053·63356	6·96
6·98	0·1539672	0·1424671	0·4664910	0·4988519	1074·91837	6·98
7·00	0·1537377	0·1422892	0·4658451	0·4980716	1096·63316	7·00

Table II. Functions of imaginary argument, and e^x

x	$e^{-x} I_0(x)$	$e^{-x} I_1(x)$	$e^x K_0(x)$	$e^x K_1(x)$	e^x	x
7·02	0·1535093	0·1421120	0·4652019	0·4972948	1118·7866	7·02
7·04	0·1532819	0·1419354	0·4645614	0·4965217	1141·3876	7·04
7·06	0·1530554	0·1417594	0·4639235	0·4957521	1164·4452	7·06
7·08	0·1528300	0·1415840	0·4632882	0·4949860	1187·9685	7·08
7·10	0·1526056	0·1414093	0·4626556	0·4942235	1211·9671	7·10
7·12	0·1523822	0·1412352	0·4620255	0·4934644	1236·4504	7·12
7·14	0·1521597	0·1410617	0·4613980	0·4927087	1261·4284	7·14
7·16	0·1519382	0·1408889	0·4607731	0·4919565	1286·9109	7·16
7·18	0·1517177	0·1407166	0·4601507	0·4912077	1312·9083	7·18
7·20	0·1514982	0·1405450	0·4595308	0·4904623	1339·4308	7·20
7·22	0·1512796	0·1403739	0·4589134	0·4897202	1366·4891	7·22
7·24	0·1510620	0·1402035	0·4582985	0·4889814	1394·0940	7·24
7·26	0·1508453	0·1400337	0·4576861	0·4882459	1422·2565	7·26
7·28	0·1506295	0·1398644	0·4570761	0·4875137	1450·9880	7·28
7·30	0·1504147	0·1396958	0·4564686	0·4867848	1480·2999	7·30
7·32	0·1502007	0·1395277	0·4558634	0·4860591	1510·2040	7·32
7·34	0·1499877	0·1393603	0·4552607	0·4853365	1540·7121	7·34
7·36	0·1497756	0·1391934	0·4546604	0·4846172	1571·8366	7·36
7·38	0·1495644	0·1390271	0·4540625	0·4839010	1603·5898	7·38
7·40	0·1493541	0·1388613	0·4534669	0·4831880	1635·9844	7·40
7·42	0·1491447	0·1386962	0·4528736	0·4824780	1669·0335	7·42
7·44	0·1489362	0·1385316	0·4522827	0·4817712	1702·7502	7·44
7·46	0·1487285	0·1383676	0·4516941	0·4810674	1737·1481	7·46
7·48	0·1485218	0·1382041	0·4511077	0·4803667	1772·2408	7·48
7·50	0·1483158	0·1380412	0·4505237	0·4796689	1808·0424	7·50
7·52	0·1481108	0·1378789	0·4499419	0·4789742	1844·5673	7·52
7·54	0·1479066	0·1377171	0·4493624	0·4782825	1881·8300	7·54
7·56	0·1477032	0·1375559	0·4487851	0·4775937	1919·8455	7·56
7·58	0·1475007	0·1373952	0·4482101	0·4769079	1958·6290	7·58
7·60	0·1472990	0·1372350	0·4476372	0·4762249	1998·1959	7·60
7·62	0·1470981	0·1370754	0·4470665	0·4755449	2038·5621	7·62
7·64	0·1468981	0·1369164	0·4464981	0·4748678	2079·7438	7·64
7·66	0·1466988	0·1367579	0·4459317	0·4741935	2121·7574	7·66
7·68	0·1465004	0·1365999	0·4453676	0·4735220	2164·6198	7·68
7·70	0·1463028	0·1364424	0·4448056	0·4728534	2208·3480	7·70
7·72	0·1461060	0·1362855	0·4442457	0·4721876	2252·9596	7·72
7·74	0·1459100	0·1361291	0·4436879	0·4715245	2298·4724	7·74
7·76	0·1457148	0·1359732	0·4431322	0·4708642	2344·9046	7·76
7·78	0·1455203	0·1358179	0·4425786	0·4702066	2392·2748	7·78
7·80	0·1453267	0·1356630	0·4420271	0·4695518	2440·6020	7·80
7·82	0·1451338	0·1355087	0·4414776	0·4688997	2489·9054	7·82
7·84	0·1449417	0·1353549	0·4409302	0·4682502	2540·2048	7·84
7·86	0·1447503	0·1352016	0·4403848	0·4676034	2591·5204	7·86
7·88	0·1445597	0·1350488	0·4398414	0·4669593	2643·8726	7·88
7·90	0·1443699	0·1348965	0·4393001	0·4663178	2697·2823	7·90
7·92	0·1441808	0·1347447	0·4387607	0·4656789	2751·7710	7·92
7·94	0·1439924	0·1345934	0·4382234	0·4650426	2807·3605	7·94
7·96	0·1438048	0·1344426	0·4376880	0·4644089	2864·0730	7·96
7·98	0·1436179	0·1342923	0·4371545	0·4637777	2921·9311	7·98
8·00	0·1434318	0·1341425	0·4366230	0·4631491	2980·9580	8·00

Table II. Functions of imaginary argument, and e^x

x	$e^{-x} I_0(x)$	$e^{-x} I_1(x)$	$e^x K_0(x)$	$e^x K_1(x)$	e^x	x
8·02	0·1432464	0·1339932	0·4360935	0·4625230	3041·1773	8·02
8·04	0·1430617	0·1338443	0·4355658	0·4618994	3102·6132	8·04
8·06	0·1428777	0·1336960	0·4350401	0·4612783	3165·2901	8·06
8·08	0·1426944	0·1335481	0·4345163	0·4606597	3229·2332	8·08
8·10	0·1425118	0·1334007	0·4339944	0·4600436	3294·4681	8·10
8·12	0·1423299	0·1332538	0·4334743	0·4594299	3361·0207	8·12
8·14	0·1421488	0·1331073	0·4329562	0·4588186	3428·9179	8·14
8·16	0·1419683	0·1329613	0·4324398	0·4582097	3498·1866	8·16
8·18	0·1417885	0·1328158	0·4319254	0·4576033	3568·8547	8·18
8·20	0·1416094	0·1326708	0·4314127	0·4569992	3640·9503	8·20
8·22	0·1414309	0·1325262	0·4309019	0·4563974	3714·5024	8·22
8·24	0·1412532	0·1323821	0·4303929	0·4557981	3789·5403	8·24
8·26	0·1410761	0·1322384	0·4298857	0·4552010	3866·0941	8·26
8·28	0·1408997	0·1320952	0·4293803	0·4546063	3944·1944	8·28
8·30	0·1407239	0·1319524	0·4288766	0·4540139	4023·8724	8·30
8·32	0·1405488	0·1318101	0·4283748	0·4534238	4105·1600	8·32
8·34	0·1403744	0·1316683	0·4278747	0·4528359	4188·0897	8·34
8·36	0·1402006	0·1315269	0·4273763	0·4522504	4272·6948	8·36
8·38	0·1400274	0·1313859	0·4268797	0·4516670	4359·0089	8·38
8·40	0·1398549	0·1312454	0·4263848	0·4510859	4447·0667	8·40
8·42	0·1396830	0·1311053	0·4258917	0·4505070	4536·9035	8·42
8·44	0·1395118	0·1309657	0·4254002	0·4499303	4628·5550	8·44
8·46	0·1393412	0·1308265	0·4249104	0·4493559	4722·0580	8·46
8·48	0·1391712	0·1306877	0·4244224	0·4487835	4817·4499	8·48
8·50	0·1390018	0·1305494	0·4239360	0·4482134	4914·7688	8·50
8·52	0·1388331	0·1304114	0·4234513	0·4476454	5014·0538	8·52
8·54	0·1386650	0·1302740	0·4229682	0·4470795	5115·3444	8·54
8·56	0·1384975	0·1301369	0·4224868	0·4465158	5218·6812	8·56
8·58	0·1383306	0·1300003	0·4220071	0·4459542	5324·1055	8·58
8·60	0·1381642	0·1298641	0·4215289	0·4453946	5431·6596	8·60
8·62	0·1379985	0·1297283	0·4210524	0·4448372	5541·3864	8·62
8·64	0·1378334	0·1295929	0·4205776	0·4442818	5653·3298	8·64
8·66	0·1376689	0·1294579	0·4201043	0·4437285	5767·5347	8·66
8·68	0·1375050	0·1293234	0·4196326	0·4431772	5884·0466	8·68
8·70	0·1373417	0·1291892	0·4191625	0·4426280	6002·9122	8·70
8·72	0·1371789	0·1290555	0·4186940	0·4420808	6124·1791	8·72
8·74	0·1370167	0·1289222	0·4182270	0·4415356	6247·8957	8·74
8·76	0·1368551	0·1287892	0·4177616	0·4409923	6374·1116	8·76
8·78	0·1366941	0·1286567	0·4172978	0·4404511	6502·8772	8·78
8·80	0·1365336	0·1285246	0·4168355	0·4399119	6634·2440	8·80
8·82	0·1363737	0·1283929	0·4163747	0·4393746	6768·2646	8·82
8·84	0·1362144	0·1282615	0·4159155	0·4388392	6904·9926	8·84
8·86	0·1360556	0·1281306	0·4154578	0·4383058	7044·4827	8·86
8·88	0·1358974	0·1280001	0·4150016	0·4377743	7186·7907	8·88
8·90	0·1357397	0·1278699	0·4145468	0·4372448	7331·9735	8·90
8·92	0·1355826	0·1277402	0·4140936	0·4367171	7480·0892	8·92
8·94	0·1354260	0·1276108	0·4136419	0·4361913	7631·1971	8·94
8·96	0·1352700	0·1274818	0·4131917	0·4356674	7785·3575	8·96
8·98	0·1351145	0·1273532	0·4127429	0·4351454	7942·6321	8·98
9·00	0·1349595	0·1272250	0·4122955	0·4346252	8103·0839	9·00

Table II. Functions of imaginary argument, and e^x

x	$e^{-x} I_0(x)$	$e^{-x} I_1(x)$	$e^x K_0(x)$	$e^x K_1(x)$	e^x	x
9·02	0·1348051	0·1270971	0·4118497	0·4341069	8266·7771	9·02
9·04	0·1346512	0·1269697	0·4114053	0·4335904	8433·7771	9·04
9·06	0·1344978	0·1268426	0·4109623	0·4330758	8604·1507	9·06
9·08	0·1343450	0·1267159	0·4105207	0·4325629	8777·9660	9·08
9·10	0·1341927	0·1265895	0·4100806	0·4320519	8955·2927	9·10
9·12	0·1340409	0·1264636	0·4096419	0·4315427	9136·2016	9·12
9·14	0·1338896	0·1263380	0·4092045	0·4310352	9320·7651	9·14
9·16	0·1337388	0·1262127	0·4087686	0·4305295	9509·0571	9·16
9·18	0·1335885	0·1260879	0·4083341	0·4300256	9701·1528	9·18
9·20	0·1334388	0·1259634	0·4079010	0·4295234	9897·1291	9·20
9·22	0·1332895	0·1258392	0·4074692	0·4290230	10097·0643	9·22
9·24	0·1331408	0·1257154	0·4070388	0·4285243	10301·0386	9·24
9·26	0·1329925	0·1255920	0·4066098	0·4280273	10509·1333	9·26
9·28	0·1328447	0·1254689	0·4061821	0·4275321	10721·4319	9·28
9·30	0·1326975	0·1253462	0·4057558	0·4270385	10938·0192	9·30
9·32	0·1325507	0·1252239	0·4053308	0·4265467	11158·9819	9·32
9·34	0·1324044	0·1251018	0·4049071	0·4260565	11384·4082	9·34
9·36	0·1322586	0·1249802	0·4044848	0·4255680	11614·3885	9·36
9·38	0·1321133	0·1248589	0·4040638	0·4250811	11849·0148	9·38
9·40	0·1319684	0·1247379	0·4036441	0·4245960	12088·3807	9·40
9·42	0·1318240	0·1246173	0·4032257	0·4241124	12332·5822	9·42
9·44	0·1316801	0·1244970	0·4028087	0·4236305	12581·7169	9·44
9·46	0·1315367	0·1243771	0·4023929	0·4231502	12835·8844	9·46
9·48	0·1313938	0·1242575	0·4019784	0·4226716	13095·1865	9·48
9·50	0·1312513	0·1241382	0·4015651	0·4221945	13359·7268	9·50
9·52	0·1311092	0·1240193	0·4011532	0·4217191	13629·6112	9·52
9·54	0·1309677	0·1239008	0·4007425	0·4212452	13904·9476	9·54
9·56	0·1308266	0·1237825	0·4003331	0·4207730	14185·8462	9·56
9·58	0·1306859	0·1236646	0·3999249	0·4203023	14472·4193	9·58
9·60	0·1305457	0·1235470	0·3995180	0·4198332	14764·7816	9·60
9·62	0·1304060	0·1234298	0·3991123	0·4193656	15063·0499	9·62
9·64	0·1302667	0·1233128	0·3987078	0·4188996	15367·3437	9·64
9·66	0·1301278	0·1231962	0·3983046	0·4184351	15677·7847	9·66
9·68	0·1299894	0·1230800	0·3979026	0·4179721	15994·4969	9·68
9·70	0·1298514	0·1229640	0·3975018	0·4175107	16317·6072	9·70
9·72	0·1297139	0·1228484	0·3971023	0·4170508	16647·2447	9·72
9·74	0·1295768	0·1227331	0·3967039	0·4165924	16983·5414	9·74
9·76	0·1294401	0·1226181	0·3963067	0·4161355	17326·6317	9·76
9·78	0·1293039	0·1225034	0·3959107	0·4156801	17676·6529	9·78
9·80	0·1291681	0·1223891	0·3955159	0·4152261	18033·7449	9·80
9·82	0·1290328	0·1222751	0·3951223	0·4147737	18398·0507	9·82
9·84	0·1288978	0·1221613	0·3947299	0·4143227	18769·7160	9·84
9·86	0·1287633	0·1220479	0·3943386	0·4138731	19148·8894	9·86
9·88	0·1286292	0·1219348	0·3939485	0·4134250	19535·7227	9·88
9·90	0·1284955	0·1218220	0·3935596	0·4129784	19930·3704	9·90
9·92	0·1283623	0·1217096	0·3931717	0·4125332	20332·9906	9·92
9·94	0·1282294	0·1215974	0·3927851	0·4120894	20743·7443	9·94
9·96	0·1280970	0·1214855	0·3923996	0·4116471	21162·7957	9·96
9·98	0·1279650	0·1213739	0·3920152	0·4112061	21590·3125	9·98
10·00	0·1278333	0·1212627	0·3916319	0·4107666	22026·4658	10·00

Table II. Functions of imaginary argument, and e^x

x	$e^{-x} I_0(x)$	$e^{-x} I_1(x)$	$e^x K_0(x)$	$e^x K_1(x)$	e^x	x
10·02	0·1277021	0·1211517	0·3912498	0·4103284	22471·430	10·02
10·04	0·1275713	0·1210411	0·3908688	0·4098917	22925·383	10·04
10·06	0·1274409	0·1209307	0·3904889	0·4094563	23388·506	10·06
10·08	0·1273109	0·1208206	0·3901101	0·4090223	23860·986	10·08
10·10	0·1271813	0·1207109	0·3897324	0·4085897	24343·009	10·10
10·12	0·1270521	0·1206014	0·3893558	0·4081584	24834·771	10·12
10·14	0·1269233	0·1204922	0·3889803	0·4077285	25336·466	10·14
10·16	0·1267948	0·1203833	0·3886059	0·4073000	25848·297	10·16
10·18	0·1266668	0·1202747	0·3882325	0·4068727	26370·467	10·18
10·20	0·1265392	0·1201664	0·3878603	0·4064468	26903·186	10·20
10·22	0·1264119	0·1200584	0·3874891	0·4060223	27446·666	10·22
10·24	0·1262850	0·1199506	0·3871189	0·4055990	28001·126	10·24
10·26	0·1261585	0·1198432	0·3867498	0·4051771	28566·786	10·26
10·28	0·1260324	0·1197360	0·3863818	0·4047565	29143·874	10·28
10·30	0·1259067	0·1196292	0·3860149	0·4043372	29732·619	10·30
10·32	0·1257813	0·1195226	0·3856489	0·4039191	30333·258	10·32
10·34	0·1256563	0·1194162	0·3852841	0·4035024	30946·030	10·34
10·36	0·1255317	0·1193102	0·3849202	0·4030869	31571·181	10·36
10·38	0·1254075	0·1192044	0·3845574	0·4026728	32208·961	10·38
10·40	0·1252836	0·1190990	0·3841956	0·4022598	32859·626	10·40
10·42	0·1251601	0·1189938	0·3838348	0·4018482	33523·434	10·42
10·44	0·1250369	0·1188888	0·3834750	0·4014378	34200·652	10·44
10·46	0·1249141	0·1187842	0·3831163	0·4010286	34891·551	10·46
10·48	0·1247917	0·1186798	0·3827586	0·4006207	35596·408	10·48
10·50	0·1246697	0·1185757	0·3824018	0·4002140	36315·503	10·50
10·52	0·1245480	0·1184718	0·3820461	0·3998085	37049·124	10·52
10·54	0·1244266	0·1183682	0·3816913	0·3994043	37797·566	10·54
10·56	0·1243056	0·1182649	0·3813375	0·3990013	38561·128	10·56
10·58	0·1241850	0·1181619	0·3809848	0·3985995	39340·114	10·58
10·60	0·1240647	0·1180591	0·3806330	0·3981989	40134·837	10·60
10·62	0·1239448	0·1179566	0·3802821	0·3977995	40945·615	10·62
10·64	0·1238252	0·1178544	0·3799323	0·3974013	41772·771	10·64
10·66	0·1237059	0·1177524	0·3795834	0·3970043	42616·637	10·66
10·68	0·1235870	0·1176507	0·3792354	0·3966084	43477·550	10·68
10·70	0·1234685	0·1175492	0·3788884	0·3962137	44355·855	10·70
10·72	0·1233503	0·1174480	0·3785424	0·3958202	45251·903	10·72
10·74	0·1232324	0·1173471	0·3781973	0·3954279	46166·052	10·74
10·76	0·1231149	0·1172464	0·3778532	0·3950367	47098·668	10·76
10·78	0·1229977	0·1171459	0·3775100	0·3946467	48050·124	10·78
10·80	0·1228808	0·1170458	0·3771677	0·3942578	49020·801	10·80
10·82	0·1227642	0·1169458	0·3768264	0·3938701	50011·087	10·82
10·84	0·1226480	0·1168462	0·3764860	0·3934835	51021·378	10·84
10·86	0·1225322	0·1167467	0·3761465	0·3930980	52052·078	10·86
10·88	0·1224166	0·1166476	0·3758079	0·3927137	53103·600	10·88
10·90	0·1223014	0·1165487	0·3754702	0·3923305	54176·364	10·90
10·92	0·1221865	0·1164500	0·3751335	0·3919484	55270·799	10·92
10·94	0·1220719	0·1163516	0·3747976	0·3915673	56387·343	10·94
10·96	0·1219577	0·1162534	0·3744627	0·3911874	57526·443	10·96
10·98	0·1218438	0·1161554	0·3741287	0·3908086	58688·554	10·98
11·00	0·1217302	0·1160578	0·3737955	0·3904309	59874·142	11·00

Table II. Functions of imaginary argument, and e^x

x	$e^{-x} I_0(x)$	$e^{-x} I_1(x)$	$e^x K_0(x)$	$e^x K_1(x)$	e^x	x
11·02	0·1216169	0·1159603	0·3734632	0·3900543	61083·680	11·02
11·04	0·1215039	0·1158631	0·3731319	0·3896788	62317·652	11·04
11·06	0·1213912	0·1157662	0·3728014	0·3893043	63576·552	11·06
11·08	0·1212789	0·1156694	0·3724717	0·3889309	64860·883	11·08
11·10	0·1211669	0·1155730	0·3721430	0·3885586	66171·160	11·10
11·12	0·1210551	0·1154767	0·3718151	0·3881873	67507·906	11·12
11·14	0·1209437	0·1153807	0·3714881	0·3878171	68871·656	11·14
11·16	0·1208326	0·1152849	0·3711619	0·3874480	70262·956	11·16
11·18	0·1207218	0·1151894	0·3708367	0·3870799	71682·362	11·18
11·20	0·1206113	0·1150941	0·3705122	0·3867128	73130·442	11·20
11·22	0·1205011	0·1149990	0·3701886	0·3863468	74607·775	11·22
11·24	0·1203912	0·1149042	0·3698659	0·3859818	76114·952	11·24
11·26	0·1202817	0·1148096	0·3695440	0·3856178	77652·576	11·26
11·28	0·1201724	0·1147152	0·3692229	0·3852548	79221·262	11·28
11·30	0·1200634	0·1146211	0·3689027	0·3848929	80821·638	11·30
11·32	0·1199547	0·1145272	0·3685833	0·3845320	82454·343	11·32
11·34	0·1198463	0·1144335	0·3682648	0·3841721	84120·031	11·34
11·36	0·1197382	0·1143401	0·3679470	0·3838132	85819·368	11·36
11·38	0·1196303	0·1142468	0·3676301	0·3834553	87553·035	11·38
11·40	0·1195228	0·1141538	0·3673140	0·3830984	89321·723	11·40
11·42	0·1194156	0·1140610	0 3669987	0·3827425	91126·142	11·42
11·44	0·1193086	0·1139685	0·3666843	0·3823875	92967·012	11·44
11·46	0·1192020	0·1138762	0·3663706	0·3820336	94845·070	11·46
11·48	0·1190956	0·1137841	0·3660578	0·3816806	96761·068	11·48
11·50	0·1189895	0·1136922	0·3657457	0·3813286	98715·771	11·50
11·52	0·1188837	0·1136005	0·3654344	0·3809775	100709·962	11·52
11·54	0·1187782	0·1135090	0·3651240	0·3806275	102744·438	11·54
11·56	0·1186729	0·1134178	0·3648143	0·3802783	104820·013	11·56
11·58	0·1185680	0·1133268	0·3645054	0·3799302	106937·518	11·58
11·60	0·1184633	0·1132360	0·3641973	0·3795830	109097·799	11·60
11·62	0·1183589	0·1131454	0·3638900	0·3792367	111301·721	11·62
11·64	0·1182548	0·1130551	0·3635834	0·3788914	113550·165	11·64
11·66	0·1181509	0·1129649	0·3632777	0·3785470	115844·030	11·66
11·68	0·1180473	0·1128750	0·3629727	0·3782035	118184·235	11·68
11·70	0·1179440	0·1127852	0·3626684	0·3778610	120571·715	11·70
11·72	0·1178410	0·1126957	0·3623650	0·3775194	123007·425	11·72
11·74	0·1177382	0·1126064	0·3620623	0·3771787	125492·340	11·74
11·76	0·1176357	0·1125173	0·3617603	0·3768389	128027·453	11·76
11·78	0·1175335	0·1124284	0·3614591	0·3765001	130613·780	11·78
11·80	0·1174315	0·1123398	0·3611587	0·3761621	133252·353	11·80
11·82	0·1173298	0·1122513	0·3608590	0·3758251	135944·229	11·82
11·84	0·1172284	0·1121630	0·3605600	0·3754890	138690·485	11·84
11·86	0·1171272	0·1120750	0·3602618	0·3751537	141492·218	11·86
11·88	0·1170263	0·1119871	0·3599643	0·3748194	144350·551	11·88
11·90	0·1169256	0·1118995	0·3596676	0·3744859	147266·625	11·90
11·92	0·1168252	0·1118120	0·3593716	0·3741533	150241·608	11·92
11·94	0·1167251	0·1117248	0·3590763	0·3738216	153276·690	11·94
11·96	0·1166252	0·1116378	0·3587818	0·3734908	156373·085	11·96
11·98	0·1165256	0·1115509	0·3584880	0·3731608	159532·031	11·98
12·00	0·1164262	0·1114643	0·3581949	0·3728318	162754·791	12·00

Table II. Functions of imaginary argument, and e^x

x	$e^{-x} I_0(x)$	$e^{-x} I_1(x)$	$e^x K_0(x)$	$e^x K_1(x)$	e^x	x
12·02	0·1163271	0·1113779	0·3579025	0·3725035	166042·66	12·02
12·04	0·1162283	0·1112916	0·3576108	0·3721762	169396·94	12·04
12·06	0·1161296	0·1112056	0·3573199	0·3718497	172818·99	12·06
12·08	0·1160313	0·1111197	0·3570296	0·3715240	176310·16	12·08
12·10	0·1159332	0·1110341	0·3567401	0·3711992	179871·86	12·10
12·12	0·1158353	0·1109487	0·3564513	0·3708753	183505·51	12·12
12·14	0·1157377	0·1108634	0·3561631	0·3705522	187212·57	12·14
12·16	0·1156404	0·1107783	0·3558757	0·3702299	190994·52	12·16
12·18	0·1155432	0·1106935	0·3555890	0·3699085	194852·86	12·18
12·20	0·1154464	0·1106088	0·3553029	0·3695879	198789·15	12·20
12·22	0·1153497	0·1105243	0·3550176	0·3692681	202804·96	12·22
12·24	0·1152533	0·1104400	0·3547329	0·3689492	206901·89	12·24
12·26	0·1151572	0·1103559	0·3544489	0·3686311	211081·59	12·26
12·28	0·1150613	0·1102720	0·3541656	0·3683138	215345·72	12·28
12·30	0·1149656	0·1101883	0·3538830	0·3679973	219695·99	12·30
12·32	0·1148702	0·1101048	0·3536010	0·3676816	224134·14	12·32
12·34	0·1147750	0·1100215	0·3533198	0·3673667	228661·95	12·34
12·36	0·1146801	0·1099383	0·3530392	0·3670527	233281·23	12·36
12·38	0·1145853	0·1098553	0·3527592	0·3667394	237993·82	12·38
12·40	0·1144909	0·1097726	0·3524800	0·3664269	242801·62	12·40
12·42	0·1143966	0·1096900	0·3522014	0·3661152	247706·54	12·42
12·44	0·1143026	0·1096076	0·3519234	0·3658044	252710·54	12·44
12·46	0·1142088	0·1095253	0·3516461	0·3654943	257815·63	12·46
12·48	0·1141153	0·1094433	0·3513695	0·3651849	263023·85	12·48
12·50	0·1140219	0·1093614	0·3510935	0·3648764	268337·29	12·50
12·52	0·1139288	0·1092798	0·3508182	0·3645687	273758·06	12·52
12·54	0·1138360	0·1091983	0·3505435	0·3642617	279288·34	12·54
12·56	0·1137433	0·1091169	0·3502694	0·3639555	284930·34	12·56
12·58	0·1136509	0·1090358	0·3499960	0·3636500	290686·31	12·58
12·60	0·1135587	0·1089549	0·3497233	0·3633453	296558·57	12·60
12·62	0·1134668	0·1088741	0·3494512	0·3630414	302549·45	12·62
12·64	0·1133750	0·1087935	0·3491797	0·3627383	308661·35	12·64
12·66	0·1132835	0·1087131	0·3489088	0·3624359	314896·72	12·66
12·68	0·1131922	0·1086328	0·3486386	0·3621342	321258·06	12·68
12·70	0·1131011	0·1085527	0·3483690	0·3618333	327747·90	12·70
12·72	0·1130103	0·1084728	0·3481000	0·3615332	334368·85	12·72
12·74	0·1129196	0·1083931	0·3478317	0·3612337	341123·55	12·74
12·76	0·1128292	0·1083136	0·3475639	0·3609351	348014·70	12·76
12·78	0·1127390	0·1082342	0·3472968	0·3606371	355045·06	12·78
12·80	0·1126490	0·1081550	0·3470303	0·3603399	362217·45	12·80
12·82	0·1125592	0·1080760	0·3467644	0·3600434	369534·73	12·82
12·84	0·1124697	0·1079971	0·3464991	0·3597477	376999·82	12·84
12·86	0·1123803	0·1079184	0·3462345	0·3594527	384615·73	12·86
12·88	0·1122912	0·1078399	0·3459704	0·3591584	392385·48	12·88
12·90	0·1122023	0·1077616	0·3457070	0·3588648	400312·19	12·90
12·92	0·1121136	0·1076834	0·3454441	0·3585719	408399·03	12·92
12·94	0·1120251	0·1076054	0·3451818	0·3582798	416649·24	12·94
12·96	0·1119368	0·1075276	0·3449202	0·3579883	425066·11	12·96
12·98	0·1118487	0·1074499	0·3446591	0·3576976	433653·02	12·98
13·00	0·1117608	0·1073724	0·3443986	0·3574076	442413·39	13·00

Table II. Functions of imaginary argument, and e^x

x	$e^{-x} I_0(x)$	$e^{-x} I_1(x)$	$e^x K_0(x)$	$e^x K_1(x)$	e^x	x
13·02	0·1116732	0·1072951	0·3441388	0·3571182	451350·74	13·02
13·04	0·1115857	0·1072179	0·3438795	0·3568296	460468·63	13·04
13·06	0·1114985	0·1071409	0·3436208	0·3565417	469770·71	13·06
13·08	0·1114114	0·1070640	0·3433626	0·3562544	479260·71	13·08
13·10	0·1113246	0·1069874	0·3431051	0·3559679	488942·41	13·10
13·12	0·1112379	0·1069109	0·3428481	0·3556820	498819·71	13·12
13·14	0·1111515	0·1068345	0·3425917	0·3553968	508896·53	13·14
13·16	0·1110652	0·1067583	0·3423359	0·3551123	519176·92	13·16
13·18	0·1109792	0·1066823	0·3420807	0·3548285	529664·99	13·18
13·20	0·1108934	0·1066064	0·3418260	0·3545454	540364·94	13·20
13·22	0·1108077	0·1065307	0·3415719	0·3542629	551281·03	13·22
13·24	0·1107223	0·1064552	0·3413184	0·3539811	562417·65	13·24
13·26	0·1106370	0·1063798	0·3410654	0·3537000	573779·24	13·26
13·28	0·1105520	0·1063046	0·3408130	0·3534195	585370·35	13·28
13·30	0·1104671	0·1062295	0·3405611	0·3531398	597195·61	13·30
13·32	0·1103825	0·1061546	0·3403098	0·3528606	609259·77	13·32
13·34	0·1102980	0·1060798	0·3400591	0·3525821	621567·63	13·34
13·36	0·1102138	0·1060052	0·3398089	0·3523043	634124·13	13·36
13·38	0·1101297	0·1059308	0·3395593	0·3520272	646934·29	13·38
13·40	0·1100458	0·1058565	0·3393102	0·3517506	660003·22	13·40
13·42	0·1099621	0·1057824	0·3390616	0·3514748	673336·17	13·42
13·44	0·1098786	0·1057084	0·3388137	0·3511995	686938·47	13·44
13·46	0·1097953	0·1056346	0·3385662	0·3509250	700815·54	13·46
13·48	0·1097122	0·1055609	0·3383193	0·3506510	714972·96	13·48
13·50	0·1096292	0·1054874	0·3380729	0·3503777	729416·37	13·50
13·52	0·1095465	0·1054140	0·3378271	0·3501051	744151·56	13·52
13·54	0·1094639	0·1053408	0·3375818	0·3498330	759184·42	13·54
13·56	0·1093816	0·1052677	0·3373371	0·3495616	774520·96	13·56
13·58	0·1092994	0·1051948	0·3370928	0·3492909	790167·32	13·58
13·60	0·1092174	0·1051221	0·3368491	0·3490207	806129·76	13·60
13·62	0·1091356	0·1050495	0·3366060	0·3487512	822414·66	13·62
13·64	0·1090540	0·1049770	0·3363633	0·3484823	839028·54	13·64
13·66	0·1089725	0·1049047	0·3361212	0·3482140	855978·04	13·66
13·68	0·1088912	0·1048325	0·3358796	0·3479463	873269·94	13·68
13·70	0·1088102	0·1047605	0·3356385	0·3476793	890911·17	13·70
13·72	0·1087293	0·1046886	0·3353980	0·3474128	908908·77	13·72
13·74	0·1086485	0·1046169	0·3351579	0·3471470	927269·94	13·74
13·76	0·1085680	0·1045453	0·3349184	0·3468818	946002·04	13·76
13·78	0·1084876	0·1044739	0·3346794	0·3466172	965112·54	13·78
13·80	0·1084074	0·1044026	0·3344409	0·3463532	984609·11	13·80
13·82	0·1083274	0·1043315	0·3342029	0·3460897	1004499·53	13·82
13·84	0·1082476	0·1042605	0·3339654	0·3458269	1024791·77	13·84
13·86	0·1081679	0·1041896	0·3337285	0·3455647	1045493·94	13·86
13·88	0·1080885	0·1041189	0·3334920	0·3453031	1066614·32	13·88
13·90	0·1080092	0·1040484	0·3332560	0·3450420	1088161·36	13·90
13·92	0·1079300	0·1039779	0·3330206	0·3447816	1110143·67	13·92
13·94	0·1078511	0·1039077	0·3327856	0·3445217	1132570·06	13·94
13·96	0·1077723	0·1038375	0·3325511	0·3442624	1155449·50	13·96
13·98	0·1076937	0·1037675	0·3323171	0·3440037	1178791·12	13·98
14·00	0·1076153	0·1036977	0·3320836	0·3437456	1202604·28	14·00

Table II. Functions of imaginary argument, and e^x

x	$e^{-x} I_0(x)$	$e^{-x} I_1(x)$	$e^x K_0(x)$	$e^x K_1(x)$	e^x	x
14·02	0·1075370	0·1036279	0·3318506	0·3434881	1226898·5	14·02
14·04	0·1074589	0·1035584	0·3316181	0·3432311	1251683·5	14·04
14·06	0·1073810	0·1034889	0·3313861	0·3429747	1276969·2	14·06
14·08	0·1073032	0·1034196	0·3311546	0·3427189	1302765·7	14·08
14·10	0·1072256	0·1033505	0·3309235	0·3424637	1329083·3	14·10
14·12	0·1071482	0·1032814	0·3306930	0·3422090	1355932·5	14·12
14·14	0·1070710	0·1032126	0·3304629	0·3419549	1383324·2	14·14
14·16	0·1069939	0·1031438	0·3302333	0·3417013	1411269·2	14·16
14·18	0·1069169	0·1030752	0·3300042	0·3414484	1439778·7	14·18
14·20	0·1068402	0·1030067	0·3297755	0·3411959	1468864·2	14·20
14·22	0·1067636	0·1029384	0·3295474	0·3409441	1498537·2	14·22
14·24	0·1066872	0·1028702	0·3293197	0·3406927	1528809·7	14·24
14·26	0·1066109	0·1028021	0·3290924	0·3404420	1559693·7	14·26
14·28	0·1065348	0·1027342	0·3288657	0·3401918	1591201·6	14·28
14·30	0·1064589	0·1026663	0·3286394	0·3399421	1623346·0	14·30
14·32	0·1063831	0·1025987	0·3284136	0·3396930	1656139·7	14·32
14·34	0·1063075	0·1025311	0·3281882	0·3394444	1689596·0	14·34
14·36	0·1062321	0·1024637	0·3279633	0·3391964	1723728·1	14·36
14·38	0·1061568	0·1023965	0·3277389	0·3389489	1758549·7	14·38
14·40	0·1060817	0·1023293	0·3275149	0·3387020	1794074·8	14·40
14·42	0·1060067	0·1022623	0·3272914	0·3384555	1830317·5	14·42
14·44	0·1059319	0·1021954	0·3270684	0·3382097	1867292·4	14·44
14·46	0·1058572	0·1021287	0·3268458	0·3379643	1905014·2	14·46
14·48	0·1057827	0·1020621	0·3266236	0·3377195	1943498·0	14·48
14·50	0·1057084	0·1019956	0·3264019	0·3374752	1982759·3	14·50
14·52	0·1056342	0·1019292	0·3261807	0·3372315	2022813·7	14·52
14·54	0·1055602	0·1018630	0·3259599	0·3369883	2063677·2	14·54
14·56	0·1054863	0·1017969	0·3257396	0·3367455	2105366·2	14·56
14·58	0·1054126	0·1017309	0·3255197	0·3365034	2147897·5	14·58
14·60	0·1053391	0·1016650	0·3253002	0·3362617	2191287·9	14·60
14·62	0·1052657	0·1015993	0·3250812	0·3360206	2235554·8	14·62
14·64	0·1051924	0·1015337	0·3248626	0·3357799	2280716·0	14·64
14·66	0·1051193	0·1014682	0·3246445	0·3355398	2326789·6	14·66
14·68	0·1050464	0·1014029	0·3244268	0·3353002	2373793·8	14·68
14·70	0·1049736	0·1013377	0·3242096	0·3350611	2421747·6	14·70
14·72	0·1049009	0·1012726	0·3239928	0·3348226	2470670·2	14·72
14·74	0·1048284	0·1012076	0·3237764	0·3345845	2520581·0	14·74
14·76	0·1047561	0·1011428	0·3235604	0·3343469	2571500·1	14·76
14·78	0·1046839	0·1010780	0·3233449	0·3341098	2623447·9	14·78
14·80	0·1046119	0·1010135	0·3231298	0·3338733	2676445·1	14·80
14·82	0·1045400	0·1009490	0·3229152	0·3336372	2730512·8	14·82
14·84	0·1044682	0·1008846	0·3227010	0·3334017	2785672·8	14·84
14·86	0·1043966	0·1008204	0·3224872	0·3331666	2841947·2	14·86
14·88	0·1043252	0·1007563	0·3222738	0·3329320	2899358·3	14·88
14·90	0·1042539	0·1006923	0·3220608	0·3326979	2957929·2	14·90
14·92	0·1041827	0·1006284	0·3218483	0·3324644	3017683·4	14·92
14·94	0·1041117	0·1005647	0·3216362	0·3322313	3078644·6	14·94
14·96	0·1040408	0·1005011	0·3214245	0·3319987	3140837·4	14·96
14·98	0·1039701	0·1004376	0·3212132	0·3317665	3204286·5	14·98
15·00	0·1038995	0·1003742	0·3210024	0·3315349	3269017·4	15·00

Table II. Functions of imaginary argument, and e^x

x	$e^{-x} I_0(x)$	$e^{-x} I_1(x)$	$e^x K_0(x)$	$e^x K_1(x)$	e^x	x
15·02	0·1038291	0·1003109	0·3207919	0·3313037	3335055·9	15·02
15·04	0·1037588	0·1002478	0·3205819	0·3310731	3402428·5	15·04
15·06	0·1036887	0·1001847	0·3203723	0·3308429	3471162·1	15·06
15·08	0·1036186	0·1001218	0·3201631	0·3306132	3541284·2	15·08
15·10	0·1035488	0·1000590	0·3199543	0·3303839	3612822·9	15·10
15·12	0·1034791	0·0999964	0·3197459	0·3301552	3685806·8	15·12
15·14	0·1034095	0·0999338	0·3195379	0·3299269	3760265·0	15·14
15·16	0·1033400	0·0998714	0·3193303	0·3296990	3836227·4	15·16
15·18	0·1032707	0·0998090	0·3191231	0·3294717	3913724·4	15·18
15·20	0·1032016	0·0997468	0·3189164	0·3292448	3992786·8	15·20
15·22	0·1031325	0·0996847	0·3187100	0·3290184	4073446·5	15·22
15·24	0·1030636	0·0996228	0·3185040	0·3287924	4155735·6	15·24
15·26	0·1029949	0·0995609	0·3182985	0·3285670	4239687·0	15·26
15·28	0·1029263	0·0994991	0·3180933	0·3283419	4325334·3	15·28
15·30	0·1028578	0·0994375	0·3178885	0·3281174	4412711·9	15·30
15·32	0·1027895	0·0993760	0·3176841	0·3278933	4501854·6	15·32
15·34	0·1027213	0·0993146	0·3174801	0·3276696	4592798·1	15·34
15·36	0·1026532	0·0992533	0·3172766	0·3274464	4685578·8	15·36
15·38	0·1025853	0·0991921	0·3170734	0·3272237	4780233·7	15·38
15·40	0·1025175	0·0991310	0·3168705	0·3270014	4876800·9	15·40
15·42	0·1024498	0·0990701	0·3166681	0·3267796	4975318·8	15·42
15·44	0·1023823	0·0990092	0·3164661	0·3265582	5075826·9	15·44
15·46	0·1023149	0·0989485	0·3162644	0·3263372	5178365·4	15·46
15·48	0·1022476	0·0988879	0·3160632	0·3261168	5282975·3	15·48
15·50	0·1021805	0·0988274	0·3158623	0·3258967	5389698·5	15·50
15·52	0·1021135	0·0987670	0·3156618	0·3256771	5498577·6	15·52
15·54	0·1020466	0·0987067	0·3154617	0·3254580	5609656·2	15·54
15·56	0·1019799	0·0986465	0·3152619	0·3252392	5722978·8	15·56
15·58	0·1019133	0·0985864	0·3150626	0·3250210	5838590·7	15·58
15·60	0·1018468	0·0985265	0·3148636	0·3248031	5956538·0	15·60
15·62	0·1017805	0·0984666	0·3146650	0·3245857	6076868·1	15·62
15·64	0·1017143	0·0984069	0·3144668	0·3243687	6199628·9	15·64
15·66	0·1016482	0·0983472	0·3142689	0·3241522	6324869·8	15·66
15·68	0·1015822	0·0982877	0·3140714	0·3239361	6452640·6	15·68
15·70	0·1015164	0·0982283	0·3138743	0·3237204	6582992·6	15·70
15·72	0·1014507	0·0981690	0·3136776	0·3235052	6715977·9	15·72
15·74	0·1013851	0·0981097	0·3134812	0·3232903	6851649·6	15·74
15·76	0·1013197	0·0980506	0·3132852	0·3230759	6990062·1	15·76
15·78	0·1012544	0·0979916	0·3130896	0·3228620	7131270·7	15·78
15·80	0·1011892	0·0979328	0·3128943	0·3226484	7275332·0	15·80
15·82	0·1011241	0·0978740	0·3126994	0·3224353	7422303·4	15·82
15·84	0·1010592	0·0978153	0·3125049	0·3222226	7572243·9	15·84
15·86	0·1009944	0·0977567	0·3123107	0·3220103	7725213·4	15·86
15·88	0·1009297	0·0976983	0·3121169	0·3217985	7881273·0	15·88
15·90	0·1008651	0·0976399	0·3119235	0·3215870	8040485·3	15·90
15·92	0·1008007	0·0975816	0·3117304	0·3213760	8202913·9	15·92
15·94	0·1007363	0·0975235	0·3115376	0·3211654	8368623·7	15·94
15·96	0·1006722	0·0974654	0·3113453	0·3209552	8537681·1	15·96
15·98	0·1006081	0·0974075	0·3111533	0·3207454	8710153·7	15·98
16·00	0·1005441	0·0973496	0·3109616	0·3205360	8886110·5	16·00

Table III. Functions of order one-third

x	$J_{1/3}(x)$	$Y_{1/3}(x)$	$\lvert H^{(1)}_{1/3}(x)\rvert$	$\arg H^{(1)}_{1/3}(x)$	$e^x K_{1/3}(x)$	x
0	0·0000000	− ∞	∞	− 90°	∞	0
0·02	+ 0·2412455	− 3·8181574	3·8257712	− 86° 23′ 4″·72	5·8973367	0·02
0·04	+ 0·3038819	− 2·9641628	2·9796989	− 84° 8′ 47″·63	4·5650965	0·04
0·06	+ 0·3477275	− 2·5398832	2·5635758	− 82° 12′ 15″·40	3·9129445	0·06
0·08	+ 0·3825227	− 2·2665744	2·2986264	− 80° 25′ 14″·29	3·4996127	0·08
0·10	+ 0·4117819	− 2·0682566	2·1088503	− 78° 44′ 23″·54	3·2048056	0·10
0·12	+ 0·4372223	− 1·9140102	1·9633131	− 77° 7′ 57″·21	2·9795927	0·12
0·14	+ 0·4598264	− 1·7884275	1·8465950	− 75° 34′ 51″·16	2·7996089	0·14
0·16	+ 0·4802143	− 1·6828031	1·7499806	− 74° 4′ 23″·06	2·6511003	0·16
0·18	+ 0·4988049	− 1·5917752	1·6680991	− 72° 36′ 3″·21	2·5256038	0·18
0·20	+ 0·5158967	− 1·5118289	1·5974279	− 71° 9′ 29″·88	2·4175728	0·20
0·22	+ 0·5317088	− 1·4405408	1·5355364	− 69° 44′ 26″·57	2·3231916	0·22
0·24	+ 0·5464087	− 1·3761797	1·4806867	− 68° 20′ 40″·45	2·2397331	0·24
0·26	+ 0·5601271	− 1·3174682	1·4315952	− 66° 58′ 1″·30	2·1651865	0·26
0·28	+ 0·5729677	− 1·2634392	1·3872889	− 65° 36′ 20″·81	2·0980307	0·28
0·30	+ 0·5850148	− 1·2133449	1·3470145	− 64° 15′ 32″·17	2·0370894	0·30
0·32	+ 0·5963375	− 1·1665964	1·3101777	− 62° 55′ 29″·69	1·9814363	0·32
0·34	+ 0·6069935	− 1·1227224	1·2763020	− 61° 36′ 8″·57	1·9303301	0·34
0·36	+ 0·6170312	− 1·0813409	1·2450002	− 60° 17′ 24″·70	1·8831690	0·36
0·38	+ 0·6264920	− 1·0421378	1·2159537	− 58° 59′ 14″·57	1·8394587	0·38
0·40	+ 0·6354112	− 1·0048529	1·1888973	− 57° 41′ 35″·13	1·7987884	0·40
0·42	+ 0·6438195	− 0·9692681	1·1636083	− 56° 24′ 23″·72	1·7608136	0·42
0·44	+ 0·6517435	− 0·9351991	1·1398978	− 55° 7′ 38″·01	1·7252429	0·44
0·46	+ 0·6592067	− 0·9024892	1·1176047	− 53° 51′ 15″·93	1·6918274	0·46
0·48	+ 0·6662297	− 0·8710041	1·0965902	− 52° 35′ 15″·65	1·6603536	0·48
0·50	+ 0·6728308	− 0·8406278	1·0767342	− 51° 19′ 35″·54	1·6306366	0·50
0·52	+ 0·6790265	− 0·8112601	1·0579319	− 50° 4′ 14″·14	1·6025156	0·52
0·54	+ 0·6848313	− 0·7828134	1·0400917	− 48° 49′ 10″·14	1·5758501	0·54
0·56	+ 0·6902585	− 0·7552112	1·0231329	− 47° 34′ 22″·35	1·5505163	0·56
0·58	+ 0·6953202	− 0·7283861	1·0069839	− 46° 19′ 49″·69	1·5264049	0·58
0·60	+ 0·7000271	− 0·7022788	0·9915813	− 45° 5′ 31″·20	1·5034188	0·60
0·62	+ 0·7043893	− 0·6768367	0·9768685	− 43° 51′ 25″·98	1·4814718	0·62
0·64	+ 0·7084159	− 0·6520129	0·9627948	− 42° 37′ 33″·23	1·4604863	0·64
0·66	+ 0·7121152	− 0·6277661	0·9493146	− 41° 23′ 52″·19	1·4403931	0·66
0·68	+ 0·7154951	− 0·6040589	0·9363868	− 40° 10′ 22″·19	1·4211296	0·68
0·70	+ 0·7185627	− 0·5808580	0·9239742	− 38° 57′ 2″·60	1·4026393	0·70
0·72	+ 0·7213248	− 0·5581337	0·9120431	− 37° 43′ 52″·84	1·3848710	0·72
0·74	+ 0·7237876	− 0·5358591	0·9005629	− 36° 30′ 52″·38	1·3677782	0·74
0·76	+ 0·7259570	− 0·5140100	0·8895054	− 35° 18′ 0″·72	1·3513186	0·76
0·78	+ 0·7278387	− 0·4925646	0·8788453	− 34° 5′ 17″·40	1·3354533	0·78
0·80	+ 0·7294377	− 0·4715032	0·8685590	− 32° 52′ 42″·00	1·3201469	0·80
0·82	+ 0·7307591	− 0·4508080	0·8586249	− 31° 40′ 14″·12	1·3053670	0·82
0·84	+ 0·7318076	− 0·4304628	0·8490233	− 30° 27′ 53″·39	1·2910835	0·84
0·86	+ 0·7325877	− 0·4104530	0·8397359	− 29° 15′ 39″·47	1·2772690	0·86
0·88	+ 0·7331037	− 0·3907653	0·8307458	− 28° 3′ 32″·02	1·2638979	0·88
0·90	+ 0·7333598	− 0·3713877	0·8220374	− 26° 51′ 30″·76	1·2509467	0·90
0·92	+ 0·7333600	− 0·3523093	0·8135962	− 25° 39′ 35″·39	1·2383936	0·92
0·94	+ 0·7331080	− 0·3335201	0·8054086	− 24° 27′ 45″·66	1·2262184	0·94
0·96	+ 0·7326077	− 0·3150111	0·7974623	− 23° 16′ 1″·32	1·2144022	0·96
0·98	+ 0·7318627	− 0·2967741	0·7897454	− 22° 4′ 22″·12	1·2029275	0·98
1·00	+ 0·7308764	− 0·2788016	0·7822472	− 20° 52′ 47″·84	1·1917780	1·00

To compute functions of order − 1/3, increase the phase by 60°.

Table III. Functions of order one-third

x	$J_{1/3}(x)$	$Y_{1/3}(x)$	$\lvert H^{(1)}_{1/3}(x) \rvert$	arg $H^{(1)}_{1/3}(x)$	$e^x K_{1/3}(x)$	x
1·02	+ 0·7296524	− 0·2610869	0·7749574	− 19° 41′ 18″·29	1·1809384	1·02
1·04	+ 0·7281940	− 0·2436239	0·7678666	− 18° 29′ 53″·27	1·1703945	1·04
1·06	+ 0·7265045	− 0·2264069	0·7609657	− 17° 18′ 32″·58	1·1601329	1·06
1·08	+ 0·7245872	− 0·2094308	0·7542466	− 16° 7′ 16″·06	1·1501411	1·08
1·10	+ 0·7224452	− 0·1926912	0·7477012	− 14° 56′ 3″·54	1·1404073	1·10
1·12	+ 0·7200818	− 0·1761839	0·7413222	− 13° 44′ 54″·86	1·1309205	1·12
1·14	+ 0·7175000	− 0·1599051	0·7351026	− 12° 33′ 49″·87	1·1216703	1·14
1·16	+ 0·7147030	− 0·1438514	0·7290360	− 11° 22′ 48″·44	1·1126469	1·16
1·18	+ 0·7116937	− 0·1280198	0·7231162	− 10° 11′ 50″·43	1·1038412	1·18
1·20	+ 0·7084752	− 0·1124076	0·7173372	− 9° 0′ 55″·70	1·0952444	1·20
1·22	+ 0·7050506	− 0·0970123	0·7116936	− 7° 50′ 4″·15	1·0868482	1·22
1·24	+ 0·7014229	− 0·0818317	0·7061802	− 6° 39′ 15″·66	1·0786451	1·24
1·26	+ 0·6975950	− 0·0668639	0·7007921	− 5° 28′ 30″·11	1·0706275	1·26
1·28	+ 0·6935699	− 0·0521072	0·6955245	− 4° 17′ 47″·41	1·0627885	1·28
1·30	+ 0·6893506	− 0·0375600	0·6903730	− 3° 7′ 7″·45	1·0551215	1·30
1·32	+ 0·6849400	− 0·0232209	0·6853336	− 1° 56′ 30″·14	1·0476204	1·32
1·34	+ 0·6803413	− 0·0090889	0·6804020	− 0° 45′ 55″·39	1·0402790	1·34
1·36	+ 0·6755573	+ 0·0048372	0·6755746	0° 24′ 36″·88	1·0330918	1·36
1·38	+ 0·6705909	+ 0·0185581	0·6708477	1° 35′ 6″·76	1·0260535	1·38
1·40	+ 0·6654453	+ 0·0320747	0·6662179	2° 45′ 34″·33	1·0191588	1·40
1·42	+ 0·6601234	+ 0·0453875	0·6616819	3° 55′ 59″·67	1·0124030	1·42
1·44	+ 0·6546281	+ 0·0584971	0·6572366	5° 6′ 22″·83	1·0057813	1·44
1·46	+ 0·6489626	+ 0·0714038	0·6528790	6° 16′ 43″·91	0·9992894	1·46
1·48	+ 0·6431297	+ 0·0841081	0·6486062	7° 27′ 2″·94	0·9929231	1·48
1·50	+ 0·6371326	+ 0·0966101	0·6444156	8° 37′ 20″·02	0·9866783	1·50
1·52	+ 0·6309743	+ 0·1089100	0·6403046	9° 47′ 35″·18	0·9805512	1·52
1·54	+ 0·6246578	+ 0·1210079	0·6362706	10° 57′ 48″·50	0·9745381	1·54
1·56	+ 0·6181862	+ 0·1329039	0·6323113	12° 8′ 0″·02	0·9686354	1·56
1·58	+ 0·6115625	+ 0·1445980	0·6284245	13° 18′ 9″·81	0·9628399	1·58
1·60	+ 0·6047900	+ 0·1560900	0·6246079	14° 28′ 17″·91	0·9571482	1·60
1·62	+ 0·5978715	+ 0·1673799	0·6208594	15° 38′ 24″·37	0·9515574	1·62
1·64	+ 0·5908104	+ 0·1784675	0·6171771	16° 48′ 29″·23	0·9460644	1·64
1·66	+ 0·5836096	+ 0·1893528	0·6135590	17° 58′ 32″·56	0·9406663	1·66
1·68	+ 0·5762725	+ 0·2000354	0·6100034	19° 8′ 34″·38	0·9353606	1·68
1·70	+ 0·5688020	+ 0·2105152	0·6065083	20° 18′ 34″·74	0·9301444	1·70
1·72	+ 0·5612014	+ 0·2207919	0·6030722	21° 28′ 33″·69	0·9250154	1·72
1·74	+ 0·5534739	+ 0·2308653	0·5996933	22° 38′ 31″·26	0·9199712	1·74
1·76	+ 0·5456226	+ 0·2407351	0·5963702	23° 48′ 27″·48	0·9150093	1·76
1·78	+ 0·5376509	+ 0·2504011	0·5931013	24° 58′ 22″·40	0·9101276	1·78
1·80	+ 0·5295619	+ 0·2598629	0·5898852	26° 8′ 16″·05	0·9053239	1·80
1·82	+ 0·5213588	+ 0·2691204	0·5867204	27° 18′ 8″·47	0·9005961	1·82
1·84	+ 0·5130449	+ 0·2781733	0·5836056	28° 27′ 59″·68	0·8959423	1·84
1·86	+ 0·5046236	+ 0·2870212	0·5805395	29° 37′ 49″·72	0·8913605	1·86
1·88	+ 0·4960979	+ 0·2956640	0·5775209	30° 47′ 38″·62	0·8868489	1·88
1·90	+ 0·4874713	+ 0·3041014	0·5745485	31° 57′ 26″·41	0·8824057	1·90
1·92	+ 0·4787471	+ 0·3123332	0·5716212	33° 7′ 13″·12	0·8780291	1·92
1·94	+ 0·4699285	+ 0·3203591	0·5687379	34° 16′ 58″·77	0·8737176	1·94
1·96	+ 0·4610189	+ 0·3281790	0·5658974	35° 26′ 43″·39	0·8694694	1·96
1·98	+ 0·4520215	+ 0·3357927	0·5630987	36° 36′ 27″·01	0·8652832	1·98
2·00	+ 0·4429398	+ 0·3432000	0·5603409	37° 46′ 9″·65	0·8611573	2·00

$J_{-1/3}(2{\cdot}00) = 0{\cdot}5603409 \times \cos 97^\circ\, 46'\, 9''{\cdot}65 = -0{\cdot}0757500.$
$Y_{-1/3}(2{\cdot}00) = 0{\cdot}5603409 \times \sin 97^\circ\, 46'\, 9''{\cdot}65 = +0{\cdot}5551971.$

Table III. Functions of order one-third

x	$J_{1/3}(x)$	$Y_{1/3}(x)$	$\lvert H^{(1)}_{1/3}(x) \rvert$	$\arg H^{(1)}_{1/3}(x)$	$e^x K_{1/3}(x)$	x
2·02	+ 0·4337771	+ 0·3504008	0·5576229	38° 55′ 51″·34	0·8570902	2·02
2·04	+ 0·4245367	+ 0·3573949	0·5549437	40° 5′ 32″·10	0·8530808	2·04
2·06	+ 0·4152219	+ 0·3641824	0·5523025	41° 15′ 11″·95	0·8491275	2·06
2·08	+ 0·4058363	+ 0·3707631	0·5496984	42° 24′ 50″·91	0·8452290	2·08
2·10	+ 0·3963830	+ 0·3771371	0·5471306	43° 34′ 29″·00	0·8413842	2·10
2·12	+ 0·3868655	+ 0·3833043	0·5445981	44° 44′ 6″·25	0·8375917	2·12
2·14	+ 0·3772872	+ 0·3892647	0·5421002	45° 53′ 42″·67	0·8338505	2·14
2·16	+ 0·3676514	+ 0·3950185	0·5396362	47° 3′ 18″·29	0·8301592	2·16
2·18	+ 0·3579615	+ 0·4005657	0·5372051	48° 12′ 53″·12	0·8265169	2·18
2·20	+ 0·3482210	+ 0·4059065	0·5348065	49° 22′ 27″·18	0·8229225	2·20
2·22	+ 0·3384331	+ 0·4110411	0·5324394	50° 32′ 0″·49	0·8193748	2·22
2·24	+ 0·3286012	+ 0·4159696	0·5301033	51° 41′ 33″·06	0·8158730	2·24
2·26	+ 0·3187288	+ 0·4206923	0·5277974	52° 51′ 4″·91	0·8124159	2·26
2·28	+ 0·3088193	+ 0·4252096	0·5255212	54° 0′ 36″·06	0·8090028	2·28
2·30	+ 0·2988759	+ 0·4295216	0·5232740	55° 10′ 6″·52	0·8056325	2·30
2·32	+ 0·2889021	+ 0·4336289	0·5210551	56° 19′ 36″·32	0·8023043	2·32
2·34	+ 0·2789012	+ 0·4375318	0·5188641	57° 29′ 5″·46	0·7990173	2·34
2·36	+ 0·2688766	+ 0·4412307	0·5167002	58° 38′ 33″·95	0·7957706	2·36
2·38	+ 0·2588316	+ 0·4447262	0·5145631	59° 48′ 1″·81	0·7925634	2·38
2·40	+ 0·2487696	+ 0·4480187	0·5124521	60° 57′ 29″·06	0·7893949	2·40
2·42	+ 0·2386939	+ 0·4511090	0·5103666	62° 6′ 55″·70	0·7862643	2·42
2·44	+ 0·2286079	+ 0·4539975	0·5083063	63° 16′ 21″·75	0·7831710	2·44
2·46	+ 0·2185149	+ 0·4566849	0·5062705	64° 25′ 47″·23	0·7801140	2·46
2·48	+ 0·2084181	+ 0·4591720	0·5042589	65° 35′ 12″·14	0·7770928	2·48
2·50	+ 0·1983209	+ 0·4614595	0·5022709	66° 44′ 36″·50	0·7741066	2·50
2·52	+ 0·1882266	+ 0·4635482	0·5003061	67° 54′ 0″·31	0·7711547	2·52
2·54	+ 0·1781384	+ 0·4654389	0·4983640	69° 3′ 23″·59	0·7682366	2·54
2·56	+ 0·1680595	+ 0·4671325	0·4964442	70° 12′ 46″·35	0·7653515	2·56
2·58	+ 0·1579933	+ 0·4686300	0·4945462	71° 22′ 8″·59	0·7624989	2·58
2·60	+ 0·1479429	+ 0·4699324	0·4926698	72° 31′ 30″·34	0·7596781	2·60
2·62	+ 0·1379115	+ 0·4710406	0·4908144	73° 40′ 51″·60	0·7568886	2·62
2·64	+ 0·1279023	+ 0·4719557	0·4889797	74° 50′ 12″·38	0·7541297	2·64
2·66	+ 0·1179186	+ 0·4726788	0·4871653	75° 59′ 32″·68	0·7514009	2·66
2·68	+ 0·1079633	+ 0·4732111	0·4853708	77° 8′ 52″·52	0·7487017	2·68
2·70	+ 0·0980398	+ 0·4735538	0·4835959	78° 18′ 11″·92	0·7460315	2·70
2·72	+ 0·0881509	+ 0·4737081	0·4818402	79° 27′ 30″·86	0·7433898	2·72
2·74	+ 0·0783000	+ 0·4736754	0·4801034	80° 36′ 49″·37	0·7407762	2·74
2·76	+ 0·0684899	+ 0·4734569	0·4783851	81° 46′ 7″·45	0·7381900	2·76
2·78	+ 0·0587238	+ 0·4730540	0·4766850	82° 55′ 25″·11	0·7356309	2·78
2·80	+ 0·0490046	+ 0·4724682	0·4750028	84° 4′ 42″·36	0·7330983	2·80
2·82	+ 0·0393353	+ 0·4717009	0·4733382	85° 13′ 59″·21	0·7305919	2·82
2·84	+ 0·0297189	+ 0·4707537	0·4716909	86° 23′ 15″·66	0·7281111	2·84
2·86	+ 0·0201583	+ 0·4696281	0·4700605	87° 32′ 31″·72	0·7256555	2·86
2·88	+ 0·0106564	+ 0·4683256	0·4684469	88° 41′ 47″·39	0·7232247	2·88
2·90	+ 0·0012161	+ 0·4668480	0·4668496	89° 51′ 2″·69	0·7208183	2·90
2·92	− 0·0081598	+ 0·4651970	0·4652685	91° 0′ 17″·63	0·7184359	2·92
2·94	− 0·0174685	+ 0·4633741	0·4637033	92° 9′ 32″·20	0·7160770	2·94
2·96	− 0·0267073	+ 0·4613813	0·4621537	93° 18′ 46″·41	0·7137413	2·96
2·98	− 0·0358733	+ 0·4592203	0·4606194	94° 28′ 0″·28	0·7114285	2·98
3·00	− 0·0449638	+ 0·4568930	0·4591002	95° 37′ 13″·81	0·7091381	3·00

To compute functions of order −1/3, increase the phase by 60°.

Table III. Functions of order one-third

x	$J_{1/3}(x)$	$Y_{1/3}(x)$	$\lvert H^{(1)}_{1/3}(x)\rvert$	$\arg H^{(1)}_{1/3}(x)$	$e^x K_{1/3}(x)$	x
3·02	− 0·0539763	+ 0·4544013	0·4575959	96° 46′ 26″·99	0·7068697	3·02
3·04	− 0·0629080	+ 0·4517471	0·4561061	97° 55′ 39″·85	0·7046231	3·04
3·06	− 0·0717564	+ 0·4489323	0·4546308	99° 4′ 52″·38	0·7023978	3·06
3·08	− 0·0805188	+ 0·4459590	0·4531696	100° 14′ 4″·59	0·7001936	3·08
3·10	− 0·0891928	+ 0·4428292	0·4517223	101° 23′ 16″·49	0·6980101	3·10
3·12	− 0·0977759	+ 0·4395451	0·4502888	102° 32′ 28″·08	0·6958470	3·12
3·14	− 0·1062656	+ 0·4361086	0·4488687	103° 41′ 39″·37	0·6937040	3·14
3·16	− 0·1146595	+ 0·4325221	0·4474619	104° 50′ 50″·36	0·6915807	3·16
3·18	− 0·1229552	+ 0·4287877	0·4460682	106° 0′ 1″·05	0·6894769	3·18
3·20	− 0·1311505	+ 0·4249076	0·4446874	107° 9′ 11″·46	0·6873922	3·20
3·22	− 0·1392429	+ 0·4208840	0·4433192	108° 18′ 21″·58	0·6853264	3·22
3·24	− 0·1472303	+ 0·4167194	0·4419636	109° 27′ 31″·42	0·6832792	3·24
3·26	− 0·1551105	+ 0·4124159	0·4406202	110° 36′ 40″·99	0·6812503	3·26
3·28	− 0·1628813	+ 0·4079761	0·4392890	111° 45′ 50″·29	0·6792394	3·28
3·30	− 0·1705405	+ 0·4034022	0·4379697	112° 54′ 59″·32	0·6772463	3·30
3·32	− 0·1780862	+ 0·3986968	0·4366621	114° 4′ 8″·09	0·6752708	3·32
3·34	− 0·1855162	+ 0·3938622	0·4353661	115° 13′ 16″·61	0·6733124	3·34
3·36	− 0·1928286	+ 0·3889010	0·4340816	116° 22′ 24″·87	0·6713711	3·36
3·38	− 0·2000215	+ 0·3838156	0·4328083	117° 31′ 32″·89	0·6694465	3·38
3·40	− 0·2070929	+ 0·3786087	0·4315461	118° 40′ 40″·66	0·6675385	3·40
3·42	− 0·2140411	+ 0·3732827	0·4302948	119° 49′ 48″·19	0·6656467	3·42
3·44	− 0·2208642	+ 0·3678404	0·4290543	120° 58′ 55″·48	0·6637710	3·44
3·46	− 0·2275605	+ 0·3622843	0·4278244	122° 8′ 2″·54	0·6619111	3·46
3·48	− 0·2341283	+ 0·3566170	0·4266049	123° 17′ 9″·37	0·6600668	3·48
3·50	− 0·2405659	+ 0·3508413	0·4253958	124° 26′ 15″·97	0·6582379	3·50
3·52	− 0·2468718	+ 0·3449599	0·4241969	125° 35′ 22″·35	0·6564241	3·52
3·54	− 0·2530444	+ 0·3389754	0·4230080	126° 44′ 28″·51	0·6546254	3·54
3·56	− 0·2590821	+ 0·3328906	0·4218290	127° 53′ 34″·46	0·6528413	3·56
3·58	− 0·2649836	+ 0·3267083	0·4206598	129° 2′ 40″·19	0·6510719	3·58
3·60	− 0·2707474	+ 0·3204313	0·4195001	130° 11′ 45″·72	0·6493168	3·60
3·62	− 0·2763722	+ 0·3140623	0·4183500	131° 20′ 51″·03	0·6475758	3·62
3·64	− 0·2818568	+ 0·3076042	0·4172093	132° 29′ 56″·15	0·6458489	3·64
3·66	− 0·2871997	+ 0·3010598	0·4160778	133° 39′ 1″·06	0·6441357	3·66
3·68	− 0·2924000	+ 0·2944320	0·4149554	134° 48′ 5″·78	0·6424361	3·68
3·70	− 0·2974564	+ 0·2877236	0·4138420	135° 57′ 10″·30	0·6407500	3·70
3·72	− 0·3023678	+ 0·2809376	0·4127375	137° 6′ 14″·63	0·6390771	3·72
3·74	− 0·3071333	+ 0·2740767	0·4116417	138° 15′ 18″·78	0·6374173	3·74
3·76	− 0·3117518	+ 0·2671440	0·4105546	139° 24′ 22″·73	0·6357703	3·76
3·78	− 0·3162224	+ 0·2601423	0·4094760	140° 33′ 26″·51	0·6341362	3·78
3·80	− 0·3205442	+ 0·2530746	0·4084058	141° 42′ 30″·10	0·6325146	3·80
3·82	− 0·3247164	+ 0·2459438	0·4073440	142° 51′ 33″·52	0·6309054	3·82
3·84	− 0·3287383	+ 0·2387529	0·4062904	144° 0′ 36″·76	0·6293085	3·84
3·86	− 0·3326092	+ 0·2315048	0·4052448	145° 9′ 39″·83	0·6277236	3·86
3·88	− 0·3363283	+ 0·2242025	0·4042073	146° 18′ 42″·72	0·6261508	3·88
3·90	− 0·3398952	+ 0·2168489	0·4031776	147° 27′ 45″·45	0·6245897	3·90
3·92	− 0·3433091	+ 0·2094471	0·4021558	148° 36′ 48″·01	0·6230403	3·92
3·94	− 0·3465698	+ 0·2020000	0·4011416	149° 45′ 50″·41	0·6215023	3·94
3·96	− 0·3496766	+ 0·1945106	0·4001351	150° 54′ 52″·65	0·6199758	3·96
3·98	− 0·3526292	+ 0·1869819	0·3991361	152° 3′ 54″·73	0·6184605	3·98
4·00	− 0·3554274	+ 0·1794168	0·3981444	153° 12′ 56″·65	0·6169562	4·00

$J_{-1/3}(4{\cdot}00) = 0{\cdot}3981444 \times \cos 213^\circ\ 12'\ 56''{\cdot}65 = -0{\cdot}3330932.$
$Y_{-1/3}(4{\cdot}00) = 0{\cdot}3981444 \times \sin 213^\circ\ 12'\ 56''{\cdot}65 = -0{\cdot}2181008.$

Table III. Functions of order one-third

x	$J_{1/3}(x)$	$Y_{1/3}(x)$	$\lvert H^{(1)}_{1/3}(x) \rvert$	$\arg H^{(1)}_{1/3}(x)$	$e^x K_{1/3}(x)$	x
4·02	− 0·3580707	+ 0·1718183	0·3971601	154° 21′ 58″·42	0·6154630	4·02
4·04	− 0·3605591	+ 0·1641895	0·3961831	155° 31′ 0″·04	0·6139805	4·04
4·06	− 0·3628923	+ 0·1565333	0·3952132	156° 40′ 1″·51	0·6125087	4·06
4·08	− 0·3650702	+ 0·1488527	0·3942503	157° 49′ 2″·83	0·6110475	4·08
4·10	− 0·3670927	+ 0·1411506	0·3932945	158° 58′ 4″·00	0·6095967	4·10
4·12	− 0·3689599	+ 0·1334301	0·3923455	160° 7′ 5″·03	0·6081563	4·12
4·14	− 0·3706718	+ 0·1256940	0·3914034	161° 16′ 5″·91	0·6067260	4·14
4·16	− 0·3722285	+ 0·1179455	0·3904679	162° 25′ 6″·66	0·6053058	4·16
4·18	− 0·3736302	+ 0·1101873	0·3895391	163° 34′ 7″·26	0·6038956	4·18
4·20	− 0·3748770	+ 0·1024224	0·3886169	164° 43′ 7″·73	0·6024952	4·20
4·22	− 0·3759693	+ 0·0946539	0·3877012	165° 52′ 8″·07	0·6011045	4·22
4·24	− 0·3769073	+ 0·0868845	0·3867919	167° 1′ 8″·27	0·5997234	4·24
4·26	− 0·3776914	+ 0·0791172	0·3858890	168° 10′ 8″·34	0·5983518	4·26
4·28	− 0·3783220	+ 0·0713550	0·3849923	169° 19′ 8″·28	0·5969897	4·28
4·30	− 0·3787997	+ 0·0636006	0·3841018	170° 28′ 8″·09	0·5956368	4·30
4·32	− 0·3791248	+ 0·0558570	0·3832175	171° 37′ 7″·78	0·5942930	4·32
4·34	− 0·3792981	+ 0·0481270	0·3823392	172° 46′ 7″·34	0·5929584	4·34
4·36	− 0·3793201	+ 0·0404134	0·3814669	173° 55′ 6″·78	0·5916327	4·36
4·38	− 0·3791916	+ 0·0327191	0·3806006	175° 4′ 6″·10	0·5903159	4·38
4·40	− 0·3789131	+ 0·0250469	0·3797400	176° 13′ 5″·29	0·5890079	4·40
4·42	− 0·3784856	+ 0·0173995	0·3788853	177° 22′ 4″·37	0·5877086	4·42
4·44	− 0·3779098	+ 0·0097798	0·3780363	178° 31′ 3″·33	0·5864178	4·44
4·46	− 0·3771866	+ 0·0021904	0·3771930	179° 40′ 2″·18	0·5851356	4·46
4·48	− 0·3763170	− 0·0053659	0·3763553	180° 49′ 0″·91	0·5838617	4·48
4·50	− 0·3753019	− 0·0128864	0·3755231	181° 57′ 59″·52	0·5825961	4·50
4·52	− 0·3741423	− 0·0203684	0·3746963	183° 6′ 58″·03	0·5813387	4·52
4·54	− 0·3728394	− 0·0278093	0·3738750	184° 15′ 56″·43	0·5800895	4·54
4·56	− 0·3713941	− 0·0352065	0·3730591	185° 24′ 54″·72	0·5788483	4·56
4·58	− 0·3698078	− 0·0425573	0·3722485	186° 33′ 52″·90	0·5776151	4·58
4·60	− 0·3680815	− 0·0498592	0·3714431	187° 42′ 50″·97	0·5763897	4·60
4·62	− 0·3662167	− 0·0571096	0·3706429	188° 51′ 48″·94	0·5751721	4·62
4·64	− 0·3642144	− 0·0643061	0·3698478	190° 0′ 46″·81	0·5739622	4·64
4·66	− 0·3620762	− 0·0714460	0·3690578	191° 9′ 44″·58	0·5727599	4·66
4·68	− 0·3598033	− 0·0785270	0·3682729	192° 18′ 42″·24	0·5715652	4·68
4·70	− 0·3573972	− 0·0855466	0·3674929	193° 27′ 39″·81	0·5703779	4·70
4·72	− 0·3548595	− 0·0925024	0·3667178	194° 36′ 37″·27	0·5691980	4·72
4·74	− 0·3521915	− 0·0993921	0·3659476	195° 45′ 34″·64	0·5680255	4·74
4·76	− 0·3493949	− 0·1062133	0·3651822	196° 54′ 31″·91	0·5668601	4·76
4·78	− 0·3464712	− 0·1129637	0·3644216	198° 3′ 29″·09	0·5657019	4·78
4·80	− 0·3434221	− 0·1196411	0·3636657	199° 12′ 26″·17	0·5645508	4·80
4·82	− 0·3402493	− 0·1262432	0·3629145	200° 21′ 23″·16	0·5634067	4·82
4·84	− 0·3369544	− 0·1327677	0·3621679	201° 30′ 20″·06	0·5622696	4·84
4·86	− 0·3335393	− 0·1392127	0·3614258	202° 39′ 16″·87	0·5611393	4·86
4·88	− 0·3300057	− 0·1455758	0·3606883	203° 48′ 13″·58	0·5600158	4·88
4·90	− 0·3263554	− 0·1518551	0·3599553	204° 57′ 10″·21	0·5588991	4·90
4·92	− 0·3225903	− 0·1580485	0·3592267	206° 6′ 6″·75	0·5577890	4·92
4·94	− 0·3187124	− 0·1641539	0·3585026	207° 15′ 3″·21	0·5566856	4·94
4·96	− 0·3147234	− 0·1701695	0·3577827	208° 23′ 59″·58	0·5555886	4·96
4·98	− 0·3106254	− 0·1760932	0·3570672	209° 32′ 55″·86	0·5544982	4·98
5·00	− 0·3064205	− 0·1819232	0·3563559	210° 41′ 52″·06	0·5534141	5·00

To compute functions of order −1/3, increase the phase by 60°.

Table III. Functions of order one-third

x	$J_{1/3}(x)$	$Y_{1/3}(x)$	$\lvert H^{(1)}_{1/3}(x) \rvert$	arg $H^{(1)}_{1/3}(x)$	$e^x K_{1/3}(x)$	x
5·02	−0·3021105	−0·1876576	0·3556489	211° 50′ 48″·18	0·5523364	5·02
5·04	−0·2976976	−0·1932947	0·3549460	212° 59′ 44″·22	0·5512650	5·04
5·06	−0·2931838	−0·1988326	0·3542473	214° 8′ 40″·17	0·5501998	5·06
5·08	−0·2885714	−0·2042696	0·3535527	215° 17′ 36″·05	0·5491408	5·08
5·10	−0·2838623	−0·2096041	0·3528621	216° 26′ 31″·84	0·5480878	5·10
5·12	−0·2790589	−0·2148343	0·3521756	217° 35′ 27″·56	0·5470410	5·12
5·14	−0·2741632	−0·2199588	0·3514930	218° 44′ 23″·20	0·5460001	5·14
5·16	−0·2691776	−0·2249760	0·3508144	219° 53′ 18″·76	0·5449651	5·16
5·18	−0·2641042	−0·2298843	0·3501397	221° 2′ 14″·25	0·5439360	5·18
5·20	−0·2589454	−0·2346822	0·3494688	222° 11′ 9″·67	0·5429127	5·20
5·22	−0·2537034	−0·2393685	0·3488018	223° 20′ 5″·01	0·5418952	5·22
5·24	−0·2483807	−0·2439417	0·3481386	224° 29′ 0″·27	0·5408834	5·24
5·26	−0·2429794	−0·2484004	0·3474791	225° 37′ 55″·46	0·5398773	5·26
5·28	−0·2375020	−0·2527435	0·3468234	226° 46′ 50″·59	0·5388767	5·28
5·30	−0·2319509	−0·2569696	0·3461714	227° 55′ 45″·64	0·5378818	5·30
5·32	−0·2263285	−0·2610776	0·3455230	229° 4′ 40″·62	0·5368923	5·32
5·34	−0·2206371	−0·2650664	0·3448782	230° 13′ 35″·53	0·5359082	5·34
5·36	−0·2148793	−0·2689349	0·3442370	231° 22′ 30″·37	0·5349296	5·36
5·38	−0·2090575	−0·2726820	0·3435994	232° 31′ 25″·14	0·5339563	5·38
5·40	−0·2031741	−0·2763068	0·3429653	233° 40′ 19″·85	0·5329883	5·40
5·42	−0·1972317	−0·2798083	0·3423346	234° 49′ 14″·49	0·5320256	5·42
5·44	−0·1912327	−0·2831855	0·3417075	235° 58′ 9″·07	0·5310681	5·44
5·46	−0·1851797	−0·2864378	0·3410837	237° 7′ 3″·58	0·5301157	5·46
5·48	−0·1790751	−0·2895642	0·3404634	238° 15′ 58″·02	0·5291685	5·48
5·50	−0·1729216	−0·2925640	0·3398464	239° 24′ 52″·40	0·5282263	5·50
5·52	−0·1667216	−0·2954366	0·3392328	240° 33′ 46″·72	0·5272892	5·52
5·54	−0·1604777	−0·2981813	0·3386224	241° 42′ 40″·97	0·5263570	5·54
5·56	−0·1541924	−0·3007974	0·3380154	242° 51′ 35″·16	0·5254298	5·56
5·58	−0·1478684	−0·3032845	0·3374116	244° 0′ 29″·30	0·5245074	5·58
5·60	−0·1415082	−0·3056420	0·3368109	245° 9′ 23″·37	0·5235899	5·60
5·62	−0·1351143	−0·3078695	0·3362135	246° 18′ 17″·38	0·5226772	5·62
5·64	−0·1286894	−0·3099667	0·3356193	247° 27′ 11″·33	0·5217693	5·64
5·66	−0·1222361	−0·3119330	0·3350282	248° 36′ 5″·22	0·5208661	5·66
5·68	−0·1157569	−0·3137683	0·3344401	249° 44′ 59″·06	0·5199676	5·68
5·70	−0·1092543	−0·3154723	0·3338552	250° 53′ 52″·84	0·5190737	5·70
5·72	−0·1027311	−0·3170448	0·3332733	252° 2′ 46″·56	0·5181844	5·72
5·74	−0·0961898	−0·3184857	0·3326945	253° 11′ 40″·22	0·5172997	5·74
5·76	−0·0896330	−0·3197948	0·3321186	254° 20′ 33″·83	0·5164195	5·76
5·78	−0·0830632	−0·3209721	0·3315457	255° 29′ 27″·38	0·5155438	5·78
5·80	−0·0764830	−0·3220176	0·3309758	256° 38′ 20″·88	0·5146725	5·80
5·82	−0·0698951	−0·3229313	0·3304088	257° 47′ 14″·33	0·5138056	5·82
5·84	−0·0633019	−0·3237134	0·3298446	258° 56′ 7″·72	0·5129432	5·84
5·86	−0·0567061	−0·3243639	0·3292834	260° 5′ 1″·06	0·5120850	5·86
5·88	−0·0501102	−0·3248832	0·3287250	261° 13′ 54″·34	0·5112312	5·88
5·90	−0·0435167	−0·3252714	0·3281694	262° 22′ 47″·57	0·5103816	5·90
5·92	−0·0369283	−0·3255288	0·3276166	263° 31′ 40″·75	0·5095362	5·92
5·94	−0·0303473	−0·3256557	0·3270667	264° 40′ 33″·88	0·5086951	5·94
5·96	−0·0237764	−0·3256526	0·3265194	265° 49′ 26″·96	0·5078581	5·96
5·98	−0·0172181	−0·3255199	0·3259749	266° 58′ 19″·98	0·5070252	5·98
6·00	−0·0106747	−0·3252580	0·3254331	268° 7′ 12″·96	0·5061964	6·00

$J_{-1/3}(6{\cdot}00) = 0{\cdot}3254331 \times \cos 328^\circ\ 7'\ 12''{\cdot}96 = +0{\cdot}2763443.$

$Y_{-1/3}(6{\cdot}00) = 0{\cdot}3254331 \times \sin 328^\circ\ 7'\ 12''{\cdot}96 = -0{\cdot}1718736.$

Table III. Functions of order one-third

x	$J_{1/3}(x)$	$Y_{1/3}(x)$	$\lvert H^{(1)}_{1/3}(x) \rvert$	$\arg H^{(1)}_{1/3}(x)$	$e^x K_{1/3}(x)$	x
6·02	− 0·0041490	− 0·3248675	0·3248940	269° 16′ 5″·89	0·5053717	6·02
6·04	+ 0·0023568	− 0·3243490	0·3243576	270° 24′ 58″·77	0·5045509	6·04
6·06	+ 0·0088402	− 0·3237031	0·3238238	271° 33′ 51″·60	0·5037342	6·06
6·08	+ 0·0152987	− 0·3229304	0·3232926	272° 42′ 44″·39	0·5029215	6·08
6·10	+ 0·0217298	− 0·3220317	0·3227640	273° 51′ 37″·12	0·5021126	6·10
6·12	+ 0·0281313	− 0·3210077	0·3222380	275° 0′ 29″·81	0·5013077	6·12
6·14	+ 0·0345007	− 0·3198593	0·3217146	276° 9′ 22″·46	0·5005066	6·14
6·16	+ 0·0408356	− 0·3185872	0·3211936	277° 18′ 15″·05	0·4997094	6·16
6·18	+ 0·0471337	− 0·3171924	0·3206753	278° 27′ 7″·60	0·4989160	6·18
6·20	+ 0·0533927	− 0·3156758	0·3201594	279° 36′ 0″·11	0·4981263	6·20
6·22	+ 0·0596103	− 0·3140384	0·3196460	280° 44′ 52″·57	0·4973404	6·22
6·24	+ 0·0657842	− 0·3122813	0·3191350	281° 53′ 44″·99	0·4965582	6·24
6·26	+ 0·0719122	− 0·3104053	0·3186265	283° 2′ 37″·36	0·4957796	6·26
6·28	+ 0·0779920	− 0·3084118	0·3181204	284° 11′ 29″·69	0·4950048	6·28
6·30	+ 0·0840213	− 0·3063018	0·3176167	285° 20′ 21″·98	0·4942335	6·30
6·32	+ 0·0899981	− 0·3040764	0·3171154	286° 29′ 14″·22	0·4934659	6·32
6·34	+ 0·0959202	− 0·3017371	0·3166164	287° 38′ 6″·42	0·4927018	6·34
6·36	+ 0·1017854	− 0·2992849	0·3161198	288° 46′ 58″·58	0·4919413	6·36
6·38	+ 0·1075917	− 0·2967212	0·3156255	289° 55′ 50″·70	0·4911843	6·38
6·40	+ 0·1133370	− 0·2940474	0·3151335	291° 4′ 42″·77	0·4904308	6·40
6·42	+ 0·1190192	− 0·2912648	0·3146438	292° 13′ 34″·80	0·4896807	6·42
6·44	+ 0·1246363	− 0·2883748	0·3141564	293° 22′ 26″·80	0·4889341	6·44
6·46	+ 0·1301863	− 0·2853790	0·3136712	294° 31′ 18″·75	0·4881909	6·46
6·48	+ 0·1356673	− 0·2822787	0·3131883	295° 40′ 10″·67	0·4874510	6·48
6·50	+ 0·1410775	− 0·2790756	0·3127076	296° 49′ 2″·54	0·4867145	6·50
6·52	+ 0·1464147	− 0·2757711	0·3122291	297° 57′ 54″·38	0·4859814	6·52
6·54	+ 0·1516774	− 0·2723669	0·3117527	299° 6′ 46″·18	0·4852516	6·54
6·56	+ 0·1568635	− 0·2688647	0·3112786	300° 15′ 37″·94	0·4845250	6·56
6·58	+ 0·1619714	− 0·2652659	0·3108066	301° 24′ 29″·66	0·4838017	6·58
6·60	+ 0·1669992	− 0·2615725	0·3103367	302° 33′ 21″·34	0·4830817	6·60
6·62	+ 0·1719453	− 0·2577860	0·3098690	303° 42′ 12″·98	0·4823648	6·62
6·64	+ 0·1768080	− 0·2539082	0·3094033	304° 51′ 4″·59	0·4816511	6·64
6·66	+ 0·1815856	− 0·2499409	0·3089398	305° 59′ 56″·16	0·4809406	6·66
6·68	+ 0·1862766	− 0·2458860	0·3084783	307° 8′ 47″·70	0·4802333	6·68
6·70	+ 0·1908793	− 0·2417452	0·3080189	308° 17′ 39″·20	0·4795290	6·70
6·72	+ 0·1953922	− 0·2375205	0·3075615	309° 26′ 30″·66	0·4788279	6·72
6·74	+ 0·1998139	− 0·2332136	0·3071062	310° 35′ 22″·08	0·4781298	6·74
6·76	+ 0·2041429	− 0·2288267	0·3066528	311° 44′ 13″·47	0·4774347	6·76
6·78	+ 0·2083778	− 0·2243615	0·3062015	312° 53′ 4″·83	0·4767427	6·78
6·80	+ 0·2125171	− 0·2198201	0·3057522	314° 1′ 56″·15	0·4760537	6·80
6·82	+ 0·2165596	− 0·2152044	0·3053048	315° 10′ 47″·44	0·4753677	6·82
6·84	+ 0·2205041	− 0·2105165	0·3048593	316° 19′ 38″·70	0·4746846	6·84
6·86	+ 0·2243491	− 0·2057584	0·3044159	317° 28′ 29″·92	0·4740045	6·86
6·88	+ 0·2280935	− 0·2009321	0·3039743	318° 37′ 21″·11	0·4733273	6·88
6·90	+ 0·2317362	− 0·1960398	0·3035347	319° 46′ 12″·26	0·4726530	6·90
6·92	+ 0·2352760	− 0·1910836	0·3030969	320° 55′ 3″·38	0·4719815	6·92
6·94	+ 0·2387118	− 0·1860655	0·3026610	322° 3′ 54″·47	0·4713130	6·94
6·96	+ 0·2420426	− 0·1809877	0·3022271	323° 12′ 45″·52	0·4706472	6·96
6·98	+ 0·2452674	− 0·1758524	0·3017949	324° 21′ 36″·55	0·4699843	6·98
7·00	+ 0·2483853	− 0·1706616	0·3013646	325° 30′ 27″·54	0·4693242	7·00

To compute functions of order − 1/3, increase the phase by 60°.

Table III. Functions of order one-third

x	$J_{1/3}(x)$	$Y_{1/3}(x)$	$\lvert H^{(1)}_{1/3}(x)\rvert$	$\arg H^{(1)}_{1/3}(x)$	$e^x K_{1/3}(x)$	x
7·02	+ 0·2513952	− 0·1654177	0·3009362	326° 39′ 18″·50	0·4686668	7·02
7·04	+ 0·2542964	− 0·1601228	0·3005095	327° 48′ 9″·42	0·4680122	7·04
7·06	+ 0·2570881	− 0·1547791	0·3000847	328° 57′ 0″·32	0·4673604	7·06
7·08	+ 0·2597694	− 0·1493888	0·2996617	330° 5′ 51″·19	0·4667113	7·08
7·10	+ 0·2623395	− 0·1439541	0·2992404	331° 14′ 42″·03	0·4660648	7·10
7·12	+ 0·2647979	− 0·1384774	0·2988209	332° 23′ 32″·83	0·4654211	7·12
7·14	+ 0·2671439	− 0·1329609	0·2984032	333° 32′ 23″·61	0·4647800	7·14
7·16	+ 0·2693769	− 0·1274068	0·2979872	334° 41′ 14″·36	0·4641415	7·16
7·18	+ 0·2714962	− 0·1218174	0·2975730	335° 50′ 5″·07	0·4635057	7·18
7·20	+ 0·2735015	− 0·1161950	0·2971604	336° 58′ 55″·76	0·4628725	7·20
7·22	+ 0·2753921	− 0·1105419	0·2967496	338° 7′ 46″·42	0·4622419	7·22
7·24	+ 0·2771678	− 0·1048604	0·2963405	339° 16′ 37″·05	0·4616138	7·24
7·26	+ 0·2788281	− 0·0991527	0·2959330	340° 25′ 27″·65	0·4609884	7·26
7·28	+ 0·2803727	− 0·0934213	0·2955273	341° 34′ 18″·22	0·4603654	7·28
7·30	+ 0·2818013	− 0·0876683	0·2951232	342° 43′ 8″·77	0·4597450	7·30
7·32	+ 0·2831136	− 0·0818961	0·2947207	343° 51′ 59″·28	0·4591271	7·32
7·34	+ 0·2843096	− 0·0761070	0·2943199	345° 0′ 49″·77	0·4585116	7·34
7·36	+ 0·2853889	− 0·0703033	0·2939207	346° 9′ 40″·23	0·4578986	7·36
7·38	+ 0·2863516	− 0·0644874	0·2935231	347° 18′ 30″·66	0·4572881	7·38
7·40	+ 0·2871974	− 0·0586615	0·2931272	348° 27′ 21″·07	0·4566801	7·40
7·42	+ 0·2879266	− 0·0528279	0·2927328	349° 36′ 11″·45	0·4560744	7·42
7·44	+ 0·2885390	− 0·0469890	0·2923400	350° 45′ 1″·80	0·4554712	7·44
7·46	+ 0·2890347	− 0·0411470	0·2919488	351° 53′ 52″·12	0·4548703	7·46
7·48	+ 0·2894138	− 0·0353042	0·2915592	353° 2′ 42″·42	0·4542718	7·48
7·50	+ 0·2896766	− 0·0294630	0·2911711	354° 11′ 32″·70	0·4536757	7·50
7·52	+ 0·2898232	− 0·0236256	0·2907845	355° 20′ 22″·95	0·4530819	7·52
7·54	+ 0·2898538	− 0·0177943	0·2903995	356° 29′ 13″·17	0·4524905	7·54
7·56	+ 0·2897689	− 0·0119713	0·2900160	357° 38′ 3″·37	0·4519013	7·56
7·58	+ 0·2895686	− 0·0061589	0·2896341	358° 46′ 53″·54	0·4513145	7·58
7·60	+ 0·2892534	− 0·0003594	0·2892536	359° 55′ 43″·69	0·4507299	7·60
7·62	+ 0·2888237	+ 0·0054250	0·2888746	361° 4′ 33″·81	0·4501476	7·62
7·64	+ 0·2882799	+ 0·0111920	0·2884971	362° 13′ 23″·91	0·4495676	7·64
7·66	+ 0·2876227	+ 0·0169396	0·2881211	363° 22′ 13″·98	0·4489898	7·66
7·68	+ 0·2868525	+ 0·0226654	0·2877465	364° 31′ 4″·03	0·4484142	7·68
7·70	+ 0·2859699	+ 0·0283672	0·2873734	365° 39′ 54″·05	0·4478408	7·70
7·72	+ 0·2849756	+ 0·0340430	0·2870018	366° 48′ 44″·05	0·4472697	7·72
7·74	+ 0·2838702	+ 0·0396905	0·2866315	367° 57′ 34″·03	0·4467007	7·74
7·76	+ 0·2826545	+ 0·0453076	0·2862627	369° 6′ 23″·98	0·4461339	7·76
7·78	+ 0·2813292	+ 0·0508922	0·2858954	370° 15′ 13″·91	0·4455692	7·78
7·80	+ 0·2798952	+ 0·0564422	0·2855294	371° 24′ 3″·82	0·4450067	7·80
7·82	+ 0·2783532	+ 0·0619554	0·2851648	372° 32′ 53″·70	0·4444463	7·82
7·84	+ 0·2767042	+ 0·0674298	0·2848017	373° 41′ 43″·56	0·4438880	7·84
7·86	+ 0·2749490	+ 0·0728635	0·2844399	374° 50′ 33″·40	0·4433318	7·86
7·88	+ 0·2730886	+ 0·0782542	0·2840794	375° 59′ 23″·22	0·4427777	7·88
7·90	+ 0·2711241	+ 0·0836001	0·2837204	377° 8′ 13″·01	0·4422257	7·90
7·92	+ 0·2690564	+ 0·0888991	0·2833627	378° 17′ 2″·78	0·4416757	7·92
7·94	+ 0·2668867	+ 0·0941493	0·2830063	379° 25′ 52″·53	0·4411278	7·94
7·96	+ 0·2646159	+ 0·0993488	0·2826513	380° 34′ 42″·25	0·4405819	7·96
7·98	+ 0·2622454	+ 0·1044956	0·2822976	381° 43′ 31″·96	0·4400381	7·98
8·00	+ 0·2597762	+ 0·1095878	0·2819453	382° 52′ 21″·64	0·4394962	8·00

$J_{-1/3}(8{\cdot}00) = 0{\cdot}2819453 \times \cos 442°\ 52'\ 21''{\cdot}64 = +0{\cdot}0349823.$

$Y_{-1/3}(8{\cdot}00) = 0{\cdot}2819453 \times \sin 442°\ 52'\ 21''{\cdot}64 = +0{\cdot}2797667.$

Table III. Functions of order one-third

x	$J_{1/3}(x)$	$Y_{1/3}(x)$	$\lvert H^{(1)}_{1/3}(x) \rvert$	$\arg H^{(1)}_{1/3}(x)$	$e^x K_{1/3}(x)$	x
8·02	+ 0·2572095	+ 0·1146236	0·2815942	384° 1′ 11″·30	0·4389564	8·02
8·04	+ 0·2545467	+ 0·1196012	0·2812445	385° 10′ 0″·94	0·4384185	8·04
8·06	+ 0·2517890	+ 0·1245187	0·2808961	386° 18′ 50″·56	0·4378827	8·06
8·08	+ 0·2489377	+ 0·1293743	0·2805489	387° 27′ 40″·16	0·4373487	8·08
8·10	+ 0·2459942	+ 0·1341664	0·2802030	388° 36′ 29″·74	0·4368168	8·10
8·12	+ 0·2429598	+ 0·1388931	0·2798584	389° 45′ 19″·29	0·4362867	8·12
8·14	+ 0·2398360	+ 0·1435528	0·2795151	390° 54′ 8″·83	0·4357586	8·14
8·16	+ 0·2366242	+ 0·1481438	0·2791730	392° 2′ 58″·35	0·4352324	8·16
8·18	+ 0·2333259	+ 0·1526645	0·2788322	393° 11′ 47″·84	0·4347081	8·18
8·20	+ 0·2299425	+ 0·1571133	0·2784927	394° 20′ 37″·32	0·4341857	8·20
8·22	+ 0·2264758	+ 0·1614886	0·2781543	395° 29′ 26″·78	0·4336652	8·22
8·24	+ 0·2229271	+ 0·1657888	0·2778172	396° 38′ 16″·21	0·4331466	8·24
8·26	+ 0·2192981	+ 0·1700125	0·2774813	397° 47′ 5″·63	0·4326298	8·26
8·28	+ 0·2155904	+ 0·1741581	0·2771467	398° 55′ 55″·03	0·4321148	8·28
8·30	+ 0·2118057	+ 0·1782243	0·2768132	400° 4′ 44″·41	0·4316017	8·30
8·32	+ 0·2079456	+ 0·1822096	0·2764809	401° 13′ 33″·76	0·4310905	8·32
8·34	+ 0·2040118	+ 0·1861127	0·2761498	402° 22′ 23″·10	0·4305810	8·34
8·36	+ 0·2000061	+ 0·1899322	0·2758200	403° 31′ 12″·42	0·4300733	8·36
8·38	+ 0·1959301	+ 0·1936668	0·2754912	404° 40′ 1″·73	0·4295675	8·38
8·40	+ 0·1917857	+ 0·1973152	0·2751637	405° 48′ 51″·01	0·4290634	8·40
8·42	+ 0·1875747	+ 0·2008763	0·2748373	406° 57′ 40″·27	0·4285611	8·42
8·44	+ 0·1832989	+ 0·2043488	0·2745121	408° 6′ 29″·52	0·4280605	8·44
8·46	+ 0·1789600	+ 0·2077315	0·2741880	409° 15′ 18″·74	0·4275617	8·46
8·48	+ 0·1745601	+ 0·2110234	0·2738651	410° 24′ 7″·95	0·4270646	8·48
8·50	+ 0·1701008	+ 0·2142234	0·2735433	411° 32′ 57″·14	0·4265693	8·50
8·52	+ 0·1655842	+ 0·2173304	0·2732227	412° 41′ 46″·32	0·4260757	8·52
8·54	+ 0·1610121	+ 0·2203434	0·2729031	413° 50′ 35″·48	0·4255838	8·54
8·56	+ 0·1563865	+ 0·2232615	0·2725847	414° 59′ 24″·61	0·4250936	8·56
8·58	+ 0·1517093	+ 0·2260837	0·2722674	416° 8′ 13″·73	0·4246051	8·58
8·60	+ 0·1469824	+ 0·2288092	0·2719512	417° 17′ 2″·83	0·4241183	8·60
8·62	+ 0·1422079	+ 0·2314370	0·2716361	418° 25′ 51″·92	0·4236331	8·62
8·64	+ 0·1373878	+ 0·2339664	0·2713221	419° 34′ 40″·99	0·4231496	8·64
8·66	+ 0·1325240	+ 0·2363966	0·2710092	420° 43′ 30″·04	0·4226678	8·66
8·68	+ 0·1276185	+ 0·2387270	0·2706973	421° 52′ 19″·08	0·4221876	8·68
8·70	+ 0·1226734	+ 0·2409567	0·2703866	423° 1′ 8″·10	0·4217090	8·70
8·72	+ 0·1176907	+ 0·2430852	0·2700768	424° 9′ 57″·10	0·4212321	8·72
8·74	+ 0·1126725	+ 0·2451118	0·2697682	425° 18′ 46″·08	0·4207568	8·74
8·76	+ 0·1076208	+ 0·2470360	0·2694606	426° 27′ 35″·04	0·4202831	8·76
8·78	+ 0·1025377	+ 0·2488572	0·2691541	427° 36′ 23″·99	0·4198109	8·78
8·80	+ 0·0974252	+ 0·2505751	0·2688486	428° 45′ 12″·93	0·4193404	8·80
8·82	+ 0·0922855	+ 0·2521890	0·2685441	429° 54′ 1″·85	0·4188715	8·82
8·84	+ 0·0871206	+ 0·2536987	0·2682407	431° 2′ 50″·75	0·4184041	8·84
8·86	+ 0·0819327	+ 0·2551038	0·2679383	432° 11′ 39″·63	0·4179383	8·86
8·88	+ 0·0767237	+ 0·2564039	0·2676369	433° 20′ 28″·50	0·4174740	8·88
8·90	+ 0·0714958	+ 0·2575988	0·2673365	434° 29′ 17″·36	0·4170113	8·90
8·92	+ 0·0662512	+ 0·2586882	0·2670371	435° 38′ 6″·20	0·4165501	8·92
8·94	+ 0·0609918	+ 0·2596720	0·2667388	436° 46′ 55″·03	0·4160905	8·94
8·96	+ 0·0557199	+ 0·2605500	0·2664414	437° 55′ 43″·83	0·4156323	8·96
8·98	+ 0·0504375	+ 0·2613221	0·2661450	439° 4′ 32″·62	0·4151757	8·98
9·00	+ 0·0451467	+ 0·2619882	0·2658496	440° 13′ 21″·40	0·4147206	9·00

To compute functions of order $-1/3$, increase the phase by 60°.

Table III. Functions of order one-third

x	$J_{1/3}(x)$	$Y_{1/3}(x)$	$\lvert H^{(1)}_{1/3}(x) \rvert$	$\arg H^{(1)}_{1/3}(x)$	$e^x K_{1/3}(x)$	x
9·02	+ 0·0398497	+ 0·2625482	0·2655552	441° 22′ 10″·17	0·4142670	9·02
9·04	+ 0·0345485	+ 0·2630023	0·2652618	442° 30′ 58″·91	0·4138148	9·04
9·06	+ 0·0292452	+ 0·2633504	0·2649693	443° 39′ 47″·64	0·4133642	9·06
9·08	+ 0·0239420	+ 0·2635927	0·2646778	444° 48′ 36″·36	0·4129150	9·08
9·10	+ 0·0186408	+ 0·2637293	0·2643873	445° 57′ 25″·06	0·4124673	9·10
9·12	+ 0·0133439	+ 0·2637603	0·2640977	447° 6′ 13″·75	0·4120210	9·12
9·14	+ 0·0080532	+ 0·2636861	0·2638090	448° 15′ 2″·42	0·4115762	9·14
9·16	+ 0·0027709	+ 0·2635068	0·2635213	449° 23′ 51″·08	0·4111328	9·16
9·18	− 0·0025010	+ 0·2632227	0·2632346	450° 32′ 39″·73	0·4106908	9·18
9·20	− 0·0077604	+ 0·2628342	0·2629487	451° 41′ 28″·36	0·4102503	9·20
9·22	− 0·0130053	+ 0·2623417	0·2626639	452° 50′ 16″·97	0·4098112	9·22
9·24	− 0·0182336	+ 0·2617456	0·2623799	453° 59′ 5″·57	0·4093735	9·24
9·26	− 0·0234434	+ 0·2610463	0·2620968	455° 7′ 54″·16	0·4089372	9·26
9·28	− 0·0286326	+ 0·2602443	0·2618147	456° 16′ 42″·74	0·4085023	9·28
9·30	− 0·0337991	+ 0·2593402	0·2615334	457° 25′ 31″·30	0·4080687	9·30
9·32	− 0·0389411	+ 0·2583346	0·2612531	458° 34′ 19″·84	0·4076366	9·32
9·34	− 0·0440566	+ 0·2572281	0·2609737	459° 43′ 8″·37	0·4072058	9·34
9·36	− 0·0491435	+ 0·2560212	0·2606951	460° 51′ 56″·89	0·4067764	9·36
9·38	− 0·0541999	+ 0·2547148	0·2604175	462° 0′ 45″·40	0·4063483	9·38
9·40	− 0·0592239	+ 0·2533095	0·2601407	463° 9′ 33″·89	0·4059216	9·40
9·42	− 0·0642137	+ 0·2518062	0·2598648	464° 18′ 22″·37	0·4054962	9·42
9·44	− 0·0691672	+ 0·2502055	0·2595898	465° 27′ 10″·83	0·4050722	9·44
9·46	− 0·0740826	+ 0·2485083	0·2593157	466° 35′ 59″·28	0·4046495	9·46
9·48	− 0·0789581	+ 0·2467156	0·2590424	467° 44′ 47″·72	0·4042281	9·48
9·50	− 0·0837918	+ 0·2448282	0·2587700	468° 53′ 36″·15	0·4038080	9·50
9·52	− 0·0885819	+ 0·2428471	0·2584984	470° 2′ 24″·56	0·4033893	9·52
9·54	− 0·0933266	+ 0·2407732	0·2582277	471° 11′ 12″·96	0·4029718	9·54
9·56	− 0·0980241	+ 0·2386075	0·2579579	472° 20′ 1″·35	0·4025557	9·56
9·58	− 0·1026727	+ 0·2363512	0·2576889	473° 28′ 49″·72	0·4021408	9·58
9·60	− 0·1072706	+ 0·2340052	0·2574207	474° 37′ 38″·08	0·4017272	9·60
9·62	− 0·1118161	+ 0·2315707	0·2571534	475° 46′ 26″·43	0·4013149	9·62
9·64	− 0·1163076	+ 0·2290489	0·2568869	476° 55′ 14″·77	0·4009038	9·64
9·66	− 0·1207433	+ 0·2264409	0·2566212	478° 4′ 3″·09	0·4004940	9·66
9·68	− 0·1251217	+ 0·2237479	0·2563563	479° 12′ 51″·41	0·4000855	9·68
9·70	− 0·1294411	+ 0·2209712	0·2560923	480° 21′ 39″·71	0·3996782	9·70
9·72	− 0·1336999	+ 0·2181120	0·2558290	481° 30′ 28″·00	0·3992721	9·72
9·74	− 0·1378966	+ 0·2151716	0·2555666	482° 39′ 16″·27	0·3988673	9·74
9·76	− 0·1420297	+ 0·2121514	0·2553050	483° 48′ 4″·53	0·3984637	9·76
9·78	− 0·1460977	+ 0·2090527	0·2550442	484° 56′ 52″·78	0·3980614	9·78
9·80	− 0·1500990	+ 0·2058768	0·2547842	486° 5′ 41″·02	0·3976602	9·80
9·82	− 0·1540322	+ 0·2026253	0·2545250	487° 14′ 29″·25	0·3972603	9·82
9·84	− 0·1578960	+ 0·1992996	0·2542665	488° 23′ 17″·47	0·3968616	9·84
9·86	− 0·1616889	+ 0·1959010	0·2540089	489° 32′ 5″·67	0·3964641	9·86
9·88	− 0·1654096	+ 0·1924311	0·2537520	490° 40′ 53″·87	0·3960677	9·88
9·90	− 0·1690567	+ 0·1888915	0·2534959	491° 49′ 42″·05	0·3956726	9·90
9·92	− 0·1726290	+ 0·1852836	0·2532406	492° 58′ 30″·22	0·3952786	9·92
9·94	− 0·1761252	+ 0·1816090	0·2529860	494° 7′ 18″·37	0·3948858	9·94
9·96	− 0·1795441	+ 0·1778693	0·2527323	495° 16′ 6″·52	0·3944942	9·96
9·98	− 0·1828845	+ 0·1740662	0·2524792	496° 24′ 54″·66	0·3941037	9·98
10·00	− 0·1861452	+ 0·1702011	0·2522270	497° 33′ 42″·78	0·3937144	10·00

$J_{-1/3}(10{\cdot}00) = 0{\cdot}2522270 \times \cos 557^\circ\ 33'\ 42''{\cdot}78 = -0{\cdot}2404711.$

$Y_{-1/3}(10{\cdot}00) = 0{\cdot}2522270 \times \sin 557^\circ\ 33'\ 42''{\cdot}78 = -0{\cdot}0761059.$

Table III. Functions of order one-third

x	$J_{1/3}(x)$	$Y_{1/3}(x)$	$\lvert H^{(1)}_{1/3}(x) \rvert$	$\arg H^{(1)}_{1/3}(x)$	$e^x K_{1/3}(x)$	x
10·02	− 0·1893250	+ 0·1662759	0·2519755	498° 42′ 30″·90	0·3933263	10·02
10·04	− 0·1924230	+ 0·1622921	0·2517247	499° 51′ 19″·00	0·3929393	10·04
10·06	− 0·1954380	+ 0·1582514	0·2514747	501° 0′ 7″·09	0·3925534	10·06
10·08	− 0·1983690	+ 0·1541556	0·2512254	502° 8′ 55″·17	0·3921687	10·08
10·10	− 0·2012150	+ 0·1500064	0·2509769	503° 17′ 43″·24	0·3917851	10·10
10·12	− 0·2039750	+ 0·1458056	0·2507291	504° 26′ 31″·30	0·3914026	10·12
10·14	− 0·2066482	+ 0·1415548	0·2504820	505° 35′ 19″·35	0·3910213	10·14
10·16	− 0·2092336	+ 0·1372559	0·2502357	506° 44′ 7″·39	0·3906411	10·16
10·18	− 0·2117305	+ 0·1329107	0·2499901	507° 52′ 55″·41	0·3902619	10·18
10·20	− 0·2141379	+ 0·1285209	0·2497452	509° 1′ 43″·43	0·3898839	10·20
10·22	− 0·2164551	+ 0·1240885	0·2495010	510° 10′ 31″·43	0·3895070	10·22
10·24	− 0·2186814	+ 0·1196151	0·2492576	511° 19′ 19″·42	0·3891311	10·24
10·26	− 0·2208160	+ 0·1151028	0·2490148	512° 28′ 7″·41	0·3887564	10·26
10·28	− 0·2228584	+ 0·1105533	0·2487728	513° 36′ 55″·39	0·3883827	10·28
10·30	− 0·2248078	+ 0·1059685	0·2485314	514° 45′ 43″·35	0·3880101	10·30
10·32	− 0·2266637	+ 0·1013503	0·2482908	515° 54′ 31″·30	0·3876386	10·32
10·34	− 0·2284255	+ 0·0967006	0·2480509	517° 3′ 19″·25	0·3872681	10·34
10·36	− 0·2300928	+ 0·0920212	0·2478116	518° 12′ 7″·18	0·3868987	10·36
10·38	− 0·2316649	+ 0·0873142	0·2475731	519° 20′ 55″·11	0·3865304	10·38
10·40	− 0·2331416	+ 0·0825814	0·2473352	520° 29′ 43″·03	0·3861631	10·40
10·42	− 0·2345224	+ 0·0778247	0·2470980	521° 38′ 30″·93	0·3857968	10·42
10·44	− 0·2358069	+ 0·0730460	0·2468615	522° 47′ 18″·82	0·3854316	10·44
10·46	− 0·2369948	+ 0·0682473	0·2466257	523° 56′ 6″·70	0·3850674	10·46
10·48	− 0·2380858	+ 0·0634306	0·2463905	525° 4′ 54″·58	0·3847043	10·48
10·50	− 0·2390797	+ 0·0585977	0·2461560	526° 13′ 42″·44	0·3843422	10·50
10·52	− 0·2399763	+ 0·0537506	0·2459222	527° 22′ 30″·30	0·3839811	10·52
10·54	− 0·2407753	+ 0·0488913	0·2456891	528° 31′ 18″·14	0·3836210	10·54
10·56	− 0·2414768	+ 0·0440217	0·2454566	529° 40′ 5″·98	0·3832620	10·56
10·58	− 0·2420805	+ 0·0391437	0·2452248	530° 48′ 53″·81	0·3829039	10·58
10·60	− 0·2425864	+ 0·0342593	0·2449936	531° 57′ 41″·62	0·3825468	10·60
10·62	− 0·2429945	+ 0·0293704	0·2447631	533° 6′ 29″·43	0·3821908	10·62
10·64	− 0·2433049	+ 0·0244790	0·2445332	534° 15′ 17″·23	0·3818357	10·64
10·66	− 0·2435175	+ 0·0195870	0·2443040	535° 24′ 5″·02	0·3814816	10·66
10·68	− 0·2436325	+ 0·0146963	0·2440754	536° 32′ 52″·80	0·3811285	10·68
10·70	− 0·2436501	+ 0·0098090	0·2438474	537° 41′ 40″·57	0·3807764	10·70
10·72	− 0·2435703	+ 0·0049268	0·2436201	538° 50′ 28″·33	0·3804253	10·72
10·74	− 0·2433934	+ 0·0000518	0·2433935	539° 59′ 16″·08	0·3800751	10·74
10·76	− 0·2431198	− 0·0048141	0·2431674	541° 8′ 3″·83	0·3797259	10·76
10·78	− 0·2427495	− 0·0096692	0·2429420	542° 16′ 51″·56	0·3793777	10·78
10·80	− 0·2422830	− 0·0145113	0·2427172	543° 25′ 39″·29	0·3790304	10·80
10·82	− 0·2417207	− 0·0193387	0·2424930	544° 34′ 27″·00	0·3786841	10·82
10·84	− 0·2410629	− 0·0241495	0·2422695	545° 43′ 14″·71	0·3783387	10·84
10·86	− 0·2403100	− 0·0289417	0·2420466	546° 52′ 2″·41	0·3779942	10·86
10·88	− 0·2394626	− 0·0337136	0·2418242	548° 0′ 50″·10	0·3776507	10·88
10·90	− 0·2385212	− 0·0384633	0·2416025	549° 9′ 37″·78	0·3773082	10·90
10·92	− 0·2374863	− 0·0431888	0·2413814	550° 18′ 25″·46	0·3769665	10·92
10·94	− 0·2363584	− 0·0478886	0·2411610	551° 27′ 13″·12	0·3766258	10·94
10·96	− 0·2351382	− 0·0525606	0·2409411	552° 36′ 0″·78	0·3762860	10·96
10·98	− 0·2338264	− 0·0572030	0·2407218	553° 44′ 48″·43	0·3759472	10·98
11·00	− 0·2324236	− 0·0618143	0·2405031	554° 53′ 36″·07	0·3756092	11·00

To compute functions of order − 1/3, increase the phase by 60°.

Table III. Functions of order one-third

x	$J_{1/3}(x)$	$Y_{1/3}(x)$	$\lvert H^{(1)}_{1/3}(x) \rvert$	arg $H^{(1)}_{1/3}(x)$	$e^x K_{1/3}(x)$	x
11·02	− 0·2309306	− 0·0663924	0·2402850	556° 2′ 23″·69	0·3752722	11·02
11·04	− 0·2293480	− 0·0709358	0·2400675	557° 11′ 11″·31	0·3749360	11·04
11·06	− 0·2276768	− 0·0754426	0·2398506	558° 19′ 58″·93	0·3746008	11·06
11·08	− 0·2259176	− 0·0799111	0·2396342	559° 28′ 46″·54	0·3742665	11·08
11·10	− 0·2240715	− 0·0843397	0·2394185	560° 37′ 34″·13	0·3739330	11·10
11·12	− 0·2221392	− 0·0887266	0·2392033	561° 46′ 21″·72	0·3736005	11·12
11·14	− 0·2201217	− 0·0930702	0·2389887	562° 55′ 9″·31	0·3732688	11·14
11·16	− 0·2180199	− 0·0973689	0·2387747	564° 3′ 56″·88	0·3729380	11·16
11·18	− 0·2158348	− 0·1016210	0·2385613	565° 12′ 44″·44	0·3726081	11·18
11·20	− 0·2135675	− 0·1058249	0·2383484	566° 21′ 32″·00	0·3722791	11·20
11·22	− 0·2112189	− 0·1099790	0·2381361	567° 30′ 19″·55	0·3719509	11·22
11·24	− 0·2087902	− 0·1140819	0·2379244	568° 39′ 7″·09	0·3716236	11·24
11·26	− 0·2062824	− 0·1181319	0·2377132	569° 47′ 54″·62	0·3712972	11·26
11·28	− 0·2036967	− 0·1221276	0·2375026	570° 56′ 42″·15	0·3709716	11·28
11·30	− 0·2010343	− 0·1260674	0·2372926	572° 5′ 29″·67	0·3706469	11·30
11·32	− 0·1982963	− 0·1299499	0·2370831	573° 14′ 17″·18	0·3703230	11·32
11·34	− 0·1954839	− 0·1337736	0·2368741	574° 23′ 4″·68	0·3700000	11·34
11·36	− 0·1925985	− 0·1375373	0·2366658	575° 31′ 52″·17	0·3696779	11·36
11·38	− 0·1896412	− 0·1412393	0·2364579	576° 40′ 39″·66	0·3693565	11·38
11·40	− 0·1866134	− 0·1448785	0·2362506	577° 49′ 27″·14	0·3690360	11·40
11·42	− 0·1835164	− 0·1484535	0·2360439	578° 58′ 14″·61	0·3687164	11·42
11·44	− 0·1803515	− 0·1519629	0·2358377	580° 7′ 2″·08	0·3683975	11·44
11·46	− 0·1771202	− 0·1554055	0·2356320	581° 15′ 49″·53	0·3680795	11·46
11·48	− 0·1738238	− 0·1587801	0·2354269	582° 24′ 36″·98	0·3677623	11·48
11·50	− 0·1704636	− 0·1620853	0·2352223	583° 33′ 24″·42	0·3674460	11·50
11·52	− 0·1670413	− 0·1653202	0·2350182	584° 42′ 11″·86	0·3671304	11·52
11·54	− 0·1635582	− 0·1684833	0·2348147	585° 50′ 59″·29	0·3668157	11·54
11·56	− 0·1600158	− 0·1715738	0·2346117	586° 59′ 46″·70	0·3665018	11·56
11·58	− 0·1564157	− 0·1745904	0·2344092	588° 8′ 34″·11	0·3661886	11·58
11·60	− 0·1527593	− 0·1775320	0·2342072	589° 17′ 21″·52	0·3658763	11·60
11·62	− 0·1490482	− 0·1803977	0·2340058	590° 26′ 8″·92	0·3655648	11·62
11·64	− 0·1452839	− 0·1831865	0·2338049	591° 34′ 56″·31	0·3652541	11·64
11·66	− 0·1414681	− 0·1858973	0·2336044	592° 43′ 43″·69	0·3649441	11·66
11·68	− 0·1376024	− 0·1885292	0·2334046	593° 52′ 31″·07	0·3646350	11·68
11·70	− 0·1336883	− 0·1910814	0·2332052	595° 1′ 18″·44	0·3643266	11·70
11·72	− 0·1297276	− 0·1935528	0·2330063	596° 10′ 5″·81	0·3640190	11·72
11·74	− 0·1257217	− 0·1959428	0·2328079	597° 18′ 53″·16	0·3637122	11·74
11·76	− 0·1216725	− 0·1982505	0·2326101	598° 27′ 40″·51	0·3634062	11·76
11·78	− 0·1175816	− 0·2004750	0·2324127	599° 36′ 27″·85	0·3631009	11·78
11·80	− 0·1134507	− 0·2026158	0·2322159	600° 45′ 15″·18	0·3627964	11·80
11·82	− 0·1092815	− 0·2046720	0·2320195	601° 54′ 2″·51	0·3624927	11·82
11·84	− 0·1050756	− 0·2066430	0·2318237	603° 2′ 49″·84	0·3621897	11·84
11·86	− 0·1008350	− 0·2085282	0·2316283	604° 11′ 37″·15	0·3618875	11·86
11·88	− 0·0965611	− 0·2103269	0·2314335	605° 20′ 24″·46	0·3615861	11·88
11·90	− 0·0922560	− 0·2120386	0·2312391	606° 29′ 11″·76	0·3612854	11·90
11·92	− 0·0879212	− 0·2136627	0·2310452	607° 37′ 59″·05	0·3609854	11·92
11·94	− 0·0835585	− 0·2151988	0·2308518	608° 46′ 46″·34	0·3606862	11·94
11·96	− 0·0791698	− 0·2166464	0·2306589	609° 55′ 33″·62	0·3603878	11·96
11·98	− 0·0747568	− 0·2180050	0·2304664	611° 4′ 20″·90	0·3600901	11·98
12·00	− 0·0703214	− 0·2192744	0·2302745	612° 13′ 8″·17	0·3597931	12·00

$J_{-1/3}(12{\cdot}00) = 0{\cdot}2302745 \times \cos 672° \; 13' \; 8''{\cdot}17 = +0{\cdot}1547365.$

$Y_{-1/3}(12{\cdot}00) = 0{\cdot}2302745 \times \sin 672° \; 13' \; 8''{\cdot}17 = -0{\cdot}1705373.$

Table III. Functions of order one-third

x	$J_{1/3}(x)$	$Y_{1/3}(x)$	$\lvert H^{(1)}_{1/3}(x) \rvert$	arg $H^{(1)}_{1/3}(x)$	$e^x K_{1/3}(x)$	x
12·02	− 0·0658652	− 0·2204540	0·2300830	613° 21′ 55″·43	0·3594968	12·02
12·04	− 0·0613901	− 0·2215437	0·2298920	614° 30′ 42″·69	0·3592013	12·04
12·06	− 0·0568980	− 0·2225430	0·2297015	615° 39′ 29″·93	0·3589065	12·06
12·08	− 0·0523905	− 0·2234519	0·2295114	616° 48′ 17″·17	0·3586125	12·08
12·10	− 0·0478696	− 0·2242700	0·2293219	617° 57′ 4″·41	0·3583191	12·10
12·12	− 0·0433371	− 0·2249971	0·2291327	619° 5′ 51″·64	0·3580265	12·12
12·14	− 0·0387947	− 0·2256333	0·2289441	620° 14′ 38″·86	0·3577346	12·14
12·16	− 0·0342443	− 0·2261782	0·2287559	621° 23′ 25″·08	0·3574434	12·16
12·18	− 0·0296877	− 0·2266320	0·2285682	622° 32′ 13″·29	0·3571529	12·18
12·20	− 0·0251267	− 0·2269945	0·2283809	623° 41′ 0″·50	0·3568631	12·20
12·22	− 0·0205632	− 0·2272658	0·2281941	624° 49′ 47″·70	0·3565741	12·22
12·24	− 0·0159989	− 0·2274458	0·2280078	625° 58′ 34″·89	0·3562857	12·24
12·26	− 0·0114356	− 0·2275347	0·2278219	627° 7′ 22″·08	0·3559980	12·26
12·28	− 0·0068753	− 0·2275326	0·2276365	628° 16′ 9″·26	0·3557110	12·28
12·30	− 0·0023196	− 0·2274397	0·2274515	629° 24′ 56″·43	0·3554248	12·30
12·32	+ 0·0022296	− 0·2272560	0·2272670	630° 33′ 43″·60	0·3551392	12·32
12·34	+ 0·0067705	− 0·2269819	0·2270829	631° 42′ 30″·76	0·3548543	12·34
12·36	+ 0·0113014	− 0·2266176	0·2268993	632° 51′ 17″·92	0·3545700	12·36
12·38	+ 0·0158205	− 0·2261634	0·2267161	634° 0′ 5″·07	0·3542865	12·38
12·40	+ 0·0203259	− 0·2256196	0·2265333	635° 8′ 52″·21	0·3540036	12·40
12·42	+ 0·0248159	− 0·2249866	0·2263510	636° 17′ 39″·35	0·3537215	12·42
12·44	+ 0·0292889	− 0·2242647	0·2261692	637° 26′ 26″·48	0·3534400	12·44
12·46	+ 0·0337429	− 0·2234544	0·2259877	638° 35′ 13″·61	0·3531591	12·46
12·48	+ 0·0381763	− 0·2225562	0·2258067	639° 44′ 0″·73	0·3528790	12·48
12·50	+ 0·0425874	− 0·2215705	0·2256262	640° 52′ 47″·84	0·3525995	12·50
12·52	+ 0·0469744	− 0·2204979	0·2254460	642° 1′ 34″·95	0·3523206	12·52
12·54	+ 0·0513356	− 0·2193390	0·2252664	643° 10′ 22″·05	0·3520424	12·54
12·56	+ 0·0556694	− 0·2180943	0·2250871	644° 19′ 9″·15	0·3517649	12·56
12·58	+ 0·0599741	− 0·2167645	0·2249082	645° 27′ 56″·24	0·3514881	12·58
12·60	+ 0·0642479	− 0·2153502	0·2247298	646° 36′ 43″·33	0·3512118	12·60
12·62	+ 0·0684894	− 0·2138521	0·2245518	647° 45′ 30″·41	0·3509363	12·62
12·64	+ 0·0726967	− 0·2122710	0·2243743	648° 54′ 17″·49	0·3506614	12·64
12·66	+ 0·0768684	− 0·2106077	0·2241971	650° 3′ 4″·56	0·3503871	12·66
12·68	+ 0·0810028	− 0·2088628	0·2240204	651° 11′ 51″·62	0·3501135	12·68
12·70	+ 0·0850983	− 0·2070373	0·2238441	652° 20′ 38″·68	0·3498405	12·70
12·72	+ 0·0891534	− 0·2051320	0·2236682	653° 29′ 25″·73	0·3495681	12·72
12·74	+ 0·0931665	− 0·2031477	0·2234927	654° 38′ 12″·78	0·3492964	12·74
12·76	+ 0·0971361	− 0·2010854	0·2233176	655° 46′ 59″·82	0·3490254	12·76
12·78	+ 0·1010607	− 0·1989460	0·2231429	656° 55′ 46″·86	0·3487549	12·78
12·80	+ 0·1049388	− 0·1967305	0·2229687	658° 4′ 33″·89	0·3484851	12·80
12·82	+ 0·1087869	− 0·1944399	0·2227948	659° 13′ 20″·92	0·3482159	12·82
12·84	+ 0·1125496	− 0·1920752	0·2226214	660° 22′ 7″·94	0·3479473	12·84
12·86	+ 0·1162795	− 0·1896374	0·2224484	661° 30′ 54″·96	0·3476794	12·86
12·88	+ 0·1199571	− 0·1871278	0·2222757	662° 39′ 41″·97	0·3474120	12·88
12·90	+ 0·1235811	− 0·1845472	0·2221035	663° 48′ 28″·97	0·3471453	12·90
12·92	+ 0·1271502	− 0·1818970	0·2219317	664° 57′ 15″·97	0·3468792	12·92
12·94	+ 0·1306629	− 0·1791782	0·2217602	666° 6′ 2″·96	0·3466137	12·94
12·96	+ 0·1341180	− 0·1763920	0·2215892	667° 14′ 49″·95	0·3463488	12·96
12·98	+ 0·1375142	− 0·1735397	0·2214186	668° 23′ 36″·94	0·3460846	12·98
13·00	+ 0·1408503	− 0·1706224	0·2212483	669° 32′ 23″·92	0·3458209	13·00

To compute functions of order − 1/3, increase the phase by 60°.

Table III. Functions of order one-third

x	$J_{1/3}(x)$	$Y_{1/3}(x)$	$\lvert H^{(1)}_{1/3}(x)\rvert$	$\arg H^{(1)}_{1/3}(x)$	$e^x K_{1/3}(x)$	x
13·02	+ 0·1441250	− 0·1676415	0·2210785	670° 41′ 10″·89	0·3455578	13·02
13·04	+ 0·1473371	− 0·1645982	0·2209090	671° 49′ 57″·86	0·3452954	13·04
13·06	+ 0·1504854	− 0·1614938	0·2207399	672° 58′ 44″·82	0·3450335	13·06
13·08	+ 0·1535688	− 0·1583297	0·2205712	674° 7′ 31″·78	0·3447722	13·08
13·10	+ 0·1565861	− 0·1551071	0·2204029	675° 16′ 18″·74	0·3445115	13·10
13·12	+ 0·1595363	− 0·1518276	0·2202350	676° 25′ 5″·69	0·3442514	13·12
13·14	+ 0·1624183	− 0·1484923	0·2200675	677° 33′ 52″·63	0·3439919	13·14
13·16	+ 0·1652310	− 0·1451029	0·2199003	678° 42′ 39″·57	0·3437330	13·16
13·18	+ 0·1679735	− 0·1416606	0·2197335	679° 51′ 26″·50	0·3434747	13·18
13·20	+ 0·1706447	− 0·1381670	0·2195671	681° 0′ 13″·43	0·3432169	13·20
13·22	+ 0·1732437	− 0·1346234	0·2194011	682° 9′ 0″·35	0·3429597	13·22
13·24	+ 0·1757696	− 0·1310315	0·2192355	683° 17′ 47″·27	0·3427031	13·24
13·26	+ 0·1782214	− 0·1273926	0·2190702	684° 26′ 34″·18	0·3424471	13·26
13·28	+ 0·1805984	− 0·1237083	0·2189053	685° 35′ 21″·09	0·3421917	13·28
13·30	+ 0·1828996	− 0·1199802	0·2187408	686° 44′ 8″·00	0·3419368	13·30
13·32	+ 0·1851244	− 0·1162097	0·2185766	687° 52′ 54″·90	0·3416825	13·32
13·34	+ 0·1872718	− 0·1123985	0·2184128	689° 1′ 41″·79	0·3414288	13·34
13·36	+ 0·1893413	− 0·1085481	0·2182494	690° 10′ 28″·68	0·3411756	13·36
13·38	+ 0·1913320	− 0·1046601	0·2180864	691° 19′ 15″·57	0·3409230	13·38
13·40	+ 0·1932433	− 0·1007361	0·2179237	692° 28′ 2″·45	0·3406709	13·40
13·42	+ 0·1950745	− 0·0967777	0·2177613	693° 36′ 49″·33	0·3404194	13·42
13·44	+ 0·1968252	− 0·0927866	0·2175994	694° 45′ 36″·20	0·3401685	13·44
13·46	+ 0·1984946	− 0·0887643	0·2174378	695° 54′ 23″·06	0·3399181	13·46
13·48	+ 0·2000822	− 0·0847125	0·2172765	697° 3′ 9″·92	0·3396683	13·48
13·50	+ 0·2015875	− 0·0806329	0·2171157	698° 11′ 56″·78	0·3394190	13·50
13·52	+ 0·2030101	− 0·0765272	0·2169551	699° 20′ 43″·63	0·3391702	13·52
13·54	+ 0·2043496	− 0·0723969	0·2167949	700° 29′ 30″·48	0·3389221	13·54
13·56	+ 0·2056054	− 0·0682438	0·2166351	701° 38′ 17″·33	0·3386744	13·56
13·58	+ 0·2067772	− 0·0640696	0·2164757	702° 47′ 4″·17	0·3384273	13·58
13·60	+ 0·2078647	− 0·0598759	0·2163166	703° 55′ 51″·00	0·3381808	13·60
13·62	+ 0·2088676	− 0·0556645	0·2161578	705° 4′ 37″·83	0·3379347	13·62
13·64	+ 0·2097855	− 0·0514370	0·2159994	706° 13′ 24″·65	0·3376892	13·64
13·66	+ 0·2106183	− 0·0471951	0·2158413	707° 22′ 11″·47	0·3374443	13·66
13·68	+ 0·2113658	− 0·0429407	0·2156836	708° 30′ 58″·29	0·3371999	13·68
13·70	+ 0·2120278	− 0·0386752	0·2155262	709° 39′ 45″·11	0·3369560	13·70
13·72	+ 0·2126040	− 0·0344006	0·2153692	710° 48′ 31″·92	0·3367126	13·72
13·74	+ 0·2130946	− 0·0301184	0·2152125	711° 57′ 18″·72	0·3364698	13·74
13·76	+ 0·2134993	− 0·0258305	0·2150561	713° 6′ 5″·52	0·3362275	13·76
13·78	+ 0·2138181	− 0·0215384	0·2149001	714° 14′ 52″·31	0·3359857	13·78
13·80	+ 0·2140510	− 0·0172440	0·2147445	715° 23′ 39″·10	0·3357444	13·80
13·82	+ 0·2141981	− 0·0129489	0·2145891	716° 32′ 25″·89	0·3355037	13·82
13·84	+ 0·2142594	− 0·0086548	0·2144341	717° 41′ 12″·67	0·3352635	13·84
13·86	+ 0·2142350	− 0·0043635	0·2142795	718° 49′ 59″·45	0·3350237	13·86
13·88	+ 0·2141251	− 0·0000766	0·2141251	719° 58′ 46″·22	0·3347845	13·88
13·90	+ 0·2139298	+ 0·0042041	0·2139712	721° 7′ 32″·99	0·3345459	13·90
13·92	+ 0·2136494	+ 0·0084770	0·2138175	722° 16′ 19″·76	0·3343077	13·92
13·94	+ 0·2132840	+ 0·0127404	0·2136642	723° 25′ 6″·52	0·3340700	13·94
13·96	+ 0·2128339	+ 0·0169926	0·2135112	724° 33′ 53″·28	0·3338329	13·96
13·98	+ 0·2122994	+ 0·0212319	0·2133585	725° 42′ 40″·03	0·3335962	13·98
14·00	+ 0·2116809	+ 0·0254567	0·2132061	726° 51′ 26″·78	0·3333600	14·00

$J_{-1/3}(14{\cdot}00) = 0{\cdot}2132061 \times \cos 786^\circ\ 51'\ 26''{\cdot}78 = +0{\cdot}0837943.$

$Y_{-1/3}(14{\cdot}00) = 0{\cdot}2132061 \times \sin 786^\circ\ 51'\ 26''{\cdot}78 = +0{\cdot}1960494.$

Table III. Functions of order one-third

x	$J_{1/3}(x)$	$Y_{1/3}(x)$	$\lvert H^{(1)}_{1/3}(x)\rvert$	$\arg H^{(1)}_{1/3}(x)$	$e^x K_{1/3}(x)$	x
14·02	+ 0·2109787	+ 0·0296652	0·2130541	728° 0′ 13″·52	0·3331244	14·02
14·04	+ 0·2101933	+ 0·0338559	0·2129024	729° 9′ 0″·26	0·3328892	14·04
14·06	+ 0·2093250	+ 0·0380272	0·2127510	730° 17′ 47″·00	0·3326546	14·06
14·08	+ 0·2083743	+ 0·0421773	0·2126000	731° 26′ 33″·73	0·3324204	14·08
14·10	+ 0·2073417	+ 0·0463046	0·2124493	732° 35′ 20″·46	0·3321867	14·10
14·12	+ 0·2062277	+ 0·0504076	0·2122989	733° 44′ 7″·18	0·3319536	14·12
14·14	+ 0·2050330	+ 0·0544847	0·2121488	734° 52′ 53″·90	0·3317209	14·14
14·16	+ 0·2037580	+ 0·0585342	0·2119990	736° 1′ 40″·62	0·3314887	14·16
14·18	+ 0·2024034	+ 0·0625547	0·2118495	737° 10′ 27″·33	0·3312570	14·18
14·20	+ 0·2009699	+ 0·0665445	0·2117004	738° 19′ 14″·04	0·3310257	14·20
14·22	+ 0·1994581	+ 0·0705021	0·2115516	739° 28′ 0″·75	0·3307950	14·22
14·24	+ 0·1978687	+ 0·0744259	0·2114031	740° 36′ 47″·45	0·3305648	14·24
14·26	+ 0·1962026	+ 0·0783145	0·2112549	741° 45′ 34″·14	0·3303350	14·26
14·28	+ 0·1944604	+ 0·0821664	0·2111070	742° 54′ 20″·83	0·3301057	14·28
14·30	+ 0·1926429	+ 0·0859801	0·2109594	744° 3′ 7″·52	0·3298769	14·30
14·32	+ 0·1907510	+ 0·0897541	0·2108121	745° 11′ 54″·21	0·3296485	14·32
14·34	+ 0·1887856	+ 0·0934869	0·2106652	746° 20′ 40″·89	0·3294206	14·34
14·36	+ 0·1867475	+ 0·0971773	0·2105185	747° 29′ 27″·57	0·3291932	14·36
14·38	+ 0·1846376	+ 0·1008236	0·2103721	748° 38′ 14″·24	0·3289663	14·38
14·40	+ 0·1824569	+ 0·1044246	0·2102261	749° 47′ 0″·91	0·3287398	14·40
14·42	+ 0·1802063	+ 0·1079789	0·2100804	750° 55′ 47″·58	0·3285138	14·42
14·44	+ 0·1778868	+ 0·1114852	0·2099349	752° 4′ 34″·24	0·3282883	14·44
14·46	+ 0·1754995	+ 0·1149420	0·2097898	753° 13′ 20″·90	0·3280632	14·46
14·48	+ 0·1730454	+ 0·1183482	0·2096449	754° 22′ 7″·55	0·3278386	14·48
14·50	+ 0·1705256	+ 0·1217023	0·2095004	755° 30′ 54″·20	0·3276145	14·50
14·52	+ 0·1679410	+ 0·1250032	0·2093562	756° 39′ 40″·85	0·3273908	14·52
14·54	+ 0·1652930	+ 0·1282497	0·2092122	757° 48′ 27″·49	0·3271676	14·54
14·56	+ 0·1625825	+ 0·1314404	0·2090686	758° 57′ 14″·13	0·3269448	14·56
14·58	+ 0·1598109	+ 0·1345743	0·2089252	760° 6′ 0″·77	0·3267225	14·58
14·60	+ 0·1569792	+ 0·1376501	0·2087822	761° 14′ 47″·40	0·3265006	14·60
14·62	+ 0·1540886	+ 0·1406666	0·2086394	762° 23′ 34″·03	0·3262792	14·62
14·64	+ 0·1511404	+ 0·1436229	0·2084969	763° 32′ 20″·65	0·3260583	14·64
14·66	+ 0·1481359	+ 0·1465178	0·2083548	764° 41′ 7″·27	0·3258378	14·66
14·68	+ 0·1450763	+ 0·1493501	0·2082129	765° 49′ 53″·89	0·3256177	14·68
14·70	+ 0·1419629	+ 0·1521190	0·2080713	766° 58′ 40″·51	0·3253981	14·70
14·72	+ 0·1387970	+ 0·1548233	0·2079300	768° 7′ 27″·12	0·3251789	14·72
14·74	+ 0·1355800	+ 0·1574621	0·2077889	769° 16′ 13″·73	0·3249602	14·74
14·76	+ 0·1323131	+ 0·1600344	0·2076482	770° 25′ 0″·33	0·3247419	14·76
14·78	+ 0·1289979	+ 0·1625393	0·2075077	771° 33′ 46″·93	0·3245240	14·78
14·80	+ 0·1256355	+ 0·1649758	0·2073676	772° 42′ 33″·53	0·3243066	14·80
14·82	+ 0·1222275	+ 0·1673432	0·2072277	773° 51′ 20″·12	0·3240896	14·82
14·84	+ 0·1187753	+ 0·1696405	0·2070881	775° 0′ 6″·71	0·3238731	14·84
14·86	+ 0·1152803	+ 0·1718669	0·2069488	776° 8′ 53″·30	0·3236570	14·86
14·88	+ 0·1117439	+ 0·1740217	0·2068097	777° 17′ 39″·88	0·3234413	14·88
14·90	+ 0·1081677	+ 0·1761041	0·2066710	778° 26′ 26″·46	0·3232261	14·90
14·92	+ 0·1045530	+ 0·1781133	0·2065325	779° 35′ 13″·04	0·3230112	14·92
14·94	+ 0·1009014	+ 0·1800487	0·2063943	780° 43′ 59″·61	0·3227969	14·94
14·96	+ 0·0972143	+ 0·1819095	0·2062564	781° 52′ 46″·18	0·3225829	14·96
14·98	+ 0·0934933	+ 0·1836952	0·2061187	783° 1′ 32″·75	0·3223694	14·98
15·00	+ 0·0897400	+ 0·1854051	0·2059813	784° 10′ 19″·31	0·3221562	15·00

To compute functions of order −1/3, increase the phase by 60°.

Table III. Functions of order one-third

x	$J_{1/3}(x)$	$Y_{1/3}(x)$	$\lvert H^{(1)}_{1/3}(x) \rvert$	$\arg H^{(1)}_{1/3}(x)$	$e^x K_{1/3}(x)$	x
15·02	+ 0·0859558	+ 0·1870386	0·2058442	785° 19′ 5″·87	0·3219435	15·02
15·04	+ 0·0821423	+ 0·1885953	0·2057074	786° 27′ 52″·42	0·3217313	15·04
15·06	+ 0·0783010	+ 0·1900745	0·2055709	787° 36′ 38″·97	0·3215194	15·06
15·08	+ 0·0744336	+ 0·1914758	0·2054346	788° 45′ 25″·52	0·3213080	15·08
15·10	+ 0·0705416	+ 0·1927988	0·2052986	789° 54′ 12″·07	0·3210970	15·10
15·12	+ 0·0666265	+ 0·1940430	0·2051628	791° 2′ 58″·61	0·3208864	15·12
15·14	+ 0·0626899	+ 0·1952080	0·2050273	792° 11′ 45″·15	0·3206762	15·14
15·16	+ 0·0587336	+ 0·1962935	0·2048921	793° 20′ 31″·69	0·3204664	15·16
15·18	+ 0·0547589	+ 0·1972992	0·2047572	794° 29′ 18″·22	0·3202570	15·18
15·20	+ 0·0507677	+ 0·1982247	0·2046225	795° 38′ 4″·75	0·3200481	15·20
15·22	+ 0·0467614	+ 0·1990697	0·2044881	796° 46′ 51″·28	0·3198395	15·22
15·24	+ 0·0427416	+ 0·1998342	0·2043540	797° 55′ 37″·80	0·3196314	15·24
15·26	+ 0·0387101	+ 0·2005178	0·2042201	799° 4′ 24″·32	0·3194237	15·26
15·28	+ 0·0346684	+ 0·2011203	0·2040865	800° 13′ 10″·84	0·3192164	15·28
15·30	+ 0·0306181	+ 0·2016418	0·2039531	801° 21′ 57″·35	0·3190094	15·30
15·32	+ 0·0265609	+ 0·2020820	0·2038200	802° 30′ 43″·86	0·3188029	15·32
15·34	+ 0·0224983	+ 0·2024408	0·2036872	803° 39′ 30″·37	0·3185968	15·34
15·36	+ 0·0184321	+ 0·2027184	0·2035546	804° 48′ 16″·87	0·3183911	15·36
15·38	+ 0·0143638	+ 0·2029145	0·2034223	805° 57′ 3″·37	0·3181857	15·38
15·40	+ 0·0102950	+ 0·2030294	0·2032902	807° 5′ 49″·87	0·3179808	15·40
15·42	+ 0·0062274	+ 0·2030630	0·2031584	808° 14′ 36″·36	0·3177763	15·42
15·44	+ 0·0021626	+ 0·2030154	0·2030269	809° 23′ 22″·85	0·3175721	15·44
15·46	− 0·0018978	+ 0·2028867	0·2028956	810° 32′ 9″·34	0·3173684	15·46
15·48	− 0·0059522	+ 0·2026772	0·2027646	811° 40′ 55″·83	0·3171650	15·48
15·50	− 0·0099990	+ 0·2023869	0·2026338	812° 49′ 42″·31	0·3169621	15·50
15·52	− 0·0140366	+ 0·2020162	0·2025032	813° 58′ 28″·79	0·3167595	15·52
15·54	− 0·0180634	+ 0·2015652	0·2023730	815° 7′ 15″·27	0·3165573	15·54
15·56	− 0·0220777	+ 0·2010343	0·2022429	816° 16′ 1″·74	0·3163555	15·56
15·58	− 0·0260781	+ 0·2004237	0·2021132	817° 24′ 48″·21	0·3161541	15·58
15·60	− 0·0300630	+ 0·1997338	0·2019836	818° 33′ 34″·68	0·3159531	15·60
15·62	− 0·0340307	+ 0·1989651	0·2018544	819° 42′ 21″·15	0·3157524	15·62
15·64	− 0·0379798	+ 0·1981178	0·2017253	820° 51′ 7″·61	0·3155522	15·64
15·66	− 0·0419086	+ 0·1971924	0·2015965	821° 59′ 54″·07	0·3153523	15·66
15·68	− 0·0458157	+ 0·1961894	0·2014680	823° 8′ 40″·52	0·3151528	15·68
15·70	− 0·0496995	+ 0·1951093	0·2013397	824° 17′ 26″·98	0·3149537	15·70
15·72	− 0·0535585	+ 0·1939526	0·2012117	825° 26′ 13″·43	0·3147549	15·72
15·74	− 0·0573911	+ 0·1927199	0·2010839	826° 34′ 59″·88	0·3145565	15·74
15·76	− 0·0611960	+ 0·1914118	0·2009563	827° 43′ 46″·32	0·3143586	15·76
15·78	− 0·0649716	+ 0·1900289	0·2008290	828° 52′ 32″·76	0·3141609	15·78
15·80	− 0·0687165	+ 0·1885717	0·2007019	830° 1′ 19″·20	0·3139637	15·80
15·82	− 0·0724292	+ 0·1870411	0·2005751	831° 10′ 5″·63	0·3137668	15·82
15·84	− 0·0761082	+ 0·1854377	0·2004485	832° 18′ 52″·06	0·3135703	15·84
15·86	− 0·0797522	+ 0·1837622	0·2003221	833° 27′ 38″·49	0·3133741	15·86
15·88	− 0·0833597	+ 0·1820154	0·2001960	834° 36′ 24″·92	0·3131784	15·88
15·90	− 0·0869294	+ 0·1801980	0·2000701	835° 45′ 11″·34	0·3129830	15·90
15·92	− 0·0904599	+ 0·1783110	0·1999445	836° 53′ 57″·76	0·3127879	15·92
15·94	− 0·0939498	+ 0·1763550	0·1998191	838° 2′ 44″·18	0·3125932	15·94
15·96	− 0·0973977	+ 0·1743311	0·1996939	839° 11′ 30″·60	0·3123989	15·96
15·98	− 0·1008025	+ 0·1722400	0·1995690	840° 20′ 17″·01	0·3122050	15·98
16·00	− 0·1041627	+ 0·1700828	0·1994442	841° 29′ 3″·42	0·3120114	16·00

$J_{-1/3}(16\cdot00) = 0\cdot1994442 \times \cos 901° \, 29' \, 3''\cdot42 = -0\cdot1993773.$

$Y_{-1/3}(16\cdot00) = 0\cdot1994442 \times \sin 901° \, 29' \, 3''\cdot42 = -0\cdot0051662.$

Table IV. Values of $J_n(x)$

x	$J_2(x)$	$J_3(x)$	$J_4(x)$	$J_5(x)$	x
0·1	+ 0·0012490	+ 0·0000208	+ 0·0000003	—	0·1
0·2	+ 0·0049834	+ 0·0001663	+ 0·0000042	+ 0·0000001	0·2
0·3	+ 0·0111659	+ 0·0005593	+ 0·0000210	+ 0·0000006	0·3
0·4	+ 0·0197347	+ 0·0013201	+ 0·0000661	+ 0·0000026	0·4
0·5	+ 0·0306040	+ 0·0025637	+ 0·0001607	+ 0·0000081	0·5
0·6	+ 0·0436651	+ 0·0043997	+ 0·0003315	+ 0·0000199	0·6
0·7	+ 0·0587869	+ 0·0069297	+ 0·0006101	+ 0·0000429	0·7
0·8	+ 0·0758178	+ 0·0102468	+ 0·0010330	+ 0·0000831	0·8
0·9	+ 0·0945863	+ 0·0144340	+ 0·0016406	+ 0·0001487	0·9
1·0	+ 0·1149035	+ 0·0195634	+ 0·0024766	+ 0·0002498	1·0
1·1	+ 0·1365642	+ 0·0256945	+ 0·0035878	+ 0·0003987	1·1
1·2	+ 0·1593490	+ 0·0328743	+ 0·0050227	+ 0·0006101	1·2
1·3	+ 0·1830267	+ 0·0411358	+ 0·0068310	+ 0·0009008	1·3
1·4	+ 0·2073559	+ 0·0504977	+ 0·0090629	+ 0·0012901	1·4
1·5	+ 0·2320877	+ 0·0609640	+ 0·0117681	+ 0·0017994	1·5
1·6	+ 0·2569678	+ 0·0725234	+ 0·0149952	+ 0·0024524	1·6
1·7	+ 0·2817389	+ 0·0851499	+ 0·0187902	+ 0·0032746	1·7
1·8	+ 0·3061435	+ 0·0988020	+ 0·0231965	+ 0·0042936	1·8
1·9	+ 0·3299257	+ 0·1134234	+ 0·0282535	+ 0·0055385	1·9
2·0	+ 0·3528340	+ 0·1289432	+ 0·0339957	+ 0·0070396	2·0
2·1	+ 0·3746236	+ 0·1452767	+ 0·0404526	+ 0·0088284	2·1
2·2	+ 0·3950587	+ 0·1623255	+ 0·0476471	+ 0·0109369	2·2
2·3	+ 0·4139146	+ 0·1799789	+ 0·0555957	+ 0·0133973	2·3
2·4	+ 0·4309800	+ 0·1981148	+ 0·0643070	+ 0·0162417	2·4
2·5	+ 0·4460591	+ 0·2166004	+ 0·0737819	+ 0·0195016	2·5
2·6	+ 0·4589729	+ 0·2352938	+ 0·0840129	+ 0·0232073	2·6
2·7	+ 0·4695615	+ 0·2540453	+ 0·0949836	+ 0·0273876	2·7
2·8	+ 0·4776855	+ 0·2726986	+ 0·1066687	+ 0·0320690	2·8
2·9	+ 0·4832271	+ 0·2910926	+ 0·1190335	+ 0·0372756	2·9
3·0	+ 0·4860913	+ 0·3090627	+ 0·1320342	+ 0·0430284	3·0
3·1	+ 0·4862070	+ 0·3264428	+ 0·1456177	+ 0·0493448	3·1
3·2	+ 0·4835277	+ 0·3430664	+ 0·1597218	+ 0·0562380	3·2
3·3	+ 0·4780317	+ 0·3587689	+ 0·1742754	+ 0·0637169	3·3
3·4	+ 0·4697226	+ 0·3733889	+ 0·1891991	+ 0·0717854	3·4
3·5	+ 0·4586292	+ 0·3867701	+ 0·2044053	+ 0·0804420	3·5
3·6	+ 0·4448054	+ 0·3987627	+ 0·2197990	+ 0·0896796	3·6
3·7	+ 0·4283297	+ 0·4092251	+ 0·2352786	+ 0·0994854	3·7
3·8	+ 0·4093043	+ 0·4180256	+ 0·2507362	+ 0·1098400	3·8
3·9	+ 0·3878547	+ 0·4250437	+ 0·2660587	+ 0·1207178	3·9
4·0	+ 0·3641281	+ 0·4301715	+ 0·2811291	+ 0·1320867	4·0
4·1	+ 0·3382925	+ 0·4333147	+ 0·2958266	+ 0·1439079	4·1
4·2	+ 0·3105347	+ 0·4343943	+ 0·3100286	+ 0·1561363	4·2
4·3	+ 0·2810592	+ 0·4333470	+ 0·3236110	+ 0·1687200	4·3
4·4	+ 0·2500861	+ 0·4301265	+ 0·3364501	+ 0·1816009	4·4
4·5	+ 0·2178490	+ 0·4247040	+ 0·3484230	+ 0·1947147	4·5
4·6	+ 0·1845931	+ 0·4170686	+ 0·3594094	+ 0·2079912	4·6
4·7	+ 0·1505730	+ 0·4072280	+ 0·3692925	+ 0·2213550	4·7
4·8	+ 0·1160504	+ 0·3952085	+ 0·3779603	+ 0·2347252	4·8
4·9	+ 0·0812915	+ 0·3810551	+ 0·3853066	+ 0·2480168	4·9
5·0	+ 0·0465651	+ 0·3648312	+ 0·3912324	+ 0·2611405	5·0

Table IV. Values of $J_n(x)$

x	$J_0(x)$	$J_1(x)$	$J_2(x)$	$J_3(x)$	$J_4(x)$	$J_5(x)$	x
1	+0·765198	+0·440051	+0·114903	+0·019563	+0·002477	+0·000250	1
2	+0·223891	+0·576725	+0·352834	+0·128943	+0·033996	+0·007040	2
3	−0·260052	+0·339059	+0·486091	+0·309063	+0·132034	+0·043028	3
4	−0·397150	−0·066043	+0·364128	+0·430171	+0·281129	+0·132087	4
5	−0·177597	−0·327579	+0·046565	+0·364831	+0·391232	+0·261141	5
6	+0·150645	−0·276684	−0·242873	+0·114768	+0·357642	+0·362087	6
7	+0·300079	−0·004683	−0·301417	−0·167556	+0·157798	+0·347896	7
8	+0·171651	+0·234636	−0·112992	−0·291132	−0·105357	+0·185775	8
9	−0·090334	+0·245312	+0·144847	−0·180935	−0·265471	−0·055039	9
10	−0·245936	+0·043473	+0·254630	+0·058379	−0·219603	−0·234062	10
11	−0·171190	−0·176785	+0·139048	+0·227348	−0·015040	−0·238286	11
12	+0·047689	−0·223447	−0·084930	+0·195137	+0·182499	−0·073471	12

x	$J_6(x)$	$J_7(x)$	$J_8(x)$	$J_9(x)$	$J_{10}(x)$	$J_{11}(x)$	x
1	+0·000021	+0·000002	—	—	—	—	1
2	+0·001202	+0·000175	+0·000022	+0·000002	—	—	2
3	+0·011394	+0·002547	+0·000493	+0·000084	+0·000013	+0·000002	3
4	+0·049088	+0·015176	+0·004029	+0·000939	+0·000195	+0·000037	4
5	+0·131049	+0·053376	+0·018405	+0·005520	+0·001468	+0·000351	5
6	+0·245837	+0·129587	+0·056532	+0·021165	+0·006964	+0·002048	6
7	+0·339197	+0·233584	+0·127971	+0·058921	+0·023539	+0·008335	7
8	+0·337576	+0·320589	+0·223455	+0·126321	+0·060767	+0·025597	8
9	+0·204317	+0·327461	+0·305067	+0·214881	+0·124694	+0·062217	9
10	−0·014459	+0·216711	+0·317854	+0·291856	+0·207486	+0·123117	10
11	−0·201584	+0·018376	+0·224972	+0·308856	+0·280428	+0·201014	11
12	−0·243725	−0·170254	+0·045095	+0·230381	+0·300476	+0·270412	12

x	$J_{12}(x)$	$J_{13}(x)$	$J_{14}(x)$	$J_{15}(x)$	$J_{16}(x)$	$J_{17}(x)$	x
4	+0·000006	+0·000001	—	—	—	—	4
5	+0·000076	+0·000015	+0·000003	—	—	—	5
6	+0·000545	+0·000133	+0·000030	+0·000006	+0·000001	—	6
7	+0·002656	+0·000770	+0·000205	+0·000051	+0·000012	+0·000002	7
8	+0·009624	+0·003275	+0·001019	+0·000293	+0·000078	+0·000019	8
9	+0·027393	+0·010830	+0·003895	+0·001286	+0·000393	+0·000112	9
10	+0·063370	+0·028972	+0·011957	+0·004508	+0·001567	+0·000506	10
11	+0·121600	+0·064295	+0·030369	+0·013009	+0·005110	+0·001856	11
12	+0·195280	+0·120148	+0·065040	+0·031613	+0·013991	+0·005698	12

Table IV. Values of $J_n(x)$ and $Y_n(x)$

x	$J_{18}(x)$	$J_{19}(x)$	$J_{20}(x)$	x
7	+ 0·000001	—	—	7
8	+ 0·000005	+ 0·000001	—	8
9	+ 0·000030	+ 0·000007	+ 0·000002	9
10	+ 0·000152	+ 0·000043	+ 0·000012	10
11	+ 0·000628	+ 0·000199	+ 0·000059	11
12	+ 0·002152	+ 0·000759	+ 0·000251	12

x	$Y_0(x)$	$Y_1(x)$	$Y_2(x)$	$Y_3(x)$	$Y_4(x)$	x
6	− 0·2881947	− 0·1750103	+ 0·2298579	+ 0·3282489	+ 0·0983910	6
7	− 0·0259497	− 0·3026672	− 0·0605266	+ 0·2680806	+ 0·2903100	7
8	+ 0·2235215	− 0·1580605	− 0·2630366	+ 0·0265422	+ 0·2829432	8
9	+ 0·2499367	+ 0·1043146	− 0·2267557	− 0·2050949	+ 0·0900258	9
10	+ 0·0556712	+ 0·2490154	− 0·0058681	− 0·2513627	− 0·1449495	10
11	− 0·1688473	+ 0·1637055	+ 0·1986120	− 0·0914830	− 0·2485118	11
12	− 0·2252373	− 0·0570992	+ 0·2157208	+ 0·1290061	− 0·1512177	12

x	$Y_5(x)$	$Y_6(x)$	$Y_7(x)$	$Y_8(x)$	$Y_9(x)$	x
6	− 0·1970609	− 0·4268259	− 0·6565908	− 1·1052194	− 2·2906609	6
7	+ 0·0637022	− 0·1993068	− 0·4053710	− 0·6114352	− 0·9921953	7
8	+ 0·2564011	+ 0·0375581	− 0·2000639	− 0·3876699	− 0·5752760	8
9	+ 0·2851178	+ 0·2267718	+ 0·0172446	− 0·1999469	− 0·3727057	9
10	+ 0·1354030	+ 0·2803526	+ 0·2010200	+ 0·0010755	− 0·1992993	10
11	− 0·0892528	+ 0·1673728	+ 0·2718414	+ 0·1786071	− 0·0120492	11
12	− 0·2298179	− 0·0402973	+ 0·1895207	+ 0·2614047	+ 0·1590189	12

x	$Y_{10}(x)$	$Y_{11}(x)$	$Y_{12}(x)$	$Y_{13}(x)$	x
6	− 5·7667633	− 16·9318836	− 56·3168097	− 208·3353554	6
7	− 1·9399240	− 4·5504447	− 12·3614737	− 37·8317507	7
8	− 0·9067010	− 1·6914765	− 3·7448595	− 9·5431018	8
9	− 0·5454645	− 0·8394376	− 1·5064942	− 3·1778801	9
10	− 0·3598142	− 0·5203290	− 0·7849097	− 1·3634543	10
11	− 0·1983240	− 0·3485399	− 0·4987558	− 0·7396546	11
12	− 0·0228763	− 0·1971461	− 0·3385583	− 0·4799704	12

Table IV. Values of $Y_n(x)$

x	$Y_0(x)$	$Y_1(x)$	$Y_2(x)$	$Y_3(x)$	x
0·1	− 1·5342387	− 6·4589511	− 127·6447832	− 5099·3323786	0·1
0·2	− 1·0811053	− 3·3238250	− 32·1571446	− 639·8190662	0·2
0·3	− 0·8072736	− 2·2931051	− 14·4800940	− 190·7748150	0·3
0·4	− 0·6060246	− 1·7808720	− 8·2983357	− 81·2024845	0·4
0·5	− 0·4445187	− 1·4714724	− 5·4413708	− 42·0594943	0·5
0·6	− 0·3085099	− 1·2603913	− 3·8927946	− 24·6915728	0·6
0·7	− 0·1906649	− 1·1032499	− 2·9614776	− 15·8194791	0·7
0·8	− 0·0868023	− 0·9781442	− 2·3585582	− 10·8146466	0·8
0·9	+ 0·0056283	− 0·8731266	− 1·9459096	− 7·7753605	0·9
1·0	+ 0·0882570	− 0·7812128	− 1·6506826	− 5·8215176	1·0
1·1	+ 0·1621632	− 0·6981196	− 1·4314715	− 4·5072313	1·1
1·2	+ 0·2280835	− 0·6211364	− 1·2633108	− 3·5898996	1·2
1·3	+ 0·2865354	− 0·5485197	− 1·1304119	− 2·9296706	1·3
1·4	+ 0·3378951	− 0·4791470	− 1·0223908	− 2·4419696	1·4
1·5	+ 0·3824489	− 0·4123086	− 0·9321938	− 2·0735414	1·5
1·6	+ 0·4204269	− 0·3475780	− 0·8548994	− 1·7896705	1·6
1·7	+ 0·4520270	− 0·2847262	− 0·7869991	− 1·5670362	1·7
1·8	+ 0·4774317	− 0·2236649	− 0·7259482	− 1·3895534	1·8
1·9	+ 0·4968200	− 0·1644058	− 0·6698787	− 1·2458651	1·9
2·0	+ 0·5103757	− 0·1070324	− 0·6174081	− 1·1277838	2·0
2·1	+ 0·5182937	− 0·0516786	− 0·5675115	− 1·0292956	2·1
2·2	+ 0·5207843	+ 0·0014878	− 0·5194317	− 0·9459092	2·2
2·3	+ 0·5180754	+ 0·0522773	− 0·4726169	− 0·8742197	2·3
2·4	+ 0·5104147	+ 0·1004889	− 0·4266740	− 0·8116122	2·4
2·5	+ 0·4980704	+ 0·1459181	− 0·3813358	− 0·7560555	2·5
2·6	+ 0·4813306	+ 0·1883635	− 0·3364356	− 0·7059567	2·6
2·7	+ 0·4605035	+ 0·2276324	− 0·2918869	− 0·6600575	2·7
2·8	+ 0·4359160	+ 0·2635454	− 0·2476693	− 0·6173586	2·8
2·9	+ 0·4079118	+ 0·2959401	− 0·2038152	− 0·5770644	2·9
3·0	+ 0·3768500	+ 0·3246744	− 0·1604004	− 0·5385416	3·0
3·1	+ 0·3431029	+ 0·3496295	− 0·1175355	− 0·5012882	3·1
3·2	+ 0·3070533	+ 0·3707113	− 0·0753587	− 0·4649097	3·2
3·3	+ 0·2690920	+ 0·3878529	− 0·0340296	− 0·4291009	3·3
3·4	+ 0·2296153	+ 0·4010153	+ 0·0062760	− 0·3936317	3·4
3·5	+ 0·1890219	+ 0·4101884	+ 0·0453714	− 0·3583353	3·5
3·6	+ 0·1477100	+ 0·4153918	+ 0·0830632	− 0·3230993	3·6
3·7	+ 0·1060743	+ 0·4166744	+ 0·1191551	− 0·2878581	3·7
3·8	+ 0·0645032	+ 0·4141147	+ 0·1534519	− 0·2525864	3·8
3·9	+ 0·0233759	+ 0·4078200	+ 0·1857626	− 0·2172943	3·9
4·0	− 0·0169407	+ 0·3979257	+ 0·2159036	− 0·1820221	4·0
4·1	− 0·0560946	+ 0·3845940	+ 0·2437015	− 0·1468365	4·1
4·2	− 0·0937512	+ 0·3680128	+ 0·2689954	− 0·1118267	4·2
4·3	− 0·1295959	+ 0·3483938	+ 0·2916395	− 0·0771012	4·3
4·4	− 0·1633365	+ 0·3259707	+ 0·3115049	− 0·0427844	4·4
4·5	− 0·1947050	+ 0·3009973	+ 0·3284816	− 0·0090137	4·5
4·6	− 0·2234600	+ 0·2737452	+ 0·3424796	+ 0·0240631	4·6
4·7	− 0·2493876	+ 0·2445013	+ 0·3534308	+ 0·0562908	4·7
4·8	− 0·2723038	+ 0·2135652	+ 0·3612893	+ 0·0875092	4·8
4·9	− 0·2920546	+ 0·1812467	+ 0·3660328	+ 0·1175556	4·9
5·0	− 0·3085176	+ 0·1478631	+ 0·3676629	+ 0·1462672	5·0

Table IV. Values of $Y_n(x)$

x	$Y_4(x)$	$Y_5(x)$	$Y_6(x)$	$Y_7(x)$	x
0·1	− 305832·2979	− 24461484·502	− 2445842617·9	− 293476652667	0·
0·2	− 19162·4148	− 765856·775	− 38273676·3	− 2295654722	0·
0·3	− 3801·0162	− 101169·657	− 3368520·9	− 134639666	0·
0·4	− 1209·7389	− 24113·576	− 601629·7	− 18024776	0·
0·5	− 499·2726	− 7946·301	− 158426·8	− 3794296	0·
0·6	− 243·02293	− 3215·6142	− 53350·547	− 1063795·33	0·
0·7	− 132·63406	− 1499·9983	− 21295·913	− 363572·80	0·
0·8	− 78·75129	− 776·6983	− 9629·977	− 143672·96	0·
0·9	− 49·88983	− 435·6898	− 4791·108	− 63445·75	0·
1·0	− 33·27842	− 260·4059	− 2570·780	− 30588·96	1·
1·1	− 23·153427	− 163·88133	− 1466·6768	− 15836·229	1·
1·2	− 16·686187	− 107·65135	− 880·4084	− 8696·433	1·
1·3	− 12·391145	− 73·32353	− 551·6360	− 5018·701	1·
1·4	− 9·443193	− 51·51913	− 358·5506	− 3021·772	1·
1·5	− 7·361972	− 37·19031	− 240·5734	− 1887·397	1·
1·6	− 5·856365	− 27·492154	− 165·96960	− 1217·2798	1·
1·7	− 4·743717	− 20·756338	− 117·35239	− 807·6135	1·
1·8	− 3·905897	− 15·969987	− 84·81625	− 549·4717	1·
1·9	− 3·264432	− 12·499113	− 62·52037	− 382·366	1·
2·0	− 2·765943	− 9·935989	− 46·91400	− 271·5480	2·
2·1	− 2·3733331	− 8·011973	− 35·77892	− 196·43901	2·
2·2	− 2·0603205	− 6·546165	− 27·69498	− 144·51734	2·
2·3	− 1·8079562	− 5·414324	− 21·73258	− 107·97306	2·
2·4	− 1·6023566	− 4·529576	− 17·27088	− 81·82481	2·
2·5	− 1·4331973	− 3·830176	− 13·88751	− 62·82986	2·
2·6	− 1·2926953	− 3·271567	− 11·290256	− 48·83731	2·
2·7	− 1·1749076	− 2·821150	− 9·273796	− 38·39572	2·
2·8	− 1·0752421	− 2·454762	− 7·691764	− 30·50994	2·
2·9	− 0·9901112	− 2·154277	− 6·438430	− 24·48750	2·
3·0	− 0·9166828	− 1·905946	− 5·436470	− 19·83994	3·
3·1	− 0·8526997	− 1·6992271	− 4·628678	− 16·218237	3·
3·2	− 0·7963470	− 1·5259577	− 3·972271	− 13·370058	3·
3·3	− 0·7461539	− 1·3797571	− 3·434928	− 11·110890	3·
3·4	− 0·7009203	− 1·2555924	− 2·991999	− 9·304403	3·
3·5	− 0·6596606	− 1·1494603	− 2·624512	− 7·848866	3·
3·6	− 0·6215621	− 1·0581497	− 2·3177427	− 6·667659	3·
3·7	− 0·5859520	− 0·9790651	− 2·0601699	− 5·702567	3·
3·8	− 0·5522725	− 0·9100926	− 1·8427079	− 4·908985	3·
3·9	− 0·5200615	− 0·8494985	− 1·6581398	− 4·252470	3·
4·0	− 0·4889368	− 0·7958514	− 1·5006918	− 3·706224	4·
4·1	− 0·4585842	− 0·7479619	− 1·3657130	− 3·249247	4·
4·2	− 0·4287478	− 0·7048359	− 1·2494327	− 2·864972	4·
4·3	− 0·3992226	− 0·6656385	− 1·1487739	− 2·540242	4·
4·4	− 0·3698472	− 0·6296652	− 1·0612100	− 2·264544	4·
4·5	− 0·3404998	− 0·5963194	− 0·9846543	− 2·029425	4·
4·6	− 0·3110929	− 0·5650943	− 0·9173730	− 1·8280526	4·
4·7	− 0·2815701	− 0·5355591	− 0·8579174	− 1·6548682	4·
4·8	− 0·2519027	− 0·5073471	− 0·8050705	− 1·5053290	4·
4·9	− 0·2220872	− 0·4801469	− 0·7578045	− 1·3757009	4·
5·0	− 0·1921423	− 0·4536948	− 0·7152474	− 1·2628988	5·

Table IV. Values of $Y_n(x)$

x	$Y_8(x)$	$Y_9(x)$	$Y_{10}(x)$	x
0·1	− 41084285308[2]	− 6573192208278[3]	− 11831335132045[5]	0·1
0·2	− 1606575569[2]	− 12850308895[3]	− 11563671430[5]	0·2
0·3	− 62798159[2]	− 334788875[3]	− 200810527[5]	0·3
0·4	− 6302655[2]	− 25192597[3]	− 11330366[5]	0·4
0·5	− 1060819[2]	− 3390825[3]	− 1219636[5]	0·5
0·6	− 24768540·5	− 659430617	− 197581500[2]	0·6
0·7	− 7250160·1	− 165354373	− 42447194[2]	0·7
0·8	− 2504646·8	− 49949263	− 11213538[2]	0·8
0·9	− 982142·7	− 17396869	− 3469552[2]	0·9
1·0	− 425674·6	− 6780205	− 1216180[2]	1·0
1·1	− 200085·33	− 2894495·9	− 47164393	1·1
1·2	− 100577·97	− 1332343·2	− 19884570	1·2
1·3	− 53495·91	− 653392·5	− 8993478	1·3
1·4	− 29859·17	− 338225·9	− 4318759	1·4
1·5	− 17375·13	− 183447·3	− 2183993	1·5
1·6	− 10485·229	− 103635·01	− 1155408·6	1·6
1·7	− 6533·582	− 60684·92	− 636012·7	1·7
1·8	− 4188·852	− 36684·77	− 362658·9	1·8
1·9	− 2754·916	− 22816·93	− 213405·5	1·9
2·0	− 1853·922	− 14559·83	− 129184·5	2·0
2·1	− 1273·8144	− 9508·814	− 80230·30	2·1
2·2	− 891·9608	− 6342·471	− 51000·98	2·2
2·3	− 635·4947	− 4312·860	− 33117·32	2·3
2·4	− 460·0405	− 2985·112	− 21928·30	2·4
2·5	− 337·9597	− 2100·112	− 14782·85	2·5
2·6	− 251·67985	− 1499·9618	− 10132·671	2·6
2·7	− 189·81514	− 1086·4347	− 7053·083	2·7
2·8	− 144·85794	− 797·2497	− 4980·319	2·8
2·9	− 111·77710	− 592·2137	− 3564·032	2·9
3·0	− 87·14989	− 444·9595	− 2582·607	3·0
3·1	− 68·61497	− 337·92355	− 1893·5218	3·1
3·2	− 54·52173	− 259·23861	− 1403·6955	3·2
3·3	− 43·70218	− 200·77848	− 1051·4532	3·3
3·4	− 35·32025	− 156·90853	− 795·3720	3·4
3·5	− 28·77095	− 123·67548	− 607·2744	3·5
3·6	− 23·612043	− 98·27476	− 467·7617	3·6
3·7	− 19·517110	− 78·69575	− 363·3271	3·7
3·8	− 16·243027	− 63·48271	− 284·4645	3·8
3·9	− 13·607138	− 51·57168	− 224·4160	3·9
4·0	− 11·471092	− 42·17814	− 178·3306	4·0
4·1	− 9·729277	− 34·71866	− 142·69412	4·1
4·2	− 8·300474	− 28·75588	− 114·93902	4·2
4·3	− 7·121782	− 23·95941	− 93·17343	4·3
4·4	− 6·144157	− 20·07785	− 75·99248	4·4
4·5	− 5·329114	− 16·91853	− 62·34502	4·5
4·6	− 4·646265	− 14·332870	− 51·43888	4·6
4·7	− 4·071477	− 12·205480	− 42·67291	4·7
4·8	− 3·585472	− 10·446246	− 35·58795	4·8
4·9	− 3·172769	− 8·984362	− 29·83101	4·9
5·0	− 2·820869	− 7·763883	− 25·12911	5·0

The numbers in [ ] are the numbers of digits between the last digits given and the decimal oints. For example, the integral part of Y_{10} (0·1) is a number containing 19 digits of which the rst 14 are given.

Table IV. Values of $e^{-x}I_n(x)$

x	$e^{-x}I_2(x)$	$e^{-x}I_3(x)$	$e^{-x}I_4(x)$	$e^{-x}I_5(x)$	x
0·1	0·0011320	0·0000189	0·0000002	—	0·1
0·2	0·0041073	0·0001368	0·0000034	0·0000001	0·2
0·3	0·0083969	0·0004191	0·0000157	0·0000005	0·3
0·4	0·0135860	0·0009027	0·0000450	0·0000018	0·4
0·5	0·0193521	0·0016043	0·0001000	0·0000050	0·5
0·6	0·0254458	0·0025257	0·0001886	0·0000113	0·6
0·7	0·0316770	0·0036585	0·0003182	0·0000222	0·7
0·8	0·0379022	0·0049877	0·0004948	0·0000394	0·8
0·9	0·0440151	0·0064938	0·0007233	0·0000647	0·9
1·0	0·0499388	0·0081553	0·0010069	0·0000999	1·0
1·1	0·0556193	0·0099497	0·0013479	0·0001468	1·1
1·2	0·0610206	0·0118547	0·0017471	0·0002072	1·2
1·3	0·0661209	0·0138486	0·0022045	0·0002826	1·3
1·4	0·0709088	0·0159110	0·0027189	0·0003746	1·4
1·5	0·0753811	0·0180231	0·0032885	0·0004843	1·5
1·6	0·0795406	0·0201679	0·0039110	0·0006129	1·6
1·7	0·0833947	0·0223299	0·0045834	0·0007611	1·7
1·8	0·0869539	0·0244955	0·0053023	0·0009298	1·8
1·9	0·0902306	0·0266527	0·0060642	0·0011192	1·9
2·0	0·0932390	0·0287912	0·0068654	0·0013298	2·0
2·1	0·0959939	0·0309022	0·0077019	0·0015615	2·1
2·2	0·0985103	0·0329781	0·0085701	0·0018142	2·2
2·3	0·1008034	0·0350127	0·0094659	0·0020879	2·3
2·4	0·1028881	0·0370010	0·0103857	0·0023819	2·4
2·5	0·1047787	0·0389387	0·0113259	0·0026960	2·5
2·6	0·1064892	0·0408227	0·0122829	0·0030293	2·6
2·7	0·1080327	0·0426507	0·0132534	0·0033813	2·7
2·8	0·1094217	0·0444207	0·0142344	0·0037511	2·8
2·9	0·1106680	0·0461318	0·0152228	0·0041380	2·9
3·0	0·1117825	0·0477833	0·0162159	0·0045409	3·0
3·1	0·1127758	0·0493750	0·0172112	0·0049590	3·1
3·2	0·1136572	0·0509071	0·0182063	0·0053913	3·2
3·3	0·1144358	0·0523802	0·0191910	0·0058369	3·3
3·4	0·1151197	0·0537949	0·0201876	0·0062947	3·4
3·5	0·1157167	0·0551523	0·0211700	0·0067638	3·5
3·6	0·1162339	0·0564535	0·0221447	0·0072431	3·6
3·7	0·1166776	0·0576999	0·0231102	0·0077318	3·7
3·8	0·1170540	0·0588928	0·0240654	0·0082288	3·8
3·9	0·1173686	0·0600338	0·0250090	0·0087333	3·9
4·0	0·1176265	0·0611243	0·0259400	0·0092443	4·0
4·1	0·1178323	0·0621661	0·0268576	0·0097611	4·1
4·2	0·1179905	0·0631607	0·0277610	0·0102826	4·2
4·3	0·1181048	0·0641096	0·0286495	0·0108082	4·3
4·4	0·1181791	0·0650147	0·0295227	0·0113371	4·4
4·5	0·1182166	0·0658774	0·0303800	0·0118685	4·5
4·6	0·1182204	0·0666994	0·0312212	0·0124017	4·6
4·7	0·1181933	0·0674822	0·0320458	0·0129361	4·7
4·8	0·1181380	0·0682274	0·0328538	0·0134711	4·8
4·9	0·1180568	0·0689364	0·0336449	0·0140060	4·9
5·0	0·1179519	0·0696107	0·0344190	0·0145403	5·0

Table IV. Values of $K_n(x)$

x	$K_0(x)$	$K_1(x)$	$K_2(x)$	$K_3(x)$	x
0·1	2·4270690	9·8538448	199·5039646	7990·0124305	0·1
0·2	1·7527039	4·7759725	49·5124293	995·0245583	0·2
0·3	1·3724601	3·0559920	21·7457403	292·9991958	0·3
0·4	1·1145291	2·1843544	12·0363013	122·5473670	0·4
0·5	0·9244191	1·6564411	7·5501836	62·0579095	0·5
0·6	0·7775221	1·3028349	5·1203052	35·4382031	0·6
0·7	0·6605199	1·0502835	3·6613300	21·9721690	0·7
0·8	0·5653471	0·8617816	2·7198012	14·4607876	0·8
0·9	0·4867303	0·7165336	2·0790271	9·9566542	0·9
1·0	0·4210244	0·6019072	1·6248389	7·1012628	1·0
1·1	0·3656024	0·5097600	1·2924388	5·2095375	1·1
1·2	0·3185082	0·4345924	1·0428289	3·9106886	1·2
1·3	0·2782476	0·3725475	0·8513976	2·9922325	1·3
1·4	0·2436551	0·3208359	0·7019921	2·3265275	1·4
1·5	0·2138056	0·2773878	0·5836560	1·8338037	1·5
1·6	0·1879548	0·2406339	0·4887471	1·4625018	1·6
1·7	0·1654963	0·2093625	0·4118051	1·1783157	1·7
1·8	0·1459314	0·1826231	0·3488460	0·9578363	1·8
1·9	0·1288460	0·1596602	0·2969093	0·7847324	1·9
2·0	0·1138939	0·1398659	0·2537598	0·6473854	2·0
2·1	0·1007837	0·1227464	0·2176851	0·5373847	2·1
2·2	0·0892690	0·1078968	0·1873570	0·4485459	2·2
2·3	0·0791399	0·0949824	0·1617334	0·3762579	2·3
2·4	0·0702173	0·0837248	0·1399880	0·3170382	2·4
2·5	0·0623476	0·0738908	0·1214602	0·2682271	2·5
2·6	0·0553983	0·0652840	0·1056168	0·2277714	2·6
2·7	0·0492554	0·0577384	0·0920246	0·1940711	2·7
2·8	0·0438200	0·0511127	0·0803290	0·1658685	2·8
2·9	0·0390062	0·0452864	0·0702383	0·1421668	2·9
3·0	0·0347395	0·0401564	0·0615105	0·1221704	3·0
3·1	0·0309547	0·0356341	0·0539444	0·1052398	3·1
3·2	0·0275950	0·0316429	0·0473718	0·0908577	3·2
3·3	0·0246106	0·0281169	0·0416512	0·0786032	3·3
3·4	0·0219580	0·0249990	0·0366633	0·0681323	3·4
3·5	0·0195989	0·0222394	0·0323071	0·0591618	3·5
3·6	0·0174996	0·0197950	0·0284968	0·0514581	3·6
3·7	0·0156307	0·0176280	0·0251593	0·0448273	3·7
3·8	0·0139659	0·0157057	0·0222321	0·0391079	3·8
3·9	0·0124823	0·0139993	0·0196614	0·0341649	3·9
4·0	0·0111597	0·0124835	0·0174014	0·0298849	4·0
4·1	0·0099800	0·0111363	0·0154123	0·0261727	4·1
4·2	0·0089275	0·0099382	0·0136599	0·0229477	4·2
4·3	0·0079880	0·0088722	0·0121146	0·0201416	4·3
4·4	0·0071491	0·0079233	0·0107506	0·0176965	4·4
4·5	0·0063999	0·0070781	0·0095457	0·0155631	4·5
4·6	0·0057304	0·0063250	0·0084804	0·0136993	4·6
4·7	0·0051321	0·0056538	0·0075380	0·0120691	4·7
4·8	0·0045972	0·0050552	0·0067036	0·0106415	4·8
4·9	0·0041189	0·0045212	0·0059643	0·0093900	4·9
5·0	0·0036911	0·0040446	0·0053089	0·0082918	5·0

Table IV. Values of $K_n(x)$

x	$K_4(x)$	$K_5(x)$	$K_6(x)$	$K_7(x)$	x
0·1	479600·2498	38376010·00	3838080599·8	460608047990	0·1
0·2	29900·2492	1197004·99	59880149·8	3594005995	0·2
0·3	5881·7297	157139·12	5243852·5	209911239	0·3
0·4	1850·2468	37127·48	930037·3	27938248	0·4
0·5	752·2451	12097·98	242711·8	5837182	0·5
0·6	359·502336	4828·8027	80839·547	1621619·74	0·6
0·7	191·994207	2216·1917	31851·875	548248·34	0·7
0·8	111·175708	1126·2179	14188·899	213959·70	0·8
0·9	68·456722	618·4609	6940·244	93155·05	0·9
1·0	44·232416	360·9606	3653·838	44207·02	1·0
1·1	29·708098	221·26843	2041·2393	22489·333	1·1
1·2	20·596272	141·21917	1197·4227	12115·446	1·2
1·3	14·661702	93·21809	731·7239	6847·593	1·3
1·4	10·672824	63·31409	462·9164	4031·169	1·4
1·5	7·918871	44·06778	301·7041	2457·700	1·5
1·6	5·973129	31·328146	201·77404	1544·6334	1·6
1·7	4·570567	22·686864	138·02271	996·9648	1·7
1·8	3·541634	16·698431	96·31069	658·7697	1·8
1·9	2·775011	12·468991	68·40128	444·4771	1·9
2·0	2·195916	9·431049	49·35116	305·5380	2·0
2·1	1·7530699	7·215746	36·113765	213·58012	2·1
2·2	1·4106641	5·578234	26·766271	151·57608	2·2
2·3	1·1432756	4·352869	20·068791	109·05961	2·3
2·4	0·9325836	3·425650	15·206127	79·45628	2·4
2·5	0·7652054	2·716884	11·632743	58·55405	2·5
2·6	0·6312432	2·1700581	8·977621	43·60523	2·6
2·7	0·5232937	1·7445711	6·984668	32·78754	2·7
2·8	0·4357615	1·4109012	5·474694	24·87388	2·8
2·9	0·3643764	1·1473430	4·320732	19·02623	2·9
3·0	0·3058512	0·9377736	3·431763	14·66483	3·0
3·1	0·2576343	0·7701024	2·7418356	11·383660	3·1
3·2	0·2177299	0·6351824	2·2026750	8·895214	3·2
3·3	0·1845662	0·5260364	1·7786158	6·993730	3·3
3·4	0·1568967	0·4373011	1·4430764	5·530512	3·4
3·5	0·1337274	0·3648244	1·1760828	4·397108	3·5
3·6	0·1142604	0·3053701	0·9625106	3·513739	3·6
3·7	0·0978523	0·2563998	0·7908246	2·821236	3·7
3·8	0·0839814	0·2159108	0·6521676	2·275387	3·8
3·9	0·0722228	0·1823141	0·5396949	1·842914	3·9
4·0	0·0622288	0·1543425	0·4480852	1·498598	4·0
4·1	0·0537139	0·1309802	0·3731778	1·2232080	4·1
4·2	0·0464423	0·1114092	0·3117023	1·0019872	4·2
4·3	0·0402191	0·0949678	0·2610745	0·8235478	4·3
4·4	0·0348822	0·0811187	0·2192429	0·6790539	4·4
4·5	0·0302965	0·0694236	0·1845713	0·5616138	4·5
4·6	0·0263491	0·0595239	0·1557490	0·4658258	4·6
4·7	0·0229453	0·0511250	0·1317219	0·3874361	4·7
4·8	0·0200054	0·0439839	0·1116385	0·3230800	4·8
4·9	0·0174623	0·0378998	0·0948088	0·2700847	4·9
5·0	0·0152591	0·0327063	0·0806716	0·2263181	5·0

Table IV. Values of $K_n(x)$

x	$K_8(x)$	$K_9(x)$	$K_{10}(x)$	x
0·1	644889647992[2]	10318694975920[3]	18574295846304[5]	0·1
0·2	2516402998[2]	20134817990[3]	18123852594[5]	0·2
0·3	98011017[2]	522935335[3]	313859212[5]	0·3
0·4	9787687[2]	39178686[3]	17640197[5]	0·4
0·5	1636838[2]	5243719[3]	1889376[5]	0·5
0·6	37918633·59	1012785182	304214741[2]	0·6
0·7	10996818·69	251904104	64885309[2]	0·7
0·8	3758483·72	75383634	16998902[2]	0·8
0·9	1456018·75	25977933	5210147[2]	0·9
1·0	622552·12	10005041	1807133[2]	1·0
1·1	288269·12	4215494·70	69269092	1·1
1·2	142544·29	1912706·01	28833134	1·2
1·3	74475·03	923463·36	12860891	1·3
1·4	40774·60	470026·62	6083974	1·4
1·5	23240·24	250353·61	3027484	1·5
1·6	13717·316	138717·80	1574292·56	1·6
1·7	8348·321	79569·40	850847·84	1·7
1·8	5220·075	47059·44	475814·46	1·8
1·9	3343·496	28600·23	274293·04	1·9
2·0	2188·117	17810·48	162482·40	2·0
2·1	1459·9812	11337·247	98636·38	2·1
2·2	991·3413	7361·331	61220·41	2·2
2·3	683·9099	4866·694	38771·08	2·3
2·4	478·7011	3270·797	25009·68	2·4
2·5	339·5354	2231·581	16406·92	2·5
2·6	243·77501	1543·7592	10931·338	2·6
2·7	176·99414	1081·6417	7387·939	2·7
2·8	129·84408	766·8400	5059·530	2·8
2·9	96·17151	549·6277	3507·654	2·9
3·0	71·86762	397·9588	2459·620	3·0
3·1	54·15191	290·87739	1743·1174	3·1
3·2	41·11923	214·49139	1247·6333	3·2
3·3	31·44899	159·47366	901·3053	3·3
3·4	24·21577	119·48709	656·7945	3·4
3·5	18·76452	90·17775	482·5358	3·5
3·6	14·627050	68·52285	357·2413	3·6
3·7	11·465773	52·40296	266·3991	3·7
3·8	9·035174	40·31822	200·0162	3·8
3·9	7·155283	31·19792	151·1457	3·9
4·0	5·693179	24·27131	114·9141	4·0
4·1	4·549986	18·979250	87·87352	4·1
4·2	3·651659	14·913071	67·56482	4·2
4·3	2·942393	11·771986	52·22047	4·3
4·4	2·379869	9·333122	40·56082	4·4
4·5	1·931814	7·430286	31·65296	4·5
4·6	1·5734796	5·938798	24·812255	4·6
4·7	1·2857868	4·764583	19·533125	4·7
4·8	1·0539552	3·836264	15·439946	4·8
4·9	0·8664794	3·099405	12·252049	4·9
5·0	0·7143624	2·512278	9·758563	5·0

The numbers in [ ] are the numbers of digits between the last digits given and the decimal points. For example, the integral part of K_{10} (0·1) is a number containing 19 digits of which the first 14 are given.

Table V. Values of $J_{\pm(n+\frac{1}{2})}(x)$

x	$J_{\frac{1}{2}}(x)$	$J_{-\frac{1}{2}}(x)$	$J_{\frac{3}{2}}(x)$	$J_{-\frac{3}{2}}(x)$	$J_{\frac{5}{2}}(x)$	$J_{-\frac{5}{2}}(x)$	x
1	+ 0·671397	+ 0·431099	+ 0·240298	− 1·102496	+ 0·049497	+ 2·876388	1
2	+ 0·513016	− 0·234786	+ 0·491294	− 0·395623	+ 0·223925	+ 0·828221	2
3	+ 0·065008	− 0·456049	+ 0·477718	+ 0·087008	+ 0·412710	+ 0·369041	3
4	− 0·301921	− 0·260766	+ 0·185286	+ 0·367112	+ 0·440885	− 0·014568	4
5	− 0·342168	+ 0·101218	− 0·169651	+ 0·321925	+ 0·240377	− 0·294372	5
6	− 0·091016	+ 0·312761	− 0·327930	+ 0·038889	− 0·072950	− 0·332205	6
7	+ 0·198129	+ 0·227356	− 0·199052	− 0·230608	− 0·283437	− 0·128524	7
8	+ 0·279093	− 0·041045	+ 0·075931	− 0·273962	− 0·250619	+ 0·143781	8
9	+ 0·109608	− 0·242326	+ 0·254504	− 0·082683	− 0·024773	+ 0·269886	9
10	− 0·137264	− 0·211709	+ 0·197983	+ 0·158435	+ 0·196659	+ 0·164179	10
11	− 0·240569	+ 0·001064	− 0·022934	+ 0·240472	+ 0·234314	− 0·066647	11
12	− 0·123589	+ 0·194364	− 0·204663	+ 0·107392	+ 0·072423	− 0·221212	12
13	+ 0·092980	+ 0·200812	− 0·193660	− 0·108427	− 0·137671	− 0·175790	13
14	+ 0·211241	+ 0·029158	− 0·014070	− 0·213323	− 0·214256	+ 0·016554	14
15	+ 0·133968	− 0·156506	+ 0·165437	− 0·123534	− 0·100880	+ 0·181212	15
16	− 0·057428	− 0·191025	+ 0·187436	+ 0·069367	+ 0·092573	+ 0·178019	16
17	− 0·186045	− 0·053248	+ 0·042305	+ 0·189178	+ 0·193511	+ 0·019864	17
18	− 0·141233	+ 0·124181	− 0·132027	+ 0·134334	+ 0·119229	− 0·146570	18
19	+ 0·027435	+ 0·180980	− 0·179536	− 0·036960	− 0·055782	− 0·175144	19
20	+ 0·162881	+ 0·072807	− 0·064663	− 0·166521	− 0·172580	− 0·047829	20
21	+ 0·145672	− 0·095367	+ 0·102303	− 0·141131	− 0·131058	+ 0·115528	21
22	− 0·001506	− 0·170103	+ 0·170034	+ 0·009238	+ 0·024692	+ 0·168843	22
23	− 0·140786	− 0·088648	+ 0·082527	+ 0·144640	+ 0·151550	+ 0·069782	23
24	− 0·147489	+ 0·069085	− 0·075230	+ 0·144611	+ 0·138086	− 0·087161	24
25	− 0·021120	+ 0·158173	− 0·159018	+ 0·014793	+ 0·002038	− 0·159948	25
26	+ 0·119324	+ 0·101229	− 0·096639	− 0·123217	− 0·130474	− 0·087011	26
27	+ 0·146854	− 0·044859	+ 0·050298	− 0·145193	− 0·141266	+ 0·060991	27
28	+ 0·040849	− 0·145147	+ 0·146606	− 0·035665	− 0·025141	+ 0·148969	28
29	− 0·098326	− 0·110835	+ 0·107444	+ 0·102148	+ 0·109441	+ 0·100268	29
30	− 0·143930	+ 0·022470	− 0·027268	+ 0·143181	+ 0·141203	− 0·036788	30
31	− 0·057900	+ 0·131087	− 0·132954	+ 0·053672	+ 0·045034	− 0·136281	31
32	+ 0·077777	+ 0·117665	− 0·115235	− 0·081454	− 0·088581	− 0·110029	32
33	+ 0·138882	− 0·001844	+ 0·006053	− 0·138826	− 0·138331	+ 0·014465	33
34	+ 0·072398	− 0·116115	+ 0·118244	− 0·068982	− 0·061964	+ 0·122202	34
35	− 0·057748	− 0·121878	+ 0·120228	+ 0·061230	+ 0·068053	+ 0·116630	35
36	− 0·131887	− 0·017017	+ 0·013353	+ 0·132360	+ 0·133000	+ 0·005987	36
37	− 0·084414	+ 0·100400	− 0·102682	+ 0·081700	+ 0·076088	− 0·107025	37
38	+ 0·038360	+ 0·123619	− 0·122610	− 0·041613	− 0·048040	− 0·120334	38
39	+ 0·123138	+ 0·034067	− 0·030910	− 0·124012	− 0·125516	− 0·024528	39
40	+ 0·094001	− 0·084139	+ 0·086489	− 0·091898	− 0·087514	+ 0·091031	40
41	− 0·019766	− 0·123031	+ 0·122549	+ 0·022766	+ 0·028733	+ 0·121365	41
42	− 0·112839	− 0·049245	+ 0·046558	+ 0·114011	+ 0·116164	+ 0·041101	42
43	− 0·101207	+ 0·067544	− 0·069898	+ 0·099636	+ 0·096331	− 0·074495	43
44	+ 0·002129	+ 0·120267	− 0·120218	− 0·004863	− 0·010326	− 0·119958	44
45	+ 0·101208	+ 0·062483	− 0·060234	− 0·102596	− 0·105223	− 0·055643	45
46	+ 0·106088	− 0·050842	+ 0·053148	− 0·104983	− 0·102622	+ 0·057689	46
47	+ 0·014382	− 0·115491	+ 0·115797	− 0·011925	− 0·006991	+ 0·116253	47
48	− 0·088476	− 0·073722	+ 0·071879	+ 0·090012	+ 0·092968	+ 0·068096	48
49	− 0·108712	+ 0·034263	− 0·036481	+ 0·108013	+ 0·106479	− 0·040876	49
50	− 0·029606	+ 0·108885	− 0·109477	+ 0·027428	+ 0·023037	− 0·110530	50

Table V. Values of $J_{\pm(n+\frac{1}{2})}(x)$

x	$J_{\frac{7}{2}}(x)$	$J_{-\frac{7}{2}}(x)$	$J_{\frac{9}{2}}(x)$	$J_{-\frac{9}{2}}(x)$	x
1	+ 0·007186	− 13·279444	+ 0·000807	+ 90·079718	1
2	+ 0·068518	− 1·674928	+ 0·015887	+ 5·034028	2
3	+ 0·210132	− 0·702076	+ 0·077598	+ 1·269137	3
4	+ 0·365820	− 0·348902	+ 0·199300	+ 0·625147	4
5	+ 0·410029	− 0·027552	+ 0·333663	+ 0·332945	5
6	+ 0·267139	+ 0·237949	+ 0·384612	+ 0·054598	6
7	− 0·003403	+ 0·322411	+ 0·280034	− 0·193887	7
8	− 0·232568	+ 0·184099	+ 0·047122	− 0·304868	8
9	− 0·268267	− 0·067254	− 0·183879	− 0·217577	9
10	− 0·099653	− 0·240524	− 0·266416	+ 0·004188	10
11	+ 0·129440	− 0·210178	− 0·151943	+ 0·200397	11
12	+ 0·234840	− 0·015220	+ 0·064567	+ 0·230091	12
13	+ 0·140709	+ 0·176039	+ 0·213437	+ 0·081000	13
14	− 0·062450	+ 0·207411	+ 0·183031	− 0·120260	14
15	− 0·199063	+ 0·063130	+ 0·007984	− 0·210673	15
16	− 0·158507	− 0·124998	− 0·161920	− 0·123323	16
17	+ 0·014610	− 0·195020	− 0·187495	+ 0·060438	17
18	+ 0·165146	− 0·093620	− 0·055005	+ 0·182978	18
19	+ 0·164856	+ 0·083050	+ 0·116519	+ 0·144546	19
20	+ 0·021518	+ 0·178478	+ 0·180111	− 0·014639	20
21	− 0·133507	+ 0·113625	+ 0·086555	− 0·153403	21
22	− 0·164423	− 0·047611	− 0·077008	− 0·153694	22
23	− 0·049581	− 0·159810	− 0·166640	− 0·021144	23
24	+ 0·103998	− 0·126452	− 0·107753	+ 0·124043	24
25	+ 0·159426	+ 0·017196	+ 0·042601	+ 0·155133	25
26	+ 0·071548	+ 0·139950	+ 0·149737	+ 0·049333	26
27	− 0·076458	+ 0·133898	+ 0·121443	− 0·095706	27
28	− 0·151096	+ 0·009064	− 0·012633	− 0·151235	28
29	− 0·088575	− 0·119436	− 0·130821	− 0·071438	29
30	+ 0·050802	− 0·137049	− 0·129349	+ 0·068766	30
31	+ 0·140218	− 0·031691	− 0·013372	+ 0·143437	31
32	+ 0·101394	+ 0·098646	+ 0·110760	+ 0·088450	32
33	− 0·027012	+ 0·136634	+ 0·132602	− 0·043448	33
34	− 0·127357	+ 0·051012	+ 0·035744	− 0·132704	34
35	− 0·110507	− 0·077891	− 0·090154	− 0·101062	35
36	+ 0·005119	− 0·133192	− 0·132005	+ 0·019912	36
37	+ 0·112964	− 0·067238	− 0·054717	+ 0·119745	37
38	+ 0·116289	+ 0·057447	+ 0·069461	+ 0·109751	38
39	+ 0·014818	+ 0·127156	+ 0·128176	+ 0·001705	39
40	− 0·097428	+ 0·080519	+ 0·070464	− 0·105122	40
41	− 0·119045	− 0·037567	− 0·049057	− 0·114951	41
42	− 0·032729	− 0·118904	− 0·121619	− 0·021284	42
43	+ 0·081099	− 0·090974	− 0·083128	+ 0·089305	43
44	+ 0·119045	+ 0·018494	+ 0·029265	+ 0·117016	44
45	+ 0·048542	+ 0·108779	+ 0·112774	+ 0·038744	45
46	− 0·064303	+ 0·098712	+ 0·092836	− 0·072710	46
47	− 0·116541	− 0·000443	− 0·010367	− 0·116187	47
48	− 0·062195	− 0·082918	− 0·102038	− 0·056004	48
49	+ 0·047346	− 0·103842	− 0·099715	+ 0·055710	49
50	+ 0·111781	− 0·016375	− 0·007388	+ 0·112833	50

Table V. Values of $J_{\pm(n+\frac{1}{2})}(x)$

x	$J_{\frac{11}{2}}(x)$	$J_{-\frac{11}{2}}(x)$	$J_{\frac{13}{2}}(x)$	$J_{-\frac{13}{2}}(x)$	x
1	+ 0·000074	− 797·438019	+ 0·000006	+ 8681·738496	1
2	+ 0·002973	− 20·978200	+ 0·000467	+ 110·346069	2
3	+ 0·022661	− 3·105334	+ 0·005493	+ 10·117087	3
4	+ 0·082606	− 1·057678	+ 0·027866	+ 2·283448	4
5	+ 0·190564	− 0·571750	+ 0·085579	+ 0·924903	5
6	+ 0·309779	− 0·319846	+ 0·183316	+ 0·531787	6
7	+ 0·363446	− 0·073127	+ 0·291096	+ 0·308802	7
8	+ 0·285580	+ 0·158877	+ 0·345551	+ 0·086412	8
9	+ 0·084388	+ 0·284832	+ 0·287020	− 0·130550	9
10	− 0·140121	+ 0·236755	+ 0·112283	− 0·264618	10
11	− 0·253757	+ 0·046217	− 0·101814	− 0·246614	11
12	− 0·186414	− 0·157348	− 0·235447	− 0·085855	12
13	+ 0·007055	− 0·232116	− 0·207468	+ 0·115406	13
14	+ 0·180113	− 0·130102	− 0·041513	+ 0·222482	14
15	+ 0·203854	+ 0·063274	+ 0·141509	+ 0·164272	15
16	+ 0·067428	+ 0·194373	+ 0·208276	− 0·010308	16
17	− 0·113872	+ 0·163023	+ 0·113813	− 0·165924	17
18	− 0·192649	+ 0·002131	− 0·062725	− 0·184280	18
19	− 0·109663	− 0·151520	− 0·180008	− 0·056824	19
20	+ 0·059532	− 0·171891	− 0·147369	+ 0·109179	20
21	+ 0·170603	− 0·047880	+ 0·002808	+ 0·178483	21
22	+ 0·132919	+ 0·110486	+ 0·143468	+ 0·098451	22
23	− 0·015626	+ 0·168084	+ 0·159167	− 0·059244	23
24	− 0·144405	+ 0·079936	+ 0·041567	− 0·160681	24
25	− 0·144089	− 0·073044	− 0·106000	− 0·122994	25
26	− 0·019716	− 0·157027	− 0·158079	+ 0·017102	26
27	+ 0·116939	− 0·101996	− 0·073801	+ 0·137260	27
28	+ 0·147035	+ 0·039548	+ 0·070397	+ 0·135698	28
29	+ 0·047975	+ 0·141606	+ 0·149019	+ 0·017726	29
30	− 0·089606	+ 0·116419	+ 0·096493	− 0·111453	30
31	− 0·144100	− 0·009951	− 0·037760	− 0·139905	31
32	− 0·070242	− 0·123523	− 0·134906	− 0·045989	32
33	+ 0·063176	− 0·124785	− 0·111543	+ 0·085043	33
34	+ 0·136818	− 0·015884	+ 0·008521	+ 0·137843	34
35	+ 0·087324	+ 0·103879	+ 0·117599	+ 0·068414	35
36	− 0·038120	+ 0·128214	+ 0·120357	− 0·059088	36
37	− 0·126274	+ 0·038110	+ 0·017176	− 0·131075	37
38	− 0·099837	− 0·083440	− 0·098362	− 0·085598	38
39	+ 0·014761	− 0·127550	− 0·124012	+ 0·034271	39
40	+ 0·113283	− 0·056866	− 0·039312	+ 0·120760	40
41	+ 0·108276	+ 0·062800	+ 0·078107	+ 0·098102	41
42	+ 0·006668	+ 0·120424	+ 0·123365	− 0·010256	42
43	− 0·098498	+ 0·072282	+ 0·057931	− 0·107796	43
44	− 0·113059	− 0·042429	− 0·057530	− 0·106408	44
45	− 0·025987	− 0·116528	− 0·119127	− 0·010259	45
46	+ 0·082467	− 0·084486	− 0·073116	+ 0·092913	46
47	+ 0·114556	+ 0·022691	+ 0·037178	+ 0·110876	47
48	+ 0·043062	+ 0·093419	+ 0·111907	+ 0·034596	48
49	− 0·065662	+ 0·093609	+ 0·084974	− 0·076725	49
50	− 0·113110	− 0·003933	− 0·017496	− 0·111958	50

Table V. Values of $J_{n+\frac{1}{2}}(x)$

x	$J_{\frac{15}{2}}(x)$	$J_{\frac{17}{2}}(x)$	$J_{\frac{19}{2}}(x)$	$J_{\frac{21}{2}}(x)$	$J_{\frac{23}{2}}(x)$	$J_{\frac{25}{2}}(x)$	x
2	+ 0·000063	+ 0·000008	+ 0·000001	—	—	—	2
3	+ 0·001140	+ 0·000207	+ 0·000034	+ 0·000005	+ 0·000001	—	3
4	+ 0·007957	+ 0·001974	+ 0·000434	+ 0·000086	+ 0·000015	+ 0·000002	4
5	+ 0·031941	+ 0·010243	+ 0·002887	+ 0·000727	+ 0·000165	+ 0·000034	5
6	+ 0·087406	+ 0·035199	+ 0·012324	+ 0·003827	+ 0·001069	+ 0·000272	6
7	+ 0·177161	+ 0·088535	+ 0·037852	+ 0·014205	+ 0·004763	+ 0·001446	7
8	+ 0·275940	+ 0·171837	+ 0·089213	+ 0·040045	+ 0·015904	+ 0·005680	8
9	+ 0·330196	+ 0·263308	+ 0·167162	+ 0·089590	+ 0·041882	+ 0·017442	9
10	+ 0·286089	+ 0·316850	+ 0·252556	+ 0·163007	+ 0·089759	+ 0·043438	10
11	+ 0·133432	+ 0·283766	+ 0·305116	+ 0·243253	+ 0·159277	+ 0·089780	11
12	− 0·068653	+ 0·149630	+ 0·280630	+ 0·294700	+ 0·235095	+ 0·155899	12
13	− 0·214523	− 0·040059	+ 0·162139	+ 0·277030	+ 0·285372	+ 0·227859	13
14	− 0·218661	− 0·192766	− 0·015412	+ 0·171849	+ 0·273186	+ 0·276957	14
15	− 0·081213	− 0·222722	− 0·171205	+ 0·005862	+ 0·179412	+ 0·269236	15
16	+ 0·101797	− 0·112842	− 0·221691	− 0·150416	+ 0·024269	+ 0·185304	16
17	+ 0·200906	+ 0·063457	− 0·137449	− 0·217076	− 0·130704	+ 0·040241	17
18	+ 0·147348	+ 0·185515	+ 0·027861	− 0·156106	− 0·209985	− 0·112207	18
19	− 0·013500	+ 0·169350	+ 0·165024	− 0·004326	− 0·169805	− 0·201228	19
20	− 0·155322	+ 0·030877	+ 0·181568	+ 0·141612	− 0·032875	− 0·179418	20

x	$J_{\frac{27}{2}}(x)$	$J_{\frac{29}{2}}(x)$	$J_{\frac{31}{2}}(x)$	$J_{\frac{33}{2}}(x)$	$J_{\frac{35}{2}}(x)$	$J_{\frac{37}{2}}(x)$	x
5	+ 0·000007	+ 0·000001	—	—	—	—	5
6	+ 0·000063	+ 0·000014	+ 0·000003	+ 0·000001	—	—	6
7	+ 0·000402	+ 0·000103	+ 0·000024	+ 0·000005	+ 0·000001	—	7
8	+ 0·001846	+ 0·000551	+ 0·000152	+ 0·000039	+ 0·000009	+ 0·000002	8
9	+ 0·006568	+ 0·002261	+ 0·000718	+ 0·000212	+ 0·000058	+ 0·000015	9
10	+ 0·018837	+ 0·007421	+ 0·002683	+ 0·000898	+ 0·000280	+ 0·000082	10
11	+ 0·044768	+ 0·020106	+ 0·008237	+ 0·003108	+ 0·001086	+ 0·000355	11
12	+ 0·089695	+ 0·045914	+ 0·021263	+ 0·009017	+ 0·003532	+ 0·001288	12
13	+ 0·152818	+ 0·089532	+ 0·046907	+ 0·022324	+ 0·009760	+ 0·003955	13
14	+ 0·221379	+ 0·149989	+ 0·089312	+ 0·047774	+ 0·023297	+ 0·010469	14
15	+ 0·269315	+ 0·215531	+ 0·147378	+ 0·089050	+ 0·048533	+ 0·024193	15
16	+ 0·265267	+ 0·262335	+ 0·210215	+ 0·144957	+ 0·088758	+ 0·049201	16
17	+ 0·189882	+ 0·261336	+ 0·255927	+ 0·205354	+ 0·142701	+ 0·088443	17
18	+ 0·054141	+ 0·193419	+ 0·257478	+ 0·250016	+ 0·200884	+ 0·140592	18
19	− 0·094968	+ 0·066273	+ 0·196122	+ 0·253715	+ 0·244541	+ 0·196755	19
20	− 0·191398	− 0·078969	+ 0·076893	+ 0·198153	+ 0·250059	+ 0·239451	20

Table V. Fresnel's integrals

x	$\frac{1}{2}\int_0^x J_{-\frac{1}{2}}(t)\,dt$	$\frac{1}{2}\int_0^x J_{\frac{1}{2}}(t)\,dt$
0·5	+ 0·550247	+ 0·092366
1·0	+ 0·721706	+ 0·247558
1·5	+ 0·779084	+ 0·415348
2·0	+ 0·753302	+ 0·562849
2·5	+ 0·670986	+ 0·665787
3·0	+ 0·561020	+ 0·711685
3·5	+ 0·452047	+ 0·700180
4·0	+ 0·368193	+ 0·642119
4·5	+ 0·325249	+ 0·556489
5·0	+ 0·328457	+ 0·465942
5·5	+ 0·372439	+ 0·391834
6·0	+ 0·443274	+ 0·349852
6·5	+ 0·522202	+ 0·347099
7·0	+ 0·590116	+ 0·381195
7·5	+ 0·631845	+ 0·441485
8·0	+ 0·639301	+ 0·512010
8·5	+ 0·612868	+ 0·575457
9·0	+ 0·560804	+ 0·617214
9·5	+ 0·496895	+ 0·628573
10·0	+ 0·436964	+ 0·608436
10·5	+ 0·395087	+ 0·563176
11·0	+ 0·380390	+ 0·504784
11·5	+ 0·395149	+ 0·447809
12·0	+ 0·434557	+ 0·405810
12·5	+ 0·488146	+ 0·388217
13·0	+ 0·542511	+ 0·398268
13·5	+ 0·584583	+ 0·432489
14·0	+ 0·604721	+ 0·481770
14·5	+ 0·598871	+ 0·533736
15·0	+ 0·569335	+ 0·575803
15·5	+ 0·524009	+ 0·598183
16·0	+ 0·474310	+ 0·596126
16·5	+ 0·432343	+ 0·570890
17·0	+ 0·407985	+ 0·529259
17·5	+ 0·406589	+ 0·481750
18·0	+ 0·427837	+ 0·439989
18·5	+ 0·465971	+ 0·413893
19·0	+ 0·511332	+ 0·409336
19·5	+ 0·552774	+ 0·426853
20·0	+ 0·580389	+ 0·461646
20·5	+ 0·587849	+ 0·504875
21·0	+ 0·573842	+ 0·545885
21·5	+ 0·542266	+ 0·574811
22·0	+ 0·501167	+ 0·584939
22·5	+ 0·460707	+ 0·574246
23·0	+ 0·430662	+ 0·545782
23·5	+ 0·418080	+ 0·506824
24·0	+ 0·425635	+ 0·467029
24·5	+ 0·451078	+ 0·436051
25·0	+ 0·487880	+ 0·421217
25·5	+ 0·526896	+ 0·425797
26·0	+ 0·558628	+ 0·448300
26·5	+ 0·575524	+ 0·482927
27·0	+ 0·573766	+ 0·521054
27·5	+ 0·554127	+ 0·553369
28·0	+ 0·521695	+ 0·572142
28·5	+ 0·484566	+ 0·573060
29·0	+ 0·451832	+ 0·556212
29·5	+ 0·431358	+ 0·525995
30·0	+ 0·427908	+ 0·489969
30·5	+ 0·442034	+ 0·456974
31·0	+ 0·470019	+ 0·434973
31·5	+ 0·504844	+ 0·429129
32·0	+ 0·537944	+ 0·440605
32·5	+ 0·561307	+ 0·466343
33·0	+ 0·569407	+ 0·499873
33·5	+ 0·560508	+ 0·532930
34·0	+ 0·537026	+ 0·557490
34·5	+ 0·504881	+ 0·567709
35·0	+ 0·472012	+ 0·561313
35·5	+ 0·446415	+ 0·540094
36·0	+ 0·434212	+ 0·509417
36·5	+ 0·438182	+ 0·476871
37·0	+ 0·457140	+ 0·450396
37·5	+ 0·486272	+ 0·436345
38·0	+ 0·518359	+ 0·437971
38·5	+ 0·545560	+ 0·454670
39·0	+ 0·561321	+ 0·482187
39·5	+ 0·561957	+ 0·513690
40·0	+ 0·547503	+ 0·541464
40·5	+ 0·521665	+ 0·558799
41·0	+ 0·490870	+ 0·561608
41·5	+ 0·462670	+ 0·549384
42·0	+ 0·443897	+ 0·525282
42·5	+ 0·439006	+ 0·495309
43·0	+ 0·449025	+ 0·466829
43·5	+ 0·471341	+ 0·446755
44·0	+ 0·500382	+ 0·439878
44·5	+ 0·529002	+ 0·447720
45·0	+ 0·550239	+ 0·468209
45·5	+ 0·559004	+ 0·496215
46·0	+ 0·553301	+ 0·524837
46·5	+ 0·534676	+ 0·547099
47·0	+ 0·507802	+ 0·557650
47·5	+ 0·479313	+ 0·554044
48·0	+ 0·456160	+ 0·537309
48·5	+ 0·443930	+ 0·511657
49·0	+ 0·445486	+ 0·483428
49·5	+ 0·460311	+ 0·459523
50·0	+ 0·484658	+ 0·445722

Table V. Fresnel's integrals

x	$\frac{1}{2}\int_0^x J_{-\frac{1}{2}}(t)\,dt$	$\frac{1}{2}\int_0^x J_{\frac{1}{2}}(t)\,dt$
0·02	+ 0·1128334	+ 0·0007522
0·04	+ 0·1595514	+ 0·0021274
0·06	+ 0·1953707	+ 0·0039078
0·08	+ 0·2255314	+ 0·0060153
0·10	+ 0·2520611	+ 0·0084044
0·12	+ 0·2759976	+ 0·0110444
0·14	+ 0·2979565	+ 0·0139124
0·16	+ 0·3183378	+ 0·0169904
0·18	+ 0·3374186	+ 0·0202639
0·20	+ 0·3554002	+ 0·0237204
0·22	+ 0·3724338	+ 0·0273496
0·24	+ 0·3886365	+ 0·0311421
0·26	+ 0·4041012	+ 0·0350898
0·28	+ 0·4189028	+ 0·0391853
0·30	+ 0·4331026	+ 0·0434218
0·32	+ 0·4467517	+ 0·0477932
0·34	+ 0·4598932	+ 0·0522937
0·36	+ 0·4725635	+ 0·0569181
0·38	+ 0·4847941	+ 0·0616612
0·40	+ 0·4966121	+ 0·0665185
0·42	+ 0·5080410	+ 0·0714853
0·44	+ 0·5191018	+ 0·0765575
0·46	+ 0·5298125	+ 0·0817309
0·48	+ 0·5401895	+ 0·0870016
0·50	+ 0·5502472	+ 0·0923658
0·52	+ 0·5599985	+ 0·0978198
0·54	+ 0·5694551	+ 0·1033602
0·56	+ 0·5786275	+ 0·1089835
0·58	+ 0·5875253	+ 0·1146863
0·60	+ 0·5961571	+ 0·1204654
0·62	+ 0·6045308	+ 0·1263176
0·64	+ 0·6126537	+ 0·1322398
0·66	+ 0·6205324	+ 0·1382290
0·68	+ 0·6281731	+ 0·1442820
0·70	+ 0·6355815	+ 0·1503961
0·72	+ 0·6427627	+ 0·1565683
0·74	+ 0·6497217	+ 0·1627958
0·76	+ 0·6564631	+ 0·1690757
0·78	+ 0·6629910	+ 0·1754054
0·80	+ 0·6693095	+ 0·1817820
0·82	+ 0·6754224	+ 0·1882030
0·84	+ 0·6813330	+ 0·1946656
0·86	+ 0·6870448	+ 0·2011673
0·88	+ 0·6925609	+ 0·2077055
0·90	+ 0·6978843	+ 0·2142775
0·92	+ 0·7030179	+ 0·2208809
0·94	+ 0·7079643	+ 0·2275131
0·96	+ 0·7127261	+ 0·2341717
0·98	+ 0·7173059	+ 0·2408543
1·00	+ 0·7217059	+ 0·2475583

Maxima and minima of Fresnel's integrals

$x = (n - \frac{1}{2})\pi$	$\frac{1}{2}\int_0^x J_{-\frac{1}{2}}(t)\,dt$
1·570796	+ 0·779893
4·712389	+ 0·321056
7·853982	+ 0·640807
10·995574	+ 0·380389
14·137167	+ 0·605721
17·278760	+ 0·404260
20·420352	+ 0·588128
23·561945	+ 0·417922
26·703538	+ 0·577121
29·845130	+ 0·427036
32·986723	+ 0·569413
36·128316	+ 0·433666
39·269908	+ 0·563631
42·411501	+ 0·438767
45·553093	+ 0·559088
48·694686	+ 0·442848

$x = n\pi$	$\frac{1}{2}\int_0^x J_{\frac{1}{2}}(t)\,dt$
3·141593	+ 0·713972
6·283185	+ 0·343415
9·424778	+ 0·628940
12·566371	+ 0·387969
15·707963	+ 0·600361
18·849556	+ 0·408301
21·991149	+ 0·584942
25·132741	+ 0·420516
28·274334	+ 0·574957
31·415927	+ 0·428877
34·557519	+ 0·567822
37·699112	+ 0·435059
40·840704	+ 0·562398
43·982297	+ 0·439868
47·123890	+ 0·558096
50·265482	+ 0·443747

Table VI. Functions of equal order and argument

n	$J_n(n)$	$n^{\frac{1}{3}} J_n(n)$	$J_n'(n)$	$n^{\frac{2}{3}} J_n'(n)$	n
1	0·4400506	0·4400506	0·3251471	0·3251471	1
2	0·3528340	0·4445430	0·2238908	0·3554045	2
3	0·3090627	0·4457456	0·1770285	0·3682342	3
4	0·2811291	0·4462646	0·1490424	0·3755633	4
5	0·2611405	0·4465441	0·1300918	0·3803908	5
6	0·2458369	0·4467152	0·1162502	0·3838497	6
7	0·2335836	0·4468293	0·1056130	0·3864704	7
8	0·2234550	0·4469100	0·0971341	0·3885364	8
9	0·2148806	0·4469696	0·0901865	0·3902143	9
10	0·2074861	0·4470153	0·0843696	0·3916089	10
11	0·2010140	0·4470512	0·0794142	0·3927897	11
12	0·1952802	0·4470800	0·0751323	0·3938047	12
13	0·1901489	0·4471036	0·0713880	0·3946882	13
14	0·1855174	0·4471233	0·0680806	0·3954655	14
15	0·1813063	0·4471399	0·0651336	0·3961557	15
16	0·1774532	0·4471540	0·0624879	0·3967734	16
17	0·1739079	0·4471662	0·0600969	0·3973300	17
18	0·1706299	0·4471768	0·0579234	0·3978347	18
19	0·1675857	0·4471861	0·0559374	0·3982948	19
20	0·1647478	0·4471943	0·0541141	0·3987163	20
21	0·1620927	0·4472015	0·0524332	0·3991041	21
22	0·1596009	0·4472080	0·0508777	0·3994624	22
23	0·1572555	0·4472138	0·0494332	0·3997946	23
24	0·1550422	0·4472191	0·0480874	0·4001035	24
25	0·1529484	0·4472239	0·0468301	0·4003917	25
26	0·1509633	0·4472282	0·0456522	0·4006614	26
27	0·1490774	0·4472321	0·0445460	0·4009143	27
28	0·1472823	0·4472358	0·0435048	0·4011521	28
29	0·1455706	0·4472391	0·0425226	0·4013762	29
30	0·1439359	0·4472422	0·0415942	0·4015877	30
31	0·1423721	0·4472450	0·0407151	0·4017879	31
32	0·1408742	0·4472476	0·0398812	0·4019775	32
33	0·1394373	0·4472500	0·0390889	0·4021576	33
34	0·1380567	0·4472523	0·0383350	0·4023288	34
35	0·1367305	0·4472544	0·0376165	0·4024918	35
36	0·1354531	0·4472564	0·0369309	0·4026472	36
37	0·1342222	0·4472583	0·0362758	0·4027956	37
38	0·1330349	0·4472600	0·0356491	0·4029374	38
39	0·1318885	0·4472616	0·0350489	0·4030732	39
40	0·1307805	0·4472632	0·0344734	0·4032033	40
41	0·1297089	0·4472646	0·0339210	0·4033281	41
42	0·1286716	0·4472660	0·0333904	0·4034479	42
43	0·1276667	0·4472673	0·0328800	0·4035631	43
44	0·1266925	0·4472685	0·0323888	0·4036738	44
45	0·1257473	0·4472697	0·0319156	0·4037805	45
46	0·1248297	0·4472708	0·0314594	0·4038833	46
47	0·1239383	0·4472718	0·0310192	0·4039824	47
48	0·1230719	0·4472728	0·0305941	0·4040781	48
49	0·1222291	0·4472738	0·0301833	0·4041705	49
50	0·1214090	0·4472747	0·0297861	0·4042599	50

For values of n exceeding 50, the following approximations may be used with seven-figure accuracy:

$$J_n(n) \sim \frac{0{\cdot}44730\,73184}{n^{\frac{1}{3}}}\left[1 - \frac{1}{225n^2}\right] - \frac{0{\cdot}00586\,92885}{n^{\frac{5}{3}}}\left[1 - \frac{1213}{14625n^2}\right],$$

$$J_n'(n) \sim \frac{0{\cdot}41085\,01939}{n^{\frac{2}{3}}}\left[1 + \frac{23}{3150n^2}\right] - \frac{0{\cdot}08946\,14637}{n^{\frac{4}{3}}}\left[1 - \frac{947}{69300n^2}\right]$$

Table VI. Functions of equal order and argument

n	$-Y_n(n)$	$-n^{\frac{1}{3}} Y_n(n)$	$Y_n'(n)$	$n^{\frac{2}{3}} Y_n'(n)$	n
1	0·7812128	0·7812128	0·8694698	0·8694698	1
2	0·6174081	0·7778855	0·5103757	0·8101709	2
3	0·5385416	0·7767114	0·3781412	0·7865654	3
4	0·4889368	0·7761387	0·3069147	0·7733765	4
5	0·4536948	0·7758072	0·2615525	0·7647843	5
6	0·4268259	0·7755941	0·2297650	0·7586672	6
7	0·4053710	0·7754469	0·2060642	0·7540520	7
8	0·3876699	0·7753399	0·1876060	0·7504241	8
9	0·3727057	0·7752590	0·1727588	0·7474840	9
10	0·3598142	0·7751961	0·1605149	0·7450441	10
11	0·3485399	0·7751458	0·1502159	0·7429809	11
12	0·3385583	0·7751049	0·1414121	0·7412092	12
13	0·3296303	0·7750711	0·1337852	0·7396683	13
14	0·3215755	0·7750426	0·1271029	0·7383135	14
15	0·3142546	0·7750184	0·1211915	0·7371112	15
16	0·3075580	0·7749976	0·1159184	0·7360358	16
17	0·3013982	0·7749796	0·1111803	0·7350670	17
18	0·2957040	0·7749638	0·1068955	0·7341890	18
19	0·2904173	0·7749499	0·1029987	0·7333887	19
20	0·2854894	0·7749374	0·0994367	0·7326559	20
21	0·2808800	0·7749266	0·0961658	0·7319817	21
22	0·2765546	0·7749168	0·0931499	0·7313591	22
23	0·2724839	0·7749079	0·0903586	0·7307820	23
24	0·2686456	0·7748999	0·0877663	0·7302453	24
25	0·2650095	0·7748925	0·0853514	0·7297446	25
26	0·2615652	0·7748859	0·0830953	0·7292763	26
27	0·2582933	0·7748798	0·0809819	0·7288371	27
28	0·2551791	0·7748742	0·0789973	0·7284242	28
29	0·2522100	0·7748690	0·0771295	0·7280352	29
30	0·2493744	0·7748642	0·0753678	0·7276680	30
31	0·2466622	0·7748598	0·0737029	0·7273206	31
32	0·2440643	0·7748557	0·0721267	0·7269914	32
33	0·2415724	0·7748519	0·0706318	0·7266790	33
34	0·2391794	0·7748483	0·0692116	0·7263820	34
35	0·2368784	0·7748450	0·0678605	0·7260991	35
36	0·2346635	0·7748419	0·0665732	0·7258295	36
37	0·2325292	0·7748389	0·0653451	0·7255720	37
38	0·2304705	0·7748362	0·0641718	0·7253259	38
39	0·2284828	0·7748336	0·0630496	0·7250904	39
40	0·2265620	0·7748312	0·0619751	0·7248647	40
41	0·2247042	0·7748289	0·0609450	0·7246483	41
42	0·2229059	0·7748267	0·0599565	0·7244405	42
43	0·2211637	0·7748246	0·0590071	0·7242407	43
44	0·2194748	0·7748227	0·0580942	0·7240486	44
45	0·2178364	0·7748208	0·0572157	0·7238636	45
46	0·2162458	0·7748191	0·0563695	0·7236853	46
47	0·2147007	0·7748174	0·0555539	0·7235134	47
48	0·2131988	0·7748158	0·0547671	0·7233475	48
49	0·2117381	0·7748143	0·0540074	0·7231873	49
50	0·2103166	0·7748128	0·0532735	0·7230324	50

For values of n exceeding 50, the following approximations may be used with seven-figure accuracy:

$$-Y_n(n) \sim \frac{0{\cdot}77475\,90021}{n^{\frac{1}{3}}}\left[1 - \frac{1}{225n^2}\right] + \frac{0{\cdot}01016\,59059}{n^{\frac{5}{3}}}\left[1 - \frac{1213}{14625n^2}\right],$$

$$Y_n'(n) \sim \frac{0{\cdot}71161\,34100}{n^{\frac{2}{3}}}\left[1 + \frac{23}{3150n^2}\right] + \frac{0{\cdot}15495\,18004}{n^{\frac{4}{3}}}\left[1 - \frac{947}{69300n^2}\right];$$

The coefficients are numerically equal to $\sqrt{3}$ times the corresponding coefficients in $J_n(n)$ and $J_n'(n)$.

Table VII. Zeros, $j_{0,n}$, $y_{0,n}$, $j_{1,n}$, $y_{1,n}$, of $J_0(x)$, $Y_0(x)$, $J_1(x)$, $Y_1(x)$

n	$j_{0,n}$	$y_{0,n}$	$j_{1,n}$	$y_{1,n}$	n
1	2·4048256	0·8935770	3·8317060	2·1971413	1
2	5·5200781	3·9576784	7·0155867	5·4296811	2
3	8·6537279	7·0860511	10·1734681	8·5960059	3
4	11·7915344	10·2223450	13·3236919	11·7491548	4
5	14·9309177	13·3610975	16·4706301	14·8974421	5
6	18·0710640	16·5009224	19·6158585	18·0434023	6
7	21·2116366	19·6413097	22·7600844	21·1880689	7
8	24·3524715	22·7820280	25·9036721	24·3319426	8
9	27·4934791	25·9229577	29·0468285	27·4752950	9
10	30·6346065	29·0640303	32·1896799	30·6182865	10
11	33·7758202	32·2052041	35·3323076	33·7610178	11
12	36·9170984	35·3464523	38·4747662	36·9035553	12
13	40·0584258	38·4877567	41·6170942	40·0459446	13
14	43·1997917	41·6291045	44·7593190	43·1882181	14
15	46·3411884	44·7704866	47·9014609	46·3303993	15
16	49·4826099	47·9118963	51·0435352	49·4725057	16
17	52·6240518	51·0533286	54·1855536	52·6145508	17
18	55·7655108	54·1947794	57·3275254	55·7565449	18
19	58·9069839	57·3362457	60·4694578	58·8984962	19
20	62·0484692	60·4777252	63·6113567	62·0404111	20
21	65·1899648	63·6192158	66·7532267	65·1822951	21
22	68·3314693	66·7607160	69·8950718	68·3241522	22
23	71·4729816	69·9022246	73·0368952	71·4659861	23
24	74·6145006	73·0437403	76·1786996	74·6077996	24
25	77·7560256	76·1852624	79·3204872	77·7495953	25
26	80·8975559	79·3267901	82·4622599	80·8913753	26
27	84·0390908	82·4683228	85·6040194	84·0331412	27
28	87·1806298	85·6098598	88·7457671	87·1748947	28
29	90·3221726	88·7514008	91·8875043	90·3166370	29
30	93·4637188	91·8929453	95·0292318	93·4583692	30
31	96·6052680	95·0344930	98·1709507	96·6000923	31
32	99·7468199	98·1760436	101·3126618	99·7418072	32
33	102·8883743	101·3175968	104·4543658	102·8835147	33
34	106·0299309	104·4591523	107·5960633	106·0252153	34
35	109·1714896	107·6007100	110·7377548	109·1669097	35
36	112·3130503	110·7422697	113·8794408	112·3085985	36
37	115·4546127	113·8838313	117·0211219	115·4503820	37
38	118·5961766	117·0253944	120·1627983	118·5919607	38
39	121·7377421	120·1669592	123·3044705	121·7336349	39
40	124·8793089	123·3085253	126·4461387	124·8753051	40

Table VII. Zeros, $j_{2,n}$, $y_{2,n}$, $j_{3,n}$, $y_{3,n}$, of $J_2(x)$, $Y_2(x)$, $J_3(x)$, $Y_3(x)$

n	$j_{2,n}$	$y_{2,n}$	$j_{3,n}$	$y_{3,n}$	n
1	5·1356223	3·3842418	6·3801619	4·5270247	1
2	8·4172441	6·7938074	9·7610231	8·0975538	2
3	11·6198412	10·0234780	13·0152007	11·3964667	3
4	14·7959518	13·2099868	16·2234640	14·6230726	4
5	17·9598195	16·3789666	19·4094148	17·8184543	5
6	21·1169971	19·5390400	22·5827295	20·9972845	6
7	24·2701123	22·6939559	25·7481667	24·1662357	7
8	27·4205736	25·8456137	28·9083508	27·3287998	8
9	30·5692045	28·9950804	32·0648524	30·4869896	9
10	33·7165195	32·1430023	35·2186707	33·6420494	10
11	36·8628565	35·2897939	38·3704724	36·7947910	11
12	40·0084467	38·4357335	41·5207197	39·9457672	12
13	43·1534538	41·5810149	44·6697431	43·0953675	13
14	46·2979967	44·7257771	47·8177857	46·2438744	14
15	49·4421641	47·8701227	50·9650299	49·3914980	15
16	52·5860235	51·0141287	54·1116156	52·5383976	16
17	55·7296271	54·1578545	57·2576516	55·6846964	17
18	58·8730158	57·3013461	60·4032241	58·8304911	18
19	62·0162224	60·4446401	63·5484022	61·9758587	19
20	65·1592732	63·5877658	66·6932417	65·1208612	20
21	68·3021898	66·7307471	69·8377884	68·2655491	21
22	71·4449899	69·8736034	72·9820804	71·4099642	22
23	74·5876882	73·0163509	76·1261492	74·5541409	23
24	77·7302971	76·1590031	79·2700214	77·6981084	24
25	80·8728269	79·3015713	82·4137195	80·8418910	25
26	84·0152867	82·4440651	85·5572629	83·9855095	26
27	87·1576839	85·5864927	88·7006678	87·1289817	27
28	90·3000252	88·7288612	91·8439487	90·2723230	28
29	93·4423160	91·8711766	94·9871177	93·4155465	29
30	96·5845614	95·0134441	98·1301857	96·5586637	30
31	99·7267657	98·1556685	101·2731621	99·7016848	31
32	102·8689327	101·2978536	104·4160552	102·8446186	32
33	106·0110655	104·4400031	107·5588722	105·9874728	33
34	109·1531673	107·5821201	110·7016197	109·1302542	34
35	112·2952406	110·7242073	113·8443033	112·2729691	35
36	115·4372877	113·8662672	116·9869284	115·4156229	36
37	118·5793107	117·0083021	120·1294994	118·5582204	37
38	121·7213115	120·1503138	123·2720205	121·7007659	38
39	124·8632917	123·2923041	126·4144954	124·8432635	39
40	128·0052530	126·4342746	129·5569276	127·9857167	40

Table VII. Zeros, $j_{4,n}$, $y_{4,n}$, $j_{5,n}$, $y_{5,n}$, of $J_4(x)$, $Y_4(x)$, $J_5(x)$, $Y_5(x)$

n	$j_{4,n}$	$y_{4,n}$	$j_{5,n}$	$y_{5,n}$	n
1	7·5883427	5·6451479	8·7714838	6·7471838	1
2	11·0647095	9·3616206	12·3386042	10·5971767	2
3	14·3725367	12·7301445	15·7001741	14·0338041	3
4	17·6159660	15·9996271	18·9801339	17·3470864	4
5	20·8269330	19·2244290	22·2177999	20·6028990	5
6	24·0190195	22·4248106	25·4303411	23·8265360	6
7	27·1990878	25·6102671	28·6266183	27·0301349	7
8	30·3710077	28·7858937	31·8117167	30·2203357	8
9	33·5371377	31·9546867	34·9887813	33·4011056	9
10	36·6990011	35·1185295	38·1598686	36·5749725	10
11	39·8576273	38·2786681	41·3263833	39·7436277	11
12	43·0137377	41·4359606	44·4893191	42·9082482	12
13	46·1678535	44·5910182	47·6493998	46·0696791	13
14	49·3203607	47·7442881	50·8071652	49·2285437	14
15	52·4715514	50·8961052	53·9630266	52·3853121	15
16	55·6216509	54·0467255	57·1173028	55·5403458	16
17	58·7708357	57·1963482	60·2702451	58·6939271	17
18	61·9192462	60·3451302	63·4220540	61·8462803	18
19	65·0669953	63·4931972	66·5728919	64·9975855	19
20	68·2141749	66·6406512	69·7228912	68·1479890	20
21	71·3608607	69·7875753	72·8721613	71·2976113	21
22	74·5071155	72·9340384	76·0207934	74·4465520	22
23	77·6529918	76·0800980	79·1688641	77·5948946	23
24	80·7985341	79·2258022	82·3164380	80·7427095	24
25	83·9437799	82·3711919	85·4635703	83·8900562	25
26	87·0887615	85·5163019	88·6103082	87·0369859	26
27	90·2335065	88·6611620	91·7566925	90·1835423	27
28	93·3780390	91·8057980	94·9027585	93·3297633	28
29	96·5223797	94·9502321	98·0485369	96·4756819	29
30	99·6665468	98·0944839	101·1940546	99·6213268	30
31	102·8105563	101·2385704	104·3393353	102·7667232	31
32	105·9544223	104·3825064	107·4843998	105·9118934	32
33	109·0981571	107·5263053	110·6292667	109·0568569	33
34	112·2417718	110·6699788	113·7739523	112·2016312	34
35	115·3852762	113·8135372	116·9184713	115·3462317	35
36	118·5286792	116·9569899	120·0628368	118·4906725	36
37	121·6719886	120·1003451	123·2070606	121·6349657	37
38	124·8152114	123·2436104	126·3511534	124·7791228	38
39	127·9583541	126·3867924	129·4951246	127·9231536	39
40	131·1014225	129·5298972	132·6389830	131·0670674	40

Table VII. Zeros, $j_{1/3,n}$, $y_{1/3,n}$, of $J_{1/3}(x)$, $Y_{1/3}(x)$; with zeros, s_n, d_n, of $J_{-1/3}(x)+J_{1/3}(x)$, $J_{-1/3}(x)-J_{1/3}(x)$

[NOTE. The last two functions are equal to

$$\sqrt{3}\cdot\{J_{1/3}(x)\cos 30^\circ - Y_{1/3}(x)\sin 30^\circ\},\quad \sqrt{3}\cdot\{J_{1/3}(x)\cos 120^\circ - Y_{1/3}(x)\sin 120^\circ\}$$

respectively.]

n	$j_{1/3,n}$	$y_{1/3,n}$	s_n	d_n	n
1	2·9025862	1·3530196	2·3834466	0·8477186	1
2	6·0327471	4·4657883	5·5101956	3·9441020	2
3	9·1705067	7·6012412	8·6473577	7·0782997	3
4	12·3101938	10·7402128	11·7868429	10·2169407	4
5	15·4506490	13·8803575	14·9272068	13·3569532	5
6	18·5914863	17·0210330	18·0679953	16·4975630	6
7	21·7325412	20·1619929	21·2090210	19·6384856	7
8	24·8737314	23·3031228	24·3501925	22·7795923	8
9	28·0150117	26·4443623	27·4914601	25·9208165	9
10	31·1563549	29·5856767	30·6327941	29·0621201	10
11	34·2977437	32·7270444	33·7741762	32·2034801	11
12	37·4391666	35·8684514	36·9155941	35·3448813	12
13	40·5806158	39·0098884	40·0570394	38·4863138	13
14	43·7220857	42·1513485	43·1985061	41·6277704	14
15	46·8635719	45·2928269	46·3399899	44·7692461	15
16	50·0050715	48·4343202	49·4814874	47·9107371	16
17	53·1465821	51·5758256	52·6229964	51·0522406	17
18	56·2881019	54·7173410	55·7645147	54·1937545	18
19	59·4296294	57·8588648	58·9060410	57·3352769	19
20	62·5711634	61·0003956	62·0475740	60·4768067	20
21	65·7127030	64·1419325	65·1891127	63·6183427	21
22	68·8542475	67·2834747	68·3306564	66·7598840	22
23	71·9957961	70·4250213	71·4722044	69·9014299	23
24	75·1373484	73·5665718	74·6137562	73·0429798	24
25	78·2789040	76·7081259	77·7553112	76·1845333	25
26	81·4204625	79·8496829	80·8968692	79·3260899	26
27	84·5620234	82·9912426	84·0384298	82·4676492	27
28	87·7035867	86·1328048	87·1799926	85·6092109	28
29	90·8451519	89·2743691	90·3215576	88·7507749	29
30	93·9867191	92·4159353	93·4631244	91·8923408	30
31	97·1282878	95·5575032	96·6046929	95·0339085	31
32	100·2698581	98·6990728	99·7462629	98·1754777	32
33	103·4114297	101·8406437	102·8878343	101·3170485	33
34	106·5530025	104·9822160	106·0294070	104·4586205	34
35	109·6945765	108·1237894	109·1709808	107·6001938	35
36	112·8361516	111·2653639	112·3125557	110·7417681	36
37	115·9777275	114·4069394	115·4541315	113·8833435	37
38	119·1193044	117·5485159	118·5957082	117·0249197	38
39	122·2608821	120·6900931	121·7372858	120·1664969	39
40	125·4024605	123·8316712	124·8788641	123·3080748	40

Table VIII. Integrals of functions of order zero

x	$\frac{1}{2}\int_0^x J_0(t)\,dt$	$\frac{1}{2}\int_0^x Y_0(t)\,dt$
0·02	+ 0·0099997	− 0·0320078
0·04	+ 0·0199973	− 0·0551846
0·06	+ 0·0299910	− 0·0750205
0·08	+ 0·0399787	− 0·0926801
0·10	+ 0·0499583	− 0·1087153
0·12	+ 0·0599280	− 0·1234500
0·14	+ 0·0698858	− 0·1370979
0·16	+ 0·0798295	− 0·1498103
0·18	+ 0·0897573	− 0·1617001
0·20	+ 0·0996672	− 0·1728544
0·22	+ 0·1095571	− 0·1833430
0·24	+ 0·1194252	− 0·1932224
0·26	+ 0·1292695	− 0·2025397
0·28	+ 0·1390880	− 0·2113348
0·30	+ 0·1488788	− 0·2196416
0·32	+ 0·1586399	− 0·2274894
0·34	+ 0·1683694	− 0·2349040
0·36	+ 0·1780654	− 0·2419080
0·38	+ 0·1877260	− 0·2485215
0·40	+ 0·1973493	− 0·2547624
0·42	+ 0·2069333	− 0·2606471
0·44	+ 0·2164763	− 0·2661901
0·46	+ 0·2259764	− 0·2714049
0·48	+ 0·2354316	− 0·2763037
0·50	+ 0·2448403	− 0·2808977
0·52	+ 0·2542004	− 0·2851974
0·54	+ 0·2635103	− 0·2892124
0·56	+ 0·2727682	− 0·2929516
0·58	+ 0·2819722	− 0·2964234
0·60	+ 0·2911206	− 0·2996358
0·62	+ 0·3002117	− 0·3025960
0·64	+ 0·3092437	− 0·3053111
0·66	+ 0·3182150	− 0·3077877
0·68	+ 0·3271238	− 0·3100319
0·70	+ 0·3359684	− 0·3120498
0·72	+ 0·3447472	− 0·3138471
0·74	+ 0·3534586	− 0·3154290
0·76	+ 0·3621010	− 0·3168010
0·78	+ 0·3706727	− 0·3179678
0·80	+ 0·3791722	− 0·3189344
0·82	+ 0·3875979	− 0·3197054
0·84	+ 0·3959484	− 0·3202852
0·86	+ 0·4042220	− 0·3206781
0·88	+ 0·4124174	− 0·3208884
0·90	+ 0·4205330	− 0·3209201
0·92	+ 0·4285674	− 0·3207771
0·94	+ 0·4365192	− 0·3204634
0·96	+ 0·4443870	− 0·3199827
0·98	+ 0·4521694	− 0·3193386
1·00	+ 0·4598652	− 0·3185347

Maxima and minima of $\frac{1}{2}\int_0^x J_0(t)\,dt$ and $\frac{1}{2}\int_0^x Y_0(t)\,dt$

$x = j_{0,n}$	$\frac{1}{2}\int_0^x J_0(t)\,dt$
2·4048256	+ 0·7352208
5·5200781	+ 0·3344230
8·6537279	+ 0·6340842
11·7915344	+ 0·3845594
14·9309177	+ 0·6028269
18·0710640	+ 0·4064156
21·2116366	+ 0·5864441
24·3524715	+ 0·4192836
27·4934791	+ 0·5759911
30·6346065	+ 0·4279931
33·7758202	+ 0·5685888
36·9170984	+ 0·4343856
40·0584258	+ 0·5629957
43·1997917	+ 0·4393331
46·3411884	+ 0·5585784
49·4826099	+ 0·4433085

$x = y_{0,n}$	$\frac{1}{2}\int_0^x Y_0(t)\,dt$
0·8935770	− 0·3209291
3·9576784	+ 0·1920149
7·0860511	− 0·1474447
10·2223450	+ 0·1237411
13·3610975	− 0·1085949
16·5009224	+ 0·0978827
19·6413097	− 0·0898033
22·7820280	+ 0·0834339
25·9229577	− 0·0782474
29·0640303	+ 0·0739188
32·2052041	− 0·0702357
35·3464523	+ 0·0670523
38·4877567	− 0·0642652
41·6291045	+ 0·0617985
44·7704866	− 0·0595953
47·9118963	+ 0·0576118

For values of x between 0 and 16, the integrals may be calculated with the help of Table I from the formulae (cf. § 10·74)

$$\tfrac{1}{2}\int_0^x J_0(t)\,dt = \tfrac{1}{4}\pi x\{J_0(x)\mathbf{H}_0'(x) + J_1(x)\mathbf{H}_0(x)\},\quad \tfrac{1}{2}\int_0^x Y_0(t)\,dt = \tfrac{1}{4}\pi x\{Y_0(x)\mathbf{H}_0'(x) + Y_1(x)\mathbf{H}_0(x)\}.$$

BIBLIOGRAPHY*

ADAMOFF, A.

On the asymptotic representation of the cylinder functions $J_\nu(z)$ and $J_\nu'(z)$ for large values of the modulus of z. *Petersburg, Ann. Inst. polyt.* 1906, pp. 239—265. [*Jahrbuch über die Fortschritte der Math.* 1907, pp. 492—493.]

AICHI, K.

Note on the Function $K_m(x)$, the Solution of the Modified Bessel's Equation. *Proc. Phys. Math. Soc. of Japan*, (3) II. (1920), pp. 8—19.

AIREY, J. R.

The Roots of the Neumann and Bessel Functions (Dec. 29, 1910). *Proc. Phys. Soc.* XXIII. (1911), pp. 219—224.

The Vibrations of Circular Plates and their Relation to Bessel Functions (Feb. 15, 1911). *Proc. Phys. Soc.* XXIII. (1911), pp. 225—232.

The Oscillations of Chains and their Relation to Bessel and Neumann Functions. *Phil. Mag.* (6) XXI. (1911), pp. 736—742.

Tables of Neumann Functions $G_n(x)$ and $Y_n(x)$. *Phil. Mag.* (6) XXII. (1911), pp. 658—663.

The Asymptotic expansions of Bessel and other functions. *Archiv der Math. und Phys.* (3) XX. (1913), pp. 240—244.

The Vibrations of Cylinders and Cylindrical Shells. *Archiv der Math. und Phys.* (3) XX. (1913), pp. 289—294.

Tables of the Neumann functions or Bessel functions of the second kind. *Archiv der Math. und Phys.* (3) XXII. (1914), pp. 30—43.

Bessel and Neumann Functions of Equal Order and Argument. *Phil. Mag.* (6) XXXI. (1916), pp. 520—528.

The Roots of Bessel and Neumann Functions of High Order. *Phil. Mag.* (6) XXXII. (1916), pp. 7—14.

Bessel Functions of Equal Order and Argument. *Phil. Mag.* (6) XXXII. (1916), pp. 237—238.

The Numerical Calculation of the Roots of the Bessel Function $J_n(x)$ and its first derivative $J_n'(x)$. *Phil. Mag.* (6) XXXIV. (1917), pp. 189—195.

The Addition Theorem of the Bessel Functions of Zero and Unit Orders. *Phil. Mag.* (6) XXXVI. (1918), pp. 234—242.

The Lommel-Weber Ω Function and its Application to the Problem of Electric Waves on a Thin Anchor Ring (Dec. 7, 1917). *Proc. Royal Soc.* XCIV. A (1918), pp. 307—314.

Bessel Functions of small Fractional Order and their application to problems of Elastic Stability. *Phil. Mag.* (6) XLI. (1921), pp. 200—205.

AIRY, SIR GEORGE B.

On the Diffraction of an Object-glass with Circular Aperture (Nov. 24, 1834). *Trans. Camb. Phil. Soc.* V. (1835), pp. 283—291.

On the Intensity of Light in the neighbourhood of a Caustic (May 2, 1836; March 26, 1838). *Trans. Camb. Phil. Soc.* VI. (1838), pp. 379—402.

On the Diffraction of an Annular Aperture (Dec. 4, 1840). *Phil. Mag.* (3) XVIII. (1841), pp. 1—10.

Supplement to a Paper, On the Intensity of Light in the neighbourhood of a Caustic (March 24, 1848). *Trans. Camb. Phil. Soc.* VIII. (1849), pp. 595—599.

AKIMOFF, M.

Transcendantes de Fourier-Bessel à plusieurs variables (July 10, 1916; June 25, 1917; Dec. 24, 1917). *Comptes Rendus*, CLXIII. (1916), pp. 26—29; CLXV. (1917), pp. 23—25, 1100—1103.

* In the case of a few inaccessible memoirs, references are given to abstracts in the *Jahrbuch über die Fortschritte der Math.* or elsewhere.

ALDIS, W. S.

Tables for the Solution of the Equation

$$\frac{d^2y}{dx^2}+\frac{1}{x}\frac{dy}{dx}-\left(1+\frac{n^2}{x^2}\right)y=0$$

(June 16, 1898). *Proc. Royal Soc.* LXIV. (1899), pp. 203—223.

On the numerical computation of the functions $G_0(x)$, $G_1(x)$ and $J_n(x\sqrt{i})$ (June 15, 1899). *Proc. Royal Soc.* LXVI. (1900), pp. 32—43.

ALEXANDER, P.

Expansion of Functions in terms of Linear, Cylindric, Spherical and Allied Functions (Dec. 20, 1886). *Trans. Edinburgh Royal Soc.* XXXIII. (1888), pp. 313—320.

ANDING, E.

Sechsstellige Tafeln der Besselschen Funktionen imaginären Arguments (Leipzig, 1911).

ANGER, C. T.*

Untersuchungen über die Function I_k^h mit Anwendungen auf das Kepler'sche Problem. *Neueste Schriften der Naturforschenden der Ges. in Danzig*, v. (1855), pp. 1—29.

ANISIMOV, V. A.

The generalised form of Riccati's equation. *Proceedings of Warsaw University*, 1896, pp. 1—33. [*Jahrbuch über die Fortschritte der Math.* 1896, p. 256.]

APPELL, P. É.

Sur l'inversion approchée de certaines intégrales réelles et sur l'extension de l'équation de Kepler et des fonctions de Bessel (April 6, 1915). *Comptes Rendus*, CLX. (1915), pp. 419—423.

AUTONNE, L.

Sur la nature des intégrales algébriques de l'équation de Riccati (May 7, 1883). *Comptes Rendus*, XCVI. (1883), pp. 1354—1356.

Sur les intégrales algébriques de l'équation de Riccati (Feb. 13, 1899). *Comptes Rendus*, CXXVIII. (1899), pp. 410—412.

BACH, D.

De l'intégration par les séries de l'équation

$$\frac{d^2y}{dx^2}-\frac{n-1}{x}\frac{dy}{dx}=y.$$

Ann. sci. de l'École norm. sup. (2) III. (1874), pp. 47—68.

BAEHR, G. F. W.

Sur les racines des équations $\int_0^\pi \cos(x\cos\omega)\,d\omega=0$ et $\int_0^\pi \cos(x\cos\omega)\sin^2\omega\,d\omega=0$ (April, 1872). *Archives Néerlandaises*, VII. (1872), pp. 351—358.

BALL, L. DE.

Ableitung einiger Formeln aus der Theorie der Bessel'schen Functionen (June 6, 1891). *Astr. Nach.* CXXVIII. (1891), col. 1—4.

BARNES, E. W.

On the homogeneous linear difference equation of the second order with linear coefficients. *Messenger*, XXXIV. (1905), pp. 52—71.

On Functions defined by simple types of Hypergeometric Series (March 12, 1906). *Trans. Camb. Phil. Soc.* XX. (1908), pp. 253—279.

The asymptotic Expansion of Integral Functions defined by generalised Hypergeometric Series (Dec. 3, 1906). *Proc. London Math. Soc.* (2) V. (1907), pp. 59—116.

BASSET, A. B.

On a method of finding the potentials of circular discs by means of Bessel's functions (May 10, 1886). *Proc. Camb. Phil. Soc.* V. (1886), pp. 425—443.

On the Potentials of the surfaces formed by the revolution of Limaçons and Cardioids about their axes (Oct. 25, 1886). *Proc. Camb. Phil. Soc.* VI. (1889), pp. 2—19.

A Treatise on Hydrodynamics (2 vols.) (Cambridge, 1888).

On the Radial Vibrations of a Cylindrical Elastic Shell (Dec. 12, 1889). *Proc. London Math. Soc.* XXI. (1891), pp. 53—58.

On a Class of Definite Integrals connected with Bessel's Functions (Nov. 13, 1893). *Proc. Camb. Phil. Soc.* VIII. (1895), pp. 122—128.

* See also under Bourget and Cauchy.

BATEMAN, H.

Certain definite integrals connected with the Legendre and Bessel functions. *Messenger*, XXXIII. (1904), pp. 182—188.

A generalisation of the Legendre polynomial (Jan. 1, 1905). *Proc. London Math. Soc.* (2) III. (1905), pp. 111—123.

The inversion of a definite integral (Nov. 8, 1906). *Proc. London Math. Soc.* (2) IV. (1906), pp. 461—498.

On an expansion of an arbitrary function into a series of Bessel functions. *Messenger*, XXXVI. (1907), pp. 31—37.

The Solution of Linear Differential Equations by means of Definite Integrals (Jan. 25, 1909). *Trans. Camb. Phil. Soc.* XXI. (1912), pp. 171—196.

The History and Present State of the Theory of Integral Equations. *British Association Report*, 1910, pp. 345—424.

Notes on integral equations. *Messenger*, XLI. (1912), pp. 94—101, 180—184.

Some equations of mixed differences occurring in the Theory of Probability and the related expansions in series of Bessel's functions. *Proc. Int. Congress of Math.* I. (Cambridge, 1912), pp. 291—294.

Electrical and Optical Wave-motions (Cambridge, 1915).

BAUER, G.

Von den Coefficienten der Reihen von Kugelfunctionen einer Variablen. *Journal für Math.* LVI. (1859), pp. 101—121.

Bemerkungen über Reihen nach Kugelfunctionen und insbesondere auch über Reihen, welche nach Producten oder Quadraten von Kugelfunctionen fortschreiten mit Anwendung auf Cylinderfunctionen (July 3, 1875). *Münchener Sitzungsberichte*, V. (1875), pp. 247—272.

BECKER, J.

Die Riccatische Differential-Gleichung. *Programm Karlsbad*, 1908 (25 pp.). [*Jahrbuch über die Fortschritte der Math.* 1908, p. 395.]

BELTRAMI, E.

Intorno ad un teorema di Abele e ad alcune sue applicazioni. *R. Ist. Lombardo Rendiconti*, (2) XIII. (1880), pp. 327—337.

Intorno ad alcune serie trigonometriche (June 17, 1880). *R. Ist. Lombardo Rendiconti*, (2) XIII. (1880), pp. 402—413.

Sulle funzioni cilindriche (Jan. 14, 1881). *Atti della R. Accad. delle Sci. di Torino*, XVI. (1880—81), pp. 201—205.

Sulla teoria delle funzioni potenziali simmetriche (April 28, 1881). *Bologna Memorie*, (4) II. (1880), pp. 461—505.

BERNOULLI, DANIEL.

Correspondence with Leibniz 1697—1704. [Published in *Leibnizens Ges. Werke, Dritte Folge* (*Mathematik*), III. (Halle, 1855).]

Notata in praecedens schediasma Ill. Co. Jacobi Riccati, *Actorum Eruditorum quae Lipsiae publicantur Supplementa*, VIII. (1724), pp. 73—75.

Solutio problematis Riccatiani propositi in Act. Lips. Suppl. Tom. VIII. p. 73. *Acta Eruditorum publicata Lipsiae*, 1725, pp. 473—475.

Exercitationes quaedam mathematicae (Venice, 1724), pp. 77—80.

Theoremata de oscillationibus corporum filo flexili connexorum et catenae verticaliter suspensae. *Comm. Acad. Sci. Imp. Petrop.* VI. (1732—33) [1738], pp. 108—122.

Demonstrationes Theorematum suorum de oscillationibus corporum filo flexili connexorum et catenae verticaliter suspensae. *Comm. Acad. Sci. Imp. Petrop.* VII. (1734—35) [1740], pp. 162—179.

BERNOULLI, JOHN.

Methodus generalis construendi omnes aequationes differentiales primi gradus. *Acta Eruditorum publicata Lipsiae*, 1694, pp. 435—437. [*Opera*, I. Lausanne and Geneva, 1742, p. 124.]

BERNOULLI, NICHOLAS (the younger).

Correspondence with Goldbach. See under Fuss.

BESSEL, F. W.

Analytische Auflösung der Keplerschen Aufgabe (July 2, 1818). *Berliner Abh.* 1816—17 [1819], pp. 49—55. [*Abhandlungen*, herausgegeben von R. Engelmann, I. (1875), pp. 17—20.]

Ueber die Entwickelung der Functionen zweier Winkel u und u' in Reihen welche nach den Cosinussen und Sinussen der Vielfachen von u und u' fortgehen (June 21, 1821). *Berliner Abh.* 1820—21 [1822], pp. 56—60. [*Abhandlungen*, II. (1876), pp. 362—364.]

Untersuchung des Theils der planetarischen Störungen welcher aus der Bewegung der Sonne entsteht (Jan. 29, 1824). *Berliner Abh.* 1824 [1826], pp. 1—52. [*Abhandlungen*, I. (1875), pp. 84—109.]

Beitrag zu den Methoden die Störungen der Kometen zu berechnen (Sept. 24, 1836). *Astr. Nach.* XIV. (1837), col. 1—48. [*Abhandlungen*, I. (1875), pp. 29—54.]

BINET, J. P. M.

Note sur l'intégrale $\int_a^y y^{2i} dy\, e^{-\frac{p}{y^2} - qy^2}$ prise entre des limites arbitraires (May 24, 1841). *Comptes Rendus*, XII. (1841), pp. 958—962.

BÔCHER, M.

On Bessel's functions of the second kind (Jan. 1892). *Annals of Math.* VI. (1892), pp. 85—90.

On some applications of Bessel's functions with pure imaginary index (Feb. 11, 1892). *Annals of Math.* VI. (1892), pp. 137—160*.

On certain methods of Sturm and their application to the roots of Bessel's functions (Feb. 1897). *Bulletin American Math. Soc.* III. (1897), pp. 205—213.

An elementary proof that Bessel's functions of the zeroth order have an infinite number of real roots (Feb. 25, 1899). *Bulletin American Math. Soc.* V. (1899), pp. 385—388.

Non-oscillatory linear differential equations of the second order (Feb. 4, 1901). *Bulletin American Math. Soc.* VII. (1901), pp. 333—340.

BÖHMER, P. E.

Über die Zylinderfunktionen (Nov. 26, 1913). *Sitz. der Berliner Math. Ges.* XIII. (1913), pp. 30—36.

BOHREN, A.

Über das Airysche Integral† (Oct. 6, 1902). *Bern Mittheilungen*, 1902, pp. 236—239.

BOOLE, G.

On the transformation of Definite Integrals. *Camb. Math. Journal*, III. (1843), pp. 216—224.

On a general method in analysis (Jan. 18, 1844). *Phil. Trans. of the Royal Soc.* 1844, pp. 225—282.

A Treatise on Differential Equations (London, 1872).

BOURGET, J.

Note sur une formule de M. Anger (Aug. 7, 1854). *Comptes Rendus*, XXXIX. (1854), p. 283.

Mémoire sur les nombres de Cauchy et leur application à divers problèmes de mécanique céleste. *Journal de Math.* (2) VI. (1861), pp. 33—54.

Mémoire sur le mouvement vibratoire des membranes circulaires (June 5, 1865). *Ann. sci. de l'École norm. sup.* III. (1866), pp. 55—95.

BRAJTZEW, J. R.

Über die Fourier-Besselschen Funktionen und deren Anwendung zur Auffindung asymptotischer Darstellungen von Integralen der linearen Differentialgleichungen mit rationalen Koeffizienten. *Warschau Polyt. Inst. Nach.* 1902, nos. 1, 2. [*Jahrbuch über die Fortschritte der Math.* 1903, pp. 575—577.]

BRASSINNE, E.

Sur diverses équations différentielles du premier ordre analogues à l'équation de Ricatti (sic). *Mém. de l'Acad. R. des Sci. de Toulouse*, (3) IV. (1848), pp. 234—236.

Sur des équations différentielles qui se rattachent à l'équation de Riccati. *Journal de Math.* XVI. (1851), pp. 255—256.

BRENKE, W. C.

Summation of a series of Bessel's functions by means of an integral (Nov. 27, 1909). *Bull. American Math. Soc.* XVI. (1910), pp. 225—230.

BRIDGEMAN, P. W.

On a Certain Development in Bessel's Functions (July 22, 1908). *Phil. Mag.* (6) XVI. (1908), pp. 947—948.

* See also *ibid.* p. 136.

† This paper, which might have been mentioned in § 6·4, contains the formula § 6·4 (1); but the author does not give the formula § 6·4 (2).

BRUNS, H.

Ueber die Beugungsfigur des Heliometer-Objectives (Oct. 15, 1882). *Astr. Nach.* CIV. (1883), col. 1—8.

BRYAN, G. H.

On the waves on a viscous rotating cylinder (June 4, 1888). *Proc. Camb. Phil. Soc.* VI. (1889), pp. 248—264.

Wave Motion and Bessel's Functions. *Nature*, LXXX. (1909), p. 309.

BURKHARDT, H. F. K. L.

Trigonometrische Reihen und Integrale (bis etwa 1850). *Encyklopädie der Math. Wiss.* II. 1 (Leipzig, 1904—16), pp. 819—1354.

BUTTERWORTH, S.

On the Evaluation of Certain Combinations of the Ber, Bei and Allied Functions (May 30, 1913). *Proc. Phys Soc.* XXV. (1913), pp. 294—297.

CAILLER, C.

Sur les fonctions de Bessel. *Archives des Sci.* (*Soc. Helvétique*), (4) XIV. (1902), pp. 347—350.

Note sur une opération analytique et son application aux fonctions de Bessel (March, 1904). *Mém. de la Soc. de phys. et d'histoire naturelle de Genève*, XXXIV. (1902—5), pp. 295—368.

CALLANDREAU, O.

Calcul des transcendantes de Bessel

$$J_n(a) = \frac{\left(\frac{a}{2}\right)^n}{1.2 \ldots n}\left[1 - \frac{\left(\frac{a}{2}\right)^2}{1.(n+1)} + \frac{\left(\frac{a}{2}\right)^4}{1.2.(n+1)(n+2)} - \ldots\right],$$

pour les grandes valeurs de a au moyen de séries sémiconvergentes. *Bulletin des Sci. Math.* (2) XIV. (1890), pp. 110—114.

Sur le calcul des polynômes $X_n(\cos\theta)$ de Legendre pour les grandes valeurs de n. *Bulletin des Sci. Math.* (2) XV. (1891), pp. 121—124.

CARLINI, F.

*Ricerche sulla convergenza della serie che serva alla soluzione del problema di Keplero** (Milan, 1817).

CARSLAW, H. S.

Some Multiform Solutions of the Partial Differential Equations of Physical Mathematics and their Applications (Nov. 10, 1898). *Proc. London Math. Soc.* XXX. (1899), pp. 121—161.

The Green's function for a wedge of any angle and other Problems in the Conduction of Heat (Oct. 30, 1909). *Proc. London Math. Soc.* (2) VIII. (1910), pp. 365—374.

The Scattering of Sound Waves by a Cone. *Math. Ann.* LXXV. (1914), pp. 133—147.

The Green's function for the equation $\nabla^2 u + k^2 u = 0$ (April 28, 1913; March 20, 1916). *Proc. London Math. Soc.* (2) XIII. (1914), pp. 236—257; (2) XVI. (1917), pp. 84—93.

The Theory of the Conduction of Heat (London, 1921).

CATALAN, E. C.

Note sur l'intégrale $\int_0^\infty \frac{\cos ax\,dx}{(1+x^2)^n}$ (Feb. 1840). *Journal de Math.* V. (1840), pp. 110—114. [Reprinted, *Mém. de la Soc. R. des Sci. de Liège*, (2) XII. (1885), pp. 26—31.]

Sur l'équation de Riccati (March 4, 1871). *Bulletin de l'Acad. R. de Belgique*, (2) XXXI. (1871), pp. 68—73.

Note sur l'équation $xy'' + ky' - xy = 0$ par M. C. Le Paige (Rapport de M. Catalan). *Bulletin de l'Acad. R. de Belgique*, (2) XLI. (1876), pp. 935—939.

Application d'une formule de Jacobi (Nov. 1868). *Mém. de la Soc. R. des Sci. de Liège*, (2) XII. (1885), pp. 312—316.

CAUCHY, A. L.

Résumé d'un mémoire sur la mécanique céleste et sur un nouveau calcul appelé calcul des limites (Lu à l'Acad. de Turin, Oct. 11, 1831). *Exercices d'Analyse*, II. (Paris, 1841), pp. 48—112. [*Oeuvres*, (2) XII. (1916), pp. 48—112.]

Mémoire sur la convergence des séries (Nov. 11, 1839). *Comptes Rendus*, IX. (1839), pp. 587—588. [*Oeuvres*, (1) IV. (1884), pp. 518—520.]

* Translated into German by Jacobi, *Astr. Nach.* XXX. (1850), col. 197—254. [*Ges. Math. Werke*, VII. (1891), pp. 189—245.]

Considérations nouvelles sur la théorie des suites et sur les lois de leur convergence (April 20, 1840). *Comptes Rendus*, x. (1840), pp. 640—656. [*Oeuvres*, (1) v. (1885), pp. 180—198.]

Méthode simple et générale pour la détermination numérique des coefficients que renferme le développement de la fonction perturbatrice (Sept. 14, 1840). *Comptes Rendus*, xi. (1840), pp. 453—475. [*Oeuvres*, (1) v. (1885), pp. 288—310.]

Note sur le développement de la fonction perturbatrice (Sept. 21, 1840). *Comptes Rendus*, xi. (1840), pp. 501—511. [*Oeuvres*, (1) v. (1885), pp. 311—321.]

Méthodes propres à simplifier le calcul des inégalités périodiques et séculaires des mouvements des planètes (Jan. 11, 1841). *Comptes Rendus*, xii. (1841), pp. 84—101. [*Oeuvres*, (1) vi. (1888), pp. 16—34.]

Note sur une transcendante que renferme le développement de la fonction perturbatrice relative au système planétaire (Oct. 4, 1841). *Comptes Rendus*, xiii. (1841), pp. 682—687. [*Oeuvres*, (1) vi. (1888), pp. 341—346.]

Note sur la substitution des anomalies excentriques aux anomalies moyennes, dans le développement de la fonction perturbatrice (Oct. 25, 1841). *Comptes Rendus*, xiii. (1841), pp. 850—854. [*Oeuvres*, (1) vi. (1888), pp. 354—359.]

Nouveau Mémoire sur le calcul des inégalités des mouvements planétaires (April 8, 1844). *Comptes Rendus*, xviii. (1844), pp. 625—643. [*Oeuvres*, (1) viii. (1893), pp. 168—188.

Sur la transformation des fonctions implicites en moyennes isotropiques, et sur leurs développements en séries trigonométriques (May 22, 1854). *Comptes Rendus*, xxxviii. (1854), pp. 910—913. [*Oeuvres*, (1) xii. (1900), pp. 148—151.]

Sur la transformation des variables qui déterminent les mouvements d'une planète ou même d'une comète en fonction explicite du temps, et sur le développement de ses fonctions en séries convergentes (June 5, 1854). *Comptes Rendus*, xxxviii. (1854), pp. 990—993. [*Oeuvres*, (1) xii. (1900), pp. 160—164.]

Sur la résolution des équations et sur le développement de leurs racines en séries convergentes (June 26, 1854). *Comptes Rendus*, xxxviii. (1854), pp. 1104—1107. [*Oeuvres*, (1) xii. (1900), pp. 167—170.]

Sur une formule de M. Anger et sur d'autres formules analogues (July 15, 1854). *Comptes Rendus*, xxxix. (1854), pp. 129—135. [*Oeuvres*, (1) xii. (1900), pp. 171—177.]

CAYLEY, A.

Sur quelques formules du calcul intégral. *Journal de Math.* xii. (1847), pp. 231—240. [*Collected Papers*, i. (1889), pp. 309—316.]

On Riccati's equation (Sept. 29, 1868). *Phil. Mag.* (4) xxxvi. (1868), pp. 348—351. [*Collected Papers*, vii. (1894), pp. 9—12.]

Note on the integration of certain differential equations by series. *Messenger* (Old Series), v. (1869), pp. 77—82. [*Collected Papers*, viii. (1895), pp. 458—462.]

*Proc. London Math. Soc.** v. (1874), pp. 123—124. [*Collected Papers*, ix. (1896), pp. 19—20.]

CHALLIS, H. W.

Extension of the Solution of Riccati's Equation (Oct. 5, 1864). *Quarterly Journal*, vii. (1866), pp. 51—53.

CHAPMAN, S.

On the general theory of summability with applications to Fourier's and other series. *Quarterly Journal*, xliii. (1911), pp. 1—52.

CHESSIN, A. S.

Note on the General Solution of Bessel's Equation. *American Journal of Math.* xvi. (1894), pp. 186—187.

On the expression of Bessel's Functions in Form of Definite Integrals. *Johns Hopkins Univ. Circulars*, xiv. (1895), pp. 20—21.

Note on Cauchy's Numbers. *Annals of Math.* x. (1896), pp. 1—2.

On the relation between Cauchy's numbers and Bessel's Functions (July 1, 1898). *Annals of Math.* xii. (1899), pp. 170—174.

On some relations between Bessel functions of the first and of the second kind (Oct. 20, 1902). *Trans. Acad. Sci. of St Louis*, xii. (1902), pp. 99—108.

Sur l'équation de Bessel avec second membre (Oct. 27, 1902). *Comptes Rendus*, cxxxv. (1902), pp. 678—679.

Sur une classe d'équations différentielles rédactibles à l'équation de Bessel (May 11, 1903). *Comptes Rendus*, cxxxvi. (1903), pp. 1124—1126.

* See under Lord Rayleigh.

CHREE, C.

Longitudinal vibrations of a circular bar. *Quarterly Journal*, XXI. (1886), pp. 287—298.

On the Coefficients in certain Series of Bessel's Functions. *Phil. Mag.* (6) XVII. (1909), pp. 329—331.

CHRISTOFFEL, E. B.

Zur Abhandlung: "Ueber die Zähler und Nenner der Näherungswerte von Kettenbrüchen" pag. 231 des vorigen Bandes* (March, 1860). *Journal für Math.* LVIII. (1861), pp. 90—92.

CINELLI, M.

Diffrazione per aperture fatte sopra superfici curve. *Il Nuovo Cimento*, (4) I. (1895), pp. 141—155.

CLEBSCH, R. F. A.

Ueber die Reflexion an einer Kugelfläche (Oct. 30, 1861). *Journal für Math.* LXI. (1863), pp. 195—262.

CLIFFORD, W. K.

On Bessel's Functions†. *Mathematical Papers* (London, 1882), pp. 346—349.

COATES, C. V.

Bessel's functions of the Second order. *Quarterly Journal*, XX. (1885), pp. 250—260.

Bessel's functions of the Second order. *Quarterly Journal*, XXI. (1886), pp. 183—192.

COCKLE, SIR JAMES.

On Linear Differential Equations of the Second Order (Dec. 24, 1861; Jan. 15, 1862; May 21, 1862). *Messenger* (Old Series), I. (1862), pp. 118—124, 164—173, 241—247.

COTTER, J. R.

A New Method of Solving Legendre's and Bessel's Equations and others of a similar type (May 27, 1907). *Proc. R. Irish Acad.*‡ XXVII. A (1909), pp. 157—161.

CRAWFORD, L.

A proof of Rodrigues' Theorem§ $\sin nx = \frac{n}{1 \cdot 3 \cdot 5 \ldots (2n-1)} \left(\frac{1}{\sin x}\frac{d}{dx}\right)^{n-1} \sin^{2n-1} x$ and some expansions derived from it (Dec. 13, 1901). *Proc. Edinburgh Math. Soc.* XX. (1902), pp. 11—15.

CRELIER, L.

Sur quelques propriétés des fonctions Bésséliennes tirées de la théorie des fractions continues (June, 1895). *Ann. di Mat.* (2) XXIV. (1896), pp. 131—163. [*Dissertation, Bern*, 1895.]

Sur la fonction Bessélienne de IIe espèce $S^n(x)$ (Dec. 1896). *Bern Mittheilungen*, 1897 [1898], pp. 61—96.

Sur les fonctions besséliennes $O^n(x)$ et $S^n(x)$ (Sept. 6, 1897; Nov. 29, 1897). *Comptes Rendus*, CXXV. (1897), pp. 421—423, 860—863.

CURTIS, A. H.

On the integration of Linear and Partial Differential Equations (Nov. 24, 1854). *Camb. and Dublin Math. Journal*, IX. (1854), pp. 272—290.

CURZON, H. E. J.

Generalisations of the Hermite functions and their connexion with Bessel functions (Nov. 10, 1913). *Proc. London Math. Soc.* (2) XIII. (1914), pp. 417—440.

DATTA, A.

On a generalisation of Neumann's Expansion in a Series of Bessel Functions (Feb. 29, 1920). *Bulletin Calcutta Math. Soc.* XI. (1921), pp. 23—34.

On an extension of Sonine's Integral in Bessel Functions|| (Jan. 6, 1921). *Bulletin Calcutta Math. Soc.* XI. (1921), pp. 221—230.

* See under Heine.

† Clifford died March 3, 1879.

‡ Not *Trans. Camb. Phil. Soc.* XXI., as stated in the *Jahrbuch über die Fortschritte der Math.* 1908, p. 383.

§ See footnote † on p. 27.

|| This paper, which deals with the corrected form, given in § 13·46 (10), of Nicholson's integral, and with various related integrals, was published after Chapter XIII had been passed for Press.

DEBYE, P.

Näherungsformeln für die Zylinderfunktionen für grosse Werte des Arguments und unbeschränkt veränderliche Werte des Index (Dec. 1908). *Math. Ann.* LXVII. (1909), pp. 535—558.

Semikonvergente Entwickelungen für die Zylinderfunktionen und ihre Ausdehnung ins Komplexe (Feb. 5, 1910). *Münchener Sitzungsberichte*, XL. (1910), no. 5.

DE LA VALLÉE POUSSIN, C. J.

Intégration de l'équation de Bessel sous forme finie (Jan. 26, 1905). *Ann. de la Soc. Sci. de Bruxelles*, XXIX. (1ère partie) (1905), pp. 140—143.

DENDY, A. AND NICHOLSON, J. W.

On the Influence of Vibrations upon the Form of Certain Sponge Spicules (May 11, 1917). *Proc. Royal Soc.* LXXXIX. B (1917), pp. 573—587.

DINI, U.

Serie di Fourier e altere rappresentazioni analitiche delle funzioni di una variabile reale (Pisa, 1880).

DINNIK, A.

Über die Darstellung einer willkürlichen Funktion durch Bessel'sche Reihe. *Kief. Polyt. Inst.* (*Engineering Section*), 1911, no. 1, pp. 83—85. [*Jahrbuch über die Fortschritte der Math.* 1911, p. 492.]

Tafeln der Besselschen Funktionen $J_{\pm\frac{1}{3}}$ und $J_{\pm\frac{2}{3}}$. *Archiv der Math. und Phys.* (3) XVIII. (1911), pp. 337—338.

Tafeln der Besselschen Funktionen $J_{\pm\frac{1}{2}}$, $J_{\pm\frac{3}{2}}$, $J_{\pm\frac{5}{2}}$. *Archiv der Math. und Phys.* (3) XX. (1913), pp. 238—240.

Tafeln der Besselschen Funktionen $J_{\pm\frac{1}{4}}$ und $J_{\pm\frac{3}{4}}$. *Archiv der Math. und Phys.* (3) XXI. (1913), pp. 324—326.

Tafeln der Besselschen Funktionen $J_{\pm\frac{1}{3}}(xi)$ und $J_{\pm\frac{2}{3}}(xi)$. *Archiv der Math. und Phys.* (3) XXII. (1914), pp. 226—227.

DIXON, A. C.

On a property of Bessel's Functions. *Messenger*, XXXII. (1903), pp. 7—8.

The expansion of x^n in Bessel's Functions. *Messenger*, XXXII. (1903), p. 8.

DONKIN, W. F.

On the Equation of Laplace's Functions, etc. (Dec. 11, 1856). *Phil. Trans. of the Royal Soc.* CXLVII. (1857), pp. 43—57.

DOUGALL, J.

The determination of Green's function by means of Cylindrical or Spherical Harmonics (March 9, 1900). *Proc. Edinburgh Math. Soc.* XVIII. (1900), pp. 33—83.

A Theorem of Sonine in Bessel Functions with two Extensions to Spherical Harmonics (Dec. 13, 1918). *Proc. Edinburgh Math. Soc.* XXXVII. (1919), pp. 33—47.

DU BOIS REYMOND, P. D. G.

Die Theorie der Fourier'schen Integrale und Formeln (June 26, 1871). *Math. Ann.* IV. (1871), pp. 362—390.

EARNSHAW, S.

Partial differential equations. An essay towards an entirely new method of integrating them (London, 1871).

ELLIS, R. L.

On the Integration of certain Differential Equations (Nov. 1840 and Feb. 1841). *Camb. Math. Journal*, II. (1841), pp. 169—177, 193—201.

On the Method of Least Squares (March 4, 1844). *Trans. Camb. Phil. Soc.* VIII. (1849), pp. 204—219.

EMDE, F.*

Zur Berechnung der reellen Nullstellen der Bessel'schen Zylinderfunktionen. *Archiv der Math. und Phys.* (3) XXIV. (1916), pp. 239—250.

ENNEPER, A.

Ueber ein bestimmtes Integral. *Math. Ann.* VI. (1873), pp. 360—365.

EPSTEIN, S. S.

Die vier Rechnungsoperationen mit Bessel'schen Functionen nebst einer geschichtlichen Einleitung (Bern, 1894, 58 pp.). [*Jahrbuch über die Fortschritte der Math.* 1893—94, pp. 845—846.]

* See also under Jahnke.

ERMAKOFF, W.

Ueber die Cylinderfunctionen (May, 1872). *Math. Ann.* V. (1872), pp. 639—640.

ESCHERICH, G. VON.

Zur Bessel'schen Differential-Gleichung. *Monatshefte für Math. und Phys.* III. (1892), p. 142.

Über eine Näherungsformel. *Monatshefte für Math. und Phys.* III. (1892), p. 234.

EULER, L.

Lettre de M. Euler à M. de la Grange (Jan. 1, 1760). *Misc. Taurinensia*, II. (1760—61), pp. 1—10.

Recherches sur l'intégration de l'équation

$$\frac{ddz}{dt^2} = aa\frac{ddz}{dx^2} + \frac{b}{x}\frac{dz}{dx} + \frac{c}{xx}z.$$

Misc. Taurinensia, III. (1762—65), pp. 60—91.

De integratione aequationum differentialium. *Novi Comm. Acad. Petrop.* VIII. (1760—61) [1763], pp. 154—169.

De resolutione aequationis $dy + ayy\,dx = bx^m dx$. *Novi Comm. Acad. Petrop.* IX. (1762—63) [1764], pp. 154—169.

De motu vibratorio tympanorum. *Novi Comm. Acad. Petrop.* X. (1764) [1766], pp. 243—260.

Institutionum Calculi Integralis, II. (Petersburg, 1769).

De oscillationibus minimis funis libere suspensi. *Acta Acad. Petrop.* V. pars 1 (1781) [1784], pp. 157—177.

De perturbatione motus chordarum ab earum pondere oriunda. *Acta Acad. Petrop.* V. pars 1 (1781) [1784], pp. 178—190.

Analysis facilis aequationem Riccatianam per fract.onem continuam resolvendi. *Mém. de l'Acad. R. des Sci. de St. Pétersbourg*, VI. (1818), p] 12—29.

FALKENHAGEN, J. H. M.

Ueber das Verhalten der Integrale einer Riccati'schen Gleichung in der Nähe einer singulären Stelle. *Nieuw Archief voor Wiskunde*, (2) VI. (1905), pp. 209—248.

FAXÉN, H.

Expansion in series of the integral $\int_y^\infty e^{-x(t \pm t^{-\mu})} t^\nu dt$ (April 14, 1920). *Arkiv för Mat. Astr. och Fysik*, XV. (1921), no. 13.

FELDBLUM, M.

The theory of Riccati's equation and applications of the function which satisfies it. *Warsaw University*, 1898, nos. 5 and 7; 1899, no. 4. [*Jahrbuch über die Fortschritte der Math.* 1898, pp. 279—280.]

FÉRIET, K. DE.

Sur les fonctions hypercylindriques (June 13, 1921). *Comptes Rendus*, CLXXII. (1921), pp. 1464—1466.

FIELDS, J. C.

A method of solving Riccati's Equation (April 8, 1886). *Johns Hopkins Univ. Circulars* VI. (1886—87), p. 29.

Solutions Analogous to Riccati's of Equations of the form $\frac{d^n y}{dx^n} = x^m y$ (May 19, 1886). *Johns Hopkins Univ. Circulars*, VI. (1886—87), pp. 29—30.

FILON, L. N. G.

On a New Mode of Expressing Solutions of Laplace's Equation in Terms of Operators involving Bessel Functions. *Phil. Mag.* (6) VI. (1903), pp. 193—213.

On the expansion of polynomials in series of functions (May 10, 1906). *Proc. London Math. Soc.* (2) IV. (1906), pp. 396—430.

FORD, W. B.

On the possibility of differentiating term-by-term the developments for an arbitrary function of one real variable in terms of Bessel functions (June, 1902). *Trans. American Math. Soc.* IV. (1903), pp. 178—184.

FORSYTH, A. R.

On linear differential equations: in particular that satisfied by the series

$$1+\frac{\alpha\beta\theta}{\gamma\epsilon}x+\frac{\alpha\,.\,\alpha+1\,.\,\beta\,.\,\beta+1\,.\,\theta\,.\,\theta+1}{1\,.\,2\,.\,\gamma\,.\,\gamma+1\,.\,\epsilon\,.\,\epsilon+1}x^2+\ldots$$

Quarterly Journal, XIX. (1883), pp. 292—337.

The expression of Bessel functions of positive order as products, and of their inverse powers as sums of rational fractions. *Messenger*, L. (1921), pp. 129—149.

FOURIER, J. B. J.

La Théorie analytique de la Chaleur (Paris, 1822). [Translated by A. Freeman, Cambridge, 1878.]

FREEMAN, A.

Note on the value of the least root of an equation allied to $J_0(z)=0$ (April 19, 1880). *Proc. Camb. Phil. Soc.* III. (1880), pp. 375—377.

FRESNEL, A. J.

Mémoire sur la diffraction de la lumière [July 29, 1818; crowned 1819]. *Mém. de l'Acad. R. des Sci.* V. (1821—22), pp. 339—476. [*Oeuvres*, I. (1866), pp. 247—382.]

FRULLANI, G.

Sopra la dipendenza fra i differenziali delle funzioni e gli Integrali definiti (Feb. 4, 1818). *Mem. soc. ital.* (*Modena*), XVIII. (1820), pp. 458—517.

FUSS, P. H.

Correspondance mathématique et physique de quelques célèbres géomètres du XVIIIème *siècle**, II. (Petersburg, 1843).

GALLOP, E. G.

The distribution of electricity on the circular disc and the spherical bowl. *Quarterly Journal*, XXI. (1886), pp. 229—256.

GASSER, A.

Ueber die Nullstellen der Besselschen Funktionen (July, 1904). *Mittheilungen der Naturf. Ges. in Bern*, 1904, pp. 92—135.

GEGENBAUER, L.

Note über die Bessel'schen Functionen zweiter Art (Feb. 8, 1872). *Wiener Sitzungsberichte*, LXV. (2) (1872), pp. 33—35.

Zur Theorie der Bessel'schen Functionen zweiter Art (July 4, 1872). *Wiener Sitzungsberichte*, LXVI. (2) (1872), pp. 220—223.

Note über bestimmte Integrale (Feb. 6, 1873). *Wiener Sitzungsberichte*, LXVII. (2) (1873), pp. 202—204.

Über die Functionen X_n^m (June 13, 1873). *Wiener Sitzungsberichte*, LXVIII. (2) (1874), pp. 357—367.

Über die Bessel'schen Functionen (March 19, 1874). *Wiener Sitzungsberichte*, LXX. (2) (1875), pp. 6—16.

Über einige bestimmte Integrale (June 18, 1874). *Wiener Sitzungsberichte*, LXX. (2) (1875), pp. 433—443.

Über einige bestimmte Integrale (June 17, 1875). *Wiener Sitzungsberichte*, LXXII. (2) (1876), pp. 343—354.

Über die Bessel'schen Functionen (June 22, 1876). *Wiener Sitzungsberichte*, LXXIV. (2) (1877), pp. 124—130.

Zur Theorie der Bessel'schen Functionen (Jan. 18, 1877). *Wiener Sitzungsberichte*, LXXV. (2) (1877), pp. 218—222.

Über die Functionen $C_n^\nu(x)$ (April 12, 1877). *Wiener Sitzungsberichte*, LXXV. (2) (1877), pp. 891—905.

Das Additionstheorem derjenigen Functionen welche bei der Entwicklung von e^{axi} nach den Näherungsnennern regulärer Kettenbrüche auftreten. *Wiener Sitzungsberichte*, LXXXV. (2) (1882), pp. 491—502.

Über die Bessel'schen Functionen (Oct. 11, 1883). *Wiener Sitzungsberichte*, LXXXVIII. (2) (1884), pp. 975—1003.

* This work contains a number of letters from Nicholas Bernoulli (the younger) to Goldbach, in which Bernoulli's solution of Riccati's equation is to be found. The reader should notice that, in Daniel Bernoulli's letters to Goldbach (*ibid.* pp. 254, 256, 259), the equation described in the table of contents as Riccati's equation is really the linear equation; Riccati's equation is mentioned on p. 260.

Zur Theorie der Functionen $C_n^\nu(x)$. *Wiener Akad. Denkschriften*, XLVIII. (1884), pp. 293—316.

Über die Bessel'schen Functionen (March 10, 1887). *Wiener Sitzungsberichte*, XCV. (2) (1887), pp. 409—410.

Einige Sätze über die Functionen $C_n^\nu(x)$. *Wiener Akad. Denkschriften*, LVII. (1890), pp. 425—480.

Über die Ringfunctionen (June 4, 1891). *Wiener Sitzungsberichte*, C. (2 *a*) (1891), pp. 745—766.

Bemerkung zu der von Herrn Elsas gegebenen Theorie der elektrischen Schwingungen in cylindrischen Drahten. *Monatshefte für Math. und Phys.* IV. (1893), pp. 379—380.

Über die zum elektromagnetischen Potentiale eines Kreisstromes associierte Function. *Monatshefte für Math. und Phys.* IV. (1893), pp. 393—401.

Eine Integralrelation. *Monatshefte für Math. und Phys.* V. (1894), pp. 53—61.

Bemerkung über die Bessel'schen Functionen. *Monatshefte für Math. und Phys.* VIII. (1897), pp. 383—384.

Notiz über die Bessel'schen Functionen erster Art. *Monatshefte für Math. und Phys.* X. (1899), pp. 189—192.

Quelques propriétés nouvelles des racines des fonctions de Bessel. *Mém. de la Soc. R. des Sci. de Liège*, (3) II. (1900), no. 3.

[Letter on a paper by H. M. Macdonald.] *Proc. London Math. Soc.* XXXII. (1901), pp. 433—436.

Über eine Relation des Herrn Hobson (May 22, 1902). *Wiener Sitzungsberichte*, CXI. (2*a*) (1902), pp. 563—572.

On integrals containing functions of Bessel. [Letter to Kapteyn.] *Proc. Section of Sci., K. Akad. van Wet. te Amsterdam*, IV. (1902), pp. 584—588.

GENOCCHI, A.

Studi intorno ai casi d' integrazione sotto forma finita. *Mem. dell' Accad. delle Sci. di Torino*, XXIII. (1866), pp. 299—362.

Sur l'équation de Riccati (Aug. 13, 1877). *Comptes Rendus*, LXXXV. (1877), pp. 391—394.

GIBSON, G. A.

A Proof of the Binomial Theorem with some Applications (Dec. 12, 1919). *Proc. Edinburgh Math. Soc.* XXXVIII. (1920), pp. 6—9.

GILBERT, L. P.

Recherches analytiques sur la diffraction de la lumière. *Mém. couronnés de l'Acad. R. des Sci. de Bruxelles*, XXXI. (1863), pp. 1—52.

GIULIANI, G.

Sopra la funzione $P^n(\cos\gamma)$ per n infinito. *Giornale di Mat.* XXII. (1884), pp. 236—239.

Sopra alcune funzioni analoghe alle funzioni cilindriche. *Giornale di Mat.* XXV. (1887) pp. 198—202.

Alcune osservazioni sopra le funzioni spheriche di ordine superiore al secondo e sopra altre funzioni che se ne possono dedurre (April, 1888). *Giornale di Mat.* XXVI. (1888), pp. 155—171.

GLAISHER, J. W. L.

On Riccati's equation (March 15, 1871). *Quarterly Journal*, XI. (1871), pp. 267—273.

On the Relations between the particular Integrals in Cayley's solution of Riccati's Equation (May 12, 1872). *Phil. Mag.* (4) XLIII. (1872), pp. 433—438.

On the Evaluation in Series of certain Definite Integrals. *British Association Report*, 1872, pp. 15—17.

Notes on definite integrals. *Messenger*, II. (1873), pp. 72—79.

On a Differential Equation allied to Riccati's (Oct. 11, 1872). *Quarterly Journal*, XII. (1873), pp. 129—137.

Sur une Propriété de la Fonction $e^{\sqrt{x}}$. *Nouvelle Corr. Math.* II. (1876), pp. 240—243, 349—350.

On a Formula of Cauchy's for the Evaluation of a class of Definite Integrals (Nov. 6, 1876). *Proc. Camb. Phil. Soc.* III. (1880), pp. 5—12.

On certain Identical Differential Relations (Nov. 9, 1876). *Proc. London Math. Soc.* VIII. (1877), pp. 47—51.

A Generalised Form of Certain Series (May 9, 1878). *Proc. London Math. Soc.* IX. (1878), pp. 197—202.

On the Solution of a Differential Equation allied to Riccati's. *British Association Report*, 1878, pp. 469—470.

Example illustrative of a point in the solution of differential equations by series. *Messenger*, VIII. (1879), pp. 20—23.

On a symbolic theorem involving repeated differentiations (May 19, 1879). *Proc. Camb. Phil. Soc.* III. (1880), pp. 269—271.

On Riccati's Equation and its Transformations and on some Definite Integrals which satisfy them (June 16, 1881). *Phil. Trans. of the Royal Soc.* 172 (1881), pp. 759—828. [*Proc. Royal Soc.* XXXII. (1881), p. 444.]

GORDAN, P.

See under Hermite.

GRAF, J. H.

Ueber die Addition und Subtraction der Argumente bei Bessel'schen Functionen nebst einer Anwendung (March, 1893). *Math. Ann.* XLIII. (1893), pp. 136—144.

Ueber einige Eigenschaften der Bessel'schen Function erster Art, insbesondere für ein grosses Argument. *Zeitschrift für Math.* XXXVIII. (1893), pp. 115—120.

Beiträge zur Auflösung von linearen Differentialgleichungen zweiter Ordnung mit linearen Coefficienten so wie von Differentialgleichungen zweiter Ordnung denen gewisse bestimmte Integrale genügen (March, 1894). *Math. Ann.* XLV. (1894), pp. 235—262.

Relations entre la fonction Bessélienne de 1re espèce et une fraction continue (May, 1894). *Ann. di Mat.* (2) XXIII. (1895), pp. 45—65.

Ableitung der Formeln für die Bessel'schen Functionen bei welchen das Argument ein Distanz darstellt (Aug. 4, 1896). *Verhandlungen der Schweiz-Naturf. Ges.* LXXIX. (1896), pp. 59—62.

Einleitung in die Theorie der Besselschen Funktionen. Von J. H. Graf und E. Gubler (2 Hefte; Bern, 1898, 1900).

Beitrag zur Auflösung von Differentialgleichungen zweiter Ordnung denen gewisse bestimmte Integrale genügen (May, 1902). *Math. Ann.* LVI. (1903), pp. 423—444.

GRAY, A.*

A Treatise on Bessel Functions. By Andrew Gray and G. B. Mathews (London, 1895).

GREENHILL, SIR A. GEORGE.

On Riccati's Equation and Bessel's Equation. *Quarterly Journal*, XVI. (1879), pp. 294—298.

On the Differential Equation of the Ellipticities of the Strata in the Theory of the Figure of the Earth (April 8, 1880). *Quarterly Journal*, XVII. (1880), pp. 203—207.

Determination of the greatest height consistent with stability that a vertical pole or mast can be made, and of the greatest height to which a tree of given proportions can grow (Feb. 7, 1881). *Proc. Camb. Phil. Soc.* IV. (1883), pp. 65—73.

The Bessel-Clifford Function (March 14, 1919). *Engineering*, CVII. (1919), p. 334.

The Bessel-Clifford Function and its applications (Aug. 11, 1919). *Phil. Mag.* (6) XXXVIII. (1919), pp. 501—528.

GRUNERT, J. A.

Beweis der Gleichung $\dfrac{d^{i-1}(1-z^2)^{i-\frac{1}{2}}}{dz^{i-1}} = (-1)^{i-1}1.3\ldots(2i-1)\dfrac{\sin ix}{i}$ für $z=\cos x$. *Archiv der Math. und Phys.* IV. (1844), pp. 104—109.

GUBLER, E.†

Die Darstellung der allgemeinen Bessel'schen Function durch bestimmte Integrale (Sept. 1888). *Zürich Vierteljahrsschrift*, XXXIII. (1888), pp. 130—172.

Verwandlung einer hypergeometrischen Reihe im Anschluss an das Integral

$$\int_0^\infty J^a(x)\, e^{-bx} x^{c-1}\, dx.$$

Inaugural-dissertation, *Zürich*, 1894 (38 pp.). [Graf and Gubler, *Einleitung in die Theorie der Besselschen Funktionen*, II. (Bern, 1900), pp. 110—135, 156.]

Ueber ein discontinuierliches Integrale (Dec. 1895). *Math. Ann.* XLVIII. (1897), pp. 37—48.

Beweis einer Formel des Herrn Sonine (Dec. 1896). *Math. Ann.* XLIX. (1897), pp. 583—584.

Ueber bestimmte Integrale mit Bessel'schen Functionen (Oct. 1902). *Zürich Vierteljahrsschrift*, XLVII. (1902), pp. 422—428.

GÜNTHER, S.

Bemerkungen über Cylinderfunctionen. *Archiv der Math. und Phys.* LVI. (1874), pp. 292—297.

* See also under Sir Joseph John Thomson.

† See also under Graf.

GWYTHER, R. F.

The employment of a geometrical construction to prove Schlömilch's series, and to aid in its development into a definite integral. *Messenger*, XXXIII. (1904), pp. 97—107.

HADAMARD, J.

Sur l'expression asymptotique de la fonction de Bessel. *Bulletin de la Soc. Math. de France*, XXXVI. (1908), pp. 77—85.

HAENTZSCHEL, E.

Ueber die functionentheoretischen Zusammenhang zwischen den Lamé'schen, Laplace'-schen und Bessel'schen Functionen. *Zeitschrift für Math.* XXXI. (1886), pp. 25—33.

Ueber die Fourier Bessel'sche Transcendente (Nov. 20, 1887). *Zeitschrift für Math.* XXXIII. (1888), pp. 185—186.

HAFEN, M.

Studien über einige Probleme der Potentialtheorie. *Math. Ann.* LXIX. (1910), pp. 517—537.

HAGUE, B.

A Note on the Graphs of the Bessel Functions of Integral Order. *Proc. Phys. Soc.* XXIX. (1917), pp. 211—214.

HALL, A.

The Besselian Function. *The Analyst*, I. (1874), pp. 81—84.

HAMILTON, SIR WILLIAM ROWAN*.

On Fluctuating Functions (June 22, 1840). *Trans. R. Irish Acad.* XIX. (1843), pp. 264—321.

On the Calculation of the Numerical Values of a certain class of Multiple and Definite Integrals (Sept. 29, 1857). *Phil. Mag.* (4) XIV. (1857), pp. 375—382.

HANKEL, H.

Die Cylinderfunctionen erster und zweiter Art (Dec. 15, 1868). *Math. Ann.* I. (1869), pp. 467—501.

Bestimmte Integrale mit Cylinderfunctionen†. *Math. Ann.* VIII. (1875), pp. 453—470.

Die Fourier'schen Reihen und Integrale für Cylinderfunctionen† (May 16, 1869). *Math. Ann.* VIII. (1875), pp. 471—494.

HANSEN, P. A.

Ermittelung der absoluten Störungen in Ellipsen von beliebiger Excentricität und Neigung, I. *Schriften der Sternwarte Seeberg* (Gotha, 1843). [Mémoire sur la détermination des perturbations absolues dans les ellipses d'une excentricité et d'une inclinaison quelconques. Par M. Hansen. Traduit de l'Allemand par M. Victor Mauvais (Paris, 1845).]

Entwickelung des Products einer Potenz des Radius Vectors mit den Sinus oder Cosinus eines Vielfachen der wahren Anomalie in Reihen. *Leipziger Abh.* II. (1855), pp. 181—281.

HANUMANTA RAO, C. V.

On a certain definite integral. *Messenger*, XLVII. (1918), pp. 134—137.

HARDY, G. H.

General theorems in contour integration: with some applications. *Quarterly Journal*, XXXII. (1901), pp. 369—384.

Notes on some points in the integral calculus, XVIII. *Messenger*, XXXV. (1906), pp. 158—166.

Further researches in the Theory of Divergent Series and Integrals (May 18, 1908). *Trans. Camb. Phil. Soc.* XXI. (1912), pp. 1—48.

On an Integral Equation (Feb. 20, 1909). *Proc. London Math. Soc.* (2) VII. (1909), pp. 445—472.

On certain definite integrals whose values can be expressed in terms of Bessel's functions. *Messenger*, XXXVIII. (1909), pp. 129—132.

On certain definite integrals considered by Airy and Stokes. *Quarterly Journal*, XLI. (1910), pp. 226—240.

Notes on some points in the integral calculus, XXVII. *Messenger*, XL. (1911), pp. 44—51.

Notes on some points in the integral calculus, XXXV. *Messenger*, XLII. (1913), pp. 89—93.

On the expression of a number as the sum of two squares. *Quarterly Journal*, XLVI. (1915), pp. 263—283.

On Dirichlet's divisor problem (April 22, 1915). *Proc. London Math. Soc.* (2) XV. (1916), pp. 1—25.

Notes on some points in the integral calculus, XLVII. *Messenger*, XLVIII. (1918), pp. 81—88.

* A letter by Hamilton on Bessel functions is published in *Sir G. G. Stokes, Memoir and Scientific Correspondence*, I. (Cambridge, 1907), pp. 131—135.

† Hankel died Aug. 29, 1873. These memoirs were composed from materials found among his papers.

HARGREAVE, C. J.

On the Solution of Linear Differential Equations (June 10, 1847). *Phil. Trans. of the Royal Soc.* 1848, pp. 31—54.

On Riccati's Equation (April 4, 1865). *Quarterly Journal*, VII. (1866), pp. 256—258.

HARGREAVES, R.

A Diffraction Problem and an Asymptotic Theorem in Bessel's Series. *Phil. Mag.* (6) XXXVI. (1918), pp. 191—199.

HARNACK, A.*

Ueber die Darstellung einer willkürlichen Function durch die Fourier-Bessel'schen Functionen (Dec. 12, 1887). *Leipziger Berichte*, XXXIX. (1887), pp. 191—214; *Math. Ann.* XXXV. (1889), pp. 41—62.

HARRIS, R. A.

On Harmonic Functions. *American Journal of Math.* XXXIV. (1912), pp. 391—420.

HARTENSTEIN, J. H.

Integration der Differentialgleichung $\frac{\partial^2 f}{\partial x^2}+\frac{\partial^2 f}{\partial y^2}=k^2 f$ für elliptische und parabolische Coordinaten. *Archiv der Math. und Phys.* (2) XIV. (1896), pp. 170—199.

HATTENDORF, K.

See under Riemann.

HAVELOCK, T. H.

Mathematical Analysis of Wave Propagation in Isotropic Space of p Dimensions (March 16, 1904). *Proc. London Math. Soc.* (2) II. (1904), pp. 122—137.

HAYASHI, T.

On a definite integral for Neumann's cylindrical function. *Nyt Tidsskrift*, XXIII. B (1912), pp. 86—90. [*Jahrbuch über die Fortschritte der Math.* 1912, p. 555.]

On the Integrals $\int_0^\pi \exp\left(x \begin{smallmatrix}\cos\\ \sin\end{smallmatrix} p\theta\right) \begin{smallmatrix}\cos\\ \sin\end{smallmatrix} q\theta\, d\theta$ (Dec. 1920). *Tôhoku Math. Journal*, XX. (1922), pp. 107—114.

HEAVISIDE, O.

Electrical Papers, I., II. (London, 1892).

On Operators in Physical Mathematics (Dec. 15, 1892; June 8, 1893). *Proc. Royal Soc.* LII. (1893), pp. 504—529; LIV. (1893), pp. 105—143.

Electromagnetic Theory†, II., III. (London, 1899, 1912).

HEINE, H. E.

Ueber die Zähler und Nenner der Näherungswerthe von Kettenbruchen‡ (Sept. 1859). *Journal für Math.* LVII. (1860), pp. 231—247.

Die *Fourier-Bessel*sche Function (June, 1868). *Journal für Math.* LXIX. (1869), pp. 128—141.

Handbuch der Kugelfunctionen: Theorie und Anwendungen (2 Bände) (Berlin, 1878, 1881).

HERMITE, C.

Sur la transcendante E_n. *Ann. di Mat.* (2) III. (1870), p. 83§.

Extrait d'une lettre de Monsieur *Ch. Hermite* à Monsieur *Paul Gordan* (June 9, 1873). *Journal für Math.* LXXVI. (1873), pp. 303—311.

Extrait d'une lettre à M. E. Jahnke (Nov. 25, 1900). *Archiv der Math. und Phys.* (3) I. (1901), pp. 20—21.

HERTZ, H.

Über die Induktion in rotierenden Kugeln. *Dissertation, Berlin*, March 15, 1880. [*Ges. Werke*, I. (Leipzig, 1895), pp. 37—134.]

Über das Gleichgewicht schwimmender elasticher Platten. *Ann. der Physik und Chemie*, (3) XXII. (1884), pp. 449—455. [*Ges. Werke*, I. (Leipzig, 1895), pp. 288—294.]

* Harnack died April 3, 1888.

† This work consists of a series of articles first published in *The Electrician*, *Nature* and elsewhere in and after 1894, with numerous additions.

‡ See also under Christoffel.

§ This note contains a statement of Carlini's formula, which Hermite apparently derived from Poisson's integral.

HERZ, N.

Bemerkungen zur Theorie der Bessel'schen Functionen (Sept. 21, 1883). *Astr. Nach.* CVII. (1884), col. 17—28.

Note, betreffend die Entwicklung der störenden Kräfte (Mar. 30, 1884). *Astr. Nach.* CVII. (1884), col. 429—432.

HILB, E.

Zur Theorie der Entwicklungen willkürlicher Funktionen nach Eigenfunktionen (Sept. 11, 1917). *Math. Zeitschrift*, I. (1918), pp. 58—69.

Über die Laplacesche Reihe (March 15, 1919; Nov. 17, 1919). *Math. Zeitschrift*, V. (1919), pp. 17—25; VIII. (1920), pp. 79—90.

HILL, C. J. D.

De radicibus rationalibus aequationis Riccatianae

$$\partial_x y + a + by + cy^2 = 0,$$

ubi a, b, c functiones rationales ipsius x (May 24, 1840). *Journal für Math.* XXV. (1843), pp. 22—37.

HOBSON, E. W.*

Systems of Spherical Harmonics (June 11, 1891). *Proc. London Math. Soc.* XXII. (1891), pp. 431—449.

On the Evaluation of a certain Surface-Integral, and its application to the Expansion, in Series, of the Potential of Ellipsoids (Jan. 12, 1893). *Proc. London Math. Soc.* XXIV. (1893), pp. 80—96.

On Bessel's Functions, and Relations connecting them with Hyper-spherical and Spherical Harmonics (Dec. 14, 1893). *Proc. London Math. Soc.* XXV. (1894), pp. 49—75.

On the most general solution of given Degree of Laplace's Equation (May 9, 1895). *Proc. London Math. Soc.* XXVI. (1895), pp. 492—494.

Note on some properties of Bessel's Functions (Jan. 14, 1897). *Proc. London Math. Soc.* XXVIII. (1897), pp. 370—375.

On the representation of a function by series of Bessel's functions (Dec. 10, 1908). *Proc. London Math. Soc.* (2) VII. (1909), pp. 359—388.

HOPF, L. UND SOMMERFELD, A. J. W.

Über komplexe Integraldarstellungen der Zylinderfunktionen. *Archiv der Math. und Phys.* (3) XVIII. (1911), pp. 1—16.

HORN, J.

Ueber lineare Differentialgleichungen mit einem veränderlichen Parameter (Dec. 30, 1898). *Math. Ann.* LII. (1899), pp. 340—362.

HURWITZ, A.

Ueber die Nullstellen der Bessel'schen Function (June 2, 1888). *Math. Ann.* XXXIII. (1889), pp. 246—266.

Ueber die Wurzeln einiger transcendenten Gleichungen. *Hamburger Mittheilungen*, II. (1890), pp. 25—31. [*Jahrbuch über die Fortschritte der Math.* 1890, p. 115.]

HYMERS, J.

Treatise on Differential Equations, and on the Calculus of Finite Differences (Cambridge 1839).

IGNATOWSKY, W. VON.

Über die Reihenentwicklungen mit Zylinderfunktionen (May 13, 1911). *Archiv der Math. und Phys.* (3) XVIII. (1911), pp. 322—327.

Über Reihen mit Zylinderfunktionen nach dem Vielfachen des Argumentes (Dec. 1913). *Archiv der Math. und Phys.* (3) XXIII. (1915), pp. 193—219.

ISELI, F.

Die Riccati'sche Gleichung. *Dissertation, Bern*, 1909 (42 pp.). [*Jahrbuch über die Fortschritte der Math.* 1909, p. 369.]

ISHERWOOD, J. G.

Tables of the Bessel Functions for pure imaginary values of the argument (April 26, 1904). *Manchester Memoirs*, XLVIII. (1903—4), no. 19.

JACKSON, F. H.

Generalised forms of the series of Bessel and Legendre. *Proc. Edinburgh Math. Soc.* XXI. (1903), pp. 65—72.

On Generalised functions of Legendre and Bessel (Nov. 16, 1903). *Trans. Edinburgh Royal Soc.* XLI. (1905), pp. 1—28.

* See also under Gegenbauer.

A generalization of Neumann's expansion of an arbitrary function in a series of Bessel functions (Nov. 26, 1903). *Proc. London Math. Soc.* (2) I. (1904), pp. 361—366.

Theorems relating to a Generalisation of the Bessel-Function (March 21, 1904). *Trans. Edinburgh Royal Soc.* XLI. (1905), pp. 105—118.

Note on a theorem of Lommel (May 13, 1904). *Proc. Edinburgh Math. Soc.* XXII. (1904), pp. 80—85.

The application of Basic numbers to Bessel's and Legendre's functions (June 2, 1904; Dec. 1, 1904; Jan. 18, 1905). *Proc. London Math. Soc.* (2) II. (1905), pp. 192—220; (2) III. (1905), pp. 1—20, 21—23.

The complete solution of the differential equation for $J_{[n]}$ (July 4, 1904). *Proc. Edinburgh Royal Soc.* XXV. (1904), pp. 273—276.

Theorems relating to a Generalization of Bessel's Function (Feb. 20, 1905). *Trans. Edinburgh Royal Soc.* XLI. (1905), pp. 399—408.

JACKSON, W. H.

On the diffraction of light produced by an opaque prism of finite angle (Nov. 11, 1903). *Proc. London Math. Soc.* (2) I. (1904), pp. 393—414.

JACOBI, C. G. J.*

Formula transformationis integralium definitorum (July 9, 1835). *Journal für Math.* XV. (1836), pp. 1—26. [*Ges. Math. Werke*, VI. (1891), pp. 86—118.]—Formule pour la transformation d'une classe d'intégrales définies, *Journal de Math.* I. (1836), pp. 195—196.

Versuch einer Berechnung der grossen Ungleichheit des Saturns nach einer strengen Entwicklung. *Astr. Nach.* XXVIII. (1849), col. 65—80, 81—94. [*Ges. Math. Werke*, VII. (1891), pp. 145—174.]

Über die annähernde Bestimmung sehr entfernter Glieder in der Entwickelung der elliptischen Coordinaten, nebst einer Ausdehnung der *Laplace*'schen Methode zur Bestimmung der Functionen grosser Zahlen. *Astr. Nach.* XXVIII. (1849), col. 257—270. [*Ges. Math. Werke*, VII. (1891), pp. 175—188.]

JAHNKE, P. R. E.†

Funktionentafeln mit Formeln und Kurven. Von E. Jahnke und F. Emde (Leipzig, 1909).

Über einige, in der elektromagnetischen Strahlungstheorie auftretende, bestimmte Integrale. *Archiv der Math. und Phys.* (3) XXIII. (1914), pp. 264—267.

JAMET, E. V.

Sur les équations anharmoniques (Sept. 13, 1901). *Comptes Rendus de l'Assoc. Française*, XXX. (Ajaccio) (1901), pp. 207—228.

Sur les équations anharmoniques. *Ann. de la Fac. des Sci. de Marseille*, XII. (1902), pp. 1—21.

JEKHOWSKY, B.

Les fonctions de Bessel de plusieurs variables exprimées par des fonctions de Bessel d'une variable (Feb. 28, 1916). *Comptes Rendus*, CLXII. (1916), pp. 318—319.

Sur la fonction génératrice des fonctions de Bessel à plusieurs variables. *Bulletin des Sci. math.* (2) XLI. (1917), pp. 58—60.

Sur les fonctions de Bessel à deux variables (May 30, 1921). *Comptes Rendus*, CLXXII. (1921), pp. 1331—1332.

JOHNSON, W. W.

On the Differential Equation

$$\frac{dy}{dx}+y^2+Py+Q=0.$$

Annals of Math. III. (1887), pp. 112—115.

JOLLIFFE, A. E.

The expansion of the square of a Bessel function in the form of a series of Bessel functions. *Messenger*, XLV. (1916), p. 16.

JULIUS, V. A.

Sur les fonctions de Bessel de deuxième espèce (Dec. 1893). *Archives Néerlandaises*, XXVIII. (1895), pp. 221—225.

Sur les ondes lumineuses sphériques et cylindriques (Dec. 1893). *Archives Néerlandaises*, XXVIII. (1895), pp. 226—244.

* See also under Carlini. † See also under Hermite.

KALÄHNE, A.

Über die Wurzeln einiger Zylinderfunktionen und gewisser aus ihnen gebildeter Gleichungen (March 1, 1906). *Zeitschrift für Math. und Phys.* LIV. (1907), pp. 55—86.

KAPTEYN, W.*

Nouvelles formules pour répresenter la fonction $J_{n-\frac{1}{2}}(x)$. *Bulletin des Sci. Math.* (2) XVI. (1892), pp. 41—44.

Over BESSEL'sche Functiën (March 12, 1892). *Nieuw Archief voor Wiskunde*, XX. (1893), pp. 116—127.

Recherches sur les fonctions de Fourier-Bessel. *Ann. sci. de l'École norm. sup.* (3) X. (1893), pp. 91—120.

Sur quelques intégrales définies contenant des fonctions de Bessel. *Archives Néerlandaises*, (2) VI. (1901), pp. 103—116.

A definite integral containing Bessel's functions (June 29, 1901). *Proc. Section of Sci., K. Akad. van Wet. te Amsterdam*, IV. (1902), pp. 102—103.

Sur un développement de M. Neumann. *Nieuw Archief voor Wiskunde*, (2) VI. (1905), pp. 49—55.

Einige Bemerkungen über Bessel'sche Functionen. *Monatshefte für Math. und Phys.* XIV. (1903), pp. 275—282.

The values of some definite integrals connected with Bessel functions (Nov. 26, 1904). *Proc. Section of Sci., K. Akad. van Wet. te Amsterdam*, VII. (1905), pp. 375—376.

On a series of Bessel functions (Dec. 24, 1904). *Proc. Section of Sci., K. Akad. van Wet. te Amsterdam*, VII. (1905), pp. 494—500.

A definite integral of Kummer (Sept. 30, 1905). *Proc. Section of Sci., K. Akad. van Wet. te Amsterdam*, VIII. (1906), pp. 350—357.

On an expansion of an arbitrary function in a series of Bessel functions. *Messenger*, XXXV. (1906), pp. 122—125.

Sur la sommation d'une série infinie. *Nieuw Archief voor Wiskunde*, (2) VII. (1907), pp. 20—25.

The quotient of two successive Bessel Functions (Dec. 30, 1905 ; Jan. 27, 1906). *Proc. Section of Sci., K. Akad. van Wet. te Amsterdam*, VIII. (1906), pp. 547—549, 640—642.

Sur le quotient de deux fonctions besséliennes successives. *Archives Néerlandaises*, (2) XI. (1906), pp. 149—168.

Recherches sur les fonctions cylindriques. *Mém. de la Soc. R. des Sci. de Liège*, (3) VI. (1906), no. 5.

Sur le calcul numérique de la série

$$\sum_{s=0}^{\infty} \frac{1}{(a^2+\beta^2 s^2)^{\frac{q}{2}}}.$$

Mém. de la Soc. R. des Sci. de Liège, (3) VI. (1906), no. 9.

On some relations between Bessel's Functions (Feb. 24, 1912). *Proc. Section of Sci., K. Akad. van Wet. te Amsterdam*, XIV. (1912), pp. 962—969.

KELVIN, LORD.

On the waves produced by a single impulse in water of any depth or in a dispersive medium (Feb. 3, 1887). *Proc. Royal Soc.* XLII. (1887), p. 80; *Phil. Mag.* (5) XXIII. (1887), pp. 252—255. [*Math. and Phys. Papers*, IV. (1910), pp. 303—306.]

Ether, Electricity and Ponderable Matter (Jan. 10, 1889). *Journal of Inst. of Electrical Engineers*, XVIII. (1890), pp. 4—37. [*Math. and Phys. Papers*, III. (1890), pp. 484—515.]

KEPIŃSKI, S.

Über die Differentialgleichung

$$\frac{\partial^2 z}{\partial x^2}+\frac{m+1}{x}\frac{\partial z}{\partial x}-\frac{n}{x}\frac{\partial z}{\partial t}=0$$

(Jan. 1905). *Math. Ann.* LXI. (1906), pp. 397—406.

Integration der Differentialgleichung

$$\frac{\partial^2 j}{\partial \xi^2}-\frac{1}{\xi}\frac{\partial j}{\partial t}=0.$$

Bull. int. de l'Acad. des Sci. de Cracovie, 1905, pp. 198—205.

* See also under Gegenbauer.

KIRCHHOFF, G.

Ueber den inducirten Magnetismus eines unbegrenzten Cylinders von weichem Eisen (June, 1853). *Journal für Math.* XLVIII. (1854), pp. 348—376.

Zur Theorie der Bewegung der Elektricität in unterseeischen oder unterirdischen Telegraphendrahten (Oct. 29, 1877). *Berliner Monatsberichte*, 1877, pp. 598—611.

Ueber die Transversalschwingungen eines Stabes von veränderlichem Querschnitt (Oct. 27, 1879). *Berliner Monatsberichte*, 1879, pp. 815—828; *Ann. der Physik und Chemie*, (3) X. (1880), pp. 501—512.

Ueber die elektrischen Strömungen in einem Kreiscylinder (April 26, 1883). *Berliner Sitzungsberichte*, 1883, pp. 519—524.

KLUYVER, J. C.

A local probability problem (Sept. 30, 1905). *Proc. Section of Sci., K. Akad. van Wet. te Amsterdam*, VIII. (1906), pp. 341—350.

An integral theorem of Gegenbauer (Feb. 27, 1909). *Proc. Section of Sci., K. Akad. van Wet. te Amsterdam*, XI. (1909), pp. 749—755.

KNESER, J. C. C. A.

Die Entwicklung willkürlicher Funktionen in Reihen die nach Bessel'schen Funktionen fortschreiten (July, 1903). *Archiv der Math. und Phys.* (3) VII. (1903), pp. 123—133.

Die Theorie der Integralgleichungen und die Darstellung willkürlicher Funktionen in der mathematischen Physik. *Math. Ann.* LXIII. (1907), pp. 477—524.

KNOCKENHAUER, K. W.

Ueber die Oerter der Maxima und Minima des gebeugten Lichtes nach den Fresnel'schen Beobachtungen. *Ann. der Physik und Chemie*, (2) XLI. (1837), pp. 103—110.

KÖNIG, J.

Ueber die Darstellung von Functionen durch unendliche Reihen (Sept. 1871). *Math. Ann.* V. (1872), pp. 310—340.

Ueber Reihenentwicklung nach Bessel'schen Functionen (1880). *Math. Ann.* XVII. (1880), pp. 85—86.

KOESTLER, W.

Beiträge zu Reihenentwickelungen nach Besselsche Zylinderfunktionen. *Dissertation, Bern*, 1907 (110 pp.). [*Jahrbuch über die Fortschritte der Math.* 1908, p. 535.]

KOPPE, M.

Die Ausbreitung einer Erschütterung an der Wellenmaschine darstellbar durch einen neuen Grenzfall der Bessel'schen Functionen. *Programm* (96). *Andreas Realgymn. Berlin*, 1899 (28 pp.). [*Jahrbuch über die Fortschritte der Math.* 1899, pp. 420—421.]

KUMMER, E. E.

Sur l'intégration de l'équation de Riccati par des intégrales définies. *Journal für Math.* XII. (1834), pp. 144—147.

Ueber die hypergeometrische Reihe

$$1+\frac{\alpha.\beta}{1.\gamma}x+\frac{\alpha(\alpha+1)\beta(\beta+1)}{1.2.\gamma(\gamma+1)}x^2+\frac{\alpha(\alpha+1)(\alpha+2)\beta(\beta+1)(\beta+2)}{1.2.3.\gamma(\gamma+1)(\gamma+2)}x^3+\dots$$

Journal für Math. XV. (1836), pp. 39—83, 127—172.

De integralibus quibusdam definitis et seriebus infinitis (April, 1837). *Journal für Math.* XVII. (1837), pp. 228—242.

Note sur l'intégration de l'équation $\dfrac{d^n y}{dx^n}=x^m.y$. *Journal für Math.* XIX. (1839), pp. 286—288.

Sur l'intégration de l'équation $\dfrac{d^n y}{dx^n}=x^m.y$. *Journal de Math.* IV. (1839), pp. 390—391.

LAGRANGE, J. L. DE.

Sur le problème de Kepler (Nov. 1, 1770). *Hist. de l'Acad. R. des Sci. de Berlin*, XXV. (1769) [1771], pp. 204—233. [*Oeuvres*, III. (1869), pp. 113—138.]

LAMB, H.

On the Oscillations of a Viscous Spheroid (Nov. 10, 1881). *Proc. London Math. Soc.* XIII. (1882), pp. 51—66.

On the Vibrations of an Elastic Sphere (May 11, 1882). *Proc. London Math. Soc.* XIII. (1882), pp. 189—212.

On Electrical Motions in a Spherical Conductor (March 14, 1883). *Phil. Trans. of the Royal Soc.* CLXXIV. (1883), pp. 519—549.

On the Induction of Electric Currents in Cylindrical and Spherical Conductors (Jan. 10, 1884). *Proc. London Math. Soc.* XV. (1884), pp. 139—149.

Note on the Induction of Electric Currents in a Cylinder placed across the Lines of Magnetic Force (June 12, 1884). *Proc. London Math. Soc.* XV. (1884), pp. 270—274.

On the Motion of a Viscous Fluid contained in a Spherical Vessel (Nov. 13, 1884). *Proc. London Math. Soc.* XVI. (1885), pp. 27—43.

A problem in Resonance illustrative of the Theory of Selective Absorption of Light (Jan. 8, 1900). *Proc. London Math. Soc.* XXXII. (1901), pp. 11—20.

Problems relating to the Impact of Waves on a Spherical Obstacle in an Elastic Medium (March, 8, 1900). *Proc. London Math. Soc.* XXXII. (1901), pp. 120—150.

On Boussinesq's Problem (Feb. 13, 1902). *Proc. London Math. Soc.* XXXIV. (1902), pp. 276—284.

On Deep-water Waves (Nov. 10, 1904). *Proc. London Math. Soc.* (2) II. (1905), pp. 371—400. [Presidential Address.]

On the theory of waves propagated vertically in the atmosphere (Dec. 6, 1908). *Proc. London Math. Soc.* (2) VII. (1909), pp. 122—141.

LAMBERT, A.

See Wangerin, A.

LANDAU, E. G. H.

Über die Gitterpunkte in einem Kreise (May 8, 1915; June 5, 1915; June 4, 1920). *Nachrichten von der K. Ges. der Wiss. zu Göttingen*, 1915, pp. 148—160, 161—171; 1920, pp. 109—134.

Zur analytischen Zahlentheorie der definiten quadratischen Formen (Über die Gitterpunkte in einem mehrdimensionalen Ellipsoid) (May 20, 1915). *Berliner Sitzungsberichte* (*Math.-phys. Klasse*), XXXI. (1915), pp. 458—476.

Über Dirichlets Teilerproblem (July 3, 1915). *Münchener Sitzungsberichte*, 1915, pp. 317—328.

LAPLACE, P. S. DE.

Mémoire sur la diminution de la durée du jour par le refroidissement de la terre. *Conn. des Tems*, 1823 [published 1820], pp. 245—257.

Traité de Mécanique Céleste, V. (Paris, 1825 and 1882).

LAURENT, PAUL MATHIEU HERMANN.

Mémoire sur les fonctions de Legendre. *Journal de Math.* (3) I. (1875), pp. 373—398.

LEBEDEFF, WERA MYLLER.

Über die Anwendung der Integralgleichungen in einer parabolischen Randwertaufgabe *Math. Ann.* LXVI. (1909), pp. 325—330.

LEBESGUE, V. A.

Remarques sur l'équation $y''+\frac{m}{x}y'+ny=0$. *Journal de Math.* XI. (1846), pp. 338—340.

LEFORT, F.

Expression numérique des intégrales définies qui se présentent quand on cherche les termes généraux du développement des coordonnées d'une planète, dans son mouvement elliptique. *Journal de Math.* XI. (1846), pp. 142—152.

LEIBNIZ, G. W.

See Daniel Bernoulli.

LE PAIGE, C.*

Note sur l'équation $xy''+ky'-xy=0$. *Bull. de l'Acad. R. de Belgique*, (2) XLI. (1876), pp. 1011—1016.

LERCH, M.

Mittheilungen aus der Integralrechnung. *Monatshefte für Math. und Phys.* I. (1890), pp. 105—112.

Betrachtungen über einige Fragen der Integralrechnung. *Rozpravy*, V. (1896), no. 23 (16 pp.). [*Jahrbuch über Fortschritte der Math.* 1896, pp. 233—234.]

* See also under Catalan.

LINDNER, P.

Die Beziehungen der begrenzten Ableitungen mit komplexen Zeiger zu den Besselschen Funktionen und ihren Nullstellen (Nov. 29, 1911). *Sitz. der Berliner Math. Ges.* XI. (1911), pp. 3—5.

LINDSTEDT, A.

Zur Theorie der Fresnel'schen Integrale (Aug. 1882). *Ann. der Physik und Chemie*, (3) XVII. (1882), pp. 720—725.

LIOUVILLE, J.

Sur la classification des transcendantes et sur l'impossibilité d'exprimer des racines de certaines équations en fonctions finies explicites des coefficients (June 8, 1835). *Journal de Math.* II. (1837), pp. 56—105; III. (1838), pp. 523—547.

Sur l'intégration d'une classe d'Équations différentielles du second ordre en quantités finies explicites (Oct. 28, 1839). *Comptes Rendus*, IX. (1839), pp. 527—530; *Journal de Math.* IV. (1839), pp. 423—456.

Remarques nouvelles sur l'équation de Riccati (Nov. 9, 1840). *Comptes Rendus*, XI. (1840), p. 729; *Journal de Math.* VI. (1841), pp. 1—13.

Sur l'intégrale $\int_0^\pi \cos i(u - x\sin u)\, du$. *Journal de Math.* VI. (1841), p. 36.

Sur une formule de M. Jacobi (March, 1841). *Journal de Math.* VI. (1841), pp. 69—73.

LIPSCHITZ, R. O. S.

Ueber ein Integral der Differentialgleichung $\frac{\partial^2 I}{\partial x^2} + \frac{1}{x}\frac{\partial I}{\partial x} + I = 0$ (July 14, 1858). *Journal für Math.* LVI. (1859), pp. 189—196.

LOBATTO, R.

Sur l'intégration des équations

$$\frac{d^n y}{dx^n} - xy = 0, \quad \frac{d^2 y}{dx^2} + abx^n y = 0$$

(April, 1837). *Journal für Math.* XVII. (1837), pp. 363—371.

LODGE, A.

Note on the Semiconvergent Series of $J_n(x)$. *British Association Report*, *York*, 1906, pp. 494—498.

LOMMEL, E. C. J. VON.

Beiträge zur Theorie der Beugung des Lichts. *Archiv der Math. und Phys.* XXXVI. (1861), pp. 385—419.

Methode zur Berechnung einer Transcendenten. *Archiv der Math und Phys.* XXXVII. (1861), pp. 349—360.

Zur Integration linearer Differentialgleichungen. *Archiv der Math. und Phys.* XL. (1863), pp. 101—126.

Studien über die Bessel'schen Functionen (Leipzig, 1868).

Integration der Gleichung $x^{m+\frac{1}{2}} \frac{\partial^{2m+1} y}{\partial x^{2m+1}} \mp y = 0$*. *Math. Ann.* II. (1870), pp. 624—635.

Ueber der Anwendung der Bessel'schen Functionen in der Theorie der Beugung. *Zeitschrift für Math. und Phys.* XV. (1870), pp. 141—169.

Zur Theorie der Bessel'schen Functionen (Dec. 1870). *Math. Ann.* III. (1871), pp. 475—487.

Zur Theorie der Bessel'schen Functionen (Jan. 1871). *Math. Ann.* IV. (1871), pp. 103—116.

Ueber eine mit den Bessel'schen Functionen verwandte Function (Aug. 1875). *Math. Ann.* IX. (1876), pp. 425—444.

Zur Theorie der Bessel'schen Functionen (Oct. 1878). *Math. Ann.* XIV. (1879), pp. 510—536.

Zur Theorie der Bessel'schen Functionen (Oct. 1879). *Math. Ann.* XVI. (1880), pp. 183—208.

Die Beugungserscheinungen einer kreisrunden Oeffnung und eines kreisrunden Schirmchens theoretisch und experimentell bearbeitet. *Münchener Abh.* XV. (1884—86), pp. 233—328.

Die Beugungserscheinungen geradlinig begrenzter Schirme. *Münchener Abh.* XV. (1884—86), pp. 531—663.

* The headings of the pages of this paper are "Zur Theorie der Bessel'schen Functionen."

LORENZ, L.

Sur le développement des fonctions arbitraires au moyen de fonctions données. *Tidsskrift for Mathematik*, 1876, pp. 129—144. [*Oeuvres Scientifiques*, II. (1899), pp. 495—513.]

Théorie de la dispersion. *Ann. der Physik und Chemie*, (3) XX. (1883), pp. 1—21. [*Oeuvres Scientifiques*, I. (1898), pp. 371—396.]

Sur la lumière refléchie et refractée par une sphère transparente. *K. Danske Videnskabernes Selskabs Skrifter*, (6) VI. (1890), pp. 1—62. [*Oeuvres Scientifiques*, I. (1898), pp. 405—502.]

LOVE, A. E. H.

The Scattering of Electric Waves by a Dielectric Sphere (Feb. 9, 1899). *Proc. London Math. Soc.* XXX. (1899), pp. 308—321.

The Transmission of Electric Waves over the Surface of the Earth (Dec. 19, 1914). *Phil. Trans. of the Royal Soc.* CCXV. A (1915), pp. 105—131.

MACDONALD, H. M.*

Note on Bessel Functions (Nov. 11, 1897). *Proc. London Math. Soc.* XXIX. (1898), pp. 110—115.

Zeroes of the Bessel Functions (April 7, 1898; Jan. 12, 1899). *Proc. London Math. Soc.* XXIX. (1898), pp. 575—584; XXX. (1899), pp. 165—179.

The Addition Theorem for the Bessel Functions (April 5, 1900). *Proc. London Math. Soc.* XXXII. (1901), pp. 152—157.

Some Applications of Fourier's Theorem (Dec. 11, 1902). *Proc. London Math. Soc.* XXXV. (1903), pp. 428—443.

The Bending of Electric Waves round a Conducting Obstacle (Jan. 21, 1903; May 12, 1903). *Proc. Royal Soc.* LXXI. (1903), pp. 251—258; LXXII. (1904), pp. 59—68.

Note on the evaluation of a certain integral containing Bessel's functions (Dec. 6, 1908). *Proc. London Math. Soc.* (2) VII. (1909), pp. 142—149.

The Diffraction of Electric Waves round a Perfectly Reflecting Obstacle (Aug. 13, 1909). *Phil. Trans. of the Royal Soc.* CCX. A (1910), pp. 113—144.

The Transmission of Electric Waves around the Earth's Surface (Jan. 10, 1914). *Proc. Royal Soc.* XC. A (1914), pp. 50—61.

Formulae for the Spherical Harmonic $P_n^{-m}(\mu)$ when $1-\mu$ is a small quantity (Feb. 6, 1914). *Proc. London Math. Soc.* (2) XIII. (1914), pp. 220—221.

[Extract from a letter to Prof. Carslaw, Oct. 17, 1912.] *Proc. London Math. Soc.* (2) XIII. (1914), pp. 239—240.

A class of diffraction problems (Jan. 14, 1915). *Proc. London Math. Soc.* (2) XIV. (1915), pp. 410—427.

The Transmission of Electric Waves around the Earth's Surface (May 23, 1916). *Proc. Royal Soc.* XCII. A (1916), pp. 433—437.

McMAHON, J.

On the Descending Series for Bessel's Functions of Both Kinds. *Annals of Math.* VIII. (1894), pp. 57—61.

On the Roots of the Bessel and certain Related Functions. *Annals of Math.* IX. (1895), pp. 23—30.

The Expression for a Rational Polynomial in a Series of Bessel functions of nth Order as required in certain Cases of Dirichlet's Problem. *Proc. American Assoc.* 1900, pp. 42—43.

MacROBERT, T. M.

The Modified Bessel Function $K_n(z)$ (Feb. 13, 1920). *Proc. Edinburgh Math. Soc.* XXXVIII. (1920), pp. 10—19.

Asymptotic Expressions for the Bessel Functions and the Fourier-Bessel Expansion (Jan. 14, 1921). *Proc. Edinburgh Math. Soc.* XXXIX. (1921), pp. 13—20.

MAGGI, G. A.

Sulla storia delle funzioni cilindriche. *Atti della R. Accad. dei Lincei*, Ser. 3 (Transunti), vol. IV. (1880), pp. 259—263.

Sopra un problema di elettrostatica (June 3, 1880). *Rend. del R. Ist. Lombardo*, (2) XIII. (1880), pp. 384—390.

MALMSTÉN, C. J.

Om Integrale $\int_0^\infty \frac{\cos ax\,dx}{(1+x^2)^n}$. *K. Svenska V. Akad. Handlingar*, LXII. (1841), pp. 65—74.

* See also under Gegenbauer.

De l'équation différentielle $y_x''+\frac{y_x'}{x}+Ax^m y=0$ (March 18, 1849). *Journal für Math.* XXXIX. (1850), pp. 108—115.

Théorèmes sur l'intégration de l'Équation $\frac{d^2y}{dx^2}+\frac{r}{x}\frac{dy}{dx}=\left(bx^m+\frac{s}{x^2}\right)y$. *Camb. and Dublin Math. Journal*, V. (1850), pp. 180—182.

MALTÉZOS, C.

Sur la chute des corps dans le vide et sur certaines fonctions transcendantes. *Nouvelles Annales de Math.* (4) XI. (1902), pp. 197—204.

MANFREDIUS, G.

De constructione aequationum differentialium primi gradus (Bologna, 1707).

MARCH, H. W.

Über die Ausbreitung der Wellen der drahtlosen Telegraphie auf der Erdkugel (Oct. 21, 1911). *Ann. der Physik und Chemie*, (4) XXXVII. (1912), pp. 29—50.

MARCOLONGO, R.

Alcuni teoremi sulle funzioni cilindriche di prima specie (April 6, 1889). *Napoli Rendiconto*, (2) III. (1889), pp. 96—99.

MARSHALL, W.

On a New Method of Computing the Roots of Bessel's Functions (Oct. 1909). *Annals of Math.* (2) XI. (1910), pp. 153—160.

MATHEWS, G. B.

See under Gray.

MAYALL, R. H. D.

On the Diffraction Pattern near the Focus of a Telescope (Feb. 22, 1897). *Proc. Camb. Phil. Soc.* IX. (1898), pp. 259—269.

MEECH, L. W.

Integration of Riccati's Equation. *Annals of Math.* I. (1886), pp. 97—103; III. (1887), pp. 47—49.

MEHLER, F. G.

Ueber die Vertheilung der statischen Electricität in einem von zwei kugelkalotten begrenzten Körper (July, 1867). *Journal für Math.* LXVIII. (1868), pp. 134—150.

Ueber eine mit den Kugel- und Cylinderfunctionen verwandte Function und ihre Anwendung in die Theorie der Electricitätsvertheilung. [*Elbing Jahresbericht*, 1870.] *Math. Ann.* XVIII. (1881), pp. 161—194.

Ueber die Darstellung einer willkürlichen Function zweier Variablen durch Cylinderfunction* (Dec. 1, 1871). *Math. Ann.* V. (1872), pp. 135—140.

Notiz über die Dirichlet'schen Integralausdrücke für die Kugelfunctionen $P^n(\cos\vartheta)$ und über eine analoge Integralform für die Cylinderfunction $J(x)$† (Dec. 2, 1871). *Math. Ann.* V. (1872), pp. 141—144.

MEISSEL, D. F. E.

Iserlohn Programm, 1862. [Nielsen, *Nouvelles Annales de Math.* (4) II. (1902), pp. 396—410.]

Tafel der Bessel'schen Functionen I_k^0 und I_k^1 von $k=0$ bis $k=15{\cdot}5$ (Nov. 8, 1888). *Berliner Abh.* 1888. (*Math. Abh.* I.)

Ueber die Besselschen Functionen J_k^0 und J_k^1. *Programm, Oberrealschule, Kiel*, 1890. [*Jahrbuch über die Fortschritte der Math.* 1890, pp. 521—522.]

Einige Entwickelungen die Bessel'schen I-Functionen betreffend (May 3, 1891). *Astr. Nach.* CXXVII. (1891), col. 359—362.

Beitrag zur Theorie der allgemeinen Bessel'schen Function (June 11, 1891). *Astr. Nach.* CXXVIII. (1891), col. 145—154.

Abgekürzte Tafel der Bessel'schen Functionen I_k^h (July 10, 1891). *Astr. Nach.* CXXVIII. (1891), col. 154—155.

Beitrag zur Theorie der Bessel'schen Functionen (Oct. 7, 1891). *Astr. Nach.* CXXVIII. (1891), col. 435—438.

* The headings of the pages of this paper are "Ueber die Cylinderfunction $J(x)$."

† The headings of the pages of this paper are "Notiz über die Functionen $P^n(\cos\vartheta)$ und $J(x)$."

Neue Entwickelungen über die Bessel'schen Functionen (Jan. 25, 1892). *Astr. Nach.* CXXIX. (1892), col. 281—284.

Weitere Entwickelungen über die Bessel'schen Functionen (May 2, 1892). *Astr. Nach.* CXXX. (1892), col. 363—368.

Über die Absoluten Maxima der Bessel'schen Functionen. *Programm, Oberrealschule, Kiel*, 1892 (11 pp.). [*Jahrbuch über die Fortschritte der Math.* 1892, pp. 476—478.]

MELLIN, R. HJ.

Abriss einer einheitlichen Theorie der Gamma und der hypergeometrischen Funktionen (June, 1909). *Math. Ann.* LXVIII. (1910), pp. 305—337.

MICHELL, J. H.

The Wave Resistance of a Ship* (Aug. 9, 1897). *Phil. Mag.* (5) XLV. (1898), pp. 106—123.

MOLINS, H.

Sur l'intégration de l'équation différentielle $\frac{d^k y}{dx^k} = ax^m y$ (March 6, 1876). *Mém. de l'Acad. des Sci. de Toulouse*, (7) VIII. (1876), pp. 167—189.

MOORE, C. N.

Note on the roots of Bessel functions. *Annals of Math.* (2) IX. (1908), pp. 156—162.

The summability of the developments in Bessel functions with applications (Sept. 11, 1908). *Trans. American Math. Soc.* X. (1909), pp. 391—435.

On the uniform convergence of the developments in Bessel functions (Oct. 30, 1909). *Trans. American Math. Soc.* XII. (1911), pp. 181—206.

A continuous function whose development in Bessel's functions is non-summable of certain orders (Sept. 4, 1916). *Bulletin American Math. Soc.* XXIV. (1918), pp. 145—149.

On the summability of the developments in Bessel's functions (Sept. 4, 1917). *Trans. American Math. Soc.* XXI. (1920), pp. 107—156.

MORTON, W. B.

The Value of the Cylinder Function of the Second Kind for Small Arguments (Oct. 25, 1900). *Nature*, LXIII. (1901), p. 29.

MURPHY, R.

On the general properties of definite integrals (May 24, 1830). *Trans. Camb. Phil. Soc.* III. (1830), pp. 429—443.

MYLLER-LEBEDEFF, W.

See under Lebedeff.

NAGAOKA, H.

Diffraction Phenomena produced by an Aperture on a Curved Surface. *Journal of the Coll. of Sci. Imp. Univ. Japan*, IV. (1891), pp. 301—322.

NEUMANN, CARL GOTTFRIED†.

Allgemeine Lösung des Problems über den stationären Temperaturzustand eines homogenen Körpers, welcher von zwei nicht concentrischen Kugelflächen begrenzt wird (Halle, 1862).

Ueber das Gleichgewicht der Wärme und das der Electricität in einem Körper, welcher von zwei nicht concentrischen Kugelflächen begrenzt wird (1862). *Journal für Math.* LXII. (1863), pp. 36—49.

Ueber die Entwickelung beliebig gegebener Functionen nach den *Bessel*schen Functionen (March 28, 1867). *Journal für Math.* LXVII. (1867), pp. 310—314.

Theorie der Bessel'schen Functionen. Ein Analogon zur Theorie der Kugelfunctionen (Leipzig, 1867).

Ueber die Entwicklungen einer Function nach Quadraten und Producten der Fourier-Bessel'schen Functionen (Nov. 1869.) *Leipziger Berichte*, XXI. (1869), pp. 221—256. [*Math. Ann.* III. (1871), pp. 581—610.]

Ueber Producte und Quadrate der Bessel'schen Functionen. *Math. Ann.* II. (1870), p. 192.

Über die nach Kreis-, Kugel- und Cylinder-functionen fortschreitenden Entwickelungen (Leipzig, 1881).

Ueber gewisse particulare Integrale der Differentialgleichung $\Delta F = F$ insbesondere über die Entwickelung dieser particularen Integrale nach Kugelfunctionen (March 8, 1886). *Leipziger Berichte*, XXXVIII. (1886), pp. 75—82.

* This paper contains some Tables of Bessel functions which were computed by **B. A. Smith.**

† See also under Schläfli.

NEUMANN, F. E.

Eine Verallgemeinerung der Zylinderfunktionen. *Dissertation, Halberstadt*, 1909 (25 pp.). [*Jahrbuch über die Fortschritte der Math.* 1909, p. 575.]

NICHOLSON, J. W.*

The Asymptotic Expansions of Bessel Functions of High Order. *Phil. Mag.* (6) XIV. (1907), pp. 697—707.

On Bessel Functions of Equal Argument and Order. *Phil. Mag.* (6) XVI. (1908), pp. 271—279.

On the Inductance of Two Parallel Wires (June 12, 1908). *Phil. Mag.* (6) XVII. (1909), pp. 255—275.

On the Relation of Airy's Integral to the Bessel Functions. *Phil. Mag.* (6) XVIII. (1909), pp. 6—17.

The Asymptotic Expansions of Bessel Functions. *Phil. Mag.* (6) XIX. (1910), pp. 228—249.

On the Bending of Electric Waves round a Large Sphere, I. *Phil. Mag.* (6) XIX. (1910), pp. 516—537.

The Approximate Calculation of Bessel Functions of Imaginary Argument. *Phil. Mag.* (6) XX. (1910), pp. 938—943.

The Scattering of Light by a Large Conducting Sphere (March 10, 1910). *Proc. London Math. Soc.* (2) IX. (1911), pp. 67—80.

Notes on Bessel Functions. *Quarterly Journal*, XLII. (1911), pp. 216—224.

The products of Bessel functions. *Quarterly Journal*, XLIII. (1912), pp. 78—100.

The pressure of radiation on a cylindrical obstacle (Dec. 14, 1911). *Proc. London Math. Soc.* (2) XI. (1913), pp. 104—126.

The Lateral Vibrations of Bars of Variable Section (May 11, 1917). *Proc. Royal Soc.* XCIII. A (1917), pp. 506—519.

Generalisation of a theorem due to Sonine. *Quarterly Journal*, XLVIII. (1920), pp. 321—329.

A Problem in the Theory of Heat Conduction (Sept. 29, 1921). *Proc. Royal Soc.* C. A (1922), pp. 226—240.

NICOLAS, J.

Étude des fonctions de Fourier (première et deuxième espèce). *Ann. sci. de l'École norm. sup.* (2) XI. (1882), supplément, pp. 3—90.

NIELSEN, N.†

Sur le produit de deux fonctions cylindriques (July 30, 1898). *Math. Ann.* LII. (1899), pp. 228—242.

Udviklinger efter Cylinderfunktioner (Aug. 28, 1898). *Nyt Tidsskrift*, IX. B (1898), pp. 73—83.

Sur le développement de zéro en fonctions cylindriques (March 25, 1899). *Math. Ann.* LII. (1899), pp. 582—587.

Flertydige Udviklinger efter Cylinderfunktioner. *Nyt Tidsskrift*, X. B (1899), pp. 73—81.

Note sur les développements schloemilchiens en série de fonctions cylindriques (Nov. 15, 1899). *Oversigt K. Danske Videnskabernes Selskabs*, 1899, pp. 661—665.

Note supplémentaire relative aux développements schloemilchiens en série de fonctions cylindriques (June 23, 1900). *Oversigt K. Danske Videnskabernes Selskabs*, 1900, pp. 55—60.

Sur une classe de polynômes qui se présentent dans la théorie des fonctions cylindriques (Feb. 26, 1900; Jan. 8, 1901). *Ann. di Mat.* (3) V. (1901), pp. 17—31; (3) VI. (1901), pp. 331—340.

Évaluation nouvelle des intégrales indéfinies et des séries infinies contenant une fonction cylindrique (May 23, 1900). *Ann. di Mat.* (3) VI. (1901), pp. 43—115.

Sur une classe de séries infinies analogues à celles de Schlömilch selon les fonctions cylindriques (Aug. 17, 1900). *Ann. di Mat.* (3) VI. (1901), pp. 301—329.

Recherche sur les séries des fonctions cylindriques dues à MM. C. Neumann et W. Kapteyn. *Ann. sci. de l'École norm. sup.* (3) XVIII. (1901), pp. 39—75.

Note sur la convergence d'une série neumannienne de fonctions cylindriques (Feb. 10, 1901). *Math. Ann.* LV. (1902), pp. 493—496.

Recherches sur une classe de séries infinies analogue à celle de M. W. Kapteyn (April 13, 1901). *Oversigt K. Danske Videnskabernes Selskabs*, 1901, pp. 127—146.

Sur les séries de factorielles (Jan. 20, 1902). *Comptes Rendus*, CXXXIV. (1902), pp. 157—160.

Théorie nouvelle des séries asymptotiques obtenues pour les fonctions cylindriques et pour les fonctions analogues (March 16, 1902). *Oversigt K. Danske Videnskabernes Selskabs*, 1902, pp. 117—177.

* See also under Datta and under Dendy. † See also under Sonine.

Équations différentielles linéaires obtenues pour le produit de deux fonctions cylindriques. *Nouvelles Annales de Math.* (4) II. (1902), pp. 396—410.

Handbuch der Theorie der Cylinderfunktionen (Leipzig, 1904).

Note sur les séries de fonctions bernoulliennes (Dec. 14, 1903). *Math. Ann.* LIX. (1904), pp. 103—109.

Sur une intégrale définie (June 5, 1904). *Math. Ann.* LIX. (1904), pp. 89—102.

Sur quelques propriétés nouvelles des fonctions cylindriques (April 1, 1906). *Atti della R. Accad. dei Lincei*, (5) XV. (1906), pp. 490—497.

Recherches sur quelques généralisations d'une identité intégrale d'Abel (Sept. 17, 1906). *K. Danske Videnskabernes Selskabs Skrifter*, (7) V. (1910), pp. 1—37.

Sur les séries des fonctions cylindriques. *Journ. für Math.* CXXXII. (1907), pp. 127—146.

Sur quelques propriétés fondamentales des fonctions sphériques (Dec. 20, 1906). *Ann. di Mat.* (3) XIV. (1908), pp. 69—90.

Über die Verallgemeinerung einiger von F. und C. Neumann gegebenen nach Kugel- und Zylinderfunktionen fortschreitenden Reihenentwickelungen (Jan. 26, 1908). *Leipziger Berichte*, LXI. (1909), pp. 33—61.

Über das Produkt zweier Zylinderfunktionen. *Monatshefte für Math. und Phys.* XIX. (1908), pp. 164—170.

Über den Legendre-Besselschen Kettenbruch (July 4, 1908). *Münchener Sitzungsberichte*, XXXVIII. (1908), pp. 85—88.

NIEMÖLLER, F.

Ueber Schwingungen einer Saite deren Spannung eine stetige Function der Zeit ist. *Zeitschrift für Math.* XXV. (1880), pp. 44—48.

Formeln zur numerischen Berechnung des allgemeinen Integrals der Bessel'schen Differentialgleichungen. *Zeitschrift für Math.* XXV. (1880), pp. 65—71.

OLBRICHT, R.

Studien über die Kugel- und Cylinderfunctionen (June 2, 1886). *Nova Acta Acad. Caes. Leop. (Halle)*, 1888, pp. 1—48.

OLTRAMARE, G.

Note sur l'intégrale

$$\int_0^\infty \frac{\cos yx}{(a^2+b^2x^2)^n}\,dx.$$

Comptes Rendus de l'Assoc. Française, XXIV. (1895), part I. p. 182; part II. pp. 167—171.

ONO, A.

On the First Root of Bessel Functions of Fractional Order (April 2, 1921). *Phil. Mag.* (6) XLII. (1921), pp. 1020—1021.

ORR, W. McF.

On Divergent (or Semiconvergent) Hypergeometric Series (May 16, 1898; April 3, 1899). *Trans. Camb. Phil. Soc.* XVII. (1899), pp. 171—199, 283—290.

On the product $J_m(x)J_n(x)$ (May 15, 1899). *Proc. Camb. Phil. Soc.* X. (1900), pp. 93—100.

Extensions of Fourier's and the Bessel-Fourier Theorems (Dec. 14, 1908). *Proc. R. Irish Acad.* XXVII. A (1910), pp. 205—248.

OSEEN, C. W.

Neue Lösung des Sommerfeldschen Diffraktionsproblems (Jan. 10, 1912). *Arkiv för Mat. Astr. och Fysik*, VII. (1912), no. 40.

OTTI, H.

Eigenschaften Bessel'scher Funktionen II^{ter} Art. *Bern Mittheilungen*, 1898 [1899], pp. 61—96.

PAOLI, P.

Sopra gl' integrali definiti (Oct. 8, 1827). *Mem. di Mat. e di Fis. della Soc. Italiana delle Sci. (Modena)*, XX. (1828), pp. 161—182.

Sull' integrazione dell' equazione

$$\frac{d^2y}{dx^2}+\left(1-\frac{i(i+1)}{x^2}\right)y=0$$

(Oct. 8, 1827). *Mem. di Mat. e di Fis. della Soc. Italiana delle Sci.* XX. (1828), pp. 183—188.

PARSEVAL, M. A.

Mémoire sur les séries et sur l'intégration complète d'une équation aux différences partielles linéaires du second ordre à coefficiens constans (Le 16 germinal, an 7—April 5, 1799). *Mém. présentés à l'Inst. par divers savans*, I. (1805), pp. 639—648.

PEARSON, K.

On the solution of some differential equations by Bessel's functions. *Messenger*, IX. (1880), pp. 127—131.

The Problem of the Random Walk. *Nature*, LXXII. (1905), pp. 294, 342.

A mathematical theory of random migration. *Drapers Company Research Memoirs*, Biometric Series, III. (1906).

PEIRCE, B. O.

See under Willson.

PÉRÈS, J.

Sur les fonctions de Bessel de plusieurs variables (August 9, 1915). *Comptus Rendus*, CLXI. (1915), pp. 168—170.

PERRON, O.

Über die Kettenbruchentwicklung des Quotienten zweier Bessel'schen Functionen (Dec. 7, 1907). *Münchener Sitzungsberichte*, XXXVII. (1907), pp. 483—504.

PETZVAL, J.

Integration der linearen Differential-Gleichungen, I. (Vienna, 1851).

PICARD, C. É.

Application de la théorie des complexes linéaires à l'étude des surfaces et des courbes gauches. *Ann. sci. de l'Ecole norm. sup.* (2) VI. (1877), pp. 329—366.

PINCHERLE, S.

Sopra alcuni sviluppi in serie per funzioni analitiche (Jan. 12, 1882). *Bologna Memorie*, (4) III. (1881), pp. 151—180.

Alcuni teoremi sopra gli sviluppi in serie per funzioni analitiche (March 23, 1882). *Rend. del R. Ist. Lombardo*, (2) XV. (1882), pp. 224—225.

Della trasformazione di Laplace e di alcune sue applicazioni (Feb. 27, 1887). *Bologna Memorie*, (4) VIII. (1887), pp. 125—143.

PLANA, G. A. A.

Note sur l'intégration de l'équation $\frac{d^2y}{dx^2}+gx^my=0$ (Jan. 20, 1822). *Mem. della R. Accad delle Sci. di Torino*, XXVI. (1821), pp. 519—538.

Recherches analytiques sur la découverte de la loi de pesanteur des planètes vers le Soleil et sur la théorie de leur mouvement elliptique (June 20, 1847). *Mem. della R. Accad. delle Sci. di Torino*, (2) X. (1849), pp. 249—332.

POCHHAMMER, L.

Ueber die lineare Differentialgleichung zweiter Ordnung mit linearen Coefficienten. *Math. Ann.* XXXVI. (1890), pp. 84—96.

Ueber einige besondere Fälle der linearen Differentialgleichung zweiter Ordnung mit linearen Coefficienten (Sept. 1890). *Math. Ann.* XXXVIII. (1891), pp. 225—246.

Ueber eine binomische lineare Differentialgleichung n^{ter} Ordnung (Sept. 1890). *Math. Ann.* XXXVIII. (1891), pp. 247—262.

Ueber die Differentialgleichung der allgemeineren F-Reihe (Jan. 1891). *Math. Ann.* XXXVIII. (1891), pp. 586—597.

Ueber eine specielle lineare Differentialgleichung 2ter Ordnung mit linearen Coefficienten (May, 1891). *Math. Ann.* XLI. (1893), pp. 174—178.

Ueber die Differentialgleichungen der Reihen $\mathfrak{F}(\rho, \sigma; x)$ und $\mathfrak{F}(\rho, \sigma, \tau; x)$ (June, 1891). *Math. Ann.* XLI. (1893), pp. 197—218.

POISSON, S. D.

Mémoire sur les intégrales définies. *Journal de l'École Polytechnique*, IX. (cahier 16) (1813), pp. 215—246.

Sur une nouvelle manière d'exprimer les coordonnées des planètes dans le mouvement elliptique. *Connaissance des Tems*, 1825 [1822], pp. 379—385.

Mémoire sur l'Intégration des équations linéaires aux différences partielles. *Journal de l'École Polytechnique*, XII. (cahier 19) (1823), pp. 215—248.

Suite du Mémoire sur les intégrales définies et sur la sommation des séries inséré dans les précedens volumes de ce Journal. *Journal de l'École Polytechnique*, XII. (cahier 19) (1823), pp. 404—509.

Mémoire sur le calcul des variations (Nov. 10, 1831). *Mém. de l'Acad. R. des Sci.* XII. (1833), pp. 223—331.

Sur le développement des coordonnées d'une planète dans son mouvement elliptique et la fonction perturbatrice de ce mouvement. *Connaissance des Tems*, 1836 [1833], pp. 3—31.

La Théorie de la Chaleur (Paris, 1835).

PORTER, M. B.

Note on the roots of Bessel's Functions. *Bulletin American Math. Soc.* IV. (1898), pp. 274—275.

On the Roots of the Hypergeometric and Bessel's Functions (June 3, 1897). *American Journal of Math.* XX. (1898), pp. 193—214.

On the roots of functions connected by a linear recurrent relation of the second order (Feb. 1901). *Annals of Math.* (2) III. (1902), pp. 55—70.

PUISEUX, V.

Sur la convergence des séries qui se présentent dans la théorie du mouvement elliptique des planètes. *Journal de Math.* XIV. (1849), pp. 33—39.

Seconde note sur la convergence des séries du mouvement elliptique. *Journal de Math.* XIV. (1849), pp. 242—246.

PURSER, F.

On the application of Bessel's functions to the elastic equilibrium of a homogeneous isotropic cylinder (Nov. 11, 1901). *Trans. R. Irish Acad.* XXXII. (1902), pp. 31—60.

Some applications of Bessel's functions to Physics (May 14, 1906). *Proc. R. Irish Acad.* XXVI. A (1907), pp. 25—66.

RAFFY, L.

Une leçon sur l'équation de Riccati. *Nouv. Ann. de Math.* (4) II. (1902), pp. 529—545.

RAMANUJAN, S.

A class of definite integrals. *Quarterly Journal*, XLVIII. (1920), pp. 294—310.

RAWSON, R.

On Cognate Riccatian Equations (1876). *Messenger*, VII. (1878), pp. 69—72.

Note on a transformation of Riccati's equation. *Messenger*, XII. (1883), pp. 34—36.

RAYLEIGH (J. W. STRUTT), LORD.

On the Vibrations of a Gas contained within a Rigid Spherical Envelope (March 14, 1872). *Proc. London Math. Soc.* IV. (1873), pp. 93—103.

Investigation of the Disturbance produced by a Spherical Obstacle on the Waves of Sound (Nov. 14, 1872). *Proc. London Math. Soc.* IV. (1873), pp. 253—283.

Notes on Bessel's Functions. *Phil. Mag.* (4) XLIV. (1872), pp. 328—344.

Note on the Numerical Calculation of the Roots of Fluctuating Functions (June 11, 1874). *Proc. London Math. Soc.* V. (1874), pp. 112—194. [*Scientific Papers*, I. (1899), pp. 190—195.]

The Theory of Sound (2 vols. London, 1877, 1878; 2nd edition, 1894, 1896).

On the Relation between the Functions of Laplace and Bessel (Jan. 10, 1878). *Proc. London Math. Soc.* IX. (1878), pp. 61—64. [*Scientific Papers*, I. (1899), pp. 338—341.]

On Images formed without Reflection or Refraction. *Phil. Mag.* (5) XI. (1881), pp. 214—218. [*Scientific Papers*, I. (1899), pp. 513—517.]

On the Electromagnetic Theory of Light. *Phil. Mag.* (5) XII. (1881), pp. 81—101. [*Scientific Papers*, I. (1899), pp. 518—536.]

On the Vibrations of a Cylindrical Vessel containing Liquid. *Phil. Mag.* (5) XV. (1883), pp. 385—389. [*Scientific Papers*, II. (1900), pp. 208—211.]

On Point-, Line-, and Plane Sources of Sound (June 14, 1888). *Proc. London Math. Soc.* XIX. (1889), pp. 504—507. [*Scientific Papers*, III. (1902), pp. 44—46.]

On the Theory of Optical Images, with special reference to the Microscope. *Phil. Mag.* (5) XLII. (1896), pp. 167—195. [*Scientific Papers*, IV. (1904), pp. 235—260.]

On the Passage of Waves through Apertures in Plane Screens, and Allied Problems. *Phil. Mag.* (5) XLIII. (1897), pp. 259—272. [*Scientific Papers*, IV. (1904), pp. 283—296.]

On the Bending of Waves round a Spherical Obstacle (May 23, 1903). *Proc. Roy. Soc.* LXXII. (1903), pp. 40—41. [*Scientific Papers*, V. (1912), pp. 112—114.]

On the Acoustic Shadow of a Sphere (Jan. 21, 1904). *Phil. Trans. of the Royal Soc.* CCIII. A (1904), pp. 87—110. [*Scientific Papers*, V. (1912), pp. 149—161.]

On the Open Organ Pipe Problem in two Dimensions. *Phil. Mag.* (6) VIII. (1904), pp. 481—487. [*Scientific Papers*, V. (1912), pp. 206—211.]

On the Experimental Determination of the Ratio of the Electrical Units. *Phil. Mag.* (6) XII. (1906), pp. 97—108. [*Scientific Papers*, V. (1912), pp. 330—340.]

On the Light dispersed from Fine Lines ruled upon Reflecting Surfaces. *Phil. Mag.* (6) XIV. (1907), pp. 350—359. [*Scientific Papers*, V. (1912), pp. 410—418.]

The Problem of the Whispering Gallery. *Phil. Mag.* (6) XX. (1910), pp. 1001—1004. [*Scientific Papers*, V. (1912), pp. 617—620.]

Note on Bessel's Functions as applied to the Vibrations of a Circular Membrane. *Phil. Mag.* (6) XXI. (1911), pp. 53—58. [*Scientific Papers*, VI. (1920), pp. 1—5.]

On a Physical Interpretation of Schlömilch's Theorem in Bessel's Functions. *Phil. Mag.* (6) XXI. (1911), pp. 567—571. [*Scientific Papers*, VI. (1920), pp. 22—25.]

Problems in the Conduction of Heat. *Phil. Mag.* (6) XXII. (1911), pp. 381—396. [*Scientific Papers*, VI. (1920), pp. 51—64.]

On the propagation of Waves through a stratified Medium (Jan. 11, 1912). *Proc. Royal Soc.* LXXXVI. A (1912), pp. 207—226. [*Scientific Papers*, VI. (1920), pp. 111—120.]

Electrical Vibrations on a thin Anchor Ring (June 27, 1912). *Proc. Royal Soc.* LXXXVII. A (1912), pp. 193—202. [*Scientific Papers*, VI. (1920), pp. 111—120.]

Further Applications of Bessel's Functions of High Order to the Whispering Gallery and Allied Problems. *Phil. Mag.* (6) XXVII. (1914), pp. 100—109. [*Scientific Papers*, VI. (1920), pp. 211—219.]

Further remarks on the Stability of Viscous Fluid Motion. *Phil. Mag.* (6) XXVIII. (1914), pp. 609—619. [*Scientific Papers*, VI. (1920), pp. 266—275.]

On the Stability of the simple Shearing Motion of a Viscous Incompressible Fluid. *Phil. Mag.* (6) XXX. (1915), pp. 329—338. [*Scientific Papers*, VI. (1920), pp. 341—349.]

On Legendre's Function* $P_n(\theta)$, when n is great and θ has any value (April 27, 1916). *Proc. Royal Soc.* XCII. A (1916), pp. 433—437. [*Scientific Papers*, VI. (1920), pp. 393—397.]

On Convection Currents in a horizontal Layer of Fluid, when the higher Temperature is on the under side. *Phil. Mag.* (6) XXXII. (1916), pp. 529—546. [*Scientific Papers*, VI. (1920), pp. 432—446.]

On the Problem of Random Vibrations and of Random Flights in one, two, or three dimensions. *Phil. Mag.* (6) XXXVII. (1919), pp. 321—347. [*Scientific Papers*, VI. (1920), pp. 604—626.]

REINECK, A.

Die Verwandtschaft zwischen Kugelfunktionen und Besselschen Funktionen. *Dissertation, Bern* (Halle, 1907). (72 pp.) [*Jahrbuch über die Fortschritte der Math.* 1907, p. 495.]

RICCATI, COUNT J. F.

Animadversationes in aequationes differentiales secundi gradus. *Actorum Eruditorum quae Lipsiae publicantur Supplementa*, VIII. (1724), pp. 66—73.

Com. Jacobi Riccati Appendix ad Animadversationes in aequationes differentiales secundi gradus, editus in Actis Eruditorum quae Lipsiae publicantur Tomo VIII. Supplementorum, Sectione II. p. 66. *Acta Eruditorum publicata Lipsiae*, 1723, pp. 502—510.

RIEMANN, G. F. B.

Zur Theorie der Nobili'schen Farbenringe (March 28, 1855). *Ann. der Physik und Chemie*, (2) XCV. (1855), pp. 130—139.

Partielle Differentialgleichungen und deren Anwendung auf physikalische Fragen. Von Bernhard Riemann. Für den Druck bearbeitet und herausgegebenen von Karl Hattendorf (Brunswick, 1876).

RÖHRS, J. H.

Spherical and Cylindric Motion in Viscous Fluid (May 14, 1874). *Proc. London Math. Soc.* V. (1874), pp. 125—139.

RUDSKI, P.

Ueber eine Klasse transcendenter Gleichungen. *Prace Mat. Fiz.* III. (1892), pp. 69—81. [*Jahrbuch über die Fortschritte der Math.* 1892, pp. 107—108.]

Note sur la situation des racines des fonctions transcendantes $J_{n+\frac{1}{2}}(x)=0$ (May 10, 1891). *Mém. de la Soc. R. des Sci. de Liège*, (2) XVIII. (1895), no. 3.

RUSSELL, A.

The Effective Resistance and Inductance of a Concentric Main, and Methods of Computing the Ber and Bei and Allied Functions (Jan. 22, 1909). *Phil. Mag.* (6) XVII. (1909), pp. 524--552.

* This is the function which other writers denote by the symbol $P_n(\cos\theta)$.

RUTGERS, J. G.

Over die bepaalde integraal $\int_0^1 e^{-q^2 z} z^{p-1} dz$. *Nieuw Archief voor Wiskunde*, (2) VI. (1905), pp. 368—373.

Over eene reeks met Besselsche Functies. *Nieuw Archief voor Wiskunde*, (2) VII. (1907), pp. 88—90.

Over reeksen van Besselsche functies en daarmede samenhangende bepaalde integralen, waarin Besselsche functies voorkomen. *Nieuw Archief voor Wiskunde*, (2) VII. (1907), pp. 164—181.

Sur les fonctions cylindriques de première espèce. *Nieuw Archief voor Wiskunde*, (2) VII. (1907), pp. 385—405.

Over eenige toepassingen der Fourier'sche ontwikkeling van een willekeurige functie naar Bessel'sche functies. *Nieuw Archief voor Wiskunde*, VIII. (1909), pp. 375—380.

RYBCZYŃSKI, W. VON.

Über die Ausbreitung der Wellen der drahtlosen Telegraphie auf der Erdkugel (March 5, 1913). *Ann. der Physik und Chemie*, (4) XLI. (1913), pp. 191—208.

SASAKI, S.

On the roots of the equation $\frac{Y_n(kr)}{J_n(kr)} - \frac{Y_n'(ka)}{J_n'(ka)} = 0$. *Tôhoku Math. Journal*, V. (1914), pp. 45—47.

SAVIDGE, H. G.

Tables of the ber and bei and ker and kei Functions with further formulae for their Computation (Nov. 12, 1909). *Phil. Mag.* (6) XIX. (1910), pp. 49—58.

SCHAFHEITLIN, P.*

Ueber die Darstellung der hypergeometrische Reihe durch ein bestimmtes Integral (Feb. 1887). *Math. Ann.* XXX. (1887), pp. 157—178.

Ueber eine Integraldarstellung der hypergeometrischen Reihen (Jan. 1888). *Math. Ann.* XXXI. (1888), p. 156.

Ueber die *Gauss*-sche und *Bessel*-sche Differentialgleichung und eine neue Integralform der letzteren (Feb. 16, 1894). *Journal für Math.* CXIV. (1895), pp. 31—44.

Die Nullstellen der *Bessel*schen Functionen. *Journal für Math.* CXXII. (1900), pp. 299—321.

Über die Nullstellen der Besselschen Funktionen zweiter Art (Jan. 10, 1901). *Archiv der Math. und Phys.* (3) I. (1901), pp. 133—137.

Über den Verlauf der Besselschen Funktionen (May 18, 1904). *Berliner Sitzungsberichte*, III. (1904), pp. 83—85.

Die Lage der Nullstellen der Besselschen Funktionen zweiter Art (June 27, 1906). *Berliner Sitzungsberichte*, V. (1906), pp. 82—93.

Über den Verlauf der Besselschen Funktionen zweiter Art. *Jahresbericht der Deutschen Math. Vereinigung*, XVI. (1907), pp. 272—279.

Die Theorie der Besselschen Funktionen. Von Paul Schafheitlin (Leipzig und Berlin, 1908; *Math. Phys. Schriften für Ingenieure und Studierende*, 4).

Beziehungen zwischen dem Integrallogarithmus und den Besselschen Funktionen (Feb. 24, 1909). *Berliner Sitzungsberichte*, VIII. (1909), pp. 62—67.

Die semikonvergenten Reihen für die Besselschen Funktionen (Sept. 1, 1909). *Jahresbericht der Deutschen Math. Vereinigung*, XIX. (1910), pp. 120—129.

SCHEIBNER, W.

On the asymptotic values of the coefficients in the development of any power of the radius vector according to the mean anomaly (March, 1856). *Gould's Astronomical Journal*, IV. (1856), pp. 177—182. [*Math. Ann.* XVII. (1880), pp. 531—544.]

Ueber die asymptotische Werthe der Coefficienten in den nach der mittleren Anomalie vorgenommenen Entwickelungen (May 31, 1856). *Leipziger Berichte*, VIII. (1856), pp. 40—64. [*Math. Ann.* XVII. (1880), pp. 545—560.]

SCHERK, H. F.

Über die Integration der Gleichung $\frac{d^n y}{dx^n} = (\alpha + \beta x) y$. *Journal für Math.* X. (1833), pp. 92—97.

* See also under Sonine.

SCHLÄFLI, L.

Sulle relazioni tra diversi integrali definiti che giovano ad esprimere la soluzione generale della equazione di Riccati. *Ann. di Mat.* (2) I. (1868), pp. 232—242.

Einige Bemerkungen zu Herrn Neumann's Untersuchungen über die Bessel'schen Functionen (May 4, 1870). *Math. Ann.* III. (1871), pp. 134—149. [With note by C. N. p. 149.]

Sopra un teorema di Jacobi recato a forma più generale ed applicata alla funzione cilindrica (Aug. 1871). *Ann. di Mat.* (2) V. (1873), pp. 199—205.

Sull' uso delle linee lungo le quali il valore assoluto di una funzione è constante (Oct. 4, 1872). *Ann. di Mat.* (2) VI. (1875), pp. 1—20.

Ueber die Convergenz der Entwicklung einer arbiträren Function $f(x)$ nach den Bessel'schen Functionen $J^a(\beta_1 x)$, $J^a(\beta_2 x)$, $J^a(\beta_3 x)$, ..., wo β_1, β_2, β_3, ... die positiven Wurzeln der Gleichung $J^a(\beta)$ vorstellen (Jan. 17, 1876). *Math. Ann.* X. (1876), pp. 137—142.

SCHLÖMILCH, O. X.

Note sur la variation des constantes arbitraires d'une intégrale définie. *Journal für Math.* XXXIII. (1846), pp. 268—280.

Analytische Studien, II. (Leipzig, 1848).

Ueber die Bessel'schen Function. *Zeitschrift für Math. und Phys.* II. (1857), pp. 137—165.

SCHÖNHOLZER, J. J.

Ueber die Auswerthung bestimmte Integrale mit Hülfe von Veränderungen des Integrationsweges (*Dissertation, Bern*, 1877). [Graf and Gubler, *Einleitung in die Theorie der Bessel'schen Funktionen*, II. (Bern, 1900).]

SCHOTT, G. A.

Electromagnetic Radiation (Cambridge, 1912).

SCHWARZSCHILD, K.

Die Beugung und Polarisation des Lichts durch einen Spalt. *Math. Ann.* LV. (1902), pp. 177—247.

SCHWERD, F. M.

Die Beugungserscheinungen aus den Fundamentalgesetzen der Undulationstheorie (Mannheim, 1835).

SEARLE, J. H. C.

On the propagation of waves in an atmosphere of varying density. *Quarterly Journal*, XXXIX. (1908), pp. 51—66.

SEGAR, H. W.

On the roots of certain continuants. *Messenger*, XXII. (1893), pp. 171—181.

SERRET, J. A.

Note sur quelques formules de calcul intégral. *Journal de Math.* VIII. (1843), pp. 1—27.

Mémoire sur l'intégration d'une équation différentielle à l'aide des différentielles à indices quelconques (Sept. 4, 1843). *Comptes Rendus*, XVII. (1843), pp. 458—475; *Journal de Math.* IX. (1844), pp. 193—216.

SHARPE, H. J.

On the Reflection of Sound at the Surface of a Paraboloid (Nov. 1874). *Quarterly Journal*, XV. (1877), pp. 1—8.

On a differential equation. *Messenger*, X. (1881), pp. 174—185; XI. (1882), pp. 41—44.

On a transcendental differential equation. *Messenger*, XI. (1882), pp. 56—63.

On a differential equation. *Messenger*, XIII. (1884), pp. 66—79.

Note on Legendre's coefficients. *Quarterly Journal*, XXIV. (1890), pp. 383—386.

On The Reflection of Sound at a Paraboloid (June 20, 1899). *Proc. Camb. Phil. Soc.* X. (1900), pp. 101—136.

SHEPPARD, W. F.

On some expressions of a function of a single variable in terms of Bessel's functions. *Quarterly Journal*, XXIII. (1889), pp. 223—260.

SIACCI, F.

Sulla integrazione di una equazione differenziale e sulla equazione di Riccati (April 13, 1901). *Napoli Rendiconto*, (3) VII. (1901), pp. 139—143.

SIBIRANI, F.

Sopra l' equazione di Riccati. *Riv. fis. mat.* XIX. (1909), pp. 216—220. [*Jahrbuch über die Fortschritte der Math.* 1909, p. 369.]

SIEMON, P.

Ueber die Integrale einer nicht homogenen Differentialgleichung zweiter Ordnung. *Programm, Luisenschule, Berlin,* 1890, 22 pp. [*Jahrbuch über die Fortschritte der Math.* 1890, pp. 340—342.]

SMITH, B. A.*

Table of Bessel's Functions Y_0 and Y_1. *Messenger,* XXVI. (1897), pp. 98—101.

Arched Dams. *Proc. American Soc. of Civil Engineers,* XLVI. (1920), pp. 375—425.

SMITH, CLARA E.

A theorem of Abel and its application to the development of a function in terms of Bessel's functions (June 8, 1906). *Trans. American Math. Soc.* VIII. (1907), pp. 92—106.

SMITH, O. A.

Quelques relations intégrales entre les fonctions sphériques et cylindriques. *Giornale di Mat.* (2) XII. (1905), pp. 365—373.

SOMMERFELD, A. J. W.†

Die willkürlichen Functionen in der mathematischen Physik. *Dissertation, Königsberg,* 1891 (75 pp.). [*Jahrbuch über die Fortschritte der Math.* 1891, pp. 519—523.]

Zur analytischen Theorie der Wärmeleitung. *Math. Ann.* XLV. (1894), pp. 263—277.

Mathematische Theorie der Diffraction (Summer, 1895). *Math. Ann.* XLVII. (1896), pp. 317—374.

Über die Ausbreitung der Wellen der drahtlosen Telegraphie (Jan. 15, 1909). *Ann. der Physik und Chemie,* (4) XXVIII. (1909), pp. 665—736.

Die Greensche Funktion der Schwingungsgleichung. *Jahresbericht der Deutscher Math. Vereinigung,* XXI. (1912), pp. 309—353.

SONINE, N. J.

On the resolution of functions into infinite series (Jan. 17/29, 1870). *Mathematical Collection published by the Moscow Math. Soc.* [Математическій Сборникъ], v. (1870), pp. 271—302, 323—382.

Recherches sur les fonctions cylindriques et le développement des fonctions continues en séries (Aug. 1879). *Math. Ann.* XVI. (1880), pp. 1—80.

Sur les fonctions cylindriques (Oct. 24, 1887). *Math. Ann.* XXX. (1887), pp. 582—583. [Note on a paper by Schafheitlin.]

Sur les fonctions cylindriques. (Extrait d'une Lettre adressée à M. Niels Nielsen, à Copenhague, May 6, 1904.) *Math. Ann.* LIX. (1904), pp. 529—552.

SPITZER, S.

Integral der Differentialgleichung $xy'' - y = 0$. *Zeitschrift für Math. und Phys.* II. (1857), pp. 165—170.

Entwickelung von $e^{\lambda x + \mu/x}$ in unendliche Reihen. *Zeitschrift für Math. und Phys.* III. (1858), pp. 244—246.

Darstellung des unendlichen Kettenbruches

$$2x+1+\cfrac{1}{2x+3+\cfrac{1}{2x+5+\cfrac{1}{2x+7+\dots}}}$$

in geschlossener Form. *Archiv der Math. und Phys.* XXX. (1858), pp. 331—334.

Ueber die Integration der Differentialgleichung

$$x^m \frac{d^n y}{dx^n} = \pm y$$

(1859). *Journal für Math.* LVII. (1860), pp. 82—87.

STEARN, H.

On some cases of the Varying Motion of a Viscous Fluid. *Quarterly Journal,* XVII. (1880), pp. 90—104.

* See also under Michell. † See also under Hopf.

STEINER, L.

Intensitäts-verhältnisse der Beugungserscheinung durch eine kreisförmige Öffnung (June 19, 1893). *Math. und Naturwiss. Berichte aus Ungarn*, XI. (1892—93) [1894], pp 362—373.

STEINTHAL, A. E.

On the Solution of the Equation

$$(1-x^2)\frac{d^2u}{dx^2}-2x\frac{du}{dx}+n(n+1)u=0.$$

Quarterly Journal, XVIII. (1882), pp. 330—345.

STEPHENSON, A.

An extension of the Fourier method of expansion in sine series. *Messenger*, XXXIII. (1904), pp. 70—77.

A more general case of expansion in sine series. *Messenger*, XXXIII. (1904), pp. 178—182.

On Expansion in Bessel's Functions (June, 1907). *Phil. Mag.* (6) XIV. (1907), pp. 547—549.

STERN, M. A.

Ueber die Anwendung der Sturm'schen Methode auf transcendente Gleichungen. *Journal für Math.* XXXIII. (1846), pp. 363—365.

STIELTJES, T. J.

Recherches sur quelques séries sémi-convergentes. *Ann. sci. de l'École norm. sup.* (3) III. (1886), pp. 201—258.

STOKES, SIR GEORGE GABRIEL.

On the numerical Calculation of a Class of Definite Integrals and Infinite Series (March 11, 1850). *Trans. Camb. Phil. Soc.* IX. (1856), pp. 166—187. [*Math. and Phys. Papers*, II. (1883), pp. 329—357.]

On the Effect of the Internal Friction of Fluids on the Motion of Pendulums (Dec. 9, 1850). *Trans. Camb. Phil. Soc.* IX. (1856), pp. [8]—[106]. [*Math. and Phys. Papers*, III. (1901), pp. 1—141.]

On the Discontinuity of Arbitrary Constants which appear in Divergent Developments (May 11, 1857). *Trans. Camb. Phil. Soc.* X. (1864), pp. 106—128. [*Math. and Phys. Papers*, IV. (1904), pp. 77—109.]

Supplement to a paper on the Discontinuity of Arbitrary Constants which appear in Divergent Developments (May 25, 1868). *Trans. Camb. Phil. Soc.* XI. (1871), pp. 412—425. [*Math. and Phys. Papers*, IV. (1904), pp. 283—298.]

On the Communication of Vibration from a Vibrating Body to a surrounding Gas* (June 18, 1868). *Phil. Trans. of the Royal Soc.* CLVIII. (1868) [1869], pp. 447—463. [*Math. and Phys. Papers*, IV. (1904), pp. 299—324.]

Smith's Prize Examination Papers, Feb. 1853 and Jan. 29, 1867. [*Math. and Phys. Papers*, V. (1905), pp. 319, 347.]

Note on the Determination of Arbitrary Constants which appear as Multipliers of Semi-convergent Series (June 3, 1889). *Proc. Camb. Phil. Soc.* VI. (1889), pp. 362—366. [*Math. and Phys. Papers*, V. (1905), pp. 221—225.]

On the Discontinuity of Arbitrary Constants that appear as Multipliers of Semi-convergent Series† (April 23, 1902). *Acta Math.* XXVI. (1902), pp. 393—397. [*Math. and Phys. Papers*, V. (1905), pp. 283—287.]

STRUTT, J. W.

See Rayleigh (Lord).

STRUVE, H.

Ueber den Einfluss der Diffraction an Fernröhren auf Lichtscheiben (May 25, 1882). *Mém. de l'Acad. Imp. des Sci. de St. Pétersbourg*, (7) XXX. (1882), no. 8.

Beitrag zur Theorie der Diffraction an Fernröhren (Aug. 1882). *Ann. der Physik und Chemie*, (3) XVII. (1882), pp. 1008—1016.

STURM, J. C. F

Sur les Équations différentielles linéaires du second ordre (Sept. 28, 1833). *Journal de Math.* I. (1836), pp. 106—186.

* An abstract will be found in *Proc. Royal Soc.* XVI. (1868), pp. 470—471.

† This, the last published paper which Stokes wrote, contains a summary of his researches composed for the Abel centenary volume.

SUCHAR, P. J.

Sur les équations différentielles linéaires réciproques du seco. d ordre (Nov. 18, 1903). *Bull. de la Soc. Math. de France*, XXXII. (1904), pp. 103—116.

SVANBERG, A. F.

De integralibus definitis disquisitiones. *Nova Acta R. Soc. Sci. Upsala*, X. (1832), pp. 231—288.

TAKEUCHI, T.

On the integral $\int_0^\pi e^{x \cos p\theta} \sin q\theta, \cos q\theta d\theta$ (May 17, 1920). *Tôhoku Math. Journal*, XVIII. (1920), pp. 295—296.

THEISINGER, L.

Bestimmte Integrale. *Monatshefte für Math. und Phys.* XXIV. (1913), pp. 328—346.

THOMAE, J.

Die Brauchbarkeit der Bessel-Fourierschen Reihe. *Berichte der Math. Seminar zu Jena*, 1912—13, pp. 8—10. [*Int. Catalogue of Sci. Lit.* XIII. (1913), p. 100.]

THOMSON, SIR JOSEPH JOHN.

Note on $\int_0^\infty \frac{\cos sx}{(a^2+x^2)^{\frac{1}{2}(2p+1)}} dx$. *Quarterly Journal*, XVIII. (1882), pp. 377—381.

Recent Researches in Electricity and Magnetism (Oxford, 1893). [Review by A. Gray, *Nature*, XLIX. (1894), pp. 357—359.]

THOMSON, W.

See Kelvin (Lord).

TODHUNTER, I.

Elementary Treatise on Laplace's Functions, Lamé's Functions and Bessel's Functions (London, 1875).

TURRIÈRE, E.

Une application géométrique de la série considérée par Airy dans la diffraction des ouvertures circulaires. *Nouvelles Annales de Math.* (4) IX. (1909), pp. 433—441.

UNFERDINGER, F.

Über die beiden Integrale $\int e^{\sin x} \frac{\cos}{\sin} (nx - \cos x)\, dx$ (April 16, 1868). *Wiener Sitzungsberichte*, LVII. (1868), pp. 611—620.

VALEWINK, G. C. A.

Over asymptotische ontwikkelingen (*Dissertation, Haarlem*, pp. 138). [*Jahrbuch über die Fortschritte der Math.* 1905, p. 328.]

VAN VLECK, E. B.

On the Roots of Bessel- and *P*-functions. *American Journal of Math.* XIX. (1897), pp. 75—85.

VERDET, É.

Leçons d'Optique Physique, I. (Paris, 1869).

VESSIOT, E.

Sur quelques équations différentielles ordinaires du second ordre. *Annales de la Fac. des Sci. de Toulouse*, IX. (1895), no. 6.

VOLTERRA, V.

Sopra alcune questioni di inversione di integrali definiti. *Ann. di Mat.* (2) XXV. (1897), pp. 139—178.

VORONOI, G.

Sur une fonction transcendante et ses applications à la sommation de quelques séries. *Ann. sci. de l'École norm. sup.* (3) XXI. (1904), pp. 207—268, 459—534.

Sur le développement, à l'aide des fonctions cylindriques, des sommes doubles

$$\Sigma f(pm^2+2qmn+rn^2)$$

où $pm^2+2qmn+rn^2$ est une forme positive à coefficients entiers. *Verh. des dritten Int. Kongresses in Heidelberg* (1904), pp. 241—245.

WAGNER, C.

Beiträge zur Entwicklung der Bessel'schen Function I. Art. *Bern Mittheilungen*, 1894, pp. 204—266.

Über die Darstellung einiger bestimmten Integrale durch Bessel'sche Funktionen (June 12, 1895; Aug. 5, 1895). *Bern Mittheilungen*, 1895, pp. 115—119; 1896, pp. 53—60.

WALKER, G. T.

Some formulae for transforming the origin of reference of Bessel's functions. *Messenger*, XXV. (1896), pp. 76—80.

WALKER, G. W.

The scattering of electromagnetic waves by a sphere. *Quarterly Journal*, XXXI. (1900), pp. 36—49.

WALKER, J.

The Analytical Theory of Light (Cambridge, 1904).

WALLENBURG, G.

Ueber *Riccati*'sche Differentialgleichungen höherer Ordnung. *Journal für Math.* CXXI. (1900), pp. 196—199.

Die Differentialgleichungen deren allgemeines Integral eine lineare gebrochene Function der willkürlichen Constanten ist. *Journal für Math.* CXXI. (1900), pp. 210—217.

Sur l'équation différentielle de Riccati du second ordre (Dec. 14, 1903). *Comptes Rendus*, CXXXVII. (1903), pp. 1033—1035.

WANGERIN, A.

Cylinderfunktionen oder Bessel'sche Funktionen. *Encyclopädie der Math. Wiss.* Bd. II. Teil 1 (Leipzig, 1904—16), pp. 742—757. [*Encyclopédie des Sci. Math.* Tome II. vol. 5 (Paris et Leipzig, 1914, pp. 209—229), translation by A. Lambert.]

WATSON, G. N.

On a certain difference equation of the second order. *Quarterly Journal*, XLI. (1910), pp. 50—55.

Bessel Functions and Kapteyn Series (April 26, 1916). *Proc. London Math. Soc.* (2) XVI. (1917), pp. 150—174.

Bessel Functions of Equal Order and Argument (June 17, 1916). *Phil. Mag.* (6) XXXII. (1916), pp. 232—237.

Simple types of Kapteyn Series. *Messenger*, XLVI. (1917), pp. 150—157.

Bessel functions of equal order and argument (Nov. 13, 1916). *Proc. Camb. Phil. Soc.* XIX. (1918), pp. 42—48.

The limits of applicability of the Principle of Stationary Phase (Nov. 22, 1916). *Proc. Camb. Phil. Soc.* XIX. (1918), pp. 49—55.

Bessel functions of large order (June 14, 1917). *Proc. Camb. Phil. Soc.* XIX. (1918), pp. 96—110.

The Zeros of Bessel Functions (Aug. 17, 1917). *Proc. Royal Soc.* XCIV. A (1918), pp. 190—206.

Bessel Functions of Equal Order and Argument. *Phil. Mag.* (6) XXXV. (1918), pp. 364—370.

The Diffraction of Electric Waves by the Earth (May 29, 1918). *Proc. Royal Soc.* XCV. A (1919), pp. 83—99.

The Transmission of Electric Waves round the Earth (Jan. 13, 1919). *Proc. Royal Soc.* XCV. A (1919), pp. 546—563.

On Nielsen's functional equations. *Messenger*, XLVIII. (1919), pp. 49—53.

The zeros of Lommel's polynomials (May 15, 1919). *Proc. London Math. Soc.* (2) XIX. (1921), pp. 266—272.

WEBB, H. A.

The expansion of an arbitrary function in a series of Bessel functions. *Messenger*, XXXIII. (1904), pp. 55—58.

On the Convergence of Infinite Series of Analytic Functions (Nov. 10, 1904). *Phil. Trans. of the Royal Soc.* CCIV. A (1904), pp. 481—497.

WEBER, H.

Ueber einige bestimmte Integrale (Jan. 1868). *Journal für Math.* LXIX. (1869), pp. 222—237.

Ueber die Integration der partiellen Differentialgleichung: $\frac{\partial^2 u}{\partial x^2} + \frac{\partial^2 u}{\partial y^2} + k^2 u = 0$ (July, 1868). *Math. Ann.* I. (1869), pp. 1—36.

Ueber die *Bessel*schen Functionen und ihre Anwendung auf die Theorie der elektrischen Ströme (May, 1872). *Journal für Math.* LXXV. (1873), pp. 75—105.

Ueber die stationären Strömungen der Electricität in Cylindern (Sept. 1872). *Journal für Math.* LXXVI. (1873), pp. 1—20.

Ueber eine Darstellung willkürlicher Functionen durch Bessel'sche Functionen (Oct. 1872). *Math. Ann.* VI. (1873), pp. 146—161.

Zur Theorie der Bessel'schen Functionen (July, 1890). *Math. Ann.* XXXVII. (1890), pp. 404—416.

WEBER, H. F.

Die wahre Theorie der Fresnel'schen Interferenz-Erscheinungen. *Zürich Vierteljahrsschrift*, XXIV. (1879), pp. 33—76; *Ann. der Physik und Chemie*, (3) VIII. (1879), pp. 407—444.

WEBSTER, A. G.

Application of a definite integral involving Bessel's functions to the self-inductance of solenoids (Dec. 29, 1905). *Bulletin American Math. Soc.* XIV. (1907), pp. 1—6.

WENDT, CÄCILIE.

Eine Verallgemeinerung des Additionstheoremes der Bessel'schen Functionen erster Art. *Monatshefte für Math. und Phys.* XI. (1900), pp. 125—131.

WEYL, H.

Singuläre Integralgleichungen (April, 1908). *Math. Ann.* LXVI. (1909), pp. 273—324.

WEYR, E.

Zur Integration der Differential-Gleichungen erster Ordnung. *Abh. böhm. Ges. Wiss. (Prag)*, (6) VIII. (1875—76), Math. Mem. 1.

WHEWELL, W.

Of the Intrinsic Equation of a Curve and its Application (Feb. 12, 1849). *Trans. Camb. Phil. Soc.* VIII. (1849), pp. 659—671.

Second Memoir on the Intrinsic Equation of a Curve and its Application (April 15, 1850). *Trans. Camb. Phil. Soc.* IX. (1856), pp. 150—156.

WHIPPLE, F. J. W.

Diffraction by a wedge and kindred problems (Nov. 8, 1915). *Proc. London Math. Soc.* (2) XVI. (1917), pp. 94—111.

WHITE, F. P.

The Diffraction of Plane Electromagnetic Waves by a Perfectly Reflecting Sphere (June 9, 1921). *Proc. Royal Soc.* C. A (1922), pp. 505—525.

WHITEHEAD, C. S.

On the functions ber x, bei x, ker x, kei x. *Quarterly Journal*, XLII. (1911), pp. 316—342.

WHITTAKER, E. T.

On the General Solution of Laplace's Equation and the Equation of Wave Motions and on an undulatory explanation of Gravity. *Monthly Notices of the R. A. S.* LXII. (1902), pp. 617—620.

On the partial differential equations of mathematical physics. *Math. Ann.* LVII. (1903), pp. 333—355.

On a new Connexion of Bessel Functions with Legendre Functions (Nov. 13, 1902). *Proc. London Math. Soc.* XXXV. (1903), pp. 198—206.

WIGERT, S.

Sur quelques fonctions arithmétiques (March, 1913). *Acta Math.* XXXVII. (1914), pp. 113—140.

WILLIAMSON, B.

On the Solution of certain Differential Equations (March 5, 1856). *Phil. Mag.* (4) XI (1856), pp. 364—371.

WILLSON, R. W. AND PEIRCE, B. O.

Table of the first forty roots of the Bessel equation $J_0(x)=0$ with the corresponding values of $J_1(x)$. *Bulletin American Math. Soc.* III. (1897), pp. 153—155.

WILTON, J. R.

A continued fraction solution of the linear differential equation of the second order. *Quarterly Journal*, XLVI. (1915), pp. 318—334.

WIRTINGER, W.

Zwei Bemerkungen zu Airy's Theorie des Regenbogens. *Berichte des natur.-med. Vereins in Innsbruck*, XXIII. (1897), pp. 7—15.

WORMS DE ROMILLY, P.

Note sur l'intégration de l'équation

$$\frac{d^2 V}{dx^2}+\frac{\mu+1}{x}\frac{dV}{dx}+V=0.$$

Journal de Math. (3) IV. (1878), pp. 177—186.

YOUNG, W. H.

On infinite integrals involving a generalisation of the sine and cosine functions (Oct. 6, 1911). *Quarterly Journal*, XLIII. (1911), pp. 161—177.

On series of Bessel functions (Dec. 6, 1917). *Proc. London Math. Soc.* (2) XVIII. (1920) pp. 163—200.

ZELINSKIJ, I. I.

On the integration of Riccati's equation (Jan. 27, 1890). *Proc. Phys. Math. Section, Naturalists' Soc. Imp. Univ. Kazan*, VIII. (1890), pp. 337—342.

INDEX OF SYMBOLS

[The numbers refer to the pages on which the symbols are defined.]

LIST OF AUTHORS QUOTED

[*The numbers refer to the pages. References are not given to entries in the bibliography, pp.* 753—788.]

GENERAL INDEX

[*The numbers refer to the pages.*]